Matrix and Linear Algebra

AF646807

Matrix and Linear Algebra
Aided with MATLAB

Third Edition

KANTI BHUSHAN DATTA
Former Professor
Department of Electrical Engineering
Indian Institute of Technology Kharagpur

PHI Learning Private Limited
Delhi-110092
2026

In fond memory of ***Shri Asoke K. Ghosh*** *(October 1942 – February 2024), Founder Chairman and Managing Director of PHI Learning, whose vision endlessly inspires.*

The Legacy Continues . . . .

Published by Pushpita Ghosh, PHI Learning Private Limited, Rimjhim House, 111, Patparganj Industrial Estate, Delhi-110092 and Printed by Gopsons Printers Pvt. Ltd., A-14, Sector-60, Noida, Gautambudh Nagar, Noida, Uttar Pradesh-201301.

₹1295.00

MATRIX AND LINEAR ALGEBRA—Aided with MATLAB, Third Edition
Kanti Bhushan Datta

© 2026 by PHI Learning Private Limited, Delhi. Previous editions © 2008, 1991.

All right reserved. This publication is protected by copyright and may not be reproduced, in whole or in part, in any form or by any electronic, mechanical, or other means, now known or hereafter invented, including photocopying, recording, or any information storage and retrieval system, OR to train artificial intelligence systems without permission in writing from the publisher.

ISBN-978-81-203-5286-5 (Print Book)
ISBN-978-93-5443-121-0 (e-Book)

The export rights of the book are vested solely with the publisher.

To my daughter
Somantika
to provide inspiration for strengthening her
skill in mathematics

Contents

Preface

The third edition witnesses twenty-six years of survival of *Matrix and Linear Algebra*. Some of the important revisions warranted us to bring out this edition, the foremost one being a detailed exposition of image and data analysis. The main features of revisions are summarized in what follows.

Examples related to rank and invertibility are updated in Chapter 5. In Chapter 6, apart from inserting two important theorems related to *ker* and *range* of a linear transformation, a new Section 6.7 *Linear Transformation*: *Dilation, Rotation, Reflection and Projection* is added with structural properties that highlight the significance of these transformations in practical applications. In Chapter 8, again a new Section 8.9 *Simultaneous Diagonalization* is introduced to expound the theory of simultaneous diagonalization of two diagonalizable matrices with numerical illustrations. In Chapter 9, the Section 9.6 *Generalized Inverse of a Matrix* is fully revised related to singular values and singular vectors elaborating its properties, emphasizing its applications in shear analysis in Engineering and stressing its importance in orthogonal projection on a subspace. The *Polar Decomposition of a Matrix* embodying the representation of a matrix similar to a complex number is added at the end of Section 12.3 *Square Root of a Matrix A*, sin *A*, cos *A*, ln *A* visualizing its usefulness in theoretical analysis. In Chapter 13, Section 13.7.2 *Image Compression using SVD* outlines a new feature of SVD in transmission of (colored) images with extensive saving in time and storage space. The result of a numerical experimentation with a black and white picture 'lena_bw.gif' of Swedish girl Lena Söderberg, the first lady of the internet is provided, and the MATLAB codes for a black and white as well as for a colour image are listed for the convenience of the reader. Finally, Section 13.7.3 contains a comprehensive and thorough account of PCA method of data analysis and PCA method of image compression with a list of MATLAB codes, two major additions of the present edition.

As, course material, this book can be used in many ways. For an elementary course, one can choose Chapters 1–3, Sections 4.1–4.4; 5.1–5.3, 5.6; 7.1, 7.2, 7.4–7.6, skipping the related linear transformation portions. Last but not least, Chapter 13, the most important part from the application point of view, outlines numerical linear algebra. These topics may form a forty-hour lecture course of one semester supported by homework and tutorials. The remaining chapters and sections may form a second semester advanced course on matrix and linear algebra for those students who are pursuing M.Sc. in Mathematics or Ph.D. programmes.

The book now contains very rich ingredients of Linear Algebra that are prerequisites for an advanced course in Signal Analysis. The related topics in this connection are: vector spaces, linear dependence and independence of vectors, Gram–Schmidt process of orthogonalization (Chapter 5); linear transformations, range space and null space, matrix of a linear transformation, special matrices such as permutation matrix, Toeplitz matrix, Vandermonde matrix, circulant matrix, triangular matrix, unitary and orthogonal matrices, and some of their applications, orthogonal projections on a subspace (Chapter 6); eigenvalues and eigenvectors, linear independence of eigenvectors, diagonalization of a matrix, Gershgorin circle theorem, and computation of eigenvalues and eigenvectors (Chapter 7); singular value decomposition (SVD) and principal components analysis (PCA), effective rank of a matrix, matrix structure from SVD, SVD and pseudo-inverse, applications of SVD and PCA to image compression and data analysis (Chapters 9 and 13); numerical linear algebra, LU factorization, Cholesky factorization, and QR factorization (Chapter 13).

KANTI BHUSHAN DATTA
e-mail: attadbitnak@yahoo.com

Preface to the First Edition

Most of the textbooks are an outgrowth of a series of lectures delivered or courses given to students and/or professionals over a period of time. This book, on the other hand, is the outcome of self-teaching, in which the author is both the teacher and the student. In this process of self-teaching and self-learning, there is a need for an evaluation of correct inconsistencies and removal of the inherent lacunae. One of the major objectives of writing this text is therefore to invite such a self-imposed evaluation from the learned readers, whose valuable suggestions may improve the content of this volume in the next edition.

This compact and self-contained volume, which has drawn extensively from the classic and contemporary books on matrices and linear transformations, in written with a view to creating, in a single volume, an ideal source for the study of Linear Algebra.

The title of this book is a shortened form of Matrix Algebra and Linear Algebra. A linear space or a vector space V over a field $\mathbb{F}$ (see Chapter 5) is called an algebra L if it is closed under a second operation called multiplication such that if $\mathbf{x}$, $\mathbf{y}$, $\mathbf{z}$ are vectors in V and α is a scalar in $\mathbb{F}$, then

(a) $\mathbf{xy} \in L$,

(b) multiplication is associative: $\mathbf{x}(\mathbf{yz}) = (\mathbf{xy})\mathbf{z}$,

(c) multiplication is distributive

$$\mathbf{x}(\mathbf{y} + \mathbf{z}) = \mathbf{xy} + \mathbf{xz},$$

$$(\mathbf{x} + \mathbf{y})\mathbf{z} = \mathbf{xz} + \mathbf{yz},$$

(d) $\alpha(\mathbf{xy}) = (\alpha\mathbf{x})\mathbf{y} = \mathbf{x}(\alpha\mathbf{y})$.

The axiom (d) shows the interaction of the scalar elements with the operation of multiplication. In commutative algebra, multiplication is commutative, i.e., if $\mathbf{x}$ and $\mathbf{y}$ are any two elements, then $\mathbf{xy} = \mathbf{yx}$. The set of all square matrices constitutes *noncommutative matrix algebra*. The set of all linear transformations $\mathscr{L}: V \to V$ satisfies all the properties of an algebra (see Chapter 6) and forms what is known as linear algebra. Such linear algebra is, in general, noncommutative.

Today, in almost all disciplines of Science, Engineering and Technology, an extensive course on the Theory of Matrices is an indispensable part of the curriculum. The modern trend is to offer this course based on the teaching of Linear Algebra. As a part of the curriculum in the discipline of mathematics, this course is taught in a much more sophisticated fashion by considering first the modules M over a ring R, in contrast with vector spaces over a field, and then endomorphisms on M over R in place of linear transformations over a field. Such a refined area in mathematics is generally covered in Abstract Algebra, with which we are not concerned at all.

After the basic structure of Linear Algebra has been established on a firm foundation, computation with linear transformations is shown to be equivalent to computation with matrices by a proper selection of a basis. In such a treatment, the emphasis is on Linear Algebra, and the theory of matrices takes virtually a secondary role in the process of learning. Consequently, the beginners suffer considerably in understanding the computational aspect of Linear Algebra. This computational part is, in fact, much more important from the engineering point of view.

Teaching Matrix Algebra on its own right is too restrictive, because it is only by going deep into Linear Algebra that one can provide a strong structural foundation to the theory of matrices. To strike a balance between these two allied disciplines, and to amalgamate them subsequently in this book, a thorough acquaintance with matrices, determinants, inverse of a matrix, rank and equivalence, and vector spaces is first developed, and then by abstracting the properties of Matrix Algebra contained in the preceding discussion, the notion of a linear transformation is introduced. The link of Matrix Algebra with Linear Algebra is shown through the notion of *matrix representation of a linear transformation*, after which the conceptual development of the structural components like eigenvalues, eigenvectors, and characteristic equation of a matrix as well as a linear transformation proceeds in a synchronous fashion. The role played by matrices in defining bilinear, quadratic and Hermitian forms and bilinear functions and their matrix representation provides ample evidence that Matrix Algebra and Linear Algebra can be understood well only by combining them. For a better understanding of either of them, they should be developed simultaneously. Vector norms and matrix norms, the normal forms of a matrix and a linear transformation, and the function of a matrix are discussed in great depth. Today, as digital computers which employ finite arithmetic are used extensively in scientific computation, special numerical techniques should be developed for computing with matrices. This important aspect is therefore given sufficient coverage in the book.

The first six chapters and a few sections in Chapters 7 and 8 dealing with characteristic equation, eigenvalues and eigenvectors, quadratic forms and definite matrices would form the core of a course for most of the students pursuing a first course on the subject. In fact, these topics would cover the syllabus of the Honours course in Mathematics at the undergraduate level of most Indian universities. The remaining topics may be included for a second course on the subject or may be used by masters and research students in Mathematics, Physics, Economics and Statistics, and undergraduate and masters students of engineering, for their core courses or electives, or for reference purposes.

The coverage of the book is quite extensive; a good collection of problems is included at the end of each chapter and answers to almost all the problems are provided. These will be of great help to those diligent students who wish to use this book for self-study. A large number of solved examples besides a few diagrams illustrating the concepts introduced have also been provided.

The seed of this book, which was supposed to form the nucleus of a voluminous book on Engineering Mathematics, was sown in the Department of Applied Physics, Calcutta University, where the author was a teacher during 1965–1982. The first chapter of the Engineering Mathematics book was on Matrix and Linear Algebra. The standard books on these topics dealing with the basic principles and their varied applications were of considerable assistance in gathering new ideas and presenting these in a single volume such as this. The author is, indeed, indebted to his distinguished teachers and former colleagues in this department, whose inspiration and guidance encouraged him through the initial stages of the book. It was only in the screne academic environment of the Indian Institute of Technology Kharagpur that this book could have taken its final shape. The learned and erudite colleages of the author at the Department of Electrical Engineering at the Institute and his students encouraged and motivated him and strengthened his conviction in completing this book within a reasonably short time. The author is extremely grateful to them for such encouragement. The author also wishes to thank Mr. Subrata Kumar Giri for his excellent typing of the manuscript. Finally the author expresses his sincere thanks to his wife, Dr. Surekha Datta, who provided a proper, indeed ideal, environment for writing this book, showed remarkable patience, and proved to be a source of great encouragement.

KANTI BHUSHAN DATTA

CHAPTER

1

Matrix Algebra

Half the battle in mathematics is the invention of a good notation. —P.S. Laplace

In the development of mathematics, the theory of determinants predated that of matrices by more than one and a half century. Determinants originated from (i) finding the numerators and denominators of fractions which express the values of unknowns in a set of linear equations, or (ii) finding the resultant of elimination of n quantities from $(n+1)$ linear equations. To understand this, let us consider the two simultaneous equations

$$a_{11}x_1 + a_{12}x_2 = b_1, \quad a_{21}x_1 + a_{22}x_2 = b_2. \tag{1.1}$$

Eliminating x_2 from the first equation, we have

$$a_{11}x_1 + a_{12}(b_2 - a_{21}x_1)/a_{22} = b_1.$$

Therefore,
$$x_1 = (b_1a_{22} - b_2a_{12})/(a_{11}a_{22} - a_{12}a_{21}). \tag{1.2}$$

Similarly,
$$x_2 = (b_2a_{11} - b_1a_{21})/(a_{11}a_{22} - a_{12}a_{21}). \tag{1.3}$$

On examining (1.2) and (1.3), which represent the solution of (1.1), it is evident that x_1 and x_2 appear as a fraction with the same denominator

$$a_{11}a_{22} - a_{12}a_{21}.$$

This expression is called the *determinant* of the system (1.1) and is denoted by

$$\begin{vmatrix} a_{11} & a_{12} \\ a_{21} & a_{22} \end{vmatrix} = a_{11}a_{22} - a_{12}a_{21}.$$

On examining the numerators of x_1 and x_2 in (1.2) and (1.3), it is evident that they are also determinants, i.e.,

$$\begin{vmatrix} b_1 & a_{12} \\ b_2 & a_{22} \end{vmatrix} = b_1a_{22} - b_2a_{12}, \qquad \begin{vmatrix} a_{11} & b_1 \\ a_{21} & b_2 \end{vmatrix} = b_2a_{11} - b_1a_{21}.$$

The Chinese, possibly as early as 1100 B.C., solved equations of the form (1.1) by a rule equivalent to the usual method by determinants [14].

On the other hand, to get the *resultant* of a set of *n linear equations* involving $(n - 1)$ variables, we eliminate the variables step by step. Thus the resultant of the equations

$$a_{11}x_1 + a_{12}x_2 + a_{13} = 0, \quad a_{21}x_1 + a_{22}x_2 + a_{23} = 0, \quad a_{31}x_1 + a_{32}x_2 + a_{33} = 0 \tag{1.4}$$

is obtained by removing the variables x_1 and x_2 to give

$$a_{11}a_{22}a_{33} - a_{12}a_{21}a_{33} + a_{13}a_{21}a_{32} - a_{13}a_{22}a_{31} + a_{12}a_{23}a_{31} - a_{11}a_{23}a_{32} = 0. \tag{1.5}$$

Equation (1.5) can also be written as

$$\begin{vmatrix} a_{11} & a_{12} & a_{13} \\ a_{21} & a_{22} & a_{23} \\ a_{31} & a_{32} & a_{33} \end{vmatrix}.$$

Gottfried Wilhelm Leibniz (1646–1716) is most frequently credited with the discovery of the determinants in 1693. He arrived at the expression (1.5) by eliminating x_1 and x_2 from (1.4), used for it a new notation which was numerical in character and appearance, and defined a rule for forming each term from the coefficients a_{ij} in (1.4) and for determining its sign, even though the connection of the new notation with the determinant with which we are familiar today was not given. Some historians believe the Japanese mathematician Seki Kowa (1642–1708) knew of its existence prior to 1683. Determinants remained forgotten until, in 1750, Gabriel Cramer (1704–52), the Swiss mathematician, rediscovered them while working on the analysis of curves. He provided a rule (i) for forming the terms of the common denominator of the fractions which express the values of the unknowns in a set of linear equations, (ii) for determining the sign of any individual term in the said common denominator, and (iii) for obtaining the numerators from the expression for the common denominator.

It was Alexandre-Theophile Vandermonde (1735–96) who first, in 1771, recognized determinants as independent functions, improved the notation, isolated them as objects of independent study, and gave a systematic account of what little was then known to the world. He may therefore be called the *formal discoverer of determinants* [14, 179]. Pierre-Simon de Laplace (1749–1827) gave the general method of expanding a determinant in 1772, and Joseph-Louis Lagrange (1736–1813) discovered useful identities in 1773. A major step forward in the development of determinant was taken in 1801 by Carl Friedrich Gauss (1777–1855). He introduced a new terminology to call $b^2 - ac$ the *determinant of the form* $ax^2 + 2bx + c$, which is perfectly identical in meaning and usage with the modern term *discriminant,* and denoted it by (a, b, c).

After this base work was done, a strong foundation for modern determinant theory was laid by Jacques P.M. Binet (1786–1856) and Augustin-Louis Cauchy (1789–1857) of France, Carl Gustav Jacob Jacobi (1804–51) of Germany, followed by James Joseph Sylvester (1814–97), Arthur Cayley (1821–95) of England, and Karl Weierstrass (1815–97) and Leopold Kronecker (1823–91) of Germany.

In a memoir presented to the Académie on November 30, 1812, Cauchy presented before the mathematicians of that time an almost exhaustive treatise on the theory of general determinants and used the term *determinant* in its present sense [137]. Subsequent researchers shifted their focus from the determinant theory to the more general theory of matrices.

In mathematics, the purpose of a *transformation* such as

$$x_1 = a_{11}y_1 + a_{12}y_2, \qquad x_2 = a_{21}y_1 + a_{22}y_2, \tag{1.6}$$

is to reduce an unsolved problem in variables x_1 and x_2 to a simpler one than the original. What Cayley did in 1858 is as given below. In addition to the transformation (1.6), he considered another transformation

$$y_1 = b_{11}z_1 + b_{12}z_2, \qquad y_2 = b_{21}z_1 + b_{22}z_2 \tag{1.7}$$

and consequently, (1.6) becomes

$$\begin{aligned} x_1 &= (a_{11}b_{11} + a_{12}b_{21})\, z_1 + (a_{11}b_{12} + a_{12}b_{22})\, z_2, \\ x_2 &= (a_{21}b_{11} + a_{22}b_{21})\, z_1 + (a_{21}b_{12} + a_{22}b_{22})\, z_2. \end{aligned} \tag{1.8}$$

The transformation from variables x_1, x_2 to variables z_1, z_2 through y_1, y_2 as is evident from (1.8) is not easy to recollect. If, however, the coefficients in the transformations shown in Equations (1.6) and (1.7) are separated and arranged schematically in terms of arrays, or *matrices* as Cayley called them, then everything becomes straightforward. Thus, if

$$A = \begin{bmatrix} a_{11} & a_{12} \\ a_{21} & a_{22} \end{bmatrix}, \qquad B = \begin{bmatrix} b_{11} & b_{12} \\ b_{21} & b_{22} \end{bmatrix}, \tag{1.9}$$

then,

$$AB = \begin{bmatrix} a_{11}b_{11} + a_{12}b_{21}, & a_{11}b_{12} + a_{12}b_{22} \\ a_{21}b_{11} + a_{22}b_{21}, & a_{21}b_{12} + a_{22}b_{22} \end{bmatrix}, \tag{1.10}$$

in which the element $a_{11}b_{11}+a_{12}b_{21}$ is constituted by the first row elements of A and the first column elements of B etc., and are easy to remember. This discovery of matrices illustrated the power and suggestiveness of a well-devised notation and paved the way for a vast and indispensible branch of mathematics with a multitude of practical applications.

With the foregoing historical account of the origin of determinants and matrices, let us now make a brief appraisal of the key notion of matrix discussed in this book. In keeping with the perspective that Cayley introduced the term *matrix* to square out some notational bottleneck in the transformation of variables, in present-day applied mathematics it is used as a shortened notation in linear simultaneous equations in an overwhelmingly large majority of cases.

In *operations research*, an important aspect is the *transportation problem.* A simple variety of it is now exemplified.

A company has plants in Ahmedabad and Mumbai, and warehouses in Kolkata and Delhi. Ahmedabad is abbreviated as *A*, Mumbai as *B*, Kolkata as *C* and Delhi as *D*. In *A* and *B* the company has a certain commodity available, each of 4 and 6 units respectively, which should be transported to *C* and *D*, where 7 and 3 units of the commodity are required. The transport cost per unit from each plant to each warehouse is given in Table 1.1.

Let x_1, x_2, x_3 and x_4 be the quantities transported from A to C, A to D, B to C, and B to D, respectively. Then the total transport cost, f, is

$$f = 8x_1 + 2x_2 + 6x_3 + 3x_4,$$

where the variables would satisfy the constraints

$$1 \cdot x_1 + 1 \cdot x_2 + 0 \cdot x_3 + 0 \cdot x_4 = 4,$$

$$0 \cdot x_1 + 0 \cdot x_2 + 1 \cdot x_3 + 1 \cdot x_4 = 6,$$

$$1 \cdot x_1 + 0 \cdot x_2 + 1 \cdot x_3 + 0 \cdot x_4 = 7,$$

$$0 \cdot x_1 + 1 \cdot x_2 + 0 \cdot x_3 + 1 \cdot x_4 = 3.$$

Table 1.1 Transportation problem

		Warehouses C	Warehouses D
Plants	Quantities→	7	3
	Quantities ↓	Cost per unit	
A	4	8	2
B	6	6	3

The problem is to determine x_1, x_2, x_3 and x_4 by solving the above equations such that f is minimized. One very often detaches the coefficients of the variables x_1, x_2, x_3 and x_4 and writes the above system of equations in a compact form as

$$\begin{bmatrix} 1 & 1 & 0 & 0 \\ 0 & 0 & 1 & 1 \\ 1 & 0 & 1 & 0 \\ 0 & 1 & 0 & 1 \end{bmatrix} \begin{bmatrix} x_1 \\ x_2 \\ x_3 \\ x_4 \end{bmatrix} = \begin{bmatrix} 4 \\ 6 \\ 7 \\ 3 \end{bmatrix}$$

On the other hand, in statistical input-output analysis of a national economy known as the *Leontief system,* we have the following problem: The flow of goods and services between all individual sectors of a national economy over a year is described in the input-output analysis (Table 1.2) in an oversimplified two-industry case [52],

Table 1.2 Input-output analysis in Leontief system

Industries	Inputs to agriculture	Inputs to manufacturing	Final demand	Total outputs
Agriculture	$x_{11} = 50$	$x_{12} = 150$	$c_1 = 50$	$y_1 = 250$
Manufacturing	$x_{21} = 75$	$x_{22} = 100$	$c_2 = 25$	$y_2 = 200$
Labour services	$x_{31} = 125$	$x_{32} = 25$	0	$y_3 = 150$

The first row above shows that of the total 250 units (millions of tons) of agriculture output per year, 50 units are required in agriculture itself (e.g., oats used to feed horses that pull harvesters of wheat and oats), 150 units in manufacturing production and the remaining 50 units go directly to households and government for consumption. Similarly, the second row shows allocation of 75, 100 and 25 units of manufactured goods per year to agriculture, manufacturing and final consumption. The third row shows that 150 units (thousands of man-years) are allocated as inputs to the two industries in the respective amounts (125 and 25). On the other hand, the first column shows that an input of 50 units of agricultural products, 75 units of manufactured goods and 125 units of labour are required for an agricultural output of 250 units. Likewise, the second column describes the cost structure of manufacturing industry. It is evident that we can derive, from Table 1.2, the equations

$$x_{11} + x_{12} + c_1 = y_1, \quad x_{21} + x_{22} + c_2 = y_2, \quad x_{31} + x_{32} = y_3.$$

Denoting by $a_{ij} = x_{ij}/y_j$ = min input of commodity i per unit output of commodity j, $i = 1, 2, 3, j = 1, 2$, the above equations become

$$(1 - a_{11})y_1 - a_{12}y_2 = c_1,$$

$$-a_{21}y_1 + (1 - a_{22})y_2 = c_2,$$

$$-a_{31}y_1 - a_{32}y_2 + y_3 = 0,$$

which in compact matrix form can be written as

$$\begin{bmatrix} 1-a_{11} & -a_{12} & 0 \\ -a_{21} & 1-a_{22} & 0 \\ -a_{31} & -a_{32} & 1 \end{bmatrix} \begin{bmatrix} y_1 \\ y_2 \\ y_3 \end{bmatrix} = \begin{bmatrix} c_1 \\ c_2 \\ 0 \end{bmatrix}.$$

The problem under consideration is to compute y_1, y_2, and y_3, given the consumptions c_1 and c_2.

It is not always that a matrix notation is used to write, in compact form, a large number of simultaneous equations.

Even students whose major area of study is not mathematics, such as the students of *psychology* and *social sciences*, are becoming more and more dependent on mathematical techniques. Psychologists often set out their quantitative observations or conclusions in the form of a table of numbers [69, 92] (see Tables 1.3 and 1.4).

Table 1.3 Test results of Group *A*

	Intelligence	Spatial aptitude	Verbal aptitude
Person 1	120	35	10
Person 2	95	40	15
Person 3	115	25	12
Person 4	100	30	9

A *psychological space* is a representation in which the individuals and/or stimuli are mapped into points, and the relations among the points in the space reflect the responses, the interaction of individuals and the stimuli.

Table 1.4 Weights by which to multiply test results for best prediction of job success

	Mathematical aptitude	Spatial aptitude	Verbal aptitude
Scientist	4	2	2
Engineer	3	4	3
Politician	1	0	5

The set of numbers in both Tables 1.3 and 1.4 can be shown as in Tables 1.5 and 1.6, by surrounding them with square brackets. These are called *matrices.*

Table 1.5 Test results as a matrix *A*

$$A = \begin{bmatrix} 120 & 35 & 10 \\ 95 & 40 & 15 \\ 115 & 25 & 12 \\ 100 & 30 & 9 \end{bmatrix}$$

Table 1.6 Weights as a matrix *B*

$$B = \begin{bmatrix} 4 & 2 & 2 \\ 3 & 4 & 3 \\ 1 & 0 & 5 \end{bmatrix}$$

If the table contains hundreds of such entries, it is economical to carry out any computation among the numbers with the help of the associated matrix. Measurement data matrices, as given above, are transformed by elementary computation to make a comparative study of the results which are important from the point of view of psychology. This results in a variety of matrices such as *dominance matrix, proximity matrix*, and *residual matrix*, in mathematical psychology. The matrix symbol is used here for a compact and efficient notation, as against the shortened notation, to write a large number of simultaneous equations in engineering problems.

Apart from the above applications in operations research, economics and psychology, matrices are now indispensable in all branches of engineering, physical and social sciences, business management, statistics, and modern control systems. As Cayley formulated the general theory of matrices as a pure algebraic system, he is generally acclaimed as its founding father.

This chapter introduces some common types of matrices, their properties and operations.

1.1 DEFINITION OF A MATRIX

Definition 1.1 *Matrix.* A scheme of elements a_{ij} set out in m rows and n columns such as

$$A = \begin{bmatrix} a_{11} & a_{12} & \cdots & a_{1n} \\ a_{21} & a_{22} & \cdots & a_{2n} \\ \vdots & \vdots & & \vdots \\ a_{m1} & a_{m2} & \cdots & a_{mn} \end{bmatrix} \quad \begin{matrix} \text{(column } \downarrow \text{ above } a_{12}\text{)} \\ \leftarrow \text{row} \end{matrix} \tag{1.11}$$

is called a *matrix A of order m by n* $(m \times n)$. □

The first subscript i, ranging from 1 to m, identifies the rows; and the second one j, ranging from 1 to n, identifies the columns. When $m = n$, the matrix is square, and is called a *matrix of order n* or an *n-square matrix.* A matrix is generally denoted by capital letters such as A in (1.11) or alternatively by $[a_{ij}]_{m \times n}$, where the subscript $m \times n$ specifies its order and may be omitted if it is obvious from the context. For example, B, a matrix of order $m \times n$, is also written as $[b_{jk}]$. A square matrix A of order n is denoted by $A = [a_{ij}]_n$.

The elements of a matrix may be real or complex numbers. If real, it is called a *real matrix.* These real or complex numbers constitute what is commonly known as a *field* $\mathbb{F}$. Thus, if it is said that A is a matrix over a field $\mathbb{F}$ of complex numbers $\mathbb{C}$ (respectively real numbers $\mathbb{R}$), then it implies that the elements of A are complex (respectively real) numbers. A matrix in which all elements are equal to zero is referred to as a *zero* or *null matrix* and is denoted by 0. The elements a_{11}, a_{22}, ... in A are called the *main diagonal elements.*

A rectangular matrix consisting of a single column such as

$$\mathbf{x}: = \begin{bmatrix} x_1 \\ x_2 \\ \vdots \\ x_n \end{bmatrix} \quad \text{or} \quad \mathbf{x}: = \text{col}\ [x_1, x_2, ..., x_n] \tag{1.12}$$

is called a *vector* or an *n-vector,* where n denotes the number of elements. Analogously, a matrix having a single row

$$\mathbf{y}: = (y_1, y_2, ..., y_n) \quad \text{or} \quad \mathbf{y}: = [y_1, y_2, ..., y_n] \tag{1.13}$$

is referred to as a *row vector.* These vectors are denoted by small bold letters such as **x, y,** etc. The symbol $(y_1, y_2, ..., y_n)$ is also used to denote the *coordinates of a point* as in coordinate geometry.

1.2 OPERATIONS ON MATRICES

Definition 1.2 *Equality of Matrices.* Two matrices are said to be *equal* if and only if they are of the same order and have all the corresponding elements equal. Thus, matrices $[a_{ij}]_{m \times n}$ and $[b_{ij}]_{m \times n}$ are equal if $a_{ij} = b_{ij}$ for all values of i and j. This generates mn ordinary equations between their elements. □

Definition 1.3 *Addition.* Two matrices A and B are conformable for addition if they are of the same order and the sum $A + B$ on addition is obtained by adding the corresponding elements in both the matrices. Thus, the sum $C = [c_{ij}]_{m\times n}$ of two matrices $A = [a_{ij}]_{m\times n}$ and $B = [b_{ij}]_{m \times n}$ is

$$[c_{ij}]_{m \times n} = [a_{ij} + b_{ij}]_{m \times n}$$

and is written as $C = A + B$. □

Definition 1.4 *Scalar Multiple of a Matrix.* The product of matrix $A = [a_{ij}]_{m\times n}$ by a scalar α (a real or a complex number) is the matrix $C = [c_{ij}]_{m \times n}$ whose elements are α times the corresponding elements in A and is denoted by αA or $A\alpha$. Thus

$$\alpha A = A\alpha = C,$$

i.e.

$$\alpha[a_{ij}]_{m \times n} = [\alpha\, a_{ij}]_{m \times n} = [c_{ij}]_{m \times n}.$$ □

Definition 1.5 *Negative of a Matrix.* The negative of a matrix $A = [a_{ij}]_{m \times n}$ is defined by $-A = [-a_{ij}]_{m \times n}$ and is obtained by changing the sign of each element a_{ij} of A. □

Definition 1.6 *Subtraction.* The difference $A - B$ of two matrices A and B of the same order is defined by

$$A - B = A + (-1)\,B \qquad \square$$

and is called the *subtraction* of B from A.

Example 1.1 Let

$$A = \begin{bmatrix} a & b & c \\ d & e & f \end{bmatrix}, \quad B = \begin{bmatrix} r & s & t \\ u & v & w \end{bmatrix}$$

Then the sum

$$A + B = \begin{bmatrix} a+r & b+s & c+t \\ d+u & e+v & f+w \end{bmatrix}.$$

If $\alpha = -1$, the scalar product αB is

$$(-1)B = \begin{bmatrix} -r & -s & -t \\ -u & -v & -w \end{bmatrix}$$

and the difference

$$A - B = A + (-1)B = \begin{bmatrix} a-r & b-s & c-t \\ d-u & e-v & f-w \end{bmatrix}$$

If $A = B$, then

$$a = r, \quad b = s, \quad c = t, \quad d = u, \quad e = v, \quad f = w.$$

The operations of addition and scalar multiplication possess the properties discussed in Section 1.2.1, where A, B, C are arbitrary matrices of the same order, and α, β are scalars.

1.2.1 Properties of Matrix Addition and Scalar Multiplication

Property I $\quad A + B = B + A,$ (commutativity)
Property II $\quad (A + B) + C = A + (B + C),$ (associativity)
Property III $\quad \alpha(A + B) = \alpha A + \alpha B,$
Property IV $\quad (\alpha + \beta)A = \alpha A + \beta A,$
Property V $\quad (\alpha\beta)A = \alpha(\beta A).$

Thus the matrix addition is *commutative* and *associative*; and the scalar multiplication of a matrix is *distributive* with respect to matrix addition (Property III).

These properties are not difficult to visualize and it is, therefore, left to the reader to prove them (see Problem 1.4).

We have seen that two successive transformations of variables represented by (1.6) and (1.7) have a compact form given by A and B in (1.9). Under these transformations, the old variables x_1 and x_2 are given in terms of new ones z_1 and z_2 as represented by (1.8) whose compact form appears as AB in (1.10) and represents the matrix product AB of A and B in this order. This development suggests the following definition:

Definition 1.7 *Multiplication of Matrices.* The multiplication of a row vector $\mathbf{y}$ into a column vector $\mathbf{x}$, having the same number of elements as given by (1.13) and (1.12) respectively, is defined to be the 1×1 matrix

$$\mathbf{yx} = [y_1, y_2, \ldots, y_n] \begin{bmatrix} x_1 \\ x_2 \\ \vdots \\ x_n \end{bmatrix} = [y_1x_1 + y_2x_2 + \cdots + y_nx_n]_{1\times 1}.$$

This operation also contains the rule for the multiplication of two matrices of any order.

Two matrices A and B defined by

$$A = \begin{bmatrix} a_{11} & a_{12} & \cdots & a_{1j} & \cdots & a_{1n} \\ a_{21} & a_{22} & \cdots & a_{2j} & \cdots & a_{2n} \\ \vdots & \vdots & & \vdots & & \vdots \\ a_{i1} & a_{i2} & \cdots & a_{ij} & \cdots & a_{in} \\ \vdots & \vdots & & \vdots & & \vdots \\ a_{m1} & a_{m2} & \cdots & a_{mj} & \cdots & a_{mn} \end{bmatrix} \leftarrow i\text{th row},$$

$$B = \begin{bmatrix} b_{11} & b_{12} & \cdots & b_{1j} & \cdots & b_{1p} \\ b_{21} & b_{22} & \cdots & b_{2j} & \cdots & b_{2p} \\ \vdots & \vdots & & \vdots & & \vdots \\ b_{i1} & b_{i2} & \cdots & b_{ij} & \cdots & b_{ip} \\ \vdots & \vdots & & \vdots & & \vdots \\ b_{n1} & b_{n2} & \cdots & b_{nj} & \cdots & b_{np} \end{bmatrix} \quad (\downarrow j\text{th column})$$

are said to be *conformable* for multiplication in the order AB if the number of columns of the first matrix A is equal to the number of rows of the second matrix B. The product AB is the matrix C defined by

$$C = \begin{bmatrix} c_{11} & c_{12} & \cdots & \cdot & \cdots & c_{1p} \\ c_{21} & c_{22} & \cdots & \cdot & \cdots & c_{2p} \\ \vdots & \vdots & & \vdots & & \vdots \\ \cdot & \cdot & & c_{ij} & & \cdot \\ \vdots & \vdots & & \vdots & & \vdots \\ c_{m1} & c_{m2} & \cdots & \cdot & \cdots & c_{mp} \end{bmatrix} \leftarrow i\text{th row} \quad (j\text{th column} \downarrow)$$

in which the element c_{ij} in the ith row and jth column is obtained by multiplying the elements in the ith row of A into the elements of the jth column of B, and summing the products so obtained. Therefore,

$$c_{ij} = \sum_{k=1}^{n} a_{ik} b_{kj}.$$

The number of rows and the number of columns of C are equal to the number of rows of A and the number of columns of B respectively. Thus,

$$\underset{m \times n}{A} \quad \underset{n \times p}{B} = \underset{m \times p}{C}$$

Example 1.2

$$\underset{2\times 3}{\begin{bmatrix} r & s & t \\ u & v & w \end{bmatrix}} \underset{3\times 2}{\begin{bmatrix} a & d \\ b & e \\ c & f \end{bmatrix}} = \underset{2\times 2}{\begin{bmatrix} (ra + sb + tc) & (rd + se + tf) \\ (ua + vb + wc) & (ud + ve + wf) \end{bmatrix}}.$$

The direction in which the product of corresponding elements is to be obtained is shown by arrows.

1.2.2 Properties of Matrix Multiplication

In defining the properties of matrix multiplication below, the matrices A, B and C are assumed to be of compatible dimensions (i.e., orders) for the operations in which they appear.

Property I Matrix multiplication is not commutative in general. That is, $AB \neq BA$ in general, even if the matrices are conformable for multiplication in a reverse order. If A is of order $n \times m$ and B of $m \times n$, the order n of the product AB is different from that m of BA. This property is true even if A and B are of the same order.

Example 1.3 Let

$$A = \begin{bmatrix} 1 & 0 & 3 \\ 2 & 3 & 1 \end{bmatrix}, \quad B = \begin{bmatrix} 3 & 0 \\ 2 & 1 \\ 0 & 3 \end{bmatrix}.$$

Then we have

$$AB = \begin{bmatrix} 3 & 9 \\ 12 & 6 \end{bmatrix},$$

which is not equal to

$$BA = \begin{bmatrix} 3 & 0 & 9 \\ 4 & 3 & 7 \\ 6 & 9 & 3 \end{bmatrix}.$$

Again, let

$$A = \begin{bmatrix} 1 & 4 \\ 3 & 2 \end{bmatrix}, \quad B = \begin{bmatrix} 4 & 3 \\ 1 & 2 \end{bmatrix}.$$

Then we have, by matrix multiplication,

$$AB = \begin{bmatrix} 8 & 11 \\ 14 & 13 \end{bmatrix}, \quad BA = \begin{bmatrix} 13 & 22 \\ 7 & 8 \end{bmatrix}$$

which are not equal. The above property does not mean that in no case $AB = BA$. For the matrices given by

$$A = \begin{bmatrix} 2 & 1 \\ -1 & 2 \end{bmatrix}, \quad B = \begin{bmatrix} 4 & 1 \\ -1 & 4 \end{bmatrix},$$

we have,

$$AB = \begin{bmatrix} 7 & 6 \\ -6 & 7 \end{bmatrix} = BA.$$

In the product AB, A is said to *premultiply* B or B is said to *postmultiply* A; and if $AB = BA$, then A and B are said to be *commutative* or to *commute*. A and B are said to be *anticommutative* or to *anticommute* if $AB = -BA$ (see Problem 1.11).

Property II From $AB = 0$, *it does not follow that either* $A = 0$ or $B = 0$. Here 0's are null matrices of appropriate order.

Example 1.4 For the matrices A and B given by

$$A = \begin{bmatrix} 1 & 1 & 2 \\ 2 & 2 & 4 \\ 1 & 1 & 2 \end{bmatrix}, \quad B = \begin{bmatrix} -1 & 1 & 1 \\ 1 & -1 & -7 \\ 0 & 0 & 3 \end{bmatrix},$$

we have,

$$AB = \begin{bmatrix} 0 & 0 & 0 \\ 0 & 0 & 0 \\ 0 & 0 & 0 \end{bmatrix},$$

a null matrix, even though A or B is not a null matrix.

Property III *The relation* $AB = AC$ *or* $BA = CA$ *does not imply that* $B = C$. That is, the cancellation law does not hold in general as in scalar algebra.

Example 1.5 For the matrices

$$A = \begin{bmatrix} 1 & 2 & 3 \\ 1 & 1 & 2 \\ -1 & 4 & 3 \end{bmatrix}, \quad B = \begin{bmatrix} 1 & 2 & 3 \\ 1 & 1 & -1 \\ 2 & 2 & 2 \end{bmatrix}, \quad C = \begin{bmatrix} 2 & 3 & 4 \\ 2 & 2 & 0 \\ 1 & 1 & 1 \end{bmatrix},$$

we have, by direct multiplication,

$$AB = \begin{bmatrix} 9 & 10 & 7 \\ 6 & 7 & 6 \\ 9 & 8 & -1 \end{bmatrix} = AC, \text{ although } B \neq C.$$

Property IV *The matrix multiplication is associative.* That is,

$$(AB)C = A(BC).$$

Property V *The multiplication of matrices is distributive with respect to addition.* That is,

$$A(B + C) = AB + AC, \qquad (B + C)A = BA + CA.$$

For verification of Properties IV and V, see Problem 1.8.

Definition 1.8 *Transpose of a Matrix.* The transpose of an $m \times n$ matrix $A = [a_{ij}]_{m \times n}$ is an $n \times m$ matrix

$$A^T = \left[a_{ij}^T\right]_{n \times m} = \left[a_{ji}\right]_{n \times m},$$

where $a_{ij}^T = a_{ji}$ and is obtained by interchanging the rows and columns of A. Alternatively, a prime ($'$) is used to denote the transpose. Thus, A' represents A^T.

Example 1.6 The transpose of the matrix

$$A = \begin{bmatrix} a_{11} & a_{12} \\ a_{21} & a_{22} \\ a_{31} & a_{32} \end{bmatrix}$$

is

$$A^T = \begin{bmatrix} a_{11}^T & a_{12}^T & a_{13}^T \\ a_{21}^T & a_{22}^T & a_{23}^T \end{bmatrix} = \begin{bmatrix} a_{11} & a_{21} & a_{31} \\ a_{12} & a_{22} & a_{32} \end{bmatrix}.$$

1.2.3 Properties of Matrix Transposition

If A and B are two matrices of compatible dimensions for the operations in which they appear and α is a scalar, then

Property I $(A^T)^T = A,$

Property II $(A + B)^T = A^T + B^T,$

Property III $(\alpha A)^T = \alpha A^T,$

Property IV $(AB)^T = B^T A^T.$

To prove Property IV, we note that if

$$A = [a_{ik}]_{m \times n}, \qquad B = [b_{kj}]_{n \times p}$$

then

$$B^T A^T = \left[b_{ik}^T\right]_{p\times n} \left[a_{kj}^T\right]_{n\times m} = \left[\sum_{k=1}^{n} b_{ik}^T\, a_{kj}^T\right]_{p\times m} = \left[\sum_{k=1}^{n} a_{jk}\, b_{ki}\right]_{p\times m}, \tag{1.14}$$

the last step follows from the definition of a transpose. Also,

$$AB = \left[a_{ik}\right]_{m\times n} \left[b_{kj}\right]_{n\times p} = \left[\sum_{k=1}^{n} a_{ik}\, b_{kj}\right]_{m\times p},$$

which on being transposed (i.e., on interchanging the subscripts i and j) gives

$$(AB)^T = \left[\sum_{k=1}^{n} a_{jk}\, b_{ki}\right]_{p\times m}. \tag{1.15}$$

Property IV is verified by comparing (1.14) and (1.15).

The remaining properties can be proved similarly (see Problem 1.12).

Example 1.7 Let

$$A = \begin{bmatrix} 1 & 2 & 0 \\ 0 & 3 & 1 \end{bmatrix}, \quad B = \begin{bmatrix} 1 & 2 & 0 \\ 2 & 3 & 2 \\ 0 & 2 & 1 \end{bmatrix}.$$

Then, by direct computation,

$$B^T A^T = \begin{bmatrix} 5 & 6 \\ 8 & 11 \\ 4 & 7 \end{bmatrix} = (AB)^T.$$

1.3 SYMMETRIC, HERMITIAN AND TRIANGULAR MATRICES

Definition 1.9 *Symmetric and Skew-symmetric Matrices. A square matrix* $A = [a_{ij}]_n$ is said to be *symmetric* if $A = A^T$, i.e., $a_{ij} = a_{ji}$, and *skew-symmetric* if $A = -A^T$, i.e. $a_{ij} = -a_{ji}$, where i and j vary from 1 to n. The matrices

$$\begin{bmatrix} a & h & g \\ h & b & f \\ g & f & c \end{bmatrix}, \quad \begin{bmatrix} 0 & h & g \\ -h & 0 & f \\ -g & -f & 0 \end{bmatrix}$$

are respectively symmetric and skew-symmetric. It can be easily shown that all the diagonal elements of a skew-symmetric matrix must be zero (see Problem 1.16). In a symmetric matrix, all the elements placed symmetrically about the main diagonal are equal, and in a skew-symmetric one, they differ by a multiple of –1.

Definition 1.10 *Hermitian and Skew-Hermitian Matrices.* If the elements of a matrix A are complex, then by replacing these elements with their conjugates, a matrix known as the *conju-*

gate of A is obtained, and is denoted by $\bar{A}$. The transpose of $\bar{A}$ denoted by $(\bar{A})^T$ or A^* is referred to as the *transposed conjugate or tranjugate of A*.

Example 1.8 Let

$$A = \begin{bmatrix} 1 & 1-i4 \\ 1+i4 & 2 \end{bmatrix}.$$

Then,

$$\bar{A} = \begin{bmatrix} 1 & 1+i4 \\ 1-i4 & 2 \end{bmatrix}, \quad A^* = (\bar{A})^T = \begin{bmatrix} 1 & 1-i4 \\ 1+i4 & 2 \end{bmatrix}.$$

A square matrix A is called *Hermitian* if $A = A^*$, i.e., $a_{ij} = \bar{a}_{ji}$; and *skew-Hermitian* if $A = -A^*$, i.e., $a_{ij} = -\bar{a}_{ji}$. The matrix A in Example 1.8 is a Hermitian matrix, and the matrix

$$A = \begin{bmatrix} 3i & 2+i \\ -2+i & i \end{bmatrix}$$

is an example of a skew-Hermitian matrix. It can be easily verified that the diagonal elements of a Hermitian matrix are real and those in a skew-Hermitian matrix are imaginary (see Problem 1.16). The symmetric matrix is a particular case of a Hermitian matrix if all the elements are real.

Definition 1.11 *Diagonal, Scalar and Unit Matrices.* In the square matrix A given by

$$A = \begin{bmatrix} a_{11} & a_{12} & \cdots & a_{1n-1} & a_{1n} \\ a_{21} & a_{22} & \cdots & a_{2n-1} & a_{2n} \\ \vdots & \vdots & & \vdots & \vdots \\ a_{n1} & a_{n2} & \cdots & a_{nn-1} & a_{nn} \end{bmatrix}, \tag{1.16}$$

the elements $a_{11}, a_{22}, ..., a_{nn}$ ($a_{1n}, a_{2n-1}, ..., a_{n1}$) are said to constitute the main diagonal (the secondary diagonal or antidiagonal) of A and the elements a_{ij} for which $j = i + k$ ($j = i - k$) are said to constitute the kth superdiagonal (subdiagonal) of A. For example, $a_{12}, a_{23}, ..., a_{n-1n}$ constitute the first superdiagonal and $a_{21}, a_{32}, ..., a_{nn-1}$ constitute the first subdiagonal of the square matrix A. These definitions of the main diagonal, superdiagonal and subdiagonal are also applicable when A is not necessarily square. A square matrix A, in which all the elements except those along the main diagonal are zero, is called a *diagonal matrix*. The matrix D given by

$$D = \begin{bmatrix} d_1 & 0 & \cdots & 0 \\ 0 & d_2 & \cdots & 0 \\ \vdots & \vdots & & \vdots \\ 0 & 0 & \cdots & d_n \end{bmatrix} \tag{1.17}$$

is a diagonal matrix and is often represented by diag $[d_1, d_2, ..., d_n]$ or diag $(d_1, d_2, ..., d_n)$ or diag $[d_i]$. If in (1.17), all the diagonal elements are equal, it is referred to as a *scalar matrix* S_n, and if these elements are equal to 1, we have a *unit matrix* or an *identity matrix* I_n of order n. Thus,

$$S_n = \begin{bmatrix} \alpha & 0 & \cdots & 0 \\ 0 & \alpha & \cdots & 0 \\ \vdots & \vdots & & \vdots \\ 0 & 0 & \cdots & \alpha \end{bmatrix}_n, \qquad I_n = \begin{bmatrix} 1 & 0 & \cdots & 0 \\ 0 & 1 & \cdots & 0 \\ \vdots & \vdots & & \vdots \\ 0 & 0 & \cdots & 1 \end{bmatrix}_n$$

are respectively a scalar and a unit matrix. The parts played by S_n and I_n are respectively similar to those of α and 1 in scalar algebra. For example, if A is a matrix of order $m \times n$, then

$$\alpha A = S_m A = A S_n, \qquad A = I_m A = A I_n$$

which are similar to the operations: $\alpha a = a\alpha$ and $a = 1 \cdot a = a \cdot 1$ in scalar algebra. The subscript n is omitted, when from the context of the problem, the order of S or I becomes evident.

Definition 1.12 *Upper and Lower Triangular Matrices.* An upper (or a lower) triangular matrix has all elements below (or above) the main diagonal equal to zero and is denoted by U (or L). Symbolically,

$$U = \left[a_{ij}\right], \quad a_{ij} = 0 \quad \text{if } i > j \quad \text{or} \quad L = \left[a_{ij}\right], \quad a_{ij} = 0 \quad \text{if } i < j.$$

In other words,

$$U = \begin{bmatrix} a_{11} & a_{12} & \cdots & a_{1n} \\ 0 & a_{22} & \cdots & a_{2n} \\ \vdots & \vdots & & \vdots \\ 0 & 0 & \cdots & a_{nn} \end{bmatrix}, \qquad L = \begin{bmatrix} a_{11} & 0 & \cdots & 0 \\ a_{21} & a_{22} & \cdots & 0 \\ \vdots & \vdots & & \vdots \\ a_{n1} & a_{n2} & \cdots & a_{nn} \end{bmatrix}$$

for square matrices. Similar structural representation can be given for rectangular matrices (see Problem 1.27). If $a_{11} = \cdots = a_{nn} = 0$, it is called a *strictly upper* (or *lower*) *triangular* matrix.

1.4 POWERS AND TRACE OF A SQUARE MATRIX

Definition 1.13 *Power of a Square Matrix.* We define zero power and pth power of a square matrix A by

$$A^0 = I, \qquad A^p = A \cdot A \cdot \cdots \cdot A \qquad (p = 1, 2, \ldots) \qquad p \text{ times}.$$

Example 1.9 If

$$A = \begin{bmatrix} i & 1 \\ 1 & -i \end{bmatrix},$$

then

$$A^2 = A \cdot A = \begin{bmatrix} i & 1 \\ 1 & -i \end{bmatrix} \begin{bmatrix} i & 1 \\ 1 & -i \end{bmatrix} = \begin{bmatrix} 0 & 0 \\ 0 & 0 \end{bmatrix}.$$

Definition 1.14 *Trace of a Square Matrix.* The sum of the main diagonal elements of a square matrix A is called the *trace* of A and is denoted by tr A. In view of (1.16), tr A is written explicitly as

$$\text{tr}\, A = a_{11} + a_{22} + \cdots + a_{nn}.$$

If A and B are of the same order, then it follows that

$$\text{tr}\,(A + B) = \text{tr}\, A + \text{tr}\, B.$$

Let A be of order $m \times n$ and B of order $n \times m$. Then

$$\text{tr}\, AB = \text{tr}\, BA.$$

The proof follows by observing that the iith diagonal element c_{ii} of AB is

$$c_{ii} = \sum_{j=1}^{n} a_{ij}\, b_{ji},$$

$$\text{tr}\, AB = \sum_{i=1}^{m} c_{ii} = \sum_{i=1}^{m} \sum_{j=1}^{n} a_{ij}\, b_{ji} = \sum_{j=1}^{n} \left(\sum_{i=1}^{m} b_{ji} a_{ij} \right) = \sum_{j=1}^{n} d_{jj} = \text{tr}\, BA,$$

where d_{jj} is the jjth diagonal element of BA.

1.5 DIFFERENTIATION AND INTEGRATION OF A MATRIX

The elements of a matrix A may be functions of a variable, say, t. This functional dependence of A on t is shown by writing A and its elements as $A(t)$ and $a_{ij}(t)$. If from the context the dependence is obvious, the usual notations A and a_{ij} are retained.

A variation in t by the amount δt would cause variations in the elements of A to the tune of δa_{ij}, the variation in the matrix being denoted by δA. Thus,

$$\delta A = \left[\delta a_{ij} \right].$$

The derivative of A with respect to t denoted by $\dot{A}$ or dA/dt is now defined by

$$\dot{A} = \frac{dA}{dt} = \lim_{\delta t \to 0} \frac{\delta A}{\delta t} = \left[\lim_{\delta t \to 0} \frac{\delta a_{ij}}{\delta t} \right] = \left[\dot{a}_{ij} \right].$$

Hence, the elements in $\dot{A}$ are derivatives of the elements a_{ij} in A.

The operator D is sometimes used to denote a derivative operator. A matrix A, raised to the power k within parenthesis, designates its kth order derivative. So,

$$D^k A = A^{(k)}.$$

The following are some results arising from matrix multiplication, assuming that matrices A, A_i's and B have properties for the indicated operations:

Result I $\quad \dfrac{d}{dt}(A+B)=\dfrac{dA}{dt}+\dfrac{dB}{dt},$

Result II $\quad \dfrac{d}{dt}(AB)=\dfrac{dA}{dt}B+A\dfrac{dB}{dt},$

Result III $\quad \dfrac{d}{dt}A^3=\dot{A}A^2+A\dot{A}A+A^2\dot{A},$

Result IV $\quad \dfrac{d}{dt}(A_1A_2\cdots A_n)=\dot{A}_1A_2\cdots A_n+A_1\dot{A}_2\cdots A_n+\cdots+A_1A_2\cdots\dot{A}_n.$

The integration with respect to t between the limits t_0 and t of a matrix $A(t) = [a_{ij}(t)]$ is defined as one having the elements

$$\int_{t_0}^{t} a_{ij}(t)\,dt,$$

assuming the elements in $A(t)$ to be integrable.

We thus see that, operations on matrices depend on the operations of the elements constituting them. These elements are very often scalars, i.e., real and complex numbers, functions, polynomials and sometimes matrices. We have used mostly scalar elements; and noted that although these scalar elements are always commutative in this multiplication, and the product of two nonzero scalars never results in a zero value, the corresponding operations with matrices do not exactly follow these rules.

1.6 FIELD AND MATRIX OVER AN ARBITRARY FIELD

We have defined a matrix over a field $\mathbb{F}$ of real and complex numbers. In fact, the elements of a matrix may be from any arbitrary field whose formal definition at this stage is, therefore, very pertinent. A nonempty collection $\mathbb{F}$ of real or complex numbers, which does not consist of the number zero alone, is called a *field* if addition, subtraction, multiplication and division of any two elements of $\mathbb{F}$ again belong to $\mathbb{F}$; division by zero is excluded. More precisely, we have the following definition.

Definition 1.15 *Field.* Let $\mathbb{F}$ be a nonempty collection or a set of elements in which two compositions called *addition* and *multiplication* and denoted by + and · respectively are defined. Also, for any two elements α and β in $\mathbb{F}$, $\alpha+\beta$ and $\alpha\cdot\beta$ are uniquely defined elements in $\mathbb{F}$, i.e., $\mathbb{F}$ is closed under the above two compositions. Such a set $\mathbb{F}$ is called a *field* if the following axioms are satisfied whenever α, β and γ are any elements in $\mathbb{F}$.

(i) Addition and multiplication are commutative, i.e., $\alpha+\beta=\beta+\alpha$ and $\alpha\cdot\beta=\beta\cdot\alpha$.

(ii) Addition and multiplication are associative, i.e., $(\alpha+\beta)+\gamma=\alpha+(\beta+\gamma)$ and $(\alpha\cdot\beta)\cdot\gamma = \alpha\cdot(\beta\cdot\gamma)$.

(iii) $\mathbb{F}$ contains a zero element 0 and a unit element $1 \neq 0$ such that $\alpha+0=\alpha$ and $\alpha\cdot 1 = \alpha$. The elements 0 and 1 are called *identity elements* for + and · respectively.

(iv) $\mathbb{F}$ contains inverses for addition and multiplication, i.e., corresponding to α and γ in $\mathbb{F}$ there exist β and δ in $\mathbb{F}$ such that $\alpha + \beta = 0$ and $\gamma \cdot \delta = 1$. β is called an *additive inverse* of α, and δ a *multiplicative inverse* of γ.

(v) Multiplication is distributive with respect to addition, i.e., $\alpha \cdot (\beta + \gamma) = \alpha \cdot \beta + \alpha \cdot \gamma$.

Example 1.10 (a) The set of all rational numbers of the form α/β, where α and β are integers and $\beta \neq 0$ with the usual definition of addition and multiplication, is a field.

(b) The set of all real numbers with usual definition of addition and multiplication is a field, and is denoted by $\mathbb{R}$.

(c) The set of all complex numbers of the form $\alpha + i\beta$, where α and β are real and $i^2 = -1$, is a field and is denoted by $\mathbb{C}$.

(d) The set of all numbers of the form $\alpha + \beta\sqrt{2}$, where α and β are rational numbers, is a field with addition and multiplication as in $\mathbb{R}$.

(e) The set of all scalar matrices of order n over a field $\mathbb{F}$ defined in (a) to (d) is itself a field.

(f) The set of matrices

$$\begin{bmatrix} \alpha & -\beta \\ \beta & \alpha \end{bmatrix},$$

where α and β are real numbers, is a field.

(g) The number of elements in a field is called the *order* of the field. If the order is infinite, the field is called *infinite*, of which (a) to (f) are examples; and if the order is finite, the field is called *finite*. To provide an example of a finite field, let p be a prime number and let us denote by $\mathbb{Z}_p$ the set

$$\mathbb{Z}_p = \{0, 1, 2, ..., p - 1\}.$$

In this set the compositions + and $\cdot$, called *addition modulo* (p) and *multiplication modulo* (p), are defined as follows. If α and β are two members of $\mathbb{Z}_p$, then

$$\alpha + \beta = \rho, \qquad \alpha \cdot \beta = \delta,$$

where ρ is the least positive remainder $0 \le \rho < p$, obtained by dividing the ordinary sum of α and β by p, and δ is the least positive remainder $0 \le \delta < p$ obtained by dividing the ordinary product of α and β by p. The reader can verify that $\mathbb{Z}_p$ is a field with 0 and 1 as identity elements, $p-\alpha$ as additive inverse of any element α. The multiplicative inverse cannot be given explicitly. If, for instance, $p = 5$, addition and multiplication modulo 5 among the elements of $\mathbb{Z}_5$ can be presented in a tabular form as follows:

Addition modulo 5 in $\mathbb{Z}_5$

	0	1	2	3	4
0	0	1	2	3	4
1	1	2	3	4	0
2	2	3	4	0	1
3	3	4	0	1	2
4	4	0	1	2	3

Multiplication modulo 5 in $\mathbb{Z}_5$

	0	1	2	3	4
0	0	0	0	0	0
1	0	1	2	3	4
2	0	2	4	1	3
3	0	3	1	4	2
4	0	4	3	2	1

The multiplicative inverses of 1, 2, 3, 4 in $\mathbb{Z}_5$ are respectively 1, 3, 2, 4.

The field $\mathbb{Z}_2$ finds widest application in Computer Science, and matrices over $\mathbb{Z}_2$ and other finite fields lie in the core of Algebraic Coding Theory.

(h) The set of positive integers, the set of integers, and the set of imaginary numbers of the form $i\beta$, β real are not fields. This is true also of $\mathbb{Z}_m$ when m is not a prime number. For example, in these sets, Axiom (iv) is not satisfied.

Example 1.11 Obtain the sum and product of the following two matrices over the finite field $\mathbb{Z}_2 = \{0, 1\}$:

$$A = \begin{bmatrix} 1 & 0 \\ 1 & 1 \end{bmatrix}, \quad B = \begin{bmatrix} 1 & 1 \\ 0 & 1 \end{bmatrix}.$$

Solution : We recall that in $\mathbb{Z}_2$, $0 + 1 = 1 + 0 = 1$, $0 + 0 = 1 + 1 = 0$. Therefore,

$$A + B = \begin{bmatrix} 0 & 1 \\ 1 & 0 \end{bmatrix}, \quad AB = \begin{bmatrix} 1 & 1 \\ 1 & 0 \end{bmatrix}.$$

1.7 MATRIX OPERATIONS WITH MATLAB

If MATLAB software is run on a computer, one gets the prompt >> in the command window. To insert small matrices, one has to put brackets around the data, with the list of elements being separated by blanks or commas, and the rows separated with semicolons. With the MATLAB command.

```
>> A = [1 2 3; 4 5 6; 7 8 19]
```

MATLAB will respond with

```
A =
    1   2   3
    4   5   6
    7   8   19
```

Similarly, for a row vector

```
>> v = [3 2 4 1]
```

MATLAB responds with

```
v =
    3 2 4 1
```

and for a column vector

```
>> u = [2; 3; 4; 1]
```

MATLAB responds with

```
u =
    2
    3
    4
    1
```

Once a matrix is entered, it is automatically remembered in the MATLAB workspace. A matrix is known as a *magic square* if the sums of any row, column and diagonal elements are equal. The general command to invoke this matrix is magic (n), ($n > 0$, $n \neq 2$), which is an $n \times n$ matrix constructed from the integers 1 through n^2, with equal column, row and diagonal sums. Thus,

```
>> B = magic (3)
```

has the MATLAB response

```
B =
    8  1  6
    3  5  7
    4  9  2
```

Similarly, eye (n) generates an identity matrix of order n, ones (n) — an $n \times n$ matrix with all elements equal to 1, and zeros (n) — an $n \times n$ matrix with all elements equal to 0. However, ones (m, n) and zeros (m, n) generate the above matrices of order $m \times n$.

The : operator is used to create regularly spaced vectors:

```
>> x = [-1.5 : 0.7 : 1]
   x =
      -1.5000    -0.8000    -0.1000    0.6000
>> y = [2.4 : -0.9 : -0.5]
   y =
      2.4000    1.5000    0.6000    -0.3000
```

In the MATLAB response, successive elements in x are incremented by 0.7, so that the last element is less than 1, while in y the successive elements are decremented by 0.9, so that the last element is greater than –0.5. The other uses of : operator are ($k > i, j$)

A (:, j) is the jth column of A

A (i, :) is the ith row of A

A (:, j : k) is A (:, j), A (:, j + 1), ..., A (:, k)

A (i : k, :) is A (i, :), A (i + 1, :), ..., A (k, :)

Let u = [3 2 4 7] and v = [3 8 5 6] be two vectors. Then

```
>> T = toeplitz (u, v)
```

```
T =
    3  8  5  6
    2  3  8  5
    4  2  3  8
    7  4  2  3
```

is a *Toeplitz matrix* whose super-, sub- and main-diagonal elements are equal.

Let

$$u = [3\ 2\ 4\ 7] \quad \text{and} \quad v = [7\ 8\ 5\ 6].$$

Then

```
>> H = hankel (u, v)
H =
    3  2  4  7
    2  4  7  8
    4  7  8  5
    7  8  5  6
```

which is a finite *Hankel matrix* $H = \left[s_{i+j-2}\right]_{i,\,j=1}^{2n-2}$ with $n = 4$. For (2, 3) element, put $i = 2$ and $j = 3$.

A *Pascal matrix* is generated with $a_{i1} = a_{1j} = 1$ and $a_{ij} = a_{i,\,j-1} + a_{i-1,\,j}$. The statement

```
>> P = pascal (4)
```

is responded by MATLAB with

```
P =
    1  1   1   1
    1  2   3   4
    1  3   6  10
    1  4  10  20
```

The statements *A* + *B*, *A* – *B*, *A***B*, and *A*^*k*, where *k* is a positive integer, after MATLAB prompt >>, produce the addition, subtraction, multiplication of two matrices and matrix product of *A* with itself *k* times; and *A*′ the transpose of *A*. Thus

```
>> A + B

ans =   9   3   9
        7  10  13
       11  17  21

>> A – B

ans =
       –7   1  –3
        1   0  –1
        3  –1  17
```

```
>> A * B
ans =
        26    38    26
        71    83    71
       156   218   136
>> A ^ 2
ans =
        30    36    72
        66    81   156
       172   206   430
>> A'
ans =
        1    4    7
        2    5    8
        3    6   19
```

MATLAB usually shows accuracy upto 4 decimal places:

```
>>1/2560
ans =
       3.9063 e - 004
```

However, if one uses the commands

```
>> format long
>> 1/2560
```

The response is

```
ans =
       3.906250000000000 e - 004
```

which gives an accuracy upto 15 decimal places. With the format short command one can go back to 4 decimal places of accuracy. MATLAB can also display values in rational form. The commands

```
>> format rat
>> 3.906250000000000 e - 004
```

respond with

```
ans =
       1/2560
```

More details are available in the on-line help manual of MATLAB.

PROBLEMS

1.1 For the matrices

$$A=\begin{bmatrix}1 & 0 & 0\\ 0 & 3 & -1\\ 0 & -1 & 3\end{bmatrix},\qquad P=\begin{bmatrix}1 & 0 & 0\\ 0 & 1 & 1\\ 0 & 1 & -1\end{bmatrix},\qquad Q=\frac{1}{2}\begin{bmatrix}2 & 0 & 0\\ 0 & 1 & 1\\ 0 & 1 & -1\end{bmatrix},$$

perform the following operations:
(a) PA, (b) AQ, (c) PQ, (d) $(PA)Q$, (e) $P(AQ)$,
(f) $Q^TA^TP^T$, (g) tr A, (h) tr PAQ, and (i) tr PQA.

1.2 Given $A = \begin{bmatrix} 1 & 0 & 2 \\ 3 & 1 & 2 \end{bmatrix}$, $B = \begin{bmatrix} 1 & 0 \\ 2 & 3 \\ 1 & 0 \end{bmatrix}$,

compute
(a) AB, (b) BA, (c) tr AB, and (d) tr BA.

1.3 Compute the following products:

(a) $\begin{bmatrix} 1-\lambda & 1-\lambda+\lambda^2 \\ -1 & 1+\lambda \\ 1+\lambda & \lambda \end{bmatrix} \begin{bmatrix} 1+\lambda+\lambda^2 & -1 \\ 1+\lambda & 1 \end{bmatrix}$,

(b) $\begin{bmatrix} 2+i & i & 1 \\ i & 1 & 1-i \end{bmatrix} \begin{bmatrix} 1 & 1+3i \\ i & 3 \\ -i & 1 \end{bmatrix}$.

1.4 Prove the properties of matrix addition and scalar multiplication I-V in Section 1.2.1.

1.5 Determine x and y, if

(a) $\begin{bmatrix} x & 1 \\ 2 & y \end{bmatrix} \begin{bmatrix} 1 \\ 2 \end{bmatrix} = \begin{bmatrix} -1 \\ 2 \end{bmatrix}$, (b) $\begin{bmatrix} a \\ b \end{bmatrix} \begin{bmatrix} a & b \end{bmatrix} = \begin{bmatrix} 4 & x \\ y & 1 \end{bmatrix}$.

1.6 For a lower triangular matrix L and an upper triangular matrix U of order three with unit diagonal elements, determine P such that
(a) $LP = PL = I$, (b) $UP = PU = I$.

1.7 Let $\mathbf{x} = [x_1\ x_2\ \cdots\ x_n]$ and $\mathbf{y} = [y_1\ y_2\ \cdots\ y_n]$ be two row vectors. Obtain the products $\mathbf{x}\mathbf{y}^T$ and $\mathbf{x}^T\mathbf{y}$.

1.8 If A, B and C are three matrices of compatible dimensions, then prove the associative and distributive laws of multiplication: $(AB)C = A(BC)$ and $A(B + C) = AB + AC$.

1.9 Explain why the relations

$$(A + B)^2 = A^2 + 2AB + B^2, \qquad (A + B)(A - B) = A^2 - B^2,$$

do not hold, in general, in matrix algebra. When does the equality hold? If A and B commute, show that

$$(A + B)^n = \sum_{i=0}^{n} {}^nC_i A^{n-1} B^i.$$

1.10 Show that the following matrices are commutative:

(a) $A=\begin{bmatrix} x & y \\ -y & x \end{bmatrix}$, $B=\begin{bmatrix} u & v \\ -v & u \end{bmatrix}$;

(b) $A_i=\begin{bmatrix} \cos\theta_i & -\sin\theta_i \\ \sin\theta_i & \cos\theta_i \end{bmatrix}$, $i=1,2.$

1.11 Show that the following matrices are anticommutative with each other:

$$\sigma_x=\begin{bmatrix} 0 & 1 \\ 1 & 0 \end{bmatrix},\quad \sigma_y=\begin{bmatrix} 0 & -i \\ i & 0 \end{bmatrix},\quad \sigma_z=\begin{bmatrix} 1 & 0 \\ 0 & -1 \end{bmatrix},$$

where $i^2 = -1$. These matrices are known as *Pauli spin matrices* in quantum mechanics.

1.12 Prove properties I-III of matrix transposition in Section 1.2.3.

1.13 Show by direct multiplication that $(AB)^T = B^T A^T$, where

(a) $A=\begin{bmatrix} 1 & 3 \\ 0 & 2 \\ 2 & 1 \end{bmatrix}$, $B=\begin{bmatrix} 2 & 1 & 3 \\ 0 & 3 & 0 \end{bmatrix}$;

(b) $A^T=\begin{bmatrix} 2 \\ 0 \\ 1 \end{bmatrix}$, $B=\begin{bmatrix} 1 & 0 & 1 \\ 0 & 2 & 0 \\ 3 & 0 & 3 \end{bmatrix}$.

1.14 Generalize Properties II and IV of matrix transposition in Section 1.2.3.

1.15 Show that any square matrix can be represented uniquely as a sum of a symmetric and a skew-symmetric matrix.

[*Hint*: $A + A^T = A^S$, and $A - A^T = A^{SS}$, where S ≡ symmetric, SS ≡ skew-symmetric.]

1.16 Prove that the diagonal elements of a Hermitian matrix are real and those of a skew-Hermitian matrix imaginary. Hence, show that diagonal elements of a skew-symmetric matrix are all zeros.

1.17 A square matrix with complex elements can be uniquely represented as the sum of a Hermitian and a skew-Hermitian matrix.

1.18 Use mathematical induction and associative law of multiplication to show that, if A is a square matrix, and p and q are positive integers, then

$$A^p A^q = A^{p+q}, \qquad (A^p)^q = A^{pq}.$$

1.19 For three square matrices A, B and C of order n, if $BC = CB$, show that

$$\text{tr } ABC = \text{tr } ACB = \text{tr } BAC = \text{tr } BCA = \text{tr } CAB = \text{tr } CBA.$$

1.20 Let $\mathbf{e}_i$ be an n-vector having 1 in the ith row and zeros elsewhere, and A a matrix of order n. Compute (a) $\mathbf{e}_i^T A$, (b) $A\mathbf{e}_i$, (c) $\mathbf{e}_i^T A\mathbf{e}_i$, (d) $\mathbf{e}_i^T \mathbf{e}_j$, (e) $\mathbf{e}_i\mathbf{e}_j^T$, (f) $\mathbf{e}_i^T A\mathbf{e}_j$.

1.21 Prove that if A and B are symmetric and of the same order, then AB is symmetric if A and B commute.

1.22 Let $\mathbf{e}_i$ be an n-vector having 1 in the ith row and zeros elsewhere, and E_{ij} be a square matrix of order n with 1 in the ijth position and zero in all others. Compute

(a) $E_{ij}E_{ij}$, (b) $E_{ij}E_{kl}$, (c) $E_{ij}\,\mathbf{e}_j$, (d) $E_{ij}\,\mathbf{e}_k$, $j \neq k$, (e) $\mathbf{e}_k^T\,E_{ij}$.

1.23 Perform the following operations on the given matrices A, B, P and Q:

(a) $\dot{A}, \dot{B}, \dot{P}, \dot{Q}$; (b) $D(AB), D(PQ)$;

(c) $(DA)B, B(DA), (DB)A$; (d) $D(AB)^T, D(PQ^T)$;

$$A = \begin{bmatrix} \cos t & \sin t \\ -\sin t & \cos t \end{bmatrix}, \qquad B = \begin{bmatrix} \cos t & \sin t \\ \sin t & -\cos t \end{bmatrix},$$

$$P = \begin{bmatrix} e^{-at} & t \\ t^{-1} & e^{at} \end{bmatrix}, \qquad Q = \begin{bmatrix} e^{at} & -t \\ -t^{-1} & e^{-at} \end{bmatrix};$$

where a is a constant, t is a variable parameter, and D is a derivative operator.

1.24 Integrate (a) A between the limits 0 and π, and (b) P between the limits ½ and 1, when A and P are given in Problem 1.23.

1.25 Show that the postmultiplication of a matrix $A = [a_{ij}]_{m \times n}$ by $D_n = \text{diag}\,[d_1, d_2, \ldots, d_n]$ is equal to a matrix obtained from A by multiplying its jth column by d_j and premultiplication of A by D_m is a matrix whose ith row is the ith row of A multiplied by d_i.

1.26 Show that $A(\alpha\mathbf{x}) = \alpha A\mathbf{x}, \quad A(\alpha\mathbf{x} + \beta\mathbf{y}) = \alpha A\mathbf{x} + \beta A\mathbf{y}$,

where α and β are scalars, $\mathbf{x}$ and $\mathbf{y}$ are n-vectors, and A an $m \times n$ matrix.

1.27 For a rectangular matrix A of order $m \times n$, write down its structural representation if it is upper triangular when (a) $m > n$, and (b) $m < n$. Repeat this example for a lower triangular matrix A.

1.28 Prove that the product of any two conformable upper (or lower) triangular matrices is an upper (or lower) triangular matrix.

1.29 Prove that the following matrices are commutative, where λ and μ are scalars but m and n are integers:

(a) $(\lambda I - A)$ and A,

(b) $(\lambda I - A)$ and $(\mu I - A)$,

(c) $(\lambda I - A)^n$ and A,

(d) $(\lambda I - A)^n$ and A^m,

(e) $(\lambda I - A)^m$ and $(\mu I - A)^n$.

1.30 Prove that a square matrix A commutes with a diagonal matrix D if and only if D is a scalar matrix.

1.31 Prove that any two diagonal matrices of the same order commute.

1.32 Show that if D is a diagonal matrix, then $AD^p = D^pA$ if and only if $AD = DA$.

1.33 If $AB = \lambda B$, where A is square and λ is a scalar, then prove that $A^pB = \lambda^pB$ for any positive integer p.

1.34 A square matrix A is called *nilpotent of degree or index p* if $A^p = 0$, a null matrix, but $A^{p-1} \neq 0$ for some positive integer p. Prove that the following matrices are nilpotent and find the associated index:

(a) $\begin{bmatrix} xy & y^2 \\ -x^2 & -xy \end{bmatrix}$,

(b) a strictly (upper or lower) triangular matrix of order n,

(c) $\left[\begin{array}{c|ccc} 0 & & I_{n-1} & \\ \vdots & & & \\ \hline 0 & \cdots & & 0 \end{array}\right]_n$,

(d) $\left[\begin{array}{ccc|c} 0 & & \cdots & 0 \\ \hline & I_{n-1} & & \vdots \\ & & & 0 \end{array}\right]_n$.

1.35 Find all diagonal matrices D of order n which are idempotent, i.e., $D^2 = D$.

1.36 A matrix is called a *Toeplitz matrix* if its entries are constant along each (main, super- and sub-) diagonal. For examples,

$$\begin{bmatrix} a_{11} & a_{12} & a_{13} & a_{14} \\ a_{21} & a_{22} & a_{12} & a_{13} \\ a_{31} & a_{21} & a_{33} & a_{12} \\ a_{41} & a_{31} & a_{21} & a_{44} \end{bmatrix} \quad \text{and} \quad \begin{bmatrix} a_0 & 0 & 0 & 0 \\ a_1 & a_0 & 0 & 0 \\ a_2 & a_1 & a_0 & 0 \\ a_3 & a_2 & a_1 & a_0 \end{bmatrix}$$

are respectively a square and a triangular Toeplitz matrix. A Toeplitz matrix is symmetric about its secondary or northeast-southwest diagonal. Matrices which are symmetric about its secondary diagonal are called *persymmetric matrices.* If T is a Toeplitz matrix with real elements of order n, then show that

$$T = E\,T'\,E, \quad \text{where} \quad E = [\mathbf{e}_n, \ldots, \mathbf{e}_1]$$

and $\mathbf{e}_i$ is an n-vector with 1 in the ith row. The matrix E is called an *exhange matrix.*

CHAPTER

2

Determinants

An important attribute in the study of matrix algebra is the concept of determinant, ascribed to a square matrix. In this section we shall introduce this important notion, the various properties of a determinant, and the methods for its computation.

2.1 PERMUTATION AND INVERSION

To each square matrix A we assign a unique and well-defined scalar, formulated according to a given rule called the *determinant* of A and denoted by det A. Before prescribing the rules of formulation we need to define the term *inversion* in connection with permutation of positive integers.

The *permutation* of the integers from 1 to n is written as $j_1\ j_2,\ ...,\ j_n$, where each of the integers can occur only once among the j's. For example, in the permutation 4132 of the integers 1 to 4, $j_1 = 4$, $j_2 = 1$, $j_3 = 3$, $j_4 = 2$.

The number of inversions associated with the integer k in the permutation of integers from 1 to n is the number of smaller integers occurring after k. The total number of *inversions in a permutation* is the sum of inversions for each integer. Thus, 4132 has a total number of four inversions because 4 is followed by three smaller integers 1, 3, 2. The inversions associated with it are therefore 3, and similarly the inversions associated with 1 are zero, and those associated with 3 are one. A permutation is said to be *even* or *odd* according to whether the number of inversions in it is an even or an odd number. For example, the permutation 4132 is even.

2.2 DETERMINANT, COFACTOR AND MINOR

Definition 2.1 *Determinant.* The determinant associated with a square matrix

$$A = \begin{bmatrix} a_{11} & a_{12} & \cdots & a_{1n} \\ a_{21} & a_{22} & \cdots & a_{2n} \\ \vdots & \vdots & & \vdots \\ a_{n1} & a_{n2} & \cdots & a_{nn} \end{bmatrix} \tag{2.1}$$

is a scalar (a real number or a complex number) defined by

$$\det A: = \sum_{(j)} (-1)^{c(j)} a_{1j_1} a_{2j_2} \ldots a_{nj_n}, \tag{2.2}$$

where the summation extends over all possible permutations $j_1\, j_2 \cdots j_n\, [\equiv (j)]$ of n second (i.e., column) subscripts in a's and $c(j)$ is the total number of inversions in each permutation. □

The first subscript of a's in (2.2) runs over 1 to n in natural order and the second subscript is a permutation $j_1\, j_2 \cdots j_n$ of the integers from 1 to n, each of which can occur only once among the j's. Recalling that the first subscript identifies the row and the second one the column, it is evident that in (2.2), each product $a_{1j_1} a_{2j_2} \cdots a_{nj_n}$ is constituted by taking one and only one element from each row and each column of A, and has a positive or a negative sign depending on whether $c(j)$ is even or odd. The total number of permutations $j_1\, j_2 \cdots j_n$ being $n!$, suggests that the expansion (2.2) of a determinant contains $n!$ terms.

The determinant of a matrix is also denoted by vertical bars. For example,

$$\begin{vmatrix} a & b \\ c & d \end{vmatrix} = \det A,$$

where

$$A = \begin{bmatrix} a & b \\ c & d \end{bmatrix}.$$

Example 2.1 To compute det A, where

$$A = \begin{bmatrix} a_{11} & a_{12} & a_{13} \\ a_{21} & a_{22} & a_{23} \\ a_{31} & a_{32} & a_{33} \end{bmatrix}, \tag{2.3}$$

we have, in view of (2.2), the relation

$$\det A = (-1)^{c(123)} a_{11}a_{22}a_{33} + (-1)^{c(132)} a_{11}a_{23}a_{32} + (-1)^{c(213)} a_{12}a_{21}a_{33}$$
$$+ (-1)^{c(231)} a_{12}a_{23}a_{31} + (-1)^{c(312)} a_{13}a_{21}a_{32} + (-1)^{c(321)} a_{13}a_{22}a_{31},$$

where in each term, the row subscripts of the elements are kept in the natural order and the column subscripts are permutations of 1, 2, 3 — the number of inversions in which determines $c(j)$. In this respect, we note that $c(1\ 2\ 3)$ = number of inversions in 1 2 3 = 0, $c(1\ 3\ 2)$ = number of inversions in 1 3 2 = 1, and so on. Therefore, we have

$$\begin{aligned}\det \mathrm{A} &= (-1)^0\, a_{11}a_{22}a_{33} + (-1)^1\, a_{11}a_{23}a_{32} + (-1)^1\, a_{12}a_{21}a_{33} \\ &\quad + (-1)^2\, a_{12}a_{23}a_{31} + (-1)^2\, a_{13}a_{21}a_{32} + (-1)^3\, a_{13}a_{22}a_{31} \\ &= a_{11}a_{22}a_{33} - a_{11}a_{23}a_{32} - a_{12}a_{21}a_{33} \\ &\quad + a_{12}a_{23}a_{31} + a_{13}a_{21}a_{32} - a_{13}a_{22}a_{31}. \end{aligned} \tag{2.4}$$

If the first two columns of A are adjoined to its right, then the terms in (2.4) can be obtained as the product of diagonal elements with the assigned signs as shown in Figure 2.1.

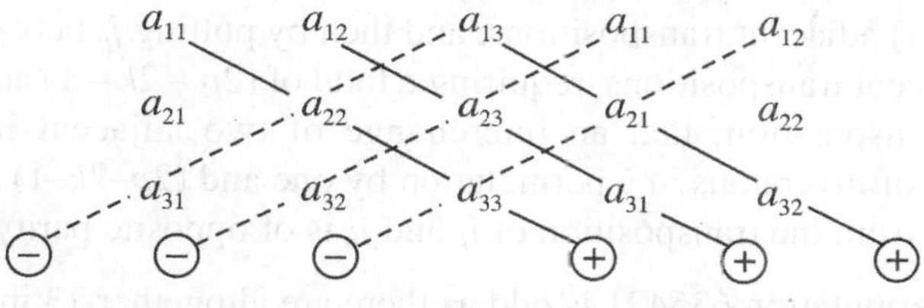

Figure 2.1 Row expansion of a third order determinant.

This rule is only applicable for third order determinants. □

Each product in (2.2) is obtained by keeping the row subscripts in the natural order and performing permutation among the column subscripts, and hence it is known as the *row expansion* of det A.

Definition 2.2 *Column Expansion of* det A. Here, det A is obtained by keeping the column subscripts in a natural order and making all possible permutations among the row subscripts, yielding

$$\det A = \sum_{(i)} (-1)^{r(i)}\, a_{i_1 1} a_{i_2 2} \cdots a_{i_n n}, \tag{2.5}$$

where the summation extends over all $n!$ permutations $i_1\ i_2 \cdots i_n [\equiv (i)]$ of the integers 1 to n and $r(i)$ is the number of inversions in the permutation $i_1\ i_2 \cdots i_n$. □

Example 2.2 If A is given as in (2.3), by column expansion of a determinant, we have

$$\begin{aligned}\det A &= (-1)^{r(123)}a_{11}a_{22}a_{33} + (-1)^{r(132)}a_{11}a_{32}a_{23} \\ &\quad + (-1)^{r(213)}a_{21}a_{12}a_{33} + (-1)^{r(231)}a_{21}a_{32}a_{13} \\ &\quad + (-1)^{r(312)}a_{31}a_{12}a_{23} + (-1)^{r(321)}a_{31}a_{22}a_{13} \\ &= (-1)^0 a_{11}a_{22}a_{33} + (-1)^1 a_{11}a_{32}a_{23} + (-1)^1 a_{21}a_{12}a_{33} \\ &\quad + (-1)^2 a_{21}a_{32}a_{13} + (-1)^2 a_{31}a_{12}a_{23} + (-1)^3 a_{31}a_{22}a_{13} \\ &= a_{11}a_{22}a_{33} - a_{11}a_{32}a_{23} - a_{21}a_{12}a_{33} \\ &\quad + a_{21}a_{32}a_{13} + a_{31}a_{12}a_{23} - a_{31}a_{22}a_{13}, \end{aligned} \tag{2.6}$$

where we have used, for example, $r(321)$ = number of inversions in 321 = 3, ... □

Two permutations are said to have the *same parity* if they are both even or odd. If one is even and the other is odd or vice versa, they are said to have *opposite parity.*

Theorem 2.1 If two integers, say, j_k and j_p $(k < p)$, in a given permutation

$$j_1 j_2 \cdots j_{k-1} \underset{j_p}{\uparrow} j_k j_{k+1} \cdots j_{p-1} \overset{j_k}{\downarrow} j_p \cdots j_n$$

are interchanged (i.e., transposed), then a permutation of opposite parity is obtained.

Proof : The transposition of j_k and j_p can be accomplished first by putting j_k between j_{p-1} and j_p by $(p - k - 1)$ adjacent transpositions, and then by putting j_p between j_{k-1} and j_{k+1} by an additional $(p - k)$ adjacent transpositions, requiring a total of $(2p - 2k - 1)$ adjacent transpositions. Since an adjacent transposition, i.e., an interchange of two adjacent integers, increases or decreases the number of inversions in a permutation by one and $(2p-2k-1)$ is an odd number, the permutation resulting from the transposition of j_k and j_p is of opposite parity. □

Example 2.3 The permutation 635421 is odd as there are altogether 13 inversions. By adjacent transposition of 3 and 5 the above permutation transforms to 653421, in which the number of inversions increases by 1 to 14. Again, by transposing 5 and 1, we get 631425 having 8 number of inversions. Thus by transposition, adjacent or not, the given odd permutation changes to an even one, i.e., to one of opposite parity. □

Theorem 2.2 The row expansion (2.2) and the column expansion (2.5) give the same algebraic expression for det A.

Proof : Referring to Example 2.1, we observe that the term $a_{12}a_{23}a_{31}$ in the row expansion (2.4), if rearranged, is seen to occur in the column expansion (2.6) as $a_{31}a_{12}a_{23}$. The latter is obtained from the former by an equal number of transpositions in the row and column subscripts as shown in Table 2.1, the column subscripts being in natural order. This way of obtaining an element in the column expansion introduces the same number of inversions in the row subscripts, as previously existed among the column subscripts. Extending this notion to an nth order case, we can say a product $a_{1j_1}\ a_{2j_2}\ \cdots\ a_{nj_n}$ in row expansion (2.2), if rearranged, can be shown to occur as a product $a_{i_1 1}\ a_{i_2 2}\ \cdots\ a_{i_n n}$ in column expansion (2.5), and the number of inversions in the column subscripts in the former product is the same as that in row subscripts in the latter product, showing that the two products under consideration should have the same sign. □

Table 2.1 Column expansion from row expansion

	Existing	Step 1	Transposed Step 2
Row subscripts	1 2 3	1 3 2	3 1 2
Column subscripts	2 3 1	2 1 3	1 2 3
	$a_{12}a_{23}a_{31}$	$a_{12}a_{31}a_{23}$	$a_{31}a_{12}a_{23}$

Example 2.4 Referring to Example 2.1, we can write det A by factoring out the elements in the second row as

$$\begin{aligned}\det A &= a_{21}(a_{32}a_{13} - a_{12}a_{33}) + a_{22}(a_{11}a_{33} - a_{31}a_{13}) + a_{23}(a_{12}a_{31} - a_{11}a_{32})\\ &= a_{21}A_{21} + a_{22}A_{22} + a_{23}A_{23},\end{aligned}$$

where

$$A_{21} = (a_{32}a_{13} - a_{12}a_{33}), \quad A_{22} = (a_{11}a_{33} - a_{31}a_{13}), \quad A_{23} = (a_{12}a_{31} - a_{11}a_{32}).$$

The elements A_{21}; A_{22} and A_{23} are called *cofactors* of a_{21}, a_{22} and a_{23} respectively. □

Definition 2.3 *Cofactor.* In the expansion of det A the factor A_{ij}, which is left out when a_{ij} is factored out, is called the *cofactor* of the element *aij* in the square matrix A. □

Theorem 2.3 The cofactor A_{ij} of the element a_{ij} in A is $(-1)^{i+j}$ times the determinant of the submatrix of order $(n - 1)$ obtained by deleting the ith row and the jth column from $A = [a_{ij}]_n$.

Proof : The row expansion of det A is

$$\det A = \sum_{(j)} \varepsilon_{j_1 \cdots j_{i-1}\, j_i\, j_{i+1} \cdots j_n}\, a_{1j_1} \cdots a_{i-1,\, j_{i-1}}\, a_{ij_i}\, a_{i+1,\, j_{i+1}} \cdots a_{nj_n}, \tag{2.7}$$

where the summation extends over the permutations $j_1\, j_2\, \ldots\, j_n$ of the integers 1 to n and

$$(-1)^{c(j)} =: \varepsilon_{j_1 j_2 \ldots j_n} = +1 \text{ or } -1$$

according as the permutation $j_1\, j_2\, \ldots\, j_n$ is even or odd.

As a's are scalars, bringing a_{ij_i} to the first position in (2.7), we have

$$\det A = \sum_{(j)} \varepsilon_{j_1 \cdots j_{i-1}\, j_i\, j_{i+1} \cdots j_n}\, a_{ij_i}\, (a_{1j_1} \cdots a_{i-1,\, j_{i-1}}\, a_{i+1,\, j_{i+1}} \cdots a_{nj_n}).$$

The epsilon symbol (ε) determines sign, and transferring j_i to the first position introduces $i - 1$ number of inversions, and therefore

$$\det A = \sum_{(j)} (-1)^{i-1}\, \varepsilon_{j_i j_1 \cdots j_{i-1}\, j_{i+1} \cdots j_n}\, a_{ij_i}\, (a_{1j_1} \cdots a_{i-1,\, j_{i-1}}\, a_{i+1,\, j_{i+1}} \cdots a_{nj_n})$$

$$= \sum_{j_i = 1}^{n} \sum (-1)^{i-1}\, \varepsilon_{j_i j_1 \cdots j_{i-1}\, j_{i+1} \cdots j_n}\, a_{ij_i}\, (a_{1j_1} \cdots a_{i-1,\, j_{i-1}}\, a_{i+1,\, j_{i+1}} \cdots a_{nj_n}), \tag{2.8}$$

where the second summation is over the permutation of $j_1 \ldots j_{i-1}\, j_{i+1} \ldots j_n$. When $j_i = j$, the coefficient of a_{ij} (termed the cofactor of a_{ij}) is

$$A_{ij} = \sum (-1)^{i-1}\, \varepsilon_{j\, j_1 \cdots j_{i-1}\, j_{i+1} \cdots j_n}\, a_{1j_1} \cdots a_{i-1,\, j_{i-1}}\, a_{i+1,\, j_{i+1}} \cdots a_{nj_n} \tag{2.9}$$

or, as j associated with ε contains $j - 1$ number of inversions,

$$A_{ij} = \sum (-1)^{i+j}\, \varepsilon_{j_1 \cdots j_{i-1}\, j_{i+1} \cdots j_n}\, a_{1j_1} \ldots a_{i-1,\, j_{i-1}}\, a_{i+1,\, j_{i+1}} \ldots a_{nj_n},$$

where the summation is over permutation $j_1 \ldots j_{i-1}\, j_{i+1} \ldots j_n$ of integers from $1, \ldots, j - 1, j + 1, \ldots, n$. The product a_{ij}'s in the above sum is the same as can be obtained from the matrix A with the ith row and the jth column deleted. □

Definition 2.4 *Minor or Subdeterminant.* The determinant of the submatrix of order $(n - 1)$, obtained by striking out the ith row and the jth column, is defined to be the *minor* (or *subdeterminant*) M_{ij} of a_{ij} in the matrix $A = [a_{ij}]_n$. In view of Theorem 2.3,

$$A_{ij} = (-1)^{i+j}\, M_{ij}.$$ □

Example 2.5 Find the minors and the cofactors of the matrix

$$A = \begin{bmatrix} 2 & 3 & 4 \\ 4 & 3 & 1 \\ 1 & 2 & 4 \end{bmatrix}.$$

Solution We have,

$$M_{11} = \begin{vmatrix} 3 & 1 \\ 2 & 4 \end{vmatrix} = 10 = A_{11}, \qquad M_{21} = \begin{vmatrix} 3 & 4 \\ 2 & 4 \end{vmatrix} = 4 = -A_{21}.$$

Similarly,

$$M_{31} = -9 = A_{31}, \qquad M_{12} = 15 = -A_{12}, \qquad M_{22} = 4 = A_{22},$$
$$M_{32} = -14 = -A_{32}, \qquad M_{13} = 5 = A_{13}, \qquad M_{23} = 1 = -A_{23},$$
$$M_{33} = -6 = A_{33}.$$

Observe that when the sum of row and column subscripts is odd, the minor and the corresponding cofactor are of opposite signs, and when even, they have the same sign. □

Example 2.6 For the third order matrix

$$\begin{bmatrix} a_{11} & a_{12} & a_{13} \\ a_{21} & a_{22} & a_{23} \\ a_{31} & a_{32} & a_{33} \end{bmatrix},$$

deleting the second row and third column in which the element a_{23} is located, we have a sub-matrix of order 2 constituted by the elements a_{11}, a_{12}, a_{31}, and a_{32}. Therefore, the *cofactor* A_{23} is

$$A_{23} = (-1)^{2+3} \det \begin{bmatrix} a_{11} & a_{12} \\ a_{31} & a_{32} \end{bmatrix} = (-1)^{2+3} M_{23},$$

where M_{23} is the minor of a_{23}. □

An alternative expansion of det A is obtained by substituting (2.9) into (2.8) and choosing j in place of j_i as

$$\det A = \sum_{j=1}^{n} a_{ij} A_{ij} = a_{i1} A_{i1} + a_{i2} A_{i2} + \cdots + a_{in} A_{in}, \quad i = 1, 2, ..., n \tag{2.10}$$

which is called the *expansion of* det A *in terms of the elements in the ith row.* Similarly, considering the column expansion, we can arrive at the relation

$$\det A = a_{1j}A_{1j} + a_{2j}A_{2j} + \cdots + a_{nj}A_{nj}, \quad j = 1, 2, ..., n \tag{2.11}$$

which is referred to as the *expansion of* det A *in terms of the elements in the jth column.*

Example 2.7 Let

$$A = \begin{bmatrix} a_{11} & a_{12} \\ a_{21} & a_{22} \end{bmatrix}$$

The cofactors of a_{11} and a_{21} are respectively $A_{11} = (-1)^{1+1}a_{22} = a_{22}$ and $A_{21} = (-1)^{2+1}a_{12} = -a_{12}$. Therefore, det A obtained by the first column expansion is

$$\det A = a_{11}A_{11} + a_{21}A_{21} = a_{11}a_{22} - a_{21}a_{12}. \qquad \square$$

From this example, we can conclude that det A of a second order matrix A is obtained as a difference of the products of main diagonal and secondary diagonal elements.

Example 2.8 Either row or column expansion can be employed to obtain the determinant of a diagonal matrix (see (1.17))

$$A = \text{diag}\,[a_{11}, a_{22}, \ldots, a_{nn}] = \begin{bmatrix} a_{11} & 0 & \cdots & 0 \\ 0 & a_{22} & \cdots & 0 \\ \vdots & \vdots & & \vdots \\ 0 & 0 & \cdots & a_{nn} \end{bmatrix}.$$

Employing row expansion (2.10) for $i = 1$, since all but a_{11} in the first row are zero, we have

$$\det A = a_{11}A_{11},$$

where

$$A_{11} = \det \text{diag}\,[a_{22}, \ldots, a_{nn}].$$

Applying (2.10) again, we have $A_{11} = a_{22}\Delta_{22}$, where $\Delta_{22} = \det \text{diag}\,[a_{33}, \ldots, a_{nn}]$. Therefore, det A becomes, by substitution of A_{11},

$$\det A = a_{11}a_{22}\Delta_{22}.$$

Continuing thus we finally get

$$\det A = a_{11}a_{22} \ldots a_{nn}.$$

Also, noting that $A_{11} = \det A/a_{11}$, the cofactor of any diagonal element can be easily shown to be $A_{ii} = \det A/a_{ii}$. On the other hand, $A_{ij} = 0$, $i \neq j$, since any minor determinant M_{ij} (e.g., M_{12}) has one column (e.g., first column) consisting of zero elements. □

Example 2.9 As in Example 2.8, we can use either row or column expansion to show that the determinant of an upper triangular or a lower triangular matrix, viz.,

$$U = \begin{bmatrix} a_{11} & a_{12} & \cdots & a_{1n} \\ 0 & a_{22} & \cdots & a_{2n} \\ \vdots & \vdots & & \vdots \\ 0 & 0 & \cdots & a_{nn} \end{bmatrix} \quad \text{or} \quad L = \begin{bmatrix} b_{11} & 0 & \cdots & 0 \\ b_{21} & b_{22} & \cdots & 0 \\ \vdots & \vdots & & \vdots \\ b_{n1} & b_{n2} & \cdots & b_{nn} \end{bmatrix}$$

is the product of diagonal elements, i.e., det $U = a_{11}a_{22} \ldots a_{nn}$ and det $L = b_{11}b_{22} \ldots b_{nn}$. □

2.3 PROPERTIES OF DETERMINANTS

Property I: The determinant of a matrix A with any two rows or any two columns transposed is –det A.

For the transposition of kth and pth rows ($k < p$) we have, from (2.7), the relation

$$\det A = \sum_{(j)} \varepsilon_{j1\cdots jk\cdots jp\cdots jn} a_{1j1} \cdots a_{kjk} \cdots a_{pjp} \cdots a_{njn}$$

$$= \sum_{(j)} \varepsilon_{j1\cdots jk\cdots jp\cdots jn} a_{1j1} \cdots a_{pjp} \cdots a_{kjk} \cdots a_{njn}.$$

The rearrangement in the last step is possible as a_{ij}'s are scalars. The column subscripts of a_{ij}'s and the subscripts of ε can be brought into correspondence by transposing j_k and j_p to ε which, by Theorem 2.1, introduces a negative sign yielding

$$\det A = -\sum_{(j)} \varepsilon_{j1\cdots jp\cdots jk\cdots jn} a_{1j1} \cdots a_{pjp} \cdots a_{kjk} \cdots a_{njn}.$$

Observing the row subscripts on a_{ij}'s, it is not difficult to see that the expression under the summation is the determinant of the matrix A with the kth and pth rows interchanged. The proof for columns is similar.

Example 2.10 Let

$$A = \begin{bmatrix} a & h & g \\ h & b & f \\ g & f & c \end{bmatrix}.$$

Then,

$$\det \begin{bmatrix} g & h & a \\ f & b & h \\ c & f & g \end{bmatrix} = -\det A$$

because there is an interchange between the first and third columns of A. □

Property II: The determinant of a matrix A is equal to that of its transpose A^T, i.e., $\det A = \det A^T$, but $\det A^* = \overline{\det A}$, where a bar represents a complex conjugate (see Definition 1.10).

Since the rows of A are the columns of A^T, the row expansion of A must equal the column expansion of A^T. Hence the property follows, since column expansion and row expansion of a determinant are equal.

Example 2.11 Let

$$P = \begin{bmatrix} a & b & c \\ d & e & f \\ g & h & k \end{bmatrix}. \quad (2.12)$$

Then,

$$\begin{vmatrix} a & d & g \\ b & e & h \\ c & f & k \end{vmatrix} = \begin{vmatrix} a & b & c \\ d & e & f \\ g & h & k \end{vmatrix}.$$

The cofactors of the first row elements of P are, respectively,

$$A = (ek - fh), \qquad B = -(dk - fg), \qquad C = (dh - eg) \tag{2.13}$$

which are the same as those of the first column elements of P^T. Therefore,

$$\det P = a(ek - fh) - b(dk - fg) + c(dh - eg) = \det P^T. \qquad \square$$

Property III: If all the elements in a row or in a column of the matrix A are zero, then det $A = 0$.

This property is a direct consequence of (2.10) and (2.11).

Property IV: The determinant of a matrix having two identical rows or columns is zero. Det A remains unaltered if the two identical rows or columns are interchanged. Observing this we have, in view of Property I, the relation

$$\det A = -\det A, \text{ i.e., } \det A = 0.$$

Example 2.12 Let A have the same elements in the first and third columns, i.e.,

$$A = \begin{bmatrix} a & d & a \\ b & e & b \\ c & f & c \end{bmatrix}.$$

Then, expanding det A in terms of first column elements by (2.11), we get

$$\det A = a(ec - bf) - b(dc - af) + c(db - ae) = 0. \qquad \square$$

Property V: If all the elements in a row or in a column of a matrix A are multiplied by a constant scalar k, then the determinant of the resultant matrix is k det A.

This property is a direct result of relations (2.10) and (2.11).

Example 2.13 For the matrix A given in Example 2.10, we have

$$\begin{vmatrix} a & h & g \\ h & b & f \\ kg & kf & kc \end{vmatrix} = \begin{vmatrix} a & kh & g \\ h & kb & f \\ g & kf & c \end{vmatrix} = k \begin{vmatrix} a & h & g \\ h & b & f \\ g & f & c \end{vmatrix}. \qquad \square$$

Property VI: If all the elements in the ith row (resp jth column) of a matrix A are replaced correspondingly by the elements $c_1, c_2, ..., c_n$, then the determinant of the resultant matrix is

$$c_1A_{i1} + c_2A_{i2} + \cdots + c_nA_{in} \text{ (resp } c_1A_{1j} + c_2A_{2j} + \cdots + c_nA_{nj}).$$

The conclusion follows directly from (2.10) and (2.11).

Property VII: The sum of the products obtained by multiplying the elements of a row or a column into the cofactors of the corresponding elements of a different row or column is always zero. That is,

$$a_{k1}A_{i1} + a_{k2}A_{i2} + \cdots + a_{kn}A_{in} = 0, \qquad i \neq k; \tag{2.14}$$

$$a_{1k}A_{1j} + a_{2k}A_{2j} + \cdots + a_{nk}A_{nj} = 0, \qquad j \neq k. \tag{2.15}$$

A comparison with (2.10) shows that the left-hand side of (2.14) is a result of the expansion of det A when ith row elements in the matrix A are replaced by kth row elements. This indicates that the ith and kth row elements of A are identical, thus verifying relation (2.14) by virtue of Property IV. A similar proof is in order for (2.15).

Example 2.14 Let a matrix P be given by (2.12) and the cofactors of first row elements by (2.13). Multiplying these cofactors correspondingly by the second row elements, we obtain

$$dA + eB + fC = d(ek - fh) - e(dk - fg) + f(dh - eg) = 0. \qquad \square$$

Property VIII: If each element in the kth row (or kth column) in matrix A is a sum of p terms, then det A can be expressed as the sum of determinants of p matrices which have, in the kth row (or kth column) as elements respectively, the first, second, ..., pth term of the corresponding elements in the kth row (or kth column) of A, all other rows (or columns) being the same throughout.

This property again is easily proved by (2.10) and (2.11).

Example 2.15 Let a matrix P be given by

$$P = \begin{bmatrix} a & b & c \\ d+r & e+s & f+t \\ g & h & k \end{bmatrix}.$$

Denoting the cofactors of second row elements by D, E, F, we have

$$\begin{aligned} \det P &= (d + r)\, D + (e + s)\, E + (f + t)\, F \\ &= (dD + eE + fF) + (rD + sE + tF), \end{aligned}$$

or, using Property VI,

$$\begin{vmatrix} a & b & c \\ d+r & e+s & f+t \\ g & h & k \end{vmatrix} = \begin{vmatrix} a & b & c \\ d & e & f \\ g & h & k \end{vmatrix} + \begin{vmatrix} a & b & c \\ r & s & t \\ g & h & k \end{vmatrix}. \qquad \square$$

Property IX: The determinant of a matrix is left unchanged if a constant multiple of the elements of any row or column is added to the corresponding elements of a different row or column.

Let k times the elements of jth row be added to the corresponding elements of the ith row in a matrix A. Then, expanding det A in terms of ith row elements, we have, by virtue of (2.10), the determinant of the resultant matrix as:

$$\begin{aligned} &(a_{i1} + ka_{j1})A_{i1} + (a_{i2} + ka_{j2})A_{i2} + \cdots + (a_{in} + ka_{jn})A_{in} \\ &= (a_{i1}A_{i1} + a_{i2}A_{i2} + \cdots + a_{in}A_{in}) + k(a_{j1}A_{i1} + a_{j2}A_{i2} + \cdots + a_{jn}A_{in}) \\ &= \det A + k \cdot 0 \quad \text{[by (2.10) and (2.14)]} \\ &= \det A. \end{aligned}$$

Example 2.16 For a multiple *m*, we have

$$\begin{vmatrix} a & b & c \\ d & e & f \\ g & h & k \end{vmatrix} = \begin{vmatrix} a+mc & b & c \\ d+mf & e & f \\ g+mk & h & k \end{vmatrix} = \begin{vmatrix} a & b & c \\ d+mg & e+mh & f+mk \\ g & h & k \end{vmatrix}.$$ □

Property X: The conjugate of the determinant of a matrix A over a complex field $\mathbb{C}$ is equal to the determinant of its conjugate $\bar{A}$, i.e., $\overline{\det A} = \det \bar{A}$ (see Section 1.3). This property follows directly from the expansion [(2.2) or (2.5)] of a determinant.

2.4 EVALUATION OF DETERMINANTS

There exist many methods for the evaluation of the determinant of a square matrix *A*. We shall describe four such methods: three for a square matrix and one for the product of two rectangular matrices.

Method I: Using properties of determinants. By repeated application of Property IX, the given matrix *A* is transformed into a new matrix *B* having all but one of the elements in a given row (or column) reduced to zero. If b_{ij} is the nonzero element and B_{ij} the corresponding cofactor, then

$$\det A = \det B = b_{ij}B_{ij}.$$

The problem thus becomes one of evaluation of a determinant B_{ij} of order $(n - 1)$, which can be treated similarly until we arrive at a determinant whose computation becomes easy at sight.

Example 2.17

$$\begin{vmatrix} 2 & 2 & 3 & 3 \\ 2 & 1 & 4 & 2 \\ 3 & 2 & 3 & 2 \\ 3 & 1 & 2 & 4 \end{vmatrix} = \begin{vmatrix} -2 & 2 & -5 & -1 \\ 0 & 1 & 0 & 0 \\ -1 & 2 & -5 & -2 \\ 1 & 1 & -2 & 2 \end{vmatrix} = 1\,(-1)^{2+2} \begin{vmatrix} -2 & -5 & -1 \\ -1 & -5 & -2 \\ 1 & -2 & 2 \end{vmatrix}$$

$$= 1 \cdot 1\,(-1)^2 \begin{vmatrix} 2 & 5 & 1 \\ 1 & 5 & 2 \\ 1 & -2 & 2 \end{vmatrix} = \begin{vmatrix} 2 & 5 & 1 \\ 1 & 5 & 2 \\ 0 & -7 & 0 \end{vmatrix}$$

$$= (-7) \cdot (-1)^{3+2} \begin{vmatrix} 2 & 1 \\ 1 & 2 \end{vmatrix} = 21.$$

□

The first step follows by adding –2, –4 and –2 times the second column to the first, third and fourth columns, respectively; the second step from the expansion (2.10) in terms of the second row, i.e., 1 times its cofactor; the third step by factoring out the constant multiple

(–1) of the first and second rows in view of Property V (see Example 2.13); the fourth step from subtracting the second row from the last one; and the fifth step from expansion in terms of the last row elements. □

Before stepping into the second method, we should be acquainted with a few more terms and their definitions.

Definition 2.5 *Minor Determinant of Order p.* Referring to (2.1), let $1 \le i_1 < i_2 \dots < i_p \le n$ and $1 \le j_1 < j_2 \dots < j_p \le n$, $1 \le p \le n$ be p row and column indices selected from the integers 1 to n and arranged in order of magnitude. Then

$$A\begin{pmatrix} i_1\ i_2 \dots i_p \\ j_1\ j_2 \dots j_p \end{pmatrix} = \begin{vmatrix} a_{i_1 j_1} & \cdots & a_{i_1 j_p} \\ \vdots & & \vdots \\ a_{i_p j_1} & \cdots & a_{i_p j_p} \end{vmatrix}$$

is defined as the *minor determinant of order p of A*.

If $p = n - 1$, *minor of order p reduces* to *minors of the elements* a_{ij} as defined in Definition 2.4, and the minor of order n is det A itself. □

We know that p rows can be selected from n rows in ${}^nC_p$ ways and there are ${}^nC_p$ ways to select p columns from these p rows. Therefore, in a matrix of order n, there are altogether $({}^nC_p)^2$ number of minors of order p.

Definition 2.6 *Complementary Minors.* Referring to the matrix A as defined in (2.1), let $1 \le i_1 < i_2 < \dots < i_p \le n$ be p row indices and let $1 \le j_1 < j_2 < \dots < j_p \le n$ be p column indices selected from among the integers 1 to n and arranged in increasing magnitude. Let the *remaining* row and column indices arranged in order of magnitude be denoted by $i_{p+1} < i_{p+2} < \dots < i_n$ and $j_{p+1} < j_{p+2} < \dots < j_n$, respectively. By this division of row and column indices, we define a pair of determinants:

$$A\begin{pmatrix} i_1\ i_2 \cdots i_p \\ j_1\ j_2 \cdots j_p \end{pmatrix} = \begin{vmatrix} a_{i_1 j_1} & a_{i_1 j_2} & \cdots & a_{i_1 j_p} \\ a_{i_2 j_1} & a_{i_2 j_2} & \cdots & a_{i_2 j_p} \\ \vdots & \vdots & & \vdots \\ a_{i_p j_1} & a_{i_p j_2} & \cdots & a_{i_p j_p} \end{vmatrix} \tag{2.16}$$

and

$$A\begin{pmatrix} i_{p+1}\ i_{p+2} \cdots i_n \\ j_{p+1}\ j_{p+2} \cdots j_n \end{pmatrix} = \begin{vmatrix} a_{i_{p+1} j_{p+1}} & \cdots & a_{i_{p+1} j_n} \\ \vdots & & \vdots \\ a_{i_n j_{p+1}} & \cdots & a_{i_n j_n} \end{vmatrix} \tag{2.17}$$

which are called the *complementary minors* of A. □

In the definition of complementary minors, the division of row (resp column) indices is such that there exists no common index in the two divisions.

Example 2.18 Let A be a matrix of order 5:

$$A = \begin{bmatrix} a_{11} & a_{12} & a_{13} & a_{14} & a_{15} \\ a_{21} & a_{22} & a_{23} & a_{24} & a_{25} \\ a_{31} & a_{32} & a_{33} & a_{34} & a_{35} \\ a_{41} & a_{42} & a_{43} & a_{44} & a_{45} \\ a_{51} & a_{52} & a_{53} & a_{54} & a_{55} \end{bmatrix}. \tag{2.18}$$

Then

$$A\begin{pmatrix} 1\ 3\ 5 \\ 2\ 4\ 5 \end{pmatrix} = \begin{vmatrix} a_{12} & a_{14} & a_{15} \\ a_{32} & a_{34} & a_{35} \\ a_{52} & a_{54} & a_{55} \end{vmatrix},$$

$$A\begin{pmatrix} 2 & 4 \\ 1 & 3 \end{pmatrix} = \begin{vmatrix} a_{21} & a_{23} \\ a_{41} & a_{43} \end{vmatrix}, \qquad A\begin{pmatrix} 2 & 3 \\ 1 & 5 \end{pmatrix} = \begin{vmatrix} a_{21} & a_{25} \\ a_{31} & a_{35} \end{vmatrix}$$

are minors of A of orders 3, 2, and 2, respectively; $A\begin{pmatrix} 1 & 3 & 5 \\ 2 & 4 & 5 \end{pmatrix}$ and $A\begin{pmatrix} 2 & 4 \\ 1 & 3 \end{pmatrix}$ are complementary minors of A. Observe that the first and the third are not complementary minors since all the row (resp column) indices in the divisions 1 3 5 (resp 2 4 5) and 2 3 (resp 1 5) are not distinct because 3 (resp 5) is present in both. □

Definition 2.7 *Algebraic Complements.* If

$$i_1 + i_2 + \cdots + i_p + j_1 + j_2 + \cdots + j_p = \alpha, \tag{2.19}$$

$$i_{p+1} + i_{p+2} + \cdots + i_n + j_{p+1} + j_{p+2} + \cdots + j_n = \beta, \tag{2.20}$$

then

$$(-1)^{\alpha} A\begin{pmatrix} i_1\ i_2 \cdots i_p \\ j_1\ j_2 \cdots j_p \end{pmatrix}$$

is referred to as the *algebraic complement* of

$$A\begin{pmatrix} i_{p+1}\ i_{p+2} \cdots i_n \\ j_{p+1}\ j_{p+2} \cdots j_n \end{pmatrix}$$

and

$$(-1)^{\beta} A\begin{pmatrix} i_{p+1}\ i_{p+2} \cdots i_n \\ j_{p+1}\ j_{p+2} \cdots j_n \end{pmatrix}$$

as the algebraic complement of $\quad A\begin{pmatrix} i_1\ i_2 \cdots i_p \\ j_1\ j_2 \cdots j_p \end{pmatrix}$,

where the row and column indices are as defined in Definition 2.6. Since, from (2.19) and (2.20), $\alpha + \beta = n(n + 1)$ is an even number, α and β are either both even or both odd. Hence, for sign determination in algebraic complements, β can be used in place of α and vice versa.

In the particular case when $p = n - 1$, $i_n = i$ and $j_n = j$, we have $\beta = i + j$, and

$$(-1)^{\alpha} A\begin{pmatrix} i_1\, i_2 \cdots i_{n-1} \\ j_1\, j_2 \cdots j_{n-1} \end{pmatrix} = (-1)^{i+j} M_{ij}$$

becomes the cofactor of

$$A\begin{pmatrix} i_n \\ j_n \end{pmatrix} = a_{ij}. \qquad \square$$

Example 2.19 Referring to the matrix A defined by (2.18), we see that

$$(-1)^{1+2+5+2+3+4} A\begin{pmatrix} 1 & 2 & 5 \\ 2 & 3 & 4 \end{pmatrix} = -\begin{vmatrix} a_{12} & a_{13} & a_{14} \\ a_{22} & a_{23} & a_{24} \\ a_{52} & a_{53} & a_{54} \end{vmatrix}$$

is the algebraic complement of

$$A\begin{pmatrix} 3 & 4 \\ 1 & 5 \end{pmatrix} = \begin{vmatrix} a_{31} & a_{35} \\ a_{41} & a_{45} \end{vmatrix}$$

and

$$-\begin{vmatrix} a_{31} & a_{35} \\ a_{41} & a_{45} \end{vmatrix}$$

is the algebraic complement of

$$\begin{vmatrix} a_{12} & a_{13} & a_{14} \\ a_{22} & a_{23} & a_{24} \\ a_{52} & a_{53} & a_{54} \end{vmatrix}. \qquad \square$$

The following method is a generalization of evaluation of determinants by (2.10).

Method II: *Laplace expansion.* (a) Select any p rows (or columns) of the matrix A as given in (2.1).

(b) Form all possible minors of order p from these p rows (or columns).

(c) Multiply each of these minors into its algebraic complement.

Then det A is obtained by adding all the products derived in (c).

Example 2.20 Evaluate

$$\det A = \begin{vmatrix} 1 & 2 & 3 & 4 \\ 2 & 3 & 4 & 1 \\ 3 & 4 & 0 & 2 \\ 4 & 1 & 0 & 3 \end{vmatrix}$$

using minors of the first two rows.

Solution : Here p rows are the first two rows from which ${}^4C_2 = 6$ minors of order two can be formed by selecting columns: 1, 2; 1, 3; 1, 4; 2, 3; 2, 4 and 3, 4. Multiplying these minors into the corresponding algebraic complements with sign determined by α in place of β (see Definition 2.7), we have

$$\det A = \begin{vmatrix} 1 & 2 \\ 2 & 3 \end{vmatrix} \cdot (-1)^{1+2+1+2} \cdot \begin{vmatrix} 0 & 2 \\ 0 & 3 \end{vmatrix} + \begin{vmatrix} 1 & 3 \\ 2 & 4 \end{vmatrix} \cdot (-1)^{1+2+1+3} \cdot \begin{vmatrix} 4 & 2 \\ 1 & 3 \end{vmatrix}$$

$$+ \begin{vmatrix} 1 & 4 \\ 2 & 1 \end{vmatrix} \cdot (-1)^{1+2+1+4} \cdot \begin{vmatrix} 4 & 0 \\ 1 & 0 \end{vmatrix} + \begin{vmatrix} 2 & 3 \\ 3 & 4 \end{vmatrix} \cdot (-1)^{1+2+2+3} \cdot \begin{vmatrix} 3 & 2 \\ 4 & 3 \end{vmatrix}$$

$$+ \begin{vmatrix} 2 & 4 \\ 3 & 1 \end{vmatrix} \cdot (-1)^{1+2+2+4} \cdot \begin{vmatrix} 3 & 0 \\ 4 & 0 \end{vmatrix} + \begin{vmatrix} 3 & 4 \\ 4 & 1 \end{vmatrix} \cdot (-1)^{1+2+3+4} \cdot \begin{vmatrix} 3 & 4 \\ 4 & 1 \end{vmatrix}$$

$$= 0 + (-2)(-1) \cdot 10 + 0 + (-1)\, 1.1 + 0 + (-13) \cdot 1 \cdot (-13)$$

$$= 20 - 1 + 169 = 188.$$ □

Before describing the Binet-Cauchy formula for the evaluation of determinants, we shall define two more terms which are of no relevance here, but very important in the study of positive definite matrices (see Chapter 8).

Definition 2.8 *Principal Minor and Leading Principal Minor.* Selecting identical row and column indices in Definition 2.5, we have

$$A = \begin{pmatrix} i_1\, i_2 \cdots i_p \\ i_1\, i_2 \cdots i_p \end{pmatrix} = \begin{vmatrix} a_{i_1 i_1} & \cdots & a_{i_1 i_p} \\ \vdots & & \vdots \\ a_{i_p i_1} & \cdots & a_{i_p i_p} \end{vmatrix},$$

which is defined as the *principal minor* of order p of A. Choosing $i_1 = 1, i_2 = 2, \ldots, i_p = p$, the principal minor becomes

$$\begin{vmatrix} a_{11} & \cdots & a_{1p} \\ \vdots & & \vdots \\ a_{p1} & \cdots & a_{pp} \end{vmatrix}, \qquad 1 \le p \le n,$$

which is termed the *leading principal minor* of order p of A. □

Example 2.21 For A given by (2.18),

$$\begin{vmatrix} a_{11} & a_{13} & a_{15} \\ a_{31} & a_{33} & a_{35} \\ a_{51} & a_{53} & a_{55} \end{vmatrix}$$

is a principal minor of order three, while the leading principal minor of order three is

$$\begin{vmatrix} a_{11} & a_{12} & a_{13} \\ a_{21} & a_{22} & a_{23} \\ a_{31} & a_{32} & a_{33} \end{vmatrix},$$

which is also a principal minor. A leading principal minor is a principal minor but not conversely. In general, there is only one leading principal minor of order p, but the number of principal minors of order p is ${}^nC_p$. □

Method III: *Binet-Cauchy formula*. The determinant of a square matrix obtained from the product of two rectangular matrices

$$A = \begin{bmatrix} a_{11} & a_{12} & \cdots & a_{1n} \\ \vdots & \vdots & & \vdots \\ a_{m1} & a_{m2} & \cdots & a_{mn} \end{bmatrix}, \quad B = \begin{bmatrix} b_{11} & \cdots & b_{1m} \\ b_{21} & \cdots & b_{2m} \\ \vdots & & \vdots \\ b_{n1} & \cdots & b_{nm} \end{bmatrix}$$

is

$$\det AB = \sum_{(j)} \begin{vmatrix} a_{1j_1} & \cdots & a_{1j_m} \\ \vdots & & \vdots \\ a_{mj_1} & \cdots & a_{mj_m} \end{vmatrix} \begin{vmatrix} b_{j_11} & \cdots & b_{j_1m} \\ \vdots & & \vdots \\ b_{j_m1} & \cdots & b_{j_mm} \end{vmatrix} \quad \text{if } m \le n,$$

$$= 0 \quad \text{if} \quad m > n \tag{2.21}$$

where the summation is over all possible permutation of the indices

$$1 \le j_1 < j_2 < \ldots < j_m \le n.$$

Before providing a proof of the above formula, we shall take up an elementary example to understand the basic processes involved in working it out.

Example 2.22 For $m = 2$ and $n = 3$ in A and B of the Binet-Cauchy formula, we have, by direct multiplication and by using Properties V and VIII of a determinant, the relation

$$\det AB = \det \begin{bmatrix} a_{11} & a_{12} & a_{13} \\ a_{21} & a_{22} & a_{23} \end{bmatrix} \begin{bmatrix} b_{11} & b_{12} \\ b_{21} & b_{22} \\ b_{31} & b_{32} \end{bmatrix}$$

$$= \begin{vmatrix} (a_{11}b_{11} + a_{12}b_{21} + a_{13}b_{31}) & (a_{11}b_{12} + a_{12}b_{22} + a_{13}b_{32}) \\ (a_{21}b_{11} + a_{22}b_{21} + a_{23}b_{31}) & (a_{21}b_{12} + a_{22}b_{22} + a_{23}b_{32}) \end{vmatrix}$$

$$= \begin{vmatrix} a_{11} & a_{11} \\ a_{21} & a_{21} \end{vmatrix} b_{11}b_{12} + \begin{vmatrix} a_{11} & a_{12} \\ a_{21} & a_{22} \end{vmatrix} b_{11}b_{22} + \begin{vmatrix} a_{11} & a_{13} \\ a_{21} & a_{23} \end{vmatrix} b_{11}b_{32}$$

$$+ \begin{vmatrix} a_{12} & a_{11} \\ a_{22} & a_{21} \end{vmatrix} b_{21}b_{12} + \begin{vmatrix} a_{12} & a_{12} \\ a_{22} & a_{22} \end{vmatrix} b_{21}b_{22} + \begin{vmatrix} a_{12} & a_{13} \\ a_{22} & a_{23} \end{vmatrix} b_{21}b_{32}$$

$$+ \begin{vmatrix} a_{13} & a_{11} \\ a_{23} & a_{21} \end{vmatrix} b_{31}b_{12} + \begin{vmatrix} a_{13} & a_{12} \\ a_{23} & a_{22} \end{vmatrix} b_{31}b_{22} + \begin{vmatrix} a_{13} & a_{13} \\ a_{23} & a_{23} \end{vmatrix} b_{31}b_{32}$$

$$= 0 + \begin{vmatrix} a_{11} & a_{12} \\ a_{21} & a_{22} \end{vmatrix} (b_{11}b_{22} - b_{12}b_{21}) + \begin{vmatrix} a_{11} & a_{13} \\ a_{21} & a_{23} \end{vmatrix} (b_{11}b_{32} - b_{12}b_{31})$$

$$+ 0 + \begin{vmatrix} a_{12} & a_{13} \\ a_{22} & a_{23} \end{vmatrix} (b_{21}b_{32} - b_{31}b_{22}) + 0$$

$$= \begin{vmatrix} a_{11} & a_{12} \\ a_{21} & a_{22} \end{vmatrix} \begin{vmatrix} b_{11} & b_{12} \\ b_{21} & b_{22} \end{vmatrix} + \begin{vmatrix} a_{11} & a_{13} \\ a_{21} & a_{23} \end{vmatrix} \begin{vmatrix} b_{11} & b_{12} \\ b_{31} & b_{32} \end{vmatrix}$$

$$+ \begin{vmatrix} a_{12} & a_{13} \\ a_{22} & a_{23} \end{vmatrix} \begin{vmatrix} b_{21} & b_{22} \\ b_{31} & b_{32} \end{vmatrix}$$

which is, in fact, obtainable by applying directly the Binet-Cauchy formula. The reader is asked to try a simple case when $m > n$, e.g., $m = 3$, $n = 2$. □

Proof of Binet-Cauchy Formula : We have

$$\det AB = \begin{vmatrix} \sum\limits_{\alpha_1=1}^{n} a_{1\alpha_1} b_{\alpha_1 1} & \cdots & \sum\limits_{\alpha_m=1}^{n} a_{1\alpha_m} b_{\alpha_m m} \\ \vdots & & \vdots \\ \sum\limits_{\alpha_1=1}^{n} a_{m\alpha_1} b_{\alpha_1 1} & \cdots & \sum\limits_{\alpha_m=1}^{n} a_{m\alpha_m} b_{\alpha_m m} \end{vmatrix}$$

$$= \sum_{\alpha_1, \ldots, \alpha_m = 1}^{n} \begin{vmatrix} a_{1\alpha_1} b_{\alpha_1 1} & \cdots & a_{1\alpha_m} b_{\alpha_m m} \\ \vdots & & \vdots \\ a_{m\alpha_1} b_{\alpha_1 1} & \cdots & a_{m\alpha_m} b_{\alpha_m m} \end{vmatrix}$$

$$= \sum_{\alpha_1, \ldots, \alpha_m = 1}^{n} \begin{vmatrix} a_{1\alpha_1} & \cdots & a_{1\alpha_m} \\ \vdots & & \vdots \\ a_{m\alpha_1} & \cdots & a_{m\alpha_m} \end{vmatrix} b_{\alpha_1 1} b_{\alpha_2 2} \cdots b_{\alpha_m m}. \tag{2.22}$$

In case $m > n$, we observe that the number of columns in matrix A is n, which is less than the m number of columns in the determinants in (2.22) constituted by a_{ij}'s. Therefore, if $m > n$, it is evident that at least two columns of the determinants in (2.22) are identical, i.e., they have the same value for the column indices α's. This reduces det AB to zero.

Considering the case $m \le n$, we find that (2.22) is a sum of n^m determinants which divide into two classes. To one class belong such determinants where each column of A is repeated upto a maximum of m times, implying that at least two α's are identical in (2.22). The determinants in this class are therefore equal to zero (cf. Example 2.22).

There are some determinants which belong to the other class. Each of these determinants has the same m columns of A, but their column positions are interchanged in (2.22) in all possible ways. Since in this case the m α's are distinct but fixed integers and $m!$ permutations are possible among these m α's, these determinants form a group of $m!$ members. Again, since m distinct α's can be selected in ${}^nC_m$ ways from n α's, the determinants in this class are divided into ${}^nC_m$ groups, each group being represented by $m!$ determinants. For any such group, we arrange the m column indices α's of each determinant in increasing order, $1 \le j_1 < j_2 < \ldots <$

$j_m \leq n$. As this operation implies interchange among columns of this determinant and a possible change of its sign, we multiply it by $r(\alpha) = \varepsilon_{\alpha_1\alpha_2\cdots\alpha_m}$ to keep its sign intact. Assembling the determinants of this group, we get

$$\sum_{(\alpha)} \varepsilon_{\alpha_1\alpha_2\cdots\alpha_m} \begin{vmatrix} a_{1j_1} & \cdots & a_{1j_m} \\ \vdots & & \vdots \\ a_{mj_1} & \cdots & a_{mj_m} \end{vmatrix} b_{\alpha_1 1} b_{\alpha_2 2} \cdots b_{\alpha_m m}$$

$$= \begin{vmatrix} a_{1j_1} & \cdots & a_{1j_m} \\ \vdots & & \vdots \\ a_{mj_1} & \cdots & a_{mj_m} \end{vmatrix} \sum_{(\alpha)} \varepsilon_{\alpha_1\alpha_2\cdots\alpha_m} b_{\alpha_1 1} b_{\alpha_2 2} \cdots b_{\alpha_m m}$$

where the summation extends over $m!$ permutations of α's among the fixed set of integers $j_1 j_2 \ldots j_m$. In view of (2.5), the above summation represents column expansion of a determinant. Consequently, we have, for this group, a product of two determinants given by

$$\begin{vmatrix} a_{1j_1} & \cdots & a_{1j_m} \\ \vdots & & \vdots \\ a_{mj_1} & \cdots & a_{mj_m} \end{vmatrix} \begin{vmatrix} b_{j_1 1} & \cdots & b_{j_1 m} \\ \vdots & & \vdots \\ b_{j_m 1} & \cdots & b_{j_m m} \end{vmatrix}.$$

As already pointed out, since there are ${}^nC_m$ groups in the second class, we have, summing the products of all these groups, the Binet-Cauchy formula (2.21). □

A direct consequence of the Binet-Cauchy formula is the following corollary.

Corollary 2.1 Let A and B be two matrices of order n. Then

$$\det AB = \det A \det B.$$ □

Example 2.23 Applying the Binet-Cauchy formula, the determinant Δ of the product

$$\begin{bmatrix} 1 & 2 & -1 \\ 2 & 1 & 1 \end{bmatrix} \begin{bmatrix} 2 & -1 \\ -1 & 2 \\ 1 & 1 \end{bmatrix}$$

is

$$\Delta = \begin{vmatrix} 1 & 2 \\ 2 & 1 \end{vmatrix} \begin{vmatrix} 2 & -1 \\ -1 & 2 \end{vmatrix} + \begin{vmatrix} 1 & -1 \\ 2 & 1 \end{vmatrix} \begin{vmatrix} 2 & -1 \\ 1 & 1 \end{vmatrix} + \begin{vmatrix} 2 & -1 \\ 1 & 1 \end{vmatrix} \begin{vmatrix} -1 & 2 \\ 1 & 1 \end{vmatrix}$$

$$= -3 \cdot 3 + 3 \cdot 3 + 3 \cdot -3 = -9.$$ □

Method IV: *Pivotal condensation.* By computing the determinants of only second order matrices, we can evaluate the determinant of an n-square matrix in $(n - 1)$ successive steps: in each step the original matrix is reduced in order by one.

To initiate the procedure, we have to select any nonzero element a_{ij} which is termed the *pivot element*. We select three more elements a_{ik}, a_{qj} and a_{qk}, of which a_{ik} is in the same row and a_{qj} is in the same column as a_{ij}, so that these four elements in their respective places form a second order matrix (see Figure 2.2).

$$\text{(a)}\ \begin{bmatrix} a_{11} & a_{12} & \cdots & a_{1j} & \cdots & a_{1k} & \cdots & a_{1n} \\ \vdots & \vdots & & \vdots & & \vdots & & \vdots \\ a_{i1} & a_{i2} & \cdots & a_{ij} & \cdots & a_{ik} & \cdots & a_{in} \\ \vdots & \vdots & & \vdots & & \vdots & & \vdots \\ a_{q1} & a_{q2} & \cdots & a_{qj} & \cdots & a_{qk} & \cdots & a_{qn} \\ \vdots & \vdots & & \vdots & & \vdots & & \vdots \\ a_{n1} & a_{n2} & \cdots & a_{nj} & \cdots & a_{nk} & \cdots & a_{nn} \end{bmatrix},$$

$$\text{(b)}\ \begin{bmatrix} a_{qk} & \cdots & a_{qj} \\ \vdots & & \vdots \\ a_{ik} & \cdots & a_{ij} \end{bmatrix}, \quad \text{(c)}\ \begin{bmatrix} a_{ik} & \cdots & a_{ij} \\ \vdots & & \vdots \\ a_{qk} & \cdots & a_{qj} \end{bmatrix}, \quad \text{(d)}\ \begin{bmatrix} a_{qj} & \cdots & a_{qk} \\ \vdots & & \vdots \\ a_{ij} & \cdots & a_{ik} \end{bmatrix}.$$

Figure 2.2 Computation of second order matrices in pivotal condensation: a_{ij} is the pivot element.

Form all possible second order matrices, including the pivot element in each case. The original n-square matrix is now replaced by $1/(a_{ij})^{n-2}$ times an $(n-1)$ square matrix with elements as determinants of these second order matrices. When a_{ij} is on the main diagonal, each determinant takes the position $(q-1, k-1)$ if a_{ij} is above a_{qk} (see Figure 2.2(a)) or (q, k) if a_{ij} is below a_{qk} (see Figure 2.2(b)). On the other hand, when a_{ij} is on a secondary diagonal, each determinant takes the position $(q-1, k)$ if a_{ij} is above a_{qk} (Figure 2.2(c)) or $(q, k-1)$, if a_{ij} is below a_{qk} (see Figure 2.2(d)). This process is repeated until the original matrix is reduced to a second order matrix from which the required determinant can be written down just by seeing it.

To prove the above computational steps, we take the case of a third order matrix A given by

$$A = \begin{bmatrix} a_{11} & a_{12} & a_{13} \\ a_{21} & a_{22} & a_{23} \\ a_{31} & a_{32} & a_{33} \end{bmatrix}.$$

If we choose any nonzero element, say, a_{23} as the pivot element, then we can write

$$\det A = \frac{1}{a_{23}^2} \begin{bmatrix} a_{23}a_{11} & a_{23}a_{12} & a_{23}a_{13} \\ a_{21} & a_{22} & a_{23} \\ a_{23}a_{31} & a_{23}a_{32} & a_{23}a_{33} \end{bmatrix}.$$

Now subtracting the second row a_{13} times from the first and a_{33} times from the third, we have

$$\det A = \frac{1}{a_{23}^2}\begin{vmatrix} a_{23}a_{11} - a_{13}a_{21} & a_{23}a_{12} - a_{13}a_{22} & 0 \\ a_{21} & a_{22} & a_{23} \\ a_{23}a_{31} - a_{33}a_{21} & a_{23}a_{32} - a_{33}a_{22} & 0 \end{vmatrix}$$

$$= \frac{1}{a_{23}}\begin{Vmatrix} \begin{vmatrix} a_{11} & a_{13} \\ a_{21} & a_{23} \end{vmatrix} & \begin{vmatrix} a_{12} & a_{13} \\ a_{22} & a_{23} \end{vmatrix} \\ \begin{vmatrix} a_{21} & a_{23} \\ a_{31} & a_{33} \end{vmatrix} & \begin{vmatrix} a_{22} & a_{23} \\ a_{32} & a_{33} \end{vmatrix} \end{Vmatrix},$$

which shows that in a third order case for evaluating determinants, the matrix can be replaced by $1/(a_{23})^{n-2}$ times an $(n - 1)$ square matrix with elements as the determinants of second order matrices. The extension of the proof to a higher order matrix case is now obvious.

It can be shown that the number of multiplications required in evaluation of determinants of an nth order matrix by this method is $(2n^3 - 3n^2 + n)/3$, which is less than $n!\,(n - 1)$ needed by (2.2) or (2.5) when $n > 2$. For instance, if $n = 5$, then (2.2) or (2.5) needs 480 multiplications while pivotal condensation requires 60 only.

Example 2.24 To evaluate the determinant

$$\Delta = \begin{vmatrix} 1 & 2 & 3 & 4 \\ 2 & 3 & 4 & 1 \\ 3 & 4 & 1 & 2 \\ 4 & 1 & 2 & 3 \end{vmatrix}$$

by pivotal condensation, we choose $a_{11} = 1$ as the pivot element and hence get, by the above method, the relation

$$\Delta = \frac{1}{1^2}\begin{Vmatrix} \begin{vmatrix} 1 & 2 \\ 2 & 3 \end{vmatrix} & \begin{vmatrix} 1 & 3 \\ 2 & 4 \end{vmatrix} & \begin{vmatrix} 1 & 4 \\ 2 & 1 \end{vmatrix} \\ \begin{vmatrix} 1 & 2 \\ 3 & 4 \end{vmatrix} & \begin{vmatrix} 1 & 3 \\ 3 & 1 \end{vmatrix} & \begin{vmatrix} 1 & 4 \\ 3 & 2 \end{vmatrix} \\ \begin{vmatrix} 1 & 2 \\ 4 & 1 \end{vmatrix} & \begin{vmatrix} 1 & 3 \\ 4 & 2 \end{vmatrix} & \begin{vmatrix} 1 & 4 \\ 4 & 3 \end{vmatrix} \end{Vmatrix} = \begin{vmatrix} -1 & -2 & -7 \\ -2 & -8 & -10 \\ -7 & -10 & -13 \end{vmatrix}.$$

Again choosing $a_{11} = -1$ as the pivot element, we obtain

$$\Delta = \frac{1}{(-1)}\begin{Vmatrix} \begin{vmatrix} -1 & -2 \\ -2 & -8 \end{vmatrix} & \begin{vmatrix} -1 & -7 \\ -2 & -10 \end{vmatrix} \\ \begin{vmatrix} -1 & -2 \\ -7 & -10 \end{vmatrix} & \begin{vmatrix} -1 & -7 \\ -7 & -13 \end{vmatrix} \end{Vmatrix} = -\begin{vmatrix} 4 & -4 \\ -4 & -36 \end{vmatrix}.$$

Hence, $\Delta = -(-144 - 16) = 160$.

```
In MATLAB,
>> det (Δ)
responds with
ans =
     160
```

PROBLEMS

2.1 How many inversions are there in the following permutations?

(a) 5 3 1 6 4 2,

(b) 7 3 4 1 5 6 2,

(c) 9 1 2 8 7 4 6 5 3,

(d) $n, n-1, ..., 3, 2, 1,$

(e) $1, 3, 5, 7, ..., 2n-1, 2, 4, 6, ..., 2n,$

(f) $2, 4, 6, 8, ..., 2n, 1, 3, 5, ..., 2n-1,$

(g) $3, 6, 9, ..., 3n, 1, 4, 7, ..., 3n-2, 2, 5, ..., 3n-1,$

(h) $1, 4, 7, ..., 3n-2, 2, 5, ..., 3n-1, 3, 6, ..., 3n.$

2.2 Determine i and k in each of the following cases so that the permutation is even:

(a) 2 1 4 i 6 8 k 5, (b) i 4 2 3 k 6, (c) 3 8 5 i k 1 7 6,

(d) 4 i 3 8 k 1 5 7, (e) 9 5 1 i 6 8 k 2 4, (f) 5 7 3 i 1 4 k.

2.3 Determine i and k in each of the following cases so that the permutation is odd:

(a) 1 5 7 i 9 8 k 3 6, (b) 6 1 8 i 2 7 3 k, (c) 7 i 2 6 1 5 k,

(d) 2 5 8 i 1 6 7 k, (e) 5 2 i k 1 8 6 3, (f) i 8 2 6 5 1 3 k.

2.4 Determine $c(j)$ as in Definition 2.1, and hence the sign associated with each of the following terms in the expansion of a third order determinant:

(a) $a_{11}\, a_{22}\, a_{33}$, (b) $a_{12}\, a_{23}\, a_{31}$, (c) $a_{13}\, a_{21}\, a_{32}$,

(d) $a_{13}\, a_{22}\, a_{31}$, (e) $a_{11}\, a_{23}\, a_{32}$, (f) $a_{12}\, a_{21}\, a_{33}$.

2.5 Determine $r(i)$ as in Definition 2.2, and hence the sign associated with each of the following terms in the column expansion of a third order determinant:

(a) $a_{21}\, a_{32}\, a_{13}$, (b) $a_{31}\, a_{12}\, a_{23}$, (c) $a_{31}\, a_{22}\, a_{13}$,

(d) $a_{11}\, a_{32}\, a_{23}$, (e) $a_{21}\, a_{12}\, a_{33}$, (f) $a_{11}\, a_{22}\, a_{33}$.

2.6 Formulate a rule as in Fig. 2.1 for the column expansion of a third order determinant.

2.7 Counting the number of inversions, determine the sign of each of the following terms in the row expansion of an nth order determinant of $[a_{ij}]$:

(a) The product of principal diagonal elements,

(b) The product of secondary diagonal elements.

2.8 Which of the following products actually appear in the expansion of a fifth order determinant? Determine their sign.

(a) $a_{45}\ a_{32}\ a_{14}\ a_{51}\ a_{23}$, (b) $a_{14}\ a_{34}\ a_{25}\ a_{41}\ a_{52}$,

(c) $a_{13}\ a_{21}\ a_{34}\ a_{42}\ a_{55}$, (d) $a_{32}\ a_{43}\ a_{14}\ a_{25}\ a_{51}$,

(e) $a_{31}\ a_{23}\ a_{43}\ a_{14}\ a_{52}$, (f) $a_{23}\ a_{35}\ a_{41}\ a_{12}\ a_{54}$.

2.9 Evaluate the determinants of the following second order matrices:

(a) $\begin{bmatrix} 2 & 1 \\ 3 & 2 \end{bmatrix}$, (b) $\begin{bmatrix} \cos\alpha & \sin\alpha \\ -\sin\beta & \cos\beta \end{bmatrix}$, (c) $\begin{bmatrix} \cosh t & \sinh t \\ -\sinh t & \cosh t \end{bmatrix}$,

(d) $\begin{bmatrix} 1 & \log_x y \\ \log_y x & 1 \end{bmatrix}$, (e) $\begin{bmatrix} \lambda^2-\lambda+1 & \lambda^3 \\ 1 & \lambda+1 \end{bmatrix}$,

(f) $\begin{bmatrix} 1 & \omega^2 \\ -\omega & \omega \end{bmatrix}$, $\omega^3 = 1$ but $\omega \neq 1$;

(g) $\begin{bmatrix} 1+i & 1 \\ i & i \end{bmatrix}$, $i^2 = -1$.

2.10 Determine the minors and cofactors of all the elements for the matrices given in Problem 2.9.

2.11 With the help of Figure 2.1, evaluate the determinants of the following third order matrices:

(a) $\begin{bmatrix} 1 & 3 & 2 \\ 3 & 2 & 1 \\ 2 & 1 & 3 \end{bmatrix}$, (b) $\begin{bmatrix} a & c & b \\ c & b & a \\ b & a & c \end{bmatrix}$,

(c) $\begin{bmatrix} 1 & 1 & \omega \\ 1 & 1 & \omega^2 \\ \omega^2 & \omega & 1 \end{bmatrix}$, $\omega^3 = 1$ but $\omega \neq 1$;

(d) $\begin{bmatrix} 1-i & 0 & 1 \\ 0 & i & 1+i \\ 1+i & -i & 1 \end{bmatrix}$, $i^2 = -1$; (e) $\begin{bmatrix} -1 & 6 & 3 \\ 1 & 2 & 0 \\ -4 & 2 & -5 \end{bmatrix}$.

2.12 Determine the minors and cofactors of all the elements of the matrices in Problem 2.11.

2.13 Use only the definition of a determinant to show that

(a) $$\begin{vmatrix} a & \alpha & 0 & 0 & 0 \\ b & \beta & 0 & 0 & 0 \\ c & \gamma & 0 & 0 & 0 \\ d & \delta & x & y & z \\ e & \varepsilon & u & v & w \end{vmatrix} = 0,$$

(b) $$\begin{vmatrix} 0 & 0 & \cdots & 0 & 1 \\ 0 & 0 & \cdots & 1 & 0 \\ \vdots & \vdots & & \vdots & \vdots \\ 1 & 0 & \cdots & 0 & 0 \end{vmatrix}_n = (-1)^{\alpha},$$

where $\alpha = \dfrac{n(n-1)}{2}$.

2.14 Show that

$$\sum_{j=1}^{n} c_j A_{ij} \text{ (resp } \sum_{i=1}^{n} c_i A_{ij})$$

is the determinant of a matrix $A = [a_{ij}]$ obtained by replacing all the elements in its ith row (resp jth column) by the corresponding elements $c_1, c_2, \ldots, c_n$.

2.15 Show by induction that the determinant of the *Vandermonde matrix*

$$V(\lambda_1, \lambda_2, \ldots, \lambda_n) = \begin{bmatrix} 1 & 1 & \cdots & 1 \\ \lambda_1 & \lambda_2 & \cdots & \lambda_n \\ \vdots & \vdots & & \vdots \\ \lambda_1^{n-1} & \lambda_2^{n-1} & \cdots & \lambda_n^{n-1} \end{bmatrix}$$

is

$$\prod_{1 \le i < j \le n} (\lambda_j - \lambda_i).$$

2.16 Compute the values of the following determinants by using the properties of a determinant, pivotal condensation, and Laplace expansion:

(a) $$\begin{vmatrix} 5 & 6 & 7 \\ 6 & 7 & 8 \\ 6 & 5 & 4 \end{vmatrix},$$

(b) $$\begin{vmatrix} \cos\theta & \cos(\theta - 120°) & \cos(\theta + 120°) \\ -\sin\theta & -\sin(\theta - 120°) & -\sin(\theta + 120°) \\ \tfrac{1}{2} & \tfrac{1}{2} & \tfrac{1}{2} \end{vmatrix},$$

(c) $\begin{vmatrix} 1 & -1 & 1 & -1 \\ 2 & 1 & -1 & 1 \\ 3 & 2 & 1 & -1 \\ 4 & 3 & 2 & 1 \end{vmatrix}$,

(d) $\begin{vmatrix} 1 & 3 & 2 & 0 \\ 4 & 2 & 1 & 0 \\ 3 & 0 & 3 & -1 \\ 2 & 0 & 1 & -4 \end{vmatrix}$,

(e) $\begin{vmatrix} 1 & -2 & 2 & -1 & 3 \\ 2 & -1 & 1 & 4 & 1 \\ 0 & 0 & -2 & 0 & 0 \\ 2 & 0 & 2 & 2 & 3 \\ 1 & 0 & 1 & 3 & 1 \end{vmatrix}$.

2.17 Prove the following identities without evaluating the determinant:

(a) $\begin{vmatrix} 1 & a & bc \\ 1 & b & ca \\ 1 & c & ab \end{vmatrix} = \begin{vmatrix} 1 & a & a^2 \\ 1 & b & b^2 \\ 1 & c & c^2 \end{vmatrix}$,

(b) $\begin{vmatrix} a+b & b+c & c+a \\ b+c & c+a & a+b \\ c+a & a+b & b+c \end{vmatrix} = 2\begin{vmatrix} a & b & c \\ b & c & a \\ c & a & b \end{vmatrix}$,

(c) $\begin{vmatrix} 0 & a & b & c \\ a & 0 & c & b \\ b & c & 0 & a \\ c & b & a & 0 \end{vmatrix} = \begin{vmatrix} 0 & 1 & 1 & 1 \\ 1 & 0 & c^2 & b^2 \\ 1 & c^2 & 0 & a^2 \\ 1 & b^2 & a^2 & 0 \end{vmatrix}$,

(d) $\begin{vmatrix} 0 & x^2 & y^2 & z^2 \\ x^2 & 0 & c^2 & b^2 \\ y^2 & c^2 & 0 & a^2 \\ z^2 & b^2 & a^2 & 0 \end{vmatrix} = \begin{vmatrix} 0 & ax & by & cz \\ ax & 0 & cz & by \\ by & cz & 0 & ax \\ cz & by & ax & 0 \end{vmatrix}$,

(e) $\begin{vmatrix} bcd & cda & dab & abc \\ a & b & c & d \\ a^2 & b^2 & c^2 & d^2 \\ a^3 & b^3 & c^3 & d^3 \end{vmatrix} = \begin{vmatrix} 1 & 1 & 1 & 1 \\ a^2 & b^2 & c^2 & d^2 \\ a^3 & b^3 & c^3 & d^3 \\ a^4 & b^4 & c^4 & d^4 \end{vmatrix}$.

2.18 Prove the following identities:

(a) $$\begin{vmatrix} 1 & 1 & 1 \\ a & b & c \\ bc & ca & ab \end{vmatrix} = (a-b)(b-c)(c-a),$$

(b) $$\begin{vmatrix} 1 & 1 & 1 \\ b+c & c+a & a+b \\ b^2+c^2 & c^2+a^2 & a^2+b^2 \end{vmatrix} = (a-b)(b-c)(c-a),$$

(c) $$\begin{vmatrix} a & b & c \\ a^2 & b^2 & c^2 \\ b+c & c+a & a+b \end{vmatrix} = (a-b)(b-c)(c-a)(a+b+c),$$

(d) $$\begin{vmatrix} 1 & 1 & 1 \\ a & b & c \\ a^3 & b^3 & c^3 \end{vmatrix} = (a-b)(b-c)(c-a)(a+b+c),$$

(e) $$\begin{vmatrix} 1 & 1 & 1 & 1 \\ a & b & c & d \\ a^2 & b^2 & c^2 & d^2 \\ a^3 & b^3 & c^3 & d^3 \end{vmatrix} = (a-b)(a-c)(a-d)(b-c)(b-d)(c-d),$$

(f) $$\begin{vmatrix} 1 & 1 & 1 & 1 \\ 1 & 1+a & 1 & 1 \\ 1 & 1 & 1+b & 1 \\ 1 & 1 & 1 & 1+c \end{vmatrix} = abc,$$

(g) $$\begin{vmatrix} x & a & b & c \\ a & x & b & c \\ a & b & x & c \\ a & b & c & x \end{vmatrix} = (x-a)(x-b)(x-c)(x+a+b+c).$$

2.19 Let $A(t) = [a_{ij}(t)]$ be a matrix with time varying elements. Show that

$$\frac{d}{dt}\det A(t) = \sum_{i=1}^{n} \det A_i(t)$$

where $A_i(t)$ is the matrix $A(t)$ with the ith row (or column) elements being replaced by their respective derivatives.

2.20 Compute all the principal minors and leading principal minors of the matrices associated with Problem 2.16.

2.21 Employ the Binet-Cauchy formula to evaluate the determinants of the square of the matrices associated with Problem 2.16.

2.22 Apply the Binet-Cauchy formula to compute the determinants of the following products:

(a) $\begin{bmatrix} 1 & 2 & 3 & 4 \\ 2 & 1 & 4 & 3 \end{bmatrix}\begin{bmatrix} 1 & 1 \\ 0 & 2 \\ 3 & 0 \\ 2 & 4 \end{bmatrix}$,

(b) $\begin{bmatrix} 1 & 2 & 3 & 3 \\ 2 & 3 & 1 & 2 \\ 3 & 1 & 2 & 1 \end{bmatrix}\begin{bmatrix} 1 & 4 & 1 \\ 4 & 2 & 4 \\ 3 & 4 & 3 \\ 1 & 2 & 1 \end{bmatrix}$.

2.23 Let

$$A = \begin{bmatrix} x_1 & x_2 & \cdots & x_n \\ y_1 & y_2 & \cdots & y_n \end{bmatrix}.$$

Computing det (AA^T) in two different ways, prove the *Schwarz inequality* for real variables, viz.

$$|(x_1y_1 + \cdots + x_ny_n)| \le (x_1^2 + \cdots + x_n^2)^{1/2}\,(y_1^2 + \cdots + y_n^2)^{1/2}.$$

2.24 Show by induction that the determinant of the n-matrix

$$A = \begin{bmatrix} c & 1 & 0 & 0 & \cdots & 0 & 0 & 0 \\ 1 & c & 1 & 0 & \cdots & 0 & 0 & 0 \\ 0 & 1 & c & 1 & \cdots & 0 & 0 & 0 \\ \vdots & \vdots & \vdots & \vdots & & \vdots & \vdots & \vdots \\ 0 & 0 & 0 & 0 & \cdots & 1 & c & 1 \\ 0 & 0 & 0 & 0 & \cdots & 0 & 1 & c \end{bmatrix}_n, c = 2\cos\theta \qquad \text{(Rayleigh)}$$

which arises in the study of transverse vibration of strings by Lagrange's method, is

$$\det A = \frac{\sin\,(n+1)\theta}{\sin\,\theta}.$$

2.25 Show by row or column expansion that the determinant of an upper and a lower triangular matrix is the product of the diagonal elements.

2.26 If A, B and C are square matrices of order n, then prove that det (ABC) = det A det B det C. Extend this result to the case of r number of matrices.

2.27 Use Laplace expansion to show that

$$\det\begin{bmatrix} A_1 & 0 \\ 0 & A_2 \end{bmatrix} = \det A_1 \det A_2 = \det\begin{bmatrix} A_1 & 0 \\ B_1 & A_2 \end{bmatrix} = \det\begin{bmatrix} A_1 & C_1 \\ 0 & A_2 \end{bmatrix},$$

where A_1 and A_2 are square matrices of any order, 0's are null matrices, and B_1 and C_1 are matrices of appropriate order. Extend this result to the case of n matrices $A_1, ..., A_n$. Can you extend this result to an infinite number of matrices?

2.28 Prove that if A_1 and A_2 are of order m and n respectively, then

$$\det\begin{bmatrix} A_3 & A_1 \\ A_2 & 0 \end{bmatrix} = (-1)^{(m+n+1)m} \det A_1 \det A_2.$$

2.29 Prove that the determinant of each of the four companion matrices given in (7.92) and (7.93) is equal to $(-1)^n a_0$.

2.30 If A and B are $m \times n$ and $n \times m$ matrices respectively and I_n an identity matrix of order n, then show that

$$\det\begin{bmatrix} A & 0 \\ -I_n & B \end{bmatrix} = \det\begin{bmatrix} A & AB \\ -I_n & 0 \end{bmatrix}.$$

2.31 Prove that the determinant of a
  (i) skew-symmetric matrix of odd order is equal to zero,
  (ii) skew-symmetric matrix is a perfect square,
  (iii) Hermitian matrix is a real number, and
  (iv) det A^*A is real if A is any matrix over $\mathbb{C}$.

2.32 The matrix of order n given by

$$A_n = \begin{bmatrix} 1 & 1 & 0 & \cdots & 0 & 0 \\ -1 & 1 & 1 & \cdots & 0 & 0 \\ 0 & -1 & 1 & \cdots & 0 & 0 \\ \vdots & \vdots & \vdots & & \vdots & \vdots \\ 0 & 0 & 0 & \cdots & -1 & 1 \end{bmatrix}$$

is known as the *Fibonacci matrix.* If $a_n = \det A_n$, show that $a_n = a_{n-1} + a_{n-2}$, $n \geq 3$ and det A_n is the nth term of the sequence 1, 2, 3, 5, 8, 13, ...

2.33 Define a tridiagonal matrix A_n of order n whose diagonal elements are $a_1, ..., a_n$, first superdiagonal elements $b_1, ..., b_{n-1}$ and first subdiagonal elements $c_1, ..., c_{n-1}$. Show that $\det A_n = a_n \det A_{n-1} - b_{n-1} c_{n-1} \det A_{n-2}$.

2.34 Show that the determinant of a matrix A of order n whose main diagonal elements are equal to zero except the nth one which is r and the remaining elements are equal to x, is

$$(-x)^{n-1} [(n-1)x - (n-2)r].$$

2.35 Prove that

(a) $$\begin{vmatrix} 0 & a_{12} & 0 & 0 & 0 \\ 0 & 0 & 0 & a_{24} & 0 \\ 0 & 0 & a_{33} & a_{34} & 0 \\ a_{41} & a_{42} & 0 & 0 & 0 \\ 0 & 0 & 0 & a_{54} & a_{55} \end{vmatrix} = a_{12}\, a_{24}\, a_{33}\, a_{41}\, a_{55},$$

(b) $$\begin{vmatrix} 1-n & 1 & \cdots & 1 \\ 1 & 1-n & \cdots & 1 \\ \vdots & \vdots & & \vdots \\ 1 & 1 & \cdots & 1-n \end{vmatrix}_n = 0,$$

(c) $$\begin{vmatrix} 1 & 1 & 1 & \cdots & 1 \\ 1 & 2 & 2^2 & \cdots & 2^{n-1} \\ 1 & 3 & 3^2 & \cdots & 3^{n-1} \\ \vdots & \vdots & \vdots & & \vdots \\ 1 & n & n^2 & \cdots & n^{n-1} \end{vmatrix} = 1!\,2!\,3!\cdots(n-1)!.$$

2.36 Show that if $A = aI$, a is a scalar and

$$B = \begin{bmatrix} b_1 & b_2 & \cdots & b_{n-1} & b_n \\ -1 & 0 & \cdots & 0 & 0 \\ 0 & -1 & \cdots & 0 & 0 \\ \vdots & \vdots & & \vdots & \vdots \\ 0 & 0 & \cdots & -1 & 0 \end{bmatrix},$$

then $\det(A + B) = a^n + a^{n-1} b_1 + a^{n-2} b_2 + \cdots + b_n$.

CHAPTER

3

Inverse of a Matrix

An important notion about the operation of scalars, which are real or complex numbers, is the division of one scalar by another nonzero scalar. After our acquaintance with three of the four basic operations, *viz.* addition, subtraction and multiplication of matrices in Chapter 1, we now turn to the definition of the final basic operation of division. This division operation in the case of a square matrix is implemented through what is known as the inverse of a matrix, the nonzero condition of a scalar for such division operation to be valid being that the determinant associated with the square matrix must not be zero. This concept of inverse of a matrix will be introduced in this chapter, which will be accompanied by important notions such as singularity of a matrix and adjoint of a matrix. Important properties of inverse operations are discussed and evaluation of matrix inversions by the method of partitioning introduced.

3.1 SINGULAR MATRIX: ADJOINT AND INVERSE OF A MATRIX

Definition 3.1 *Singular and Nonsingular Matrix.* A square matrix A is said to be *singular* if $\det A = 0$; if $\det A \neq 0$, it is referred to as *nonsingular.* □

Example 3.1 Let us define

$$A = \begin{bmatrix} 1 & 2 & 3 \\ 3 & -1 & 2 \\ 2 & 0 & 2 \end{bmatrix}, \qquad B = \begin{bmatrix} 2 & 4 \\ 1 & 3 \end{bmatrix}.$$

By direct computation, it is easily seen that $\det A = 0$ and $\det B = 2 \neq 0$. Hence A is a singular matrix while B is nonsingular.

Definition 3.2 *Adjoint of a Matrix.* Let A_{ij} be the cofactor of any element a_{ij} in a matrix A of order n. Then the matrix formed with these cofactors placed in a transposed position is defined as the *adjoint* or *adjugate* of A and is denoted by adj A. □

Thus

$$\text{adj } A = [A_{ij}]^T = \begin{bmatrix} A_{11} & A_{21} & \cdots & A_{n1} \\ A_{12} & A_{22} & \cdots & A_{n2} \\ \vdots & \vdots & & \vdots \\ A_{1n} & A_{2n} & \cdots & A_{nn} \end{bmatrix}, \tag{3.1}$$

Recalling Definition 2.4, the adjoint of a third order matrix A in terms of minors has the representation

$$\text{adj } A = [(-1)^{i+j} M_{ij}]^T = \begin{bmatrix} +M_{11} & -M_{21} & +M_{31} \\ -M_{12} & +M_{22} & -M_{32} \\ +M_{13} & -M_{23} & +M_{33} \end{bmatrix},$$

where the alternate positive and negative sign pattern in the columns may be noted. Irrespective of the order of the matrix, the first column starts with a plus, the second with a minus, and so on.

Example 3.2 For the third order matrix given in Example 2.5, the adjoint is

$$\text{adj } A = \begin{bmatrix} 10 & -4 & -9 \\ -15 & 4 & 14 \\ 5 & -1 & -6 \end{bmatrix}.$$ □

Recalling Property VII in Section 2.3, for the expansion of a determinant in terms of a row (2.14) or a column (2.15), we obtain from (3.1) the relation

$$A \bullet \text{adj } A = \begin{bmatrix} a_{11} & a_{12} & \cdots & a_{1n} \\ a_{21} & a_{22} & \cdots & a_{2n} \\ \vdots & \vdots & & \vdots \\ a_{n1} & a_{n2} & \cdots & a_{nn} \end{bmatrix} \begin{bmatrix} A_{11} & A_{21} & \cdots & A_{n1} \\ A_{12} & A_{22} & \cdots & A_{n2} \\ \vdots & \vdots & & \vdots \\ A_{1n} & A_{2n} & \cdots & A_{nn} \end{bmatrix}$$

$$= \begin{bmatrix} \det A & 0 & \cdots & 0 \\ 0 & \det A & \cdots & 0 \\ \vdots & \vdots & & \vdots \\ 0 & 0 & \cdots & \det A \end{bmatrix} = \begin{bmatrix} A_{11} & A_{21} & \cdots & A_{n1} \\ A_{12} & A_{22} & \cdots & A_{n2} \\ \vdots & \vdots & & \vdots \\ A_{1n} & A_{2n} & \cdots & A_{nn} \end{bmatrix}$$

$$\times \begin{bmatrix} a_{11} & a_{12} & \cdots & a_{1n} \\ a_{21} & a_{22} & \cdots & a_{2n} \\ \vdots & \vdots & & \vdots \\ a_{n1} & a_{n2} & \cdots & a_{nn} \end{bmatrix} = \text{adj } A \cdot A,$$

or,

$$A \cdot \text{adj } A = \text{adj } A \cdot A = \det A \cdot I_n. \tag{3.2}$$

If A is nonsingular, dividing (3.2) by det A, we get

$$AR = RA = I_n, \qquad R = \text{adj } A/\det A. \tag{3.3}$$

Definition 3.3 *Inverse.* For a nonsingular matrix A of order n, there exists a matrix R of the same order such that

$$AR = RA = I_n.$$

Then R is defined as the *reciprocal* or *inverse* of the matrix A and is denoted by A^{-1}. □

Relation (3.3) suggests that, given A, its inverse is obtained as

$$A^{-1} = \frac{\text{adj } A}{\det A} = \frac{1}{\det A}\begin{bmatrix} A_{11} & A_{21} & \cdots & A_{n1} \\ A_{12} & A_{22} & \cdots & A_{n2} \\ \vdots & \vdots & & \vdots \\ A_{1n} & A_{2n} & \cdots & A_{nn} \end{bmatrix}, \tag{3.4}$$

which implies that the inverse of a square matrix A can be computed by dividing all the elements of adj A by det A. A^{-1} is unique and exists only when A is square and nonsingular. A singular matrix has an adjoint but no inverse since division by det $A = 0$ is not possible.

That the matrix multiplication is not commutative implies, in general, that for any matrix B, $A^{-1}B$ and BA^{-1} are different. Therefore, division by a matrix A cannot be defined in the form B/A, in which there is an ambiguity whether A predivides or postdivides B.

Example 3.3 Find the inverse of P, where

$$P = \begin{bmatrix} a & b \\ c & d \end{bmatrix}.$$

Solution If the cofactors of the elements are written by their capital letters, then

$$A = d, \quad B = -c, \quad C = -b, \quad D = a,$$
$$\det P = \Delta = ad - bc.$$

Therefore,

$$P^{-1} = \frac{1}{\Delta}\begin{bmatrix} A & C \\ B & D \end{bmatrix} = \frac{1}{ad - bc}\begin{bmatrix} d & -b \\ -c & a \end{bmatrix}.$$ □

This example illustrates that the inverse of a second order matrix is obtained by inspection, i.e., by interchanging the main diagonal elements, reversing the sign of the secondary diagonal elements, and then dividing each element by the determinant of the matrix.

Example 3.4 Find the inverse of the matrix A in Example 2.5. From the cofactors computed there,

$$\det A = a_{11}A_{11} + a_{12}A_{12} + a_{13}A_{13} = 2\cdot 10 + 3\cdot(-15) + 4\cdot 5 = -5.$$

Therefore, from the adj A computed in Example 3.2, we have

$$A^{-1} = \frac{\text{adj } A}{\det A} = -\frac{1}{5}\begin{bmatrix} 10 & -4 & -9 \\ -15 & 4 & 14 \\ 5 & -1 & -6 \end{bmatrix}.$$ □

Example 3.5 To obtain the inverse of a diagonal matrix

$$A = [a_{ij}] = \text{diag}\,[a_{11}, a_{12}, ..., a_{nn}],$$

we note from Example 2.8 that

$$A_{ii} = \frac{\det A}{a_{ii}}; \qquad A_{ij} = 0, \qquad i \neq j.$$

Therefore,

$$A^{-1} = \text{diag}\left[\frac{1}{a_{11}}, \frac{1}{a_{22}}, \ldots, \frac{1}{a_{nn}}\right]. \qquad \square$$

Definition 3.4 *Adjoint Determinant.* The determinant of adj A of a square matrix A is defined as the *adjoint determinant* of A. □

From (3.2), observing that $(\det A)I_n$ is a diagonal matrix with each diagonal element equal to det A, we have, by Corollary 2.1 and Example 2.8, the relation

$$\det A \cdot \det\,(\text{adj}\,A) = (\det A)^n.$$

Hence, it follows that

$$\det\,(\text{adj}\,A) = (\det A)^{n-1}.$$

3.2 IMPORTANT PROPERTIES OF MATRIX INVERSION

Property I: For two nonsingular matrices A and B of order n,

$$(AB)^{-1} = B^{-1}A^{-1}. \tag{3.5}$$

This property follows from the observation that

$$(AB)(B^{-1}A^{-1}) = A(BB^{-1})A^{-1} = AI_nA^{-1} = AA^{-1} = I_n,$$
$$(B^{-1}A^{-1})(AB) = B^{-1}I_nB = I_n,$$

and from *Definition 3.3.*

Property II: If A is nonsingular, then $AB = AC$ implies $B = C$.
By premultiplying both sides of $AB = AC$ by A^{-1}, the result follows.

Property III: If A is nonsingular, then

(i) $(A^{-1})^{-1} = A$, (ii) $(A^T)^{-1} = (A^{-1})^T$,

(iii) $(\bar{A})^{-1} = \overline{(A^{-1})}$, and (iv) $(A^*)^{-1} = (A^{-1})^*$.

All the properties follow from $AA^{-1} = I_n$: (i) postmultiplying both sides by $(A^{-1})^{-1}$, (ii) taking the transpose of both sides and postmultiplying by $(A^T)^{-1}$, (iii) and (iv) follow similarly.

Property IV: If A is nonsingular and n is a positive integer, define $(A^{-1})^n = A^{-n}$. Then for all integral values of p and q, $A^pA^q = A^{p+q}$ and $(A^p)^q = A^{pq}$.

The proof follows from mathematical induction and the associative law of multiplication, and is given in Example 3.9.

Property V: If A and B are nonsingular and commutative, so too are : (i) A^{-1} and B^{-1}, (ii) A^{-1} and B, and (iii) A and B^{-1}. In addition, if A and B are symmetric, so too are the matrix products in (i), (ii) and (iii).

Since $AB = BA$, result (i) follows from (3.5), result (ii) from $A^{-1}(AB)A^{-1}$, and result (iii) from $B^{-1}(BA)B^{-1}$. The symmetry follows obviously from Property IV of matrix transposition and Property III (ii) above.

Property VI: If A and B are of order n, then (i) adj AB = adj B adj A, and (ii) adj (adj A) = $(\det A)^{n-2} A$ if $\det A \neq 0$.

The proof is left as an exercise to the reader (see Problem 3.9).

3.3 INVERSE OF A MATRIX BY PARTITIONING

Sometimes it is convenient to group the elements in a matrix into rectangular blocks called *submatrices* and treat these submatrices as elements in matrix operations. This process of treating the original matrix as a matrix whose elements are submatrices is known as the *partitioning of matrices*. For example, let A and B be two third order matrices given by

$$A = \left[\begin{array}{cc|c} a & b & c \\ d & e & f \\ \hline g & h & k \end{array}\right], \qquad B = \left[\begin{array}{cc|c} r & s & t \\ u & v & w \\ \hline x & y & z \end{array}\right],$$

where the elements are partitioned into rectangular blocks shown by dashed lines. The submatrices are, therefore,

$$\alpha_{11} = \begin{bmatrix} a & b \\ d & e \end{bmatrix}, \qquad \beta_{11} = \begin{bmatrix} r & s \\ u & v \end{bmatrix}, \qquad \alpha_{12} = \begin{bmatrix} c \\ f \end{bmatrix},$$

$$\beta_{12} = \begin{bmatrix} t \\ w \end{bmatrix}, \qquad \alpha_{21} = [g \quad h], \qquad \beta_{21} = [x \quad y],$$

$$\alpha_{22} = [k], \qquad \beta_{22} = [z].$$

After partitioning, A and B become, respectively,

$$A = \begin{bmatrix} \alpha_{11} & \alpha_{12} \\ \alpha_{21} & \alpha_{22} \end{bmatrix}, \qquad B = \begin{bmatrix} \beta_{11} & \beta_{12} \\ \beta_{21} & \beta_{22} \end{bmatrix},$$

where the elements α_{ij}'s and β_{ij}'s are submatrices as specified above.

Here subdivisions of columns and rows in A are identical with those in B. We define two matrices A and B of the same order to be *identically partitioned* if they are subdivided into equal number of submatrices and if the corresponding submatrices α_{ij} and β_{ij} are of the same order. In terms of submatrices, the sum

$$A + B = \begin{bmatrix} \alpha_{11} & \alpha_{12} \\ \alpha_{21} & \alpha_{22} \end{bmatrix} + \begin{bmatrix} \beta_{11} & \beta_{12} \\ \beta_{21} & \beta_{22} \end{bmatrix} = \begin{bmatrix} \alpha_{11} + \beta_{11} & \alpha_{12} + \beta_{12} \\ \alpha_{21} + \beta_{21} & \alpha_{22} + \beta_{22} \end{bmatrix}.$$

Two matrices of the same order are conformable for addition after partitioning if they are identically partitioned. The sum $\alpha_{ij} + \beta_{ij}$ is obtained in the same way as for matrix addition.

The matrices A and B are conformable for multiplication after partitioning as AB if the subdivisions of columns in A into groups are identical with those of rows in B. The sub-divisions of rows in A and columns in B are immaterial. Thus we obtain the following product according to the ordinary rule of matrix multiplication if we treat the submatrices as if they are scalar elements:

$$AB = \overset{\rightarrow\, p\, \leftarrow}{\left[\begin{array}{c|cc} a & b & c \\ d & e & f \\ \hline g & h & k \end{array}\right]} \left[\begin{array}{c|c} x & u \\ \hline y & v \\ z & w \end{array}\right] \begin{matrix} \downarrow \\ p \\ \uparrow \end{matrix} = \begin{bmatrix} \alpha_{11} & \alpha_{12} \\ \alpha_{21} & \alpha_{22} \end{bmatrix} \begin{bmatrix} \beta_{11} & \beta_{12} \\ \beta_{21} & \beta_{22} \end{bmatrix}$$

$$= \begin{bmatrix} \alpha_{11}\beta_{11} + \alpha_{12}\beta_{21} & \alpha_{11}\beta_{12} + \alpha_{12}\beta_{22} \\ \alpha_{21}\beta_{11} + \alpha_{22}\beta_{21} & \alpha_{21}\beta_{12} + \alpha_{22}\beta_{22} \end{bmatrix},$$

where the submatrices α_{ij} and β_{ij} are easy to identify.

Example 3.6 Obtain the following matrix product using the notion of partitioning:

$$\left[\begin{array}{c|cc|cc} 1 & 0 & 0 & 2 & 3 \\ 2 & 0 & 0 & 3 & 1 \\ \hline 3 & 2 & 1 & 0 & 0 \end{array}\right] \left[\begin{array}{cc} 2 & 3 \\ \hline 1 & 2 \\ 3 & 1 \\ \hline 2 & 2 \\ 1 & 1 \end{array}\right].$$

Solution Specifying the partitioning by dashed lines, for identical partitioning of columns of the first matrix and rows of the second matrix, we have

$$\left[\begin{array}{c|cc|cc} 1 & 0 & 0 & 2 & 3 \\ 2 & 0 & 0 & 3 & 1 \\ \hline 3 & 2 & 1 & 0 & 0 \end{array}\right] \left[\begin{array}{cc} 2 & 3 \\ \hline 1 & 2 \\ 3 & 1 \\ \hline 2 & 2 \\ 1 & 1 \end{array}\right]$$

$$= \begin{bmatrix} \begin{bmatrix} 1 \\ 2 \end{bmatrix} [2 \quad 3] + \begin{bmatrix} 0 & 0 \\ 0 & 0 \end{bmatrix} \begin{bmatrix} 1 & 2 \\ 3 & 1 \end{bmatrix} + \begin{bmatrix} 2 & 3 \\ 3 & 1 \end{bmatrix} \begin{bmatrix} 2 & 2 \\ 1 & 1 \end{bmatrix} \\ [3][2 \quad 3] + [2 \quad 1] \begin{bmatrix} 1 & 2 \\ 3 & 1 \end{bmatrix} + [0 \quad 0] \begin{bmatrix} 2 & 2 \\ 1 & 1 \end{bmatrix} \end{bmatrix}$$

$$= \begin{bmatrix} \begin{bmatrix} 2 & 3 \\ 4 & 6 \end{bmatrix} + \begin{bmatrix} 7 & 7 \\ 7 & 7 \end{bmatrix} \\ [6 \quad 9] + [5 \quad 5] \end{bmatrix} = \begin{bmatrix} 9 & 10 \\ 11 & 13 \\ 11 & 14 \end{bmatrix}.$$

Definition 3.5 *Quasi-diagonal Matrix.* In a partitioned matrix, all the elements are submatrices. A partitioned matrix with all the off-diagonal elements equal to null matrices is called a *quasi-diagonal* matrix. □

Let

$$A=\left[\begin{array}{cc|ccc} 2 & -1 & 0 & 0 & 0 \\ 1 & 0 & 0 & 0 & 0 \\ \hline 0 & 0 & 3 & 1 & 3 \\ 0 & 0 & 0 & 2 & 2 \\ 0 & 0 & 4 & 3 & 1 \end{array}\right], \quad \alpha=\begin{bmatrix} 2 & -1 \\ 1 & 0 \end{bmatrix}, \quad \beta=\begin{bmatrix} 3 & 1 & 3 \\ 0 & 2 & 2 \\ 4 & 3 & 1 \end{bmatrix}.$$

Then

$$A=\begin{bmatrix} \alpha & 0 \\ 0 & \beta \end{bmatrix},$$

where 0's are null matrices of appropriate order, is a quasi-diagonal matrix and is also written as diag (α, β).

The principle of partitioning offers a simple method of obtaining the inversion of a matrix. In order to formulate the basic equations leading to the inverse of a matrix using the concept of partitioning, let us consider n equations in n unknown $x_1, x_2, \ldots, x_n$:

$$\begin{aligned} a_{11}x_1 + a_{12}x_2 + \cdots + a_{1n}x_n &= y_1, \\ a_{21}x_1 + a_{22}x_2 + \cdots + a_{2n}x_n &= y_2, \\ &\vdots \\ a_{n1}x_1 + a_{n2}x_2 + \cdots + a_{nn}x_n &= y_n, \end{aligned}$$

where a_{ij}'s are known coefficients and $y_1, \ldots, y_n$ have prescribed values. In matrix notation these equations are

$$\begin{array}{r} k \text{ rows} \\ \\ (n-k) \text{ rows} \end{array} \left[\begin{array}{ccc|ccc} a_{11} & \cdots & a_{1k} & \cdots & a_{1n} \\ \vdots & & \vdots & & \vdots \\ a_{k1} & \cdots & a_{kk} & \cdots & a_{kn} \\ \hline \vdots & & \vdots & & \vdots \\ a_{n1} & \cdots & a_{nk} & \cdots & a_{nn} \end{array}\right] \left[\begin{array}{c} x_1 \\ \vdots \\ x_k \\ \hline \vdots \\ x_n \end{array}\right] \begin{array}{l} k \\ \\ (n-k) \end{array} = \left[\begin{array}{c} y_1 \\ \vdots \\ y_k \\ \hline \vdots \\ y_n \end{array}\right] \begin{array}{l} k \\ \\ (n-k) \end{array}, \tag{3.6}$$

$$k \text{ cols.} \qquad (n-k) \text{ cols.}$$

which in terms of submatrices is

$$\begin{bmatrix} \alpha_{11} & \alpha_{12} \\ \alpha_{21} & \alpha_{22} \end{bmatrix} \begin{bmatrix} \mathbf{x}_1 \\ \mathbf{x}_2 \end{bmatrix} = \begin{bmatrix} \mathbf{y}_1 \\ \mathbf{y}_2 \end{bmatrix}. \tag{3.7}$$

We denote in (3.6) the coefficient matrix by A, and the column vector on the left by $\mathbf{x}$, and on the right by $\mathbf{y}$. Thus, both (3.6) and (3.7) can be written as $A\mathbf{x} = \mathbf{y}$. Assuming A to be nonsingular, premultiplying both sides of $A\mathbf{x} = \mathbf{y}$ by A^{-1}, we have $\mathbf{x} = A^{-1}\mathbf{y}$. The problem of finding the inverse of A is thus equivalent to expressing the vector $\mathbf{x}$ in terms of the vector $\mathbf{y}$, when the coefficient matrix of $\mathbf{y}$ is the required inverse.

□ Carrying out the multiplication in (3.7) and by definition of equality of two matrices, we have

$$\alpha_{11}\mathbf{x}_1 + \alpha_{12}\mathbf{x}_2 = \mathbf{y}_1, \tag{3.8a}$$

$$\alpha_{21}\mathbf{x}_1 + \alpha_{22}\mathbf{x}_2 = \mathbf{y}_2. \tag{3.8b}$$

Choosing the partitioning in such a way that all the inverses appearing in the following relations exist, we get from (3.8a) the equation

$$\mathbf{x}_1 = \alpha_{11}^{-1}\mathbf{y}_1 - \alpha_{11}^{-1}\alpha_{12}\mathbf{x}_2. \tag{3.9}$$

Putting $\mathbf{x}_1$ into (3.8b), we obtain the result

$$\mathbf{x}_2 = (\alpha_{22} - \alpha_{21}\alpha_{11}^{-1}\alpha_{12})^{-1}(\mathbf{y}_2 - \alpha_{21}\alpha_{11}^{-1}\mathbf{y}_1), \tag{3.10}$$

which in turn eliminates $\mathbf{x}_2$ in (3.9), giving thereby

$$\mathbf{x}_1 = \alpha_{11}^{-1}\mathbf{y}_1 - \alpha_{11}^{-1}\alpha_{12}(\alpha_{22} - \alpha_{21}\alpha_{11}^{-1}\alpha_{12})^{-1}(\mathbf{y}_2 - \alpha_{21}\alpha_{11}^{-1}\mathbf{y}_1). \tag{3.11}$$

Let
$$\alpha_{22} - \alpha_{21}\alpha_{11}^{-1}\alpha_{12} = \beta. \tag{3.12}$$

Then (3.11) and (3.10) respectively become

$$\mathbf{x}_1 = (\alpha_{11}^{-1} + \alpha_{11}^{-1}\alpha_{12}\beta^{-1}\alpha_{21}\alpha_{11}^{-1})\mathbf{y}_1 - \alpha_{11}^{-1}\alpha_{12}\beta^{-1}\mathbf{y}_2,$$

$$\mathbf{x}_2 = -\beta^{-1}\alpha_{21}\alpha_{11}^{-1}\mathbf{y}_1 + \beta^{-1}\mathbf{y}_2,$$

or,

$$\begin{bmatrix} \mathbf{x}_1 \\ \hline \mathbf{x}_2 \end{bmatrix} = \left[\begin{array}{c|c} \alpha_{11}^{-1} + (\alpha_{11}^{-1}\alpha_{12})\beta^{-1}(\alpha_{21}\alpha_{11}^{-1}) & -\alpha_{11}^{-1}\alpha_{12}\beta^{-1} \\ \hline -\beta^{-1}\alpha_{21}\alpha_{11}^{-1} & \beta^{-1} \end{array}\right] \begin{bmatrix} \mathbf{y}_1 \\ \hline \mathbf{y}_2 \end{bmatrix}.$$

Therefore, according to the discussion after (3.7), we have

$$A^{-1} = \left[\begin{array}{c|c} \alpha_{11}^{-1} + (\alpha_{11}^{-1}\alpha_{12})\beta^{-1}(\alpha_{21}\alpha_{11}^{-1}) & -\alpha_{11}^{-1}\alpha_{12}\beta^{-1} \\ \hline -\beta^{-1}\alpha_{21}\alpha_{11}^{-1} & \beta^{-1} \end{array}\right], \tag{3.13}$$

where β is given by (3.12) and α_{ij}'s are as in (3.7). It is evident from (3.13) that inversion of a given order matrix is obtained in terms of inversion of lower order matrices, thus reducing computational labour. Particularly, if *A is symmetric,* then α_{ii}'s too are symmetric, and computational steps can be further reduced by putting $\gamma = \alpha_{11}^{-1}\alpha_{12}$, so that $\gamma^T = \alpha_{21}\alpha_{11}^{-1}$. Also,

$$\beta = \alpha_{22} - \alpha_{21}\alpha_{11}^{-1}\alpha_{12} = \alpha_{22} - \alpha_{21}\gamma,$$

which is symmetric.

In view of the foregoing relations, the inverse (3.13) assumes the following *convenient symmetrical form*:

$$A^{-1} = \begin{bmatrix} \alpha_{11}^{-1} + \gamma\beta^{-1}\gamma^T & (-\gamma\beta^{-1}) \\ (-\gamma\beta^{-1})^T & \beta^{-1} \end{bmatrix}. \tag{3.14a}$$

In case the inverses occurring in (3.13) or (3.14a) do not exist, new choices of α_{ij}'s should be made. In particular, if either α_{12} or α_{21} is zero, A^{-1} takes a very simple form, viz. from (3.12) and (3.13),

$$A^{-1} = \begin{bmatrix} \alpha_{11}^{-1} & 0 \\ -\alpha_{22}^{-1}\alpha_{21}\alpha_{11}^{-1} & \alpha_{22}^{-1} \end{bmatrix} \quad \text{or} \quad \begin{bmatrix} \alpha_{11}^{-1} & -\alpha_{11}^{-1}\alpha_{12}\alpha_{22}^{-1} \\ 0 & \alpha_{22}^{-1} \end{bmatrix}. \tag{3.14b}$$

If β is singular, but $\alpha = \alpha_{11} - \alpha_{12}\,\alpha_{22}^{-1}\,\alpha_{21}$ and α_{22} are nonsingular, then

$$A^{-1} = \left[\begin{array}{c|c} \alpha^{-1} & -\alpha^{-1}\alpha_{12}\alpha_{22}^{-1} \\ \hline -\alpha_{22}^{-1}\,\alpha_{21}\,\alpha^{-1} & \alpha_{22}^{-1} + (\alpha_{22}^{-1}\alpha_{21})\alpha^{-1}(\alpha_{12}\alpha_{22}^{-1}) \end{array}\right].$$

Example 3.7 Obtain the inverse of the following matrix by partitioning:

$$A = \begin{bmatrix} 2 & 1 & 4 \\ 4 & 3 & 1 \\ 1 & 2 & 4 \end{bmatrix}$$

Solution Choose

$$\alpha_{11} = \begin{bmatrix} 2 & 1 \\ 4 & 3 \end{bmatrix}, \quad \alpha_{12} = \begin{bmatrix} 4 \\ 1 \end{bmatrix}, \quad \alpha_{21} = [1 \quad 2], \quad \alpha_{22} = [4].$$

Then, in view of Example 3.3, by inspection,

$$\alpha_{11}^{-1} = \frac{1}{2}\begin{bmatrix} 3 & -1 \\ -4 & 2 \end{bmatrix}$$

and

$$\alpha_{21}\alpha_{11}^{-1} = \frac{1}{2}[1 \quad 2]\begin{bmatrix} 3 & -1 \\ -4 & 2 \end{bmatrix} = \frac{1}{2}[-5 \quad 3],$$

$$\alpha_{11}^{-1}\alpha_{12} = \frac{1}{2}\begin{bmatrix} 3 & -1 \\ -4 & 2 \end{bmatrix}\begin{bmatrix} 4 \\ 1 \end{bmatrix} = \frac{1}{2}\begin{bmatrix} 11 \\ -14 \end{bmatrix},$$

$$\beta = \alpha_{22} - \alpha_{21}\alpha_{11}^{-1}\alpha_{12} = [4] - \frac{1}{2}[-5 \quad 3]\begin{bmatrix} 4 \\ 1 \end{bmatrix} = \left[\frac{25}{2}\right].$$

Therefore,

$$\beta^{-1} = \left[\frac{2}{25}\right], \tag{3.15a}$$

$$\beta^{-1}\alpha_{21}\alpha_{11}^{-1} = \frac{1}{25}[-5 \quad 3], \tag{3.15b}$$

$$\alpha_{11}^{-1}\alpha_{12}\beta^{-1} = \frac{1}{25}\begin{bmatrix} 11 \\ -14 \end{bmatrix}, \tag{3.15c}$$

$$\alpha_{11}^{-1} + (\alpha_{11}^{-1}\alpha_{12})\beta^{-1}(\alpha_{21}\alpha_{11}^{-1}) = \frac{1}{2}\begin{bmatrix} 3 & -1 \\ -4 & 2 \end{bmatrix} + \frac{1}{2}\begin{bmatrix} 11 \\ -14 \end{bmatrix} \times \frac{2}{25} \times \frac{1}{2}[-5 \quad 3]$$

$$= \frac{1}{2}\begin{bmatrix} 3 & -1 \\ -4 & 2 \end{bmatrix} + \frac{1}{50}\begin{bmatrix} -55 & 33 \\ 70 & -42 \end{bmatrix} = \frac{1}{25}\begin{bmatrix} 10 & 4 \\ -15 & 4 \end{bmatrix}. \tag{3.15d}$$

Substituting the values from (3.15a)–(3.15d) into (3.13), the inverse of A is

$$A^{-1} = \frac{1}{25}\begin{bmatrix} 10 & 4 & -11 \\ -15 & 4 & 14 \\ 5 & -3 & 2 \end{bmatrix}. \qquad \square$$

If the matrix A for inversion is symmetric, then instead of following the above procedure, it is advisable to use (3.14a) which would entail less computation as shown below.

Example 3.8 Find the inverse of the following symmetric matrix by partitioning:

$$A = \begin{bmatrix} 1 & 4 & 2 \\ 4 & 2 & 1 \\ 2 & 1 & 3 \end{bmatrix}.$$

Solution Choosing

$$\alpha_{11} = \begin{bmatrix} 1 & 4 \\ 4 & 2 \end{bmatrix}, \quad \alpha_{12} = \alpha_{21}^T = \begin{bmatrix} 2 \\ 1 \end{bmatrix}, \quad \alpha_{22} = [3],$$

we have,

$$\alpha_{11}^{-1} = -\frac{1}{14}\begin{bmatrix} 2 & -4 \\ -4 & 1 \end{bmatrix},$$

$$\gamma = \alpha_{11}^{-1}\alpha_{12} = -\frac{1}{14}\begin{bmatrix} 2 & -4 \\ -4 & 1 \end{bmatrix}\begin{bmatrix} 2 \\ 1 \end{bmatrix} = -\frac{1}{14}\begin{bmatrix} 0 \\ -7 \end{bmatrix} = \frac{1}{2}\begin{bmatrix} 0 \\ 1 \end{bmatrix},$$

$$\beta = \alpha_{22} - \alpha_{21}\gamma = [3] - [2 \quad 1]\begin{bmatrix} 0 \\ \frac{1}{2} \end{bmatrix} = \left[\frac{5}{2}\right].$$

Therefore,

$$\gamma\beta^{-1} = \begin{bmatrix} 0 \\ \frac{1}{2} \end{bmatrix}\left[\frac{2}{5}\right] = \begin{bmatrix} 0 \\ \frac{1}{5} \end{bmatrix},$$

$$\alpha_{11}^{-1} + \gamma\beta^{-1}\gamma^T = -\frac{1}{14}\begin{bmatrix} 2 & -4 \\ -4 & 1 \end{bmatrix} + \frac{1}{2}\begin{bmatrix} 0 \\ 1 \end{bmatrix}\left[\frac{2}{5}\right]\cdot\frac{1}{2}\begin{bmatrix} 0 & 1 \end{bmatrix}$$

$$= -\frac{1}{14}\begin{bmatrix} 2 & -4 \\ -4 & 1 \end{bmatrix} + \frac{1}{10}\begin{bmatrix} 0 & 0 \\ 0 & 1 \end{bmatrix} = \frac{1}{35}\begin{bmatrix} -5 & 10 \\ 10 & 1 \end{bmatrix}.$$

Hence by (3.14a), the inverse is

$$A^{-1} = \frac{1}{35}\begin{bmatrix} -5 & 10 & 0 \\ 10 & 1 & -7 \\ 0 & -7 & 14 \end{bmatrix}$$ □

In MATLAB to evaluate the inverse of a square matrix A, one has to enter inv (A). Thus

```
>> inv (A)
ans =
     -0.1429        0.2857        0
      0.2857        0.0286       -0.2000
      0            -0.2000        0.4000
```

However, to get it in a rational form, we enter

```
>> format rat
>> inv (A)
ans =

     -1/7       2/7      0
      2/7       1/35    -1/5
      0        -1/5      2/5
```

□

Two methods, viz. (i) inverse from the adjoint, and (ii) inverse by partitioning, are described in this chapter. The other two, viz. (i) inverse by step-by-step reduction of $[A; I]$ and (ii) inverse from elementary matrices, are introduced in Chapter 4.

We shall conclude this chapter with an important example of general interest.

Example 3.9 Prove that for all (positive or negative) integral values of p and q,

$$\text{(i) } A^pA^q = A^{p+q}, \qquad \text{(ii) } (A^p)^q = A^{pq}$$

when A is nonsingular.

Solution We define for a positive integer n,

$$A^0 := I, \qquad A^n := AA\cdots A, \qquad n \text{ times}; \tag{3.16}$$

$$A^{-n} := (A^{-1})^n := A^{-1}A^{-1}\cdots A^{-1}, \qquad n \text{ times}. \tag{3.17}$$

(a) *p and q are positive integers.* We have, by definition,

$$A^n A = AA \cdots AA, \qquad n + 1 \text{ times} = A^{n+1}. \tag{3.18}$$

Following the process of mathematical induction, suppose that for a positive integer $m > 1$,

$$A^n A^{m-1} = A^{n+m-1}. \tag{3.19}$$

Then, in view of (3.18), we have

$$A^m = A^{m-1} A \tag{3.20}$$

and hence by the associative law of matrix multiplication,

$$\begin{aligned} A^n A^m &= (A^n A^{m-1})A = A^{n+m-1} A \quad \text{[by (3.19)]} \\ &= A^{n+m} \qquad \text{[by (3.20)]}. \end{aligned} \tag{3.21}$$

So (i) is true when $p = n$ and $q = m$ are positive integers. Again, by (3.21),

$$(A^n)^2 = A^n A^n = A^{n+n} = A^{2n}.$$

Suppose now that for a positive integer $m > 1$,

$$(A^n)^{m-1} := A^n A^n \cdots A^n, \qquad (m - 1) \text{ times} = A^{n(m-1)}. \tag{3.22}$$

Then, by (3.22) and (3.21),

$$(A^n)^m = \underbrace{A^n A^n \cdots A^n}_{(m-1)\text{ times}} A^n = (A^n)^{m-1} A^n = A^{n(m-1)} A^n = A^{nm}.$$

So (ii) holds when $p = n$ and $q = m$ are positive integers.

(b) *p and q are negative integers.* For a positive integer n, by definition, from (3.17) we have

$$A^{-n} A^{-1} = A^{-1} A^{-1} \cdots A^{-1} A^{-1}, \qquad (n + 1) \text{ times} = A^{-n-1}. \tag{3.23}$$

Following mathematical induction, let us suppose that

$$A^{-n} A^{-(m-1)} = A^{-n-m+1} \tag{3.24}$$

for a positive integer, $m > 1$. Then, by (3.23), $A^{-m} = A^{-(m-1)} A^{-1}$, and hence in view of (3.24) and the associative law of matrix multiplication, we have

$$\begin{aligned} A^{-n} A^{-m} &= (A^{-n} A^{-m+1})A^{-1} = A^{-n-m+1} A^{-1} \\ &= A^{-n-m} \qquad \text{[by (3.23)]}. \end{aligned} \tag{3.25}$$

So (i) is true when $p = -n$ and $q = -m$ are negative integers. We know that

$$(A^{-1})^{-1} = A \qquad (A^{-2})^{-1} = (A^{-1}A^{-1})^{-1} = AA = A^2$$

where we have used (3.17) and Property I of Section 3.2. Suppose that

$$(A^{-(n-1)})^{-1} = A^{n-1}. \tag{3.26}$$

Then, by (3.17) and (3.26),

$$(A^{-n})^{-1} = (A^{-(n-1)} A^{-1})^{-1} = AA^{n-1} = A^n.$$

Consequently,

$$(A^{-n})^{-m} = (A^{-n})^{-1} \cdots (A^{-n})^{-1}, \quad m \text{ times}$$
$$= A^{n} \cdots A^{n}, \quad m \text{ times} = A^{nm}$$

by repeated application of (3.21). Thus (ii) is true when $p = -n$ and $q = -m$ are negative integers.

(c) *p is a positive and q is a negative integer.* Now for positive integers m and n,

$$A^n A^{-m} = \underbrace{A \cdots A}_{n \text{ times}} \underbrace{A^{-1} \cdots A^{-1}}_{m \text{ times}}$$

$$= \begin{cases} A \cdots A, & n - m \text{ times} \quad \text{if } n \geq m; \\ A^{-1} \cdots A^{-1}, & m - n \text{ times} \quad \text{if } m > n; \end{cases}$$

$$= \begin{cases} A^{n-m}, & \text{if } n \geq m; \\ (A^{-1})^{m-n} = A^{n-m}, & \text{if } m > n. \end{cases}$$

Thus (i) is true if $p = n$ is a positive and $q = -m$ a negative integer. Again, from (3.16), Property I of Section 3.2 and (3.17),

$$(A^n)^{-1} = (A \cdots A)^{-1} = A^{-1} \cdots A^{-1}, \quad n \text{ times} = A^{-n}$$

Suppose now that for a positive integer $m > 1$,

$$(A^n)^{-(m-1)} = A^{-n(m-1)} \tag{3.27}$$

Hence, by (3.27) and (3.25),

$$(A^n)^{-m} = (A^n)^{-1} \cdots (A^n)^{-1}, \quad m \text{ times}$$
$$= (A^n)^{-(m-1)} (A^n)^{-1}$$
$$= A^{-n(m-1)} A^{-n} = A^{-nm}$$

which shows that (ii) is true when $p = n$ is a positive integer and $q = -m$ a negative integer.

(d) *p is a negative and q is a positive integer.* This is left as an exercise to the reader.

(e) *p = 0 and q is any integer (or q = 0 and p is any integer).* This case is obvious because

$$A^0 A^q = IA^q = A^q = A^{0+q}, \qquad (A^0)^q = I^q = I = A^0 = A^{0q}. \qquad \square$$

PROBLEMS

3.1 Test whether the following matrices are singular or not:

(a) $\begin{bmatrix} 1 & 4 & 0 \\ 3 & 6 & 5 \\ 1 & 2 & 3 \end{bmatrix}$, (b) $\begin{bmatrix} 4 & 2 & 2 \\ 3 & 3 & 2 \\ 2 & 4 & 2 \end{bmatrix}$, (c) $\begin{bmatrix} 1+i & 2 & -2 \\ 0 & i+2 & -1 \\ i & 1 & 0 \end{bmatrix}$, $i^2 = -1$;

(d) $\begin{bmatrix} \sigma^2 & \sigma & 1 \\ \sigma & \sigma^2 & \sigma \\ 1 & \sigma & 1 \end{bmatrix}$, (e) $\begin{bmatrix} 1+\sigma & \sigma & \sigma \\ 1 & 1 & 1 \\ \sigma & \sigma-1 & \sigma \end{bmatrix}$, (f) $\begin{bmatrix} 1 & \omega & \omega^2 \\ \omega & \omega & 1 \\ 1 & \omega^2 & \omega^2 \end{bmatrix}$, $\omega^3 = 1$.

3.2 Determine the adjoint of the matrices in Problem 3.1.

3.3 Determine the matrix which on premultiplication (or postmultiplication) of the matrix in Problem 3.1(a) produces a diagonal matrix. Make the same computation for the remaining matrices in Problem 3.1.

3.4 Write by inspection the inverse of the following second order matrices:

(a) $\begin{bmatrix} 2 & 5 \\ 1 & 3 \end{bmatrix}$, (b) $\begin{bmatrix} 1+i & i \\ i & i \end{bmatrix}$, (c) $\begin{bmatrix} \sigma^2+1 & \sigma-1 \\ \sigma+1 & 1 \end{bmatrix}$,

(d) $\begin{bmatrix} 1 & \omega \\ \omega & -\omega \end{bmatrix}$, $\omega^3 = 1$; (e) $\begin{bmatrix} \cos\theta & \sin\theta \\ -\sin\theta & \cos\theta \end{bmatrix}$, (f) $\begin{bmatrix} s & \sigma \\ \sigma & s \end{bmatrix}$.

3.5 Compute the inverse of the matrices in Problems 3.1(a), 3.1(c), and 3.1(e).

3.6 Find the solution to the following equations by using a matrix inverse:

(a) $2x + y + 3z = 3$, $2y + z = 2$, $x + y + 2z = 1$;

(b) $2x + y - z = 2$, $x + 2y + 3z = 1$, $2x + 3y + 4z = 1$;

(c) $2x - y + 3z = 33$, $y = -2$, $2x + y + z = 11$;

(d) $x + z = 2$, $-x + 2z = 1$, $y + 3z = 3$.

3.7 Prove that, if $PAQ = I$, where all the four matrices are of order n, then A^{-1} exists and $A^{-1} = QP$.

3.8 If A and B are nonsingular, commutative and symmetric, prove that the following matrix products are symmetric:

(a) $A^{-1}B^{-1}$, (b) $A^{-1}B$, (c) AB^{-1}.

3.9 Prove that

(a) adj AB = adj B adj A,

(b) adj (adj A) = (det A)^{n-2}A, if det $A \neq 0$;

(c) A adj A = (adj A) $A = 0$, if A is singular.

3.10 Obtain the following matrix product by partitioning:

(a) $\left[\begin{array}{cc|c} 1 & 0 & 0 \\ 0 & 1 & 0 \\ \hline 0 & 0 & 1 \\ 2 & 3 & 4 \end{array}\right] \left[\begin{array}{cc} 1 & 0 \\ 0 & 1 \\ \hline 1 & 2 \end{array}\right]$, (b) $\left[\begin{array}{c|cc|c} 1 & 2 & 3 & 4 \\ \hline 2 & 0 & 0 & 3 \\ 3 & 0 & 0 & 2 \end{array}\right] \left[\begin{array}{ccc} 1 & 2 & 1 \\ \hline 2 & 3 & 0 \\ 2 & 1 & 0 \\ \hline 1 & 2 & 3 \end{array}\right]$.

3.11 Use Laplace expansion to show that:

(a) det AB = det $A\cdot$ det B, where A and B are square matrices of the same order;

(b) $\det\begin{bmatrix} A_1 & 0 \\ 0 & A_2 \end{bmatrix} = \det A_1 \cdot \det A_2,$

$$\det\begin{bmatrix} A_1 & & & \\ & A_2 & & \\ & & \ddots & \\ & & & A_n \end{bmatrix} = \det A_1 \cdot \det A_2 \cdot \cdots \cdot \det A_n,$$

where $A_1, A_2, ..., A_n$ are square matrices not necessarily of the same order.

(c) $\det\begin{bmatrix} A_{11} & A_{12} \\ 0 & A_{22} \end{bmatrix} = \det\begin{bmatrix} A_{11} & 0 \\ A_{21} & A_{22} \end{bmatrix} = \det A_{11}\cdot\det A_{22}.$

3.12 (a) If A_{ij} is the cofactor of a_{ij} in the matrix $A = [a_{ij}]_n$, show that

$$\det\begin{bmatrix} A & \mathbf{p} \\ \mathbf{q}^{\mathbf{T}} & 0 \end{bmatrix} = \det\begin{bmatrix} 0 & \mathbf{q}^{\mathbf{T}} \\ \mathbf{p} & A \end{bmatrix} = -\sum_{i=1}^{n}\sum_{j=1}^{n} p_i q_j A_{ij}$$

$$= -\mathbf{q}^T (\text{adj } A)\mathbf{p},$$

where $\mathbf{p}^T = [p_1, p_2, \ldots, p_n]$, $q^T = [q_1, q_2, \ldots, q_n]$;

(b) $\det\begin{bmatrix} A & \mathbf{p} \\ \mathbf{q}^T & a \end{bmatrix} = \det\begin{bmatrix} a & \mathbf{q}^{\mathbf{T}} \\ \mathbf{p} & A \end{bmatrix} = a \det A - \mathbf{q}^T (\text{adj } A)\mathbf{p},$

where a is a scalar, **p** and **q** are given in (a).

3.13 Use the principle of (i) adjoint and (ii) partitioning separately to find the inversion of the following matrices:

(a) $\begin{bmatrix} \cos\theta & -\sin\theta & 1 \\ \cos(\theta-120°) & -\sin(\theta-120°) & 1 \\ \cos(\theta+120°) & -\sin(\theta+120°) & 1 \end{bmatrix},$

(b) $\begin{bmatrix} 1 & 0 & 3 \\ 0 & 1 & 2 \\ 5 & 2 & 4 \end{bmatrix},$ (c) $\begin{bmatrix} 0 & 3 & 2 & 1 \\ 3 & 1 & 1 & 0 \\ 2 & 1 & 4 & 2 \\ 1 & 0 & 2 & 3 \end{bmatrix},$ (d) $\begin{bmatrix} 1 & 0 & 0 & 1 \\ 0 & 1 & 0 & -2 \\ 0 & 0 & 1 & 3 \\ 4 & 3 & 2 & 8 \end{bmatrix},$

(e) $\begin{bmatrix} 1 & 3 & 2 & 5 \\ 3 & 1 & 0 & 0 \\ 2 & 0 & 1 & 0 \\ 5 & 0 & 0 & 1 \end{bmatrix},$ (f) $\begin{bmatrix} a & 1 & 2 & 0 \\ 0 & b & 0 & 2 \\ 0 & 0 & c & 1 \\ 0 & 0 & 0 & d \end{bmatrix},$ $a, b, c, d \neq 0;$

$$\text{(g)}\ \begin{bmatrix} 2 & 1 & 1 \\ 1 & 1 & 2 \\ 0 & 0 & 4 \end{bmatrix}, \qquad \text{(h)}\ \begin{bmatrix} 3 & 2 & 0 \\ 1 & 1 & 0 \\ 4 & 1 & 2 \end{bmatrix}, \qquad \text{(i)}\ \begin{bmatrix} 0 & 0 & 1 & 2 \\ 1 & -2 & 2 & 8 \\ 1 & -1 & 3 & 14 \\ 0 & 1 & 2 & 7 \end{bmatrix}.$$

3.14 Determine the inverse of the Vandermonde matrix in Problem 2.15 for $n = 3$, and hence extend to the nth order case if λ_i's are distinct.

3.15 Prove that by using the principle of matrix partitioning,

$$\text{(a)}\ \begin{bmatrix} A_1 & 0 \\ 0 & A_2 \end{bmatrix}^{-1} = \begin{bmatrix} A_1^{-1} & 0 \\ 0 & A_2^{-1} \end{bmatrix},$$

$$\text{(b)}\ \begin{bmatrix} A_1 & & & \\ & A_2 & & \\ & & \ddots & \\ & & & A_n \end{bmatrix}^{-1} = \begin{bmatrix} A_1^{-1} & & & \\ & A_2^{-1} & & \\ & & \ddots & \\ & & & A_n^{-1} \end{bmatrix}$$

where $A_1, A_2, ..., A_n$ are square nonsingular matrices, not necessarily of the same order.

3.16 Show that

$$\text{(a)}\ \det(A^{-1}) = \frac{1}{\det A}, \qquad \text{(b)}\ \frac{d}{dt} A^{-1} = -A^{-1} \frac{dA}{dt} A^{-1}.$$

3.17 (a) If r is a scalar, and $\mathbf{a}$ and $\mathbf{b}$ are n-column and n-row vectors respectively, then

$$\begin{array}{l} \qquad\ \ \begin{array}{c|c} n\ \text{cols.} & 1\ \text{col.} \end{array} \\ \det \left[\begin{array}{c|c} I & \mathbf{a} \\ \hline \mathbf{b} & \mathbf{r} \end{array}\right] \begin{array}{l} n\ \text{rows} \\ \hline 1\ \text{row} \end{array} = r - \mathbf{ba}, \end{array}$$

$$\det \begin{bmatrix} A & A\mathbf{a} \\ \mathbf{a}^T A & r \end{bmatrix} = \det A \cdot (r - \mathbf{a}^T A\mathbf{a}) = \det \begin{bmatrix} r & \mathbf{a}^T A \\ A\mathbf{a} & A \end{bmatrix}.$$

(b) Show that

$$\det \begin{bmatrix} A & -\mathbf{b} \\ \mathbf{c}^T & \alpha \end{bmatrix} = (\alpha + \mathbf{c}^T A^{-1}\mathbf{b}) \det A,$$

where A is $n \times n$, $\mathbf{b}$ and $\mathbf{c}$ are n-vectors, and α is a scalar.

3.18 Prove that

$$\det S = \det \begin{bmatrix} A & B \\ C & D \end{bmatrix} = \det A \det(D - CA^{-1}B) = \det D \det(A - BD^{-1}C),$$

where A and D are nonsingular. The matrix $D - CA^{-1}B$ (resp $A - BD^{-1}C$) is known as the *Schur complement* of A (resp D) in S.
[*Hint*: Subtract CA^{-1} times the first row from the second, etc.]

3.19 (a) Show that if A and D are nonsingular, then the inverse of

$$\left[\begin{array}{c|c} A & 0 \\ \hline C & D \end{array}\right]$$

is

$$\left[\begin{array}{c|c} A^{-1} & 0 \\ \hline -D^{-1}CA^{-1} & D^{-1} \end{array}\right].$$

(b) Prove that if A and C are nonsingular, then

$$\left[\begin{array}{c|c} A & B \\ \hline 0 & C \end{array}\right]^{-1} = \frac{1}{\det A \cdot \det C}\left[\begin{array}{c|c} \operatorname{adj} A \det C & -(\operatorname{adj} A)\, B(\operatorname{adj} C) \\ \hline 0 & \operatorname{adj} C \det A \end{array}\right].$$

3.20 Let A and B be two $n \times n$ matrices, and let there exist a nonsingular matrix P such that AP and BP are both diagonal and nonsingular. Prove that any row of A is a scalar multiple of the corresponding row of B and vice versa. State the corresponding results for columns.

3.21 Let A be an n-square matrix with all elements equal to a. Show that

(a) $\det (A + \lambda I) = \lambda^{n-1} (na + \lambda)$,

(b) $$(A + \lambda I)^{-1} = \frac{1}{\lambda(na+\lambda)}\begin{bmatrix} (n-1)a+\lambda & -a & \cdots & -a \\ -a & (n-1)a+\lambda & \cdots & -a \\ \vdots & \vdots & & \vdots \\ -a & -a & \cdots & (n-1)a+\lambda \end{bmatrix},$$

3.22 Prove that

(a) $$\begin{bmatrix} 1 & a_{12} & a_{13} & \cdots & a_{1n} \\ & 1 & a_{23} & \cdots & a_{2n} \\ & & & \ddots & \vdots \\ & & & & a_{n-1,n} \\ & & & & 1 \end{bmatrix}^{-1} = \begin{bmatrix} 1 & b_{12} & b_{13} & \cdots & b_{1n} \\ & 1 & b_{23} & \cdots & b_{2n} \\ & & & \ddots & \vdots \\ & & & & b_{n-1,n} \\ & & & & 1 \end{bmatrix}$$

where

$$\begin{bmatrix} b_{1j} \\ b_{2j} \\ \vdots \\ b_{j-1,j} \end{bmatrix} = -\begin{bmatrix} 1 & b_{12} & \cdots & b_{1\,j-1} \\ & 1 & \cdots & b_{2\,j-1} \\ & & \ddots & \vdots \\ & & & b_{j-2,j-1} \\ & & & 1 \end{bmatrix}\begin{bmatrix} a_{1j} \\ a_{2j} \\ \vdots \\ a_{j-1,j} \end{bmatrix}$$

with $j = 2, \ldots, n$;

$$\text{(b)}\quad \begin{bmatrix} a_1 & 0 & 0 & \cdots & 0 \\ a_2 & 1 & 0 & \cdots & 0 \\ a_3 & 0 & 1 & \cdots & 0 \\ \vdots & \vdots & \vdots & & \vdots \\ a_n & 0 & 0 & \cdots & 1 \end{bmatrix}^{-1} = \begin{bmatrix} \dfrac{1}{a_1} & 0 & 0 & \cdots & 0 \\ -\dfrac{a_2}{a_1} & 1 & 0 & \cdots & 0 \\ -\dfrac{a_3}{a_1} & 0 & 1 & \cdots & 0 \\ \vdots & \vdots & \vdots & & \vdots \\ -\dfrac{a_n}{a_1} & 0 & 0 & \cdots & 1 \end{bmatrix}, \quad a_1 \neq 0;$$

$$\text{(c)}\quad \begin{bmatrix} 0 & 1 & 0 & \cdots & 0 \\ 0 & 0 & 1 & \cdots & 0 \\ \vdots & \vdots & & & \vdots \\ 0 & 0 & 0 & \cdots & 1 \\ -a_0 & -a_1 & -a_2 & \cdots & -a_{n-1} \end{bmatrix}^{-1} = \begin{bmatrix} -\dfrac{a_1}{a_0} & -\dfrac{a_2}{a_0} & \cdots & -\dfrac{a_{n-1}}{a_0} & -\dfrac{1}{a_0} \\ 1 & 0 & \cdots & 0 & 0 \\ 0 & 1 & \cdots & 0 & 0 \\ \vdots & \vdots & & \vdots & \vdots \\ 0 & 0 & \cdots & 1 & 0 \end{bmatrix}, \quad a_0 \neq 0.$$

3.23 Find A^{-1} if (a) $A^2 = -I$, (b) $A^2 - A + I = 0$.

3.24 Prove that:
(a) If A is nonsingular, $AB = AC$ implies $B = C$.
(b) If det $(I - A) \neq 0$, then $(I - A)^{-1} = I + A + A^2 + \ldots$.
(c) $(I - A)^{-1}$ and $(I + A)$ commute.

3.25 Show that AB and BA are nonsingular if A and B are nonsingular n-square matrices; AB and BA are singular if at least one of the n-square matrices A and B is singular.

3.26 Determine the inverse of the tridiagonal matrix given in Problem 2.24.

3.27 Given A, a matrix of order n, $\mathbf{x}$ and $\mathbf{y}$ n-column vectors, prove that det $(A + \mathbf{x}\mathbf{y}^T) =$ det $A + \mathbf{y}^T$ (adj $A)\mathbf{x}$.

3.28 If P is invertible, then show that tr $P^{-1}AP =$ tr A, where A is any $n \times n$ matrix (see Definition 1.14).

3.29 Prove that:

(a) $\det \begin{bmatrix} C & 0 \\ 0 & D \end{bmatrix} = \det C \cdot \det D$;

(b) det $(I_n + \mathbf{a}\mathbf{b}^T) = 1 + \mathbf{b}^T\mathbf{a}$;
(c) det $(I_n + AB) =$ det $(I_m + BA)$;
where $\mathbf{a}$ and $\mathbf{b}$ are n-vectors, A and B are matrices of order $n \times m$ and $m \times n$ respectively, and C and D are any square matrices.

3.30 If det $(I_n + AB) =$ det $(I_m + BA) \neq 0$ (see Problem 3.29), where A and B are matrices of order $n \times m$ and $m \times n$ respectively, then show that
(a) $(I_n + AB)^{-1} A = A(I_m + BA)^{-1}$,
(b) $(I_n + AB)^{-1} = I_n - A(I_m + BA)^{-1}B$.

3.31 Show that $(I + \mathbf{ab}^T)^{-1} = I - (1 + \mathbf{b}^T\mathbf{a})^{-1}\mathbf{ab}^T$, where $\mathbf{a} = [a_1, a_2, ..., a_n]^T$ and $\mathbf{b} = [b_1, b_2, ..., b_n]^T$.

3.32 *Matrix Inversion Lemma.* If A, C and $C^{-1} + D A^{-1} B$ are nonsingular square matrices, then

$$(A + BCD)^{-1} = A^{-1} - A^{-1}B\,(C^{-1} + DA^{-1}B)^{-1}\,DA^{-1}.$$

3.33 Find the inverse of the following tridiagonal matrices:

(a) $\begin{bmatrix} 1 & 1 \\ 1 & 2 \end{bmatrix}$, (b) $\begin{bmatrix} 1 & 1 & 0 \\ 1 & 2 & 1 \\ 0 & 1 & 2 \end{bmatrix}$, (c) $\begin{bmatrix} 1 & 1 & 0 & 0 \\ 1 & 2 & 1 & 0 \\ 0 & 1 & 2 & 1 \\ 0 & 0 & 1 & 2 \end{bmatrix}$,

(d) $\begin{bmatrix} 1 & 1 & 0 & \cdots & 0 & 0 \\ 1 & 2 & 1 & \cdots & 0 & 0 \\ 0 & 1 & 2 & \cdots & 0 & 0 \\ \vdots & \vdots & \vdots & & \vdots & \vdots \\ 0 & 0 & 0 & \cdots & 2 & 1 \\ 0 & 0 & 0 & \cdots & 1 & 2 \end{bmatrix}_n$.

3.34 Show that if the order is even

$$\begin{bmatrix} 0 & -1 & -1 & -1 & \cdots & -1 \\ 1 & 0 & -1 & -1 & \cdots & -1 \\ 1 & 1 & 0 & -1 & \cdots & -1 \\ \vdots & \vdots & \vdots & \vdots & & \vdots \\ 1 & 1 & 1 & 1 & \cdots & 0 \end{bmatrix}^{-1} = \begin{bmatrix} 0 & 1 & -1 & 1 & \cdots & 1 \\ -1 & 0 & 1 & -1 & \cdots & -1 \\ 1 & -1 & 0 & 1 & \cdots & 1 \\ \vdots & \vdots & \vdots & \vdots & & \vdots \\ -1 & 1 & -1 & 1 & \cdots & 0 \end{bmatrix}.$$

3.35 Let A be a matrix whose diagonal elements are all equal to u but off-diagonal elements are all equal to v. Show that the inverse of A is of the same form as A, i.e. all the diagonal elements of A^{-1} are equal and all its off-diagonal elements are also equal.

3.36 In computing the inverse by partitioning, if β is singular but $\alpha = \alpha_{11} - \alpha_{12}\,\alpha_{22}^{-1}\,\alpha_{21}$ and α_{22} are nonsingular, then prove that (3.13) is modified to

$$A^{-1} = \left[\begin{array}{c|c} \alpha^{-1} & -\alpha^{-1}(\alpha_{12}\,\alpha_{22}^{-1}) \\ \hline -\alpha_{22}^{-1}\,\alpha_{21}\,\alpha^{-1} & \alpha_{22}^{-1} + (\alpha_{22}^{-1}\,\alpha_{21})\alpha^{-1}(\alpha_{12}\,\alpha_{22}^{-1}) \end{array}\right].$$

If A is symmetric, then show that

$$A^{-1} = \left[\begin{array}{c|c} \alpha^{-1} & (-\delta\alpha^{-1})^T \\ \hline -\delta\alpha^{-1} & \alpha_{22}^{-1} + \delta\alpha^{-1}\,\delta^T \end{array}\right],$$

where $\delta = \alpha_{22}^{-1}\,\alpha_{21}$.

CHAPTER 4

Rank and Equivalence

In this chapter, the concept of a *submatrix* introduced in Section 3.3 is generalized in an obvious way. This will be followed by formal definitions of rank of a matrix, elementary row and column transformations, elementary and permutation matrices, equivalence, and normal forms.

4.1 SUBMATRIX: RANK

Recalling the definition of matrix A which is nothing but a set-up of elements arranged in rectangular blocks, we can form a new matrix by taking some, if not all, of the elements of A, such that the ith row (jth column) elements of this new matrix belong to only the kth row (rth column) elements of the parent matrix A, and, moreover, occur in the same order. This process is equivalent to striking out some rows and some columns from A. The matrix thus formed having elements which are subset of the elements in A is termed a *submatrix* of A. For example,

$$\begin{bmatrix} 4 & 3 \\ 2 & 5 \end{bmatrix}$$

is a submatrix of order 2 of

$$A = \begin{bmatrix} 1 & \textcircled{4} & 0 & \textcircled{3} \\ 3 & 1 & 2 & \textcircled{1} \\ 2 & \textcircled{2} & 1 & 5 \end{bmatrix},$$

which is obtained by striking out first and third columns, and second row of A, whereas

$$\begin{bmatrix} 4 & 3 \\ 2 & 1 \end{bmatrix}$$

is not, because either $[3\ 1]^T$ or $[4\ 2]^T$ may not belong to the same column in A, or the elements in [2 1] may not belong to the same row (see the elements marked by a circle). Again,

$$\begin{bmatrix} 1 & 0 & 3 \\ 3 & 2 & 1 \end{bmatrix}$$

is a submatrix of order 2 × 3, and the whole matrix A is a submatrix of itself. Only square submatrices play a role in the definition of a rank given below. Note also that square submatrices can always be defined even though the parent matrix is nonsquare.

Definition 4.1 *Rank.* An $m \times n$ matrix A is said to have a *rank* r if it has at least one submatrix of order r which is nonsingular but all submatrices of order greater than r are singular. It is denoted by $\rho(A)$ or rank A. The determinants of square submatrices in A are also referred to as *subdeterminants* of A. □

The rank r of an $m \times n$ matrix A can at the most be equal to the smaller of the numbers m and n, but it may be less. An $n \times n$ matrix A has rank $r < n$ iff det $A = 0$ and rank $r = n$ iff det $A \neq 0$. Owing to Definition 3.1, we can say that an n-square matrix A is singular iff its rank $r < n$, and nonsingular iff $r = n$.

Example 4.1 Find rank A, where A is given by

$$A = \begin{bmatrix} 1 & 2 & 3 \\ 2 & 4 & 6 \end{bmatrix}.$$

Solution In A, there are three submatrices of order 2, viz.

$$\begin{bmatrix} 1 & 2 \\ 2 & 4 \end{bmatrix}, \begin{bmatrix} 1 & 3 \\ 2 & 6 \end{bmatrix}, \begin{bmatrix} 2 & 3 \\ 4 & 6 \end{bmatrix},$$

all the determinants of which are equal to zero. But among the submatrices of order one, viz. [1], [2], [3], [2], [4], [6], there is at least one nonzero single element and hence rank A is 1. □

It is important to note that only a null matrix can have *zero rank.*

Example 4.2 Find rank A, where A is given by

$$A = \begin{bmatrix} 2 & 1 & 1 \\ 4 & 2 & 2 \\ 1 & 2 & 2 \end{bmatrix}.$$

Solution Here, det $A = 0$, but there is a submatrix of order 2, i.e., $\begin{bmatrix} 4 & 2 \\ 1 & 2 \end{bmatrix}$, whose determinant is not zero. Hence, rank $A = 2$. □

4.2 ELEMENTARY TRANSFORMATIONS

The process of ascertaining the rank of a matrix by computing determinants is indeed a laborious task except for lower order matrices. An alternative way to achieve this is by means of elementary transformations which are now defined.

Definition 4.2 *Elementary row (resp column) transformations* of a matrix are obtained by subjecting it to the following three operations:

(i) Interchange of any two rows (resp columns).
(ii) Multiplication of any row (resp column) by a nonzero scalar.
(iii) Addition to any row (resp column) a suitable multiple of any other row (resp column).

These operations are defined as *elementary row* (resp *column*) *operations*. The adjective *elementary* may be dropped if there is no ambiguity from the context of description. □

Comparing the elementary operations with the properties of determinants described in Section 2.3, we observe that after elementary transformations, a singular matrix remains singular and the determinant of a nonsingular matrix is altered to the extent of a nonzero scalar multiple. Also, the determinants of the submatrices of all orders in any $m \times n$ matrix A are affected similarly. Therefore, elementary transformations do not affect the rank of a matrix.

$$\text{Let} \qquad \text{(a)} \begin{bmatrix} 0 & 0 & 1 \\ 0 & 1 & 0 \\ 1 & 0 & 0 \end{bmatrix}, \quad \text{(b)} \begin{bmatrix} 1 & 0 & 0 \\ 0 & 1 & 0 \\ 0 & 0 & \alpha \end{bmatrix}, \quad \text{(c)} \begin{bmatrix} 1 & \alpha & 0 \\ 0 & 1 & 0 \\ 0 & 0 & 1 \end{bmatrix} \tag{4.1}$$

be three matrices obtained from a unit matrix of order 3 by elementary row transformations RT, viz.

RT (a) by interchanging rows 1 and 3;
RT (b) by multiplying the third row with a scalar α; and
RT (c) by adding α times the second row to the first row.

Such matrices obtained from a unit matrix by elementary operations are referred to as *left elementary matrices*. The matrix (a), which is obtained by interchanging rows (or columns), is also known as *a permutation matrix*. Premultiplying a matrix of order 3 × 2, i.e.,

$$\begin{bmatrix} a_1 & b_1 \\ a_2 & b_2 \\ a_3 & b_3 \end{bmatrix} \tag{4.2}$$

by the left elementary matrices (4.1), we obtain

$$\text{(a)} \begin{bmatrix} 0 & 0 & 1 \\ 0 & 1 & 0 \\ 1 & 0 & 0 \end{bmatrix} \begin{bmatrix} a_1 & b_1 \\ a_2 & b_2 \\ a_3 & b_3 \end{bmatrix} = \begin{bmatrix} a_3 & b_3 \\ a_2 & b_2 \\ a_1 & b_1 \end{bmatrix},$$

$$\text{(b)} \begin{bmatrix} 1 & 0 & 0 \\ 0 & 1 & 0 \\ 0 & 0 & \alpha \end{bmatrix} \begin{bmatrix} a_1 & b_1 \\ a_2 & b_2 \\ a_3 & b_3 \end{bmatrix} = \begin{bmatrix} a_1 & b_1 \\ a_2 & b_2 \\ \alpha a_3 & \alpha b_3 \end{bmatrix},$$

$$\text{(c)} \begin{bmatrix} 1 & \alpha & 0 \\ 0 & 1 & 0 \\ 0 & 0 & 1 \end{bmatrix} \begin{bmatrix} a_1 & b_1 \\ a_2 & b_2 \\ a_3 & b_3 \end{bmatrix} = \begin{bmatrix} a_1 + \alpha a_2 & b_1 + \alpha b_2 \\ a_2 & b_2 \\ a_3 & b_3 \end{bmatrix},$$

where we see that matrix (4.2) is subjected to the same elementary row transformations RT (a), RT (b), RT (c), respectively, as the unit matrix given in (4.1). Similarly, we obtain elementary column transformations of a matrix A by postmultiplying it with a matrix known as *right elementary matrix* which is obtained by subjecting a unit matrix of proper order to the same column transformations which we intend to perform on A. Elementary matrices are always nonsingular.

We may look upon the matrices in (4.1) as derived from a third order unit matrix by column transformations, viz.,

CT (a) by interchanging columns 1 and 3;
CT (b) by multiplying the third column with α; and
CT (c) by adding α times the first column to the second.

Postmultiplying a matrix of order 2 × 3, viz.

$$A = \begin{bmatrix} a_1 & a_2 & a_3 \\ b_1 & b_2 & b_3 \end{bmatrix} \tag{4.3}$$

by the right elementary matrices (4.1), we obtain

$$\text{(a)} \quad \begin{bmatrix} a_1 & a_2 & a_3 \\ b_1 & b_2 & b_3 \end{bmatrix} \begin{bmatrix} 0 & 0 & 1 \\ 0 & 1 & 0 \\ 1 & 0 & 0 \end{bmatrix} = \begin{bmatrix} a_3 & a_2 & a_1 \\ b_3 & b_2 & b_1 \end{bmatrix},$$

$$\text{(b)} \quad \begin{bmatrix} a_1 & a_2 & a_3 \\ b_1 & b_2 & b_3 \end{bmatrix} \begin{bmatrix} 1 & 0 & 0 \\ 0 & 1 & 0 \\ 0 & 0 & \alpha \end{bmatrix} = \begin{bmatrix} a_1 & a_2 & \alpha a_3 \\ b_1 & b_2 & \alpha b_3 \end{bmatrix},$$

$$\text{(c)} \quad \begin{bmatrix} a_1 & a_2 & a_3 \\ b_1 & b_2 & b_3 \end{bmatrix} \begin{bmatrix} 1 & \alpha & 0 \\ 0 & 1 & 0 \\ 0 & 0 & 1 \end{bmatrix} = \begin{bmatrix} a_1 & a_2 + \alpha a_1 & a_3 \\ b_1 & b_2 + \alpha b_1 & b_3 \end{bmatrix},$$

which shows that column transformations, CT (a), CT (b), CT (c) are realized in (4.3) as in the unit matrix. In the above discussion "left" is used to attribute premultiplying and "right" to postmultiplying elementary matrices even though they may happen to be the same.

We now conclude the foregoing observation in the following definition.

Definition 4.3 To effect elementary row (resp column) transformations on a matrix A, we have to perform first the same transformations on a unit matrix of proper order getting a left (resp right) elementary matrix P(resp Q), and then obtain the desired row (resp column) transformations of A by premultiplying (resp postmultiplying) it with the left (resp right) elementary matrix P (resp Q).

The left (resp right) elementary matrices are designated by H (resp K) and the concise notations for elementary operations on these matrices are:

H_{ij} (resp K_{ij}) : interchange of the ith and jth rows (resp columns).
H_i (α) (resp K_i (α)) : multiplication of the ith row (resp column) by scalar α.
H_i (α) (resp K_i (α)) : addition of α times the jth row (resp column) to the ith row (resp column).

The inverse of an elementary transformation is an operation which undoes the effect of elementary transformation and can therefore be achieved by premultiplying or postmultiplying the associated elementary matrix by a similar elementary matrix. Thus the inverse of H_{ij} obtained by interchanging the *i*th and *j*th rows of a unit matrix is an elementary matrix of the same form because, by interchanging the same two rows, H_{ij} is reverted to its parent unit matrix. $H_i(\alpha)$ obtained by multiplying the *i*th row of a unit matrix by α is put back to the unit matrix by multiplying its *i*th row by $1/\alpha$ and, therefore, it has an inverse $H_i(1/\alpha)$. Finally, $H_{ij}(\alpha)$ obtained by adding α times the *j*th row to the *i*th row of a unit matrix is transformed back to the unit matrix by adding $-\alpha$ times the *j*th row to the *i*th row and, therefore, has an inverse $H_{ij}(-\alpha)$. Similar arguments are applicable to right elementary matrices. Also, by direct computation, the following results associated with inversion of elementary matrices can be easily verified:

$$H_{ij}^{-1} = H_{ij}, \quad H_i(\alpha)^{-1} = H_i(1/\alpha), \quad H_{ij}^{-1}(\alpha) = H_{ij}(-\alpha);$$

$$K_{ij}^{-1} = K_{ij}, \quad K_i(\alpha)^{-1} = K_i(1/\alpha), \quad K_{ij}^{-1}(\alpha) = K_{ij}(-\alpha).$$

The readers are asked to verify the above results by taking a third order unit matrix. For example,

$$H_{31}(\alpha) = \begin{bmatrix} 1 & 0 & 0 \\ 0 & 1 & 0 \\ \alpha & 0 & 1 \end{bmatrix}, \qquad H_{31}(\alpha)^{-1} = \begin{bmatrix} 1 & 0 & 0 \\ 0 & 1 & 0 \\ -\alpha & 0 & 1 \end{bmatrix} = H_{31}(-\alpha),$$

as direct computation shows $H_{31}(\alpha)H_{31}(-\alpha) = I$.

We have seen how to carry out single row operation such as RT(a) or RT(b) on the matrix (4.2) by premultiplication with an elementary matrix. To effect both RT(a) and RT(b) successively, we need to premultiply (4.2) by (4.1a) and (4.1b) in succession:

$$\overset{E_2}{\begin{bmatrix} 1 & 0 & 0 \\ 0 & 1 & 0 \\ 0 & 0 & \alpha \end{bmatrix}} \overset{E_1}{\begin{bmatrix} 0 & 0 & 1 \\ 0 & 1 & 0 \\ 1 & 0 & 0 \end{bmatrix}} \begin{bmatrix} a_1 & b_1 \\ a_2 & b_2 \\ a_3 & b_3 \end{bmatrix} = \begin{bmatrix} 1 & 0 & 0 \\ 0 & 1 & 0 \\ 0 & 0 & \alpha \end{bmatrix} \begin{bmatrix} a_3 & b_3 \\ a_2 & b_2 \\ a_1 & b_1 \end{bmatrix}$$

$$= \begin{bmatrix} a_3 & b_3 \\ a_2 & b_2 \\ \alpha a_1 & \alpha b_1 \end{bmatrix} = \overset{E}{\begin{bmatrix} 0 & 0 & 1 \\ 0 & 1 & 0 \\ \alpha & 0 & 0 \end{bmatrix}} \begin{bmatrix} a_1 & b_1 \\ a_2 & b_2 \\ a_3 & b_3 \end{bmatrix},$$

which imposes on (4.2) both the row operations RT(a) and RT(b). *We can avoid multiplication to get E as the product of E_2E_1*, because E can be obtained by subjecting a unit matrix of order 3 to operations denned by RT(a) and RT(b) successively, which we intend to perform on the given matrix (4.2) as shown below:

$$\begin{bmatrix} 1 & 0 & 0 \\ 0 & 1 & 0 \\ 0 & 0 & 1 \end{bmatrix} \xrightarrow{\text{by RT(a)}} \begin{bmatrix} 0 & 0 & 1 \\ 0 & 1 & 0 \\ 1 & 0 & 0 \end{bmatrix} \xrightarrow{\text{by RT(b)}} \begin{bmatrix} 0 & 0 & 1 \\ 0 & 1 & 0 \\ \alpha & 0 & 0 \end{bmatrix} = E.$$

Thus the product of a number of left (resp right) elementary matrices can be obtained virtually by manipulation on the rows (resp columns). Similarly, a product of the form $H_{23}(1)H_1(2)K_{23}$ can be obtained without resorting to matrix multiplication, because the left elementary matrices occur to the left of K_{23} to be multiplied. However, $K_{23}H_{23}(1)H_1(2)$ needs multiplication which can be avoided by nothing that $K_{23} = H_{23}$ and, consequently $K_{23}H_{23}(1)H_1(2) = H_{23}H_{23}(1)H_1(2)$. The case with right elementary matrices can be handled in a reverse order, i.e., if these matrices $E_1\ E_2 \ldots E_m$ occur to the right of a matrix A, then without executing multiplication operation, the product E can be obtained simply by manipulation on columns of E_1.

4.3 EQUIVALENCE AND NORMAL FORM

Definition 4.4 *Equivalence.* Two matrices A and B are said to be *equivalent* if one can be obtained from the other by a sequence of elementary row and column operations. In other words, A is said to be equivalent to B if there exist nonsingular matrices P and Q such that $PAQ = B$. It follows that $UBV = A$, where $U = P^{-1}$ and $V = Q^{-1}$ are nonsingular. So, if A is equivalent to B, then B is also equivalent to A. The equivalence is denoted by ~, i.e., A is equivalent to B by $A \sim B$. In particular, if $Q = I$, then $PA = B$, and A is said to be *row equivalent* to B. If $P = I$, then $AQ = B$, and A is said to be *column equivalent* to B. □

The equivalence in Definition 4.4 can be used to partition the set of all, say, $m \times n$ matrices into equivalent classes. The property of all the matrices belonging to one equivalent class S is that they satisfy, as a pair, a *relation* of the form $PAQ = B$, where A and B are in S and P and Q are nonsingular. This relation is known as an *equivalence relation*, having the following properties:

(i) Reflexive : $A \sim A$,
(ii) Symmetric : if $A \sim B$ then $B \sim A$,
(iii) Transitive : if $A \sim B$ and $B \sim C$ then $A \sim C$,

for all matrices A, B, C in the class S. But matrices belonging to different equivalent classes do not satisfy properties (i)-(iii). Conversely, any relation satisfying these three properties is referred to as equivalence relation.

Let A, B, C be three matrices in an equivalent class S. Then

(i) $IAI = A$ shows that $A \sim A$.
(ii) If for nonsingular matrices P and Q we have $PAQ = B$, then $P^{-1}BQ^{-1} = A$, which implies that if $A \sim B$, then $B \sim A$.
(iii) Moreover, if $B \sim C$, then for nonsingular matrices R and T, we have $RBT = C$ which, when coupled with the relation $A \sim B$, i.e. $PAQ = B$ in (ii), gives $(RP)A(QT) = C$. This shows that if $A \sim B$ and $B \sim C$, then $A \sim C$.

Thus we have shown that all matrices in the equivalent class S have the property of symmetry, reflexivity and transitivity.

Example 4.3 Transform the matrix A in Example 4.2 by elementary operations to a simple form.

Solution Subjecting A to elementary row operations, viz. (i) adding –2 times the first row to the second and third rows, (ii) dividing the third row by –3, and (iii) interchanging the second and third rows, we get

$$A = \begin{bmatrix} 2 & 1 & 1 \\ 4 & 2 & 2 \\ 1 & 2 & 2 \end{bmatrix} \sim \begin{bmatrix} 2 & 1 & 1 \\ 0 & 0 & 0 \\ -3 & 0 & 0 \end{bmatrix} \sim \begin{bmatrix} 2 & 1 & 1 \\ 0 & 0 & 0 \\ 1 & 0 & 0 \end{bmatrix} \sim \begin{bmatrix} 2 & 1 & 1 \\ 1 & 0 & 0 \\ 0 & 0 & 0 \end{bmatrix}. \tag{4.4}$$

Again, subjecting the last matrix in (4.4) to column operations, i.e., (i) adding –2 and –1 times the second column to the first and third columns, and (ii) interchanging the first and second columns, we obtain

$$\begin{bmatrix} 2 & 1 & 1 \\ 1 & 0 & 0 \\ 0 & 0 & 0 \end{bmatrix} \sim \begin{bmatrix} 0 & 1 & 0 \\ 1 & 0 & 0 \\ 0 & 0 & 0 \end{bmatrix} \sim \begin{bmatrix} 1 & 0 & 0 \\ 0 & 1 & 0 \\ 0 & 0 & 0 \end{bmatrix} \sim \left[\begin{array}{c|c} I_2 & 0 \\ \hline 0 & 0 \end{array}\right] =: B. \tag{4.5}$$

From this it is evident that rank $A = 2$. If we perform the same row and column operations on a unit matrix of third order, we obtain, respectively,

$$P = \begin{bmatrix} 1 & 0 & 0 \\ \frac{2}{3} & 0 & -\frac{1}{3} \\ -2 & 1 & 0 \end{bmatrix}, \quad Q = \begin{bmatrix} 0 & 1 & 0 \\ 1 & -2 & -1 \\ 0 & 0 & 1 \end{bmatrix}.$$

Therefore,

$$\begin{bmatrix} 1 & 0 & 0 \\ \frac{2}{3} & 0 & -\frac{1}{3} \\ -2 & 1 & 0 \end{bmatrix} \begin{bmatrix} 2 & 1 & 1 \\ 4 & 2 & 2 \\ 1 & 2 & 2 \end{bmatrix} \begin{bmatrix} 0 & 1 & 0 \\ 1 & -2 & -1 \\ 0 & 0 & 1 \end{bmatrix} = \left[\begin{array}{cc|c} 1 & 0 & 0 \\ 0 & 1 & 0 \\ \hline 0 & 0 & 0 \end{array}\right],$$

i.e., $PAQ = B$.

Since P as well as Q is made of a product of elementary matrices and elementary matrices are nonsingular, they are therefore nonsingular. So, $A = P^{-1}BQ^{-1}$, i.e.,

$$\begin{bmatrix} 2 & 1 & 1 \\ 4 & 2 & 2 \\ 1 & 2 & 2 \end{bmatrix} = \begin{bmatrix} 1 & 0 & 0 \\ 2 & 0 & 1 \\ 2 & -3 & 0 \end{bmatrix} \begin{bmatrix} 1 & 0 & 0 \\ 0 & 1 & 0 \\ 0 & 0 & 0 \end{bmatrix} \begin{bmatrix} 2 & 1 & 1 \\ 1 & 0 & 0 \\ 0 & 0 & 1 \end{bmatrix}.$$

Hence, A and B are equivalent. □

It is a useful observation that as elementary transformations do not alter the rank of a matrix, the order of unit submatrix I_2 in B of (4.5) refers to rank A. This observation leads to the following definition.

Definition 4.5 *Normal Form.* By a sequence of elementary transformations, a matrix A of order $m \times n$ can be reduced to any of the forms

$$\left[\begin{array}{c|c} I_r & 0_{r,n-r} \\ \hline 0_{m-r,r} & 0_{m-r,n-r} \end{array}\right], \qquad r < m, n; \qquad (4.6a)$$

$$[I_r \mid 0], \qquad r = m < n; \qquad (4.6b)$$

$$\left[\begin{array}{c} I_r \\ \hline 0 \end{array}\right], \qquad r = n < m; \qquad (4.6c)$$

$$[I_r], \qquad r = m = n; \qquad (4.6d)$$

known as the normal form of A, where r is rank A, 0's are null matrices of appropriate order, or as indicated. Because of Definition 4.3, there exist matrices P and Q of order m and n respectively, such that PAQ has any of the above forms. □

The elementary operations to obtain the normal form NF of a given nonzero $m \times n$ matrix A which was calculated for a third order matrix in Example 4.3 are the following:

NF (a) By interchanging rows as well as columns, bring a nonzero element, preferably 1, in the (1, 1) position of A; call this element a_{11}.

NF (b) Divide all the first row or column elements by a_{11} if already $a_{11} \neq 1$. Therefore, now A has 1 in (1, 1) position.

NF (c) Subtract suitable multiples of the first row from the remaining rows, so that the first column elements of A are zero but for the one in the first row.

NF (d) Subtracting suitable multiples of the first column from other columns, reduce the first row elements of all columns except for the one in the first column to zero. The matrix A is now reduced to

$$\left[\begin{array}{c|ccc} 1 & 0 & \cdots & 0 \\ \hline 0 & a_{22} & \cdots & a_{2n} \\ \vdots & \vdots & B & \vdots \\ 0 & a_{m2} & \cdots & a_{mn} \end{array}\right]$$

NF (e) Repeat the foregoing steps for the lower right submatrix B by first bringing a nonzero element, preferably 1, in the position (2, 2) if $a_{22} = 0$ already.

NF (f) Continue the above procedure down the diagonal elements until one of the forms in (4.6) is obtained (see Example 4.5).

In Sections 4.4 and 4.5, we shall describe two methods for obtaining the inverse of a square matrix which rely extensively on elementary operations.

4.4 INVERSE BY STEP-BY-STEP REDUCTION OF [A; I]

The solution of the matrix equation (3.6), viz.

$$A\mathbf{x} = \mathbf{y}, \tag{4.7}$$

where A is nonsingular, is given by

$$\mathbf{x} = A^{-1}\mathbf{y}. \tag{4.8}$$

If we take into consideration only the coefficient matrices of **x** and **y**, the process of solution transforms $[A; I]$ in (4.7) to $[I; A^{-1}]$ in (4.8). By the sequence of *row transformations* as in Definition 4.2, $[A; I]$ can be reduced to $[I; B]$ (see (4.6d)), in which case B defines A^{-1}. The *elementary row operations* (i), (ii), (iii), when applied to $[A; I]$, correspond to (i) interchange of any two equations, (ii) multiplication of any equation by a nonzero scalar, and (iii) addition of a suitable multiple of one equation to any other, in the simultaneous equations described by (4.7).

Example 4.4 Considering A as given below, we have

$$[A;I] = \left[\begin{array}{ccc|ccc} 2 & 1 & 3 & 1 & 0 & 0 \\ 0 & 2 & 1 & 0 & 1 & 0 \\ 1 & 1 & 2 & 0 & 0 & 1 \end{array}\right] \sim \left[\begin{array}{ccc|ccc} 1 & 1 & 2 & 0 & 0 & 1 \\ 0 & 2 & 1 & 0 & 1 & 0 \\ 2 & 1 & 3 & 1 & 0 & 0 \end{array}\right]$$

$$\sim \left[\begin{array}{ccc|ccc} 1 & 1 & 2 & 0 & 0 & 1 \\ 0 & 2 & 1 & 0 & 1 & 0 \\ 0 & -1 & -1 & 1 & 0 & -2 \end{array}\right] \sim \left[\begin{array}{ccc|ccc} 1 & 1 & 2 & 0 & 0 & 1 \\ 0 & 1 & 1 & -1 & 0 & 2 \\ 0 & 2 & 1 & 0 & 1 & 0 \end{array}\right]$$

$$\sim \left[\begin{array}{ccc|ccc} 1 & 0 & 1 & 1 & 0 & -1 \\ 0 & 1 & 1 & -1 & 0 & 2 \\ 0 & 0 & -1 & 2 & 1 & -4 \end{array}\right] \sim \left[\begin{array}{ccc|ccc} 1 & 0 & 0 & 3 & 1 & -5 \\ 0 & 1 & 0 & 1 & 1 & -2 \\ 0 & 0 & 1 & -2 & -1 & 4 \end{array}\right]$$

by (i) interchanging the first and third rows, (ii) subtracting twice the first row from the third row, (iii) interchanging the second and third rows and multiplying the second row by –1, (iv) subtracting 1 and 2 times respectively the second row from the first and third rows, and (v) adding the third row to-the first and second rows, and then multiplying it by –1. Therefore,

$$A^{-1} = \begin{bmatrix} 2 & 1 & 3 \\ 0 & 2 & 1 \\ 1 & 1 & 2 \end{bmatrix}^{-1} = \begin{bmatrix} 3 & 1 & -5 \\ 1 & 1 & -2 \\ -2 & -1 & 4 \end{bmatrix}. \qquad \square$$

4.5 INVERSE FROM ELEMENTARY MATRICES

Let a nonsingular matrix A be transformed to I (see (4.6d)) by a sequence of elementary row and column operations NF(a)-NF(f). The elementary row operations on A are implemented by premultiplying it with H_1, H_2, ..., H_r and column operations by postmultiplying A with, K_1, K_2, ..., K_p so that

$$H_r \cdots H_2 H_1 A K_1 K_2 \cdots K_p = I.$$

Since elementary matrices are nonsingular, it follows that

$$A = (H_r \cdots H_2 H_1)^{-1} (K_1 K_2 \cdots K_p)^{-1} \tag{4.9}$$

or,

$$A^{-1} = (K_1 K_2 \cdots K_p)(H_r \cdots H_2 H_1) = PQ,$$

where

$$P = K_1 K_2 \cdots K_p, \quad Q = H_r \cdots H_2 H_1.$$

In view of our observations at the end of Section 4.2, we need not multiply elementary matrices to get P and Q, which can be obtained from a unit matrix by imposing on it separately the elementary column and row transformations, respectively, subjected to A.

As any nonsingular matrix A has the representation shown in (4.9), where inverses of H_i's and K_i's are elementary matrices, we can state that every nonsingular matrix can be decomposed into a product of elementary matrices.

Example 4.5 Take A as given in Example 4.4. We have, by elementary operations,

$$A = \begin{bmatrix} 2 & 1 & 3 \\ 0 & 2 & 1 \\ 1 & 1 & 2 \end{bmatrix} \sim \begin{bmatrix} 1 & 1 & 2 \\ 0 & 2 & 1 \\ 2 & 1 & 3 \end{bmatrix} \sim \begin{bmatrix} 1 & 1 & 2 \\ 0 & 2 & 1 \\ 0 & -1 & -1 \end{bmatrix}$$

$$\sim \begin{bmatrix} 1 & 0 & 0 \\ 0 & 2 & 1 \\ 0 & -1 & -1 \end{bmatrix} \sim \begin{bmatrix} 1 & 0 & 0 \\ 0 & 2 & 1 \\ 0 & 1 & 1 \end{bmatrix} \sim \begin{bmatrix} 1 & 0 & 0 \\ 0 & 1 & 1 \\ 0 & 2 & 1 \end{bmatrix}$$

$$\sim \begin{bmatrix} 1 & 0 & 0 \\ 0 & 1 & 1 \\ 0 & 0 & -1 \end{bmatrix} \sim \begin{bmatrix} 1 & 0 & 0 \\ 0 & 1 & 0 \\ 0 & 0 & -1 \end{bmatrix} \sim \begin{bmatrix} 1 & 0 & 0 \\ 0 & 1 & 0 \\ 0 & 0 & 1 \end{bmatrix}$$

The above elementary operations are described by (see Section 4.2)

(i) H_{13}, (ii) H_{31} (–2), (iii) $K_{21}(-1)K_{31}(-2)$, (iv) H_3 (–1),
(v) H_{23}, (vi) H_{32} (–2), (vii) K_{32} (–1), (viii) H_3 (–1),

in the order of their execution on A. Therefore, only (iii) and (vii) constitute P and the remaining ones Q. So,

$$P = \begin{bmatrix} 1 & -1 & 0 \\ 0 & 1 & 0 \\ 0 & 0 & 1 \end{bmatrix} \begin{bmatrix} 1 & 0 & -2 \\ 0 & 1 & 0 \\ 0 & 0 & 1 \end{bmatrix} \begin{bmatrix} 1 & 0 & 0 \\ 0 & 1 & -1 \\ 0 & 0 & 1 \end{bmatrix} = \begin{bmatrix} 1 & -1 & -1 \\ 0 & 1 & -1 \\ 0 & 0 & 1 \end{bmatrix}$$

and observing that Q is a product of H_i's in the reverse order, we have

$$Q=\begin{bmatrix}1&0&0\\0&1&0\\0&0&-1\end{bmatrix}\begin{bmatrix}1&0&0\\0&1&0\\0&-2&1\end{bmatrix}\begin{bmatrix}1&0&0\\0&0&1\\0&1&0\end{bmatrix}\begin{bmatrix}1&0&0\\0&1&0\\0&0&-1\end{bmatrix}$$

$$\times\begin{bmatrix}1&0&0\\0&1&0\\-2&0&1\end{bmatrix}\begin{bmatrix}0&0&1\\0&1&0\\1&0&0\end{bmatrix}=\begin{bmatrix}0&0&1\\-1&0&2\\-2&-1&4\end{bmatrix}.$$

Hence in view of (4.9),

$$A^{-1}=PQ=\begin{bmatrix}3&1&-5\\1&1&-2\\-2&-1&4\end{bmatrix}.$$

If we subject a third order unit matrix to the column operations (iii) and (vii) in succession, then we obtain P without the need to compute the product as shown above. Similarly, with the aid of row operations (viii), (vi), (v), (iv), (ii), (i) executed in this order on a third order unit matrix, we get Q. □

4.6 ROW-EQUIVALENT AND COLUMN-EQUIVALENT CANONICAL FORM

If we exclude column operations in obtaining the normal form as described in Section 4.3, we can derive a different type of normal form known as *row-equivalent canonical form* or *row echelon normal form* by elementary row operations alone.

Definition 4.6 *Row-Equivalent Canonical* (*or Row Echelon Normal*) *Form.* Any nonzero $m \times n$ matrix A of rank r is row-equivalent to a canonical matrix C in which

(a) one or more elements of each of the first r rows are nonzero while all other rows have only zero elements;

(b) in the ith row ($i = 1, 2, ..., r$) the first nonzero element is 1 which occurs in the j_ith column;

(c) $j_1 < j_2 < \ldots < j_r$ which are called *indices* of the form;

(d) the j_ith column ($i = 1, 2, ..., r$) has only one nonzero element which is 1 located in the ith row; these columns are elementary m vectors (see (5.4)) $\mathbf{e}_1, \mathbf{e}_2, ..., \mathbf{e}_r$; and

(e) rank A = rank $C = r$. □

Example 4.6 Obtain the row-equivalent canonical form of the given matrix A.

$$A=\begin{bmatrix}0&1&2&-1&2\\0&2&4&0&14\\0&-1&-2&4&13\end{bmatrix}\sim\begin{bmatrix}0&1&2&-1&2\\0&0&0&2&10\\0&0&0&3&15\end{bmatrix}$$

$$\sim \begin{bmatrix} 0 & 1 & 2 & -1 & 2 \\ 0 & 0 & 0 & 1 & 5 \\ 0 & 0 & 0 & 3 & 15 \end{bmatrix} \sim \begin{bmatrix} 0 & 1 & 2 & 0 & 7 \\ 0 & 0 & 0 & 1 & 5 \\ 0 & 0 & 0 & 0 & 0 \end{bmatrix} = C.$$

Here, $j_1 = 2, j_2 = 4$. We can easily confirm that the structure of row-equivalent canonical form C conforms to the stipulations (a)–(e). From this canonical form, rank A is immediately read as 2. □

The following sequence of row operations can be of much help to obtain the row-equivalent canonical form in a systematic manner.

RC (a) Let j_1 be the first column in A with nonzero elements. By interchanging rows, if necessary, bring a nonzero element in the first row of column j_1. Call this element a_{1j_1}.

RC (b) Divide all nonzero elements in the first row by a_{1j_1}. So, A has 1 in $(1, j_1)$ position.

RC (c) Subtract suitable multiples of the first row from *the remaining ones* so that the j_1th column elements in these rows are equal to zero.

RC (d) Let j_2 be the first column with nonzero elements below the first row. Obviously, $j_1 < j_2$. By interchanging of rows, bring, if needed, a nonzero element to be called a_{2j_2} on the second row of this column.

RC (e) Repeat steps RC (b) and RC (c), replacing the "first row" by the "second row" and $1j_1$ by $2j_2$.

RC (f) Continue this process for the remaining columns until at the rth step all rows below row r have only zero elements.

An alternative term for the row-equivalent canonical form is *reduced row echelon form* (RREF). The MATLAB macro for it is rref(A) for a matrix A. Thus, for A given in Example 4.6,

```
>> rref (A)
```

responds with the matrix C as given in that example.

To compute the inverse of a nonsingular matrix A, we use

```
>> rref ([A eye(size (A))])
```

which for the matrix A given in Example 4.4 responds with the row echelon form and A^{-1} given in that example. Here, size (A) evaluates the order of A which is 3 and eye (3) is the identity matrix of order 3.

Theorem 4.1 The row-equivalent canonical (RC) form of a matrix A is unique.

Proof Let A be an $m \times n$ matrix of rank r, and B and C its two RC forms, presumably different. Hence B and C are of the same rank r and, therefore, have the same r number of nonzero rows. Denote the indices (see Definition 4.6) of B and C, by

$$j_1 < j_2 < \cdots < j_r; \quad k_1 < k_2 < \cdots < k_r,$$

respectively. Since $B \sim A$ and $C \sim A$, we have, by symmetry and transitivity of row-equivalence (see Section 4.3), $B \sim C$. Therefore, there exists a nonsingular matrix P such that (see Definition 4.4)

$$B = PC. \tag{4.10}$$

Denoting the ith row of B and C by $\mathbf{b}_i$ and $\mathbf{c}_i$, respectively, we have, from (4.10), the relation

$$\mathbf{b}_i = \alpha_{i1}\mathbf{c}_1 + \alpha_{i2}\mathbf{c}_2 + \cdots + \alpha_{ir}\mathbf{c}_r, \quad 1 \le i \le r; \tag{4.11}$$

where $[\alpha_{i1}\ \alpha_{i2}\ . \ . \ . \ \alpha_{im}]$ is the ith row of P. We shall first prove that

$$j_1 = k_1, j_2 = k_2, ..., j_r = k_r \tag{4.12}$$

by induction. Since the first $k_1 - 1$ elements of $\mathbf{c}_1, \mathbf{c}_2, ..., \mathbf{c}_r$ are zero (see Definition 4.6), it follows, from (4.11) when $i = 1$, that the first k_1–1 elements of $\mathbf{b}_1$ are also zero. Hence $j_1 \ge k_1$. Writing $C = P^{-1}B$ from (4.10), we can likewise show that $k_1 \ge j_1$. Therefore, $j_1 = k_1$. Suppose now that $j_1 = k_1, \cdot\cdot\cdot, \ \ j_{r-1} = k_{r-1}$. So $j_r > j_{r-1} = k_{r-1}$. Comparing both sides of (4.11) when $i = r$, we obtain

$$\alpha_{r1} = \cdot\cdot\cdot = \alpha_{rr-1} = 0,$$

so the first k_r–1 elements of $\mathbf{b}_r$ are zero and $j_r \ge k_r$. Employing $C = P^{-1}B$ again we can show similarly that $k_r \ge j_r$. Hence $j_r = k_r$.

Now, we shall prove that $\mathbf{b}_i = \mathbf{c}_i$, $1 \le i \le r$. Since the j_ith element of $\mathbf{b}_i$ is 1 and the elements in columns j_p, $p \ne i$, $1 \le p \le r$ are zero, we have, in view of (4.12), comparing both sides of (4.11),

$$\alpha_{ip} = 0, p \ne i, 1 \le p \le r \text{ and } \alpha_{ii} = 1.$$

Hence, from (4.11), $\mathbf{b}_i$, $= \mathbf{c}_i$. □

A very easy consequence of Theorem 4.1 is the following corollary.

Corollary 4.1 The RC form of a nonsingular matrix A of order n is I_n. □

Remark 4.1 In computing the inverse with "step-by-step reduction of $[A; I]$", essentially we have to determine the RC form of $[A; I]$. Since A is nonsingular, it is transformed into I and since RC form is unique, so is A^{-1} (see Section 4.4).

Corollary 4.2 If A and B are row-equivalent, then they have the same RC form.

Proof Let C be the RC form of A, and B be row-equivalent to A. Then there exist nonsingular matrices P and Q such that

$$PA = C \quad A = QB$$

Hence $PQB = C$, which shows that C is the RC form of B. □

Remark 4.2 Similarly, we can obtain *column-equivalent canonical* (CC) form by elementary column operations, which is exactly the same as obtaining the row-equivalent canonical form C of A^T and transposing C. The rank of A can be read from C. The CC form of A is unique.

4.7 PROPERTIES OF RANK

We shall now study the effect of matrix multiplication on the rank of a matrix, which will be studied in further detail from the viewpoint of range space and null space in Section 5.7.

Property I: The rank of a matrix A remains unchanged by premultiplication (resp postmultiplication) with a nonsingular matrix, i.e., $\rho(PA) = \rho(A) = \rho(AQ)$, where P and Q are nonsingular.

We observe that in view of (4.9) any nonsingular matrix can be expressed as a product of elementary matrices. Premultiplication (resp postmultiplication) of A with these elementary matrices are nothing but elementary row (resp column) operations and thus keeps the rank of A unaltered. A direct consequence of Property I is the following property.

Property II: If P and Q are nonsingular, then $\rho(PAQ) = \rho(A)$.

Property III: Two $m \times n$ matrices A and B are equivalent iff they have the same rank.

If A and B are equivalent, then there exist nonsingular matrices P and Q such that

$$B = PAQ.$$

Hence, by Property II, $\rho(A) = P(B)$. On the other hand, if $\rho(A) = \rho(B) = r$, then there exist nonsingular matrices $P, Q, \hat{P}, \hat{Q}$ such that

$$PAQ = N = \left[\begin{array}{c|c} I_r & 0 \\ \hline 0 & 0 \end{array}\right] = \hat{P}B\hat{Q}. \tag{4.13}$$

Therefore,

$$B = \hat{P}^{-1}PAQ\hat{Q}^{-1}$$

which shows that A and B are equivalent.

Property IV: The rank of the product of two matrices A and B cannot exceed the rank of either factor, i.e.,

$$\rho(AB) \leq \min\,[\rho(A), \rho(B)] \tag{4.14}$$

Let the $m \times p$ matrix A be of rank r. Therefore, owing to (4.13), we can write

$$A = P^{-1}NQ^{-1}.$$

Let B be a $p \times n$ matrix; then consider the rank of AB:

$$\rho(AB) = \rho(P^{-1}NQ^{-1}B) = \rho(NQ^{-1}B).$$

The last step follows from Property I. In view of the structure of N in (4.13), it is easily verified that the first r rows of $NQ^{-1}B$ are the same as those of $Q^{-1}B$, but the remaining $m - r$ rows in $NQ^{-1}B$ are zeros. So, taking into consideration Property I, we can write

$$\rho(AB) \leq \rho(B).$$

Similarly,

$$\rho(AB) \leq \rho(A).$$

These two results jointly prove (4.14).

Property V: If $AB = 0$, where A is an $m \times p$ matrix of rank r and B is any $p \times n$ matrix satisfying the above relation, then $\rho(B) \leq p - r$.

In view of (4.13), from $AB = 0$ it follows that $P^{-1}NQ^{-1}B = 0$, i.e., $NQ^{-1}B = 0$, which implies that $Q^{-1}B$ has first the r zero rows and remaining $p - r$ rows arbitrary. Hence $\rho(B) \leq p - r$ by Property I.

4.8 RIGHT INVERSE AND LEFT INVERSE OF A MATRIX

We have defined the inverse of an n-square matrix A in Chapter 3 and have seen that the matrix must be nonsingular, i.e., it must be of full rank n for the inverse to exist. The basic notion in the definition of an inverse is the existence of another matrix B which on multiplication into A gives an identity matrix. On the other hand, if A is not a square but an $m \times n$, $m \neq n$ matrix, it is sometimes possible to find an $n \times m$ matrix B which on premultiplication or postmultiplication into A gives rise to an identity matrix. This aspect may be clarified in the following example.

Example 4.7 Let a 2×3 matrix A be given by

$$A = \begin{bmatrix} 1 & 1 & 1 \\ 2 & 3 & 4 \end{bmatrix}. \tag{4.15}$$

Then for the matrix

$$C = \begin{bmatrix} 3 & -1 \\ -2 & 1 \\ 0 & 0 \end{bmatrix}$$

we have,

$$AC = \begin{bmatrix} 1 & 0 \\ 0 & 1 \end{bmatrix} = I_2.$$

There are also other matrices, e.g.,

$$\hat{C} = \begin{bmatrix} 0 & 0 \\ 4 & -1 \\ -3 & 1 \end{bmatrix}$$

such that $A\hat{C} = I_2$. On the other hand,

$$CA = \begin{bmatrix} 1 & 0 & -1 \\ 0 & 1 & 2 \\ 0 & 0 & 0 \end{bmatrix}, \quad \hat{C}A = \begin{bmatrix} 0 & 0 & 0 \\ 2 & 1 & 0 \\ -1 & 0 & 1 \end{bmatrix},$$

which are not unit matrices. Thus AC (resp $A\hat{C}$) is an identity matrix while CA (resp $\hat{C}A$) is not.

Again, defining

$$A = \begin{bmatrix} 1 & 2 \\ 2 & 1 \\ 3 & 2 \end{bmatrix},$$

we find that if

$$B = \begin{bmatrix} 0 & 2 & -1 \\ 0 & -3 & 2 \end{bmatrix},$$

then $BA = I_2$. But,

$$AB = \begin{bmatrix} 0 & -4 & 3 \\ 0 & 1 & 0 \\ 0 & 0 & 1 \end{bmatrix}$$

is not an identity matrix. □

This example illustrates the fact that for any matrix A there does not exist a matrix R which on premultiplying and postmultiplying A gives in both cases the same identity matrix as in Definition 3.3. This leads to the following definition.

Definition 4.7 *Left and Right Inverse.* For a given $m \times n$ matrix A, if there exists an $n \times m$ matrix B such that $BA = I_n$, then B is defined as the *left inverse* of A, On the other hand, if there exists an $n \times m$ matrix C such that $AC = I_m$, then C is defined as the *right inverse* of A. □

Theorem 4.2 An $m \times n$ matrix A has a right inverse iff rank $A = m$ and a left inverse iff rank $A = n$.

Proof Let A have a right inverse. So there exists a matrix C such that $AC = I_m$. The necessity will be proved by contradiction. Suppose $x = \rho(A) < m$. Since $\rho(AC) = m$, taking into consideration (4.14), we have

$$m \le \min\,[\rho(A),\ \rho(C)].$$

Setting $y = \rho(C)$, we see that if $x \le y$ then $m \le x < m$ and if $x > y$, then $m \le y < x < m$, leading to contradictory conditions. Hence, $\rho(A) = m$, because $\rho(A) > m$ is not possible (see Definition 4.1).

Sufficiency will now be proved constructively. If rank $A = m$, then there is an $m \times m$ submatrix $\hat{A}$ formed by some m columns of A which is nonsingular. Also, there exists an $n \times m$ matrix R such that $AR = \hat{A}$. For example, let

$$A = \begin{bmatrix} a & b & c \\ x & y & z \end{bmatrix}, \quad \hat{A} = \begin{bmatrix} a & c \\ x & z \end{bmatrix}.$$

Then,

$$R = \begin{bmatrix} 1 & 0 \\ 0 & 0 \\ 0 & 1 \end{bmatrix}.$$

It now immediately follows that $C = R(AR)^{-1}$ is a right inverse of A because $AC = I_m$ by direct multiplication. The theorem is therefore established. □

Remark 4.3 The right or left inverse, if it exists, is not unique. In fact, a matrix may possess an infinite number of right or left inverses. For example, the matrix A given by (4.15) in Example 4.7 has right inverses of the form

$$\begin{bmatrix} \alpha+3 & \beta-1 \\ -2(1+\alpha) & 1-2\beta \\ \alpha & \beta \end{bmatrix},$$

where α and β are any real numbers. In fact, putting $\alpha = 0$, $\beta = 0$, the right inverse C, and putting $\alpha = -3$, $\beta = 1$, the right inverse $\hat{C}$ are obtained. This class of right inverses can be obtained by solving the matrix equation

$$\begin{bmatrix} 1 & 1 & 1 \\ 2 & 3 & 4 \end{bmatrix} \begin{bmatrix} x & u \\ y & v \\ z & w \end{bmatrix} = \begin{bmatrix} 1 & 0 \\ 0 & 1 \end{bmatrix}$$

for x, y, z; u, v, w. If both a left inverse B and a right inverse C of A exist, then $B = C$ which is unique and is defined as the inverse of A. Indeed, from $BA = I$, $AC = I$, we have $BAC = C$ or $B \cdot I = C$. For another left inverse $\hat{B}$, we can show $\hat{B} = C$; so $B = \hat{B}$, and the inverse is unique.

PROBLEMS

4.1 Determine all the possible square submatrices of the following matrices:

(a) $\begin{bmatrix} 3 & 3 & 1 \\ 5 & 3 & 2 \\ 1 & 2 & -1 \end{bmatrix}$, (b) $\begin{bmatrix} 1 & 2 & 0 & 1 \\ 2 & -3 & 7 & 9 \\ 1 & 4 & -2 & -1 \end{bmatrix}$

(c) $\begin{bmatrix} 2 & 3 & 1 & 1 \\ 1 & -1 & 3 & -2 \\ 1 & 0 & 2 & -1 \\ 3 & 5 & 1 & 2 \end{bmatrix}$, (d) $\begin{bmatrix} 2 & 3 \\ 3 & 2 \\ 1 & 4 \\ 5 & 3 \end{bmatrix}$.

4.2 Determine the rank of the matrices given in Problem 4.1.

4.3 Determine the left elementary matrices which would premultiply the matrices in Problem 4.1 to give

(a) $\begin{bmatrix} 3 & 3 & 1 \\ 1 & 2 & -1 \\ 2 & 0 & 1 \end{bmatrix}$, (b) $\begin{bmatrix} 1 & 2 & 0 & 1 \\ 0 & -7 & 7 & 7 \\ -1 & 0 & -2 & -3 \end{bmatrix}$,

(c) $\begin{bmatrix} 1 & 0 & 2 & -1 \\ 0 & 3 & -3 & 3 \\ 2 & -1 & 5 & -3 \\ -1 & -1 & -1 & 0 \end{bmatrix}$, (d) $\begin{bmatrix} 1 & -1 \\ 2 & 3 \\ 0 & 5 \\ 1 & -3 \end{bmatrix}$.

4.4 Determine the right elementary matrices which would postmultiply the matrices in Problem 4.1 to give

(a) $\begin{bmatrix} 0 & 2 & 1 \\ 2 & 0 & 1 \\ -1 & 3 & 0 \end{bmatrix}$, (b) $\begin{bmatrix} 2 & 0 & 0 & 1 \\ 4 & -7 & 7 & 9 \\ 2 & 2 & -2 & -1 \end{bmatrix}$,

(c) $\begin{bmatrix} 2 & 3 & 0 & 1 \\ 1 & -1 & 5 & -2 \\ 1 & 0 & 3 & -1 \\ 3 & 5 & -1 & 2 \end{bmatrix}$, (d) $\begin{bmatrix} 2 & 1 \\ -2 & -6 \\ 0 & -1 \\ -4 & -11 \end{bmatrix}$.

4.5 Without computing the matrix product, determine a third order matrix P which is expressed as a multiplication of the following elementary matrices:

(a) $H_{23}H_{21}(2)K_{12}(-1)$, (b) $H_1(2)K_3(2)K_{23}(-1)K_3\left(\frac{1}{2}\right)$, (c) $H_2(2)H_{31}(-2)H_{23}(1)K_{23}(-1)$.

4.6 Without doing matrix multiplication, determine the inverse of a third order matrix P which is a product of elementary matrices as given below:

(a) $K_{12}(-1)H_{21}(2)H_{23}$, (b) $K_3\left(\frac{1}{2}\right)K_{23}(-1)K_3(2)H_1(2)$, (c) $K_{23}(-1)H_{23}(1)H_{31}(-2)H_2(2)$.

4.7 By determining P and Q in Definition 4.4, show that the matrices in Problem 4.1 are equivalent to

(a) $[I_3]$, (b) $\left[\begin{array}{c|c} I_2 & 0 \\ \hline 0 & 0 \end{array}\right]$, (c) $\left[\begin{array}{c|c} I_2 & 0 \\ \hline 0 & 0 \end{array}\right]$, (d) $\begin{bmatrix} I_2 \\ \hline 0 \end{bmatrix}$.

4.8 By the method of (a) step-by-step reduction and (b) elementary matrices, obtain the inverse of the matrices given in Problem 3.13.

4.9 Express the following matrices as a product of elementary matrices:

(a) $\begin{bmatrix} 1 & 1 \\ 1 & 2 \end{bmatrix}$, (b) $\begin{bmatrix} 3 & 6 \\ 4 & 2 \end{bmatrix}$, (c) $\begin{bmatrix} 2 & 0 & 3 \\ 0 & 1 & 2 \\ 1 & 0 & 3 \end{bmatrix}$, (d) $\begin{bmatrix} 3 & 1 & 0 \\ 6 & 3 & 1 \\ 0 & 1 & 3 \end{bmatrix}$.

4.10 Obtain the row echelon normal form of the following matrices:

(a) $\begin{bmatrix} 2 & 6 & 4 & 2 \\ -1 & -6 & -3 & 0 \\ 2 & 6 & 6 & 6 \end{bmatrix}$, (b) $\begin{bmatrix} 1 & 1+i & -i \\ 0 & i & 1+2i \\ 1 & 1+2i & 1+i \end{bmatrix}$, (c) $\begin{bmatrix} 1 & 3 & 1 & 2 \\ 2 & 6 & 2 & 5 \\ -1 & -3 & -1 & -2 \end{bmatrix}$.

4.11 Find an appropriate inverse, if it exists, of each of the following matrices:

(a) $\begin{bmatrix} 1 & 2 \\ 2 & 1 \\ 3 & 2 \end{bmatrix}$, (b) $\begin{bmatrix} 3 & 2 \\ 2 & 3 \\ 5 & 4 \end{bmatrix}$, (c) $\begin{bmatrix} 3 & 3 \\ 1 & 1 \\ 3 & 3 \end{bmatrix}$, (d) $\begin{bmatrix} 1 & 2 & 1 & 0 \\ 0 & 0 & 2 & 0 \end{bmatrix}$.

4.12 Prove that for an n-square matrix A, (a) if rank $A < n - 1$, then adj $A = 0$; (b) if rank $A = n - 1$, then rank adj $A = 1$.

4.13 Using the structure of normal forms given in (4.6), prove that the ranks of A, A^T, $\bar{A}$ and A^* are the same.

4.14 Prove that (a) AB and BA are nonsingular if A and B are nonsingular n-square matrices; (b) AB and BA are singular if at least one of the n-square matrices A and B is singular.

4.15 Using MATLAB, solve Problem 4.10.

CHAPTER

5

Vector Space

In this chapter, a formal definition of a vector space, whose elements are called *vectors*, is given and its connection with vectors in vector algebra is brought out clearly. The structure of a vector space depends on the notion of (i) linear independence of vectors, (ii) basis, and (iii) dimension which are then introduced. These structural constituents are now employed to describe vector subspaces and direct sum of subspaces. After these qualitative structures, some quantitative aspects such as norm and orthogonality of vectors of a vector space are presented under inner product space, followed by a detailed description of the Gram-Schmidt process for the construction of an orthonormal basis. Based on the foundation of a vector space thus established, the existence and methods of solutions of linear simultaneous equations are then described. The concepts of null space and range space, and some important results on the rank of a product of two matrices including Sylvester's inequality, are outlined in a comprehensive manner. Finally, the computational methods for testing the dependence or independence of a set of vectors are critically discussed.

5.1 VECTOR SPACE

In the following discussion and elsewhere in the text, we refer to the class of real numbers or the class of complex numbers as *scalars*.

Definition 5.1 *Vector Space.* A *vector space* or a *linear space* denoted by V has elements or members **x, y, z, ...,** called *vectors*, satisfying the following axioms:

To each pair of vectors **x, y** in V there corresponds in V a unique third vector called their sum and written as $\mathbf{x} + \mathbf{y}$ (i.e., V is closed under addition) in such a manner that:

1. addition is commutative: $\mathbf{x} + \mathbf{y} = \mathbf{y} + \mathbf{x}$;
2. addition is associative: $\mathbf{x} + (\mathbf{y} + z) = (\mathbf{x} + \mathbf{y}) + z$;

3. there exists a unique vector $\boldsymbol{\theta}$ in V called the zero vector or the additive identity, such that $\mathbf{x} + \boldsymbol{\theta} = \mathbf{x}$ for every vector $\mathbf{x}$; and
4. each vector $\mathbf{x}$ in V can be associated with a unique vector $-\mathbf{x}$ in V called the additive inverse such that $\mathbf{x} + (-\mathbf{x}) = \boldsymbol{\theta}$.

   Any scalar α and any vector $\mathbf{x}$ in V can be combined by an operation called *scalar multiplication* to give a unique vector $\alpha\mathbf{x}$ in V (i.e., V is closed under scalar multiplication) in such a manner that:

5. multiplication by scalars α, β is associative, e.g., $\alpha(\beta\mathbf{x}) = (\alpha\beta)\mathbf{x}$;
6. there exists a unique nonzero scalar 1 (called one) such that $1\mathbf{x} = \mathbf{x}$ for every vector $\mathbf{x}$;
7. multiplication by scalars is distributive with respect to vector addition, e.g., $\alpha(\mathbf{x} + \mathbf{y}) = \alpha\mathbf{x} + \alpha\mathbf{y}$; and
8. multiplication by vectors is distributive with respect to scalar addition, e.g., $(\alpha + \beta)\mathbf{x} = \alpha\mathbf{x} + \beta\mathbf{x}$. □

The definition of a vector space thus needs a class of scalars to be termed a *field* $\mathbb{F}$ *of scalars**, and the vector space V is said to be over the field $\mathbb{F}$. If the field is finite, we say that V is a vector space over a finite field $\mathbb{F}$ (see Definition 1.15). In this text we shall be concerned principally with either real or complex numbers as elements of $\mathbb{F}$. An aggregate or collection of elements such as $\mathbf{x}, \mathbf{y}, \mathbf{z}$ in a vector space V is referred to as a *set* and is denoted by $\{\mathbf{x}, \mathbf{y}, \mathbf{z}\}$. Also, $\mathbf{x}$ is an element of the vector space V and is denoted by $\mathbf{x} \in V$ which is read as $\mathbf{x}$ belongs to (or is a member of) V. There are four compositions involved in the definition of a vector space: (i) addition in V, (ii) scalar multiplication of elements of V with elements in $\mathbb{F}$, (iii) addition in $\mathbb{F}$, and (iv) multiplication in $\mathbb{F}$. For example, in axiom 8, $\alpha + \beta$ is an element of $\mathbb{F}$ which multiplies into $\mathbf{x}$, an element of V to give the sum of $\alpha\mathbf{x}$ and $\beta\mathbf{x}$ which are elements of V. Axioms 5–8 show the interaction between the elements of $\mathbb{F}$ and V. If $\mathbb{F}$ is the set of real numbers, V is termed a *real vector space*, whereas if $\mathbb{F}$ is the set of complex numbers, it is called a *complex vector space*.

In Euclidean plane (Figure 5.1(a)), from the viewpoint of coordinate geometry, a point P is described by the coordinates (x_1, x_2), and the *directed line segment* $\overrightarrow{OP}$ having a magnitude and a direction is known as a vector. Multiplication of $\overrightarrow{OP}$ by a scalar α is defined as $\overrightarrow{OQ} = \alpha\overrightarrow{OP}$, where Q has the coordinates $(\alpha x_1, \alpha x_2)$. If A and B are the points (x_1, x_2) and (y_1, y_2), then completing the parallelogram $OACB$ (Figure 5.1(b)), we define $\overrightarrow{OC}$ as *the sum of* $\overrightarrow{OA}$ *and* $\overrightarrow{OB}$. By geometric construction, the coordinates of C are $(x_1 + y_1, x_2 + y_2)$. Again, if α and β are any two real numbers, X and Y are the points (x_1, x_2) and (y_1, y_2), respectively, *a composite operation* $\alpha\overrightarrow{OX} + \beta\overrightarrow{OY}$ gives a resulting directed line segment as $\overrightarrow{OX_1} + \overrightarrow{OY_1} = \overrightarrow{OZ}$, as in Figure 5.1(c), and we can find geometrically the coordinates of Z as $(\alpha x_1 + \beta y_1, \alpha x_2 + \beta x_2)$. Finally, if we define the zero element as the origin and the negative of $\overrightarrow{OP}$ (Figure 5.1(a)) as $\overrightarrow{OP'}$

*Small Greek letters are generally used to denote scalars and small bold Latin letters vectors. The symbol for a zero vector is $\boldsymbol{\theta}$.

where P' is the point $(-x_1, -x_2)$, then it is not difficult to show that axioms 1 to 8 are satisfied by the foregoing set of directed line segments in vector algebra which can truly be termed vectors.

On the other hand, we can call a column matrix

$$\begin{bmatrix} x_1 \\ x_2 \end{bmatrix},$$

a vector, meaning that (x_1, x_2) are the coordinates of P in $\overrightarrow{OP}$. Similarly, scalar multiplication

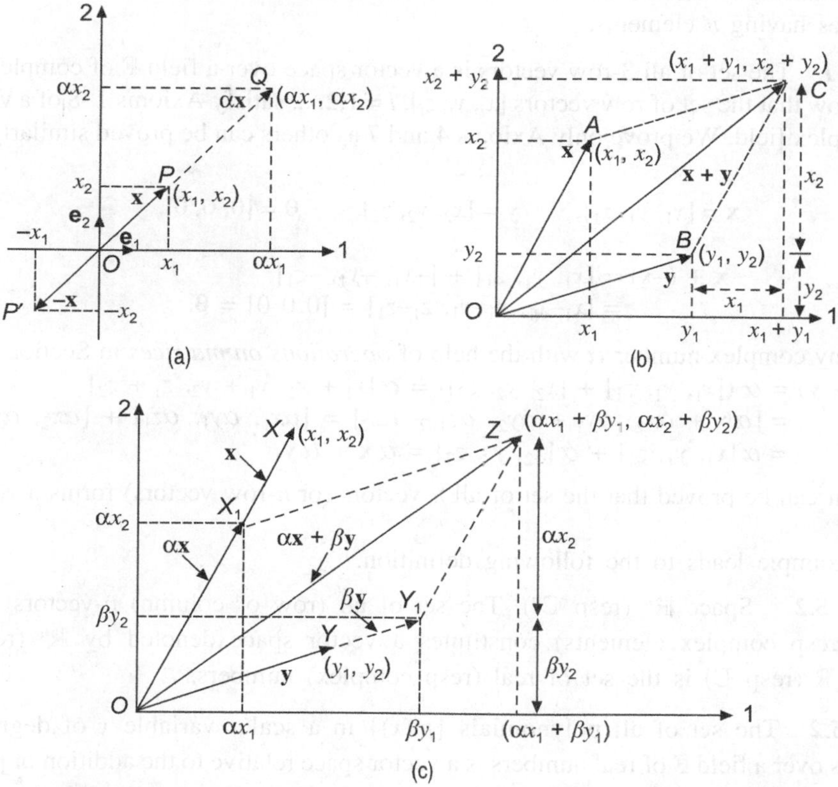

Figure 5.1 (a) Scalar multiplication and negative of a vector **x**; (b) sum of two vectors **x** and **y**; (c) composite operation of two vectors **x** and **y**.

(Figure 5.1(a)), addition (Figure 5.1(b)), and composite operation (Figure 5.1(c)) of two directed line segments conform to (or induce) similar operations on column matrices, viz.,

$$\alpha \begin{bmatrix} x_1 \\ x_2 \end{bmatrix} = \begin{bmatrix} \alpha x_1 \\ \alpha x_2 \end{bmatrix} \quad \text{(scalar multiplication)} \qquad (5.1a)$$

$$\begin{bmatrix} x_1 \\ x_2 \end{bmatrix} + \begin{bmatrix} y_1 \\ y_2 \end{bmatrix} = \begin{bmatrix} x_1 + y_1 \\ x_2 + y_2 \end{bmatrix} \quad \text{(addition of vectors)} \qquad (5.1b)$$

$$\alpha\begin{bmatrix} x_1 \\ x_2 \end{bmatrix} + \beta\begin{bmatrix} y_1 \\ y_2 \end{bmatrix} = \begin{bmatrix} \alpha x_1 + \beta y_1 \\ \alpha x_2 + \beta y_2 \end{bmatrix} \quad \text{(composite operation)} \qquad (5.1c)$$

where (5.la) represents $\overrightarrow{OQ}$, i.e., coordinates of Q, (5.1b) $\overrightarrow{OC}$, and (5.1c) $\overrightarrow{OZ}$.

The foregoing discussion shows that column matrices can be termed vectors in the sense of directed line segments. Stated differently, we can call a point in an n-coordinate space as well as a directed line segment joining the origin to that point a vector. Similar arguments apply to 2×1 row matrices which we call *row vectors.* This notion of a row vector can be extended to row matrices having n elements.

Example 5.1 The set of all 3-row vectors is a vector space over a field $\mathbb{F}$ of complex numbers. Here we show that the set of row vectors $[x_i, y_i, z_i]$, $i = 1,2, \ldots$ satisfy Axioms 1–8 of a vector space over a complex field. We prove only Axioms 4 and 7 as others can be proved similarly.

Let

$$\mathbf{x} = [x_1, y_1, z_1], \qquad \mathbf{y} = [x_2, y_2, z_2], \qquad \boldsymbol{\theta} = [0, 0, 0].$$

Then

$$\begin{aligned}\mathbf{x} + (-\mathbf{x}) &= [x_1, y_1, z_1] + [-x_1, -y_1, -z_1] \\ &= [x_1 - x_1, y_1 - y_1, z_1 - z_1] = [0\ 0\ 0] = \boldsymbol{\theta}.\end{aligned}$$

Also, for any complex number α with the help of *operations on matrices* in Section 1.2, we get

$$\begin{aligned}\alpha\,(\mathbf{x} + \mathbf{y}) &= \alpha\,([x_1, y_1, z_1] + [x_2, y_2, z_2]) = \alpha\,[x_1 + x_2, y_1 + y_2, z_1 + z_2] \\ &= [\alpha x_1 + \alpha x_2, \alpha y_1 + \alpha y_2, \alpha z_1 + \alpha z_2] = [\alpha x_1, \alpha y_1, \alpha z_1] + [\alpha x_2, \alpha y_2, \alpha z_2] \\ &= \alpha\,[x_1, y_1, z_1] + \alpha\,[x_2, y_2, z_2] = \alpha\,\mathbf{x} + \alpha\,\mathbf{y}.\end{aligned}$$

Similarly, it can be proved that the set of all n-vectors (or n-row vectors) forms a vector space. □

This example leads to the following definition.

Definition 5.2 Space $\mathbb{R}^n$ (resp $\mathbb{C}^n$). The set of all (row or column) n-vectors having real elements (resp complex elements) constitutes a vector space denoted by $\mathbb{R}^n$ (resp $\mathbb{C}^n$). In particular, $\mathbb{R}$ (resp $\mathbb{C}$) is the set of real (resp complex) numbers. □

Example 5.2 The set of all polynomials $\{p_i(x)\}$ in a scalar variable x of degree $\le n$ with coefficients over a field $\mathbb{F}$ of real numbers is a vector space relative to the addition of polynomials and the product of a polynomial with a scalar as the two vector space compositions.

Let $p_i(x) = \alpha_{i0} + \alpha_{i1}x + \cdots + \alpha_{in}x^n$, where α_{ij}'s are in $\mathbb{R}$, be an element in the set and scalar 0 be the zero element. Then, it is not difficult to check that all the Axioms 1-8 of a vector space are satisfied by the set of polynomials $\{p_i(x)\}$. This vector space of polynomials with degree $\le n$ is denoted by $\mathbb{P}_n(x)$ over $\mathbb{R}$ or over $\mathbb{C}$. □

Example 5.3 The set of all real valued continuous (differentiable or integrable) functions defined in some interval $[a, b]$, where a and b are real, is a vector space V, which can be shown by defining the following operations for any two functions f and g and a scalar α. Thus

$$(f + g)(x) := f(x) + g(x), \qquad (\alpha f)(x) := \alpha f(x),$$

where the first relation defines addition $f + g$ of the elements in V and the second, multiplication of an element f in V with a scalar α in $\mathbb{F}$. This space is denoted by $C([a, b])$. □

Example 5.4 The set of all $m \times n$ matrices over a field $\mathbb{F}$ forms a vector space over $\mathbb{F}$ under operations of addition and scalar multiplication as defined in Definitions 1.3 and 1.4, respectively. The verification of Axioms 1, 2, and 5–8 in Definition 5.1 follows directly from Properties I–V of the above operations in Section 1.2.1. With the null matrix defined as in Section 1.1 and the negative of a matrix as in Definition 1.5, it is easy to verify Axioms 3 and 4. □

Example 5.5 The set $\mathbb{C}$ of all complex numbers, e.g., the numbers of the form $a + ib$, where a and b are real and $i = \sqrt{-1}$ forms a vector space over $\mathbb{R}$ under addition and scalar multiplication defined by

$$(a + ib) + (c + id) = (a + c) + i(b + d) \qquad \alpha(a + ib) = \alpha a + i\alpha b.$$

Identifying the zero and negative elements as

$$0 + i0 \quad \text{and} \quad -(a + ib) = -a - ib,$$

it is easily seen that all vector space axioms are satisfied by $\mathbb{C}$. □

Example 5.6 The set V of all polynomials $\{p(x)\}$ of degree n in a variable x over $\mathbb{R}$ is *not a vector space.* This is because if $p(x)$ and $q(x)$ are two polynomials in this set V, then $p(x) + q(x)$ may have a degree less than n, and does not, therefore, belong to V.

Example 5.7 The set V of all 2-row vectors defined by $[x, \log x]$, where $x > 0$ is a real number, is not a vector space. This is because the zero element $[0\ 0]$ does not belong to V, and since for $x, y > 0$, $[x, \log x] + [y, \log y] \neq [x + y, \log(x + y)]$ and $\alpha[x, \log x] \neq [\alpha x, \log \alpha x]$, V is not closed both under addition and scalar multiplication.

5.2 LINEAR DEPENDENCE, BASIS AND DIMENSION

We shall be mainly concerned with column matrices (vectors) or row matrices (row vectors) as the elements of a vector space, even though other vector spaces will show up by way of examples. With this remark in view, we now turn our attention to understand properly the notion of linear independence of vectors constituting a vector space in order to know more about its structure.

Definition 5.3 *Linear Dependence.* A set of vectors $\mathbf{x}_1, \mathbf{x}_2, ..., \mathbf{x}_n$ in a vector space V over a field $\mathbb{F}$ is said to be *linearly dependent* if there exist scalars $\alpha_1, \alpha_2, ..., \alpha_n$, *not all zero*, in $\mathbb{F}$ such that

$$\sum_{i=1}^{n} \alpha_i \mathbf{x}_i = \alpha_1 \mathbf{x}_1 + \alpha_2 \mathbf{x}_2 + \cdots + \alpha_n \mathbf{x}_n = \boldsymbol{\theta}. \tag{5.2}$$

If, on the other hand, (5.2) implies that $\alpha_i = 0$ for each i, then the set of vectors $\mathbf{x}_1, \mathbf{x}_2, ..., \mathbf{x}_n$ is said to be *linearly independent.* □

Example 5.8 The vectors

$$\mathbf{x}_1 = \begin{bmatrix} 1 \\ -1 \\ 0 \end{bmatrix}, \quad \mathbf{x}_2 = \begin{bmatrix} 0 \\ 1 \\ -1 \end{bmatrix}, \quad \mathbf{x}_3 = \begin{bmatrix} 0 \\ 0 \\ 1 \end{bmatrix}, \quad \mathbf{x}_4 = \begin{bmatrix} 1 \\ 2 \\ 3 \end{bmatrix}$$

of space $\mathbb{R}^3$ satisfy the equation

$$\mathbf{x}_1 + 3\mathbf{x}_2 + 6\mathbf{x}_3 - \mathbf{x}_4 = \boldsymbol{\theta} \tag{5.3}$$

and are, therefore, linearly dependent. But, if we consider any three of these vectors, say, $\mathbf{x}_1$, $\mathbf{x}_3$ and $\mathbf{x}_4$, then $\alpha_1\mathbf{x}_1 + \alpha_3\mathbf{x}_3 + \alpha_4\mathbf{x}_4 = \boldsymbol{\theta}$ can hold iff $\alpha_1 = \alpha_3 = \alpha_4 = 0$ (see Problem 3.6(d)). The vectors $\mathbf{x}_1$, $\mathbf{x}_3$, and $\mathbf{x}_4$ are, therefore, independent. □

In view of (5.3), we can write $\mathbf{x}_4 = \mathbf{x}_1 + 3\mathbf{x}_2 + 6\mathbf{x}_3$, which shows that in case of n dependent vectors, we can express any vector $\mathbf{x}_i$ (here $\mathbf{x}_4$) having a nonzero coefficient in terms of other $n - 1$ (here 3) vectors on which $\mathbf{x}_i$ (here $\mathbf{x}_4$) is dependent, so to say. Hence, the attribute *dependent*. But no such description is possible for independent vectors.

In geometrical terms, two vectors in $\mathbb{R}^2$ are dependent iff they are coillinear (e.g., vectors $\overrightarrow{OP}$ and $\overrightarrow{OQ}$ in Figure 5.1(a) are linearly dependent because $\overrightarrow{OQ} = \alpha\overrightarrow{OP}$, i.e., $\overrightarrow{OQ}$ depends on $\overrightarrow{OP}$ and conversely) and in $\mathbb{R}^3$ three vectors are dependent iff they are coplanar.

Example 5.9 (a) *Elementary Vectors.* We call an n-vector in $\mathbb{R}^n$ having 1 in the ith row, zeros elsewhere, an ith *elementary vector*, and denote it by $\mathbf{e}_i$, i.e.,

$$\mathbf{e}_1 = \begin{bmatrix}1\\0\\0\\\vdots\\0\end{bmatrix}, \quad \mathbf{e}_2 = \begin{bmatrix}0\\1\\0\\\vdots\\0\end{bmatrix}, \quad \ldots, \quad \mathbf{e}_n = \begin{bmatrix}0\\\vdots\\0\\0\\1\end{bmatrix}. \tag{5.4}$$

The 2-vectors $\mathbf{e}_1 = [1\ 0]^T$ and $\mathbf{e}_2 = [0\ 1]^T$ in $\mathbb{R}^2$ (Figure 5.1(a)) are linearly independent because, by putting $\mathbf{x}_1 = \mathbf{e}_1$, $\mathbf{x}_2 = \mathbf{e}_2$, $n = 2$ in (5.2), we get $\alpha_1 = \alpha_2 = 0$. Similarly, the n-vectors $\mathbf{e}_i$, $i = 1, 2, \ldots, n$, in $\mathbb{R}^n$ given by (5.4) are linearly independent. Indeed, by putting $\mathbf{x}_i = \mathbf{e}_i$, $i = 1, 2, \ldots, n$, in (5.2), we get, in view of (5.4), the relation $\alpha_1\mathbf{e}_1 + \alpha_2\mathbf{e}_2 + \cdots + \alpha_n\mathbf{e}_n = [\alpha_1\ \alpha_2 \cdots \alpha_n]^T = [0\ 0 \cdots 0]^T$. Thus, by Definition 1.2, $\alpha_1 = \alpha_2 = \cdots = \alpha_n = 0$.

(b) The zero vector $\boldsymbol{\theta} = [0\ 0 \cdots 0]$ in $\mathbb{R}^n$ is linearly dependent, which follows by (5.2) with $n = 1$, $\mathbf{x}_1 = \boldsymbol{\theta}$, and so α_1 is arbitrary. Similarly, a single vector $\mathbf{x} \neq \boldsymbol{\theta}$ in V is linearly independent.

(c) The vectors $\mathbf{x}_1, \ldots, \mathbf{x}_n$, $\boldsymbol{\theta}$ are linearly dependent, because for any scalar $\alpha \neq 0$,

$$0\mathbf{x}_1 + \cdots + 0\mathbf{x}_n + \alpha\boldsymbol{\theta} = \boldsymbol{\theta}.$$

When the elements in $\mathbb{R}^n$ are row vectors, the elementary vectors in it are still denoted by $\mathbf{e}_i$, $i = 1, 2, \ldots, n$. □

Example 5.10 Let the elements of a vector space be two continuous functions f_1 and f_2 defined by

$$f_1(t) = t, \qquad t \text{ in } [-1, 1];$$

$$f_2(t) = \begin{cases} t, & t \text{ in } [0,1]; \\ -t, & t \text{ in } [-1,0]. \end{cases}$$

We observe that by choosing $\alpha_1 = 1$ and $\alpha_2 = -1$, $\alpha_1 f_1(t) + \alpha_2 f_2(t) = 0$ for all t in [0, 1]. Hence, f_1 and f_2 are linearly dependent on [0, 1]. They are also linearly dependent on the interval [–1, 0]. However, f_1 and f_2 are linearly independent on [–1, 1]. We see therefore that although the functions are linearly independent on an interval, they need not be linearly independent on a subinterval. □

Definition 5.4 *Linear Combination.* Let $\mathbf{x}$ be a vector in a vector space V over a field $\mathbb{F}$. If there exist vectors $\mathbf{x}_1, \mathbf{x}_2, \ldots, \mathbf{x}_n$ in V and scalars $\alpha_1, \alpha_2, \ldots, \alpha_n$ in $\mathbb{F}$ such that

$$\mathbf{x} = \alpha_1\mathbf{x}_1 + \alpha_2\mathbf{x}_2 + \cdots + \alpha_n\mathbf{x}_n, \tag{5.5}$$

then $\mathbf{x}$ is said to be a linear combination of (or a vector generated by) the vectors $\mathbf{x}_1$, $\mathbf{x}_2, \ldots, \mathbf{x}_n$. □

Example 5.11 The vector $\mathbf{x} = [4\ 1\ 3]^T$ in $\mathbb{R}^3$ is a linear combination of the vectors $\mathbf{x}_1$, $\mathbf{x}_2$, $\mathbf{x}_3$ and $\mathbf{x}_4$ given in Example 5.8, because

$$\mathbf{x} = 3\mathbf{x}_1 + 2\mathbf{x}_2 + 2\mathbf{x}_3 + \mathbf{x}_4.$$

Alternatively, we can represent $\mathbf{x}$ as

$$\mathbf{x} = 5\mathbf{x}_1 + 8\mathbf{x}_2 + 14\mathbf{x}_3 - \mathbf{x}_4$$

In fact, $\mathbf{x}$ admits of infinitely many representations in terms of $\mathbf{x}_1$, $\mathbf{x}_2$, $\mathbf{x}_3$ and $\mathbf{x}_4$ which, it might be observed, are linearly dependent. But representation of $\mathbf{x}$ in terms of $\mathbf{x}_1$, $\mathbf{x}_2$ and $\mathbf{x}_3$ which are linearly independent is *unique*, i.e.,

$$\mathbf{x} = 4\mathbf{x}_1 + 5\mathbf{x}_2 + 8\mathbf{x}_3$$

is a unique representation of $\mathbf{x}$ in terms of $\mathbf{x}_1$, $\mathbf{x}_2$ and $\mathbf{x}_3$ (see Corollary 5.1). Apart from being unique, the last representation of $\mathbf{x}$ is in terms of less number of vectors compared to the other two representations given above, where there are, therefore, some redundant vectors. □

Again in Figure 5.1(c), the vector $\overrightarrow{OZ}$ is a linear combination of the vectors $\overrightarrow{OX}$ and $\overrightarrow{OY}$.

Definition 5.5 *Basis and Dimension.* In a vector space V over a field $\mathbb{F}$, a basis is a set of *linearly independent* vectors $\{\mathbf{x}_1, \mathbf{x}_2, \ldots, \mathbf{x}_n\}$ given in a definite order such that *every vector* $\mathbf{x}$ in V is a *linear combination of the vectors in this set.* The number n of independent vectors in the basis is referred to as the dimension of the vector space, and a vector space of dimension n is attributed as *finite dimensional* and written as V_n. The vector space V_n is said to be *spanned* or *generated* by the basis vectors $\{\mathbf{x}_1, \mathbf{x}_2, \ldots, \mathbf{x}_n\}$ as any vector $\mathbf{x}$ in V_n can be obtained as a linear combination of the vectors $\mathbf{x}_i$'s. The coefficients $(\alpha_1, \alpha_2, \ldots, \alpha_n)$ in the linear combination of the vector $\mathbf{x}$ (see (5.5)) are called the *coordinates* of $\mathbf{x}$, and are written as $\mathbf{u} = [\alpha_1, \alpha_2, \ldots, \alpha_n]^T$, when we say that $\mathbf{x}$ has the representation $\mathbf{u}$ or, equivalently, $\mathbf{x} \equiv \mathbf{u}$ with respect to the basis $\{\mathbf{x}_1, \mathbf{x}_2, \ldots, \mathbf{x}_n\}$. If a vector space is of dimension n, in short it is written as $\dim V = n$. In other words, a *dimension* in a vector space V is the *maximum number* of *linearly independent vectors* in the space V, while the *set of these independent vectors* is called *a basis* for that space. □

Thus, a set of vectors is a possible candidate for a basis in a vector space V if the set satisfies two things: *first*, the set of vectors is linearly independent; *second*, every vector in V can be expressed as a linear combination of these vectors in this set.

Other notations for vector spaces are U and W.

It is important to note that the vectors in a basis must be in a particular order and, as a matter of fact, the basis $\{\mathbf{x}_1, \mathbf{x}_2, \ldots, \mathbf{x}_n\}$ is considered to be different from the basis $\{\mathbf{x}_2, \mathbf{x}_1, \ldots, \mathbf{x}_n\}$. This is because the coordinates of a vector $\mathbf{x}$ in V_n in terms of these bases are, respectively, $(\alpha_1, \alpha_2, \ldots, \alpha_n)$ and $(\alpha_2, \alpha_1, \ldots, \alpha_n)$ which are obviously different if $\alpha_1 \neq \alpha_2$. Therefore, our subsequent discussion, a basis will always imply an ordered basis in the sense given above.

Example 5.12 Referring to Figure 5.1(a), we consider the vector space V of all *directed line segments* on the Euclidean plane. Any vector $\overrightarrow{OP}$ having (x_1, x_2) *as the coordinates* of P, from the viewpoint of coordinate geometry, is according to our convention represented by

$$\begin{bmatrix} x_1 \\ x_2 \end{bmatrix} = x_1 \begin{bmatrix} 1 \\ 0 \end{bmatrix} + x_2 \begin{bmatrix} 0 \\ 1 \end{bmatrix} = x_1\mathbf{e}_1 + x_2\mathbf{e}_2 , \tag{5.6}$$

where $\mathbf{e}_1$ and $\mathbf{e}_2$ are represented by the *directed line segments* Oe_1 and Oe_2, respectively, and are linearly independent, because they are not collinear. Thus, the set $\{\mathbf{e}_1, \mathbf{e}_2\}$ (i.e., Oe_1 and Oe_2) is a basis for the vectors of *directed line segments* on the Euclidean plane.

On the other hand, if we look upon $[x_1\ x_2]^T$ as any element of $\mathbb{R}^2$, then its representation by (5.6) is obvious regarding $\mathbf{e}_1$ and $\mathbf{e}_2$ to be elements of $\mathbb{R}^2$ which are linearly independent as shown in Example 5.9. Thus, space $\mathbb{R}^2$ is generated by $\mathbf{e}_1$ and $\mathbf{e}_2$. The *coordinates* of $[x_1, x_2]^T$ with respect to the basis $\{\mathbf{e}_1, \mathbf{e}_2\}$, following Definition 5.5, are x_1 and x_2, and may, therefore, be compared with the notion of coordinates in coordinate geometry mentioned above [see Remark 5.1 and Example 5.37].

We have already observed in Example 5.9 that the elementary vectors $\mathbf{e}_i$ of order $n \times 1$, $i = 1, 2, \ldots, n$, are linearly independent in $\mathbb{R}^n$, or $\mathbb{C}^n$. We can follow the above arguments of $\mathbb{R}^2$ to say that $\mathbb{R}^n$ or $\mathbb{C}^n$ is generated by $\mathbf{e}_i$'s given in (5.4). In other words, $\mathbf{e}_i$'s are a set of basis vectors in $\mathbb{R}^n$ or $\mathbb{C}^n$.

Finally, it should be kept in mind that the dimension of a vector space need not always be finite [see Example 5.14]. In this text, however, we shall concentrate only on finite dimensional vector spaces. □

Example 5.13 *Elementary Basis Matrices.* Let us define E_{ij} to be a matrix of order $m \times n$ in which the element in the ijth position is 1 and all others are equal to zero:

$$E_{ij} = \begin{bmatrix} 0 & 0 & \cdots & 0 & \cdots & 0 \\ 0 & 0 & \cdots & 0 & \cdots & 0 \\ \vdots & \vdots & & \vdots & & \vdots \\ 0 & 0 & \cdots & 1 & \cdots & 0 \\ \vdots & \vdots & & \vdots & & \vdots \\ 0 & 0 & \cdots & 0 & \cdots & 0 \end{bmatrix}_{m \times n} \begin{matrix} \\ \\ \\ \leftarrow i\text{th row} \\ \\ \\ \end{matrix} \quad i = 1, \ldots, m; \quad j = 1, \ldots, n. \tag{5.7}$$

$\uparrow$ jth column

For any scalars α_{ij} $(i = 1, \ldots, m;\ j = 1, \ldots, n)$, the relation

$$\sum_{i=1}^{m} \sum_{j=1}^{n} \alpha_{ij} E_{ij} = 0$$

implies that $\alpha_{ij} = 0$ for all i, j in view of Definition 1.2 and Example 1.1. Hence, the set of

E_{ij}'s given by (5.7) is linearly independent. Moreover, any $m \times n$ matrix A given by (1.11) can be expressed as a linear combination of E_{ij}'s, viz.

$$A = [a_{ij}]_{m\times n} = \sum_{i=1}^{m} \sum_{j=1}^{n} a_{ij} E_{ij}.$$

Thus, the set of E_{ij}'s given by (5.7) forms a basis for the vector space of $m \times n$ matrices with real elements to be designated by $\mathbb{R}^{m\times n}$ or with complex elements to be designated by $\mathbb{C}^{m\times n}$, both of which are designated, in general, by $V_{m\times n}$. □

Example 5.14 The set $\{1, x, x^2, \ldots, x^n\}$ is a basis which is called a *natural basis* for the vector space of polynomials $\mathbb{P}_n(x)$ over $\mathbb{R}$ or $\mathbb{C}$. The dimension of the space $\mathbb{P}_n(x)$ is, therefore, $n + 1$. On the other hand, $\{1, x, ..., x^n, x^{n+1}, \ldots\}$ is a natural basis for the vector space of polynomials of any arbitrary degree designated by $\mathbb{P}(x)$ over $\mathbb{R}$ or over $\mathbb{C}$ whose dimension is consequently infinite. □

As a matter of fact, there are many distinct sets of basis vectors in a vector space, and it can be proved that the number of elements in these basis are the same. Further, given a vector $\mathbf{x}$ and a basis $\mathbf{x}_1, \mathbf{x}_2, ..., \mathbf{x}_n$, the coefficients of the linear combination defined by (5.5) are unique. The first of these results can be embodied in a theorem which is now stated.

Theorem 5.1 In a finite dimensional vector space V_n, every basis has exactly n elements.

The proof of this important theorem relies on the following lemma.

Lemma 5.1 Let $\{\mathbf{x}_1, \mathbf{x}_2, \ldots, \mathbf{x}_n\}$ be a basis in the vector space V_n over a field $\mathbb{F}$. Then any set of vectors $\{\mathbf{y}_1, \mathbf{y}_2, \ldots, \mathbf{y}_m\}$, $m > n$, are linearly dependent in V_n over $\mathbb{F}$.

Proof We shall prove the theorem by contradiction. Suppose that $\{\mathbf{y}_1, \mathbf{y}_2, ..., \mathbf{y}_m\}$ are linearly independent. Then there exist scalars $\alpha_1, \alpha_2, \ldots, \alpha_n$ in $\mathbb{F}$ such that

$$\mathbf{y}_1 = \alpha_1\mathbf{x}_1 + \alpha_2\mathbf{x}_2 + \cdots + \alpha_n\mathbf{x}_n.$$

Since $\mathbf{y}_1 \neq \boldsymbol{\theta}$ by assumption (see Example 5.9(c)), some $\alpha_i \neq 0$ which, by renumbering of x_i's without any loss of generality, can be taken as $\alpha_1 \neq 0$. Hence,

$$\mathbf{x}_1 = (\mathbf{y}_1 - \alpha_2\mathbf{x}_2 - \cdots - \alpha_n\mathbf{x}_n)/\alpha_1 \tag{5.8}$$

We can thus replace $\mathbf{x}_1$ in the basis vectors by the right-hand side of (5.8), which means that the vector space V can now be generated by $\{\mathbf{y}_1, \mathbf{x}_2, ..., \mathbf{x}_n\}$. This act of replacement of $\mathbf{x}_1$ by $\mathbf{y}_1$ can be continued step by step for other $\mathbf{x}_i$'s until all the basis vectors $\{\mathbf{x}_1, \mathbf{x}_2, ..., \mathbf{x}_n\}$ are replaced by $\{\mathbf{y}_1, \mathbf{y}_2, ..., \mathbf{y}_n\}$, which then would generate the vector space V. This will now be proved by induction.

At the kth step $1 \leq k < n$, let the vectors

$$\{\mathbf{y}_1, ..., \mathbf{y}_k, \mathbf{x}_{k+1}, ..., \mathbf{x}_n\} \tag{5.9}$$

generate V, allowing renumbering of $\mathbf{x}_i$'s if necessary. Then there exist scalars $a_1, ..., a_k, a_{k+1}, \ldots, a_n$ in $\mathbb{F}$ such that

$$\mathbf{y}_{k+1} = a_1\mathbf{y}_1 + \cdots + a_k\mathbf{y}_k + a_{k+1}\mathbf{x}_{k+1} + \cdots + a_n\mathbf{x}_n.$$

Since $\mathbf{y}_i$'s are linearly independent, all of $a_{k+1}, ..., a_n$ are not zero. Allowing necessary renumbering of $\mathbf{x}_{k+1}, ..., \mathbf{x}_n$, we may assume that $a_{k+1} \neq 0$. So,

$$\mathbf{x}_{k+1} = (\mathbf{y}_{k+1} - a_1\mathbf{y}_1 - \cdots - a_k\mathbf{y}_k - a_{k+2}\mathbf{x}_{k+2} - \cdots - a_n\mathbf{x}_n)/a_{k+1} \tag{5.10}$$

and if the right-hand side of (5.10) replaces $\mathbf{x}_{k+1}$ in (5.9), the generating set of vectors for V becomes

$$\{\mathbf{y}_1, ..., \mathbf{y}_k, \mathbf{y}_{k+1}, \mathbf{x}_{k+2}, ..., \mathbf{x}_n\}.$$

Hence by induction it is proved that $\{\mathbf{y}_1, ..., \mathbf{y}_n\}$ generates V. Therefore, since $m > n$ and $\mathbf{y}_m$ is in V, we have

$$\mathbf{y}_m = b_1\mathbf{y}_1 + \cdots + b_n\mathbf{y}_n$$

for some scalars b_i's in $\mathbb{F}$, which contradicts the assumption that $\mathbf{y}_i$'s are linearly independent. Hence, $\{\mathbf{y}_1, \mathbf{y}_2, ..., \mathbf{y}_m)$, $m > n$, must be linearly dependent. □

Proof of Theorem 5.1 Let $\{\mathbf{x}_1, ..., \mathbf{x}_n\}$ and $\{\mathbf{y}_1, ..., \mathbf{y}_m\}$ be two basis of the vector space V. By Lemma 5.1, $n < m$ is not possible because this would imply that $\{\mathbf{y}_1, ..., \mathbf{y}_m\}$ are linearly dependent. Conversely, for a similar reason, $n > m$ is not possible. Hence $m = n$. □

A direct consequence of Theorem 5.1 is the following corollary.

Corollary 5.1 Every vector $\mathbf{x}$ in a vector space V_n over $\mathbb{F}$ can be uniquely represented by (5.5) in terms of the basis vectors $\{\mathbf{x}_1, \mathbf{x}_2, ..., \mathbf{x}_n\}$. In such representation, $\alpha_1, \alpha_2, ..., \alpha_n$ in (5.5) are unique (see Example 5.11).

Proof Suppose there are two ways of representing the vector $\mathbf{x}$ in V_n given by

$$\mathbf{x} = \alpha_1\mathbf{x}_1 + \alpha_2\mathbf{x}_2 + \cdots + \alpha_n\mathbf{x}_n = a_1\mathbf{x}_1 + a_2\mathbf{x}_2 + \cdots + a_n\mathbf{x}_n,$$

where α_i's and a_i's are scalars in $\mathbb{F}$. Then

$$(\alpha_1 - a_1)\mathbf{x}_1 + (\alpha_2 - a_2)\mathbf{x}_2 + \cdots + (\alpha_n - a_n)\mathbf{x}_n = \boldsymbol{\theta}.$$

Since $\{\mathbf{x}_1, \mathbf{x}_2, ..., \mathbf{x}_n\}$ are linearly independent,

$$\alpha_1 - a_1 = \alpha_2 - a_2 = \cdots = \alpha_n - a_n = 0.$$

Therefore,

$$\alpha_1 = a_1, \quad \alpha_2 = a_2, ..., \quad \alpha_n = a_n.$$ □

An important property of an n-dimensional vector space V_n is that any set of n linearly independent vectors forms a basis of this space. This result is a consequence of Lemma 5.1 and is given in the following corollary.

Corollary 5.2 Any n linearly independent vectors $\mathbf{y}_1, \mathbf{y}_2, ..., \mathbf{y}_n$ in an n-dimensional vector space V_n form a basis of V_n.

Proof Because of Definition 5.5, we have to prove that any vector $\mathbf{x}$ in V_n can be expressed as a linear combination of $\mathbf{y}_i$'s. In view of Lemma 5.1, the vectors $\{\mathbf{y}_1, \mathbf{y}_2, ..., \mathbf{y}_n, \mathbf{x}\}$ are linearly dependent and hence by Definition 5.3 there exist scalars $\alpha_1, \alpha_2, ..., \alpha_n, \alpha$, *not all zero*, such that

$$\alpha_1\mathbf{y}_1 + \alpha_2\mathbf{y}_2 + \cdots + \alpha_n\mathbf{y}_n + \alpha\mathbf{x} = \boldsymbol{\theta},$$

where $\alpha \neq 0$. Indeed, if $\alpha = 0$ in view of independence of $\mathbf{y}_i$'s, all α_i's are zero. Hence, $\mathbf{x}$ can be expressed as a linear combination of $\mathbf{y}_i$'s.

Remark 5.1 In fact, a basis in a vector space can be thought of as a yardstick of our measurement. Different yardsticks give different values on measuring the same quantity. For instance, the diameter of a given circle is, say, 100 inches in the FPS system while it is 2.54 m in the MKS system. Similar is the effect of different basis in a vector space. To elucidate, if with respect to the basis $\{\mathbf{e}_1, \mathbf{e}_2\}$ (see (5.6)), the coordinates of P are (2, 2) because

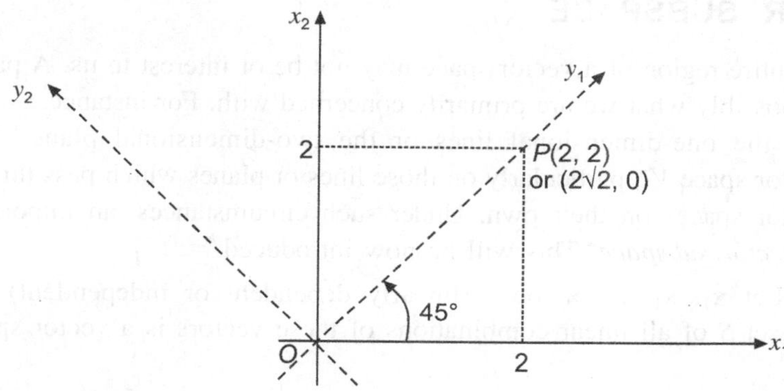

Figure 5.2 Change of basis.

$[2, 2]^T = 2\mathbf{e}_1 + 2\mathbf{e}_2$, then with respect to the basis vectors (see Figure 5.2),

$$\mathbf{a}_1 = \frac{1}{\sqrt{2}}\begin{bmatrix}1\\1\end{bmatrix}, \qquad \mathbf{a}_2 = \frac{1}{\sqrt{2}}\begin{bmatrix}1\\-1\end{bmatrix};$$

P is represented by $2\sqrt{2}\mathbf{a}_1 + 0\mathbf{a}_2$, i.e., by coordinates $(2\sqrt{2}, 0)$, if we agree (see Definition 5.5) to call the coefficients of the linear combination of basis vectors representing the vector $\overrightarrow{OP}$ as its coordinates. Thus, $\overrightarrow{OP}$ is the vector $[2, 2]^T$ with respect to the basis $\{\mathbf{e}_1, \mathbf{e}_2\}$ and the vector $[2\sqrt{2}, 0]^T$ relative to the basis $\{\mathbf{a}_1, \mathbf{a}_2\}$. Indeed, it is easy to see that $(2\sqrt{2}, 0)$ are the coordinates of P with respect to a new coordinate system (y_1, y_2) obtained by rotating the old system (x_1, x_2) through an angle 45° (see Figure 5.2), implying that representation of a point in $\mathbb{R}^n$ with a new basis is the same as representing it in a new coordinate system.

Definition 5.6 *Natural Basis*. Among the distinct sets of basis in $\mathbb{R}^n$ or $\mathbb{C}^n$, the one formed by the set of elementary vectors $\mathbf{e}_i$'s (see (5.4)) of order $n \times 1$ (or of order $1 \times n$ if the elements of $\mathbb{R}^n$ or $\mathbb{C}^n$ are n-row vectors), $i = 1, 2, ..., n$, is referred to as a *natural basis* or a *standard basis*. Similarly, the set of $m \times n$ matrices E_{ij} defined by (5.7) is referred to as a *natural* or a *standard basis* for the vector space $\mathbb{R}^{m \times n}$ or $\mathbb{C}^{m \times n}$ of all $m \times n$ matrices with real or complex elements.

□

Although we have familiarized ourselves in this section with the notion of linear dependence of a set of vectors, we propose to defer to Section 5.8 the testing procedure, on performing which, one can ensure whether any given set is independent or not. The reason for this is that, some understanding of such testing procedure calls for our acquaintance with the solution of linear simultaneous equations, which is therefore to be completed before undertaking such computational study.

5.3 VECTOR SUBSPACE

Sometimes the entire region of a vector space may not be of interest to us. A part of the entire vector space is possibly what we are primarily concerned with. For instance, our attention may be focussed on the one-dimensional lines or the two-dimensional planes V_2 in a three-dimensional vector space V_3, particularly on those lines or planes which pass through the origin constituting vector spaces on their own. Under such circumstances, an important concept to reckon with is *vector subspace*. This will be now introduced.

Theorem 5.2 Let $\mathbf{x}_1, \mathbf{x}_1, \ldots, \mathbf{x}_r$ be r (linearly dependent or independent) vectors in V_n, $r \leq n$. Then the set S of all linear combinations of these vectors is a vector space.

Proof Let

$$\mathbf{x} = \alpha_1\mathbf{x}_1 + \cdots + \alpha_r\mathbf{x}_r, \qquad \mathbf{y} = \beta_1\mathbf{x}_1 + \cdots + \beta_r\mathbf{x}_r, \qquad \mathbf{z} = \gamma_1\mathbf{x}_1 + \cdots + \gamma_r\mathbf{x}_r,$$

and consider $\boldsymbol{\theta}$ to be the zero vector as in V_n. These elements $\mathbf{x}$, $\mathbf{y}$ and $\mathbf{z}$ in S satisfy Axioms 1–8 of Definition 5.1, as can be easily verified by direct computation. Hence S is a vector space. □

Definition 5.7 *Vector Subspace.* The vector space generated (alternatively spanned) by any r number of (linearly dependent or independent) vectors $\mathbf{x}_1, \mathbf{x}_2, \ldots, \mathbf{x}_r$ of V_n (see Definition 5.5), $r \leq n$, is defined as a *subspace* of V_n and is denoted by S_r if $\mathbf{x}_1, \mathbf{x}_2, \ldots, \mathbf{x}_r$ are linearly independent. If $r < n$, then S_r is a proper subspace of V_n. That S_r is a (proper) subspace of V_n is shown symbolically by ($S_r \subset V_n$ or $V_n \supset S_r$) $S_r \subseteq V_n$ or $V_n \supseteq S_r$, and two subspaces $\overline{S}_r$ and S_r are equal iff $\overline{S}_r \subseteq S_r$ and $\overline{S}_r \supseteq S_r$. If $S_r \subseteq V_n$ (resp $S_r \subset V_n$), then evidently, dim $S_r \leq$ dim V_n (resp dim $S_r <$ dim V_n). □

A subspace S_r in V_n is, in fact, a vector space in its own right and, therefore, the elements in S_r must satisfy the Axioms 1–8 of Definition 5.1 as proved in Theorem 5.2. Consequently, S_r must contain $\boldsymbol{\theta}$ as its zero element.

Example 5.15 Let $\mathbf{e}_1 = [1\ 0\ 0]^T$ and $\mathbf{e}_2 = [0\ 1\ 0]^T$ be two vectors in $\mathbb{R}^3$ which are linearly independent, as proved in Example 5.9, and can be represented as in Figure 5.3. The vector space generated by $\mathbf{e}_1$ and $\mathbf{e}_2$ (see Theorem 5.2) is the x_1x_2-plane which is a proper subspace of the three-dimensional $x_1x_2x_3$-space. In Definition 5.7, $V_n = \mathbb{R}^3$, $S_r = x_1x_2$-plane, $n = 3$, $r = 2$, $S_r \subset V_n$, dim $S_r = 2 <$ dim $V_n = 3$, and $\mathbf{e}_1$ and $\mathbf{e}_2$ form a basis of S_2. □

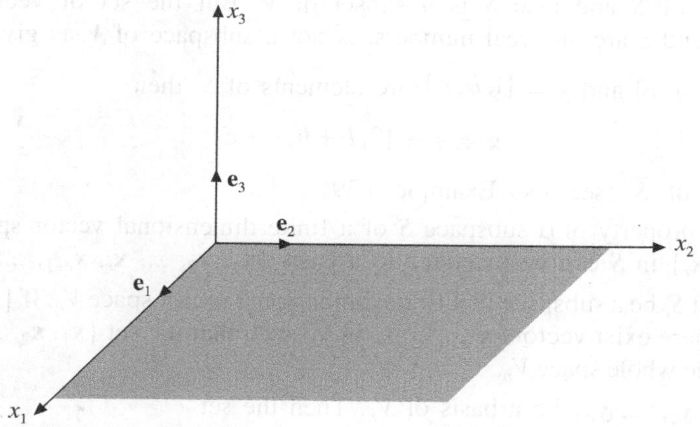

Figure 5.3 A vector subspace: the x_1x_2–plane, a vector subspace of the three-dimensional $x_1x_2x_3$–space.

Example 5.16 The set of all $n \times n$ diagonal matrices is a proper subspace S_n of the vector space $V_{n \times n}$ formed by the set of all $n \times n$ matrices. The matrices E_{ii}, $i = 1, 2, ..., n$, form a basis of the subspace S_n (see Example 5.13). □

Example 5.17 The set of all polynomials generated by $\{1, x^2, x^4\}$ over $\mathbb{R}$ forms a proper subspace of the vector space spanned by $\{1, x, x^2, x^3, x^4\}$ over $\mathbb{R}$. □

Sometimes we come across a vector subspace S in which the linearly independent vectors generating S are not known from the context, but the fact that the elements of S are also the elements of a vector space V (i.e., S is a subset of V) is known. In this case, one can show whether S is a subspace or not by checking all the axioms of a vector space in Definition 5.1. More simply, since it is known that S is a subset of a vector space V, only two vector space axioms need to be satisfied in view of the following theorem.

Theorem 5.3 If S is a subset of a vector space V, i.e., the elements of S are also the elements of V, then S is also a subspace of V iff (i) $\mathbf{x} + \mathbf{y}$ is an element of S whenever $\mathbf{x}$ and $\mathbf{y}$ are in S. (ii) $\alpha\mathbf{x}$ is an element of S whenever $\mathbf{x}$ is in S and α is a scalar.

Proof : (a) *Necessity.* If S is a subspace then, obviously, the given conditions (i) and (ii) hold (see Definition 5.1), since S is a vector space in its own right.

(b) *Sufficiency.* Supposing that conditions (i) and (ii) are true, we shall show that S is a vector space. To this effect we need to verify only Axioms 3 and 4 of Definition 5.1. Others, as they are true for the elements of V, are also true for the elements of S. If we choose $\alpha = -1$, then by the given condition (ii) $\alpha\mathbf{x} = (-1)\mathbf{x} = -\mathbf{x}$ (see Problem 5.2(e)) is an element of S which is Axiom 4. Again, by the given condition (i), $\mathbf{x} + (-\mathbf{x}) = \boldsymbol{\theta}$ is an element of S, which is Axiom 3. □

Example 5.18 The set S of 3-row vectors defined by $[x, 0, 0]$, where x is a real number, is a subspace of the vector space V of all 3-row vectors $[x, y, z]$, see Example 5.1. This is because if $\mathbf{x} = [a, 0, 0]$ and $\mathbf{y} = [b, 0, 0]$ are any two elements in S, and α is any scalar, then,

$$\mathbf{x} + \mathbf{y} = [a + b, 0, 0], \qquad \alpha\mathbf{x} = [\alpha a, 0, 0]$$

are also elements of S and also S is a subset of V. But the set of vectors $\overline{S}$ given by $[1, y, z]$, where y and z are any real numbers, is not a subspace of V as given above. This is because if $\mathbf{x} = [1, b, c]$ and $\mathbf{y} = [1, \overline{b}, \overline{c}]$ are elements of S, then

$$\mathbf{x} + \mathbf{y} = [2, b+\overline{b}, c+\overline{c}]$$

is not an element of $\overline{S}$ (see also Example 5.39). □

An important property of a subspace S of a finite dimensional vector space V is that any basis $\{\mathbf{x}_1, \mathbf{x}_2, ..., \mathbf{x}_r\}$ in S can be extended to a basis $\{\mathbf{x}_1, \mathbf{x}_2, ..., \mathbf{x}_r, \mathbf{x}_{r+1}, ..., \mathbf{x}_n\}$ of V.

Theorem 5.4 Let S_r be a subspace of a finite dimensional vector space V_n. If $\{\mathbf{x}_1, \mathbf{x}_2, ..., \mathbf{x}_r)$ is the basis of S_r, then there exist vectors $\mathbf{x}_{r+1}, ..., \mathbf{x}_n$ in V_n such that the set $\{\mathbf{x}_1, \mathbf{x}_2, ..., \mathbf{x}_r; \mathbf{x}_{r+1}, ..., \mathbf{x}_n\}$ forms a basis for the whole space V_n.

Proof : Let $\{\mathbf{y}_1, \mathbf{y}_2, ..., \mathbf{y}_n\}$ be a basis of V_n. Then the set

$$\{\mathbf{x}_1, \mathbf{x}_2, ..., \mathbf{x}_r; \mathbf{y}_1, \mathbf{y}_2, ..., \mathbf{y}_n\} \tag{5.11}$$

obviously spans the whole space V_n. From this set, a set of n linearly independent vectors can be selected in the following way: we choose the vectors $\mathbf{x}_1, \mathbf{x}_2, ..., \mathbf{x}_r$ which are linearly independent. If $r = n$, then we are done. Otherwise, for $r < n$, we pick up $\mathbf{y}_1$, and if it is linearly independent of the preceding vectors in the set (5.11), we retain it; else we reject it. Next, we pick up $\mathbf{y}_2$; if it is independent of the preceding vectors in (5.11), we retain it; else we reject it. Proceeding in this way, we choose a set of m linearly independent vectors. Obviously, $m \geq n$. By Lemma 5.1, only the equality holds, and by Corollary 5.2 the selected n linearly independent vectors containing $\mathbf{x}_1, \mathbf{x}_2, ..., \mathbf{x}_r$ form a basis of V_n.

Example 5.19 Let the subspace S_2 of $\mathbb{R}^3$ be generated by $\mathbf{x}_1 = [1, 0, 0]^T$ and $\mathbf{x}_2 = [0,1, 1]^T$. To extend this basis of S_2 to a basis of the whole space $\mathbb{R}^3$, we choose the natural basis $\mathbf{e}_i$, $i = 1, 2, 3$ in $\mathbb{R}^3$, and form a set consisting of the vectors

$$\begin{bmatrix}1\\0\\0\end{bmatrix}, \begin{bmatrix}0\\1\\1\end{bmatrix}, \begin{bmatrix}1\\0\\0\end{bmatrix}, \begin{bmatrix}0\\1\\0\end{bmatrix}, \begin{bmatrix}0\\0\\1\end{bmatrix}$$

from which we find that $\mathbf{x}_1, \mathbf{x}_2, \mathbf{e}_2$, or $\mathbf{x}_1, \mathbf{x}_2, \mathbf{e}_3$ form a basis of $\mathbb{R}^3$. Either of these is an extension of the basis $\mathbf{x}_1, \mathbf{x}_2$ of S_2 to that of the whole space $\mathbb{R}^3$ (see Example 5.38).

5.3.1 Vector Space as a Direct Sum of Subspaces

Referring to Figure 5.1(a), we see that the axes $O1$ and $O2$ are two proper subspaces in $\mathbb{R}^2$. The vector $\overrightarrow{OP}$ in it can be represented as the sum of $x_1\mathbf{e}_1 + x_2\mathbf{e}_2$, where $x_1\mathbf{e}_1$ and $x_2\mathbf{e}_2$ are respectively vectors in subspaces $O1$ and $O2$, and $\mathbf{e}_1$ and $\mathbf{e}_2$ are unit vectors defined in (5.4). This representation of $\overrightarrow{OP}$ in $\mathbb{R}^2$ as the sum of $x_1\mathbf{e}_1$ in subspace $O1$ and $x_2\mathbf{e}_2$ in subspace $O2$ is unique in view of Corollary 5.1. Extending this idea to a general vector space V, we get the following definition.

Definition 5.8 A vector space V is said to be the *direct sum* of two subspaces U and W written as

$$V = U \oplus W$$

if every vector **y** in V has a unique representation of the form

$$\mathbf{y} = \mathbf{x} + \mathbf{z},$$

where **x** and **z** are in U and W respectively.

Example 5.20 Considering the three-dimensional space $V \equiv \mathbb{R}^3$, let

$$\mathbf{y} = \begin{bmatrix} x_1 \\ x_2 \\ x_3 \end{bmatrix}, \quad \mathbf{x} = \begin{bmatrix} x_1 \\ x_2 \\ 0 \end{bmatrix}, \quad \mathbf{z} = \begin{bmatrix} 0 \\ 0 \\ x_3 \end{bmatrix}$$

(see Figure 5.4).

Because the third coordinate is always zero, **x** lies on the plane $1O2$, i.e., by changing x_1 and x_2, **x** describes the subspace U represented by the plane $1O2$. On the other hand, z always represents a straight line in the vertically upward direction $O3$, i.e., z describes the subspace W represented by $O3$. Moreover, any vector **y** in $V \equiv \mathbb{R}^3$ can be written as

$$\mathbf{y} = \begin{bmatrix} x_1 \\ x_2 \\ x_3 \end{bmatrix} = \begin{bmatrix} x_1 \\ x_2 \\ 0 \end{bmatrix} + \begin{bmatrix} 0 \\ 0 \\ x_3 \end{bmatrix} = \mathbf{x} + \mathbf{z}.$$

This is the only way to divide **y** into two parts, one lying in subspace U and the other in W. In other words, V is the direct sum of the subspaces U and W, and is written as

$$V = U \oplus W.$$

□

As a consequence of Theorem 5.2 and the foregoing discussion, we have the following corollary.

Corollary 5.3 If a vector **x** in a vector space V_n is expressed as a linear combination of r independent vectors $\mathbf{x}_1$, $\mathbf{x}_2$, ..., $\mathbf{x}_r$ ($r \le n$) in V_n, then this representation is unique. More-

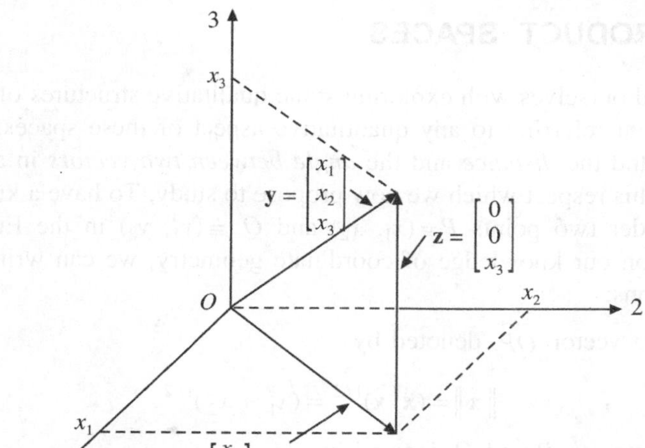

Figure 5.4 Direct sum of subspaces.

over, if $\mathbf{x}_1, \mathbf{x}_2, ..., \mathbf{x}_r, \mathbf{x}_{r+1}$ are linearly dependent, then the vector space spanned by them and that by $\mathbf{x}_1, \mathbf{x}_2, ..., \mathbf{x}_r$ are the same.

Proof The first part follows as in Corollary 5.1.

For the second part, let S be the vector space spanned by $\mathbf{x}_1\ \mathbf{x}_2 \cdots \mathbf{x}_r$ denoted by $V^r{}_n$, and $\bar{S}$ be the vector space spanned by $\mathbf{x}_1\ \mathbf{x}_2 \cdots \mathbf{x}_r\ \mathbf{x}_{r+1}$. Since $\mathbf{x}_1\ \mathbf{x}_2 \cdots \mathbf{x}_r\ \mathbf{x}_{r+1}$ are dependent, by Definition 5.3, there exist scalars $\bar{\alpha}_1, \bar{\alpha}_2, ..., \bar{\alpha}_{r+1}$ in $\mathbb{F}$ not all zero, such that

$$\bar{\alpha}_1\mathbf{x}_1 + \bar{\alpha}_2\mathbf{x}_2 + \cdots + \bar{\alpha}_r\mathbf{x}_r + \bar{\alpha}_{r+1}\mathbf{x}_{r+1} = \mathbf{\theta},$$

where $\bar{\alpha}_{r+1} \neq 0$; indeed, if $\bar{\alpha}_{r+1} = 0$, then all $\bar{\alpha}_i$'s are zero because $\mathbf{x}_1, \mathbf{x}_2, ..., \mathbf{x}_r$ are linearly independent. Hence for some scalars $\alpha_1, \alpha_2, ..., \alpha_r$ in $\mathbb{F}$, we have

$$\mathbf{x}_{r+1} = \alpha_1\mathbf{x}_1 + \alpha_2\mathbf{x}_2 + \cdots + \alpha_r\mathbf{x}_r.$$

Let $\mathbf{x}$ be a vector in S. So there exist in $\mathbb{F}$ scalars $a_1, a_2, ..., a_{r+1}$ satisfying

$$\mathbf{x} = a_1\mathbf{x}_1 + a_2\mathbf{x}_2 + \cdots + a_{r+1}\mathbf{x}_{r+1}.$$

In view of the last relation, this equation becomes

$$\mathbf{x} = (a_1 + \alpha_1 a_{r+1})\mathbf{x}_1 + (a_2 + \alpha_2 a_{r+2})\mathbf{x}_2 + \cdots + (a_r + \alpha_r a_{r+1})\mathbf{x}_r$$

which shows that $\mathbf{x} \in S$, and hence $\bar{S} \subseteq S$.

On the other hand, if $\mathbf{x} \in S$, then for some scalars $c_1, c_2, ..., c_r$ in $\mathbb{F}$,

$$\mathbf{x} = c_1\mathbf{x}_1 + c_2\mathbf{x}_2 + \cdots + c_r\mathbf{x}_r + 0\bullet\mathbf{x}_{r+1}$$

which shows that $\mathbf{x} \in \bar{S}$ as well. That is, $S \subseteq \bar{S}$. Thus, $S = \bar{S} = V_n^r$. □

The reader is referred to Section 11.1 for further results on the direct sum of subspaces.

5.4 INNER PRODUCT SPACES

So far, we contended ourselves with exploring some qualitative structures of a vector space and their elements without referring to any quantitative aspect of these spaces. The notion of the *length of a vector*, and the *distance* and the *angle between two vectors* in a vector space is of vital importance in this respect which we now propose to study. To have a knowledge about this aspect, let us consider two points $P \equiv (x_1, x_2)$ and $Q \equiv (y_1, y_2)$ in the Euclidean plane (see Figure 5.5). Based on our knowledge of coordinate geometry, we can write, from Figure 5.5, the following relations:

(i) Length of a vector $\overrightarrow{OP}$ denoted by

$$\|\mathbf{x}\| = (\mathbf{x}^T\mathbf{x})^{1/2} = (x_1^2 + x_2^2)^{1/2}. \tag{5.12}$$

(ii) Distance between P and Q denoted by

$$d(\mathbf{x}, \mathbf{y}) = \|\mathbf{x} - \mathbf{y}\| = [(x_1 - y_1)^2 + (x_2 - y_2)^2]^{1/2}. \tag{5.13}$$

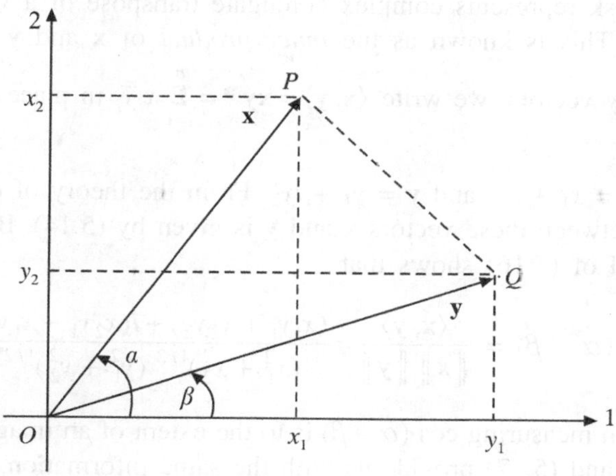

Figure 5.5 $\alpha - \beta$ is the angle between vectors $\mathbf{x} = [x_1 \quad x_2]^T$ and $\mathbf{y} = [y_1 \quad y_2]^T$; $\|\mathbf{x} - \mathbf{y}\|$ is the distance between P and Q.

(iii) The cosine of the angle between $\overrightarrow{OP}$ and $\overrightarrow{OQ}$ denoted by

$$\cos(\alpha - \beta) = \cos\alpha\cos\beta + \sin\alpha\sin\beta = \frac{x_1y_1 + x_2y_2}{(x_1^2 + x_2^2)^{1/2}(y_1^2 + y_2^2)^{1/2}} \tag{5.14}$$

$$= \frac{\langle \mathbf{x}, \mathbf{y} \rangle}{\|\mathbf{x}\| \|\mathbf{y}\|}, \tag{5.15}$$

where

$$\mathbf{x} = [x_1, x_2]^T, \qquad \mathbf{y} = [y_1, y_2]^T, \qquad \langle \mathbf{x}, \mathbf{y} \rangle = \mathbf{y}^T\mathbf{x}.$$

In vector algebra, if (x_1, x_2, x_3) and (y_1, y_2, y_3) are two vectors $\mathbf{x}$ and $\mathbf{y}$ and $\mathbf{x}\cdot\mathbf{y}$ is their dot product, the angle θ between them is expressed by

$$\cos\theta = \frac{\langle \mathbf{x}, \mathbf{y} \rangle}{\|\mathbf{x}\| \|\mathbf{y}\|},$$

where $\qquad \|\mathbf{x}\|^2 = x_1^2 + x_2^2 + x_3^2$ and $\|\mathbf{y}\|^2 = y_1^2 + y_2^2 + y_3^2$.

This definition of length and angle can be extended to n-dimensional vectors in $\mathbb{R}^n$ in an obvious way. But it shows some anomalies in $\mathbb{C}^n$. In $\mathbb{C}$, consider $x = i$; then by (5.12), $|x| = (i^2)^{1/2} = i$ (*single vertical bars* | | *are used for scalars*, and double bars for vectors *to denote length*) which contradicts the notion that a length is always positive. This contradiction can be avoided by defining

$$\langle \mathbf{x}, \mathbf{y} \rangle = \mathbf{y} * \mathbf{x} = \sum_{i=1}^{n} x_i \overline{y}_i \tag{5.16}$$

in $\mathbb{C}^n$ where the asterisk represents complex conjugate transpose of a vector, and a bar only a complex conjugate. This is known as the *inner product* of **x** and **y**. When, however, the elements in $\mathbb{C}^n$ are row vectors, we write $\langle \mathbf{x}, \mathbf{y} \rangle = \mathbf{x}\mathbf{y}^* = \sum_{i=1}^{n} x_i \bar{y}_i$ in place of (5.16) as the inner product of **x** and **y**.

Again in $\mathbb{C}$, let $\mathbf{x} = x_1 + ix_2$ and $\mathbf{y} = y_1 + iy_2$. From the theory of complex numbers, we know that the angle between these vectors **x** and **y** is given by (5.14). But actual computation by (5.15) with the aid of (5.16) shows that

$$\cos(\alpha - \beta) = \frac{\langle \mathbf{x}, \mathbf{y} \rangle}{\|\mathbf{x}\| \, \|\mathbf{y}\|} = \frac{(x_1 y_1 + x_2 y_2) + i(x_2 y_1 - x_1 y_2)}{(x_1^2 + x_2^2)^{1/2} \, (y_1^2 + y_2^2)^{1/2}}, \tag{5.17}$$

where the inaccuracy in measuring $\cos(\alpha - \beta)$ is to the extent of an imaginary term. But, when $(\alpha - \beta) = 90°$, (5.14) and (5.17) provide us with the same information, i.e.,

$$\cos(\alpha - \beta) = 0 \quad \text{iff } \langle \mathbf{x}, \mathbf{y} \rangle = 0$$

which is defined either by $\mathbf{y}^T\mathbf{x} = 0$ or by $\mathbf{y}^*\mathbf{x} = 0$, so that, according to the inner product defined in (5.16), (5.15) can only tell us when two vectors are perpendicular. Thus, the concept of length of the elements in any vector space is contained in the inner product of two vectors defined in (5.16) which, although, cannot give an accurate measurement of the angle between two vectors, can at least guarantee when these vectors are at right angles. From a practical point of view, it is important to be able to know when two vectors are at right angles. The foregoing discussion motivates us to offer the following definition for a general vector space *V*.

Definition 5.9 *Inner or Scalar Product—Inner product space.* Let **x**, **y**, **z** be any three elements in an arbitrary vector space *V* over $\mathbb{F}$. The inner product or scalar product of two vectors **x** and **y** in *V* denoted by $\langle \mathbf{x}, \mathbf{y} \rangle$ is defined as a scalar in $\mathbb{F}$ satisfying the following axioms:

1. $\langle \mathbf{x}, \mathbf{x} \rangle =: \|\mathbf{x}\|^2 \geq 0$; $\langle \mathbf{x}, \mathbf{x} \rangle = 0$ iff $\mathbf{x} = \boldsymbol{\theta}$,
2. $\langle \mathbf{x}, \mathbf{y} \rangle = \overline{\langle \mathbf{y}, \mathbf{x} \rangle}$,
3. $\langle \alpha\mathbf{x} + \beta\mathbf{y}, \mathbf{z} \rangle = \alpha \langle \mathbf{x}, \mathbf{z} \rangle + \beta \langle \mathbf{y}, \mathbf{z} \rangle$,

where α and β are scalars in $\mathbb{F}$. The *inner product space* is a vector space in which the vectors satisfy Axioms 1–3. With an inner product, the space $\mathbb{R}^n$ is called the *Euclidean space* and the space $\mathbb{C}^n$ the *unitary space*. More generally, a real vector space *V*, for which a real inner product is defined, is called a *Euclidean space*. A complex vector space *V*, for which a complex inner product is defined, is called *a unitary space*. Moreover, if *A* is a matrix of order *n* over $\mathbb{C}$, then for vectors **x** and **y** in $\mathbb{C}^n$, we have, in view of (5.16), $\langle \mathbf{x}, A\mathbf{y} \rangle = (A\mathbf{y})^*\mathbf{x} = \mathbf{y}^* A^* \mathbf{x} = \langle A^*\mathbf{x}, \mathbf{y} \rangle$. □

Example 5.21 (a) Let us consider the vector space $V = \mathbb{C}^n$. Let the inner product of any two vectors

$$\mathbf{x} = [x_1 \; x_2 \; . \; . \; . \; x_n]^T \quad \text{and} \quad \mathbf{y} = [y_1 \; y_2 \; . \; . \; . \; y_n]^T$$

in $\mathbb{C}^n$ be defined by (5.16). Then

$$\langle \mathbf{x}, \mathbf{x} \rangle = |x_1|^2 + |x_2|^2 + \cdots + |x_n|^2 \geq 0,$$

because each term is nonnegative and $\langle \mathbf{x}, \mathbf{x} \rangle = 0$ iff

$$|x_1|^2 = |x_2|^2 = \cdots = |x_n|^2 = 0,$$

i.e., iff $x_1 = x_2 = \cdots = x_n = 0$, i.e., iff $\mathbf{x} = \boldsymbol{\theta}$. Again,

$$\langle \mathbf{x}, \mathbf{y} \rangle = \sum_{i=1}^{n} x_i \bar{y}_i = \text{complex conjugate of } \sum_{i=1}^{n} y_i \bar{x}_i = \overline{\langle \mathbf{y}, \mathbf{x} \rangle}.$$

Suppose that

$$\mathbf{z} = [z_1 \; z_2 \; \cdot \; \cdot \; \cdot \; z_n]^T$$

is another element of $\mathbb{C}^n$ and α and β are two scalars in $\mathbb{C}$. Then

$$\langle \alpha\mathbf{x} + \beta\mathbf{y}, \mathbf{z} \rangle = \sum_{i=1}^{n} (\alpha x_i + \beta y_i)\bar{z}_i = \alpha \sum_{i=1}^{n} x_i \bar{z}_i + \beta \sum_{i=1}^{n} y_i \bar{z}_i = \alpha \langle \mathbf{x}, \mathbf{z} \rangle + \beta \langle \mathbf{y}, \mathbf{z} \rangle.$$

Thus we find that all the three axioms in Definition 5.9 are satisfied by the elements of $\mathbb{C}^n$ which is an inner product space known as *unitary space*.

(b) Consider the vector space $V = C([a, b])$ of all continuous real-valued functions on the interval $[a, b]$ with $a \neq b$ (see Example 5.3). The inner product in this space can be defined by

$$\langle f, g \rangle = \int_a^b f(t)g(t)\, dt,$$

where f and g are elements of V. Then it can be shown that all the axioms in Definition 5.9 are satisfied by the elements of V. Hence, the space V is an inner product space known as the *Euclidean space of continuous functions*.

(c) Consider the space of all polynomials of degree $\leq n$, viz., $\mathbb{P}_n(t)$ over $\mathbb{C}$ (see Example 5.2). If p and q are any two elements in $\mathbb{P}_n(t)$, then

$$\langle p, q \rangle = \int_a^b p(t)q(t)\, dt$$

for any scalars a, b in $\mathbb{C}$, $a \neq b$, defines an inner product. By varying the interval $[a, b]$, many different inner product functions can be defined with the same set of elements in $\mathbb{P}_n(t)$. In view of the discussion in Example 5.21(b), each such inner product transforms $\mathbb{P}_n(t)$ into an inner product space.

(d) The set of continuous complex-valued functions defined on the interval $[0, 2\pi]$ with the inner product

$$\langle f, g\rangle = \int_0^{2\pi} f(t)\overline{g(t)}\, dt,$$

where f and g are two elements in the set, is an inner product space. We know that every complex-valued function f can be written as

$$f = f_1 + if_2, \qquad i = \sqrt{-1}.$$

Therefore,

$$\langle f, f\rangle = \int_0^{2\pi} [f_1^2 + f_2^2]\, dt$$

which is greater than zero iff $f \neq 0$ on some subinterval of $[0, 2\pi]$, and is zero iff $f = 0$. It is easily verified that other axioms of Definition 5.9 are satisfied by the elements of the above set which, therefore, constitutes the *unitary space of continuous functions* denoted by $I([0, 2\pi])$. □

An important result in inner product space is Schwarz inequality which has extensive application in mathematical analysis. Before obtaining its derivation, we shall present an important lemma needed in this respect.

Lemma 5.2 Let $\mathbf{x}, \mathbf{y}, \mathbf{z}$ be any three vectors in an inner product space. Then there holds, for scalars α and β, the equality

$$\langle \mathbf{x}, \alpha\mathbf{y} + \beta\mathbf{z}\rangle = \bar{\alpha}\,\langle \mathbf{x}, \mathbf{y}\rangle + \bar{\beta}\,\langle \mathbf{x}, \mathbf{z}\rangle. \tag{5.18}$$

Proof Applying first, axiom 2, then 3, and finally 2 of an inner product (Definition 5.9) to the left-hand side of (5.18), we get

$$\langle \overline{\alpha\mathbf{y}} + \overline{\beta\mathbf{z}}, \overline{\mathbf{x}}\rangle = \bar{\alpha}\,\langle \overline{\mathbf{y}}, \overline{\mathbf{x}}\rangle + \bar{\beta}\,\langle \overline{\mathbf{z}}, \overline{\mathbf{x}}\rangle = \bar{\alpha}\,\langle \mathbf{x}, \mathbf{y}\rangle + \bar{\beta}\,\langle \mathbf{x}, \mathbf{z}\rangle. \qquad \square$$

Theorem 5.5 *Schwarz Inequality.* If $\mathbf{x}$ and $\mathbf{y}$ are two vectors in an inner product space, then there holds the inequality

$$\left[\langle \mathbf{x}, \mathbf{y}\rangle \overline{\langle \mathbf{x}, \mathbf{y}\rangle}\right]^{1/2} = |\langle \mathbf{x}, \mathbf{y}\rangle| \leq \|\mathbf{x}\|\,\|\mathbf{y}\|. \tag{5.19}$$

Proof We note that when $\mathbf{y} = \boldsymbol{\theta}$, the inequality is trivially true. Assume therefore that $\mathbf{y} \neq \boldsymbol{\theta}$, and define a vector $\mathbf{z}$ and a scalar α as

$$\mathbf{z} = \frac{\mathbf{y}}{\|\mathbf{y}\|}, \qquad \alpha = \langle \mathbf{x}, \mathbf{z}\rangle = \left\langle \mathbf{x}, \frac{\mathbf{y}}{\|\mathbf{y}\|}\right\rangle = \frac{1}{\|\mathbf{y}\|}\langle \mathbf{x}, \mathbf{y}\rangle$$

such that $\langle \mathbf{z}, \mathbf{z}\rangle = 1$. Now, using Axioms 1 and 3 of Definition 5.9, we have

$$0 \leq \langle \mathbf{x} - \alpha\mathbf{z}, \mathbf{x} - \alpha\mathbf{z}\rangle = \langle \mathbf{x}, \mathbf{x} - \alpha\mathbf{z}\rangle - \alpha\,\langle \mathbf{z}, \mathbf{x} - \alpha\mathbf{z}\rangle$$

$$= \langle \mathbf{x}, \mathbf{x}\rangle - \bar{\alpha}\,\langle \mathbf{x}, \mathbf{z}\rangle - \alpha\,\langle \mathbf{z}, \mathbf{x}\rangle + \alpha\bar{\alpha}\langle \mathbf{z}, \mathbf{z}\rangle$$

$$= \|\mathbf{x}\|^2 - \bar{\alpha}\alpha - \alpha\bar{\alpha} + \alpha\bar{\alpha}$$

$$= \|\mathbf{x}\|^2 - |\alpha|^2,$$

where the third step follows from (5.18). By substituting for α and rearranging, the result (5.19) follows.

An easy consequence of Theorem 5.5 is the following corollary.

Corollary 5.4 *Schwarz Inequality.* If **x** and **y** are two vectors in an Euclidean space $\mathbb{R}^n$, then there holds the inequality

$$-1 \le \frac{\langle \mathbf{x}, \mathbf{y} \rangle}{\|\mathbf{x}\| \|\mathbf{y}\|} \le 1. \tag{5.20}$$

□

It might be recalled that the notion of an angle of the elements in a vector space was explained by arguments and ideas fixed on a two-dimensional Euclidean space $\mathbb{R}^2$ (Figure 5.5). In this case the angle between any two vectors **x** and **y** as given by (5.15) is known to be bounded by

$$-1 \le \cos(\alpha - \beta) = \frac{\langle \mathbf{x}, \mathbf{y} \rangle}{\|\mathbf{x}\| \|\mathbf{y}\|} \le 1. \tag{5.21}$$

The notion of angle will be considered to be fruitfully extended to a general vector space *V*, should relation (5.21) hold also in this case. In view of the Schwarz inequality (5.19) or (5.20), this indeed is the case. In $\mathbb{R}^n$ or $\mathbb{C}^n$, the length of a vector $\mathbf{x} = [x_1\ x_2\ \cdots x_n]^T$ (see 5.12) is termed *norm*, and in view of (5.16), is precisely expressed by

$$\text{Norm of } \mathbf{x}: = \|\mathbf{x}\| = \left[\langle \mathbf{x}, \mathbf{x} \rangle\right]^{1/2}$$

$$= (x_1\bar{x}_1 + x_2\bar{x}_2 + \cdots + x_n\bar{x}_n)^{1/2}$$

$$= \left(|x_1|^2 + |x_2|^2 + \cdots + |x_n|^2\right)^{1/2} \quad \text{in } \mathbb{C}^n \tag{5.22}$$

$$= \left(x_1^2 + x_2^2 + \cdots + x_n^2\right)^{1/2} \text{ in } \mathbb{R}^n, \tag{5.23}$$

and

$$\text{Norm of } x_i: = |x_i| = (x_i\bar{x}_i)^{1/2}. \tag{5.24}$$

Drawing analogy with the distance between two points on the Euclidean plane, the distance between the vectors $\mathbf{x} = [x_1\ x_2\ \cdots\ x_n]^T$ and $\mathbf{y} = [y_1\ y_2\ \cdots\ y_n]^T$ in $\mathbb{R}^n$ or $\mathbb{C}^n$ can be expressed by extending representation given by (5.13), as

$$\text{Distance between } \mathbf{x} \text{ and } \mathbf{y}: = \left(|x_1 - y_1|^2 + \cdots + |x_n - y_n|^2\right)^{1/2} \quad \text{in } \mathbb{C}^n$$

$$= \left\{(x_1 - y_1) + \cdots + (x_n - y_n)^2\right\}^{1/2} \quad \text{in } \mathbb{R}^n.$$

The foregoing discussion motivates us to offer the following definition for an inner product space.

Definition 5.10 *Length of a Vector and Distance between two vectors.* Let V be an inner product space. For a vector $\mathbf{x}$ in V, we define the length or norm of $\mathbf{x}$ as

$$\text{Norm of } \mathbf{x}: = \| \mathbf{x} \| = \sqrt{\langle \mathbf{x}, \mathbf{x} \rangle}$$

and for each ordered pair of vectors $\mathbf{x}$ and $\mathbf{y}$ in V, we define the distance between $\mathbf{x}$ and $\mathbf{y}$ as

$$\text{Distance between } \mathbf{x} \text{ and } \mathbf{y}: = d(\mathbf{x}, \mathbf{y}): = \| \mathbf{x} - \mathbf{y} \|. \quad \square$$

Again in V_n, particularly in $\mathbb{C}^n$, precise measurement of angle between any two vectors $\mathbf{x}$ and $\mathbf{y}$ by extending (5.15) is beset with difficulties as already pointed out. Therefore, one remains satisfied by knowing the condition when $\mathbf{x}$ and $\mathbf{y}$ are at right angles. As stated already, this condition is $\langle \mathbf{x}, \mathbf{y} \rangle = 0$. In view of this, we have the following important notion of an orthogonal set.

Definition 5.11 *Orthogonal Set.* Two vectors $\mathbf{x}_1$ and $\mathbf{x}_2$ in an inner product space V are said to be orthogonal if $\langle \mathbf{x}_1, \mathbf{x}_2 \rangle = 0$. A set of vectors $\{\mathbf{x}_1, \mathbf{x}_2, ..., \mathbf{x}_n\}$ in V is called an orthogonal set if each vector of the set is orthogonal to every other vector in the set, i.e., if $\langle \mathbf{x}_i, \mathbf{x}_j \rangle = 0, i \neq j$.

Example 5.22 The vectors $\mathbf{x} = [1\ i\ 0]^T$ and $\mathbf{y} = [i\ 1\ 0]^T$ are orthogonal. The set of vectors $\mathbf{x}$, $\mathbf{y}$ and $\mathbf{z} = [0\ 0\ 2 + 3i]^T$ is an orthogonal set. $\square$

Definition 5.12 *Unit Vector, Normalized Vector.* A vector in an inner product space V is called a *unit vector* if its norm is equal to one. A nonzero vector is said to be normalized if it is divided by its norm. After normalizing $\mathbf{x}$, we have $\mathbf{x}/ \| \mathbf{x} \|$ which has also a norm equal to one. Thus a *normalized vector* is also a unit vector. $\square$

Example 5.23 (a) Find $\langle \mathbf{x}, \mathbf{y} \rangle, \langle \mathbf{y}, \mathbf{x} \rangle, \| \mathbf{x} \|, \| \mathbf{y} \|$, the normalized $\mathbf{x}$ and $\mathbf{y}$, where

$$\mathbf{x} = \begin{bmatrix} 2 \\ i \\ 3 \end{bmatrix} \text{ and } \mathbf{y} = \begin{bmatrix} 1+2i \\ 2i \\ 1 \end{bmatrix}.$$

Solution we have

$$\langle \mathbf{x}, \mathbf{y} \rangle = \bar{y}_1 x_1 + \bar{y}_2 x_2 + \bar{y}_3 x_3 = (1 - 2i)2 + (-2i)i + 1 \cdot 3 = 7 - 4i,$$

$$\langle \mathbf{y}, \mathbf{x} \rangle = 2(1 + 2i) - i(2i) + 3 \cdot 1 = 7 + 4i.$$

So,

$$\langle \mathbf{x}, \mathbf{y} \rangle = \overline{\langle \mathbf{y}, \mathbf{x} \rangle},$$

$$\| \mathbf{x} \| = (\bar{x}_1 x_1 + \bar{x}_2 x_2 + \bar{x}_3 x_3)^{1/2} = (2 \cdot 2 + (-i)i + 3 \cdot 3)^{1/2} = \sqrt{14},$$

$$\| \mathbf{y} \| = \{(1 - 2i)(1 + 2i) + (-2i)(2i) + 1 \cdot 1\}^{1/2} = \sqrt{10}.$$

The normalized **x** and **y** are, respectively,

$$\frac{\mathbf{x}}{\|\mathbf{x}\|} = \begin{bmatrix} \frac{2}{\sqrt{14}} \\ \frac{i}{\sqrt{14}} \\ \frac{3}{\sqrt{14}} \end{bmatrix} \quad \text{and} \quad \frac{\mathbf{y}}{\|\mathbf{y}\|} = \begin{bmatrix} \frac{1+2i}{\sqrt{10}} \\ \frac{2i}{\sqrt{10}} \\ \frac{1}{\sqrt{10}} \end{bmatrix}$$

The distance between **x** and **y** is

$$d(\mathbf{x}, \mathbf{y}) = \|\mathbf{x} - \mathbf{y}\| = (|x_1 - y_1|^2 + |x_2 - y_2|^2 + |x_3 - y_3|^2)^{1/2}$$

$$= (|1 - 2i|^2 + |-i|^2 + |2|^2)^{1/2} = \sqrt{10}. \qquad \square$$

Definition 5.13 *Orthonormal Set.* A set of vectors $\{\mathbf{x}_1, \mathbf{x}_2, ..., \mathbf{x}_m\}$ in an inner product space V is said to be orthonormal if it is an orthogonal set, and moreover, if each vector in the set is a unit vector, i.e., $\langle \mathbf{x}_i, \mathbf{x}_j \rangle = \delta_{ij}$, where δ_{ij} is a Kronecker delta (see (6.51)). $\square$

Example 5.23(b) The set of vectors

$$\mathbf{e}_1 = \begin{bmatrix} 1 \\ 0 \\ 0 \end{bmatrix}, \quad \mathbf{e}_2 = \begin{bmatrix} 0 \\ 1 \\ 0 \end{bmatrix}, \quad \text{and} \quad \mathbf{e}_3 = \begin{bmatrix} 0 \\ 0 \\ 1 \end{bmatrix}$$

is an orthonormal set with respect to the inner product defined by (5.16). $\square$

Example 5.23(c) Let us consider the set of sine and cosine functions $\cos n\theta$ and $\sin m\theta$, where n and m are integers in the vector space C ([0, 2π]), see Example 5.21(b), and define an inner product in it via

$$\langle f, g \rangle = \int_0^{2\pi} f(\theta) g(\theta)\, d\theta.$$

Then,

$$\|\cos n\theta\|^2 = \int_0^{2\pi} \cos^2 n\theta \, d\theta = \begin{cases} \pi, & \text{if } n \neq 0; \\ 2\pi, & \text{if } n = 0. \end{cases}$$

Similarly,

$$\|\sin n\theta\|^2 = \pi, \; n \neq 0.$$

Moreover,

$$\langle \cos n\theta, \sin m\theta \rangle = \int_0^{2\pi} \cos n\theta \sin m\theta \, d\theta = 0.$$

Also,

$$\langle \cos n\theta, \cos m\theta \rangle = \langle \sin n\theta, \sin m\theta \rangle = 0, \qquad m \neq n.$$

Therefore, the set of vectors

$$\left\{\frac{1}{\sqrt{2\pi}}, \frac{\cos\theta}{\sqrt{\pi}}, \ldots, \frac{\cos n\theta}{\sqrt{\pi}}, \ldots; \frac{\sin\theta}{\sqrt{\pi}}, \ldots, \frac{\sin n\theta}{\sqrt{\pi}}, \ldots\right\}$$

is an orthonormal set in the space $C([0, 2\pi])$ with respect to the inner product defined above. It can be shown that $C([0, 2\pi])$ is an *infinite dimensional* space. □

Example 5.23(d) The set of functions $\{\exp(in\theta)/\sqrt{2\pi}\}$, where $i = \sqrt{-1}$ and n is an integer, in the vector space $I([0, 2\pi])$, see Example 5.21(d), is an orthonormal set. Indeed,

$$\langle \exp(in\theta), \exp(im\theta)\rangle = \int_0^{2\pi} \exp(in\theta)\exp(-im\theta)\, d\theta$$

$$= \begin{cases} 0, & m \neq n; \\ 2\pi, & m = n. \end{cases}$$ □

Theorem 5.6 Every orthonormal (or orthogonal) set of vectors $\{\mathbf{x}_1, \mathbf{x}_2, \ldots, \mathbf{x}_m\}$ in an inner product space V is linearly independent.

Proof Let

$$c_1\mathbf{x}_1 + c_2\mathbf{x}_2 + \cdots + c_m\mathbf{x}_m = \boldsymbol{\theta}. \tag{5.25}$$

Taking the inner product on both sides of (5.25) with $\mathbf{x}_1, \mathbf{x}_2, \ldots, \mathbf{x}_m$, we find successively $c_1 = 0$, $c_2 = 0$, ..., $c_m = 0$, using Definition 5.13 and Axiom 3 of Definition 5.9. □

5.5 ORTHONORMAL BASIS AND GRAM-SCHMIDT PROCESS OF ORTHOGONALIZATION

In an n-dimensional vector space V_n, there are n number of linearly independent basis vectors, and by forming the inner product among them, we are able to say whether or not these vectors are mutually orthogonal. If they are orthogonal, we can find out their norms to see whether each of them has a unit norm. When the basis vectors are orthogonal, each having a unit norm, they are said to constitute an *orthonormal basis.* It is pleasant to work with an orthonormal basis, because representation of any vector or the proof of any theorem with its aid becomes very simple. Given a set of n linearly independent vectors in an inner product space V_n, it is always possible to construct an orthonormal basis by what is known as the *Gram-Schmidt process.*

With the above idea in view, let us therefore consider a set of m linearly independent vectors $\mathbf{x}_1, \mathbf{x}_2, \ldots, \mathbf{x}_m$ in an inner product space V_n. We shall now describe a procedure whereby we can generate a set of m orthogonal vectors $\mathbf{y}_1, \mathbf{y}_2, \ldots, \mathbf{y}_m$. In view of Theorem 5.6, these vectors $\mathbf{y}_i$'s are then linearly independent. To this effect, define the $\mathbf{y}_i$'s by

$$\begin{aligned} \mathbf{y}_1 &= \mathbf{x}_1, \\ \mathbf{y}_2 &= a_{21}\mathbf{y}_1 + \mathbf{x}_2, \\ \mathbf{y}_3 &= a_{31}\mathbf{y}_1 + a_{32}\mathbf{y}_2 + \mathbf{x}_3, \\ &\vdots \\ \mathbf{y}_m &= a_{m1}\mathbf{y}_1 + a_{m2}\mathbf{y}_2 + \cdots + a_{mm-1}\mathbf{y}_{m-1} + \mathbf{x}_m, \end{aligned} \tag{5.26}$$

where a_{ij}'s are scalars to be determined by requiring that each $\mathbf{y}_i$ is orthogonal to every other vector in the set. Because of this stipulation, $\mathbf{y}_1$ and $\mathbf{y}_2$ are orthogonal, and hence from the second equation in (5.26), we can write

$$\langle \mathbf{y}_2, \mathbf{y}_1 \rangle = a_{21} \langle \mathbf{y}_1, \mathbf{y}_1 \rangle + \langle \mathbf{x}_2, \mathbf{y}_1 \rangle = 0 \tag{5.27}$$

which gives

$$a_{21} = -\frac{\langle \mathbf{x}_2, \mathbf{y}_1 \rangle}{\langle \mathbf{y}_1, \mathbf{y}_1 \rangle}$$

Consequently,

$$\mathbf{y}_2 = \mathbf{x}_2 - \frac{\langle \mathbf{x}_2, \mathbf{y}_1 \rangle}{\langle \mathbf{y}_1, \mathbf{y}_1 \rangle} \mathbf{y}_1.$$

Again, by requiring that $\mathbf{y}_3$ be orthogonal to $\mathbf{y}_1$ and $\mathbf{y}_2$, we have, from the third equation in (5.26), the relations

$$\langle \mathbf{y}_3, \mathbf{y}_1 \rangle = a_{31} \langle \mathbf{y}_1, \mathbf{y}_1 \rangle + a_{32} 0 + \langle \mathbf{x}_3, \mathbf{y}_1 \rangle = 0,$$

$$\langle \mathbf{y}_3, \mathbf{y}_2 \rangle = a_{31} 0 + a_{32} \langle \mathbf{y}_2, \mathbf{y}_2 \rangle + \langle \mathbf{x}_3, \mathbf{y}_2 \rangle = 0,$$

where we have used the condition (5.27). Solving these equations, we get

$$a_{31} = -\frac{\langle \mathbf{x}_3, \mathbf{y}_1 \rangle}{\langle \mathbf{y}_1, \mathbf{y}_1 \rangle} \text{ and } a_{32} = -\frac{\langle \mathbf{x}_3, \mathbf{y}_2 \rangle}{\langle \mathbf{y}_2, \mathbf{y}_2 \rangle}$$

which, when substituted in (5.26), give us

$$\mathbf{y}_3 = \mathbf{x}_3 - \frac{\langle \mathbf{x}_3, \mathbf{y}_1 \rangle}{\langle \mathbf{y}_1, \mathbf{y}_1 \rangle} \mathbf{y}_1 - \frac{\langle \mathbf{x}_3, \mathbf{y}_2 \rangle}{\langle \mathbf{y}_2, \mathbf{y}_2 \rangle} \mathbf{y}_2.$$

Now, by induction, let

$$\mathbf{y}_{m-1} = \mathbf{x}_{m-1} - \frac{\langle \mathbf{x}_{m-1}, \mathbf{y}_1 \rangle}{\langle \mathbf{y}_1, \mathbf{y}_1 \rangle} \mathbf{y}_1 - \frac{\langle \mathbf{x}_{m-1}, \mathbf{y}_2 \rangle}{\langle \mathbf{y}_2, \mathbf{y}_2 \rangle} \mathbf{y}_2 - \cdots - \frac{\langle \mathbf{x}_{m-1}, \mathbf{y}_{m-2} \rangle}{\langle \mathbf{y}_{m-2}, \mathbf{y}_{m-2} \rangle} \mathbf{y}_{m-2}$$

so that $\mathbf{y}_1$, $\mathbf{y}_2$, ..., $\mathbf{y}_{m-1}$ are mutually orthogonal. By requiring that $\mathbf{y}_m$ be orthogonal to these vectors $\mathbf{y}_i$'s, $i = 1, 2, ..., m - 1$, the last equation in (5.26) gives, successively,

$$a_{m1} = -\frac{\langle \mathbf{x}_m, \mathbf{y}_1 \rangle}{\langle \mathbf{y}_1, \mathbf{y}_1 \rangle}, \ldots, a_{mm-1} = -\frac{\langle \mathbf{x}_m, \mathbf{y}_{m-1} \rangle}{\langle \mathbf{y}_{m-1}, \mathbf{y}_{m-1} \rangle}$$

and hence

$$\mathbf{y}_m = \mathbf{x}_m - \frac{\langle \mathbf{x}_m, \mathbf{y}_1 \rangle}{\langle \mathbf{y}_1, \mathbf{y}_1 \rangle} \mathbf{y}_1 - \frac{\langle \mathbf{x}_m, \mathbf{y}_2 \rangle}{\langle \mathbf{y}_2, \mathbf{y}_2 \rangle} \mathbf{y}_2 - \cdots - \frac{\langle \mathbf{x}_m, \mathbf{y}_{m-1} \rangle}{\langle \mathbf{y}_{m-1}, \mathbf{y}_{m-1} \rangle} \mathbf{y}_{m-1}.$$

The set of vectors $\mathbf{y}_1, \mathbf{y}_2, ..., \mathbf{y}_m$ thus generated is mutually orthogonal because these vectors are forced to be so in the process of their determination. We have therefore established the following theorem.

Theorem 5.7 For a set of m linearly independent vectors $\mathbf{x}_1, \mathbf{x}_2, ..., \mathbf{x}_m$ in an inner product space V_n, there exists a set of orthogonal vectors $\mathbf{y}_1, \mathbf{y}_2, ..., \mathbf{y}_m$ defined by

$$\mathbf{y}_1 = \mathbf{x}_1;$$

$$\mathbf{y}_k = \mathbf{x}_k - \sum_{i=1}^{k-1} \frac{\langle \mathbf{x}_k, \mathbf{y}_i \rangle}{\langle \mathbf{y}_i, \mathbf{y}_i \rangle} \mathbf{y}_i, \quad k = 2, \ldots, m. \tag{5.28}$$

□

With the help of this result, we shall now prove an extremely important theorem.

Theorem 5.8 *Gram-Schmidt Orthonormalization.* There exists an orthonormal set of vectors $\boldsymbol{\varepsilon}_1, \boldsymbol{\varepsilon}_2, \ldots, \boldsymbol{\varepsilon}_m$ associated with a linearly independent set of vectors $\mathbf{x}_1, \mathbf{x}_2, ..., \mathbf{x}_m$ in an inner product space V_n.

Proof By Theorem 5.7, it is possible to construct an orthogonal set of vectors $\mathbf{y}_1, \mathbf{y}_2, ..., \mathbf{y}_m$, each one of which is a linear combination of $\mathbf{x}_i$'s. If we choose

$$\boldsymbol{\varepsilon}_i = \frac{\mathbf{y}_i}{\|\mathbf{y}_i\|} \tag{5.29}$$

we get an orthonormal set of vectors. □

We shall now offer a physical interpretation of the orthogonality concept contained in the Gram-Schmidt process by introducing the notion of *orthogonal projection*. With this object in view, let us recall that the two vectors $\mathbf{x}$ and $\mathbf{y}$ in V_n are orthogonal if $\langle \mathbf{x}, \mathbf{y} \rangle = 0$.

Now, as a result of Axiom 3 of Definition 5.9 and (5.18), if $\mathbf{x}$ and $\mathbf{y}$ are orthogonal, we have

$$\begin{aligned}\langle \mathbf{x}+\mathbf{y}, \mathbf{x}+\mathbf{y} \rangle &= \langle \mathbf{x}, \mathbf{x}+\mathbf{y} \rangle + \langle \mathbf{y}, \mathbf{x}+\mathbf{y} \rangle = \langle \mathbf{x}, \mathbf{x} \rangle + \langle \mathbf{x}, \mathbf{y} \rangle \\ &\quad + \langle \mathbf{y}, \mathbf{x} \rangle + \langle \mathbf{y}, \mathbf{y} \rangle = \langle \mathbf{x}, \mathbf{x} \rangle + \langle \mathbf{y}, \mathbf{y} \rangle.\end{aligned}$$

Therefore, using (5.22),

$$\|\mathbf{x}+\mathbf{y}\|^2 = \|\mathbf{x}\|^2 + \|\mathbf{y}\|^2$$

which is similar to the Pythagorean theorem in Euclidean geometry (see Figure 5.6(a)), i.e., in the right-angled triangle OAB on the plane, $OA^2 + AB^2 = OB^2$. This formula can be extended-

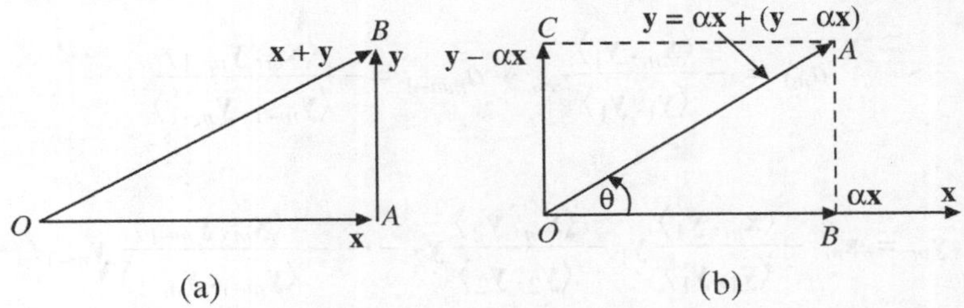

Figure 5.6 (a) Vectors $\mathbf{x}$ and $\mathbf{y}$ are orthogonal; (b) $\alpha \mathbf{x}$ is orthogonal projection of $\mathbf{y}$ on $\mathbf{x}$.

in a natural way to the case when there are a set of orthogonal vectors, $\mathbf{x}_1, \mathbf{x}_2, ..., \mathbf{x}_m$ in V_n, i.e.,

$$\left\langle \sum_{i=1}^{m} \mathbf{x}_i, \sum_{i=1}^{m} \mathbf{x}_i \right\rangle = \sum_{i=1}^{m} \langle \mathbf{x}_i, \mathbf{x}_i \rangle.$$

Definition 5.14 *Orthogonal Projection.* Let $\mathbf{x}$ and $\mathbf{y}$ be two vectors in V_n. Then, the vector $\alpha\mathbf{x}$, where α is a nonzero scalar, is called the *orthogonal projection* (OP) of the vector $\mathbf{y}$ on the vector $\mathbf{x}$ if $\alpha\mathbf{x}$ and $\mathbf{y} - \alpha\mathbf{x}$ are orthogonal, i.e.,

$$\langle \alpha\mathbf{x}, \mathbf{y} - \alpha\mathbf{x} \rangle = 0. \tag{5.30}$$

□

Applying (5.18) and axioms of Definition 5.9 to (5.30), we get

$$\alpha \langle \mathbf{x}, \mathbf{y} \rangle - \alpha\bar{\alpha} \langle \mathbf{x}, \mathbf{x} \rangle = 0$$

from which we have

$$\alpha = \frac{\langle \mathbf{y}, \mathbf{x} \rangle}{\langle \mathbf{x}, \mathbf{x} \rangle}.$$

So,

$$\text{Orthogonal projection of } \mathbf{y} \text{ on } \mathbf{x} = \frac{\langle \mathbf{y}, \mathbf{x} \rangle}{\langle \mathbf{x}, \mathbf{x} \rangle}\mathbf{x}. \tag{5.31}$$

Referring to Figure 5.6(b), if θ denotes the angle between $\mathbf{x}$ and $\mathbf{y}$ in $\mathbb{R}^n$ and is given by (5.15), where $\langle \mathbf{x}, \mathbf{y} \rangle$ is defined by (5.16), then we have, for the orthogonal projection of $\overrightarrow{OA}$ on $\mathbf{x}$,

$$\alpha\mathbf{x} = \overrightarrow{OB} = \frac{\langle \mathbf{y}, \mathbf{x} \rangle}{\langle \mathbf{x}, \mathbf{x} \rangle}\mathbf{x} = \boldsymbol{\varepsilon} \| \mathbf{y} \| \cos\theta, \qquad \boldsymbol{\varepsilon} = \mathbf{x}/\| \mathbf{x} \|, \tag{5.32}$$

where the final result is obtained on simplification by using (5.15) and $\theta = \alpha - \beta$. The expression (5.32) shows that finding the orthogonal projection of one vector $\mathbf{y}$ on another vector $\mathbf{x}$ in $\mathbb{R}^n$ is strikingly similar to resolution of a vector along two perpendicular directions on the plane.

It is important to note that the vector $\overrightarrow{OC}$ (see Figure 5.6(b)), which can be written as

$$\overrightarrow{OC} = \mathbf{y} - \text{orthogonal projection of } \mathbf{y} \text{ on } \mathbf{x} \tag{5.33}$$

is a component of $\mathbf{y}$ having the property of being perpendicular to $\mathbf{x}$.

With the help of the principle of orthogonal projection, we are now ready to offer a physical explanation why the vectors $\mathbf{y}_1, \mathbf{y}_1, ..., \mathbf{y}_m$, as determined in Theorem 5.7, are orthogonal. By referring to Figure 5.7(a), we see that, in view of (5.33), the relation

$$\begin{aligned}\overrightarrow{OC} &= \mathbf{x}_2 - \text{orthogonal projection of } \mathbf{x}_2 \text{ on } \mathbf{y}_1 \\ &= \mathbf{x}_2 - \frac{\langle \mathbf{x}_2, \mathbf{y}_1 \rangle}{\langle \mathbf{y}_1, \mathbf{y}_1 \rangle}\mathbf{y}_1\end{aligned}$$

is perpendicular to $\mathbf{y}_1$ and is selected as $\mathbf{y}_2$ so that $<\mathbf{y}_1, \mathbf{y}_2> = 0$. In fact, from the given set $\mathbf{x}_1, \mathbf{x}_2, ..., \mathbf{x}_m$, we

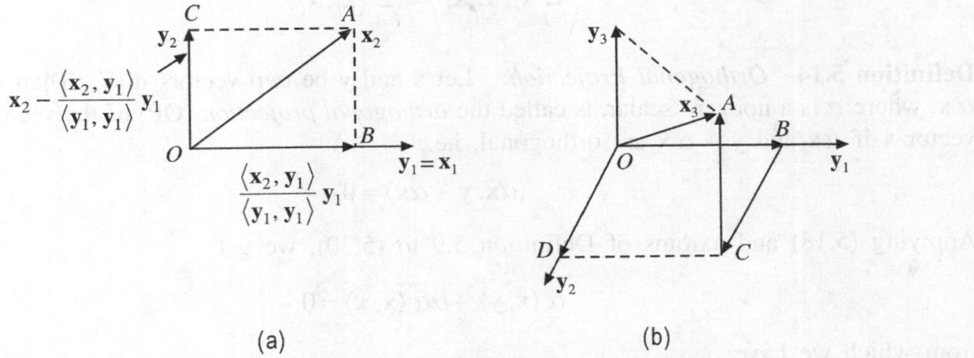

Figure 5.7 (a) Vector *OB* is an orthogonal projection of $\mathbf{x}_2$ on $\mathbf{y}_1$; (b) vector *OB* is an orthogonal projection of $\mathbf{x}_3$ on $\mathbf{y}_1$ and vector *OD* is an orthogonal projection of $\mathbf{x}_3$ on $\mathbf{y}_2$.

select $\mathbf{y}_1 = \mathbf{x}_1$. Now, there are two orthogonal components of $\mathbf{x}_2$: $\overrightarrow{OB}$ along $\mathbf{y}_1 = \mathbf{x}_1$, which we reject, and $\overrightarrow{OC}$ orthogonal to $\mathbf{y}_1$, which we select to be $\mathbf{y}_2$. This component $\overrightarrow{OC}$ exists as a nonzero orthogonal component because $\mathbf{x}_1$ and $\mathbf{x}_2$ are independent.

Again, by repeated application of (5.33), the given $\mathbf{x}_3$ is resolved into $\overrightarrow{OB}$ and $\overrightarrow{BA}$ (see Figure 5.7(b)), of which $\overrightarrow{OB}$ along $\mathbf{y}_1$ is rejected and $\overrightarrow{BA}$ is again resolved into $\overrightarrow{BC}$ and $\overrightarrow{CA}$, of which $\overrightarrow{BC}$ along $\mathbf{y}_2$ is rejected and $\overrightarrow{CA}$ is chosen as $\mathbf{y}_3$. Hence, by (5.31), we have

$$\mathbf{y}_3 = \overrightarrow{CA} = \mathbf{x}_3 - \overrightarrow{OB} - \overrightarrow{BC} = \mathbf{x}_3 - \overrightarrow{OB} - \overrightarrow{OD}$$

$$= \mathbf{x}_3 - [\text{OP of } \mathbf{x}_3 \text{ on } \mathbf{y}_1 + \text{OP of } \mathbf{x}_3 \text{ on } \mathbf{y}_2]$$

$$= \mathbf{x}_3 - \frac{\langle \mathbf{x}_3, \mathbf{y}_1 \rangle}{\langle \mathbf{y}_1, \mathbf{y}_1 \rangle}\mathbf{y}_1 - \frac{\langle \mathbf{x}_3, \mathbf{y}_2 \rangle}{\langle \mathbf{y}_2, \mathbf{y}_2 \rangle}\mathbf{y}_2$$

and it is easy to verify by direct computation that

$$\langle \mathbf{y}_3, \mathbf{y}_2 \rangle = 0 \text{ and } \langle \mathbf{y}_3, \mathbf{y}_2 \rangle = 0$$

and, therefore, $\mathbf{y}_1$, $\mathbf{y}_2$ and $\mathbf{y}_3$ are mutually orthogonal.

Thus the orthogonal vectors $\mathbf{y}_1, \mathbf{y}_2, ..., \mathbf{y}_{m-1}$ having been selected from the given independent vectors $\mathbf{x}_1, \mathbf{x}_2, ..., \mathbf{x}_{m-1}$, a new (and the last) $\mathbf{x}_m$ is picked up, its components along $\mathbf{y}_1, \mathbf{y}_2, ..., \mathbf{y}_{m-1}$ are rejected while the component of $\mathbf{x}_m$ perpendicular to all $\mathbf{y}_1, \mathbf{y}_2, ..., \mathbf{y}_{m-1}$ is selected to be $\mathbf{y}_m$. This perpendicular component exists because $\mathbf{x}_1, \mathbf{x}_2, ..., \mathbf{x}_m$ are independent. In this way, it is possible to select a set of m orthogonal vectors $\mathbf{y}_1, \mathbf{y}_2, ..., \mathbf{y}_m$ from the given independent set of m vectors $\mathbf{x}_1, \mathbf{x}_2, ..., \mathbf{x}_m$.

Example 5.24(a) Use the Gram-Schmidt orthonormalization process to determine an orthonormal basis in $\mathbb{R}^3$ for the given set of independent vectors

$$\mathbf{x}_1 = [1\ 0\ 1]^T, \qquad \mathbf{x}_2 = [-1\ 1\ 0]^T, \qquad \mathbf{x}_3 = [-3,\ 2\ 0]^T.$$

Solution Using (5.28), we have

$$\mathbf{y}_1 = \mathbf{x}_1 = [1\ 0\ 1]^T, \qquad \langle \mathbf{y}_1, \mathbf{y}_1 \rangle = 2;$$

$$\mathbf{y}_2 = \mathbf{x}_2 - \frac{\langle \mathbf{x}_2, \mathbf{y}_1 \rangle}{\langle \mathbf{y}_1, \mathbf{y}_1 \rangle}\mathbf{y}_1 = [-1\,1\,0]^T - \frac{-1}{2}[1\,0\,1]^T = \frac{1}{2}[-1\,2\,1]^T,$$

$$\langle \mathbf{y}_2, \mathbf{y}_2 \rangle = \frac{3}{2};$$

$$\mathbf{y}_3 = \mathbf{x}_3 - \frac{\langle \mathbf{x}_3, \mathbf{y}_1 \rangle}{\langle \mathbf{y}_1, \mathbf{y}_1 \rangle}\mathbf{y}_1 - \frac{\langle \mathbf{x}_3, \mathbf{y}_2 \rangle}{\langle \mathbf{y}_2, \mathbf{y}_2 \rangle}\mathbf{y}_2 = [-3\,2\,0]^T - \frac{-3}{2}[1\,0\,1]^T$$

$$-\frac{7/2}{3/2}\cdot\frac{1}{2}[-1\ 2\ 1]^T = \frac{1}{3}[-1\ -1\ 1]^T.$$

Therefore, by (5.29), the orthonormal basis is

$$\varepsilon_1 = \frac{\mathbf{y}_1}{\|\mathbf{y}_1\|} = \frac{1}{\sqrt{2}}[1\ 0\ 1]^T,$$

$$\varepsilon_2 = \frac{\mathbf{y}_2}{\|\mathbf{y}_2\|} = \frac{1}{2}\sqrt{\frac{2}{3}}[-1\,2\,1]^T = \frac{1}{\sqrt{6}}[-1\,2\,1]^T,$$

$$\varepsilon_3 = \frac{\mathbf{y}_3}{\|\mathbf{y}_3\|} = \frac{1}{3}\sqrt{3}\,[-1\,-1\,1]^T = \frac{1}{\sqrt{3}}[-1\,-1\,1]^T.$$ □

Example 5.24(b) The set $\{\mathbf{x}_1 = 1,\ \mathbf{x}_2 = 1 + t,\ \mathbf{x}_3 = t + t^2\}$ is given as a basis of $\mathbb{P}_2(t)$ over $\mathbb{R}$. Define the inner product as

$$\langle p, q \rangle = \int_{-1}^{1} p(t)q(t)\,dt,$$

where $p(t)$ and $q(t)$ are elements of $\mathbb{P}_2(t)$. Construct an orthonormal basis of $\mathbb{P}_2(t)$ from the given set.

Solution Following (5.28), choose $\mathbf{y}_1 = \mathbf{x}_1 = 1$. Then

$$\mathbf{y}_2 = 1 + t - \frac{\langle 1+t, 1 \rangle}{\langle 1, 1 \rangle}1$$

Now,

$$\langle 1, 1 \rangle = \int_{-1}^{1} dt = 2, \text{ and } \langle 1+t, 1 \rangle = \int_{-1}^{1} (1+t)\,dt = 2.$$

Therefore,

$$\mathbf{y}_2 = 1 + t - (2/2) = t.$$

Similarly,

$$\mathbf{y}_3 = t + t^2 - \frac{\langle t+t^2, 1\rangle}{\langle 1, 1\rangle} 1 - \frac{\langle t+t^2, t\rangle}{\langle t, t\rangle} t$$

$$= t + t^2 - \frac{2/3}{2} 1 - \frac{2/3}{2/3} t = t^2 - \frac{1}{3}$$

Therefore, the set $\left\{1, t, t^2 - \frac{1}{3}\right\}$ is an orthogonal basis in $\mathbb{P}_2(t)$. Again, $\|\mathbf{y}_1\| = \sqrt{2}$, $\|\mathbf{y}_2\| = \sqrt{6}/3$, $\|\mathbf{y}_3\| = 2\sqrt{10}/15$. Consequently, the set

$$\left\{\frac{\sqrt{2}}{2}, \frac{\sqrt{6}}{2} t, \frac{\sqrt{10}}{4}(3t^2 - 1)\right\}$$

is an orthonormal basis of $\mathbb{P}_2(t)$. □

Example 5.24(c) If the columns of $X = [\mathbf{x}_1\ \mathbf{x}_2\ \ldots\ \mathbf{x}_m]$, $\mathbf{x}_i \in \mathbb{R}^n$ and $m \le n$ are linearly independent, then prove that ∃ a nonsingular upper triangular matrix $R_0 \in \mathbb{R}^{m\times m}$ and a matrix $Q_0 \in \mathbb{R}^{n\times m}$ with $Q_0^T Q_0 = I_m$ such that $X = Q_0R_0$. Also show that $X = QR$, where Q is orthogonal and R is upper triangular.

Solution If we define $X = [\mathbf{x}_1\ \mathbf{x}_2\ \ldots\ \mathbf{x}_m]$, and $U_0 = [\mathbf{y}_1\ \mathbf{y}_2\ \cdots\ \mathbf{y}_m]$, from (5.26) we can write $X = U_0V_0$, where

$$V_0 = \begin{bmatrix} 1 & -a_{21} & -a_{31} & \cdots & -a_{m1} \\ 0 & 1 & -a_{32} & \cdots & -a_{m2} \\ \vdots & \vdots & \vdots & & \vdots \\ 0 & 0 & 0 & \cdots & -a_{mm-1} \\ 0 & 0 & 0 & \cdots & 1 \end{bmatrix}$$

Appending a set of $n - m$ number of orthogonal vectors $\mathbf{y}_{m+1} \cdots \mathbf{y}_n$ as columns to U_0, each of which is also orthogonal to $\mathbf{y}_i$, $i = 1, 2, \ldots, m$ we get $X = UV$, where

$$U = [U_0\ \vdots\ \mathbf{y}_{m+1}\ \cdots\ \mathbf{y}_n], \quad V = \begin{bmatrix} V_0 \\ \cdots \\ 0 \end{bmatrix} \begin{matrix} m \\ \cdots \\ n-m \end{matrix}$$

Let us define, $\alpha_i = \|\mathbf{y}_i\|$, $i = 1, 2, \ldots, n$; and $\varepsilon_i = \mathbf{y}_i/\alpha_i$. Then, $X = QR$, where $Q := [\varepsilon_1 \ldots \varepsilon_n]$ is orthogonal and ith row of R is α_i times the corresponding row of V. Alternatively, $X = Q_0R_0$, where $Q_0 = [\varepsilon_1\ \ldots\ \varepsilon_m]$ such that $Q_0^T Q_0 = I_m$ and ith row of R_0 is α_i times the corresponding row of V_0, for all $i = 1, 2, \ldots, m$.

5.6 LINEAR SIMULTANEOUS EQUATIONS: CRAMER'S RULE

The solution of linear simultaneous equations and its existence are the main concern of this section. Before setting ourselves to this task, we shall prove a remarkable property of any matrix which relates to its number of independent rows and columns, because this concept of row or column independence of a matrix will be needed for our main purpose. With this objective in view, we now present the following three lemmas:

Lemma 5.3 In a vector space V_n, a vector $\mathbf{x}$ is a linear combination of the independent vectors $\mathbf{x}_1, \mathbf{x}_2, ..., \mathbf{x}_n$ iff $\mathbf{x}, \mathbf{x}_1, ..., \mathbf{x}_n$ are linearly dependent.

Proof Let $\mathbf{x}$ be a linear combination of the given independent vectors. Therefore, $\mathbf{x} = \alpha_1\mathbf{x}_1 + \alpha_2\mathbf{x}_2 + \cdots + \alpha_n\mathbf{x}_n$, i.e., $\mathbf{x} - (\alpha_1\mathbf{x}_1 + \cdots + \alpha_n\mathbf{x}_n) = \boldsymbol{\theta}$, which proves the necessity. On the other hand, if $\mathbf{x}, \mathbf{x}_1, ..., \mathbf{x}_n$ are linearly dependent, then $\alpha\,\mathbf{x} + \alpha_1\mathbf{x}_1 + \cdots + \alpha_n\mathbf{x}_n = \boldsymbol{\theta}$, where $\alpha \neq 0$. Otherwise, because of linear independence of $\mathbf{x}_i$'s, all α_i's are zero. Hence, $\mathbf{x} = -\dfrac{1}{\alpha}(\alpha_1\mathbf{x}_1 + \cdots + \alpha_n\mathbf{x}_n)$, which proves the sufficiency. □

Lemma 5.4 Under elementary column (resp row) transformations, the number of independent columns (resp rows) in any matrix A of order $m \times n$ is invariant.

Proof Let $A = [\mathbf{a}_1, \mathbf{a}_2, \mathbf{a}_3, ..., \mathbf{a}_n]$, where m-vectors $\mathbf{a}_i$'s represent columns of A. Obviously, an interchange of columns does not alter the number of independent columns.

Now let us select $\mathbf{a}_1$. If $\mathbf{a}_2$ is dependent on $\mathbf{a}_1$, then subtracting a suitable scalar multiple of $\mathbf{a}_1$ from $\mathbf{a}_2$, the second column is reduced to zero. If $\mathbf{a}_2$ is not dependent, select $\mathbf{a}_2$ so that $\mathbf{a}_1$ and $\mathbf{a}_2$ are linearly independent. Let $\mathbf{a}_1$ and $\mathbf{a}_2$ be independent. If $\mathbf{a}_1, \mathbf{a}_2, \mathbf{a}_3$ are dependent, then by Lemma 5.3, $\mathbf{a}_3$ is a linear combination of $\mathbf{a}_1$ and $\mathbf{a}_2$ so that this combination when subtracted from $\mathbf{a}_3$ reduces column 3 of A to zero. If not, select $\mathbf{a}_3$ to be independent. In this way, subtracting the linear combination of independent vectors preceding a vector to be selected, we can pick up a set $\mathbf{a}_{i_1}, \mathbf{a}_{i_2}, ..., \mathbf{a}_{i_r}$ of r independent vectors from among the columns of A. Any other way of selection gives the same number of independent vectors since this number is invariant under interchange of columns.

Without loss of generality, let $\mathbf{a}_1, \mathbf{a}_2, ..., \mathbf{a}_r$, $r \leq n$, be linearly independent columns of A. Under elementary column transformation (ii) (see Definition 4.2), let $\mathbf{a}_p$ be multiplied by $\alpha \neq 0$. Then, for $1 \leq p \leq r$, $\alpha_1\mathbf{a}_1 + \cdots + \alpha\alpha_p\mathbf{a}_p + \cdots + \alpha_r\mathbf{a}_r = \boldsymbol{\theta}$ implies $\alpha_1 = 0, ..., \alpha\alpha_p = 0$, i.e., $\alpha_p = 0, ..., \alpha_r = 0$. So, $\mathbf{a}_1, ..., \mathbf{a}_p, ..., \mathbf{a}_r$ remain linearly independent. Again for $r < p \leq n$, $\mathbf{a}_p$ is linearly dependent, and so, $\mathbf{a}_p = \alpha_1\mathbf{a}_1 + \cdots + \alpha_r\mathbf{a}_r$ shows that $\alpha\,\mathbf{a}_p$, $\alpha \neq 0$, remains linearly dependent.

Applying column transformation (iii) (see Definition 4.2), i.e., adding $\alpha \neq 0$ times $\mathbf{a}_p$ to $\mathbf{a}_k$ and for linear independence, we have from $\alpha_1\mathbf{a}_1 + \cdots + \alpha_k(\mathbf{a}_k + \alpha\mathbf{a}_p) + \cdots + \alpha_p\mathbf{a}_p + \cdots + \alpha_r\mathbf{a}_r = \boldsymbol{\theta}$, $\alpha_1 = 0, ..., \alpha_k = 0, ..., \alpha\alpha_k + \alpha_p = 0$, i.e., $\alpha_p = 0, ..., \alpha_r = 0$ which shows that independence of $\mathbf{a}_1, \mathbf{a}_2, ..., \mathbf{a}_r$ does not change under elementary column transformation (iii). For rows the proof is similar. □

Example 5.25 Let

$$A = \begin{bmatrix} a_{11} & a_{12} & a_{13} \\ a_{21} & a_{22} & a_{23} \\ a_{31} & a_{32} & a_{33} \end{bmatrix}.$$

Adding α times first column to the second (elementary column transformation of type (iii), see Definition 4.2), we get

$$A \sim A_1 = \begin{bmatrix} a_{11} & a_{12} + \alpha a_{11} & a_{13} \\ a_{21} & a_{22} + \alpha a_{21} & a_{23} \\ a_{31} & a_{32} + \alpha a_{31} & a_{33} \end{bmatrix}.$$

We now show that the number of independent rows in A and A_1 are same. For scalars α_1, α_2, α_3, considering A_1, the relation

$$\alpha_1[a_{11}\ a_{12} + \alpha a_{11}\ a_{13}] + \alpha_2[a_{21}\ a_{22} + \alpha a_{21}\ a_{23}] + \alpha_3[a_{31}\ a_{32} + \alpha a_{31}\ a_{33}] = \boldsymbol{\theta}$$

implies, as $\alpha \neq 0$, by property of matrix equality (see Section 1.2), and in view of the fact that $\boldsymbol{\theta}$ is a 1×3 null matrix, that

$$\alpha_1[a_{11}\ a_{12}\ a_{13}] + \alpha_2[a_{21}\ a_{22}\ a_{23}] + \alpha_3[a_{31}\ a_{32}\ a_{33}] = \boldsymbol{\theta}.$$

The equations which govern the dependent rows in A and A_1 are, therefore, the same. This shows that under elementary column transformation (iii), see Definition 4.2, row dependence of a matrix does not change. This is true for the remaining columns as well as for row transformations. In view of this observation, we now have the following lemma.

Lemma 5.5 If a matrix A over a field $\mathbb{F}$ is row-equivalent to B, then:

(a) LC (linear combination) of certain columns of A are zero iff the same LC of same numbered columns of B are zero.

(b) Certain columns of A are independent (resp dependent) iff the same numbered columns of B are independent (resp dependent).

Proof (a) Since A and B are row-equivalent, by Definition 4.4, there exists a nonsingular matrix P such that $PA = B$. Let $\mathbf{a}_1$, $\mathbf{a}_2$, ..., $\mathbf{a}_r$ be any r number of columns of A satisfying

(i) $$\alpha_1\mathbf{a}_1 + \alpha_2\mathbf{a}_2 + \cdots + \alpha_r\mathbf{a}_r = \boldsymbol{\theta}$$

where, α_1, α_2, ..., α_r are elements of $\mathbb{F}$. Premultiplying both sides of (i) by P, we get

(ii) $$\alpha_1\mathbf{b}_1 + \alpha_2\mathbf{b}_2 + \cdots + \alpha_r\mathbf{b}_r = \boldsymbol{\theta}$$

where, $\mathbf{b}_1$, $\mathbf{b}_2$, ..., $\mathbf{b}_r$ are correspondingly numbered columns of B as can be seen from $PA = B$. Conversely, if (ii) is true, we have (i).

(b) This result is a direct consequence of the correspondence between (i) and (ii). □

Remark 5.2 Lemma 5.5 asserts that under elementary row transformations (i.e., under row-equivalence) of a matrix A, its column dependence (i.e., the number of independent columns in A and/or linear combination of dependent columns in A) does not change. By employing A^T, a dual version of this lemma can be stated since, under elementary column transformations, the row dependence of a matrix is invariant. So, we have the following corollary.

Corollary 5.5 Under elementary column (resp row) transformations, the row (resp column) dependence of a matrix is invariant. □

Theorem 5.9 The number of independent columns or rows of a matrix is equal to its rank.

Proof By elementary transformations (see Definition 4.2), let an $m \times n$ matrix A transform to any of the normal forms N (see Definition 4.5). By virtue of Lemma 5.4 and Corollary 5.5, the numbers of independent rows and columns in A and N are the same. Hence the theorem, as the number of independent rows or columns in N is the same as its rank r. □

This theorem has the following corollaries which can be easily verified.

Corollary 5.6 The rank of a matrix A is equal to that of its transpose or transposed conjugate. □

Corollary 5.7 In $\mathbb{C}^n$ the number of independent elements in a set of vectors is equal to the rank of the matrix having these vectors as columns. □

Corollary 5.8 A square matrix A is nonsingular iff all its rows or columns are linearly independent.

Proof It follows by rewriting Theorem 5.9 as: number of independent columns or rows in A = rank $A = n$. □

Linear simultaneous equations. A set of m linear simultaneous equations in n unknowns $x_1, x_2, \ldots, x_n$, viz.,

$$\begin{aligned} a_{11}x_1 + a_{12}x_2 + \cdots + a_{1n}x_n &= b_1 \\ a_{21}x_1 + a_{22}x_2 + \cdots + a_{2n}x_n &= b_2 \\ &\vdots \\ a_{m1}x_1 + a_{m2}x_2 + \cdots + a_{mn}x_n &= b_m \end{aligned} \tag{5.34}$$

can be written in the form

$$A\mathbf{x} = \mathbf{b}, \tag{5.35}$$

where

$$A = \begin{bmatrix} a_{11} & a_{12} & \cdots & a_{1n} \\ a_{21} & a_{22} & \cdots & a_{2n} \\ \vdots & \vdots & & \vdots \\ a_{m1} & a_{m2} & \cdots & a_{mn} \end{bmatrix}, \quad \mathbf{x} = \begin{bmatrix} x_1 \\ x_2 \\ \vdots \\ x_n \end{bmatrix}, \quad \mathbf{b} = \begin{bmatrix} b_1 \\ b_2 \\ \vdots \\ b_m \end{bmatrix}. \tag{5.36}$$

Equation (5.35) is said to be *homogeneous* if $\mathbf{b} = \boldsymbol{\theta}$, and *inhomogeneous* if $\mathbf{b} \neq \boldsymbol{\theta}$. Any vector $\mathbf{x}$ satisfying $A\mathbf{x} = \mathbf{b}$ is called a *solution* of (5.35). If no such vector exists, then the set of equations (5.34) or (5.35) is said to be *inconsistent.* A solution $\mathbf{x}$ of homogeneous equation $A\mathbf{x} = \boldsymbol{\theta}$ is referred to as *trivial* or *nontrivial* according as $\mathbf{x} = \boldsymbol{\theta}$ or $\mathbf{x} \neq \boldsymbol{\theta}$.

Theorem 5.10 The set of linear algebraic equations $A\mathbf{x} = \mathbf{b}$ in (5.35) has a solution iff rank A is equal to the rank of $m \times (n + 1)$ *augmented matrix* $B = [A, \mathbf{b}]$.

Proof (a) *Necessity.* Denoting by $\mathbf{a}_i = [a_{1i}, \ldots, a_{mi}]^T$, $i = 1, \ldots, n$, the ith column of A, if $\mathbf{x} = [x_1, \ldots, x_n]$ is a solution, from (5.35) we get

$$x_1\mathbf{a}_1 + \cdots + x_n\mathbf{a}_n = \mathbf{b} \tag{5.37}$$

implying **b** in B is a linear combination of the columns in A and is, therefore, dependent on them. Thus, $[A, \mathbf{b}]$ has the same number of independent columns as A. Hence by Theorem 5.9, rank A = rank $[A, \mathbf{b}]$.

(b) *Sufficiency.* If rank A = rank $[A, \mathbf{b}]$, by Lemma 5.3, **b** is a linear combination of the columns of A such as (5.37), and the coefficients of this combination constitute the solution vector **x**. □

Remark 5.3 From Theorem 5.10, it follows that for $A\mathbf{x} = \mathbf{b}$ to be *consistent*, the number of independent equations = rank A = rank $[A, \mathbf{b}] \le$ number of columns in A = number of unknowns.

Corollary 5.9 The set of homogeneous equations $A\mathbf{x} = \boldsymbol{\theta}$ is always consistent and has a unique trivial solution if A is square and nonsingular.

Proof Consistency is an easy consequence of Theorem 5.10, where $\mathbf{b} = \boldsymbol{\theta}$. Since all the columns $\mathbf{a}_1, \ldots, \mathbf{a}_n$ in A are linearly independent, from (5.37), where $\mathbf{b} = \boldsymbol{\theta}$, we conclude $x_1 = \cdots = x_n = 0$, which is unique [see Definition 5.3]. □

Example 5.26 The equation

$$\underset{A}{\begin{bmatrix} 1 & 0 & 3 \\ -1 & 2 & -2 \\ 1 & 0 & 3 \end{bmatrix}} \underset{\mathbf{x}}{\begin{bmatrix} x_1 \\ x_2 \\ x_3 \end{bmatrix}} = \underset{\mathbf{b}}{\begin{bmatrix} 1 \\ 2 \\ 3 \end{bmatrix}}$$

has no solution. Because here rank A = 2, while rank $[A, \mathbf{b}]$ = 3. □

Example 5.27 The equation

$$\begin{bmatrix} 1 & 0 & 3 \\ -1 & 2 & -2 \\ 0 & 2 & 1 \end{bmatrix} \begin{bmatrix} x_1 \\ x_2 \\ x_3 \end{bmatrix} = \begin{bmatrix} 0 \\ 0 \\ 0 \end{bmatrix} \tag{5.38}$$

has a solution; in fact, infinitely many solutions. Writing these equations explicitly, we obtain

$$\begin{aligned} x_1 + 3x_3 &= 0, \\ -x_1 + 2x_2 - 2x_3 &= 0, \\ 2x_2 + x_3 &= 0. \end{aligned}$$

Subtracting the second equation from the third, we get

$$\begin{aligned} x_1 + 3x_3 &= 0, \\ -x_1 + 2x_2 - 2x_3 &= 0, \\ x_1 + 3x_3 &= 0, \end{aligned} \tag{5.39}$$

where the first and last equations are the same, showing that all the equations in (5.38) are not independent. Therefore, solving the first two equations in (5.39), we obtain

$$x_1 = -3x_3, \qquad x_2 = -(½)x_3$$

Let $x_3 = -k$ be any scalar. Then the solution of (5.38) becomes $x_1 = 3k$, $x_2 = 0.5k$, $x_3 = -k$, giving an infinite number of values for **x** by varying k. Note that here r = rank A = 2. Two equations are therefore independent, and $r = 2$ number of variables can be expressed as a linear combination of remaining $n - r = 1$ number of variables. In this problem, in the set of solutions which are a constant multiple of a single vector of the form $k[3, 0.5, -1]^T$, there is only one linearly independent vector. Indeed, it can be shown that the number of independent solution vectors for $A\mathbf{x} = \boldsymbol{\theta}$ is $n - r$ which is the number of variables selected arbitrarily (see Theorem 5.16). □

Corollary 5.10 The set of linear homogeneous equations $A\mathbf{x} = \boldsymbol{\theta}$, where A is an $m \times n$ matrix and $\mathbf{x}$ an n-vector, has an infinitely many solutions iff rank $A < n$, the number of unknowns.

Proof Let rank $A = r < n$ and, consequently, r of the equations $A\mathbf{x} = \boldsymbol{\theta}$ are linearly independent. If $r < m$, the remaining $m - r$ number of equations can be reduced to zero by subtracting from them a suitable linear combination of the independent equations (see Example 5.27). Without any loss of generality, we consider the first r equations to be independent which are written as

$$\begin{bmatrix} a_{11} & \cdots & a_{1r} \\ \vdots & & \vdots \\ a_{r1} & \cdots & a_{rr} \end{bmatrix} \begin{bmatrix} x_1 \\ \vdots \\ x_r \end{bmatrix} + \begin{bmatrix} a_{1r+1} & \cdots & a_{1n} \\ \vdots & & \vdots \\ a_{rr+1} & \cdots & a_{rn} \end{bmatrix} \begin{bmatrix} x_{r+1} \\ \vdots \\ x_n \end{bmatrix} = \boldsymbol{\theta},$$

or

$$C\mathbf{u} + D\mathbf{v} = \boldsymbol{\theta}, \tag{5.40}$$

where

$$C = \begin{bmatrix} a_{11} & \cdots & a_{1r} \\ \vdots & & \vdots \\ a_{r1} & \cdots & a_{rr} \end{bmatrix}, \quad D = \begin{bmatrix} a_{1r+1} & \cdots & a_{1n} \\ \vdots & & \vdots \\ a_{rr+1} & \cdots & a_{rn} \end{bmatrix},$$

$$\mathbf{u} = \begin{bmatrix} x_1 \\ \vdots \\ x_r \end{bmatrix}, \quad \mathbf{v} = \begin{bmatrix} x_{r+1} \\ \vdots \\ x_n \end{bmatrix}, \tag{5.41}$$

and the variables are so arranged that C is nonsingular. It is easy to see that $\mathbf{x}^T = [\mathbf{u}^T, \mathbf{v}^T]$, where $\mathbf{u} = -C^{-1}D\mathbf{v}$ and $\mathbf{v}$ is any arbitrary $(n - r)$ – vector, is a solution of (5.40) and defines a set of infinite number of solution vectors, one solution vector for each choice of $\mathbf{v}$. This proves the sufficiency.

To prove the necessity, assume that $A\mathbf{x} = \boldsymbol{\theta}$ has infinitely many solutions but rank $A = r = n$. Considering only n independent equations, by Corollary 5.9, $A\mathbf{x} = \boldsymbol{\theta}$ has a unique trivial solution which contradicts the assumptions of many solutions. □

A direct consequence of Corollary 5.10 is the following corollary.

Corollary 5.11 The set of linear homogeneous equations $A\mathbf{x} = \boldsymbol{\theta}$, where A is an $n \times n$ matrix and $\mathbf{x}$ an n-vector, has an infinite number of nontrivial solutions iff rank $A < n$, i.e., iff det $A = 0$. □

Theorem 5.11 If $\mathbf{u}$ is a particular solution of $A\mathbf{x} = \mathbf{b}$ and $\mathbf{v}$ is any solution of its homogeneous part $A\mathbf{x} = \boldsymbol{\theta}$, then a class of solutions for $A\mathbf{x} = \mathbf{b}$ is described by $\mathbf{u} + \alpha\,\mathbf{v}$, where α is a scalar.

Proof We have $A(\mathbf{u} + \alpha\,\mathbf{v}) = A\mathbf{u} + \alpha A\mathbf{v} = \mathbf{b} + \alpha \bullet \theta = \mathbf{b}$. Hence, $\mathbf{u} + \alpha\,\mathbf{v}$ is a solution vector of $A\mathbf{x} = \mathbf{b}$. □

Theorem 5.12 The solution of $A\mathbf{x} = \mathbf{b}$ is unique iff A is square and nonsingular.

Proof Let A be of order $m \times n$. For the solution to exist, the number of independent equations must be less than or equal to the number of unknowns by Theorem 5.10, i.e., $m \le n$ (see Remark 5.3). Let $\mathbf{x} = [x_1, ..., x_n]^T$ and $\mathbf{y} = [y_1, ..., y_n]^T$ be two solutions of $A\mathbf{x} = \mathbf{b}$ and let $\mathbf{a}_1, ..., \mathbf{a}_n$ represent the columns of A. So

$$\mathbf{b} = x_1\mathbf{a}_1 + \cdots + x_n\mathbf{a}_n = y_1\mathbf{a}_1 + \cdots + y_n\mathbf{a}_n.$$

Subtracting, we have

$$(x_1 - y_1)\,\mathbf{a}_1 + \cdots + (x_n - y_n)\,\mathbf{a}_n = \boldsymbol{\theta}. \tag{5.42}$$

Since the solution of $A\mathbf{x} = \mathbf{b}$ is unique, $x_i = y_i$. Hence (5.42) implies that $\mathbf{a}_1, ..., \mathbf{a}_n$ are linearly independent. Therefore, by Theorem 5.9, rank $A = n$. and hence $m = n$.

Conversely, since A^{-1} exists, $A^{-1}\mathbf{b}$ is a solution of $A\mathbf{x} = \mathbf{b}$. To prove that the solution is unique, let $\mathbf{x}$ and $\mathbf{y}$ be any two solutions. Then $A(\mathbf{x} - \mathbf{y}) = \boldsymbol{\theta}$, which by Corollary 5.9 implies $\mathbf{x} = \mathbf{y}$, as A is square and nonsingular. □

Theorem 5.13 A consistent set of linear inhomogeneous equations $A\mathbf{x} = \mathbf{b}$, where A is an $m \times n$ matrix, $\mathbf{x}$ an n-vector and $\mathbf{b}$ an m-vector has (a) a unique solution iff $m = n$ and A is nonsingular, and (b) an infinitely many solutions iff rank $A = r < n$, the number of unknowns.

Proof (a) It is only a restatement of Theorem 5.12.

(b) Since $A\mathbf{x} = \mathbf{b}$ is consistent, r = rank A = rank $[A, \mathbf{b}]$ and r number of equations in $A\mathbf{x} = \mathbf{b}$ are independent. The remaining $(m - r)$ equations can be reduced to zero by subtracting from them suitable linear combinations of r independent equations. So, without loss of generality, we consider A to be of order $r \times n$ in $A\mathbf{x} = \mathbf{b}$, whose solution according to Theorem 5.11 is in the form $\mathbf{u} + \alpha\,\mathbf{v}$. By Corollary 5.10, $\mathbf{v}$ is an infinite number of vectors, iff $r < n$. □

Example 5.28 Prove that the vector space $\mathbb{C}^n$ has dimension n.

Solution We have to prove that in $\mathbb{C}^n$ (a) any $(n + 1)$ number of n-vectors $\mathbf{x}_1, \mathbf{x}_2, ..., \mathbf{x}_{n+1}$ are linearly dependent, and (b) there is at least one set of n number of independent vectors.

To prove (a) let $\alpha_1, \alpha_2, ..., \alpha_{n-1}$ be scalars such that

$$\alpha_1\mathbf{x}_1 + \alpha_2\mathbf{x}_2 + \cdots + \alpha_{n+1}\mathbf{x}_{n+1} = X\mathbf{a} = \boldsymbol{\theta} \tag{5.43}$$

where X is an $n \times (n + 1)$ matrix $[\mathbf{x}_1, \mathbf{x}_1, ..., \mathbf{x}_{n+1}]$ and $\mathbf{a}$ is a vector with elements $\alpha_1, \alpha_2, ..., \alpha_{n+1}$. By Corollary 5.10, (5.43) has a nontrivial solution for α_i's, and hence $\mathbf{x}_i$'s are linearly dependent.

For (b), we observe that the elementary n-vectors $\mathbf{e}_1, \mathbf{e}_2, ..., \mathbf{e}_n$ (see (5.4)) are n numbers of linearly independent vectors in $\mathbb{C}^n$. □

Theorem 5.14 The set of all solutions $\mathbf{x}$ of the linear simultaneous equations $A\mathbf{x} = \boldsymbol{\theta}$, where A is an $m \times n$ matrix and $\mathbf{x}$ an n-vector, constitutes a vector space.

Proof Let $\mathbf{x}_1$, $\mathbf{x}_2$, and $\mathbf{x}_3$ be any three solutions of $A\mathbf{x} = \boldsymbol{\theta}$. Referring to Definition 5.1, we have, in view of the Distributive law of matrix multiplication (as described in Property V, Section 1.2.2).

$$A(\mathbf{x}_1 + \mathbf{x}_2) = A\mathbf{x}_1 + A\mathbf{x}_2 = \boldsymbol{\theta} + \boldsymbol{\theta} = \boldsymbol{\theta},$$

$$A(\alpha\mathbf{x}_1) = \alpha A\mathbf{x}_1 = \alpha\boldsymbol{\theta} = \boldsymbol{\theta},$$

so that $\mathbf{x}_1 + \mathbf{x}_1$ and $\alpha\,\mathbf{x}_1$ are solutions of $A\mathbf{x} = \boldsymbol{\theta}$. Recognizing the solutions as vectors, Axioms 1 to 8 in Definition 5.1 can be easily verified, assuming zero vector $\boldsymbol{\theta}$ to be an element of $\mathbb{C}^n$, i.e., V_n. Alternatively, applying Theorem 5.3, the set of all solutions is seen to form a vector subspace of V_n. □

Thus, existence and uniqueness of solutions of linear homogeneous equations $A\mathbf{x} = \boldsymbol{\theta}$ are related to the rank and nonsingularity of A, and results embracing them are contained in Corollaries 5.9–5.11. On the other hand, solutions of inhomogeneous equations $A\mathbf{x} = \mathbf{b}$ are related to the properties of A and $\mathbf{b}$, and the results concerning them are described by Theorems 5.10–5.13. In most of the cases, explicit formulae for the nontrivial solutions cannot be stated, except for the case when a unique solution of $A\mathbf{x} = \mathbf{b}$ exists as given in Theorem 5.12. In this situation, the solution of $A\mathbf{x} = \mathbf{b}$ can be expressed elegantly by what is popularly known as *Cramer's Rule.*

Theorem 5.15 *Cramer's Rule.* Let a set of linear simultaneous equations be described by

$$A\mathbf{x} = \mathbf{b}, \tag{5.44}$$

where $A = [a_{ij}]$ is an n-square nonsingular matrix, $\mathbf{x}$ and $\mathbf{b}$ are n-vectors having elements x_i's and b_i's. Then a unique solution of (5.44) is given by

$$x_i = \det A_i/\det A,$$

where A_i is a matrix obtained from A by replacing its ith column elements by the elements of $\mathbf{b}$.

Proof By Theorem 5.12, a unique solution to (5.44) exists. Premultiplying both sides of (5.44) by A^{-1} and using (3.4), we get

$$\mathbf{x} = A^{-1}\mathbf{b}, \quad \text{or} \quad \begin{bmatrix} x_1 \\ x_2 \\ \vdots \\ x_n \end{bmatrix} = \frac{1}{\det A}\begin{bmatrix} A_{11} & A_{21} & \cdots & A_{n1} \\ A_{12} & A_{22} & \cdots & A_{n2} \\ \vdots & \vdots & & \vdots \\ A_{1n} & A_{2n} & \cdots & A_{nn} \end{bmatrix}\begin{bmatrix} b_1 \\ b_2 \\ \vdots \\ b_n \end{bmatrix}, \tag{5.45}$$

where A_{ij}'s are cofactors of a_{ij}'s. From (5.45), the ith element x_i of $\mathbf{x}$ is

$$x_i = \frac{1}{\det A}(b_1A_{1i} + b_2A_{2i} + \cdots + b_nA_{ni})$$

$$= \frac{\det A_i}{\det A}, \qquad i = 1, 2, ..., n; \tag{5.46}$$

where, by Property VI of determinants in Section 2.3, the expression in the numerator on the right-hand side of (5.46) is the determinant of a matrix A_i which is obtained from A by replacing its ith column elements by the elements of $\mathbf{b}$. □

Example 5.29 Find the solution of the following set of equations:

$$\underset{A}{\begin{bmatrix} 1 & 1 & 0 \\ 2 & -1 & 2 \\ 3 & 1 & 0 \end{bmatrix}}\underset{\mathbf{x}}{\begin{bmatrix} x_1 \\ x_2 \\ x_3 \end{bmatrix}} = \underset{\mathbf{b}}{\begin{bmatrix} 3 \\ -2 \\ 3 \end{bmatrix}}.$$

Solution Since $\det A = 4$ is not zero, we can use Cramer's rule. Here, A_1, A_2 and A_3 are

$$A_1 = \begin{bmatrix} 1 & 1 & 0 \\ -2 & -1 & 2 \\ 3 & 1 & 0 \end{bmatrix}, \quad A_2 = \begin{bmatrix} 1 & 3 & 0 \\ 2 & -2 & 2 \\ 3 & 3 & 0 \end{bmatrix} \text{ and } A_3 = \begin{bmatrix} 1 & 1 & 3 \\ 2 & -1 & -2 \\ 3 & 1 & 3 \end{bmatrix},$$

and hence $\det A_1 = 0$, $\det A_2 = 12$ and $\det A_3 = 2$. Therefore,

$$x_1 = \frac{\det A_1}{\det A} = 0, \quad x_2 = \frac{\det A_2}{\det A} = 3, \quad x_3 = \frac{\det A_3}{\det A} = \frac{1}{2}. \quad \square$$

We have already remarked that, in general, no explicit formula for the solution of $A\mathbf{x} = \mathbf{b}$ exists. However, powerful computational algorithms are available for solving these equations. We shall now describe, with the help of an example, the most famous of these algorithms known as *Gaussian Elimination.* An improved version of this algorithm paying regard to numerical accuracy is outlined in Chapter 13.

Example 5.30 *Gaussian Elimination for the Solution of Linear Simultaneous Equations.* Obtain a general solution to the following simultaneous equations where we count the variables in the order *x, y, z, w*:

$$\begin{aligned} z + 2w &= 8, \\ x - 3y + 2z - w &= 2, \\ 2x - 6y + 5z &= 12, \\ 3x - 9y + 8z + 4w &= 31. \end{aligned} \tag{5.47}$$

Solution A general method of solving (5.47) is as follows:

(a) We bring to the first position an equation whose coefficient of the first variable x is not zero; then we subtract a suitable scalar multiple of the first equation from the following equations so that coefficients of x are zero in them.

(b) Leaving out the first equation from among the remaining equations, we bring to the second position an equation in which the coefficient of the *first variable among the variables in these equations* is not zero. Subtracting a suitable scalar multiple of the second equation from the following equations, the coefficient of the first *variable* is made zero in them as in step (a).

(c) Keeping aside the first and the second equation, we treat the remaining equations as in step (a) and continue this process.

This method of solving linear simultaneous equations is known as *Gaussian Elimination.*

Returning to equations (5.47), we do the following:

(i) Interchange the first and the second equation, then subtract twice the first equation from the third and thrice from the fourth to get, respectively,

$$\begin{aligned} x - 3y + 2z - w &= 2, \\ z + 2w &= 8, \\ 2x - 6y + 5z &= 12, \\ 3x - 9y + 8z + 4w &= 31; \end{aligned}$$

$$\begin{aligned} x - 3y + 2z - w &= 2, \\ z + 2w &= 8, \\ z + 2w &= 8, \\ 2z + 7w &= 25. \end{aligned} \tag{5.48}$$

This procedure eliminates the first variable x from all equations except for the first equation in (5.48). Leaving out the first equation, among the remaining equations the first variable z is present in the second equation.

(ii) We, therefore, subtract once the second equation from the third and twice from the fourth; then we interchange the third and fourth equations to get

$$\begin{aligned} x - 3y + 2z - w &= 2, \\ z + 2w &= 8, \\ w &= 3, \\ 0 &= 0, \end{aligned} \tag{5.49}$$

in which the variable z has been eliminated from the last two equations.

We now start from the last nonzero equation in (5.49) which gives us $w = 3$. This value, when substituted in the preceding equation, gives $z = 2$. Coming to the first equation, we find that any arbitrary value can be assigned to x or y. Putting $y = \alpha$, an arbitrary parameter, we have $x = 3\alpha + 1$. As we proceed in the reverse direction in solving the final set of equations (5.49), this process is known as *back substitution.*

Therefore, the solution of (5.47) is finally given by

$$x = 3\alpha + 1, \qquad y = \alpha, \qquad z = 2, \text{ and } w = 3,$$

where α is an arbitrary parameter. □

The intrinsic idea behind the Gaussian Elimination is to reduce, at each step, the number of equations as well as the number of unknowns involved. For instance, in Example 5.30, though we have started with four equations in four variables, in the next step we have reduced them to two equations in two unknowns [see the last two equations in (5.48)] and, finally, in one equation in one unknown [see the last nonzero equation in (5.49)].

It is not hard to see that the operations performed by employing equations in Gaussian Elimination to solve $A\mathbf{x} = \mathbf{b}$ are nothing but elementary row operations effected on the augmented matrix $B{:}= [A \mid \mathbf{b}]$. Therefore, there is no point in carrying along the variables at every step, and hence the following *revised version of Gaussian Elimination* is useful in practice.

(i) If the element in position (1, 1) of B is zero, by interchanging rows, bring in this position a nonzero element; otherwise go to step (ii).

(ii) Subtract a suitable scalar multiple of the first row from the following ones so that all elements in the first column of B except the one in position (1, 1) are zero.

(iii) Leaving aside the first row of B, in the matrix B_1 formed by the remaining rows, let j be the first nonzero column counted from the left. By interchanging rows, bring a nonzero element in the position (1, j) of B_1. Now subtract a suitable multiple of the first row of B_1 from the following rows to reduce to zero all the elements in column j of B_1 except the one in position (1, j).

(iv) Again leaving aside the first and the second rows of B, we form a matrix B_2 by the remaining rows and transform it as the matrix B_1 in (iii).

(v) Similarly, we successively construct matrices B_3, ..., B_{m-2}, and in each case we transform the matrix as B_1 in step (iii).

It is important to note that, after transformation of B in each step above, the transformed matrix is still denoted by B. After these row transformations, $[A \mid \mathbf{b}]$ becomes equivalent to $[\tilde{A} \mid \tilde{\mathbf{b}}]$, from which it is easy to verify whether rank A is equal to rank $[A \mid \mathbf{b}]$ or not. Indeed, if the number of nonzero rows in $\tilde{A}$ is equal to that in $[\tilde{A} \mid \tilde{\mathbf{b}}]$, then rank A = rank $[A \mid \mathbf{b}]$, and the system of simultaneous equations has a solution. So the algorithmic procedure known as Gaussian Elimination does also suggest whether or not the given set of equations is consistent, and if consistent, it gives immediately the solutions by back substitution.

Example 5.31 The augmented matrix associated with the linear simultaneous equations given in Examples 5.30 is

$$B := [A \mid \mathbf{b}] = \left[\begin{array}{cccc|c} 0 & 0 & 1 & 2 & 8 \\ 1 & -3 & 2 & -1 & 2 \\ 2 & -6 & 5 & 0 & 12 \\ 3 & -9 & 8 & 4 & 31 \end{array}\right]$$

We now apply the following steps to B:

(i) Interchange the first and the second row.

(ii) Subtract twice the first row from the third and thrice from the fourth.

(iii) Subtract once the second row from the third and twice from the fourth.

(iv) Interchange third and fourth rows.

Thus, we get successively,

$$\text{(a)} \quad \left[\begin{array}{cccc|c} 1 & -3 & 2 & -1 & 2 \\ 0 & 0 & 1 & 2 & 8 \\ 2 & -6 & 5 & 0 & 12 \\ 3 & -9 & 8 & 4 & 31 \end{array}\right]$$

$$\text{(b)} \quad \left[\begin{array}{cccc|c} 1 & -3 & 2 & -1 & 2 \\ \hline 0 & 0 & 1 & 2 & 8 \\ 0 & 0 & 1 & 2 & 8 \\ 0 & 0 & 2 & 7 & 25 \end{array}\right] \left.\vphantom{\begin{array}{c} 0 \\ 0 \\ 0 \end{array}}\right\} B_1$$

$$(c)\ \left[\begin{array}{cccc|c} 1 & -3 & 2 & -1 & 2 \\ 0 & 0 & 1 & 2 & 8 \\ \hline 0 & 0 & 0 & 0 & 0 \\ 0 & 0 & 0 & 3 & 9 \end{array}\right] \Big\} B_2$$

$$(d)\ \left[\begin{array}{cccc|c} 1 & -3 & 2 & -1 & 2 \\ 0 & 0 & 1 & 2 & 8 \\ 0 & 0 & 0 & 3 & 9 \\ 0 & 0 & 0 & 0 & 0 \end{array}\right] = [\tilde{A} \mid \tilde{\mathbf{b}}]$$

Since the number of nonzero rows in $\tilde{A}$ and $[\tilde{A} \mid \tilde{\mathbf{b}}]$ is the same, the given set of simultaneous equations is consistent. Now, the equations corresponding to (d) above are obviously those given by (5.49), from which the solution is obtained as before. □

A variation of Gaussian Elimination is the following. Referring to (5.47), a general method of solving these equations is:

(i) Consider the first equation of (5.47) and then find the first variable whose coefficient is not zero (here it is z); next subtract a suitable multiple of the first equation from the following ones so that the coefficients of z are zero in them.

(ii) Now consider the second equation in the set obtained after step (i), and determine the first variable whose coefficient is not zero (here it is x); then subtract a suitable multiple of this equation from the following ones to reduce to zero the coefficients of x in them.

(iii) Consider successively third, fourth, . . . equations and treat them as in step (ii).

For the set of simultaneous equations (5.47), the above steps give, in succession,

(a)
$$\begin{aligned} z + 2w &= 8, \\ x - 3y - 5w &= -14, \\ 2x - 6y - 10w &= -28, \\ 3x - 9y - 12w &= -33; \end{aligned}$$

(b)
$$\begin{aligned} z + 2w &= 8, \\ x - 3y - 5w &= -14, \\ 0 &= 0, \\ 3w &= 9. \end{aligned}$$

Hence, the solution is $w = 3$, $z = 2$, $y = \alpha$ (= an arbitrary parameter), $x = 3\alpha + 1$, as before.

The foregoing procedure can be carried out with the help of the augmented matrix B: = $[A \mid \mathbf{b}]$, which avoids carrying the variables at every step. The revised version of this algorithm is therefore stated as follows:

(i) Consider the first row of B.

(ii) Find the first nonzero element a_{1j_1}, counting from the left in this row, and subtract a suitable multiple of this row from the following rows to reduce the elements of column j_1 in them to zero.

(iii) Next, consider the second row of B so reduced and treat it as in step (ii).

(iv) The third, fourth, . . . rows are treated similarly. In this process, if any row is zero, pass on to the following row.

By the elementary row operations as in the above algorithmic procedure, the augmented matrix $[A \mid \mathbf{b}]$ is transformed into $[\tilde{A} \mid \tilde{\mathbf{b}}]$. As already indicated, if the numbers of nonzero rows in $\tilde{A}$ and $[\tilde{A} \mid \tilde{\mathbf{b}}]$ are equal, then rank A = rank $[A \mid \mathbf{b}]$, and the solution to the linear simultaneous equations having the augmented matrix $[A \mid \mathbf{b}]$ exists.

These steps, when applied to B in Example 5.31, give

$$\text{(a)} \quad \left[\begin{array}{rrrr|r} 0 & 0 & 1 & 2 & 8 \\ 1 & -3 & 0 & -5 & -14 \\ 2 & -6 & 0 & -10 & -28 \\ 3 & -9 & 0 & -12 & -33 \end{array}\right]$$

$$\text{(b)} \quad \left[\begin{array}{rrrr|r} 0 & 0 & 1 & 2 & 8 \\ 1 & -3 & 0 & -5 & -14 \\ 0 & 0 & 0 & 0 & 0 \\ 0 & 0 & 0 & 3 & 9 \end{array}\right] =: [\tilde{A} \mid \tilde{\mathbf{b}}]$$

in which (b) suggests that (5.47) is consistent and gives (5.49), and hence the solutions as noted earlier.

Remark 5.4 If $A\mathbf{x} = \mathbf{b}$ is consistent and r = rank A, then without any loss of generality, we can assume A to be of order $r \times n$ as observed in the proof of Theorem 5.13. Referring to the proof of Corollary 5.10, we can write the solution of $A\mathbf{x} = \mathbf{b}$ as $\mathbf{u} = C^{-1}(\mathbf{b} - D\mathbf{v})$, where $\mathbf{v}$ is any arbitrary vector having $n - r$ elements. Hence, in the solution of $A\mathbf{x} = \mathbf{b}$, n–rank $A = n - r$ number of variables from $x_1, x_2, \ldots, x_n$ can be assigned arbitrarily. If $\mathbf{b} = \boldsymbol{\theta}$, the vector space generated by the solution vectors $\mathbf{x}$ has dimension $n - r$ (see Theorem 5.16).

5.7 RANK AND NULLITY: SYLVESTER'S INEQUALITY

The effect on the rank of a matrix when it is premultiplied or postmultiplied by a nonsingular matrix as also the rank of the product of two or more matrices were considered in Section 4.7. The basic tool for such a study was elementary row or column transformations.

Introducing the concepts of range space and null space, we shall prove here some new results on the rank of sum or product of two or more matrices. It will also be shown how these concepts enable us to give sharp proofs of some of the results of Section 4.7.

Definition 5.15 *Null Space and Range Space.* Let A be an $m \times n$ matrix and $\mathbf{x}$ an n-vector. The set of all solutions of $A\mathbf{x} = \boldsymbol{\theta}$ constitutes a vector space (see Theorem 5.14) which is defined as the *null space* of A and is denoted by $\mathbb{N}(A)$. The set of all vectors $\mathbf{y}$ in V_m such that $A\mathbf{x} = \mathbf{y}$, $\mathbf{y} \neq \boldsymbol{\theta}$, for some $\mathbf{x}$ in V_n is defined as the *range space* of A and is denoted by $\mathbb{R}(A)$. □

Since $\mathbf{y} = x_1\mathbf{a}_1 + \cdots + x_n\mathbf{a}_n$, where $\mathbf{a}_i$'s are columns in A, (see (5.35) and (5.36) where now $\mathbf{y} = \mathbf{b}$), any vector in $\mathbb{R}(A)$ is a linear combination of the columns in A, and thus in view of Corollary 5.3, $\mathbb{R}(A)$ is generated by the linearly independent columns in A. Therefore, the range space $\mathbb{R}(A)$ is also known as the column space of A and has dimension in view of Theorem 5.9 equal to the rank A.

Example 5.32 In $\mathbb{R}^3$, the null space of [1 –3 2] is obtained from

$$\underset{A}{\begin{bmatrix} 1 & -3 & 2 \end{bmatrix}} \underset{\mathbf{x}}{\begin{bmatrix} x_1 \\ x_2 \\ x_3 \end{bmatrix}} = 0$$

a the set of all vectors $[x_1\ x_2\ x_3]^T$ orthogonal to the vector $\overrightarrow{OP} = [1\ -3\ 2]^T$, and hence lie in a *plane AOB* perpendicular to $\overrightarrow{OP}$ (see Figure 5.8(a)). For the equation in $\mathbb{R}^3$ given by

$$\underset{\tilde{A}}{\begin{bmatrix} 0 & 3 & 3 \\ 0 & -2 & 3 \end{bmatrix}} \underset{\mathbf{x}}{\begin{bmatrix} x_1 \\ x_2 \\ x_3 \end{bmatrix}} = \boldsymbol{\theta},$$

any vector in $\mathbb{N}(\tilde{A})$ is orthogonal to both $\overrightarrow{OP} = [0\ 3\ 3]^T$ and $\overrightarrow{OQ} = [0\ -2\ 3]^T$. Hence, $\mathbb{N}(\tilde{A})$ is a straight line perpendicular to the plane containing $\overrightarrow{OP}$ and $\overrightarrow{OQ}$ (see Fig. 5.8b). Here, both $\overrightarrow{OP}$ and $\overrightarrow{OQ}$ are in the plane x_2Ox_3, hence, $\mathbb{N}(\tilde{A})$ is represented by the Ox_1 axis.

In the first case, the range space $\mathbb{R}(A)$ is a set of real numbers, while in the second case, it is a set of straight lines in a plane generated by the columns of $\tilde{A}$. Again, in the first case, the dimension of $\mathbb{R}(A)$ is 1 and that of $\mathbb{N}(A)$ is 2, so that dim $\mathbb{R}(A)$ + dim $\mathbb{N}(A)$ = 3. In the second case, dim $\mathbb{R}(\tilde{A}) = 2$ and dim $\mathbb{N}(\tilde{A}) = 1$, but the sum remains 3; the space under consideration is $\mathbb{R}^3$ which is three-dimensional. This finding is true, in general, which we shall substantiate in the following theorem and in Corollary 5.12.

Theorem 5.16 The rank of a matrix A in $V_{m \times n}$ is r iff dim $\mathbb{N}(A)$ is $n - r$.

Proof : (a) *Necessity*: Let rank $A = r$. Following Corollary 5.10, we note that any solution of $A\mathbf{x} = \boldsymbol{\theta}$ is in the form

$$\mathbf{x} = \begin{bmatrix} \mathbf{u} \\ \mathbf{w} \end{bmatrix},$$

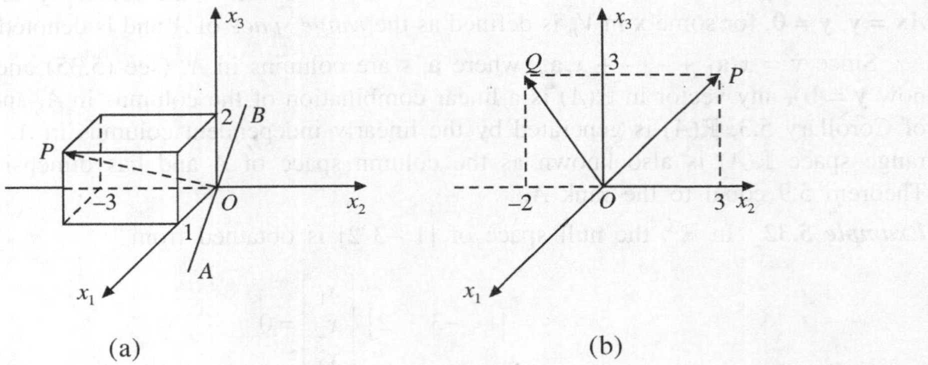

Figure 5.8 (a) Plane *AOB* of dim 2 orthogonal to vector $\overrightarrow{OP}$ is a null space of [1 −3 2];
(b) Ox_1 of dim 1 is a null space of $\tilde{A}$ in Example 5.32.

where $\mathbf{u} = -C^{-1}D\mathbf{w}$ is an r-vector and $\mathbf{w}$ is an arbitrary $(n - r)$-vector. For all $i = 1, 2, ..., n - r$, let $\mathbf{e}_i$ be elementary vectors in V_{n-r} such that

$$\mathbf{w} = \sum_{i=1}^{n-r} w_i \mathbf{e}_i,$$

where w_i are components of $\mathbf{w}$, and let $\mathbf{b}_i = -C^{-1}D\mathbf{e}_i$. Then

$$\mathbf{x} = \begin{bmatrix} \mathbf{u} \\ \mathbf{w} \end{bmatrix} = \begin{bmatrix} -C^{-1}D \\ I_{n-r} \end{bmatrix} \mathbf{w} = \sum_{i=1}^{n-r} w_i \begin{bmatrix} \mathbf{b}_i \\ \mathbf{e}_i \end{bmatrix} = \sum_{i=1}^{n-r} w_i \mathbf{y}_i, \tag{5.50}$$

where

$$\mathbf{y}_i = \begin{bmatrix} \mathbf{b}_i \\ \mathbf{e}_i \end{bmatrix}.$$

The $n - r$ number of vectors $\mathbf{y}_i$ are linearly independent by Corollary 5.7, and any solution of $A\mathbf{x} = \boldsymbol{\theta}$ can always be represented in terms of these $n - r$ number of vectors, as is evident from (5.50). Therefore, the set of $\mathbf{y}_i$'s, $i = 1, 2, ..., n - r$, is a basis of $\mathbb{N}(A)$ and dimension of $\mathbb{N}(A)$ is $n - r$ [see Example 5.41(b)].

(b) *Sufficiency* Let dim $\mathbb{N}(A) = n - r$. Then there exist $(n - r)$ number of independent vectors $\mathbf{x}_i$'s in V_n such that $A\mathbf{x}_i = \boldsymbol{\theta}$ for all $i = r + 1, ..., n$. Denoting by $\mathbf{x}_k$'s, $k = 1, ..., r$, the remaining independent vectors in V_n so that $\mathbf{x}_i$'s and $\mathbf{x}_k$'s form a basis in V_n, we have $A\mathbf{x}_k = \mathbf{y}_k \neq \boldsymbol{\theta}$ for $k = 1, ..., r$, where $\mathbf{y}_k$'s are in the range space $\mathbb{R}(A)$ of A. Let

$$\alpha_1\mathbf{y}_1 + \cdots + \alpha_r\mathbf{y}_r = \boldsymbol{\theta}. \tag{5.51}$$

Therefore, substituting for $\mathbf{y}_k$'s, we get

$$A(\alpha_1\mathbf{x}_1 + \cdots + \alpha_r\mathbf{x}_r) = \boldsymbol{\theta}, \quad \text{or} \quad A\mathbf{x} = \boldsymbol{\theta},$$

where $\mathbf{x} = \alpha_1\mathbf{x}_1 + \cdots + \alpha_r\mathbf{x}_r$ is not in $\mathbb{N}(A)$. Hence, $\mathbf{x}$ must be equal to zero. Since $\mathbf{x}_k$'s are linearly independent, $\alpha_1 = \cdots = \alpha_r = 0$. Thus, in view of (5.51), $\mathbf{y}_k$'s are also linearly independent. Therefore, $r = \dim \mathbb{R}(A) = \text{rank } A$. □

Literally interpreted, Theorem 5.16 implies that in the solution of linear simultaneous equations defined by $A\mathbf{x} = \boldsymbol{\theta}$, the number of linearly independent solution $\mathbf{x}$ is (n–rank A). See Example 5.27.

Definition 5.16 *Nullity* The dimension of a null space of a matrix A is called the nullity of A and is denoted by null A. □

An easy consequence of Theorem 5.16 is the following corollary.

Corollary 5.12 Let A be an $m \times n$ matrix. Then rank A + null $A = n$. □

Example 5.33 Referring to Example 5.32 (see Figure 5.8(a)), we have already observed that the dimension of $\mathbb{N}(A)$, which is a plane, is 2 and the dimension of $\mathbb{R}(A)$ is 1; so rank A + null $A = n = 3$. Again, in Figure 5.8(b), the dimension of $\mathbb{N}(\tilde{A})$, which is a straight line, is 1, and the dimension of $\mathbb{R}(\tilde{A})$ is 2, with $n = 3$, thereby verifying the truth of Corollary 5.12. Again, the coefficient matrix in Example 5.27 has rank 2, and the dimension of the null space, i.e., the space spanned by the solution vectors, is 1; therefore, rank A + null $A = 3$. □

Theorem 5.17 If A and B are two matrices respectively in $V_{m \times n}$ and $V_{n \times p}$, then rank $(AB) \le \min$ (rank A, rank B).

Proof Let $\mathbf{x}$ be in V_p. Then $B\mathbf{x} = \boldsymbol{\theta}$ implies that $AB\mathbf{x} = \boldsymbol{\theta}$. Hence, $\mathbb{N}(B) \subseteq \mathbb{N}(AB)$. Therefore, in view of Definition 5.7, $\dim \mathbb{N}(B) \le \dim \mathbb{N}(AB)$. By Theorem 5.16, p – rank $B \le p$ – rank (AB), or

$$\text{rank } (AB) \le \text{rank } B. \tag{5.52}$$

Similarly, we can show rank $(B'A') \le$ rank A'. Hence, by Corollary 5.6,

$$\text{rank } (AB) \le \text{rank } A. \tag{5.53}$$

Combining (5.52) and (5.53), we get the assertion of the theorem. □

Corollary 5.13 If P and Q are nonsingular matrices in $V_{m \times m}$ and $V_{n \times n}$ respectively, then rank (PA) = rank A and rank (AQ) = rank A, where A is in $V_{m \times n}$.

Proof By Theorem 5.17, rank $(PA) \le$ rank A, and rank A = rank $(P^{-1}PA) \le$ rank (PA), which proves rank (PA) = rank A. For the second assertion, we have rank (AQ) = rank $(Q'A')$ = rank A' = rank A, by using Corollary 5.6 and the first assertion of this corollary. □

With the help of this corollary we can easily prove the following corollary.

Corollary 5.14 If P and Q are nonsingular, then rank A = rank (PAQ), where A is in $V_{m \times n}$, P in $V_{m \times m}$ and Q in $V_{n \times n}$. □

Theorem 5.18 If the matrices A and B are in $V_{m \times n}$, then rank $(A + B) \le$ rank A + rank B.

Proof Let rank $A = r$ and rank $B = p$. Let $\mathbf{x}_1, \ldots, \mathbf{x}_r$ and $\mathbf{y}_1, \ldots, \mathbf{y}_p$ be the independent columns of A and B, respectively. The columns $\mathbf{z}_1, \ldots, \mathbf{z}_n$ of $(A + B)$ which span a vector space W, say, can be expressed as a linear combination of $\mathbf{x}_i$'s and $\mathbf{y}_i$'s and, therefore, for all $k = 1, \ldots, n$, $\mathbf{z}_k$ is in W implies that $\mathbf{z}_k$ too is in vector space V spanned by $\mathbf{x}_i$'s and $\mathbf{y}_i$'s. So, W is a subspace of V. Hence, in view of Definition 5.7,

$$\dim W \le \dim V \le r + p,$$

because the number of independent vectors among $\mathbf{x}_i$'s and $\mathbf{y}_i$'s is less than or equal to $r + p$. Therefore,

$$\text{rank } (A + B) \le \text{rank } A + \text{rank } B.$$

Theorem 5.19 Let A and B be two matrices in $V_{m\times n}$ and $V_{n\times p}$, respectively. Prove that the number of linearly independent vectors $\mathbf{y} = B\mathbf{x}$ satisfying $A\mathbf{y} = \mathbf{\theta}$ is rank B – rank (AB).

Proof Let

$$\dim \mathbb{N}(B) = p - \text{rank } B = q \text{ and } \dim \mathbb{N}(AB) = p - \text{rank } (AB) = k.$$

There exist therefore q number of linearly independent vectors: $\mathbf{x}_1, ..., \mathbf{x}_q$ in V_p satisfying $B\mathbf{x} = \mathbf{\theta}$ which is a basis of $\mathbb{N}(B)$. Now every vector satisfying $B\mathbf{x} = \mathbf{\theta}$ also satisfies $AB\mathbf{x} = \mathbf{\theta}$. So, $\mathbb{N}(B)$ is a subspace of $\mathbb{N}(AB)$, and hence $k \ge q$ in view of Definition 5.7. From this it follows that in V_p there are $(k - q)$ number of independent vectors $\mathbf{x}_{q+1}, ..., \mathbf{x}_k$ such that $\mathbf{x}_1, \mathbf{x}_2, ..., \mathbf{x}_k$ is a basis of $\mathbb{N}(AB)$ and $B\mathbf{x}_i \ne \mathbf{\theta}$, $i = q + 1, ..., k$. We shall show that $B\mathbf{x}_{q+1}, ..., B\mathbf{x}_k$ are linearly independent. Assume that for scalars $\alpha_{q+1}, ..., \alpha_k$

$$\alpha_{q+1}B\mathbf{x}_{q+1} + \cdots + \alpha_k B\mathbf{x}_k = B(\alpha_{q+1}\mathbf{x}_{q+1} + \cdots + \alpha_k\mathbf{x}_k) = \mathbf{\theta} \tag{5.54}$$

which implies that the vector within parentheses is in $\mathbb{N}(B)$ and, consequently, a linear combination of $\mathbf{x}_1, ..., \mathbf{x}_q$. Therefore, there are scalars $\alpha_1, ..., \alpha_q$ such that

$$\alpha_{q+1}\mathbf{x}_{q+1} + \cdots + \alpha_k\mathbf{x}_k = \alpha_1\mathbf{x}_1 + \cdots + \alpha_q\mathbf{x}_q.$$

Since $\mathbf{x}_1, ..., \mathbf{x}_q, \mathbf{x}_{q+1}, ..., \mathbf{x}_k$ are linearly independent, $\alpha_1, ..., \alpha_q, \alpha_{q+1}, ..., \alpha_k$ are all zero. As a result, it follows from (5.54) that $B\mathbf{x}_{q+1}, ..., B\mathbf{x}_k$ are linearly independent. Hence, the number of independent vectors $\mathbf{y} = B\mathbf{x}$ satisfying $A\mathbf{y} = \mathbf{\theta}$ is

$$k - q = \text{rank } B - \text{rank } (AB).$$

In other words,

$$\dim \mathbb{N}(AB) = k = q + \text{rank } B - \text{rank } (AB).$$

□

Theorem 5.20 *Sylvester's Inequality.* If the matrices A and B are in $V_{m\times n}$ and $V_{n\times p}$ respectively, then

$$\text{rank } A + \text{rank } B - n \le \text{rank } (AB) \le \min (\text{rank } A, \text{rank } B)$$

Proof We observe that n – rank A is dim $\mathbb{N}(A)$, i.e., the number of independent vectors $\mathbf{y}$ satisfying $A\mathbf{y} = \mathbf{\theta}$. Again, by Theorem 5.19, rank B – rank (AB) is the number of independent vectors $\mathbf{y} = B\mathbf{x}$ satisfying $A\mathbf{y} = \mathbf{\theta}$ and forms a subspace of $\mathbb{N}(A)$. Hence,

$$\text{rank } B - \text{rank } (AB) \le n - \text{rank } A,$$

or

$$\text{rank } A + \text{rank } B - n \le \text{rank } (AB).$$

This result along with Theorem 5.17 proves the assertion of this theorem. □

Sylvester's inequality is a powerful tool for establishing the rank of the product or the sum of two or more matrices. It is a special case of the following theorem:

Theorem 5.21 If A, B and C are in $V_{m\times n}$, $V_{n\times p}$ and $V_{p\times k}$ respectively, then

$$\text{rank } (AB) + \text{rank } (BC) \le \text{rank } B + \text{rank } (ABC)$$

Proof We note that the number of independent vectors $BC\mathbf{x}$ such that $ABC\mathbf{x} = \boldsymbol{\theta}$ is $\{\text{rank } (BC) - \text{rank } (ABC)\}$ and span a vector space V, and the number of independent vectors $B\mathbf{x}$ satisfying $AB\mathbf{x} = \boldsymbol{\theta}$ is $\{\text{rank } B - \text{rank } (AB)\}$ and span a vector space W. Since V is a subspace of W,

$$\text{rank } (BC) - \text{rank } (ABC) \le \text{rank } B - \text{rank } (AB). \qquad \square$$

Example 5.34 (a) Let P be a matrix of order $m \times n$ and Q be of order $n \times m$ such that $PQ = I_m$. Then rank P = rank $Q = m$.

The result will be proved by contradiction. Suppose rank $P < m$. Then by Sylvester's inequality (or specifically by Theorem 5.17), we have

$$m = \text{rank } (PQ) \le \text{rank } P < m$$

which is a contradiction. Therefore, rank $P = m$. Similarly, rank $Q = m$.

(b) For a matrix $A \in \mathbb{C}^{m\times n}$, prove that rank $(A^* A)$ = rank $(A A^*)$ = rank A.

By (5.32), rank $(A^* A)$ rank A. To get the reverse inequality, we note that $A^* A\mathbf{x} = 0$ implies that $\mathbf{x}^* A^* A\mathbf{x} = 0$, i.e., $\|A\mathbf{x}\|^2 = 0$. Hence, $A\mathbf{x} = 0$ by Definition 5.9. Therefore, $\mathbb{N}(A^*A) \subseteq \mathbb{N}(A)$ and by Definition 5.7, null $(A^* A) \le$ null A. Since by Corollary 5.12, rank $(A^* A)$ +null $(A^* A) = n$ = rank A +null A, and as a result rank $A \le$ rank $(A^* A)$. Hence, rank $(A^* A)$ = rank A. Switching the roles of A and A^* in the above arguments, we can similarly prove that rank (AA^*) = rank A.

(c) Let $V = R \oplus N$, where R and N are subspaces of $V = \mathbb{R}^n$ or $\mathbb{C}^n$. It is given that $\{\mathbf{u}_1,\ldots,\mathbf{u}_r\}$ is a basis of R and $\{\mathbf{v}_1,\ldots,\mathbf{v}_r\}$ is that of $N^\perp$. If $A = [\mathbf{u}_1,\ldots,\mathbf{u}_r]$ and $B = [\mathbf{v}_1,\ldots,\mathbf{v}_r]$, prove that $B^* A$ is invertible.

We assume that $\{\mathbf{u}_{r+1},\ldots,\mathbf{u}_n\} \notin R$ is a basis of N and $\{\mathbf{v}_1,\ldots,\mathbf{v}_r\} \notin R^\perp$ is a basis of $N^\perp$. Then, it is evident $\{\mathbf{u}_1,\ldots,\mathbf{u}_r; \mathbf{u}_{r+1},\ldots\mathbf{u}_n\}$ is a basis of V. Hence, $B^*[\mathbf{u}_{r+1}\cdots\mathbf{u}_n] = [\boldsymbol{\theta}\cdots\boldsymbol{\theta}]_{r\times(n-r)}$ and the set $\{\mathbf{u}_{r+1},\ldots,\mathbf{u}_n\}$ generates the null space of B^*. If $\mathbf{z} \in \mathbb{N}(B^*A)$, then $B^* A\mathbf{z} = \boldsymbol{\theta}$ and $A\mathbf{z} \in \mathbb{N}(B^*)$. Therefore, $A\mathbf{z} = \alpha_{r+1}\mathbf{u}_{r+1} + \cdots + \alpha_n\mathbf{u}_n$ for some scalars $\alpha_{r+1},\ldots,\alpha_n$. If we denote the elements of $\mathbf{z}$ by $\{-\alpha_1,\ldots,-\alpha_r\}$ then $A\mathbf{z} = -(\alpha_1\mathbf{u}_1 + \cdots + \alpha_r\mathbf{u}_r)$, and consequently,

$$\alpha_1\mathbf{u}_1 + \cdots + \mathbf{u}_r\alpha_r + \alpha_{r+1}\mathbf{u}_{r+1} + \cdots + \alpha_n\mathbf{u}_n = \boldsymbol{\theta}.$$

Since, $\mathbf{u}_i$s are linearly independent, all α_is are zero and $\mathbf{z} = \boldsymbol{\theta}$. It follows, therefore, ker $B^*A = \{\boldsymbol{\theta}\}$ and rank $B^*A = r$. This proves that B^*A is invertible. $\square$

5.8 COMPUTATION OF LINEAR DEPENDENCE AND INDEPENDENCE OF VECTORS

The notion of linear dependence and independence of vectors was explained in detail in Section 5.2. Given a set of vectors, one must carry out some computations before finally ensuring oneself whether the set under test is linearly dependent or independent. Or one may be asked to choose a basis of the vector space generated by a given set of vectors and to express a vector in the set in terms of independent ones. Or one may encounter the problem as to whether two or more given sets of vectors span the same subspace S or not; if yes, find the basis of S. With this purpose in mind, we shall now set ourselves to the task of establishing some theorems which would answer these questions.

Let us designate a set of vectors in V_m as

$$\mathbf{x}_1 = \begin{bmatrix} x_{11} \\ x_{21} \\ \vdots \\ x_{m1} \end{bmatrix}, \quad \mathbf{x}_2 = \begin{bmatrix} x_{12} \\ x_{22} \\ \vdots \\ x_{m2} \end{bmatrix}, \quad \ldots, \quad \mathbf{x}_n = \begin{bmatrix} x_{1n} \\ x_{2n} \\ \vdots \\ x_{mn} \end{bmatrix}, \tag{5.55}$$

and the matrix A formed with them as

$$A = [\mathbf{x}_1, \mathbf{x}_2, \ldots, \mathbf{x}_n] = \begin{bmatrix} x_{11} & x_{12} & \cdots & x_{1n} \\ x_{21} & x_{22} & \cdots & x_{2n} \\ \vdots & \vdots & & \vdots \\ x_{m1} & x_{m2} & \cdots & x_{mn} \end{bmatrix}. \tag{5.56}$$

In view of Theorem 5.9, it is evident that the number of independent vectors in $\mathbf{x}_1$, $\mathbf{x}_2$, ..., $\mathbf{x}_n$ is equal to the rank of A.

Theorem 5.22 A sufficient condition for the set of vectors $\mathbf{x}_1$, $\mathbf{x}_2$, ..., $\mathbf{x}_n$ in V_m defined in (5.55) to be linearly dependent is that $n > m$.

Proof The maximum number of linearly independent vectors in V_m is m by Definition 5.5. Any number n of vectors greater than m are therefore linearly dependent. □

Theorem 5.23 A necessary and sufficient condition for the set of vectors $\mathbf{x}_1$, $\mathbf{x}_2$, ..., $\mathbf{x}_n$ in V_m, to be linearly dependent is that rank $A < n$, where A is given by (5.56). Then the number of independent vectors is equal to the rank A.

Proof Let $\mathbf{x}_1$, $\mathbf{x}_2$, ..., $\mathbf{x}_n$ be linearly dependent. Then there exist scalars α_1, α_2, ..., α_n not all zeros such that

$$\alpha_1\mathbf{x}_1 + \alpha_2\mathbf{x}_2 + \cdots + \alpha_n\mathbf{x}_n = \boldsymbol{\theta}, \quad \text{or} \quad A\mathbf{u} = \boldsymbol{\theta},$$

where

$$\mathbf{u} = [\alpha_1, \alpha_2, \ldots, \alpha_n]^T,$$

from which it follows that n columns of A are linearly dependent. In other words, by Theorem 5.9, the number of independent columns, i.e., the rank of A is less than n.

Conversely, assume that rank $A < n$. Then by Theorem 5.9, n number of columns of A are linearly dependent, i.e., the number of independent vectors $\mathbf{x}_i$'s is equal to rank A. □

Theorem 5.24 A necessary and sufficient condition for the set of vectors $\mathbf{x}_1, \mathbf{x}_2, ..., \mathbf{x}_n$ in V_m to be linearly independent is that rank $A = n$, where A is given by (5.56).

Proof Let $\mathbf{x}_1, \mathbf{x}_2, ..., \mathbf{x}_n$ be linearly independent. Then in A there are n number of independent columns, and hence by Theorem 5.9, rank $A = n$.

Conversely, suppose rank $A = n$ which implies (by Theorem 5.9) that A has n number of independent columns, i.e., $\mathbf{x}_1, \mathbf{x}_2, ..., \mathbf{x}_n$ are linearly independent. □

A new matrix, known as the Gramian matrix, will now be introduced in terms of which, criteria pertaining to dependency or independency of a set of vectors can be formulated. To this effect, let $\mathbf{x}_1, \mathbf{x}_2, ..., \mathbf{x}_n$ be a set of vectors, over a real or a complex field $\mathbb{F}$ in V_m. Then the matrix defined by

$$G = A^*A = \begin{bmatrix} \langle \mathbf{x}_1, \mathbf{x}_1 \rangle, & \langle \mathbf{x}_2, \mathbf{x}_1 \rangle, & \cdots, & \langle \mathbf{x}_n, \mathbf{x}_1 \rangle \\ \langle \mathbf{x}_1, \mathbf{x}_2 \rangle, & \langle \mathbf{x}_2, \mathbf{x}_2 \rangle, & \cdots, & \langle \mathbf{x}_n, \mathbf{x}_2 \rangle \\ \vdots & & & \\ \langle \mathbf{x}_1, \mathbf{x}_n \rangle, & \langle \mathbf{x}_2, \mathbf{x}_n \rangle, & \cdots, & \langle \mathbf{x}_n, \mathbf{x}_n \rangle \end{bmatrix} \tag{5.57}$$

is called the *Gramian matrix*, where A is defined by (5.56). It is evident that $\mathbf{x}_i$'s are mutually orthogonal (see Definition 5.11) iff G is diagonal.

The following lemma plays a crucial part in the proof of the next two theorems.

Lemma 5.6 Let $\mathbf{x}_1, \mathbf{x}_2, ..., \mathbf{x}_n$ be a set of vectors in V_m over a field (real or complex) $\mathbb{F}$, and $\alpha_1, \alpha_2, ..., \alpha_n$ be scalars (real or complex numbers). Then

$$\alpha_1 \langle \mathbf{x}_1, \mathbf{x}_j \rangle \langle\ \rangle + \alpha_2 \langle \mathbf{x}_2, \mathbf{x}_j \rangle + \cdots + \alpha_n \langle \mathbf{x}_n, \mathbf{x}_j \rangle = 0 \tag{5.58}$$

for $j = 1,2, ..., n$, implies that

$$\mathbf{y} := \alpha_1\mathbf{x}_1 + \alpha_2\mathbf{x}_2 + \cdots + \alpha_n\mathbf{x}_n = \boldsymbol{\theta}.$$

Proof Taking the help of Axiom 3 in Definition 5.9, it follows from (5.58) that

$$\langle \alpha_1\mathbf{x}_1 + \alpha_2\mathbf{x}_2 + \cdots + \alpha_n\mathbf{x}_n, \mathbf{x}_j \rangle = 0, \quad \text{or} \quad \langle \mathbf{y}, \mathbf{x}_j \rangle = 0 \tag{5.59}$$

for $j = 1, 2, ..., n$. In view of (5.18) and (5.59), we now have

$$\langle \mathbf{y}, \mathbf{y} \rangle = \langle \mathbf{y}, \alpha_1\mathbf{x}_1 + \alpha_2\mathbf{x}_2 + \cdots + \alpha_n\mathbf{x}_n \rangle = 0.$$

Hence by Axiom 1 of Definition 5.9, $\mathbf{y} = \boldsymbol{\theta}$. □

Theorem 5.25 A set of vectors $\mathbf{x}_1, \mathbf{x}_2, ..., \mathbf{x}_n$ in V_m over a field $\mathbb{F}$ are linearly dependent iff the determinant of their Gramian matrix G is zero.

Proof If $\mathbf{x}_1, \mathbf{x}_2, ..., \mathbf{x}_n$ are linearly dependent, then there exist in $\mathbb{F}$ scalars $\alpha_1, \alpha_2, ..., \alpha_n$ *not all zero* such that

$$\mathbf{y} := \alpha_1\mathbf{x}_1 + \alpha_2\mathbf{x}_2 + \cdots + \alpha_n\mathbf{x}_n = \boldsymbol{\theta}. \tag{5.60}$$

Taking into account (5.60), the scalar product of $\mathbf{y}$ and $\mathbf{x}_j$, in view of Axiom 3 of Definition 5.9 and (5.57), can be written as

$$\langle \mathbf{y}, \mathbf{x}_j \rangle = \alpha_1 \langle \mathbf{x}_1, \mathbf{x}_j \rangle + \alpha_2 \langle \mathbf{x}_2, \mathbf{x}_j \rangle + \cdots + \alpha_n \langle \mathbf{x}_n, \mathbf{x}_j \rangle = 0 \tag{5.61}$$

for $j = 1, 2, ..., n$, or

$$G\mathbf{u} = \mathbf{\theta}, \tag{5.62}$$

where $\mathbf{u} = [\alpha_1, \alpha_2, ..., \alpha_n]^T$. Since $\mathbf{u}$ is nontrivial, by Corollary 5.11, det $G = 0$.

Conversely, if det $G = 0$, by Theorem 5.9, the number of independent columns in G is less than n. Therefore, there exists an n-vector $\mathbf{u} = [\alpha_1, \alpha_2, ..., \alpha_n]^T \neq \mathbf{\theta}$ such that (5.62), and also (5.61), hold. By Lemma 5.6, we therefore have

$$\alpha_1\mathbf{x}_1 + \alpha_2\mathbf{x}_2 + \cdots + \alpha_n\mathbf{x}_n = \mathbf{\theta},$$

because $\mathbf{u} \neq \theta$, $\mathbf{x}_1, \mathbf{x}_2, ..., \mathbf{x}_n$ are linearly dependent. □

This theorem cannot ascertain how many independent vectors there are in the given set while the former Theorem 5.23 can.

Theorem 5.26 A set of vectors $\mathbf{x}_1, \mathbf{x}_2, ..., \mathbf{x}_n$ in V_m over a field $\mathbb{F}$ are linearly independent iff the determinant of their Gramian matrix is positive.

Proof Let det $G > 0$. Then for an n-vector, $\mathbf{u} = [\alpha_1, \alpha_2, ..., \alpha_n]^T$, $G\mathbf{u} = \mathbf{\theta}$ implies $\alpha_1 = \alpha_2 = \cdots = \alpha_n = 0$ by Corollary 5.9. Now as a result of (5.57), from $G\mathbf{u} = \mathbf{\theta}$, it follows that

$$\alpha_1 \langle \mathbf{x}_1, \mathbf{x}_j \rangle + \alpha_2 \langle \mathbf{x}_2, \mathbf{x}_j \rangle + \cdots + \alpha_n \langle \mathbf{x}_n, \mathbf{x}_j \rangle = 0$$

for $j = 1, 2, ..., n$. By Lemma 5.6,

$$\alpha_1\mathbf{x}_1 + \alpha_2\mathbf{x}_2 + \cdots + \alpha_n\mathbf{x}_n = \mathbf{\theta}.$$

Hence, $G\mathbf{u} = 0$ or, equivalently, $\alpha_1\mathbf{x}_1 + \alpha_2\mathbf{x}_2 + \cdots + \alpha_n\mathbf{x}_n = \mathbf{\theta}$ implies that all α_i's are zero; so $\mathbf{x}_1, \mathbf{x}_2, ..., \mathbf{x}_n$ are linearly independent.

Let $\mathbf{x}_1, \mathbf{x}_2, ..., \mathbf{x}_n$ be linearly independent, which define A in (5.56). Then there exist scalars $\alpha_1, \alpha_2, ..., \alpha_n$, not all zero, such that

$$\alpha_1\mathbf{x}_1 + \alpha_2\mathbf{x}_2 + \cdots + \alpha_n\mathbf{x}_n = A\mathbf{u} \neq \mathbf{\theta}, \tag{5.63}$$

where $\mathbf{u} = [\alpha_1, \alpha_2, ..., \alpha_n]^T$. On premultiplication of the left-hand side of (5.63) by its transposed conjugate and taking into consideration (5.57), we have

$$\mathbf{u}^*A^*A\mathbf{u} = \mathbf{u}^*G\mathbf{u} > 0$$

in which the left-hand side represents a positive definite Hermitian form, G being a Hermitian matrix. Therefore, by Corollary 8.7, det $G > 0$. □

Theorems 5.22, 5.23, and 5.25 contain the criteria for testing whether or not a given set of vectors are linearly dependent. Theorems 5.24 and 5.26 contain the criteria whether or not these vectors are linearly independent. In case the set is linearly dependent, Theorem 5.23 tells us how to ascertain the number of independent vectors in the set. Under these circumstances, if it is required to express any vector as a linear combination of the independent vectors (see Definition 5.4 and Example 5.11), the following theorem offers a constructive procedure of doing this.

Theorem 5.27 Let r be the number of independent vectors in the set $\mathbf{x}_1, \mathbf{x}_2, ..., \mathbf{x}_n$ in V_m. Then each of the remaining $(n - r)$ vectors can be expressed as a linear combination of these r independent vectors.

Proof Obviously, $n > r$, and let $n \le m$ and form the matrix A given in (5.56). Since A has r independent columns, rank $A = r$ by Theorem 5.9. If the minor of order r at the upper left-hand corner of A is zero, we *interchange rows as well as columns* so that such minor M_r of order r at this position is not zero. Reindexing the rows and columns in their natural order, we have

$$M_r = \begin{vmatrix} x_{11} & x_{12} & \cdots & x_{1r} \\ x_{21} & x_{22} & \cdots & x_{2r} \\ \vdots & \vdots & & \vdots \\ x_{r1} & x_{r2} & \cdots & x_{rr} \end{vmatrix} \neq 0$$

so that the first r columns in A so rearranged are linearly independent.

Since $m \ge n > r$ and rank $A = r$, minor M_{r+1} of order $r + 1$ exists, which has a zero value. So, with the help of the remaining rows and columns, we form the minor M_{r+1} given by

$$M_{r+1} = \begin{vmatrix} x_{11} & x_{12} & \cdots & x_{1r} & x_{1j} \\ x_{21} & x_{22} & \cdots & x_{2r} & x_{2j} \\ \vdots & \vdots & & \vdots & \vdots \\ x_{r1} & x_{r2} & \cdots & x_{rr} & x_{rj} \\ x_{i1} & x_{i2} & \cdots & x_{ir} & x_{ij} \end{vmatrix} = 0,$$

where $i \ge 1$, $j > r$. Let $c_1, c_2, ..., c_r, c_{r+1} = M_r$ be the cofactors of $x_{i1}, x_{i2}, ..., x_{ir}, x_{ij}$, respectively, and as a consequence of Property VII in Section 2.3 and that $M_{r+1} = 0$, we can write

$$c_1 x_{k1} + c_2 x_{k2} + \cdots + c_r x_{kr} + c_{r+1} x_{kj} = 0 \tag{5.64}$$

for $k = 1, 2, ..., m$; $j = r + 1, ..., n$, because i and j were arbitrary in our previous selection. For a fixed j if we consider m equations (5.64) by varying k from 1 to m, we get

$$c_1 \mathbf{x}_1 + c_2 \mathbf{x}_2 + \cdots + c_r \mathbf{x}_r + c_{r+1} \mathbf{x}_j = \boldsymbol{\theta}, \qquad j = r + 1, ..., n. \tag{5.65}$$

Since $c_{r+1} = M_r \neq 0, n - r$ number of vectors $\mathbf{x}_j$, $j = r + 1, ..., n$, can be expressed as a linear combination of r independent vectors $\mathbf{x}_1, \mathbf{x}_2, ..., \mathbf{x}_r$.

If $n > m$, $n - m$ rows are added to A (these rows may be one of the rows of A) without affecting its rank. The foregoing procedure is therefore applicable, as the number of rows in A is now equal to the number of columns. □

Example 5.35 Show that the following vectors are linearly dependent:

$$\mathbf{x}_1 = [1 \quad 2 \quad 1 \quad 1]^T, \qquad \mathbf{x}_2 = [0 \quad 1 \quad 0 \quad 0]^T, \qquad \mathbf{x}_3 = [3 \quad 0 \quad 3 \quad 3]^T,$$

and express the dependent vector as a combination of linearly independent vectors.

Solution (a) *Rank Test* Here, the matrix A in (5.56) is

$$A = \begin{bmatrix} 1 & 0 & 3 \\ 2 & 1 & 0 \\ 1 & 0 & 3 \\ 1 & 0 & 3 \end{bmatrix} \sim \begin{bmatrix} 1 & 0 & 0 \\ 0 & 1 & 0 \\ 0 & 0 & 0 \\ 0 & 0 & 0 \end{bmatrix}$$

by elementary transformations as in Example 4.3. Hence, rank $A = 2$ which is less than the number of given vectors. Therefore, by Theorem 5.23, $\mathbf{x}_1$, $\mathbf{x}_2$, $\mathbf{x}_3$ are linearly dependent and two vectors of them are independent. Instead of a normal form, we can reduce A to row-echelon normal form to decide on rank A (see Section 4.6).

(b) *Gramian Matrix Test* The Gramian matrix formed by the given vectors is

$$G = A^T A = \begin{bmatrix} 7 & 2 & 9 \\ 2 & 1 & 0 \\ 9 & 0 & 27 \end{bmatrix}$$

which is singular. Therefore, the given set of vectors is linearly dependent. But the number of independent vectors cannot be ascertained by this test.

By rank test, the number of independent vectors is 2. Since the upper left-hand corner minor in A is not zero, the vectors $\mathbf{x}_1$ and $\mathbf{x}_2$ are linearly independent by Theorem 5.9. We now designate

$$M_2 = \begin{vmatrix} 1 & 0 \\ 2 & 1 \end{vmatrix}, \qquad M_3 = \begin{vmatrix} 1 & 0 & 3 \\ 2 & 1 & 0 \\ 1 & 0 & 3 \end{vmatrix}.$$

The cofactors of the third row elements in M_3 are

$$c_1 = -3, \qquad c_2 = 6, \qquad c_3 = 1.$$

Therefore, with the help of (5.65), we have

$$-3\mathbf{x}_1 + 6\mathbf{x}_2 + \mathbf{x}_3 = \boldsymbol{\theta} \quad \text{which gives} \quad \mathbf{x}_3 = 3\mathbf{x}_1 - 6\mathbf{x}_2.$$

In fact, any two vectors of the given set are linearly independent, and any one vector can be expressed in terms of the remaining two. □

Example 5.36 Choose a basis of the vector space spanned by

$$\mathbf{x}_1 = [1 \ 2]^T, \qquad \mathbf{x}_2 = [2 \ 4]^T, \qquad \mathbf{x}_3 = [1 \ 3]^T, \qquad \mathbf{x}_4 = [2 \ 3]^T,$$

and express the remaining vectors in terms of the basis vectors.

Solution Here the matrix A in (5.56) is

$$A = \begin{matrix} \mathbf{x}_1 & \mathbf{x}_2 & \mathbf{x}_3 & \mathbf{x}_4 \end{matrix} \\ A=\begin{bmatrix} 1 & 2 & 1 & 2 \\ 2 & 4 & 3 & 3 \end{bmatrix}$$

and, therefore, $m = 2$, $n = 4$. Since rank $A = 2$, by Theorem 5.9, A has two independent columns. Since the upper left-hand corner minor of order two in A is zero, we interchange $\mathbf{x}_2$ and $\mathbf{x}_3$ to get such nonzero minor. This implies that $\mathbf{x}_1$ and $\mathbf{x}_3$ are linearly independent and form a basis for the vector space spanned by $\mathbf{x}_i$'s. Moreover, we augment A by appending to it its first row twice to get a transformed A as (where $\mathbf{x}_i$'s are columns above the dashed line):

$$\begin{array}{cccc} \mathbf{x}_1 & \mathbf{x}_3 & \mathbf{x}_2 & \mathbf{x}_4 \end{array} \\ \left[\begin{array}{cccc} 1 & 1 & 2 & 2 \\ 2 & 3 & 4 & 3 \\ \hdashline 1 & 1 & 2 & 2 \\ 1 & 1 & 2 & 2 \end{array}\right]$$

from which we construct the following two minors;

$$\text{(a)} \begin{array}{|ccc|} \mathbf{x}_1 & \mathbf{x}_3 & \mathbf{x}_2 \\ 1 & 1 & 2 \\ 2 & 3 & 4 \\ \hdashline 1 & 1 & 2 \end{array}, \qquad \text{(b)} \begin{array}{|ccc|} \mathbf{x}_1 & \mathbf{x}_3 & \mathbf{x}_4 \\ 1 & 1 & 2 \\ 2 & 3 & 3 \\ \hdashline 1 & 1 & 2 \end{array}.$$

Hence the cofactors of the last row elements are, respectively,

(a) –2, 0, 1, (b) –3, 1, 1.

Therefore, by (5.65),

(a) $-2\mathbf{x}_1 + \mathbf{x}_2 = \boldsymbol{\theta}$, (b) $-3\mathbf{x}_1 + \mathbf{x}_3 + \mathbf{x}_4 = \boldsymbol{\theta}$

which give

$$\mathbf{x}_2 = 2\mathbf{x}_1, \qquad \mathbf{x}_4 = 3\mathbf{x}_1 - \mathbf{x}_3. \qquad \square$$

Among a given set of vectors, we can express the dependent vectors in terms of independent basis vectors by Theorem 5.27 and as illustrated by Examples 5.35 and 5.36. The difficult task in this method is to bring to the upper left-hand corner a nonzero minor of order r as no easy systematic procedure seems to be available. The following two methods have no such concomitant demerits and therefore are useful for our purpose.

5.8.1 Gaussian Elimination Method

From the matrix $A = [\mathbf{x}_1\ \mathbf{x}_2\ .\ .\ .\ \mathbf{x}_n]$ constructed by the given m-vectors $\mathbf{x}_i$'s [see (5.55) and (5.56)], the number r of linearly independent vectors which form a basis of the vector space S generated by $\mathbf{x}_1$, $\mathbf{x}_2$, ..., $\mathbf{x}_n$ can be selected by the following procedure.

Step 1 Let $x_{i_1 1}$ be the first nonzero element in $\mathbf{x}_1$, counting from the top. Adding suitable multiples of $\mathbf{x}_1$ to the remaining columns, the elements in their i_1th rows can be reduced to zero.

Step 2 If now $x_{i_2 2}$ is the first nonzero element from the top in the i_2th row of second column of A so reduced in step 1, then this column can be similarly employed to reduce to zero the i_2th row elements of all columns except for the first and the second.

Step 3 This process is continued till all the columns are exhausted.

Step 4 At any step if a zero column is encountered, one has to pass to the next column for one's consideration.

Step 5 The indices $i_1 i_2 \ldots i_r$ are all distinct and correspond to r linearly independent columns in A. To justify this, observe that in the foregoing procedure, only the elementary transformation (iii) in Definition 4.2 is used in a restricted sense, i.e., addition of a suitable multiple of any column to a column following it. Since in the foregoing procedure *no interchange of columns* is permitted, there exists a nonsingular *upper triangular* matrix Q such that

$$AQ = B, \tag{5.66}$$

where B is the matrix transformed from A by the column operations effected in Gaussian Elimination and, therefore, has r nonzero columns.

The nonzero columns of B designated by $j_1, j_2, \ldots, j_r$ are linearly independent. Indeed, the j_rth column has a nonzero element on the i_rth row and zero elements on the rows $i_1 \cdots i_{r-1}$; the j_{r-1}th column has a nonzero element on the i_{r-1}th row and zero elements on the rows $i_1 \cdots i_{r-2}$, and so on. Interchanging rows and columns of the matrix formed by these nonzero columns of B, we have an upper triangular matrix with nonzero diagonal elements. So B has r independent columns as stipulated. Now by Lemma 5.4, the number of linearly independent columns in A is r in view of (5.66).

Again, since the nonzero columns in B were used in the column transformation of A, in any column p of the upper triangular matrix Q, the elements in the rows $j_1, j_2, \ldots, j_k$ $(j_1 < j_2 < \cdots < j_k < p,\ k \le r)$ are, in general, nonzero, the element in the row p is 1 and the elements in the remaining rows are zero (see Example 5.37). Consequently, the columns in B can be represented by

$$\mathbf{b}_p = \mathbf{a}_{j_1} q_{j_1 p} + \mathbf{a}_{j_2} q_{j_2 p} + \cdots + \mathbf{a}_{j_k} q_{j_k p} + \mathbf{a}_p.$$

When p is other than $j_1 j_2 \ldots j_r$, $\mathbf{b}_p = \mathbf{0}$, and the columns $\mathbf{a}_p$'s in A can be written as a linear combination of $\mathbf{a}_{j_1}, \mathbf{a}_{j_2}, \ldots, \mathbf{a}_{j_r}$. On the other hand, considering p to be successively equal to $j_1, j_2, \ldots, j_r$, we have

$$[\mathbf{b}_{j_1} \mathbf{b}_{j_2} \ldots \mathbf{b}_{j_r}] = [\mathbf{a}_{j_1} \mathbf{a}_{j_2} \ldots \mathbf{a}_{j_r}] \begin{bmatrix} 1 & x & \cdots & x \\ & 1 & \cdots & x \\ & & \ddots & \vdots \\ & & & 1 \end{bmatrix}$$

which shows that $\mathbf{a}_{j_1}, \mathbf{a}_{j_2}, \ldots, \mathbf{a}_{j_r}$ are linearly independent by Property I in Section 4.7 and Theorem 5.9, because $\mathbf{b}_{j_1}, \mathbf{b}_{j_2}, \ldots, \mathbf{b}_{j_r}$ are so where x's represent the parameters q_{j_k}'s. By virtue of Definition 5.5, it is evident that $\mathbf{a}_{j_1}, \mathbf{a}_{j_2}, \ldots, \mathbf{a}_{j_r}$ form a basis of the vector space S generated by $\mathbf{x}_1, \mathbf{x}_2, \ldots, \mathbf{x}_n$.

Example 5.37 Choose a basis of the vector space S generated by the set of vectors

$$\mathbf{x}_1 = [1\ 2\ 1]^T, \quad \mathbf{x}_2 = [2\ 4\ 2]^T, \quad \mathbf{x}_3 = [1\ 2\ 2]^T, \quad \mathbf{x}_4 = [2\ 4\ 0]^T,$$

and express the dependent vectors in terms of the basis vectors.

Solution We form the matrix

$$A = [\mathbf{x}_1\ \mathbf{x}_2\ \mathbf{x}_3\ \mathbf{x}_4]$$

and following the procedure given in Gaussian Elimination, we transform it into a matrix B. Therefore, we can find an upper triangular matrix Q (see Example 4.3) such that

$$\underset{A}{\begin{array}{cccc} \mathbf{x}_1 & \mathbf{x}_2 & \mathbf{x}_3 & \mathbf{x}_4 \end{array} \atop \begin{bmatrix} 1 & 2 & 1 & 2 \\ 2 & 4 & 2 & 4 \\ 1 & 2 & 2 & 0 \end{bmatrix}} \underset{Q}{\begin{bmatrix} 1 & -2 & -1 & -4 \\ & 1 & 0 & 0 \\ & & 1 & 2 \\ & & & 1 \end{bmatrix}} = \underset{B}{\begin{bmatrix} 1 & 0 & 0 & 0 \\ 2 & 0 & 0 & 0 \\ 1 & 0 & 1 & 0 \end{bmatrix}}$$

Corresponding to nonzero columns of B, we choose $\mathbf{x}_1$ and $\mathbf{x}_3$ to be the basis vectors. The columns in the product AQ which equate to zero columns in B give

$$\mathbf{x}_2 = 2\mathbf{x}_1, \quad \mathbf{x}_4 = 4\mathbf{x}_1 - 2\mathbf{x}_3.$$

Other choices of basis rearranging $\mathbf{x}_i$'s are also possible. Such choices are $(\mathbf{x}_1, \mathbf{x}_4)$, $(\mathbf{x}_3, \mathbf{x}_4)$, $(\mathbf{x}_2, \mathbf{x}_3)$, $(\mathbf{x}_2, \mathbf{x}_4)$ which show that, except for the pair $(\mathbf{x}_1, \mathbf{x}_2)$, all other pairs form a basis of S. □

Example 5.38 To solve Example 5.19 by Gaussian Elimination (re-read the proof of Theorem 5.4 at this stage), we set $A = [\mathbf{x}_1\ \mathbf{x}_2\ \mathbf{e}_1\ \mathbf{e}_2\ \mathbf{e}_3]$ and then determine an upper triangular Q and B such that

$$\underset{A}{\begin{array}{ccccc} \mathbf{x}_1 & \mathbf{x}_2 & \mathbf{e}_1 & \mathbf{e}_2 & \mathbf{e}_3 \end{array} \atop \begin{bmatrix} 1 & 0 & 1 & 0 & 0 \\ 0 & 1 & 0 & 1 & 0 \\ 0 & 1 & 0 & 0 & 1 \end{bmatrix}} \underset{Q}{\begin{bmatrix} 1 & 0 & -1 & 0 & 0 \\ & 1 & 0 & -1 & -1 \\ & & 1 & 0 & 0 \\ & & & 1 & 1 \\ & & & & 1 \end{bmatrix}} = \underset{B}{\begin{bmatrix} 1 & 0 & 0 & 0 & 0 \\ 0 & 1 & 0 & 0 & 0 \\ 0 & 1 & 0 & -1 & 0 \end{bmatrix}}$$

from which it follows that $\mathbf{x}_1, \mathbf{x}_2, \mathbf{e}_2$ form an extension of the basis of S_2 to that of the whole space $\mathbb{R}^3$. □

Lemma 5.7 If the columns (resp rows) in each of A and B are linearly independent and generate the same subspace, then there exists a nonsingular matrix P (resp R) such that $AP = B$ (resp $RA = B$).

Proof Since the columns of B are in the subspace generated by those of A, we can find a matrix of appropriate order satisfying

$$AP = B. \tag{5.67}$$

Conversely, the columns of A are in the subspace generated by those of B. Therefore, there exists a matrix Q such that $A = BQ$. Hence,

$$A(I - PQ) = 0 \quad \text{and} \quad B(QP - I) = 0. \tag{5.68}$$

Since the columns in each of A and B are linearly independent, we have from (5.68) the relation

$$PQ = QP = I$$

which shows that Q is the inverse of P and P is nonsingular in (5.67).

The proof of the part in parenthesis of the Lemma follows similarly by considering the transpose of A and B. □

Sometimes it is important to know whether two given sets of vectors in V_n span the same space or not; if the answer is yes, then find a basis for the space. Making use of Gaussian Elimination, a solution to the above problem can be given easily. To this end, let A and C be formed respectively by the two given sets and subject the compound matrix $[A \mid C]$ to Gaussian Elimination. If $[A \mid C] \sim [B \mid 0]$, where $A \sim B$ (here ~ stands for column equivalence, see Definition 4.4, which is obtained with restricted column operations (iii), see Definition 4.2, as defined in Gaussian Elimination method in this section), then columns of C are linearly dependent on those of A. We now have the following theorem.

Theorem 5.28 Two sets of vectors forming the columns of A and C span the same vector space iff

$$\text{(a) rank } A = \text{rank } C \text{ and (b) } [A \,\vdots\, C] \sim [B \,\vdots\, 0], \tag{5.69}$$

where $A \sim B$.

Proof Without any loss of generality, assume that each of A and C has full column rank.

(a) *Necessity* Suppose that the columns of A and C span the same space. Then by Lemma 5.7, there exists a nonsingular matrix P such that $AP = C$. Hence, in view of Property III of Section 4.7, rank A = rank C. Also,

$$[A \,\vdots\, C]\left[\begin{array}{c|c} I & -P \\ \hline 0 & I \end{array}\right] = [A \,\vdots\, 0].$$

So (5.69b) holds as well.

(b) *Sufficiency* Now suppose (5.69) is true. Then by Theorem 5.9, it follows from (5.69a) that A and C have the same number of columns which are linearly independent. From (5.69b) it follows that

$$[A \,\vdots\, C]\left[\begin{array}{c|c} P & -R \\ \hline 0 & I \end{array}\right] = [B \,\vdots\, 0], \tag{5.70}$$

where P is nonsingular and R is a matrix of appropriate order. From (5.70) we therefore have $AR = C$ and, by Sylvester's inequality, R is nonsingular because A and C have the same number of independent columns. This shows that the columns of A and C span the same space. □

Example 5.39 Show that in $V_3 = \mathbb{R}^3$ the set of

$$\mathbf{x}_1 = [3\ 1\ 0]^T, \quad \mathbf{x}_2 = [0\ 2\ 3]^T, \quad \mathbf{x}_3 = [1\ 1\ 1]^T$$

spans the same vector space as the set

$$[-1\ \ -3\ \ -4]^T, \quad [7\ \ 11\ \ 13]^T, \quad [6\ \ 8\ \ 9]^T, \quad [3\ \ -11\ \ -18]^T.$$

Solution By Gaussian elimination, it is easy to see that

$$\bar{A} := \begin{bmatrix} 3 & 0 & 1 \\ 1 & 2 & 1 \\ 0 & 3 & 1 \end{bmatrix} \sim \begin{bmatrix} 3 & 0 & 0 \\ 1 & 2 & 0 \\ 0 & 3 & 0 \end{bmatrix},$$

$$\bar{C} := \begin{bmatrix} -1 & 7 & 6 & 3 \\ -3 & 11 & 8 & -11 \\ -4 & 13 & 9 & -18 \end{bmatrix} \sim \begin{bmatrix} -1 & 0 & 0 & 0 \\ -3 & -10 & 0 & 0 \\ -4 & -15 & 0 & 0 \end{bmatrix}.$$

Hence, the first two columns of $\bar{A}$ as well as $\bar{C}$ are linearly independent. Choose

$$A := \begin{bmatrix} 3 & 0 \\ 1 & 2 \\ 0 & 3 \end{bmatrix}, \qquad C := \begin{bmatrix} -1 & 7 \\ -3 & 11 \\ -4 & 13 \end{bmatrix}$$

and using restricted column operation [Definition 4.2(iii) with restriction], it can be easily seen that

$$[A \,\vdots\, C] \sim [A \,\vdots\, 0]$$

and rank A = rank C. Hence the two given sets span the same vector subspace S_2 of dimension 2 in V_3. Moreover, if we choose $\mathbf{x}_1$ and $\mathbf{x}_2$ as the basis of S_2, then we can show that $3\mathbf{x}_3 = \mathbf{x}_1 + \mathbf{x}_2$. The vector subspace S_2 is contained in the plane π through the straight lines OP (representing $\mathbf{x}_1$) and OQ (representing $\mathbf{x}_2$). The vector $\mathbf{x}_3$ lies in π and $3\mathbf{x}_3 = \overrightarrow{OT} = \overrightarrow{OP} + \overrightarrow{OQ}$, as shown in Figure 5.9. □

5.8.2 RC (or RREF) Method (Based on Row-equivalent Canonical Form)

By the given set of vectors $\mathbf{x}_1, \mathbf{x}_2, ..., \mathbf{x}_n$ in V_m, we form the matrix A as in (5.56) which is then transformed into RC form B (see Section 4.6). Obviously, the elementary m-vectors $\mathbf{e}_1, \mathbf{e}_2, ..., \mathbf{e}_r$ in the columns $j_1, j_2, ..., j_r$ of B are linearly independent and the dependent vectors can be expressed as a linear combination of these independent vectors merely by eye inspection. So, similarly numbered columns of A are linearly independent by Lemma 5.5, and the remaining columns in A can be expressed by an identical LC (in respect of B's) of the independent vectors of A (see Proof (a) of Lemma 5.5).

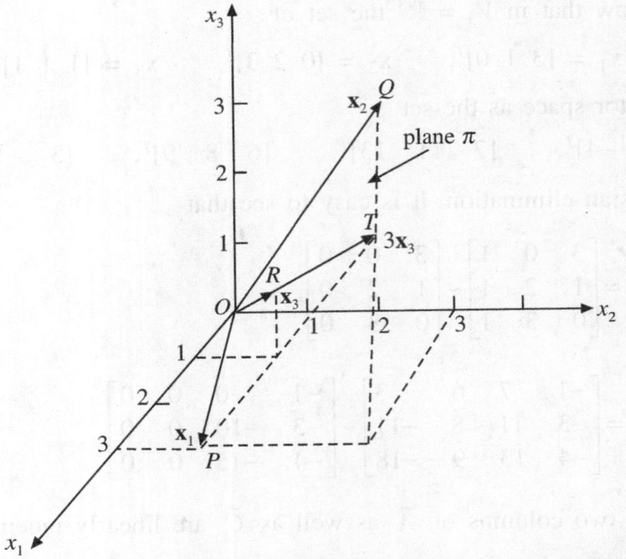

Figure 5.9 Plane π which is represented by *OPTQ* is a vector subspace in R^3.

Example 5.40 Choose a basis of the vector space generated by the vectors

$$\mathbf{x}_1 = [1 \quad 2 \quad 1]^T, \qquad \mathbf{x}_2 = [2 \quad 1 \quad 1]^T, \qquad \mathbf{x}_3 = [3 \quad 3 \quad 2]^T,$$
$$\mathbf{x}_4 = [1 \quad 1 \quad 1]^T, \qquad \mathbf{x}_5 = [5 \quad 4 \quad 3]^T,$$

and express the dependent vectors in terms of the basis vectors.

Solution We have

$$\begin{array}{cc} & \begin{array}{ccccc} \mathbf{x}_1 & \mathbf{x}_2 & \mathbf{x}_3 & \mathbf{x}_4 & \mathbf{x}_5 \end{array} \quad \begin{array}{ccccc} \mathbf{y}_1 & \mathbf{y}_2 & \mathbf{y}_3 & \mathbf{y}_4 & \mathbf{y}_5 \end{array} \\ A := & \begin{bmatrix} 1 & 2 & 3 & 1 & 5 \\ 2 & 1 & 3 & 1 & 4 \\ 1 & 1 & 2 & 1 & 3 \end{bmatrix} \sim \begin{bmatrix} 1 & 0 & 1 & 0 & 1 \\ 0 & 1 & 1 & 0 & 2 \\ 0 & 0 & 0 & 1 & 0 \end{bmatrix} =: B. \end{array}$$

In the light of Lemma 5.5, we now draw the following conclusions by inspecting the columns of B:

(i) $\mathbf{x}_4$ is linearly independent of all other vectors, since $\mathbf{y}_4$ is so in B.
(ii) $\mathbf{x}_3 = \mathbf{x}_1 + \mathbf{x}_2$, $\mathbf{x}_5 = \mathbf{x}_1 + 2\mathbf{x}_2$, since $\mathbf{y}_3 = \mathbf{y}_1 + \mathbf{y}_2$ and $\mathbf{y}_5 = \mathbf{y}_1 + 2\mathbf{y}_2$.
(iii) $\mathbf{x}_1$, $\mathbf{x}_2$, and $\mathbf{x}_4$ form a basis in V_3.
(iv) From (ii), since $\mathbf{x}_1 = 2\mathbf{x}_3 - \mathbf{x}_5$, $\mathbf{x}_2 = -\mathbf{x}_3 + \mathbf{x}_5$, we can replace $\mathbf{x}_1$ and $\mathbf{x}_2$ by $\mathbf{x}_3$ and $\mathbf{x}_5$ so that $\mathbf{x}_3$, $\mathbf{x}_4$ and $\mathbf{x}_5$ form also a basis of V_n.
(v) No other combination is a basis. □

Example 5.41 Determine a basis for: (a) the range space, and (b) the null space of A given by

$$A = \begin{bmatrix} 1 & 2 & 3 & 1 & 5 \\ 2 & 1 & 3 & 1 & 4 \\ 1 & 1 & 2 & 1 & 3 \end{bmatrix}.$$

Solution (a) The basis for the range space can be computed as in Example 5.40 from which it is apparent that rank $A = 3$.

(b) To compute the basis of null space of A, we follow the constructive necessity proof of Theorem 5.16. So, we interchange the third and fourth columns of A to get

$$AT = \begin{bmatrix} 1 & 2 & 1 & 3 & 5 \\ 2 & 1 & 1 & 3 & 4 \\ 1 & 1 & 1 & 2 & 3 \end{bmatrix},$$

where

$$T = \begin{bmatrix} 1 & \cdot & \cdot & \cdot & \cdot \\ & 1 & \cdot & \cdot & \cdot \\ & & \cdot & 1 & \cdot \\ & & 1 & \cdot & \cdot \\ & & & & 1 \end{bmatrix}$$

so that we can define a nonsingular C, C^{-1} and a matrix D by

$$C = \begin{bmatrix} 1 & 2 & 1 \\ 2 & 1 & 1 \\ 1 & 1 & 1 \end{bmatrix}, \quad C^{-1} = \begin{bmatrix} 0 & 1 & -1 \\ 1 & 0 & -1 \\ -1 & -1 & 3 \end{bmatrix} \text{ and } D = \begin{bmatrix} 3 & 5 \\ 3 & 4 \\ 2 & 3 \end{bmatrix}.$$

We next define

$$[\mathbf{b}_1 \ \mathbf{b}_2] = -C^{-1} D = \begin{bmatrix} -1 & -1 \\ -1 & -2 \\ 0 & 0 \end{bmatrix}$$

and from the proof (necessity) of Theorem 5.16,

$$\mathbf{y}_1 = [-1 \ \ -1 \ \ 0 \ \ 1 \ \ 0]^T \text{ and } \mathbf{y}_2 = [-1 \ -2 \ \ 0 \ \ 0 \ \ 1]^T$$

form a basis for the null space of AT. Hence, the vectors

$$T\mathbf{y}_1 = [-1 \ \ -1 \ \ 1 \ \ 0 \ \ 0]^T \text{ and } T\mathbf{y}_2 = [-1 \ \ -2 \ \ 0 \ \ 0 \ \ 1]^T$$

form a basis for the null space of A. □

We shall now describe how, based on the RC method, we can check whether or not the two given sets of vectors generate the same subspace. To begin with, let us present the following lemma.

Lemma 5.8 If A is row-equivalent (resp column equivalent) to B, then the rows (resp columns) of A span the same subspace as the rows (resp columns) of B.

Proof We shall not prove that part of the Lemma which is given in parentheses because this is similar to the proof given below.

We denote by S_a the subspace generated by the rows of A and by S_b that generated by the rows of B. If A is row-equivalent to B, then there exists a nonsingular matrix P such that (see Definition 4.4) $PA = B$, which means that each row of B is an LC of the rows in A. Hence $S_b \subseteq S_a$. Also, we have $A = P^{-1}B$, which shows that each row of A is an LC of the rows of B. So $S_a \subseteq S_b$. From these two set inclusions, it follows easily that $S_a = S_b$. □

Theorem 5.29 Let S be the subspace generated by the rows of a matrix A, and let C be its RC form. Then, (a) the nonzero rows of C form a basis of S, and (b) the rows of another matrix B with RC form D span the same subspace S iff the nonzero rows of D constitute the same basis as the nonzero rows of C.

(a) From the structure of the RC form C, its nonzero rows are linearly independent (see Section 4.6). Moreover, by Lemma 5.8, the rows of A span the same space S as the nonzero rows of C, which therefore, form a basis of S (see Definition 5.5).

(b) *Necessity.* Since A and B span the same subspace we have, denoting by ~ row-equivalence,

$$X := \begin{bmatrix} A \\ B \end{bmatrix} \sim \begin{bmatrix} A \\ 0 \end{bmatrix} \sim \begin{bmatrix} C \\ 0 \end{bmatrix}, \tag{5.71a}$$

$$Y := \begin{bmatrix} B \\ A \end{bmatrix} \sim \begin{bmatrix} B \\ 0 \end{bmatrix} \sim \begin{bmatrix} D \\ 0 \end{bmatrix}, \tag{5.71b}$$

where zeros are zero matrices of appropriate order. This shows that the RC forms of X and Y are, respectively, $[C^T\ 0]^T$ and $[D^T\ 0]^T$. Since Y can be obtained from X by an interchange of rows, we can write $X = QY$ for some nonsingular matrix Q (see Section 4.2). For the row-equivalence shown in (5.7la), we can find a nonsingular matrix P such that

$$PX = \begin{bmatrix} C \\ 0 \end{bmatrix} = PQY$$

which shows that $[C^T\ 0]^T$ is also an RC form of Y. Since by Theorem 4.1 the RC form of a matrix is unique, $C = D$.

Sufficiency is obvious. □

Remark 5.5 The foregoing theorem shows that if we have two sets of vectors represented by the rows of A and B, respectively, and if the nonzero rows in the RC form of A are the same as those of the RC form of B, then the subspaces generated by the above two sets are identical. This implies that, whatever be the spanning set for a subspace S, the same basis can always be determined.

Example 5.42 Show that in $V_3 = \mathbb{R}^3$ the set of vectors $[3\ 0\ 1]^T$, $[1\ 2\ 1]^T$, $[0\ 3\ 1]^T$ does not span the same space as the set

$$[7\ \ 2\ \ 13]^T \qquad [-1\ \ 7\ \ -4]^T \qquad [2\ \ 3\ \ 3]^T \qquad [1\ \ 27\ \ -6]^T$$

Solution We now construct a matrix A with rows as the first set of vectors and another matrix B with rows as the second set of vectors. These matrices A and B are

$$A = \begin{bmatrix} 3 & 0 & 1 \\ 1 & 2 & 1 \\ 0 & 3 & 1 \end{bmatrix} \qquad B = \begin{bmatrix} 7 & 2 & 13 \\ -1 & 7 & -4 \\ 2 & 3 & 3 \\ 1 & 27 & -6 \end{bmatrix}$$

which have respectively the RC forms C and D given by

$$A \sim C = \begin{bmatrix} 1 & 0 & 1/3 \\ 0 & 1 & 1/3 \\ 0 & 0 & 0 \end{bmatrix}, \qquad B \sim D = \begin{bmatrix} 1 & 0 & 33/17 \\ 0 & 1 & -5/17 \\ 0 & 0 & 0 \\ 0 & 0 & 0 \end{bmatrix}.$$

As the nonzero rows in these RC forms are different, the two given sets span different subspaces in $\mathbb{R}^3$. □

Sometimes, instead of requiring to pick up all the independent vectors in a set and to express the dependent vectors in terms of the independent vectors, one may be supplied with a set of independent vectors in V_m and then be asked to express a given vector in terms of the independent ones. Such problems can obviously be formulated befitting Theorem 5.27. Nonetheless, an alternative but direct attack on such questions more often than not pays a high dividend.

Let $\mathbf{x}_1 \cdots \mathbf{x}_n$ be n independent vectors in V_m, $n \le m$. We have to express an unknown vector $\mathbf{x}$ in V_m in terms of these $\mathbf{x}_i$'s. In other words, determine scalars $\alpha_1 \cdots \alpha_n$ satisfying $\alpha_1\mathbf{x}_1 + \cdots + \alpha_n\mathbf{x}_n = \mathbf{x}$, i.e.

$$[\mathbf{x}_1 \cdots \mathbf{x}_n] \begin{bmatrix} \alpha_1 \\ \vdots \\ \alpha_n \end{bmatrix} = \mathbf{x}, \tag{5.72}$$

the solution of which, if it exists, can be obtained by applying Theorems 5.10–5.13 and 5.15 or by Gaussian Elimination (see Examples 5.30 and 5.31).

Example 5.43 Find the coordinates (see Definition 5.5) of the vector $\mathbf{x} = [5, -1, 2]^T$ with respect to the basis $\mathbf{x}_1 = [1\ 4\ 2]^T$, $\mathbf{x}_2 = [4\ 2\ 1]^T$, $\mathbf{x}_3 = [2\ 1\ 3]^T$.

Solution Let α_1, α_2, α_3 be the coordinates of $\mathbf{x}$ with respect to the basis $\mathbf{x}_1$, $\mathbf{x}_2$, $\mathbf{x}_3$. Then (5.72) becomes

$$\begin{bmatrix} 1 & 4 & 2 \\ 4 & 2 & 1 \\ 2 & 1 & 3 \end{bmatrix} \begin{bmatrix} \alpha_1 \\ \alpha_2 \\ \alpha_3 \end{bmatrix} = \begin{bmatrix} 5 \\ -1 \\ 2 \end{bmatrix}$$

which, when solved by determining the inverse of the coefficient matrix (see Example 3.8), gives

$$\begin{bmatrix} \alpha_1 \\ \alpha_2 \\ \alpha_3 \end{bmatrix} = \begin{bmatrix} 1 & 4 & 2 \\ 4 & 2 & 1 \\ 2 & 1 & 3 \end{bmatrix}^{-1} \begin{bmatrix} 5 \\ -1 \\ 2 \end{bmatrix}$$

$$= \frac{1}{35} \begin{bmatrix} -5 & 10 & 0 \\ 10 & 1 & -7 \\ 0 & -7 & 14 \end{bmatrix} \begin{bmatrix} 5 \\ -1 \\ 2 \end{bmatrix} = \begin{bmatrix} -1 \\ 1 \\ 1 \end{bmatrix}.$$ □

5.9 MATLAB METHODS IN VECTOR SPACES

MATLAB methods in vector spaces are outlined below under the following six cateogries:

Case I We are given a set of linearly independent (LI) vectors $\mathbf{x}_1, \mathbf{x}_2, ..., \mathbf{x}_n$ in V_m; we form a matrix $A = [\mathbf{x}_1, \mathbf{x}_2, ..., \mathbf{x}_n]$ with the given vectors as columns. To express any other vector $\mathbf{x}$ in V_m as a linear combination (LC) of these vectors, we obtain the reduced row echelon form (RREF) of $B = [A\ \mathbf{x}]$, using the MATLAB command rref (B). The elements in the last column of rref (B) which corresponds to the linearly dependent (LD) vector are the coefficients of the linear combination (see Example 5.40). The vectors $\mathbf{x}_i$'s are LI if rref (A) has the structure $[I_r]$ or $\begin{bmatrix} I_r \\ 0 \end{bmatrix}$.

Example 5.44 Express $\mathbf{x} = [10\ 2\ 3]^T$ as an LC of the independent vectors $\mathbf{x}_1 = [2\ 4\ 1]^T$, $\mathbf{x}_2 = [1\ 3\ 2]^T$, $\mathbf{x}_3 = [4\ 1\ 4]^T$.

Solution Let

$$\begin{array}{cc} & \begin{array}{cccc} \mathbf{x}_1 & \mathbf{x}_2 & \mathbf{x}_3 & \mathbf{x} \end{array} \\ A = & \begin{bmatrix} 2 & 1 & 4 & 10 \\ 4 & 3 & 1 & 2 \\ 1 & 2 & 4 & 3 \end{bmatrix}. \end{array}$$

Then

```
>> rref (A)

ans =

     1     0     0     3
     0     1     0    -4
     0     0     1     2
```

Hence, the elements in the last column are the coefficients of the LC, so that

$$\mathbf{x} = 3\mathbf{x}_1 - 4\mathbf{x}_2 + 2\mathbf{x}_3.$$

Case II If it is required to pick up the LI vectors from the set $\mathbf{x}_1, \mathbf{x}_2, ..., \mathbf{x}_n$, then the matrix $A = [\mathbf{x}_1, \mathbf{x}_2, ..., \mathbf{x}_n]$ is transformed to RREF and the vectors $\mathbf{x}_j$, whose corresponding columns in RREF are elementary vectors $\mathbf{e}_i$, are LI. The remaining vectors $\mathbf{x}_k$, $k \neq j$, are LD. The coefficients in the column in RREF corresponding to $\mathbf{x}_k$ are the coefficients of LC for $\mathbf{x}_k$ in terms of LI vectors preceding it.

Example 5.45 Find the vectors which are LD in the set of columns of

$$\begin{array}{cc} & \begin{array}{ccccc} \mathbf{x}_1 & \mathbf{x}_2 & \mathbf{x}_3 & \mathbf{x}_4 & \mathbf{x}_5 \end{array} \\ A = & \begin{bmatrix} -1 & 2 & 1 & 2 & 1 \\ 2 & -4 & -3 & 1 & 2 \\ 1 & -2 & 1 & 3 & 6 \\ 3 & -6 & 1 & 1 & 8 \end{bmatrix} \end{array}$$

and express them as LC of the independent vectors.

Solution The RREF of A is

```
>> rref (A)
ans =

   1  -2   0   0   2
   0   0   1   0   1
   0   0   0   1   1
   0   0   0   0   0
```

Hence, $\mathbf{x}_1$, $\mathbf{x}_3$ and $\mathbf{x}_4$ are LI, and $\mathbf{x}_2$ and $\mathbf{x}_5$ are LD, which are expressed as LC of LI vectors by

$$\mathbf{x}_2 = -2\mathbf{x}_1, \qquad \mathbf{x}_5 = 2\mathbf{x}_1 + \mathbf{x}_3 + \mathbf{x}_4.$$

Case III MATLAB command Z = null (A) computes an orthonormal basis, which is the set of columns in Z, for the null space of A obtained from a singular value decomposition so that $AZ = 0$, a null matrix, and $Z'Z = I$, an identity matrix. On the other hand, Z = null (A, 'r') computes a rational basis for the null space obtained from the reduced row echelon form. The command size (Z, 2) gives the nullity of A.

Example 5.46 Let a matrix A be described by

$$A = \begin{matrix} \mathbf{x}_1 & \mathbf{x}_2 & \mathbf{x}_3 & \mathbf{x}_4 \end{matrix} \\ \begin{bmatrix} 2 & 3 & 1 & 1 \\ 1 & -1 & 3 & -2 \\ 1 & 0 & 2 & -1 \\ 3 & 5 & 1 & 2 \end{bmatrix}.$$

Determine the null space of A in orthonormal as well as rational basis.

Solution With the statement

```
>> Z = null (A)
Z =

    0.8159    0.0323
   -0.4241    0.3918
   -0.3918   -0.4241
    0.0323   -0.8159
```

the columns of which are orthonormal and generate the null space of A, so that $Z'Z = I_2$. On the other hand,

```
>> Z = null (A, 'r')
Z =
   -2    1
    1   -1
    1    0
    0    1
```

the columns of which are not orthonormal and generate the null space of A. As

```
>> rref (A)
```

responds with

```
ans =

     1   0   2  -1
     0   1  -1   1
     0   0   0   0
     0   0   0   0
```

$\mathbf{x}_1$ and $\mathbf{x}_2$ are LI and form the basis of range space of A. Moreover, it is easily concluded that $\mathbf{x}_3 = 2\mathbf{x}_1 - \mathbf{x}_2$ and $\mathbf{x}_4 = -\mathbf{x}_1 + \mathbf{x}_2$.

Case IV If the vectors to be tested for linear independence are polynomials, then associate with each polynomial a column of coefficients. Any missing terms in the polynomial are associated with a zero coefficient. The coefficient of the highest-power term is taken as the first entry of the column, the coefficient of the next-highest-power term as the second entry and so on. The linear independence and linear combination problems are now solved as in Cases I and II.

Example 5.47 Test if the following polynomials are LI:

$$t^2 + 2t + 2, \qquad 2t^2 + 1, \qquad t^2 + 3t.$$

Solution The polynomials $t^2 + 2t + 2$, $2t^2 + 1$ and $t^2 + 3t$ correspond respectively to the columns of

$$A = \begin{bmatrix} 1 & 2 & 1 \\ 2 & 0 & 3 \\ 2 & 1 & 0 \end{bmatrix}.$$

```
>> rref (A)
ans =
     1   0   0
     0   1   0
     0   0   1
```

which shows that the given polynomials are LI.

Case V If the vectors are $m \times n$ matrices, N in number, then with each such matrix AJ, a column vector BJ is formed by stringing together its columns one after the other. In MATLAB this transformation is done by using

```
>> BJ = reshape(AJ, m * n, 1)
```

J = 1, 2, ..., N. Then the command rref is used:

```
>> rref ([B1 B2 ··· BN])
```

If the response of RREF is of the form $[I_r]$ or $[I_r \mid 0]^T$ then the set of matrices are LI, otherwise not.

Example 5.48 Test if the following set of four 2×2 matrices are LI or not:

$$A_1 = \begin{bmatrix} 1 & 2 \\ 1 & 1 \end{bmatrix}, \quad A_2 = \begin{bmatrix} 2 & 3 \\ 1 & 2 \end{bmatrix}, \quad A_3 = \begin{bmatrix} 2 & 1 \\ 1 & 1 \end{bmatrix}, \quad A_4 = \begin{bmatrix} 3 & 2 \\ 1 & 1 \end{bmatrix}.$$

Solution With commands

```
>> A = [reshape(A1, 4, 1) reshape(A2, 4, 1) reshape(A3, 4, 1) reshape(A4, 4, 1)]
A =

     1   2   2   3
     1   1   1   1
     2   3   1   2
     1   2   1   1

>> rref (A)
ans =

     1   0   0   0
     0   1   0   0
     0   0   1   0
     0   0   0   1
```

and from the above results we conclude that the set of four 2×2 matrices are LI as rref (A) = I_4.

Case VI The general solution $\mathbf{x}$ to a system of linear equations $A\mathbf{x} = \mathbf{b}$ is the sum of (i) $Z_\mathbf{q}$ where Z = null (A) or null (A, 'r') and $\mathbf{q}$ is a vector with a number of arbitrary parameters equal to the number of columns of Z, i.e. nullity (A); and (ii) a particular solution $\mathbf{p}$ to the non-homogeneous system $A\mathbf{x} = \mathbf{b}$ implying $A\mathbf{p} = \mathbf{b}$. Therefore, a general solution is $\mathbf{x} = \mathbf{p} + Z_\mathbf{q}$. Indeed, $A\mathbf{x} = A\,(\mathbf{p} + Z_\mathbf{q}) = A\mathbf{p} = \mathbf{b}$ since AZ is a null matrix.

Example 5.49 Solve $A\mathbf{x} = \mathbf{b}$ if

$$A = \begin{array}{c} \begin{matrix} \mathbf{x}_1 & \mathbf{x}_2 & \mathbf{x}_3 & \mathbf{x}_4 \end{matrix} \\ \begin{bmatrix} -1 & 2 & 1 & -1 \\ 2 & -4 & -3 & 1 \\ 1 & -2 & 1 & 3 \\ 3 & -6 & 1 & 7 \end{bmatrix} \end{array}, \quad \mathbf{b} = \begin{bmatrix} 1 \\ -5 \\ 5 \\ 9 \end{bmatrix}.$$

Solution Using MATLAB

```
>> Z = null (A, 'r')
Z =
     2  -2
     1   0
     0  -1
     0   1
```

```
>> rref ([A  b])
ans =

   1  -2   0   2   2
   0   0   1   1   3
   0   0   0   0   0
   0   0   0   0   0
```

which shows that $\mathbf{b} = 2\mathbf{x}_1 + 3\mathbf{x}_3$, as a result $\mathbf{p} = [2\ 0\ 3\ 0]^T$ is a particular solution. Therefore,

$$\mathbf{x} = \begin{bmatrix} 2 \\ 0 \\ 3 \\ 0 \end{bmatrix} + \alpha \begin{bmatrix} 2 \\ 1 \\ 0 \\ 0 \end{bmatrix} + \beta \begin{bmatrix} -2 \\ 0 \\ -1 \\ 1 \end{bmatrix}$$

is a general solution, α and β being arbitrary parameters.

PROBLEMS

Wherever a prime occurs in the following problems, it denotes a transpose (see Definition 1.8).

5.1 Show that the following sets of vectors form a vector space over a field $\mathbb{C}$:

(a) The set of n-vectors $[x_1, x_2, ..., x_n]^T$ over $\mathbb{C}$.
(b) The set of vectors $1, x, x^2, ..., x^n$; x is a complex variable.
(c) The set of vectors $1, 1 + x, x^2, ..., x^n$; x is a complex variable.
(d) The set consisting of the single element 0.
(e) The set of all $n \times m$ matrices over $\mathbb{C}$.

5.2 Using the axioms of a vector space, prove that:

(a) $\boldsymbol{\theta} + \mathbf{x} = \mathbf{x}$, (b) $\mathbf{x} + \mathbf{z} = \mathbf{y} + \mathbf{z}$ implies $\mathbf{x} = \mathbf{y}$,
(c) $\alpha \cdot \boldsymbol{\theta} = \boldsymbol{\theta}$, (d) $0 \cdot \mathbf{x} = \boldsymbol{\theta}$,
(e) $(-1)\ \mathbf{x} = -\mathbf{x}$, (f) $\alpha\ \mathbf{x} = \boldsymbol{\theta}$ implies $\alpha = \boldsymbol{\theta}$ and/or $\mathbf{x} = \Theta$,
(g) $\alpha\ \mathbf{x} = \beta\ \mathbf{x}$ and $\mathbf{x} \neq \boldsymbol{\theta}$ implies $\alpha = \beta$, and
(h) $\alpha\ \mathbf{x} = \alpha\ \mathbf{y}$ and $\alpha \neq 0$ implies $\mathbf{x} = \mathbf{y}$.

5.3 Prove that the following sets of vectors over $\mathbb{F}$ form a vector space over the same field by showing that all the axioms in Definition 5.1 are satisfied:

(a) The set of vectors $[x_1, x_2]'$ such that $x_1 + x_2 = 0$.
(b) The set of vectors $[x_1, x_2, x_3]'$ such that $x_3 = 0$.
(c) The set of vectors $[x_1, x_2, x_3, x_4]'$ such that $x_1 = x_3$, $x_2 = x_4$.

5.4 Prove that the following sets of vectors over a field $\mathbb{F}$ do not form a vector space over the same field by showing which of the axioms in Definition 5.1 are not satisfied:

(a) The set of vectors $[x_1, x_2]'$ such that $x_2 = 1$.
(b) The set of vectors $[x_1, x_2]'$ such that x_1 and x_2 are integers.
(c) The set of vectors $[x_1, x_2]'$ such that $x_1 + x_2 = 1$.

5.5 Find t, for which the following vectors are linearly independent:

(a) $[\cos t, \sin t]'$, $[-\sin t, \cos t]'$,

(b) $[\cos t, \sin t]'$, $[\sin t, \cos t]'$,

(c) $[e^{\alpha t}, \alpha e^{\alpha t}]'$, $[e^{\beta t}, \beta e^{\beta t}]'$.

5.6 Prove that all the basis in a finite dimensional vector space have the same number of vectors.

5.7 (a) Prove that the following sets of vectors form a basis:

(i) $[1, -1\ 0]'$, $[0\ 1\ -1]'$, $[0\ 0\ 1]'$ in $\mathbb{R}^3$.

(ii) $[1\ 1\ 1]'$, $[0\ 1\ 1]'$, $[0\ 0\ 1]'$ in $\mathbb{R}^3$.

(iii) $[1\ 1\ \ldots\ 1]'$, $[0\ 1\ \ldots\ 1]'$, ..., $[0\ 0\ \ldots\ 0\ 1]'$ in $\mathbb{C}^n$.

(iv) The set of vectors forming the columns of an upper (or a lower) triangular matrix of order n in $\mathbb{C}^n$, with nonzero diagonal elements.

(v) The elementary n-vectors $\mathbf{e}_i$, $i = 1, 2, ..., n$ in $\mathbb{C}^n$.

(b) Prove that any vector $[a_1\ a_2\ a_3]'$ cannot be represented in terms of the vectors $\mathbf{x}_1 = [1\ 1\ 1]$, $\mathbf{x}_2 = [0\ 1\ 2]$, $\mathbf{x}_3 = [0\ 2\ 4]$, and that $\mathbf{x}_i$'s do not form a basis in $\mathbb{R}^3$.

5.8 In $\mathbb{C}^3$ find a basis containing (a) the elementary vector $\mathbf{e}_i$, (b) the vectors $[1\ 1\ 0]'$, $[1\ 2\ 1]'$, and (c) the vectors $[i\ 1\ 0]'$, $[1, -i\ 1]'$.

5.9 Show that the set of vectors $\mathbf{x}_1 = [1\ 3\ 3]'$, $\mathbf{x}_2 = [-1\ 1\ 2]'$, $\mathbf{x}_3 = [1\ 1\ 1]'$ forms a basis in $\mathbb{R}^3$. With respect to this basis, find the representation (see Definition 5.5) and the coordinates of the following vectors:

(a) $[2\ 2\ 3]'$, (b) $[-1\ 0\ -1]'$, (c) $[5\ 7\ 2]'$, (d) $[2i, 2(i + 1), 2(i - 1)]'$, (e) $[1\ 0\ 0]'$.

5.10 Show that if in $\mathbb{R}^2$, $P \equiv (x_1, x_2)$ is a point with respect to the elementary basis $\mathbf{e}_1$, $\mathbf{e}_2$, then $(x_1 \cos \theta + x_2 \sin \theta, -x_1 \sin \theta + x_2 \cos \theta)$ is the same point with respect to the basis $[\cos \theta\ \ \sin \theta]'$, $[-\sin \theta\ \ \cos \theta]'$.

5.11 Prove that the following sets of vectors over $\mathbb{R}$ form a vector subspace of $\mathbb{R}^3$ and give a geometric construction of the subspace and the vector space:

(a) The set of vectors $[x_1, x_2, x_3]$ such that $x_1 + x_2 + x_3 = 0$.

(b) The set of vectors $[x_1, x_2, x_3]$ such that $x_2 = 0$.

(c) The set of vectors $[x_1, x_2, x_3]$ such that $x_1 = x_2$.

5.12 Determine the dimension r of the vector subspace S_r of V_3 generated by the following vectors (see Example 5.39) and give, if possible, geometric construction of the subspaces:

(a) $[1\ -1\ 1]'$,$[-1\ 1\ -1]'$, $[2\ -2\ 2]'$,

(b) $[1\ 2\ 3]'$, $[2\ 3\ 5]'$, $[1\ 1\ 2]'$,

(c) $[1\ 0\ 1]'$, $[1 + i\ i\ 1 + 2i]'$, $[-i\ 1 + 2i\ 1 + i]'$,

(d) $[1\ 2\ 1]'$, $[-1\ -2\ -1]'$, $[1\ 3\ 3]'$, $[2\ 3\ 2]'$,

(e) $[-1\ 1\ 1]'$, $[2\ 1\ 2]'$, $[3\ 3\ 1]'$.

5.13 Compute (a) $\langle \mathbf{x}, \mathbf{y} \rangle$, (b) $\langle \mathbf{y}, \mathbf{x} \rangle$, (c) $\| \mathbf{x} \|$, $\| \mathbf{y} \|$, (d) $\langle \alpha\mathbf{x}, \beta\mathbf{y} \rangle$, (e) $\| \mathbf{x} - \mathbf{y} \|$, where

(i) $\mathbf{x} = [2, -1, 1, -2]'$, $\mathbf{y} = [1\ -1\ 1\ -1]'$, $\alpha = 1 + i$, $\beta = 1 - i$

(ii) $\mathbf{x} = [i\ 1\ -1]'$, $\mathbf{y} = [1\ -i\ 1]$, $\alpha = 1$, $\beta = 2i$

(iii) $\mathbf{x} = [1\ -1\ 2]'$, $\mathbf{y} = [1\ -i\ 1 + i\ 0]$, $\alpha = 1 + i$, $\beta = 1$.

5.14 Find a vector orthogonal to the vectors

(a) $[1\ 1\ 0]'$, $[1\ -2\ 1]'$

(b) $[i\ 1\ -1]'$, $[1\ -i\ 1]'$

(c) $[1\ 3\ 1]'$, $[2\ 1\ 1]'$

[Hint: Use (3.3), where A has the first two columns as the given vectors to determine the last row of R].

5.15 Resolve the vector y into two perpendicular components, one along **x**, where

(a) $\mathbf{x} = [2\ 1\ 2]$, $\mathbf{y} = [1\ 2\ 1]$;

(b) $\mathbf{x} = [1\ 2\ 0\ 1]$, $\mathbf{y} = [1\ 2\ 2\ 4]$;

(c) $\mathbf{x} = [i\ -1\ i\ 1]$, $\mathbf{y} = [1, i + 1\ 0]$;

(d) $\mathbf{x} = [1\ 2 + i\ i\ 1\ -i]$, $\mathbf{y} = [1 + i\quad 2 + i,\quad -5, -6]$;

(e) $\mathbf{x} = [1\ 2\ 1\ 2]$, $\mathbf{y} = [2\ -3\ 0\ 2]$.

5.16 Test if the following sets of vectors are linearly independent; if they are, construct an orthonormal set of vectors with the help of them.

(a) $[2\ 1\ 2]$, $[1\ 2\ 1]$;

(b) $[i\ -1\ i\ 1]$, $[1\ i + 1\ 0]$;

(c) $[1\ 2\ 0\ 1]$, $[1\ 2\ 2\ 4]$, $[2, -1, -1, 0]$;

(d) $[1\ 2\ -1\ 0]$, $[1\ 2\ 0\ -1]$;

(e) $[1\quad 2 + i\quad i\quad 1 - i]$, $[1 + i\quad 2 + i, -5, -6]$;

(f) $[1\ 2\ 3\ 4]$, $[1\ -1\ -2\ -3]$, $[5\ 4\ 5\ 6]$.

5.17 (a) If $\mathbf{x}$ is in $\mathbb{C}^n$ and $\langle \mathbf{x}, \mathbf{y} \rangle = 0$ for every $\mathbf{y}$ in $\mathbb{C}^n$, then prove that $\mathbf{x} = \boldsymbol{\theta}$.

(b) If A is an n-square matrix over $\mathbb{C}$ and $\langle A\mathbf{x}, \mathbf{y} \rangle = 0$ for all $\mathbf{x}$ and $\mathbf{y}$ in $\mathbb{C}$, then prove that $A = 0$.

5.18 Show that the following sets of vectors form a basis in V_3 and then find an orthonormal basis using the Gram-Schmidt process:

(a) $[1\ 0\ 0]'$, $[1\ 1\ 0]'$, $[1\ 1\ 1]'$;

(b) $[1\ 0\ 1]'$, $[1\ 1\ 0]'$, $[0\ 1\ 1]'$;

(c) $[1\ 1\ 1]'$, $[1\ 0\ -1]'$,$[1\ -2\ 1]'$;

(d) $[1\ 2\ 1]'$, $[2\ 1\ 2]'$, $[1\ 0\ -1]'$;

(e) $[i\ 0\ 1]'$, $[i\ -1\ i\ 1]'$, $[1\ i + 1\ 0]'$;

(f) $[2i\ 1, -2i]'$, $[1\ 2i\ 0]'$, $[1\ 2\ -1]'$.

5.19 Show that the following sets of vectors form a basis in V_4 and then determine an orthogonal basis:

(a) $[1\ 0\ 0\ 0]'$, $[1\ 1\ 0\ 0]'$, $[1\ 1\ 1\ 0]'$, $[1\ 1\ 1\ 1]'$;

(b) $[1\ -1\ 1\ -1]'$, $[1\ 2\ -1\ 0]'$, $[1\ 2\ 0\ -1]'$, $[1\ 0\ 2\ -1]'$;

(c) $[2\ -1\ -1\ 0]'$, $[1\ 2\ 0\ 1]'$, $[0\ -1\ 0\ 2]'$, $[1\ 2\ 2\ 4]'$;

(d) $[1\ 2\ 0\ 0]'$, $[1\ 0\ 2\ 0]'$, $[1\ 0\ 0\ 2]'$, $[0\ 1\ 2\ 0]'$.

5.20 Use the Gram-Schmidt process and Problem 5.14 to determine an orthonormal basis in $\mathbb{R}^3$ with the help of the following vectors:

(a) $[1\ 1\ 1]'$, $[-1\ 1\ 0]'$;

(b) $[i+1\ 1\ 0]'$, $[i-1\ 0\ 1]'$;

(c) $[2\ 1\ -1]'$, $[i\ i+1\ 1]'$.

5.21 Find the solution of the following simultaneous equations, if consistent, by the Gaussian Elimination method:

(a) $2x_1 + 2x_2 + x_3 = 8,$
$x_1 - x_2 = -2,$
$x_2 + x_3 = 0.$

(b) $x_1 + x_3 = 11,$
$x_2 = -2,$
$2x_1 + x_3 = 13.$

(c) $x_1 + x_2 + x_3 = 2,$
$x_1 + 2x_2 + 3x_3 = 3,$
$2x_1 + 3x_2 + 4x_3 = 4.$

(d) $x_1 + x_2 + x_3 = 2,$
$x_1 + 2x_2 + 3x_3 = 3,$
$2x_1 + 3x_2 + 4x_3 = 5.$

(e) $x_1 + x_2 - x_3 = 2,$
$x_1 + 2x_2 + 3x_3 = 1,$
$x_1 + x_2 + x_3 = 0.$

(f) $2x_1 - x_2 + 3x_3 = 2,$
$x_1 + x_2 = 4,$
$2x_1 + x_2 + x_3 = 6.$

(g) $2x_1 - x_2 = 3,$
$x_1 + 2x_2 + x_3 = 3,$
$-x_1 + x_3 = 3.$

(h) $2x_1 - x_2 + 3x_3 = 5,$
$6x_1 + x_2 - x_3 = 3,$
$2x_1 + x_2 + x_3 = -1.$

(i) $x_1 + 3x_3 = 1,$
$-x_1 + 2x_2 - 2x_3 = 2,$
$x_1 + 3x_3 = 3.$

(j) $x_1 + 2x_2 = 1,$
$3x_1 + x_2 = 2,$
$2x_1 + 4x_2 = 2,$
$x_1 + x_2 = 1.$

(k) $2x_1 + 7x_2 + x_3 + 3x_4 = -1,$
$x_1 + 3x_2 + x_3 + 9x_4 = -2,$
$5x_2 - 6x_4 = 0,$
$2x_1 + 2x_2 = 2.$

(l) $x_1 + 2x_2 - x_3 + 2x_4 = 1,$
$3x_1 + x_3 + 4x_4 = -1,$
$x_1 - x_2 + x_3 + x_4 = -1.$

(m) $3x_1 - 2x_2 + 5x_3 + x_4 = 1,$
$x_1 + x_2 - 3x_3 = 2,$
$2x_1 + x_2 + x_3 + x_4 = -1,$
$x_1 + 3x_3 + 2x_4 = 3.$

(n) $3x_1 + x_2 + x_3 = 3,$
$3x_1 + 2x_2 + 2x_3 = 5,$
$2x_1 - 3x_2 - 2x_3 = 1.$

(o) $x_1 - x_2 + x_3 - x_4 = 2,$
$x_2 + x_4 = -2,$
$x_1 - x_3 = -3,$
$x_2 - x_4 = 4.$

5.22 Solve the simultaneous equations in Problem 5.21 by Cramer's rule, if they are consistent.

5.23 Find a so that $A\mathbf{x} = \boldsymbol{\theta}$, where A is given below, has a nontrivial solution:

(a) $\begin{bmatrix} 1 & 1 & a \\ 1 & a & 1 \\ a & 1 & 1 \end{bmatrix}$, (b) $\begin{bmatrix} 1 & 0 & a \\ 2 & 1 & 0 \\ 3 & 2 & a \end{bmatrix}$, (c) $\begin{bmatrix} 1 & a & 0 & 0 \\ a & 1 & 0 & 0 \\ 1 & -a & 1 & a \\ -a & 1 & a & 1 \end{bmatrix}$.

5.24 Let $\mathbf{x} = A\mathbf{y}$, where A is nonsingular. Then $\mathbf{x}$ is not a zero vector iff $\mathbf{y}$ is not a zero vector.

5.25 If B is not a null matrix and A is nonsingular, then AB is not a null matrix.

5.26 Determine the dimension of range space and nullity of A, where A is given by

$$\text{(a)} \begin{bmatrix} 1 & 2 & -1 & 2 \\ 3 & 0 & 1 & 4 \\ 1 & -1 & 1 & 1 \end{bmatrix}, \quad \text{(b)} \begin{bmatrix} 1 & 2 & 1 & 1 \\ 2 & -3 & 7 & 9 \\ 1 & 4 & -2 & -1 \end{bmatrix}, \quad \text{(c)} \begin{bmatrix} 2 & 3 & 1 & 1 \\ 1 & -1 & 3 & -2 \\ 1 & 0 & 2 & -1 \\ 3 & 5 & 1 & 2 \end{bmatrix}$$

5.27 Determine a basis for the range space and null space of the matrices given in Problem 5.26.

5.28 Prove Sylvester's inequality with the help of Theorems 5.17 and 5.21.

5.29 If the n-square matrix A is *idempotent*, i.e., $A = A^2$, then $\rho(A) + \rho(I - A) = n$.

5.30 If A and B are in $V_{m \times n}$ and $V_{n \times p}$ respectively, then $\rho(A) + \rho(B) \leq n$ if $AB = 0$.

5.31 (a) If $A = A^2$, $AB = BA = 0$, then $\rho(A + B) = \rho(A) + \rho(B)$.

(b) Let A be a matrix of order $n \times p$, $n \leq p$. Show that A is of full rank iff AA^* is nonsingular.

5.32 Prove that the Gramian matrix test cannot ascertain the number of independent vectors in a set.

5.33 Denoting the following vectors by $\mathbf{x}_1$, $\mathbf{x}_2$, $\mathbf{x}_3$ and $\mathbf{x}_4$, respectively, determine a linearly independent set among them and express each of the remaining vectors as a linear combination of this set:

(a) $[1\ 2\ 1]'$, $[2\ 1\ 3]'$, $[1\ 5\ 0]'$, $[2\ {-5},\ 5]'$;

(b) $[2\ 3\ 1\ 0]'$, $[3\ 2\ {-1}\ 1]'$, $[1\ 4\ 3\ {-1}]'$, $[0\ 5\ 5\ {-2}]'$;

(c) $[4\ 2\ 1\ {-2}\ 3]'$, $[2\ 1\ 3\ 2\ {-1}]'$, $[2\ 1\ {-2}\ {-4}\ 4]'$, $[2\ 1\ {-7},\ {-10},\ 9]'$;

(d) $[2\ 5\ 3\ 1]'$, $[4,\ 11,\ 7,\ 2]'$, $[0\ 1\ 1\ 0]'$, $[2\ 4\ 2\ 1]'$.

5.34 Denoting the vectors in each case of (a)–(d) in Problem 5.12 by $\mathbf{x}_1$, $\mathbf{x}_2$, $\mathbf{x}_3$, ..., respectively, determine the set of independent vectors and express each of the remaining vectors as a linear combination of this set, using (a) Theorem 5.27, (b) the Gaussian Elimination method, and (c) the *RC* method.

5.35 A set $\mathbf{x}_j \pm i\mathbf{y}_j$, $i = \sqrt{-1}$, $j = 1, \ldots, n$, of $2n$-vectors in $\mathbb{C}^{2n}$ is linearly independent iff the set $\mathbf{x}_j$, $\mathbf{y}_j$, $j = 1, \ldots, n$, of $2n$-vectors in $\mathbb{R}^{2n}$ is linearly independent.

5.36 Let A be an $m \times n$ matrix with coefficients in a field $\mathbb{F}$. Then: (a) rank $A = n$ iff rank $A^*A = n$ or, equivalently, det $A^*A \neq 0$; (b) rank $A = m$ iff rank $AA^* = m$, or equivalently, det $AA^* \neq 0$.

5.37 Explain why in $\mathbb{R}^3$ the vector $[1\ 1\ 1]^T$ cannot be expressed in terms of $\mathbf{x}_1 = [1\ 2\ 1]^T$, $\mathbf{x}_2 = [1\ 1\ 0]^T$ and $\mathbf{x}_3 = [-1\ 1\ 2]^T$.

5.38 Prove that the determinant of a Gramian matrix is nonnegative.

5.39 Verify that the set of complex numbers $a + ib$ constitutes a vector space over the field of real numbers.

5.40 Show that the set of all $n \times n$ symmetric matrices forms a vector subspace S in $\mathbb{C}^{n\times n}$ and that its dimension is $n(n + 1)/2$. Write down a basis for S when $n = 2$, 3, and 4.

5.41 Show that the set of all $n \times n$ skew-symmetric matrices form a vector subspace S in $\mathbb{C}^{n\times n}$ and that its dimension is $n(n - 1)/2$. Write down a basis for S when (a) $n = 2$ and (b) $n = 3$.

5.42 Show that the set of all $n \times n$ upper (resp lower) triangular matrices forms a vector subspace S in $\mathbb{C}^{n\times n}$ and that its dimension is $n(n + 1)/2$. Exhibit a basis for S when (a) $n = 3$ and (b) $n = 4$.

5.43 Show that the vectors

$$\begin{bmatrix} 0 & 1 \\ 1 & 1 \end{bmatrix} \quad \begin{bmatrix} 1 & 0 \\ 1 & 1 \end{bmatrix} \quad \begin{bmatrix} 1 & 1 \\ 0 & 1 \end{bmatrix} \quad \begin{bmatrix} 1 & 1 \\ 1 & 0 \end{bmatrix}$$

form a basis of $V_{2\times 2}$. Find the coordinates (see Remark 5.1) of $A = \begin{bmatrix} 3 & 0 \\ 2 & 1 \end{bmatrix}$ with respect to the above basis.

5.44 Show that the vectors 1, $1 + x$, $x + x^2$ form a basis of $\mathbb{P}_2(x)$, and find the coordinates of $2 + 3x + x^2$ with respect to the above basis.

5.45 Determine the coordinates of the vector $[1\ 2\ {-1}]^T$ with respect to the basis given in Problem 5.18.

5.46 Let $A = [\mathbf{e}_{j_1}\ \mathbf{e}_{j_2} \ldots \mathbf{e}_{j_n}]$, where $\mathbf{e}_r$ is an rth elementary vector, be a permutation matrix; $j_1 j_2 \ldots j_n$ is a permutation of the integers 1 to n. Show that $A^{-1} = A^T$.

5.47 For a matrix linear transformation $A_{m\times n} : V_n \rightarrow W_m$ over a complex field $\mathbb{C}$, prove that $\ker A = [\text{range } A^*]^{\perp}$ and $V_n = \ker A \oplus \text{range } A^*$.

CHAPTER

6

Linear Transformation and Matrices

In this chapter we shall acquaint ourselves with the notion of a linear transformation (or mapping or function) and its various properties. The sole aim of such a study is to show how a linear transformation can be represented by a matrix. In the course of this study, the striking similarity between kernel or null space, rank, nullity and other features of linear transformations and matrices would not escape the notice of an astute reader. Abstracting two basic properties of a matrix, viz., (i) property of additivity and (ii) property of homogeneity, the definition of a linear transformation is given.

Linear transformations are classified into (i) *one-to-one* or *injective*, (ii) *onto* or *surjective*, (iii) *one-to-one* and *onto* or *bijective*.

An injective linear transformation is represented by a matrix which is left invertible, a surjective linear transformation by a right invertible matrix, and a bijective transformation by an invertible matrix. The matrix of a linear transformation is uniquely determined for a particular selection of the basis. But, for different choices of the bases the same linear transformation can be represented by different matrices. These matrices pertaining to a single linear transformation are defined as similar or equivalent.

6.1 LINEAR TRANSFORMATION

On recalling linear simultaneous equations (5.35), i.e.,

$$A\mathbf{x} = \mathbf{b} \tag{6.1}$$

where A, $\mathbf{x}$ and $\mathbf{b}$ are given by (5.36), we note that $\mathbf{x}$ is an element of the n-dimensional vector space V_n (see Definitions 5.2 and 5.5) which on premultiplication by the matrix A is transformed into an element **b** of the m-dimensional vector space. That is, for each element of V_n, the matrix A assigns one and only one element of W_m. Here, the matrix A which carries out such assignments is called a mapping (or function or transformation) because A maps (or transforms) an element **x** of V_n into an element **b** of W_m. This notion is represented as

$$A: V_n \to W_m: \mathbf{x} \mapsto A\mathbf{x} = \mathbf{b}$$

which means that A is a mapping from V_n to W_m indicated by the arrow, i.e., A sends an element **x** in V_n into an element **b** in W_m shown by a special arrow. For a different element **y** in V_n under the same mapping or transformation by the matrix A, there exists an element **c** in W_m satisfying

$$A\mathbf{y} = \mathbf{c}. \tag{6.2}$$

An important property of such a transformation by the matrix A is that any linear combination $\alpha\mathbf{x} + \beta\mathbf{y}$ (α and β are scalars, see Definition 5.4) of the vectors **x** and **y** in V_n is transformed by A into *the same linear combination* $\alpha\,\mathbf{b} + \beta\,\mathbf{c}$ of the transformed vectors **b** and **c** in W_m. In other words, taking (6.1) and (6.2) into account, we get

$$A(\alpha\mathbf{x} + \beta\mathbf{y}) = \alpha A\mathbf{x} + \beta A\mathbf{y} = \alpha\mathbf{b} + \beta\mathbf{c} \tag{6.3}$$

which we can easily verify using the definition of a scalar multiple of a matrix in Section 1.2, and the distributive law of matrix multiplication. Applying (6.3) to the cases (a) $\alpha = \beta = 1$ and (b) $\beta = 0$, we have respectively,

(a) $A(\mathbf{x} + \mathbf{y}) = A\mathbf{x} + A\mathbf{y}$: property of additivity,
(b) $A(\alpha\mathbf{x}) = \alpha A\mathbf{x}$: property of homogeneity,

where in (a), addition is preserved and in (b), scalar multiplication is preserved by the matrix A. The transformation by the matrix A satisfying properties (a) and (b) as defined here is termed as a *linear transformation* or a *linear mapping* or *a linear function.* Observe carefully the attribute *linear.* Conversely, we have the following definition.

Definition 6.1 *Linear Transformation (or mapping or function).* Given two vector spaces V_n and W_m, a transformation (or mapping or function) L from V_n to W_m is a correspondence that associates to each element **x** in V_n a unique element $\mathbf{b} = L\mathbf{x}$ in W_m. Any transformation (or mapping or function) L from a vector space V_n to another vector space W_m is said to be linear if for any vectors **x** and **y** in V_n there are assigned vectors $L\mathbf{x}$ and $L\mathbf{y}$ in W_m satisfying the relations

$$L(\mathbf{x} + \mathbf{y}) = L\mathbf{x} + L\mathbf{y} \quad \text{(property of additivity)} \tag{6.4}$$

$$L(\alpha\mathbf{x}) = \alpha\, L\mathbf{x} \quad \text{(property of homogeneity)} \tag{6.5}$$

both of which can be written together as

$$L(\alpha\mathbf{x} + \beta\mathbf{y}) = \alpha L\mathbf{x} + \beta L\mathbf{y} \quad \text{(the principle of superposition)} \tag{6.6}$$

where α and β are any scalars. If L is a linear transformation from V_n to W_m, then symbolically, we write

$$L : V_n \to W_m : \mathbf{x} \mapsto L\mathbf{x} \tag{6.7}$$

which is read as *L is a linear transformation* (or *mapping* or *function*) *from* V_n *to* W_m (first part) *which sends an element* **x** in V_n *to an element* $L\mathbf{x}$ in W_m (second part). A similar reading is valid, omitting the attribute *linear*. Sometimes, (6.7) is written alternatively as

$$V_n \xrightarrow{L} W_m$$

Very often, the description *homomorphism* from a vector space V_n into a vector space W_m instead of a linear transformation, is used in the literature to designate such linear mapping. □

If the two vector spaces W_m and V_n in (6.7) are the same V_n, then L is said to be a linear transformation (or mapping or function) on V_n. In this book, we prefer to use the term *linear transformation* without alluding to *linear mapping* or *linear function*, side by side.

A vector space V is defined over a field $\mathbb{F}$ of scalars (see Definition 5.1). Similarly, in defining a linear transformation, reference must be made to the underlying field which contains the scalars α and β in (6.6) of Definition 6.1. Stated more specifically, L is a linear transformation from a vector space V_n over a field $\mathbb{F}$ to another vector space W_m over $\mathbb{F}$. The vector spaces V_n and W_m must also be defined over the same field $\mathbb{F}$. For the sake of brevity, such reference to a field $\mathbb{F}$ would not be explicitly made although it is always implicit in a linear transformation under discussion.

In what follows we give further definitions applicable to a transformation which need not necessarily be linear.

Definition 6.2 *Domain, Co-domain, Range, and Image of a Transformation* (see Figure 6.1). Let L be a transformation $L : V_n \to W_m$, and $\mathbf{x}$ an element in V_n. Then $\mathbf{b} = L\mathbf{x}$ is called the image of $\mathbf{x}$ under L. If S_r is a subspace (see Definition 5.7) in V_n, then denote

$$L(S_r) = \text{the set of all elements } L\mathbf{x} \text{ in } W_m \tag{6.8}$$

where $\mathbf{x}$ is in S_r. We call $L(S_r)$ the image of S_r under L.

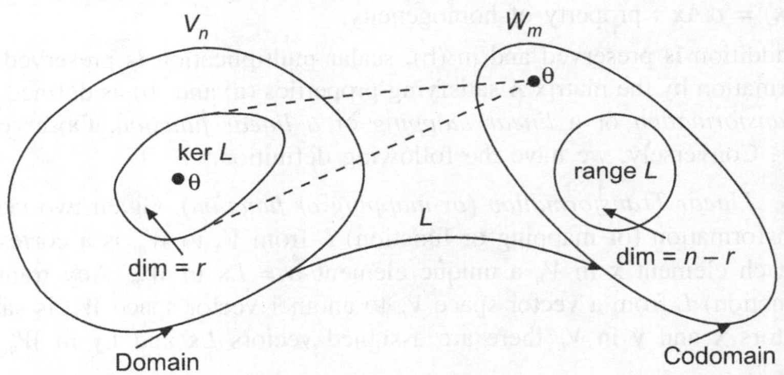

Figure 6.1 Diagram showing domain, codomain, range, and kernel of a linear transformation $L : V_n \to W_m$.

Moreover, the vector space V_n is called the *domain* of L and $L(V_n)$ (see (6.8)) is called the *range* of L, and W_m is called the *codomain* of L. □

Observe that every element $\mathbf{b}$ in the range of L is the image of some element $\mathbf{x}$ in V_n and that the range of a linear transformation L is a subspace of W_m. Indeed, if $\mathbf{b}$ and $\mathbf{c}$ are in $L(V_n)$, then $L\mathbf{x} = \mathbf{b}$ and $L\mathbf{y} = \mathbf{c}$ for some $\mathbf{x}$ and $\mathbf{y}$ in V_n. Hence, $\mathbf{b} + \mathbf{c} = L\mathbf{x} + L\mathbf{y} = L(\mathbf{x} + \mathbf{y})$ is in $L(V_n)$. Again for any scalar α in $\mathbb{F}$, $\alpha\mathbf{b} = \alpha L\mathbf{x} = \mathrm{L}(\alpha\mathbf{x})$ is in $L(V_n)$. Since $L(V_n)$ is a subset of W_m, by Theorem 5.3, $L(V_n)$ = range of L, is a subspace of W_m.

Definition 6.3 *One-to-one and Onto Transformation.* A transformation $L : V_n \to W_m$ is said to be *onto* if the range of L is the entire vector space W_m.

Again, if every element $\mathbf{b}$ in $L(V_n)$, which is the range of L lying in W_m, is an image of exactly one element of the domain V_n under the transformation, then L is called a *one-to-one transformation.* To be more specific, a transformation L is one-to-one if distinct elements of V_n are transformed into distinct elements of W_m. Stated mathematically, this means that if $\mathbf{x}$ and $\mathbf{y}$ are in V_n, then $L\mathbf{x} = L\mathbf{y}$ implies that $\mathbf{x} = \mathbf{y}$. Alternatively, a transformation which is onto is called *surjective,* a one-to-one transformation *injective,* and the one which is both one-to-one and onto is called *bijective.* A linear transformation is also called *homomorphism;* and a one-to-one and onto linear transformation is also termed *isomorphism.* □

Example 6.1 *Projection* The transformation $L: \mathbb{R}^3 \to \mathbb{R}^2$ defined by

$$L\mathbf{x} = L\begin{bmatrix} x_1 \\ x_2 \\ x_3 \end{bmatrix} = \begin{bmatrix} x_1 \\ x_2 \end{bmatrix} \tag{6.9}$$

is a linear transformation and is known as a *projection.* This is because, for scalars α and β, $\mathbf{x}$ and $\mathbf{y}$ in $\mathbb{R}^3$ we have, by (6.9),

$$L(\alpha\mathbf{x} + \beta\mathbf{y}) = L\begin{bmatrix} \alpha x_1 + \beta y_1 \\ \alpha x_2 + \beta y_2 \\ \alpha x_3 + \beta y_3 \end{bmatrix} = \begin{bmatrix} (\alpha x_1 + \beta y_1) \\ \alpha x_2 + \beta y_2 \end{bmatrix}$$

$$= \alpha\begin{bmatrix} x_1 \\ x_2 \end{bmatrix} + \beta\begin{bmatrix} y_1 \\ y_2 \end{bmatrix} = \alpha L\mathbf{x} + \beta L\mathbf{y}$$

satisfying (6.6). Referring to Figure 6.2(a), we see that $\mathbf{x}$ (point P or the vector OP) is an element in three-dimensional space $\mathbb{R}^3$, while the image $L\mathbf{x}$ of $\mathbf{x}$ (point Q or the vector OQ) lies in x_1x_2-plane $\mathbb{R}^2$. The domain of L is $\mathbb{R}^3$, and it can be verified that the range $L(\mathbb{R}^3)$ of L is the whole of $\mathbb{R}^2$. In fact, any point $[y_1, y_2]^T$ in $\mathbb{R}^2$ is the image of $[y_1, y_2, y_3]^T$ in $\mathbb{R}^3$. Hence L is an *onto* transformation. Let $S_2 = [x_1, x_1, x_3]^T$ be a subspace in $\mathbb{R}^3$ represented by a plane passing through OR (the line $x_1 = x_2$, $x_3 = 0$) and perpendicular to the plane 102. Then $L(S_2) = [x_1, x_1]^T$ is the line OR which is the *image* of S_2 under L. The transformation L here is *not* one-to-one because the image of P: (x_1, x_2, x_3) and that of P' : $(x_1, x_2, \tilde{x}_3)$, $x_3 \neq \tilde{x}_3$, (see Figure 6.2(a)) are the same

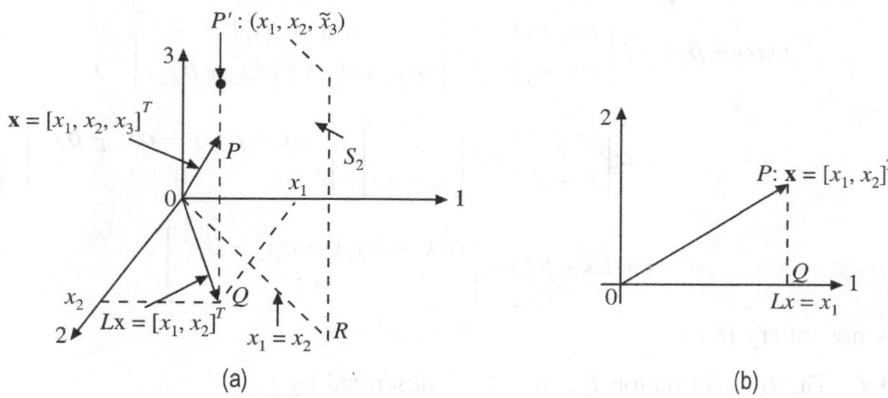

Figure 6.2 Linear onto transformation: projection.

Q: (x_1, x_2). The transformation L projects the three-dimensional vector space $\mathbb{R}^3$ onto the two-dimensional plane $\mathbb{R}^2$, in particular, OP to OQ (see Figure 6.2(a)). Similarly, $\tilde{L}: \mathbb{R}^n \to \mathbb{R}^m$, $m < n$ defined by $\tilde{L}[x_1\ x_2 \ldots x_n]^T = [x_1 x_2 \ldots x_m]^T$ is a linear transformation and a projection of $\mathbb{R}^n$ onto $\mathbb{R}^m$, which for $m = 1$ and $n = 2$ is shown in Figure 6.2(b). □

Example 6.2 Let $\mathbb{P}_n(t)$ be the set of all polynomials in the variable t and of order $\leq n$, i.e., any element $p(t)$ in $\mathbb{P}_n(t)$ is

$$p(t) = a_0 + a_1 t + a_2 t^2 + \cdots + a_n t^n.$$

Since 1, t, ..., t^n are linearly independent, they form a basis of $\mathbb{P}_n(t)$. The elements in $\mathbb{P}_n(t)$ satisfy the properties 1–8 of the vector space in Definition 5.1 (see Example 5.2). Now, $D = d/dt$ is a linear transformation $D: \mathbb{P}_n(t) \to \mathbb{P}_{n-1}(t)$ because, we note that

$$Dp(t) = a_1 + 2a_2 t + \cdots + na_n t^{n-1} \tag{6.10}$$

and if $q(t)$ is another element in $\mathbb{P}_n(t)$ defined by

$$q(t) = b_0 + b_1 t + b_2 t^2 + \cdots + b_n t^n,$$

then for any scalars α and β,

$$\begin{aligned} D(\alpha p + \beta q) &= D[(\alpha a_0 + \beta b_0) + (\alpha a_1 + \beta b_1)t + (\alpha a_2 + \beta b_2)t^2 + \cdots + (\alpha a_n + \beta b_n)t^n] \\ &= (\alpha a_1 + \beta b_1) + 2(\alpha a_2 + \beta b_2)t + \cdots + n(\alpha a_n + \beta b_n)t^{n-1} \\ &= \alpha(a_1 + 2a_2 t + \cdots + na_n t^{n-1}) + \beta(b_1 + 2b_2 t + \cdots + nb_n t^{n-1}) \\ &= \alpha\, Dp + \beta\, Dq \end{aligned}$$

in view of (6.10), and similar expression for $Dq\ (t)$. This transformation D is again onto but not one-to-one, because $p(t)$ and $p(t) + a$ constant, have the same image under D. □

Example 6.3 The transformation $L: \mathbb{R}^2 \to \mathbb{R}^2$ defined by

$$L\mathbf{x} = L[x_1\ x_2]^T = [x^2_1,\ x_1 + x_2]^T$$

is not linear. This is so because, for scalars α and β and elements $\mathbf{x} = [x_1\ x_2]^T$ and $\mathbf{y} = [y_1\ y_2]^T$ in $\mathbb{R}^2$, we have

$$L(\alpha x + \beta y) = L\begin{bmatrix} \alpha x_1 + \beta y_1 \\ \alpha x_2 + \beta y_2 \end{bmatrix} = \begin{bmatrix} (\alpha x_1 + \beta y_1)^2 \\ (\alpha x_1 + \beta y_1) + (\alpha x_2 + \beta y_2) \end{bmatrix}$$

$$= \alpha\begin{bmatrix} x_1^2 \\ x_1 + x_2 \end{bmatrix} + \beta\begin{bmatrix} y_1^2 \\ y_1 + y_2 \end{bmatrix} + \begin{bmatrix} (\alpha x_1 + \beta y_1)^2 - \alpha x_1^2 - \beta y_1^2 \\ 0 \end{bmatrix}$$

$$= \alpha L\mathbf{x} + \beta L\mathbf{y} + \begin{bmatrix} (\alpha x_1 + \beta y_1)^2 - \alpha x_1^2 - \beta y_1^2 \\ 0 \end{bmatrix}$$

which does not satisfy (6.6). □

Example 6.4 The transformation $L : \mathbb{R}^2 \to \mathbb{R}^2$ described by

$$L\mathbf{x} = L[x_1\ x_2]^T = [x_1 - x_2,\ x_1 + x_2]^T \tag{6.11}$$

satisfies (6.6) as can be easily verified and is therefore linear. Again, any point $[y_1, y_2]^T$ in $W_m = \mathbb{R}^2$ is the image of the point

$$\begin{bmatrix} x_1 \\ x_2 \end{bmatrix} = \begin{bmatrix} \frac{1}{2}(y_1 + y_2) \\ \frac{1}{2}(y_2 - y_1) \end{bmatrix}$$

in $V_n = \mathbb{R}^2$, and hence the transformation is onto. The transformation is also one-to-one because if $x = [x_1, x_2]^T$ and $\mathbf{y} = [y_1, y_2]^T$ are any two points in $V_n = \mathbb{R}^2$, then in view of (6.11), $Lx = Ly$ implies that

$$\begin{bmatrix} x_1 - x_2 \\ x_1 + x_2 \end{bmatrix} = \begin{bmatrix} y_1 - y_2 \\ y_1 + y_2 \end{bmatrix}$$

which on solving gives $x_1 = y_1$ and $x_2 = y_2$, i.e., $\mathbf{x} = \mathbf{y}$. Thus the linear transformation L defined by (6.11) is bijective; in other words, L is an *isomorphism* on $\mathbb{R}^2$. □

Example 6.5 (a) The translation of coordinate axes on the plane $\mathbb{R}^2$ defined by (see Figure 6.3(a))

$$L\mathbf{x} = L[x_1\, x_2]^T = \mathbf{y} = [y_1\, y_2]^T = [x_1 - \alpha,\ x_2 - \beta]^T$$

is not a linear transformation on $\mathbb{R}^2$, as can be easily verified,

(b) The rotation of coordinate axes on the plane $\mathbb{R}^2$ defined by

$$\begin{aligned} L\mathbf{x} &= L[x_1\ x_2]^T = \mathbf{y} = [y_1\ y_2]^T \\ &= [x_1 \cos\theta + x_2 \sin\theta,\ -x_1 \sin\theta + x_2 \cos\theta]^T \end{aligned}$$

is a linear *bijective transformation* (i.e., an *isomorphism)* on $\mathbb{R}^2$, as can be easily verified.

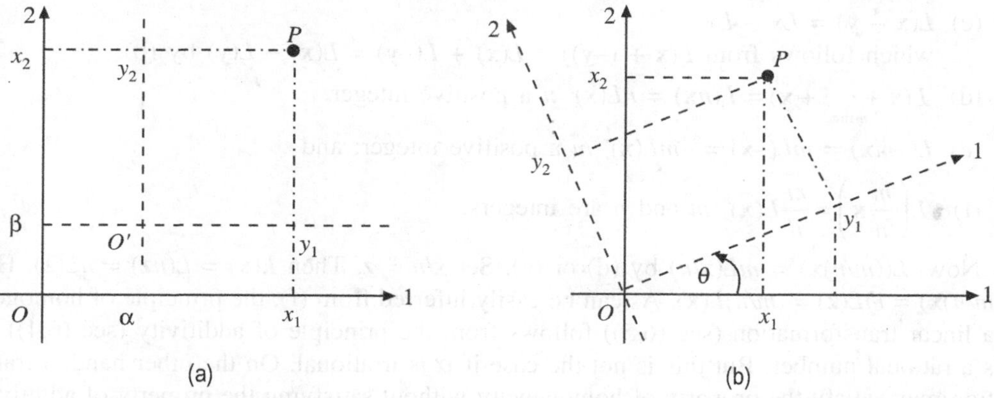

Figure 6.3 (a) Translation of coordinate axes; (b) rotation of coordinate axes.

(c) The transformation defined by

$$L: V_{2\times2} \to \mathbb{P}_2(x): A = \begin{bmatrix} a & b \\ c & d \end{bmatrix} \mapsto a + (b + c)x + dx^2$$

is linear, Indeed, if

$$B = \begin{bmatrix} \tilde{a} & \tilde{b} \\ \tilde{c} & \tilde{d} \end{bmatrix}$$

is any other element in $V_{2\times 2}$, then

$L(\alpha A + \beta B) = (\alpha a + \beta \tilde{a}) + \alpha (b + c)x + \beta (\tilde{b} + \tilde{c})x + (\alpha d + \beta \tilde{d})x^2 = \alpha LA + \beta LB.$ □

Definition 6.4 *Identity and Zero (or Null) Transformations.* An identity transformation $I : V_n \to V_n : \mathbf{x} \mapsto \mathbf{x}$ is defined by $I(\mathbf{x}) = \mathbf{x}$ for all $\mathbf{x}$ in V_n, which sends an element $\mathbf{x}$ of V_n onto itself. A zero (or null) transformation $\Theta : V_n \to W_m : \mathbf{x} \mapsto \boldsymbol{\theta}$ defined by $\Theta(\mathbf{x}) = \boldsymbol{\theta}$ for all $\mathbf{x}$ in V_n, sends every element $\mathbf{x}$ in V_n to the zero-element $\boldsymbol{\theta}$ in W_m. □

It can be easily shown that an identity transformation and a zero transformation are linear. Moreover, an identity transformation is both injective and surjective, i.e., it is bijective. Sometimes we write I_V in place of $I : V \to V$ to underline the associated vector space V.

6.2 PROPERTIES OF LINEAR TRANSFORMATIONS

Let $L : V_n \to W_m$ be a linear transformation, and $\mathbf{x}$, $\mathbf{y}$, $\mathbf{z}$ be elements in V_n, and $\boldsymbol{\theta}$ be a zero element either in V_n or W_m. The zero elements in V_n and W_m are not distinguished by symbols even though they can be distinguished from the context. We now have

(a) $L(\boldsymbol{\theta}) = \boldsymbol{\theta}$,
which follows from subtracting $L(\boldsymbol{\theta})$ so that from both sides of
$L(\boldsymbol{\theta}) = L(\boldsymbol{\theta} + \boldsymbol{\theta}) = L(\boldsymbol{\theta}) + L(\boldsymbol{\theta})$
$L(\boldsymbol{\theta}) - L(\boldsymbol{\theta}) = L(\boldsymbol{\theta}) + L(\boldsymbol{\theta}) - L(\boldsymbol{\theta})$, i.e., $L(\boldsymbol{\theta}) = \boldsymbol{\theta}$.

(b) $L(-\mathbf{x}) = -L(\mathbf{x})$.
We note that $L(\boldsymbol{\theta}) = L(\mathbf{x} - \mathbf{x}) = L(\mathbf{x}) + L(-\mathbf{x})$. Hence, in view of (a), $L(-\mathbf{x}) = L(\boldsymbol{\theta}) - L(\mathbf{x}) = \boldsymbol{\theta} - L(\mathbf{x}) = -L(\mathbf{x})$.

(c) $L(\mathbf{x} - \mathbf{y}) = L\mathbf{x} - L\mathbf{y}$
which follows from $L(\mathbf{x} + (-\mathbf{y})) = L(\mathbf{x}) + L(-\mathbf{y}) = L(\mathbf{x}) - L(\mathbf{y})$ by (b).

(d) $L(\underbrace{\mathbf{x} + \cdots + \mathbf{x}}_{n\text{-times}}) = L(n\mathbf{x}) = nL(\mathbf{x})$, n a positive integer.

(e) $L(-m\mathbf{x}) = mL(-\mathbf{x}) = -mL(\mathbf{x})$, m a positive integer; and

(f) $L\left(\dfrac{m}{n}\mathbf{x}\right) = \dfrac{m}{n}L(\mathbf{x})$, m and n are integers.

Now, $L((m/n)\mathbf{x}) = mL(\mathbf{x}/n)$ by (d) or (e). Set $\mathbf{x}/n = \mathbf{z}$. Then $L(\mathbf{x}) = L(n\mathbf{z}) = nL(\mathbf{z})$. Hence, $L((m/n)\mathbf{x}) = mL(\mathbf{z}) = (m/n)L(\mathbf{x})$. As can be easily inferred from (f), the principle of homogeneity of a linear transformation (see (6.5)) follows from the principle of additivity (see (6.4)) when α is a rational number. But this is not the case if α is irrational. On the other hand, a transformation may satisfy the property of homogeneity without satisfying the property of additivity as in the following example:

$$L[x(t)] = \begin{cases} \dfrac{x^2(t)}{x(t-1)}, & x(t-1) \neq 0; \\ 0, & x(t-1) = 0; \end{cases}$$

for all t.

Let us now consider the set of all linear transformations from $V_n \to W_m$ which we denote by $\mathcal{L}(V_n, W_m)$, or simply by $\mathcal{L}$, if the underlying vector spaces V_n and W_m are clear from the context. If L and $\bar{L}$ are two elements in $\mathcal{L}$ and α is any scalar, then we define addition of two linear transformations and their scalar multiplication as

$$(L+\bar{L})\mathbf{x} = L\mathbf{x} + \bar{L}\mathbf{x}, \quad (\alpha L)\mathbf{x} = L(\alpha \mathbf{x}), \tag{6.12}$$

where $\mathbf{x}$ is an element of V_n.

By putting $\alpha = 1$ and $\alpha = -1$ successively in (6.12), we have, by Problem 5.2(e) and Property (b) of a linear transformation above, the relations

$$1L = L, \qquad (-L)\mathbf{x} = -L\mathbf{x}.$$

The latter defines the negative of a linear transformation. Also,

$$(L + \Theta)\mathbf{x} = L\mathbf{x} + \Theta\mathbf{x} = L\mathbf{x} + \boldsymbol{\theta} = L\mathbf{x},$$
$$(L + (-L))\mathbf{x} = L\mathbf{x} + (-L)\mathbf{x} = L\mathbf{x} - L\mathbf{x} = \boldsymbol{\theta} = \Theta\mathbf{x}$$

for any $\mathbf{x}$. Hence,

$$L + \Theta = L \text{ and } L + (-L) = \Theta.$$

The elements in W_m are commutative and associative, by virtue of which we can prove that linear transformations also have the same properties. This proves Axioms 1 – 2 of a vector space. The remaining Axioms 5 – 8 follow from the definition of a scalar multiplication of a linear transformation in (6.12). *It is thus shown that* (see Example 5.3) *that $\mathcal{L}$ satisfies all the axioms of Definition 5.1 and so it is a vector space.* Furthermore, the sum $L+\bar{L}$ of any two linear transformations in $\mathcal{L}$ is again linear. In other words, for $\mathbf{x}$ and $\mathbf{y}$ in V_n and α scalar, we can show with the aid of (6.12) that

$$(L+\bar{L})(\mathbf{x}+\mathbf{y}) = (L+\bar{L})\mathbf{x} + (L+\bar{L})\mathbf{y},$$

$$(L+\bar{L})(\alpha\mathbf{x}) = \alpha(L+\bar{L})\mathbf{x}.$$

Definition 6.5 *Equality of Two Linear Transformations.* Two linear transformations L and $\bar{L}$ from $V_n \to W_m$ are defined as equal iff $L\mathbf{x} = \bar{L}\mathbf{x}$ for every $\mathbf{x}$ in V_n. □

A linear transformation $L : V_n \to W_m$ is completely determined if it is specified at the basis elements of V_n. In other words, we have the following theorem, which plays a crucial role in linear algebra.

Theorem 6.1 Let $\{\mathbf{x}_1, ..., \mathbf{x}_n\}$ be a basis in V_n and $\{\mathbf{z}_1, ..., \mathbf{z}_n\}$ be the given elements in W_m. Then, there is one and only one linear transformation $L : V_n \to W_m$ for which

$$L\mathbf{x}_i = \mathbf{z}_i, \quad i = 1, ..., n. \tag{6.13}$$

Proof For any element $\mathbf{x}$ in V_n, there are unique scalars α_i's such that

$$\mathbf{x} = \alpha_1\mathbf{x}_1 + \cdots + \alpha_n\mathbf{x}_n. \tag{6.14}$$

Now, define the desired linear transformation by

$$L\mathbf{x} := \alpha_1\mathbf{z}_1 + \cdots + \alpha_n\mathbf{z}_n \tag{6.15}$$

in terms of the coordinates α_i's of $\mathbf{x}$ and the given $\mathbf{z}_i$'s. It can be easily checked that L is linear. Indeed, if

$$\mathbf{y} = \beta_1\mathbf{x}_1 + \cdots + \beta_n\mathbf{x}_n$$

is another element in V_n, then

$$L\mathbf{y} = \beta_1\mathbf{z}_1 + \cdots + \beta_n\mathbf{z}_n, \quad \mathbf{x}+\mathbf{y} = (\alpha_1+\beta_1)\mathbf{x}_1 + \cdots + (\alpha_n+\beta_n)\mathbf{x}_n, \quad \alpha\mathbf{x} = \alpha\alpha_1\mathbf{x}_1 + \cdots + \alpha\alpha_n\mathbf{x}_n.$$

Therefore,

$$\begin{aligned} L(\mathbf{x}+\mathbf{y}) &= (\alpha_1+\beta_1)\mathbf{z}_1 + \cdots + (\alpha_n+\beta_n)\mathbf{z}_n \\ &= (\alpha_1\mathbf{z}_1 + \cdots + \alpha_n\mathbf{z}_n) + (\beta_1\mathbf{z}_1 + \cdots + \beta_n\mathbf{z}_n) = L\mathbf{x} + L\mathbf{y} \end{aligned}$$

and

$$L(\alpha\mathbf{x}) = \alpha\alpha_1\mathbf{z}_1 + \cdots + \alpha\alpha_n\mathbf{z}_n = \alpha L\mathbf{x}.$$

Hence, there exists a linear transformation defined by (6.15) as required in the theorem.

To prove that L is unique, assume that $\bar{L}$ is another linear transformation satisfying the condition $\bar{L}\mathbf{x}_i = \mathbf{z}_i$ in the theorem. Then, in view of (6.14) and (6.15), we have

$$\bar{L}\mathbf{x} = \alpha_1\bar{L}\mathbf{x}_1 + \cdots + \alpha_n\bar{L}\mathbf{x}_n = \alpha_1\mathbf{z}_1 + \cdots + \alpha_n\mathbf{z}_n = L\mathbf{x}$$

for every $\mathbf{x}$ in V_n. Therefore, $\bar{L} = L$ in view of Definition 6.5. □

However, if the set $\{\mathbf{x}_1, \mathbf{x}_2, \ldots, \mathbf{x}_n\}$ is not linearly independent, then the representation (6.14) is not unique and an infinite number of linear transformations (6.15) exists. Such a nonunique set of linear transformations may define the left or the right inverse K of a given linear transformation. See Definition 6.10 and Example 6.13.

Example 6.6 Let $\mathbf{x}_1 = (1, 1, -1)$, $\mathbf{x}_2 = (4, 1, 1)$, $\mathbf{x}_3 = (1, -1, 2)$ be a basis of $\mathbb{R}^3$ and let $L : \mathbb{R}^3 \to \mathbb{R}^2$ be the linear transformation such that $L\mathbf{x}_1 = (1, 0)$, $L\mathbf{x}_2 = (0, 1)$, $L\mathbf{x}_3 = (1, 1)$. Find L.

Solution Let $\mathbf{y} = (y_1, y_2, y_3)$ be any element of $\mathbb{R}^3$. We have to find the expression for $L(y_1, y_2, y_3)$. Now there exist scalars α_1, α_2, and α_3 such that

$$\mathbf{y} = \alpha_1\mathbf{x}_1 + \alpha_2\mathbf{x}_2 + \alpha_3\mathbf{x}_3.$$

Therefore,

$$\begin{aligned} (y_1, y_2, y_3) &= \alpha_1(1 \quad 1 \quad -1) + \alpha_2(4 \quad 1 \quad 1) + a_3(1 \quad -1 \quad 2) \\ &= (\alpha_1, \alpha_2, \alpha_3)\begin{bmatrix} 1 & 1 & -1 \\ 4 & 1 & 1 \\ 1 & -1 & 2 \end{bmatrix}. \end{aligned}$$

Postmultiplying both sides by the inverse of the coefficient matrix,

$$(\alpha_1, \alpha_2, \alpha_3) = (y_1, y_2, y_3)\begin{bmatrix} 3 & -1 & 2 \\ -7 & 3 & -5 \\ -5 & 2 & -3 \end{bmatrix}.$$

Hence,

$$\alpha_1 = 3y_1 - 7y_2 - 5y_3, \quad \alpha_2 = -y_1 + 3y_2 + 2y_3, \quad \alpha_3 = 2y_1 - 5y_2 - 3y_3.$$

Therefore,

$$\begin{aligned} L\mathbf{y} = L(\alpha_1\mathbf{x}_1 + \alpha_2\mathbf{x}_2 + \alpha_3\mathbf{x}_3) &= \alpha_1 L\mathbf{x}_1 + \alpha_2 L\mathbf{x}_2 + \alpha_3 L\mathbf{x}_3 \\ &= \alpha_1(1\ 0) + \alpha_2(0\ 1) + \alpha_3(1\ 1) \\ &= (5y_1 - 12y_2 - 8y_3,\ y_1 - 2y_2 - y_3). \end{aligned}$$

□

Theorem 6.2 The dimension of the vector space formed by the set of all linear transformations $\mathcal{L}(V_n, W_m)$ is $(\dim V_n \times \dim W_m)$.

Proof We denote a basis of V_n by $\{\mathbf{x}_1, ..., \mathbf{x}_n\}$ and that of W_m by $\{\mathbf{y}_1, ..., \mathbf{y}_m\}$. In view of Theorem 6.1, there exist unique linear transformations $E_{ij} : V_n \to W_m$ in L defined by

$$E_{ij}\mathbf{x}_k = \mathbf{\theta}, \qquad k \neq i;$$

$$E_{ij}\mathbf{x}_i = \mathbf{y}_j, \qquad i = 1, ..., n; \qquad j = 1, ..., m. \tag{6.16}$$

To prove the theorem, we have to show that (i) E_{ij}'s span $\mathcal{L}(V_n, W_m)$, and (ii) E_{ij}'s are linearly independent.

Let $A : V_n \to W_m$ be any linear transformation in $\mathcal{L}$. Then the element $A\mathbf{x}_i$ in W_m can be expressed as

$$A\mathbf{x}_i = \sum_{j=1}^{m} a_{ij}\mathbf{y}_j. \tag{6.17}$$

Let us now define a linear transformation in $\mathcal{L}$ by

$$B := \sum_{i=1}^{n}\sum_{j=1}^{m} a_{ij}E_{ij}. \tag{6.18}$$

Then, in view of (6.16) and (6.17), we have

$$B\mathbf{x}_k = \sum_{i=1}^{n}\sum_{j=1}^{m} a_{ij}E_{ij}\mathbf{x}_k = \sum_{j=1}^{m} a_{kj}E_{kj}\mathbf{x}_k = \sum_{j=1}^{m} a_{kj}\mathbf{y}_j = A\mathbf{x}_k. \tag{6.19}$$

Hence $A = B$ by Definition 6.5, which implies that any transformation A can be expressed as a linear combination of E_{ij}'s.

Again from (6.18), by virtue of (6.19), $B = \Theta$ implies that

$$\sum_{j=1}^{m} a_{kj}\mathbf{y}_j = \theta.$$

From this equation, it follows that $a_{kj} = 0$ ($k = 1, ..., n$; $j = 1, ..., m$) since y_j's are linearly independent. Therefore, E_{ij}'s form a basis of $\mathcal{L}(V_n, W_m)$. □

An important notion in the study of a linear transformation is its *kernel* which literally means central (or essential) part or core. We shall now see in the following that this meaning is worthy of the role kernel plays in determining the structure of linear transformations.

Definition 6.6 *Ker L or Null Space of L.* The kernel (or null space) of a linear transformation $L : V_n \to W_m$ is the set of elements in V_n whose image under L is the zero element of W_m and is denoted by ker L. In view of (6.8), this implies L (ker L) = $\boldsymbol{\theta}$ (see Definition 5.15 and Figure 6.1).

Definition 6.7 *Rank and Nullity of L.* The rank of a linear transformation $L : V_n \to W_m$ is defined by

$$\text{rank } L = \dim L(V_n),$$

and the nullity (abbreviated as null) by (see Definition 5.16 and Figure 6.1)

$$\text{nullity } L = \dim \ker L.$$ □

When a matrix is regarded as a linear transformation (see (6.3)), we can define its null space or kernel, rank and nullity as above. Alternatively, they can be given as in Definitions 5.15 and 5.16 which are exemplified in Example 5.32. The following examples are given to clarify these notions for a general linear transformation.

Example 6.7 The ker L is a subspace. If x and y are any two elements in ker L and α is a scalar, then $L(\mathbf{x} + \mathbf{y}) = L\mathbf{x} + L\mathbf{y} = \boldsymbol{\theta} + \boldsymbol{\theta} = \boldsymbol{\theta}$, $L(\alpha\mathbf{x}) = \alpha L\mathbf{x} = \alpha\boldsymbol{\theta} = \boldsymbol{\theta}$. So, $\mathbf{x} + \mathbf{y}$ and $\alpha\mathbf{x}$ are in ker L. Therefore, ker L is a subspace in view of Theorem 5.3. □

Example 6.8 Let V be the vector space of real-valued functions in the variable t having derivatives of all orders (see Example 5.3), and let $L = D - aI_V : V \to V$, where $D = d/dt$; a, a real number and I_V the identity transformation (see Definition 6.4), be a linear transformation. Suppose that $f(t)$ is an element of V. Then, ker L is the solution of $Lf(t) = (D - aI_V)f(t) = 0$ which gives ker $L = c \exp(at)$, where c is a constant. □

Example 6.9 Let $L : \mathbb{R}^3 \to \mathbb{R}$ be a linear transformation defined by

$$L(x_1, x_2, x_3) = x_1 - 3x_2 + 2x_3.$$

Then ker L, obtained by solving

$$x_1 - 3x_2 + 2x_3 = 0, \tag{6.20}$$

is the set of all vectors (x_1, x_2, x_3) orthogonal to (1, –3, 2), see Example 5.32, Figure 5.8(a). We observe that the range space of L consists of real numbers: if $\mathbf{x} = (x_1, x_2, x_3) = (2, 1, 1)$, then $L\mathbf{x} = 1$ and is of dimension 1. The ker L is generated by vectors of the form $(3\alpha - 2\beta, \alpha, \beta)$ which is a solution of (6.20) where α and β are arbitrary real numbers and this set of vectors has dimension 2. This is because, in the matrix

$$\begin{bmatrix} 3a - 2b & a & b \\ 3c - 2d & c & d \\ 3e - 2f & e & f \end{bmatrix}$$

formed with three vectors in the set as rows, the first column is a linear combination of the remaining two columns. □

Theorem 6.3 A linear transformation $L : V_n \to W_m$ is injective iff ker $L = \boldsymbol{\theta}$, i.e., iff null $L = 0$.

Proof (a) *Necessity.* Let $\mathbf{x}$ be any element in ker L. Then $L\mathbf{x} = \boldsymbol{\theta} = L\boldsymbol{\theta}$. Since L is, injective (see Definition 6.3), $\mathbf{x} = \boldsymbol{\theta}$.

(b) *Sufficiency.* Suppose ker $L = \boldsymbol{\theta}$ and $\mathbf{x}$ and $\mathbf{y}$ are any two elements in V_n. Then $L\mathbf{x} = L\mathbf{y}$ implies that $L(\mathbf{x} - \mathbf{y}) = \boldsymbol{\theta}$. Hence, $\mathbf{x} - \mathbf{y} = \boldsymbol{\theta}$ because ker $L = \boldsymbol{\theta}$. Therefore, $L\mathbf{x} = L\mathbf{y}$ implies that $\mathbf{x} = \mathbf{y}$. Thus by Definition 6.3, L is injective. □

Theorem 6.4 Let $L : V_n \to W_m$ be an injective linear transformation and $\{\mathbf{x}_1, ..., \mathbf{x}_k\}$ be the set of linearly independent vectors in V_n. Then, the set $\{L\mathbf{x}_1, ..., L\mathbf{x}_k\}$ is also linearly independent.

Proof For scalars $\alpha_1, ..., \alpha_k$, from

$$\alpha_1 L\mathbf{x}_1 + \cdots + \alpha_k L\mathbf{x}_k = L(\alpha_1\mathbf{x}_1 + \cdots + \alpha_k\mathbf{x}_k) = \boldsymbol{\theta} = L\boldsymbol{\theta}$$

we have, since L is injective,

$$\alpha_1\mathbf{x}_1 + \cdots + \alpha_k\mathbf{x}_k = \boldsymbol{\theta}.$$

Hence, $\alpha_1 = \cdots = \alpha_k = 0$, since $\mathbf{x}_1, ..., \mathbf{x}_k$ are linearly independent. □

Theorem 6.5 Let $L : V_n \to W_m$ be a linear transformation. Then,

$$\dim V_n = \dim \ker L + \dim \text{range } L,$$

i.e.,

$$\dim V_n = \text{null } L + \text{rank } L,$$

where range $L = L(V_n)$, see Definition 6.2.

Proof Let the set $\{\mathbf{x}_1, ..., \mathbf{x}_k\}$, $k \le n$, be the basis of ker L such that

$$L\mathbf{x}_i = \boldsymbol{\theta}, \qquad i = 1, ..., k. \tag{6.21}$$

This basis can be extended to a basis $\{\mathbf{x}_1, ..., \mathbf{x}_k, \mathbf{x}_{k+1}, ..., \mathbf{x}_n\}$ of V_n, since ker L is a subspace of V_n (see Example 6.7 and Theorem 5.4). The theorem will be proved if we can show that the set of vectors $\{L\mathbf{x}_{k+1}, ..., L\mathbf{x}_n\}$ is a basis for the range L. Any element $\mathbf{x}$ in V_n can be expressed in terms of its basis elements $\mathbf{x}_i$'s as

$$\mathbf{x} = \alpha_1\mathbf{x}_1 + \cdots + \alpha_k\mathbf{x}_k + \alpha_{k+1}\mathbf{x}_{k+1} + \cdots + \alpha_n\mathbf{x}_n. \tag{6.22}$$

Therefore, by virtue of (6.21) we have from (6.22) the relation

$$L\mathbf{x} = \alpha_{k+1}L\mathbf{x}_{k+1} + \cdots + \alpha_n L\mathbf{x}_n, \tag{6.23}$$

which shows that any vector $L\mathbf{x}$ in range L is spanned by the set $\{L\mathbf{x}_{k+1}, ..., L\mathbf{x}_n\}$. To show that these vectors are linearly independent, suppose that

$$\alpha_{k+1}L\mathbf{x}_{k+1} + \cdots + \alpha_n L\mathbf{x}_n = \boldsymbol{\theta}.$$

Then,

$$L(\alpha_{k+1}\mathbf{x}_{k+1} + \cdots + \alpha_n\mathbf{x}_n) = \boldsymbol{\theta}$$

which shows that $\alpha_{k+1}\mathbf{x}_{k+1} + \cdots + \alpha_n\mathbf{x}_n$, is in ker L and can therefore be expressed in terms of its basis vectors $\mathbf{x}_1, ..., \mathbf{x}_k$. That is, there are scalars $\alpha_1, ..., \alpha_k$ such that

$$\alpha_{k+1}\mathbf{x}_{k+1} + \cdots + \alpha_n\mathbf{x}_n = -\alpha_1\mathbf{x}_1 - \cdots - \alpha_k\mathbf{x}_k.$$

Therefore,

$$\alpha_1 \mathbf{x}_1 + \cdots + \alpha_n \mathbf{x}_n = \boldsymbol{\theta}. \tag{6.24}$$

Since the set $\{\mathbf{x}_1, ..., \mathbf{x}_n\}$ is a basis of V_n, (6.24) implies that $\alpha_1 = \cdots = \alpha_k = \alpha_{k+1} = \cdots = \alpha_n = 0$. Hence, $\{L\mathbf{x}_{k+1}, ..., L\mathbf{x}_n\}$ is a linearly independent set, which, as already proved, spans range L and as a result forms its basis. □

Remark 6.1 The proof of Theorem 6.5 suggests a method of finding the basis for the range space $L(V_n)$ of a linear transformation $L : V_n \to W_m$. Indeed, if $\{\mathbf{x}_1, \mathbf{x}_2, ..., \mathbf{x}_n\}$ is a basis of V_n, then the set $\{L\mathbf{x}_1, L\mathbf{x}_2, ..., L\mathbf{x}_n\}$ spans W_m from which a basis for W_m can be selected by, e.g., the Gaussian elimination method (see Example 6.10(d)).

Remark 6.2 For an injective or one-to-one linear transformation $L : V_n \to W_m$, $\ker L = \boldsymbol{\theta}$. Hence by Theorem 6.5, L is injective iff $\dim V_n = \text{rank } L = \dim \text{range } L$. On the other hand, by definition, a linear transformation is onto iff $\dim W_m = \text{rank } L = \dim \text{range } L$.

Example 6.10 (a) In Example 6.9, $\dim \mathbb{R}^3 = \dim V_n = 3$, $\dim \ker L = 2$, and dim range $L = 1$.

(b) For identity transformation $I_V : V_n \to V_n$, rank $I_V = n$, null $I_V = 0$, but for zero-transformation $\Theta : V_n \to W_m$, rank $\Theta = 0$, null $\Theta = n$.

(c) Find a linear transformation $L : \mathbb{R}^4 \to \mathbb{R}^3$ for which the null space is spanned by $\mathbf{x}_1 = [1\ 1\ 1\ 1]^T$, $\mathbf{x}_2 = [1\ 0\ 0\ 1]^T$, and the image space of L is spanned by $\mathbf{y}_1 = [1\ 1\ 0]^T$ and $\mathbf{y}_2 = [1\ 0\ 1]^T$.

Solution Let the set $\{\mathbf{x}_1, \mathbf{x}_2, \mathbf{x}_3, \mathbf{x}_4\}$, where $\mathbf{x}_3 = [0\ 0\ 1\ 0]^T = \mathbf{e}_3$ and $\mathbf{x}_4 = [0\ 0\ 0\ 1]^T = \mathbf{e}_4$, be a basis of $\mathbb{R}^4$. By the constructive proof of Theorem 6.5, the set $\{L\mathbf{x}_3, L\mathbf{x}_4\}$ forms a basis of the range space of L. Therefore, the condition provided in the example gives $L\mathbf{x}_1 = \boldsymbol{\theta}$, $L\mathbf{x}_2 = \boldsymbol{\theta}$, $L\mathbf{x}_3 = \mathbf{y}_1$, and $L\mathbf{x}_4 = \mathbf{y}_2$, where $\boldsymbol{\theta} = [0\ 0\ 0]^T$. Any element $[\alpha\ \beta\ \gamma\ \delta]^T$ in $\mathbb{R}^4$ can be expressed in terms of the basis vectors $\mathbf{x}_i$'s as

$$[\alpha\ \beta\ \gamma\ \delta]^T = \beta\,\mathbf{x}_1 + (\alpha - \beta)\,\mathbf{x}_2 + (\gamma - \beta)\,\mathbf{x}_3 + (\delta - \alpha)\,\mathbf{x}_4$$

Hence, as a result of Theorem 6.1 and the foregoing relations, the required linear transformation L is given by

$$\begin{aligned} L[\alpha\ \beta\ \gamma\ \delta\ ]^T &= \beta\boldsymbol{\theta} + (\alpha - \beta)\boldsymbol{\theta} + (\gamma - \beta\ )\mathbf{y}_1 + (\delta - \alpha)\mathbf{y}_2 \\ &= [\gamma + \delta - \alpha - \beta,\ \gamma - \beta,\ \delta - \alpha]^T. \end{aligned}$$

Here, the basis of $\mathbb{R}^4$ is so selected that the right bottom corner submatrix in $X = [\mathbf{x}_1\ \mathbf{x}_2\ \mathbf{e}_3\ \mathbf{e}_4]$ is invertible for the convenience of inversion of X.

(d) Consider the linear transformation (see Example 6.5(c))

$$L : V_{2\times 2} \to \mathbb{P}_2(x) : A = \begin{bmatrix} a & b \\ c & d \end{bmatrix} \mapsto a + (b + c)x + dx^2.$$

Since E_{11}, E_{12}, E_{21}, E_{22} form a basis of $V_{2\times2}$, the set $\{LE_{11} = 1, LE_{12} = LE_{21} = x, LE_{22} = x^2\}$ generates the range space of L as a result of Remark 6.1. Hence rank $L = \dim \mathbb{P}_2(x) = 3$ and

L is onto. By Theorem 6.5, null $L = 4 - 3 = 1$ and hence in view of Theorem 6.3, L is not one-to-one. □

Remark 6.3 In the foregoing discussion, the dimensions n and m of the underlying vector spaces V_n and W_m are specifically mentioned as the subscripts. Sometimes it is convenient to write the vector spaces as V and W without explicitly referring to their dimensions as subscripts. Without any major difficulty, the previous definitions, theorems and discussion can be adapted to this changed notation.

Because of the fact that $L(V)$ is a subspace of W for a linear transformation $L : V \to W$, rank $L \le \dim W$ (see Definitions 6.2 and 6.7) and by virtue of Theorem 6.5, we have rank $L \le \dim V$. Linear transformations for which rank $L = \dim V = \dim W$ are of particular interest to us. This gives rise to Definition 6.8.

Definition 6.8 *Nonsingular Linear Transformation.* A linear transformation $L : V \to W$ is said to be nonsingular iff it is bijective, i.e., iff it is one-to-one and onto; otherwise, it is called singular. □

Theorem 6.6 A linear transformation $L : V \to W$ is nonsingular iff the set $\{L\mathbf{x}_1, ..., L\mathbf{x}_n\}$ is a basis of W whenever the set $\{\mathbf{x}_1, ..., \mathbf{x}_n)$ is a basis of V.

Proof (a) *Necessity.* Since L is injective, by Theorem 6.4, the set $\{L\mathbf{x}_1, ..., L\mathbf{x}_n\}$ is linearly independent. Since L is surjective, for any element $\mathbf{y}$ in W, there exists an element $\mathbf{x} = \alpha_1\mathbf{x}_1 + \cdots + \alpha_n\mathbf{x}_n$ in V_n such that

$$\mathbf{y} = L\mathbf{x} = \alpha_1 L\mathbf{x}_1 + \cdots + \alpha_n L\mathbf{x}_n.$$

Therefore, the set $\{L\mathbf{x}_1, ..., L\mathbf{x}_n\}$ is a basis of W (see Definition 5.5).

(b) *Sufficiency.* Let $\{\mathbf{x}_1, ..., \mathbf{x}_n\}$ be a basis of V. Then $\{L\mathbf{x}_1, ..., L\mathbf{x}_n\}$ is also a basis of W. For any two elements $\mathbf{x} = \alpha_1\mathbf{x}_1 + \cdots + \alpha_n\mathbf{x}_n$ and $\tilde{\mathbf{x}} = \tilde{\alpha}_1\mathbf{x}_1 + \cdots + \tilde{\alpha}_n\mathbf{x}_n$ such that $L\mathbf{x} = L\tilde{\mathbf{x}}$, we have $L(\mathbf{x} - \tilde{\mathbf{x}}) = \theta$. Therefore,

$$L[(\alpha_1 - \tilde{\alpha}_1)\mathbf{x}_1 + \cdots + (\alpha_n - \tilde{\alpha}_n)\mathbf{x}_n] = \theta,$$

which, as L is linear, gives us

$$(\alpha_1 - \tilde{\alpha}_1)L\mathbf{x}_1 + \cdots + (\alpha_n - \tilde{\alpha}_n)L\mathbf{x}_n = \boldsymbol{\theta}.$$

This implies $\alpha_1 - \tilde{\alpha}_1 = 0, \ldots, \alpha_n - \tilde{\alpha}_n = 0$. Hence, $\mathbf{x} = \tilde{\mathbf{x}}$, and L is injective.

Again, any element $\mathbf{y}$ in W can be expressed as

$$\mathbf{y} = \alpha_1 L\mathbf{x}_1 + \cdots + \alpha_n L\mathbf{x}_n.$$

Since L is a linear transformation, we have

$$\mathbf{y} = L(\alpha_1\mathbf{x}_1 + \cdots + \alpha_n\mathbf{x}_n) = L\mathbf{x},$$

i.e., every element $\mathbf{y}$ in W is the image of some element $\mathbf{x}$ in V. Therefore, L is surjective. □

Definition 6.9 *Composite Transformation or Product of Transformations.* Let $K : U \to V$ and $L : V \to W$ be two linear transformations. Then, a composite transformation from $U \to W$, or a product of two transformations K and L is denoted by $L \cdot K$ and is defined by

$$L \cdot K : U \to W : \mathbf{x} \mapsto (L \cdot K)\mathbf{x} = L(K\mathbf{x})$$

for all $\mathbf{x}$ in U. A composition of three and more transformations may be similarly defined. □

Example 6.11 Let two transformations $L : \mathbb{R}^3 \to \mathbb{R}^2$ and $K : \mathbb{R}^2 \to \mathbb{R}^3$ be defined by

$$L(x, y, z) = (x,\ z), \qquad K(x, y) = (0,\ x,\ y).$$

Then,

$$(K \cdot L)(x, y, z) = K(x, z) = (0, x, z),$$

$$(L \cdot K)(x, y) = L(0, x, y) = (0, y).$$

So, the two composite transformations $K \cdot L$ and $L \cdot K$ are, in general, not equal. In other words, the product of two transformations K and L is not generally commutative. This fact is analogous to the product of two matrices (see Example 1.3). For the transformation $\tilde{L} : \mathbb{R}^3 \to \mathbb{R}^2 : (x, y, z) \mapsto (0, x)$, we have $(\tilde{L} \cdot K)(x, y) = \tilde{L}(0, x, y) = (0, 0)$ so that $\tilde{L} \cdot K$ is the zero transformation Θ, but $\tilde{L} \neq \Theta$ and $K \neq \Theta$. Hence, a product of nonzero transformations can be the zero transformation. □

Definition 6.10 *Inverse Transformations.* For a linear transformation $L : V \to W$ if there exists a linear transformation $K : W \to V$ implying

$$V \xrightarrow{L} W \xrightarrow{K} V$$

such that

$$K \cdot L = I_V \tag{6.25}$$

then K is called the *left inverse* of L. On the other hand, if

$$W \xrightarrow{K} V \xrightarrow{L} W$$

ensures that

$$L \cdot K = I_W \tag{6.26}$$

then K is called the *right inverse* of L. If both (6.25) and (6.26) are satisfied, then K is said to be the *inverse* of L. Alternatively, L is called respectively the right inverse, left inverse and inverse of K. □

Example 6.12 Let L and K be two linear transformations $\mathbb{R} \to \mathbb{R}$, where $\mathbb{R}$ is the set of real numbers, defined by $Lx = x + 5$, $Kx = x - 5$, where x is in $\mathbb{R}$. Then $(K \cdot L)x = K(x + 5) = x = Ix$ and $(L \cdot K)x = L(x - 5) = x = Ix$. Therefore, $K \cdot L = L \cdot K = I$, and L has an inverse which is K; conversely, K has an inverse which is L.

Example 6.13 *Right Inverse.* Let L be a linear transformation: $\mathbb{R}^3 \to \mathbb{R}^2$ defined by $L(x, y, z) = (2x + 3y + z, x + 2y + 2z)$. Then $K : \mathbb{R}^2 \to \mathbb{R}^3$ defined by $K(x, y) = (2x - 3y, -x + 2y, 0)$ is a right inverse of L. This is because

$$(L \cdot K)(x, y) = L(2x - 3y, -x + 2y, 0) = (x, y).$$

If L is given, we may be asked to find its right inverse $K : \mathbb{R}^2 \to \mathbb{R}^3$ such that $(L \cdot K)(u\ v) = (u\ v)$. To do this, we define $K(u\ v) = (x\ y\ z)$. As L is known, we get

$$u = 2x + 3y + z, \quad v = x + 2y + 2z.$$

Solving these equations, we obtain

$$x = 2u - 3v + 4\alpha, \qquad y = -u + 2v - 3\alpha, \qquad z = \alpha = \text{arbitrary}.$$

Thus, there are an infinite number of right inverses of L given by

$$K(u\ v) = (2u - 3v + 4\alpha, -u + 2v - 3\alpha, \alpha),$$

α being arbitrary. □

Theorem 6.7 A linear transformation $L : V \to W$ has an inverse iff it is bijective.

Proof (a) *Necessity.* Suppose that $K : W \to V$ is the inverse of L so that (6.25) and (6.26) are satisfied. Let $\mathbf{x}$ and $\mathbf{y}$ be two elements of V such that $L\mathbf{x} = L\mathbf{y}$. Then

$$(K \cdot L)\mathbf{x} = K(L\mathbf{x}) = K(L\mathbf{y}) = (K \cdot L)\mathbf{y}$$

Since (6.25) is true, from above, $\mathbf{x} = \mathbf{y}$; so L is injective. On the other hand, let $\mathbf{z}$ be any element of W so that $K\mathbf{z} = \mathbf{x}$ is in V. In view of (6.26), we have

$$(L \cdot K)\mathbf{z} = \mathbf{z}.$$

Therefore, $L\mathbf{x} = \mathbf{z}$, where $\mathbf{x} = K\mathbf{z}$ is in V. This means that every element $\mathbf{z}$ of W is the image of some element $\mathbf{x}$ in V under L. Hence L is surjective.

(b) *Sufficiency.* Since L is surjective, for every element $\mathbf{y}$ of W, there is an element $\mathbf{x}$ in V such that $L\mathbf{x} = \mathbf{y}$. Because L is injective this element $\mathbf{x}$ is uniquely determined by $\mathbf{y}$. This correspondence of $\mathbf{y}$ in W to $\mathbf{x}$ in V can be defined by the transformation $K\mathbf{y} = \mathbf{x}$. Therefore,

$$K(L\mathbf{x}) = (K \cdot L)\mathbf{x} = \mathbf{x} \quad \text{and} \quad L(K\mathbf{y}) = (L \cdot K)\mathbf{y} = \mathbf{y}$$

which imply that K is an inverse of L. □

Corollary 6.1 A linear transformation $L : V \to W$ is invertible iff it is nonsingular.

Proof The proof follows from Definition 6.8 and Theorem 6.7. □

Theorem 6.8 A linear transformation $L : V \to W$ has

(a) a left inverse iff it is injective, and
(b) a right inverse iff it is surjective.

Proof (a) *Necessity.* It follows from the necessity part of Theorem 6.7.

Sufficiency. Since L is injective, for each $\mathbf{y}$ in W, there is at most one $\mathbf{x}$ in V with $\mathbf{y} = L\mathbf{x}$. If $\hat{\mathbf{x}}$ is an arbitrary element in V, then the function $K : W \to V$ defined by

$$K\mathbf{y} = \mathbf{x} \text{ with } L\mathbf{x} = \mathbf{y}, \quad \text{when } \mathbf{y} \text{ is in } L(V);$$
$$= \hat{\mathbf{x}}, \quad \text{otherwise};$$

returns each elements $L\mathbf{x}$ to $\mathbf{x}$. So, $K(L\mathbf{x}) = \mathbf{x}$ for every $\mathbf{x}$ implying that $K \cdot L = I_V$ and that K is the left inverse of L. In fact, L has many left inverses, one for each choice of $\hat{\mathbf{x}}$.

(b) *Necessity* It follows from the Necessity part of Theorem 6.7.

Sufficiency since $L : V \to W$ is surjective, to each $\mathbf{y}$ in W there is at least one $\mathbf{x}$ in V such that

$$L\mathbf{x} = \mathbf{y}.$$

Choose one such $\mathbf{x}$ for each $\mathbf{y}$ and define

$$K : W \to V$$

by letting $K\mathbf{y} = \mathbf{x}$, Then $L(K\mathbf{y}) = \mathbf{y}$; so, $L \cdot K = \mathrm{I}_W$. Since $L\mathbf{x} = \mathbf{y}$ may hold for more than one value of $\mathbf{x}$, there exist infinitely many right inverses for L.

Example 6.14 (a) *Inverse.* Let $L : \mathbb{R}^2 \to \mathbb{R}^2$ be a linear transformation defined by

$$L(x, y) = (2x + y, 3x + y)$$

Now ker L is obtained by solving $2x + y = 0$ and $3x + y = 0$ which give $x = 0$, $y = 0$. By Theorem 6.3, L is injective.

Let (u, v) be any given element in $\mathbb{R}^2$. Then

$$(u, v) = (2x + y, 3x + y)$$

gives the solution as

$$x = v - u \quad \text{and} \quad y = 3u - 2v. \tag{6.27}$$

Thus, any element (u, v) in the co-domain is the image of some element (x, y) given by (6.27) in the domain of L. So, L is surjective. Hence, L has an inverse by Theorem 6.7.

The inverse of L is a linear transformation K satisfying

$$(K \cdot L)(x, y) = K(2x + y, 3x + y) = (x, y)$$

By choosing the elementary vectors $\mathbf{e}_1$ and $\mathbf{e}_2$ as a basis in $\mathbb{R}^2$, we have, by Theorem 6.6, since L is nonsingular (see Definition 6.8), that the vectors

$$L(1\ 0) = (2\ 3), \quad L\ (0\ 1) = (1\ 1)$$

form a basis of $\mathbb{R}^2$. Our problem is now to determine, as in Example 6.6, a linear transformation K satisfying

$$K(2\ 3) = (1\ 0) \quad \text{and} \quad K(1\ 1) = (0\ 1).$$

Any element (u, v) in $\mathbb{R}^2$ can be expressed in terms of basis vectors (2 3) and (1 1) as (see Example 6.6)

$$(u\ \ v) = (v - u, 3u - 2v)\begin{bmatrix} 2 & 3 \\ 1 & 1 \end{bmatrix}.$$

By Theorem 6.1, the required linear transformation K is

$$K(u, v) = (v - u)(1\ 0) + (2u - 2v)(0\ 1) = (v - u, 3u - 2v).$$

(b) *Left Inverse.* Let a linear transformation $L : \mathbb{R}^2 \to \mathbb{R}^3$ be defined by $L(x, y) = (x - y, 0.5x, -0.5x + y)$. Since $L(x, y) = 0$ gives $x = y = 0$, we have $\ker L = 0$, and L is injective by Theorem 6.3. By Theorem 6.8, L is left invertible. To find the left inverse $K : \mathbb{R}^3 \to \mathbb{R}^2$ satisfying $(K \cdot L)(x, y) = (x, y)$, we define $K(u, v, w) = (x, y)$. And as L is known, equating the left sides of the above two equations, we have

$$u = x - y, \qquad v = 0.5x \quad \text{and} \quad w = -0.5x + y.$$

Hence, solving $x = 2v$, $y = 2v - u = v + w$, and so $K(u, v, w) = (2v, 2v - u)$ defines an infinite number of left inverses for every w or $K(u, v, w) = (2v, w + v)$ for every u.

6.3 MATRIX OF A LINEAR TRANSFORMATION

We shall now study how explicit computations with linear transformations can be translated into computations with matrices having scalar elements. In view of this, let $L : V \to W$ be a linear transformation between two arbitrary vector spaces V and W over the field $\mathbb{F}$ in which $X := \{\mathbf{x}_1, \mathbf{x}_2, ..., \mathbf{x}_n\}$ and $Y := \{\mathbf{y}_1, \mathbf{y}_2, ..., \mathbf{y}_m\}$ are respectively two given bases. Since every vector $L\mathbf{x}_i$ $(i = 1, 2, ..., n)$ is in W, it is a linear combination of the basis vectors $\mathbf{y}_1, \mathbf{y}_2, ..., \mathbf{y}_m$ in W and can, therefore, be written as

$$L\mathbf{x}_i = a_{1i}\mathbf{y}_1 + a_{2i}\mathbf{y}_2 + \cdots + a_{mi}\mathbf{y}_m, \tag{6.28}$$

in which the elements a_{ij}'s are uniquely determined scalars in $\mathbb{F}$ as a result of Corollary 5.1. Now any vector $\mathbf{x} = \alpha_1\mathbf{x}_1 + \alpha_2\mathbf{x}_2 + \ldots + \alpha_n\mathbf{x}_n$ in V is transformed into a vector $\mathbf{y}$ in W by L so that we can write

$$\begin{aligned}
\mathbf{y} = L\mathbf{x} &= \alpha_1 L\mathbf{x}_1 + \alpha_2 L\mathbf{x}_2 + \cdots + \alpha_n L\mathbf{x}_n \\
&= \alpha_1(a_{11}\mathbf{y}_1 + \alpha_{21}\mathbf{y}_2 + \cdots + \alpha_{m1}\mathbf{y}_m) + \alpha_2(\alpha_{12}\mathbf{y}_1 + a_{22}\mathbf{y}_2 + \cdots + \alpha_{m2}\mathbf{y}_m) \\
&\quad + \cdots + \alpha_n(a_{1n}\mathbf{y}_1 + a_{2n}\mathbf{y}_2 + \cdots + a_{mn}\mathbf{y}_m) \\
&= (\alpha_1 a_{11} + \alpha_2 a_{12} + \cdots + \alpha_n a_{1n})\mathbf{y}_1 + \cdots + (\alpha_1 a_{m1} + \alpha_2 a_{m2} + \cdots + \alpha_n a_{mn})\mathbf{y}_m \\
&= \beta_1\mathbf{y}_1 + \cdots + \beta_m\mathbf{y}_m,
\end{aligned} \tag{6.29}$$

where

$$\beta_i = \alpha_1 a_{i1} + \cdots + \alpha_n a_{in} \tag{6.30}$$

and α_i's are scalars in $\mathbb{F}$. So it appears from (6.29) that computations with the linear transformation $L : V \to W$, as far as assignment of vectors $\mathbf{x}$ to $\mathbf{y} = L\mathbf{x}$ is concerned, can be done if the coordinates $(\beta_1, \beta_2, ..., \beta_m)$ of $\mathbf{y}$ with respect to the given basis Y, are known. Writing (6.30) explicitly, we have

$$\begin{bmatrix} \beta_1 \\ \beta_2 \\ \vdots \\ \beta_m \end{bmatrix} = \begin{bmatrix} a_{11} & a_{12} & \cdots & a_{1n} \\ a_{21} & a_{22} & \cdots & a_{2n} \\ \vdots & \vdots & & \vdots \\ a_{m1} & a_{m2} & \cdots & a_{mn} \end{bmatrix} \begin{bmatrix} \alpha_1 \\ \alpha_2 \\ \vdots \\ \alpha_n \end{bmatrix}, \qquad (6.31)$$

which reveals that the determination of the coordinates ($\beta_1, \beta_2, ..., \beta_m$) of **y** can be accomplished if we know the matrix $A = [a_{ij}]$ and the coordinates ($\alpha_1, \alpha_2, ..., \alpha_n$) of **x** with respect to the given basis X. For a known linear transformation L, the elements a_{ij} in the matrix $A = [a_{ij}]$ are uniquely determined by relations (6.28) with reference to the given bases X and Y, and so are the coordinates of the given vector **x**.

Thus, given a linear transformation $L : V \to W$ with respective bases X and Y, and an element **x** in V, one can compute the vector $\mathbf{y} = L\mathbf{x}$ in W by first determining the matrix $A = [a_{ij}]$ whose elements a_{ij} are uniquely determined by (6.28) with respect to the bases X and Y and the coordinate vector $\mathbf{u} = [\alpha_1\ \alpha_2\ \cdots\ \alpha_n]^T$ of **x** with respect to X. Having determined A and **u**, the coordinate vector $\mathbf{v} = [\beta_1\ \beta_2\ \cdots\ \beta_m]^T$ of **y** is then obtained by making use of (6.31) which when put into (6.29) specifies $\mathbf{y} = L\mathbf{x}$ completely. So the computation of the linear transformation $\mathbf{y} = L\mathbf{x}$ is greatly assisted by the associated matrix relation $\mathbf{v} = A\mathbf{u}$ and these relations $\mathbf{y} = L\mathbf{x}$ and $\mathbf{v} = \mathbf{Au}$ are very similar in nature.

Provided that the bases X and Y are kept unaltered, the matrix A defined by the coefficients a_{ij}'s in (6.28) is unique for a given linear transformation L. The rectangular array of mn elements a_{ij}'s uniquely determined as (6.28) is defined by *the matrix* $A = [a_{ij}]$ *of the given linear transformation* L with respect to the specified bases X and Y.

Conversely, given any arbitrary $m \times n$ matrix A, the bases X and Y in vector spaces V and W respectively, there is a unique linear transformation $L : V \to W$, whose action on $\mathbf{x}_1, \mathbf{x}_2, ..., \mathbf{x}_n$ is defined by (6.28) as a result of Theorem 6.1.

The results pertaining to the foregoing discussion can be summarized in the following theorem.

Theorem 6.9 Let $L : V \to W$ be a linear transformation between any two arbitrary vector spaces V and W with bases X: = $\{\mathbf{x}_1, \mathbf{x}_2, ..., \mathbf{x}_n\}$ and Y: = $\{\mathbf{y}_1, \mathbf{y}_2, ..., \mathbf{y}_m\}$. If $\mathbf{u} = \{\alpha_1\ \alpha_2\ \cdots\ \alpha_n\}^T$ is the coordinate vector of an element **x** in V, $\mathbf{v} = [\beta_1\ \beta_2\ \cdots\ \beta_m]^T$ that of an element **y** in W, and A is a matrix representation of L, all with respect to the given bases, then

$$L\mathbf{x} = \mathbf{y} \quad \text{iff } A\mathbf{u} = \mathbf{v}.$$

Proof (a) *Sufficiency.* Suppose $A\mathbf{u} = \mathbf{v}$. By definition, $\mathbf{x} = \alpha_1\mathbf{x}_1 + \alpha_2\mathbf{x}_2 + \cdots + \alpha_n\mathbf{x}_n$ and $\mathbf{y} = \beta_1\mathbf{y}_1 + \beta_2\mathbf{y}_2 + \cdots + \beta_m\mathbf{y}_m$. From the latter relation, we retrace the steps in (6.29) to get $\mathbf{y} = L\mathbf{x}$.

(b) *Necessity.* Assuming $L\mathbf{x} = \mathbf{y}$, it follows from (6.30) that $A\mathbf{u} = \mathbf{v}$.
Thus the vector equation $\mathbf{y} = L\mathbf{x}$ corresponds to the matrix equation $\mathbf{v} = A\mathbf{u}$ and vice versa. □

Example 6.15 Let A, the matrix representation of a linear transformation $L : \mathbb{R}^3 \to \mathbb{R}^2$ with respect to the bases $\mathbf{x}_1 = [5\ 1\ 3]^T$, $\mathbf{x}_2 = [3\ 2\ 2]^T$, $\mathbf{x}_3 = [1\ 2\ 1]^T$ and $\mathbf{y}_1 = [-1\ 1]^T$, $\mathbf{y}_2 = [2\ 1]^T$, be given by

$$A = \begin{bmatrix} 2 & -1 & 1 \\ 1 & 2 & -2 \end{bmatrix}.$$

Find the linear transformation L.

Solution We have

$$L[\mathbf{x}_1\ \mathbf{x}_2\ \mathbf{x}_3] = [\mathbf{y}_1\ \mathbf{y}_2]A = \begin{bmatrix} 0 & 5 & -5 \\ 3 & 1 & -1 \end{bmatrix}. \tag{6.32}$$

The coordinates $(\alpha_1, \alpha_2, \alpha_3)$ of any vector $\mathbf{x} = [x_1, x_2, x_3]^T$ in $\mathbb{R}^3$ with respect to the given basis vectors [see (5.72) and Example 5.43] can be determined to give

$$\alpha_1 = -2x_1 - x_2 + 4x_3, \qquad \alpha_2 = 5x_1 + 2x_2 - 9x_3, \qquad \alpha_3 = -4x_1 - x_2 + 7x_3.$$

Observing that $\mathbf{x} = \alpha_1\mathbf{x}_1 + \alpha_2\mathbf{x}_2 + \alpha_3\mathbf{x}_3$, we have, with the help of (6.32), the relation

$$L\mathbf{x} = \alpha_1 L\mathbf{x}_1 + \alpha_2 L\mathbf{x}_2 + \alpha_3 L\mathbf{x}_3 = \begin{bmatrix} 5(9x_1 + 3x_2 - 16x_3) \\ 3x_1 - 4x_3 \end{bmatrix},$$

which defines the linear transformation L. □

Example 6.16 Let A, the matrix representation of the linear transformation $L : \mathbb{P}_1(t) \to \mathbb{P}_2(t)$ with respect to the bases $\{1 - t, t\}$ and $\{1, t, 1 + t + t^2\}$, be given by

$$A = \begin{bmatrix} 2 & -1 \\ -1 & -1 \\ 1 & 1 \end{bmatrix}.$$

Determine L.

Solution We have

$$L[1 - t, t] = [1, t, 1 + t + t^2]A = [3 + t^2, t^2].$$

Now, the coordinates of any vector $p(t) = a_0 + a_1 t$ in $\mathbb{P}_1(t)$ with respect to the given basis are $(a_0, a_0 + a_1)$ because

$$p(t) = a_0 + a_1 t = a_0(1 - t) + (a_0 + a_1)t.$$

Hence, taking into consideration the expression for $L[1 - t, t]$, we have

$$\begin{aligned} Lp(t) &= a_0 L(1 - t) + (a_0 + a_1)L(t) \\ &= a_0(3 + t^2) + (a_0 + a_1)t^2 = 3a_0 + (2a_0 + a_1)t^2, \end{aligned}$$

which is the desired linear transformation. □

Corollary 6.2 Let $L : V \to W$ be a linear transformation whose matrix representation with respect to the bases $X: = \{\mathbf{x}_1, \mathbf{x}_2, ..., \mathbf{x}_n\}$ and $Y: = \{\mathbf{y}_1, \mathbf{y}_2, ..., \mathbf{y}_m\}$ is A. Prove that L is invertible iff A is invertible.

Proof Let $\mathbf{u} = [\alpha_1\ \alpha_2 \cdots \alpha_n]^T$ and $\mathbf{v} = [\beta_1\ \beta_2 \cdots \beta_m]^T$ be the coordinate vectors of $\mathbf{x}$ in V and of $\mathbf{y}$ in W respectively, with respect to the given bases.

(a) *Necessity.* If $L\mathbf{x} = \mathbf{y} = \boldsymbol{\theta}$ and $\mathbf{x} = \boldsymbol{\theta}$, then $\mathbf{v} = \boldsymbol{\theta}$ and $\mathbf{u} = \boldsymbol{\theta}$, because the coordinate vector $\mathbf{u}$ or $\mathbf{v}$ of a zero element in V or W is $\boldsymbol{\theta}$; also, by Theorem 6.9 $L\mathbf{x} = \mathbf{y}$ iff $A\mathbf{u} = \mathbf{v}$. If we now suppose that L is one-to-one, i.e., if $L\mathbf{x} = \boldsymbol{\theta}$ implies $\mathbf{x} = \boldsymbol{\theta}$, then from the foregoing it follows that $A\mathbf{u} = \boldsymbol{\theta}$ implies $\mathbf{u} = \boldsymbol{\theta}$. So A is one-to-one. Next suppose that L is onto. Therefore, for every $\mathbf{y}$ in W there exists a $\mathbf{x}$ in V such that $L\mathbf{x} = \mathbf{y}$. This implies by Theorem 6.9 that for every $\mathbf{v}$ there exists a $\mathbf{u}$ such that $A\mathbf{u} = \mathbf{v}$. Hence A is onto. Thus, if L is invertible, so also is A.

(b) *Sufficiency.* It follows by retracing the above arguments. □

As a consequence of Corollary 6.2 and Theorem 4.2, a linear transformation L is one-to-one, iff its matrix representation A is left invertible (i.e., has full column rank) and onto iff A is right invertible (i.e., has full row rank). Indeed, if A is left invertible and B is its left inverse then $A\mathbf{x} = \boldsymbol{\theta}$ implies $BA\mathbf{x} = B\boldsymbol{\theta}$, i.e., $\mathbf{x} = \boldsymbol{\theta}$. However, if A is right invertible and C is its right inverse then for each $\mathbf{y}$, the equation $A\mathbf{x} = \mathbf{y}$ has a solution $\mathbf{x} = C\mathbf{y}$.

In the foregoing discussion, a matrix is so construed as to be tied up with a linear transformation and the bases of the underlying vector spaces V and W, although it can exist in its own right, as already discussed. It is important to note that the coefficients $a_{1i}, a_{2i}, ..., a_{mi}$ in the expression (6.28) for $L\mathbf{x}_i$ change their horizontal disposition to take up a vertical position in the ith column of the associated matrix $[a_{ij}]$ in (6.31).

In any vector space there may be a number of different bases, and consequently, the matrices $[a_{ij}]$ associated with a given linear transformation in respect of different bases are different.

Any linear transformation can be associated with a matrix, and a matrix defined as a rectangular array of elements satisfies (6.3), and is, therefore, a linear transformation. We can thus use the words *matrix* and *linear transformation* interchangeably. Matrices associated with linear transformations from one vector space V into itself, i.e., with $L : V \to V$, are square matrices. In such cases we shall say that A is a matrix associated with a linear transformation L on a vector space V with respect to a basis $\{\mathbf{x}_1, \mathbf{x}_2, ..., \mathbf{x}_n\}$ or, more concisely, A is a matrix with respect to the basis $\{\mathbf{x}_1, \mathbf{x}_2, ..., \mathbf{x}_n\}$. We shall always select the same basis of the two underlying vector spaces V while considering linear transformation L from V into itself although two different bases may be selected if it is desired.

Remark 6.4 An important vector space V of interest is $\mathbb{C}^n$ or $\mathbb{R}^n$, where the elements are column n-vectors. Another vector space which will play a significant role in the clarification of basic concepts in linear algebra is the vector space $\mathbb{P}_n(t)$ of polynomials in the variable t of degree $\leq n$. In these cases the expression (6.28) can be rewritten as

$$L[\mathbf{x}_1\ \ \mathbf{x}_2\ \cdots\ \mathbf{x}_n]: = [L\mathbf{x}_1\ \ L\mathbf{x}_2\ \cdots\ L\mathbf{x}_n]$$

$$= [\mathbf{y}_1\ \ \mathbf{y}_2 \cdots \mathbf{y}_m]\begin{bmatrix} a_{11} & a_{12} & \cdots & a_{1n} \\ a_{21} & a_{22} & \cdots & a_{2n} \\ \vdots & \vdots & & \vdots \\ a_{m1} & a_{m2} & \cdots & a_{mn} \end{bmatrix}. \tag{6.33}$$

Therefore,

$$L\mathbf{V} = \mathbf{W}A,$$

where $\mathbf{V}: = [\mathbf{x}_1\ \ \mathbf{x}_2\ \cdots\ \mathbf{x}_n]$ and $\mathbf{W}: = [\mathbf{y}_1\ \ \mathbf{y}_2\ \cdots\ \mathbf{y}_n]$ are called *basis matrices* in V_n and W_m respectively. Concentrating on a linear transformation $L : \mathbb{R}^n \to \mathbb{R}^m$ such that $L = \mathbf{W}A\mathbf{V}^{-1}$ and choosing natural bases in these spaces, we have $\mathbf{V} = I_n$ and $\mathbf{W} = I_m$; consequently, $L\mathbf{x} = A\mathbf{x}$ for each element $\mathbf{x}$ in $\mathbb{R}^n$. However, for arbitrary bases in the vector spaces we have specified, this transformation through the matrix A is executed by $L\mathbf{x} = \mathbf{W}A\mathbf{V}^{-1}\mathbf{x}$ (see Example 6.17). The reason why representation of the above form is not defined for a general vector space will be clear if we consider the elements of V_m, viz., $\mathbf{y}_i$'s in (6.33), to be row-vectors.

6.3.1 Matrix of an Identity and a Zero Transformation

The matrix of an identity transformation I_V: $V \to V$ (see Definition 6.4) is obtained by noting that in place of (6.28) we now have

$$I_V\mathbf{x}_i = 1 \cdot \mathbf{x}_i, \quad i = 1, 2, ..., n,$$

which shows that $a_{ii} = 1$, and $a_{ij} = 0$ $(i \neq j)$ in (6.28). Thus, the matrix of I_V turns out to be I_n, a unit matrix of order n where $n = \dim V$. So, irrespective of bases, the matrix of an identity transformation I_V is a unit matrix. Similarly, it can be shown that a zero transformation $\Theta : V \to W$, see Definition 6.4, can be represented by a null matrix of order $m \times n$, which is not dependent on the bases of V and W.

Example 6.17 A linear transformation L on $\mathbb{R}^3$ is defined as

$$L\mathbf{x} = \begin{bmatrix} x_1 + 2x_2 \\ 3x_3 \\ x_2 - 2x_3 \end{bmatrix}, \qquad \mathbf{x} = \begin{bmatrix} x_1 \\ x_2 \\ x_3 \end{bmatrix}.$$

Find the matrix A associated with L with respect to the following bases:

$$\text{(a)} \quad \mathbf{e}_1 = \begin{bmatrix} 1 \\ 0 \\ 0 \end{bmatrix}, \qquad \mathbf{e}_2 = \begin{bmatrix} 0 \\ 1 \\ 0 \end{bmatrix}, \qquad \mathbf{e}_3 = \begin{bmatrix} 0 \\ 0 \\ 1 \end{bmatrix}.$$

$$\text{(b)} \quad \mathbf{a}_1 = \begin{bmatrix} 1 \\ 1 \\ 1 \end{bmatrix}, \qquad \mathbf{a}_2 = \begin{bmatrix} 0 \\ 1 \\ 1 \end{bmatrix}, \qquad \mathbf{a}_3 = \begin{bmatrix} 0 \\ 0 \\ 1 \end{bmatrix}.$$

Solution (a) Here (see Example 5.43),

$$L\mathbf{e}_1 = \begin{bmatrix} 1 \\ 0 \\ 0 \end{bmatrix} = 1 \cdot \mathbf{e}_1 + 0 \cdot \mathbf{e}_2 + 0 \cdot \mathbf{e}_3,$$

$$L\mathbf{e}_2 = \begin{bmatrix} 2 \\ 0 \\ 1 \end{bmatrix} = 2\mathbf{e}_1 + 0 \cdot \mathbf{e}_2 + 1 \cdot \mathbf{e}_3,$$

$$L\mathbf{e}_3 = \begin{bmatrix} 0 \\ 3 \\ -2 \end{bmatrix} = 0 \cdot \mathbf{e}_1 + 3\mathbf{e}_2 - 2\mathbf{e}_3.$$

Therefore,

$$A = \begin{bmatrix} 1 & 2 & 0 \\ 0 & 0 & 3 \\ 0 & 1 & -2 \end{bmatrix}.$$

In this example, the matrix A plays the same role as the linear transformation L in the sense that $L\mathbf{x} = A\mathbf{x}$. This is so because, A is the matrix representation of the linear transformation L on a natural basis.

(b) In this case, (see Example 5.43),

$$L\mathbf{a}_1 = \begin{bmatrix} 3 \\ 3 \\ -1 \end{bmatrix} = 3\begin{bmatrix} 1 \\ 1 \\ 1 \end{bmatrix} + 0\begin{bmatrix} 0 \\ 1 \\ 1 \end{bmatrix} - 4\begin{bmatrix} 0 \\ 0 \\ 1 \end{bmatrix} = 3\mathbf{a}_1 + 0 \cdot \mathbf{a}_2 - 4\mathbf{a}_3,$$

$$L\mathbf{a}_2 = \begin{bmatrix} 2 \\ 3 \\ -1 \end{bmatrix} = 2\mathbf{a}_1 + \mathbf{a}_2 - 4\mathbf{a}_3, \qquad L\mathbf{a}_3 = \begin{bmatrix} 0 \\ 3 \\ -2 \end{bmatrix} = 0 \cdot \mathbf{a}_1 + 3\mathbf{a}_2 - 5\mathbf{a}_3$$

Therefore,

$$A = \begin{bmatrix} 3 & 2 & 0 \\ 0 & 1 & 3 \\ -4 & -4 & -5 \end{bmatrix}.$$ □

In this example, however, $L\mathbf{x} = \mathbf{V}A\mathbf{V}^{-1}\mathbf{x}$, where $\mathbf{V} = [\mathbf{a}_1\ \mathbf{a}_2\ \mathbf{a}_3]$, because the basis in which A is represented is not a natural one (see Remark 6.4).

Example 6.18 Consider the vector space $\mathbb{P}_n(t)$ of polynomials and the linear transformation D: $\mathbb{P}_n(t) \to \mathbb{P}_{n-1}(t)$, as given in Example 6.2. Observing that $\{1, t, ..., t^n\}$ is a basis of $\mathbb{P}_n(t)$ and $\{1, t, ..., t^{n-1}\}$ is that of $\mathbb{P}_{n-1}(t)$, we can write, as a result of (6.28), the equations

$$
\begin{aligned}
D1 &= 0 \cdot 1 + 0 \cdot t + \cdots + 0 \cdot t^{n-1}, \\
Dt &= 1 \cdot 1 + 0 \cdot t + \cdots + 0 \cdot t^{n-1}, \\
&\vdots \\
Dt^n &= 0 \cdot 1 + 0 \cdot t + \cdots + n \cdot t^{n-1}.
\end{aligned}
$$

Therefore, the matrix A associated with the linear transformation D is

$$
A = \begin{bmatrix} 0 & 1 & 0 & \cdots & 0 \\ 0 & 0 & 2 & \cdots & 0 \\ \vdots & \vdots & \vdots & & \vdots \\ 0 & 0 & 0 & \cdots & n \end{bmatrix}_{n+1 \times n}
$$

(see also Example 6.31). □

6.3.2 Matrix of the Sum of Two Linear Transformations and a Scalar Multiple of a Linear Transformation

Referring to (6.12), if L and $\hat{L}$ are the elements of $\mathcal{L}(V_n, W_m)$ with a basis $\{\mathbf{x}_1, \ldots, \mathbf{x}_n\} =: X$ in V_n and a basis $\{\mathbf{y}_1, \ldots, \mathbf{y}_m\} =: Y$ in W_m, and α is scalar, then we have, in view of (6.28), the relations

$$
\begin{aligned}
(L + \hat{L})\mathbf{x}_i &= L\mathbf{x}_i + \hat{L}\mathbf{x}_i = (a_{1i}\mathbf{y}_1 + \cdots + a_{mi}\mathbf{y}_m) \\
&\quad + (\hat{a}_{1i}\mathbf{y}_1 + \cdots + \hat{a}_{mi}\mathbf{y}_m), \qquad i = 1, \ldots, n, \\
(\alpha L)\mathbf{x}_i &= L(\alpha \mathbf{x}_i) = \alpha L\mathbf{x}_i = (\alpha a_{1i}\mathbf{y}_1 + \cdots + \alpha a_{mi}\mathbf{y}_m), \qquad i = 1, \ldots, n,
\end{aligned}
$$

where a_{ij}'s and $\hat{a}_{ij}$'s are unique scalars for the given basis Y, and determine the matrices $A = [a_{ij}]$ and $\hat{A} = [\hat{a}_{ij}]$ of L and $\hat{L}$ respectively. Hence, the equations corresponding to (6.30) for $L + \hat{L}$ and αL now become

$$
\begin{aligned}
\beta_i &= \alpha_1(a_{i1} + \hat{a}_{i1}) + \cdots + \alpha_n(a_{in} + \hat{a}_{in}), \qquad i = 1, \ldots, m, \\
\beta_i &= \alpha_1(\alpha a_{i1}) + \cdots + \alpha_n(\alpha a_{in}), \qquad i = 1, \ldots, m,
\end{aligned}
$$

respectively, from which it follows that (i) the matrix associated with the sum $L + \hat{L}$ of two linear transformations L and $\hat{L}$ is the sum $A + \hat{A}$ of their individual matrices and (ii) the matrix associated with α times a linear transformation L is α times its associated matrix A.

6.3.3 Matrix of a Composite Transformation

We now consider two linear transformations $K : U \to V$ and $L : V \to W$, and the bases in these spaces U, V, and W are respectively denoted by $Z = \{\mathbf{z}_1, \ldots, \mathbf{z}_p\}$, $X = \{\mathbf{x}_1, \ldots, \mathbf{x}_n\}$, and $Y =$

$\{\mathbf{y}_1, ..., \mathbf{y}_m\}$. Denote by $A = [a_{ki}]_{n\times p}$ and $B = [b_{ik}]_{m\times n}$ the matrices of these linear transformations K and L respectively, with respect to the above bases. Then we have

$$K\mathbf{z}_j = a_{1j}\mathbf{x}_1 + \cdots + a_{kj}\mathbf{x}_k + \cdots + a_{nj}\mathbf{x}_n,$$

$$L\mathbf{x}_k = b_{1k}\mathbf{y}_1 + \cdots + b_{ik}\mathbf{y}_k + \cdots + b_{mk}\mathbf{y}_m,$$

Therefore, because of linearity of K and L,

$$(L\cdot K)\mathbf{z}_j = L(K\mathbf{z}_j) = \sum_{k=1}^{n} a_{kj} L\mathbf{x}_k = \sum_{k=1}^{n}\sum_{i=1}^{m} a_{kj} b_{ik}\mathbf{y}_i = \sum_{i=1}^{m} c_{ij}\mathbf{y}_i,$$

where

$$c_{ij} = \sum_{k=1}^{n} b_{ik} a_{kj},$$

which shows that the matrix of the composite transformation $L\cdot K$ is $C = [c_{ij}] = BA$ in view of Definition 1.7. In particular, the matrix of L^2 is A^2 and that of L^k, the composite of k number of like transformations on V, is A^k.

6.3.4 Matrix of an Inverse Transformation

Let $K : W \to V$ be the inverse of a linear transformation $L : V \to W$ which are related by (6.25) and (6.26). By Theorem 6.7, L is bijective, and as a result of Definition 6.8 and Theorem 6.6, it obviously follows that dim V = dim W. If $X := \{\mathbf{x}_1, \mathbf{x}_2, ..., \mathbf{x}_n\}$ and $Y := \{\mathbf{y}_1, \mathbf{y}_2 ..., \mathbf{y}_n\}$ are the bases of V and W respectively, and $A = [a_{ij}]$ and $B = [b_{ij}]$ are the matrix representations of L, and K respectively with respect to these bases, then we have, in view of (6.28), the relations

$$L\mathbf{x}_k = a_{1k}\mathbf{y}_1 + \cdots + a_{ik}\mathbf{y}_i + \cdots + a_{nk}\mathbf{y}_n,$$

$$K\mathbf{y}_j = b_{1j}\mathbf{x}_1 + \cdots + b_{kj}\mathbf{x}_k + \cdots + b_{nj}\mathbf{x}_n.$$

Hence, since L is linear,

$$(L\cdot K)\mathbf{y}_j = L(K\mathbf{y}_j) = \sum_{k=1}^{n} b_{kj} L\mathbf{x}_k = \sum_{k=1}^{n} b_{kj}\sum_{i=1}^{n} a_{ik}\mathbf{y}_i = \sum_{i=1}^{n}\left(\sum_{k=1}^{n} a_{ik} b_{kj}\right)\mathbf{y}_i.$$

So the matrix of $L\cdot K$ with respect to the given basis Y is AB (see Section 6.3.3). Again since L is right invertible, $L\cdot K = I_W$ (see Definition 6.10) which shows that the matrix of $L\cdot K$ is alternatively given by I_n (see section 6.3.1) with respect to the same basis Y. As the matrix of a linear transformation is unique with respect to a given basis, we must have $AB = I_n$. On the contrary, when L is considered to be left invertible, we get, $BA = I_n$. So the matrix A of a linear transformation L has an inverse equal to the matrix B of its inverse transformation K.

We have already observed that the set of all linear transformations $\mathcal{L}(V_n, W_m)$ is a vector space and since an $m \times n$ matrix A is a linear transformation (see (6.3)) $A : V_n \to W_m$, the set of all such $m \times n$ matrices form an mn dimensional vector space as a result of Theorem 6.2.

If the linear transformation $L : V \to W$ is invertible and A is a matrix of L with respect to bases X and Y, then the matrix of L^{-1} is A^{-1}. Consequently, it follows from $L\mathbf{x} = \mathbf{y}$ iff

$A\mathbf{u} = \mathbf{v}$ that $L^{-1}\,\mathbf{y} = \mathbf{x}$ iff $A^{-1}\,\mathbf{v} = \mathbf{u}$, where $\mathbf{u}$ and $\mathbf{v}$ are coordinates with respect to bases X and Y respectively.

For example, let $L : \mathbb{P}(t) \to \mathbb{P}(t)$ over $\mathbb{R}$ defined by $Lf(t) = f(t) + f'(t) + tf''(t)$. To find the matrix representation of L with respect to a standard basis $\{1, t, t^2\}$, we note that

$$L(1) = 1 = 1 \cdot 1 + 0 \cdot t + 0 \cdot t^2,$$
$$L(t) = t + 1 = 1 \cdot 1 + 1 \cdot t + 0 \cdot t^2,$$
$$L(t^2) = t^2 + 4t = 0 \cdot 1 + 4 \cdot t + 1 \cdot t^2.$$

So,

$$A = \begin{bmatrix} 1 & 1 & 0 \\ 0 & 1 & 4 \\ 0 & 0 & 1 \end{bmatrix}.$$

Since A is invertible so also L. The matrix representation of L^{-1} is

$$A^{-1} = \begin{bmatrix} 1 & -1 & 4 \\ 0 & 1 & -4 \\ 0 & 0 & 1 \end{bmatrix}.$$

Let $\mathbf{y} = g(t) = a_0 + a_1 t + a_2 t^2$ with coordinates v $= [a_0\ a_1\ a_2]^T$. To find $L^{-1}\,g(t)$ we note that

$$\mathbf{u} = A^{-1}\mathbf{v} = [a_0 - a_1 + 4a_2,\ a_1 - 4a_2,\ a_2]^T$$

$\mathbf{x} = (a_0 - a_1 + 4a_2) + (a_1 - 4a_2)t + a_2 t^2$. So, $L^{-1}\,\mathbf{y} = \mathbf{x}$, where $\mathbf{x}$ and $\mathbf{y}$ are as given above.

6.4 CHANGE OF BASIS

We have already explained (see Figure 5.2) that if with respect to a basis $\{\mathbf{e}_1, \mathbf{e}_2\}$ a vector has the representation $[2\ 2]^T$, then with respect to a different basis $\{\mathbf{a}_1 = (1/2)[\sqrt{2}, \sqrt{2}]^T, \mathbf{a}_2 = (1/2)[\sqrt{2}, -\sqrt{2}]^T\}$, the same vector has the representation $[2\sqrt{2},\ 0]^T$. Sometimes it is imperative to change the basis in representing a linear transformation L, because relative to this new basis, the representation of L may become very much simplified. We shall, therefore, turn our attention to establish one important result concerning the matrix representation of a linear transformation when the basis is changed. This aspect has an important bearing on our study of canonical forms of matrices in Section 10.6.

Theorem 6.10 If two sets of vectors $X = \{\mathbf{x}_1, \mathbf{x}_2, ..., \mathbf{x}_n\}$ and $\hat{X} = \{\hat{\mathbf{x}}_1, \hat{\mathbf{x}}_2, ..., \hat{\mathbf{x}}_n\}$ are the bases of a vector space V_n, then there exists a nonsingular matrix $B = [b_{ij}]$ such that

$$\hat{\mathbf{x}}_i = b_{1i}\mathbf{x}_1 + b_{2i}\mathbf{x}_2 + \cdots + b_{ni}\mathbf{x}_n, \qquad i = 1, 2, \ldots, n. \tag{6.34}$$

The nonsingular matrix B is defined as a *(coordinate) transformation matrix* in the vector space V_n.

Proof Suppose that X and $\hat{X}$ are two bases in V_n. Then $\hat{\mathbf{x}}_i$'s ($i = 1, 2, ..., n$) can be expressed as a linear combination of $\mathbf{x}_1, \mathbf{x}_2, ..., \mathbf{x}_n$, i.e.,

$$\hat{\mathbf{x}}_i = b_{1i}\mathbf{x}_1 + b_{2i}\mathbf{x}_2 + \cdots + b_{ni}\mathbf{x}_n, \quad i = 1, 2, \ldots, n,$$

where b_{ij}'s are scalars. Define

$$B = [\mathbf{b}_1 \; \mathbf{b}_2 \; \cdots \; \mathbf{b}_n],$$

where

$$\mathbf{b}_i = [b_{1i} \; b_{2i} \; \cdots \; b_{ni}]^T \tag{6.35}$$

is an n-vector. To show that B is nonsingular, for scalars $\alpha_1 \; \cdots \; \alpha_n$ we write, with the help of (6.34),

$$\alpha_1\hat{\mathbf{x}}_1 + \cdots + \alpha_n\hat{\mathbf{x}}_n = \alpha_1(b_{11}\mathbf{x}_1 + \cdots + b_{n1}\mathbf{x}_n) + \cdots + \alpha_n(b_{1n}\mathbf{x}_1 + \cdots + b_{nn}\mathbf{x}_n). \tag{6.36}$$

Now, if

$$\alpha_1\mathbf{b}_1 + \cdots + \alpha_n\mathbf{b}_n = \boldsymbol{\theta},$$

which in view of (6.35) implies

$$\alpha_1 b_{i1} + \cdots + \alpha_n b_{in} = 0, \quad i = 1, 2, ..., n. \tag{6.37}$$

Then, inserting (6.37) in (6.36), we have

$$\alpha_1\hat{\mathbf{x}}_1 + \cdots + \alpha_n\hat{\mathbf{x}}_n = \boldsymbol{\theta}. \tag{6.38}$$

Since $\hat{\mathbf{x}}_1, \ldots, \hat{\mathbf{x}}_n$ are linearly independent from (6.38), it follows that $\alpha_1 = \cdots = \alpha_n = 0$, and (6.37) now implies that $\mathbf{b}_1, ..., \mathbf{b}_n$ are linearly independent. By Corollary 5.7, B is nonsingular. □

Remark 6.5 If the vector space V_n consists of n-vectors or is the space $\mathbb{P}_{n-1}(t)$, then (6.34) can be written compactly as (see Remark 6.4)

$$\hat{\mathbf{V}} = \mathbf{V}B,$$

where $\mathbf{V} := [\mathbf{x}_1 \; \mathbf{x}_2 \; \cdots \; \mathbf{x}_n]$ and $\hat{\mathbf{V}} = [\hat{\mathbf{x}}_1\hat{\mathbf{x}}_2 \cdots \hat{\mathbf{x}}_n]$ are called bases matrices in $\mathbf{V}$ and $\hat{\mathbf{V}}$ respectively. Specifically, if $\mathbf{U}$ is a basis matrix in V_n, its columns will represent a basis for the space V_n.

Theorem 6.11 Suppose that $X := \{\mathbf{x}_1, ..., \mathbf{x}_n\}$ and $\hat{X} = \{\hat{\mathbf{x}}_1, ..., \hat{\mathbf{x}}_n\}$ are two bases of a vector space V_n such that

$$\hat{\mathbf{x}}_i = b_{1i}\mathbf{x}_1 + \cdots + b_{ni}\mathbf{x}_n \tag{6.39}$$

where $B = [b_{ij}]$ is a nonsingular matrix of order n. If $\mathbf{u} = [\alpha_1 \cdots \alpha_n]^T$ and $\hat{\mathbf{u}} = [\hat{\alpha}_1 \cdots \hat{\alpha}_n]^T$ are the representations of a vector $\mathbf{x}$ in V_n with respect to X and $\hat{X}$ respectively, then $\hat{\mathbf{u}} = B^{-1}\mathbf{u}$.

Proof Since $\mathbf{u}$ and $\hat{\mathbf{u}}$ are the coordinates of the same vector $\mathbf{x}$, we have, in view of Corollary 5.1 and (6.39), the relation

$$\alpha_1 \mathbf{x}_1 + \cdots + \alpha_n \mathbf{x}_n = \hat{\alpha}_1 \hat{\mathbf{x}}_1 + \cdots + \hat{\alpha}_n \hat{\mathbf{x}}_n = \sum_{i=1}^{n} \hat{\alpha}_i \sum_{j=1}^{n} b_{ji} \mathbf{x}_j = \sum_{j=1}^{n} \left(\sum_{i=1}^{n} \hat{\alpha}_i b_{ji} \right) \mathbf{x}_j.$$

Hence,

$$\alpha_j = \sum_{i=1}^{n} \hat{\alpha}_i b_{ji}$$

i.e., $\mathbf{u} = B\hat{\mathbf{u}}$ (see Example 6.21). □

Theorem 6.12 Suppose that A is an $m \times n$ matrix of the linear transformation $L : V_n \to W_m$ with respect to the bases $X : = \{\mathbf{x}_1, ..., \mathbf{x}_n\}$ and $Y : = \{\mathbf{y}_1, ..., \mathbf{y}_m\}$. If $\hat{A}$ is an $m \times n$ matrix of L with respect to different bases $\hat{X} := \{\hat{\mathbf{x}}_1, ..., \hat{\mathbf{x}}_n\}$ and $\hat{Y} := \{\hat{\mathbf{y}}_1, ..., \hat{\mathbf{y}}_m\}$, then there exist nonsingular matrices B and C of order n and m respectively, such that $\hat{A} = C^{-1}AB$.

Proof If $A = [a_{ki}]$ is the matrix of L with respect to bases X and Y, we have, in view of (6.28), the relation

$$L\mathbf{x}_i = a_{1i}\mathbf{y}_1 + a_{2i}\mathbf{y}_2 + \cdots + a_{mi}\mathbf{y}_m.$$

Similarly, for $\hat{A} = [\hat{a}_{ij}]$, we can write

$$L\hat{\mathbf{x}}_j = \hat{a}_{1j}\hat{\mathbf{y}}_1 + \hat{a}_{2j}\hat{\mathbf{y}}_2 + \cdots + \hat{a}_{mj}\hat{\mathbf{y}}_m.$$

As a result of Theorem 6.10, there exist nonsingular coordinate transformation matrices $B = [b_{ij}]$ and $C = [c_{ki}]$ satisfying

$$\hat{\mathbf{x}}_j = b_{1j}\mathbf{x}_1 + b_{2j}\mathbf{x}_2 + \cdots + b_{nj}\mathbf{x}_n,$$
$$\hat{\mathbf{y}}_i = c_{1i}\mathbf{y}_1 + c_{2i}\mathbf{y}_2 + \cdots + c_{mi}\mathbf{y}_m.$$

Hence,

$$L\hat{\mathbf{x}}_j = \sum_{i=1}^{m} \hat{a}_{ij}\hat{\mathbf{y}}_i = \sum_{i=1}^{m} \hat{a}_{ij} \sum_{k=1}^{m} c_{ki}\mathbf{y}_k = \sum_{k=1}^{m} \left(\sum_{i=1}^{m} c_{ki}\hat{a}_{ij} \right) \mathbf{y}_k$$

Alternatively, since L is linear, we have

$$L\hat{\mathbf{x}}_j = \sum_{i=1}^{n} b_{ij} L\mathbf{x}_i = \sum_{i=1}^{n} b_{ij} \sum_{k=1}^{m} a_{ki}\mathbf{y}_k = \sum_{k=1}^{m} \left(\sum_{i=1}^{n} a_{ki} b_{ij} \right) \mathbf{y}_k$$

which shows that (see Definition 1.7) $C\hat{A} = AB$. □

Remark 6.6 In Theorem 6,12, let the vector spaces V_n and W_m be $\mathbb{C}^n$ and $\mathbb{C}^m$ respectively, which consist of n- and m-vectors or $\mathbb{P}_{n-1}(t)$ and $\mathbb{P}_{m-1}(t)$. Let $\mathbf{V}: = [\mathbf{x}_1 \ \ \mathbf{x}_2 \ \cdots \ \mathbf{x}_n]$ and $\hat{\mathbf{V}} := [\hat{\mathbf{x}}_1 \ \ \hat{\mathbf{x}}_2 \cdots \hat{\mathbf{x}}_n]$ be the old and new basis matrices in V, and $\mathbf{W}: = [\mathbf{y}_1 \ \ \mathbf{y}_2 \ \cdots \ \mathbf{y}_m]$ and $\hat{\mathbf{W}} := [\hat{\mathbf{y}}_1 \ \ \hat{\mathbf{y}}_2 \ldots \hat{\mathbf{y}}_m]$ be those in W_m. Then, it is easy to see that $\hat{\mathbf{V}} = \mathbf{V}B$ and $\hat{\mathbf{W}} = \mathbf{W}C$ (see Remark 6.5), where B is called the coordinate transformation matrix in V_n and C that in W_m.

As a direct consequence of Theorem 6.12, we have the following corollary.

Corollary 6.3 If a linear transformation $L : V \to V$ has matrices A and $\hat{A}$ with respect to two different bases of V, then there exist a nonsingular matrix B such that $\hat{A} = B^{-1}AB$. □

Theorem 6.11 says that if in V_n an element $\mathbf{x}$ is given, of which $\mathbf{u}$ is the coordinate vector relative to a basis X, then relative to a different basis $\hat{X}$, the coordinate vector $\hat{\mathbf{u}}$ of the same element $\mathbf{x}$ can be obtained if the coordinate transformation matrix B connecting X and $\hat{X}$ is known. Thus, $\mathbf{u} = B\hat{\mathbf{u}}$ by Theorem 6.11. Similarly, if $\mathbf{y}$ is an element in W_m and $\mathbf{v}$ and $\hat{\mathbf{v}}$ are its coordinate vectors with respect to bases Y and $\hat{Y}$ respectively, then $\mathbf{v} = C\hat{\mathbf{v}}$ when C is the coordinate transformation matrix connecting Y and $\hat{Y}$. On the other hand, Theorem 6.12 tells us that if a matrix A relative to the bases X and Y is given, then the matrix $\hat{A}$ relative to different bases $\hat{X}$ and $\hat{Y}$ under the same linear transformation L can be determined if the coordinate transformation matrix B connecting X and $\hat{X}$, and C connecting Y and $\hat{Y}$ are known. Thus, $C\hat{A} = AB$ by Theorem 6.12. Figure 6.4 provides a visual aid to the computational steps involved in the above process.

As already pointed out, there may be many matrices which represent a single linear transformation L on a vector space V although with respect to different bases. Corollary 6.3 shows how these matrices are related to each other by a nonsingular matrix. These matrices associated with a single linear transformation are called *similar* which is now formally defined.

Definition 6.11 *Similar Matrices.* Let A and $\hat{A}$ be two matrices of order n. Then A is said to be similar to $\hat{A}$ if there exists a nonsingular matrix B such that $B^{-1}AB = \hat{A}$. From this similarity relation, it follows now that $C^{-1}\hat{A}C = A$, where $C = B^{-1}$ is nonsingular. Hence,

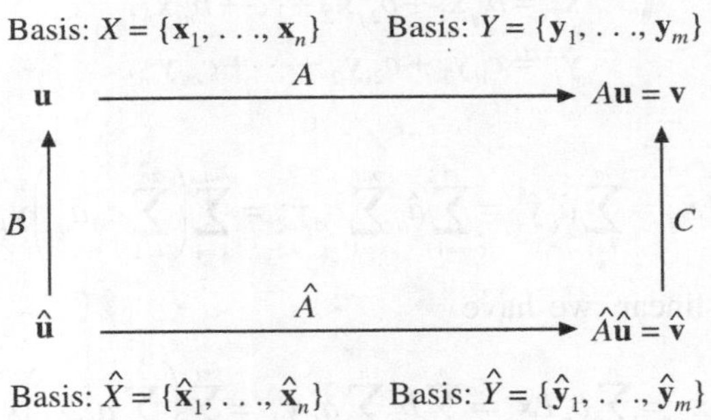

Figure 6.4 A linear transformation L becomes a matrix A or $\hat{A}$ with the selection of a basis: $AB = C\hat{A}$ where B and C are coordinate transformation matrices.

if A is similar to $\hat{A}$, then $\hat{A}$ too is similar to A. Thus, when $B^{-1}AB = \hat{A}$ holds for a nonsingular matrix B, A and $\hat{A}$ are said to be similar. □

One common property for the class of similar matrices which is evident from Definition 6.11 is that they have the same determinant, i.e., det $\hat{A}$ = det B^{-1} det A det B = det A (see Problem 3.16(a)). Other criteria such as the necessary and sufficient conditions for two matrices to be similar will be given in Section 10.6.

There is another class of matrices representing the same linear transformation $L : V_n \to W_m$ with respect to different pairs of bases as shown in Theorem 6.12. These are known as *equivalent matrices.* These matrices were introduced in Definition 4.4 and a criterion for equivalence was established in Section 4.7, Property III.

Example 6.19 In Example 6.17 with respect to the bases $\hat{X} := \{\mathbf{e}_1 \;\; \mathbf{e}_2 \;\; \mathbf{e}_3\}$ and $X := \{\mathbf{a}_1 \;\; \mathbf{a}_2 \;\; \mathbf{a}_3\}$, the matrices associated with the linear transformation $L : \mathbb{R}^3 \to \mathbb{R}^3$ are defined as

$$\hat{A} = \begin{bmatrix} 1 & 2 & 0 \\ 0 & 0 & 3 \\ 0 & 1 & -2 \end{bmatrix}$$

and

$$A = \begin{bmatrix} 3 & 2 & 0 \\ 0 & 1 & 3 \\ -4 & -4 & -5 \end{bmatrix}$$

respectively (see Fig. 6.4), where $V_n = W_m = \mathbb{R}^3, X = Y, \hat{X} = \hat{Y}$ (i.e., the corresponding elements in X and Y are equal; so also in $\hat{X}$ and $\hat{Y}$) and $B = C$. Now,

$$\mathbf{e}_1 = \mathbf{a}_1 - \mathbf{a}_2, \qquad \mathbf{e}_2 = \mathbf{a}_2 - \mathbf{a}_3, \qquad \mathbf{e}_3 = \mathbf{a}_3.$$

Therefore, in view of Theorem 6.10 and Remark 6.5,

$$[\mathbf{e}_1 \;\; \mathbf{e}_2 \;\; \mathbf{e}_3] = [\mathbf{a}_1 \;\; \mathbf{a}_2 \;\; \mathbf{a}_3] \begin{bmatrix} 1 & 0 & 0 \\ -1 & 1 & 0 \\ 0 & -1 & 1 \end{bmatrix}$$

according to which we define

$$B := \begin{bmatrix} 1 & 0 & 0 \\ -1 & 1 & 0 \\ 0 & -1 & 1 \end{bmatrix}$$

It is easy to check that $\hat{A} = B^{-1}AB$. The matrices A and $\hat{A}$ are therefore similar. □

Example 6.20 A linear transformation $L : \mathbb{R}^3 \to \mathbb{R}^2$ is given by

$$L\mathbf{x} = \begin{bmatrix} x_1 + x_2 + x_3 \\ 2x_2 + x_3 \end{bmatrix}, \quad \mathbf{x} = \begin{bmatrix} x_1 \\ x_2 \\ x_3 \end{bmatrix}.$$

The basis matrices in $\mathbb{R}^3$ are

$$\mathbf{V} = [\mathbf{x}_1 \;\; \mathbf{x}_2 \;\; \mathbf{x}_3] = \begin{bmatrix} 1 & 0 & 0 \\ 0 & 1 & 0 \\ 0 & 0 & 1 \end{bmatrix},$$

$$\hat{\mathbf{V}} = [\hat{\mathbf{x}}_1 \;\; \hat{\mathbf{x}}_2 \;\; \hat{\mathbf{x}}_3] = \begin{bmatrix} 2 & 0 & 1 \\ 2 & 1 & 0 \\ 1 & 0 & 1 \end{bmatrix},$$

and those in $\mathbb{R}^2$ are

$$\mathbf{W} = [\mathbf{y}_1 \;\; \mathbf{y}_2] = \begin{bmatrix} 2 & 1 \\ 1 & 1 \end{bmatrix} \text{ and } \hat{\mathbf{W}} = [\hat{\mathbf{y}}_1 \;\; \hat{\mathbf{y}}_2] = \begin{bmatrix} 1 & 1 \\ 0 & 1 \end{bmatrix}.$$

Find the matrix A with respect to the bases defined by $\mathbf{V}$ and $\mathbf{W}$, and $\hat{A}$ with respect to $\hat{V}$ and $\hat{\mathbf{W}}$ of the linear transformation L. Determine nonsingular matrices B and C such that $\hat{A} = C^{-1}AB$.

Solution As a result of Remark 6.4, we have the relations

$$L\mathbf{V} = \begin{bmatrix} 1 & 1 & 1 \\ 0 & 2 & 1 \end{bmatrix} = \begin{bmatrix} 2 & 1 \\ 1 & 1 \end{bmatrix} \begin{bmatrix} 1 & -1 & 0 \\ -1 & 3 & 1 \end{bmatrix} = \mathbf{W}A,$$

$$L\hat{\mathbf{V}} = \begin{bmatrix} 5 & 1 & 2 \\ 5 & 2 & 1 \end{bmatrix} = \begin{bmatrix} 1 & 1 \\ 0 & 1 \end{bmatrix} \begin{bmatrix} 0 & -1 & 1 \\ 5 & 2 & 1 \end{bmatrix} = \hat{\mathbf{W}}\hat{A}$$

from which we get

$$A = \begin{bmatrix} 1 & -1 & 0 \\ -1 & 3 & 1 \end{bmatrix}, \qquad \hat{A} = \begin{bmatrix} 0 & -1 & 1 \\ 5 & 2 & 1 \end{bmatrix}.$$

The decomposition of $L\mathbf{V}$ into $\mathbf{W}A$ and $L\hat{\mathbf{V}}$ into $\hat{\mathbf{W}}\hat{A}$ is carried out, taking the help of the procedure given in Example 5.43. As hinted in the proof of Theorem 6.12, the matrices B and C are determined satisfying $\hat{\mathbf{V}} = \mathbf{V}B$ and $\hat{\mathbf{W}} = \mathbf{W}C$ (see also Remark 6.6). Therefore,

$$B = \begin{bmatrix} 2 & 0 & 1 \\ 2 & 1 & 0 \\ 1 & 0 & 1 \end{bmatrix} \text{ and } C = \begin{bmatrix} 1 & 0 \\ -1 & 1 \end{bmatrix}.$$

□

Example 6.21 A vector in $\mathbb{R}^3$ has a coordinate vector $\mathbf{u} = [1 \quad 2 \quad -1]^T$ with respect to a basis defined by

$$\mathbf{V} = [\mathbf{x}_1 \quad \mathbf{x}_2 \quad \mathbf{x}_3] = \begin{bmatrix} 1 & 2 & 1 \\ -1 & 0 & 4 \\ 1 & 0 & 1 \end{bmatrix}.$$

Find its coordinate vector with respect to the basis defined by

$$\hat{\mathbf{V}} = [\hat{\mathbf{x}}_1 \quad \hat{\mathbf{x}}_2 \quad \hat{\mathbf{x}}_3] = \begin{bmatrix} 1 & 1 & 0 \\ 0 & 1 & 1 \\ 1 & 0 & 1 \end{bmatrix}.$$

Solution Since the basis matrix $\hat{\mathbf{V}}$ is nonsingular, solving $\hat{\mathbf{V}} = \mathbf{V}B$, we get (see Theorem 6.11 and Remark 6.5)

$$B^{-1} = \hat{\mathbf{V}}^{-1}\mathbf{V} = \begin{bmatrix} 1 & 1 & 0 \\ 0 & 1 & 1 \\ 1 & 0 & 1 \end{bmatrix}^{-1} \begin{bmatrix} 1 & 2 & 1 \\ -1 & 0 & 4 \\ 1 & 1 & 0 \end{bmatrix} = \frac{1}{2} \begin{bmatrix} 3 & 3 & -3 \\ -1 & 1 & 5 \\ -1 & -1 & 3 \end{bmatrix}.$$

Therefore,

$$\hat{\mathbf{u}} = B^{-1}\mathbf{u} = [6, -2, -3]^T. \qquad \square$$

6.5 ORTHOGONAL AND UNITARY TRANSFORMATIONS

Let us consider two sets of coordinate systems (x, y, z) and (u, v, w). The former is fixed in earth while the latter in a rigid body and is capable of rotation about the three axes x, y and z. Let the coordinate system (u, v, w) be rotated anticlockwise through an angle θ in the xy-plane about the z-axis (Figure 6.5(a)) assuming that the w-axis is coincident with the z-axis and that both

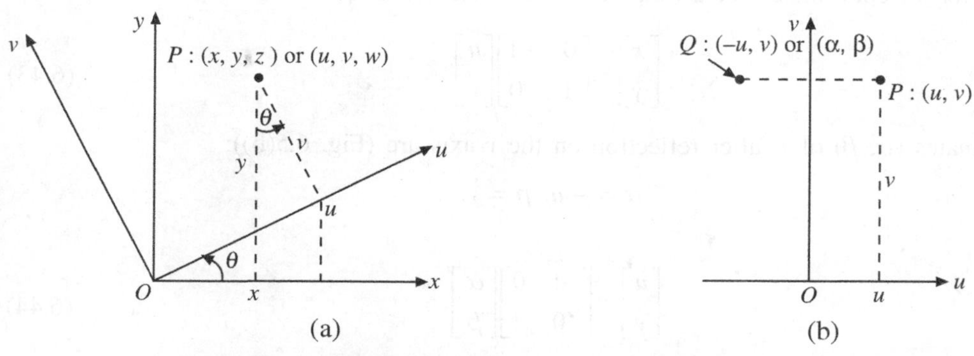

Figure 6.5 Transformation of coordinates: (a) rotation; (b) reflection about vertical axis.

the systems have the same origin O. The position (x, y, z) and (u, v, w) of a point P in the rigid body in the two coordinate systems are related by

$$x = u\cos\theta - v\sin\theta, \quad y = u\sin\theta + v\cos\theta, \quad z = w. \tag{6.40}$$

This is a particular case of a general motion of a rigid body which is important to scientists associated with the study of rocket flights and attitude control of satellites orbiting round the earth and also in robotics. Using matrix notation, equations in (6.40) can be written as

$$\begin{bmatrix} x \\ y \\ z \end{bmatrix} = \begin{bmatrix} \cos\theta & -\sin\theta & 0 \\ \sin\theta & \cos\theta & 0 \\ 0 & 0 & 1 \end{bmatrix} \begin{bmatrix} u \\ v \\ w \end{bmatrix},$$

where the matrix P transforming one set of coordinates to another is

$$P = \begin{bmatrix} \cos\theta & -\sin\theta & 0 \\ \sin\theta & \cos\theta & 0 \\ 0 & 0 & 1 \end{bmatrix} \tag{6.41}$$

which is known as *the coordinate transformation matrix.*

A characteristic property of transformation matrix P in (6.41) which distinguishes a class of real matrices is $PP^T = P^TP = I$.

Definition 6.12 *Orthogonal Matrix.* A square matrix P over a real field $\mathbb{R}$ is said to be orthogonal if

$$PP^T = P^TP = I. \tag{6.42}$$

Two properties of an orthogonal matrix, as is obvious from (6.42), are

(a) $P^{-1} = P^T$, (b) $\det P = \pm 1$. □

Example 6.22 The transformation matrix P in (6.41) is orthogonal and $\det P = 1$, as can be easily verified. Let us consider two two-dimensional coordinate systems (x, y) and (u, v). The transformation of coordinates for a rotation of 90° from (6.40) is

$$\begin{bmatrix} x \\ y \end{bmatrix} = \begin{bmatrix} 0 & -1 \\ 1 & 0 \end{bmatrix} \begin{bmatrix} u \\ v \end{bmatrix}. \tag{6.43}$$

The coordinates (α, β) of P after reflection on the v-axis are (Fig. 6.5(b)):

$$\alpha = -u, \ \beta = v.$$

Therefore,

$$\begin{bmatrix} u \\ v \end{bmatrix} = \begin{bmatrix} -1 & 0 \\ 0 & 1 \end{bmatrix} \begin{bmatrix} \alpha \\ \beta \end{bmatrix}. \tag{6.44}$$

Substituting (6.44) into (6.43), we obtain

$$\begin{bmatrix} x \\ y \end{bmatrix} = \begin{bmatrix} 0 & -1 \\ -1 & 0 \end{bmatrix} \begin{bmatrix} \alpha \\ \beta \end{bmatrix},$$

where the transformation matrix

$$Q = \begin{bmatrix} 0 & -1 \\ -1 & 0 \end{bmatrix}$$

is orthogonal but has det $P = -1$. □

Definition 6.13 *Proper and Improper Orthogonal Matrix.* An orthogonal matrix P with det $P = 1$ is called *proper* and with det $P = -1$ *improper.* □

Definition 6.12 can be extended to matrices over the field of complex numbers $\mathbb{C}$ when they are called *unitary matrices.*

Definition 6.14 *Unitary Matrix.* A square matrix U over a field $\mathbb{C}$ of complex numbers is called unitary if (see Section 1.3)

$$UU^* = U^*U = I. \quad \square \tag{6.45}$$

Unitary matrices are analogues of complex numbers z of absolute value 1, i.e., $z\bar{z} = \bar{z}z = 1$ which implies $|z|^2 = 1$.

On taking the determinants of both sides of (6.45), we have (see Section 1.3, and Properties II and X in Section 2.3)

$$\det U \det \mathbf{U}^* = 1, \quad \text{or} \quad \det U\ \overline{(\det U^T)} = 1, \quad \text{or} \quad |\det U| = 1.$$

Thus, two properties of a unitary matrix which follow automatically from (6.45) are:

(a) $U^{-1} = U^*$, (b) det $U = e^{i\alpha}$, α real.

Example 6.23 A simple example of a unitary matrix is

$$U = \frac{1}{\sqrt{2}} \begin{bmatrix} 1 & i \\ i & 1 \end{bmatrix}$$

in which det U = 1, i.e., $\alpha = 0$ in (b). □

The properties of orthogonal and unitary matrices are strikingly similar. Hence, the properties of the former will be sufficiently developed while those of the latter will be mentioned in passing. For applications of orthogonal and unitary transformations, see Section 13.3.

Definition 6.15 *Orthogonal and Unitary Transformation.* A linear transformation of coordinates in V_n defined by $\mathbf{y} = A\mathbf{x}$ is called orthogonal if the transformation matrix A is orthogonal and is called unitary if A is unitary. □

Definition 6.16 *Isometry.* A linear transformation A on V_n which preserves the length of a vector is called an isometry (*iso*: same; *metry*: length). That is, if $\mathbf{y} = A\mathbf{x}$, where $\mathbf{x}$, $\mathbf{y}$ are any elements in V_n and A is isometry, then $\| \mathbf{x} \| = \| \mathbf{y} \|$ (see Definition 5.9(1)). □

Theorem 6.13 A linear transformation $\mathbf{y} = A\mathbf{x}$ is $\mathbb{R}^n$ is isometry iff A is orthogonal.

Proof (a) *Sufficiency.* Let $\mathbf{y} = A\mathbf{x}$, where $\mathbf{x}$ and $\mathbf{y}$ are in V_n and A is orthogonal. Then, since $AA^T = I$, we have (see Definition 5.9)

$$\| \mathbf{y} \|^2 = \langle \mathbf{y}, \mathbf{y} \rangle = \langle A\mathbf{x}, A\mathbf{x} \rangle = \mathbf{x}^T A^T A\mathbf{x} = \mathbf{x}^T\mathbf{x} = \| \mathbf{x} \|^2$$

So, A is an isometry in $\mathbb{R}^n$.

(b) *Necessity.* Conversely, let A be an isometry in $\mathbb{R}^n$. Then, for elements $\mathbf{x}$ and $\mathbf{y}$ in $\mathbb{R}^n$ such that $A\mathbf{x} = \mathbf{y}$, we have

$$\|\mathbf{x}\|^2 = \|\mathbf{y}\|^2.$$

Therefore,

$$\mathbf{x}^T\mathbf{x} = \mathbf{x}^T A^T A\mathbf{x}. \tag{6.46}$$

If $\mathbf{a}_j$ denotes the jth column of A, then $A\mathbf{e}_j = \mathbf{a}_j$, where $\mathbf{e}_j$'s are elementary vectors (see (5.4)). Putting in (6.46), $\mathbf{x} = \mathbf{e}_j$, we get $1 = \mathbf{a}_j{}^T\mathbf{a}_j$ for all $j = 1, ..., n$. Again, putting $\mathbf{x} = \mathbf{e}_i + \mathbf{e}_j$, $i \neq j$, in (6.46), since $A(\mathbf{e}_i + \mathbf{e}_j) = \mathbf{a}_i + \mathbf{a}_j$, we obtain

$$2 = (\mathbf{a}_i + \mathbf{a}_j)^T(\mathbf{a}_i + \mathbf{a}_j), \quad i \neq j.$$

Hence, $\mathbf{a}_i{}^T\mathbf{a}_j = 0$, $i \neq j$. So, we have $\mathbf{a}_j{}^T\mathbf{a}_j = 1$ and $\mathbf{a}_i{}^T\mathbf{a}_j = 0$, $i \neq j$, which implies that $A^TA = I$. Therefore, A is orthogonal. □

In a similar fashion we can prove the following corollary.

Corollary 6.4 A linear transformation $\mathbf{y} = \mathrm{U}\mathbf{x}$ in $\mathbb{C}^n$ is isometry iff U is unitary. □

Corollary 6.5 A linear transformation A in $\mathbb{R}^n$ is orthogonal iff the distance between any two vectors $\mathbf{x}$ and $\mathbf{y}$ in it is invariant under this transformation, i.e.,

$$\|\mathbf{x} - \mathbf{y}\| = \|A(\mathbf{x} - \mathbf{y})\|.$$

Proof Putting $\mathbf{z} = \mathbf{x} - \mathbf{y}$, we note that the result to be proved is a direct consequence of Theorem 6.13. □

Corollary 6.6 A linear transformation U in $\mathbb{C}^n$ is unitary iff the distance between any two vectors $\mathbf{x}$ and $\mathbf{y}$ in it is invariant under this transformation, i.e.,

$$\| \mathbf{x} - \mathbf{y} \| = \| \mathrm{U}\,(\mathbf{x} - \mathbf{y}) \|.$$

The results of Theorem 6.13 and Corollary 6.5 admit of a good geometrical interpretation. Since, the rotation of coordinates can be represented by an orthogonal transformation (see (6.41)), we see that the distance of the point P in a rigid body from the origin O is the same in whatever coordinate system, e.g. xy or uv, it is measured (this is Theorem 6.13); so also the distance PQ between any two points in it is invariant (this is Corollary 6.5), see Figure 6.6.

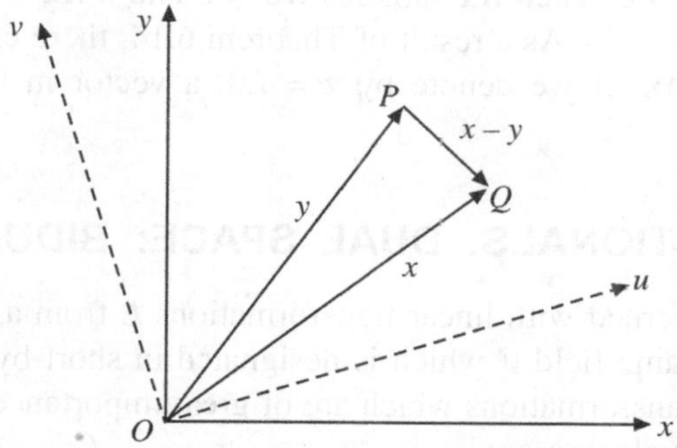

Figure 6.6 Invariance of length and distance under an orthogonal transformation.

Theorem 6.14 Let L be a transformation (see Definition 6.1) in $\mathbb{R}^n$ such that $L\boldsymbol{\theta} = \boldsymbol{\theta}$. If for any pair of elements $\mathbf{x}$, $\mathbf{y}$ in $\mathbb{R}^n$,

$$\| L\mathbf{x} - L\mathbf{y} \| = \| \mathbf{x} - \mathbf{y} \| \tag{6.47}$$

then $L\mathbf{x} = A\mathbf{x}$, where A is an orthogonal matrix.

Proof Putting $\mathbf{y} = \boldsymbol{\theta}$ in (6.47) since $L\boldsymbol{\theta} = \boldsymbol{\theta}$, we have

$$\| L\mathbf{x} \| = \| \mathbf{x} \|. \tag{6.48}$$

Again, writing both sides of (6.47) explicitly with the help of Definition 5.9 and substituting from (6.48), we have, for any $\mathbf{x}$, $\mathbf{y}$ in $\mathbb{R}^n$, the relation

$$\begin{aligned} \| L\mathbf{x} - L\mathbf{y} \|^2 &= \langle L\mathbf{x} - L\mathbf{y}, L\mathbf{x} - L\mathbf{y} \rangle = \| L\mathbf{x} \|^2 - \langle L\mathbf{x}, L\mathbf{y} \rangle - \langle L\mathbf{y}, L\mathbf{x} \rangle + \| L\mathbf{y} \|^2 \\ &= \| \mathbf{x} \|^2 - \langle \mathbf{x}, \mathbf{y} \rangle - \langle \mathbf{y}, \mathbf{x} \rangle + \| \mathbf{y} \|^2. \end{aligned}$$

Therefore,

$$\langle L\mathbf{x}, L\mathbf{y} \rangle = \langle \mathbf{x}, \mathbf{y} \rangle, \tag{6.49}$$

where we have used the fact that $\langle L\mathbf{x}, L\mathbf{y} \rangle = \langle L\mathbf{y}, L\mathbf{x} \rangle$ in $\mathbb{R}^n$.

Let $\mathbf{x} = \mathbf{e}_i$, $\mathbf{y} = \mathbf{e}_j$, where $\mathbf{e}_i$ and $\mathbf{e}_j$ are elementary vectors in $\mathbb{R}^n$ (see (5.4)). Then putting these relations in (6.49), we have

$$(L\mathbf{e}_i, L\mathbf{e}_j) = (\mathbf{e}_i, \mathbf{e}_j) = \delta_{ij}, \tag{6.50}$$

where

$$\delta_{ij} = \begin{cases} 0, & i \neq j; \\ 1, & i = j; \end{cases} \tag{6.51}$$

is called the *Kronecker delta*. Let $L\mathbf{e}_i, = \mathbf{a}_i$ for all $i = 1, ..., n$, where $\mathbf{a}_i$, is a vector in $\mathbb{R}^n$. Then the matrix of the transformation L is $[\mathbf{a}_1, \mathbf{a}_2, ..., \mathbf{a}_n]$, where in view of (6.50), $\mathbf{a}_i^T\mathbf{a}_j = \delta_{ij}$. This implies that $A^TA = I$, i.e., A is orthogonal. □

Corollary 6.7 If L is a transformation in $\mathbb{R}^n$ such that for any pair of elements $\mathbf{x}$, $\mathbf{y}$ in $\mathbb{R}^n$, $\| L\mathbf{x} - L\mathbf{y} \| = \| \mathbf{x} - \mathbf{y} \|$, then there exists a vector $\mathbf{z}$ and an orthogonal transformation A in $\mathbb{R}^n$ connected by $L\mathbf{x} = A\mathbf{x} + \mathbf{z}$.

Proof Define $K\mathbf{x} := L\mathbf{x} - L\boldsymbol{\theta}$. Then $K\mathbf{x}$ satisfies $K\boldsymbol{\theta} = \boldsymbol{\theta}$ and $\| K\mathbf{x} - K\mathbf{y} \| = \| (L\mathbf{x} - L\boldsymbol{\theta}) - (L\mathbf{y} - L\boldsymbol{\theta}) \| = \| L\mathbf{x} - L\mathbf{y} \| = \| \mathbf{x} - \mathbf{y} \|$. As a result of Theorem 6.14, there exists an orthogonal matrix A in $\mathbb{R}^n$ such that $K\mathbf{x} = A\mathbf{x}$. If we denote by $\mathbf{z} = L\boldsymbol{\theta}$, a vector in $\mathbb{R}^n$, then $L\mathbf{x} = K\mathbf{x} + L\boldsymbol{\theta} = A\mathbf{x} + \mathbf{z}$. □

6.6 LINEAR FUNCTIONALS: DUAL SPACE: BIDUAL SPACE

So far, we have been concerned with linear transformations L from a vector space V to another vector space W over the same field $\mathbb{F}$ which is designated in short by $L : V \to W$ over $\mathbb{F}$. Two particular cases of these transformations which are of great importance for a clear understanding of general topics in linear algebra are

(a) The case where L is a linear transformation from V into itself
(b) The case where L is a linear transformation from V into $\mathbb{F}$.

In this section, our prime concern is to study linear transformations belonging to class (b) which, as a matter of fact, take up scalar values and are also sometimes properly termed as scalar-valued linear transformations. A detailed study of transformations belonging to class (a) will be carried out in Chapter 7. We have already established in Section 6.2 that the set of all linear transformations $L : V \to W$ over $\mathbb{F}$ is a vector space over the same field. The foregoing discussion is a prelude to the following definition.

Definition 6.17 *Linear Functional and Dual Space.* A linear transformation f from a vector space V over a field $\mathbb{F}$ into this field $\mathbb{F}$ of scalars ($f: V \to \mathbb{F}$) is called a *linear functional* on the space V. The set of all linear functionals on V is a vector space defined as the *dual space* of V and is denoted by V^*. □

As noted in the definition, a linear functional is designated by small letters f, h, etc., as against capital letters L, K, etc. for linear transformations. We shall now provide examples of some typical linear functionals encountered in linear algebra.

Example 6.24 Let V be the set of all $n \times n$ matrices. Then $f : V \to \mathbb{F} : A \mapsto \operatorname{tr} A$ (see Definition 1.14) is a linear functional.

Example 6.25 Let V be the vector space of all real (or complex) valued functions integrable over the interval $a \le t \le b$ (see Example 5.3). Then, the transformation $f : V \to \mathbb{R}$ (or $\mathbb{C}$) defined by

$$x(t) \mapsto \int_a^b x(t)\,dt$$

is a linear functional on V. This mapping f assigns to each integrable function $x(t)$ a real (or a complex) number on the interval $a \le t \le b$.

Example 6.26 Let V be the vector space of all real (or complex) valued functions integrable over the interval $a \le t \le b$, and let $p(t)$ be a fixed function in V. Then the transformation $f : V \to \mathbb{R}$ (or $\mathbb{C}$) defined by

$$x(t) \mapsto \int_a^b x(t)p(t)dt$$

is a linear functional on V. In particular, if $0 \le t \le 2\pi$ and $p(t) = \exp(-int)$, then the linear functional f_n (n = an integer) defined by

$$f_n(x) = \frac{1}{2\pi}\int_0^{2\pi} x(t)\exp(-int)dt$$

is known as the nth Fourier coefficient of x.

Example 6.27 Let $\{\mathbf{x}_1, \mathbf{x}_2, ..., \mathbf{x}_n\}$ be a basis of the n-dimensional vector space V over $\mathbb{F}$. Any vector $\mathbf{x}$ in V can be represented by $\mathbf{x} = \alpha_1\mathbf{x}_1 + \alpha_2\mathbf{x}_2 + \cdots + \alpha_n\mathbf{x}_n$, where α_i's are scalars in $\mathbb{F}$. We now consider *a fixed vector* $\mathbf{z}$ in V and represent it by $\mathbf{z} = c_1\mathbf{x}_2 + c_2\mathbf{x}_2 + \cdots + c_n\mathbf{x}_n$, where c_i's are scalars in $\mathbb{F}$. Denote by $\mathbf{w} = [c_1 \; c_2 \; \cdots \; c_n]^T$ and $\mathbf{u} = [\alpha_1 \; \alpha_2 \; \cdots \; \alpha_n]^T$, the coordinate vectors of $\mathbf{z}$ and $\mathbf{x}$ respectively. Then, the linear transformation $f_\mathbf{z} : V \to \mathbb{F}$ defined by

$$f_\mathbf{z}(\mathbf{x}) = c_1\alpha_1 + c_2\alpha_2 + \cdots + c_n\alpha_n = \mathbf{w}^T\mathbf{u} \tag{6.52}$$

is a linear functional on V. In fact, $f_\mathbf{z}(\mathbf{x})$ is obtained as an inner product of coordinate vectors of $\mathbf{z}$ and $\mathbf{x}$. Restricting $\mathbf{z}$ to be the basis vectors $\mathbf{x}_i$'s, we get from (6.52), n number of linear functional $f_{\mathbf{x}_i}$ $(i = 1, 2, ..., n)$ in V^* given by

$$f_{\mathbf{x}_i}(\mathbf{x}) = \alpha_i, \tag{6.53}$$

because now the elementary vector $\mathbf{e}_i = [0 \cdots 1 \cdots 0]^T$, see (5.4), is the coordinate vector of $\mathbf{x}_i$ with respect to the basis vectors $\{\mathbf{x}_1, \mathbf{x}_2, ..., \mathbf{x}_n\}$ and should replace $\mathbf{w}$ in (6.52). This functional may be regarded as a linear transformation on V that maps each vector in V onto its ith coordinate relative to the selected basis. These linear functionals are very important and would assume a dominant role in our study. Taking into account (6.53), a noteworthy property of these functionals is, obviously,

$$f_{\mathbf{x}_i}(\mathbf{x}_j) = \delta_{ij}; \quad i, j = 1, 2, ..., n; \tag{6.54}$$

where δ_{ij} is the Kronecker delta (see (6.51)), because $\mathbf{e}_j$ is the coordinate vector of $\mathbf{x}_j$ with respect to the basis $\{\mathbf{x}_1, \mathbf{x}_2, ..., \mathbf{x}_n\}$, and $\mathbf{u} = \mathbf{e}_j$ and $\mathbf{w} = \mathbf{e}_i$, in (6.52). Moreover, if $\boldsymbol{\theta}$ is a zero vector in V, then

$$f_{\boldsymbol{\theta}}(\mathbf{x}) = 0, \tag{6.55}$$

the scalar zero in $\mathbb{F}$, because the coordinate vector of $\boldsymbol{\theta}$ is the null vector $[0 \;\; 0 \cdots 0]^T$ to replace $\mathbf{w}$ in (6.52). So, $f_{\boldsymbol{\theta}}$ transforms any vector $\mathbf{x}$ in V to the zero scalar 0 and therefore, defines a zero (or null) functional in V^*. □

Now we shall state and prove a very crucial theorem on dual basis.

Theorem 6.15 *Dual Basis.* Let $\{\mathbf{x}_1, \mathbf{x}_2, ..., \mathbf{x}_n\}$ be a basis of the n-dimensional vector space V over a field $\mathbb{F}$. Then the unique linear functionals $f_{\mathbf{x}_1}, f_{\mathbf{x}_2}, ..., f_{\mathbf{x}_n}$ defined by (6.54) form a basis of the dual space V^* defined as the dual basis of $\{\mathbf{x}_1, \mathbf{x}_2, ..., \mathbf{x}_n\}$, and any element f in V^* can be expressed as

$$f = f(\mathbf{x}_1)f_{\mathbf{x}_1} + \cdots + f(\mathbf{x}_n)f_{\mathbf{x}_n} \tag{6.56}$$

and for each vector $\mathbf{x}$ in V, we have

$$\mathbf{x} = f_{\mathbf{x}_1}(\mathbf{x})\mathbf{x}_1 + f_{\mathbf{x}_2}(\mathbf{x})\mathbf{x}_2 + \cdots + f_{\mathbf{x}_n}(\mathbf{x})\mathbf{x}_n. \tag{6.57}$$

Proof We have to prove (see Definition 5.5) that

(a) $f_{\mathbf{x}_i}$'s are linearly independent in V^*

(b) Any vector f in V^* can be expressed as a linear combination of $f_{\mathbf{x}_i}$'s.

(a) Suppose that for scalars $\alpha_1, \alpha_2, ..., \alpha_n$ in $\mathbb{F}$ we have

$$\alpha_1 f_{\mathbf{x}_1} + \alpha_2 f_{\mathbf{x}_2} + \cdots + \alpha_n f_{\mathbf{x}_n} = f_{\boldsymbol{\theta}}.$$

Then,

$$(\alpha_1 f_{\mathbf{x}_1} + \alpha_2 f_{\mathbf{x}_2} + \cdots + \alpha_n f_{\mathbf{x}_n})\mathbf{x}_i = f_{\boldsymbol{\theta}}(\mathbf{x}_i),$$

which by taking into account (6.54) and (6,55), implies that

$$\alpha_i = 0, \qquad (i = 1, 2, ..., n)$$

Therefore, $f_{\mathbf{x}_i}$'s are linearly independent in V^*.

(b) Suppose now that f is any linear functional in V^*. Then, we can find n scalars $a_1\ a_2\ \cdots\ a_n$ in $\mathbb{F}$ satisfying

$$f(\mathbf{x}_i) = a_i, \qquad (i = 1, 2, ..., n), \tag{6.58}$$

because f is presumably a known linear functional. For any vector

$$\mathbf{x} = \alpha_1\mathbf{x}_1 + \alpha_2\mathbf{x}_2 + \cdots + \alpha_n\mathbf{x}_n \tag{6.59}$$

in V, because of the linearity of f, and using (6.58) (6.53) and (6.12) respectively, we get

$$\begin{aligned} f(\mathbf{x}) &= \alpha_1 f(\mathbf{x}_1) + \alpha_2 f(\mathbf{x}_2) + \cdots + \alpha_n f(\mathbf{x}_n) \\ &= \alpha_1 a_1 + \alpha_2 a_2 + \cdots + \alpha_n a_n \\ &= a_1 f_{\mathbf{x}_1}(\mathbf{x}) + a_2 f_{\mathbf{x}_2}(\mathbf{x}) + \cdots + a_n f_{\mathbf{x}_n}(\mathbf{x}) \\ &= (a_1 f_{\mathbf{x}_1} + a_2 f_{\mathbf{x}_2} + \cdots + a_n f_{\mathbf{x}_n})(\mathbf{x}) \end{aligned} \tag{6.60}$$

Since $\mathbf{x}$ is arbitrary, and a combination of linear functionals is again a linear functional, we have, by Definition 6.5, the relation

$$f = a_1 f_{\mathbf{x}_1} + a_2 f_{\mathbf{x}_2} + \cdots + a_n f_{\mathbf{x}_n} \tag{6.61}$$

which proves (b).

Substituting (6.58) into (6.61), and (6.53) into (6.59), we get (6.56) and (6.57) respectively. Because of Theorem 6.1, the linear functional $f_{\mathbf{x}_i}$ is uniquely defined by (6.54). □

Remark 6.7 Equation (6.57) reassures the physical interpretation of a dual basis already given: that each element $f_{\mathbf{x}_i}$ in the dual basis $\{f_{\mathbf{x}_1}, f_{\mathbf{x}_2}, ..., f_{\mathbf{x}_n}\}$ is precisely a function which assigns to each vector $\mathbf{x}$ in V the ith coordinate of $\mathbf{x}$ relative to the given basis $\{\mathbf{x}_1, \mathbf{x}_2, ..., \mathbf{x}_n\}$.

Since $\mathbf{u} = [\alpha_1 \ \alpha_2 \cdots \alpha_n]^T$ is the coordinate vector of an element $\mathbf{x}$ in V and $\mathbf{a} = [a_1 \ a_2 \cdots a_n]^T$ is that for f in V^*, (6.60) shows that every linear functional on V has the form $f(\mathbf{x}) = \mathbf{a}^T\mathbf{u}$.

Definition 6.18 *Dual Basis.* For any basis $\{\mathbf{x}_1, \mathbf{x}_2, ..., \mathbf{x}_n\}$ of a vector space V over a field $\mathbb{F}$, there exists a unique basis $\{f_1, f_2, ..., f_n\}$ of V^* such that $f_i(\mathbf{x}_j) = \delta_{ij}$, $(i, j = 1, 2, \ldots, n)$, as substantiated by Theorem 6.15. The basis $\{f_1, f_2, ..., f_n\}$ of V^* is said to be dual to the given basis $\{\mathbf{x}_1, \mathbf{x}_2, ..., \mathbf{x}_n\}$ of V. □

Example 6.28 Determine a dual basis $\{f_{\mathbf{x}_1}, f_{\mathbf{x}_2}\}$ in V^* of the basis $\mathbf{x}_1 = [3 \ 2]^T$, $\mathbf{x}_2 = [1 \ 1]^T$ in $V = \mathbb{R}^2$.

Solution We have to determine explicitly the functional $f_{\mathbf{x}_1}$ and $f_{\mathbf{x}_2}$. To this end we note that the coordinate vector $[\alpha_1 \ \alpha_2]^T$ of any element $\mathbf{x} = [x_1 \ x_2]^T$ in $\mathbb{R}^2$ is obtained by solving $\mathbf{x} = \alpha_1\mathbf{x}_1 + \alpha_2\mathbf{x}_2$, i.e.,

$$\begin{bmatrix} x_1 \\ x_2 \end{bmatrix} = [\mathbf{x}_1\mathbf{x}_2]\begin{bmatrix} \alpha_1 \\ \alpha_2 \end{bmatrix} = \begin{bmatrix} 3 & 1 \\ 2 & 1 \end{bmatrix}\begin{bmatrix} \alpha_1 \\ \alpha_2 \end{bmatrix}$$

which gives

$$\alpha_1 = x_1 - x_2, \qquad \alpha_2 = -2x_1 + 3x_2.$$

From (6.53), we immediately get

$$f_{\mathbf{x}_1}(\mathbf{x}) = x_1 - x_2, \quad f_{\mathbf{x}_2}(\mathbf{x}) = -2x_1 + 3x_2.$$

Example 6.29 Determine a dual basis $\{f_{\mathbf{x}_1}, f_{\mathbf{x}_2}, f_{\mathbf{x}_3}\}$ in V^* of the basis $\{\mathbf{x}_1 = 1, \mathbf{x}_2 = 1 + t, \mathbf{x}_3 = t + t^2\}$ in $\mathbb{P}_2(t)$ over $\mathbb{R}$.

Solution As in Example 6.28, the coordinate vector $[\alpha_1 \ \alpha_2 \ \alpha_3]^T$ of any element $\mathbf{x}(t) = a_0 + a_1t + a_2t^2$ in $\mathbb{P}_2(t)$ (see Definition 5.5) is obtained by solving

$$\begin{aligned} \mathbf{x}(t) &= a_0 + a_1t + a_2t^2 = \alpha_1\mathbf{x}_1 + \alpha_2\mathbf{x}_2 + \alpha_3\mathbf{x}_3 \\ &= (\alpha_1 + \alpha_2) + (\alpha_2 + \alpha_3)t + \alpha_3t^2 \end{aligned}$$

which, on equating the coefficients of various powers of t on both sides, gives

$$a_0 = \alpha_1 + \alpha_2, \qquad a_1 = \alpha_2 + \alpha_3, \qquad a_2 = \alpha_3.$$

Hence,

$$\alpha_1 = a_0 - a_1 + a_2, \qquad \alpha_2 = a_1 - a_2, \qquad \alpha_3 = a_2.$$

Consequently, employing (6.53), we obtain the dual basis defined by

$$f_{\mathbf{x}_1}(\mathbf{x}) = a_0 - a_1 + a_2, \quad f_{\mathbf{x}_2}(\mathbf{x}) = a_1 - a_2, \quad f_{\mathbf{x}_3}(\mathbf{x}) = a_2.$$

6.6.1 Linear Transformation and Transpose of a Matrix: Dual Space

In Section 6.3, we have discussed in detail as to how a linear transformation $L : V \to W$ can be represented with respect to chosen bases by a matrix A. Our task now would be to segregate those linear transformations whose matrix representation is given by the transpose of A, i.e., by

A^T. With this object in our mind, we shall now define a new linear functional on V in the following way. With the given L and for any linear functional $g : W \rightarrow \mathbb{F}$, we define the new linear functional $L^T g : V \rightarrow \mathbb{F}$ as

$$L^T g = g \cdot L \tag{6.62}$$

which is represented by the commutative diagram in Figure 6.7. Since both L and g are linear, the composite mapping $g \cdot L$ and hence $L^T g$ are linear. We also note that g is an element of W^* and $L^T g$ is that of V^*. Consequently, we can consider L^T to be a transformation from W^* to V^* (see Fig. 6.7(b)) which, as a matter of fact, is linear and defined as the transpose of L.

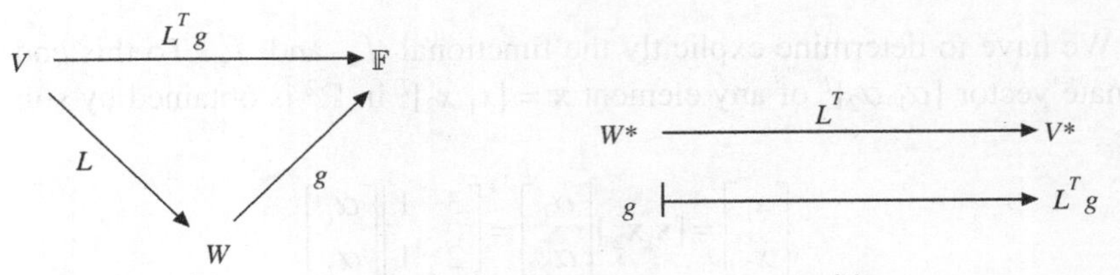

Figure 6.7 Definition of (a) the functional $L^T g$ and (b) the transpose L^T of a linear transformation L.

Definition 6.19: *Transpose of a Linear Transformation.* Let V and W be two vector spaces over a field $\mathbb{F}$ with respective dual spaces V^* and W^*. If L is a linear transformation: $V \rightarrow W$, then the transpose L^T of L is a linear transformation: $W^* \rightarrow V^*$ defined by

$$L^T g = g \cdot L$$

for every element g in W^*. □

The transpose L^T of L as defined above is linear. Indeed, in view of (6.62) and (6.12), for any linear functional f and g in W^*, scalars α and β in $\mathbb{F}$, and any $\mathbf{x}$ in V, we have:

$$L^T(\alpha f + \beta g)\mathbf{x} = [(\alpha f + \beta g) \cdot L]\mathbf{x} = (\alpha f + \beta g)\ (L\mathbf{x}) = \alpha f(L\mathbf{x}) + \beta g\ (L\mathbf{x})$$

$$= \alpha (f \cdot L)\mathbf{x} + \beta (g \cdot L)\ \mathbf{x} = [\alpha L^T f + \beta L^T g]\mathbf{x},$$

which shows that, because $\mathbf{x}$ is arbitrary, L^T satisfies

$$L^T(\alpha f + \beta g) = \alpha L^T f + \beta L^T g,$$

which is the principle of superposition (see Definition 6.1).

This linear transformation L^T from a vector space W^* to another vector space V^* can be represented by a matrix as expounded in Section 6.3, which we now want to investigate. Although this matrix of L^T depends on the bases of the relevant vector spaces, a judicious choice however gives A^T as the matrix of L^T whenever A is the matrix of L. This important result is contained in the following theorem.

Theorem 6.16 Suppose that $\{f_1, f_2, ..., f_n\}$ in V^* is a dual basis of the basis $\{\mathbf{x}_1, \mathbf{x}_2, ..., \mathbf{x}_n\}$ in V and that $\{g_1, g_2, ..., g_m\}$ in W^* is a dual basis of that $\{\mathbf{y}_1, \mathbf{y}_2, ..., \mathbf{y}_m\}$ in W. If A is the matrix of the linear transformation $L : V \rightarrow W$ with respect to the above bases, then A^T is the matrix of $L^T : W^* \rightarrow V^*$ with respect to the above dual bases.

Proof By definition of a dual basis and the transpose of a linear transformation, we have

$$f_i(\mathbf{x}_k) = \delta_{ik}, \qquad (i, k = 1, 2, \ldots, n); \tag{6.63}$$

$$g_k(\mathbf{y}_j) = \delta_{kj}, \qquad (k, j = 1, 2, \ldots, m); \tag{6.64}$$

$$L^T g_j = g_j \cdot L, \qquad (j = 1, 2, \ldots, m). \tag{6.65}$$

Let $A = [a_{ij}]$ of order $m \times n$ be the matrix of L with respect to the given bases and $B = [b_{ij}]$ of order $n \times m$ that of L^T with respect to the given dual bases. Then (see (6.28)),

$$L\mathbf{x}_i = a_{1i}\mathbf{y}_1 + a_{2i}\mathbf{y}_2 + \cdots + a_{mi}\mathbf{y}_m, \quad (i = 1, 2, \ldots, n); \tag{6.66}$$

$$L^T g_j = b_{1j}f_1 + b_{2j}f_2 + \cdots + b_{nj}f_n, \quad (j = 1, 2, \ldots, m). \tag{6.67}$$

Hence, making use of (6.65)–(6.67), we have

$$(b_{1j}f_1 + b_{2j}f_2 + \cdots + b_{nj}f_n)\mathbf{x}_i = L^T g_j \mathbf{x}_i = (g_j \cdot L)\mathbf{x}_i = g_j(L\mathbf{x}_i)$$
$$= g_j(a_{1i}\mathbf{y}_1 + a_{2i}\mathbf{y}_2 + \cdots + a_{mi}\mathbf{y}_m).$$

Therefore, in view of the linearity of g_j and using (6.63) and (6.64), the above equation yields

$$b_{ij} = a_{ji},$$

which shows that the matrix of L^T with respect to the given dual bases is $B = A^T$. □

***Example* 6.30** *Transpose of a Linear Transformation.* Find the transpose of the linear transformation

$$L : \mathbb{R}^3 \to \mathbb{R}^2 : \mathbf{x} = [x_1 \;\; x_2 \;\; x_3]^T \mapsto L\mathbf{x} = \begin{bmatrix} 2x_1 - x_2 + 2x_3 \\ x_1 + 2x_2 - x_3 \end{bmatrix},$$

when the bases in $\mathbb{R}^3$ and $\mathbb{R}^2$ are respectively (see Section 6.3) given by the columns in the basis matrices

$$P = [\mathbf{x}_1 \;\; \mathbf{x}_2 \;\; \mathbf{x}_3] = \begin{bmatrix} 2 & 1 & 0 \\ 0 & 2 & 1 \\ 1 & 0 & 0 \end{bmatrix}, \qquad Q = [\mathbf{y}_1 \;\; \mathbf{y}_2] = \begin{bmatrix} 3 & 1 \\ 2 & 1 \end{bmatrix}. \tag{6.68}$$

Solution To determine the transformation L^T

$$(\mathbb{R}^2)^* \xrightarrow{L^T} (\mathbb{R}^3)^*$$

we note that $L\mathbf{x}$ is an element of $\mathbb{R}^2$. From

$$L\mathbf{x} = \begin{bmatrix} 2x_1 - x_2 + 2x_3 \\ x_1 + 2x_2 - x_3 \end{bmatrix} = Q\begin{bmatrix} x_1 - 3x_2 + 3x_3 \\ -x_1 + 8x_2 - 7x_3 \end{bmatrix},$$

it follows that the coordinates of $L\mathbf{x}$ with respect to the basis defined by Q are $(x_1 - 3x_2 + 3x_3, -x_1 + 8x_2 - 7x_3)$. Hence, if in $(\mathbb{R}^2)^*$, $\{g_1, g_2\}$ is a dual basis of the columns in Q and $g = a_1g_1 + a_2g_2$ is any element, where a_1, a_2 are scalars in $\mathbb{R}$, then taking cognizance of (6.62) and (6.53), we have

$$L^Tg(\mathbf{x}) = (a_1g_1 + a_2g_2)\,(L\mathbf{x}) = a_1(x_1 - 3x_2 + 3x_3) + a_2(-x_1 + 8x_2 - 7x_3)$$
$$= [a_1 - a_2,\ -3a_1 + 8a_2,\ 3a_1 - 7a_2]\mathbf{x}. \qquad (6.69)$$

Therefore, the required transformation L^T is

$$L^Tg = [a_1 - a_2,\ -3a_1 + 8a_2,\ 3a_1 - 7a_2].$$

Now, if in $(\mathbb{R}^3)^*$ the dual basis of the columns in P is $\{f_1, f_2, f_3,\}$, then we write

$$L^Tg_1 = b_1f_1 + b_2f_2 + b_3f_3, \qquad L^Tg_2 = c_1f_1 + c_2f_2 + c_3f_3, \qquad (6.70)$$

where b_i's and c_i's are scalars in $\mathbb{R}$, which in view of (6.54) are expressed as

$$b_i = L^Tg_1(\mathbf{x}_i), \qquad c_i = L^Tg_2(\mathbf{x}_i), \quad i = 1, 2, 3.$$

Putting $\mathbf{x} = \mathbf{x}_1$ and $g = g_1$ in (6.69) (when $a_1 = 1, a_2 = 0$), and using (6.68), we have,

$$L^Tg_1(\mathbf{x}_1) = [1\ {-3}\ 3]\mathbf{x}_1 = 5 = b_1.$$

Similarly, we determine $b_2 = -5$, $b_3 = -3$, $c_1 = -9$, $c_2 = 15$, $c_3 = 8$. Hence, with the help of (6.70), we have the matrix B of L^T with respect to the specified dual basis (see Section 6.3) as

$$B = \begin{bmatrix} 5 & -9 \\ -5 & 15 \\ -3 & 8 \end{bmatrix}.$$

Again, we can easily verify that

$$L[\mathbf{x}_1\ \ \mathbf{x}_2\ \ \mathbf{x}_3] = \begin{bmatrix} 6 & 0 & -1 \\ 1 & 5 & 2 \end{bmatrix} = [\mathbf{y}_1\ \ \mathbf{y}_2]\begin{bmatrix} 5 & -5 & -3 \\ -9 & 15 & 8 \end{bmatrix}$$

which shows that the matrix A of L with respect to the given bases defined by the columns in P and Q is

$$A = \begin{bmatrix} 5 & -5 & -3 \\ -9 & 15 & 8 \end{bmatrix} = B^T.$$

□

Example 6.31 Consider the linear transformation L (see Example 6.5(c)) over $\mathbb{R}$ defined by

$$L : \mathbb{R}^{2\times2} \to \mathbb{P}_2(t) : B = \begin{bmatrix} a & b \\ c & d \end{bmatrix} \mapsto a + (b + c)t + dt^2.$$

The matrices

$$B_{11} = \begin{bmatrix} 0 & 1 \\ 1 & 1 \end{bmatrix}, \quad B_{12} = \begin{bmatrix} 1 & 0 \\ 1 & 1 \end{bmatrix}, \quad B_{21} = \begin{bmatrix} 1 & 1 \\ 0 & 1 \end{bmatrix}, \quad B_{22} = \begin{bmatrix} 1 & 1 \\ 1 & 0 \end{bmatrix}$$

form a basis of the vector space $\mathbb{R}^{2\times2}$. This is because, if

$$\alpha_1B_{11} + \alpha_2B_{12} + \alpha_3B_{21} + \alpha_4B_{22} = 0$$

for scalars α_i's in $\mathbb{R}$, then it follows that $\alpha_1 = \alpha_2 = \alpha_3 = \alpha_4 = 0$. The polynomials $\{1 + t, t + t^2, t^2\}$ form a basis of $\mathbb{P}_2(t)$ because $a_1(1 + t) + a_2(t + t^2) + a_3t^2 = 0$ for all values of t, where the scalars a_i's are in $\mathbb{R}$, implies that $a_1 = a_2 = a_3 = 0$. Since

$$LB_{11} = 2t + t^2 = 0\cdot(1 + t) + 2\cdot(t + t^2) + (-1)\cdot t^2,$$

$$LB_{12} = 1 + t + t^2 = 1\cdot(1 + t) + 0\cdot(t + t^2) + 1\cdot t^2 = LB_{21},$$

$$LB_{22} = 1 + 2t = 1\cdot(1 + t) + 1\cdot(t + t^2) + (-1)\cdot t^2,$$

the matrix A of L with respect to the given bases is

$$A = \begin{bmatrix} 0 & 1 & 1 & 1 \\ 2 & 0 & 0 & 1 \\ -1 & 1 & 1 & -1 \end{bmatrix}.$$

We note that

$$LB = a + (b + c)t + dt^2 = a(1 + t) + (b + c - a)\,(t + t^2) + (a - b - c + d)t^2$$

has coordinates $(a, b + c - a, a - b - c + d)$ with respect to the given basis in $\mathbb{P}_2(t)$. To determine the transformation L^T:

$$[\mathbb{P}_2(t)]^* \xrightarrow{L^T} (\mathbb{R}^{2\times 2})^*,$$

if in $[\mathbb{P}_2(t)]^*$ we have $\{g_1, g_2, g_3\}$ as a dual basis of $\{1 + t, t + t^2, t^2\}$, then for a given element

$$g = a_1g_1 + a_2g_2 + a_3g_3 \tag{6.71}$$

where a_1, a_2, a_3 are scalars in $\mathbb{R}$, we obtain

$$L^Tg(B) = (a_1g_1 + a_2g_2 + a_3g_3)(LB) = a_1a + a_2(b + c - a) + a_3(a - b - c + d), \tag{6.72}$$

So, the transpose L^T of L is defined as

$$L^Tg = f : \mathbb{R}^{2\times 2} \to \mathbb{R}$$

such that

$$f(B) = a_1a + a_2(b + c - a) + a_3(a - b - c + d)$$

when g is given by (6.71).

We note that L^Tg_i's, $i = 1, 2, 3$, are elements of $(\mathbb{R}^{2\times 2})^*$ in which $\{f_{11}, f_{12}, f_{21}, f_{22}\}$ is a dual basis of $\{B_{11}, B_{12}, B_{21}, B_{22}\}$ satisfying

$$f_{ij}(B_{pq}) = \begin{cases} 1, & \text{when } i = p, j = q; \\ 0, & \text{otherwise, } i, j, p, q = 1, 2. \end{cases} \tag{6.73}$$

Hence, we can express $L^T g_1$ as

$$L^T g_1 = c_1 f_{11} + c_2 f_{12} + c_3 f_{21} + c_4 f_{22}. \tag{6.74}$$

Allowing the functional in (6.74) to operate successively on B_{11}, B_{12}, B_{21}, B_{22}, we have, using (6.71)–(6.73),

$$L^T g_1(B_{11}) = c_1 = 0, \quad L^T g_1(B_{12}) = c_2 = 1, \quad L^T g_1(B_{21}) = c_3 = 1, \quad L^T g_1(B_{22}) = c_4 = 1.$$

Consequently,

$$L^T g_1 = 0 \cdot f_{11} + 1 \cdot f_{12} + 1 \cdot f_{21} + 1 \cdot f_{22}.$$

Similarly,

$$L^T g_2 = 2 \cdot f_{11} + 0 \cdot f_{12} + 0 \cdot f_{21} + 1 \cdot f_{22},$$

$$L^T g_3 = -1 \cdot f_{11} + 1 \cdot f_{12} + 1 \cdot f_{21} - 1 \cdot f_{22}.$$

It follows that the matrix of L^T is A^T. □

6.6.2 Bidual Space

If we concentrate on a fixed vector **x** in V, then for every linear functional f in the dual space V^*, we have a scalar $f(\mathbf{x})$ in $\mathbb{F}$. This correspondence of every element in V^* to an element in $\mathbb{F}$ by the fixed element **x** defines a mapping

$$\phi_{\mathbf{x}} : V^* \to \mathbb{F} : f \mapsto \phi_{\mathbf{x}}(f) = f(\mathbf{x}). \tag{6.75}$$

This mapping is linear. Indeed, for scalars α, β in $\mathbb{F}$ and linear functionals f, g in V^*, we have

$$\phi_{\mathbf{x}}(\alpha f + \beta g) = (\alpha f + \beta g)(\mathbf{x}) = \alpha f(\mathbf{x}) + \beta g(\mathbf{x}) = \alpha \phi_{\mathbf{x}}(f) + \beta f_{\mathbf{x}}(g)$$

which shows that for a fixed **x**, $\phi_{\mathbf{x}}$ is linear.

If we now keep f as a fixed element in V^*, then each element **x** in V again determines such a mapping $\phi_{\mathbf{x}} : V^* \to \mathbb{F} : f \mapsto \phi_{\mathbf{x}}(f) = f(\mathbf{x})$. All such mappings $\phi_{\mathbf{x}}$'s for different f's and **x**'s constitute the dual space of V^* which is denoted by V^{**}, and is called the *second dual space* or *bidual space* of V.

Example 6.32 Let the mapping $\phi_{\mathbf{x}} : (\mathbb{R}^3)^* \to \mathbb{R}$ when $\mathbf{x} = [1\ 1\ 2]^T$ is a fixed vector in $\mathbb{R}^3$ be defined as

$$\phi_{\mathbf{x}}(f) = [a\ b\ c]\mathbf{x} = a + b + 2c,$$

where a, b, c are reals. Different choices of $[a\ b\ c]$ define different linear functionals. In particular,

$$\phi_{\mathbf{x}}(f_k) = [k, -2k, k]\mathbf{x} = k,$$

where k is real, maps the linear functional f_k's onto the real number k. On the other hand, if $f = f_1$, a fixed linear functional and $\mathbf{x} = \mathbf{x}_p = [p\ p\ 2p]^T$, p real, then

$$\phi_{\mathbf{x}_p}(f_1) = [1, -2, 1]\mathbf{x}_p = p$$

maps the linear functional f_1 onto the real number p for every $\mathbf{x}_p$.

Here $\phi_{\mathbf{x}}$ or $\phi_{\mathbf{x}_p}$, are elements of $(\mathbb{R}^3)^{**}$, the *second dual space* of $\mathbb{R}$ or *dual space* of $(\mathbb{R}^3)^*$. □

The correspondence of each element $\mathbf{x}$ in V to an element $\phi_{\mathbf{x}}$ in V^{**} as defined by (6.75) is a mapping

$$\psi : V \to V^{**} : \mathbf{x} \mapsto \psi(\mathbf{x}) = \phi_{\mathbf{x}}, \tag{6.76}$$

which is linear. Indeed, if $\mathbf{x}$ and $\mathbf{y}$ are elements in V, then

$$\begin{aligned}\psi(\alpha\mathbf{x} + \beta\mathbf{y})(f) = \phi_{\alpha\mathbf{x}+\beta\mathbf{y}}(f) &= f(\alpha\mathbf{x} + \beta\mathbf{y}) = \alpha f(\mathbf{x}) + \beta f(\mathbf{y})\\ &= \alpha\phi_{\mathbf{x}}(f) + \beta\phi_{\mathbf{y}}(f) = (\alpha\phi_{\mathbf{x}} + \beta\phi_{\mathbf{y}})(f)\\ &= [\alpha\psi(\mathbf{x}) + \beta\psi(\mathbf{y})](f)\end{aligned}$$

for any f in V^* and any scalars α and β in $\mathbb{F}$. Hence,

$$\psi(\alpha\mathbf{x} + \beta\mathbf{y}) = \alpha\psi(\mathbf{x}) + \beta\psi(\mathbf{y})$$

which shows that the mapping ψ is linear.

We shall now prove that the vector spaces V and V^{**} are *isomorphic*, i.e., the map ψ is an *isomorphism*. The following Lemma will prove instrumental for the above purpose.

Lemma 6.1 Let V be a vector space of dimension n and let $\mathbf{x}$ be an element in V. If $\phi_{\mathbf{x}}(f) = 0$, where $\phi_{\mathbf{x}}$ is defined by (6.75) for all f in V^*, then $\mathbf{x} = \mathbf{\theta}$.

Proof We shall prove the Lemma by contradiction. Suppose that $\phi_{\mathbf{x}}(f) = f(\mathbf{x}) = 0$ for all f in V^*, but $\mathbf{x} \neq \mathbf{\theta}$. Then we may choose a basis $\{\mathbf{x}_1 = \mathbf{x}, \mathbf{x}_2, ..., \mathbf{x}_n\}$ in V, of which $\{f_{\mathbf{x}_1}, f_{\mathbf{x}_2} ..., f_{\mathbf{x}_n})$ is the dual basis. Then $f_{\mathbf{x}}(\mathbf{x}) = 1 \neq 0$ which is a contradiction. □

Theorem 6.17 If V is a finite dimensional vector space over a field $\mathbb{F}$, then it is isomorphic to its second dual space V^{**}.

Proof We need to prove that the mapping ψ defined by (6.76) is an isomorphism. The mapping ψ is one-to-one. To see this, suppose that $\psi(\mathbf{x}) = \phi_{\mathbf{x}}$ = zero functional on V^* for some $\mathbf{x}$ in V. Then $\psi(\mathbf{x})(f) = \phi_{\mathbf{x}}(f) = f(\mathbf{x}) = 0$, the zero scalar in $\mathbb{F}$ for every f in V^*. By Lemma 6.1, $\mathbf{x} = \mathbf{\theta}$. Also, in view of (6.75) and Theorem 6.2, dim V = dim V^* = dim V^{**}, which shows that ψ is onto. Therefore, ψ is an isomorphism. □

An important application of the bidual space V^{**} resolves the pertinent question in the study of dual space whether every basis of the vector space V^* is a dual of some basis in V. This answer is contained in the following corollary.

Corollary 6.8 Let V be a finite dimensional vector space over the field $\mathbb{F}$ with dual space V^*. Then, every basis for V^* is the dual of some basis of V.

Proof Let $\{f_1, f_2, ..., f_n\}$ be a basis of V^*. By Theorem 6.15, there is a dual basis $\{\phi_1, \phi_2, ..., \phi_n\}$ in V^{**} of the above basis defined by

$$\phi_i(f_j) = \delta_{ij} = f_i(\mathbf{x}_i).$$

Hence, it follows that $\{\mathbf{x}_1, \mathbf{x}_2, ..., \mathbf{x}_n\}$ is a basis for V and that $\{f_1, f_2, ..., f_n\}$ is a dual of this basis.

Example 6.33 With respect to the basis $\{\mathbf{x}_1, \mathbf{x}_2\}$, any vector $\mathbf{x}$ in $\mathbb{R}^2$ is represented by $\mathbf{x} = \alpha_1\mathbf{x}_1 + \alpha_2\mathbf{x}_2$. If $\{f_1, f_2\}$ is the dual basis of $\{\mathbf{x}_1, \mathbf{x}_2\}$ in $\mathbb{R}^2$, then

$$f_1(\mathbf{x}) = \alpha_1 \quad \text{and} \quad f_2(\mathbf{x}) = \alpha_2. \tag{6.77}$$

In fact, $\{f_1, f_2\}$ forms a basis of V^*, the vector space of linear functionals $f : \mathbb{R}^2 \to \mathbb{R}$. Let us define any such linear functional by

$$f(\mathbf{x}) = c_1\alpha_1 + c_2\alpha_2$$

which becomes, as a result of (6.77), $f(\mathbf{x}) = (c_1f_1 + c_2f_2)(\mathbf{x})$. Hence, $f = c_1f_1 + c_2f_2$. The dual basis of $\{f_1, f_2\}$ designated by $\{\phi_1, \phi_2\}$ is defined by

$$\phi_1(f) = f(\mathbf{x}_1) = (c_1f_1 + c_2f_2)(\mathbf{x}_1) = c_1,$$

$$\phi_2(f) = f(\mathbf{x}_2) = (c_1f_1 + c_2f_2)(\mathbf{x}_2) = c_2.$$

Example 6.34 Let us consider linear functionals, $f : \mathbb{P}_2(t) \to \mathbb{R}$ over $\mathbb{R}$, defined by $f(a_0 + a_1t + a_2t^2) = a_0 + a_1 + a_2$, where a_0, a_1, a_2 are elements of $\mathbb{R}$ and where we have tacitly assumed that the basis in $\mathbb{P}_2(t)$ is $\{\mathbf{x}_1 = 1, \mathbf{x}_2 = t, \mathbf{x}_3 = t^2\}$. Obviously, the dual basis of $\{1, t, t^2\}$ is $\{f_1, f_2, f_3\}$ defined by

$$f_1(\mathbf{x}) = a_0, \quad f_2(\mathbf{x}) = a_1, \quad f_3(\mathbf{x}) = a_2,$$

which in turn has a dual basis $\{\phi_1, \phi_2, \phi_3\}$ in $\mathbb{P}_2{}^{**}(t)$ given by

$$\phi_1(f) = f(\mathbf{x}_1) = 1, \quad \phi_2(f) = f(\mathbf{x}_2) = 1, \quad \phi_3(f) = f(\mathbf{x}_3) = 1.$$

6.6.3 Adjoint of a Linear Transformation

We shall now consider linear transformations L on the inner product space V over $\mathbb{C}$ in which there exists, as we have seen in Section 5.5, a set of orthonormal basis vectors. In this inner product spaces if A is the matrix of L with respect to a given orthonormal basis, an obvious question is: What is the linear transformation on V which is represented with respect to the given basis by A^*, the transposed conjugate or tranjugate of A? The answer is given by introducing the *adjoint of the linear transformation L*, which is unique and is denoted by L^*, and by showing that A^* is the matrix of L^* with respect to the given orthonormal basis. Over the real field $\mathbb{R}$, this matrix of L^* turns out to be A^T. In Section 6.6.2 we have seen that A^T is also the matrix of the transpose L^T of L with respect to the dual basis of the given orthonormal basis. Are, therefore, L^T and L^* the same over the real field $\mathbb{R}$? It is justified that even though L^T and L^* are not the same, they give rise to the same matrix over $\mathbb{R}$ and a connection between these two linear transformations is established.

Lemma 6.2 Let f be a given linear functional on a finite dimensional inner product space V over a field $\mathbb{F}$. Then there exists a unique vector $\mathbf{y}$ in V such that $f(\mathbf{x}) = \langle \mathbf{x}, \mathbf{y} \rangle$ for all $\mathbf{x}$ in V.

Proof Suppose that $\{\mathbf{x}_1, \mathbf{x}_2, ..., \mathbf{x}_n\}$ is an orthonormal basis of V. We choose the vector $\mathbf{y}$ in V satisfying

$$\mathbf{y} = \overline{f(\mathbf{x}_1)}\,\mathbf{x}_1 + \overline{f(\mathbf{x}_2)}\,\mathbf{x}_2 + \cdots + \overline{f(\mathbf{x}_n)}\,\mathbf{x}_n. \tag{6.78}$$

If we define $g(\mathbf{x}) = \langle \mathbf{x}, \mathbf{y} \rangle$ for any vector **x** in V, then taking cognizance of (5.18), we have

$$g(\mathbf{x}_i) = \left\langle \mathbf{x}_i, \overline{f(\mathbf{x}_1)}\,\mathbf{x}_1 + \overline{f(\mathbf{x}_2)}\,\mathbf{x}_2 + \cdots + \overline{f(\mathbf{x}_n)}\,\mathbf{x}_n \right\rangle = f(\mathbf{x}_i), \qquad (i = 1, 2, ..., n).$$

Since the functionals g and f agree on the basis vectors, we have, by Theorem 6.1, $f = g$ which proves that $f(\mathbf{x}) = \langle \mathbf{x}, \mathbf{y} \rangle$.

To show that **y** is unique, suppose there is another vector **z** in V such that

$$f(\mathbf{x}) = \langle \mathbf{x}, \mathbf{y} \rangle = \langle \mathbf{x}, \mathbf{z} \rangle.$$

Then, $\langle \mathbf{x}, \mathbf{y} \rangle - \langle \mathbf{x}, \mathbf{z} \rangle = \langle \mathbf{x}, \mathbf{y} - \mathbf{z} \rangle = 0$ for all **x** in V. In particular, $\langle \mathbf{y} - \mathbf{z}, \mathbf{y} - \mathbf{z} \rangle = 0$, giving **y** = **z** by Axiom 1 of Definition 5.9. Hence, **y** is unique. □

The foregoing Lemma, which is a special case of the famous *Riesz representation theorem*, gives a simple characterization of all linear functionals on the inner product space. The existence of an orthonormal basis in such a space is instrumental in establishing the above simple characterization which enables us to write a linear functional on an inner product space as an inner product. On any arbitrary vector space, however, such a linear functional is expressed in terms of the dual basis of the space as substantiated in Theorem 6.15.

Example 6.35 Suppose that the set of vectors

$$\mathbf{x}_1 = \frac{1}{2}[1 - i, 1 + i, 0]^T, \qquad \mathbf{x}_2 = \frac{1}{2}[-1 + i, 1 + i, 0]^T, \qquad \mathbf{x}_3 = [0\ 0\ 1]^T$$

is an orthonormal basis in $\mathbb{C}^3$. We define a linear functional f on $\mathbb{C}^3$ by

$$f(\mathbf{x}_1) = 1 + i, \qquad f(\mathbf{x}_2) = 1 - i, \qquad f(\mathbf{x}_3) = 1,$$

which specify f completely as a result of Theorem 6.1. Determine a vector **y** such that $f(\mathbf{x}) = \langle \mathbf{x}, \mathbf{y} \rangle$ for all **x** in $\mathbb{C}^3$.

Solution Making use of (6.78), the required **y** can be written as

$$\mathbf{y} = (1 - i)\mathbf{x}_1 + (1 + i)\mathbf{x}_2 + \mathbf{x}_3 = [-i - 1,\ 1 + i,\ 1]^T.$$ □

Example 6.36 Let L be a linear transformation on a finite dimensional inner product space V over $\mathbb{F}$ in which **y** is a fixed element. Show that the function $f : V \to \mathbb{F}$ defined by $f(\mathbf{x}) = \langle L\mathbf{x}, \mathbf{y} \rangle$ for any **x** in V is linear.

Solution For scalars α and β in $\mathbb{F}$, and for any two elements **x** and **z** in V, we have

$$f(\alpha\mathbf{x} + \beta\mathbf{z}) = \langle L(\alpha\mathbf{x} + \beta\mathbf{z}), \mathbf{y} \rangle = \langle \alpha L\mathbf{x} + \beta L\mathbf{z}, \mathbf{y} \rangle$$

$$= \alpha\langle L\mathbf{x}, \mathbf{y} \rangle + \beta\langle L\mathbf{z}, \mathbf{y} \rangle = \alpha f(\mathbf{x}) + \beta f(\mathbf{z}).$$

Hence, f is a linear functional. □

Lemma 6.3 Let L be a linear transformation on a finite dimensional inner product space V. Then, there exists a unique linear transformation L^* on V such that

$$\langle L\mathbf{x}, \mathbf{y}\rangle = \langle \mathbf{x}, L^*\mathbf{y}\rangle$$

for all $\mathbf{x}$ and $\mathbf{y}$ in V.

Proof Suppose that $\mathbf{y}$ is a fixed element in V. In the light of Example 6.36, we can define a linear transformation f on V by

$$f(\mathbf{x}) = \langle L\mathbf{x}, \mathbf{y}\rangle$$

for any $\mathbf{x}$ in V. By Lemma 6.2, there exists a unique vector $\mathbf{z}$ in V such that

$$f(\mathbf{x}) = \langle L\mathbf{x}, \mathbf{y}\rangle = \langle \mathbf{x}, \mathbf{z}\rangle \tag{6.79}$$

for all $\mathbf{x}$ in V. Since $f(\mathbf{x})$ depends on $\mathbf{y}$, this unique vector $\mathbf{z}$ constructed by Lemma 6.2 must be a function of $\mathbf{y}$ which we designate by

$$\mathbf{z} = L^*\mathbf{y}. \tag{6.80}$$

Thus, we obtain a transformation L^* on V whose connection with L, in view of (6.79) and (6.80), is given by

$$\langle L\mathbf{x}, \mathbf{y}\rangle = \langle \mathbf{x}, L^*\mathbf{y}\rangle.$$

Since for a given $\mathbf{y}$, one can determine $\mathbf{z}$ uniquely by Lemma 6.2, L^* defined by (6.80) is unique.

We shall now show that L^* is linear. For any vectors $\mathbf{u}$ and $\mathbf{v}$ in V, and scalars α and β in $\mathbb{F}$, using (5.18), we have

$$\begin{aligned}
\langle \mathbf{x}, L^*(\alpha\mathbf{u} + \beta\mathbf{v})\rangle &= \langle L\mathbf{x}, \alpha\mathbf{u} + \beta\mathbf{v}\rangle = \bar{\alpha}\langle L\mathbf{x}, \mathbf{u}\rangle + \bar{\beta}\langle L\mathbf{x}, \mathbf{v}\rangle \\
&= \bar{\alpha}\langle \mathbf{x}, L^*\mathbf{u}\rangle + \bar{\beta}\langle \mathbf{x}, L^*\mathbf{v}\rangle \\
&= \langle \mathbf{x}, \alpha L^*\mathbf{u}\rangle + \langle \mathbf{x}, \beta L^*\mathbf{v}\rangle \\
&= \langle \mathbf{x}, \alpha L^*\mathbf{u} + \beta L^*\mathbf{v}\rangle.
\end{aligned}$$

Hence,

$$L^*(\alpha\,\mathbf{u} + \beta\,\mathbf{v}) = \alpha L^*\mathbf{u} + \beta L^*\mathbf{v}.$$

So, L^* defined by (6.80) is a unique linear transformation. □

Definition 6.20 *Adjoint of a Linear Transformation.* Let L be a linear transformation on an inner product space V. If for any $\mathbf{x}$, $\mathbf{y}$ in V, there exists a linear transformation L^* on V such that

$$\langle L\mathbf{x}, \mathbf{y}\rangle = \langle \mathbf{x}, L^*\mathbf{y}\rangle,$$

then L^* is called the *adjoint* of L. □

For the linear transformation $L : V_n \to W_m$, if there exists a linear transformation $L^* : W_m \to V_n$ such that $\langle L\mathbf{x}, \mathbf{y}\rangle = \langle \mathbf{x}, L^*\mathbf{y}\rangle$ then L^* is defined as adjoint of L. See Theorem 6.19.

We have just seen that, in a finite dimensional inner product space, the adjoint L^* does always exist. On the contrary, in the infinite dimensional space, in some cases, it exists while in others it does not. The foregoing definition is appropriate to both the cases.

Example 6.37 Let a linear transformation on $\mathbb{C}^3$ be defined by

$$L\mathbf{x} = (x_1 + ix_2,\ x_2 + ix_3,\ x_3 + ix_1), \tag{6.81}$$

where $\mathbf{x} = (x_1, x_2, x_3)$ is in $\mathbb{C}^3$. Determine the adjoint of L.

Solution If L^* is the adjoint of L, then for any $\mathbf{x} = (x_1, x_2, x_3)$ and $\mathbf{y} = (y_1, y_2, y_3)$ in $\mathbb{C}^3$, we can write, in view of Definition 6.20,

$$f(\mathbf{x}) = \langle L\mathbf{x}, \mathbf{y}\rangle = \langle \mathbf{x}, L^*\mathbf{y}\rangle = \langle \mathbf{x}, \mathbf{z}\rangle, \tag{6.82}$$

where $L^*\mathbf{y} = \mathbf{z} = (z_1, z_2, z_3)$ can be determined by (6.78) in the constructive proof of Lemma 6.2. Choosing the row elementary vectors $\mathbf{e}_1$, $\mathbf{e}_2$, $\mathbf{e}_3$ in $\mathbb{C}^3$ as the orthonormal basis, we write, following (6.78),

$$\mathbf{z} = \overline{f(\mathbf{e}_1)}\,\mathbf{e}_1 + \overline{f(\mathbf{e}_2)}\,\mathbf{e}_2 + \overline{f(\mathbf{e}_3)}\,\mathbf{e}_3. \tag{6.83}$$

By virtue of (6.81) and (6.82), it follows easily that

$$f(\mathbf{x}) = \langle L\mathbf{x}, \mathbf{y}\rangle = \langle (x_1 + ix_2, x_2 + ix_3, x_3 + ix_1), (y_1, y_2, y_3)\rangle$$

and hence,

$$\begin{aligned} \overline{f(\mathbf{e}_1)} &= \overline{\langle (1, 0, i), (y_1, y_2, y_3)\rangle} = y_1 - iy_3, \\ \overline{f(\mathbf{e}_2)} &= y_2 - iy_1, \qquad \overline{f(\mathbf{e}_3)} = y_3 - iy_2. \end{aligned} \tag{6.84}$$

Therefore, by (6.83) and (6.84),

$$L^*\mathbf{y} = \mathbf{z} = (y_1 - iy_3,\ y_2 - iy_1,\ y_3 - iy_2).$$ □

Example 6.38 Determine the adjoint of the linear transformation (see Problem 6.5(a))

$$L : \mathbb{P}_2(t) \to \mathbb{P}_2(t) : p(t) \mapsto t\dot{p}(t) + p(t)$$

over $\mathbb{R}$ with respect to the inner product

$$\langle p(t), q(t)\rangle = \int_{-1}^{1} p(t)q(t)\,dt.$$

Solution The orthonormal basis in $\mathbb{P}_2(t)$ is (see Example 5.24(b))

$$\left\{ \mathbf{x}_1(t) = \frac{\sqrt{2}}{2}, \quad \mathbf{x}_2(t) = \frac{\sqrt{6}\,t}{2}, \quad \mathbf{x}_3(t) = \frac{\sqrt{10}(3t^2-1)}{4} \right\}. \tag{6.85}$$

If $p(t)$ and $q(t)$ are two arbitary elements in $\mathbb{P}_2(t)$, then by definition of the adjoint of L, we can write

$$f(p) = \langle Lp, q \rangle = \langle p, L^* q \rangle, \tag{6.86}$$

where $f : \mathbb{P}_2(t) \to \mathbb{R}$ is a linear functional. Following the constructive proof of Lemma 6.2, the adjoint L^* is explicitly written as

$$L^*q = : h(t) = f(\mathbf{x}_1)\mathbf{x}_1 + f(\mathbf{x}_2)\mathbf{x}_2 + f(\mathbf{x}_3)\mathbf{x}_3. \tag{6.87}$$

Denoting $q(t) = \alpha_0 + \alpha_1 t + \alpha_2 t^2$, where α_0, α_1, α_2 are real, making use of (6.85) and (6.86), and using the definition of L given in the problem, we successively get

$$f(\mathbf{x}_1) = \langle L\mathbf{x}_1, q \rangle = \frac{1}{\sqrt{2}} \int_{-1}^{1} (\alpha_0 + \alpha_1 t + \alpha_2 t^2)\, dt = \frac{\sqrt{2}\,(3\alpha_0 + \alpha_2)}{3},$$

$$f(\mathbf{x}_2) = \langle L\mathbf{x}_2, q \rangle = \int_{-1}^{1} t\sqrt{6}\,(\alpha_0 + \alpha_1 t + \alpha_2 t^2)\, dt = \frac{2\sqrt{6}}{3}\alpha_1$$

$$f(\mathbf{x}_3) = \langle L\mathbf{x}_3, q \rangle = \left(\sqrt{10}/4\right) \int_{-1}^{1} (9t^2 - 1)(\alpha_0 + \alpha_1 t + \alpha_2 t^2)\, dt = \sqrt{10}\left(\frac{15\alpha_0 + 11\alpha_2}{15}\right)$$

Inserting these expressions in (6.87), finally we obtain

$$L^*(\alpha_0 + \alpha_1 t + \alpha_2 t^2) = \frac{-3(\alpha_0 + \alpha_2)}{2} + 2\alpha_1 t + \frac{(15\alpha_0 + 11\alpha_2)t^2}{2},$$

which is the required adjoint of L. □

Theorem 6.18 Let $\{\mathbf{x}_1, \mathbf{x}_2, ..., \mathbf{x}_n\}$ be an orthonormal basis in a finite dimensional inner product space V over $\mathbb{F}$ on which L is a linear transformation. If A is the matrix of L with respect to the above basis, then A^* is the matrix of L^* with respect to the same basis.

Proof We note that the vectors in the orthonormal basis satisfy the relation $\langle \mathbf{x}_i, \mathbf{x}_j \rangle = \delta_{ij}$. Suppose that $\mathbf{x}$ and $\mathbf{y}$ are any elements in V, and $A = [a_{ij}]$ is the matrix of L with respect to the given orthonormal basis. Then, we can write

$$\begin{aligned} \mathbf{x} &= \alpha_1\mathbf{x}_1 + \alpha_2\mathbf{x}_2 + \cdots + \alpha_n\mathbf{x}_n, \\ \mathbf{y} &= \beta_1\mathbf{x}_1 + \beta_2\mathbf{x}_2 + \cdots + \beta_n\mathbf{x}_n, \\ L\mathbf{x}_j &= a_{1j}\mathbf{x}_1 + a_{2j}\mathbf{x}_2 + \cdots + a_{nj}\mathbf{x}_n, \end{aligned} \tag{6.88}$$

where α_i's and β_i's are scalars in $\mathbb{F}$. Hence, making use of Definition 5.9 and (5.18), we can write, successively,

$$\langle \mathbf{x}, L^*\mathbf{y}\rangle = \langle L\mathbf{x}, \mathbf{y}\rangle = \langle \sum_{j=1}^{n} \alpha_j L\mathbf{x}_j, \sum_{k=1}^{n} \beta_k \mathbf{x}_k\rangle = \langle \sum_{j=1}^{n} \alpha_j \sum_{i=1}^{n} a_{ij}\mathbf{x}_i, \sum_{k=1}^{n} \beta_k \mathbf{x}_k\rangle$$

$$= \sum_{i=1}^{n}\sum_{j=1}^{n} \alpha_j a_{ij} \langle \mathbf{x}_i, \sum_{k=1}^{n} \beta_k \mathbf{x}_k\rangle = \sum_{i=1}^{n}\sum_{j=1}^{n} \alpha_j a_{ij}\bar{\beta}_i$$

$$= \sum_{i=1}^{n}\sum_{j=1}^{n} \overline{\beta_i a_{ij}} \langle \sum_{k=1}^{n} \alpha_k \mathbf{x}_k, \mathbf{x}_j\rangle = \langle \mathbf{x}, \sum_{i=1}^{n}\sum_{j=1}^{n} \beta_i \bar{a}_{ij}\mathbf{x}_j\rangle.$$

Since L^* is unique, therefore, it follows that

$$L^*\mathbf{y} = \sum_{i=1}^{n}\sum_{j=1}^{n} \beta_i \bar{a}_{ij}\mathbf{x}_j.$$

Setting $\mathbf{y} = \mathbf{x}_i$ in the above expression, because of (6.88), we have,

$$L^*\mathbf{x}_i = \sum_{j=1}^{n} \bar{a}_{ij}\mathbf{x}_j,$$

which in view of (6.28) ensures that the matrix of L^* is A^*.

Theorem 6.18 is also valid for $L : \mathbb{C}^n \to \mathbb{C}^m$. □

Example 6.39 Let $\mathbf{x} = [x_1\ x_2\ x_3]$ and $\mathbf{y} = [y_1\ y_2\ y_3]$ be two elements in $\mathbb{C}^3$ in which the elements are row vectors. The adjoint of the linear transformation L on $\mathbb{C}^3$ defined by

$$L\mathbf{x} = [x_1 + ix_2, x_2 + ix_3, x_3 + ix_1]$$

is

$$L^*\mathbf{y} = [y_1 - iy_3, y_2 - iy_1, y_3 - iy_2]$$

as determined in Example 6.37. With respect to the natural orthonormal basis $\{\mathbf{e}_1, \mathbf{e}_2, \mathbf{e}_3\}$, it is easily seen that

$$L\mathbf{e}_1 = 1\cdot\mathbf{e}_1 + i\cdot\mathbf{e}_3; \quad L\mathbf{e}_2 = i\cdot\mathbf{e}_1 + 1\cdot\mathbf{e}_2; \quad L\mathbf{e}_3 = i\cdot\mathbf{e}_2 + 1\cdot\mathbf{e}_3;$$
$$L^*\mathbf{e}_1 = 1\cdot\mathbf{e}_1 - i\cdot\mathbf{e}_2; \quad L^*\mathbf{e}_2 = 1\cdot\mathbf{e}_2 - i\cdot\mathbf{e}_3; \quad L^*\mathbf{e}_3 = -i\cdot\mathbf{e}_1 + 1\cdot\mathbf{e}_3.$$

Hence, the matrices A and A^* of L and L^* are respectively,

$$A = \begin{bmatrix} 1 & i & 0 \\ 0 & 1 & i \\ i & 0 & 1 \end{bmatrix}$$

and

$$A^* = \begin{bmatrix} 1 & 0 & -i \\ -i & 1 & 0 \\ 0 & -i & 1 \end{bmatrix}$$

which show that the matrix of L^* is transposed conjugate of that of L with respect to the same orthonormal basis. □

Example 6.40 Referring to Example 6.38, we consider the same linear transformation L and its adjoint L^* as calculated there in terms of the orthonormal basis (6.85) with respect to the specified inner product. For an element $p(t) = \alpha_0 + \alpha_1 t + \alpha_2 t^2$ in $\mathbb{P}_2(t)$, it is shown that (see the definition of L in Example 6.38),

$$Lp(t) = \alpha_0 + 2\alpha_1 t + 3\alpha_2 t^2 = \sqrt{2}\,(\alpha_0 + \alpha_2)\mathbf{x}_1 + \frac{2\sqrt{6}}{3}\alpha_1\mathbf{x}_2 + \frac{2\sqrt{10}}{5}\alpha_2\mathbf{x}_3;$$

$$L^*p(t) = \frac{(6\alpha_0 + 2\alpha_2)\sqrt{2}}{6}\mathbf{x}_1 + \frac{4\alpha_1}{\sqrt{6}}\mathbf{x}_2 + \frac{30\alpha_0 + 22\alpha_2}{3\sqrt{10}}\mathbf{x}_3.$$

We note that, if

$$p(t) = \mathbf{x}_1(t) = 1/\sqrt{2}, \quad \text{then} \quad \alpha_0 = 1/\sqrt{2}, \alpha_1 = \alpha_2 = 0;$$

$$p(t) = \mathbf{x}_2(t) = \sqrt{6}\,t/2, \quad \text{then} \quad \alpha_0 = \alpha_2 = 0, \ \alpha_1 = \sqrt{6}/2;$$

$$p(t) = \mathbf{x}_3(t) = \sqrt{10}\,(3t^2 - 1)/4, \quad \text{then} \quad \alpha_0 = -\sqrt{10}/4, \ \alpha_1 = 0, \ \alpha_2 = 3\sqrt{10}/4.$$

Consequently,

$$L[\mathbf{x}_1\ \mathbf{x}_2\ \mathbf{x}_3] = [\mathbf{x}_1\ \mathbf{x}_2\ \mathbf{x}_3]\begin{bmatrix} 1 & 0 & \sqrt{5} \\ 0 & 2 & 0 \\ 0 & 0 & 3 \end{bmatrix},$$

$$L^*[\mathbf{x}_1\ \mathbf{x}_2\ \mathbf{x}_3] = [\mathbf{x}_1\ \mathbf{x}_2\ \mathbf{x}_3]\begin{bmatrix} 1 & 0 & 0 \\ 0 & 2 & 0 \\ \sqrt{5} & 0 & 3 \end{bmatrix}$$

which show that the matrix of L^* is the transpose of that of L. □

From what we have just now established it is clear that, if L is a linear transformation on the inner product space V over the real field $\mathbb{R}$ represented by a matrix A with respect to an orthonormal basis, then its adjoint L^* is also a linear transformation on V represented by the matrix A^T with respect to the same basis. From our earlier study we know that, the tranpose L^T is a linear transformation on the dual space V^* and can be represented by the same A^T with respect to the dual basis of the above orthonormal basis. It may be argued that L^* acts on V while L^T on V^*, and consequently, L^* and L^T are different indeed. How is it then that they give rise to the same matrix A^T over the real field $\mathbb{R}$? In linear algebra, the same matrix may be associated with different linear transformations, and there is no point, therefore, in the above objection.

We shall now take up the interesting question posed above, and show that, L^* and L^T have identical mapping properties.

It is ensured by Lemma 6.2 that any linear functional f on a finite dimensional inner product space V over a field $\mathbb{F}$ can be expressed as $f(\mathbf{x}) = \langle \mathbf{x}, \mathbf{y} \rangle$ for a unique vector $\mathbf{y}$ in V. This $\mathbf{y}$ is, in fact, a function of f as corroborated by (6.78). So, we can define a function $\phi : V^* \to V$ such that, $\phi(f) = \mathbf{y}$ and that $f(\mathbf{x}) = \langle \mathbf{x}, \phi(f) \rangle$ for all $\mathbf{x}$ in V. The function ϕ is, as will be shown now, an isomorphism. If for another element g in V^*, $\phi(f) = \phi(g)$, then $f(\mathbf{x}) = g(\mathbf{x})$ for all $\mathbf{x}$ in V. Hence, in view of Definition 6.5, $f = g$ and ϕ is one-to-one by Definition 6.3. Also, ϕ is onto. This is because, for every $\mathbf{z}$ in V we can define a corresponding linear functional g in V^* by

$$g(\mathbf{x}) = \langle \mathbf{x}, \mathbf{z} \rangle.$$

It is obvious that g is a function ψ of $\mathbf{z}$ and that $\mathbf{z} = \phi(g)$; so there is a mapping $\psi : V \rightarrow V^*$ such that $\psi(\mathbf{z}) = g$, giving $\mathbf{z} = \phi(g) = (\phi \cdot \psi)(\mathbf{z})$, which holds for all $\mathbf{z}$. Hence, $\phi \cdot \psi = I_V$, and by Theorem 6.8(b), ϕ is onto. Finally, for f and g in V^* and α in $\mathbb{C}$, it is easy to prove that

$$\phi(f + g) = \phi(f) + \phi(g), \tag{6.89}$$

$$\phi(\alpha f) = \bar{\alpha}\phi(f). \tag{6.90}$$

Indeed,

$$\begin{aligned}(f + g)(\mathbf{x}) &= \langle \mathbf{x}, \phi(f + g) \rangle = f(\mathbf{x}) + g(\mathbf{x}) \\ &= \langle \mathbf{x}, \phi(f) \rangle + \langle \mathbf{x}, \phi(g) \rangle = \langle \mathbf{x}, \phi(f) + \phi(g) \rangle, \\ (\alpha f)(\mathbf{x}) &= \langle \mathbf{x}, \phi(\alpha f) \rangle = f(\alpha \mathbf{x}) = \langle \alpha \mathbf{x}, \phi(f) \rangle = \langle \mathbf{x}, \bar{\alpha}\, \phi(f) \rangle\end{aligned}$$

from which follow (6.89) and (6.90). From the foregoing discussion, it follows obviously that over the real field $\mathbb{R}$, ϕ is an isomorphism on $V^* \rightarrow V$ because in this case, $\phi(\alpha f) = \alpha \phi(f)$. We recall that, by Definition 6.19 for the transpose of a linear transformation L, and by Definition 6.20 for the adjoint of a linear transformation L, we have respectively,

$$L^T f\mathbf{x} = f \cdot L\mathbf{x}$$

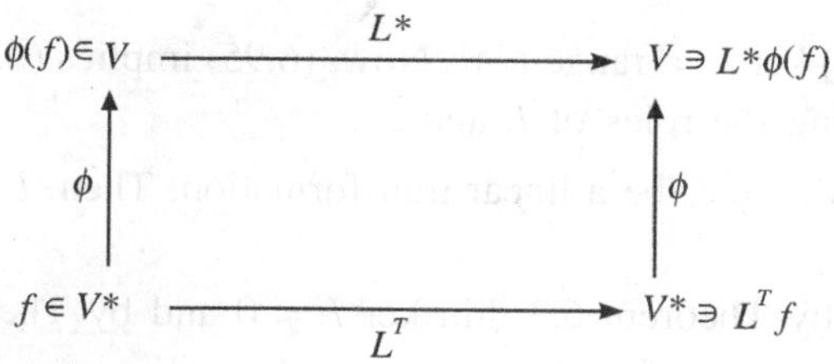

Figure 6.8 Mapping properties of L^* and L^T over $\mathbb{R}$.

for all $\mathbf{x}$ in V and all f in V^*, and

$$\langle L\mathbf{x}, \mathbf{y} \rangle = \langle \mathbf{x}, L^* \mathbf{y} \rangle$$

for all $\mathbf{x}$ and $\mathbf{y}$ in V. Hence, we get

$$\langle \mathbf{x}, \phi(L^T f) \rangle = L^T f(\mathbf{x}) = (f \cdot L)(\mathbf{x}) = f(L\mathbf{x}) = \langle L\mathbf{x}, \phi(f) \rangle = \langle \mathbf{x}, L^* \phi(f) \rangle,$$

from which it follows that

$$\phi(L^T f) = L^* \phi(f). \tag{6.91}$$

Equation (6.91) shows that, to every element f in V^* which is mapped to $L^T f$ by L^T, there is a corresponding element $\phi(f)$ in V, which is mapped to $L^*\phi(f)$ by L^* (see Figure 6.8). The isomorphism ϕ ensures that V and V^* have the same number of elements and relation (6.91) ensures that L^* and L^T have identical mapping properties.

Theorem 6.19 Let $L : V_n \to W_m$ be a linear transformation. Then, $\ker L = (\text{range } L^*)^{\perp}$ and

$$V_n = \ker L \oplus \text{range } L^*, \tag{6.92}$$

$$W_m = \ker L^* \oplus \text{range } L. \tag{6.93}$$

In particular, $\dim V_n = \dim \ker L + \dim \text{range } L^*$ and $\dim W_m = \dim \ker L^* + \dim \text{range } L$.

Proof By definition $L^* : W_m \to V_n$ and if $\mathbf{z} \in \text{range } L^*$ then $\exists\ \mathbf{y} \in W_m$ such that $L^*\mathbf{y} = \mathbf{z}$. For any element $\mathbf{x} \in \ker L$, $L\mathbf{x} = 0$ and $\langle \mathbf{x}, \mathbf{z}\rangle = \langle \mathbf{x}, L^*\mathbf{y}\rangle = \langle L\mathbf{x}, \mathbf{y}\rangle = 0$. Hence,

$$\ker L \subseteq (\text{range } L^*)^{\perp} \tag{6.94}$$

a space every element of which is orthogonal to any vector of range L^* in view of Definition 5.11. Next, let $\mathbf{v} \in V_n$ and

$$\mathbf{v} = \mathbf{x} + \mathbf{z}, \tag{6.95}$$

where $\mathbf{z} \in M = \text{range } L^* \subseteq V_n$ and $\mathbf{x} \in M^{\perp} \subseteq V_n$. Therefore, every $\mathbf{y} \in W_m$

$$\langle L\mathbf{x}, \mathbf{y}\rangle = \langle \mathbf{x}, L^*\mathbf{y}\rangle = 0.$$

Hence, $L\mathbf{x} = 0$ which means that $\mathbf{x} \in \ker L$. So,

$$(\text{range } L^*)^{\perp} = M^{\perp} \subseteq \ker L$$

and from (6.94) and (6.96) $\ker L = (\text{range } L^*)^{\perp}$. Now, (6.95) implies (6.92). The second assertion (6.93) follows by switching the roles of L and L^*. □

Theorem 6.20 Let $L : V_n \to V_n$ be a linear transformation. Then, L is injective if and only if L is surjective.

Proof If L is injective, by Theorem 6.3 $\dim \ker L = 0$ and by Theorem 6.5

$$\dim V_n = \dim \ker L + \dim \text{range } L = \dim \text{range } L.$$

Hence, by Definition 5.7 $V_n = \text{range } L$ and L is onto. Conversely, if L is onto, $\dim \text{range } L = n = \dim V_n$ and from $\dim V_n = \dim \ker L + \dim \text{range } L$, it follows that $\dim \ker L = 0$. Hence, L is injective by Theorem 6.3. □

6.7 LINEAR TRANSFORMATION: DILATION, ROTATION, REFLECTION AND PROJECTION

Dilation If the linear transformation L is a matrix of order 2, then $|\det A|$ is a scale factor that represents a magnification of area produced by L. Let us consider the effect of transformation given by

$$A = \begin{bmatrix} a & b \\ c & d \end{bmatrix}$$

on a unit square with vertices at (0,0), (1,0), (0,1) and (1,1) and of area 1. If we post-multiply A with $[1\ 1]^T$ then we find $[a + b\ c + d]^T$ implying that vertex (1,1) of the square is transformed to $(a + b, c + d)$. Similarly, other vertices (0,0), (1,0), and (0,1) are, respectively, transformed to (0,0), (a, c), and (b, d) as shown in Figure 6.9(a) changing the square into a parallelogram. Considering the indicated boundaries shown by dotted lines, it is obvious that its area is $(a+b)(c+d) - 2[bc + \frac{1}{2}bd + \frac{1}{2}ac] = ad - bc$ which is det A. So, the unit square is transformed to a parallelogram with a dilation of its area |det A| times.

It can be be proved that if $(x_{1j}, x_{2j},\ldots x_{nj})$, $j = 1, 2,\ldots, n$ are the coordinates of the vertices of n adjacent sides of an n-dimensional parallelepiped, then its volume is |det A|, where A is the matrix with columns formed with these coordinates. So, a unit hypercube in $\mathbb{R}^n$ is dilated | dat A | times by a transformation A whose columns are $[x_{1j}\ x_{2j}\ldots x_{nj}]'$, $j = 1, 2,\ldots, n$. Depending on the ordering of the columns in A, det A may turn out to be positive or negative so we take the absolute value of det A.

If we set $a = d = m$ in A but $b = c = 0$ then $A[x\ y]^T = [mx\ my]^T$ implying that the length of the segment OP is magnified det $A = m$ times to OP' as shown in Figure 6.9(b). However, if we choose $A = [0\ -1;\ 1\ 0]$ then the point $(x, 0)$ is rotated counter-clockwise through 90° to $(0, x)$ as shown in Figure 6.9(c).

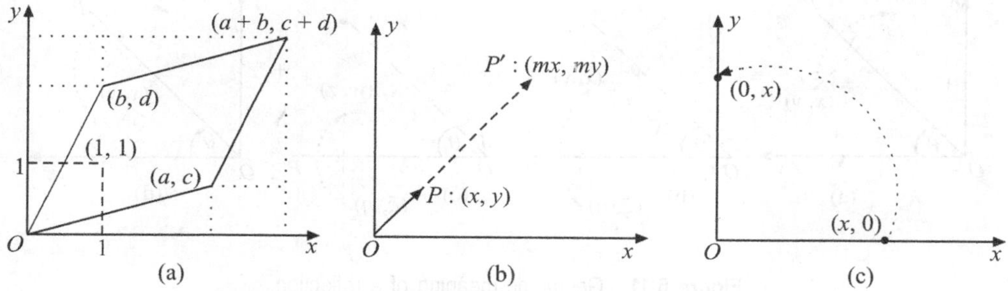

Figure 6.9 Geometric meaning of a transformation.

The above examples illustrate that a linear transformation represents dilation, stretching and rotation of an object that is further supported by polar decomposition in Section 12.3.

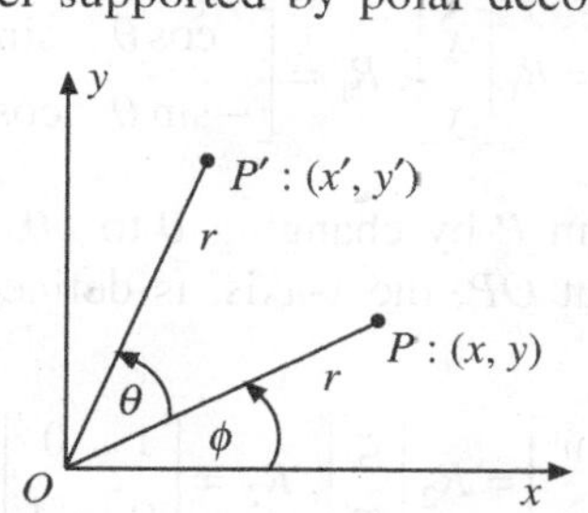

Figure 6.10 Geometric meaning of a rotation.

Rotation Let $P:(x, y)$ be a point in the plane with $OP = r$ making an angle ϕ with the x-axis. Then, $x = r\cos\phi$ and $y = r\sin\phi$ as shown in Figure 6.10. Let the image of P after an *anti-clockwise* rotation θ about O be $P':(x', y')$. Therefore,

$$x' = r\cos(\phi+\theta) = x\cos\theta - y\sin\theta,$$

$$y' = r\sin(\phi+\theta) = x\sin\theta + y\cos\theta.$$

Thus,

$$\begin{bmatrix} x' \\ y' \end{bmatrix} = P\begin{bmatrix} x \\ y \end{bmatrix}, P = \begin{bmatrix} \cos\theta & -\sin\theta \\ \sin\theta & \cos\theta \end{bmatrix}. \tag{6.96}$$

For a *clockwise* rotation θ, we have to use $-\theta$ above. The rotation represented by P is a linear transformation and is orthogonal.

Reflection Examining Figure 6.11, we may conclude that a reflection of a point (x, y), Figure 6.11(a), in a plane about the line OP making an angle θ with the x-axis represented by $y = x\tan\theta$, can be looked upon as the composition of three simple transformations:

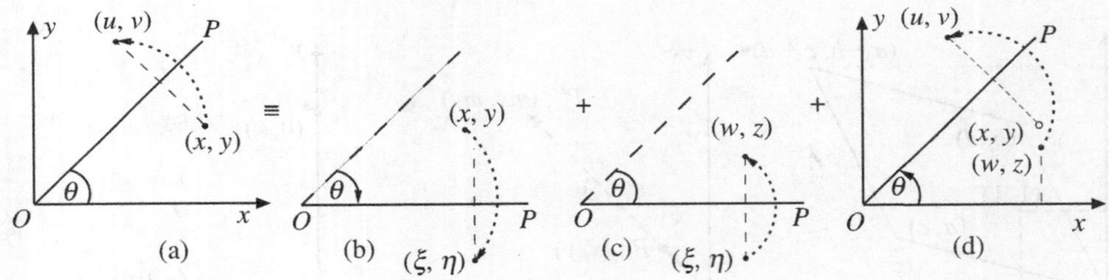

Figure 6.11 Geometric meaning of a reflection.

1. Rotation of the plane *clockwise* through θ so that OP is now mapped onto the x-axis and (x, y) into (ξ, η) as shown in Figure 6.11(b). Call this transformation R_1.

$$\begin{bmatrix} \xi \\ \eta \end{bmatrix} = R_1\begin{bmatrix} x \\ y \end{bmatrix}, R_1 = \begin{bmatrix} \cos\theta & \sin\theta \\ -\sin\theta & \cos\theta \end{bmatrix},$$

where R_1 is obtained from P by changing θ to $-\theta$.

2. Reflection of (ξ, η) about OP, the x-axis, is defined as $w = \xi$ and $z = -\eta$. Call this transformation R_2:

$$\begin{bmatrix} w \\ z \end{bmatrix} = R_2\begin{bmatrix} \xi \\ \eta \end{bmatrix}, R_2 = \begin{bmatrix} 1 & 0 \\ 0 & -1 \end{bmatrix}.$$

3. Rotation of OP back through θ anticlockwise about O so OP now takes the original position and (w, z) is mapped into (u, v), the final image of (x, y) under reflection about OP. Call this transformation R_3.

$$\begin{bmatrix} u \\ v \end{bmatrix} = R_3 \begin{bmatrix} w \\ z \end{bmatrix}, \ R_3 = \begin{bmatrix} \cos\theta & -\sin\theta \\ \sin\theta & \cos\theta \end{bmatrix}.$$

Finally,

$$\begin{bmatrix} u \\ v \end{bmatrix} = R_3 \begin{bmatrix} w \\ z \end{bmatrix} = R_3 R_2 \begin{bmatrix} \xi \\ \eta \end{bmatrix} = R_3 R_2 R_1 \begin{bmatrix} x \\ y \end{bmatrix}$$

$$= \begin{bmatrix} \cos\theta & -\sin\theta \\ \sin\theta & \cos\theta \end{bmatrix} \begin{bmatrix} 1 & 0 \\ 0 & -1 \end{bmatrix} \begin{bmatrix} \cos\theta & \sin\theta \\ -\sin\theta & \cos\theta \end{bmatrix} \begin{bmatrix} x \\ y \end{bmatrix}$$

$$= \begin{bmatrix} \cos 2\theta & \sin 2\theta \\ \sin 2\theta & -\cos 2\theta \end{bmatrix} \begin{bmatrix} x \\ y \end{bmatrix}.$$

The transformation matrix for reflection in line $y = x \tan\theta$

$$Q = \begin{bmatrix} \cos 2\theta & \sin 2\theta \\ \sin 2\theta & -\cos 2\theta \end{bmatrix} \tag{6.97}$$

is orthogonal and represents a linear transformation.

However, for reflection in line $y = x \tan\theta + c$, $c \neq 0$, we use translation defined by

$$\begin{bmatrix} x' \\ y' \end{bmatrix} = \begin{bmatrix} x \\ y \end{bmatrix} + \begin{bmatrix} 0 \\ -c \end{bmatrix},$$

so in (x', y') coordinates the line $y' = x' \tan\theta$ passes through the origin. So, the reflection of (x', y') in the line $y' = x' \tan\theta$ is

$$\begin{bmatrix} u' \\ v' \end{bmatrix} = \begin{bmatrix} \cos 2\theta & \sin 2\theta \\ \sin 2\theta & -\cos 2\theta \end{bmatrix} \begin{bmatrix} x' \\ y' \end{bmatrix},$$

where (u', v') is in $x'y'$-coordinate frame. Reverting back to xy-coordinate frame, the point of reflection (u, v) is $u' = u$ and $v' = v - c$, i.e.,

$$\begin{bmatrix} u \\ v - c \end{bmatrix} = \begin{bmatrix} \cos 2\theta & \sin 2\theta \\ \sin 2\theta & -\cos 2\theta \end{bmatrix} \begin{bmatrix} x \\ y - c \end{bmatrix}. \tag{6.98}$$

For reflection in a vertical line, $x = a$, the change of coordinates to be used is defined by $x' = x - a$ and $y' = y$. Since, now $\theta = \pi/2$, proceeding as above the reflection transformation in line $x = a$ is

$$\begin{bmatrix} u' \\ v' \end{bmatrix} = \begin{bmatrix} -1 & 0 \\ 0 & 1 \end{bmatrix} \begin{bmatrix} x' \\ y' \end{bmatrix}.$$

In original coordinate system, the above becomes

$$\begin{bmatrix} u-a \\ v \end{bmatrix} = \begin{bmatrix} -1 & 0 \\ 0 & 1 \end{bmatrix} \begin{bmatrix} x-a \\ y \end{bmatrix}.$$

where (u, v) is in xy-coordinate frame.

Similarly, for rotation about (a, b), where both a and b are not zero, we use translation defined by

$$\begin{bmatrix} x' \\ y' \end{bmatrix} = \begin{bmatrix} x \\ y \end{bmatrix} - \begin{bmatrix} a \\ b \end{bmatrix},$$

so in coordinates (x', y') the center of rotation is the origin in which the transformation for rotation about the origin anticlockwise through θ is P as given above:

$$\begin{bmatrix} u' \\ v' \end{bmatrix} = \begin{bmatrix} \cos\theta & -\sin\theta \\ \sin\theta & \cos\theta \end{bmatrix} \begin{bmatrix} x' \\ y' \end{bmatrix},$$

where (u', v') is the image of (x', y') under *anti-clockwise* rotation about the origin in $x'y'$-coordinate frame. Reverting back to xy-coordinate frame

$$\begin{bmatrix} u-a \\ v-b \end{bmatrix} = \begin{bmatrix} \cos\theta & -\sin\theta \\ \sin\theta & \cos\theta \end{bmatrix} \begin{bmatrix} x-a \\ y-b \end{bmatrix}, \tag{6.99}$$

where (u, v) is the image of (x, y) under *anti-clockwise* rotation about (a, b).

Example 6.41 (a) Find the reflection of the point (1,1) in the line $y = \sqrt{3}x + 1$. (b) Find the rotation of the point (3,2) about (2,3) through an angle of $\pi/4$.

Solution (a) Here, $(x, y) = (1,1)$, $c = 1$ and the angle of the line is $\theta = \pi/3$ obtained from $\tan\theta = \sqrt{3}$. The desired point of reflection (u, v) from (6.98) is

$$\begin{bmatrix} u \\ v \end{bmatrix} = \begin{bmatrix} \cos 2\theta & \sin 2\theta \\ \sin 2\theta & -\cos 2\theta \end{bmatrix} \begin{bmatrix} x \\ y-c \end{bmatrix} + \begin{bmatrix} 0 \\ c \end{bmatrix}$$

$$= \frac{1}{2}\begin{bmatrix} -1 & \sqrt{3} \\ \sqrt{3} & 1 \end{bmatrix} \begin{bmatrix} 1 \\ 1-1 \end{bmatrix} + \begin{bmatrix} 0 \\ 1 \end{bmatrix} = \frac{1}{2}\begin{bmatrix} -1 \\ \sqrt{3}+2 \end{bmatrix}.$$

(b) The rotation of the point $(x,y) = (3,2)$ about $(a,b) = (2,3)$ through angle of $\theta = \pi/4$ is (u, v) and is obtained from (6.99) as

$$\begin{bmatrix} u \\ v \end{bmatrix} = \begin{bmatrix} \cos\theta & -\sin\theta \\ \sin\theta & \cos\theta \end{bmatrix} \begin{bmatrix} x-a \\ y-b \end{bmatrix} + \begin{bmatrix} a \\ b \end{bmatrix}$$

$$= \frac{1}{\sqrt{2}} \begin{bmatrix} 1 & -1 \\ 1 & 1 \end{bmatrix} \begin{bmatrix} 3-2 \\ 2-3 \end{bmatrix} + \begin{bmatrix} 2 \\ 3 \end{bmatrix} = \begin{bmatrix} 2+\sqrt{2} \\ 3 \end{bmatrix}.$$

□

Projection The subspaces U and W of $\mathbb{R}^n$ are said to be orthogonal if for all $u \in U$ and $w \in W$, $\langle u, w \rangle = 0$. We denote this by $U \perp W$. Alternatively, we also write $W = U^{\perp}$, or $U = W^{\perp}$. Let

$$u = \begin{bmatrix} 1 \\ 1 \\ 1 \end{bmatrix}, w_1 = \begin{bmatrix} 1 \\ -1 \\ 0 \end{bmatrix}, \text{ and } w_2 = \begin{bmatrix} 0 \\ 1 \\ -1 \end{bmatrix}.$$

The subspace U generated by $\{\mathbf{u}\}$ and the subspace W generated by $\{\mathbf{w}_1, \mathbf{w}_2\}$ are orthogonal.

Definition 6.21 Let V be the direct sum of two subspaces U and $W : V = U \oplus W$. A linear transformation $P : V \rightarrow V$ is called a projection on U along W if for $v = u + w$ with $u \in U$ and $w \in W$, we have $P(v) = u$. If $U \perp W$, we say P is an *orthogonal* projection otherwise P is called an *oblique* projection. □

According to the above definition, $P(\boldsymbol{\theta} + w) = P(w) = P(\boldsymbol{\theta}) = \boldsymbol{\theta}$ and $P(u + \boldsymbol{\theta}) = P(u) = u$. Consequently, $P(V) = P(U) = U$ and $P(W) = \mathbb{N}(P)$, the null space of P. The range space U and null space W of P are orthogonal for an orthogonal projection.

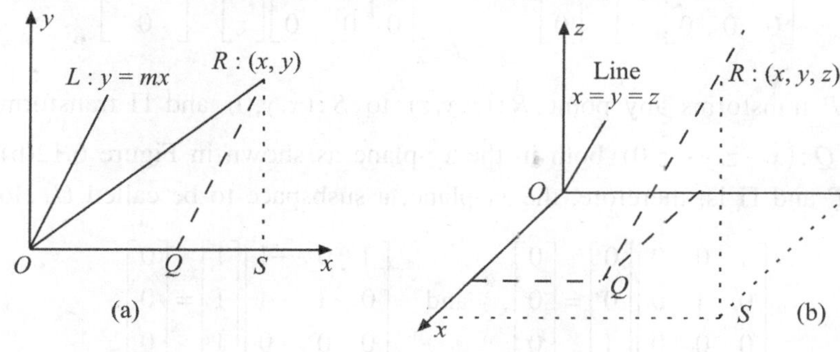

Figure 6.12 Projection transformation.

Example 6.42 (a) Let P and Π be defined by

$$P = \begin{bmatrix} 1 & 0 \\ 0 & 0 \end{bmatrix}, \text{ and } \Pi = \begin{bmatrix} 1 & -\frac{1}{m} \\ 0 & 0 \end{bmatrix}, m \neq 0.$$

Since,

$$\begin{bmatrix}1 & 0\\0 & 0\end{bmatrix}\begin{bmatrix}x\\y\end{bmatrix}=\begin{bmatrix}x\\0\end{bmatrix}, \quad \text{and} \quad \begin{bmatrix}1 & -\frac{1}{m}\\0 & 0\end{bmatrix}\begin{bmatrix}x\\y\end{bmatrix}=\begin{bmatrix}x-\frac{y}{m}\\0\end{bmatrix},$$

the projection P transforms any point $R:(x,y)$ to $S:(x,0)$ and Π transforms any point $R:(x,y)$ to $Q:(x-(y/m),0)$ as shown in Figure 6.12(a). The range space of both P and Π is, therefore, the x-axis, a subspace to be called U. However, as

$$\begin{bmatrix}1 & 0\\0 & 0\end{bmatrix}\begin{bmatrix}0\\1\end{bmatrix}=\begin{bmatrix}0\\0\end{bmatrix}, \quad \text{and} \quad \begin{bmatrix}1 & -\frac{1}{m}\\0 & 0\end{bmatrix}\begin{bmatrix}1\\m\end{bmatrix}=\begin{bmatrix}0\\0\end{bmatrix},$$

the null space of P is y-axis generated by $[0\ \ 1]'$ a subspace to be called W and that of Π is a subspace along the line $y = mx$ generated by $[1\ \ m]'$ to be called Ω.

The two-dimensional space V of the xy-plane can be represented as $V = U \oplus W$ or $V = U \oplus \Omega$. Therefore, $P(V) = \Pi(V) = U$, W is a null space of P and Ω, a null space of Π as explained above. Since, U and W are orthogonal, P is an orthogonal projection while Π is an oblique projection.

(b) Let P and Π be defined by

$$P=\begin{bmatrix}1 & 0 & 0\\0 & 1 & 0\\0 & 0 & 0\end{bmatrix} \quad \text{and} \quad \Pi=\begin{bmatrix}1 & 0 & -1\\0 & 1 & -1\\0 & 0 & 0\end{bmatrix}.$$

Since,

$$\begin{bmatrix}1 & 0 & 0\\0 & 1 & 0\\0 & 0 & 0\end{bmatrix}\begin{bmatrix}x\\y\\z\end{bmatrix}=\begin{bmatrix}x\\y\\0\end{bmatrix}, \quad \text{and} \quad \begin{bmatrix}1 & 0 & -1\\0 & 1 & -1\\0 & 0 & 0\end{bmatrix}\begin{bmatrix}x\\y\\z\end{bmatrix}=\begin{bmatrix}x-z\\y-z\\0\end{bmatrix},$$

the projection P transforms any point $R:(x,y,z)$ to $S:(x,y,0)$ and Π transforms any point $R:(x,\ y,\ z)$ to $Q:(x-z,y-z,0)$ both in the xy-plane as shown in Figure 6.12(b). The range space of both P and Π is, therefore, the xy-plane, a susbspace to be called U. However, as

$$\begin{bmatrix}1 & 0 & 0\\0 & 1 & 0\\0 & 0 & 0\end{bmatrix}\begin{bmatrix}0\\0\\1\end{bmatrix}=\begin{bmatrix}0\\0\\0\end{bmatrix}, \quad \text{and} \quad \begin{bmatrix}1 & 0 & -1\\0 & 1 & -1\\0 & 0 & 0\end{bmatrix}\begin{bmatrix}1\\1\\1\end{bmatrix}=\begin{bmatrix}0\\0\\0\end{bmatrix},$$

the null space of P is the z-axis generated by $[0\,0\,1]'$, a subspace to be called W and that of Π is the subspace along the line $x = y = z$ generated by $[1\,1\,1]'$ to be called Ω.

The three-dimensional space V of $\mathbb{R}^3$ can be represented as $V = U \oplus W$ or $V = U \oplus \Omega$. Therefore, $P(V) = \Pi(V) = U$, W is a null space of P and Ω, a null space of Π as explained above. Since, U and W are orthogonal, P is an orthogonal projection while Π is an oblique projection.

Theorem 6.21 Let P be a linear transformation on a vector space V. Then, P is an orthogonal projection if and only if it is self-adjoint, i.e., $P = P^*$.

Proof Suppose P is an orthogonal projection on a subspace U of V. Any $\mathbf{v}_1$ and $\mathbf{v}_2$ of V can be written as the sum of vectors in U and $W = U^\perp$:

$$\mathbf{v}_i = \mathbf{u}_i + \mathbf{w}_i.$$

Then,

$$\langle P\mathbf{v}_1, \mathbf{v}_2\rangle = \langle \mathbf{u}_1, \mathbf{u}_2 + \mathbf{w}_2\rangle = \langle \mathbf{u}_1, \mathbf{u}_2\rangle + \langle \mathbf{u}_1, \mathbf{w}_2\rangle = \langle \mathbf{u}_1, \mathbf{u}_2\rangle,$$

$$\langle \mathbf{v}_1, P\mathbf{v}_2\rangle = \langle \mathbf{u}_1 + \mathbf{w}_1, \mathbf{u}_2\rangle = \langle \mathbf{u}_1, \mathbf{u}_2\rangle + \langle \mathbf{w}_1, \mathbf{u}_2\rangle = \langle \mathbf{u}_1, \mathbf{u}_2\rangle,$$

so, $\langle P\mathbf{v}_1, \mathbf{v}_2\rangle = \langle \mathbf{v}_1, P\mathbf{v}_2\rangle$ for all $\mathbf{v}_1$, $\mathbf{v}_2 \in V$. Hence, $P = P^*$.

Conversely, if $P = P^*$, by Theorem 6.19, ker P = (range $P^*)^\perp$ = (range $P)^\perp$, this means that P is an orthogonal projection as desired. □

From the above discussion, properties of a projection P are summarized below:

Properties of Projection P

Property I Projection P is idempotent $P^2 = P$.

Property II Range $(P) = U$ and null $(P) = W$.

Property III Eigenvalues of P are 1 or 0.

Property IV If P is orthogonal, then $P = P^*$.

For Property III, see Problem 7.70.

Theorem 6.22 Let R and N be subspaces of $V = \mathbb{R}^n$ or $\mathbb{C}^n$ such that $V = R \oplus N$, where any element of N need not be orthogonal to R. It is given that $\{\mathbf{u}_1, \ldots, \mathbf{u}_r\}$ is a basis of R and $\{\mathbf{v}_1, \ldots, \mathbf{v}_r\}$ is that of $N^\perp$. Define $A := [\mathbf{u}_1, \ldots, \mathbf{u}_r]$ and $B := [\mathbf{v}_1, \ldots, \mathbf{v}_r]$, where the vectors are expressed as the coordinates of the standard basis. Then the matrix in the standard basis of the projection on R parallel to N is

$$P = A(B^*A)^{-1}B^*. \tag{6.100}$$

Proof Let $\mathbf{x} \in V$. If $P\mathbf{x} \in R$, then for some scalars $\{y_1, \ldots, y_r\}$

$$P\mathbf{x} = \sum_{j=1}^{r} y_j \mathbf{u}_j = [\mathbf{u}_1 \cdots \mathbf{u}_r] \begin{bmatrix} y_1 \\ \vdots \\ y_r \end{bmatrix} = A\mathbf{y},$$

where $\mathbf{y} \in R$ represents the coordinates with respect to the basis $\{\mathbf{u}_1, \ldots, \mathbf{u}_r\}$ of the projection $\mathbf{x}$ on R. Since $\mathbf{x} - P\mathbf{x} \in N = (N^\perp)^\perp$, this means that $\mathbf{x} - P\mathbf{x} \perp N^\perp$, therefore,

$$\langle \mathbf{v}_k, \mathbf{x} - P\mathbf{x}\rangle = 0, \forall k = 1, \ldots, r.$$

Hence,

$$0 = \langle \mathbf{v}_k, \mathbf{x} - P\mathbf{x}\rangle = \langle \mathbf{v}_k, \mathbf{x}\rangle - \langle \mathbf{v}_k, P\mathbf{x}\rangle = \langle \mathbf{v}_k, \mathbf{x}\rangle - \langle \mathbf{v}_k, \sum_{j=1}^{r} y_j \mathbf{u}_j\rangle,$$

so,

$$0 = \langle \mathbf{v}_k, \mathbf{x}\rangle - \sum_{j=1}^{r} \overline{y}_j \langle \mathbf{v}_k, \mathbf{u}_j\rangle, \forall k = 1, \ldots, r.$$

Consequently,

$$\mathbf{x}^*[\mathbf{v}_1\cdots\mathbf{v}_r]=[\overline{y}_1\cdots\overline{y}_r]\begin{bmatrix}\mathbf{u}_1^*\mathbf{v}_1 & \cdots & \mathbf{u}_1^*\mathbf{v}_r\\ \vdots & & \vdots\\ \mathbf{u}_r^*\mathbf{v}_1 & \cdots & \mathbf{u}_r^*\mathbf{v}_r\end{bmatrix}=\mathbf{y}^*\begin{bmatrix}\mathbf{u}_1^*\\ \vdots\\ \mathbf{u}_r^*\end{bmatrix}[\mathbf{v}_1\cdots\mathbf{v}_r]=\mathbf{y}^*A^*B.$$

Hence,

$$B^*A\mathbf{y}=B^*\mathbf{x},\quad \text{i.e.,}\quad \mathbf{y}=(B^*A)^{-1}B^*\mathbf{x}.$$

Therefore,

$$P\mathbf{x}=A\mathbf{y}=A(B^*A)^{-1}B^*\mathbf{x},$$

and $P = A(B^* A)^{-1} B^*$. It is obvious that, $P^2=(A(B^*A)^{-1}B^*)(A(B^*A)^{-1}B^*)=A(B^*A)^{-1}\ B^*=P.$

So P is a projection. Now, as $A\mathbf{x}\in R$, so $PA\mathbf{x}=AA(B^*A)^{-1}B^*\mathbf{x}=A\mathbf{x}$. By construction for $\forall\,\mathbf{x}\in V,\ \mathbf{x}-P\mathbf{x}\in N$ and $P(A\mathbf{x}-PA\mathbf{x})=P(A\mathbf{x}-A\mathbf{x})=\theta$, the null space. This shows that P is a projection with R as the range space and N as the null space. □

Corollary 6.9 Let R and N be *two orthogonal subspaces* of $V=\mathbb{R}^n$ or $\mathbb{C}^n$ such that $V=R\oplus N$. Define $A:=[\mathbf{u}_1,\ldots,\mathbf{u}_r]$, where $\{\mathbf{u}_1,\ldots,\mathbf{u}_r\}$ is a basis of R. Then the matrix of the orthogonal transformation on R is $P=A(A^*A)^{-1}A^*$.

Proof In the proof of Theorem 6.22, we note that as $R=N^{\perp}$, the basis of $N^{\perp}$ is also $\{\mathbf{u}_1,\ldots,\mathbf{u}_r\}$ and hence $B=A$. Setting $B=A$ in (6.100) we get the desired result. In view of Example 5.34(c), A^*A is invertible. □

Example 6.43 (a) Let the range space R of a projection Π be generated by $\mathbf{u}=[1\ 0]'$ while the null space N by $\mathbf{n}=[1\ m]'$, $m\neq 0$. Find Π. (b) If the range space R of a projection Π is generated by $\mathbf{u}_1=[1\ 0\ 0]'$ and $\mathbf{u}_2=[0\ 1\ 0]'$ and the null space N by $\mathbf{n}=[1\ 1\ 1]'$, find Π.

Solution (a) By solving $[x\ y]\ \mathbf{n}=0$, we get $\mathbf{v}=[-m\ 1]'$. Therefore, $N^{\perp}=\text{span}\ \{\mathbf{v}\}$. Defining $A=[\mathbf{u}]$ and $B=[\mathbf{v}]$, we get,

$$B^*A=\begin{bmatrix}-m & 1\end{bmatrix}\begin{bmatrix}1\\0\end{bmatrix}=-m,(B^*A)^{-1}=-\tfrac{1}{m},$$

$$\Pi=A(B^*A)^{-1}B^*=-\begin{bmatrix}1\\0\end{bmatrix}\frac{1}{m}[-m\quad 1]=\begin{bmatrix}1 & -\frac{1}{m}\\ 0 & 0\end{bmatrix}$$

which gives us the oblique projection with range space $R=\text{span}\ \{\mathbf{u}\}$ and null space $N=\text{span}\ \{\mathbf{n}\}$ as explained in Example 6.41(a). Similarly, the orthogonal projection with $R=\text{span}\ \{\mathbf{u}\}$ is

$$P=A(A^*A)^{-1}A^*=\begin{bmatrix}1 & 0\\ 0 & 0\end{bmatrix}$$

as explained in Example 6.42(a).

(b) Two linearly independent solutions, by solving

$$[x \quad y \quad z]\begin{bmatrix}1\\1\\1\end{bmatrix} = 0,$$

are $\mathbf{v}_1 = [1\ 0 - 1]'$ and $\mathbf{v}_2 = [1\ -1\ 0]'$. Therefore, $N^{\perp} = \text{span}\ \{\mathbf{v}_1, \mathbf{v}_2\}$. Defining A = $[\mathbf{u}_1\ \mathbf{u}_2]$ and $B = [\mathbf{v}_1\ \mathbf{v}_2]$, we get,

$$B^* A = \begin{bmatrix}1 & 0 & -1\\1 & -1 & 0\end{bmatrix}\begin{bmatrix}1 & 0\\0 & 1\\0 & 0\end{bmatrix} = \begin{bmatrix}1 & 0\\1 & -1\end{bmatrix} = (B^* A)^{-1},$$

$$\Pi = A(B^* A)^{-1} B^* = \begin{bmatrix}1 & 0\\0 & 1\\0 & 0\end{bmatrix}\begin{bmatrix}1 & 0\\1 & -1\end{bmatrix}\begin{bmatrix}1 & 0 & -1\\1 & -1 & 0\end{bmatrix} = \begin{bmatrix}1 & 0 & -1\\0 & 1 & -1\\0 & 0 & 0\end{bmatrix},$$

which gives us the oblique projection with range space $R = \text{span}\ \{\mathbf{u}_1, \mathbf{u}_2\}$ and null space $N = \text{span}\ \{\mathbf{n}\}$ as explained in Example 6.42(b). Similarly, the orthogonal projection with $R = \text{span}\{\mathbf{u}_1, \mathbf{u}_2\}$, as explained in Example 6.42(b), is

$$P = A(A^* A)^{-1} A^* = \begin{bmatrix}1 & 0 & 0\\0 & 1 & 0\\0 & 0 & 0\end{bmatrix}.$$

□

PROBLEMS

6.1 Show that the following transformation $L : \mathbb{C}^2 \to \mathbb{C}^2$ preserves scalar multiplication but not addition: $L(x, y) = (x^2/y, x)$.

6.2 Show that the following transformations are linear:

(a) $L : V_3 \to V_3 : (x, y, z) \mapsto (x, y, z)$.

(b) $L : V_3 \to V_3 : (x, y, z) \mapsto (-x, -y, -z)$.

(c) $L : \mathbb{C} \to \mathbb{C} : a + ib \mapsto a - ib,\ i = \sqrt{-1}$.

(d) $L : V \to V : u(t) \mapsto a\ddot{u}(t) + b\dot{u}(t) + cu(t)$ when V is the vector space of functions having derivatives of all orders and a, b, c are arbitrary scalars.

(e) $L : V \to \mathbb{R} : u(t) \mapsto \int_0^1 u(t)\,dt$, where V is the vector space of continuous real valued functions defined on $\mathbb{R}$.

(f) $L : \mathbb{P}(t) \to \mathbb{R} : p(t) \mapsto \int_0^1 p(x)\,dx$.

(g) $L : \mathbb{P}(x) \to \mathbb{R}(x) : p(x) \mapsto p(x) \sin t$, t is real.

(h) $L : \mathbb{P}_n(t) \to \mathbb{P}_{n+1}(t) : p(t) \mapsto tp(t)$.

(i) $L : \mathbb{C}^n \to \mathbb{C}^n : (x_1\ x_2\ \cdots\ x_n) \mapsto (x_{j_1}\ x_{j_2}\ \cdots\ x_{j_n})$, where $j_1 j_2 \cdots j_n$ is a permutation of the integers from 1 to n (see Section 2.1).

6.3 Show that the following transformations are nonlinear:

(a) $L : \mathbb{R} \to \mathbb{R} : x \mapsto t \sin x$, t is real.

(b) $L : V \to V : x(t) \mapsto a\dot{x}^2$, where a is a constant and V is defined in Problem 6.2(d).

(c) $L : V \to V : u(t) \mapsto a\ddot{u}(t) + b\dot{u}^2(t) + cu(t)$, where a, b, c are real constants and V is defined in (b).

(d) $L : V \to V$: $u(t) \mapsto a\ddot{u}(t) + b\dot{u}(t) + cu^2(t)$, where a, b, c are real constants and V is defined in (b).

(e) $L : V \to V : u(x, y, z) \mapsto u_x^2 + u_y^2 + u_z^2$, where V is the vector space of differentiable functions of all orders with respect to x, y, z and $u_x = \partial u / \partial x$, etc.

(f) $L : \mathbb{R} \to \mathbb{R} : x \mapsto \int_0^x u(t)\,dt$, where $u(t)$ is a continuous function.

6.4 Test if the following transformations on $\mathbb{R}^3$ are linear or not. Give, wherever possible, a geometrical interpretation of the transformation:

(a) $L(x, y, z) = (x, y, 0)$,

(b) $L(x, y, z) = (x, 1, 0)$,

(c) $L(x, y, z) = (x + 1, y, z)$,

(d) $L(x, y, z) = (-x, y, z)$,

(e) $L(x, y, z) = (x - y, y - z, z - x)$.

6.5 Show that the following transformations are linear:

(a) $L : \mathbb{P}_3(x) \to \mathbb{P}_3(x) : p(x) \mapsto xp'(x) + p(x)$.

(b) $L : \mathbb{P}_2(x) \to \mathbb{P}_3(x) : p(x) \mapsto p(x) + 6 \int_0^x p(t)\,dt$.

(c) $L : \mathbb{P}_2(x) \to V_{2\times 2} : a + bx + cx^2 \mapsto \begin{bmatrix} a & a+b \\ b+c & c+a \end{bmatrix}$.

(d) $L : V_{2\times 2} \to \mathbb{P}_3(x) : \begin{bmatrix} a & b \\ c & d \end{bmatrix} \mapsto (a-b)+(b-c)x+(c-d)x^2+(d-a)x^3$.

6.6 For the given transformations $L : \mathbb{R}^3 \to \mathbb{R}^3$ and subspace S_r, $r < 3$ of $\mathbb{R}^3$, find (i) the image of S_r, and (ii) the range of L. Give a geometrical interpretation in each case.

(a) $L(x, y, z) = (x, 1, 1)$, S_r is the y-axis.

(b) $L(x, y, z) = (x, y, 1)$, S_r, is the y-axis.

(c) $L(x, y, z) = (x - y, y - z, z - x)$, S_r is the xy-plane.

(d) $L(x, y, z) = (x, y, 0)$, S_r, is yz-plane.

What is the co-domain in each case?

6.7 Find a linear transformation L in each case below, which transforms

(a) the row vectors (1, 1, 1), (1, 1, 0), (1, 0, 0) in $\mathbb{R}^3$ to (2, 1), (2, 1) (2 1) in $\mathbb{R}^2$;

(b) the row vectors (2 3), (3 2) in $\mathbb{R}^2$ to (1, 2) (2 3) in $\mathbb{R}^2$;

(c) the vectors $[1\ 2\ 3]^T$, $[2\ 1\ 3]^T$, $[3\ 1\ 2]^T$ in $\mathbb{R}^3$ to $[1\ 0\ 2]^T$, $[0\ 1\ 2]^T$, $[1\ 2\ 0]^T$ in $\mathbb{R}^3$;

(d) the elementary vectors $\mathbf{e}_1$, $\mathbf{e}_2$, $\mathbf{e}_3$, $\mathbf{e}_4$ in $\mathbb{R}^4$ to $[1\ 2\ 3]^T$, $[1\ 1\ 2]^T$, $[1\ 2\ 2]^T$, $[2\ 2\ 2]$ in $\mathbb{R}^3$;

(e) the vectors $[2\ {-1}\ 3]^T$, $[4\ {-3}\ 0]^T$, $[3\ {-2}\ 1]^T$ in $\mathbb{R}^3$ to $[3\ 1\ 2]^T$, $[2\ 1\ 4]^T$ $[2\ 1\ 3]^T$ in $\mathbb{R}^3$;

(f) the row vectors (3 –1 –2), (1 1 0), (–2 0 2) in $\mathbb{R}^3$ to twice the elementary vectors $2\mathbf{e}_1$, $2\mathbf{e}_2$, $2\mathbf{e}_3$ in $\mathbb{R}^3$.

(g) the vector $[0\ 3]^T$, $[-1\ 5]^T$ to $[3\ 1]^T$, $[1\ 3]^T$ in $\mathbb{R}^2$;

(h) the row vectors (–3 –5 9), (–4 –7 12), (1 1 –2) in $\mathbb{R}^3$ to (2 3), (1 2), (–1 –1) in $\mathbb{R}^2$;

(i) the row vectors (1 1 1), (–1 1 –1), (1 1 2) in $\mathbb{R}^3$ to (1 1), (1 1), (1 0) in $\mathbb{R}^2$.

Which of the above transformations are nonsingular?

6.8 Prove that the set of linear transformations with addition and scalar multiplication defined by (6.12) forms a vector space.

6.9 A linear operator $L : \mathbb{R}^3 \to \mathbb{R}^4$ transforms $\mathbf{x}_1 = [2\ {-1}\ 3]^T$, $\mathbf{x}_2 = [4\ {-3}\ 0]^T$, $\mathbf{x}_3 = [3\ {-2}\ 1]^T$ to $\mathbf{z}_1 = [1\ 0\ 1\ 0]^T$, $\mathbf{z}_2 = [2\ 1\ 0\ {-1}]^T$ $\mathbf{z}_3 = [-1\ 2\ 1\ {-2}]^T$ respectively. Find its value at

(a) $\mathbf{e}_1$, $\mathbf{e}_2$, $\mathbf{e}_3$ in $\mathbb{R}^3$, (b) $[2\ 3\ 1]^T$, (c) $[0\ 0\ 0]^T$.

6.10 Determine ker L and nullity L when L is given below and give a geometrical interpretation of ker L.

(a) $L(x\ y\ z) = (x - y\quad z - 1)$.

(b) $L(x\ y\ z) = (x - y,\ y - z,\ z - x)$.

(c) $L(x\ y\ z) = (x^2 - 2xz + z^2,\ x - y,\ y - 1)$.

(d) $L(x\ y\ z) = (x + y - z)$.

(e) $L(x\ y\ z) = (x + z \quad y + z)$.

(f) $L(x\ y) = A\mathbf{x}$, where A is a nonsingular $n \times n$ matrix.

6.11 Determine the kernel of the following maps:

(a) $L(x,\ y) = (\cos x,\ \sin y)$,

(b) $Lx = (e^x - 1\ x)$,

(c) $Lx = (\log x,\ x - 1)$,

(d) $L(x\ y) = (\sin x \quad \cos y\ x)$,

(e) $L(x\ y) = (x^2 + y^2 - 1 \quad x + y - 1)$.

6.12 Let V be the vector space of all real valued functions in the variable t having derivatives of all orders. Determine ker L of the linear operator $L : V \rightarrow V$ when

(a) $L = D = \dfrac{d}{dt}$; (b) $L = D^2 = \dfrac{d^2}{dt^2}$;

(c) $L = D^n = \dfrac{d^n}{dt^n}$; (d) $L = D^2 + (\alpha + \beta)D + \alpha\beta\, I_V$,

where α and β are scalars.

6.13 Let V be the vector space of all $n \times n$ matrices and W the vector space of all $n \times n$ upper triangular matrices. Show that the map $L : V \rightarrow W$ is linear. Determine ker L and nullity L.

6.14 Show that the map $L : V_{n\times n} \rightarrow V_{n\times n} : A \mapsto (A + A^T)/2$ is linear, ker L is the vector space of all $n \times n$ skew-symmetric matrices and nullity $L = n(n - 1)/2$.

6.15 Show that the map $L : V_{n\times n} \rightarrow V_{n\times n} : A \mapsto (A - A^T)/2$ is linear; ker L is the vector space of all $n \times n$ symmetric matrices, and nullity $L = n(n + 1)/2$.

6.16 Test whether or not the following transformations are onto:

(a) $L(x\ y\ z) = (x - y \quad y - z \quad z - x)$;

(b) $L(x\ y) = (x \cos\theta - y \sin\theta,\ x \sin\theta + y \cos\theta)$;

(c) $L(x\ y\ z) = (x - y - z, \quad y - x - z, \quad z - x - y)$;

(d) $L(x\ y\ z) = (x - y,\ y - z)$;

(e) $L(x\ y\ z) = (x - y,\ y)$;

(f) $L(x\ y) = (x\ x - y \quad x + y)$.

Which of the transformations are one-to-one?

6.17 (a) If V and W are vector spaces of the same dimension, then the linear map $L : V \to W$ is one-to-one iff it is onto

(b) Prove that a linear transformation L on V_n is nonsingular iff L is one-to-one or iff L is onto.

6.18 Find a linear transformation $L : \mathbb{R}^4 \to \mathbb{R}^4$ for which the null space is spanned by $\mathbf{x}_1 = [2\ {-2}\ 1\ {-2}]^T$, $\mathbf{x}_2 = [-3\ 4\ 3\ 1]^T$, and the range space by $[3\ 2\ 1\ 0]^T$, $[0\ 1\ 2\ 3]^T$.

6.19 Find a linear transformation $L : \mathbb{R}^4 \to \mathbb{R}^3$ for which the null space is spanned by $\mathbf{x}_1 = [2\ 1\ {-1}\ {-2}]^T$ and image space by $[\cos\theta\ \sin\theta\ 0]^T$, $[-\sin\theta\ \cos\theta\ 0]^T$, and $[0\ 0\ 1]^T$, where θ is an arbitrary real number.

6.20 Find a linear transformation $L : \mathbb{R}^4 \to \mathbb{R}^4$ whose null space is the same as the range space.

6.21 Justify that there is no nontrivial linear transformation $L : \mathbb{R}^n \to \mathbb{R}^n$, where n is an odd integer for which the null space is the same as the range space.

6.22 Find a linear transformation $L : \mathbb{R}^4 \to \mathbb{R}^4$ for which the range space and null space are complementary but have the same dimension.

6.23 Find bases of (i) null space, and (ii) range space of the linear transformations L given in Problem 6.7.

6.24 Two transformations K and L are given below. In each case find the composite transformation $L{\cdot}K$ and $K{\cdot}L$ wherever it is defined.

(a) $L(x\ y\ z) = (y\ \ x + y - z)$, $K(x, y) = (x - y\ \ y\ x)$;

(b) $L(x\ y) = (y\ \ x + y\ \ x)$, $K(x\ y\ z) = (x - y\ \ y - z)$;

(c) $L(x\ y) = (x\cos\theta - y\sin\theta\ \ x\sin\theta + y\cos\theta)$;

$K(x\ y) = (x\cos\theta + y\sin\theta, -x\sin\theta + y\cos\theta)$;

(d) $L\mathbf{x} = \mathbf{x}^T A^{-1}\mathbf{a}$, $K\mathbf{x} = A^T\mathbf{x}$, where A is a nonsingular matrix of order n and $\mathbf{a}$ is a constant n-vector.

6.25 Find the composite transformation $(L{\cdot}K{\cdot}M)\ (x\ y\ z)$ when

(a) $L(x\ y) = (x - y\ \ 0\ \ x + y)$, $K(x\ y\ z) = (x + y\ \ x + y + z)$,

$M(x\ y\ z) = (x\ y - x\ \ x - y + z)$;

(b) $L(x\ y\ z) = (x - y\ \ y - z\ \ z - x)$, $K(x\ y\ z) = (x + y\ \ y + z\ \ z + x)$,

$M(x\ y\ z) = (x - y - z\ \ y - x - z\ \ z - x - y)$.

6.26 A linear transformation L is given below. Find another linear transformation K such that $(K{\cdot}L)\mathbf{x} = (L{\cdot}K)\mathbf{x} = \mathbf{x}$.

(a) $L[x\ y]^T = [-y\ \ 3x + 5y]^T$, $\mathbf{x} = [x\ y]^T$;

(b) $L[x\ y\ z]^T = [-x - 2y + z, x + y, x]^T$, $\mathbf{x} = [x\ y\ z]^T$;

(c) $L[x\ y\ z]^T = [3x + y - 2z, -x + y, -2x + 2z]^T$, $\mathbf{x} = [x\ y\ z]^T$.

6.27 (a) Find a linear transformation L which transforms the vectors $[1\ -1\ 1]^T$, $[1\ 1\ -1]^T$, $[-1\ 1\ 1]^T$ in $\mathbb{R}^3$ to $[1\ -1]^T$, $[-1\ 1]^T$, $[1\ 1]^T$ in $\mathbb{R}^2$. Show that $[u\ v]^T$ in $\mathbb{R}^2$ is the image of infinite number of vectors $[x\ y\ z]^T$ in $\mathbb{R}^3$. Interpret the operator L geometrically.

(b) Prove that the transformation L in (a) is right invertible (i.e., onto but not one-to-one). Find another transformation $K : \mathbb{R}^2 \to \mathbb{R}^3$ such that $(L{\cdot}K)\mathbf{x} = \mathbf{x}$, where $\mathbf{x} = [u\ v]^T$, and show that infinite number of such transformations K exist.

6.28 Show that the following transformations are onto but not one-to-one. Find the right inverse in each case.

(a) $L[x\ y\ z\ w]^T = [x + y + z + 2w,\ 2x + y + 2z + 2w,\ 3x + 2y + 2z + 2w]^T$.

(b) $L[x\ y\ z]^T = [5y + 3z,\ x + 6y + 4z]^T$.

(c) $L(x\ y\ z) = (y \quad x + y - z)$.

6.29 Show that the following transformations L are one-to-one. Find the left inverse K of L in each case.

(a) $L(x\ y) = (x \quad x - y \quad x + y)$.

(b) $L(x\ y\ z) = (x - y - z \quad y - x - z \quad z - x - y)$.

(c) $L(x\ y\ z) = (x + y + z \quad x \quad y - z)$.

6.30 Determine a linear transformation L which maps every vector in $\mathbb{R}^3$ over $\mathbb{R}$ onto a vector of a subspace S generated by

(a) $[1\ 1\ 0]^T$, $[1\ 0\ 1]^T$; (b) $[1\ 0\ 1]^T$; (c) $[1\ 1\ 1]^T$, $[1\ 1\ 0]^T$.

6.31 Let D and T be linear transformations on $\mathbb{P}(t)$ defined by $Dp(t) = \dot{p}(t)$ and $Tp(t) = tp(t)$. Prove that $DT - TD$ is an identity transformation.

6.32 Let L and K be operators on $\mathbb{P}(t)$ defined by $L(a_0 + a_1t + \cdots + a_nt) = a_1 + a_2t + \cdots + a_nt^{n-1}$ and $K(a_0 + a_1t + \cdots + a_nt^n) = a_0t + a_1t^2 + \cdots + a_nt^{n+1}$. Show that L and K are linear and that $L{\cdot}K = I_{\mathbb{P}(t)}$ and $K{\cdot}L \neq I_{\mathbb{P}(t)}$. Does the operator L have an inverse? Show that K has infinitely many left inverses.

6.33 Does a solution always exist in Example 6.14(b).

6.34 A linear transformation on $\mathbb{R}^3$ is given by

$$L\mathbf{x} = [-x - y + z \quad x - 4y + z \quad 2x - 5y]^T,$$

where $\mathbf{x} = [x\ y\ z]^T$ is in $\mathbb{R}^3$. Determine the matrix representation of L with respect to the following basis matrices:

$$\text{(a) } [\mathbf{e}_1\mathbf{e}_2\mathbf{e}_3], \qquad \text{(b) } \begin{bmatrix} 1 & 1 & -1 \\ 0 & 1 & -1 \\ 0 & 0 & 1 \end{bmatrix}, \qquad \text{(c) } \begin{bmatrix} 1 & 2 & 1 \\ 0 & 1 & 0 \\ 2 & 0 & 1 \end{bmatrix}.$$

6.35 Consider the vector space $\mathbb{C}^2$ over the real field $\mathbb{R}$. Show that $[1\ 0]^T$, $[i\ 0]^T$, $[0\ 1]^T$, $[0\ i]^T$ form a basis of $\mathbb{C}^2$. Define a linear transformation

$$L:\mathbb{C}^2 \to \mathbb{C}^2 : \begin{bmatrix} a+ib \\ c+id \end{bmatrix} \mapsto \begin{bmatrix} a-ib \\ c-id \end{bmatrix}$$

over $\mathbb{R}$ and find its matrix representation with respect to the above basis.

6.36 Let $L : \mathbb{C} \to \mathbb{C} : a + ib \mapsto a - ib$ be a linear transformation where $\mathbb{C}$ is a vector space over $\mathbb{R}$. Find the matrix representation of L with respect to the basis $\{1, i\}$.

6.37 Find the matrix representation of the linear transformation L on $\mathbb{C}^n$ defined by $L\mathbf{x} = [x_{j1} \ x_{j2} \ \cdots \ x_{jn}]^T$ with respect to the elementary vectors $\{\mathbf{e}_1, \mathbf{e}_2, ..., \mathbf{e}_n\}$, where $j_1 j_2 ..., j_n$ is the permutation of the integers from 1 to n and $\mathbf{x} = [x_1 \ x_2 \ \cdots \ x_n]^T$.

6.38 Let $\mathbb{P}(t)$ be the set of polynomials over $\mathbb{C}$ and let an integral operator S on $\mathbb{P}(t)$ be defined by (see Example 5.14) $S : \mathbb{P}(t) \to \mathbb{P}(t) : Sp(t) = S(a_0 + a_1 t + \cdots + a_n t^n) \mapsto a_0 t + a_1 t^2/2 + \cdots + a_n t^{n+1}/n+1$. Show that

$$Sp(t) = \int_0^t p(\tau)\, d\tau$$

and that S is linear. Find the matrix representation of S with respect to the basis $\{1, t, t^2, ...\}$.

6.39 Show that the transformation defined by

$$Q : \mathbb{P}(t) \to \mathbb{P}(t) : p(t) \mapsto q(t)p(t),$$

where $q(t)$ is a given polynomial not necessarily in $\mathbb{P}(t)$ with complex coefficients, is linear.

(a) Find the matrix representation of $L : \mathbb{P}_n(t) \to \mathbb{P}_{n+1}(t) : p(t) \mapsto tp(t)$ with respect to the standard basis.

(b) Is L injective?

(c) Is L surjective?

6.40 Find the matrix representation of the derivative operator $D : \mathbb{P}_n(t) \to \mathbb{P}_{n-1}(t)$ (see Example 6.2) with respect to the bases $\{1, t, ..., t^n\}$ and $\{1, 1 + t, t + t^2, ..., t^{n-2} + t^{n-1}\}$.

6.41 Show that the transformation $L : \mathbb{P}_n(t) \to \mathbb{P}_n(t) : p(t) = a_0 + a_1 t + \cdots + a_n t^n \mapsto p(t + 1) = a_0 + a_1(t + 1) + \cdots + a_n(t + 1)^n$ is linear and can be represented as

$$L = 1 + \frac{D}{1!} + \frac{D^2}{2!} + \cdots + \frac{D^n}{n!}, \text{ where } \quad D^i = \frac{d^i}{dt^i}.$$

Find the matrix representation with respect to the basis $\{1, t, ..., t^n\}$.

6.42 Let $V := V_{2\times 2}$ be the vector space of all 2×2 matrices over a field $\mathbb{F}$ and let E_{11}, E_{12}, E_{21} and E_{22} be a basis of $V_{2\times 2}$ (see Examples 5.4 and 5.13). Find the matrix representation of the linear transformation $L : V \to V : X \mapsto BX$, where

$$B = \begin{bmatrix} 1 & -1 \\ -1 & 1 \end{bmatrix}.$$

Show that L is neither one-to-one nor onto.

6.43 Find the matrix representation of the linear transformation L given in Problem 6.42 when

$$B = \begin{bmatrix} 1 & -1 \\ 1 & 1 \end{bmatrix}.$$

Show that L is now one-to-one and onto.

6.44 Let A, the matrix representation of a linear transformation $L : \mathbb{R}^3 \to \mathbb{R}^3$ with respect to the basis $[5\ 1\ 3]^T$, $[3\ 2\ 2]^T$, and $[1\ 2\ 1]^T$ be given by

$$A = \begin{bmatrix} 3 & 2 & -2 \\ -1 & 0 & 1 \\ 2 & 1 & -1 \end{bmatrix}.$$

Find the linear transformation L.

6.45 Let A, the matrix representation of the linear transformation $L : \mathbb{P}_3(t) \to \mathbb{P}_3(t)$ with respect to the bases $\{1 - t,\ t - t^2,\ t^2 - t^3,\ t^3\}$ and $\{1,\ 1 + t,\ t + t^2,\ t^2 + t^3\}$ be given by

$$A = \begin{bmatrix} 1 & -1 & 1 & 1 \\ 1 & 1 & -1 & 1 \\ 1 & 1 & 1 & -1 \\ -1 & 1 & 1 & 1 \end{bmatrix}.$$

Determine L.

6.46 A transformation L is defined by (see Examples 5.2 and 5.13)

$$L : V_{2\times 2} \to \mathbb{P}_2(x) : \begin{bmatrix} a & b \\ c & d \end{bmatrix} \mapsto (a+d) + (a+b)x + (b+c)x^2 .$$

Show that L is linear and find the matrix representation of L with respect to bases $\{E_{11},\ E_{12},\ E_{21},\ E_{22}\}$ and $\{1,\ x,\ x^2\}$.

6.47 Show that the map $L : \mathbb{C}^n \to \mathbb{C}$ defined by $L\mathbf{x} = x_1 + x_2 + \cdots + x_n$, when $\mathbf{x} = [x_1\ x_2\ \cdots\ x_n]^T$, is linear and its kernel has the dimension $n - 1$. What is the matrix representation of L with respect to bases $\{\mathbf{e}_1, \mathbf{e}_2, ..., \mathbf{e}_n\}$ and $\{1\}$?

6.48 Show that the map $L : \mathbb{C}^{n\times n} \to \mathbb{C}$ defined by $LA = a_{11} + a_{22} + \cdots + a_{nn}$, where $A = [a_{ij}]$ is an $n \times n$ matrix over $\mathbb{C}$, is linear and its kernel has the dimension $n^2 - 1$. What is the matrix representation of L with respect to the bases E_{ij}'s and 1?

6.49 Find the matrix of the sum of two linear transformations L and K given by

(a) $L[x\ y]^T = [x + y\quad x\quad x - y]^T$, $K[x\ y]^T = [x + 2y\quad x - y\quad y]^T$ with respect to the bases represented respectively by the columns of

$$\begin{bmatrix} 1 & 1 \\ 0 & 1 \end{bmatrix} \text{ and } \begin{bmatrix} 1 & 1 & 1 \\ 0 & 1 & 1 \\ 0 & 0 & 1 \end{bmatrix};$$

(b) $L[x\ y\ z]^T = [x + y \quad y + z \quad z + x]^T$, $K[x\ y\ z]^T = [x - y \quad y - z \quad z - x]^T$ with respect to the basis $\{[1\ 0\ 1]^T, [0\ 1\ 1]^T, [1\ 1\ 0]^T\}$.

6.50 The matrix of a linear transformation L, on $\mathbb{C}^2$ with respect to the bases $[1\ 1]^T$, $[1\ 0]^T$ is

$$\begin{bmatrix} 2 & 3 \\ 1 & 2 \end{bmatrix}.$$

Find the matrix of L with respect to
(a) the basis $\{[1\ 0]^T, [0\ 1]^T\}$;
(b) the basis $\{[1\ -1]^T, [1\ 1]^T\}$.

6.51 The matrix of a linear transformation L on $\mathbb{C}^3$ with respect to the elementary basis $\{\mathbf{e}_1, \mathbf{e}_2, \mathbf{e}_3\}$ is

$$\begin{bmatrix} 3 & 2 & -3 \\ -2 & 1 & 1 \\ 1 & 2 & 1 \end{bmatrix}.$$

Find the matrix of L with respect to

(a) the basis $\{[0\ 1\ 1]^T, [1\ 1\ 0]^T, [1\ 0\ 1]^T\}$, and
(b) the basis $\{[1\ 1\ -1]^T, [-1\ 1\ 1]^T, [1\ -1\ 1]^T\}$.

6.52 The matrix of a linear transformation $L : \mathbb{C}^3 \to \mathbb{C}^2$ with respect to the bases $\{[0\ 1\ 1]^T, [1\ 1\ 0]^T, [1\ 0\ 1]^T\}$ and $\{[1\ 0]^T, [1\ 1]^T\}$ is

$$\begin{bmatrix} 2 & -1 & 3 \\ 1 & 2 & -2 \end{bmatrix}.$$

Find the matrix of L with respect to the bases
(a) $\{[1\ 1\ -1]^T, [-1\ 1\ 1]^T, [1\ -1\ 1]^T\}$ and $\{[1\ -1]^T, [1\ 1]^T\}$;
(b) $\{[1\ 0\ 0]^T, [0\ 1\ 0]^T, [0\ 0\ 1]^T\}$ and $\{[1\ 0]^T, [0\ 1]^T\}$.

6.53 Show that if A and B are orthogonal, then (a) A^T, (b) A^{-1}, and (c) AB are orthogonal.

6.54 Prove that if U and V are unitary, so also U^T, U^{-1}, $\bar{U}$, UV.

6.55 Prove that for an orthogonal matrix P (a) det $P = \pm 1$ and (b) $P^{-1} = P^T$.

6.56 Prove that the columns (resp rows) of an $n \times n$ unitary matrix U form an orthogonal basis in V_n.

6.57 If an orthogonal matrix U is proper (resp improper), then each of its elements is equal to its cofactor (resp negative of its cofactor).

6.58 Show that the coordinates of the point P with respect to xy- and uv-frames in Figure 6.13 are related by

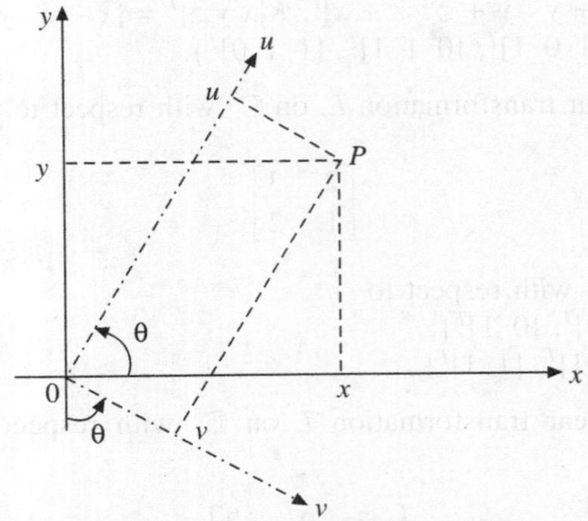

Figure 6.13 Improper orthogonal transformation.

$$\begin{bmatrix} x \\ y \end{bmatrix} = \begin{bmatrix} \cos\theta & \sin\theta \\ \sin\theta & -\cos\theta \end{bmatrix} \begin{bmatrix} u \\ v \end{bmatrix}.$$

Prove also that the above transformation matrix is improper orthogonal and can be interpreted as composition of two orthogonal transformations, viz., rotation of coordinate axes (proper orthogonal transformation) and reflection about the coordinate axis u (see Figure 6.5 without z or w coordinate), i.e.,

$$\begin{bmatrix} \cos\theta & \sin\theta \\ \sin\theta & -\cos\theta \end{bmatrix} = \begin{bmatrix} \cos\theta & -\sin\theta \\ \sin\theta & \cos\theta \end{bmatrix} \begin{bmatrix} 1 & 0 \\ 0 & -1 \end{bmatrix}.$$

6.59 Prove Corollary 6.4.

6.60 Find the transpose of the following linear transformations and its matrix with respect to the dual of the given basis:

(a) $L : \mathbb{R}^2 \to \mathbb{R}^2 : \begin{bmatrix} x_1 \\ x_2 \end{bmatrix} \mapsto \begin{bmatrix} x_1 - x_2 \\ x_1 + x_2 \end{bmatrix}$; basis: elementary vectors in $\mathbb{R}^2$

(b) L same as in (a); basis : $\{[3 \quad 2]^T, [1 \quad 1]^T\}$.

(c) $L : \mathbb{R}^2 \to \mathbb{R}^3 : \mathbf{x} = \begin{bmatrix} x_1 \\ x_2 \end{bmatrix} \mapsto L\mathbf{x} = \begin{bmatrix} 2x_1 - x_2 \\ 3x_1 \\ x_1 + 2x_2 \end{bmatrix}$;

$$\text{basis matrices}: \begin{bmatrix} 2 & 1 \\ 1 & 1 \end{bmatrix}; \begin{bmatrix} 1 & 2 & 1 \\ 0 & 1 & 0 \\ 1 & 0 & 2 \end{bmatrix}.$$

6.61 Given a linear transformation $L : \mathbb{R}^3 \to \mathbb{R}^2 : (x\ y\ z) \mapsto (2x - y - z,\ x - 2y + z)$ and a functional $g : \mathbb{R}^2 \to \mathbb{R} : (uv) \mapsto u - 2v$. Use the definition of transpose of a linear transformation to compute $L^T g$.

6.62 Determine the transpose of the linear transformation L defined by

$$L : \mathbb{P}_2(t) \to \mathbb{R}^{2\times2} : a + bt + ct^2 \mapsto \begin{bmatrix} a & b/2 \\ b/2 & c \end{bmatrix}$$

and its matrix representation when the bases are as given in Example 6.31.

6.63 Let $L : V \to W$ be a linear transformation between finite dimensional inner product spaces V and W.

(a) Prove that there exists a unique linear transformation $L^* : W \to V$ such that $\langle L\mathbf{x}, y\rangle = \langle \mathbf{x}, L^* y\rangle$ for all $\mathbf{x}$ in V and $\mathbf{y}$ in W.

(b) Let $\{\mathbf{x}_1, \mathbf{x}_2, ..., \mathbf{x}_n\}$ and $\{\mathbf{y}_1, \mathbf{y}_2, ..., \mathbf{y}_m\}$ be orthonormal bases for V and W respectively, and let A be the matrix of L with respect to these bases. Show that A^* is equal to the matrix of L^* with respect to the above bases.

6.64 If the linear transformation $L : U \to V$ is surjective and 237 that $K : V \to W$ is injective, then prove that rank $(K{\cdot}L)$ = dim (V).

6.65 Prove that $L^T g = g{\cdot}L$ in (6.62) is a linear map.

6.66 Find the projection P on the range space $\{\mathbf{u}\}$ along the space $\{\mathbf{v}\}$ and give a geometrical interpretation of the projection.

(a) $\mathbf{u} = [1\ 1]^T$; $\mathbf{v} = [-1\ 1]^T$; (b) $\mathbf{u} = [a\ b]^T$; $\mathbf{v} = [-b\ a]^T$;

(c) $\mathbf{u} = [1\ 1]^T$; $\mathbf{v} = [0\ 1]^T$; (d) $\mathbf{u} = [2\ 1]^T$; $\mathbf{v} = \frac{1}{2}[2 - \sqrt{3}\ 1 + 2\sqrt{3}]^T$.

6.67 Find the projection P on the range space generated by $\mathbf{u} = [1\ 1\ 1]^T$ along the space generated by $\mathbf{v}_1 = [1\ -1\ 0]^T$ and $\mathbf{v}_2 = [0\ 1\ -1]^T$ and give a geometrical interpretation of the projection.

6.68 Find the projection P on the range space generated by $\mathbf{u}_1 = [1\ -1\ 0]^T$ and $\mathbf{u}_2 = [0\ 1\ -1]^T$ along the space generated by $\mathbf{v} = [1\ 1\ 1]^T$ and give a geometrical interpretation of the projection.

6.69 Find the projection P on the range space generated by $\mathbf{u}_1 = [1\ -1\ 0]^T$ and $\mathbf{u}_2 = [0\ 1\ -1]^T$ along the space generated by $\mathbf{v} = [1\ -1\ 1]^T$ and give a geometrical interpretation of the projection.

CHAPTER

7

Eigenvalues, Eigenvectors and the Characteristic Equation

In this chapter we shall discuss mainly square matrices A and throughout the ensuing discussion any new facts and developments will be based on the determination of a vector $\mathbf{x}$ (to be called *characteristic vector* or *eigenvector*) and a scalar λ (to be called *characteristic value* or *eigenvalue*) such that $A\mathbf{x} = \lambda\mathbf{x}$. Based on these concepts of eigenvalues and eigenvectors, we shall indicate the conditions on A under which a nonsingular matrix P can be selected such that $P^{-1}AP$ is diagonal, i.e., A is similar to a diagonal matrix. It can be shown that Hermitian and real symmetric matrices can always be reduced by similarity transformation to a diagonal form.

On the other hand, any linear transformation L from a vector space on itself can be represented by a square matrix. The set of all matrices representing L was shown in Chapter 6 to be similar and to have the same determinant although these matrices of L are different. In this chapter we show how suitable bases can be selected for some types of linear transformations such that their matrix representation is as simple as possible in having a diagonal form. Matrices which cannot be transformed into a strictly diagonal form or linear transformations whose matrix representation is not possible to be diagonal but can be transformed into or represented by quasidiagonal normal forms or Jordan's canonical forms will be studied in detail in Chapters 10 and 11.

7.1 EIGENVALUES, EIGENVECTORS AND THE CHARACTERISTIC EQUATION OF A MATRIX

Let us consider a simple second order differential equation, viz.

$$\ddot{x} + 3\dot{x} + 2x = 0 \tag{7.1}$$

which has a solution of the form $c_1e^{-t} + c_2e^{-2t}$, c_1 and c_2 are arbitrary constants. The character of the solution is determined by the indices of the exponential functions, viz. –1 and –2, which are in fact a solution for λ of the equation

$$\lambda^2 + 3\lambda + 2 = 0 \tag{7.2}$$

obtained by substituting $x = e^{\lambda t}$ in (7.1). These indices would force the solution to decay in course of time. The roots of (7.2) are therefore the guiding factors for the behaviour of the modes of the system described by (7.1). On the other hand, forming the equations for a simple parallel inductor-resistor circuit, we have, when $e(t) = 0$ (see Figure 7.1),

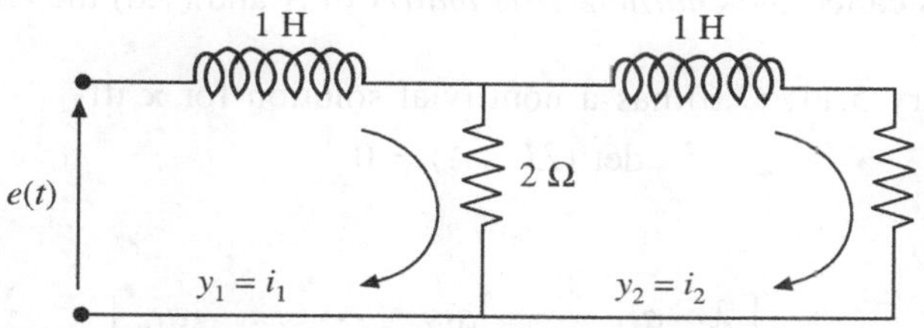

Figure 7.1 A parallel inductor-resistor circuit.

$$\dot{y}_1 + 2y_1 - 2y_2 = 0, \qquad \dot{y}_2 + 5y_2 - 2y_1 = 0,$$

or as a vector-matrix equation

$$\dot{\mathbf{y}}(t) = A\mathbf{y}(t), \tag{7.3}$$

where

$$\mathbf{y}(t) = \begin{bmatrix} y_1(t) \\ y_2(t) \end{bmatrix}, \qquad A = \begin{bmatrix} -2 & 2 \\ 2 & -5 \end{bmatrix}.$$

A solution to (7.3) may be assumed to be of the form

$$\mathbf{y}(t) = \mathbf{x}e^{\lambda t}, \tag{7.4}$$

where $\mathbf{x}$ is a *nonzero*, constant 2-vector and λ is a constant scalar which are still unknown. Substituting (7.4) into (7.3), we get

$$\lambda\mathbf{x}e^{\lambda t} = A\mathbf{x}e^{\lambda t}$$

from which we have

$$\lambda\mathbf{x} = A\mathbf{x}. \tag{7.5}$$

In many practical cases, a great number of problems can be formulated in the form (7.5). The solution (7.4) of the loop currents in the above circuit or of the system described by (7.3), where A is an $n \times n$ matrix, therefore becomes: given a matrix A of order n, determine a nonzero vector $\mathbf{x}$ and a scalar λ such that (7.5) is satisfied. This problem is known as the *characteristic value* or *eigenvalue problem.* In fact, denoting $y_1 = x$, $y_2 = \dot{x} = \dot{y}_1$, the solution to (7.1) can be transformed into that of a characteristic value problem defined by (7.5) pertaining to $\dot{y}_1 = y_2$, $\dot{y}_2 = \ddot{x} = -2y_1 - 3y_2$.

Rewriting (7.5), we get

$$(\lambda I - A)\mathbf{x} = \boldsymbol{\theta}, \tag{7.6}$$

where in general,

$$A = \begin{bmatrix} a_{11} & \cdots & a_{1n} \\ \vdots & & \vdots \\ a_{n1} & \cdots & a_{nn} \end{bmatrix}, \quad \mathbf{x} = \begin{bmatrix} x_1 \\ \vdots \\ x_n \end{bmatrix}. \tag{7.7}$$

In (7.6), $(\lambda I - A)$ is called the *characteristic matrix* of A and (7.6) the *characteristic matrix equation.*

In view of Corollary 5.11, (7.6) has a nontrivial solution for $\mathbf{x}$ iff

$$\det(\lambda I - A) = 0$$

or, writing explicitly,

$$\phi(\lambda) = \det \begin{bmatrix} \lambda - a_{11} & -a_{12} & \cdots & -a_{1n} \\ -a_{21} & \lambda - a_{22} & \cdots & -a_{2n} \\ \vdots & \vdots & & \vdots \\ -a_{n1} & -a_{n2} & \cdots & \lambda - a_{nn} \end{bmatrix} = 0. \tag{7.8}$$

By expansion of the determinant, it is evident that $\phi(\lambda)$ is a polynomial of degree n in λ and is referred to as a *characteristic polynomial* of the matrix A. Equation (7.8) is known as a *characteristic equation*, and n number of its roots *characteristic* (or *eigen-* or *proper* or *latent*) *values* or *roots of A*. The term *spectrum of a matrix* is often used to imply the set of characteristic values of A and is denoted by spec(A).

Denoting by λ_i any characteristic root of A, we note that (7.8) is satisfied for this λ_i and (7.6), i.e. $(\lambda_i I - A)\mathbf{x} = \boldsymbol{\theta}$ has nontrivial solutions for $\mathbf{x}$, say, $\mathbf{x}_i$. Any such solution $\mathbf{x}_i$ is called the *characteristic* (*eigen-*, or *proper* or *latent*) *vector* of A associated with the characteristic value λ_i; consequently, we have

$$A\mathbf{x}_i = \lambda_i \mathbf{x}_i. \tag{7.9}$$

Example 7.1 Determine the characteristic roots and the characteristic vectors of the matrix

$$A = \begin{bmatrix} 1 & 2 \\ 1 & 0 \end{bmatrix}.$$

Solution The characteristic polynomial of A by (7.8) is written as

$$\phi(\lambda) = \det \begin{bmatrix} \lambda - 1 & -2 \\ -1 & \lambda \end{bmatrix} = \lambda^2 - \lambda - 2 = (\lambda - 2)(\lambda + 1).$$

Solving $\phi(\lambda) = 0$, we get $\lambda_1 = 2$, and $\lambda_2 = -1$ as the characteristic roots of A. For $\lambda = \lambda_1 = 2$, denoting $\mathbf{x}_1 = [x_1\ x_2]^T$, from (7.9) we have

$$-x_1 + 2x_2 = 0, \qquad x_1 - 2x_2 = 0 \tag{7.10}$$

of which only one equation is independent. Therefore, assigning an arbitrary value to x_1, e.g., $x_1 = 2$, we have (see Example 5.27)

$$\mathbf{x}_1 = \begin{bmatrix} 2 \\ 1 \end{bmatrix}$$

as a characteristic vector of A associated with $\lambda_1 = 2$ satisfying $A\mathbf{x}_1 = \lambda_1\mathbf{x}_1$. Similarly, a characteristic vector corresponding to $\lambda_2 = -1$ is computed as

$$\mathbf{x}_2 = \begin{bmatrix} 1 \\ -1 \end{bmatrix}$$

which satisfies $A\mathbf{x}_2 = \lambda_2\mathbf{x}_2$.

Indeed, any nonzero scalar (k) multiple of $\mathbf{x}_1$ satisfies (7.10) and is therefore a characteristic vector of A, thus giving infinite number of characteristic vectors $k\mathbf{x}_1$ corresponding to λ_1. Similar arguments hold for λ_2 as well.

This example can be employed to offer a geometrical interpretation of (7.5) or (7.6). In this example, from (7.9) it is evident that $A\mathbf{x}_1 = 2\mathbf{x}_1$ (or $A\mathbf{x}_2 = -\mathbf{x}_2$), see Figure 7.2, and hence $A\mathbf{x}_1$ and $\mathbf{x}_1$ (or $A\mathbf{x}_2$ and $\mathbf{x}_2$) are linearly dependent. But if $\mathbf{x}$ is not a characteristic vector, then $A\mathbf{x}$ and $\mathbf{x}$ are linearly independent. So if $\mathbf{x}$ is a characteristic vector of a matrix A, then $\mathbf{x}$ and $A\mathbf{x}$ point in the same direction, i.e., they lie on the same line. □

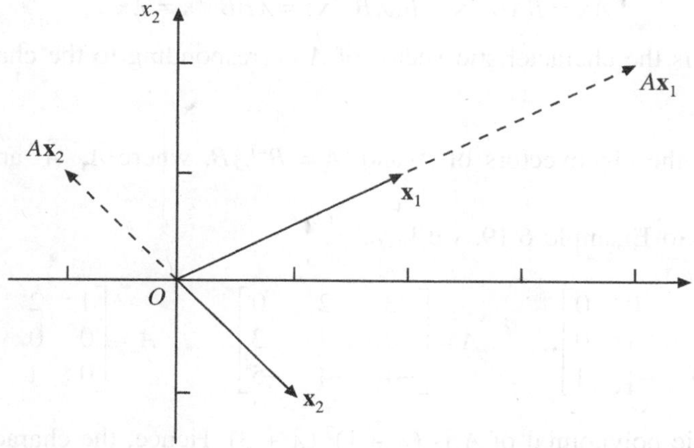

Figure 7.2 Interpretation of $A\mathbf{x} = l\mathbf{x}$.

Example 7.2 If λ is a characteristic root of a matrix A, then show that λ^k is the characteristic root of A^k.

Solution This is so because, in view of (7.5), we have

$$A^k\mathbf{x} = A^{k-1}A\mathbf{x} = A^{k-1}\lambda\mathbf{x} = \lambda A^{k-2}\lambda\mathbf{x} = \cdots = \lambda^k\mathbf{x},$$

i.e., $A^k\mathbf{x} = \lambda^k\mathbf{x}$, which shows that λ^k is the characteristic root of A^k by definition. □

We recall that similar matrices are representations of a single linear transformation on a vector space with respect to different bases. Moreover, the matrices A and $\hat{A}$ are similar if there exists a nonsingular B such that $B^{-1}AB = \hat{A}$ (see Section 6.4). Regarding characteristic values and characteristic vectors, we have the following theorem:

Theorem 7.1 Two similar matrices A and $\hat{A} = B^{-1}AB$ (where B is nonsingular) have the same characteristic polynomial and hence the same characteristic roots. Moreover, $\mathbf{x}$ is a characteristic vector of A corresponding to the characteristic root λ iff $B^{-1}\mathbf{x}$ is a characteristic vector of $\hat{A}$ corresponding to the same characteristic root λ of A.

Proof We have, in view of Problems 2.26 and 3.16(a), the relation

$$\begin{aligned}\det(\lambda I - \hat{A}) &= \det(\lambda I - B^{-1}AB) = \det[B^{-1}(\lambda I - A)B] \\ &= \det B^{-1}\cdot\det(\lambda I - A)\cdot\det B = \det(\lambda I - A)\end{aligned}$$

which proves that the characteristic polynomials of A and $\hat{A}$ are the same and so these matrices have the same characteristic roots.

Suppose now that $\mathbf{x}$ is a characteristic vector of A associated with λ, i.e., $A\mathbf{x} = \lambda\mathbf{x}$. Then

$$\hat{A}B^{-1}\mathbf{x} = B^{-1}A\mathbf{x} = B^{-1}\lambda\mathbf{x} = \lambda B^{-1}\mathbf{x} \tag{7.11}$$

which shows that $B^{-1}\mathbf{x}$ is a characteristic vector of $\hat{A}$ associated with its characteristic root λ. On the contrary, if (7.11) holds, then

$$A\mathbf{x} = B\hat{A}B^{-1}\mathbf{x} = B(\lambda B^{-1}\mathbf{x}) = \lambda BB^{-1}\mathbf{x} = \lambda\mathbf{x}$$

which shows that $\mathbf{x}$ is the characteristic vector of A corresponding to the characteristic root λ. □

Example 7.3 Find the eigenvectors of A and $\hat{A} = B^{-1}AB$, where A, $\hat{A}$ and B are given in Example 6.19.

Solution Referring to Example 6.19, we have

$$B = \begin{bmatrix} 1 & 0 & 0 \\ -1 & 1 & 0 \\ 0 & -1 & 1 \end{bmatrix}, \quad A = \begin{bmatrix} 3 & 2 & 0 \\ 0 & 1 & 3 \\ -4 & -4 & -5 \end{bmatrix}, \quad \hat{A} = \begin{bmatrix} 1 & 2 & 0 \\ 0 & 0 & 3 \\ 0 & 1 & -2 \end{bmatrix}.$$

The characteristic polynomial of A is $(\lambda - 1)^2(\lambda + 3)$. Hence, the characteristics roots are 1, 1, –3, and the computation of associated characteristic vectors gives $\mathbf{x} = \alpha\,[1, -1, 0]^T$ and $\mathbf{y} = \beta\,[1, -3, 4]^T$, where α and β are arbitrary scalars. The characteristic roots of $\hat{A}$ are also 1, 1, –3, but the associated characteristic vectors are $B^{-1}\mathbf{x} = \alpha[1\ 0\ 0]^T$ and $B^{-1}\mathbf{y} = \beta\,[1, -2, 2]^T$. □

Two matrices may have the same characteristic equation but they may not be similar. For example, the matrices

$$A = \begin{bmatrix} 1 & 0 \\ 2 & 1 \end{bmatrix}, \qquad B = \begin{bmatrix} 1 & 0 \\ 0 & 1 \end{bmatrix}$$

have the characteristic equation. But there is no nonsingular matrix P such that $B = PAP^{-1}$.

Example 7.4 (*Eigenvector method*) Determine a solution of the differential equation

$$\ddot{x}(t) + 3\dot{x}(t) + 2x(t) = 0 \tag{7.12}$$

subject to the initial conditions $x(0) = 1$, $\dot{x}(0) = 1$.

Solution Put $y_1 = x$ and $y_2 = \dot{y}_1$. Then (7.12) can be written as

$$\dot{\mathbf{y}}(t) = A\mathbf{y}(t), \tag{7.13}$$

where

$$\mathbf{y} = \begin{bmatrix} y_1 \\ y_2 \end{bmatrix}, \quad A\mathbf{y} = \begin{bmatrix} 0 & 1 \\ -2 & -3 \end{bmatrix}. \tag{7.14}$$

Assuming the solution of (7.13) to be of the form $\mathbf{y}(t) = \mathbf{x} \exp(\lambda t)$, where $\mathbf{x} = [x_1\ x_2]^T$ is a nonzero constant 2-vector and λ is a constant scalar to be determined, we have from (7.13) the relation $(\lambda I - A)\mathbf{x} = \mathbf{\theta}$. By Corollary 5.11, det $(\lambda I - A) = 0$ which gives $\lambda^2 + 3\lambda + 2 = 0$. Therefore, $\lambda = -1$ and -2, for which the solution of $(\lambda I - A)\mathbf{x} = \mathbf{\theta}$ becomes $\mathbf{x} = \alpha\,[1, -1]^T$ and $\mathbf{x} = \alpha\,[1, -2]^T$ respectively, where α and β are arbitrary scalars. Hence the solution of (7.13) is

$$\mathbf{y}(t) = \alpha\,[1, -1]^T \exp(-t) + \beta\,[1, -2]^T \exp(-2t),$$

since a general solution of (7.13) is a linear combination of all solutions in the form $\mathbf{x} \exp(\lambda t)$. Imposing the initial conditions, we have $\alpha = 3$ and $\beta = -2$. Hence, $y_1(t) = x(t) = 3 \exp(-t) - 2 \exp(-2t)$. □

7.1.1 Eigenvalues and Eigenvectors of a Linear Transformation

For the linear transformation $L: V \rightarrow V$, if there exists a nonzero vector $\mathbf{x}$ in V such that

$$L\mathbf{x} = \lambda\mathbf{x}, \qquad \mathbf{x} \neq \mathbf{\theta} \tag{7.15}$$

for some scalar λ, then we call λ a *characteristic root* (or *eigenvalue* or *proper value* or *latent root*) of L and $\mathbf{x}$ the *characteristic* (or *eigen-*, or *proper* or *latent*) *vector* of L. If $\mathbf{u} = [\alpha_1\alpha_2 \cdots \alpha_n]^T$ is the coordinate vector of $\mathbf{x}$ and A is the matrix of L, both with respect to a given basis $X = \{\mathbf{x}_1, \mathbf{x}_2, \ldots, \mathbf{x}_n\}$ in V, then by Theorem 6.9, (7.15) implies and is implied by

$$A\mathbf{u} = \lambda\mathbf{u}, \qquad \mathbf{u} \neq \mathbf{\theta} \qquad \text{or} \qquad (\lambda I - A)\mathbf{u} = \mathbf{\theta}, \qquad \mathbf{u} \neq \mathbf{\theta}, \tag{7.16}$$

where

$$\mathbf{x} = \alpha_1\mathbf{x}_1 + \alpha_2\mathbf{x}_2 + \cdots + \alpha_n\mathbf{x}_n. \tag{7.17}$$

The validity of (7.16) demands that det $(\lambda I - A) = 0$ (as stipulated in the last section) whose solution gives n number of characteristic roots or eigenvalues of L. Thus the eigenvalue problem of a linear transformation can be reformulated as an eigenvalue problem of its matrix A with respect to a convenient basis. The eigenvalues of L are exactly those of its matrix A with respect to any basis X and the eigenvector $\mathbf{x}$ of L is related to the eigenvector $\mathbf{u} = [\alpha_1\alpha_2 \cdots \alpha_n]^T$ of A by (7.17).

If, however, A is the matrix of a linear transformation L on $\mathbb{R}^n$ with respect to the standard basis $\{\mathbf{e}_1, \mathbf{e}_2, \ldots, \mathbf{e}_n\}$, then in (7.17) $\mathbf{x}_i = \mathbf{e}_i$, $i = 1, 2, \ldots, n$. Thus, $\mathbf{u} = \mathbf{x}$, and the eigenvectors $\mathbf{x}$ of L and $\mathbf{u}$ of its matrix A turn out to be the same.

We have already observed that the class of matrices associated with a linear transformation L is similar. That is, if A and $\hat{A}$ are any two matrices in this class, then there exists a nonsingular matrix B satisfying $B^{-1}AB = \hat{A}$ and $\det A = \det \hat{A}$, see Section 6.4. This prompts us to define the determinant of a linear transformation L on V by $\det L := \det A$, where A is the matrix of L with respect to any basis X in V. Moreover, the identity

$$\det(\lambda I - \hat{A}) = \det[B^{-1}(\lambda I - A)B] = \det(\lambda I - A)$$

suggests that we can define the characteristic polynomial of a linear transformation L by $\det(\lambda I_V - L) := \det(\lambda I - A)$, where A is the matrix of L with respect to any basis.

Example 7.5 Determine the eigenvalues and eigenvectors of the linear transformation L on $\mathbb{R}^3$ given by

$$L[x_1\ x_2\ x_3]^T = [x_1 + 2x_2,\ 3x_3,\ x_2 - 2x_3]^T.$$

Solution The matrix of L with respect to the standard basis $\{\mathbf{e}_1, \mathbf{e}_2, \mathbf{e}_3\}$ in $\mathbb{R}^3$ is (see Example 6.17)

$$\hat{A} = \begin{bmatrix} 1 & 2 & 0 \\ 0 & 0 & 3 \\ 0 & 1 & -2 \end{bmatrix}.$$

So the characteristic equation for L is

$$\det(\lambda I - \hat{A}) = (\lambda - 1)^2(\lambda + 3) = 0.$$

Hence the eigenvalues of L are 1, 1, and –3. Solving $(\lambda I - \hat{A})\mathbf{u} = \boldsymbol{\theta}$ for $\mathbf{u}$ when $\lambda = 1$ and –3, we get, respectively,

$$\mathbf{u}_1 = \alpha[1\ 0\ 0]^T, \qquad \mathbf{u}_2 = \beta[1, -2, 2]^T$$

as solutions where α and β are arbitrary scalars, which are the eigenvectors of L corresponding to 1 and –3, respectively.

It is interesting to note that with respect to the basis $\mathbf{x}_1 = [1\ 1\ 1]^T$, $\mathbf{x}_2 = [0\ 1\ 1]^T$, and $\mathbf{x}_3 = [0\ 0\ 1]^T$, the matrix of L is

$$A = \begin{bmatrix} 3 & 2 & 0 \\ 0 & 1 & 3 \\ -4 & -4 & -5 \end{bmatrix}$$

(see Example 6.17) whose eigenvalues obtained from $\det(\lambda I - A) = 0$ are 1, 1, and –3, as before, but the associated eigenvectors are given by $\mathbf{v}_1 = \alpha[1, -1, 0]^T$ and $\mathbf{v}_2 = \beta[1, -3, 4]^T$, where α and β are arbitrary scalars. These are, however, not the eigenvectors of L since they do not satisfy $L\mathbf{x} = \lambda\mathbf{x}$ for $\lambda = 1$ and –3, respectively. They are, as a matter of fact, related to the eigenvectors $\mathbf{z}_1$ and $\mathbf{z}_2$ of L by (7.17). We define $C := [\mathbf{x}_1\mathbf{x}_2\mathbf{x}_3]$. Then (7.17) can be written as $\mathbf{x} = C\mathbf{u}$. Consequently, we have, after substituting the values of $\mathbf{x}_1$, $\mathbf{x}_2$, $\mathbf{x}_3$, and (i) $\mathbf{u} = \mathbf{v}_1$ and (ii) $\mathbf{u} = \mathbf{v}_2$ by turn in (7.17), the relations

(i) $\mathbf{z}_1 = C\mathbf{v}_1 = \alpha(1 \cdot \mathbf{x}_1 - 1 \cdot \mathbf{x}_2 + 1 \cdot \mathbf{x}_3) = \alpha[1\ 0\ 0]^T$

(ii) $\mathbf{z}_2 = C\mathbf{v}_2 = \beta(1 \cdot \mathbf{x}_1 - 3\mathbf{x}_2 + 4\mathbf{x}_3) = \beta[1, -2, 2]^T$

as the eigenvectors of L as before. □

Example 7.6 A linear transformation L: $\mathbb{P}_2(t) \to \mathbb{P}_2(t)$ is defined as

$$L(a_0 + a_1 t + a_2 t^2) = -(a_0 + 2a_1 + a_2) + (2a_0 + 3a_1)t - 2(a_0 + a_1)t^2.$$

Find the eigenvalues and eigenvectors of L.

Solution Since, $L(1) = -1 + 2t - 2t^2$, $L(t) = -2 + 3t - 2t$ and $L(t^2) = -1 + 0.t - 0.t^2$, the matrix of L with respect to the standard basis $\{1, t, t^2\}$ is (see Section 6.3)

$$A = \begin{bmatrix} -1 & -2 & -1 \\ 2 & 3 & 0 \\ -2 & -2 & 0 \end{bmatrix}.$$

The eigenvalues of L are obtained by solving det $(\lambda I - A) = \lambda^3 - 2\lambda^2 - \lambda + 2 = 0$ which are 1, –1, and 2. The corresponding eigenvectors of A are, respectively, $\mathbf{u} = \alpha[1, -1, 0]^T$, $\mathbf{v} = \beta[2, -1, 2]^T$, and $\mathbf{w} = \gamma[1, -2, 1]^T$. For this example in (7.17) $\mathbf{x}_1 = 1$, $\mathbf{x}_2 = t$, and $\mathbf{x}_3 = t^2$. So defining $\mathbf{p} = [1\ t\ t^2]^T$, (7.17) can be written as

$$\mathbf{x} = \mathbf{p}^T\mathbf{u}.$$

Hence, the corresponding eigenvectors of L are

$$\mathbf{x} = \mathbf{p}^T\mathbf{u} = \alpha(1 - t), \qquad \mathbf{y} = \mathbf{p}^T\mathbf{v} = \beta(2 - t + 2t^2),$$
$$\mathbf{z} = \mathbf{p}^T\mathbf{w} = \gamma(1 - 2t + t^2),$$

where α, β and γ are arbitrary scalars. □

7.2 PROPERTIES OF EIGENVECTORS ASSOCIATED WITH DISTINCT EIGENVALUES

For the sake of brevity, the roots of the characteristic equation of a matrix are henceforth called by the alternative designation *eigenvalues* and the characteristic vector by *eigenvector*. The characteristic polynomial of an n-square matrix as determined by (7.8) is also of degree n. So the solution of the characteristic equation would give rise to n number of eigenvalues which may be all real or all complex or both. If the matrix A is originally defined over a real field $\mathbb{F}$ which is indeed so in most of the cases, its eigenvalues may not be in $\mathbb{F}$. We shall therefore assume in the following that the field $\mathbb{F}$ over which A is defined is such that all the eigenvalues lie also in the same field. Such a field is called an *algebraically closed field.* This field is such that any polynomial in the variable λ, say, with coefficients in this field, has also roots in the same field.

The eigenvectors $\mathbf{x}$ defined with the help of (7.5) are sometimes called *right eigenvectors,* being positioned as they are in (7.5) to the right side of A. In this section, we shall study another kind of eigenvectors arising naturally with the transpose A^T of a square matrix A, which are termed *left eigenvectors* as a result of their position to the left of A in the characteristic matrix equation (7.6).

Theorem 7.2 The eigenvectors associated with distinct eigenvalues of an n-square matrix A are linearly independent.

Proof We shall prove the theorem by mathematical induction. By assuming that $(n - 1)$ number of eigenvectors $\mathbf{x}_1, \mathbf{x}_2, ..., \mathbf{x}_{n-1}$ are linearly independent, we shall show by the method of contradiction that the eigenvectors $\mathbf{x}_1, \mathbf{x}_2, ..., \mathbf{x}_n$ are also linearly independent. Let there exist constants $\alpha_1, \alpha_2, ..., \alpha_n$ not all zero such that

$$\alpha_1\mathbf{x}_1 + \cdots + \alpha_{n-1}\mathbf{x}_{n-1} + \alpha_n\mathbf{x}_n = \boldsymbol{\theta}. \tag{7.18}$$

Here $\alpha_n \neq 0$, because in the opposite case as $\mathbf{x}_1, ..., \mathbf{x}_{n-1}$ are assumed to be linearly independent by induction hypothesis, $\alpha_1 = \cdots = \alpha_{n-1} = 0$, and evidently the theorem is proved.

So when $\alpha_n \neq 0$, premultiplying both sides of (7.18) by A and observing that when λ_i's are distinct, $A\mathbf{x}_i = \lambda_i\mathbf{x}_i$, where $\mathbf{x}_i$ is an eigenvector corresponding to λ_i, we have

$$\alpha_1\lambda_1\mathbf{x}_1 + \cdots + \alpha_{n-1}\lambda_{n-1}\mathbf{x}_{n-1} + \alpha_n\lambda_n\mathbf{x}_n = \boldsymbol{\theta}. \tag{7.19}$$

If $\lambda_n \neq 0$, from (7.18) and (7.19), we have

$$-\alpha_n\mathbf{x}_n = \alpha_1\mathbf{x}_1 + \cdots + \alpha_{n-1}\mathbf{x}_{n-1} = \frac{1}{\lambda_n}(\alpha_1\lambda_1\mathbf{x}_1 + \cdots + \alpha_{n-1}\lambda_{n-1}\mathbf{x}_{n-1}),$$

or

$$(\lambda_n - \lambda_1)\alpha_1\mathbf{x}_1 + \cdots + (\lambda_n - \lambda_{n-1})\alpha_{n-1}\mathbf{x}_{n-1} = \boldsymbol{\theta}.$$

Since $\mathbf{x}_1, ..., \mathbf{x}_{n-1}$ are linearly independent by induction hypothesis, by Definition 5.3, $(\lambda_n - \lambda_i)\alpha_i = 0$, $i = 1, ..., n - 1$. Again, since $\lambda_1, ..., \lambda_n$ are distinct, $\alpha_i = 0$, $i = 1, ..., n - 1$, which, when put in (7.18), gives $\alpha_n\mathbf{x}_n = \boldsymbol{\theta}$, where $\mathbf{x}_n$ is not a zero vector. Therefore, $\alpha_n = 0$, which is a contradiction.

If $\lambda_n = 0$, from (7.19) since λ_i's are distinct and so no λ_i $(i = 1, ..., n - 1)$ is zero, we have $\alpha_1 = \cdots = \alpha_{n-1} = 0$ because $\mathbf{x}_1, ..., \mathbf{x}_{n-1}$ are linearly independent by induction hypothesis. These values of α_i, $i = 1, ..., n - 1$, when substituted into (7.18) gives $\alpha_n = 0$. This is contrary to our assumption that not all α_i's are zero. □

Example 7.7 Check, whether or not, the eigenvectors of A given in Example 7.1 are linearly independent.

Solution The eigenvectors associated with the distinct eigenvalues 2 and –1 of the matrix A in Example 7.1 are

$$\mathbf{x}_1 = \begin{bmatrix} 2 \\ 1 \end{bmatrix}, \qquad \mathbf{x}_2 = \begin{bmatrix} 1 \\ -1 \end{bmatrix}.$$

The matrix $[\mathbf{x}_1\ \mathbf{x}_2]$ is nonsingular, and therefore, $\mathbf{x}_1$ and $\mathbf{x}_2$ are linearly independent. □

Theorem 7.3 An nth order matrix A with distinct eigenvalues $\lambda_1, ..., \lambda_n$ is similar (see Definition 6.11) to a diagonal matrix D with these eigenvalues as diagonal elements.

Proof Let $\mathbf{x}_1, ..., \mathbf{x}_n$ be the eigenvectors of A associated with its distinct eigenvalues $\lambda_1, ..., \lambda_n$ such that

$$A\mathbf{x}_i = \lambda_i\mathbf{x}_i, \qquad i = 1, ..., n,$$

which when written in an alternative way give us

$$A[\mathbf{x}_1 \cdots \mathbf{x}_n] = [\mathbf{x}_1 \cdots \mathbf{x}_n]\,\text{diag}\,(\lambda_1, \ldots, \lambda_n).$$

Therefore,

$$AP = PD, \tag{7.20}$$

where

$$P = [\mathbf{x}_1 \cdots \mathbf{x}_n] \tag{7.21}$$

is a nonsingular matrix of order n by Theorem 7.2 and Theorem 5.24, and

$$D = \text{diag}\,(\lambda_1, \ldots, \lambda_n). \tag{7.22}$$

The theorem follows from premultiplication of both sides of (7.20) by P^{-1}, giving

$$P^{-1}AP = \text{diag}\,(\lambda_1, \ldots, \lambda_n). \quad \square$$

Definition 7.1 *Modal Matrix and Spectral Matrix.* The matrix P in (7.21) formed by the linearly independent eigenvectors of a matrix A is called a *modal matrix* for A, and the diagonal matrix D in (7.22) having the eigenvalues of A as the diagonal elements is called a *spectral matrix* for A. □

Definition 7.2 *Diagonalizable Matrix.* A matrix A is said to be diagonalizable over $\mathbb{F}$ iff there exists a nonsingular matrix P over $\mathbb{F}$ such that $P^{-1}AP$ is a diagonal matrix D over $\mathbb{F}$. □

Example 7.8 Find a nonsingular matrix P such that $P^{-1}AP$ is diagonal, where A is given in Example 7.1.

Solution The matrix P formed by the eigenvectors associated with distinct eigenvalues 2, and –1 of A given in Example 7.1 is

$$P = \begin{bmatrix} 2 & 1 \\ 1 & -1 \end{bmatrix}.$$

Therefore,

$$P^{-1}AP = \frac{1}{3}\begin{bmatrix} 1 & 1 \\ 1 & -2 \end{bmatrix}\begin{bmatrix} 1 & 2 \\ 1 & 0 \end{bmatrix}\begin{bmatrix} 2 & 1 \\ 1 & -1 \end{bmatrix}$$

$$= \frac{1}{3}\begin{bmatrix} 2 & 2 \\ -1 & 2 \end{bmatrix}\begin{bmatrix} 2 & 1 \\ 1 & -1 \end{bmatrix} = \begin{bmatrix} 2 & 0 \\ 0 & -1 \end{bmatrix}$$

is a diagonal matrix having as diagonal elements the eigenvalues of A. □

Example 7.9 Show that a real matrix having distinct complex eigenvalues cannot be transformed by a nonsingular transformation to a diagonal form with real elements along the digonal.

Solution Let $\mathbf{x} + i\mathbf{y}$, where $\mathbf{x}$ and $\mathbf{y}$ are 2-vectors, be the complex eigenvector associated with the complex eigenvalue $\lambda = a + ib$ of a second order real matrix A. By Problem 7.4, $\bar{\lambda} = a - ib$ is the remaining eigenvalue and $\mathbf{x} - i\mathbf{y}$ the corresponding eigenvector. We have

$$A(\mathbf{x} + i\mathbf{y}) = (a + ib)\,(\mathbf{x} + i\mathbf{y}),$$

in which equating real and imaginary parts we have

$$A\mathbf{x} = a\mathbf{x} - b\mathbf{y}, \qquad A\mathbf{y} = b\mathbf{x} + a\mathbf{y}.$$

Hence,

$$A[\mathbf{x}\ \mathbf{y}] = [A\mathbf{x}\ A\mathbf{y}] = [\mathbf{x}\ \mathbf{y}]\begin{bmatrix} a & b \\ -b & a \end{bmatrix}. \tag{7.23}$$

Now by Theorem 7.2, $\mathbf{x} + i\mathbf{y}$ and $\mathbf{x} - i\mathbf{y}$ are linearly independent; so also $\mathbf{x}$ and $\mathbf{y}$ by Problem 5.35. Therefore, $P = [\mathbf{x}\ \mathbf{y}]$ is nonsingular which helps us to write (7.23) as

$$P^{-1}AP = \begin{bmatrix} a & b \\ -b & a \end{bmatrix}$$ □

This example illustrates that a real matrix having distinct complex eigenvalues does not have a diagonal form with real elements (see Problem 7.15) of the same type as for real eigenvalues.

When a matrix is diagonalizable, we can compute its *higher power* very easily. To see this, we first find a nonsingular matrix P such that $P^{-1}AP = \text{diag}\,[\lambda_1, \lambda_2, ..., \lambda_n] =: D$. By successive computation, we get

$$D^2 = \text{diag}\,[\lambda_1^2, \lambda_2^2, \ldots, \lambda_n^2] = P^{-1}APP^{-1}AP = P^{-1}A^2P,$$
$$\vdots$$
$$D^k = \text{diag}\,[\lambda_1^k, \lambda_2^k, \ldots, \lambda_n^k] = P^{-1}A^kP, \tag{7.24}$$

where k is any positive integer. As D^k admits of easy computation, the kth power of A can be determined, as is evident from (7.24) by

$$A^k = PD^kP^{-1}.$$

Example 7.10 Compute A^{20} if

$$A = \begin{bmatrix} 1 & 2 \\ 1 & 0 \end{bmatrix}.$$

Solution As given in Example 7.8, we have

$$P = \begin{bmatrix} 2 & 1 \\ 1 & -1 \end{bmatrix}$$

such that $P^{-1}AP = \text{diag}\,[2, -1] =: D$. Therefore,

$$A^{20} = PD^{20}P^{-1} = \frac{1}{3}\begin{bmatrix} 2 & 1 \\ 1 & -1 \end{bmatrix}\begin{bmatrix} 2 & 0 \\ 0 & -1 \end{bmatrix}^{20}\begin{bmatrix} 1 & 1 \\ 1 & -2 \end{bmatrix}$$

$$= \frac{1}{3}\begin{bmatrix} 2^{21}+1 & 2^{21}-2 \\ 2^{20}-1 & 2^{20}+2 \end{bmatrix}.$$ □

7.2.1 Left Eigenvector and Right Eigenvector

As a result of Property II of determinants in Section 2.3, over the real field $\mathbb{R}$ we can write from (7.8) det $(\lambda I - A)$ = det $(\lambda I - A^T)$, which implies that a matrix A and its transpose A^T have the same eigenvalues. However, the eigenvectors of A and A^T corresponding to the same eigenvalue are, in general, different. Let $\mathbf{y} = [y_1 \ldots y_n]^T$ be the eigenvector of A^T corresponding to the eigenvalue λ. Then (7.5) becomes

$$A^T\mathbf{y} = \lambda\mathbf{y},$$

transposing both sides of which gives

$$\mathbf{y}^T A = \lambda\mathbf{y}^T, \tag{7.25}$$

where $\mathbf{y}$ is called a *left eigenvector* of A corresponding to the eigenvalue λ. In the light of this definition, $\mathbf{x}$ in (7.5) may be termed the *right eigenvector* of A associated with the eigenvalue λ. When there is no chance of confusion, we shall call these right eigenvectors without any attribute only *eigenvectors*. As in Theorem 7.2, we can prove the following theorem.

Theorem 7.4 The left eigenvectors associated with distinct eigenvalues of a square matrix A are linearly independent. □

Definition 7.3 *Biorthogonal Sets.* Two sets of vectors $\{\mathbf{x}_1, ..., \mathbf{x}_m\}$ and $(\mathbf{y}_1, ..., \mathbf{y}_m\}$ in V_n are said to be biorthogonal if $\langle \mathbf{x}_i, \mathbf{y}_j \rangle = \delta_{ij}$ (see (6.51)), $i, j = 1, ..., m$. □

This definition gives, in particular, $\langle \mathbf{x}_i, \mathbf{x}_j \rangle = \delta_{ij}$ when $\mathbf{x}_j = \mathbf{y}_j$. It might be recalled that the set of vectors $\{\mathbf{x}_1, ..., \mathbf{x}_m\}$ satisfying this condition is called orthonormal (see Definition 5.13).

Theorem 7.5 There exists over the real field $\mathbb{R}$ a set of left eigenvectors $\{\mathbf{y}_1, ..., \mathbf{y}_n\}$ of an n-square matrix A which are biorthogonal to the set of right eigenvectors $\{\mathbf{x}_1, ..., \mathbf{x}_n\}$ when the eigenvalues of A are distinct.

Proof Since $\mathbf{y}_i$ is a left eigenvector of A corresponding to λ_i, we have, in view of (7.25), the equations

$$\mathbf{y}_i^T A = \lambda_i \mathbf{y}_i^T, \qquad i = 1, ..., n. \tag{7.26}$$

If we define

$$Q = [\mathbf{y}_1 \cdots \mathbf{y}_n], \tag{7.27}$$

then (7.26) can be written as

$$Q^T A = DQ^T, \tag{7.28}$$

where D is given by (7.22). Likewise, for the right eigenvectors $\mathbf{x}_i$'s, (7.20) is satisfied from which, since P is nonsingular (see Theorem 7.2), we can write

$$P^{-1}A = DP^{-1}. \tag{7.29}$$

Comparing (7.28) and (7.29), we see that the rows of P^{-1} define a set of left eigenvectors of A. So, if we choose the rows of P^{-1} to define the set of left eigenvectors so that $Q^T = P^{-1}$, then we get

$$Q^TP = I, \qquad PQ^T = I.$$

Writting $Q^T P = I$ explicitly with the help of (7.27) and (7.21), we get

$$\begin{bmatrix} \mathbf{y}_1^T \\ \vdots \\ \mathbf{y}_n^T \end{bmatrix} [\mathbf{x}_1 \cdots \mathbf{x}_n] = I.$$

Hence (see (5.16)), $\langle \mathbf{x}_i, \mathbf{y}_j \rangle = \delta_{ij}, i, j = 1, ..., n.$ □

Alternatively, Theorem 7.5 can be reformulated as follows.

Theorem 7.6 Over the real field $\mathbb{R}$, two sets of n number of eigenvectors associated respectively with A and its transpose A^T can be so selected that they are biorthogonal when the n-square matrix A has distinct eigenvalues. If P and Q are constructed respectively by these eigenvectors of A and A^T, then obviously $Q^T P = I$ and $Q^T AP = \text{diag}\ (\lambda_1, \lambda_2, ..., \lambda_n)$, where λ_i's are distinct eigenvalues of A. □

Example 7.11 Compute a biorthogonal system of left and right eigenvectors associated with the matrix A in Example 7.1.

Solution The square matrix A given in Example 7.1 has distinct eigenvalues 2 and –1, and the respective right eigenvectors $\mathbf{x}_1 = [2\ 1]^T$ and $\mathbf{x}_2 = [1\ –1]^T$. The left eigenvectors are obtained by solving

$$[y_1\ y_2] \begin{bmatrix} 1 & 2 \\ 1 & 0 \end{bmatrix} = \lambda_i [y_1\ y_2], \qquad i = 1, 2;$$

for $\lambda_1 = 2$ and $\lambda_2 = -1$. The solutions are, respectively,

$$\mathbf{y}_1 = \begin{bmatrix} \alpha \\ \alpha \end{bmatrix}$$

and

$$\mathbf{y}_2 = \begin{bmatrix} \beta \\ -2\beta \end{bmatrix},$$

where α and β are real numbers. Now (see Example 7.8),

$$P^{-1} = \frac{1}{3}\begin{bmatrix} 1 & 1 \\ 1 & -2 \end{bmatrix}.$$

So, if we select $\alpha = \beta = 1/3$, then the rows of P^{-1} can define the left eigenvectors of A which are $\mathbf{y}_1 = (1/3)\,[1\ \ 1]^T$ and $\mathbf{y}_2 = (1/3)\,[1, -2]^T$. It can be easily checked that the biorthogonality condition of Definition 7.3 is satisfied. □

When the matrix A is defined over a complex field $\mathbb{C}$, results analogous to Theorem 7.5 can be stated and proved. We shall now state the corresponding result, but pursue a slightly different method for the proof.

Theorem 7.7 Let λ_i and $\mathbf{x}_i$ $(i = 1, ..., n)$ be the distinct eigenvalues and the associated eigenvectors of an n-square matrix A over a complex field $\mathbb{C}$. Show that $\bar{\lambda}_i$'s are the eigenvalues of A^* and that the corresponding eigenvectors $\mathbf{y}_i$'s can be so chosen that they are biorthogonal to $\mathbf{x}_i$'s (see Definition 7.3).

Proof Since A has distinct eigenvalues $\lambda_1, ..., \lambda_n$, there exists a nonsingular matrix P such that

$$AP = PD, \tag{7.29}$$

where $D = \text{diag}\ (\lambda_1, ..., \lambda_n)$. Taking the transposed conjugate of both sides of (7.20) and with some matrix manipulation, we have

$$A^*(P^*)^{-1} = (P^*)^{-1}\ \text{diag}\ (\bar{\lambda}_1, ..., \bar{\lambda}_n). \tag{7.30}$$

Therefore, $\bar{\lambda}_i$'s are the eigenvalues of A^* with associated linearly independent eigenvectors $\mathbf{y}_i$'s which are the columns of $(P^*)^{-1}$, because from (7.30) we can write

$$A^*\mathbf{y}_j = \bar{\lambda}_j \mathbf{y}_j. \tag{7.31}$$

Designating the columns of P by $\mathbf{x}_i$'s we have, from (7.20), the relation

$$A\mathbf{x}_i = \lambda_i \mathbf{x}_i. \tag{7.32}$$

Taking the inner product with $\mathbf{x}_i$ of both sides of (7.31) and with $\mathbf{y}_j$ of both sides of (7.32), we get

$$\langle \mathbf{x}_i, A^*\mathbf{y}_j\rangle = \langle \mathbf{x}_i, \bar{\lambda}_j \mathbf{x}_j\rangle, \quad \text{or} \quad \langle A\mathbf{x}_i, \mathbf{y}_j\rangle = \lambda_j\langle \mathbf{x}_i, \mathbf{y}_j\rangle; \tag{7.33}$$

and

$$\langle A\mathbf{x}_i, \mathbf{y}_j\rangle = \langle \lambda_i\mathbf{x}_i, \mathbf{y}_j\rangle = \lambda_i\langle \mathbf{x}_i, \mathbf{y}_j\rangle, \tag{7.34}$$

where we have used (5.16), (5.18) and Definition 5.9. It follows now from (7.33) and (7.34) that

$$(\lambda_i - \lambda_j)\ \langle \mathbf{x}_i, \mathbf{y}_j\rangle = 0.$$

Hence, if $\lambda_i \neq \lambda_j$, then $\langle \mathbf{x}_i, \mathbf{y}_j\rangle = 0$ which implies that $\mathbf{x}_i$'s and $\mathbf{y}_i$'s are biorthogonal as $\mathbf{x}_i$'s are already selected by satisfying $\langle \mathbf{x}_i, \mathbf{y}_i\rangle = 1$, because $\mathbf{y}_i$'s are columns of $(P^*)^{-1}$.

Taking the tranjugate of both sides of (7.31), we get

$$\mathbf{y}_j^*\ A = \lambda_j \mathbf{y}_j^*$$

and in tune with the definition of the left eigenvector over a real field $\mathbb{R}$ we call $\mathbf{y}_j$ as the left eigenvector of A associated with λ_j. □

Following the proof of Theorem 7.7 we can now restate Theorem 7.7.

Theorem 7.8 If an n-square matrix A over a complex field $\mathbb{C}$ has distinct eigenvalues and P is constructed by the associated eigenvectors, then the eigenvectors of A^* which become eventually left eigenvectors of A can be so chosen that if Q is the matrix formed by them, then $Q^*P = I$ and $Q^*AP = \text{diag}\ (\lambda_1, \lambda_2, ..., \lambda_n)$, where λ_i's are the eigenvalues of A. In other words, any matrix over a complex field $\mathbb{C}$ having distinct eigenvalues can be characterized by a *biorthogonal system of left and right eigenvectors.* □

7.2.2 Diagonalizable Linear Transformation

We know that the set of all matrices associated with a single linear transformation L in respect of different bases are similar and therefore have the same eigenvalues, see Theorem 7.1. If the matrix A of L on V has distinct eigenvalues, it is sometimes important to find a basis $X = \{\mathbf{x}_1, \mathbf{x}_2, ..., \mathbf{x}_n\}$ in V, in terms of which the matrix of L has a diagonal form. With this background we can offer the following general definition.

Definition 7.4 *Diagonalizable Linear Transformation.* A linear transformation L on V is said to be diagonalizable if there exists a basis X in V with respect to which the matrix of L is a diagonal matrix diag $[\lambda_1, \lambda_2, ..., \lambda_n]$. □

The following theorem contains an artifice by virtue of which a basis can be selected to get a diagonal representation of a linear transformation L if its eigenvalues are distinct.

Theorem 7.9 A linear transformation L on V with distinct eigenvalues is diagonalizable.

Proof Let $A = [a_{ij}]$ be the matrix of L with respect to the basis $X = \{\mathbf{x}_1, \mathbf{x}_2, ..., \mathbf{x}_n\}$ in V so that

$$L\mathbf{x}_k = \sum_{j=1}^{n} a_{jk}\,\mathbf{x}_j, \qquad k = 1, 2, ..., n.$$

Since A has distinct eigenvalues $\lambda_1, \lambda_2, ..., \lambda_n$, we construct its modal matrix P (see Definition 7.1) such that by (7.20),

$$AP = P \text{ diag } [\lambda_1, \lambda_2, ..., \lambda_n].$$

Consequently, if

$$\mathbf{p}_i = [p_{1i}\ p_{2i}\ \cdots\ p_{ni}]^T$$

is the ith column of P, then $A\mathbf{p}_i = \lambda_i\mathbf{p}_i$, from which it follows that

$$\sum_{k=1}^{n} a_{jk}\, p_{ki} = \lambda_i p_{ji}, \qquad j = 1, 2, ..., n.$$

If we now define

$$\mathbf{y}_i = p_{1i}\mathbf{x}_1 + p_{2i}\mathbf{x}_2 + \cdots + p_{ni}\mathbf{x}_n,$$

then

$$L\mathbf{y}_i = \sum_{k=1}^{n} p_{ki} L\mathbf{x}_k = \sum_{k=1}^{n} p_{ki} \sum_{j=1}^{n} a_{jk}\,\mathbf{x}_j = \sum_{j=1}^{n} \Big(\sum_{k=1}^{n} a_{jk}\, p_{ki}\Big)\mathbf{x}_j$$

$$= \sum_{j=1}^{n} \lambda_i p_{ji}\,\mathbf{x}_j = \lambda_i \mathbf{y}_i.$$

This shows that with respect to the basis $\{\mathbf{y}_1, \mathbf{y}_2, ..., \mathbf{y}_n\}$, the matrix of L is diag $[\lambda_1, \lambda_2, ..., \lambda_n]$. □

It is easy to see that the vectors $\mathbf{y}_i$'s in the foregoing theorem are eigenvectors of L as can be inferred in the light of Section 7.1.1. Hence the following corollary easily follows.

Corollary 7.1 A linear transformation L on V is diagonalizable iff there exists a basis for V consisting of the eigenvectors of L. □

Example 7.12 Let us consider the linear transformation $L: \mathbb{P}_2(t) \to \mathbb{P}_2(t): (a_0 + a_1t + a_2t^2) \mapsto -(a_0 + 2a_1 + a_2) + (2a_0 + 3a_1)t - 2(a_0 + a_1)t^2$ as defined in Example 7.6. Find a basis in $\mathbb{P}_2(t)$, so that L can be represented by a diagonal matrix with respect to this basis.

Solution The matrices $\mathbf{p}$ and P as given in Example 7.6 are

$$\mathbf{p} = [1\ t\ t^2]^T, \qquad P = \begin{bmatrix} 1 & 2 & 1 \\ -1 & -1 & -2 \\ 0 & 2 & 1 \end{bmatrix}.$$

Hence, with respect to the basis defined by the columns of $\mathbf{p}^TP = [1 - t,\ 2 - t + 2t^2,\ 1 - 2t + t^2]$, L is represented by a diagonal matrix. To check, we have

$$L[1 - t,\ 2 - t + 2t^2,\ 1 - 2t + t^2] = [1 - t,\ 2 - t + 2t^2,\ 1 - 2t + t^2]\ \text{diag}\ [1, -1, 2]. \quad \square$$

7.3 MATRIX POLYNOMIAL AND LAMBDA MATRIX

A polynomial in the scalar variable x having coefficients which are scalars is known as a *scalar polynomial.* But there are polynomials where the variable x, instead of being a scalar is an n-square matrix A, but the coefficients remain scalars. Then, these polynomials are called *matrix polynomials*. On the other hand, if matrices take the place of the scalar coefficients in the original scalar polynomial, then we have a polynomial in the scalar variable x having coefficients which are matrices, and we call it a *polynomial matrix* or *lambda matrix*. These two different types of polynomials involving matrices and their operational properties are the subject-matter of this section.

7.3.1 Matrix Polynomials

Considering x to be a scalar variable, the mth degree polynomial in x denoted by $f(x)$ is the functional relation

$$f(x) = a_mx^m + a_{m-1}x^{m-1} + \cdots + a_1x + a_0, \tag{7.35}$$

where $a_m \neq 0$, m is a positive integer, and the coefficient of a_0 is x^0. We shall call $f(x)$ as a *monic polynomial* if the leading coefficient $a_m = 1$. When all the coefficients are real, we say that the polynomial $f(x)$ is over the real field $\mathbb{R}$ and when some are complex, $f(x)$ is said to be a polynomial over the complex field $\mathbb{C}$. In general, we say that $f(x)$ is a (scalar) polynomial over a field $\mathbb{F}$ in case the coefficients a_i's are either real or complex, which is not specified.

We can associate with the polynomial (7.35) a polynomial in a (square) matrix variable A of order n by replacing x by A and noting that $A^\circ = I$, and the integral powers of A are defined by $A^k = A^p{\cdot}A^q = A^q{\cdot}A^p$, $p + q = k$, where p, q and k are integers (see Section 1.4 and Problem 1.18; Property IV in Section 3.2 and Example 3.9). This gives

$$f(A) = a_mA^m + a_{m-1}A^{m-1} + \cdots + a_1A + a_0I \tag{7.36}$$

which is called a *matrix polynomial* or a polynomial in the matrix variable A over a field $\mathbb{F}$. Since a matrix A commutes with itself, i.e., $A^pA^q = A^qA^p$, where p and q are integers, we are allowed to multiply two polynomials in A precisely in the same way as two scalar polynomials.

Thus we have the following obvious properties of a matrix polynomial associated with scalar polynomials $f(x)$, $g(x)$, $h(x)$:

(a) If $f(x) = g(x)$, then $f(A) = g(A)$.
(b) If $f(x) + g(x) = h(x)$, then $f(A) + g(A) = h(A)$.
(c) If $f(x)\, g(x) = h(x)$, then $f(A)\, g(A) = g(A)f(A) = h(A)$.
(d) If for a polynomial $f(x)$, there exist polynomials $d(x)$, $q(x)$ and $r(x)$ such that $f(x) = d(x)q(x) + r(x)$, then $f(A) = d(A)q(A) + r(A)$, where $q(x)$ is the quotient, $r(x)$ the remainder on dividing $f(x)$ by $d(x)$ and the degree of $r(x)$ is less than that of $d(x)$. Similarly, $q(A)$ is the quotient, $r(A)$ the remainder on dividing $f(A)$ by $d(A)$. If $r(x) = 0$, $d(x)$ is called the divisor of $f(x)$; similarly $d(A)$ is the divisor of $f(A)$ if $r(A) = 0$.

Example 7.13 Let $f(x) = x^2 + 3x + 2 + a$, where a is a real number, $d(x) = x + 1$, and

$$A = \begin{bmatrix} 1 & 2 \\ 1 & 0 \end{bmatrix}$$

Find the quotient $q(A)$ and the remainder $r(A)$ on dividing $f(A)$ on dividing $f(A)$ by $d(A)$.

Solution Since, $f(x) = d(x)(x + 2) + a$; so $q(x) = x + 2$ is the quotient; $r(x) = a$, the remainder; deg $r(x) = 0 < 1 =$ deg $d(x)$, where deg $r(x)$ is the degree of the polynomial $r(x)$, etc. If $a = 0$ in $f(x)$, $r(x) = 0$, and we call $(x + 1)$ the divisor of $f(x) = x^2 + 3x + 2$. Now,

$$f(A) = A^2 + 3A + (2 + a)I = \begin{bmatrix} 3 & 2 \\ 1 & 2 \end{bmatrix} + \begin{bmatrix} 3 & 6 \\ 3 & 0 \end{bmatrix}$$

$$+ (2 + a)\begin{bmatrix} 1 & 0 \\ 0 & 1 \end{bmatrix} = \begin{bmatrix} 8+a & 8 \\ 4 & 4+a \end{bmatrix},$$

$$d(A) = A + I = \begin{bmatrix} 2 & 2 \\ 1 & 1 \end{bmatrix},$$

$$q(A) = A + 2I = \begin{bmatrix} 3 & 2 \\ 1 & 2 \end{bmatrix}, \quad r(A) = aI.$$

So,

$$d(A)q(A) = \begin{bmatrix} 8 & 8 \\ 4 & 4 \end{bmatrix}$$

and hence we find that $f(A) = d(A)q(A) + r(A)$. If $a = 0$, then $r(A) = 0$, a null matrix and $d(A)$ is called the divisor of $f(A)$ and $q(A)$ the quotient. □

7.3.2 Lambda Matrix or Polynomial Matrix

In a matrix polynomial (7.36), it might be observed that the coefficients a_m, ..., a_0 are scalars but the variable A is a matrix. There is another class of polynomials involving matrices where the coefficients are matrices while the variable is a scalar. These polynomials are referred to as *lambda* or *polynomial matrices*.

Definition 7.5 *Lambda Matrix or Polynomial Matrix:* An n-square matrix $A(\lambda)$ which is a polynomial in the scalar variable λ from a field $\mathbb{F}$ represented by

$$A(\lambda) = A_m\lambda^m + A_{m-1}\lambda^{m-1} + \cdots + A_1\lambda + A_0,$$

where the leading coefficient $A_m \neq 0$, A_i's are square matrices in $V_{n\times n}$, is defined as a *polynomial* or *a lambda matrix* or simply a λ-matrix of order n and of degree m over the field $\mathbb{F}$. If the determinant of the leading coefficient is not zero, i.e., if $\det A_m \neq 0$, then the lambda matrix is said to be *regular*. ☐

Matrix polynomials are denoted by lower case letters while polynomial or lambda matrices by capital letters.

Example 7.14 Let

$$A(\lambda) = \begin{bmatrix} 3\lambda^2+1 & 3\lambda^3+\lambda^2 \\ \lambda^3+2\lambda+2 & 2 \end{bmatrix}.$$

This matrix written as

$$A(\lambda) = \begin{bmatrix} 0 & 3 \\ 1 & 0 \end{bmatrix}\lambda^3 + \begin{bmatrix} 3 & 1 \\ 0 & 0 \end{bmatrix}\lambda^2 + \begin{bmatrix} 0 & 0 \\ 2 & 0 \end{bmatrix}\lambda + \begin{bmatrix} 1 & 0 \\ 2 & 2 \end{bmatrix}$$
$$= A_3\lambda^3 + A_2\lambda^2 + A_1\lambda + A_0$$

is a regular lambda matrix of order 2 and degree 3. It is regular since $\det A_3 = -3 \neq 0$. ☐

Since the elements of a lambda matrix are polynomials, a lambda matrix is also called a *polynomial matrix*, i.e., a matrix with polynomial elements.

7.3.3 Composition of Lambda Matrices

Definition 7.6 *Equality of Two Lambda Matrices.* Two λ-matrices are said to be equal if and only if the coefficients of like powers of λ are equal. ☐

Let two λ-matrices of the same order n be given by

$$A(\lambda) = A_m\lambda^m + A_{m-1}\lambda^{m-1} + \cdots + A_1\lambda + A_0, \tag{7.37}$$

$$B(\lambda) = B_p\lambda^p + B_{p-1}\lambda^{p-1} + \cdots + B_1\lambda + B_0, \tag{7.38}$$

which are of degree m and p, respectively.

Definition 7.7 *Addition.* The addition of two component λ-matrices of the same order but of degree m and p respectively, as given by (7.37) and (7.38), is defined as

$$A(\lambda) + B(\lambda) = A_m\lambda^m + \cdots + A_{p+1}\lambda^{p+1} + (A_p + B_p)\lambda^p + \cdots$$
$$+ (A_1 + B_1)\lambda + (A_0 + B_0),$$

where it is assumed that $m > p$. The case $p \geq m$ is treated similarly. Thus, the degree of the sum of two λ-matrices does not exceed the degree of either component. ☐

Definition 7.8 *Multiplication.* As the λ-matrix by definition is a square matrix, two λ-matrices can be multiplied if and only if they are of the same order. Therefore, the two n-square λ-matrices (7.37) and (7.38) on multiplication give

$$A(\lambda)B(\lambda) = A_mB_p\lambda^{m+p} + (A_mB_{p-1} + A_{m-1}B_p)\lambda^{m+p-1} + \cdots + A_0B_0.$$

We recall that in matrix multiplication it is possible that $A_m B_p = 0$ even though $A_m \neq 0$, $B_p \neq 0$ (see Example 1.4). But, if either of the λ-matrices is regular, i.e., if det A_m or det $B_p \neq 0$, then $A_m B_p \neq 0$ whenever $A_m \neq 0$, $B_p \neq 0$. With this observation we can say that the degree of the product of two component λ-matrices is less than or equal to the sum of the degrees of the components. □

Definition 7.9 *Subtraction.* If the negative of a λ-matrix $B(\lambda)$ given in (7.38) is defined as

$$-B(\lambda) = -B_p\lambda^p - B_{p-1}\lambda^{p-1} - \cdots - B_1\lambda - B_0,$$

then we define the subtraction of two λ-matrices $A(\lambda)$ and $B(\lambda)$ as the addition of $A(\lambda)$ to $-B(\lambda)$, i.e.,

$$\begin{aligned} A(\lambda) - B(\lambda) = A_m\lambda^m + \cdots + A_{p+1}\lambda^{p+1} + (A_p - B_p)\lambda^p + \cdots \\ + (A_1 - B_1)\lambda + (A_0 - B_0), \quad \text{if } m > p; \end{aligned}$$

and similarly if $m \leq p$. □

Definition 7.10: *Right Value and Left Value of a λ-Matrix.* If in a λ-matrix $A(\lambda)$ of order n we substitute a square matrix B of order n for λ, then we get a matrix in $V_{n\times n}$. As the matrix multiplication is not commutative, the nature of this substitution must be defined; otherwise there is an ambiguity in the result. For example, a λ-matrix of degree one can be written as

$$A(\lambda) = A_1\lambda + A_0 \quad \text{or} \quad A(\lambda) = \lambda A_1 + A_0. \tag{7.39}$$

Substituting the matrix B for λ in (7.39), we have

$$A_1B + A_0 \quad \text{or} \quad BA_1 + A_0$$

which are, in general, not equal because $A_1B \neq BA_1$. Therefore, writing the λ-matrix $A(\lambda)$ in the form

$$\begin{aligned} A(\lambda) &= A_m\lambda^m + A_{m-1}\lambda^{m-1} + \cdots + A_1\lambda + A_0 \\ &= \lambda^m A_m + \lambda^{m-1}A_{m-1} + \cdots + \lambda A_1 + A_0 \end{aligned}$$

we term,

$$A_r(B) = A_mB^m + A_{m-1}B^{m-1} + \cdots + A_1B + A_0$$

and

$$A_l(B) = B^mA_m + B^{m-1}A_{m-1} + \cdots + BA_1 + A_0,$$

respectively, as the *right value* and the *left value* of $A(\lambda)$ at B. □

Example 7.15 Let $A(\lambda)$ be given as in Example 7.14. Compute $A_r(B)$ and $A_l(B)$ if

$$B = \begin{bmatrix} 1 & 2 \\ 0 & 1 \end{bmatrix}.$$

Solution We have,

$$A_r(B) = \begin{bmatrix} 0 & 3 \\ 1 & 0 \end{bmatrix}\begin{bmatrix} 1 & 6 \\ 0 & 1 \end{bmatrix} + \begin{bmatrix} 3 & 1 \\ 0 & 0 \end{bmatrix}\begin{bmatrix} 1 & 4 \\ 0 & 1 \end{bmatrix} + \begin{bmatrix} 0 & 0 \\ 2 & 0 \end{bmatrix}\begin{bmatrix} 1 & 2 \\ 0 & 1 \end{bmatrix} + \begin{bmatrix} 1 & 0 \\ 2 & 2 \end{bmatrix}$$

$$= \begin{bmatrix} 0 & 3 \\ 1 & 6 \end{bmatrix} + \begin{bmatrix} 3 & 13 \\ 0 & 0 \end{bmatrix} + \begin{bmatrix} 0 & 0 \\ 2 & 4 \end{bmatrix} + \begin{bmatrix} 1 & 0 \\ 2 & 2 \end{bmatrix} = \begin{bmatrix} 4 & 16 \\ 5 & 12 \end{bmatrix},$$

$$A_l(B) = \begin{bmatrix} 1 & 6 \\ 0 & 1 \end{bmatrix}\begin{bmatrix} 0 & 3 \\ 1 & 0 \end{bmatrix} + \begin{bmatrix} 1 & 4 \\ 0 & 1 \end{bmatrix}\begin{bmatrix} 3 & 1 \\ 0 & 0 \end{bmatrix} + \begin{bmatrix} 1 & 2 \\ 0 & 1 \end{bmatrix}\begin{bmatrix} 0 & 0 \\ 2 & 0 \end{bmatrix} + \begin{bmatrix} 1 & 0 \\ 2 & 2 \end{bmatrix}$$

$$= \begin{bmatrix} 6 & 3 \\ 1 & 0 \end{bmatrix} + \begin{bmatrix} 3 & 1 \\ 0 & 0 \end{bmatrix} + \begin{bmatrix} 4 & 0 \\ 2 & 0 \end{bmatrix} + \begin{bmatrix} 1 & 0 \\ 2 & 2 \end{bmatrix} = \begin{bmatrix} 14 & 4 \\ 5 & 2 \end{bmatrix}. \qquad \square$$

If $A(\lambda)$ and $B(\lambda)$ are two λ-matrices of the same order, then

(a) $A(\lambda) + B(\lambda) = R(\lambda)$ implies $A_r(C) + B_r(C) = R_r(C)$, $A_l(C) + B_l(C) = R_l(C)$, but

(b) $A(\lambda)B(\lambda) = R(\lambda)$, however, *does not imply* $A_r(C)B_r(C) = R_r(C)$, or $A_l(C)B_l(C) = R_l(C)$ where C is a square matrix of the same order as A and B.

As for (b), we note that for first degree λ-matrices $A(\lambda)$ and $B(\lambda)$, we have

$$R(\lambda) = (A_1\lambda + A_0)\ (B_1\lambda + B_0) = A_1B_1\lambda^2 + (A_1B_0 + A_0B_1)\lambda + A_0B_0,$$

$$A_r(C)B_r(C) = (A_1C + A_0)(B_1C + B_0) = A_1CB_1C + A_0B_1C + A_1CB_0 + A_0B_0.$$

Hence, it is evident that, in general, $R_r(C) \neq A_r(C)B_r(C)$ although $R(\lambda) = A(\lambda)B(\lambda)$.

Definition 7.11 *Right and Left Division of Lambda Matrices.* Let $A(\lambda)$, $B(\lambda)$, $Q(\lambda)$ and $R(\lambda)$ be λ-matrices of the same order n, with $A(\lambda)$ of degree m, $B(\lambda)$ regular of degree p less than or equal to m, $Q(\lambda)$ of degree $(m - p)$ and $R(\lambda) \equiv 0$ or of degree less than m. Then $Q(\lambda)$ is called *right quotient*, $R(\lambda)$ the *right remainder* of $A(\lambda)$ on dividing by $B(\lambda)$ if

$$A(\lambda) = Q(\lambda)B(\lambda) + R(\lambda) \tag{7.40}$$

and $B(\lambda)$ is called the *right divisor* of $A(\lambda)$ if $R(\lambda) \equiv 0$.

Also, $\hat{Q}(\lambda)$ is called the *left quotient*, $\hat{R}(\lambda)$ the *left remainder* of $A(\lambda)$ if

$$A(\lambda) = B(\lambda)\hat{Q}(\lambda) + \hat{R}(\lambda)$$

and $B(\lambda)$ the *left divisor* of $A(\lambda)$ if $\hat{R}(\lambda) \equiv 0$. $\square$

Theorem 7.10 Let two λ-matrices $A(\lambda)$ and $B(\lambda)$ of the same order with respective degrees m and p, $m \geq p$, be given by (7.37) and (7.38) with $B(\lambda)$ regular. Then the right quotient $Q(\lambda)$ and the right remainder $R(\lambda)$ of $A(\lambda)$ on dividing by $B(\lambda)$ (see (7.40)) exist and are unique. Similar assertion holds for the left quotient and the left remainder.

Proof To prove the existence, we shall proceed as we do in the case of long division of one polynomial by another of lower or equal degree. So we premultiply both sides by (7.38) by $A_mB_p^{-1}\lambda^{m-p}$ to get

$$A_mB_p^{-1}\lambda^{m-p}B(\lambda) = A_m\lambda^m + (A_mB_p^{-1}B_{p-1})\lambda^{m-1} + \cdots + A_mB_p^{-1}B_0\lambda^{m-p}. \tag{7.41}$$

Subtracting (7.41) from (7.37), we get

$$A(\lambda) - A_mB_p^{-1}\lambda^{m-p}B(\lambda) = (A_{m-1} - A_mB_p^{-1}B_{p-1})\lambda^{m-1} + \cdots \tag{7.42}$$

Since the first term on the right-hand side of (7.42) may be identically zero, we get from (7.42),

$$A(\lambda) - A_mB_p^{-1}\lambda^{m-p}B(\lambda) = A^{(1)}(\lambda), \tag{7.43}$$

where $A^{(1)}(\lambda)$ is an n-square λ-matrix with degree $m_1 < m$ and is represented by

$$A^{(1)}(\lambda) = A_{m_1}^{(1)}\lambda^{m_1} + \cdots + A_0^{(1)}.$$

If $m_1 \geq p$, we repeat the above process on $A^{(1)}(\lambda)$ rather than on $A(\lambda)$ to get

$$A^{(1)}(\lambda) - A_{m_1}^{(1)}B_p^{-1}\lambda^{m_1-p}B(\lambda) = A^{(2)}(\lambda), \tag{7.44}$$

where

$$A^{(2)}(\lambda) = A_{m_2}^{(2)}\lambda^{m_2} + \cdots + A_0^{(2)}$$

is again an n-square λ-matrix with degree $m_2 < m_1$. Since the degrees of $A(\lambda)$, $A^{(1)}(\lambda)$, $A^{(2)}(\lambda)$ are decreasing progressively, we finally arrive after a finite number of $(r + 1)$ steps at a polynomial $A^{(r+1)}(\lambda)$ with degree $m_{r+1} < p$ satisfying

$$A^{(r)}(\lambda) - A_{m_r}^{(r)}B_p^{-1}\lambda^{m_r-p}B(\lambda) = A^{(r+1)}(\lambda), \tag{7.45}$$

where $A_{m_r}^{(r)}$ is the leading coefficient of $A^{(r)}(\lambda)$ with degree $m_r \geq p$. Adding (7.43) to (7.45), we get

$$A(\lambda) - \{[A_m\lambda^m + A_{m_1}^{(1)}\lambda^{m_1} + \cdots + A_{m_r}^{(r)}\lambda^{m_r}]\lambda^{-p}B_p^{-1}\}B(\lambda) = A^{(r+1)}(\lambda),$$

where it is not difficult to identify the expression within the braces with the right quotient $Q(\lambda)$ and $A^{(r+1)}(\lambda)$ with the right remainder $R(\lambda)$ in (7.40).

To prove uniqueness, we assume that $Q_1(\lambda)$ is another right quotient and $R_1(\lambda)$ another right remainder so that

$$A(\lambda) = Q_1(\lambda)B(\lambda) + R_1(\lambda). \tag{7.46}$$

Subtracting (7.46) from (7.40), we obtain

$$[Q(\lambda) - Q_1(\lambda)]B(\lambda) = R_1(\lambda) - R(\lambda) \tag{7.47}$$

where the degree of λ-matrix on the right-hand side is less than that of $B(\lambda)$. So, if $Q(\lambda) \neq Q_1(\lambda)$, we arrive at a contradiction in view of Definition 7.6 regarding the equality of two λ-matrices in (7.47). Therefore, $Q(\lambda) = Q_1(\lambda)$ and, consequently, $R_1(\lambda) = R(\lambda)$.

A similar proof can be given for the left quotient and left remainder. □

The following example shows that, in general, the left quotient and right quotient, and the left remainder and right remainder are different.

Example 7.16 Determine (a) the right quotient and right remainder, and (b) the left quotient and left remainder of $A(\lambda)$ on dividing by $B(\lambda)$ when

$$A(\lambda) = \begin{bmatrix} 1 & 2 \\ 0 & 0 \end{bmatrix}\lambda^3 + \begin{bmatrix} -1 & 0 \\ 1 & 3 \end{bmatrix}\lambda^2 + \begin{bmatrix} 2 & 3 \\ -2 & 0 \end{bmatrix}\lambda + \begin{bmatrix} 0 & -1 \\ 1 & 2 \end{bmatrix}$$

$$= A_3\lambda^3 + A_2\lambda^2 + A_1\lambda + A_0, \tag{7.48}$$

$$B(\lambda) = \begin{bmatrix} 2 & 1 \\ 1 & 1 \end{bmatrix}\lambda + \begin{bmatrix} 1 & -1 \\ 0 & 2 \end{bmatrix} = B_1\lambda + B_0. \tag{7.49}$$

Solution (a) The right quotient and right remainder: By (7.41) we have

$$A_3 B_1^{-1} \lambda^{3-1} B(\lambda) = \lambda^2 \begin{bmatrix} 1 & 2 \\ 0 & 0 \end{bmatrix} \begin{bmatrix} 1 & -1 \\ -1 & 2 \end{bmatrix} B(\lambda) = \lambda^2 \begin{bmatrix} -1 & 3 \\ 0 & 0 \end{bmatrix} B(\lambda)$$

$$= \begin{bmatrix} 1 & 2 \\ 0 & 0 \end{bmatrix} \lambda^3 + \begin{bmatrix} -1 & 7 \\ 0 & 0 \end{bmatrix} \lambda^2. \tag{7.50}$$

Therefore, subtracting (7.50) from (7.48), we get

$$A(\lambda) - \begin{bmatrix} -1 & 3 \\ 0 & 0 \end{bmatrix} \lambda^2 B(\lambda) = \begin{bmatrix} 0 & -7 \\ 1 & 3 \end{bmatrix} \lambda^2 + \begin{bmatrix} 2 & 3 \\ -2 & 0 \end{bmatrix} \lambda + \begin{bmatrix} 0 & -1 \\ 1 & 2 \end{bmatrix}$$

$$=: A^{(1)}(\lambda). \tag{7.51}$$

Again applying the operation (7.41) to (7.49) and (7.51), we obtain

$$\begin{bmatrix} 0 & -7 \\ 1 & 3 \end{bmatrix} B_1^{-1} \lambda^{2-1} B(\lambda) = \lambda \begin{bmatrix} 0 & -7 \\ 1 & 3 \end{bmatrix} \begin{bmatrix} 1 & -1 \\ -1 & 2 \end{bmatrix} B(\lambda)$$

$$= \lambda \begin{bmatrix} 7 & -14 \\ -2 & 5 \end{bmatrix} B(\lambda) = \begin{bmatrix} 0 & -7 \\ 1 & 3 \end{bmatrix} \lambda^2 + \begin{bmatrix} 7 & -35 \\ -2 & 12 \end{bmatrix} \lambda. \tag{7.52}$$

Subtracting (7.52) from (7.51), we have

$$A^{(1)}(\lambda) - \lambda \begin{bmatrix} 7 & -14 \\ -2 & 5 \end{bmatrix} B(\lambda) = \begin{bmatrix} -5 & 38 \\ 0 & -12 \end{bmatrix} \lambda + \begin{bmatrix} 0 & -1 \\ 1 & 2 \end{bmatrix} =: A^{(2)}(\lambda). \tag{7.53}$$

Again applying operation (7.41) to (7.53) and (7.48), we get

$$\begin{bmatrix} -5 & 38 \\ 0 & -12 \end{bmatrix} B_1^{-1} B(\lambda) = \begin{bmatrix} -5 & 38 \\ 0 & -12 \end{bmatrix} \begin{bmatrix} 1 & -1 \\ -1 & 2 \end{bmatrix} B(\lambda)$$

$$= \begin{bmatrix} -43 & 81 \\ 12 & -24 \end{bmatrix} B(\lambda) = \begin{bmatrix} -5 & 38 \\ 0 & -12 \end{bmatrix} \lambda + \begin{bmatrix} -43 & 205 \\ 12 & -60 \end{bmatrix}. \tag{7.54}$$

Subtracting (7.54) from (7.53), we obtain

$$A^{(2)}(\lambda) - \begin{bmatrix} -43 & 81 \\ 12 & -24 \end{bmatrix} B(\lambda) = \begin{bmatrix} 43 & -206 \\ -11 & 62 \end{bmatrix}. \tag{7.55}$$

Adding (7.51), (7.53) and (7.55), we have

$$A(\lambda) - \begin{bmatrix} -1 & 3 \\ 0 & 0 \end{bmatrix} \lambda^2 B(\lambda) - \begin{bmatrix} 7 & 14 \\ -2 & 5 \end{bmatrix} \lambda B(\lambda) - \begin{bmatrix} -43 & 81 \\ 12 & -24 \end{bmatrix} B(\lambda)$$

$$= \begin{bmatrix} 43 & -206 \\ -11 & 62 \end{bmatrix}.$$

Hence,

$$Q(\lambda) = \begin{bmatrix} -1 & 3 \\ 0 & 0 \end{bmatrix}\lambda^2 + \begin{bmatrix} 7 & -14 \\ -2 & 5 \end{bmatrix}\lambda + \begin{bmatrix} -43 & 81 \\ 12 & -24 \end{bmatrix},$$

$$R(\lambda) = \begin{bmatrix} 43 & -206 \\ -11 & 62 \end{bmatrix}.$$

(b) The left quotient and left remainder. We have, from (7.48) and (7.49), the relation

$$B(\lambda)B_1^{-1}A_3\lambda^2 = B(\lambda)\begin{bmatrix} 1 & -1 \\ -1 & 2 \end{bmatrix}\begin{bmatrix} 1 & 2 \\ 0 & 0 \end{bmatrix}\lambda^2 = B(\lambda)\begin{bmatrix} 1 & 2 \\ -1 & -2 \end{bmatrix}\lambda^2$$

$$= \begin{bmatrix} 1 & 2 \\ 0 & 0 \end{bmatrix}\lambda^3 + \begin{bmatrix} 2 & 4 \\ -2 & -4 \end{bmatrix}\lambda^2 \tag{7.56}$$

Subtracting (7.56) from (7.48), we get

$$A(\lambda) - B(\lambda)\begin{bmatrix} 1 & 2 \\ -1 & -2 \end{bmatrix}\lambda^2 = \begin{bmatrix} -3 & -4 \\ 3 & 7 \end{bmatrix}\lambda^2 + \begin{bmatrix} 2 & 3 \\ -2 & 0 \end{bmatrix}\lambda + \begin{bmatrix} 0 & -1 \\ 1 & 2 \end{bmatrix}$$

$$=: A^{(1)}(\lambda). \tag{7.57}$$

Again from (7.49) and (7.57),

$$B(\lambda)B_1^{-1}\begin{bmatrix} -3 & -4 \\ 3 & 7 \end{bmatrix}\lambda = B(\lambda)\begin{bmatrix} 1 & -1 \\ -1 & 2 \end{bmatrix}\begin{bmatrix} -3 & -4 \\ 3 & 7 \end{bmatrix}\lambda = B(\lambda)\begin{bmatrix} -6 & -11 \\ 9 & 18 \end{bmatrix}\lambda$$

$$= \begin{bmatrix} -3 & -4 \\ 3 & 7 \end{bmatrix}\lambda^2 + \begin{bmatrix} -15 & -27 \\ 18 & 36 \end{bmatrix}\lambda \tag{7.58}$$

Therefore, subtracting (7.58) from (7.57), we obtain

$$A^{(1)}(\lambda) - B(\lambda)\begin{bmatrix} -6 & -11 \\ 9 & 18 \end{bmatrix}\lambda = \begin{bmatrix} 17 & 30 \\ -20 & -36 \end{bmatrix}\lambda + \begin{bmatrix} 0 & -1 \\ 1 & 2 \end{bmatrix} =: A^{(2)}(\lambda). \tag{7.59}$$

Again from (7.49) and (7.59),

$$B(\lambda)B_1^{-1}\begin{bmatrix} 17 & 30 \\ -20 & -36 \end{bmatrix} = B(\lambda)\begin{bmatrix} 1 & -1 \\ -1 & 2 \end{bmatrix}\begin{bmatrix} 17 & 30 \\ -20 & -36 \end{bmatrix}$$

$$= B(\lambda)\begin{bmatrix} 37 & 66 \\ -57 & -102 \end{bmatrix}$$

$$= \begin{bmatrix} 17 & 30 \\ -20 & -36 \end{bmatrix}\lambda + \begin{bmatrix} 94 & 168 \\ -114 & -204 \end{bmatrix}, \tag{7.60}$$

and subtracting (7.60) from (7.59), we have

$$A^{(2)}(\lambda) - B(\lambda)\begin{bmatrix} 37 & 66 \\ -57 & -102 \end{bmatrix} = \begin{bmatrix} -94 & -169 \\ 115 & 206 \end{bmatrix} \tag{7.61}$$

Finally, adding (7.57), (7.59) and (7.61), we get

$$A(\lambda) - B(\lambda)\begin{bmatrix} 1 & 2 \\ -1 & -2 \end{bmatrix}\lambda^2 - B(\lambda)\begin{bmatrix} -6 & -11 \\ 9 & 18 \end{bmatrix}\lambda - B(\lambda)\begin{bmatrix} 37 & 66 \\ -57 & -102 \end{bmatrix}$$

$$= \begin{bmatrix} -94 & -169 \\ 115 & 206 \end{bmatrix}$$

from which we obtain

$$Q(\lambda) = \begin{bmatrix} 1 & 2 \\ -1 & -2 \end{bmatrix}\lambda^2 + \begin{bmatrix} -6 & -11 \\ 9 & 18 \end{bmatrix}\lambda + \begin{bmatrix} 37 & 66 \\ -57 & -102 \end{bmatrix},$$

$$R(\lambda) = \begin{bmatrix} -94 & -169 \\ 115 & 206 \end{bmatrix}.$$

□

Theorem 7.11 *Remainder Theorem.* Suppose $A(\lambda)$, an n-square λ-matrix of degree m described by

$$A(\lambda) = A_m\lambda^m + A_{m-1}\lambda^{m-1} + \cdots + A_1\lambda + A_0 \tag{7.62}$$

and B, an n-square matrix, are given. Then there exist a unique n-square λ-matrix $Q(\lambda)$ of degree $m - 1$ represented by

$$Q(\lambda) = Q_{m-1}\lambda^{m-1} + Q_{m-2}\lambda^{m-2} + \cdots + Q_1\lambda + Q_0 \tag{7.63}$$

and an n-square matrix R such that

$$A(\lambda) = Q(\lambda)(\lambda I - B) + R, \tag{7.64}$$

where $R = A_r(B)$ is the right value of $A(\lambda)$ at B, or

$$A(\lambda) = (\lambda I - B)\hat{Q}(\lambda) + \hat{R},$$

where $\hat{R} = A_l(B)$ is the left value of $A(\lambda)$ at B and $\hat{Q}(\lambda)$ is an n-square λ-matrix of degree $(m - 1)$.

Proof Substituting in (7.64) the expressions for $A(\lambda)$ and $Q(\lambda)$ from (7.62) and (7.63) respectively, we get

$$A_m\lambda^m + A_{m-1}\lambda^{m-1} + \cdots + A_1\lambda + A_0$$

$$= (Q_{m-1}\lambda^{m-1} + Q_{m-2}\lambda^{m-2} + \cdots + Q_1\lambda + Q_0)(\lambda I - B) + R$$

in which equating coefficients of like powers of λ in view of Definition 7.6 for equality of two λ-matrices we have

$$\begin{aligned} A_m &= Q_{m-1}, \\ A_{m-1} &= -Q_{m-1}B + Q_{m-2}, \\ &\vdots \\ A_1 &= -Q_1 B + Q_0, \\ A_0 &= -Q_0 B + R. \end{aligned} \tag{7.65}$$

Hence, since A_i's and B are given, Q_i's can be determined uniquely from the first m equations. Postmultiplying equations (7.65) successively from below by I, B, B^2, ..., B^{m-1}, B^m and adding, we get

$$A_m B^m + A_{m-1} B^{m-1} + \cdots + A_1 B + A_0 = R. \tag{7.66}$$

Hence $R = A_r(B)$.

Similarly, the left quotient $\hat{Q}(\lambda)$ can be uniquely determined with $\hat{R} = A_l(B)$. □

Corollary 7.2 A λ-matrix $A(\lambda)$ is divisible on the right by $(\lambda I - B)$ if and only if $A_r(B) = 0$ and on the left if and only if $A_l(B) = 0$.

Proof The proof follows directly from Definition 7.11 and the Remainder Theorem 7.11. □

Example 7.17 Find the right quotient and right remainder of $A(\lambda)$ which is given by (7.48) on dividing by $(\lambda I - B)$ when

$$B = \begin{bmatrix} 1 & -1 \\ 0 & 2 \end{bmatrix}.$$

Solution Here, since $m = 3$, as in (7.63) we denote the right quotient by

$$Q(\lambda) = Q_2\lambda^2 + Q_1\lambda + Q_0.$$

Now from (7.65) we have, successively,

$$Q_2 = A_3 = \begin{bmatrix} 1 & 2 \\ 0 & 0 \end{bmatrix},$$

$$Q_1 = A_2 + Q_2 B = \begin{bmatrix} -1 & 0 \\ 1 & 3 \end{bmatrix} + \begin{bmatrix} 1 & 2 \\ 0 & 0 \end{bmatrix}\begin{bmatrix} 1 & -1 \\ 0 & 2 \end{bmatrix}$$

$$= \begin{bmatrix} -1 & 0 \\ 1 & 3 \end{bmatrix} + \begin{bmatrix} 1 & 3 \\ 0 & 0 \end{bmatrix} = \begin{bmatrix} 0 & 3 \\ 1 & 3 \end{bmatrix},$$

$$Q_0 = A_1 + Q_1 B = \begin{bmatrix} 2 & 3 \\ -2 & 0 \end{bmatrix} + \begin{bmatrix} 0 & 3 \\ 1 & 3 \end{bmatrix}\begin{bmatrix} 1 & -1 \\ 0 & 2 \end{bmatrix}$$

$$= \begin{bmatrix} 2 & 3 \\ -2 & 0 \end{bmatrix} + \begin{bmatrix} 0 & 6 \\ 1 & 5 \end{bmatrix} = \begin{bmatrix} 2 & 9 \\ -1 & 5 \end{bmatrix},$$

$$R = A_0 + Q_0 B = \begin{bmatrix} 0 & -1 \\ 1 & 2 \end{bmatrix} + \begin{bmatrix} 2 & 9 \\ -1 & 5 \end{bmatrix}\begin{bmatrix} 1 & -1 \\ 0 & 2 \end{bmatrix}$$

$$= \begin{bmatrix} 0 & -1 \\ 1 & 2 \end{bmatrix} + \begin{bmatrix} 2 & 16 \\ -1 & 11 \end{bmatrix} = \begin{bmatrix} 2 & 15 \\ 0 & 13 \end{bmatrix}. \tag{7.67}$$

R can also be obtained by using (7.66). Therefore, the right quotient is

$$Q(\lambda) = \begin{bmatrix} 1 & 2 \\ 0 & 0 \end{bmatrix} \lambda^2 + \begin{bmatrix} 0 & 3 \\ 1 & 3 \end{bmatrix} \lambda + \begin{bmatrix} 2 & 9 \\ -1 & 5 \end{bmatrix}$$

and the right remainder is given by (7.67). □

7.3.4 Operator Polynomial

We can associate with a polynomial $f(x)$ of the form (7.35) a polynomial in a variable which is now a linear transformation L on V. This newly defined polynomial now becomes a linear transformation $f(L)$ given by

$$f(L) = a_m L^m + a_{m-1} L^{m-1} + \cdots + a_1 L + a_0 I_V, \tag{7.68}$$

which we shall call an *operator polynomial.*

As in the case of matrices, we define $L^0 = I_V$ and integral powers of L by $L^k = L^p \cdot L^q = L^q \cdot L^p$, $p+q = k$, where p, q and k are integers (see Definition 6.9). It can be easily shown as in the matrix case that associated with three-scalar polynomials $f(x)$, $g(x)$, $h(x)$, we have

(a) if $f(x) = g(x)$, then $f(L) = g(L)$;
(b) if $f(x) + g(x) = h(x)$, then $f(L) + g(L) = h(L)$;
(c) if $f(x)\, g(x) = h(x)$, then $f(L) \cdot g(L) = h(L)$;
(d) if for a polynomial $f(x)$, there exist polynomials $d(x)$, $q(x)$ and $r(x)$ such that

$$f(x) = d(x)q(x) + r(x), \text{ then } f(L) = d(L) \cdot q(L) + r(L), \text{ where } \deg r(x) < \deg d(x).$$

Example 7.18 will familiarize the reader with manoeuvring an operator polynomial.

Example 7.18 Let $L(u; v) = (u + v, u - 2v)$ be a linear transformation on $\mathbb{R}^2$ and

$$f(x) = x^2 + 3, \qquad d(x) = x - 1 \tag{7.69}$$

be two given polynomials. Find the quotient $q(L)$ and the remainder $r(L)$ on dividing $f(L)$ by $d(L)$.

Solution Since, $f(x) = d(x)(x + 1) + 4$, so $q(x) = x + 1$ is the quotient and $r(x) = 4$ is the remainder. Now applying Definition 6.9, we have $L^2 = L \cdot L$, and so

$$L^2(u, v) = L \cdot L(u, v) = L(u + v, u - 2v) = (2u - v, -u + 5v).$$

Hence with the help of (6.12), we get

$$\begin{aligned} f(L)(u, v) &= (L^2 + 3I_V)(u, v) = L^2(u, v) + 3I_V(u, v) \\ &= (2u - v, -u + 5v) + 3(u, v) = (5u - v, -u + 8v), \qquad (7.70) \\ d(L)(u, v) &= (L - I_V)(u, v) = L(u, v) - I_V(u, v) \\ &= (u + v, u - 2v) - (u, v) = (v, u - 3v), \\ q(L)(u, v) &= (L + I_V)(u, v) = L(u, v) + I_V(u, v) \\ &= (u + v, u - 2v) + (u, v) = (2u + v, u - v), \\ r(L)(u, v) &= 4I_V(u, v) = 4\,(u, v). \end{aligned}$$

□

Let A be a matrix of a linear transformation L on V with respect to a basis X. Then, using the notion of matrix representation of (i) the sum of two linear transformations, (ii) scalar multiplication of a linear transformation, and (iii) the product of two linear transformations, as given in Sections 6.3.2 and 6.3.3, we have for the operator polynomial defined by (7.68) with respect to the basis X:

$$\text{matrix of } f(L) = \sum_{k=0}^{m} \text{matrix of } (a_k L^k) = \sum_{k=0}^{m} a_k \text{ matrix of } L^k$$

$$= \sum_{k=0}^{m} a_k A^k = f(A), \tag{7.71}$$

which shows that the matrix of $f(L)$ is $f(A)$ with respect to the basis X which is used to define A. □

Example 7.19 Determine the matrix representation of the linear transformation $L(u, v) = (u + v, u - 2v)$ on $\mathbb{R}^2$ with respect to the natural basis, and also of $f(L)$ when $f(x) = x^2 + 1$.

Solution The matrix A of linear transformation L with respect to basis $\{\mathbf{e}_1, \mathbf{e}_2\}$, which are now row-vectors in $\mathbb{R}^2$, can be found from the observation that

$$L\mathbf{e}_1 = (1, 1) = 1\cdot\mathbf{e}_1 + 1\cdot\mathbf{e}_2, \qquad L\mathbf{e}_2 = (1, -2) = 1\cdot\mathbf{e}_1 - 2\cdot\mathbf{e}_2.$$

Therefore,

$$A = \begin{bmatrix} 1 & 1 \\ 1 & -2 \end{bmatrix}. \tag{7.72}$$

Similarly, from (7.70) the matrix of $f(L)$ is

$$\begin{bmatrix} 5 & -1 \\ -1 & 8 \end{bmatrix}.$$

Now by direct computation, employing (7.72), we get

$$f(A) = A^2 + 3I = \begin{bmatrix} 2 & -1 \\ -1 & 5 \end{bmatrix} + \begin{bmatrix} 3 & 0 \\ 0 & 3 \end{bmatrix} = \begin{bmatrix} 5 & -1 \\ -1 & 8 \end{bmatrix}$$

which shows that the matrix of $f(L)$ is indeed $f(A)$. □

7.4 CHARACTERISTIC POLYNOMIAL, ANNIHILATING POLYNOMIAL AND MINIMUM POLYNOMIAL

The elementary facts which we have learnt in Section 7.3 about λ-matrices are sufficiently extensive to study in this section some aspects of a square matrix A. It is a matter of great interest to note that a characteristic matrix $(\lambda I - A)$ and its adjoint adj $(\lambda I - A)$ are both λ-matrices. If A is a matrix of order n, then adj $(\lambda I - A)$ is a regular λ-matrix of order n and degree $n - 1$.

Example 7.20 Express adj$(\lambda I - A)$ as a λ–matrix if

$$A = \begin{bmatrix} 3 & 0 & 1 \\ 2 & -1 & 3 \\ 1 & 4 & 3 \end{bmatrix}.$$

Solution Obviously

$$(\lambda I - A) = \begin{bmatrix} \lambda-3 & 0 & -1 \\ -2 & \lambda+1 & -3 \\ -1 & -4 & \lambda-3 \end{bmatrix}.$$

By direct computation, employing formula (3.1), we have

$$\text{adj } (\lambda I - A) = \begin{bmatrix} \lambda^2-2\lambda-15 & 4 & \lambda+1 \\ 2\lambda-3 & \lambda^2-6\lambda+8 & 3\lambda-7 \\ \lambda+9 & 4(\lambda-3) & \lambda^2-2\lambda-3 \end{bmatrix}$$

$$= \begin{bmatrix} 1 & 0 & 0 \\ 0 & 1 & 0 \\ 0 & 0 & 1 \end{bmatrix}\lambda^2 + \begin{bmatrix} -2 & 0 & 1 \\ 2 & -6 & 3 \\ 1 & 4 & -2 \end{bmatrix}\lambda + \begin{bmatrix} -15 & 4 & 1 \\ -3 & 8 & -7 \\ 9 & -12 & -3 \end{bmatrix}.$$

So adj $(\lambda I - A)$ associated with a third order matrix A is a regular λ-matrix of the same order, but of degree 2. □

As corroborated by the foregoing example, in general, for a matrix A of order n, adj $(\lambda I - A)$ is a regular λ-matrix of the same order n but of degree $n - 1$ and can be represented by

$$B(\lambda) := \text{adj } (\lambda I - A) = B_{n-1}\lambda^{n-1} + B_{n-2}\lambda^{n-2} + \cdots + B_1\lambda + B_0, \tag{7.73}$$

where B_i's ($i = 0, 1, \ldots, n-1$) are matrices of order n with $B_{n-1} = I$.

A characteristic polynomial $\phi(\lambda)$ associated with a square matrix A is already defined as $\phi(\lambda) := \det(\lambda I - A)$, see (7.8), and can be represented by expanding $\det(\lambda I - A)$, see Definition 2.1, in the form

$$\phi(\lambda) = a_n\lambda^n + a_{n-1}\lambda^{n-1} + \cdots + a_1\lambda + a_0 \tag{7.74}$$

which is a polynomial in λ of degree n. The matrix polynomial (see (7.36) and Section 7.3.1) associated with the characteristic polynomial $\phi(\lambda)$ is

$$\phi(A) = a_nA^n + a_{n-1}A^{n-1} + \cdots + a_1A + a_0I, \tag{7.75}$$

which is of order n and is referred to as the *characteristic matrix polynomial* of A.

An important property of the characteristic equation is contained in the Cayley-Hamilton theorem which is in its own right the most important theorem in the theory of matrices and determinants.

Theorem 7.12 *Cayley-Hamilton Theorem.* A square matrix A satisfies its own characteristic equation $\phi(\lambda) = 0$, i.e., for a square matrix A, $\phi(A) = 0$, a null matrix.

Proof We shall give two different proofs of this theorem, the first one employs the Remainder Theorem 7.11.

(a) In view of (3.2) we have

$$\det(\lambda I - A)I = \text{adj } (\lambda I - A)\,(\lambda I - A) = (\lambda I - A)\,\text{adj } (\lambda I - A), \tag{7.76}$$

which shows that the λ-matrix det $(\lambda I - A)I$ is divisible on the right as well as on the left by $(\lambda I - A)$. Since det $(\lambda I - A) = \phi(\lambda)$ is given by (7.74), we have, applying Corollary 7.2, the relation

$$(a_n\lambda^n + a_{n-1}\lambda^{n-1} + \cdots + a_1\lambda + a_0)I = 0 \text{ at } A,$$

or

$$a_nA^n + a_{n-1}A^{n-1} + \cdots + a_1A + a_0I = 0,$$

which by (7.75) implies that $\phi(A) = 0$.

(b) We note that $\phi(\lambda) := \det(\lambda I - A)$. Now, substituting from (7.73) and (7.74) into (7.76), we obtain

$$(a_n\lambda^n + a_{n-1}\lambda^{n-1} + \cdots + a_1\lambda + a_0)I = (\lambda I - A)(B_{n-1}\lambda^{n-1} + B_{n-2}\lambda^{n-2} + \cdots + B_1\lambda + B_0).$$

Hence, equating coefficients of like powers of λ on both sides, we get the following recursive relations for B_i's:

$$\begin{aligned}
\lambda^n &: a_nI = B_{n-1},\\
\lambda^{n-1} &: a_{n-1}I = B_{n-2} - AB_{n-1},\\
\lambda^{n-2} &: a_{n-2}I = B_{n-3} - AB_{n-2},\\
&\ \ \vdots\\
\lambda^1 &: a_1I = B_0 - AB_1,\\
\lambda^0 &: a_0I = -AB_0.
\end{aligned} \tag{7.77}$$

Premultiplying both sides of equations (7.77) successively by $A^n, A^{n-1}, A^{n-2}, ..., A, A^0 = I$, and adding the results, we get

$$a_nA^n + a_{n-1}A^{n-1} + \cdots + a_1A + a_0I = 0. \quad \square$$

Example 7.21 Show that the matrix A given in Example 7.1 follows the Cayley-Hamilton Theorem

Solution The characteristic polynomial of A is $\phi(\lambda) = \lambda^2 - \lambda - 2$. Now,

$$A^2 - A - 2I = \begin{bmatrix} 1 & 2 \\ 1 & 0 \end{bmatrix}^2 - \begin{bmatrix} 1 & 2 \\ 1 & 0 \end{bmatrix} - 2\begin{bmatrix} 1 & 0 \\ 0 & 1 \end{bmatrix}$$

$$= \begin{bmatrix} 3 & 2 \\ 1 & 2 \end{bmatrix} - \begin{bmatrix} 1 & 2 \\ 1 & 0 \end{bmatrix} - \begin{bmatrix} 2 & 0 \\ 0 & 2 \end{bmatrix} = \begin{bmatrix} 0 & 0 \\ 0 & 0 \end{bmatrix} = 0,$$

which shows that $\phi(A)$ is a null matrix, i.e. $\phi(A) = 0$. $\square$

Definition 7.12 *Greatest Common Divisor (GCD).* The polynomial $d(x)$ over a field $\mathbb{F}$ is said to be a common divisor of a set of polynomials $f_1(x), f_2(x), ..., f_n(x)$ over the same field $\mathbb{F}$ if $d(x)$ is a divisor of each of them. A *monic polynomial* $g(x)$ is said to be the *greatest common divisor* (GCD) of the polynomials $f_i(x)$'s if every common divisor of $f_i(x)$'s is a divisor of $g(x)$. If the GCD is one, then $f_i(x)$'s are said to be relatively prime or co-prime over the field $\mathbb{F}$. $\square$

Example 7.22 Let $f_1(x) = 2(x^2 - 1)(x + 1)$ and $f_2(x) = 4(x^2 - 1)(x - 1)$. Then $d_1(x) = 2(x + 1)$ and $d_2(x) = 2(x - 1)$ are common divisors of $f_1(x)$ and $f_2(x)$, but $d(x) = (x^2 - 1)$ is the greatest common divisor. $\square$

Example 7.23 The polynomials $x^2 + 1$ and $x^2 - 1$ are co-prime over the field $\mathbb{R}$ of real numbers. □

Definition 7.13 *Irreducible Polynomial.* Let $f(x)$ be a polynomial over a field $\mathbb{F}$, and let the only monic polynomial over $\mathbb{F}$ which divides $f(x)$ be one. Then $f(x)$ is said to be irreducible over $\mathbb{F}$. □

Example 7.24 The polynomial $f(x) = x^2 + 1$ is irreducible over $\mathbb{R}$. But over $\mathbb{C}$, the divisors of $f(x)$ are $x + i$ and $x - i$, and hence $f(x)$ is not irreducible over $\mathbb{C}$. □

Definition 7.14 *Annihilating Polynomial.* A polynomial $f(\lambda)$ (not identically zero) over a field $\mathbb{F}$ is said to be an annihilating polynomial of a square matrix A if $f(A) = 0$, a null matrix. □

It is evident that the characteristic polynomial $\phi(\lambda)$ of A is an annihilating polynomial of degree n since $\phi(A) = 0$. Consequently, it is not difficult to construct annihilating polynomials of degree higher than n by multiplying $\phi(A)$ by various powers of A. In application it is important to know how to find out an annihilating polynomial of the least degree. If this least degree is m, then obviously $0 < m \le n$.

Definition 7.15 *Minimum Polynomial.* A monic polynomial which is an annihilating polynomial of the least degree is said to be a minimum polynomial of the matrix A and is denoted by $\psi(\lambda)$. □

Theorem 7.13 Every annihilating polynomial of a square matrix A is divisible by its minimum polynomial.

Proof Let $r(\lambda)$ be the remainder and $q(\lambda)$ the quotient on dividing an annihilating polynomial $f(\lambda)$ by the minimum polynomial $\psi(\lambda)$ associated with a square matrix A. The degree of $r(\lambda)$ is lower than that of $\psi(\lambda)$. Therefore,

$$f(\lambda) = q(\lambda)\psi(\lambda) + r(\lambda). \tag{7.78}$$

Going over to the matrix polynomial (see section 7.3.1) in (7.78), we replace λ by A, and noting that $f(A) = 0$ and $\psi(A) = 0$, we have $r(A) = 0$. Since $r(\lambda)$ is of lower degree than $\psi(\lambda)$, the fact that $r(A) = 0$ contradicts that $\psi(\lambda)$ is a minimum polynomial. Consequently, $r(\lambda) = 0$, and from (7.78) we have, $f(\lambda) = q(\lambda)\psi(\lambda)$, showing that $\psi(\lambda)$ divides the annihilating polynomial $f(\lambda)$. □

The minimum polynomial of any square matrix A is unique. Suppose, on the contrary, that $\psi_1(\lambda)$ and $\psi_2(\lambda)$ are two minimum polynomials, obviously of the same degree. By Theorem 7.13, $\psi_1(\lambda)$ divides $\psi_2(\lambda)$ and $\psi_2(\lambda)$ divides $\psi_1(\lambda)$. Since $\psi_1(\lambda)$ and $\psi_2(\lambda)$ are monic, $\psi_1(\lambda) = \psi_2(\lambda)$. The next theorem gives an important relation by means of which the minimum polynomial of any square matrix can be calculated.

Theorem 7.14 Let $d_{n-1}(\lambda)$ be the greatest common divisor of all minors of order $(n - 1)$ in the matrix $(\lambda I - A)$ of order n, and let $\phi(\lambda)$ be the characteristic polynomial of A. Then the minimum polynomial $\psi(\lambda)$ of A is given by

$$\psi(\lambda) = \frac{\phi(\lambda)}{d_{n-1}(\lambda)} \tag{7.79}$$

and

$$\deg \psi(\lambda) \le \deg \phi(\lambda).$$

Proof We note that $\phi(\lambda) := \det(\lambda I - A)$ can be expressed as the sum of elements in the first row of $(\lambda I - A)$ multiplied by their corresponding cofactors. So $d_{n-1}(\lambda)$ is a divisor of $\phi(\lambda)$ in (7.79) and $\psi(\lambda)$ is a polynomial. Because $\phi(\lambda)$ and $d_{n-1}(\lambda)$ are monic, $\psi(\lambda)$ too is monic. Since $d_{n-1}(\lambda)$ is also a divisor of all the elements in adj $(\lambda I - A)$, we write

$$d_{n-1}(\lambda)R(\lambda I, A) = \text{adj}\,(\lambda I - A) \tag{7.80}$$

where the GCD of all the elements in n-square λ-matrix $R(\lambda I, A)$ is one. Now

$$(\lambda I - A)\,\text{adj}\,(\lambda I - A) = \phi(\lambda)\, I \tag{7.81}$$

which on dividing by $d_{n-1}(\lambda)$ becomes, in view of (7.79) and (7.81),

$$(\lambda I - A)R(\lambda I, A) = \psi(\lambda)I \tag{7.82}$$

By remainder Theorem 7.11, from (7.82) we have $\psi(A) = 0$, and $\psi(\lambda)$ is therefore an annihilating polynomial of A.

We now prove, by contradiction, that $\psi(\lambda)$ is a minimum polynomial. Suppose that $p(\lambda)$ is a polynomial of lower degree than $\psi(\lambda)$ such that $p(A) = 0$ and $p(\lambda)q(\lambda) = \psi(\lambda)$ for some polynomial $q(\lambda)$. By Remainder Theorem 7.11, there exists a unique λ-matrix $Q(\lambda)$ for which

$$(\lambda I - A)Q(\lambda) = p(\lambda)I,$$

or

$$(\lambda I - A)q(\lambda)Q(\lambda) = \psi(\lambda)I \tag{7.83}$$

which when compared with (7.82) gives $R(\lambda I, A) = q(\lambda)Q(\lambda)$, because of the Remainder Theorem, the quotient of $\psi(\lambda)I$ on dividing by $(\lambda I - A)$ must be unique. This contradicts the fact that the GCD of all the elements in $R(\lambda I, A)$ is one. □

Example 7.25 Find the minimum polynomial of A when

$$A = \begin{bmatrix} 1 & 0 & 1 \\ 0 & 1 & 0 \\ 0 & 0 & -1 \end{bmatrix}.$$

Solution One can easily compute

$$\lambda I - A = \begin{bmatrix} \lambda-1 & 0 & -1 \\ 0 & \lambda-1 & 0 \\ 0 & 0 & \lambda+1 \end{bmatrix},$$

$$\text{adj}\,(\lambda I - A) = \begin{bmatrix} (\lambda^2-1) & 0 & (\lambda-1) \\ 0 & \lambda^2-1 & 0 \\ 0 & 0 & (\lambda-1)^2 \end{bmatrix} \tag{7.84}$$

The GCD of all minors of the order 2 in $(\lambda I - A)$ is the same as the GCD of cofactors of all elements in $(\lambda I - A)$. This GCD is $d_2(\lambda) = (\lambda - 1)$ from adj $(\lambda I - A)$. Also, the characteristic polynomial of A is $\phi(\lambda) = (\lambda - 1)(\lambda^2 - 1)$. Hence by Theorem 7.14, the minimum polynomial $\psi(\lambda) = \lambda^2 - 1$. One can also check to see that here $A^2 - I = 0$. □

Definition 7.16 *Resolvent Matrix $R(\lambda, A)$.* The inverse of the matrix $(\lambda I - A)$, when λ is not an eigenvalue of A, is called the resolvent of the square matrix A. Symbolically we write, $R(\lambda, A) = (\lambda I - A)^{-1}$. □

Definition 7.17 *Nonderogatory and Derogatory Matrices.* A square matrix is referred to as *nonderogatory* if its characteristic and minimum polynomials are identical. Otherwise, it is called *derogatory.* □

It is not difficult to show that for any matrix A, the distinct linear factors of the minimum polynomial coincide with those of the characteristic polynomial (see Problem 7.32).

Example 7.26 Prove that the resolvent of a square matrix A can be represented by

$$(\lambda I - A)^{-1} = \{(\lambda^{n-1} + a_{n-1}\lambda^{n-2} + \cdots + a_1)I + (\lambda^{n-2} + a_{n-1}\lambda^{n-3} + \cdots + a_2)A + \cdots + (\lambda + a_{n-1})A^{n-2} + A^{n-1}\}/\phi(\lambda) \tag{7.85}$$

$$= \{(A^{n-1} + a_{n-1}A^{n-2} + \cdots + a_1 I) + (A^{n-2} + a_{n-1}A^{n-3} + \cdots + a_2 I)\lambda + \cdots + (A + a_{n-1}I)\lambda^{n-2} + I\lambda^{n-1}\}/\phi(\lambda), \tag{7.86}$$

where $a_1, ..., a_{n-1}$ are the coefficients in the characteristic polynomial $\phi(\lambda)$ of A given in (7.74).

Solution The expressions (7.85) and (7.86) can be obtained from each other by rearrangement. Alternatively, we can replace λI by A and A by λI in (7.85) to get (7.86) and vice versa.

Now $\phi(\lambda) := \det(\lambda I - A)$ and $B(\lambda) := \operatorname{adj}(\lambda I - A)$ are represented respectively by (7.74) and (7.73), and we note from (7.8) that a_n = coefficient of λ^n in $\phi(\lambda) = 1$. B_i's are now determined with the recursive relations (7.77), giving thereby

$$\begin{aligned} B_{n-1} &= I, \\ B_{n-2} &= A + a_{n-1}I, \\ B_{n-3} &= A(A + a_{n-1}I) + a_{n-2}I = A^2 + a_{n-1}A + a_{n-2}I, \\ &\vdots \\ B_k &= A^{n-k-1} + a_{n-1}A^{n-k-2} + \cdots + a_{k+1}I, \end{aligned} \tag{7.87}$$

where $k = 0, 1, ..., n - 1$. Substituting these expressions for B_i's in (7.73) and observing that $(\lambda I - A)^{-1} = B(\lambda)/\phi(\lambda)$, we get (7.86). □

Definition 7.18 *Reduced Adjoint Matrix $R(\lambda I, A)$.* The λ-matrix $R(\lambda I, A)$ defined by (7.80): $d_{n-1}(\lambda)R(\lambda I, A) = \operatorname{adj}(\lambda I - A)$, where $d_{n-1}(\lambda)$ is the GCD of all elements in $\operatorname{adj}(\lambda I - A)$, is called the reduced adjoint matrix of $(\lambda I - A)$. □

By definition, the GCD of all the elements in the reduced adjoint matrix $R(\lambda I, A)$ is one. We can express $R(\lambda I, A)$ in terms of the minimum polynomial $\psi(\lambda)$ of A by (7.82): $(\lambda I - A)R(\lambda I, A) = \psi(\lambda)I$. By the Cayley-Hamilton theorem, the characteristic polynomial $\phi(\lambda)$ is an annihilating polynomial and by Theorem (7.13), $\psi(\lambda)$ divides $\phi(\lambda)$. So the roots of $\psi(\lambda) = 0$ are also eigenvalues of A. If μ is a root of $\psi(\lambda) = 0$, from (7.82) we have

$$(\mu I - A)R(\mu I, A) = 0. \tag{7.88}$$

Now, $R(\mu I, A)$ is identically zero would mean that all the elements of $R(\lambda I, A)$ are divisible by $(\lambda - \mu)$, which is evidently not possible because the GCD of all elements in $R(\lambda I, A)$ is one. Consequently, $R(\mu I, A)$ has nonzero columns, one of which is denoted by $\mathbf{x}$. Then, we have from (7.88) the equation

$$A\mathbf{x} = \mu\mathbf{x}$$

which demonstrates that every nonzero column in $R(\mu I, A)$ is an eigenvector of A associated with the eigenvalue μ.

Example 7.27(a) For the matrix A given in Example 7.25, $\psi(\lambda) = \lambda^2 - 1$, and from (7.84) the reduced adjoint matrix $R(\lambda I, A)$ of $(\lambda I - A)$ is obtained as

$$R(\lambda I, A) = \begin{bmatrix} +1 & 0 & 1 \\ 0 & +1 & 0 \\ 0 & 0 & -1 \end{bmatrix}.$$

Now,

$$R(I, A) = \begin{bmatrix} 2 & 0 & 1 \\ 0 & 2 & 0 \\ 0 & 0 & 0 \end{bmatrix}, \qquad R(-I, A) = \begin{bmatrix} 0 & 0 & 1 \\ 0 & 0 & 0 \\ 0 & 0 & -2 \end{bmatrix}.$$

Hence, it is not difficult to see that $[1\ 0\ 0]^T$ and $[0\ 1\ 0]^T$ are two independent eigenvectors of A associated with the eigenvalue $\lambda = 1$, and $[1\ 0\ -2]^T$ is the eigenvector associated with the eigenvalue $\lambda = -1$. □

Example 7.27(b) If the minimum polynomial

$$\psi(\lambda) = \lambda^m + a_{m-1}\lambda^{m-1} + \cdots + a_1\lambda + a_0$$

of A has degree m, prove that the matrix polynomial $f(A)$ associated with

$$f(\lambda) = b_n\lambda^n + b_{n-1}\lambda^{n-1} + \cdots + b_1\lambda + b_0,$$

where $b_n \neq 0$, $n \geq m$, can be represented by one of degree less than m. Therefore, determine $f(A)$ if

$$f(\lambda) = 2\lambda^5 + 5\lambda^3 + 7\lambda, \qquad A = \begin{bmatrix} 1 & 0 & 1 \\ 0 & 1 & 0 \\ 0 & 0 & -1 \end{bmatrix}.$$

Solution Let $r(\lambda)$ be the remainder and $q(\lambda)$ the quotient on dividing $f(\lambda)$ by $\psi(\lambda)$. Then

$$f(\lambda) = q(\lambda)\psi(\lambda) + r(\lambda),$$

where $\deg r(\lambda) < \deg \psi(\lambda) = m$. Since $\psi(\lambda)$ is a minimum polynomial of A, $\psi(A) = 0$, and hence

$$f(A) = q(A)\psi(A) + r(A) = r(A)$$

in which $\deg r(A) < m$.

The minimum polynomial of A as shown in Example 7.25 is $\psi(\lambda) = \lambda^2 - 1$. Therefore,

$$f(\lambda) = 2\lambda^5 + 5\lambda^3 + 7\lambda = (2\lambda^3 + 7\lambda)(\lambda^2 - 1) + 14\lambda.$$

Since $\psi(A) = A^2 - I = 0$, it follows that

$$f(A) = 14A = 14\begin{bmatrix} 1 & 0 & 1 \\ 0 & 1 & 0 \\ 0 & 0 & -1 \end{bmatrix}.$$

□

7.4.1 Cayley-Hamilton Theorem and Minimum Polynomial for a Linear Transformation

It might be recalled that the characteristic polynomial of a linear transformation L on V is defined by det $(\lambda I_V - L)$: = det $(\lambda I - A) = \phi(\lambda)$, where A is any matrix of L and the characteristic equation by det $(\lambda I_V - L) = 0$. With this observation, by a straightforward argument, we can easily prove the Cayley-Hamilton theorem for a linear transformation L.

Corollary 7.3 A linear transformation L on V satisfies its own characteristic equation $\phi(\lambda) = 0$, i.e., for a linear transformation L on V, $\phi(L) = \Theta$, a zero transformation.

Proof Let A be a matrix of L with respect to any basis X in V, and let $\mathbf{u}$ be the coordinate vector of any element $\mathbf{x}$ in V with respect to the same basis. Then denoting det $(\lambda I - A) = \phi(\lambda)$, by Theorem 7.12 and (7.74), we have

$$\phi(A) = a_n A^n + a_{n-1} A^{n-1} + \cdots + a_1 A + a_0 I = 0, \tag{7.89}$$

a null matrix. By virtue of (7.71), the matrix of $\phi(L)$ is $\phi(A)$, and hence in view of Theorem 6.9 and (7.89), it is not difficult to deduce that

$$\phi(A)\mathbf{u} = 0 \cdot \mathbf{u} = [0\ 0\ \cdots\ 0]^T \text{ implies } \phi(L)\mathbf{x} = \boldsymbol{\theta},$$

where $\boldsymbol{\theta}$ is a null vector in V. Since $\mathbf{x}$ is an arbitrary element of V, it follows that $\phi(L) = \Theta$, a zero transformation. □

Example 7.28 Show that the linear transformation L on $\mathbb{R}^2$ defined as $L(u, v) = (u + v, u - 2v)$ satisfies Cayley-Hamilton Theorem.

Solution The matrix of L with respect to the basis $\{\mathbf{e}_1, \mathbf{e}_2\}$ in $\mathbb{R}^2$ is given by (7.72). So the characteristic polynomial of L is $\phi(\lambda) = \det(\lambda I - A) = \lambda^2 + \lambda - 3$. By direct computation we have

$$\begin{aligned}\phi(L)(u, v) &= (L^2 + L - 3I_V)(u, v)\\ &= (2u - v, -u + 5v) + (u + v, u - 2v) - 3(u, v)\\ &= (0, 0)\end{aligned}$$

for any u and v. So, $\phi(L)$ is a zero transformation, showing that L satisfies its own characteristic equation. □

The minimum polynomial $\psi(\lambda)$ of a linear transformation L on V is that the monic polynomial of least degree among monic polynomials $g(\lambda)$ such that $\psi(L) = \Theta$, a zero transformation. An important result in this regard is the following theorem.

Theorem 7.15 If A is the matrix of a linear transformation L on V, then $\psi(\lambda)$ is a minimum polynomial of L iff it is a minimum polynomial of A.

Proof To prove the necessity, suppose that $\psi(\lambda)$ is a minimum polynomial of L. Then $\psi(L) = \Theta$, a zero transformation. If X is a basis of V, then $\psi(L)X = Z$, where Z has zero vectors in all its columns. By Theorem 6.9 this implies that (see also section 7.3.4) $\psi(A)I = 0$, a null matrix. So, $\psi(\lambda)$ is an annihilating polynomial of A, which is also the minimum polynomial of A. Otherwise, if $\hat{\psi}(\lambda)$ is another polynomial of lower degree than $\psi(\lambda)$ then retracing the above steps, it follows that $\hat{\psi}(L) = \Theta$, which contradicts that $\psi(\lambda)$ is a minimum polynomial of L. The sufficiency follows easily by taking a reverse line of arguments to those already given. □

7.5 COMPUTATION OF CHARACTERISTIC POLYNOMIAL AND ADJOINT OF $(\lambda I - A)$

The characteristic equation of a matrix A is defined by (7.8), where $\phi(\lambda)$ is the characteristic polynomial which can be represented by (7.74). To obtain $\phi(\lambda)$, we must expand the determinant in (7.8) which can be done by directly taking advantage of properties of a determinant (see Section 2.4). This would involve operations with polynomials in λ. One can, however, dispense with the active participation of these polynomials by using some systematic procedures to be discussed in this section. We shall, therefore, describe computations of $\phi(\lambda)$, relying on the operations with only the scalar elements of the matrix A rather than polynomials in λ. The computation of adj $(\lambda I - A)$, see (7.73) and (7.87), can be done easily after numerical evaluation of the coefficients of $\phi(\lambda)$.

Before presenting the first result in this direction, we shall use an example to explain a crucial point to be used in proving this result.

Example 7.29 Determine the characteristic polynomial for a third order matrix $A = [a_{ij}]$.

Solution For a third order matrix A,

$$(\lambda I - A) = \begin{bmatrix} \lambda - a_{11} & -a_{12} & -a_{13} \\ -a_{21} & \lambda - a_{22} & -a_{23} \\ -a_{31} & -a_{32} & \lambda - a_{33} \end{bmatrix} = [\lambda \mathbf{e}_1 - \mathbf{c}_1 \ \lambda \mathbf{e}_2 - \mathbf{c}_2 \ \lambda \mathbf{e}_3 - \mathbf{c}_3],$$

where for $i = 1, 2, 3$, $\mathbf{c}_i = [a_{1i} \ a_{2i} \ a_{3i}]^T$ and $\mathbf{e}_i$ is an elementary vector in $\mathbb{R}^3$ given by (5.4). To compute the characteristic polynomial, we note that

$$\begin{aligned} \det(\lambda I - A) = & \,|\lambda \mathbf{e}_1 \ \lambda \mathbf{e}_2 \ \lambda \mathbf{e}_3| + |\lambda \mathbf{e}_1 \ \lambda \mathbf{e}_2, -\mathbf{c}_3| + |-\mathbf{c}_1 \ \lambda \mathbf{e}_2 \ \lambda \mathbf{e}_3| \\ & + |\lambda \mathbf{e}_1, -\mathbf{c}_2, \lambda \mathbf{e}_3| + |\lambda \mathbf{e}_1, -\mathbf{c}_2, -\mathbf{c}_3| + |-\mathbf{c}_1, \lambda \mathbf{e}_2, -\mathbf{c}_3| \\ & + |-\mathbf{c}_1, -\mathbf{c}_2, \lambda \mathbf{e}_3| + |-\mathbf{c}_1, -\mathbf{c}_2, -\mathbf{c}_3| \end{aligned}$$

is the sum of $2^3 = 8$ auxiliary determinants which must be calculated (see Property VIII, Section 2.3). For example, by repeated use of Property V of a determinant (see Section 2.3), we get

$$|\lambda \mathbf{e}_1, \lambda \mathbf{e}_2, \lambda \mathbf{e}_3| = \lambda^3 |\mathbf{e}_1 \ \mathbf{e}_2 \ \mathbf{e}_3| = \lambda^3,$$

$$|\lambda \mathbf{e}_1, \lambda \mathbf{e}_2, -\mathbf{c}_3| = \lambda^2(-1)|\mathbf{e}_1 \ \mathbf{e}_2 \ \mathbf{c}_3| = -\lambda^2 \begin{bmatrix} 1 & 0 & a_{13} \\ 0 & 1 & a_{23} \\ 0 & 0 & a_{33} \end{bmatrix}$$

$$= -\lambda^2 a_{33}, \qquad a_{33} = \text{a principal minor of order 1.}$$

Similarly,

$$|-\mathbf{c}_1 \ \lambda \mathbf{e}_2 \ \lambda \mathbf{e}_3| = -\lambda^2 a_{11}, \qquad |\lambda \mathbf{e}_1, -\mathbf{c}_2 \ \lambda \mathbf{e}_3| = -\lambda^2 a_{22}$$

which are two other principal minors of order 1;

$$|-\mathbf{c}_1, \lambda\mathbf{e}_2, -\mathbf{c}_3| = (-1)^2\lambda \begin{vmatrix} a_{11} & 0 & a_{13} \\ a_{21} & 1 & a_{23} \\ a_{31} & 0 & a_{33} \end{vmatrix}$$

$$= \lambda \begin{vmatrix} a_{11} & a_{13} \\ a_{31} & a_{33} \end{vmatrix}$$

which is λ times a principal minor of order 2. Similarly,

$$|\lambda\mathbf{e}_1, -\mathbf{c}_2, -\mathbf{c}_3| = \lambda \begin{vmatrix} a_{22} & a_{23} \\ a_{32} & a_{33} \end{vmatrix}, \qquad |-\mathbf{c}_1, -\mathbf{c}_2\ \lambda\mathbf{e}_3| = \lambda \begin{vmatrix} a_{11} & a_{12} \\ a_{21} & a_{22} \end{vmatrix}$$

which are λ times two other second order principal minors; and $|-\mathbf{c}_1, -\mathbf{c}_2, -\mathbf{c}_3| = -\det A$. Therefore,

$\det(\lambda I - A) = \lambda^3 - (a_{11} + a_{22} + a_{33})\lambda^2 +$ (sum of all principal minors of order 2)$\lambda - \det A$.

□

The important point to note here is how the auxiliary determinants can be computed with the help of principal minors of an n-square matrix A. An auxiliary determinant having elementary vectors $\mathbf{e}_i$'s in r number of columns can be computed with the help of an $(n - r)$th order principal minor of A. Check this fact when $n = 3$, as in this example.

Theorem 7.16 The characteristic polynomial of a matrix A of order n is an nth degree polynomial in λ of the form

$$\phi(\lambda) = a_n\lambda^n + a_{n-1}\lambda^{n-1} + \cdots + a_r\lambda^r + \cdots + a_1\lambda + a_0, \tag{7.74}$$

where

$$a_n = 1, \qquad a_0 = (-1)^n \det A, \tag{7.90}$$

and for all $r = 1, 2, ..., n - 1$, a_r is equal to $(-1)^{n-r}$ times the sum of ${}^nC_r$ number of principal minors of order $n - r$ of A.

Proof Denoting the columns of A by $\mathbf{c}_i$'s ($i = 1, 2, ..., n$), from (7.8) we get

$$\det(\lambda I - A) = \det(\lambda\mathbf{e}_1 - \mathbf{c}_1, \lambda\mathbf{e}_2 - \mathbf{c}_2, ..., \lambda\mathbf{e}_n - \mathbf{c}_n),$$

$\mathbf{e}_i$'s being ith elementary vectors represented by (5.4). This determinant can be expressed as the sum of 2^n determinants, of which those having λ in r number of columns are of the form

$$\det(\lambda\mathbf{e}_1, \lambda\mathbf{e}_2, ..., \lambda\mathbf{e}_r, -\mathbf{c}_{r+1}, ..., -\mathbf{c}_n) + \det(\lambda\mathbf{e}_1, \lambda\mathbf{e}_2, ..., -\mathbf{c}_r, \lambda\mathbf{e}_{r+1}, ..., -\mathbf{c}_n)$$
$$+ \cdots + \det(-\mathbf{c}_1, -\mathbf{c}_2, ..., -\mathbf{c}_{n-r}, \lambda\mathbf{e}_{n-r+1}, ..., \lambda\mathbf{e}_n)$$

with ${}^nC_r$ number of terms. With reference to Property V of a determinant (see Section 2.3) and Example 7.29, it is not difficult to see that the value of each term in the above summand is evidently $(-1)^{n-r}\lambda^r$ times the minor of A obtained by deleting those r number of columns from A which contain λ's in the determinant matrix of the term concerned and correspondingly numbered r rows, i.e., $(-1)^{n-r}\lambda^r$ times a principal minor of A of order $n - r$. The term containing

λ^n is obtained from det $(\lambda\mathbf{e}_1, \lambda\mathbf{e}_2, ..., \lambda\mathbf{e}_n) = \lambda^n$, and the constant term from det$(-\mathbf{c}_1, -\mathbf{c}_2, ..., -\mathbf{c}_n) = (-1)^n$ det A. Consequently, $a_n = 1$ and $a_0 = (-1)^n$ det A.

Thus the characteristic polynomial $\phi(\lambda)$ of an n-square matrix A is $\phi(\lambda) = \lambda^n -$ (sum of all principal minors of order one)λ^{n-1} + (sum of all principal minors of order two) $\lambda^{n-2} - \cdots + (-1)^n$ det A. □

Corollary 7.4 A matrix A of order n has eigenvalues $\lambda_1, \lambda_2, ..., \lambda_n$ and its characteristics polynomial $\phi(\lambda)$ is represented by (7.74). Show that

$$a_{n-1} = -\text{ tr } A = -(\lambda_1 + \lambda_2 + \cdots + \lambda_n)$$

$$a_0 = (-1)^n \lambda_1\lambda_2 \cdots \lambda_n, \qquad \det A = \lambda_1\lambda_2 \cdots \lambda_n.$$

Proof Indeed, from Theorem 7.16 and (1.16) we have $a_{n-1} = (-1)^{n-(n-1)}$ (sum of all principal minors of order 1 in A) $= -(a_{11} + a_{22} + \cdots + a_{nn}) = -$ tr A, where tr A is defined as in Definition 1.14.

Again, since $\lambda_1\lambda_2 \cdots \lambda_n$ are eigenvalues of A, we have

$$\phi(\lambda) = (\lambda - \lambda_1)(\lambda - \lambda_2) \cdots (\lambda - \lambda_n) \tag{7.91}$$

in which, on multiplication, the term independent of λ is easily seen to be $(-1)^n\lambda_1\lambda_2 \cdots \lambda_n$, and the coefficient of λ^{n-1} to be $-(\lambda_1 + \lambda_2 + \cdots + \lambda_n)$. Comparing (7.91) with (7.74) and then using (7.90), we have

$$a_0 = (-1)^n\lambda_1\lambda_2 \cdots \lambda_n = (-1)^n \det A,$$

$$a_{n-1} = -(\lambda_1 + \lambda_2 + \cdots + \lambda_n).$$

Hence,

$$\det A = \lambda_1\lambda_2 \cdots \lambda_n.$$ □

Example 7.30 Find the characteristic polynomial of the third order matrix A given by

$$A = \begin{bmatrix} -2 & -2 & -4 \\ 2 & 3 & 2 \\ 3 & 2 & 5 \end{bmatrix}.$$

Solution The general expression for the characteristic polynomial of a third order matrix is

$$\phi(\lambda) = a_3\lambda^3 + a_2\lambda^2 + a_1\lambda + a_0.$$

By applying Theorem 7.16 we have

$$a_3 = 1, \qquad a_0 = (-1)^3 \det A = -6.$$

Now, the sum of all principal minors of order 2 is

$$\begin{vmatrix} -2 & -2 \\ 2 & 3 \end{vmatrix} + \begin{vmatrix} -2 & -4 \\ 3 & 5 \end{vmatrix} + \begin{vmatrix} 3 & 2 \\ 2 & 5 \end{vmatrix} = -2 + 2 + 11 = 11.$$

Therefore, in view of Theorem 7.16, we have

$$a_1 = (-1)^{3-1}\, 11 = 11.$$

Finally, using Corollary 7.4, we obtain

$$a_2 = -\text{tr } A = -(-2 + 3 + 5) = -6.$$

Hence, the characteristic polynomial is

$$\phi(\lambda) = \lambda^3 - 6\lambda^2 + 11\lambda - 6.$$

The eigenvalues of A obtained by solving $\phi(\lambda) = 0$ are $\lambda_1 = 1$, $\lambda_2 = 2$, $\lambda_3 = 3$, and $\lambda_1\lambda_2\lambda_3 = 1{\cdot}2{\cdot}3 = 6$ which is the determinant of A. □

Example 7.31 If λ is an eigenvalue and $\mathbf{x}$ is the corresponding eigenvector of an invertible matrix A, then the eigenvalue and eigenvector of A^{-1} are respectively λ^{-1} and $\mathbf{x}$.

Solution By Corollary 7.4, we see that the determinant of a square matrix A is the product of its eigenvalues. Therefore, no eigenvalue of a nonsingular matrix can be zero. Hence, if A is nonsingular, and λ is one of its eigenvalues and $\mathbf{x}$ the corresponding eigenvector, then from $A\mathbf{x} = \lambda\mathbf{x}$ it follows that $A^{-1}\mathbf{x} = \lambda^{-1}\,\mathbf{x}$. This implies that A^{-1} has λ^{-1} as the eigenvalue and $\mathbf{x}$ as the corresponding eigenvector. □

Definition 7.19 *Companion Matrix.* Suppose that $\phi(\lambda)$ is an nth degree polynomial given by

$$\phi(\lambda) = \lambda^n + a_{n-1}\lambda^{n-1} + \cdots + a_1\lambda + a_0.$$

Then a matrix C with 1's along the first superdiagonal (see Definition 1.11) and negative of the coefficients $a_0, a_1, \ldots, a_{n-1}$ along the last row with remaining elements equal to zero, viz.,

$$C = \begin{bmatrix} 0 & 1 & 0 & \cdots & 0 \\ 0 & 0 & 1 & \cdots & 0 \\ \vdots & \vdots & \vdots & & \vdots \\ 0 & 0 & 0 & \cdots & 1 \\ -a_0 & -a_1 & -a_2 & \cdots & -a_{n-1} \end{bmatrix} \tag{7.92}$$

is called the *companion matrix* (alternatively *a Frobenius matrix*) of the polynomial $\phi(\lambda)$. Three other forms of the companion matrix of $\phi(\lambda)$ are

$$\begin{bmatrix} -a_{n-1} & 1 & 0 & \cdots & 0 \\ -a_{n-2} & 0 & 1 & \cdots & 0 \\ \vdots & \vdots & \vdots & & \vdots \\ -a_1 & 0 & 0 & \cdots & 1 \\ -a_0 & 0 & 0 & \cdots & 0 \end{bmatrix}, \tag{7.93a}$$

$$\begin{bmatrix} -a_{n-1} & -a_{n-2} & \cdots & -a_1 & -a_0 \\ 1 & 0 & \cdots & 0 & 0 \\ 0 & 1 & \cdots & 0 & 0 \\ \vdots & \vdots & & \vdots & \vdots \\ 0 & 0 & \cdots & 1 & 0 \end{bmatrix}, \tag{7.93b}$$

$$\begin{bmatrix} 0 & 0 & 0 & \cdots & 0 & -a_0 \\ 1 & 0 & 0 & \cdots & 0 & -a_1 \\ 0 & 1 & 0 & \cdots & 0 & -a_2 \\ \vdots & \vdots & \vdots & & \vdots & \vdots \\ 0 & 0 & 0 & \cdots & 1 & -a_{n-1} \end{bmatrix}. \tag{7.93c}$$

As a matrix of the form (7.92) accompanies any monic polynomial $\phi(\lambda)$, it is known as a *companion* or *friend* of that polynomial. □

Example 7.32 For the campanion matrix C given by (7.92), prove that

(a) the characteristic polynomial is $\phi(\lambda) = \lambda^n + a_{n-1}\lambda^{n-1} + \cdots + a_1\lambda + a_0$;

(b) the characteristic polynomial and the minimum polynomial are the same;

(c) if λ_i is an eigenvalue of C, then $\mathbf{x}_i = [1\ \lambda_i\ \cdots\ \lambda_i^{n-1}]^T$ is the associated eigenvector;

(d) if $\lambda_1, \lambda_2, ..., \lambda_n$ are distinct eigenvalues of C, then $V^{-1}CV = \text{diag}(\lambda_1, \lambda_2, ..., \lambda_n)$, where V is the Vandermonde matrix defined in Problem 2.15.

Solution (a) The characteristic polynomial of C is given by

$$\det(\lambda I - C) = \det \begin{bmatrix} \lambda & -1 & 0 & \cdots & 0 \\ 0 & \lambda & -1 & \cdots & 0 \\ \vdots & & & & \vdots \\ 0 & 0 & 0 & \cdots & -1 \\ a_0 & a_1 & a_2 & \cdots & \lambda + a_{n-1} \end{bmatrix}. \tag{7.94}$$

Adding successively λ times the second column, λ^2 times the third column $\cdots$ and λ^{n-1} times the last column to the first column in the (7.94), we have

$$\begin{vmatrix} 0 & -1 & 0 & \cdots & 0 \\ 0 & \lambda & -1 & \cdots & 0 \\ \vdots & \vdots & \vdots & & \vdots \\ 0 & 0 & 0 & \cdots & -1 \\ \phi(\lambda) & a_1 & a_2 & \cdots & \lambda + a_{n-1} \end{vmatrix}$$

which, when expanded in terms of the first column, gives

$$(-1)^{n+1}\phi(\lambda)(-1)^{n-1} = \phi(\lambda).$$

(b) From (7.94) the minor of a_0, see Definition 2.4, in $(\lambda I - C)$ is $(-1)^{n-1}$. Therefore, the GCD of all minors of order $(n-1)$ in $(\lambda I - C)$ is 1. By virtue of Theorem 7.14, the characteristic polynomial and the minimum polynomial are the same.

(c) Since λ_i is an eigenvalue of C, from its characteristic equation we have $\phi(\lambda_i) = 0$, and hence $\lambda_i^n = -(a_{n-1}\lambda_i^{n-1} + \cdots + a_1\lambda_i + a_0)$. In view of this and the fact that C is given by (7.92), by direct multiplication, we have

$$C\mathbf{x}_i = \begin{bmatrix} \lambda_i \\ \lambda_i^2 \\ \vdots \\ \lambda_i^n \end{bmatrix} = \lambda_i \mathbf{x}_i \tag{7.95}$$

which shows that $\mathbf{x}_i$ is an eigenvector of C corresponding to the eigenvalue λ_i.

(d) From (7.95) for $i = 1, ..., n$,

$$C[\mathbf{x}_1\ \mathbf{x}_2\ \cdots\ \mathbf{x}_n] = [\mathbf{x}_1\ \mathbf{x}_2\ \cdots\ \mathbf{x}_n]\ \text{diag}\ (\lambda_1, \lambda_2, ..., \lambda_n)$$

which can be written as

$$CV = V\ \text{diag}\ (\lambda_1, \lambda_2, ..., \lambda_n), \tag{7.96}$$

where

$$V = [\mathbf{x}_1\ \mathbf{x}_2 \cdots \mathbf{x}_n] = \begin{bmatrix} 1 & 1 & \cdots & 1 \\ \lambda_1 & \lambda_2 & \cdots & \lambda_n \\ \vdots & \vdots & & \vdots \\ \lambda_1^{n-1} & \lambda_2^{n-1} & \cdots & \lambda_n^{n-1} \end{bmatrix}.$$

If all the eigenvalues $\lambda_1, \lambda_2, ..., \lambda_n$ of C are distinct, V is nonsingular (see Problem 2.15). Consequently, it follows from (7.96) that

$$V^{-1}CV = \text{diag}\ (\lambda_1, \lambda_2, ..., \lambda_n).$$ □

7.5.1 Eigenvalues and Eigenvectors of Matrix Polynomials

After our study of eigenvalues and eigenvectors of a square matrix A in Section 7.1 and the introduction of matrix polynomials $f(A)$ in Section 7.3.1, we are now fully equipped to relate these eigenvalues and eigenvectors of A to those of $f(A)$. With this object in mind, we now present the following theorem.

Theorem 7.17 If $\lambda_1, \lambda_2, ..., \lambda_n$ are the eigenvalues of a square matrix A over $\mathbb{C}$ with corresponding eigenvectors $\mathbf{x}_1, \mathbf{x}_2, ..., \mathbf{x}_n$, then any matrix polynomial $f(A)$ has

(a) $f(\lambda_1), f(\lambda_2), ..., f(\lambda_n)$ as the eigenvalues;

(b) $\mathbf{x}_1, \mathbf{x}_2, ..., \mathbf{x}_n$ as the corresponding eigenvectors.

This theorem will be proved with the aid of the following lemma.

Lemma 7.1 If $p(x)$ is any polynomial, then the matrix polynomial $p(A)$ of the n-square matrix A has a determinant given by

$$\det p(A) = p(\lambda_1)\, p(\lambda_2) \cdots p(\lambda_n),$$

where λ_i's are the eigenvalues of A.

Proof The polynomial $p(x)$, say of degree r, can be resolved into linear factors over $\mathbb{C}$ to give

$$p(x) = \alpha(\mu_1 - x)\,(\mu_2 - x) \cdots (\mu_r - x), \tag{7.97}$$

where $\alpha, \mu_1, \mu_2, ..., \mu_r$ are scalars in $\mathbb{C}$. By Property (a) of matrix polynomials in Section 7.3.1, from (7.97), the associated matrix polynomial can be written as

$$p(A) = \alpha(\mu_1 I - A)\,(\mu_2 I - A) \cdots (\mu_r I - A).$$

Hence, denoting the characteristic polynomial of A by $\phi(\lambda) = (\lambda - \lambda_1)(\lambda - \lambda_2) \cdots (\lambda - \lambda_n)$ as in (7.91), we have

$$\begin{aligned}\det p(A) &= \alpha^n \det(\mu_1 I - A)\det(\mu_2 I - A) \cdots \det(\mu_r I - A) \\ &= \alpha^n \phi(\mu_1)\phi(\mu_2) \cdots \phi(\mu_r) \\ &= \prod_{i=1}^{n} \alpha(\mu_1 - \lambda_i)(\mu_2 - \lambda_i) \cdots (\mu_r - \lambda_i) \\ &= p(\lambda_1)p(\lambda_2) \cdots p(\lambda_n). \end{aligned} \tag{7.98}$$

□

Proof of Theorem 7.17 (a) Now considering the characteristic polynomial of $p(A)$, we have, by applying Lemma 7.1, the equation

$$\det[\lambda I - f(A)] = [\lambda - f(\lambda_1)][\lambda - f(\lambda_2)] \cdots [(\lambda - f(\lambda_n)],$$

where we have identified $p(A) = \lambda I - f(A)$. This shows that $f(\lambda_1), f(\lambda_2), ..., f(\lambda_n)$ are eigenvalues of $f(A)$.

(b) Assuming that $f(x)$ is of degree m, we represent the associated matrix polynomial as

$$f(A) = \alpha_m A^m + \alpha_{m-1}A^{m-1} + \cdots + \alpha_1 A + \alpha_0 I.$$

Now, if λ_i is an eigenvalue of A with associated eigenvector $\mathbf{x}_i$, then using the result of Example 7.2, it is easy to check that

$$\begin{aligned} f(A)\mathbf{x}_i &= (a_m A^m + a_{m-1}A^{m-1} + \cdots + a_1 A + a_0 I)\mathbf{x}_i \\ &= (a_m \lambda_i^m + a_{m-1}\lambda_i^{m-1} + \cdots + a_1\lambda_1 + a_0)\mathbf{x}_i \\ &= f(\lambda_i)\mathbf{x}_i \end{aligned}$$

which shows that $\mathbf{x}_i$ is an eigenvector of $f(A)$ associated with the eigenvalue $f(\lambda_i)$. Note that the eigenvectors of $f(A)$ may not all be the eigenvectors of A (see Problem 7.37). □

Theorem 7.18 Let $\lambda_1, \lambda_2, ..., \lambda_n$ be the eigenvalues of A with corresponding eigenvectors $\mathbf{x}_1, \mathbf{x}_2, ..., \mathbf{x}_n$. If $g(\lambda)$ is a polynomial such that $g(\lambda_i) \neq 0$ $(i = 1, 2, ..., n)$ and $f(\lambda)$ is any arbitrary polynomial, then (a) the eigenvalues of

$$h(A) = f(A)g^{-1}(A) = g^{-1}(A)f(A)$$

are $h(\lambda_i) = f(\lambda_i)/g(\lambda_i)$, and (b) the respective eigenvectors are $\mathbf{x}_i$ $(i = 1, 2, ..., n)$.

Proof By applying Lemma 7.1, we have

$$\det g(A) = g(\lambda_1)g(\lambda_2) \cdots g(\lambda_n)$$

which is nonzero since $g(\lambda_i) \neq 0$. Hence, $g(A)$ is invertible. Consequently,

$$\det h(A) = \det f(A)\,[\det g(A)]^{-1} = \prod_{i=1}^{n} f(\lambda_i)/g(\lambda_i) = \prod_{i=1}^{n} h(\lambda_i)$$

and, therefore,

$$\det[\lambda I - h(A)] = \prod_{i=1}^{n}[\lambda - h(\lambda_i)]$$

which proves (a). To prove (b), in view of Example 7.2, we note that $g(A)\mathbf{x}_i = g(\lambda_i)\mathbf{x}_i$ and $f(A)\mathbf{x}_i = f(\lambda_i)\mathbf{x}_i$. Therefore,

$$h(A)g(A)\mathbf{x}_i = f(A)\mathbf{x}_i,$$

or

$$h(A)\mathbf{x}_i = [f(\lambda_i)/g(\lambda_i)]\mathbf{x}_i.$$

We now consider a nonsingular matrix polynomial $g(A) = B$ having a minimum polynomial

$$\mu(\lambda) = \lambda^r + \alpha_{r-1}\lambda^{r-1} + \cdots + \alpha_1\lambda + \alpha_0,$$

where $\alpha_0 \neq 0$; otherwise, $\lambda = 0$ is an eigenvalue of $g(A)$, which contradicts that $g(A)$ is nonsingular. Consequently,

$$B^r + \alpha_{r-1}B^{r-1} + \cdots + \alpha_1 B + \alpha_0 I = 0.$$

From this equation,

$$B^{-1} = -\frac{1}{\alpha_0}(B^{r-1} + \alpha_{r-1}B^{r-2} + \cdots + \alpha_1 I) = g^{-1}(A).$$

Now, if we replace B by $g(A)$, we find that $g^{-1}(A)$ is a polynomial in A. □

If λ_i's are all nonzero, then eigenvalues of A^{-1} are $1/\lambda_i$, $(i = 1, 2, \ldots, n)$ with corresponding eigenvectors $\mathbf{x}_1, \mathbf{x}_2, \ldots, \mathbf{x}_n$. This is also a particular case of Theorem 7.18 if $g(\lambda) = \lambda$ and $f(\lambda) = 1$. Alternatively, if follows from $A\mathbf{x}_i = \lambda_i\mathbf{x}_i$ that $A^{-1}\mathbf{x}_i = \lambda_i^{-1}\mathbf{x}_i$ which proves also the same result.

7.5.2 Newton Formulae, Leverrier Method, and Faddeev Algorithm

The sum of the powers of roots of any polynomial such as $\phi(\lambda)$ is connected with its coefficients by Newton formulae.

Theorem 7.19 *Newton Formulae.* Suppose that the roots of the polynomial

$$\phi(\lambda) = \lambda^n + a_{n-1}\lambda^{n-1} + \cdots + a_1\lambda + a_0 \tag{7.99}$$

are $\lambda_1, \lambda_2, \ldots, \lambda_n$. Denote

$$s_r = \lambda_1^r + \lambda_2^r + \cdots + \lambda_n^r, \qquad r = 1, 2, \ldots, n. \tag{7.100}$$

Then,

$$s_k + a_{n-1}s_{k-1} + \cdots + a_{n-k+1}s_1 + ka_{n-k} = 0, \qquad k = 1, 2, \ldots, n. \tag{7.101}$$

Proof Since $\lambda_1, \ldots, \lambda_n$ are the roots of $\phi(\lambda)$ given in (7.99), we can write

$$\phi(\lambda) = \lambda^n + a_{n-1}\lambda^{n-1} + a_{n-2}\lambda^{n-2} + \cdots + a_k\lambda^k + \cdots + a_1\lambda + a_0 \tag{7.102}$$

$$= (\lambda - \lambda_1)\cdots(\lambda - \lambda_n). \tag{7.103}$$

(i) Differentiating (7.102) and (ii) first, taking the logarithm of (7.103) and then differentiating with respect to λ, two expressions for the derivative of $\phi(\lambda)$ are obtained:

$$\phi'(\lambda) = n\lambda^{n-1} + (n-1)a_{n-1}\lambda^{n-2} + \cdots + ka_k\lambda^{k-1} + \cdots + a_1 \tag{7.104}$$

$$= \phi(\lambda)\{(\lambda - \lambda_1)^{-1} + \cdots + (\lambda - \lambda_n)\}^{-1}. \tag{7.105}$$

For any $\mu \neq \lambda$, using (7.99), by long division we have

$$\frac{\phi(\lambda)}{\lambda-\mu} = \lambda^{n-1} + (\mu + a_{n-1})\lambda^{n-2} + (\mu^2 + a_{n-1}\mu + a_{n-2})\lambda^{n-3}$$
$$+ \cdots + (\mu^{n-k} + a_{n-1}\mu^{n-k-1} + \cdots + a_k)\lambda^{k-1}$$
$$+ \cdots + (\mu^{n-1} + a_{n-1}\mu^{n-2} + \cdots + a_2\mu + a_1) + \frac{\phi(\mu)}{\lambda-\mu}. \tag{7.106}$$

In relation (7.106), we put successively $\mu = \lambda_1, \lambda_2, \ldots, \lambda_n$ and then add the resulting expressions, to write $\phi'(\lambda)$ by (7.105) as

$$\phi'(\lambda) = n\lambda^{n-1} + (s_1 + na_{n-1})\lambda^{n-2} + (s_2 + s_1a_{n-1} + na_{n-2})\lambda^{n-3}$$
$$+ \cdots + (s_{n-k} + a_{n-1}s_{n-k-1} + \cdots + na_k)\lambda^{k-1}$$
$$+ \cdots + (s_{n-1} + a_{n-1}s_{n-2} + \cdots + a_2s_1 + na_1) \tag{7.107}$$

because $\phi(\lambda_i) = 0$. Now, equating the coefficients of like powers of λ between (7.104) and (7.107), we have successively

$$\begin{aligned}
&s_1 + a_{n-1} = 0,\\
&s_2 + s_1a_{n-1} + 2a_{n-2} = 0,\\
&\vdots\\
&s_{n-k} + a_{n-1}s_{n-k-1} + \cdots + a_{k+1}s_1 + (n-k)a_k = 0,\\
&\vdots\\
&s_{n-1} + a_{n-1}s_{n-2} + \cdots + a_2s_1 + (n-1)a_1 = 0,
\end{aligned} \tag{7.108}$$

which are (7.101) for $k = 1, 2, \ldots, n-1$. Again, since λ_i's satisfy (7.99), $\phi(\lambda_i) = 0$, i.e.,

$$\lambda_i^n + a_{n-1}\lambda_i^{n-1} + \cdots + a_1\lambda_i + a_0 = 0, \qquad i = 1, 2, \ldots, n,$$

and on adding the n equations, we get

$$s_n + a_{n-1}s_{n-1} + \cdots + a_1s_1 + na_0 = 0, \tag{7.109}$$

which is (7.101) for $k = n$. □

In view of Example 7.2 and Corollary 7.4, we observe that s_k as defined in (7.100) is

$$s_k = \operatorname{tr} A^k \tag{7.110}$$

which can be employed in (7.108) and (7.109) to determine successively the coefficients a_k's of the characteristic polynomial $\phi(\lambda)$. This procedure is known in the literature as the *method of Leverrier*.

In this method, we compute $s_1, s_2, \ldots, s_n$ as defined in (7.100) from the traces of the matrices $A, A^2, \ldots, A^n$ respectively, employing (7.110). Then, (7.108) and (7.109) are used to determine recursively the coefficients $a_{n-1}, a_{n-2}, \ldots, a_1, a_0$ of the characteristic polynomial $\phi(\lambda)$.

Example 7.33 Compute the coefficients of the characteristic polynomial of the matrix A given in Example 7.30.

Solution We have using Newton formulae

$$A = \begin{bmatrix} -2 & -2 & -4 \\ 2 & 3 & 2 \\ 3 & 2 & 5 \end{bmatrix}, \qquad A^2 = \begin{bmatrix} -12 & -10 & -16 \\ 8 & 9 & 8 \\ 13 & 10 & 17 \end{bmatrix},$$

$$A^3 = \begin{bmatrix} -44 & -38 & -52 \\ 26 & 27 & 26 \\ 45 & 38 & 53 \end{bmatrix}.$$

Therefore,

$$s_1 = \text{tr } A = -2 + 3 + 5 = 6,$$
$$s_2 = \text{tr } A^2 = -12 + 9 + 17 = 14,$$
$$s_3 = \text{tr } A^3 = -44 + 27 + 53 = 36.$$

Taking advantage of the recursive relation (7.108) and (7.109), we have successively

$$a_2 = -s_1 = -6, \qquad a_1 = -\frac{1}{2}(s_2 + s_1 a_2) = -\frac{1}{2}(14 - 6 \cdot 6) = 11,$$

$$a_0 = -\frac{1}{3}(s_3 + a_2 s_2 + a_1 s_1) = -6,$$

which are the same as obtained in Example 7.30. □

D.K. Faddeev suggested that the coefficients a_i's of the characteristic polynomial $\phi(\lambda)$, see (7.74), and B_i's of adj $(\lambda I - A)$ should be simultaneously computed, see (7.73). The coefficients a_i's are computed in terms of traces of some recursively computable matrices instead of traces of powers of A as in Leverrier method.

Theorem 7.20 *Faddeev algorithm.* The resolvent of the square matrix A written with the help of (7.73) and (7.74) is

$$(\lambda I - A)^{-1} = (\lambda^n + a_{n-1}\lambda^{n-1} + \cdots + a_1\lambda + a_0)^{-1}[B_{n-1}\lambda^{n-1} + B_{n-2}\lambda^{n-2} + \cdots + B_1\lambda + B_0],$$

in which

$$\begin{aligned} B_{n-1} &= I, & a_{n-1} &= -\text{tr }(AB_{n-1}); \\ B_{n-2} &= a_{n-1}I + AB_{n-1}, & a_{n-2} &= -\frac{1}{2}\text{tr }(AB_{n-2}); \\ &\vdots & &\vdots \\ B_k &= a_{k+1}I + AB_{k+1}, & a_k &= -\frac{1}{n-k}\text{tr }(AB_k); \\ &\vdots & &\vdots \\ B_0 &= a_1 I + AB_1, & a_0 &= -\frac{1}{n}\text{tr }(AB_0); \\ 0 &= a_0 I + AB_0, \end{aligned}$$

where the last equation checks the correctness of the computation.

Proof Since by Theorem 7.16 $a_n = 1$, from (7.77) all the expressions for B_i's on the left and also the last one can be obtained.

From Newton formulae (7.101) and (7.110), we have

$$s_{n-k} + a_{n-1}s_{n-k-1} + \cdots + a_{k+1}s_1 + (n-k)a_k = 0,$$

or

$$\begin{aligned} a_k &= -\frac{1}{(n-k)}(s_{n-k} + a_{n-1}s_{n-k-1} + \ldots + a_{k+1}s_1) \\ &= -\frac{1}{(n-k)}\operatorname{tr}(A^{n-k} + a_{n-1}A^{n-k-1} + \ldots + a_{k+1}A) \\ &= -\frac{1}{(n-k)}\operatorname{tr}(AB_k), \end{aligned}$$

where the expression for B_k given in Example 7.26 is used. □

Example 7.34 Find the resolvent of the matrix A given in Example 7.30.

Solution When $n = 3$, using Faddeev algorithm, we have

$$B_2 = I, \qquad AB_2 = A = \begin{bmatrix} -2 & -2 & -4 \\ 2 & 3 & 2 \\ 3 & 2 & 5 \end{bmatrix}.$$

Hence,

$$a_2 = -\operatorname{tr} A = -6,$$

$$B_1 = a_2 I + AB_2 = \begin{bmatrix} -8 & -2 & -4 \\ 2 & -3 & 2 \\ 3 & 2 & -1 \end{bmatrix}, \qquad AB_1 = \begin{bmatrix} 0 & 2 & 8 \\ -4 & -9 & -4 \\ -5 & -2 & -13 \end{bmatrix}.$$

Therefore,

$$a_1 = -\frac{1}{2}\operatorname{tr}(AB_1) = 11,$$

$$B_0 = a_1 I + AB_1 = \begin{bmatrix} 11 & 2 & 8 \\ -4 & 2 & -4 \\ -5 & -2 & -2 \end{bmatrix}, \qquad AB_0 = \begin{bmatrix} 6 & 0 & 0 \\ 0 & 6 & 0 \\ 0 & 0 & 6 \end{bmatrix}.$$

Thus,

$$a_0 = -\frac{1}{3}\operatorname{tr}(AB_0) = -6.$$ □

The only computational formula for $B(\lambda) = \operatorname{adj}(\lambda I - A)$ which computes it numerically is (7.87). This computation relies on the knowledge of the coefficients a_i's (see (7.74)) of the characteristic polynomial $\phi(\lambda)$ of A, and can be conveniently done by Faddeev algorithm: Theorem 7.20. We, however, note the following formulae for $B(\lambda)$ and reduced adjoint matrix $R(\lambda I, A)$, see Definition 7.18. For $\lambda \neq \mu$ in $\mathbb{F}$, since $\phi(\lambda) - \phi(\mu)$ and $\psi(\lambda) - \psi(\mu)$ are both divisible by $\lambda - \mu$, there exist polynomials $f(\lambda, \mu)$ and $g(\lambda, \mu)$ in the variables λ and μ so that

$$\phi(\lambda) - \phi(\mu) = (\lambda - \mu) f(\lambda, \mu), \tag{7.111}$$

$$\psi(\lambda) - \psi(\mu) = (\lambda - \mu) g(\lambda, \mu). \tag{7.112}$$

In these identities we replace λ and μ by the matrices (by extending Definition 7.5 to two variables) λI and A respectively, to get

$$\phi(\lambda) I = (\lambda I - A) f(\lambda I, A), \tag{7.113}$$

$$\psi(\lambda) I = (\lambda I - A) g(\lambda I, A), \tag{7.114}$$

where we used the fact that $\psi(A) = 0$ for the minimum polynomial $\psi(\lambda)$ of A and $\phi(A) = 0$. By Theorem 7.10, we note that, on dividing the λ-matrices, the quotient is unique, and hence comparing (7.113) and (7.114) with (7.81) and (7.82), we get

$$\text{adj}\,(\lambda I - A) =: B(\lambda) = f(\lambda I, A), \tag{7.115}$$

$$R(\lambda I, A) = g(\lambda I, A), \tag{7.116}$$

where f and g are given by (7.111) and (7.112), respectively.

Example 7.35 Compute the reduced adjoint matrix for the matrix A given in Example 7.25.

Solution We have $\phi(\lambda) = \det(\lambda I - A) = \lambda^3 - \lambda^2 - \lambda + 1$ and $\psi(\lambda) = \lambda^2 - 1$. Therefore,

$$f(\lambda, \mu) = (\lambda - \mu)^{-1}[\phi(\lambda) - \phi(\mu)] = \lambda^2 + \lambda\mu + \mu^2 - \lambda - \mu - 1,$$

$$g(\lambda, \mu) = (\lambda - \mu)^{-1}[\psi(\lambda) - \psi(\mu)] = \lambda + \mu.$$

Hence, by (7.115) and (7.116), we have

$$\text{adj}\,(\lambda I - A) = f(\lambda I, A) = \lambda^2 I + \lambda A + A^2 - \lambda I - A - I$$

$$= \begin{bmatrix} \lambda^2 - 1 & 0 & \lambda - 1 \\ 0 & \lambda^2 - 1 & 0 \\ 0 & 0 & (\lambda - 1)^2 \end{bmatrix},$$

$$R(\lambda I, A) = g(\lambda I, A) = \lambda I + A = \begin{bmatrix} \lambda + 1 & 0 & 1 \\ 0 & \lambda + 1 & 0 \\ 0 & 0 & \lambda - 1 \end{bmatrix}. \qquad \square$$

7.6 MULTIPLICITIES OF EIGENVALUES

Definition 7.20 *Simple and Multiple Eigenvalues.* When resolved into factors, the characteristic polynomial $\phi(\lambda)$ of an $n \times n$ matrix A gives

$$\phi(\lambda) = (\lambda - \lambda_1) \cdots (\lambda - \lambda_n),$$

where $\lambda_1, \ldots, \lambda_n$ are the eigenvalues of A. If a factor $(\lambda - \lambda_k)$ occurs once and only once, then the eigenvalue λ_k is said to be simple. On the other hand, if a factor, say $(\lambda - \lambda_r)$, occurs n_r times, we say that the matrix A has an eigenvalue λ_r of multiplicity n_r or n_r is defined as the *algebraic multiplicity* of the eigenvalue λ_r of A. In this text, the term *multiplicity of an eigenvalue* is used as synonymous with *algebraic multiplicity of an eigenvalue.*

Theorem 7.21 Let λ_i be an eigenvalue of multiplicity n_i of an n-square matrix A. Then, the rank of the matrix $(\lambda_i I - A)$ is at least $n - n_i$, i.e.,

(a) $(n - n_i) \leq \text{rank } (\lambda_i I - A) < n$,

(b) $0 < \text{null } (\lambda_i I - A) \leq n_i$.

Proof Since the characteristic polynomial (7.74) of A has a root λ_i of multiplicity n_i, we have

$$\phi(\lambda) = \det(\lambda I - A) = (\lambda - \lambda_i)^{n_i} \psi(\lambda), \tag{7.117}$$

where $\psi(\lambda)$ is a polynomial of degree $n - n_i$ and $\psi(\lambda_i) \neq 0$. Now, by virtue of (7.117), the characteristic polynomial associated with the matrix $(\lambda_i - A)$ is

$$\det\{(-\lambda_i + \lambda)I + A\} = (-1)^n \phi(\lambda_i - \lambda) = (-1)^{n+n_i} \lambda^{n_i} \psi(\lambda_i - \lambda),$$

which is a polynomial of the form

$$b_n \lambda^n + b_{n-1}\lambda^{n-1} + \cdots + b_{n_i}\lambda^{n_i},$$

where $b_{n_i} \neq 0$ since $\psi(\lambda_i) \neq 0$ and $b_{n_i} = (-1)^{n+n_i}\psi(\lambda_i)$. By Theorem 7.16, b_{n_i} is the sum of all principal minors of order $n - n_i$ in the matrix $(\lambda_i I - A)$. Therefore, at least one principal minor of order $n - n_i$ of $(\lambda_i I - A)$ is not zero.

Also, there may be nonzero principal minors of order $n - n_i + k$ contributing to the coefficient $b_{n_i - k}$ $(k = 1, ..., n_i)$, but they add together to give zero value for b_{n_i-k}. Hence, the rank of $(\lambda_i I - A)$ is at least $n - n_i$. Since $\det(\lambda_i I - A) = 0$, the first result, (a), follows. The second result, (b), is a consequence of Corollary 5.12. □

An easy consequence of this theorem is the following corollary.

Corollary 7.5 If an n-square matrix A has a simple eigenvalue λ, then rank $(\lambda I - A) = n - 1$. □

Example 7.36 Find the multiplicity of eigenvalues of the following three matrices:

(i) $A = \begin{bmatrix} 2 & 0 & 4 \\ 0 & 6 & 0 \\ 4 & 0 & 2 \end{bmatrix}$, (ii) $B = \begin{bmatrix} 2 & 2 & 3 \\ 0 & 1 & 2 \\ 0 & 0 & 1 \end{bmatrix}$,

(iii) $C = \begin{bmatrix} 1 & 2 & 3 \\ 0 & 1 & 2 \\ 0 & 0 & 1 \end{bmatrix}$.

Solution (i) The characteristic equation of A is $\lambda^3 - 10\lambda^2 + 12\lambda + 72 = 0$ which has the roots $\lambda_1 = -2$, $\lambda_2 = \lambda_3 = 6$ and rank $(\lambda_2 I - A) = 1$. Here $n = 3$, and λ_2 has a multiplicity $n_2 = 2$ and rank $(\lambda_2 I - A) = n - n_2 = 1$.

(ii) The matrix B has an eigenvalue $\lambda = 1$ of multiplicity $m = 2$. Since $n = 3$, we have $n - m = 1$, but rank $(\lambda I - B) = 2 > 1 = n - m$.

(iii) The only eigenvalue of C is $\lambda_1 = 1$ of multiplicity $n_1 = 3$. Since $n = 3$, rank $(\lambda_1 I - C) = 2 > 0 = n - n_1$. □

Theorem 7.22 The number k_i of independent eigenvectors associated with an eigenvalue λ_i of multiplicity n_i of an n-square matrix A is equal to $n -$ rank $(\lambda_i I - A)$. Also, $1 \leq k_i \leq n_i$.

Proof The eigenvectors **x** associated with the eigenvalue λ_i of matrix A are the solutions of

$$(\lambda_i I - A)\mathbf{x} = \theta.$$

By Theorem 5.16, the above equation has n – rank $(\lambda_i I - A)$ number of independent solutions for **x**; hence $k_i = n$ – rank $(\lambda_i I - A)$. Observing that det $(\lambda_i I - A) = 0$, the value of k_i is at least one, and by Theorem 7.21, $k_i \le n_i$. □

Let λ_i be an eigenvalue of A of *algebraic multiplicity* n_i. The *number of linearly independent eigenvectors* associated with λ_i is n – rank $(\lambda_i I - A)$ = null $(\lambda_i I - A) \le n_i$ (see Section 5.7) in view of Theorem 7.22 and is defined to be *the geometric multiplicity* of the eigenvalue λ_i of A.

Example 7.37 Consider the matrix C given in Example 7.36, which has the only eigenvalue $\lambda_1 = 1$ of multiplicity 3. The eigenvectors of C associated with 1 are the solutions of $(I - C)\mathbf{x} = \theta$, which are given by $\alpha\,[0\ 0\ 1]^T$, where α is any real number. The solution vectors are evidently of dimension one. So $k_1 = 1$, $n = 3$, $n_1 = 3$, and therefore the result $1 \le k_1 \le n_1$ of Theorem 7.22 is satisfied. The eigenvalue $\lambda_1 = 1$ of A is of *algebraic multiplicity* 3, but has a *geometric multiplicity* 1. □

Let n_i be the algebraic multiplicity, k_i the geometric multiplicity, and m_i multiplicity in the minimum polynomial of an eigenvalue λ_i of a matrix A. Then, $1 \le m_i \le n_i$ by Theorem 7.14 and $1 \le k_i \le n_i$ by Theorem 7.22. But no such relation exists between k_i and m_i. For the matrix diag [1, 1, 1], the characteristic polynomial is $\phi(\lambda) = (\lambda - 1)^3$, the minimum polynomial $\psi(\lambda) = (\lambda - 1)$ and $k_1 = 3$ (calculate !) for the eigenvalue $\lambda_1 = 1$. Therefore, $n_1 = 3$, $m_1 = 1$, and $k_1 > m_1$. On the other hand, for the matrix C in (iii), corresponding to the eigenvalue $\lambda_1 = 1$ (see Example 7.37), $n_1 = 3$, $m_1 = 3$ (calculate !) and $k_1 = 1$. In this case $k_1 < m_1$.

Definition 7.21 *Simple Matrix*; *Regular Eigenvalue.* The eigenvalue λ of multiplicity m of a matrix A is called *regular* if the number of independent eigenvectors (i.e., geometric multiplicity) associated with λ is equal to m (i.e., algebraic multiplicity) and a square matrix A with all regular eigenvalues is said to be a *simple matrix.* As a consequence of Theorems 7.22 and 5.16, an n-square matrix A with k distinct eigenvalues $\lambda_1, \lambda_2, ..., \lambda_k$ is simple iff

$$n - \text{rank}\,(\lambda_i I - A) = n_i, \qquad i = 1, ..., k,$$

where n_i is the multiplicity of λ_i. This is also a condition for the eigenvalue λ_i of multiplicity n_i of a matrix A to be regular. If the (algebraic) multiplicity n_i of an eigenvalue λ_i of A exceeds its geometric multiplicity k_i, then λ_i is said to be a *defective eigenvalue.* A matrix with a defective eigenvalue is referred to as a *defective matrix.* □

So, a square matrix A is *defective* if it is not simple. We thus see that a simple matrix A in $V_{n \times n}$ has n number of linearly independent eigenvectors which can be employed as a basis in V_n.

Theorem 7.23 An n-square matrix A is similar to a diagonal matrix iff all its eigenvalues are regular (i.e., iff A is simple).

Proof (a) *Sufficiency.* Suppose that A is simple, then it has n number of linearly independent eigenvectors $\mathbf{x}_1, \mathbf{x}_2, ..., \mathbf{x}_n$ associated with eigenvalues $\lambda_1, \lambda_2, ..., \lambda_n$, which are not necessarily distinct. So,

$$A[\mathbf{x}_1, \mathbf{x}_2, ..., \mathbf{x}_n] = [\mathbf{x}_1, \mathbf{x}_2, ..., \mathbf{x}_n]\ \text{diag}\ [\lambda_1, \lambda_2, ..., \lambda_n]. \tag{7.118}$$

Putting

$$P = [\mathbf{x}_1, \mathbf{x}_2, ..., \mathbf{x}_n],$$

from (7.118) we have

$$P^{-1}AP = \text{diag } [\lambda_1, \lambda_2, ..., \lambda_n].$$

(b) *Necessity.* Conversely, let A be similar to a diagonal matrix D. So there exists a nonsingular matrix X such that

$$X^{-1}AX = D, \quad \text{or} \quad AX = XD. \tag{7.119}$$

Denoting the columns of X by $\mathbf{x}_i$'s and the diagonal elements of D by d_i's, we have from (7.119), using the definition of matrix equality, the equation

$$A\mathbf{x}_i = d_i\mathbf{x}_i.$$

Hence, d_i's are the eigenvalues of A with the columns $\mathbf{x}_i$'s of X as the corresponding eigenvectors. Since X is nonsingular, $\mathbf{x}_i$'s are linearly independent and so the eigenvalues d_i's are regular. □

A weaker version of the above result is contained in Theorem 7.3.

Example 7.38 The eigenvalue $\lambda = 6$ of multiplicity 2 of the matrix A in Example 7.36 is regular. This is because of Definition 7.21, as we have here n–rank $(\lambda I - A) = 3 - 1 = 2$. Also, the eigenvectors of A associated with $\lambda = 6$ when computed as the solution of

$$(\lambda I - A)\mathbf{x} = \begin{bmatrix} 4 & 0 & -4 \\ 0 & 0 & 0 \\ -4 & 0 & 4 \end{bmatrix} \begin{bmatrix} x_1 \\ x_2 \\ x_3 \end{bmatrix} = \boldsymbol{\theta}$$

have the general form $[\alpha\ \beta\ \alpha]^T$, where α and β are real. So the eigenvector $\mathbf{x}_2 = [0\ 1\ 0]^T$ and $\mathbf{x}_3 = [1\ 0\ 1]^T$ associated $\lambda = 6$ are linearly independent, and hence $\lambda = 6$ is a regular eigenvalue of A which is also a simple matrix.

Again, one eigenvector of A corresponding to the eigenvalue $\lambda_1 = -2$ can be found to be $\mathbf{x}_1 = [1\ 0\ -1]^T$. If we now construct a nonsingular matrix P by $P = [\mathbf{x}_1\ \mathbf{x}_2\ \mathbf{x}_3]$, then it is easy to verify that $P^{-1}AP = \text{diag } (-2, 6, 6)$. □

7.7 EIGENVALUE PROBLEM FOR HERMITIAN MATRICES

It might be recalled (see Definition 1.10) that a square matrix A defined over a complex field is called Hermitian if A is equal to its transposed conjugate A^*. When all the elements of A are real, a Hermitian matrix A becomes a real symmetric matrix. Since a real symmetric matrix is a particular case of a Hermitian matrix, the properties of symmetric matrices would follow as particular cases of Hermitian matrices. In view of this, in the sequel, the theorems only for Hermitian matrices will be established, followed by mere statements of the corresponding results for real symmetric matrices.

Theorem 7.24 The eigenvalues of (a) a Hermitian matrix H are real and (b) of a skew-Hermitian matrix S are purely imaginary.

Proof (a) Let λ be an eigenvalue of H with an associated eigenvector $\mathbf{x} \neq \boldsymbol{\theta}$. Then

$$H\mathbf{x} = \lambda\mathbf{x}. \tag{7.120}$$

Premultiplying both sides of (7.120) by $\mathbf{x}^*$, the transposed conjugate of $\mathbf{x}$, we get

$$\mathbf{x}^*H\mathbf{x} = \lambda\mathbf{x}^*\mathbf{x}. \tag{7.121}$$

The transposed conjugate of both sides of this equation gives

$$\mathbf{x}^*H^*\mathbf{x} = \mathbf{x}^*H\mathbf{x} = \bar{\lambda}\,\mathbf{x}^*\mathbf{x} \tag{7.122}$$

since $H = H^*$. Substracting (7.122) from (7.121), we have

$$(\lambda - \bar{\lambda})\mathbf{x}^*\mathbf{x} = 0. \tag{7.123}$$

Since $\mathbf{x} \neq \boldsymbol{\theta}$, we must have $\lambda = \bar{\lambda}$, i.e., λ is real. □

(b) Let μ be an eigenvalue of S with an associated eigenvector $\mathbf{x} \neq \boldsymbol{\theta}$. Then $S\mathbf{x} = \mu\mathbf{x}$. Since $S = -S^*$, proceeding as in part (a), we have, in place of (7.123), the relation

$$(\mu + \bar{\mu})\mathbf{x}^*\mathbf{x} = 0.$$

Hence, $\mu + \bar{\mu} = 0$ as $\mathbf{x} \neq \boldsymbol{\theta}$. Therefore, μ is imaginary. □

A direct consequence of the foregoing theorem is the following corollary.

Corollary 7.6 The eigenvalues of a real symmetric matrix are real and of a real skew-symmetric matrix are purely imaginary. □

We have already acquainted ourselves with some basic characteristics of orthogonal and unitary matrices in Section 6.5. Presently we shall see, how these matrices play an important role in diagonalizing Hermitian and real symmetric matrices. Before pursuing this matter, however, we shall study the property of eigenvalues of these matrices.

Theorem 7.25 Suppose that P is an orthogonal matrix. Then, (a) if λ is an eigenvalue of P, so also λ^{-1}; (b) every eigenvalue of P has a unit modulus, i.e., it can be represented in the form $\lambda = e^{i\alpha}$, where $0 \le \alpha \le 2\pi$.

Proof (a) Suppose that λ is an eigenvalue of an orthogonal matrix P with the corresponding eigenvector $\mathbf{x}$. Therefore,

$$P\mathbf{x} = \lambda\mathbf{x}. \tag{7.124}$$

Since P is nonsingular (see Definition 6.12), no eigenvalue of P is zero in view of Corollary 7.4. Premultiplying both sides of (7.124) by P^T and taking into account (6.42), we have, after minor rearrangement, the relation

$$P^T\mathbf{x} = \lambda^{-1}\mathbf{x}$$

which shows that λ^{-1} is an eigenvalue of P^T. Since the eigenvalues of P and P^T are the same (see section 7.2.1), we conclude that λ^{-1} is also an eigenvalue of P.

(b) Let $\mathbf{x}$ be an eigenvector of P corresponding to the eigenvalue λ, as before. Since P is real and $P^TP = I$, applying (5.16), (7.124) and (5.18), we obtain

$$\langle\mathbf{x}, \mathbf{x}\rangle = \langle P^T P\mathbf{x}, \mathbf{x}\rangle = \mathbf{x}^T P^T P\mathbf{x} = \langle P\mathbf{x}, P\mathbf{x}\rangle = \langle\lambda\mathbf{x}, \lambda\mathbf{x}\rangle = \lambda\bar{\lambda}\,\langle\mathbf{x}, \mathbf{x}\rangle.$$

Hence, $\lambda\bar{\lambda} = |\lambda|^2 = 1$, because the eigenvector $\mathbf{x} \neq \boldsymbol{\theta}$. Therefore, λ is of the form $e^{i\alpha}$. □

In view of Theorem 7.25 we can prove, without difficulty, the following theorem.

Theorem 7.26 Suppose that λ is any eigenvalue of a unitary matrix U over a complex field $\mathbb{C}$. Then, (i) $\bar{\lambda}^{-1}$ is also an eigenvalue of U and (ii) every eigenvalue of U has a unit modulus. □

The nature of eigenvalues of a Hermitian/symmetric matrix, a skew-Hermitian/skew-symmetric matrix, and orthogonal/unitary matrix is summarized in Figure 7.3.

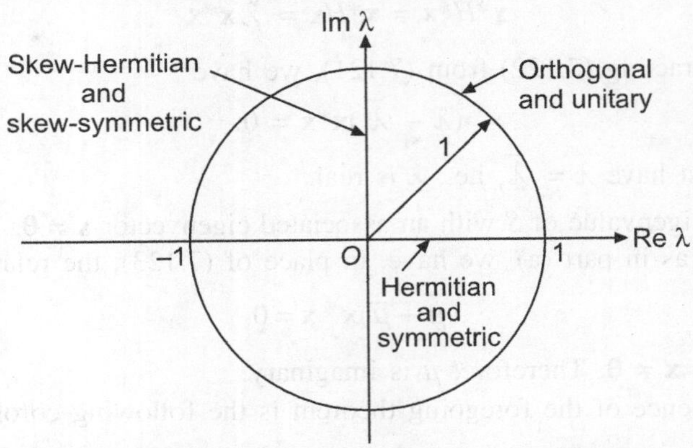

Figure 7.3 Location of eigenvalues on the complex plane for (a) Hermitian and symmetric matrices; (b) skew-Hermitian and skew-symmetric matrices; (c) orthogonal and unitary matrices.

Theorem 7.27 For every Hermitian matrix H of order n, there exists a unitary transformation S such that $S^*HS = \text{diag}(\lambda_1, ..., \lambda_n)$, where λ_i's are the eigenvalues of H not necessarily distinct and nonzero. Stated alternatively, every Hermitian matrix H is unitarily similar to a diagonal matrix whose diagonal elements are the eigenvalues of H.

Proof The theorem is obviously true when $n = 1$. Suppose that $n = 2$. Define a 2×2 matrix U whose first column is the normalized (i.e., unit), see Definition 5.12, eigenvector $\mathbf{x}_1$ $(\mathbf{x}_1^* \mathbf{x}_1 = 1)$ associated with λ_1 and the second column of U is a vector $\mathbf{x}_2$ orthonormal (see Definition 5.13) to $\mathbf{x}_1$ (i.e., $\mathbf{x}_1^*\mathbf{x}_2 = \mathbf{x}_2^*\mathbf{x}_1 = 0, \mathbf{x}_2^*\mathbf{x}_2 = 1$), making U unitary which is always available in view of the Gram-Schmidt orthonormalization process (see Theorem 5.8). Therefore,

$$H\mathbf{x}_1 = \lambda_1\mathbf{x}_1, \qquad U = [\mathbf{x}_1, \mathbf{x}_2], \qquad UU^* = U^*U = I,$$

$$U^*HU = \begin{bmatrix} \mathbf{x}_1^* \\ \mathbf{x}_2^* \end{bmatrix} H[\mathbf{x}_1, \mathbf{x}_2] = \begin{bmatrix} \mathbf{x}_1^* \\ \mathbf{x}_2^* \end{bmatrix} [\lambda_1 \mathbf{x}_1, H\mathbf{x}_2]$$

$$= \begin{bmatrix} \lambda_1 & \mathbf{x}_1^*H\mathbf{x}_2 \\ 0 & \mathbf{x}_2^*H\mathbf{x}_2 \end{bmatrix} = \begin{bmatrix} \lambda_1 & h_{12} \\ 0 & h_{22} \end{bmatrix}, \tag{7.125}$$

where $h_{12} = \mathbf{x}_1^*H\mathbf{x}_2$ and $h_{22} = \mathbf{x}_2^*H\mathbf{x}_2$. Since $U^*HU = (U^*HU)^*$ as H is Hermitian, from (7.125) we have $h_{12} = 0$. The eigenvalues of H under a unitary transformation remain unchanged (see Theorem 7.1 and note that $U^* = U^{-1}$), which means that the eigenvalues of

H are the same as those of U^*HU. So, from (7.125) we have $h_{22} = \lambda_2$, the second eigenvalue of H. We thus have

$$U^*HU = \text{diag}\,(\lambda_1, \lambda_2)$$

which is the proposition of the theorem when $n = 2$, taking $S = U$. Let us now assume that the theorem is true for order $(n-1)$ and that H is a Hermitian matrix of order n. Construct a unitary matrix U of order n whose first column is the unit eigenvector $\mathbf{x}_1$ associated with λ_1 and whose other columns are denoted by $\mathbf{x}_2 \cdots \mathbf{x}_n$. By Gram-Schmidt orthonormalization process, this is always possible.

Observing that $\mathbf{x}_1 \cdots \mathbf{x}_n$ are mutually orthonormal and $H\mathbf{x}_1 = \lambda_1\mathbf{x}_1$, we have

$$U^*HU = \begin{bmatrix} \overset{*}{\mathbf{x}}_1 \\ \overset{*}{\mathbf{x}}_2 \\ \vdots \\ \overset{*}{\mathbf{x}}_n \end{bmatrix} H[\mathbf{x}_1\ \mathbf{x}_2\ \cdots\ \mathbf{x}_n] = \begin{bmatrix} \overset{*}{\mathbf{x}}_1 \\ \overset{*}{\mathbf{x}}_2 \\ \vdots \\ \overset{*}{\mathbf{x}}_n \end{bmatrix} [\lambda_1\mathbf{x}_1, H\mathbf{x}_2, \ldots, H\mathbf{x}_n]$$

$$= \left[\begin{array}{c|ccc} \lambda_1 & \overset{*}{\mathbf{x}}_1 H\mathbf{x}_2 & \cdots & \overset{*}{\mathbf{x}}_1 H\mathbf{x}_n \\ \hline 0 & & & \\ \vdots & & G & \\ 0 & & & \end{array}\right], \tag{7.126}$$

where G is a matrix of order $n - 1$. Noting that U^*HU is Hermitian, we deduce from (7.126) that

$$\overset{*}{\mathbf{x}}_1 H\mathbf{x}_2 = \cdots = \overset{*}{\mathbf{x}}_1 H\mathbf{x}_n = 0 \tag{7.127}$$

and that G is Hermitian. As in the second order case, the eigenvalues of H are obtained from $(\lambda - \lambda_1)\det(\lambda I - G) = 0$, which shows that the remaining eigenvalues $\lambda_2 \cdots \lambda_n$ of H are the same as those of G.

By induction hypothesis, there exists a unitary matrix T of order $n - 1$ such that

$$T^*GT = \text{diag}(\lambda_2, \ldots, \lambda_n). \tag{7.128}$$

Therefore, observing that the n-square matrix P defined by

$$P = \left[\begin{array}{c|ccc} 1 & 0 & \cdots & 0 \\ \hline 0 & & & \\ \vdots & & T & \\ 0 & & & \end{array}\right]$$

is unitary, premultiplying (7.126) by P^* and post-multiplying it by P, we obtain

$$(UP)^*H(UP) = \text{diag}\,(\lambda_1, \ldots, \lambda_n), \tag{7.129}$$

where we have used (7.127) and (7.128). If we define $S = UP$, then S is unitary because $S^*S = P^*U^*UP = P^*P = I_n$. Consequently, from (7.129), $S^*HS = \text{diag}\,(\lambda_1, \ldots, \lambda_n)$, where S is unitary. □

A particular case of Theorem 7.27 is the following corollary.

Corollary 7.7 For every real symmetric matrix A, there exists an orthogonal transformation P such that P^TAP = diag $[\lambda_1, ..., \lambda_n]$, where $\lambda_1, ..., \lambda_n$ are the eigenvalues of A not necessarily distinct and nonzero. Stated alternatively, every real symmetric matrix A is orthogonally similar to a diagonal matrix whose diagonal elements are the eigenvalues of A. □

The property of a Hermltian matrix was used in the proof of Theorem 7.27 to reduce to zero the elements above the main diagonal of a transformed matrix as on the right hand side of (7.125) and (7.126). If this property were not used, by a similar induction procedure, the following corollary would have resulted.

Corollary 7.8 For every square matrix A over $\mathbb{C}$, there exists a unitary transformation S such that S^*AS is an upper (or a lower) triangular matrix (which is not unique) with the eigenvalues of A as diagonal elements. □

This result, that S^*AS is an upper (or a lower) triangular matrix, is known as *Schur decomposition* of A.

Following the proof of Theorem 7.27, the transformation of an arbitrary matrix to an upper triangular matrix is shown in Example 7.39. For a lower triangular form, one has to start with the transpose of A.

Example 7.39 Determine an orthogonal matrix S such that S^TAS is upper triangular, where

$$A = \begin{bmatrix} 0 & 1 & 1 \\ -2 & 3 & 2 \\ -3 & 3 & 4 \end{bmatrix}.$$

Solution The characteristic polynomial of A is

$$\phi(\lambda) = \det(\lambda I - A) = \lambda^3 - 7\lambda^2 + 11\lambda - 5 = (\lambda - 1)^2 (\lambda - 5).$$

Therefore, the eigenvalues of A are $\lambda_1 = 5$, $\lambda_2 = 1$, $\lambda_3 = 1$. The eigenvector associated with 1 can be computed to give $[\alpha + \beta, \alpha, \beta]^T$, where α and β are arbitrary. We choose $\mathbf{x}_1 = [1\ 1\ 0]^T$ as the eigenvector of 1 and select $\mathbf{x}_2 = \mathbf{e}_2$ and $\mathbf{x}_3 = \mathbf{e}_3$ (see (5.4)), so that $\mathbf{x}_1, \mathbf{x}_2, \mathbf{x}_3$ are linearly independent. We now apply Theorem 5.8 to get the following set of three orthogonal vectors:

$$\mathbf{y}_1 = \mathbf{x}_1 = \begin{bmatrix} 1 \\ 1 \\ 0 \end{bmatrix},$$

$$\mathbf{y}_2 = \mathbf{e}_2 - \frac{\langle \mathbf{e}_2, \mathbf{y}_1 \rangle}{\langle \mathbf{y}_1, \mathbf{y}_1 \rangle} \mathbf{y}_1 = \frac{1}{2} \begin{bmatrix} -1 \\ 1 \\ 0 \end{bmatrix},$$

$$\mathbf{y}_3 = \mathbf{e}_3 - \frac{\langle \mathbf{e}_3, \mathbf{y}_1 \rangle}{\langle \mathbf{y}_1, \mathbf{y}_1 \rangle} \mathbf{y}_1 - \frac{\langle \mathbf{e}_3, \mathbf{y}_2 \rangle}{\langle \mathbf{y}_2, \mathbf{y}_2 \rangle} \mathbf{y}_2 = \mathbf{e}_3.$$

Hence, the set of orthonormal vectors are

$$\mathbf{u}_1 = \frac{1}{\sqrt{2}} [1\,1\,0]^T, \quad \mathbf{u}_2 = \frac{1}{\sqrt{2}} [-1\,1\,0]^T, \quad \mathbf{u}_3 = \mathbf{e}_3.$$

We now define an orthogonal matrix $U = [\mathbf{u}_1\ \mathbf{u}_2\ \mathbf{u}_3]$. In the light of the proof of Theorem 7.27, we compute

$$U^TAU = \frac{1}{2}\begin{bmatrix} 1 & 1 & 0 \\ -1 & 1 & 0 \\ 0 & 0 & \sqrt{2} \end{bmatrix}\begin{bmatrix} 0 & 1 & 1 \\ -2 & 3 & 2 \\ -3 & 3 & 4 \end{bmatrix}\begin{bmatrix} 1 & -1 & 0 \\ 1 & 1 & 0 \\ 0 & 0 & \sqrt{2} \end{bmatrix}$$

$$= \left[\begin{array}{c|cc} 1 & 3 & 3/\sqrt{2} \\ \hline 0 & 2 & 1/\sqrt{2} \\ 0 & 3\sqrt{2} & 4 \end{array}\right] = \left[\begin{array}{c|c} 1 & 3 \quad 3\sqrt{2} \\ \hline 0 & \\ 0 & G \end{array}\right]$$

which corresponds to (7.126) having $\lambda_1 = 1$ at the upper left corner. In the above equation we identify G as

$$G = \begin{bmatrix} 2 & 1/\sqrt{2} \\ 3\sqrt{2} & 4 \end{bmatrix}$$

which has a characteristic polynomial det $(\lambda I - G) = \lambda^2 - 6\lambda + 5$ having eigenvalues $\lambda_2 = 1$, $\lambda_3 = 5$, which are the unused eigenvalues of A. We now repeat the foregoing procedure for G. The eigenvector of G corresponding to $\lambda_2 = 1$ can be computed to give $[1, -\sqrt{2}]^T$. We construct as before, an orthonormal set of eigenvectors:

$$\mathbf{v}_1 = \frac{1}{\sqrt{3}}\begin{bmatrix} 1 \\ -\sqrt{2} \end{bmatrix}, \qquad \mathbf{v}_2 = \frac{1}{\sqrt{3}}\begin{bmatrix} \sqrt{2} \\ 1 \end{bmatrix},$$

and hence select an orthogonal matrix $T = [\mathbf{v}_1, \mathbf{v}_2]$. In fact, we have

$$T^TGT = \frac{1}{3}\begin{bmatrix} 1 & -\sqrt{2} \\ \sqrt{2} & 1 \end{bmatrix}\begin{bmatrix} 2 & 1/\sqrt{2} \\ 3\sqrt{2} & 4 \end{bmatrix}\begin{bmatrix} 1 & \sqrt{2} \\ -\sqrt{2} & 1 \end{bmatrix} = \begin{bmatrix} 1 & -5/\sqrt{2} \\ 0 & 5 \end{bmatrix}.$$

We now define

$$P = \left[\begin{array}{c|cc} 1 & 0 & 0 \\ \hline 0 & & \\ 0 & T & \end{array}\right] = \frac{1}{\sqrt{3}}\begin{bmatrix} \sqrt{3} & 0 & 0 \\ 0 & 1 & \sqrt{2} \\ 0 & -\sqrt{2} & 1 \end{bmatrix}.$$

The required transformation is

$$S = UP = \frac{1}{\sqrt{6}}\begin{bmatrix} 1 & -1 & 0 \\ 1 & 1 & 0 \\ 0 & 0 & \sqrt{2} \end{bmatrix}\begin{bmatrix} \sqrt{3} & 0 & 0 \\ 0 & 1 & \sqrt{2} \\ 0 & -\sqrt{2} & 1 \end{bmatrix} = \frac{1}{\sqrt{6}}\begin{bmatrix} \sqrt{3} & -1 & -\sqrt{2} \\ \sqrt{3} & 1 & \sqrt{2} \\ 0 & -2 & \sqrt{2} \end{bmatrix}.$$

It can be verified that

$$S^TAS = \begin{bmatrix} 1 & \sqrt{3}/2 & 3\sqrt{6}/2 \\ 0 & 1 & -5\sqrt{2}/2 \\ 0 & 0 & 5 \end{bmatrix}$$

which is indeed an upper triangular matrix with the eigenvalues of A along the diagonal and is called the Schur decomposition of A. □

The transformation of a Hermitian matrix into a diagonal form following the method outlined in the proof of Theorem 7.27 is laborious as is evident from Example 7.39 where triangularization of a matrix is attempted. This computation for diagonal transformation of Hermitian matrices can be simplified for which we need the following Theorems 7.28–7.31.

Theorem 7.28 Let λ be an eigenvalue of multiplicity m of a Hermitian matrix H. Then, the number of linearly independent eigenvectors associated with λ is m.

Proof By Theorem 7.27, there exists a unitary matrix U such that

$$U^*HU = \text{diag}\ (\lambda_1, ..., \lambda_m; \lambda_{m+1}, ..., \lambda_n), \tag{7.130}$$

where $\lambda = \lambda_1 = \cdots = \lambda_m$, $\lambda_{m+1}, ..., \lambda_n$ are the eigenvalues of H, $\lambda_i \neq \lambda$, $i = m + 1, ..., n$, and the columns of U are the orthonormal set of vectors $\mathbf{x}_1, ..., \mathbf{x}_m; \mathbf{x}_{m+1}, ..., \mathbf{x}_n$. This ordering of eigenvalues are possible by similar transposition of columns and rows of the matrix diag $(\lambda_1, ..., \lambda_n)$ which is equivalent to premultiplication and postmultiplication of both sides of (7.130) by elementary matrices, which does not alter the unicity of U. Since $UU^* = I$, after some simple rearrangement of (7.130), we have

$$\begin{aligned} U^*(\lambda I - H)U &= \lambda I - U^*HU \\ &= \lambda I - \text{diag}\ (\lambda_1, ..., \lambda_m; \lambda_{m+1}, ..., \lambda_n) \\ &= \text{diag}\ (0, ..., 0; \lambda - \lambda_{m+1}, ..., \lambda - \lambda_n). \end{aligned}$$

Hence, as λ is different from λ_i, $i = m + 1, ..., n$, and as U is nonsingular, we have, in view of Corollary 5.14, rank $(\lambda I - H) = n - m$. By applying Theorem 7.22, it now follows that the number of linearly independent eigenvectors associated with λ is m. □

Corollary 7.9 The number of linearly independent eigenvectors associated with an eigenvalue λ of multiplicity m of a real symmetric matrix is m. □

Theorem 7.28 establishes the existence of m linearly independent eigenvectors associated with an eigenvalue λ of multiplicity m of a Hermitian matrix H, but these eigenvectors may not be orthogonal (see Example 7.40). However, these m linearly independent vectors can be chosen as orthogonal by the Gram-Schmidt process of orthogonalization as established in Theorem 7.30.

Theorem 7.29 The eigenvectors associated with distinct eigenvalues of a Hermitian matrix constitute an orthogonal set (see Definition 5.11).

Proof Suppose that H is a Hermitian matrix and λ_i and λ_j are its two distinct eigenvalues with $\mathbf{x}_i$ and $\mathbf{x}_j$ as the respective eigenvectors. Therefore, we can write

$$H\mathbf{x}_i = \lambda_i\mathbf{x}_i, \qquad H\mathbf{x}_j = \lambda_j\mathbf{x}_j.$$

Since by virtue of Theorem 7.24, λ_i and λ_j are real, taking the inner product with $\mathbf{x}_j$ and $\mathbf{x}_i$ of the above two equations respectively, we have

$$\langle H\mathbf{x}_i, \mathbf{x}_j \rangle = \langle \lambda_i\, \mathbf{x}_i, \mathbf{x}_j \rangle = \lambda_i\, \langle \mathbf{x}_i, \mathbf{x}_j \rangle, \tag{7.131}$$

$$\langle \mathbf{x}_i, H\mathbf{x}_j \rangle = \langle \mathbf{x}_i, \lambda_j\, \mathbf{x}_j \rangle, \quad \text{or} \quad \langle H\mathbf{x}_i, \mathbf{x}_j \rangle = \lambda_j\, \langle \mathbf{x}_i, \mathbf{x}_j \rangle, \tag{7.132}$$

where we have used Definition 5.9, Lemma 5.2, and the fact that $H = H^*$. By subtracting one from the other, it follows from (7.131) and (7.132) that

$$(\lambda_i - \lambda_j)\langle \mathbf{x}_i, \mathbf{x}_j \rangle = 0.$$

Since $\lambda_i \neq \lambda_j$, we have $\langle \mathbf{x}_i, \mathbf{x}_j \rangle = 0$, i.e., the eigenvectors $\mathbf{x}_i$ and $\mathbf{x}_j$ of H associated with distinct eigenvalues are orthogonal.

Corollary 7.10 The eigenvectors associated with distinct eigenvalues of a real symmetric matrix constitute an orthogonal set. □

Theorem 7.30 Suppose that a Hermitian matrix H has an eigenvalue λ of multiplicity m. Then there exists a set of m number of orthogonal (as well as orthonormal) eigenvectors associated with λ.

Proof Let λ be an eigenvalue of H having multiplicity m, and $\mathbf{x}_1, \ldots, \mathbf{x}_m$ be the corresponding eigenvectors so that

$$H\mathbf{x}_i = \lambda \mathbf{x}_i, \qquad i = 1, \ldots, m. \tag{7.133}$$

By Theorem 7.28, these eigenvectors are linearly independent but not necessarily orthogonal, and in view of Theorem 5.7, we define a set of orthogonal vectors $\mathbf{y}_i$'s by

$$\mathbf{y}_1 = \mathbf{x}_1,$$

$$\mathbf{y}_k = \mathbf{x}_k - \sum_{i=1}^{k-1} \frac{\langle \mathbf{x}_k, \mathbf{y}_i \rangle}{\langle \mathbf{y}_i, \mathbf{y}_i \rangle} \mathbf{y}_i, \qquad k = 2, \ldots, m; \tag{7.134}$$

which are also the eigenvectors of H associated with λ. Indeed, we have $H\mathbf{y}_1 = H\mathbf{x}_1 = \lambda\mathbf{x}_1 = \lambda\mathbf{y}_1$, and by mathematical induction, suppose that $H\mathbf{y}_i = \lambda\mathbf{y}_i$, $i = 1, \ldots, k-1$. Hence, using (6.3) and (7.133), from (7.134) we have

$$H\mathbf{y}_k = H\mathbf{x}_k - \sum_{i=1}^{k-1} \frac{\langle \mathbf{x}_k, \mathbf{y}_i \rangle}{\langle \mathbf{y}_i, \mathbf{y}_i \rangle} H\mathbf{y}_i = \lambda\mathbf{x}_k - \sum_{i=1}^{k-1} \frac{\langle \mathbf{x}_k, \mathbf{y}_i \rangle}{\langle \mathbf{y}_i, \mathbf{y}_i \rangle} \lambda\mathbf{y}_i = \lambda\mathbf{y}_k.$$

Therefore, $\mathbf{y}_1, \ldots, \mathbf{y}_m$ defined by (7.134) are a set of m orthogonal eigenvectors associated with the multiple eigenvalue λ. □

Corollary 7.11 A real symmetric matrix A having an eigenvalue λ of multiplicity m possesses a set of m orthogonal (as well as orthonormal) eigenvectors associated with λ. □

Theorem 7.31 A Hermitian matrix H of order n possesses an orthogonal (as well as an orthonormal) set of n eigenvectors.

Proof By constructive procedure of Theorem 7.30, we choose for every eigenvalue of multiplicity m an orthogonal set of m eigenvectors $\mathbf{x}_1, \ldots, \mathbf{x}_m$, and for a simple eigenvalue we choose an eigenvector computed in the usual way. In view of Theorem 7.29, the eigenvectors associated with distinct eigenvalues are mutually orthogonal, the set of chosen eigenvectors for multiple eigenvalues along with those computed for simple eigenvalues therefore constitute a set of orthogonal vectors. □

Corollary 7.12 An $n \times n$ real symmetric matrix A possesses an orthogonal (as well as orthonormal) set of n eigenvectors. □

An $n \times n$ Hermitian matrix H is always similar to a diagonal matrix D whose diagonal elements are the eigenvalues of H. In other words, for any Hermitian matrix H, there exists a unitary matrix U such that

$$U^*HU = D, \tag{7.135}$$

where $U^* = U^{-1}$. The existence of such a result is proved in Theorem 7.27 which does not however suggest how to construct simply the transformation matrix U. By Theorem 7.28, we know that every Hermitian matrix is simple. Also, by virtue of Theorems 7.29 and 7.30 which establish Theorem 7.31, we can construct an orthonormal set of n eigenvectors which can be employed to build up the matrix U. Since H is simple, we can apply Theorem 7.23 to show that (7.135) is satisfied with this construction of U. The orthogonal properties of eigenvectors demonstrated by Theorems 7.5 and 7.29 are the best results available for any arbitrary matrix A. On the other hand, for a real symmetric matrix A, one can construct, with the support of Corollaries 7.10–7.12, an orthogonal matrix P such that $P^TAP = D$, where D is a diagonal matrix with the eigenvalues of A as the diagonal elements.

Example 7.40 For the real symmetric matrix A given below, find an orthogonal transformation P such that P^TAP is diagonal with the eigenvalues of A as the diagonal elements:

$$A = \begin{bmatrix} 1 & 2 & 0 \\ 2 & 1 & 0 \\ 0 & 0 & -1 \end{bmatrix}.$$

Solution The characteristic equation of A is

$$\lambda^3 - \lambda^2 - 5\lambda - 3 = (\lambda + 1)^2 (\lambda - 3) = 0,$$

which shows that the eigenvalues of A are $\lambda_1 = -1$ of multiplicity 2 and $\lambda_2 = 3$ simple. By solving $(\lambda_1 I - A)\mathbf{x} = \boldsymbol{\theta}$, we see that the eigenvectors associated with λ_1 can be expressed by $[\alpha, -\alpha, \beta]^T$, where α and β are real, and hence the two independent eigenvectors corresponding to λ_1 are

$$\mathbf{x}_1 = \begin{bmatrix} 1 \\ -1 \\ 0 \end{bmatrix}, \qquad \mathbf{x}_2 = \begin{bmatrix} 1 \\ -1 \\ 1 \end{bmatrix},$$

which are not orthogonal. By Theorem 5.7 of the Gram-Schmidt orthogonalization process, we have

$$\mathbf{y}_1 = \mathbf{x}_1, \qquad \mathbf{y}_2 = \mathbf{x}_2 - \frac{\langle \mathbf{x}_2, \mathbf{y}_1 \rangle}{\langle \mathbf{y}_1, \mathbf{y}_1 \rangle}\mathbf{y}_1 = \begin{bmatrix} 0 \\ 0 \\ 1 \end{bmatrix}.$$

Therefore, $\mathbf{y}_1$ and $\mathbf{y}_2$ are two orthogonal eigenvectors associated with $\lambda_1 = -1$. By solving $(\lambda_2 I - A)\mathbf{x} = \boldsymbol{\theta}$, we get $\mathbf{y}_3 = [1 \ 1 \ 0]^T$ as an eigenvector associated with $\lambda_2 = 3$ which is automatically orthogonal to both $\mathbf{y}_1$ and $\mathbf{y}_2$ in view of Corollary 7.10. By Theorem 5.8, the orthonormal eigenvectors of the matrix A are, therefore,

$$\mathbf{z}_1 = \frac{1}{\sqrt{2}}\begin{bmatrix} 1 \\ -1 \\ 0 \end{bmatrix}, \qquad \mathbf{z}_2 = \begin{bmatrix} 0 \\ 0 \\ 1 \end{bmatrix}, \qquad \mathbf{z}_3 = \frac{1}{\sqrt{2}}\begin{bmatrix} 1 \\ 1 \\ 0 \end{bmatrix}.$$

We now define an orthogonal matrix P by $P = [\mathbf{z}_1, \mathbf{z}_2, \mathbf{z}_3]$ which satisfies $P^TP = I$, and it is easy to verify, as a result of Theorem 7.23, that $P^{-1}AP = P^TAP = \text{diag}\ (-1, -1, 3)$. □

Theorem 7.32 Let A and B be two Hermitian matrices of the same order n with A having positive eigenvalues. Then, there exists an $n \times n$ nonsingular matrix P, which is not unique, such that

$$P^*AP = I, \qquad P^*BP = \text{diag}\ (c_1, ..., c_n),$$

where c_i's are real.

Proof Since A is Hermitian, by Theorem 7.27, there exists a unitary matrix U such that

$$U^*AU = \text{diag}\ (\lambda_1, ..., \lambda_n), \tag{7.136}$$

where λ_j's are positive eigenvalues of A. We now introduce two n-vectors $\mathbf{x} = [x_1 \cdots x_n]^T$ and $\mathbf{y} = [y_1 \cdots y_n]^T$ related by $\mathbf{x} = U\mathbf{y}$. Then, by (7.136) we have

$$\mathbf{x}^*A\mathbf{x} = \mathbf{y}^*(U^*AU)\mathbf{y} = \lambda_1\bar{y}_1y_1 + \cdots + \lambda_n\bar{y}_ny_n, \tag{7.137}$$

$$\mathbf{x}^*B\mathbf{x} = \mathbf{y}^*(U^*BU)\mathbf{y}. \tag{7.138}$$

Using the transformation $\mathbf{y} = Q\mathbf{z}$, where $Q^{-1} = \text{diag}\ (\sqrt{\lambda_1}, ..., \sqrt{\lambda_n})$ and $\mathbf{z} = [z_1 \cdots z_n]^T$, we have from (7.137) and (7.138) the relations

$$\mathbf{x}^*A\mathbf{x} = \mathbf{z}^*(UQ)^*A(UQ)\mathbf{z} = \bar{z}_1z_1 + \cdots + \bar{z}_nz_n = \mathbf{z}^*I_n\mathbf{z}, \tag{7.139}$$
$$\mathbf{x}^*B\mathbf{x} = \mathbf{z}^*(UQ)^*B(UQ)\mathbf{z}.$$

Since B is Hermitian, $C = (UQ)^*B(UQ)$ too is Hermitian. Therefore, there exists a unitary matrix R by virtue of Theorem 7.27 such that

$$(UQR)^*B(UQR) = \text{diag}\ (c_1, ..., c_n), \tag{7.140}$$

where c_i's are the eigenvalues of the Hermitian matrix C, and are therefore real. Denoting by $P := UQR$, from (7.139) and (7.140) we have

$$P^*AP = (UQR)^*A(UQR) = R^*(UQ)^*A(UQ)R = R^*R = I,$$

$$P^*BP = \text{diag}\ (c_1, ..., c_n),$$

where c_i's are the eigenvalues of P^*BP. □

Corollary 7.13 Given any two real symmetric matrices A and B of order n with A having positive eigenvalues, there exists a nonsingular matrix P of order n, which is not unique, such that $P^TAP = I$ and $P^TBP = \text{diag}\ (c_1, ..., c_n)$, where c_i's, being the eigenvalues of the symmetric matrix P^TBP, are real. □

Example 7.41 Given two real symmetric matrices A and B by

$$A = \begin{bmatrix} 6 & 4 & -2 \\ 4 & 12 & -4 \\ -2 & -4 & 13 \end{bmatrix}, \qquad B = \begin{bmatrix} 1 & -1 & 1 \\ -1 & 1 & -1 \\ 1 & -1 & 1 \end{bmatrix}, \tag{7.141}$$

determine a nonsingular matrix P such that $P^TAP = I$ and P^TBP is diagonal.

Solution The characteristic polynomial of A is $\phi(\lambda) = \det(\lambda I - A) = \lambda^3 - 31\lambda^2 + 270\lambda - 648 = (\lambda - 4)(\lambda - 9)(\lambda - 18)$. The eigenvalues are therefore 4, 9, 18, .and the associated orthogonal eigenvectors are, respectively,

$$\mathbf{x}_1 = [-2\ 1\ 0]^T, \quad \mathbf{x}_2 = [2\ 4\ 5]^T, \quad \mathbf{x}_3 = [1\ 2\ -2]^T.$$

We now normalize these vectors (see Definition 5.12) and construct an orthogonal matrix with these normalized vectors as columns to get

$$U = \frac{1}{3\sqrt{5}} \begin{bmatrix} -6 & 2 & \sqrt{5} \\ 3 & 4 & 2\sqrt{5} \\ 0 & 5 & -2\sqrt{5} \end{bmatrix}.$$

It can be checked that $U^TAU = \text{diag}\ [4, 9, 18]$. We now define $Q^{-1} = \text{diag}\ [2, 3, 3\sqrt{2}]$ so that

$$(UQ)^TA(UQ) = I.$$

Computing the matrix C defined by $C = (UQ)^TB(UQ)$, we get

$$C = \frac{1}{180} \begin{bmatrix} 81 & -18 & 9\sqrt{10} \\ -18 & 4 & -2\sqrt{10} \\ 9\sqrt{10} & -2\sqrt{10} & 10 \end{bmatrix}.$$

The characteristic equation of C is $\det(\lambda I - C) = \lambda^2(\lambda - 19/36)$. Hence its eigenvalues are $\lambda_1 = 0$, $\lambda_2 = 0$, $\lambda_3 = 19/36$, and the associated eigenvectors are, respectively,

$$[-2, -4, \sqrt{10}]^T, \quad [1, -1/2, -\sqrt{10}]^T, \quad [9, -2, \sqrt{10}]^T.$$

By Theorem 7.30 we orthogonalize the first two eigenvectors associated with eigenvalues 0 and then normalize all the three eigenvectors to get

$$\mathbf{y}_1 = (1/\sqrt{30})[-2, -4, \sqrt{10}]^T, \quad \mathbf{y}_2 = (1/\sqrt{285})[2, -11, -4\sqrt{10}]^T,$$

$$\mathbf{y}_3 = (1/\sqrt{95})[9, -2, \sqrt{10}]^T,$$

as a set of three orthonormal eigenvectors. If we now define the orthogonal matrix R by $R := [\mathbf{y}_1\mathbf{y}_2\mathbf{y}_3]$, then it can be easily checked that $R^TCR = (UQR)^TB(UQR) = \text{diag}\ [0, 0, 19/36]$. Hence, the required nonsingular transformation is

$$P = UQR = \frac{1}{90\sqrt{57}} \begin{bmatrix} 15\sqrt{38} & -120 & -160\sqrt{3} \\ -15\sqrt{38} & -150 & 85\sqrt{3} \\ -30\sqrt{38} & -30 & -40\sqrt{3} \end{bmatrix}. \tag{7.142}$$

The matrix P is not unique. This is because another set of eigenvectors $\mathbf{x}_i$'s and $\mathbf{y}_i$'s would give the same result. □

We have already observed that a characteristic feature of a Hermitian matrix is that it is a *simple matrix*. In other words, an n-square Hermitian matrix has always an orthogonal set of n eigenvectors as established in Theorem 7.31. Because of this property, a Hermitian matrix can always be reduced to a diagonal matrix by a unitary transformation. Now an important question is: What is the most general class of nth order matrices A which have an orthogonal set of n eigenvectors? The answer is: If the matrices are normal. A normal matrix is, therefore, introduced in the following definition.

Definition 7.22 *Normal Matrix.* A matrix A over a complex field $\mathbb{C}$ is called normal iff $AA^* = A^*A$, i.e., iff A commutes with its tranjugate A^*. □

The commutativity relation defined above is held in common by all Hermitian, skew-Hermitian, unitary and diagonal matrices which are, therefore, normal. The following is a crucial theorem in the study of normal matrices.

Theorem 7.33 Over the complex field $\mathbb{C}$, a matrix A of order n is normal iff A has an orthogonal set of n eigenvectors.

Proof (a) *Sufficiency.* We normalize the given set of orthogonal eigenvectors (see Definition 5.12) and suppose that

$$U = [\mathbf{x}_1, \mathbf{x}_2, ..., \mathbf{x}_n]$$

has columns formed by an orthonormal set of n eigenvectors associated with eigenvalues $\lambda_1, \lambda_2, ..., \lambda_n$ of A (see Problem 7.58). Since $U^*U = I$, by (7.118) of Theorem 7.23, which holds also in our case, we have

$$A = U \operatorname{diag}(\lambda_1, \lambda_2, ..., \lambda_n)U^*.$$

Hence, by direct computation,

$$AA^* = A^*A = U \operatorname{diag}(\lambda_1\bar{\lambda}_1, \lambda_2\bar{\lambda}_2, ..., \lambda_n\bar{\lambda}_n)U^*.$$

(b) *Necessity.* Conversely, let A be normal, i.e.,

$$AA^* = A^*A. \tag{7.143}$$

By Corollary 7.8 there exists a unitary matrix S such that

$$S^*AS = V, \text{ i.e., } A = SVS^*, \tag{7.144}$$

where (see also Example 7.39) V is an upper triangular matrix represented by

$$V = \begin{bmatrix} v_{11} & v_{12} & \cdots & v_{1n} \\ 0 & v_{22} & \cdots & v_{2n} \\ \vdots & \vdots & & \vdots \\ 0 & 0 & \cdots & v_{nn} \end{bmatrix}. \tag{7.145}$$

Inserting (7.144) into (7.143), we now have

$$SVV^*S^* = SV^*VS^*, \text{ i.e., } VV^* = V^*V. \tag{7.146}$$

Computing (7.146) with the help of (7.145), we get for the element in the (1, 1) position of (7.146) the relation

$$|v_{11}|^2 + |v_{12}|^2 + \cdots + |v_{1n}|^2 = |v_{11}|^2,$$

which gives us

$$v_{12} = \cdots = v_{1n} = 0.$$

Again, considering the element of (7.146) in the position (2, 2), we have

$$|v_{22}|^2 + |v_{23}|^2 + \cdots + |v_{2n}|^2 = |v_{12}|^2 + |v_{22}|^2.$$

Hence, since $v_{12} = 0$, we have

$$v_{23} = \cdots = v_{2n} = 0.$$

Proceeding in this way, we can prove that

$$V = \text{diag } (v_{11}, v_{22}, \ldots, v_{nn}).$$

Now, (7.144) can be written as

$$AS = S \text{ diag } (v_{11}, v_{22}, \ldots, v_{nn}),$$

which shows that the columns of S constitute a set of n orthogonal eigenvectors for A. □

Theorem 7.33 proves that all normal matrices are simple. But the converse is not true; that is, all simple matrices are not normal. For example, the matrix A given in Example 7.25 is simple (see also Example 7.27(a)) but not normal. Thus simple matrices which are not normal are not diagonalizable by unitary transformation.

An easy consequence of Theorem 7.33 is the following corollary.

Corollary 7.14 A normal matrix is unitarily similar to a diagonal matrix, i.e., for a normal matrix over $\mathbb{C}$ there is a unitary matrix U such that U^*AU is diagonal. □

Example 7.42 Find a unitary transformation U for the matrix

$$A = \begin{bmatrix} 1 & -1 & 1 \\ 1 & 2 & 1 \\ -1 & 1 & 2 \end{bmatrix}$$

such that U^*AU is diagonal.

Solution The matrix A is normal as it satisfies $A^*A = AA^*$. The characteristic polynomial of A is $\phi(\lambda) = \det(\lambda I - A) = \lambda^3 - 5\lambda^2 + 9\lambda - 9$, and hence its eigenvalues and eigenvectors are, respectively, 3, $1 + \sqrt{2}i$, $1 - \sqrt{2}i$ and $\mathbf{x}_1 = [0\ 1\ 1]^T$, $\mathbf{x}_2 = [1, -i/\sqrt{2}, i/\sqrt{2}]^T$, $\mathbf{x}_3 = [1, i/\sqrt{2}, -i/\sqrt{2}]^T$. These eigenvectors are orthogonal having norms $\|\mathbf{x}_i\| = \sqrt{2}$ ($i = 1, 2, 3$). Let us, therefore, choose

$$U = (\sqrt{2})^{-1}[\mathbf{x}_1\ \mathbf{x}_2\ \mathbf{x}_3].$$

Then, it is easy to check that $UU^* = U^*U = I$ and that $U^*AU = \text{diag } [1, 1 + \sqrt{2}i, 1 - \sqrt{2}i]$. □

Without any major changes, Theorems 7.5–7.8 can be stated and proved for a simple matrix. This is because, the principal property used to prove these theorems is the existence of n linearly independent eigenvectors associated with an n-square matrix A, which is evidently true for a simple matrix. Therefore, we have the following theorem.

Theorem 7.34 Any n-square simple matrix A can be characterized by a biorthogonal system of left and right eigenvectors, which form the matrices P and Q respectively, such that

$$QP^* = Q^*P = I, \qquad Q^*AP = \text{diag}\ (\lambda_1, \lambda_2, ..., \lambda_n), \tag{7.147}$$

where λ_i's are the eigenvalues of A.

Proof Since A is simple, by Definition 7.21, it has n number of linearly independent eigenvectors $\mathbf{x}_1, \mathbf{x}_2, ..., \mathbf{x}_n$ associated respectively with its eigenvalues $\lambda_1, \lambda_2, ..., \lambda_n$. Denote

$$P: = [\mathbf{x}_1\ \mathbf{x}_2 \cdots \mathbf{x}_n].$$

Then, we have (see 7.118)

$$AP = P\ \text{diag}\ [\lambda_1, \lambda_2, ..., \lambda_n]. \tag{7.148}$$

Therefore, taking the tranjugate of both sides and rearranging, we get

$$A^*(P^*)^{-1} = (P^*)^{-1}\ \text{diag}\ [\bar{\lambda}_1, \bar{\lambda}_2, ..., \bar{\lambda}_n].$$

The eigenvectors of A^* can, therefore, be selected as the columns of $(P^*)^{-1}$ which we denote by $Q = (P^*)^{-1}$. Hence, we have a biorthogonal system of left and right eigenvectors forming respectively the matrices Q and P and satisfying

$$Q^*P = I$$

and, in view of (7.148),

$$Q^*AP = \text{diag}\ (\lambda_1, \lambda_2, ..., \lambda_n). \qquad \square$$

In Theorem 7.34, (7.147) can be written as

$$\begin{aligned} A &= P\ \text{diag}\ (\lambda_1, \lambda_2, ..., \lambda_n)Q^* \\ &= \lambda_1 P D_1 Q^* + \lambda_2 P D_2 Q^* + \cdots + \lambda_n P D_n Q^* \\ &= \lambda_1 Q_1 + \lambda_2 Q_2 + \cdots + \lambda_n Q_n, \end{aligned} \tag{7.149}$$

where

D_i = a diagonal matrix with all diagonal elements equal to zero but for the ith which is 1,

$Q_i = PD_iQ^*$.

Observing that $D_i^2 = D_i$ and $D_iD_j = 0, i \neq j$, it can be easily verified that, in view of (7.147), $Q_i^2 = Q_i, Q_iQ_j = 0,\ i \neq j$; and $Q_1 + Q_2 + \cdots + Q_n = I$, since $Q^*P = I$ implies $PQ^* = I$.

Theorem 7.34 is known as *the Spectral Theorem* of a matrix, since it describes the matrix A in terms of its spectrum, as is evident from (7.149). and is a special case of the following *Generalized Spectral Theorem.*

Theorem 7.35 If A is an n-square simple matrix, then it has a spectral resolution given by

$$f(A) = f(\lambda_1)Q_1 + \cdots + f(\lambda_n)Q_n,$$

where $f(\lambda)$ is an arbitrary polynomial and the constituent matrices Q_i's of A satisfy

(a) $Q_i^2 = Q_i$, (b) $Q_iQ_j = 0,\ i \neq j$;

(c) $Q_1 + \cdots + Q_n = I$.

Proof Let $\lambda_1, ..., \lambda_n$ be the eigenvalues of A. Since A is simple, by Theorem 7.34 there exists a biorthogonal system of right and left eigenvectors of A, represented by the columns of X and Y respectively, and given by

$$X = [\mathbf{x}_1 \ \cdots \ \mathbf{x}_n]$$

and

$$Y = [\mathbf{y}_1 \ \cdots \ \mathbf{y}_n]$$

so that

$$Y^*AX = D, \tag{7.150}$$

where

$$YX^* = Y^*X = I, \qquad D = \text{diag } (\lambda_1, ..., \lambda_n). \tag{7.151}$$

Now, premultiplying and postmultiplying (7.150) by X and Y^* respectively, and with the help of (7.151), we obtain successively

$$A = XDY^*, \qquad A^2 = XDY^*XDY^* = XD^2Y^*, ..., A^k = XD^kY^*. \tag{7.152}$$

Suppose that

$$f(\lambda) = \lambda^k + a_{k-1}\lambda^{k-1} + \cdots + a_1\lambda + a_0$$

is an arbitrary polynomial. Then, substituting from (7.152) the various powers of A in $f(A)$ and considering (7.151), we have

$$\begin{aligned} f(A) &= X[D^k + a_{k-1}D^{k-1} + \cdots + a_1D + a_0I]Y^* \\ &= X \text{ diag } [f(\lambda_1), ..., f(\lambda_n)]Y^* \\ &= f(\lambda_1)\mathbf{x}_1\mathbf{y}_1^* + \cdots + f(\lambda_n)\mathbf{x}_n\mathbf{y}_n^*. \end{aligned}$$

Setting $Q_i = \mathbf{x}_i\mathbf{y}_i^*$, we have the desired spectral resolution. The biorthogonality of the vectors $\mathbf{x}_i$'s and $\mathbf{y}_i$'s gives directly properties (a), (b) and (c). □

Corollary 7.15 Suppose that A is a matrix of order n with k distinct eigenvalues λ_i, each of multiplicity n_i, $i = 1, 2, ..., k$. Then, A is simple if and only if it admits of a spectral resolution

$$A = \lambda_1Q_1 + \cdots + \lambda_kQ_k, \qquad k \leq n; \tag{7.153}$$

where the constituent matrices Q_i's of A have rank n_i with the properties

(a) $Q_i^2 = Q_i$, (b) $Q_iQ_j = 0, \quad i \neq j$;

(c) $Q_1 + \cdots + Q_k = I$.

The decomposition of A into its constituent matrices is unique.

Proof (a) *Necessity.* Suppose that λ_i is an eigenvalue of A of multiplicity n_i, then taking $f(\lambda) = \lambda$ in Theorem 7.35, we have

$$A = \lambda_1Q_1^{(1)} + \lambda_1Q_1^{(2)} + \cdots + \lambda_1Q_1^{(n_1)} + \cdots + \lambda_kQ_k^{(n_k)}, \tag{7.154}$$

where $Q_i^{(j)}$'s are the constituent matrices of A. Defining

$$Q_i := Q_i^{(1)} + \cdots + Q_i^{(n_i)}, \qquad i = 1, 2, \ldots, k;$$

there follows from (7.154) the relation

$$A = \lambda_1 Q_1 + \cdots + \lambda_k Q_k$$

and Q_i's satisfy (a), (b) and (c) with rank n_i.

(b) *Sufficiency.* We observe that

$$\text{rank } Q_1 + \cdots + \text{rank } Q_k = n_i + \cdots + n_k = n = \text{rank } (Q_1 + \cdots + Q_k),$$

which means that the independent vectors in the columns of Q_i's remain independent in the columns of the sum $Q_1 + \cdots + Q_k$. Therefore, linearly independent vectors of Q_i and Q_j $(i \neq j)$ are mutually independent. Now, by virtue of properties (a) and (b), (7.153) gives

$$AQ_i = \lambda_i Q_i, \tag{7.155}$$

from which we choose n_i number of independent columns of Q_i to be the independent eigenvectors associated with λ_i for all $i = 1, ..., k$. Let Q be a nonsingular matrix formed by n number of these chosen eigenvectors which are mutually independent, as is evident from the foregoing observation. Then, it follows easily that

$$AQ = Q \text{ diag } [\lambda_1 I_{n_1}, ..., \lambda_k I_{n_k}], \tag{7.156}$$

i.e., A is similar to a diagonal matrix.

If the decomposition (7.153) is not unique, then it is possible to have the representation

$$A = \alpha_1 P_1 + \alpha_2 P_2 + \cdots + \alpha_m P_m, \tag{7.157}$$

where α_i's are scalars and the n-square matrices P_i's satisfy

(a) $P_i^2 = P_i$, (b) $P_i P_j = 0$, $i \neq j$;

(c) $P_1 + P_2 + \cdots + P_m = I$.

Postmultiplying (7.157) by P_i, we get

$$AP_i = \alpha_i P_i,$$

which shows that α_i is an eigenvalue and the number of independent columns of P_i equals the independent eigenvectors of A associated with α_i. Since α_i's and λ_i's both represent the eigenvalues of A, α_i can be eventually identified with λ_i giving $m = k$ in (7.157).

Let us define a polynomial $p_j(\lambda)$ in the complex variable λ by

$$p_j(\lambda) = \frac{(\lambda - \lambda_1) \cdots (\lambda - \lambda_{j-1})(\lambda - \lambda_{j+1}) \cdots (\lambda - \lambda_k)}{(\lambda_j - \lambda_1) \cdots (\lambda_j - \lambda_{j-1})(\lambda_j - \lambda_{j+1}) \cdots (\lambda_j - \lambda_k)}$$

$$= c_0 + c_1 \lambda + \cdots + c_{k-1} \lambda^{k-1} \tag{7.158}$$

which becomes, in particular,

$$p_j(\lambda_i) = \begin{cases} 0, & i \neq j; \\ 1, & i = j. \end{cases} \tag{7.159}$$

Remembering that $m = k$, from (7.153) and (7.157), we have

$$A^2 = \sum_{i=1}^{k} \lambda_i^2 P_i = \sum_{i=1}^{k} \lambda_i^2 Q_i,$$

$$\vdots$$

$$A^{k-1} = \sum_{i=1}^{k} \lambda_i^{k-1} P_i = \sum_{i=1}^{k} \lambda_i^{k-1} Q_i.$$

Hence,

$$\begin{aligned} p_j(A) &= c_0 I + c_1 A + \cdots + c_{k-1} A^{k-1} \\ &= \sum_{i=1}^{k} (c_0 + c_1\lambda_i + \cdots + c_{k-1}\lambda_i^{k-1}) P_i \\ &= \sum_{i=1}^{k} (c_0 + c_1\lambda_i + \cdots + c_{k-1}\lambda_i^{k-1}) Q_i \end{aligned}$$

from which, in view of (7.158) and (7.159), we have

$$p_j(A) = P_j = Q_j, \qquad j = 1, \ldots, k.$$

□

Remark 7.1 A matrix A, for which $A = A^2$, is said to be *idempotent*. Thus, the constituent matrices are idempotent.

Remark 7.2 The range space of Q_i is the space spanned by the linearly independent eigenvectors associated with λ_i, as is evident from (7.155), and is known as the eigenspace corresponding to λ_i. The whole eigenspace of the simple matrix A is obtained as the range space of Q which follows from (7.156) (see Theorem 11.8).

Example 7.43 Obtain a spectral resolution of the matrix

$$A = \begin{bmatrix} 7 & -6 & -6 \\ 3 & -2 & -3 \\ 2 & -2 & -1 \end{bmatrix}.$$

Solution The matrix A has eigenvalues at 1, 1, and 2. Since rank $(I - A) = 1$ (see Definition 7.21), A *is* simple. We compute the right eigenvectors of A to get

$$\mathbf{x}_1 = [1\ 1\ 0]^T, \qquad \mathbf{x}_2 = [1\ 0\ 1]^T, \qquad \mathbf{x}_3 = [6\ 3\ 2]^T,$$

with the help of which we form the matrices P and Q as

$$P = \begin{bmatrix} 1 & 1 & 6 \\ 1 & 0 & 3 \\ 0 & 1 & 2 \end{bmatrix}, \qquad Q = (P^T)^{-1} = \begin{bmatrix} -3 & -2 & 1 \\ 4 & 2 & -1 \\ 3 & 3 & -1 \end{bmatrix}.$$

As proved in Theorem 7.34, the vectors

$$\mathbf{y}_1 = [-3\ 4\ 3]^T, \qquad \mathbf{y}_2 = [-2\ 2\ 3]^T, \qquad \mathbf{y}_3 = [1\ -1\ -1]^T,$$

defined with the help of columns of Q, are the left eigenvectors of A and are biorthogonal to $\mathbf{x}_1$, $\mathbf{x}_2$, $\mathbf{x}_3$. Moreover, $Q^T AP = \text{diag}\,[1, 1, 2]$. Again defining $Q_1 = \mathbf{x}_1\mathbf{y}_1^T + \mathbf{x}_2\mathbf{y}_2^T$ and $Q_2 = \mathbf{x}_3\mathbf{y}_3^T$, as proved in Theorem 7.35, we have, by direct computation, the matrices

$$Q_1 = \begin{bmatrix} -5 & 6 & 6 \\ -3 & 4 & 3 \\ -2 & 2 & 3 \end{bmatrix}, \qquad Q_2 = \begin{bmatrix} 6 & -6 & -6 \\ 3 & -3 & -3 \\ 2 & -2 & -2 \end{bmatrix}.$$

So, A can be resolved into $A = Q_1 + 2Q_2$, as proved in Corollary 7.15. □

7.8 CONGRUENT MATRICES

In Section 7.7 we have seen that for any real symmetric matrix A and for any normal matrix A over $\mathbb{R}$, in general, there exists an orthogonal transformation P such that $P^TAP = D$, a diagonal matrix. Moreover, for a Hermitian matrix H and, in general, for a normal matrix H over $\mathbb{C}$, it is also established that there exists a unitary matrix U such that $U^*HU = D$, a diagonal matrix. The transformations of matrices which are connected by the above relations can be generalized as given in the following definition.

Definition 7.23 *Congruent and Conjunctive Matrices.* Two matrices A and B of order n over a field $\mathbb{F}$ are said to be congruent over $\mathbb{F}$, if there exists a nonsingular matrix P over $\mathbb{F}$ such that

$$B = P^TAP.$$

On the contrary, the n-square matrices A and B over $\mathbb{C}$ are said to be *conjunctive* over $\mathbb{C}$ iff there exists a nonsingular matrix Q over $\mathbb{C}$ such that

$$B = Q^*AQ.$$ □

The congruent (and also conjunctive) relation defined above is a special case of equivalence relation of Definition 4.4. Interestingly enough, these congruent (resp conjunctive) matrices arise naturally in the study of equivalent quadratic (resp Hermitian) forms in Chapters 8.

We have already seen in Section 7.7, that every symmetric matrix A over $\mathbb{F}$ is congruent over $\mathbb{F}$ to a diagonal matrix D with eigenvalues of A in the main diagonal. In fact, we have a general form of this result which is contained in the following theorem.

Theorem 7.36 Every symmetric matrix A over $\mathbb{F}$ of rank r is congruent over $\mathbb{F}$ to a diagonal matrix $D = \text{diag}\,[a_1, a_2, ..., a_r, 0, ..., 0]$ with $a_i \neq 0$ $(i = 1, 2, ..., r)$.

The following two elementary transformations will play a crucial role in proving the theorem and, consequently, we shall describe them now.

(a) *T*1. Let $A = [a_{ij}]$ be an n-square real symmetric matrix with $a_{rr} = 0$ $(r = 1, 2, ..., p-1)$, but $a_{pp} \neq 0$ $(p \le n)$. Then, for the nonsingular matrix $P = K_{1p}$ (see Section 4.2, note also $P^T = H_{1p}$), P^TAP has the nonzero element a_{pp} in the (1, 1) position. For example, let $n = p = 3$. Then, $P = K_{13}$ and $a_{33} \neq 0$. Hence,

$$P^TAP = \begin{bmatrix} 0 & 0 & 1 \\ 0 & 1 & 0 \\ 1 & 0 & 0 \end{bmatrix} \begin{bmatrix} 0 & a_{12} & a_{13} \\ a_{12} & 0 & a_{23} \\ a_{13} & a_{23} & a_{33} \end{bmatrix} \begin{bmatrix} 0 & 0 & 1 \\ 0 & 1 & 0 \\ 1 & 0 & 0 \end{bmatrix}$$

$$= \begin{bmatrix} a_{33} & a_{23} & a_{13} \\ a_{23} & 0 & a_{12} \\ a_{13} & a_{12} & 0 \end{bmatrix}$$

which has the nonzero element a_{33} in the (1, 1) position.

For a Hermitian matrix A, the procedure to transfer a nonzero diagonal element to the (1, 1) position is the same as above.

(b) $T2$. Let $A = [a_{ij}]$ be an n-square real symmetric matrix with $a_{pp} = 0$ ($p = 1, 2, ..., n$) but with an off-diagonal element $a_{ij} \neq 0$, $i \neq j$. Then, we define $P = K_{ij}(1)$ or $Q = K_{ji}(1)$ to get the nonzero element $2a_{ij}$ in the (i, i) diagonal position of P^TAP (or $2a_{ij}$ in the (j, j) diagonal position of Q^TAQ) which can be brought to the (1, 1) position by transformation of type $T1$. For example, if $n = 3$, $a_{11} = a_{22} = a_{33} = 0$, $a_{23} \neq 0$, then define $P = K_{23}(1)$ or $Q = K_{32}(1)$. Therefore,

$$P^TAP = \begin{bmatrix} 1 & 0 & 0 \\ 0 & 1 & 1 \\ 0 & 0 & 1 \end{bmatrix} \begin{bmatrix} 0 & a_{12} & a_{13} \\ a_{12} & 0 & a_{23} \\ a_{13} & a_{23} & 0 \end{bmatrix} \begin{bmatrix} 1 & 0 & 0 \\ 0 & 1 & 0 \\ 0 & 1 & 1 \end{bmatrix} = \begin{bmatrix} 0 & a_{12}+a_{13} & a_{13} \\ a_{12}+a_{13} & 2a_{23} & a_{23} \\ a_{13} & a_{23} & 0 \end{bmatrix}$$

which has the nonzero element $2a_{23}$ in the (2, 2) position. But, Q^TAQ has the nonzero element $2a_{23}$ in the (3, 3) position.

Proof of Theorem 7.36 The symmetric matrix

$$A = \left[\begin{array}{c|cccc} a_{11} & a_{12} & a_{13} & \cdots & a_{1n} \\ \hline a_{12} & a_{22} & a_{23} & \cdots & a_{2n} \\ a_{13} & a_{23} & a_{33} & \cdots & a_{3n} \\ \vdots & \vdots & \vdots & & \vdots \\ a_{1n} & a_{2n} & a_{3n} & \cdots & a_{nn} \end{array}\right]$$

has elements in the first row, which are the same as those in the first column. If all the elements of A are zero, then A is already in the required form with zero diagonal elements and we are done. Suppose, therefore, that $a_{11} \neq 0$, and define a matrix

$$P_1 = \begin{bmatrix} 1, & -a_{12}/a_{11}, & -a_{13}/a_{11}, & \ldots, & -a_{1n}/a_{11} \\ & 1 & & & \\ & & 1 & & 0 \\ & & & \ddots & \\ & 0 & & & 1 \end{bmatrix}$$

obtained from a unit matrix (see Section 4.2) such that AP_1 has all zero elements in the first row but for a_{11} in the (1, 1) position. Since, the first row elements are the same as the first column elements in A, it is not difficult to see, as can be verified by direct computation, that

$$P_1^TAP_1 = \left[\begin{array}{c|ccc} a_1 & 0 & \ldots & 0 \\ \hline 0 & & & \\ \vdots & & A_2 & \\ 0 & & & \end{array}\right],$$

which has all zero elements in the first row and the first column except the one in the (1, 1) position which we redefine by $a_1 = a_{11}$. Moreover, since A is symmetric, it is easily verified that A_2 is also symmetric. If the element in the (1, 1) position of A_2 is not zero, we can proceed as before to find a nonsingular matrix P_2 such that

$$P_2^T A_2 P_2 = \left[\begin{array}{c|ccc} a_2 & 0 & \cdots & 0 \\ \hline 0 & & & \\ \vdots & & A_3 & \\ 0 & & & \end{array}\right],$$

where A_3 is now symmetric. Now, the matrix

$$Q_2 = P_1 \left[\begin{array}{c|c} 1 & 0 \\ \hline 0 & P_2 \end{array}\right]$$

is such that

$$Q_2^T A Q_2 = \left[\begin{array}{cc|c} a_1 & & \\ & a_2 & 0 \\ \hline 0 & & A_3 \end{array}\right].$$

The foregoing procedure is pursued up to the $(k - 1)$th step until we have

$$Q_{k-1}^T A Q_{k-1} = \left[\begin{array}{ccc|cccc} a_1 & & & & & & \\ & \ddots & & & 0 & & \\ & & a_{k-1} & & & & \\ \hline & & & 0 & b_{12} & \cdots & b_{1p} \\ & 0 & & b_{12} & b_{22} & \cdots & b_{2p} \\ & & & \vdots & \vdots & & \vdots \\ & & & b_{1p} & b_{2p} & \cdots & b_{pp} \end{array}\right]$$

$$= \left[\begin{array}{c|c} a_1 \ddots a_{k-1} & 0 \\ \hline 0 & A_k \end{array}\right]$$

in which the matrix $A_k = [b_{ij}]$ of order $p = n - k + 1$ has a zero element in the (1, 1) position. By using transformations of the type $T1$ or $T2$ or both, we can bring a nonzero element in the (1, 1) position and use the foregoing procedure to reduce to zero all elements in the first column and the first row of A_k, except the one in the (1, 1) position. So, there is a nonsingular matrix P_k, such that $P_k^T A_k P_k$ = diag $[a_k, A_{k+1}]$. Defining

$$Q_k = Q_{k-1} \begin{bmatrix} I_{k-1} & 0 \\ 0 & P_k \end{bmatrix},$$

it is obvious that $Q_k^T A Q_k$ = diag $[a_1, ..., a_k, A_{k+1}]$. The foregoing steps are applied to A_{k+1} till $k = r$, when we have $P = Q_r$ and $P^T A P$ = diag $[a_1, a_2, ..., a_r, 0, ..., 0]$. □

Similarly, we can prove the following theorem.

Theorem 7.37 Every Hermitian matrix H of rank r is conjunctive to diag $[h_1, h_2, ..., h_r, 0, ..., 0]$, where h_i's $(i = 1, ..., r)$ are real. □

The following corollaries are easy consequences of the foregoing theorems, see Problem 7.63.

Corollary 7.16 An n-square symmetric matrix of rank r is congruent over $\mathbb{R}$ to the canonical form diag $[I_p, -I_{r-p}, 0, ..., 0]_n$, where $p \le r$ is a nonnegative integer.

Corollary 7.17 An n-square Hermitian matrix of rank r is conjunctive over $\mathbb{C}$ to the canonical form diag $[I_p, -I_{r-p}, 0, ..., 0]_n$, where $p \le r$ is a nonnegative integer.

The congruent (resp conjunctive) reduction to a diagonal form D of any symmetric (resp Hermitian) matrix A, and a simultaneous computation of the nonsingular transformation P can be done systematically by using the from $[A; I]$. The form $[A; I]$ is subjected to elementary row operations as envisaged in Theorem 7.36, leaving column operations to be performed only on A.

Example 7.44 Determine a nonsingular matrix P for congruent reduction of A to a diagonal form D, where

$$A = \begin{bmatrix} 1 & 2 & 1 & -2 \\ 2 & 5 & -2 & 1 \\ 1 & -2 & 17 & -1 \\ -2 & 1 & -1 & 36 \end{bmatrix}.$$

Solution We subject $[A; I]$ to elementary row transformations denoted by $H_{21}(-2)$, $H_{31}(-1)$, $H_{41}(2)$ and then column transformations $K_{21}(-2)$, $K_{31}(-1)$, $K_{41}(2)$, see Definition 4.3, to get

$$[A; I] = \left[\begin{array}{cccc|cccc} 1 & 2 & 1 & -2 & 1 & 0 & 0 & 0 \\ 2 & 5 & -2 & 1 & 0 & 1 & 0 & 0 \\ 1 & -2 & 17 & -1 & 0 & 0 & 1 & 0 \\ -2 & 1 & -1 & 36 & 0 & 0 & 0 & 1 \end{array}\right]$$

$$\sim \left[\begin{array}{cccc|cccc} 1 & 0 & 0 & 0 & 1 & 0 & 0 & 0 \\ 0 & 1 & -4 & 5 & -2 & 1 & 0 & 0 \\ 0 & -4 & 16 & 1 & -1 & 0 & 1 & 0 \\ 0 & 5 & 1 & 32 & 2 & 0 & 0 & 1 \end{array}\right].$$

Next, we subject the reduced $[A; I]$ to transformations defined by $H_{32}(4)$, $H_{42}(-5)$, $K_{32}(4)$, $K_{42}(-5)$ successively, and obtain

$$[A; I] \sim \left[\begin{array}{cccc|cccc} 1 & 0 & 0 & 0 & 1 & 0 & 0 & 0 \\ 0 & 1 & 0 & 0 & -2 & 1 & 0 & 0 \\ 0 & 0 & 0 & 21 & -9 & 4 & 1 & 0 \\ 0 & 0 & 21 & 7 & 12 & -5 & 0 & 1 \end{array}\right].$$

Since, the element in position (3, 3) is zero, we apply transformation of type T1, *viz.*, H_{34} and K_{34} to get

$$[A; I] \sim \left[\begin{array}{cccc|cccc} 1 & 0 & 0 & 0 & 1 & 0 & 0 & 0 \\ 0 & 1 & 0 & 0 & -2 & 1 & 0 & 0 \\ 0 & 0 & 7 & 21 & 12 & -5 & 0 & 1 \\ 0 & 0 & 21 & 0 & -9 & 4 & 1 & 0 \end{array}\right].$$

Finally, we apply the transformations represented by $H_{43}(-3)$ and $K_{43}(-3)$ to get

$$[A; I] \sim \left[\begin{array}{cccc|cccc} 1 & 0 & 0 & 0 & 1 & 0 & 0 & 0 \\ 0 & 1 & 0 & 0 & -2 & 1 & 0 & 0 \\ 0 & 0 & 7 & 0 & 12 & -5 & 0 & 1 \\ 0 & 0 & 0 & -63 & -45 & 19 & 1 & -3 \end{array}\right] = [D, P^T]. \tag{7.160}$$

It is interesting to note that the transformation matrix $P = K_{21}(-2)K_{31}(-1)K_{41}(2)K_{32}(4)K_{42}(-5)$ $K_{34}K_{43}(-3)$ and additional transformations $H_3(1/\sqrt{7}), H_4(1/3\sqrt{7}), K_3(1/\sqrt{7})$ and $K_4(1/3\sqrt{7})$ applied to (7.160) reduce A to the canonical form $C = \text{diag}\ [1, 1, 1, -1] = \text{diag}\ [I_3, -1]$. *So the congruent reduction of A to a diagonal D is not unique.* Observe that at every stage the reduced A is symmetric, which enables one to check the accuracy of computations at every step. □

For conjunctive transformation of a Hermitian matrix to a diagonal form, any row transformation of the type $H_{ij}(z)$ must be accompanied by a column transformation $K_{ij}(\bar{z})$, where $\bar{z}$ = complex conjugate of z.

Example 7.45 Determine a nonsingular matrix Q for conjunctive reduction of H to a diagonal form:

$$H = \begin{bmatrix} 0 & i \\ -i & 0 \end{bmatrix}.$$

Solution The matrix

$$[H; I] = \left[\begin{array}{cc|cc} 0 & i & 1 & 0 \\ -i & 0 & 0 & 1 \end{array}\right].$$

We apply the transformation of type $T2$, viz., $H_{12}(i)$ and $K_{12}(-i)$, to the above matrix to generate a nonzero element in (1, 1) position:

$$[H; I] \sim \left[\begin{array}{cc|cc} 2 & i & 1 & i \\ -i & 0 & 0 & 1 \end{array}\right].$$

Next, we apply the transformations $H_{21}(i/2)$ and $K_{21}(-i/2)$ to get

$$[H; I] \sim \left[\begin{array}{cc|cc} 2 & 0 & 1 & i \\ 0 & -1/2 & i/2 & 1/2 \end{array}\right] = [D \quad Q^*], \tag{7.161}$$

in which,

$$Q = \begin{bmatrix} 1 & -i/2 \\ -i & 1/2 \end{bmatrix}$$

reduces A to the diagonal form diag [2, –1/2]. To reduce H to a canonical form, we apply the transformations $H_1(1/\sqrt{2})$, $K_1(1/\sqrt{2})$; $H_2(\sqrt{2})$, $K_2(\sqrt{2})$ to (7.161) to get

$$[H; I] \sim \left[\begin{array}{cc|cc} 1 & 0 & 1/\sqrt{2} & i/\sqrt{2} \\ 0 & -1 & i/\sqrt{2} & 1/\sqrt{2} \end{array}\right] = [C \quad R^*],$$

which shows that the nonsingular matrix

$$R = \frac{1}{\sqrt{2}}\begin{bmatrix} 1 & -i \\ -i & 1 \end{bmatrix}$$

reduces H by conjunctive transformation to the canonical form C = diag [1, –1].

7.9 MATLAB AIDS

To compute the characteristic polynomial (CP) of an $n \times n$ matrix A, we use the command poly (A) which responds with a row vector of $n + 1$ elements that form the coefficients of CP. Let A = magic (3). Then, poly (A) responds with

```
1.0000    –15.0000    –24.0000    360.0000
```

implying that CP of A is $\phi(\lambda) = \lambda^3 - 15\lambda^2 - 24\lambda + 360$. With MATLAB command eig (A) we compute the eigenvalues and eigenvectors of A. The command E = eig (A) produces a vector containing the eigenvalues of A, and [V, D] = eig (A), a diagonal matrix D of eigenvalues and a matrix V of order n whose columns are the corresponding eigenvectors so that $AV = VA$. For A = magic (3), command [V, D] = eig (A) responds with

```
V =

   –0.5774    –0.8131    –0.3416
   –0.5774     0.4714    –0.4714
   –0.5774     0.3416     0.8131

D =

   15.0000          0          0
         0     4.8990          0
         0          0    –4.8990
```

If matrix A is defective, we use [V, D] = jordan (A) (see Chapter 11).

For the generalized eigenvalue problem ($A\mathbf{x} = \lambda B\mathbf{x}$, see Section 8.7, where A and B are interchanged), we use E = eig (A, B) which is a vector containing the generalized eigenvalues of square matrices A and B. The command [V, D] = eig [A, B] produces a diagonal matix D of generalized eigenvalues and a full matrix V whose columns are the corresponding eigenvectors, so that $AV = BVD$. For A = B = eye (3), command [V, D] = eig [A, B] responds with (justify!)

```
V =

   1  0  0
   0  1  0
   0  0  1

D =

   1  0  0
   0  1  0
   0  0  1
```

PROBLEMS

7.1 Show that the problem of determining the maximum of $Q(x, y) = ax^2 + 2bxy + cy^2$ on the circle $x^2 + y^2 = 1$ can be formulated as a characteristic value problem.

7.2 The maximum value of $Q(\mathbf{x}, \mathbf{x}) = \sum_{i=1}^{n} \sum_{j=1}^{n} a_{ij}x_i x_j$ subject to the constraint $\| \mathbf{x} \|^2 = 1$ can be determined by maximizing $Q(\mathbf{x}, \mathbf{x}) - \lambda(\| \mathbf{x} \|^2 - 1)$, where λ is a constant Lagrange multiplier. Show that the problem is equivalent to determining a nonzero vector $\mathbf{x}$ such that $A\mathbf{x} - \lambda\mathbf{x} = \boldsymbol{\theta}$.

7.3 Show that if λ is an eigenvalue of A, then $\alpha\lambda$ is that of αA (where α is a scalar) with the same eigenvector.

7.4 If $\mathbf{x}$ is an eigenvector associated with a complex eigenvalue λ of a real matrix A, then show that $\bar{\mathbf{x}}$ is the eigenvector corresponding to $\bar{\lambda}$.

7.5 Prove that the eigenvalues of diag $(d_1, d_2, ..., d_n)$ are $d_1, d_2, ..., d_n$, and the corresponding eigenvectors are $\alpha_1\mathbf{e}_1, \alpha_2\mathbf{e}_2, ..., \alpha_n\mathbf{e}_n$ where $\alpha_1, \alpha_2, ..., \alpha_n$ are arbitrary scalars.

7.6 Find the eigenvalues and the associated eigenvectors of

(a) a null matrix, (b) $\left[\begin{array}{c|c} \begin{matrix}0\\ \vdots\end{matrix} & I_{n-1} \\ \hline 0 & \cdots \quad 0 \end{array}\right]$, (c) $\begin{bmatrix} 0 & & 1 \\ & 1 & \\ & \cdot^{\cdot^{\cdot}} & \\ 1 & & 0 \end{bmatrix}_n$.

7.7 Find the eigenvalues of (a) an upper triangular matrix $A = [a_{ij}]_n$, $a_{ij} = 0$, $i > j$, and (b) a lower triangular matrix $B = [b_{ij}]_n$, $b_{ij} = 0$, $i < j$. Under what conditions do the corresponding eigenvectors form (a) an upper triangular and (b) a lower triangular matrix?

7.8 Determine the eigenvalues and eigenvectors of the following matrices:

(a) $\begin{bmatrix} 1 & 6 \\ 1 & 2 \end{bmatrix}$, (b) $\begin{bmatrix} 0 & -1 \\ 1 & 2 \end{bmatrix}$, (c) $\begin{bmatrix} 0 & 2 & 0 \\ 2 & 0 & 2 \\ 0 & 2 & 0 \end{bmatrix}$, (d) $\begin{bmatrix} 2 & 0 & 2 \\ 0 & 2 & 0 \\ 2 & 0 & 2 \end{bmatrix}$,

(e) $\begin{bmatrix} 2 & 0 & 1 \\ 1 & 0 & 1 \\ 1 & -2 & 0 \end{bmatrix}$, (f) $\begin{bmatrix} -4 & -4 & 0 \\ 10 & 9 & 1 \\ -4 & -3 & 1 \end{bmatrix}$.

7.9 Determine the solutions of the following differential equations by the eigenvector method:

(a) $\begin{bmatrix} \dot{x} \\ \dot{y} \end{bmatrix} = \begin{bmatrix} 1 & 2 \\ 2 & 1 \end{bmatrix} \begin{bmatrix} x \\ y \end{bmatrix}$; $\quad x(0) = 3, \quad y(0) = 1;$

(b) $\begin{bmatrix} \dot{x} \\ \dot{y} \end{bmatrix} = \begin{bmatrix} -1 & 2 \\ -2 & 3 \end{bmatrix} \begin{bmatrix} x \\ y \end{bmatrix}$; $x(0) = 1$, $y(0) = 2$.

Given that if A is a matrix of order 2 having an eigenvalue λ of multiplicity 2 and a single independent eigenvector **x**, then the vector **y** obtained by solving $(\lambda I - A)\mathbf{y} = \mathbf{x}$ is linearly independent of **x** and that $\alpha\mathbf{x} \exp(\lambda t) + \beta(t\mathbf{x} - \mathbf{y}) \exp(\lambda t)$ is a solution of $\dot{\mathbf{x}}(t) = A\mathbf{x}(t)$.

(c) $\ddot{x} + 5\dot{x} + 4x = 0$; $x(0) = 1$, $\dot{x}(0) = 0$.

7.10 Determine the eigenvalues and the associated eigenvectors of the following linear transformations:

(a) $L[x\ y\ z]^T = [x - y,\ y - z,\ z - x]^T$,

(b) $L[x\ y]^T = [x \cos\theta - y \sin\theta,\ x \sin\theta + y \cos\theta]^T$,

(c) $L(x, y, z) = (x - y - z,\ y - x - z,\ z - x - y)$,

(d) $L(x, y) = (x - y,\ x + y)$.

7.11 Determine the eigenvalues and associated eigenvectors of the following linear operators:

(a) $L : \mathbb{P}_2(x) \to \mathbb{P}_2(x) : p(x) \mapsto xp'(x) + p(x)$,

(b) $L : \mathbb{P}_2(x) \to \mathbb{P}_2(x) : p(x) \mapsto \dfrac{d}{dx}[(1 - x)p(x)]$,

(c) $L : V_{2\times 2} \to V_{2\times 2} : \begin{bmatrix} a & b \\ c & d \end{bmatrix} \mapsto \begin{bmatrix} d & -c \\ -b & a \end{bmatrix}$.

7.12 Can you define eigenvalues and eigenvectors for the linear operator L given by

$$L : V_{2\times 2} \to \mathbb{P}_3(x) : \begin{bmatrix} a & b \\ c & d \end{bmatrix} \mapsto (a - b) + (b - c)x + (c - d)x^2 + (d - a)x^3\ ?$$

Justify your answer.

7.13 A linear transformation L has a matrix

$$A = \begin{bmatrix} 3 & 2 & 0 \\ 0 & 1 & 3 \\ -4 & -4 & -5 \end{bmatrix}$$

with respect to the basis $[1\ 1\ 1]^T$, $[0\ 1\ 1]^T$, $[0\ 0\ 1]^T$. Find the eigenvalues and eigenvectors for L.

7.14 Find a nonsingular matrix P such that $P^{-1}AP$ is in a diagonal form, where A is given by

(a) $\begin{bmatrix} 0 & 1 \\ 1 & 0 \end{bmatrix}$, (b) $\begin{bmatrix} 0 & 0 & 1 \\ 0 & 1 & 0 \\ 1 & 0 & 0 \end{bmatrix}$, (c) $\begin{bmatrix} 0 & 2 & 0 \\ 2 & 0 & 2 \\ 0 & 2 & 0 \end{bmatrix}$,

(d) $\begin{bmatrix} 2 & 0 & 2 \\ 0 & 2 & 0 \\ 2 & 0 & 2 \end{bmatrix}$, (e) $\begin{bmatrix} 2 & 0 & 1 \\ 1 & 0 & 1 \\ 0 & -2 & 0 \end{bmatrix}$.

In each case determine the modal and the spectral matrix for A.

7.15 Let A be a real matrix of order $2n$ having distinct complex roots $\lambda_r = a_r + ib_r$, $r = 1, 2, ..., n$. Show that A is similar to diag $[A_1, A_2, ..., A_n]$, where

$$A_r = \begin{bmatrix} a_r & b_r \\ -b_r & a_r \end{bmatrix}.$$

7.16 If A is a matrix of order 3 having a real eigenvalue λ and two complex eigenvalues $a \pm ib$ with associated eigenvectors $\mathbf{x}$ and $\mathbf{y} \pm i\mathbf{z}$ respectively, then show that

$$P^{-1}AP = \begin{bmatrix} \lambda & 0 & 0 \\ 0 & a & b \\ 0 & -b & a \end{bmatrix}, \qquad P = [\mathbf{x}\ \mathbf{y}\ \mathbf{z}].$$

7.17 Find a nonsingular matrix P such that

(a) $P^{-1}\begin{bmatrix} 0 & 1 & 0 \\ 0 & 0 & 1 \\ 5 & -7 & 3 \end{bmatrix} P = \begin{bmatrix} 1 & 0 & 0 \\ 0 & 1 & 2 \\ 0 & -2 & 1 \end{bmatrix}$,

(b) $P^{-1}\begin{bmatrix} 2 & 0 & 1 \\ 0 & 0 & 1 \\ 1 & -2 & 0 \end{bmatrix} P = \begin{bmatrix} 2 & 0 & 0 \\ 0 & 0 & 1 \\ 0 & -1 & 0 \end{bmatrix}$.

7.18 Determine a basis with respect to which the matrix representation of the following linear transformations is diagonal:

(a) $L : \mathbb{P}_2(x) \to \mathbb{P}_2(x) : p(x) \mapsto \dfrac{d}{dx}[(1-x)p(x)]$,

(b) $L : \mathbb{P}_2(x) \to \mathbb{P}_2(x) : p(x) \mapsto xp'(x) + p(x)$,

(c) $L(x, y, z) = (x - y - z, y - x - z, z - x - y)$,

(d) $L : V_{2\times 2} \to V_{2\times 2} : \begin{bmatrix} a & b \\ c & d \end{bmatrix} \mapsto \begin{bmatrix} d & -c \\ -b & a \end{bmatrix}$.

7.19 Find a biorthogonal set of eigenvectors associated with the following matrices without computing any matrix inverse:

(a) $\begin{bmatrix} 1 & 6 \\ 1 & 2 \end{bmatrix}$, (b) $\begin{bmatrix} 2 & 0 & 2 \\ 0 & 2 & 0 \\ 2 & 0 & 2 \end{bmatrix}$, (c) $\begin{bmatrix} -6 & 4 & 2 \\ -14 & 9 & 4 \\ 6 & -3 & 0 \end{bmatrix}$.

7.20 Determine A^{20} if A is given by as in Problem 7.19.

7.21 Find A^{30} if

$$A = \begin{bmatrix} 1 & 3 \\ 2 & 2 \end{bmatrix}.$$

7.22 For each of the polynomials $f(x)$ and $d(x)$, find polynomials $q(x)$ and $r(x)$ such that deg $r(x) <$ deg $d(x)$:

(a) $f(x) = x^3 + 2x^2 + 2x + 3$; $d(x) = x^2 + 1$.

(b) $f(x) = x^3 - 1$; $d(x) = x^2 + x + 1$.

(c) $f(x) = 4x^4 + 4x^2 + 1$; $d(x) = 5x^5 + 3x^3 + x$.

(d) $f(x) = 6x^3 + 7x^2 + 4x + 3$; $d(x) = 2x + 1$.

7.23 For each matrix A given in Problem 7.19, determine $f(A)$, $d(A)$, $q(A)$ and $r(A)$ for polynomials given in Problem 7.22.

7.24 If $g(A)$ is nonsingular, prove that $[g(A)]^{-1} f(A) = f(A)[g(A)]^{-1}$.

7.25 A lambda matrix $A(\lambda)$ is defined by

$$A(\lambda) = \begin{bmatrix} \lambda^3 - \lambda & 3\lambda^2 - \lambda \\ -\lambda^2 + 1 & 2\lambda^3 - 1 \end{bmatrix}.$$

Determine the right value and the left value of $A(\lambda)$ at B which is given by

(a) $\begin{bmatrix} 1 & 0 \\ 0 & 2 \end{bmatrix}$, (b) $\begin{bmatrix} 2 & 0 \\ -1 & 1 \end{bmatrix}$, (c) $\begin{bmatrix} 1 & -1 \\ -1 & 1 \end{bmatrix}$.

7.26 Determine (a) the right quotient and the right remainder, and (b) the left quotient and the left remainder of $A(\lambda)$ on dividing by $B(\lambda)$ when

(i) $A(\lambda) = \begin{bmatrix} 2\lambda^2 + 1 & \lambda - 1 \\ 2\lambda + 3 & \lambda^2 + \lambda \end{bmatrix}$, $B(\lambda) = \begin{bmatrix} 2\lambda + 1 & \lambda - 1 \\ \lambda & \lambda + 2 \end{bmatrix}$;

(ii) $A(\lambda) = \begin{bmatrix} \lambda^3 & \lambda^2 \\ \lambda & 1 \end{bmatrix}$, $B(\lambda) = \begin{bmatrix} \lambda & 1 \\ 1 & \lambda \end{bmatrix}$.

7.27 Determine the right quotient and the right remainder of $A(\lambda)$ on dividing by $(\lambda I - B)$ when $A(\lambda)$ is given as in Problem 7.26 and B by

$$B = \begin{bmatrix} 2 & 1 \\ 1 & 1 \end{bmatrix}.$$

7.28 Apply the Cayley-Hamilton theorem to compute A^3 and A^4 when A is given as in Problems 7.8(c)–7.8(f).

7.29 Apply the Cayley-Hamilton theorem to compute A^{-1} when A is given by

$$\text{(a)}\ \begin{bmatrix} 2 & 1 & 1 \\ 1 & 1 & 2 \\ 0 & 0 & 4 \end{bmatrix}, \qquad \text{(b)}\ \begin{bmatrix} 1 & 3 & 2 & 5 \\ 3 & 1 & 0 & 0 \\ 2 & 0 & 1 & 0 \\ 5 & 0 & 0 & 1 \end{bmatrix}.$$

7.30 Determine the minimum polynomial of the following matrices:

$$\text{(a)}\ \begin{bmatrix} 1 & 1 & 0 \\ 0 & 1 & 0 \\ 0 & 0 & 1 \end{bmatrix}, \quad \text{(b)}\ \begin{bmatrix} a & 0 & 0 \\ 0 & a & 0 \\ 0 & 0 & a \end{bmatrix}, \quad \text{(c)}\ \begin{bmatrix} 2 & 1 & 0 & 0 \\ 0 & 2 & 0 & 0 \\ 0 & 0 & 2 & 1 \\ 0 & 0 & 0 & 2 \end{bmatrix},$$

$$\text{(d)}\ \begin{bmatrix} -1 & 0 & 0 & 3 \\ -2 & 0 & 1 & 3 \\ 2 & -1 & -2 & 3 \\ 0 & 0 & 0 & 2 \end{bmatrix}, \quad \text{(e)}\ \begin{bmatrix} 7 & -9 & -5 & -9 \\ 1 & 5 & -1 & 3 \\ 2 & -6 & 0 & -6 \\ 0 & -4 & 0 & -2 \end{bmatrix}.$$

7.31 (a) Let $\phi_a(\lambda)$, $\phi_b(\lambda)$ and $\phi_c(\lambda)$ be the characteristic polynomials of A, B and C respectively, where

$$A = \begin{bmatrix} B & 0 \\ 0 & C \end{bmatrix}$$

and $\psi_a(\lambda)$, $\psi_b(\lambda)$, and $\psi_c(\lambda)$ be the respective minimum polynomials. Prove that $\phi_a(\lambda) = \phi_b(\lambda)\phi_c(\lambda)$ and $\psi_a(\lambda) = \text{LCM of } \psi_b(\lambda) \text{ and } \psi_c(\lambda)$.

(b) Determine the characteristic polynomial and the minimum polynomial of A given by

$$A = \left[\begin{array}{ccc|cc} 0 & 1 & 0 & & \\ 0 & 0 & 1 & \multicolumn{2}{c}{0} \\ 2 & 1 & -2 & & \\ \hline & 0 & & 1 & 2 \\ & & & 3 & 2 \end{array}\right].$$

(c) Let D be a diagonal matrix in $V_{n\times n}$ with diagonal elements $d_1, \ldots, d_p$ with d_i of multiplicity n_i for all $i = 1, \ldots, p$ and $n_1 + \cdots + n_p = n$. Then show that the minimum polynomial of D is $\psi(\lambda) = (\lambda - \lambda_1) \cdots (\lambda - \lambda_p)$ where $\lambda_i = d_i$.

7.32 Prove that if A is an $n \times n$ matrix over $\mathbb{F}$ with k distinct eigenvalues λ_i of multiplicity n_i, then A has a minimum polynomial of the form

$$\psi(\lambda) = (\lambda - \lambda_1)^{m_1} \cdots (\lambda - \lambda_k)^{m_k}$$

where $1 \le m_i \le n_i$, $i = 1, \ldots, k$.

7.33 Determine the characteristic polynomial and the minimum polynomial of the following linear transformations:

(a) $L[x, y, z]^T = [2x, x + 2y, 2z]^T$,

(b) $L : \mathbb{P}_2(x) \to \mathbb{P}_2(x) : p(x) \mapsto xp'(x) + p(x)$,

(c) $L : V_{2\times 2} \to V_{2\times 2} : \begin{bmatrix} a & b \\ c & d \end{bmatrix} \mapsto \begin{bmatrix} d & -c \\ -b & a \end{bmatrix}$.

7.34 Find the reduced adjoint matrix of $(\lambda I - A)$ and adj $(\lambda I - A)$, where A is given in Problem 7.30 using (i) definitions, (ii) Faddeev algorithm, and (iii) the characteristic and minimum polynomials.

7.35 If

$$A = \begin{bmatrix} 1 & 3 & 0 \\ 2 & 2 & 0 \\ 0 & 0 & -1 \end{bmatrix},$$

then prove that

(a) $A^4 - 2A^3 - 9A - 5I = 7A^2 - 5A - 5I$,

(b) $A^8 + I = 2623A^2 + 5238A + 2617I$.

Hence, compute

(c) $A^4 - 2A^3 - 9A - 5I$, (d) $A^8 + I$.

7.36 Evaluate the matrix $A^8 + 4A^6 + 8A^4 + 4A^2 + I$ if

$$A = \begin{bmatrix} 1 & -1 \\ -1 & 1 \end{bmatrix}.$$

7.37 Let a square matrix A be defined as

$$A = \begin{bmatrix} 0 & 1 \\ 0 & 0 \end{bmatrix}$$

and let $g(\lambda) = \lambda^2 + 1$ be a polynomial. Show that $g(A)$ has $\{\mathbf{e}_1, \mathbf{e}_2\}$ as eigenvectors, of which $\mathbf{e}_1$ is an eigenvector of A while $\mathbf{e}_2$ is not.

7.38 Prove that similar matrices have the same (a) rank, (b) characteristic polynomial, (c) determinant, (d) trace, (e) eigenvalues, and (f) minimum polynomial.

7.39 Compute the characteristic polynomial of the following matrices by applying Theorem 7.16:

(a) $\begin{bmatrix} 1 & -1 & 1 \\ -1 & 1 & -1 \\ 1 & -1 & 1 \end{bmatrix}$, (b) $\begin{bmatrix} a & b & a \\ a & b & c \\ c & b & c \end{bmatrix}$, (c) $\begin{bmatrix} 1 & 2 & 3 & 4 \\ 2 & -1 & 4 & 3 \\ 3 & 4 & 1 & 2 \\ 4 & 3 & 2 & -1 \end{bmatrix}$.

7.40 It is given that $f(A) = 2A^3 + A^2 - A + 2I$ and A is

$$\text{(a)} \begin{bmatrix} 1 & 2 & 0 \\ 3 & 2 & 0 \\ 0 & 0 & 1 \end{bmatrix}, \qquad \text{(b)} \begin{bmatrix} 1 & 1 & 1 \\ 1 & 1 & 1 \\ 1 & 1 & 1 \end{bmatrix}.$$

Find in each case the eigenvalues and eigenvectors of $f(A)$.

7.41 Show that the characteristic polynomial of

$$A = \begin{bmatrix} 1 & 1 & \cdots & 1 \\ 1 & 1 & \cdots & 1 \\ \vdots & \vdots & & \vdots \\ 1 & 1 & \cdots & 1 \end{bmatrix}_n$$

is $\lambda^n - n\lambda^{n-1}$ and that A has eigenvectors $[\alpha_1, \ldots, \alpha_{n-1}, -\alpha_1 - \cdots - \alpha_{n-1}]^T$ and $[\alpha\ \alpha \cdots \alpha]^T$ corresponding to the eigenvalues $\lambda = 0$ and $\lambda = n$ respectively, where α and α_i's are arbitrary scalars.

7.42 If $\phi(\lambda)$ is the characteristic polynomial of a nonsingular matrix A of order n, then show that the characteristic polynomial of A^{-1} is

$$\frac{1}{\det A}(-\lambda)^n \phi\left(\frac{1}{\lambda}\right).$$

7.43 A matrix A is nonsingular iff in its characteristic polynomial $\phi(\lambda) = \lambda^n + a_{n-1}\lambda^{n-1} + \cdots + a_1\lambda + a_0$, $a_0 \neq 0$.

7.44 Let C be given by the alternatives (7.93a)–(7.93c) of a companion form (7.92). Prove that:

(a) In each case the characteristic polynomial is $\phi(\lambda) = \lambda^n + a_{n-1}\lambda^{n-1} + \cdots + a_1\lambda + a_0$.

(b) In each case the characteristic polynomial and the minimum polynomial are the same.

(c) If λ_i is an eigenvalue of C, then $\alpha\,[1, \lambda_i + a_{n-1}, \lambda_i^2 + a_{n-1}\lambda_i + a_{n-2}, \ldots, \lambda_i^{n-1} + a_{n-1}\lambda_i^{n-2} + \cdots + a_1]^T$, $\alpha\,[\lambda_i^{n-1}, \ldots, \lambda_i^2, \lambda_i, 1]^T$ and $\alpha\,[\lambda_i^{n-1} + a_{n-1}\lambda_i^{n-2} + \cdots + a_1, \ldots, \lambda_i^2 + a_{n-1}\lambda_i + a_{n-2}, \lambda_i + a_{n-1}, 1]^T$ are the eigenvectors of (7.93a), (7.93b) and (7.93c) respectively, corresponding to λ_i.

7.45 If the eigenvalues of A lie on the left half of the complex plane, show that the eigenvalues of $B = (\mu I - A)^{-1}(\mu I + A)$ are less than 1 in absolute value if μ, a positive real number, is not an eigenvalue of A.

7.46 If λ is not an eigenvalue of an n-square matrix A, then show that

$$1 - \mathbf{q}^*(\lambda I - A)^{-1}\,\mathbf{b} = \det(\lambda I - A - \mathbf{b}\mathbf{q}^*)/\det(\lambda I - A),$$

where $\mathbf{b}$ and $\mathbf{q}$ are n-vectors.

7.47 Find the eigenvalues and eigenvectors of the following companion matrices C, and determine a nonsingular matrix V such that $V^{-1}CV$ is diagonal. Compute also the minimum polynomial in each case.

(a) $\begin{bmatrix} 0 & 1 \\ -2 & -3 \end{bmatrix}$, (b) $\begin{bmatrix} 0 & 1 & 0 \\ 0 & 0 & 1 \\ 0 & 1 & 0 \end{bmatrix}$, (c) $\begin{bmatrix} 0 & 1 & 0 \\ 0 & 0 & 1 \\ -2 & 1 & 2 \end{bmatrix}$.

7.48 (a) Let

$$A_n = \begin{bmatrix} 2c & 1 & & & \\ 1 & 2c & 1 & & \\ & 1 & 2c & 1 & \\ & & \ddots & \ddots & \ddots \\ & & & & 1 \\ & & & 1 & 2c \end{bmatrix},$$

$D_n = \det A_n$, $c = \cos\theta$, then $D_n = \sin(n+1)\theta/\sin\theta$, and show how D_n may be evaluated if $|c| > 1$. D_n is a Chebyschev polynomial of second kind, $U_n(c)$. (b) Find the eigenvalues and eigenvectors of A_n.

7.49 Express $(A^4 + 2A^2 + I)^{-1}$ as a linear polynomial in A when

$$A = \begin{bmatrix} 1 & 2 \\ 2 & 1 \end{bmatrix}.$$

7.50 Let C be a companion matrix given by

$$C = \begin{bmatrix} 0 & 1 & 0 & \cdots & 0 \\ 0 & 0 & 1 & \cdots & 0 \\ \vdots & \vdots & \vdots & & \vdots \\ a_1 & a_2 & a_3 & \cdots & a_n \end{bmatrix},$$

which has distinct eigenvalues $\lambda_1, \lambda_2, \ldots, \lambda_n$. Equating both sides of $CP = P \operatorname{diag}(\lambda_1, \lambda_2, \ldots, \lambda_n)$, determine the nonsingular matrix $P = [p_{ij}]$ and show that $p_{ij} = \lambda_j^{i-1}\alpha_j$, ($\alpha_1, \alpha_2, \ldots, \alpha_n$ are arbitrary nonzero elements) and that a general form of eigenvectors of C is $\alpha_j[1\ \lambda_j\ \cdots\ \lambda_j^{n-1}]^T$.

7.51 Prove that the following matrices are simple:

(a) $\begin{bmatrix} 1 & 1 & \cdots & 1 \\ 1 & 1 & \cdots & 1 \\ \vdots & \vdots & & \vdots \\ 1 & 1 & \cdots & 1 \end{bmatrix}$, (b) $\begin{bmatrix} 0 & & 1 \\ & \iddots & \\ 1 & & 0 \end{bmatrix}$, (c) $\begin{bmatrix} 5 & -3 & -3 & -3 \\ 3 & -1 & -3 & -3 \\ 0 & 0 & 2 & 0 \\ -2 & 2 & 2 & 4 \end{bmatrix}$,

(d) a null matrix, (e) a diagonal matrix.

7.52 Show that the following are not simple matrices:

(a) $A = \begin{bmatrix} -1 & 3 & 0 & 3 \\ -1 & 3 & 1 & 3 \\ 1 & -1 & -2 & 0 \\ 1 & -1 & -1 & -1 \end{bmatrix}$, (b) $A = \begin{bmatrix} 7 & -9 & -5 & -9 \\ 1 & 5 & -1 & 3 \\ 2 & -6 & 0 & -6 \\ 0 & -4 & 0 & -2 \end{bmatrix}$.

7.53 Determine an orthogonal transformation to reduce A to an upper triangular form when

$$A = \begin{bmatrix} 8 & 9 & -9 & 0 \\ 0 & 2 & 0 & 2 \\ 4 & 6 & -4 & 0 \\ 0 & 0 & 0 & 3 \end{bmatrix}.$$

7.54 Determine the eigenvalues and eigenvectors of A and a unitary matrix U such that U^*AU is diagonal when A is given by

(a) $\begin{bmatrix} 1 & i & 1 \\ -i & 2 & i \\ 1 & -i & 1 \end{bmatrix}$, (b) $\begin{bmatrix} 2 & 3 & 0 & 0 \\ 3 & 2 & 0 & 0 \\ 0 & 0 & 2 & -3 \\ 0 & 0 & -3 & 2 \end{bmatrix}$, (c) $\begin{bmatrix} 5 & -2 & 0 \\ -2 & 6 & 2 \\ 0 & 2 & 7 \end{bmatrix}$,

(d) $\begin{bmatrix} 3 & 1 & 0 & 0 \\ 1 & 3 & 0 & 0 \\ 0 & 0 & 3 & -1 \\ 0 & 0 & -1 & 3 \end{bmatrix}$, (e) $\begin{bmatrix} 3 & -1 & 0 \\ -1 & 3 & 0 \\ 0 & 0 & 4 \end{bmatrix}$.

7.55 Prove that:

(a) U^*AU and A have the same eigenvalues if U is unitary.

(b) P^TAP and A have the same eigenvalues if P is orthogonal.

(c) PAQ and A have the same eigenvalues if $PQ = I$.

(d) AB and BA, where A and B need not be square, have the same eigenvalues but for some zero eigenvalues.

7.56 Determine the eigenvalues of AB and BA when

(a) $A = \begin{bmatrix} 1 & 2 \\ -1 & 2 \end{bmatrix}$, $B = \begin{bmatrix} 1 & -3 \\ 1 & 1 \end{bmatrix}$;

(b) $A = \begin{bmatrix} 1 & 2 & 1 \\ 2 & 3 & 2 \end{bmatrix}$, $B = \begin{bmatrix} 2 & -1 \\ -1 & -2 \\ 1 & 1 \end{bmatrix}$.

7.57 Test whether or not the following matrices are normal:

(a) $\begin{bmatrix} 2 & 1+i \\ 1-i & 3 \end{bmatrix}$, (b) $\begin{bmatrix} 0 & & & 1 \\ & & 1 & \\ & \cdot^{\cdot^{\cdot}} & & \\ 1 & & & 0 \end{bmatrix}_n$, (c) $\begin{bmatrix} 1 & 0 & 1 \\ 0 & 1 & 0 \\ -1 & 0 & 1 \end{bmatrix}$,

(d) $\begin{bmatrix} a & -b \\ b & a \end{bmatrix}$, a, b real; (e) $\begin{bmatrix} a & b & c \\ -b & a & d \\ -d & -c & a \end{bmatrix}$,

(f) $\begin{bmatrix} a & b & -c \\ -b & a & d \\ c & -d & a \end{bmatrix}$, (g) $\begin{bmatrix} a & b & c \\ c & a+d & d \\ b & d & a+d \end{bmatrix}$, if $b^2 = c^2$.

7.58 If $\mathbf{x}$ is an eigenvector of A associated with an eigenvalue λ, then the normalized vector (see Definition 5.12) $\mathbf{x}/\|\mathbf{x}\|$ is also an eigenvector of A associated with λ.

7.59 Two real symmetric matrices A and B are given by

$$A = \begin{bmatrix} 5 & 2 & 0 \\ 2 & 8 & 0 \\ 0 & 0 & 1 \end{bmatrix}, \qquad B = \begin{bmatrix} 1 & -1 & 1 \\ -1 & 1 & -1 \\ 1 & -1 & 1 \end{bmatrix}.$$

Determine a nonsingular matrix P such that $P^*AP = I$ and P^*BP is diagonal.

7.60 Determine a unitary transformation U such that U^*AU is diagonal when A is a normal matrix given by

(a) $\begin{bmatrix} 1 & 1 & 1 \\ -1 & 1 & 1 \\ -1 & -1 & 1 \end{bmatrix}$, (b) $\begin{bmatrix} 2 & 1 & 1 \\ -1 & 2 & -2 \\ -1 & 2 & 2 \end{bmatrix}$, (c) $\begin{bmatrix} 1 & -1 & -1 \\ -1 & 2 & 1 \\ -1 & 1 & 2 \end{bmatrix}$.

7.61 Determine the spectral resolution of A when it is given by

(a) $\begin{bmatrix} 1 & 1 & 1 & 1 \\ 1 & 1 & 1 & 1 \\ 1 & 1 & 1 & 1 \\ 1 & 1 & 1 & 1 \end{bmatrix}$, (b) $\begin{bmatrix} 5 & -3 & -3 & -3 \\ 3 & -1 & -3 & -3 \\ 0 & 0 & 2 & 0 \\ -2 & 2 & 2 & 4 \end{bmatrix}$.

7.62 Show that if λ_i's are the eigenvalues of a simple matrix A of multiplicity n_i, then

$$(\lambda I - A)^{-1} = \sum_{k=1}^{r} \frac{Q_k}{\lambda - \lambda_k},$$

where Q_k is the constituent matrix of A of rank n_k.

7.63 Prove Corollaries 7.16 and 7.17.

7.64 Find a congruent transformation P to reduce the following real symmetric matrices to a canonical form defined in Corollary 7.16.

$$\text{(a)} \begin{bmatrix} 1 & 2 & 3 \\ 2 & 3 & 1 \\ 3 & 1 & 2 \end{bmatrix}, \quad \text{(b)} \begin{bmatrix} 0 & 1 & 2 \\ 1 & 0 & 3 \\ 2 & 3 & 0 \end{bmatrix}, \quad \text{(c)} \begin{bmatrix} 0 & 0 & -2 & 3 \\ 0 & 1 & 1 & -3 \\ -2 & 1 & 1 & 1 \\ 3 & -3 & 1 & 9 \end{bmatrix}.$$

7.65 Find a conjunctive transformation U to reduce the following Hermitian matrices to a canonical form defined in Corollary 7.17:

$$\text{(a)} \begin{bmatrix} 0 & 1+i \\ 1-i & 0 \end{bmatrix}, \quad \text{(b)} \begin{bmatrix} 0 & -i & -i \\ i & 0 & -i \\ i & i & 0 \end{bmatrix}, \quad \text{(c)} \begin{bmatrix} 1 & 1+i & i \\ 1-i & 2 & 1 \\ -i & 1 & 3 \end{bmatrix}.$$

7.66 Let λ be an eigenvalue of multiplicity m of a linear transformation L on V_n. Then prove that

$$n - m \le \text{rank } (\lambda I_V - L) < n; \qquad 0 < \text{null } (\lambda I_V - L) \le m.$$

7.67 Find the sum and product of the eigenvalues of the matrices given in Problem 7.61.

7.68 A matrix of the form

$$C = \begin{bmatrix} c_0 & c_1 & c_2 & \cdots & c_{n-2} & c_{n-1} \\ c_{n-1} & c_0 & c_1 & \cdots & c_{n-3} & c_{n-2} \\ \vdots & \vdots & \vdots & \cdots & \vdots & \vdots \\ c_2 & c_3 & c_4 & \cdots & c_0 & c_1 \\ c_1 & c_2 & c_3 & \cdots & c_{n-1} & c_0 \end{bmatrix}_n$$

is known as a *circulant matrix.* Let z_k be a root of the scalar equation : $z^n = 1$ and set

$$\lambda_k = c_0 + c_1 z_k + \cdots + c_{n-1} z_k^{n-1}$$

Show that λ_k satisfies the following system of equations

$$z_k \lambda_k = c_{n-1} + c_0 z_k + \cdots + c_{n-2} z_k^{n-1},$$

$$\vdots$$

$$z_k^{n-1} \lambda_k = c_1 + c_2 z_k + \cdots + c_0 z_k^{n-1}$$

Hence, prove that λ_k is an eigenvalue of C with an associated eigenvector

$$\mathbf{x}_k = [1\; z_k \cdots z_k^{n-1}]^T$$

and that a set of n eigenvalues of C are

$$1, e^{i2\pi/n}, e^{i4\pi/n}, \ldots, e^{i2(n-1)\pi/n}.$$

7.69 Prove that if λ_i is an eigenvalue of A, then λ_i/ε is an eigenvalue of $\varepsilon^{-1}A$, where ε is a small parameter.

7.70 Show that the eigenvalues of an idempolent matrix A ($= A^2$) are either 1 or 0.

7.71 Prove that for a normal matrix A, if λ_i is an eigenvalue and $\mathbf{x}_i$ is the corresponding eigenvector, then $A^* \mathbf{x}_i = \bar{\lambda}_i \mathbf{x}_i$.

CHAPTER

8

Bilinear, Quadratic and Hermitian Forms

A bilinear form is a special type of function involving two different real or complex vector variables having the value of the function in a real or complex field $\mathbb{F}$ while a quadratic form is a particular case of this bilinear form defined for a real vector variable and a real field. A Hermitian form is a generalization of the quadratic form and is defined on a complex vector space with values in a real field $\mathbb{R}$. Appropriate nonsingular transformations of the given vector variables reduce these forms (bilinear, quadratic or Hermitian) to a simple structure and these enable one to separate structurally identical forms from the others. As the quadratic and Hermitian forms assume real values, they are classified according to their affinity for only positive, only negative, or either of positive and negative values without special preference. One must pay particular attention to these latter forms because of their dominant role in control systems (e.g., quadratic optimization problems, Lyapunov stability, etc.), statistics (e.g., in covariance of two variables, etc.), and in many branches of physical science and engineering.

Throughout this chapter we denote by V_n an n-dimensional vector space having elements which are n-vectors and whose components are in a field $\mathbb{F}$ of real or complex numbers denoted by $\mathbb{R}$ or $\mathbb{C}$.

8.1 BILINEAR FORMS

Definition 8.1 *Bilinear Forms.* Let V_m and V_n be two vector spaces over the same field $\mathbb{F}$, and let $\mathbf{x}$ and $\mathbf{y}$ be two elements of V_m and V_n respectively. Then an expression of the form

$$b(\mathbf{x}, \mathbf{y}) = \mathbf{x}^T A\mathbf{y} = \sum_{i=1}^{m} \sum_{j=1}^{n} a_{ij} x_i y_j, \tag{8.1}$$

where

$$\mathbf{x}^T = [x_1, x_2, \ldots, x_m], \qquad \mathbf{y}^T = [y_1, y_2, \ldots, y_n],$$

and $A = [a_{ij}]$ is an $m \times n$ matrix over the field $\mathbb{F}$, is called a *bilinear form* over $\mathbb{F}$. The matrix A in (8.1) is called the *matrix* of the form and rank of A is referred to as the *rank* of the form. When A is symmetric, the bilinear form (8.1) is said to be *symmetric*. The polynomial (8.1) represents a *real bilinear form* if the field $\mathbb{F}$ under consideration is real, in which case $b(\mathbf{x}, \mathbf{y})$ has a real value. In general, $b(\mathbf{x}, \mathbf{y})$ assumes values in the field $\mathbb{F}$. □

It is important to note that in (8.1) the coefficient of the product x_iy_j is the element a_{ij} of the matrix A located in the ijth position.

Example 8.1 Let

$$b(\mathbf{x},\mathbf{y}) = [x_1 x_2]\begin{bmatrix} 1 & 0 & 3 \\ -2 & 1 & 1 \end{bmatrix}\begin{bmatrix} y_1 \\ y_2 \\ y_3 \end{bmatrix}$$

$$= x_1y_1 + 3x_1y_3 - 2x_2y_1 + x_2y_2 + x_2y_3,$$

where the underlying field is assumed to be $\mathbb{R}$. This is an example of a *real bilinear form*. □

In respect of the variable $\mathbf{x}$ in V_m, the *bilinear function* $b(\mathbf{x}, \mathbf{y})$ may be looked upon as a linear mapping (see Definition 6.1) from $V_m \to \mathbb{F}$. In other words, if $\mathbf{z}$ is another vector in V_m and α, β are scalars in $\mathbb{F}$, then $b(\alpha\,\mathbf{x} + \beta\,\mathbf{z}, \mathbf{y}) = \alpha\, b(\mathbf{x}, \mathbf{y}) + \beta\, b(\mathbf{z}, \mathbf{y})$, which can be proved with the help of (8.1) when $\mathbf{z}$ is written as $\mathbf{z}^T = [z_1, z_2, \ldots, z_m]$. Again, in respect of the variable $\mathbf{y}$ in V_n, we can explain similarly that $b(\mathbf{x}, \mathbf{y})$ defines a linear map from $V_n \to \mathbb{F}$, i.e., $b(\mathbf{x}, \alpha\mathbf{y} + \beta\,\mathbf{u}) = \alpha\, b(\mathbf{x}, \mathbf{y}) + \beta\, b(\mathbf{x}, \mathbf{u})$ for any vector $\mathbf{u}$ in V_n. Because of this linear mapping properties with respect to two vector spaces V_m and V_n when considered separately, $b(\mathbf{x}, \mathbf{y})$ is known as *bilinear form.*

From the discussion in Section 4.3, we know that, by elementary row and column transformations, any matrix A can be reduced to a normal form. It might also be recalled (see Definition 4.5) that there exist nonsingular matrices P and Q of order m and n respectively such that

$$P^T AQ = \left[\begin{array}{c|c} I_r & 0 \\ \hline 0 & 0 \end{array}\right] \tag{8.2}$$

is in a normal form, where r is the rank of A. By transforming coordinates with the introduction of new variables $\mathbf{u}$ and $\mathbf{v}$ in the vector spaces V_m and V_n respectively, so that

$$\mathbf{x} = B\mathbf{u} \quad \text{and} \quad \mathbf{y} = C\mathbf{v}, \tag{8.3}$$

where B and C are nonsingular matrices of appropriate order, we introduce (8.3) in (8.1) to get the bilinear form

$$b(\mathbf{u}, \mathbf{v}) = \mathbf{u}^T(B^TAC)\mathbf{v} = \mathbf{u}^TD\mathbf{v}, \qquad D = B^TAC. \tag{8.4}$$

The bilinear forms (8.1) and (8.4) are then said to be *equivalent*.

Definition 8.2 *Equivalent Bilinear Forms*. The bilinear form $\mathbf{x}^TA\mathbf{y}$ is said to be equivalent to $\mathbf{u}^TD\mathbf{v}$, where $\mathbf{x}$ and $\mathbf{u}$ are in V_m, $\mathbf{y}$ and $\mathbf{v}$ are in V_n, A and D are matrices of the form over the same field, if there exist nonsingular matrices B and C of order m and n respectively such that

$$D = B^T AC, \tag{8.5}$$

i.e., if the matrices of the two forms A and D are equivalent (see Definition 4.4). □

It is not difficult to show that the equivalence relation as defined above is reflexive, symmetric and transitive (see Section 4.3). If we select B and C in (8.4) to be such that $\boldsymbol{B} = \boldsymbol{P}$, $C = Q$, then by virtue of (8.2), we have

$$b(\mathbf{u}, \mathbf{v}) = \mathbf{u}^T (P^T AQ)\mathbf{v} = \mathbf{u}^T \begin{bmatrix} I_r & 0 \\ 0 & 0 \end{bmatrix} \mathbf{v} = u_1 v_1 + \cdots + u_r v_r, \tag{8.6}$$

i.e., the bilinear form (8.1) is equivalent to the *canonical form* (8.6).

Example 8.2 Let a matrix of a bilinear form be given by

$$A = \begin{bmatrix} 2 & 1 & 1 \\ 4 & 2 & 2 \\ 1 & 2 & 2 \end{bmatrix}.$$

As shown in Example 4.3, we can choose

$$P^T = \begin{bmatrix} 1 & 0 & 0 \\ 2/3 & 0 & -1/3 \\ -2 & 1 & 0 \end{bmatrix} \quad \text{and} \quad Q = \begin{bmatrix} 0 & 1 & 0 \\ 1 & -2 & -1 \\ 0 & 0 & 1 \end{bmatrix}$$

so that

$$P^T AQ = \begin{bmatrix} 1 & 0 & 0 \\ 0 & 1 & 0 \\ 0 & 0 & 0 \end{bmatrix}.$$

The given bilinear form

$$[x_1\ x_2\ x_3] \begin{bmatrix} 2 & 1 & 1 \\ 4 & 2 & 2 \\ 1 & 2 & 2 \end{bmatrix} \begin{bmatrix} y_1 \\ y_2 \\ y_3 \end{bmatrix}$$

is, therefore, equivalent to

$$[u_1\ u_2\ u_3] \text{ diag } (1, 1, 0)[v_1\ v_2\ v_3]^T = u_1 v_1 + u_2 v_2.$$ □

Definition 8.3 *Cogredient Transformations and Congruent Matrices.* Let $\mathbf{x}$ and $\mathbf{y}$ in Definition 8.1 belong to the same vector space V_n over a field $\mathbb{F}$. Then, A is square and $\mathbf{x}^T A\mathbf{y}$ is called a bilinear form on the space V_n over $\mathbb{F}$. Consequently, the transformation matrices in (8.3) are the same, $B = C$ and $\mathbf{x} = B\mathbf{u}$, $\mathbf{y} = B\mathbf{v}$. Therefore, (8.4) becomes

$$b(\mathbf{u}, \mathbf{v}) = \mathbf{u}^T(B^T AB)\mathbf{v} = \mathbf{u}^T D\mathbf{v},$$

where $D = B^TAB$ and **u** and **v** are both n-vectors. Under such circumstances, i.e., when **x** and **y** are subjected to the same transformation $\mathbf{x} = B\mathbf{u}$ and $\mathbf{y} = B\mathbf{v}$, it is said that **x** and **y** are transformed *c109grediently* and that A is *congruent* to D over $\mathbb{F}$, connected as they are by $D = B^TAB$, where B is nonsingular (see Section 7.8). It can be shown that congruence is a true *equivalence* relation (see Section 4.3). □

Definition 8.4 *Contragredient Transformations.* In Definition 8.1, **x** and **y** may belong to geometrically distinct vector spaces of the same dimension n, i.e., V_n's over $\mathbb{F}$, in which case A is a square $n \times n$ matrix and **x**, **y** must be subjected to distinct linear transformations. If A is of rank n, then nonsingular transformations P and Q may be so selected that

$$b(\mathbf{u}, \mathbf{v}) = \mathbf{u}^T(P^TAQ)\mathbf{v} = u_1v_1 + \cdots + u_nv_n = \mathbf{u}^TI_n\mathbf{v}. \tag{8.7}$$

Let us now find nonsingular transformations

$$\mathbf{u} = R\tilde{\mathbf{u}} \quad \text{and} \quad \mathbf{v} = S\tilde{\mathbf{v}}, \tag{8.8}$$

where $\tilde{\mathbf{u}}$ and $\tilde{\mathbf{v}}$ are elements of two distinct V_n's over $\mathbb{F}$ and R, S are $n \times n$ matrices which leave the *canonical bilinear form* (8.7) invariant in form. Inserting (8.8) into (8.7), we get

$$\mathbf{u}^T I_n \mathbf{v} = \tilde{\mathbf{u}}^T R^T S \tilde{\mathbf{v}}^T$$

in which the canonical form (8.7) remains invariant under transformations (8.8) if

$$R^TS = I_n, \text{ i.e., } S = (R^T)^{-1}.$$

The transformations defined by $\mathbf{u} = R\tilde{\mathbf{u}}$, $\mathbf{v} = (R^T)^{-1}\tilde{\mathbf{v}}$, which would keep the canonical bilinear form (8.7) invariant in form, are referred to as *contragredient transformations.* □

Under orthogonal transformation of coordinates (see Definition 6.15) of the bilinear form $b(\mathbf{x}, \mathbf{y}) = \mathbf{x}^TA\mathbf{y}$, we have $\mathbf{x} = B\mathbf{u}$, $\mathbf{y} = B\mathbf{v}$, and $B^TB = I$. The bilinear form $b(\mathbf{x}, \mathbf{y})$ then becomes $b(\mathbf{u}, \mathbf{v}) = \mathbf{u}^TB^TAB\mathbf{v} = \mathbf{u}^TD\mathbf{v}$, where $D = B^TAB$. This is also the case for cogredient transformation but with an additional constraint $B^TB = I$. The above orthogonal transformation is again a contragredient transformation with the additional requirement that $R = S = B$. This shows that over the real field, the orthogonal transformation of coordinates, transforms a real bilinear form, both cogrediently and contragrediently (see also Section 8.8).

8.2 QUADRATIC FORMS

Definition 8.5 *Quadratic Forms.* Let $\mathbf{x} = [x_1, x_2, \ldots, x_n]^T$ be an n-vector in the vector space V_n over a field $\mathbb{F}$, and let $A = [a_{ij}]$ be an n-square matrix over $\mathbb{F}$. Then any polynomial of the form

$$q(x_1, \ldots, x_n) = \mathbf{x}^TA\mathbf{x} = \sum_{i,j=1}^{n} a_{ij}x_ix_j \tag{8.9}$$

is called a quadratic form of order n over $\mathbb{F}$ in the variables $x_1, x_2, \ldots, x_n$. A is called the matrix of the form and rank A the rank of the form. A quadratic form is said to be *real* (resp *complex*) if the matrix A is over a real field $\mathbb{R}$ (resp a complex field $\mathbb{C}$) and the variable **x** is in the vector space $\mathbb{R}^n$ (resp $\mathbb{C}^n$). □

Unless stated to the contrary, we shall confine our discussion to quadratic forms *in which the matrix A as well as the variables are considered to be real.*

Example 8.3 Let $q(x_1, x_2, x_3) = 2x_1^2 - 3x_1x_2 + 3x_2^2 + 2x_1x_3 + 4x_2x_3 + x_3^2$ be a quadratic form in the variables x_1, x_2, x_3. The matrix

$$A = \begin{bmatrix} 2 & -3 & 3 \\ 0 & 3 & 3 \\ -1 & 1 & 1 \end{bmatrix}$$

is a matrix of the form as can be verified by computing $\mathbf{x}^T A\mathbf{x}$, where $\mathbf{x}^T = [x_1, x_2, x_3]$. There are numerous ways in which this matrix A may be expressed, as the reader may verify. But there is only one way in which the matrix A can always be put into a symmetric form. This is done by writing the coefficients of the product x_ix_j, $i \neq j$, as the sum of two equal terms in the form

$$q(x_1, x_2, x_3) = 2x_1^2 - (3/2)x_1x_2 + x_1x_3 - (3/2)x_2x_1 + 3x_2^2 + 2x_2x_3 + x_3x_1 + 2x_3x_2 + x_3^2 \tag{8.10}$$

and noting that the coefficient of the product x_ix_j is a_{ij}. Consequently, the matrix of the quadratic form in this example can also be written as

$$A = \begin{bmatrix} 2 & -3/2 & 1 \\ -3/2 & 3 & 2 \\ 1 & 2 & 1 \end{bmatrix}$$

which is a symmetric matrix. □

Among the various ways in which the matrix A of a quadratic form (8.9) can be written, by the arguments put forward in this example, there is only one way in which it can be expressed uniquely by a real symmetric matrix. Indeed, given any quadratic form (8.9), we can write it as

$$q(x_1, x_2, ..., x_n) = \mathbf{x}^T A\mathbf{x} = \sum_{i,j=1}^{n} a_{ij}x_ix_j = \sum_{i,j=1}^{n} \frac{1}{2}(a_{ij} + a_{ji})x_ix_j$$

$$= \sum_{i,j=1}^{n} b_{ij}x_ix_j = \mathbf{x}^T B\mathbf{x},$$

where

$$b_{ij} = \frac{1}{2}(a_{ij} + a_{ji}) = \frac{1}{2}(a_{ji} + a_{ij}) = b_{ji}$$

and so

$$B = \frac{1}{2}(A + A^T) = B^T$$

which is a symmetric matrix of the quadratic form (8.9). Henceforth, unless otherwise stated, the matrix A of the *quadratic form* $\mathbf{x}^T A\mathbf{x}$ *will be considered to be real and symmetric.*

Definition 8.6 *Discriminant of, and Singular and Nonsingular Quadratic Forms.* By the discriminant of a quadratic form $\mathbf{x}^T A\mathbf{x}$, we mean det A. The quadratic form is said to be nonsingular if its discriminant is nonzero; otherwise, it is said to be singular. □

The substitution of $\mathbf{x} = P\mathbf{y}$ in the quadratic form $\mathbf{x}^T A\mathbf{x}$ is called a *linear transformation* of the quadratic form. The linear transformation is referred to as *real* (resp *complex*) if P is *real* (resp *complex*), and *nonsingular* (resp *singular*) if P is nonsingular (resp singular).

If a new coordinate system is introduced in the vector space V_n so that the old variable $\mathbf{x}$ is related to the new variable $\mathbf{y}$ by $\mathbf{x} = P\mathbf{y}$, where P is a nonsingular transformation matrix, then (8.9) is changed into another quadratic form, viz.

$$\mathbf{x}^T A\mathbf{x} = \mathbf{y}^T(P^T AP)\mathbf{y} = \mathbf{y}^T B\mathbf{y}, \tag{8.11}$$

where

$$B = P^T AP. \tag{8.12}$$

The matrix A of a quadratic form under the nonsingular transformation carried out in (8.11) changes to $P^T AP$ which is symmetric if A is symmetric and the discriminant changes from det A to (det P)^2 det A, but its rank, i.e., rank A = rank $P^T AP$ (see Corollary 5.14), remains invariant. When such a nonsingular transformation exists over $\mathbb{F}$ such that (8.11) holds, these quadratic forms, i.e., $\mathbf{x}^T A\mathbf{x}$ and $\mathbf{y}^T B\mathbf{y}$, are said to be equivalent over $\mathbb{F}$. Since by Definition 7.23, A is congruent to B if (8.12) holds, we can say that two quadratic forms are equivalent over a field $\mathbb{F}$ if the matrices of their form are congruent over $\mathbb{F}$. Conversely, if (8.12) holds, so does (8.11) for a nonsingular transformation defined by $\mathbf{x} = P\mathbf{y}$. Thus, $\mathbf{x}^T A\mathbf{x}$ and $\mathbf{y}^T B\mathbf{y}$ are equivalent. The following definition and theorems follow as a consequence of the foregoing discussion.

Definition 8.7 *Equivalent Quadratic Forms.* Two quadratic forms $\mathbf{x}^T A\mathbf{x}$ and $\mathbf{y}^T B\mathbf{y}$ are said to be equivalent over $\mathbb{F}$ if one can be obtained from the other by a nonsingular transformation over $\mathbb{F}$ defined by $\mathbf{x} = P\mathbf{y}$ such that (8.11) and (8.12) are satisfied. □

Theorem 8.1 The rank of a quadratic form remains invariant under a nonsingular transformation. □

Theorem 8.2 *Equivalence Theorem of Complex Quadratic Form.* Two quadratic forms over a field $\mathbb{F}$ are equivalent iff the matrices of their forms are congruent over $\mathbb{F}$. □

Example 8.4 The quadratic forms defined by

$$q(x_1, x_2, x_3) = 5x_1^2 + x_2^2 + 21x_3^2 + 4x_1x_2 - 20x_1x_3 - 8x_2x_3 = \mathbf{x}^T A\mathbf{x},$$

$$q(y_1, y_2, y_3) = y_1^2 + y_2^2 + y_3^2 = \mathbf{y}^T B\mathbf{y},$$

where

$$A = \begin{bmatrix} 5 & 2 & -10 \\ 2 & 1 & -4 \\ -10 & -4 & 21 \end{bmatrix}, \qquad B = \text{diag } (1, 1, 1)$$

are equivalent. This is so because there exists a nonsingular matrix

$$P = \begin{bmatrix} 2 & -1 & 0 \\ 0 & 2 & -1 \\ 1 & 0 & 0 \end{bmatrix} \tag{8.13}$$

such that $P^TAP = B$. The above quadratic forms are nonsingular because det A = det B = 1 is nonzero. □

This example shows that $\mathbf{x}^TA\mathbf{x}$ and $\mathbf{y}^TB\mathbf{y}$ are two equivalent quadratic forms, and A and B are two congruent matrices (see Definition 7.23) defined as they are by a relation of the form (8.12), where P is given by (8.13). Moreover, the quadratic form $\mathbf{y}^TB\mathbf{y}$ is in a very simple form in that it contains terms with the variables occurring as the squares and no cross-product terms such as y_1y_2, y_2y_3 and y_1y_3 appear. This can always be done by selecting a proper coordinate system, i.e., by choosing an appropriate nonsingular transformation matrix P.

8.3 REDUCTION OF QUADRATIC FORMS

Definition 8.8 *Diagonal and Unit Quadratic Forms.* If the matrix of a quadratic form is diagonal (respectively unity), then it is called a *diagonal* (resp *a unit) quadratic form.* For example, $\mathbf{x}^TI_n\mathbf{x} = x_1^2 + x_2^2 + \cdots + x_n^2$ is a unit and $\mathbf{x}^T$ diag $(a_1, a_2, ..., a_n)\mathbf{x} = a_1x_1^2 + a_2x_2^2 + \cdots + a_nx_n^2$ is a diagonal quadratic form. In the diagonal form, some of the a_i's may be zero. These two forms are sometimes called the *canonical representation* of a quadratic form. □

Numerous ways may be pursued to transform a quadratic form into its diagonal form. Among them we shall confine ourselves to the following two, viz., (a) orthogonal transformation and (b) Lagrange reduction. The process of transforming a given quadratic form into a diagonal form is known as *reduction to diagonal form.*

8.3.1 Orthogonal Transformation

Reduction of a quadratic form to a diagonal form by means of an orthogonal transformation is referred to as *orthogonal reduction.*

Theorem 8.3 Every quadratic form $\mathbf{x}^TA\mathbf{x}$ can be reduced to a diagonal quadratic form $\mathbf{y}^TD\mathbf{y}$ by means of an orthogonal transformation P, where the diagonal elements of D are the eigenvalues *of A.*

Proof Since A is symmetric, as a consequence of Corollary 7.7, there exists an orthogonal transformation P such that $P^TAP = D$ and $P^TP = I_n$, where D is a diagonal matrix having the eigenvalues λ_i's, $i = 1, 2, ..., n$, of A as diagonal elements, i.e., D = diag $(\lambda_1, \lambda_2, ..., \lambda_n)$. Let $\mathbf{x} = P\mathbf{y}$. Then

$$\mathbf{x}^TA\mathbf{x} = \mathbf{y}^T(P^TAP)\mathbf{y} = \mathbf{y}^TD\mathbf{y} = \lambda_1y_1^2 + \lambda_2y_2^2 + \cdots + \lambda_ny_n^2. \quad \square$$

An easy consequence of Theorem 8.3 is the following corollary.

Corollary 8.1 Every quadratic form $\mathbf{x}^TA\mathbf{x}$ of rank r can be reduced by an orthogonal transformation P to

$$\lambda_1y_1^2 + \lambda_2y_2^2 + \cdots + \lambda_ry_r^2,$$

where $\mathbf{x} = P\mathbf{y}$ and $\lambda_1, \lambda_2, ..., \lambda_r$ are the nonzero eigenvalues of A. □

Example 8.5 Let $q(x_1, x_2, x_3) = \mathbf{x}^T A\mathbf{x}$ be a real quadratic form where

$$A = \begin{bmatrix} 1 & 2 & 0 \\ 2 & 1 & 0 \\ 0 & 0 & -1 \end{bmatrix}.$$

As computed in Example 7.40, the orthogonal matrix

$$P = \begin{bmatrix} 1/\sqrt{2} & 0 & 1/\sqrt{2} \\ -1/\sqrt{2} & 0 & 1/\sqrt{2} \\ 0 & 1 & 0 \end{bmatrix}$$

transforms A to $P^T AP = \text{diag}\,(-1, -1, 3)$. Therefore, using the transformation $\mathbf{x} = P\mathbf{y}$, the given quadratic form becomes

$$q = \mathbf{y}^T P^T AP\mathbf{y} = -y_1^2 - y_2^2 + 3y_3^2. \quad \square$$

8.3.2 Lagrange Reduction

Example 8.5 shows that the higher the order of A, the larger the amount of computation needed in orthogonal reduction of a quadratic form. Consequently, the labour becomes prohibitive for a very high order matrix A. Therefore, unless orthogonality of the transformation is essential, we shall not pursue orthogonal reduction but employ what is known as Lagrange reduction. This method proceeds by forming squares successively by taking lesser and lesser number of variables among $x_1, x_2, \ldots, x_n$ of the quadratic form as elucidated in the following examples (Example 8.6 and Example 8.7).

Example 8.6 Let the quadratic form be

$$q(x_1, x_2, x_3) = 4x_1^2 + 10x_2^2 + 11x_3^2 - 4x_1x_2 + 12x_1x_3 - 12x_2x_3.$$

Collecting the terms containing x_1 and factoring out the coefficient of x_1^2, we obtain

$$q = 4\{x_1^2 - x_1(x_2 - 3x_3)\} + 10x_2^2 + 11x_3^2 - 12x_2x_3.$$

Completing a square on the term x_1 by adding and subtracting terms as required, we get

$$\begin{aligned} q &= 4\left\{x_1 - \frac{1}{2}(x_2 - 3x_3)\right\}^2 - (x_2 - 3x_3)^2 + 10x_2^2 + 11x_3^2 - 12x_2x_3 \\ &= 4\left(x_1 - \frac{1}{2}x_2 + \frac{3}{2}x_3\right)^2 + 9x_2^2 - 6x_2x_3 + 2x_3^2 \\ &= 4\left(x_1 - \frac{1}{2}x_2 + \frac{3}{2}x_3\right)^2 + 9\left(x_2^2 - \frac{2}{3}x_2x_3\right) + 2x_3^2. \end{aligned}$$

Finally, completing square on x_2, we have

$$q = 4\left(x_1 - \frac{1}{2}x_2 + \frac{3}{2}x_3\right)^2 + 9\left(x_2 - \frac{1}{3}x_3\right)^2 + x_3^2.$$

Now, put

$$y_1 = x_1 - \frac{1}{2}x_2 + \frac{3}{2}x_3,$$

$$y_2 = x_2 - \frac{1}{3}x_3,$$

$$y_3 = x_3,$$

which can be put in the form $\mathbf{y} = P\mathbf{x}$ with the transformation matrix

$$P = \begin{bmatrix} 1 & -1/2 & 3/2 \\ 0 & 1 & -1/3 \\ 0 & 0 & 1 \end{bmatrix},$$

that is nonsingular with determinant equal to one. Using this transformation, the given quadratic form reduces to the diagonal form

$$q = 4y_1^2 + 9y_2^2 + y_3^2.$$

If we choose $z_1 = 2y_1$, $z_2 = 3y_2$, $z_3 = y_3$, then we have $q = z_1^2 + z_2^2 + z_3^2$, so that the given quadratic form is reduced to a unit quadratic form (see Definition 8.8). □

In Example 8.6 above, computation of the squares would not have been possible if there were no x_1^2 term in the first step and x_2^2 term in the second step. Indeed, there might not have been any square terms on which a square could be completed. These cases can be handled by an artifice given in Example 8.7 below.

Example 8.7 We now consider the quadratic form

$$q(x_1, x_2, x_3) = x_1x_2 + 2x_1x_3 + 3x_2x_3. \tag{8.14}$$

Set

$$x_1 = y_1, \qquad x_2 = y_1 + y_2, \qquad x_3 = y_3, \tag{8.15}$$

The transformation $\mathbf{x} = P\mathbf{y}$ is nonsingular as det $P = 1$. The quadratic form q now becomes

$$q = y_1(y_1 + y_2) + 2y_1y_3 + 3(y_1 + y_2)y_3$$

$$= y_1^2 + y_1y_2 + 5y_1y_3 + 3y_2y_3.$$

Proceeding as in Example 8.6, we have

$$q = \{\, y_1^2 + y_1(y_2 + 5y_3)\} + 3y_2y_3$$

$$= \left\{ y_1 + \frac{1}{2}(y_2 + 5y_3) \right\}^2 - \frac{1}{4}(y_2^2 + 10y_2y_3 + 25y_3^2) + 3y_2y_3$$

$$= \left(y_1 + \frac{1}{2}y_2 + \frac{5}{2}y_3 \right)^2 - \frac{1}{4}(y_2^2 - 2y_2y_3) - \frac{25}{4}y_3^2$$

$$= \left(y_1 + \frac{1}{2}y_2 + \frac{5}{2}y_3 \right)^2 - \frac{1}{4}(y_2 - y_3)^2 - 6y_3^2. \tag{8.16}$$

Putting z_1, z_2 and z_3 for the square terms in (8.16) and taking (8.15) into consideration, we have

$$z_1 = y_1 + \frac{1}{2}y_2 + \frac{5}{2}y_3 = \frac{1}{2}x_1 + \frac{1}{2}x_2 + \frac{5}{2}x_3,$$

$$z_2 = y_2 - y_3 = x_2 - x_1 - x_3,$$

$$z_3 = y_3 = x_3,$$

so that the transformation $\mathbf{z} = Q\mathbf{x}$, i.e.,

$$\begin{bmatrix} z_1 \\ z_2 \\ z_3 \end{bmatrix} = \begin{bmatrix} \frac{1}{2} & \frac{1}{2} & \frac{5}{2} \\ -1 & 1 & -1 \\ 0 & 0 & 1 \end{bmatrix} \begin{bmatrix} x_1 \\ x_2 \\ x_3 \end{bmatrix}$$

with det $Q = 1$ will reduce the given quadratic form (8.14) to the diagonal form

$$q = z_1^2 - \frac{1}{4}z_2^2 - 6z_3^2.$$ □

Note that in Example 8.6 the transformation matrix P is upper triangular because at each step there are terms to form a square. As this is not the case in Example 8.7, the transformation matrix Q loses its triangular structure. The necessary and sufficient condition to reduce a quadratic form to a diagonal form by a transformation matrix which is upper triangular is given in Theorem 8.4.

In the quadratic form

$$q(x_1, \ldots, x_n) = \mathbf{x}^T A\mathbf{x} = \sum_{i,j=1}^{n} a_{ij}x_ix_j,$$

we now assume that $a_{11} \neq 0$. Therefore, observing that $a_{ij} = a_{ji}$, we have

$$q = a_{11}x_1^2 + 2x_1 \sum_{j=2}^{n} a_{1j}x_j + \sum_{i,j=2}^{n} a_{ij}x_ix_j \tag{8.17}$$

$$= a_{11}\left(x_1 + \frac{a_{12}}{a_{11}}x_2 + \cdots + \frac{a_{1n}}{a_{11}}x_n\right)^2 - a_{11}\sum_{i,j=2}^{n}\frac{a_{1i}}{a_{11}}x_i\frac{a_{1j}}{a_{11}}x_j$$

$$+ \sum_{i,j=2}^{n} a_{ij}x_ix_j = \frac{1}{a_{11}}(a_{11}x_1 + a_{12}x_2 + \cdots + a_{1n}x_n)^2 + q_1,$$

where q_1 is a quadratic form not containing x_1. To take stock of the elements in a quadratic form, we may use the crate diagram (Figure 8.1) where the coefficient a_{ij} associates itself with the variables x_i and x_j. The first summation of the right-hand side of (8.17) is accounted for by the first row and column elements except a_{11} in the crate diagram. In general, *a square in the quadratic form* $\mathbf{x}^T A\mathbf{x}$ *can be completed* if there is a nonzero diagonal element, say, a_{ij} in

	x_1	x_2	$\cdots$	x_j	$\cdots$	x_n	
x_1	a_{11}	a_{12}	$\cdots$	a_{1j}	$\cdots$	a_{1n}	← 1st row
x_2	a_{21}	a_{22}	$\cdots$	a_{2j}	$\cdots$	a_{2n}	
$\vdots$	$\vdots$	$\vdots$	$\vdots$	$\vdots$	$\vdots$	$\vdots$	
x_i	a_{i1}	a_{i2}	$\cdots$	a_{ij}	$\cdots$	a_{in}	
$\vdots$	$\vdots$	$\vdots$	$\vdots$	$\vdots$	$\vdots$	$\vdots$	
x_n	a_{n1}	a_{n2}	$\cdots$	a_{nj}	$\cdots$	a_{nn}	← *n*th row
	↑ 1st col					↑ *n*th col	

Figure 8.1 Schematic view of a quadratic form.

A in which case we can deduce as previously:

$$q = \frac{1}{a_{ii}}(a_{i1}x_1 + a_{i2}x_2 + \cdots + a_{in}x_n)^2 + q_i = \frac{1}{4a_{ii}}\left(\frac{\partial q}{\partial x_i}\right)^2 + q_i \tag{8.18}$$

where q_i is a quadratic form not containing x_i. On the other hand, if $a_{11} = a_{22} = 0$ but $a_{21} = a_{21} \neq 0$, then

$$q = 2x_1(a_{12}x_2 + a_{13}x_3 + \cdots + a_{1n}x_n)$$

$$+ 2x_3 \frac{a_{23}}{a_{12}}(a_{12}x_2 + a_{13}x_3 + \cdots + a_{1n}x_n)$$

$$\vdots$$

$$+ 2x_n \frac{a_{2n}}{a_{12}}(a_{12}x_2 + a_{13}x_3 + \cdots + a_{1n}x_n)$$

$$+ \sum_{i,j=3}^{n} a_{ij}x_{ij} - \frac{2}{a_{12}}(a_{23}x_3 + \cdots + a_{2n}x_n)(a_{13}x_3 + \cdots + a_{1n}x_n)$$

$$= \frac{2}{a_{12}}(a_{12}x_2 + a_{13}x_3 + \cdots + a_{1n}x_n)(a_{21}x_1 + a_{23}x_3 + \cdots + a_{2n}x_n) + q_{12}$$

$$= \frac{1}{2a_{12}}[\{(a_{11} + a_{12})x_1 + (a_{12} + a_{22})x_2 + \cdots + (a_{1n} + a_{2n})x_n\}^2$$

$$- \{(a_{11} - a_{21})x_1 + (a_{12} - a_{22})x_2 + \cdots + (a_{1n} - a_{2n})x_n\}^2] + q_{12},$$

where q_{12} is a quadratic form not containing the variables x_1 and x_2 and the relation $4xy = (x + y)^2 - (x - y)^2$ is used in the last step. In general, if $a_{ii} = 0$, $a_{jj} = 0$ but $a_{ij} \neq 0$, then the above equation can be written as

$$q = \frac{1}{2a_{ij}}[\{(a_{i1} + a_{j1})x_1 + (a_{i2} + a_{j2})x_2 + \cdots + (a_{in} + a_{jn})x_n\}^2$$

$$- \{(a_{i1} - a_{j1})x_1 + (a_{i2} - a_{j2})x_2 + \cdots + (a_{in} - a_{jn})x_n\}^2] + q_{ij}$$

$$= \frac{1}{8a_{ij}}\left[\left(\frac{\partial q}{\partial x_i} + \frac{\partial q}{\partial x_j}\right)^2 - \left(\frac{\partial q}{\partial x_i} - \frac{\partial q}{\partial x_j}\right)^2\right] + q_{ij}, \tag{8.19}$$

where q_{ij} is a quadratic form not containing x_i and x_j. Because the forms

$$a_{i1}x_1 + \cdots + a_{in}x_n \quad \text{and} \quad a_{j1}x_1 + \cdots + a_{jn}x_n,$$

which do not contain x_i and x_j respectively, are independent, the forms within the braces in (8.19) being the sum and difference of these forms are, therefore, linearly independent. With the help of (8.18) and (8.19), any quadratic form $\mathbf{x}^T A \mathbf{x}$ can be expressed as the sum of squares of independent forms.

Example 8.8 In the quadratic form

$$q(x_1, x_2, x_3) = x_1^2 + 4x_2^2 + 16x_3^2 + 4x_1x_2 + 8x_1x_3 + 17x_2x_3,$$

since $a_{11} = 1$, applying (8.18), where $i = 1$, we have

$$q = \frac{1}{4}(2x_1 + 4x_2 + 8x_3)^2 + x_2x_3 = (x_1 + 2x_2 + 4x_3)^2 + q_1,$$

where $q_1 = x_2x_3$. Now applying (8.19), with $i = 2$ and $j = 3$ and $a_{23} = 0.5$, we have

$$q_1 = \frac{1}{4}[(x_2 + x_3)^2 - (x_2 - x_3)^2].$$

Therefore,

$$q = (x_1 + 2x_2 + 4x_3)^2 + \frac{1}{4}(x_2 + x_3)^2 - \frac{1}{4}(x_2 - x_3)^2,$$

which can be transformed into a diagonal form by setting

$$z_1 = x_1 + 2x_2 + 4x_3, \qquad z_2 = x_2 + x_3, \qquad z_3 = x_2 - x_3.$$

The transformation matrix is nonsingular although not upper triangular. □

The following theorem gives the necessary and sufficient condition for reducing a quadratic form to a diagonal one by a nonsingular transformation matrix, which is upper triangular.

Theorem 8.4 (Jacobi) A quadratic form

$$q = \mathbf{x}^T A\mathbf{x} = \sum_{i,j=1}^{n} a_{ij}x_ix_j \tag{8.20}$$

with rank r can be transformed by a nonsingular transformation $\mathbf{x} = P\mathbf{y}$ with det $P = 1$ and P upper triangular to a diagonal form

$$q = c_1y_1^2 + c_2y_2^2 + \cdots + c_ry_r^2 \tag{8.21}$$

if and only if Δ_i's, $i = 1, 2, ..., r$, are nonzero, where

$$\begin{aligned}
y_1 &= x_1 + \alpha_{12}x_2 + \cdots + \alpha_{1r}x_r + \alpha_{1r+1}x_{r+1} + \cdots + \alpha_{1n}x_n,\\
y_2 &= x_2 + \cdots + \alpha_{2r}x_r + \alpha_{2r+1}x_{r+1} + \cdots + \alpha_{2n}x_n,\\
&\ \vdots\\
y_j &= x_j + \cdots + \alpha_{jr}x_r + \alpha_{jr+1}x_{r+1} + \cdots + \alpha_{jn}x_n,\\
&\ \vdots\\
y_r &= x_r + \alpha_{rr+1}x_{r+1} + \cdots + \alpha_{rn}x_n,\\
y_{r+1} &= x_{r+1},\\
&\ \vdots\\
y_n &= x_n,
\end{aligned} \tag{8.22}$$

are n linearly independent and real linear forms in the variables x_i's, α_{ij}'s are scalars completely determined by a_{ij}'s,

$$c_i = \Delta_i/\Delta_{i-1}, \tag{8.23}$$

and Δ_i's are the leading principal minors of A defined as

$$\Delta_0 = 1, \quad \Delta_1 = a_{11}, \quad \Delta_2 = \begin{vmatrix} a_{11} & a_{12} \\ a_{21} & a_{22} \end{vmatrix}, \quad \Delta_2 = \begin{vmatrix} a_{11} & a_{12} & a_{13} \\ a_{21} & a_{22} & a_{23} \\ a_{31} & a_{32} & a_{33} \end{vmatrix}, \ldots, \quad \Delta_n = \det A.$$

Proof (a) *Sufficiency.* This will be proved by mathematical induction. Assuming $\Delta_1 = a_{11} \neq 0$, from (8.17) we have

$$q = a_{11}\left(x_1 + \frac{a_{12}}{a_{11}}x_2 + \cdots + \frac{a_{1n}}{a_{11}}x_n\right)^2 + \sum_{i,j=2}^{n}\left(a_{ij} - \frac{a_{1i}a_{1j}}{a_{11}}\right)x_i x_j$$

$$= c_1 y_1^2 + q_2(x_2, \ldots, x_n),$$

where c_1 is given by (8.23), y_1 by (8.22) with

$$\alpha_{1p} = \frac{a_{1p}}{a_{11}}, \qquad p = 2, \ldots, n;$$

and

$$q_2(x_2, \ldots, x_n) = \sum_{i,j=2}^{n} b_{ij}x_i x_j,$$

where $\qquad b_{ij} = a_{ij} - (a_{1i}\, a_{1j} / a_{11}) = b_{ji}$

is a quadratic form in $x_2, \ldots, x_n$.

We can apply the same process of forming squares to q_2 and so on until we are able by this reduction process to arrive at the stage given by

$$q = c_1 y_1^2 + c_2 y_2^2 + \cdots + c_{p-1} y_{p-1}^2 + q_p(x_p, \ldots, x_n), \tag{8.24}$$

where y_i's are given as linearly independent forms by (8.22) and c_i's by (8.23). The existence of (8.23) implies that $\Delta_1, \ldots, \Delta_{p-1}$ are nonzero. Let

$$q_p(x_p, \ldots, x_n) = \sum_{i,j=p}^{n} \beta_{ij}x_i x_j.$$

In the next step, we shall show that if $\Delta_p \neq 0$, then $q_p(x_p, \ldots, x_n)$ contains $\beta_{pp} x_p^2$, i.e., $\beta_{pp} \neq 0$. Let us select $x_{p+1} = \cdots = x_n = 0$ and $x_1, x_2, \ldots, x_p$ not all equal to zero such that

$$\begin{aligned} y_1 &= x_1 + \alpha_{12}x_2 + \cdots + \alpha_{1p}x_p = 0, \\ y_2 &= x_2 + \alpha_{23}x_3 + \cdots + \alpha_{2p}x_p = 0, \\ &\;\;\vdots \\ y_{p+1} &= x_{p-1} \qquad\qquad + \alpha_{p-1,p}x_p = 0. \end{aligned} \tag{8.25}$$

In (8.25) if we put $x_p = 0$, then all the elements of $x_1, ..., x_{p-1}$ are equal to zero. Therefore, x_p cannot be zero if all the elements of $x_1, x_2, ..., x_p$ are not to be zero. Now, equating (8.20) and (8.24) with the above chosen x_i's and y_i's, we get

$$\sum_{i,j=1}^{p} a_{ij}x_i x_j = \beta_{pp}x_p^2,$$

which can be written as

$$[x_1 \ . \ . \ . \ x_p]\begin{bmatrix} a_{11} & \cdots & a_{1p} \\ \vdots & & \vdots \\ a_{p1} & \cdots & a_{pp}-\beta_{pp} \end{bmatrix}\begin{bmatrix} x_1 \\ \vdots \\ x_p \end{bmatrix} = 0. \tag{8.26}$$

The above equation is satisfied for nonzero values of $x_1, ..., x_p$ if the determinant of the square matrix, denoted by A_p, in (8.26) is zero. Such nonzero values can be obtained by solving $A_p[x_1 \cdots x_p]^T = \boldsymbol{\theta}$, where $\det A_p = 0$, which yields

$$\beta_{pp} = \frac{\Delta_p}{\Delta_{p-1}}.$$

So, by mathematical induction we have proved that, if the quadratic form (8.20) can be reduced to the form (8.24) at the $(p - 1)$th step, then $\beta_{pp} \neq 0$ if $\Delta_p \neq 0$, and so (8.24) can be reduced at the pth step to the form

$$q = c_1y_1^2 + c_2y_2^2 + \cdots + c_py_p^2 + q_{p+1}(x_{p+1}, ..., x_n),$$

where the coefficient of y_p^2 is $\beta_{pp} = c_p = \Delta_p/\Delta_{p-1}$, as given by (8.23) and y_p is given by y_j in (8.22) for $j = p$. Since rank $A = r$, this reduction process can be continued only up to the rth step.

The transformation P defined by (8.22), i.e., $\mathbf{x} = P\mathbf{y}$, is obviously nonsingular with $\det P = 1$.

(b) *Necessity.* These conditions are necessary because, if (8.21) holds, c_i's must exist, which demands that $\Delta_1 \cdots \Delta_{r-1}$ be different from zero. Also, for the rank of (8.21) to be r, it is necessary that Δ_r be not zero because under nonsingular transformation, the rank of the quadratic form (8.20) must be the same as that of the form (8.21). □

In view of Theorem 8.4, any quadratic form q of rank r, whose matrix has first r nonzero leading principal minors $\Delta_1, \Delta_2, ..., \Delta_r$, can be represented by

$$q = \frac{\Delta_1}{\Delta_0}y_1^2 + \frac{\Delta_2}{\Delta_1}y_2^2 + \cdots + \frac{\Delta_r}{\Delta_{r-1}}y_r^2$$

which is known as *Jacobi formula,* where $y_1, y_2, ..., y_r$ are independent forms. By setting $z_i = \Delta_i y_i$, an *alternative form* of *Jacobi formula* is given by

$$q = \frac{1}{\Delta_0\Delta_1}z_1^2 + \frac{1}{\Delta_1\Delta_2}z_2^2 + \cdots + \frac{1}{\Delta_{r-1}\Delta_r}z_r^2,$$

where $z_1, z_2, ..., z_r$ are now independent forms. □

Example 8.9 Let us now consider a quadratic form $q(x_1, x_2, x_3)$ whose matrix is

$$A = \begin{bmatrix} 4 & -2 & 6 \\ -2 & 10 & -6 \\ 6 & -6 & 11 \end{bmatrix}.$$

The leading principal minors of A are

$$\Delta_1 = 4, \qquad \Delta_2 = 36, \qquad \Delta_3 = \det A = 36.$$

By Jacobi formula, the diagonal form of q is

$$4y_1^2 + 9y_2^2 + y_3^2$$

which is the same as obtained in Example 8.6. It is shown there, how to compute an upper triangular nonsingular P to achieve this diagonal form. The alternative form of Jacobi formula, however, gives the diagonal form

$$\frac{1}{4}z_1^2 + \frac{1}{144}z_2^2 + \frac{1}{1296}z_3^2.$$ □

8.4 SYLVESTER'S LAW OF INERTIA

We have seen in Section 8.3.1 that any quadratic form is reducible to a diagonal form. The rank of any quadratic form over a field $\mathbb{F}$ remains invariant under a nonsingular transformation as a result of Theorem 8.1. This implies that, whatever nonsingular transformations are employed to reduce a quadratic form to the diagonal form

$$q = c_1y_1^2 + c_2y_2^2 + \cdots + c_ry_r^2, \tag{8.27}$$

the number of square terms present must be equal to the rank r of the form. For a quadratic form over a real field $\mathbb{R}$, the c_i's are real numbers.

There is an important invariant property, apart from the one just mentioned, possessed by a real quadratic form, which is given by Sylvester's law of inertia. To introduce this invariant property, let k number of the coefficients in (8.27) be positive and the remaining $(r–k)$ number be negative and the variables be numbered in such a way that terms with positive coefficients occur first, followed by negative ones. So, we may write (8.27) in the form

$$\alpha_1y_1^2 + \alpha_2y_2^2 + \cdots + \alpha_ky_k^2 - \alpha_{k+1}y_{k+1}^2 - \cdots - \alpha_ry_r^2, \tag{8.28}$$

where all α_i's are positive.

In view of Theorem 8.3, any quadratic form $q = \mathbf{x}^TA\mathbf{x}$, $A = A^T$ can be reduced by means of an orthogonal transformation $\mathbf{x} = P\mathbf{y}$ to the diagonal form $q = \mathbf{y}^TD\mathbf{y}$, where $D = P^TAP = \text{diag}\,[\alpha_1, \alpha_2, ..., \alpha_k, -\alpha_{k+1}, ..., -\alpha_r, 0, ..., 0]$ has k number of positive eigenvalues $\alpha_1, ..., \alpha_k$; $r - k$ number of negative eigenvalues $-\alpha_{k+1}, ..., -\alpha_r$ and $n - r$ number of zero eigenvalues. We know by virtue of Theorem 7.1 that, under any nonsingular transformation, the eigenvalues of any matrix, not necessarily symmetric or Hermitian, are invariants. This leads to the definition of inertia of a matrix A denoted by In (A) by the triple

$$(\pi(A),\ \nu(A),\ \delta(A)) =: \ \text{In}\,(A),$$

where $\pi(A)$, $\nu(A)$ and $\delta(A)$ denote, respectively, the number of eigenvalues of A with positive real parts, with negative real parts and with zero real parts. As the growth and decay of response of a system, viz. its inertia, are goverened by the eigenvalues of its mathematical model, the term *inertia* is therefore significant to study the dynamic behaviour of physical systems. The eigenvalues of a matrix A of order n can therefore be characterized in two ways: (a) its index and signature, (see Definition 8.9) or (b) its inertia, interchangeably as (a) can be derived from (b) or vice versa. Hence, in all Theorems 8.5, 8.6, 8.9, 8.10, and corollaries 8.2 – 8.3, the words *the rank r and the index k* can be replaced with *the inertia* without altering the assertions of the Theorems and Corollaries.

In Example 8.5, $q = -y_1^2 - y_2^2 + 3y_3^2$ so that $D = \text{diag}\,[-1, -1, 3]$. Therefore, $\pi(A) = 1$, $\nu(A) = 2$, $\delta(A) = 0$, while index $(A) = k = 1$, and signature $(A) = 2k - r = 2 - 3 = -1$.

Definition 8.9 *Index and Signature of a Quadratic Form.* The number of positive terms k in the diagonal form (8.28) is called the *index*, and the difference between the number of positive and the number of negative terms in it, i.e., $2k - r$ is called the *signature* of the corresponding quadratic form. □

Theorem 8.5 *Sylvester's Law of Inertia.* Under all real, nonsingular transformations, the rank r and the index k of a real quadratic form $q = \mathbf{x}^T A\mathbf{x}$ are invariants.

Proof In view of Theorem 8.1, the rank of a quadratic form q remains invariant under a nonsingular transformation. Again, by applying Definition 8.5, the rank of a diagonal form (8.28) is equal to the number of terms present in it. Therefore, the rank of q is equal to the number of terms present in its diagonal form (8.28) and is an invariant.

Suppose now that k and p are two different indices of a real quadratic form $q = \mathbf{x}^T A\mathbf{x}$ obtained by two distinct nonsingular transformations

$$\mathbf{y} = P\mathbf{x} \quad \text{and} \quad \mathbf{z} = Q\mathbf{x} \tag{8.29}$$

respectively (see Section 8.3), so that we may write

$$q \equiv \alpha_1 y_1^2 + \alpha_2 y_2^2 + \cdots + \alpha_k y_k^2 - \alpha_{k+1} y_{k+1}^2 - \cdots - \alpha_r y_r^2$$

$$\equiv \beta_1 z_1^2 + \beta_2 z_2^2 + \cdots + \beta_p z_p^2 - \beta_{p+1} z_{p+1}^2 - \cdots - \beta_r z_r^2 \tag{8.30}$$

which is an identity in the variables $x_1, \ldots, x_n$ since y_i's and z_i's are linear forms of x_i's defined by (8.29). Let $k > p$ and

$$y_{k+1} = \cdots = y_n = 0 \quad \text{and} \quad z_1 = \cdots = z_p = 0. \tag{8.31}$$

These are $n - k + p < n$ number of equations in n number of variables $x_1, \ldots, x_n$ and must, therefore, possess real nontrivial solutions. For each of these nonzero solutions, say,

$$x_1 = \xi_1, \ldots, x_n = \xi_n, \tag{8.32}$$

we have from (8.30), the relation

$$\alpha_1 y_1^2 + \alpha_2 y_2^2 + \cdots + \alpha_k y_k^2 + \beta_{p+1} z_{p+1}^2 + \cdots + \beta_r z_r^2 \equiv 0. \tag{8.33}$$

Since, all the terms on the left-hand side of (8.33) are always positive for each of the nonzero solutions (8.32), the identity (8.33) is valid iff

$$y_1 = \cdots = y_k = z_{p+1} = \cdots = z_r = 0. \tag{8.34}$$

So from (8.31) and (8.34), we have

$$y_1 = y_2 = \cdots = y_n = 0.$$

From (8.29), therefore, $P\mathbf{x} = 0$ with P nonsingular and $\mathbf{x}$ a nonzero solution given by (8.32). This is contradictory to the hypothesis that P is nonsingular. So, it is not true that $k > p$. Similarly, it can be shown that $p > k$ is also untenable. Hence $k = p$, and this proves the invariance of the index of q under nonsingular transformations. □

Corollary 8.2 If k is the index and r the rank of a real quadratic form $q = \mathbf{x}^T A\mathbf{x}$, then q can be transformed by a nonsingular transformation $\mathbf{y} = P\mathbf{x}$ to the *canonical form*

$$q = y_1^2 + y_2^2 + \cdots + y_k^2 - y_{k+1}^2 - \cdots - y_r^2.$$

Proof From Sylvester's law of inertia, we know that there exists a nonsingular transformation $\mathbf{x} = R\mathbf{z}$ which reduces $q = \mathbf{x}^T A\mathbf{x}$ to

$$q = c_1 z_1^2 + \cdots + c_k z_k^2 - c_{k+1} z_{k+1}^2 - \cdots - c_r z_r^2, \tag{8.35}$$

where k is the index and r the rank of the form and c_i's are positive real numbers. Put

$$y_1 = \sqrt{c_1} z_1, \ldots, y_r = \sqrt{c_r} z_r, \; y_{r+1} = z_{r+1}, \ldots, y_n = z_n$$

so that in the transformation $Q\mathbf{y} = \mathbf{z}$,

$$Q = \text{diag}\left[\frac{1}{\sqrt{c_1}}, \ldots, \frac{1}{\sqrt{c_1}}, 1, \ldots, 1\right]$$

is nonsingular. This transformation Q reduces (8.35) to

$$q = y_1^2 + \cdots + y_k^2 - y_{k+1}^2 - \cdots - y_r^2$$

and the matrix P transforming $q = \mathbf{x}^T A\mathbf{x}$ to the above form is $P = RQ$ because

$$\mathbf{x}^T A\mathbf{x} = \mathbf{z}^T R^T A R\mathbf{z} = \mathbf{y}^T (Q^T R^T A R Q)\mathbf{y} = \mathbf{y}^T (P^T A P)\mathbf{y}.$$ □

Example 8.10 The quadratic form q given by (8.14) has index 1 and signature –1 because its diagonal form, as worked out in Example 8.7, is

$$q = z_1^2 - \frac{1}{4} z_2^2 - 6 z_3^2.$$

By the transformation $u_1 = z_1$, $2u_2 = z_2$, $u_3 = \sqrt{6}\, z_3$, q is reduced to a *canonical form* given by

$$q = u_1^2 - u_2^2 - u_3^2.$$ □

The following corollary is an easy consequence of Theorems 8.3 and 8.5.

Corollary 8.3 If A is a real symmetric matrix, then the number of positive and negative characteristic roots are k and $r - k$ respectively, where r = rank A and k is the index of the real quadratic form associated with A. □

We know from Theorem 8.2 that two quadratic forms over a field $\mathbb{F}$ are equivalent iff they have the same rank. But a more stringent restriction may be imposed on two *real quadratic forms* so that they become equivalent.

Theorem 8.6 *Equivalence Theorem of Real Quadratic Forms.* Two real quadratic forms are equivalent iff they have the same rank and the same index.

Proof (a) *Necessity.* Let us assume that $\mathbf{x}^T A\mathbf{x}$ and $\mathbf{y}^T B\mathbf{y}$ are equivalent under a real nonsingular linear transformation $\mathbf{x} = P\mathbf{y}$. So, from (8.11) and (8.12), we have

$$B = P^T AP \tag{8.36}$$

In view of Corollary 8.2, we have nonsingular transformation $\mathbf{y} = Q\mathbf{z}$, which reduces $\mathbf{y}^T B\mathbf{y}$ to the canonical form

$$\mathbf{y}^T B\mathbf{y} = \mathbf{z}^T (Q^T BQ)\mathbf{z} = z_1^2 + z_2^2 + \cdots + z_k^2 - z_{k+1}^2 - \cdots - z_r^2 \tag{8.37}$$

where $r = \text{rank } B$ and k is the index of the associated form. Introducing (8.36) into (8.37), we have

$$\mathbf{z}^T (PQ)^T A(PQ)\mathbf{z} = z_1^2 + z_2^2 + \cdots + z_k^2 - z_{k+1}^2 - \cdots - z_r^2$$

which shows that the transformation $\mathbf{x} = PQ\mathbf{z}$ reduces $\mathbf{x}^T A\mathbf{x}$ to the same canonical form of $\mathbf{y}^T B\mathbf{y}$. Hence, these quadratic forms have the same rank r and the same index k.

(b) *Sufficiency.* We suppose that $\mathbf{x}^T A\mathbf{x}$ and $\mathbf{y}^T B\mathbf{y}$ have the same rank r and the same index k. Therefore, by Corollary 8.2, there exist nonsingular transformations $\mathbf{x} = P\mathbf{z}$ and $\mathbf{y} = Q\mathbf{z}$ such that

$$\mathbf{z}^T P^T AP\mathbf{z} = \mathbf{z}^T Q^T BQ\mathbf{z} = z_1^2 + z_2^2 + \cdots + z_k^2 - z_{k+1}^2 - \cdots - z_r^2.$$

Hence, $P^T AP = Q^T BQ$. Therefore, $B = R^T AR$, where $R = PQ^{-1}$. So, A and B are congruent and their forms are equivalent. □

8.5 HERMITIAN FORMS

So far, we have been dealing with real quadratic forms which were seen assuming only real values. On the other hand, complex quadratic forms assume, in general, complex values. For example, the value of the real quadratic form

$$q = [x_1 \ x_2]\begin{bmatrix} 1 & 2 \\ 2 & 4 \end{bmatrix}\begin{bmatrix} x_1 \\ x_2 \end{bmatrix} = (x_1 + 2x_2)^2$$

is real for real x_1, x_2, but that of the complex quadratic form

$$q = [x_1 \ x_2]\begin{bmatrix} 1 & i \\ i & 1 \end{bmatrix}\begin{bmatrix} x_1 \\ x_2 \end{bmatrix} = x_1^2 + 2ix_1x_2 + x_2^2,$$

where x_1 and x_2 are now complex variables, is, in general, complex. This shows that, the natural generalization of a real quadratic form cannot be a complex quadratic form.

There are, however, a class of forms strikingly similar to the real quadratic forms, which though defined over a complex field, give rise to only real values and, as a result, is assumed to generalize the real quadratic forms. These generalized real quadratic forms are known as the *Hermitian forms.*

Definition 8.10 *Hermitian Form.* Let $\mathbf{x} = [x_1\ x_2\ \cdots\ x_n]^T$ be an n-vector in the unitary space $\mathbf{C}^n$, and let $H = [h_{ij}]$ be a Hermitian matrix of order n over the complex field $\mathbb{C}$. Then any polynomial of the form

$$h(x_1, x_2, ..., x_n) = \mathbf{x}^*H\mathbf{x} = \sum_{i,j=1}^{n} h_{ij}\bar{x}_i x_j \tag{8.38}$$

is called a *Hermitian form* of order n. The matrix H is called the matrix of the form and rank H the rank of the form. If H is real, then a Hermitian form is called the *real Hermitian form.* Moreover, det H is defined as the *discriminant* of the Hermitian form, and a Hermitian form is called *singular* if its discriminant is zero, otherwise it is called *nonsingular.* □

From the definition it is evident that a Hermitian form always takes up a real value. Indeed, as h in (8.38) is a scalar and H a Hermitian matrix, i.e., $H = H^*$, we have

$$\bar{h} = h^* = \mathbf{x}^*H^*\mathbf{x} = \mathbf{x}^*H\mathbf{x} = h$$

which shows that h is equal to its complex conjugate, i.e., h is real for every choice of x. Moreover, we note that if $\mathbf{x}$ and H are real, then the Hermitian form (8.38) reduces to a real quadratic form. Because of the foregoing results, the properties of Hermitian forms are similar to those of real quadratic forms.

A Hermitian form remains Hermitian by a nonsingular linear transformation of coordinates defined by

$$\mathbf{x} = P\mathbf{y}, \tag{8.39}$$

where P is a nonsingular matrix over $\mathbb{C}$. Indeed, substituting (8.39) into (8.38), we get

$$h = \mathbf{x}^*H\mathbf{x} = \mathbf{y}^*(P^*HP)\mathbf{y} = \mathbf{y}^*Q\mathbf{y}, \tag{8.40}$$

where $Q = P^*HP = Q^*$. So, the reduced form (8.40) is again Hermitian. It is easily seen that, under this nonsingular transformation, the rank of a Hermitian form remains invariant.

Definition 8.11 *Diagonal and Unit Hermitian Forms.* A Hermitian form represented by

$$h = c_1\bar{x}_1x_1 + c_2\bar{x}_2x_2 + \cdots + c_n\bar{x}_nx_n,$$

where the coefficients c_i's are real as h is real, is referred to as *diagonal.* If all c_i's can be reduced to one, then a Hermitian form given by

$$h = \bar{x}_1x_1 + \bar{x}_2x_2 + \cdots + \bar{x}_nx_n = \mathbf{x}^*\mathbf{x}$$

is called a *unit Hermitian form* of order n. □

Theorem 8.7 Under a unitary transformation (see Definition 6.15), a Hermitian form $h = \mathbf{x}^*H\mathbf{x}$ can be reduced to the diagonal form

$$h = \lambda_1\bar{y}_1y_1 + \lambda_2\bar{y}_2y_2 + \cdots + \lambda_n\bar{y}_ny_n, \tag{8.41}$$

where λ_i's are the eigenvalues of H.

Proof Necessity. By Theorem 7.27, we know that for a Hermitian matrix H, there exists a unitary matrix P such that P^*HP = diag $(\lambda_1, \lambda_2, ..., \lambda_n)$, where λ_i's are the eigenvalues of H. Employing the transformation $\mathbf{x} = P\mathbf{y}$, we have

$$h = \mathbf{x}^*H\mathbf{x} = \mathbf{y}^*(P^*HP)\mathbf{y} = \mathbf{y}^* \text{ diag } (\lambda_1, \lambda_2, ..., \lambda_n)\mathbf{y}$$

$$= \lambda_1 \bar{y}_1 y_1 + \lambda_2 \bar{y}_2 y_2 + \cdots + \lambda_n \bar{y}_n y_n. \quad \square$$

Note that, in the result (8.41), $\bar{y}_i y_i$ is real for $i = 1, 2, ..., n$. So, for h to be real, the eigenvalues λ_i's must be real.

We state some theorems concerning the properties of Hermitian forms without proof, because they can be proved in the same fashion as for real quadratic forms.

Theorem 8.8 Let $h = \mathbf{x}^*H\mathbf{x}$ be a Hermitian form of rank r. Then, there exists a nonsingular transformation $\mathbf{x} = P\mathbf{y}$ which reduces h to

$$c_1 \bar{y}_1 y_1 + c_2 \bar{y}_2 y_2 + \cdots + c_r \bar{y}_r y_r, \tag{8.42}$$

where c_i's are real and are not necessarily the eigenvalues of H. Numbering the variables in (8.42) in a suitable way, the terms with positive coefficients can be made to appear first, followed by terms with negative coefficients, that is, by another singular nontransformation $\mathbf{y} = Q\mathbf{z}$, (8.42) reduces to

$$\alpha_1 \bar{z}_1 z_1 + \alpha_2 \bar{z}_2 z_2 + \cdots + \alpha_k \bar{z}_k z_k - \alpha_{k+1} \bar{z}_{k+1} z_{k+1} - \cdots - \alpha_r \bar{z}_r z_r, \tag{8.43}$$

where all α_i's are positive, or equivalently to the canonical form

$$\bar{z}_1 z_1 + \bar{z}_2 z_2 + \cdots + \bar{z}_k z_k - \bar{z}_{k+1} z_{k+1} - \cdots - \bar{z}_r y_r. \quad \square$$

To get the form (8.42), it is convenient to use Lagrange method of reduction for Hermitian forms which will now be described. As in (8.17), suppose that $h_{11} \neq 0$ in the Hermitian form (8.38) which can be rewritten as

$$h(x_1, x_2, \ldots, x_n) = h_{11}\left[\bar{x}_1 x_1 + \bar{x}_1 \sum_{j=2}^{n} \frac{h_{1j}}{h_{11}} x_j + x_1 \sum_{j=2}^{n} \frac{\bar{h}_{1j}}{h_{11}} \bar{x}_j\right] + \sum_{i,j=2}^{n} h_{ij}\bar{x}_i x_j$$

$$= h_{11}\left[\bar{x}_1 + \sum_{j=2}^{n} \frac{\bar{h}_{1j}}{h_{11}} \bar{x}_j\right]\left[x_1 + \sum_{j=2}^{n} \frac{h_{1j}}{h_{11}} x_j\right]$$

$$- h_{11}\left[\sum_{i=2}^{n} \frac{\bar{h}_{1j}}{h_{11}} \bar{x}_i\right]\left[\sum_{j=2}^{n} \frac{h_{1j}}{h_{11}} x_j\right] + \sum_{i,j=2}^{n} h_{ij}\bar{x}_i x_j$$

$$= \frac{1}{h_{11}} \left| h_{11}x_1 + h_{12}x_2 + \cdots + h_{1n}x_n \right|^2 + \sum_{i,j=2}^{n} \left(h_{ij} - \frac{h_{i1}h_{1j}}{h_{11}} \right) \bar{x}_i x_j.$$

The crate diagram (Figure 8.1) is helpful to get the various elements in the above manipulation, after replacing the left-hand side variables by $\bar{x}_1, \bar{x}_2, ..., \bar{x}_n$, respectively.

In general, a square in the Hermitian form $\mathbf{x}^*H\mathbf{x}$ can be computed if there is a nonzero diagonal element h_{ii} in H giving, as above, the relation

(a) $$h(x_1, x_2, \ldots, x_n) = \frac{1}{h_{ii}} | h_{i1}x_1 + h_{i2}x_2 + \cdots + h_{in}x_n|^2 + h_i(x_1, \ldots, x_{i-1}, x_{i+1}, \ldots, x_n),$$

where h_i is a Hermitian form not containing x_i.

If, however, $h_{11} = h_{22} = 0$ but $h_{12} \neq 0$, then a square cannot be formed in the same way as above. In this case, we note that $h_{ij} = \overline{h}_{ji}$ and

$$h(x_1, x_2, \ldots, x_n) = \overline{x}_1 (h_{11}x_1 + h_{12}x_2 + \cdots + h_{1n}x_n)$$

$$+ x_1(h_{11}\overline{x}_1 + \overline{h}_{12}\overline{x}_2 + \cdots + \overline{h}_{1n}\overline{x}_n) + h_{32}\overline{x}_3 x_2 + \overline{h}_{32}x_3\overline{x}_2 + \cdots + h_{n2}\overline{x}_n x_2 + \overline{h}_{n2}x_n\overline{x}_2 + \sum_{i,j=3}^{n} h_{ij}\overline{x}_i x_j$$

$$= \frac{1}{h_{12}}(h_{11}x_1 + h_{12}x_2 + \cdots + h_{1n}x_n)(h_{12}\overline{x}_1 + h_{22}\overline{x}_2 + h_{32}\overline{x}_3 + \cdots + h_{n2}\overline{x}_n)$$

$$+ \frac{1}{\overline{h}_{12}}(h_{11}\overline{x}_1 + \overline{h}_{12}\overline{x}_2 + \cdots + \overline{h}_{1n}\overline{x}_n)(\overline{h}_{12}x_1 + h_{22}x_2 + \overline{h}_{32}x_3 + \cdots + \overline{h}_{n2}x_n)$$

$$- \frac{1}{h_{12}}(h_{13}x_3 + \cdots + h_{1n}x_n)(h_{32}\overline{x}_3 + \cdots + h_{n2}\overline{x}_n) - \frac{1}{\overline{h}_{12}}(\overline{h}_{13}\overline{x}_3 + \cdots + \overline{h}_{1n}\overline{x}_n)(\overline{h}_{32}x_3 + \cdots + \overline{h}_{n2}x_n)$$

$$+ \sum_{i,j=3}^{n} h_{ij}\overline{x}_i x_j.$$

Hence,

(b) $$h(x_1, x_2, \ldots, x_n) = \frac{1}{2}\left| \left(h_{11} + \frac{h_{21}}{h_{21}}\right)x_1 + \left(h_{12} + \frac{h_{22}}{h_{21}}\right)x_2 + \cdots + \left(h_{1n} + \frac{h_{2n}}{h_{21}}\right)x_n\right|^2$$

$$- \frac{1}{2}\left| \left(h_{11} - \frac{h_{21}}{h_{21}}\right)x_1 + \left(h_{12} - \frac{h_{22}}{h_{21}}\right)x_2 + \cdots + \left(h_{1n} - \frac{h_{2n}}{h_{21}}\right)x_n\right|^2 + g_{12}(x_3, \ldots, x_n),$$

where we have used the relation

$$uv + \overline{u}\,\overline{v} = \frac{1}{2}\{(u+\overline{v})(\overline{u}+v) - (u-\overline{v})(\overline{u}-v)\}$$

and g_{12} is a Hermitian form not involving x_1 and x_2. In general, if $h_{ii} = h_{jj} = 0$ but $h_{ij} \neq 0$, then the Hermitian form (8.38) can be represented by

(c)
$$h(x_1, x_2, \ldots, x_n) = \frac{1}{2}\left[\left|\sum_{k=1}^{n}\left(h_{ik} + \frac{h_{jk}}{h_{ji}}\right)x_k\right|^2 - \left|\sum_{k=1}^{n}\left(h_{ik} - \frac{h_{jk}}{h_{ji}}\right)x_k\right|^2\right] + g_{ij},$$

where g_{ij} is a Hermitian form not involving x_i and x_j. By repeated application of (a) and (c), we can represent any Hermitian form as the sum of squares given in (8.42).

Example 8.11 Reduce the following Hermitian form to the canonical form (8.42) by Lagrange reduction:

$$h(x_1, x_2, x_3) = (1+i)\bar{x}_1 x_2 + (1-i)\bar{x}_2 x_1 + i\bar{x}_1 x_3 - i\bar{x}_3 x_1 + \bar{x}_2 x_3 + \bar{x}_3 x_2 + 4x_3\bar{x}_3 .$$

Solution For the Hermitian form, we determine that $h_{12} = 1 + i$, $h_{13} = i$, $h_{23} = 1$ and $h_{33} = 4$. Since $h_{11} = h_{22} = 0$ and $h_{12} \neq 0$, applying (b), we have,

$$h(x_1, x_2, x_3) = \frac{1}{2}\left\{\left| x_1 + (1+i)x_2 + \frac{1}{2}(1+3i)x_3\right|^2 - \left| -x_1 + (1+i)x_2 - \frac{1}{2}(1-i)x_3\right|^2\right\} + 3x_3\bar{x}_3 .$$

Using the transformations

$$y_1 = x_1 + (1+i)x_2 + \frac{1}{2}(1+3i)x_3,$$

$$y_2 = -x_1 + (1+i)x_2 - \frac{1}{2}(1-i)x_3, \qquad y_3 = x_3,$$

we have

$$h(y_1, y_2, y_3) = \frac{1}{2}(y_1\bar{y}_1 - y_2\bar{y}_2) + 3y_3\bar{y}_3$$

which shows that the index of h is 2 and its signature 1, see the definition below. □

Definition 8.12 *Index and Signature of a Hermitian Form.* As in the case of a real quadratic form, the number k of positive terms in the diagonal form (8.43) is defined as the index, and the difference between the number of the positive terms and the number of negative terms in (8.43), which is $2k - r$, is called the signature of the form. □

Theorem 8.9 *Law of Inertia for Hermitian Forms.* Under all nonsingular transformations, the rank r and the index k are invariants of a Hermitian form. □

Definition 8.13 *Equivalent Hermitian Forms.* Two Hermitian forms $\mathbf{x}^*H\mathbf{x}$ and $\mathbf{y}^*Q\mathbf{y}$ are said to be equivalent over $\mathbb{C}$ if one can be obtained from the other by a nonsingular transformation over $\mathbb{C}$. In other words, $\mathbf{x}^*H\mathbf{x}$ and $\mathbf{y}^*Q\mathbf{y}$ are equivalent iff there exists a nonsingular transformation $\mathbf{x} = P\mathbf{y}$ such that

$$\mathbf{x}^*H\mathbf{x} = \mathbf{y}^*(P^*HP)\mathbf{y} = \mathbf{y}^*Q\mathbf{y}.$$ □

Theorem 8.10 *Equivalence Theorem of Hermitian Forms.* Two Hermitian forms are equivalent iff they have the same rank and the same index. □

Example 8.12 Let $h(x_1, x_2, x_3)$ be a Hermitian form with a matrix of the form

$$H = \begin{bmatrix} 2 & i & 0 \\ -i & 1 & -i \\ 0 & i & 2 \end{bmatrix}.$$

Now, det $(\lambda I - H) = \lambda(\lambda - 2)(\lambda - 3)$ which gives the eigenvalues of H as $\lambda_1 = 2$, $\lambda_2 = 3$, $\lambda_3 = 0$. The normalized eigenvectors (see Definition 5.12) associated with λ_1, λ_2 and λ_3 are, respectively,

$$\mathbf{x}_1 = \frac{1}{\sqrt{2}}[1 \;\; 0 \;\; -1]^T, \quad \mathbf{x}_2 = \frac{1}{\sqrt{3}}[i \;\; 1 \;\; i]^T, \quad \mathbf{x}_3 = \frac{1}{\sqrt{6}}[-i \;\; 2 \;\; -i]^T.$$

Because of Theorem 7.31, these eigenvectors are orthonormal; if they were not so, then they could be made orthonormal by applying Theorem 5.8.

We now define a unitary matrix as

$$P = [\mathbf{x}_1 \; \mathbf{x}_2 \; \mathbf{x}_3]$$

which satisfies $P^*P = I$. Applying Theorem 7.34, we get

$$Q := P^{-1}HP = P^*HP = \text{diag } (2, 3, 0).$$

Therefore, with the transformation $\mathbf{x} = P\mathbf{y}$, the given Hermitian form reduces to

$$h = 2y_1\bar{y}_1 + 3y_2\bar{y}_2$$

which shows that the index as well as the signature of the Hermitian form is 2. Clearly, the Hermitian forms $\mathbf{x}^*H\mathbf{x}$ and $\mathbf{y}^*Q\mathbf{y}$ are equivalent. □

8.6 POSITIVE DEFINITE QUADRATIC AND HERMITIAN FORMS: POSITIVE DEFINITE MATRICES

We have seen so far that quadratic (or Hermitian) forms can be distinguished among themselves by the rank and the index which can be employed as a criterion to classify the different real quadratic or Hermitian forms: $q(x_1, ..., x_n) = \mathbf{x}^TA\mathbf{x}$ or $h(x_1, ..., x_n) = \mathbf{x}^*H\mathbf{x}$ over a real field $\mathbb{R}$ or a complex field $\mathbb{C}$. There is still another way which one can think of for such classification by observing that both the forms assume real values. For $\mathbf{x} = \boldsymbol{\theta}$, q or h is always zero and for any $\mathbf{x} \neq \boldsymbol{\theta}$, the sign of the value of q or h depends on certain structure of A or H. Therefore, a classification for the above forms may be justified, which depends on the sign of the real values which a quadratic or Hermitian form assumes for all nonzero values of the variable $\mathbf{x}$.

For clarity of exposition, we shall for the time being denote a real quadratic or Hermitian form by $q(\mathbf{x}) \equiv q(x_1, ..., x_n)$.

Definition 8.14: *Definite Quadratic and Hermitian Forms: Definite Matrices.* A real quadratic or Hermitian form $q(\mathbf{x})$ is said to be

(a) *positive definite* (*PD*)	if $q(\mathbf{x}) > 0$	when $\mathbf{x} \neq \boldsymbol{\theta}$;
(b) *negative definite* (*ND*)	if $q(\mathbf{x}) < 0$	when $\mathbf{x} \neq \boldsymbol{\theta}$;
(c) *positive semi-definite* (*PSD*)	if $q(\mathbf{x}) \geq 0$	when $\mathbf{x} \neq \boldsymbol{\theta}$;
(d) *negative semi-definite* (*NSD*)	if $q(\mathbf{x}) \leq 0$	when $\mathbf{x} \neq \boldsymbol{\theta}$; and
(e) *indefinite* (*ID*)	if $q(\mathbf{x}) \gtreqless 0$	when $\mathbf{x} \neq \boldsymbol{\theta}$;

where $\mathbf{x} \neq \boldsymbol{\theta}$ implies that there are some variables among $x_1, ..., x_n$ constituting $\mathbf{x}$ which are not zero.

The real symmetric matrix A (resp the Hermitian matrix H) is called a *positive definite matrix* (resp a *positive definite Hermitian matrix)* if the corresponding quadratic form $q(\mathbf{x}) = \mathbf{x}^TA\mathbf{x}$ (resp Hermitian form $q(\mathbf{x}) = \mathbf{x}^*H\mathbf{x}$) is positive definite and is written as $A > 0$ (resp $H > 0$). Similarly, A is called a negative definite matrix if $q(\mathbf{x}) = \mathbf{x}^T A\mathbf{x}$ is negative definite, a positive semi-definite matrix if $q(\mathbf{x}) = \mathbf{x}^T A\mathbf{x}$ is positive semi-definite, a negative semi-definite matrix if $q(\mathbf{x}) = \mathbf{x}^T A\mathbf{x}$ is negative semi-definite, and an indefinite matrix if $q(\mathbf{x}) = \mathbf{x}^T A\mathbf{x}$ is indefinite. The notation for A ND is $A < 0$, for PSD is $A \geq 0$, for NSD is $A \leq 0$ and for indefinite is $A \lesseqgtr 0$. Identical definitions and notations can be furnished in the case of a Hermitian matrix H replacing A with H. □

These terms were introduced by Gauss in his *Disquisitiones Arithmeticae.*

Example 8.13 If $\mathbf{x}$ is a vector in the space $\mathbb{R}^3$, then (i) the quadratic form $q(\mathbf{x}) = x_1^2 + 3x_2^2 + x_3^2$ is PD, (ii) $q(\mathbf{x}) = (x_1 - 2x_2)^2 + x_3^2$ is PSD, and (iii) $q(\mathbf{x}) = x_1^2 + x_2^2$ is PSD. In (ii) $q = 0$ when $x_1 = 2$, $x_2 = 1$, $x_3 = 0$, and in (iii) $q = 0$ when $x_1 = 0$, $x_2 = 0$ and $x_3 =$ any nonzero value. The result in (iii) shows that *in defining a quadratic form the space in which* $\mathbf{x}$ *lies must be mentioned.* The quadratic form $q = (x_1 - 2x_2)^2 - x_3^2$ is indefinite in $\mathbb{R}^3$ because $q = 1$ when $x_1 = 3$, $x_2 = 1$, $x_3 = 0$; $q = -1$ when $x_1 = 2$, $x_2 = 1$, $x_3 = 1$, and $q = 0$ when $x_1 = 2$, $x_2 = 1$, $x_3 = 0$. The negatives of the forms given in (i) and (ii) are respectively ND and NSD. □

Theorem 8.11 Let $q(\mathbf{x}) = q(x_1, ..., x_n)$ be a real quadratic form or a Hermitian form of order n, rank r, and signature p. Then $q(x_1, ..., x_n)$ is

(a) positive definite iff $r = p = n$;
(b) positive semi-definite iff $r = p < n$;
(c) negative definite iff $r = -p = n$;
(d) negative semi-definite iff $r = -p < n$; and
(e) indefinite iff $| p | < r$.

Proof Only part (a) for a real quadratic form will be proved as others can be proved similarly.

(i) *Necessity.* To prove the necessity, we observe that in view of Corollary 8.2, the given real quadratic form q can be transformed by a nonsingular transformation $\mathbf{y} = P\mathbf{y}$ to

$$q = y_1^2 + y_2^2 + \cdots + y_k^2 - y_{k+1}^2 - \cdots - y_r^2.$$

In this representation, all the squares along with signs must be positive for q to be PD. Otherwise, by solving $\mathbf{y} = P\mathbf{x}$, we can select values of $x_1, ..., x_n$ for which $y_1 = y_2 = \cdots = y_k = 0$ and $y_{k+1}, ..., y_r$ are not zero, giving a negative value for q. So, $p = r$ and

$$q = y_1^2 + y_2^2 + \cdots + y_r^2. \tag{8.44}$$

Moreover, $r = n$; if not, we can again choose nonzero values of x_i's for which $y_1 = \cdots = y_r = 0$ and $y_{r+1}, ..., y_n$ nonzero, giving a zero value for q.

(ii) *Sufficiency.* Conversely, if $r = p = n$, then all the n squares in the representation (8.44) are positive and, as a result, q is always positive whenever $\mathbf{x} \neq \mathbf{\theta}$. □

An easy consequence of Theorem 8.11 is the following corollary.

Corollary 8.4 If the quadratic form $\mathbf{x}^T A\mathbf{x}$ (resp Hermitian form $\mathbf{x}^* H\mathbf{x}$) is PD, then there exists a nonsingular transformation $\mathbf{x} = P\mathbf{y}$ (resp $\mathbf{x} = Q\mathbf{y}$) such that $\mathbf{x}^T A\mathbf{x} = \mathbf{y}^T\mathbf{y}$ (resp $\mathbf{x}^* H\mathbf{x} = \mathbf{y}^*\mathbf{y}$). □

Theorem 8.12 If a real quadratic form $q(\mathbf{x}) = \mathbf{x}^T A\mathbf{x}$ is positive definite, then

(a) $\det A > 0$,

(b) every principal minor of A is positive,

(c) $a_{ii} > 0$, $i = 1, 2, ..., n$, where $A = [a_{ij}]$.

Proof (a) As a result of Corollary 8.4, there exists a nonsingular transformation $\mathbf{x} = P\mathbf{y}$ which reduces the PD quadratic form $q = \mathbf{x}^T A\mathbf{x}$ to a unit diagonal form, i.e.,

$$q(\mathbf{x}) = \mathbf{x}^T A\mathbf{x} = \mathbf{y}^T(P^T A P)\mathbf{y} = \mathbf{y}^T I_n \mathbf{y}$$

so that

$$P^T A P = I_n.$$

Hence, observing that $\det P = \det P^T$, we have

$$\det A = (\det P)^{-2} > 0.$$

(b) In the form $q = \mathbf{x}^T A\mathbf{x}$ if we put some of the variables $x_1, x_2, ..., x_n$ equal to zero, we get a quadratic form in less than n number of variables with a matrix of the form corresponding to a principal minor of A. Since the obtained quadratic form must be positive definite, by the condition (a) above, this principal minor must be positive. Similarly, setting some of the variables, $x_1, x_2, ..., x_n$ in all possible ways, it is easily verified that all the principal minors of A must be positive.

(c) This result is a particular case of (b). □

Following the arguments used in the proof of Theorem 8.12, we can easily prove Corollaries 8.5 and 8.6 and Theorem 8.13 which are stated below.

Corollary 8.5 If a Hermitian form $h(\mathbf{x}) = \mathbf{x}^* H\mathbf{x}$ is PD, then (a) $\det H > 0$, (b) every principal minor of H is positive, and (c) $h_{ii} > 0$, $i = 1, 2, ..., n$, where $H = [h_{ij}]$. □

Theorem 8.13 If a real quadratic form $q(\mathbf{x}) = \mathbf{x}^T A\mathbf{x}$ is PSD, then (a) $\det A = 0$, (b) every principal minor of A is nonnegative, and (c) $a_{ii} > 0$ if x_i^2 appears in $q(\mathbf{x})$. □

Corollary 8.6 If a Hermitian form $h(\mathbf{x}) = \mathbf{x}^* H\mathbf{x}$ is PSD, then (a) $\det H = 0$, (b) every principal minor of H is nonnegative, and (c) $h_{ii} > 0$ *if* x_i^2 appears in $h(\mathbf{x})$. □

So far, we are concerned with reducing a quadratic or a Hermitian form to a diagonal one by an orthogonal transformation, the Lagrange method, or the Jacobi formula; and with characterising it with the rank, index and signature of the form. In terms of the matrix A of the form, a quadratic or a Hermitian form can be characterized by the inertia of A or by the order n, rank r and signature p of the form. We shall now show, how these forms can be classified as regards their definiteness by the principal minors, Heading principal minors and eigenvalues of the matrix A of the form.

Theorem 8.14 A real matrix A is symmetric and PD iff one of the following statements holds:

(a) P^TAP is symmetric and PD for any real and nonsingular P.
(b) $A = P^TP$ for any real and nonsingular P.

Proof of (a): (i) *Necessity.* Suppose that A is real, symmetric and PD. This implies that the real quadratic form $\mathbf{x}^TA\mathbf{x}$ is PD. Put $\mathbf{x} = P\mathbf{y}$, where $\mathbf{y} \neq \boldsymbol{\theta}$ if $\mathbf{x} \neq \boldsymbol{\theta}$ because P is nonsingular. Hence from

$$\mathbf{x}^TA\mathbf{x} = \mathbf{y}^T(P^TAP)\mathbf{y}$$

we conclude that the right-hand side is positive for all nonzero values of $\mathbf{y}$, and so $P^TAP = P^TA^TP = (P^TAP)^T$ is symmetric and PD.

(ii) *Sufficiency.* Conversely, suppose that P^TAP is symmetric and PD for any real and nonsingular P. Putting $P = I$, we must have $P^TAP = A$ to be symmetric and PD.

Proof of (b): (i) *Necessity.* We suppose that A is symmetric and PD. So, $\mathbf{x}^TA\mathbf{x}$ is PD and hence by Corollary 8.4, there exists a nonsingular transformation $\mathbf{x} = Q\mathbf{y}$ such that

$$\mathbf{x}^TA\mathbf{x} = \mathbf{y}^T(Q^TAQ)\mathbf{y} = \mathbf{y}^TI_n\mathbf{y}.$$

Therefore,

$$Q^TAQ = I_n \qquad \text{implying that} \qquad A = P^TP, \qquad P = Q^{-1}.$$

(ii) *Sufficiency.* On the other hand, let $\mathbf{x}^TA\mathbf{x} = \mathbf{x}^TP^TP\mathbf{x}$, where P is real and nonsingular. Hence, by substituting $\mathbf{y} = P\mathbf{x}$ and noting that $\mathbf{y} \neq \boldsymbol{\theta}$ iff $\mathbf{x} \neq \boldsymbol{\theta}$, we have

$$\mathbf{x}^TA\mathbf{x} = \mathbf{y}^T\mathbf{y} > 0 \quad \text{whenever } \mathbf{y} \neq \boldsymbol{\theta}.$$

Obviously, A is symmetric. Hence the theorem. □

We shall now state and establish the most important criterion for positive definiteness, known as Sylvester's criterion, which is famous for its simplicity and extensive applicability.

Theorem 8.15 *Sylvester's Criterion for Positive Definiteness.* A quadratic form

$$q(x_1, \ldots, x_n) = \mathbf{x}^T A\mathbf{x} = \sum_{i,j=1}^{n} a_{ij}x_ix_j$$

is PD iff all the leading principal minors of A are positive, i.e., iff all the determinants

$$\Delta_1 = a_{11},\ \Delta_2 = \begin{vmatrix} a_{11} & a_{12} \\ a_{21} & a_{22} \end{vmatrix}, \ldots,\ \Delta_n = \begin{vmatrix} a_{11} & \cdots & a_{1n} \\ \vdots & & \vdots \\ a_{n1} & \cdots & a_{nn} \end{vmatrix} \tag{8.45}$$

are positive.

Proof (a) *Necessity.* We assume that $q(x_1, \ldots, x_n)$ is PD. Note that

$$q(x_1, \ldots, x_p, 0, \ldots, 0) = \sum_{i,j=1}^{p} a_{ij}x_ix_j$$

is a quadratic form in the variables $x_1, \ldots, x_p$. This form must be positive definite necessitating, in view of Theorem 8.12(a), that

$$\Delta_p = \begin{vmatrix} a_{11} & \cdots & a_{1p} \\ \vdots & & \vdots \\ a_{p1} & \cdots & a_{pp} \end{vmatrix}$$

be positive; otherwise, $q(x_1, ..., x_p, 0, ..., 0)$ would be nonpositive for $x_1, ..., x_p$ not all zero, which would contradict that $q(x_1, ..., x_n)$ is PD. Therefore, for $q(x_1, ..., x_n)$ to be PD, all the determinants Δ_p, $p = 1, ..., n$, must be positive.

(b) *Sufficiency.* We observe that if the Δ_i's as given in (8.45) are positive, then the quadratic form $q(x_1, ..., x_n)$ can be transformed, by a nonsingular transformation $\mathbf{x} = P\mathbf{y}$, as a result of Theorem 8.4, into the form

$$q = \mathbf{x}^T A\mathbf{x} = \sum_{i,j=1}^{n} \frac{\Delta_i}{\Delta_{i-1}} y_i^2, \quad \Delta_0 := 1. \quad \text{(Jacobi formula)} \tag{8.46}$$

Since Δ_i's are positive, (8.46) assures that q must be positive definite.

The following theorem needs stronger conditions for a real quadratic form to be PD.

Theorem 8.16 A real quadratic form $\mathbf{x}^T A\mathbf{x}$ is PD iff all the principal minors of A are positive.

Proof If $\mathbf{x}^T A\mathbf{x}$ is PD, then all the principal minors of A are positive in view of part (b) of Theorem 8.12. Conversely, if all the principal minors of A are positive, then the leading principal minors of A must be positive. So, Sylvester's criterion for positive definiteness, i.e., Theorem 8.15, ensures that $\mathbf{x}^T A\mathbf{x}$ is PD. □

Theorems 8.14–8.16 are also valid, *mutatis mutandis,* for the positive definiteness of Hermitian forms. These results are therefore stated without proof in the following corollaries.

Corollary 8.7 A Hermitian matrix H is PD iff one of the following statements holds:

(a) P^*HP is PD for any nonsingular matrix P over $\mathbb{C}$.
(b) $H = P^*P$ for some nonsingular matrix P over $\mathbb{C}$. □

Corollary 8.8 *Sylvester's Criterion for Positive Definiteness.* A Hermitian form

$$h(x_1, ..., x_n) = \mathbf{x}^* H\mathbf{x} = \sum_{i,j=1}^{n} h_{ij}\bar{x}_i x_j$$

is PD iff all the leading principal minors of H are positive.

Corollary 8.9 A Hermitian form $\mathbf{x}^*H\mathbf{x}$ is positive definite iff all the principal minors of H are positive. □

Once the criteria for testing positive definiteness of quadratic and Hermitian forms are formulated, it is a straightforward extension to establish such criteria for negative definiteness of these forms. Because, interpreted differently, Definition 8.14(b) tells us that a quadratic or a Hermitian form $q(\mathbf{x})$ is ND if $-q(\mathbf{x})$ is PD, so the criteria of positive definiteness applied to $-q(\mathbf{x})$ would help us establish the criteria of negative definiteness for $q(\mathbf{x})$.

Theorem 8.17 The quadratic form

$$q(x_1,\ldots,x_n)=\sum_{i,j=1}^{n} a_{ij}x_ix_j$$

is ND iff

$$a_{11}<0,\quad \begin{vmatrix} a_{11} & a_{12} \\ a_{21} & a_{22} \end{vmatrix}>0,\quad \begin{vmatrix} a_{11} & a_{12} & a_{13} \\ a_{21} & a_{22} & a_{23} \\ a_{31} & a_{32} & a_{33} \end{vmatrix}<0,\ldots$$

Proof For $q(x_1, \ldots, x_n)$ to be negative definite, $-q(x_1, \ldots, x_n)$ must be PD. Applying Sylvester's criterion for positive definiteness to $-q$, we get successively

$$-a_{11}>0,\quad \begin{vmatrix} -a_{11} & -a_{12} \\ -a_{21} & -a_{22} \end{vmatrix}>0,\quad \begin{vmatrix} -a_{11} & -a_{12} & -a_{13} \\ -a_{21} & -a_{22} & -a_{23} \\ -a_{31} & -a_{32} & -a_{33} \end{vmatrix}>0,\ldots$$

From this follows the condition stated in the theorem. □

Similarly, we get the following corollary.

Corollary 8.10 The Hermitian form

$$h(x_1,\ldots,x_n)=\sum_{i,j=1}^{n} h_{ij}\bar{x}_ix_j$$

is ND iff

$$h_{11}<0,\quad \begin{vmatrix} h_{11} & h_{12} \\ h_{21} & h_{22} \end{vmatrix}>0,\quad \begin{vmatrix} h_{11} & h_{12} & h_{13} \\ h_{21} & h_{22} & h_{23} \\ h_{31} & h_{32} & h_{33} \end{vmatrix}<0,\ldots$$ □

Theorem 8.18 A real quadratic form $\mathbf{x}^T A\mathbf{x}$ is ND iff all the principal minors of A of even order are positive and those of odd order are negative. □

Corollary 8.11 A Hermitian form $\mathbf{x}^*H\mathbf{x}$ is ND iff all the principal minors of H of even order are positive and those of odd order are negative. □

Until now, we have seen how, by performing tests on the elements of the matrix A *or* H of the forms, one can ascertain some types of definiteness of quadratic and Hermitian forms. One can, however, relate definiteness of these forms directly to the eigenvalues of the matrix of the forms as given in the following theorem.

Theorem 8.19 A real quadratic form $\mathbf{x}^T A\mathbf{x}$ is

(a) PD iff all the eigenvalues of A are positive;

(b) ND iff all the eigenvalues of A are negative;

(c) PSD iff all the eigenvalues of A are nonnegative with at least one zero eigenvalue;

(d) NSD iff all the eigenvalues of A are nonpositive with at least one zero eigenvalue; and

(e) ID iff A has at least one positive and one negative eigenvalue.

Proof By Theorem 8.3, a real quadratic form associated with a real symmetric matrix A can be reduced by an orthogonal transformation $\mathbf{x} = P\mathbf{y}$ to

$$q = \mathbf{x}^T A\mathbf{x} = \lambda_1 y_1^2 + \lambda_2 y_2^2 + \cdots + \lambda_n y_n^2, \tag{8.47}$$

where λ_i's are the eigenvalues of A. Assertions (a) and (b) can then be easily established with the help of (8.47).

To prove (c), we note that in the representation (8.47), any of the λ_i's cannot be negative. If λ_i were negative, then we can select $x_1, x_2, \ldots, x_n$ with the help *of* $\mathbf{x} = P\mathbf{y}$ such that y_i is not zero but $y_i = \cdots = y_{i-1} = y_{i+1} = \cdots = y_n = 0$, for which selection we would get a negative value for q, which is not possible. So, all λ_i's must be nonnegative. Since all positive λ_i's give a PD q, at least one of the λ_i's must be zero, which is the necessity of assertion (c). Conversely, if all λ_i's are nonnegative, including at least one zero eigenvalue, then for all values of x_1, ..., x_n, we can find y_1, ..., y_n which would give nonnegative values for q. □

An exactly identical result for a Hermitian matrix may be stated.

Corollary 8.12 A Hermitian form $\mathbf{x}^*H\mathbf{x}$ is

(a) PD iff all the eigenvalues of H are positive;

(b) ND iff all the eigenvalues of H are negative;

(c) PSD iff all the eigenvalues of H are nonnegative with at least one zero eigenvalue;

(d) NSD iff all the eigenvalues of H are nonpositive with at least one zero eigenvalue; and

(e) ID iff H has at least one positive and one negative eigenvalue. □

The following Theorem 8.20 and Corollary 8.13 give the criteria of positive semi-definiteness for a quadratic and a Hermitian form, respectively.

Theorem 8.20 A quadratic form $q(\mathbf{x}) = \mathbf{x}^T A\mathbf{x}$ is PSD iff A is singular and all its principal minors are nonnegative.

Proof (a) *Necessity.* Suppose that $\mathbf{x}^T A\mathbf{x}$ is PSD. Then, by virtue of part (c) in Theorem 8.19, A must have a zero eigenvalue, and hence A is singular. If we put $n - k$ $(1 \le k \le n - 1)$ number of the variables $x_1, x_2, \ldots, x_n$ equal to zero in $q(\mathbf{x})$, we get a reduced quadratic form in k number of variables with a matrix of the form M of order k, which is one of the principal minors of A. This reduced quadratic form must be PD or PSD for $q(\mathbf{x})$ to be PSD. This requires that det $M \ge 0$ by Theorem 8.19. Thus, all the principal minors of A must be nonnegative.

(b) *Sufficiency.* Conversely, suppose that A is singular and all its principal minors are nonnegative. However, assume that

$$\mathbf{x}^T A\mathbf{x} = -\mu < 0, \qquad \mu > 0. \tag{8.48}$$

Let

$$A_m = \begin{bmatrix} a_{11} & \cdots & a_{1m} \\ \vdots & & \vdots \\ a_{m1} & \cdots & a_{mm} \end{bmatrix}, \qquad m = 1, 2, \ldots, n.$$

Now, we have

$$\det (A_m + \lambda I) = \lambda^m + a_{m-1}\lambda^{m-1} + \cdots + a_1\lambda + a_0 > 0,$$

where a_i's are the sum of $(m - i)$th order principal minors of A_m (see Theorem 7.16) if $\lambda > 0$, as the principal minors of A are nonnegative. Therefore, by Theorem 8.15, $(A_m + \lambda I)$ is PD for every $\lambda > 0$. Putting $\mu = \lambda \mathbf{x}^T\mathbf{x}$ in (8.48), we get

$$\mathbf{x}^T(A + \lambda I)\mathbf{x} = 0,$$

which is not possible because $(A + \lambda I)$ is PD. This contradicts the assumption (8.48) and, consequently, $\mathbf{x}^T A\mathbf{x} \geq 0$. Since A is singular, there are nonzero values of $\mathbf{x}$ for which $A\mathbf{x} = \boldsymbol{\theta}$, and hence $\mathbf{x}^T A\mathbf{x} = 0$. Therefore, $\mathbf{x}^T A\mathbf{x}$ is positive semi-definite. □

Similarly, one can prove the following Theorem 8.21, Corollary 8.13 and Corollary 8.14.

Corollary 8.13 A Hermitian form $h(\mathbf{x}) = \mathbf{x}^T H\mathbf{x}$ is PSD iff H is singular and all its principal minors are nonnegative. □

Theorem 8.21 A quadratic form $q(\mathbf{x}) = \mathbf{x}^T A\mathbf{x}$ is NSD iff A is singular and all principal minors of even order of A are nonnegative while those of odd order are nonpositive. □

Corollary 8.14 A Hermitian form $\mathbf{x}^* H\mathbf{x} = h(\mathbf{x})$ is NSD iff H is singular and the principal minors of even order of H are nonnegative while those of odd order are nonpositive. □

The necessary and sufficient conditions of positive semi-definiteness (or negative semi-definiteness) of a quadratic form $q(\mathbf{x}) = \mathbf{x}^T A\mathbf{x}$ (resp a Hermitian form $h(\mathbf{x}) = \mathbf{x}^* H\mathbf{x}$) cannot be given in terms of leading principal minors of A (resp H). This is erroneous as the following example shows.

Example 8.14 The leading principal minors of the matrix A associated with the quadratic form

$$\mathbf{x}^T A\mathbf{x} = [x_1\ x_2\ x_3]\begin{bmatrix} 4 & 2 & 0 \\ 2 & 1 & 0 \\ 0 & 0 & -1 \end{bmatrix}\begin{bmatrix} x_1 \\ x_2 \\ x_3 \end{bmatrix} = (2x_1 + x_2)^2 - x_3^2$$

are nonnegative, but the quadratic form is not PSD. In fact, $\mathbf{x}^T A\mathbf{x}$ here is indefinite. □

However, the sufficient conditions of positive semi-definiteness can be expressed in terms of leading principal minors as established in the following theorem.

Theorem 8.22 A quadratic form

$$q(\mathbf{x}) = \mathbf{x}^T A\mathbf{x} = \sum_{i,j=1}^{n} a_{ij}x_i x_j$$

is PSD if the leading principal minors $\Delta_1, \ldots, \Delta_{n-1}$ of A (see (8.45)) are positive and $\det A = 0$.

Proof This theorem can be proved by contradiction. Suppose that $\Delta_1, \ldots, \Delta_{n-1}$ are positive and $\det A = 0$, but

$$q(\mathbf{x}) = \mathbf{x}^T A\mathbf{x} = -\mu < 0, \tag{8.49}$$

where $\mu > 0$ and $x_1, \ldots, x_n$ are not all zero. Now, by Theorem 8.15, $q(x_1, \ldots, x_{n-1}, 0) > 0$ because the leading principal minors $\Delta_1, \ldots, \Delta_{n-1}$ of the matrix associated with this quadratic form are positive. But this contradicts assumption (8.49); so $x_n \neq 0$. Putting $\mu = \lambda x_n^2$, $\lambda > 0$ in (8.49), we have

$$\mathbf{x}^T \begin{bmatrix} a_{11} & \cdots & a_{1n} \\ \vdots & & \vdots \\ a_{n1} & \cdots & a_{nn} + \lambda \end{bmatrix} \mathbf{x} = :\mathbf{x}^T B\mathbf{x} = 0 \tag{8.50}$$

in which $\mathbf{x}$ is nonzero. Hence, $\det B = \det A + \lambda\Delta_{n-1} = \lambda\Delta_{n-1} > 0$, and as $\Delta_1, \ldots, \Delta_{n-1}$, are also positive, by Theorem 8.15, $\mathbf{x}^T B\mathbf{x}$ is PD. Consequently, (8.50) implies that $\mathbf{x} = \boldsymbol{\theta}$. This contradicts the assumption that $\mathbf{x} \neq \boldsymbol{\theta}$. So, (8.49) is not true and therefore, $\mathbf{x}^T A\mathbf{x} \geq 0$. Since A is singular, $\mathbf{x}^T A\mathbf{x}$ cannot be PD, and hence it is PSD. □

Corollary 8.15 A Hermitian form $h(\mathbf{x}) = \mathbf{x}^* H\mathbf{x}$ is PSD if $\det H = 0$ and the remaining leading principal minors H are positive. □

By Definition 8.14, a real symmetric matrix is defined as PD iff the corresponding quadratic form is PD and so on and a Hermitian matrix is PD iff the corresponding Hermitian form is PD and so on. Therefore, tests for positive definiteness of real symmetric and Hermitian matrices become identical with those for quadratic and Hermitian forms.

We conclude this section concerning the tests for definiteness of quadratic and Hermitian forms with the following theorem for indefinite forms or matrices.

Theorem 8.23 A real symmetric (or Hermitian) matrix A is indefinite iff at least one of the following conditions is satisfied:

(a) A has a negative principal minor of even order;
(b) A has a positive principal minor of odd order and a negative principal minor of odd order. □

Example 8.15 Determine the definiteness of the Hermitian form given in Example 8.12.

Solution The leading principal minors of H are respectively 2, 1 and $\det H = 0$. Other principal minors of order 2 are

$$\begin{vmatrix} 2 & 0 \\ 0 & 2 \end{vmatrix} = 4, \qquad \begin{vmatrix} 1 & -i \\ i & 2 \end{vmatrix} = 1.$$

Therefore, by Corollary 8.13, $h(x_1, x_2, x_3)$ is PSD. □

As already mentioned (see Example 8.3), there are infinite number of matrices associated with a quadratic form, but there is only one which is real symmetric. The test for definiteness is to be applied only to this real symmetric matrix. Otherwise, the test may end up with erroneous results. This is because, the derivation of these criteria for definiteness of quadratic forms critically depends on Theorem 8.3 or Corollary 8.1 which is valid for a symmetric matrix. For example, the quadratic form $q(x_1, x_2) = x_1^2 + 3x_1x_2 + 2x_2^2$ is indefinite, but if one associates the matrix

$$\begin{bmatrix} 1 & 2 \\ 1 & 2 \end{bmatrix}$$

with this form, then by Theorm 8.20 $q(x_1, x_2)$ appears to be PSD. This conclusion is, however, erroneous.

Example 8.16 Determine the definiteness of the following quadratic form:

$$q(x_1, x_2, x_3) = [x_1 \ x_2 \ x_3]\begin{bmatrix} 2 & 0 & -2 \\ 1 & 5 & 2 \\ -1 & 1 & 1 \end{bmatrix}\begin{bmatrix} x_1 \\ x_2 \\ x_3 \end{bmatrix}.$$

Solution The real symmetric matrix associated with the above quadratic form is

$$A = \begin{bmatrix} 2 & 1/2 & -3/2 \\ 1/2 & 5 & 3/2 \\ -3/2 & 3/2 & 1 \end{bmatrix}.$$

The principal minors of A of order 1 are 2, 5, 1, those of order 2 are 39/4, –1/4 and 11/4; the principal minor of order 3 is –33/4. By Theorem 8.23, $q(x_1, x_2, x_3)$ is indefinite. □

8.7 GENERALIZED EIGENVALUE PROBLEM

In the theory of small vibrations, it is important to determine a nonzero vector **x** and a scalar λ given two real symmetric or two Hermitian matrices A and B of the same order n, with A positive definite such that

$$B\mathbf{x} = \lambda A\mathbf{x},$$

which is alternatively written as

$$(\lambda A - B)\mathbf{x} = \mathbf{0}. \tag{8.51}$$

This problem is known as a *generalized eigenvalue* or *generalized characteristic value problem,* **x** is called *the eigenvector,* and λ *the eigenvalue of the* pair of matrices (A, B). By Corollary 5.10, (8.51) gives a nonzero solution for **x** iff

$$\det(\lambda A - B) = 0 \tag{8.52}$$

which is a polynomial in λ. The expression $\det(\lambda A - B)$ is known as the *characteristic polynomial,* and (8.52) is referred to as the *characteristic matrix equation* of (A, B). Since A is nonsingular (see Theorem 8.12), $\det(\lambda A - B)$ is of degree n in λ and so (8.52) has n roots. Consequently, there always exist n number of eigenvalues for the pair (A, B).

We observe that if P is a nonsingular matrix of order n, then

$$\det P^*(\lambda A - B)P = 0 \tag{8.53}$$

implies (8.52) and we can select, by applying Theorem 7.32, the nonsingular matrix P such that

$$P^*BP = \text{diag}(\lambda_1, ..., \lambda_n) =: D, \quad \text{and} \quad P^*AP = I, \tag{8.54}$$

where $\lambda_1, \lambda_2, ..., \lambda_n$ are real. Substituting (8.54) into (8.53), we get

$$\det(\lambda I - D) = (\lambda - \lambda_1)(\lambda - \lambda_2) \cdots (\lambda - \lambda_n) = 0 \tag{8.55}$$

The solutions of (8.55) for λ are the n number of real values $\lambda_1, \lambda_2, ..., \lambda_n$ which are therefore the roots of det $(\lambda A - B) = 0$, and consequently the eigenvalues of the generalized eigenvalue problem as given above. It turns out from above that, the eigenvalues of (A, B) and P^*BP are the same.

The following theorem deals with simultaneous reduction of two quadratic forms, one of which is PD and the other an arbitrary one. A corresponding result for Hermitian forms is contained in Corollary 8.16.

Theorem 8.24 Let $\mathbf{x}^T A\mathbf{x}$ be a PD and $\mathbf{x}^T B\mathbf{x}$ an arbitrary real quadratic form. Then, there exists a nonsingular transformation $\mathbf{x} = P\mathbf{y}$ such that

$$\mathbf{x}^T A\mathbf{x} = y_1^2 + y_2^2 + \cdots + y_n^2, \text{ and}$$

$$\mathbf{x}^T B\mathbf{x} = c_1 y_1^2 + c_2 y_2^2 + \cdots + c_n y_n^2,$$

where c_i's are the real eigenvalues of $P^T BP$.

Proof The assertion in the theorem follows from Corollary 7.13 by observing that A is real symmetric with positive eigenvalues and B is real symmetric. □

The corresponding result for Hermitian forms can be similarly proved as in the following corollary.

Corollary 8.16 Let $\mathbf{x}^* A\mathbf{x}$ be a PD and $\mathbf{x}^* B\mathbf{x}$ an arbitrary Hermitian form. Then, there exists a nonsingular transformation $\mathbf{x} = P\mathbf{y}$ such that

$$\mathbf{x}^* A\mathbf{x} = \bar{y}_1 y_1 + \bar{y}_2 y_2 + \cdots + \bar{y}_n y_n, \text{and}$$

$$\mathbf{x}^* B\mathbf{x} = c_1 \bar{y}_1 y_1 + c_2 \bar{y}_2 y_2 + \cdots + c_n \bar{y}_n y_n,$$

where c_i's are the real eigenvalues of P^*BP. □

An alternative statement of Theorem 8.24 and Corollary 8.16 can be given with the notion of a generalized eigenvalue as follows. This depends on the fact that the eigenvalues of (A, B) and P^*BP are the same as established above.

Theorem 8.25 There exists a nonsingular transformation $\mathbf{x} = P\mathbf{y}$ which reduces simultaneously $\mathbf{x}^T A\mathbf{x}$ (A, a PD real symmetric matrix) and $\mathbf{x}^T B\mathbf{x}$, (B, a real symmetric matrix) to

$$\mathbf{x}^T A\mathbf{x} = \bar{y}_1 y_1 + \bar{y}_2 y_2 + \cdots + \bar{y}_n y_n, \quad \text{and} \quad \mathbf{x}^T B\mathbf{x} = \lambda_1 \bar{y}_1 y_1 + \lambda_2 \bar{y}_2 y_2 + \cdots + \lambda_n \bar{y}_n y_n,$$

where λ_i's are the roots of det $(\lambda A - B) = 0$. □

Corollary 8.17 There exists a nonsingular transformation $\mathbf{x} = P\mathbf{y}$ which reduces simultaneously $\mathbf{x}^* A\mathbf{x}$ (A, a PD Hermitian matrix) and $\mathbf{x}^T B\mathbf{x}$, (B, a Hermitian matrix) to

$$\mathbf{x}^* A\mathbf{x} = \bar{y}_1 y_1 + \bar{y}_2 y_2 + \cdots + \bar{y}_n y_n \quad \text{and} \quad \mathbf{x}^T B\mathbf{x} = \lambda_1 \bar{y}_1 y_1 + \lambda_2 \bar{y}_2 y_2 + \cdots + \lambda_n \bar{y}_n y_n,$$

where λ_i's are the roots of det $(\lambda A - B) = 0$. □

It may be observed that the generalized characteristic matrix equation (8.51) reverts to a conventional characteristic matrix equation (7.6) if $A = I$. Moreover, in the study of small oscillations of a system with n degrees of freedom, the PD form $\mathbf{x}^T A\mathbf{x}$ corresponds to the kinetic energy of the system and a PD or a positive semi-definite form $\mathbf{x}^T B\mathbf{x}$ refers to the potential energy, and λ's are used to compute the normal modes of oscillations.

The computation of nonsingular transformation P referred to in Theorems 8.24 and 8.25 can be carried out by following the construction given in Theorem 7.32. This is laborious since P is computed as the product of three matrices, i.e., $P = UQR$. Of these, the determination of the unitary (or orthogonal) matrices U and R is really difficult as seen in Example 7.41. A simplification of this computational procedure is, therefore, in order which will be taken up in the sequel. In the following, we shall assume A and B to be Hermitian with A positive definite if not stated to the contrary.

Lemma 8.1 Let λ be an eigenvalue of (A, B) of multiplicity m, where A and B are Hermitian with A positive definite. Then, there exist m linearly independent vectors associated with λ.

Proof Suppose that $\mathbf{x}$ is an eigenvector associated with the eigenvalue λ of (A, B) having multiplicity m. Then,

$$(\lambda A - B)\mathbf{x} = \boldsymbol{\theta}. \tag{8.51}$$

By Theorem 7.32, there exists a nonsingular matrix P such that (see also Theorem 7.28)

$$P^*AP = I, \qquad P^*BP = \text{diag } [\lambda_1, \cdots \lambda_m; \lambda_{m+1}, \ldots, \lambda_n],$$

where $\lambda_1 = \lambda_2 = \cdots = \lambda_m = \lambda$, $\lambda_{m+1}, \ldots, \lambda_n$ are the eigenvalues of (A, B) (see Corollary 8.17). Hence, in view of Corollary 5.14 and the fact that $\lambda \neq \lambda_i$, $(i = m + 1, \ldots, n)$, we have

$$\begin{aligned}\text{rank } (\lambda A - B) &= \text{rank } P^*(\lambda A - B)P \\ &= \text{rank } (\text{diag } [0, \ldots, 0; \lambda - \lambda_{m+1}, \ldots, \lambda - \lambda_n]) = n - m,\end{aligned}$$

which implies that (8.51) has m linearly independent solutions for the eigenvector $\mathbf{x}$. □

We shall now generalize the concepts of orthogonal and orthonormal set of vectors given in Definitions 5.11 and 5.13.

Definition 8.15 *A-Orthogonal Set.* Let A be a positive definite Hermitian matrix of order n. Then, the two vectors $\mathbf{x}_1$ and $\mathbf{x}_2$ in $\mathbb{C}^n$ are said to be *orthogonal relative to* A, or in short, *A-orthogonal* if $\langle \mathbf{x}_1, A\mathbf{x}_2\rangle = 0$. A set of vectors $\mathbf{x}_1, \mathbf{x}_2, \ldots, \mathbf{x}_p$ is called an *orthogonal set relative* to A or an *A-orthogonal set* if each vector of the set is A-orthogonal to every other vector in the set, i.e., $\langle \mathbf{x}_i, A\mathbf{x}_j\rangle = 0$, $i \neq j$. When $A = I$, this definition reverts to the conventional concept of an orthogonal set given in Definition 5.11. □

Example 8.17 The set of vectors

$$\mathbf{x}_1 = [1\ 1\ 0]^T, \quad \mathbf{x}_2 = [5, -8, -13]^T, \quad \mathbf{x}_3 = [32, -17, 8]^T$$

is orthogonal relative to

$$A = \begin{bmatrix} 6 & 4 & -2 \\ 4 & 12 & -4 \\ -2 & -4 & 13 \end{bmatrix}.$$

This is because, direct computation shows

$$\langle \mathbf{x}_1, A\mathbf{x}_2\rangle = \langle \mathbf{x}_1, A\mathbf{x}_3\rangle = \langle \mathbf{x}_2, A\mathbf{x}_3\rangle = 0,$$

$$\langle \mathbf{x}_1, A\mathbf{x}_1\rangle = 26, \quad \langle \mathbf{x}_2, A\mathbf{x}_2\rangle = 2223, \quad \langle \mathbf{x}_3, A\mathbf{x}_3\rangle = 6156. \quad □$$

Example 8.18 If A is positive definite and Hermitian, then $\langle \mathbf{x}, A\mathbf{x} \rangle = 0$ if $\mathbf{x} = \boldsymbol{\theta}$. If $\mathbf{x} \neq \boldsymbol{\theta}$, then there exists a real positive number a such that $\mathbf{y} = (a)^{-1/2}\mathbf{x}$ satisfies $\langle \mathbf{y}, A\mathbf{y} \rangle = 1$.

Solution Since A is positive definite and Hermitian, there exists a nonsingular matrix P such that $P^*AP = I$ (see Problem 8.31). So, using the transformation $\mathbf{x} = P\mathbf{u}$, we have

$$\langle \mathbf{x}, A\mathbf{x} \rangle = \mathbf{x}^*A\mathbf{x} = \mathbf{u}^*P^*AP\mathbf{u} = \mathbf{u}^*\mathbf{u} = 0$$

iff $\mathbf{u} = \boldsymbol{\theta}$, i.e., iff $\mathbf{x} = \boldsymbol{\theta}$.

Because of the positive-definiteness of A, if $\mathbf{x} \neq \boldsymbol{\theta}$, then $\langle \mathbf{x}, A\mathbf{x} \rangle = a$, a positive real number. So, defining $\mathbf{y} = (a)^{-1/2}\mathbf{x}$, we have

$$\langle (a)^{-1/2}\mathbf{x}, A(a)^{-1/2}\mathbf{x} \rangle = (a)^{-1} \langle \mathbf{x}, A\mathbf{x} \rangle = 1. \qquad \square$$

Definition 8.16 *A-Orthonormal Set.* Let A be a positive definite Hermitian matrix of order n. Then, the two vectors $\mathbf{x}_1$ and $\mathbf{x}_2$ in $\mathbb{C}^n$ are said to be *orthonormal relative to A*, or in short, *A-orthonormal* if $\langle \mathbf{x}_i, A\mathbf{x}_j \rangle = \delta_{ij}$ $(i = 1, 2)$. A set of vectors $\mathbf{x}_1, \mathbf{x}_2, \ldots, \mathbf{x}_p$ is called an *orthonormal set relative to A*, or an *A-orthonormal set*, if $\langle \mathbf{x}_i, A\mathbf{x}_j \rangle = \delta_{ij}$ $(i = 1, 2, \ldots, p)$. When $A = I$, this definition conforms to the conventional notion of an orthonormal set in Definition 5.13. $\square$

Lemma 8.2 Let λ be an eigenvalue of (A, B) of multiplicity m, where A and B are Hermitian with A positive definite. Then, there exists a set of m A-orthogonal eigenvectors associated with λ.

Proof By Lemma 8.1, there exist m linearly independent eigenvectors associated with λ which are designated by $\mathbf{x}_1, \mathbf{x}_2, \ldots, \mathbf{x}_m$. We shall use the technique underlying the Gram-Schmidt process to generate m A-orthogonal vectors $\mathbf{y}_1, \mathbf{y}_2, \ldots, \mathbf{y}_m$. These $\mathbf{y}_i$'s are defined by

$$\begin{aligned}
\mathbf{y}_1 &= \mathbf{x}_1,\\
\mathbf{y}_2 &= a_{21}\mathbf{y}_1 + \mathbf{x}_2,\\
\mathbf{y}_3 &= a_{31}\mathbf{y}_1 + a_{32}\mathbf{y}_2 + \mathbf{x}_3,\\
&\vdots\\
\mathbf{y}_m &= a_{m1}\mathbf{y}_1 + a_{m2}\mathbf{y}_2 + \cdots + a_{mm-1}\mathbf{y}_{m-1} + \mathbf{x}_m,
\end{aligned} \tag{8.56}$$

where a_{ij}'s are scalars to be determined, requiring each $\mathbf{y}_i$ to be A-orthogonal to every other vector in the set. For $\mathbf{y}_2$ to be A-orthogonal to $\mathbf{y}_1$, from (8.56) we have

$$0 = \langle \mathbf{y}_2, A\mathbf{y}_1 \rangle = \langle a_{21}\mathbf{y}_1 + \mathbf{x}_2, A\mathbf{y}_1 \rangle = a_{21} \langle \mathbf{y}_1, A\mathbf{y}_1 \rangle + \langle \mathbf{x}_2, A\mathbf{y}_1 \rangle.$$

Hence,

$$a_{21} = -\langle \mathbf{x}_2, A\mathbf{y}_1 \rangle / \langle \mathbf{y}_1, A\mathbf{y}_1 \rangle$$

which specifies $\mathbf{y}_2$ completely. Suppose, in this way, we can determine the set of A-orthogonal vectors $\mathbf{y}_1, \mathbf{y}_2, \ldots, \mathbf{y}_{m-1}$. Now, taking the inner product of both sides of (8.56) with $A\mathbf{y}_1, \ldots, A\mathbf{y}_{m-1}$ in succession and requiring that $\mathbf{y}_m$ be A-orthogonal to the $\mathbf{y}_i$'s already determined, we have

$$a_{m1} = -\frac{\langle \mathbf{x}_m, A\mathbf{y}_1 \rangle}{\langle \mathbf{y}_1, A\mathbf{y}_1 \rangle}, a_{m2} = -\frac{\langle \mathbf{x}_m, A\mathbf{y}_2 \rangle}{\langle \mathbf{y}_2, A\mathbf{y}_2 \rangle}, \ldots, a_{mm-1} = -\frac{\langle \mathbf{x}_m, A\mathbf{y}_{m-1} \rangle}{\langle \mathbf{y}_{m-1}, A\mathbf{y}_{m-1} \rangle}.$$

That these $\mathbf{y}_i$'s satisfy $(\lambda A - B)\mathbf{y}_i = \boldsymbol{\theta}$ and, therefore, are eigenvectors of (A, B) can be easily verified. $\square$

Corollary 8.18 If λ is an eigenvalue of (A, B) of multiplicity m, where A and B are Hermitian with A positive definite, then there exists a set of m A-orthonormal eigenvectors associated with λ.

Proof By Lemma 8.2, there exists a set of m A-orthogonal eigenvectors designated by $\mathbf{x}_1, \mathbf{x}_2, ..., \mathbf{x}_m$, associated with λ. Hence $\langle \mathbf{x}_i, A\mathbf{x}_j\rangle = 0$, $i \neq j$, and since A is positive definite Hermitian, $\langle \mathbf{x}_i, A\mathbf{x}_i\rangle = \alpha_i$, a positive real number. Define $a_i = (\alpha_i)^{-1/2}$ and $\mathbf{y}_i = a_i\mathbf{x}_i$. Then,

$$\langle \mathbf{y}_i, A\mathbf{y}_j\rangle = \langle a_i\mathbf{x}_i, A(a_j\mathbf{x}_j)\rangle = a_i a_j \langle \mathbf{x}_i, A\mathbf{x}_j\rangle = \delta_{ij},$$

which shows that $\mathbf{y}_1, \mathbf{y}_2, ..., \mathbf{y}_m$ form an A-orthonormal set.

Lemma 8.3 Let λ and μ be two distinct eigenvalues of (A, B), where A and B are Hermitian with A positive definite, and $\mathbf{x}$ and $\mathbf{y}$ the corresponding eigenvectors. Then, $\mathbf{x}^*A\mathbf{y} = 0$ and $\mathbf{x}^*B\mathbf{y} = 0$.

Proof Suppose that λ and μ are the eigenvalues (which are real, see Theorem 7.24) with $\mathbf{x}$ and $\mathbf{y}$ as the corresponding eigenvectors. Then,

$$(\lambda A - B)\mathbf{x} = 0 \quad \text{and} \quad (\mu A - B)\mathbf{y} = 0.$$

Hence,

$$\mu\mathbf{x}^*A\mathbf{y} = \mathbf{x}^*(\mu A\mathbf{y}) = \mathbf{x}^*B\mathbf{y} = (B\mathbf{x})^*\mathbf{y} = (\lambda A\mathbf{x})^*\mathbf{y} = \lambda\mathbf{x}^*A\mathbf{y}.$$

Since $\lambda \neq \mu$, we must have $\mathbf{x}^*A\mathbf{y} = 0$, and hence $\mathbf{x}B^*\mathbf{y} = \mu\mathbf{x}^*A\mathbf{y} = 0$. □

Theorem 8.26 Let (A, B), where A and B are n-square Hermitian matrices with A positive definite, have k distinct eigenvalues λ_i, each of multiplicity m_i $(i = 1, 2, ..., k)$. Then (A, B) has n linearly independent A-orthonormal eigenvectors $\mathbf{y}_1, \mathbf{y}_2, ..., \mathbf{y}_n$, and the matrix P formed by these vectors as columns reduces A and B simultaneously to

$$P^*AP = I \quad \text{and} \quad P^*BP = \text{diag}\,[\underbrace{\lambda_1 \ldots \lambda_1}_{m_1 \text{ times}}; \ldots; \underbrace{\lambda_k \ldots \lambda_k}_{m_k \text{ times}}].$$

Proof By virtue of Corollary 8.18, we can select m_i number of A-orthonormal eigenvectors of (A, B) associated with λ_i $(i = 1, 2, ..., k)$, $m_1 + m_2 + \cdots + m_k = n$. By Lemma 8.3, all these n-eigenvectors in the set

$$\{\mathbf{y}_1, ..., \mathbf{y}_{m_1}; \mathbf{y}_{m_1+1}, ..., \mathbf{y}_{m_1+m_2}, ..., \mathbf{y}_n\}$$

form an A-orthonormal set with

$$\langle \mathbf{y}_i, A\mathbf{y}_j\rangle = \delta_{ij}. \tag{8.57}$$

To prove that these eigenvectors are all linearly independent, suppose for scalars $\alpha_1, \alpha_2, ..., \alpha_n$, we write

$$\alpha_1\mathbf{y}_1 + \alpha_2\mathbf{y}_2 + \cdots + \alpha_{m1}\mathbf{y}_{m1} + \cdots + \alpha_n\mathbf{y}_n = \boldsymbol{\theta}.$$

Taking the inner product with $A\mathbf{y}_1, A\mathbf{y}_2, ..., A\mathbf{y}_n$ of both sides of the above equation successively, we have $\alpha_1 = \alpha_2 = \cdots = \alpha_n = 0$. So, $\mathbf{y}_1, \mathbf{y}_2, ..., \mathbf{y}_n$ are linearly independent. Setting $P = [\mathbf{y}_1 \;\; \mathbf{y}_2 \;\; \cdots \;\; \mathbf{y}_n]$, we have by direct computation, by making use of (8.57), the relation

$$P^*AP = \begin{bmatrix} \mathbf{y}_1^* \\ \mathbf{y}_2^* \\ \vdots \\ \mathbf{y}_n^* \end{bmatrix} A[\mathbf{y}_1\, \mathbf{y}_2 \cdots \mathbf{y}_n] = I_n,$$

and in view of $(\lambda_i A - B)\mathbf{y}_i = \boldsymbol{\theta}$,

$$\begin{aligned} P^*BP &= P^*[\lambda_1 A\mathbf{y}_1 \cdots \lambda_1 A\mathbf{y}_{m_1} \cdots \lambda_k A\mathbf{y}_n] \\ &= P^*B\,[\mathbf{y}_1\, \mathbf{y}_2 \cdots \mathbf{y}_n] \\ &= \text{diag}\,[\lambda_1, \ldots, \lambda_1; \lambda_2, \ldots, \lambda_k]. \end{aligned}$$ □

Thus, given a Hermitian pair (A, B) with A positive definite, one can take recourse to the constructive proof of Theorem 7.32 to determine a nonsingular matrix P such that $P^*AP = I$ and $P^*BP = \text{diag}\,[\lambda_1, \lambda_2, \ldots, \lambda_n]$, where λ_i's are the eigenvalues of (A, B). Alternatively, a less laborious method would be to compute the set of A-orthonormal eigenvectors of (A, B) and construct the matrix P with these eigenvectors as columns as proved in Theorem 8.26. Since there are infinite number of eigenvectors associated with an eigenvalue (see Example 7.1), the matrix P above is not unique.

Example 8.19 Determine a nonsingular matrix P which reduces the symmetric matrices (A, B) given by (7.141), viz.,

$$A = \begin{bmatrix} 6 & 4 & -2 \\ 4 & 12 & -4 \\ -2 & -4 & 13 \end{bmatrix} \quad \text{and} \quad B = \begin{bmatrix} 1 & -1 & 1 \\ -1 & 1 & -1 \\ 1 & -1 & 1 \end{bmatrix} \tag{8.58}$$

to a unit matrix and a diagonal matrix respectively.

Solution The leading principal minors of A are 6, 56, and 648; hence, by Sylvester's criterion, A is positive definite. The characteristic equation of (A, B) is given by (8.52) which in our case is

$$\det(\lambda A - B) = 18\lambda^2\,(36\lambda - 19) = 0.$$

Hence, the eigenvalues of (A, B) are $\lambda_1 = 0$, $\lambda_2 = 0$, $\lambda_3 = 19/36$. To determine the eigenvector $\mathbf{x} = [x\ y\ z]^T$ associated with the eigenvalue 0, we have to solve (8.51) which in our case becomes

$$\begin{bmatrix} -1 & 1 & -1 \\ 1 & -1 & 1 \\ -1 & 1 & -1 \end{bmatrix}\begin{bmatrix} x \\ y \\ z \end{bmatrix} = \boldsymbol{\theta}.$$

On solving, we get $\mathbf{x}_1 = [1\ 1\ 0]^T$, $\mathbf{x}_2 = [1\ \ 0\ \ -1]$ as the two independent vectors associated with $\lambda_1 = \lambda_2 = 0$. Following Lemma 8.2, we choose $\mathbf{y}_1 = \mathbf{x}_1$ and

$$\mathbf{y}_2 = \mathbf{x}_2 - \frac{\langle \mathbf{x}_2, A\mathbf{y}_1 \rangle}{\langle \mathbf{y}_1, A\mathbf{y}_1 \rangle}\mathbf{y}_1 = \begin{bmatrix} 1 \\ 0 \\ -1 \end{bmatrix} - \frac{16}{26}\begin{bmatrix} 1 \\ 1 \\ 0 \end{bmatrix} = \frac{1}{13}\begin{bmatrix} 5 \\ -8 \\ -13 \end{bmatrix}$$

as the two, A-orthogonal vectors associated with the double root at $\lambda = 0$. Now, $\langle \mathbf{y}_1, A\mathbf{y}_1 \rangle = 26$ and $\langle \mathbf{y}_2, A\mathbf{y}_2 \rangle = 171/13$; the two A-orthonormal vectors associated with the eigenvalue 0 are (see Corollary 8.18)

$$\mathbf{u}_1 = \frac{1}{\sqrt{26}}[1 \quad 1 \quad 0]^T \quad \text{and} \quad \mathbf{u}_2 = \frac{1}{3\sqrt{247}}[5, \quad -8, \quad -13]^T.$$

The eigenvector associated $\lambda_3 = 19/36$ is obtained by solving (8.51) for $\lambda = \lambda_3$ and for A and B given in (8.58). This yields

$$\mathbf{x}_3 = [32, -17, 8]^T$$

Since $\langle \mathbf{x}_3, A\mathbf{x}_3 \rangle = 6156$, we choose

$$\mathbf{u}_3 = \frac{1}{18\sqrt{19}}[32, \quad -17, \quad 8]^T.$$

The required nonsingular transformation is, therefore,

$$P = [\mathbf{u}_1\, \mathbf{u}_2\, \mathbf{u}_3] = \frac{1}{18\sqrt{494}}\begin{bmatrix} 18\sqrt{19} & 30\sqrt{2} & 32\sqrt{26} \\ 18\sqrt{19} & -48\sqrt{2} & -17\sqrt{26} \\ 0 & -78\sqrt{2} & 8\sqrt{26} \end{bmatrix}.$$ □

8.8 BASES FOR MATRIX REPRESENTATION OF A BILINEAR FUNCTION

We shall begin this section with the abstract definition of bilinear functions and then show how by a proper selection of bases in the underlying vector spaces, these functions put in their familiar appearances in bilinear forms as seen in Sections 8.1 and the following.

Let $\mathbf{x}$ be an element in the vector space V, and $\mathbf{y}$ an element in the vector space W. Then we call the pair $(\mathbf{x}, \mathbf{y})$ an element of the vector space $V \times W$. For example, $([1, 2]^T, [1, -1, 0]^T)$ is an element of $\mathbb{R}^2 \times \mathbb{R}^3$.

Definition 8.17 *Bilinear Function.* Let V and W be two vector spaces over a field $\mathbb{F}$. A bilinear function is a mapping $f : V \times W \to \mathbb{F}$: $(\mathbf{x}, \mathbf{y}) \mapsto f(\mathbf{x}, \mathbf{y})$ which satisfies the following axioms:

(a) $f(\alpha\mathbf{w} + \beta\mathbf{x}, \mathbf{y}) = \alpha f(\mathbf{w}, \mathbf{y}) + \beta f(\mathbf{x}, \mathbf{y})$,
(b) $f(\mathbf{x}, \gamma\mathbf{y} + \delta\,\mathbf{z}) = \gamma f(\mathbf{x}, \mathbf{y}) + \delta f(\mathbf{x}, \mathbf{z})$,

for any arbitrary elements $\mathbf{w}$, $\mathbf{x}$ in V; $\mathbf{y}$, $\mathbf{z}$ in W; and $\alpha, \beta, \gamma, \delta$ in $\mathbb{F}$. □

Stated explicitly, to every pair of elements $(\mathbf{x}, \mathbf{y})$ in the vector space $V \times W$, a bilinear function f assigns an element from the field $\mathbb{F}$ in such a way that $f(\mathbf{x}, \mathbf{y})$ is linear in $\mathbf{x}$ when $\mathbf{y}$ is held fixed and is linear in $\mathbf{y}$ when $\mathbf{x}$ is held fixed as suggested by axioms (a) and (b). The elements of $\mathbb{F}$ with which we are concerned in this text are either real or complex numbers.

Example 8.20 Let V be the vector space of all continuous real functions on $\mathbb{R}$, and let $K(t, \tau)$ be a given continuous function of the real variables t, τ. Then,

$$f(x, y) = \int_a^b \int_a^b K(t,\tau)x(t)y(\tau)\,dt\,d\tau$$

where x, y are arbitrary elements in V and a, b are real numbers, is a bilinear function $f: V \times V \to \mathbb{R}$. It can be easily checked that f satisfies the axioms (a) and (b) above. □

Example 8.21 Let $p = a_0 + a_1t + a_2t^2$ and $q = b_0 + b_1t + b_2t^2$ be two elements in $\mathbb{P}_2(t)$ over the field $\mathbb{R}$ of real numbers. Then, the function defined by

$$f(p, q) = a_0b_0 + a_1b_1 + a_2b_2$$

is a bilinear function $f: \mathbb{P}_2(t) \times \mathbb{P}_2(t) \to \mathbb{R}$. □

Example 8.22 Let X and Y be two arbitrary elements in $V_{n\times m}$, the vector space V of all $n \times m$ matrices over $\mathbb{C}$ and let A be a fixed element in $V_{n\times n}$, the vector space of all $n \times n$ matrices over $\mathbb{C}$. Then, the function

$$f(X, Y) = \text{tr } X^TAY$$

is a bilinear function $f: V \times V \to \mathbb{C}$. □

With the above abstract definition of a bilinear function, we shall now show how, by selecting bases in the underlying vector spaces V and W, we can express the value $f(\mathbf{x}, \mathbf{y})$ of any such bilinear function f at $\mathbf{x}$ and $\mathbf{y}$ in terms of the coordinate vectors $\mathbf{u}$ and $\mathbf{v}$ for $\mathbf{x}$ and $\mathbf{y}$ respectively, and a matrix A associated with f. We shall represent this value by $b(\mathbf{u}, \mathbf{v}) = \mathbf{u}^TA\mathbf{v}$, to be called a *bilinear form*, which represents the bilinear function. In fact, we have $f(\mathbf{x}, \mathbf{y}) = b(\mathbf{u}, \mathbf{v})$ for the value of the bilinear function at $\mathbf{x}$ and $\mathbf{y}$. This connection conforms to the description of such forms by (8.1). The distinction between bilinear functions and bilinear forms is made for clarity of representation of these bilinear functions after selection of bases. The matrix A is called either the *matrix of the bilinear function f* with respect to the chosen bases or the *matrix of the bilinear form b*(**u**, **v**).

Theorem 8.27 Let $f: V \times W \to \mathbb{F} : (\mathbf{x}, \mathbf{y}) \mapsto f(\mathbf{x}, \mathbf{y})$ be a bilinear function, and let $\{\mathbf{x}_1, \mathbf{x}_2, ..., \mathbf{x}_n\}$ and $\{\mathbf{y}_1, \mathbf{y}_2, ..., \mathbf{y}_m\}$ be bases in V and W respectively. Then, the value of the bilinear function $f(\mathbf{x}, \mathbf{y})$ can be given by the bilinear form

$$b(\mathbf{u}, \mathbf{v}) = f(\mathbf{x}, \mathbf{y}) = \mathbf{u}^TA\mathbf{v}, \tag{8.59}$$

where $\mathbf{u} = [\alpha_1, \alpha_2, ..., \alpha_n]^T$ and $\mathbf{v} = [\beta_1, \beta_2, ..., \beta_m]^T$ are the coordinate vectors of $\mathbf{x}$ and $\mathbf{y}$ with respect to $\{\mathbf{x}_1, \mathbf{x}_2, ..., \mathbf{x}_n\}$ and $\{\mathbf{y}_1, \mathbf{y}_2, ..., \mathbf{y}_m\}$ respectively, and $A = [a_{ij}]_{n\times m}$ is given uniquely by $a_{ij} = f(\mathbf{x}_i, \mathbf{y}_j)$.

Conversely, if we have a pair of bases $\{\mathbf{x}_1, \mathbf{x}_2, ..., \mathbf{x}_n\}$ and $\{\mathbf{y}_1, \mathbf{y}_2, ..., \mathbf{y}_m\}$, and any $n \times m$ matrix A over $\mathbb{F}$, then it is easy to see that (8.59) defines a bilinear form.

Proof Suppose that $\mathbf{x}$ and $\mathbf{y}$ are arbitrary elements in V and W respectively. Then,

$$\mathbf{x} = \alpha_1\mathbf{x}_1 + \alpha_2\mathbf{x}_2 + \cdots + \alpha_n\mathbf{x}_n \quad \text{and} \quad \mathbf{y} = \beta_1\mathbf{y}_1 + \beta_2\mathbf{y}_2 + \cdots + \beta_m\mathbf{y}_m. \tag{8.60}$$

In view of axioms (a) and (b) in Definition 8.17, we can now write successively

$$\begin{aligned} f(\mathbf{x}, \mathbf{y}) &= f(\alpha_1\mathbf{x}_1 + \alpha_2\mathbf{x}_2 + \cdots + \alpha_n\mathbf{x}_n, \mathbf{y}) \\ &= \alpha_1 f(\mathbf{x}_1, \mathbf{y}) + \alpha_2 f(\mathbf{x}_2, \mathbf{y}) + \cdots + \alpha_n f(\mathbf{x}_n, \mathbf{y}), \end{aligned} \tag{8.61}$$

$$\begin{aligned} f(\mathbf{x}_i, \mathbf{y}) &= f(\mathbf{x}_i, \beta_1\mathbf{y}_1 + \beta_2\mathbf{y}_2 + \cdots + \beta_m\mathbf{y}_m) \\ &= \beta_1 f(\mathbf{x}_i, \mathbf{y}_1) + \beta_2 f(\mathbf{x}_i, \mathbf{y}_2) + \cdots + \beta_m f(\mathbf{x}_i, \mathbf{y}_m), \end{aligned} \tag{8.62}$$

where $i = 1, 2, ..., n$. Inserting (8.62) into (8.61), we finally have

$$f(\mathbf{x},\mathbf{y}) = \sum_{i=1}^{n}\sum_{j=1}^{m} \alpha_i \beta_j f(\mathbf{x}_i,\mathbf{y}_j), \tag{8.63}$$

which is of the form (8.59) after putting $a_{ij} = f(\mathbf{x}_i, \mathbf{y}_j)$.

The mn scalars $a_{ij} = f(\mathbf{x}_i, \mathbf{y}_j)$, as is obvious from (8.63), completely determine the value of $f(\mathbf{x}, \mathbf{y})$. Moreover, the matrix A of the bilinear function f is uniquely determined with the selection of bases of the underlying vector spaces.

Conversely, if we are given the bases $\{\mathbf{x}_1, \mathbf{x}_2, ..., \mathbf{x}_n\}$ and $\{\mathbf{y}_1, \mathbf{y}_2, ..., \mathbf{y}_m\}$ in V and W respectively, and an $n \times m$ matrix A over $\mathbb{F}$, then any element $\mathbf{x}$ in V and $\mathbf{y}$ in W can be represented by (8.60). So, we have

$$\mathbf{u}^T A\mathbf{v} = b(\mathbf{u}, \mathbf{v}) =: f(\mathbf{x}, \mathbf{y})$$

which can be shown to satisfy axioms (a) and (b). For example, if $\hat{\mathbf{x}} = \hat{\alpha}_1\mathbf{x}_1 + \hat{\alpha}_2\mathbf{x}_2 + \cdots + \hat{\alpha}_n\mathbf{x}_n$ is another element in V, then $\hat{\mathbf{u}} = [\hat{\alpha}_1\,\hat{\alpha}_2\,\cdots\,\hat{\alpha}_n]^T$ and

$$f(\mathbf{x}+\hat{\mathbf{x}},\mathbf{y}) := (\hat{\mathbf{u}}+\mathbf{u})^T A\mathbf{v} = \hat{\mathbf{u}}^T A\mathbf{v} + \mathbf{u}^T A\mathbf{v} = f(\hat{\mathbf{x}},\mathbf{y}) + f(\mathbf{x},\mathbf{y}).$$ □

Example 8.23 Let a bilinear form $f: \mathbb{R}^3 \times \mathbb{R}^2 \to \mathbb{R}$ be defined as

$$f(\mathbf{x}, \mathbf{y}) = 2x_1y_1 - x_1y_2 + 3x_2y_1 + x_2y_2 - 2x_3y_2 \tag{8.64}$$

where $\mathbf{x} = [x_1\ x_2\ x_3]^T$ is in $\mathbb{R}^3$ and $\mathbf{y} = [y_1\ y_2]^T$ is in $\mathbb{R}^2$. With standard bases $\{\mathbf{e}_1, \mathbf{e}_2, \mathbf{e}_3\}$ in $\mathbb{R}^3$ and $\{\hat{\mathbf{e}}_1, \hat{\mathbf{e}}_2\}$ in $\mathbb{R}^2$ (see Definition 5.6), we have the matrix of the bilinear form as

$$A = \begin{bmatrix} 2 & -1 \\ 3 & 1 \\ 0 & -2 \end{bmatrix}$$

and the coordinate vectors for $\mathbf{x}$ and $\mathbf{y}$ are respectively

$$\mathbf{u} = [x_1\ x_2\ x_3]^T \quad \text{and} \quad \mathbf{v} = [y_1\ y_2]^T.$$

From the above illustration, it is obvious that the bilinear form represented in (8.1) is with respect to standard bases. On the contrary, if we choose (distinguish carefully between the symbols $\mathbf{x}_1$ and x_1, etc.)

$$\{\mathbf{x}_1 = [1\ -1\ \ 1]^T,\ \ \mathbf{x}_2 = [0\ 1\ 1]^T,\ \ \mathbf{x}_3 = [-2\ 1\ -1]^T\},$$
$$\{\mathbf{y}_1 = [1\ 1]^T,\ \ \mathbf{y}_2 = [-1,\ 2]^T\} \tag{8.65}$$

to be the bases of V and W respectively, then for the given bilinear form, we can easily compute

$$a_{11} = f(\mathbf{x}_1, \mathbf{y}_1) = -5, \qquad a_{12} = -7,$$
$$a_{21} = 2, \qquad a_{22} = -5, \qquad a_{31} = 4, \qquad a_{32} = 11.$$

The coordinate vectors of $\mathbf{x}$ and $\mathbf{y}$ are (see Example 5.43), respectively,

$$\mathbf{u} = [\alpha_1\ \alpha_2\ \alpha_3]^T = [-x_1 - x_2 + x_3,\ (x_2 + x_3)/2, (\ -2x_1 - x_2 + x_3)/2]^T$$

$$\mathbf{v} = [\beta_1\ \beta_2]^T = \frac{1}{3}\ [2y_1 + y_2,\ -y_1 + y_2]^T. \tag{8.66}$$

So, the representation of (8.64) with respect to the bases (8.65) is

$$b(\mathbf{u},\mathbf{v}) = [\alpha_1\ \alpha_2\ \alpha_3]\begin{bmatrix} -5 & -7 \\ 2 & -5 \\ 4 & 11 \end{bmatrix}\begin{bmatrix} \beta_1 \\ \beta_2 \end{bmatrix},$$

where α_i's and β_i's are given by (8.66). □

Theorem 8.28 Let $X = \{\mathbf{x}_1, \mathbf{x}_2, ..., \mathbf{x}_n\}$ and $Y = \{\mathbf{y}_1, \mathbf{y}_2, ..., \mathbf{y}_m\}$ be the respective bases of the vector spaces V and W in terms of which A is the matrix of the bilinear function f. If $\hat{X} = \{\hat{\mathbf{x}}_1, \hat{\mathbf{x}}_2, \ldots, \hat{\mathbf{x}}_n\}$ and $\hat{Y} = \{\hat{\mathbf{y}}_1, \hat{\mathbf{y}}_2, \ldots, \hat{\mathbf{y}}_m\}$ are new bases of V and W respectively, then the new matrix of the bilinear function f is P^TAQ, where P is the coordinate transformation matrix (see Theorem 6.10) in V and Q is that in W.

Proof Suppose that $\mathbf{u} = [\alpha_1, \alpha_2, ..., \alpha_n]^T$ and $\hat{\mathbf{u}} = [\hat{\alpha}_1, \hat{\alpha}_2, ..., \hat{\alpha}_n]^T$ be the coordinate vectors of any element $\mathbf{x}$ in V with respect to bases X and $\hat{X}$ respectively; $\mathbf{v} = [\beta_1, \beta_2, ..., \beta_m]^T$ and $\hat{\mathbf{v}} = [\hat{\beta}_1, \hat{\beta}_2, \ldots, \hat{\beta}_m]^T$ be the coordinate vectors of any element $\mathbf{y}$ in W with respect to bases Y and $\hat{Y}$. Then, if P and Q are the *coordinate transformation matrices* (changing coordinate vectors $\mathbf{u}$ to $\hat{\mathbf{u}}$ and $\mathbf{v}$ to $\hat{\mathbf{v}}$, see Theorem 6.10) in V and W respectively, we have, by Theorem 6.11, $\mathbf{u} = P\hat{\mathbf{u}}$ and $\mathbf{v} = Q\hat{\mathbf{v}}$. Hence, the bilinear form (8.59) can be transformed in succession as

$$f(\mathbf{x}, \mathbf{y}) = \mathbf{u}^T A\mathbf{v} = \hat{\mathbf{u}}^T P^T AQ\hat{\mathbf{v}}$$

Therefore, it follows that the matrix of $f(\mathbf{x}, \mathbf{y})$ with respect to the new bases $\hat{X}$ and $\hat{Y}$ is P^TAQ. □

Remark 8.1 Referring to Remark 6.6, the old and new basis matrices in V, viz., $\mathbf{V} = [\mathbf{x}_1\ \mathbf{x}_2 \ldots \mathbf{x}_n]$ and $\hat{\mathbf{V}} = [\hat{\mathbf{x}}_1\ \hat{\mathbf{x}}_2 \ldots \hat{\mathbf{x}}_n]$, are related by $\hat{\mathbf{V}} = \mathbf{V}P$ and those in W, viz., $\mathbf{W} = [\mathbf{y}_1\ \mathbf{y}_2 \cdots \mathbf{y}_n]$ and $\hat{\mathbf{W}} = [\hat{\mathbf{y}}_1\ \hat{\mathbf{y}}_2 \ldots \hat{\mathbf{y}}_m]$, by $\hat{\mathbf{W}} = \mathbf{W}Q$ in Theorem 8.28 if the vector spaces V and W happen to be unitary spaces or vector spaces of polynomials.

The bilinear forms $\mathbf{u}^TA\mathbf{v}$ and $\hat{\mathbf{u}}^T B\hat{\mathbf{v}}$, where $B = P^TAQ$ associated with the bilinear function $b(\mathbf{x}, \mathbf{y})$ with respect to different bases, are by Definition 8.2, called *equivalent*. Therefore, there are infinitely many equivalent bilinear forms associated with a single bilinear function. This aspect reflects an identical notion of equivalence of matrices associated with a single linear function $L : V_n \to W_m$ expressed in Theorem 6.12 and discussed after Definition 6.11.

Example 8.24 Let

$$A = \begin{bmatrix} 5 & 3 \\ -2 & -2 \\ 2 & 1 \end{bmatrix}$$

be the matrix of a bilinear function $f : \mathbb{R}^3 \to \mathbb{R}^2$ with respect to the bases $X = \{\mathbf{x}_1, \mathbf{x}_2, \mathbf{x}_3\}$ and $Y = \{\mathbf{y}_1, \mathbf{y}_2\}$ given in Example 8.23. Determine new bases in $\mathbb{R}^3$ and $\mathbb{R}^2$ with respect to which the bilinear form assumes the canonical form (8.6).

Solution With the expedients given in Section 4.3, we can find matrices P and Q:

$$P = \begin{bmatrix} 0 & -1 & -2 \\ 0 & 0 & -1 \\ 1 & 2 & 4 \end{bmatrix}, \qquad Q = \begin{bmatrix} 1 & 1 \\ -1 & -2 \end{bmatrix},$$

such that

$$P^T AQ = \begin{bmatrix} 1 & 0 \\ 0 & 1 \\ 0 & 0 \end{bmatrix}$$

which is the matrix of the given bilinear function with respect to the bases in V and W given respectively by (see Remark 8.1) the basis matrices:

$$\hat{\mathbf{V}} = \mathbf{V}\mathbf{P} = \begin{bmatrix} 1 & 0 & -2 \\ -1 & 1 & 1 \\ 1 & 1 & -1 \end{bmatrix} \begin{bmatrix} 0 & -1 & -2 \\ 0 & 0 & -1 \\ 1 & 2 & 4 \end{bmatrix} = \begin{bmatrix} -2 & -5 & -10 \\ 1 & 3 & 5 \\ -1 & -3 & -7 \end{bmatrix},$$

$$\hat{\mathbf{W}} = \mathbf{W}Q = \begin{bmatrix} 1 & -1 \\ 1 & 2 \end{bmatrix} \begin{bmatrix} 1 & 1 \\ -1 & -2 \end{bmatrix} = \begin{bmatrix} 2 & 3 \\ -1 & -3 \end{bmatrix}.$$ □

We shall now specialize the notion of bilinear functions in Definition 8.17 to the case when $V = W = \mathbb{R}^n$ consisting of n-vectors, $\mathbb{F} = \mathbb{R}$ and $\mathbf{x} = \mathbf{y}$, which gives rise to the real quadratic forms described in Section 8.2. To do this, we need the concept of symmetric bilinear functions, which is contained in the following definition.

Definition 8.18 *Symmetric Bilinear Function.* A bilinear function $f : V \times V \to \mathbb{F}$: $(\mathbf{x}, \mathbf{y}) \mapsto f(\mathbf{x}, \mathbf{y})$ is said to be symmetric if $f(\mathbf{x}, \mathbf{y}) = f(\mathbf{y}, \mathbf{x})$, i.e., if the interchange of $\mathbf{x}$ and $\mathbf{y}$ does not alter the value of the function f. A symmetric bilinear function $f(\mathbf{x}, \mathbf{y})$ is defined as *a quadratic function* if $\mathbf{y} = \mathbf{x}$ and $\mathbb{F} = \mathbb{R}$. □

If we assume now that $V = W$ and $\{\mathbf{x}_1, \mathbf{x}_2, ..., \mathbf{x}_n\}$ is the basis of these spaces, then in (8.60)

$$\mathbf{y} = \beta_1\mathbf{x}_1 + \beta_2\mathbf{x}_2 + \cdots + \beta_n\mathbf{x}_n,$$

and consequently, (8.63) becomes

$$f(\mathbf{x},\mathbf{y}) = \sum_{i=1}^{n}\sum_{j=1}^{n} \alpha_i\beta_j f(\mathbf{x}_i, \mathbf{x}_j).$$

Moreover, if we suppose that f is symmetric, then

$$a_{ij} = f(\mathbf{x}_i, \mathbf{x}_j) = f(\mathbf{x}_j, \mathbf{x}_i) = a_{ji},$$

and as a result, in the associated bilinear form

$$b(\mathbf{u}, \mathbf{v}) = \mathbf{u}^T A\mathbf{v},$$

A is a symmetric matrix. Hence, a symmetric bilinear function $f(\mathbf{x}, \mathbf{y})$ can be represented by a bilinear form $b(\mathbf{u}, \mathbf{v})$ with a uniquely determined symmetric matrix of the form $A = [a_{ij}]$. In particular, a quadratic function $f(\mathbf{x}, \mathbf{y})$, $\mathbf{y} = \mathbf{x}$ can be obtained as a quadratic form $q(\mathbf{u}) = \mathbf{u}^T A\mathbf{u}$ from a uniquely defined symmetrix matrix A.

Theorem 8.29 Let $f : \mathbb{R}^n \times \mathbb{R}^n \to \mathbb{R} : (\mathbf{x}, \mathbf{y}) \mapsto f(\mathbf{x}, \mathbf{y})$ be a symmetric bilinear function and $X := \{\mathbf{x}_1, \mathbf{x}_2, ..., \mathbf{x}_n\}$ be a basis in $\mathbb{R}^n$. In terms of this basis, the quadratic function $f(\mathbf{x}, \mathbf{y})$, $\mathbf{y} = \mathbf{x}$ on $\mathbb{R}^n$ is represented by the quadratic form

$$q(\mathbf{u}) = \mathbf{u}^T A\mathbf{u},$$

where $\mathbf{u}$ is the coordinate vector of $\mathbf{x}$ with respect to X and

$$A = [a_{ij}],$$

where $a_{ij} = f(x_i, x_j)$, is a unique $n \times n$ real symmetric matrix. If $\hat{X} = \{\mathbf{x}_1, \mathbf{x}_2, ..., \mathbf{x}_n\}$ is another basis in $\mathbb{R}^n$, then the quadratic form representing $f(\mathbf{x}, \mathbf{y})$, $\mathbf{y} = \mathbf{x}$ is given by

$$\hat{q}(\hat{\mathbf{u}}) = \hat{\mathbf{u}}^T B\hat{\mathbf{u}} = q(\mathbf{u}),$$

where $B = P^T AP$ is real and symmetric, $\hat{\mathbf{u}}$ is the coordinate vector of $\mathbf{x}$ with respect to $\hat{X}$, and P is the coordinate transformation matrix in $\mathbb{R}^n$ (i.e., $\mathbf{u} = P\hat{\mathbf{u}}$); moreover, $[\hat{\mathbf{x}}_1\, \hat{\mathbf{x}}_2 \ldots \hat{\mathbf{x}}_n] = [\mathbf{x}_1\, \mathbf{x}_2 \cdots \mathbf{x}_n]P$. The quadratic forms q and $\hat{q}$ associated with a single bilinear function f are called *equivalent*, a necessary and sufficient condition for which is that A and B be congruent. □

Example 8.25 Let a bilinear function $f : \mathbb{R}^3 \times \mathbb{R}^3 \to \mathbb{R}$ be defined by

$$f(\mathbf{x},\mathbf{y}) = x_1y_1 + 2x_2y_2 - x_3y_3 - 3x_1y_3 - 3x_3y_1 + 2x_2y_3 + 2x_3y_2$$

where $\mathbf{x} = [x_1\ x_2\ x_3]^T$ and $\mathbf{y} = [y_1\ y_2\ y_3]^T$.

It is obvious that $f(\mathbf{x}, \mathbf{y}) = f(\mathbf{y}, \mathbf{x})$ and hence, f is symmetric. We now choose

$$\mathbf{x}_1 = [1\ {-1}\ 2], \quad \mathbf{x}_2 = [2\ {-1}\ 1], \quad \mathbf{x}_3 = [1\ 2\ {-1}]$$

as the basis of $\mathbb{R}^3$. Computing $a_{ij} = f(\mathbf{x}_i, \mathbf{x}_j)$, we have

$$A = \begin{bmatrix} -21 & -19 & 6 \\ -19 & -11 & 8 \\ 6 & 8 & 6 \end{bmatrix}.$$

The coordinate vector $[\alpha_1\ \alpha_2\ \alpha_3]^T$ of $\mathbf{x}$ is obtained by solving

$$[\alpha_1\ \alpha_2\ \alpha_3]\begin{bmatrix} 1 & -1 & 2 \\ 2 & -1 & 1 \\ 1 & 2 & -1 \end{bmatrix} = [x_1\ x_2\ x_3]$$

which gives

$$\mathbf{u} = \begin{bmatrix} \alpha_1 \\ \alpha_2 \\ \alpha_3 \end{bmatrix} = \frac{1}{6}\begin{bmatrix} -x_1 + 3x_2 + 5x_3 \\ 3x_1 - 3x_2 - 3x_3 \\ x_1 + 3x_2 + x_3 \end{bmatrix}.$$

It follows now that the quadratic function $f(\mathbf{x}, \mathbf{x})$ has the quadratic form $q(\mathbf{u}) = \mathbf{u}^T A\mathbf{u}$, where $\mathbf{u}$ and A are as defined above. □

Thus we have established and substantiated with an example, how a given quadratic function $f(\mathbf{x}, \mathbf{y})$, $\mathbf{y} = \mathbf{x}$ on $\mathbb{R}^n$ can be represented with respect to a chosen basis by a quadratic form $q(\mathbf{u}) = \mathbf{u}^T A\mathbf{u}$, where A is a real symmetric matrix. We shall now discuss the converse problem. That is, given a quadratic form $q(\mathbf{u})$, how we can determine a quadratic function $f(\mathbf{x}, \mathbf{y})$, $\mathbf{y} = \mathbf{x}$, whose value is given by the specified quadratic form.

We have discussed in detail in Section 8.2 that, given a quadratic form $q(\mathbf{u})$ as a polynomial in the elements $\alpha_1, \alpha_2, ..., \alpha_n$ of the n-vector $\mathbf{u}$, we can determine uniquely a real symmetric matrix B such that $q(\mathbf{u}) = \mathbf{u}^T B\mathbf{u}$. If we now choose a basis $\{\mathbf{x}_1, \mathbf{x}_2, ..., \mathbf{x}_n\}$ in $\mathbb{R}^n$, we can represent an element $\mathbf{x}$ in $\mathbb{R}^n$ by $\mathbf{x} = \alpha_1\mathbf{x}_1 + \alpha_2\mathbf{x}_2 + \cdots + \alpha_n\mathbf{x}_n$. Consequently, the quadratic function for the given quadratic form $q(\mathbf{u})$ is $f(\mathbf{x}, \mathbf{x}) = \mathbf{u}^T B\mathbf{u}$, where $\mathbf{x} = \alpha_1\mathbf{x}_1 + \alpha_2\mathbf{x}_2 + \cdots + \alpha_n\mathbf{x}_n$ and $\mathbf{u} = [\alpha_1\ \alpha_2\ \cdots\ \alpha_n]^T$.

Example 8.26 Let a quadratic form be given by

$$q(\mathbf{u}) = 2\alpha_1^2 + \alpha_2^2 - \alpha_3^2 + 2\alpha_1\alpha_2 - 4\alpha_1\alpha_3 - 2\alpha_2\alpha_3,$$

where

$$\mathbf{u} = [\alpha_1\ \alpha_2\ \alpha_3]^T.$$

Referring to Example 8.3, we can write $q(\mathbf{u})$ in the form

$$q(\mathbf{u}) = [\alpha_1\ \alpha_2\ \alpha_3]\begin{bmatrix} 2 & 1 & -2 \\ 1 & 1 & -1 \\ -2 & -1 & -1 \end{bmatrix}\begin{bmatrix} \alpha_1 \\ \alpha_2 \\ \alpha_3 \end{bmatrix} =: \mathbf{u}^T A\mathbf{u}, \tag{8.67}$$

where A is a real symmetric matrix. If we choose the set

$$\{\mathbf{x}_1 = [1\ -1\ 1]^T,\quad \mathbf{x}_2 = [2\ -1\ 2]^T,\quad \mathbf{x}_3 = [1\ -1\ 1]^T\}$$

as a basis in $\mathbb{R}^3$, then an element $\mathbf{x} = [x_1\ x_2\ x_3]^T$ in $\mathbb{R}^3$ can be written as

$$\mathbf{x} = [\mathbf{x}_1\ \mathbf{x}_2\ \mathbf{x}_3]\mathbf{u}.$$

Hence,

$$\mathbf{u} = [\mathbf{x}_1\ \mathbf{x}_2\ \mathbf{x}_3]^{-1}\mathbf{x}. \tag{8.68}$$

Substituting (8.68) into (8.67), we get

$$\mathbf{x}^T\begin{bmatrix} \mathbf{x}_1^T \\ \mathbf{x}_2^T \\ \mathbf{x}_3^T \end{bmatrix}^{-1}\begin{bmatrix} 2 & 1 & -2 \\ 1 & 1 & -1 \\ -2 & -1 & -1 \end{bmatrix}[\mathbf{x}_1\ \mathbf{x}_2\ \mathbf{x}_3]^{-1}\mathbf{x} = \mathbf{x}^T\begin{bmatrix} 2 & -4 & -3 \\ -4 & 5 & 6 \\ -3 & 6 & 5 \end{bmatrix}\mathbf{x} =: f(\mathbf{x}, \mathbf{x})$$

as the required quadratic function. □

We have seen in Section 6.4 that a single linear transformation on a vector space V can be associated with innumerable number of matrices with respect to different bases. This aggregate of matrices pertaining to a single linear transformation is classified as similar, and two matrices A and B in this class are similar iff $P^{-1}AP = B$ for a nonsingular matrix P. This is made specific in Definition 6.11. On the other hand, a bilinear function $f: \mathbb{R}^n \times \mathbb{R}^n \to \mathbb{R}$ has also been found to give rise to an innumerable number of quadratic forms, and consequently, with respect to different bases, innumerable number of matrices of these forms, any two matrices A and B in this class are called *congruent* as they are related by $B = P^TAP$, where P is a nonsingular matrix, see Definition 7.23. The quadratic forms associated with a single bilinear function are known to be *equivalent,* a necessary and sufficient condition for which is that the matrices of these forms must be congruent. Thus, while *similar matrices* are associated with a single linear function, *congruent matrices* are however associated with a single bilinear function f.

Since, for a symmetric matrix A, there exists a nonsingular matrix Q such that

$$Q^TAQ = D, \qquad \text{a diagonal matrix,} \tag{8.69}$$

so, given a symmetric bilinear function $f: \mathbb{R}^n \times \mathbb{R}^n \to \mathbb{R}$, we can find a basis in $\mathbb{R}^n$ such that the quadratic form q of f on $\mathbb{R}^n$ is diagonal. Indeed, if A is transformed by Q to the diagonal matrix D given by (8.69), then choosing $P = Q$ in Theorem 8.29, $\hat{q}(\hat{\mathbf{u}})$ becomes diagonal, which is the representation of f with respect to the basis defined by the columns in $[\mathbf{x}_1\ \mathbf{x}_2\ \cdots\ \mathbf{x}_n] = [\hat{\mathbf{x}}_1 \hat{\mathbf{x}}_2 \ldots \hat{\mathbf{x}}_n]Q$. Thus, we have established the following corollary.

Corollary 8.19 There exists a basis in $\mathbb{R}^n$ in terms of which the bilinear function $f: \mathbb{R}^n \times \mathbb{R}^n \to \mathbb{R}$ can be represented by a diagonal quadratic form. □

Example 8.27 Let a bilinear function $f: \mathbb{R}^3 \times \mathbb{R}^3 \to \mathbb{R}$ be defined by

$$\begin{aligned} f(\mathbf{x}, \mathbf{y}) = {} & x_1y_1 + 2x_1y_2 + 2x_2y_1 + 3x_2y_2 + 3x_1y_3 \\ & + 3x_3y_1 + x_2y_3 + x_3y_2 + 2x_3y_3, \end{aligned}$$

where $\mathbf{x} = [x_1\ x_2\ x_3]^T$ and $\mathbf{y} = [y_1\ y_2\ y_3]^T$ are the elements of $\mathbb{R}^3$. Determine a basis in $\mathbb{R}^3$ so that the associated quadratic form is diagonal.

Solution It is easily verified from definition that the bilinear function given above is symmetric. With respect to the natural basis $V = \{\mathbf{e}_1, \mathbf{e}_2, \mathbf{e}_3\}$ in $\mathbb{R}^3$, at $\mathbf{y} = \mathbf{x}$, f is represented uniquely by $\mathbf{u}^TA\mathbf{u}$ (here $\mathbf{x} = \mathbf{u}$), where A is the symmetric matrix $A = [a_{ij}]$ having the elements

$$a_{ij} = f(\mathbf{e}_i, \mathbf{e}j) = a_{ji}, \qquad i, j = 1, 2, 3$$

as in Theorem 8.29. This gives

$$A = \begin{bmatrix} 1 & 2 & 3 \\ 2 & 3 & 1 \\ 3 & 1 & 2 \end{bmatrix}.$$

Following Section 7.8, we can determine (see Problem 7.64(a)) the nonsingular matrix

$$P = \begin{bmatrix} 1 & 7a & -2 \\ 0 & -5a & 1 \\ 0 & a & 0 \end{bmatrix}, \qquad a = 1/3\sqrt{2}$$

such that $P^TAP = \text{diag}\ [1\ 1\ -1]$.

If we choose P to be the coordinate transformation matrix, then the coordinate vector of $\mathbf{x}$ with respect to a new basis is $\hat{\mathbf{u}} = P^{-1}\mathbf{u}$ (see Theorem 6.11), and consequently,

$$\mathbf{u}^T A\mathbf{u} = \hat{\mathbf{u}}^T P^T AP\hat{\mathbf{u}} = \hat{\alpha}_1^2 + \hat{\alpha}_2^2 - \hat{\alpha}_3^2,$$

where

$$\hat{\mathbf{u}} = [\hat{\alpha}_1\ \hat{\alpha}_2\ \hat{\alpha}_3],$$

and the columns in the matrix

$$[\hat{\mathbf{x}}_1\ \hat{\mathbf{x}}_2\ \hat{\mathbf{x}}_3] = [\mathbf{e}_1\ \mathbf{e}_2\ \mathbf{e}_3]P = P$$

form the new basis (see Remark 6.5). □

We have investigated in detail the class of bilinear functions which can be represented by quadratic forms. The next question is: What kind of bilinear functions are represented by Hermitian forms? To answer this question, we need the following definition.

Definition 8.19 *Conjugate Bilinear Function.* Let V and W be two vector spaces over a complex field $\mathbb{C}$. Then the function $f: V \times W \to \mathbb{C}: (\mathbf{x}, \mathbf{y}) \mapsto f(\mathbf{x}, \mathbf{y})$ is called *conjugate bilinear* if it satisfies the following axioms:

(a) $f(\alpha\,\mathbf{w} + \beta\mathbf{x}, \mathbf{y}) = \alpha f(\mathbf{w}, \mathbf{y}) + \beta f(\mathbf{x}, \mathbf{y})$,

(b) $f(\mathbf{x}, \gamma\mathbf{y} + \delta\,\mathbf{z}) = \overline{\gamma}\, f(\mathbf{x}, \mathbf{y}) + \overline{\delta}\, f(\mathbf{x}, \mathbf{z})$,

for arbitrary elements $\mathbf{w}, \mathbf{x}$ in V; $\mathbf{y}, \mathbf{z}$ in W, and $\alpha, \beta, \gamma, \delta$ in $\mathbb{C}$. When $V = W$, we call f *conjugate* or *Hermitian symmetric* if $f(\mathbf{x}, \mathbf{y}) = \overline{f(\mathbf{y}, \mathbf{x})}$, (see Definition 5.9). A conjugate bilinear function f: $V \times V \to \mathbb{C}$: $(\mathbf{x}, \mathbf{y}) \mapsto f(\mathbf{x}, \mathbf{y})$ with conjugate symmetry is called Hermitian if $\mathbf{y} = \mathbf{x}$. □

Observe that over a real field $\mathbb{R}$ this definition reverts to what is contained in Definitions 8.17 and 8.18. Moreover, a Hermitian function is always real because $f(\mathbf{x}, \mathbf{x}) = \overline{f(\mathbf{x}, \mathbf{x})}$.

We now concentrate on a Hermitian function $f(\mathbf{x}, \mathbf{y})$, $\mathbf{y} = \mathbf{x}$ as defined above and consider in V a basis given by

$$\{\mathbf{x}_1, \mathbf{x}_2, ..., \mathbf{x}_n\}. \tag{8.70}$$

Proceeding similarly as in deriving (8.63), we get

$$f(\mathbf{x}, \mathbf{x}) = \sum_{i=1}^{n} \sum_{j=1}^{n} \alpha_i \overline{\alpha}_j f(\mathbf{x}_i, \mathbf{x}_j).$$

We note that by Hermitian symmetry,

$$a_{ij} := f(\mathbf{x}_i, \mathbf{x}_j) = \overline{f(\mathbf{x}_j, \mathbf{x}_i)} = \overline{a}_{ji}$$

which shows that

$$A := [a_{ij}] = [\,\overline{a}_{ji}\,] = A^*,$$

and as a result,

$$f(\mathbf{x}, \mathbf{x}) = \sum_{j=1}^{n} \sum_{i=1}^{n} \overline{\alpha}_j \alpha_i \overline{a}_{ji} = \mathbf{u}^* \overline{A}\mathbf{u} = \mathbf{u}^* A^T \mathbf{u} =: h(\mathbf{u}),$$

where

$$\mathbf{u} = [\alpha_1 \ \alpha_2 \ \cdots \ \alpha_n]^T. \tag{8.71}$$

So, $h(\mathbf{u}) = \mathbf{u}^*H\mathbf{u}$, where $H = \bar{A}$ and $H^* = \bar{A} = H$ (i.e., H is Hermitian), is the Hermitian form of $f(\mathbf{x}, \mathbf{x})$. Since

$$\overline{h(\mathbf{u})} = \mathbf{u}^*H^*\mathbf{u} = \mathbf{u}^*H\mathbf{u} = \mathbf{h}(\mathbf{u}), \tag{8.72}$$

the value of the Hermitian form and that of the Hermitian function are real.

Conversely, if we are given

$$g(\mathbf{u}) = \sum_{i=1}^{n}\sum_{j=1}^{n} \alpha_i\bar{\alpha}_j a_{ji}, \tag{8.73}$$

which is real for given complex numbers a_{ji} and for all complex vectors $\mathbf{u}$ given by (8.71), then we can write it in a Hermitian form. Indeed,

$$g(\mathbf{u}) = \overline{g(\mathbf{u})} = \sum_{i=1}^{n}\sum_{j=1}^{n} \bar{\alpha}_i\alpha_j\bar{a}_{ji} = \sum_{i=1}^{n}\sum_{j=1}^{n} \alpha_i\bar{\alpha}_j\bar{a}_{ij}. \tag{8.74}$$

If we define

$$h_{ji} = \frac{1}{2}(\bar{a}_{ij} + a_{ji}),$$

then $\bar{h}_{ji} = h_{ij}$, and adding (8.73) and (8.74) and then dividing by 2, we get

$$g(\mathbf{u}) = \sum_{i=1}^{n}\sum_{j=1}^{n} \alpha_i\bar{\alpha}_j h_{ji} = \mathbf{u}^* H\mathbf{u},$$

where $H = [h_{ij}]$ is a Hermitian matrix, and hence $g(\mathbf{u})$ is in a Hermitian form. We now choose a basis (8.70) in V and define an element $\mathbf{x}$ in V by

$$\mathbf{x} = \alpha_1\mathbf{x}_1 + \alpha_2\mathbf{x}_2 + \cdots + \alpha_n\mathbf{x}_n. \tag{8.75}$$

Then, the Hermitian function associated with $g(\mathbf{u})$ is

$$\mathbf{u}^*H\mathbf{u}$$

with $\mathbf{u}$ and $\mathbf{x}$ given by (8.71) and (8.75) respectively. We next choose a new basis $\{\hat{\mathbf{x}}_1, \hat{\mathbf{x}}_2, \ldots, \hat{\mathbf{x}}_n\}$ different from (8.70), and if Q is a coordinate transformation matrix such that $\hat{\mathbf{u}} = Q^{-1}\mathbf{u}$, where $\mathbf{u}$ and $\hat{\mathbf{u}}$ are coordinate vectors of an element $\mathbf{x}$ in V with respect to old and new bases, then, instead of (8.72), the Hermitian form of the Hermitian function $f(\mathbf{x}, \mathbf{y})$, $\mathbf{y} = \mathbf{x}$ is given by

$$\mathbf{u}^* H\mathbf{u} = \hat{\mathbf{u}}^* Q^* HQ\hat{\mathbf{u}} = \hat{\mathbf{u}}^* C\hat{\mathbf{u}},$$

where $C = Q^*HQ$ and H are conjunctive matrices (see Definition 7.23). Thus, a Hermitian function with respect to different bases is represented by a set of innumerable number of Hermitian forms whose matrices are *conjunctive* with one another.

Example 8.28 Let a bilinear function $f: \mathbb{C}^3 \times \mathbb{C}^3 \to \mathbb{C}$ be defined by

$$f(\mathbf{x}, \mathbf{y}) = x_1\bar{y}_1 + 2x_3\bar{y}_3 - ix_1\bar{y}_2 + (1-i)x_1\bar{y}_3 + ix_2\bar{y}_1 + x_2\bar{y}_3 + (1+i)x_3\bar{y}_1 + x_3\bar{y}_2, \tag{8.75}$$

where $\mathbf{x} = [x_1\ x_2\ x_3]^T$ and $\mathbf{y} = [y_1\ y_2\ y_3]^T$ are elements of $\mathbb{C}^3$. Determine a basis in $\mathbb{C}^3$ so that the Hermitian form associated with $f(\mathbf{x}, \mathbf{y})$ is *canonical* (see Theorem 8.8).

Solution It is easily verified by computing $\overline{f(\mathbf{y},\mathbf{x})}$ that the bilinear function $f(\mathbf{y},\mathbf{x})$ given above is conjugate symmetric. With respect to the natural basis $\{\mathbf{e}_1, \mathbf{e}_2, \mathbf{e}_3\}$ in $\mathbb{C}^3$, at $\mathbf{y} = \mathbf{x}$, f is represented uniquely by $\mathbf{u}^T H\mathbf{u}$, where $\mathbf{u} = \mathbf{x}$, $H = \bar{A}$, the Hermitian matrix, and $A = [a_{ij}]$ is the matrix having elements

$$a_{ij} = f(e_i, e_j) = \bar{a}_{ji}.$$

Thus,

$$A = \begin{bmatrix} 1 & i & 1+i \\ -i & 0 & 1 \\ 1-i & 1 & 2 \end{bmatrix}$$

and the Hermitian form associated (8.75) with respect to natural basis is

$$h(\mathbf{u}) = \mathbf{u}^* H\mathbf{u},$$

where $H = \bar{A}$. Following the procedure given in the proof of Theorem 7.36, we apply successively the transformations (a) $H_{21}(-i)$, $H_{31}(-1-i)$; (b) $K_{21}(i)$, $K_{31}(-1+i)$; (c) $H_{32}(i)$, $K_{32}(-i)$; and (d) H_{23}, K_{23} to get

$$[H; I] = \left[\begin{array}{ccc|ccc} 1 & -i & 1-i & 1 & 0 & 0 \\ i & 0 & 1 & 0 & 1 & 0 \\ 1+i & 1 & 2 & 0 & 0 & 1 \end{array}\right]$$

$$\underset{\sim}{\text{(a)}} \left[\begin{array}{ccc|ccc} 1 & -i & 1-i & 1 & 0 & 0 \\ 0 & -1 & -i & -i & 1 & 0 \\ 0 & i & 0 & -1-i & 0 & 1 \end{array}\right]$$

$$\underset{\sim}{\text{(b)}} \left[\begin{array}{ccc|ccc} 1 & 0 & 0 & 1 & 0 & 0 \\ 0 & -1 & -i & -i & 1 & 0 \\ 0 & i & 0 & -1-i & 0 & 1 \end{array}\right]$$

$$\underset{\sim}{\text{(c)}} \left[\begin{array}{ccc|ccc} 1 & 0 & 0 & 1 & 0 & 0 \\ 0 & -1 & 0 & -i & 1 & 0 \\ 0 & 0 & 1 & -i & i & 1 \end{array}\right]$$

$$\underset{\sim}{\text{(d)}} \left[\begin{array}{ccc|ccc} 1 & 0 & 0 & 1 & 0 & 0 \\ 0 & 1 & 0 & -i & i & 1 \\ 0 & 0 & -1 & -i & 1 & 0 \end{array}\right] := [D \mid Q^*].$$

This gives (see Example 7.45)

$$Q^*HQ = \text{diag } [1, 1, -1] =: D.$$

Hence, if we choose (see Remark 8.1) the columns in

$$[\mathbf{e}_1\ \mathbf{e}_2\ \mathbf{e}_3]Q = Q = \begin{bmatrix} 1 & i & i \\ 0 & -i & 1 \\ 0 & 1 & 0 \end{bmatrix} \tag{8.76}$$

as the basis of $\mathbb{C}^3$, then the Hermitian canonical form associated with $f(\mathbf{x}, \mathbf{y})$ given in (8.75) can be described by using the transformation $\mathbf{u} = Q\hat{\mathbf{u}}$ as

$$\mathbf{u}^* H\mathbf{u} = \hat{\mathbf{u}}^* Q^* HQ\hat{\mathbf{u}} = \hat{\mathbf{u}}^* D\hat{\mathbf{u}} = a\bar{a} + b\bar{b} - c\bar{c}$$

where $\hat{\mathbf{u}} = Q^{-1}\mathbf{u} = [a\ b\ c]^T$ is the coordinate vector of $\mathbf{x}$ with respect to the basis defined by the columns in (8.76), see Theorem 6.11. □

8.9 SIMULTANEOUS DIAGONALIZATION

Definition 8.20 *Simultaneous Diagonalization* Two diagonalizable matrices A and B in $\mathbb{C}^{n\times n}$ are said to be simultaneously diagonalizable if there exists a similarity transformation $P \in \mathbb{C}^{n\times n}$ such that both $P^{-1}AP$ and $P^{-1}BP$ are diagonal. □

If $AB = BA$, then for an eigenvector $\mathbf{v}$ of B under an eigenvalue λ,

$$B(A\mathbf{v}) = A(B\mathbf{v}) = A(\lambda\mathbf{v}) = \lambda(A\mathbf{v}).$$

So, $A\mathbf{v}$ is again an eigenvector of B under the eigenvalue λ. This means that eigenspace E_λ of B is invariant under A, i.e., each vector $\mathbf{v}$ of E_λ is mapped by A to a scalar (different from λ, in general) multiple of $\mathbf{v}$ itself. Therefore, for any eigenbasis of B, the corresponding eigenvectors form an eigenbasis of A as well. This implies that the eigenspace E_λ of B is also the eigenspace of A. If both A and B are *diagonalizable*, then the above conclusion holds true even if the eigenvalues of B are of multiplicity more than one. Therefore, two diagonalizable matrices are simultaneously diagonalizable if they share a common eigenbasis.

Theorem 8.30 Two Hermitian matrices A and B in $\mathbb{C}^{n\times n}$ are simultaneously diagonalizable having the eigenvalues along the diagonal by a unitary similarity transformation if and only if they commute.

Proof Let V be a unitary matrix such that both $V^* AV = D_A$ and $V^* BV = D_B$ are diagonal where $V^*V = VV^* = I$, an identity matrix. Since $D_AD_B = D_BD_A$, a matrix property, we have using the above relations

$$V^* AVV^* BV = V^* BVV^* AV.$$

This implies $V^* ABV = V^* BAV$. Hence, we obtain $AB = BA$ after pre-multiplying and post-multiplying both sides by V and V^*.

Conversely, suppose $AB = BA$, where A and B are Hermitian matrices. Then, we can find a unitary matrix U such that $U^* AU = D_A$, a diagonal matrix with eigenvalues λ_is of A along the diagonal. We now define $C := U^* BU$. Explicitly,

$$U^* AU = D_A = \begin{bmatrix} \lambda_1 & 0 & \cdots & 0 \\ 0 & \lambda_2 & \cdots & 0 \\ \vdots & \vdots & \ddots & \vdots \\ 0 & 0 & \cdots & \lambda_n \end{bmatrix}, U^* BU = C = \begin{bmatrix} c_{11} & c_{12} & \cdots & c_{1n} \\ c_{21} & c_{22} & \cdots & c_{2n} \\ \vdots & \vdots & \ddots & \vdots \\ c_{n1} & c_{n2} & \cdots & c_{nn} \end{bmatrix}.$$

Since, $C^* = (U^* BU)^* = U^* BU = C$, so C is Hermitian. Again,

$$D_A C = (U^* AU)(U^* BU) = U^* ABU = U^* BAU = (U^* BU)(U^* AU) = CD_A.$$

Written explicitly,

$$D_A C = \begin{bmatrix} \lambda_1 c_{11} & \lambda_1 c_{12} & \cdots & \lambda_1 c_{1n} \\ \lambda_2 c_{21} & \lambda_2 c_{22} & \cdots & \lambda_2 c_{2n} \\ \vdots & \vdots & \ddots & \vdots \\ \lambda_n c_{n1} & \lambda_n c_{n2} & \cdots & \lambda_n c_{nn} \end{bmatrix}, CD_A = \begin{bmatrix} \lambda_1 c_{11} & \lambda_2 c_{12} & \cdots & \lambda_n c_{1n} \\ \lambda_1 c_{21} & \lambda_2 c_{22} & \cdots & \lambda_n c_{2n} \\ \vdots & \vdots & \ddots & \vdots \\ \lambda_1 c_{n1} & \lambda_2 c_{n2} & \cdots & \lambda_n c_{nn} \end{bmatrix}.$$

The condition $D_A C - CD_A = 0$, a null matrix, gives us $(\lambda_i - \lambda_j)c_{ij} = 0$. If all the λ_is are distinct, then we conclude that $c_{ij} = 0, i \neq j$. This gives us C a diagonal matrix. If the eigenvalues are not distinct, we follow in the following way.

Let A have k distinct eigenvalues λ_i, each of multiplicity n_i, $n_1 + \ldots + n_k = n$, and λ_is are contiguous along the diagonal in D_A. We would then conclude that $c_{ij} = 0$ if $\lambda_i \neq \lambda_j$. One can then write D_A and C in block matrix form:

$$D_A = \begin{bmatrix} \lambda_1 I_{n_1} & 0 & \cdots & 0 \\ 0 & \lambda_2 I_{n_2} & \cdots & 0 \\ \vdots & \vdots & \vdots & \vdots \\ 0 & 0 & \cdots & \lambda_k I_{n_k} \end{bmatrix}, C = \begin{bmatrix} C_1 & 0 & \cdots & 0 \\ 0 & C_2 & \cdots & 0 \\ \vdots & \vdots & \vdots & \vdots \\ 0 & 0 & \cdots & C_k \end{bmatrix} \tag{8.77}$$

where C_i is of order n_i. Since C_is are Hermitian, they can be diagonalized by unitary similarity transformations V_is for which $V_i^* C_i V_i$ is diagonal. Thus, we can form:

$$V = \begin{bmatrix} V_1 & 0 & \cdots & 0 \\ 0 & V_2 & \cdots & 0 \\ \vdots & \vdots & \ddots & \vdots \\ 0 & 0 & \cdots & V_k \end{bmatrix},$$

such that $V^*CV = D_B$ is diagonal. It is easily seen that $V^*D_AV = D_A$. Hence, we have succeeded in finding an invertible matrix $Q = UV$ such that $Q^* AQ = D_A$ and $Q^* BQ = D_B$. □

The above proof can be extended to establish a stronger result that two diagonalizable matrices are simultaneously diagonalizable if and only if they commute. The above generalization is based on the following Lemma.

Lemma 8.4 Let $A \in \mathbb{C}^{n\times n}$ and $B \in \mathbb{C}^{m\times m}$ be two matrices such that

$$C = \begin{bmatrix} A & 0 \\ 0 & B \end{bmatrix}$$

Then C is diagonalizable if and only if both A and B are diagonalizable.

Proof If both A and B are diagonalizable, then there exist nonsingular matrices $P_1 \in \mathbb{C}^{n\times n}$ and $P_2 \in \mathbb{C}^{m\times m}$ such that both $P_1^{-1}AP_1 = D_A$ and $P_2^{-1}BP_2 = D_B$ are diagonal matrices. Define $P = \text{diag}[P_1, P_2]$. Now, $P^{-1}CP$ is

$$\begin{bmatrix} P_1 & 0 \\ 0 & P_2 \end{bmatrix}^{-1} \begin{bmatrix} A & 0 \\ 0 & B \end{bmatrix} \begin{bmatrix} P_1 & 0 \\ 0 & P_2 \end{bmatrix} = \begin{bmatrix} P_1^{-1}AP_1 & 0 \\ 0 & P_2^{-1}BP_2 \end{bmatrix} = \begin{bmatrix} D_A & 0 \\ 0 & D_B \end{bmatrix},$$

a diagonal matrix.

Conversely, if C is diagonalizable, then there exist a non-singular matrix $P \in \mathbb{C}^{(n+m)\times(n+m)}$ such that $P^{-1}CP = \text{diag}[\lambda_1, \lambda_2, \ldots, \lambda_{n+m}] = D_C$ is diagonal. If we write $P = [\mathbf{p}_1\ \mathbf{p}_2 \ldots \mathbf{p}_{n+m}]$ with ith column as

$$\mathbf{p}_i = \begin{bmatrix} \mathbf{x}_i \\ \mathbf{y}_i \end{bmatrix} \in \mathbb{C}^{n+m},\ \mathbf{x}_i \in \mathbb{C}^n,\ \mathbf{y}_i \in \mathbb{C}^m,$$

then from $CP = PD_C$,

$$\begin{bmatrix} A & 0 \\ 0 & B \end{bmatrix} \begin{bmatrix} \mathbf{x}_i \\ \mathbf{y}_i \end{bmatrix} = \begin{bmatrix} A\mathbf{x}_i \\ B\mathbf{y}_i \end{bmatrix} = \lambda_i \begin{bmatrix} \mathbf{x}_i \\ \mathbf{y}_i \end{bmatrix}.$$

Hence, $A\mathbf{x}_i = \lambda_i\mathbf{x}_i$ and $B\mathbf{y}_i = \lambda_i\mathbf{y}_i$, $\forall i = 1, 2, \ldots, n + m$. If there were fewer than n number of linearly independent vectors in the set $\{\mathbf{x}_1, \mathbf{x}_2, \cdots, \mathbf{x}_{n+m}\} \in \mathbb{C}^n$ then the column rank, and hence the row rank of the matrix

$$[\mathbf{x}_1\ \mathbf{x}_2 \cdots \mathbf{x}_{n+m}]_{n\times(n+m)}$$

would be less than n. By a similar argument, if there were fewer than m number of linearly independent vectors in the set $\{\mathbf{y}_1, \mathbf{y}_2, \cdots, \mathbf{y}_{n+m}\} \in \mathbb{C}^m$ then the column rank, and hence the row rank of the matrix

$$[\mathbf{y}_1\ \mathbf{y}_2 \cdots \mathbf{y}_{n+m}]_{m\times(n-m)}$$

would be less than m. In either event or both, the matrix

$$P = [\mathbf{p}_1 \mathbf{p}_2 \cdots \mathbf{p}_{n+m}] = \begin{bmatrix} \mathbf{x}_1 & \mathbf{x}_2 & \cdots & \mathbf{x}_{n+m} \\ \mathbf{y}_1 & \mathbf{y}_2 & \cdots & \mathbf{y}_{n+m} \end{bmatrix}_{n+m}$$

would have row rank and hence rank less than $n + m$. This is impossible as P is invertible. Thus, there are exactly n linearly independent vectors in the set $\{\mathbf{x}_1, \mathbf{x}_2, \cdots, \mathbf{x}_{n+m}\} \in \mathbb{C}^n$ and since each of these is an eigenvector of A, the matrix A must be diagonalizable. By similar reasoning the matrix B is also diagonalizable. □

Theorem 8.31 Two diagonalizable matrices A and B in $\mathbb{C}^{n\times n}$ are simultaneously diagonalizable having the eigenvalues along the diagonal by a similarity transformation if and only if they commute.

Proof The proof follows the same steps as given for Hermitian matrices. However, as although B is diagonalizable which implies that C in (8.77) is diagonalizable, by Lemma 8.4 each of C_is is diagonalizable. This completes the proof. □

Example 8.29 Find a similarity transformation for the simultaneous diagonalization of the diagonalizable matrices A and B given by

$$A = \begin{bmatrix} 0 & 4 & 0 & 0 \\ 1 & 0 & 0 & 0 \\ 0 & 0 & 0 & 4 \\ 0 & 0 & 1 & 0 \end{bmatrix} \text{ and } B = \begin{bmatrix} 0 & 0 & 1 & 0 \\ 0 & 0 & 0 & 1 \\ 1 & 0 & 0 & 0 \\ 0 & 1 & 0 & 0 \end{bmatrix}.$$

Solution The matrices A and B commute as by direct multiplication

$$AB = \begin{bmatrix} 0 & 0 & 0 & 4 \\ 0 & 0 & 1 & 0 \\ 0 & 4 & 0 & 0 \\ 1 & 0 & 0 & 0 \end{bmatrix} = BA.$$

They can be written in a block form as

$$A = \begin{bmatrix} A_1 & 0 \\ 0 & A_2 \end{bmatrix}, \text{ where } A_1 = \begin{bmatrix} 0 & 4 \\ 1 & 0 \end{bmatrix} = A_2;$$

$$B = \begin{bmatrix} 0 & B_1 \\ B_2 & 0 \end{bmatrix}, \text{ where } B_1 = \begin{bmatrix} 1 & 0 \\ 0 & 1 \end{bmatrix} = B_2.$$

Step 1. *Diagonalize* A. The eigenvalues of A_1 or A_2 are $\lambda_1 = 2$ and $\lambda_2 = -2$. Therefore, the eigenvalues of A are the combined eigenvalues A_1 and A_2 and are $\lambda_1 = 2$ of multiplicity 2 and $\lambda_2 = -2$ again of multiplicity 2. The eigenvector of A_1 or A_2 corresponding to $\lambda_1 = 2$ is $[2\ 1]^T$ and consequently, the eigenvector of A corresponding to $\lambda_1 = 2$ of A_1 is $\mathbf{u}_1 = [2\ 1\ 0\ 0]^T$ and another eigenvector of A corresponding to $\lambda_1 = 2$ of A_2 is $\mathbf{u}_2 = [0\ 0\ 2\ 1]^T$. Similarly, the eigenvectors of A corresponding to $\lambda_2 = -2$ are $\mathbf{u}_3 = [2\ -1\ 0\ 0]^T$ and $\mathbf{u}_4 = [0\ 0\ 2\ -1]^T$. With respect to the basis $\{\mathbf{u}_1,\ldots,\mathbf{u}_4\}$, the matrix A is diagonal. If we define $U = [\mathbf{u}_1 \cdots \mathbf{u}_4]$, then

$$A[\mathbf{u}_1 \cdots \mathbf{u}_4] = [\mathbf{u}_1 \cdots \mathbf{u}_4]\begin{bmatrix} 2 & 0 & 0 & 0 \\ 0 & 2 & 0 & 0 \\ 0 & 0 & -2 & 0 \\ 0 & 0 & 0 & -2 \end{bmatrix},$$

i.e, $U^{-1}AU = \text{diag}[2, 2, -2, -2]$.

Step 2. *Compute B with respect to the basis* $\{\mathbf{u}_1,\ldots,\mathbf{u}_4\}$. Since, $B\mathbf{u}_1 = \mathbf{u}_2$, $B\mathbf{u}_2 = \mathbf{u}_1$, $B\mathbf{u}_3 = \mathbf{u}_4$ and $B\mathbf{u}_4 = \mathbf{u}_3$,

$$B[\mathbf{u}_1 \cdots \mathbf{u}_4] = [\mathbf{u}_1 \cdots \mathbf{u}_4]\begin{bmatrix} 0 & 1 & 0 & 0 \\ 1 & 0 & 0 & 0 \\ 0 & 0 & 0 & 1 \\ 0 & 0 & 1 & 0 \end{bmatrix}.$$

That is, the matrix of B with respect to the basis $\{\mathbf{u}_1,\ldots\mathbf{u}_4\}$ is the block diagonal matrix

$$C = \begin{bmatrix} 0 & 1 & 0 & 0 \\ 1 & 0 & 0 & 0 \\ 0 & 0 & 0 & 1 \\ 0 & 0 & 1 & 0 \end{bmatrix} = \begin{bmatrix} C_1 & 0 \\ 0 & C_2 \end{bmatrix}, C_1 = C_2 = \begin{bmatrix} 0 & 1 \\ 1 & 0 \end{bmatrix}.$$

Step 3. *Diagonalize the block diagonal matrix* $C = \text{diag}[C_1, C_2]$. The eigenvalues of both of C_1 and C_2 are $\lambda_1 = 1$ and $\lambda_2 = -1$, and the corresponding eigenvectors are $[1\ 1]^T$ and $[1\ -1]^T$, respectively. Therefore, the eigenvalues of C are $\{1, -1, 1, -1\}$ and the corresponding eigenvectors are $\mathbf{v}_1 = [1\ 1\ 0\ 0]^T$, $\mathbf{v}_2 = [1\ -1\ 0\ 0]^T$, $\mathbf{v}_3 = [0\ 0\ 1\ 1]^T$ and $\mathbf{v}_4 = [0\ 0\ 1\ -1]^T$, respectively, with respect to the basis $\{\mathbf{u}_1,\ldots,\mathbf{u}_4\}$. With respect to the basis $\{\mathbf{v}_1,\ldots,\mathbf{v}_4\}$, the matrix C is diagonal. If we define $V = [\mathbf{v}_1 \cdots \mathbf{v}_4]$, then

$$C[\mathbf{v}_1 \cdots \mathbf{v}_4] = [\mathbf{v}_1 \cdots \mathbf{v}_4]\begin{bmatrix} 1 & 0 & 0 & 0 \\ 0 & -1 & 0 & 0 \\ 0 & 0 & 1 & 0 \\ 0 & 0 & 0 & -1 \end{bmatrix},$$

i.e., $V^{-1}CV = \text{diag}[1, -1, 1, -1]$.

Alternatively, the eigenvalues for each of C_i are ± 1 and the corresponding eigenvectors are $[1\ 1]^T$ and $[1\ -1]^T$. Therefore,

$$V_i^{-1}C_iV_i = \frac{1}{2}\begin{bmatrix} 1 & 1 \\ 1 & -1 \end{bmatrix}\begin{bmatrix} 0 & 1 \\ 1 & 0 \end{bmatrix}\begin{bmatrix} 1 & 1 \\ 1 & -1 \end{bmatrix} = \begin{bmatrix} 1 & 0 \\ 0 & -1 \end{bmatrix}, V_i = \begin{bmatrix} 1 & 1 \\ 1 & -1 \end{bmatrix}$$

for $i = 1, 2$. Consequently, $V^{-1}CV = \text{diag}[1, -1, 1, -1]$, where $V = \text{diag}[V_1, V_2]$.

Step 4. *Revert to old basis*. The vectors $\{\mathbf{v}_1,\ldots,\mathbf{v}_4\}$ are with respect to the basis $\{\mathbf{u}_1,\ldots,\mathbf{u}_4\}$. With respect to the original basis they are

$$\tilde{\mathbf{v}}_1 = \mathbf{u}_1 + \mathbf{u}_2,\ \tilde{\mathbf{v}}_2 = \mathbf{u}_1 - \mathbf{u}_2,\ \tilde{\mathbf{v}}_3 = \mathbf{u}_3 + \mathbf{u}_4, \text{ and } \tilde{\mathbf{v}}_4 = \mathbf{u}_3 - \mathbf{u}_4.$$

Above, in $\mathbf{v}_1 = [1\ 1\ 0\ 0]^T$, the elements 1, 1, 0 and 0 are associated with $\mathbf{u}_1$, $\mathbf{u}_2$, $\mathbf{u}_3$, and $\mathbf{u}_4$, respectively, and hence, $\tilde{\mathbf{v}}_1 = \mathbf{u}_1 + \mathbf{u}_2 + 0\cdot\mathbf{u}_3 + 0\cdot\mathbf{u}_4$, and so on. Hence, the similarity transformation to diagonalize A and B simultaneously is,

$$Q = [\tilde{\mathbf{v}}_1\ \tilde{\mathbf{v}}_2\ \tilde{\mathbf{v}}_3\ \tilde{\mathbf{v}}_4] = \begin{bmatrix} 2 & 2 & 2 & 2 \\ 1 & 1 & -1 & -1 \\ 2 & -2 & 2 & -2 \\ 1 & -1 & -1 & 1 \end{bmatrix}.$$

Alternatively, the similarity transformation is $Q = UV$ and we get the matrix as given above.

It can be verified that $Q^{-1}AQ = \text{diag}[2, 2, -2, -2]$ and $Q^{-1}BQ = \text{diag}\ [1, -1, 1, -1]$. Note the order of eigenvalues along the diagonal in both of A and B and that the vectors $\{\tilde{\mathbf{v}}_1, \tilde{\mathbf{v}}_2, \tilde{\mathbf{v}}_3, \tilde{\mathbf{v}}_4\}$ generate the eigenspace of both A and B. □

PROBLEMS

8.1 Determine the matrix A of the following bilinear forms $b(\mathbf{x}, \mathbf{y})$:

(a) $2x_1y_1 + x_1y_2 - 2x_2y_1 + 3x_2y_2 - 2x_1y_3$,
(b) $x_1y_1 + 2x_1y_2 + x_1y_3 + 3x_2y_1 - 2x_2y_2 + x_3y_2 - 3x_3y_3$,
(c) $-2x_1y_1 - x_1y_2 + 2x_2y_1 - x_3y_1 + 3x_3y_2$,
(d) $3x_1y_1 + x_1y_2 + x_2y_1 - 2x_2y_2 - 4x_2y_3 - 4x_3y_2 + 3x_3y_3$.

Which of the above forms is symmetric?

8.2 Determine the transformation matrices P and Q so that the bilinear forms $\mathbf{x}^T A\mathbf{y}$ given in Problem 8.1 are equivalent to a canonical form.

8.3 Show that the equivalence relation in bilinear forms as defined in Definition 8.2 is reflexive, symmetric and transitive.

8.4 Prove that the congruence of two matrices as given in Definition 8.3 is a true equivalence relation.

8.5 Determine the co-gredient transformations $\mathbf{x} = B\mathbf{u}$ and $\mathbf{y} = B\mathbf{v}$ such that the following bilinear forms $\mathbf{x}^T A\mathbf{x}$ are congruent to $\mathbf{u}^T D\mathbf{v}$, where D is diagonal:

(a) $x_1y_1 + x_2y_2 + 3x_3y_3 + 2x_1y_2 + 2x_2y_1$
(b) $4x_1y_1 + 5(x_2y_2 + x_3y_3)/2 - 3x_2y_3/2 - 3x_3y_2/2$
(c) $8x_1y_1 + 12x_1y_3 - 2x_2y_2 + 12x_3y_1 - 2x_3y_3$

8.6 Determine the matrix B such that A is congruent to a diagonal matrix D, where A and D are given respectively by

(a) $\begin{bmatrix} 1 & 0 & -4 \\ 0 & 5 & 4 \\ -4 & 4 & 3 \end{bmatrix}$, diag [3, −3, 9];

(b) $\begin{bmatrix} 1 & 1 & 1 \\ -1 & 1 & 1 \\ -1 & -1 & 1 \end{bmatrix}$, diag $[1, \alpha, \bar{\alpha}]$, $\alpha = (1 + \sqrt{3}i)/2$.

8.7 Show that the bilinear form $\mathbf{x}^T\mathbf{y} = x_1y_1 + x_2y_2 + \cdots + x_ny_n$ is transformed into itself iff $\mathbf{x}$ and $\mathbf{y}$ are transformed contragrediently.

8.8 (a) Write down the quadratic form associated with the matrix

$$B = \begin{bmatrix} 2 & 1 & 2 \\ -3 & -3 & -1 \\ 4 & 1 & 3 \end{bmatrix}.$$

(b) Rewrite the matrix A of the form so that it is symmetric.
(c) What is the discriminant of the form?
(d) Does it depend on the matrix of the form?
(e) Give an example.

8.9 (i) Find the quadratic forms associated with A given in Problem 8.6. (ii) Determine the quadratic forms which are equivalent to those obtained in (i) but having no cross-terms.

8.10 Find an orthogonal transformation $\mathbf{x} = P\mathbf{y}$ to reduce the following quadratic forms $\mathbf{x}^TA\mathbf{x}$, to a diagonal form $\mathbf{y}^TD\mathbf{y}$, where the diagonal elements of D are the eigenvalues *of* A:

(a) $\begin{bmatrix} 5 & -2 & 0 \\ -2 & 6 & 2 \\ 0 & 2 & 7 \end{bmatrix}$, (b) $\begin{bmatrix} 3 & 1 & 0 & 0 \\ 1 & 3 & 0 & 0 \\ 0 & 0 & 3 & -1 \\ 0 & 0 & -1 & 3 \end{bmatrix}$, (c) $\begin{bmatrix} 3 & -1 & 0 \\ -1 & 3 & 0 \\ 0 & 0 & 4 \end{bmatrix}$.

8.11 Using Lagrange reduction, find a nonsingular transformation $\mathbf{y} = P\mathbf{x}$ which reduces the quadratic forms given in Problem 8.10 to diagonal. Find also the diagonal form in each case.

8.12 Using Lagrange reduction, transform to diagonal forms the following quadratic forms. Find the transformation matrix P such that $\mathbf{z} = P\mathbf{x}$ and also the diagonal form in each case:

(a) $x_1x_2 + x_2x_3 + x_3x_1$,

(b) $4x_1^2 + x_2^2 + 9x_3^2 - 4x_1x_2 + 12x_1x_3$,

(c) $(x_1 + x_2 + x_3)x_2$.

8.13 Prove that there does not exist a nonsingular transformation P which is upper triangular to reduce the quadratic forms in Problem 8.12 in a diagonal form.

8.14 Using Jacobi formula and its alternative form, reduce to diagonal forms the quadratic forms given in Problem 8.10. In each case, determine the upper triangular transformation matrix P.

8.15 Let

$$A_i(\mathbf{x}) = \sum_{j=1}^{n} a_{ij}x_j, \qquad i = 1, 2, \ldots, n;$$

$$\Delta_0 = 1, \qquad \Delta_1 = a_{11},$$

$$\Delta_j = \begin{bmatrix} a_{11} & a_{12} & \cdots & a_{1j} \\ a_{21} & a_{22} & \cdots & a_{2j} \\ \vdots & \vdots & & \vdots \\ a_{j1} & a_{j2} & \cdots & a_{jj} \end{bmatrix},$$

$$y_1 = A_1(\mathbf{x}),$$

$$y_j = \begin{bmatrix} a_{11} & \cdots & a_{1\,j-1} & A_1(\mathbf{x}) \\ a_{21} & \cdots & a_{2\,j-1} & A_2(\mathbf{x}) \\ \vdots & & \vdots & \vdots \\ a_{j1} & & a_{j\,j-1} & A_j(\mathbf{x}) \end{bmatrix} \qquad j = 2, \ldots, n;$$

and let the rank of the matrix $A = [a_{ij}]$ be r. Then prove Jacobi formula

$$q = \sum_{i,j=1}^{n} a_{ij}x_ix_j = \sum_{j=1}^{r} \frac{y_j^2}{\Delta_j \Delta_{j-1}},$$

where the linear forms y_j's (j = 1, 2, ..., r) are linearly independent.

8.16 Determine the rank, index and signature of the quadratic forms given in Problem 8.12.

8.17 Find a nonsingular transformation $\mathbf{y} = P\mathbf{x}$ which will reduce the quadratic forms in Problem 8.12 to the canonical form of Corollary 8.1.

8.18 Determine whether or not the quadratic forms $q = x_1^2 + x_2^2 + x_3^2$ and $p = y_1^2 - y_2^2 - y_3^2$ are equivalent. If yes, find the transformation which reduces p to q.

8.19 Test whether or not the quadratic forms p and q are equivalent over the real field $\mathbb{R}$.

(a) $p = x_1^2 - 2x_1x_2 + 3x_2^2;$ $\qquad q = x_1x_2 - x_2^2.$

(b) $p = x_1x_2 + x_2x_3 + x_3x_1;$ $\qquad q = 4x_1^2 + x_2^2 + 8x_1x_2 + 2x_1x_3 + 2x_2x_3.$

8.20 Let

$$H(\mathbf{x}, \mathbf{y}) = \begin{bmatrix} \mathbf{x} \\ \mathbf{y} \end{bmatrix}^* H \begin{bmatrix} \mathbf{x} \\ \mathbf{y} \end{bmatrix},$$

where

$$H = \frac{1}{2}\begin{bmatrix} A & B \\ B* & C \end{bmatrix}$$

is a Hermitian matrix, A and C being Hermitian matrices of order n and m respectively, and **x** and **y** are n- and m-vectors. Show that (a) $H(\mathbf{x}, \mathbf{y}) = H(\mathbf{u}, \mathbf{\theta}) + H(\mathbf{v}, \mathbf{y}) + \text{Re } \mathbf{u}^*(A\mathbf{v} + B\mathbf{y})$ if $\mathbf{x} = \mathbf{u} + \mathbf{v}$; and (b) $H(A_{j\omega}^{-1}\mathbf{b}y, y) = -f(j\omega) \mid y \mid^2$, where

$$A_\lambda = (\lambda I_n - A), \quad f(\lambda) = [\delta(\lambda)\delta^*(\lambda)]^{-1}\mathbf{r}^*(\lambda)H\mathbf{r}(\lambda),$$

$$\mathbf{r}(\lambda) = \begin{bmatrix} B(\lambda)\mathbf{b} \\ \delta(\lambda) \end{bmatrix}, \qquad B(\lambda) = \text{adj}\, A_\lambda, \qquad \delta(\lambda) = \det A_\lambda,$$

b, a constant $n \times 1$ vector, $m = 1$, $j = \sqrt{-1}$ and y is a scalar.

8.21 Determine whether or not the following Hermitian forms are equivalent:

(a) $2\bar{x}_1x_1 + 3i\,\bar{x}_2x_1 - 3i\,\bar{x}_1x_2 - \bar{x}_2x_2$; $\quad (2+i)\bar{x}_2x_1 + (2-i)\bar{x}_1x_2$.

(b) $\bar{x}_1x_1 + 2\bar{x}_2x_2 + (1+2i)\bar{x}_1x_2(1-2i)\bar{x}_2x_2$; $\quad i\bar{x}_1x_2 - i\bar{x}_2x_1$.

(c) $i(\bar{x}_1x_2 + \bar{x}_1x_3 + \bar{x}_2x_3 - \bar{x}_2x_1 - \bar{x}_3x_1 - \bar{x}_3x_2)$;

$\bar{x}_1x_2 + (1+i)\bar{x}_1x_2 + (1-i)\bar{x}_2x_3 + 2\bar{x}_3x_3$.

8.22 Using Lagrange reduction or otherwise, determine a linear transformation $\mathbf{y} = P\mathbf{x}$ to reduce to diagonal forms the Hermitian forms in Problem 8.21.

8.23 Justify that:

(a) $x_1^2 + 2x_3^2$ is PD in $\mathbb{R}^2$ but PSD in $\mathbb{R}^3$;

(b) $\bar{x}_1x_1 + \bar{x}_2x_2$ is PD in $\mathbb{C}^2$ but PSD in $\mathbb{C}^3$.

8.24 Applying Theorem 8.11, determine the definiteness of the following quadratic forms in $\mathbb{R}^3$:

(a) $x_1x_2 + x_1x_3 + x_2x_3$,

(b) $4x_1^2 + x_2^2 + 9x_3^2 - 4x_1x_2 + 12x_1x_3$,

(c) $(x_1 + x_2 + x_3)x_2$,

(d) $x_1^2 + 3x_2^2 + 6x_3^2 - 2x_1x_2 + 4x_1x_3$,

(e) $2x_1^2 + 4x_2^2 + 8x_3^2 + 2x_1x_2 - 2x_1x_3 - 2x_2x_3$,

(f) $-2(x_1^2 + x_2^2 + x_3^2 - 2x_2x_3)$,

(g) $-x_1^2 - 2x_2^2 - 2x_3^2 + 2x_1x_2 + 2x_2x_3$,

(h) $-2x_1^2 - 3x_2^2$.

8.25 Applying Theorem 8.11, determine the definiteness of the following Hermitian forms: $\mathbf{x}^*H\mathbf{x}$ in $\mathbb{C}^3$ if H is

(a) $\begin{bmatrix} 0 & -i & -i \\ i & 0 & -i \\ i & i & 0 \end{bmatrix}$, (b) $\begin{bmatrix} -2 & 1+2i & 0 \\ 1-2i & -4 & 0 \\ 0 & 0 & 0 \end{bmatrix}$,

(c) $\begin{bmatrix} 2 & 1-i & 2 \\ 1+i & 2 & 1+i \\ 2 & 1-i & 2 \end{bmatrix}$, (d) $\begin{bmatrix} 2 & 1-i & i \\ 1+i & 4 & 2+i \\ -i & 2-i & 8 \end{bmatrix}$,

(e) $\begin{bmatrix} -2 & 1-i & i \\ 1+i & -2 & 2+i \\ -i & 2-i & -6 \end{bmatrix}$.

8.26 Apply Sylvester's criterion to determine which of the forms in Problem 8.24 are (i) PD and (ii) ND.

8.27 Apply Sylvester's criterion to determine which of the Hermitian forms in Problem 8.25 are (i) PD and (ii) ND.

8.28 Applying Theorem 8.19, test the definiteness of the quadratic forms in Problem 8.24.

8.29 Applying Corollary 8.12, test the definiteness of the Hermitian forms in Problem 8.25.

8.30 Determine the value of a for which the following matrices are negative definite:

(a) $\begin{bmatrix} a & 1 & 2 \\ 1 & a & 3 \\ 2 & 3 & a \end{bmatrix}$, (b) $\begin{bmatrix} -2 & 1+i & i \\ 1-i & a & 1 \\ -i & 1 & -3 \end{bmatrix}$.

8.31 If A is a positive definite Hermitian matrix, then show that there exists a nonsingular matrix P such that $P^*AP = I$.

8.32 If $\mathbf{x}^*H\mathbf{x}$ is PSD, then every principal minor determinant of H is ≥ 0.

8.33 Test for definiteness of the Hermitian matrix

$$\begin{bmatrix} 1 & \omega & \omega^2 \\ \omega^2 & 1 & \omega \\ \omega & \omega^2 & 1 \end{bmatrix},$$

where $\omega^3 = 1$, $\omega \neq 1$.

8.34 A real symmetric matrix is PD iff one of the following statements holds:

(a) All its characteristic roots are positive;

(b) A^{-1} exists and is PD.

8.35 If $F(\mathbf{x}) = \mathbf{x}^*H\mathbf{x}$ is a Hermitian form, then $F(\mathbf{x}_0 e^{j\omega}) = F(\mathbf{x}_0)$, where ω is real and $j = \sqrt{-1}$.

8.36 (a) When A is PSD, $\mathbf{x}^T A\mathbf{x} = 0$ iff $A\mathbf{x} = \boldsymbol{\theta}$

(b) When A is PD, $A\mathbf{x} = \boldsymbol{\theta}$ iff $\mathbf{x} = \boldsymbol{\theta}$.

8.37 A real symmetric matrix A (resp a Hermitian matrix H) is PD iff A^p (resp H^p) is PD for every integer p.

8.38 A real symmetric matrix A of rank r is PSD iff there exists a matrix P of rank r such that $A = P^T P$.

8.39 If A is real symmetric, $\mathbf{x} = \mathbf{y} + j\mathbf{z}$, $j = \sqrt{-1}$ is complex, and $\mathbf{y}$ and $\mathbf{z}$ are real, then $\mathbf{x}^*A\mathbf{x} = \mathbf{y}^T A\mathbf{y} + \mathbf{z}^T A\mathbf{z}$.

8.40 Let A be an $n \times m$ matrix, $n \neq m$. Show that $A^* A$ is PSD if $m > n$ and AA^* is PSD if $n > m$.

8.41 If A is an n-square singular matrix, then show that $A^* A$ is PSD.

8.42 Show that a real symmetric or a Hermitian matrix $P(\omega)$ whose elements are continuous functions of ω is PD iff $P(\infty) > 0$ and $\det P(\omega) \neq 0$ $(-\infty < \omega < \infty)$.

8.43 If A is symmetric or Hermitian and has characteristic roots (distinct or not) in the negative half-plane, then (a) $(I - e^{AT})$ and (b) $A^{-1}(e^{AT} - I)$, $T > 0$ are PD.

8.44 If Q is PD, there exists no nonzero vector $\mathbf{u}$ such that $Q\mathbf{u} = \boldsymbol{\theta}$.

8.45 The rank of two matrices A and B are the same if one of the following holds:

(a) A is similar to B;

(b) A is equivalent to B;

(c) A is congruent to B.

8.46 Determine the eigenvalues and eigenvectors of the matrix pair (A, B), and hence, a nonsingular matrix P such that $P^*AP = I$ and $P^*BP =$ diagonal.

(a) $A = \begin{bmatrix} 7 & -2 & 1 \\ -2 & 10 & -2 \\ 1 & -2 & 7 \end{bmatrix}$; $\quad B = \begin{bmatrix} 3 & -1 & 1 \\ -1 & 5 & -1 \\ 1 & -1 & 3 \end{bmatrix}$.

(b) $A = \begin{bmatrix} 2 & 2 & 2 \\ 2 & 5 & 4 \\ 2 & 4 & 5 \end{bmatrix}$; $\quad B = \begin{bmatrix} 7 & -4 & -4 \\ -4 & 1 & -8 \\ -4 & -8 & 1 \end{bmatrix}$.

(c) $A = \begin{bmatrix} 2 & 3 & 1 \\ 3 & 9 & -1 \\ 1 & -1 & 2 \end{bmatrix}$; $\quad B = \begin{bmatrix} 0 & -1 & 1 \\ -1 & 1 & -1 \\ 1 & -1 & 2 \end{bmatrix}$.

8.47 Prove that the eigenvalues of (A, B), where A and B are Hermitian with A positive definite, are real.

8.48 Determine a nonsingular transformation P which transforms one of the quadratic forms to $u^2 + v^2 + w^2$ and the other to $au^2 + bv^2 + cw^2$, where a, b, c are suitable constants.

(a) $x^2 + 2y^2 + 3z^2$; $\quad -4xz + 4yz$.

(b) $x^2 + 2y^2 + 4z^2 + 2xz + 2yz$; $\quad 2xz - 2yz$.

8.49 For the bilinear functions $f: \mathbb{R}^3 \times \mathbb{R}^3 \to \mathbb{R}$ given below, determine a basis in $\mathbb{R}^3$ so that the associated quadratic form is canonical (see Corollary 8.1):

(a) $f(\mathbf{x}, \mathbf{y}) = x_1y_1 - x_1y_2 + x_1y_3 - x_2y_1 + x_2y_2 - x_2y_3 + x_3y_1 - x_3y_2 + x_3y_3$,

(b) $f(\mathbf{x}, \mathbf{y}) = x_1(y_2 + y_3) + x_2(y_1 + y_3) + x_3(y_1 + y_2)$,

(c) $f(\mathbf{x}, \mathbf{y}) = x_1(2y_1 + y_2 - y_3) + x_2(y_1 + 4y_2 - y_3) - x_3(y_1 + y_2 - 8y_3)$,

where $\mathbf{x} = [x_1\ x_2\ x_3]^T$ and $\mathbf{y} = [y_1\ y_2\ y_3]^T$.

8.50 For the bilinear functions $f: \mathbb{C}^3 \times \mathbb{C}^3 \to \mathbb{C}$ given below, determine a basis in $\mathbb{C}^3$ such that the associated Hermitian form is canonical:

(a) $f(\mathbf{x},\mathbf{y}) = ix_1(\bar{y}_2 + \bar{y}_3) + ix_2(\bar{y}_3 - \bar{y}_1) - ix_3(\bar{y}_1 + \bar{y}_2)$,

(b) $f(\mathbf{x},\mathbf{y}) = x_1\bar{y}_1 + 2x_2\bar{y}_2 + (1+2i)x_2\bar{y}_1 + (1-2i)x_1\bar{y}_2 + x_3\bar{y}_3$,

(c) $f(\mathbf{x},\mathbf{y}) = x_1(\bar{y}_1 - i\bar{y}_2 + \bar{y}_3) + x_2(i\bar{y}_1 + 2\bar{y}_2 - i\bar{y}_3) + x_3(\bar{y}_1 + i\bar{y}_2 + \bar{y}_3)$,

where $i = \sqrt{-1}$, $\mathbf{x} = [x_1\ x_2\ x_3]^T$ and $\mathbf{y} = [y_1\ y_2\ y_3]^T$.

8.51 Let A and B be PD such that $\mathbf{x}^*A\mathbf{x} = \mathbf{x}^*B\mathbf{x}$ for all $\mathbf{x} \in \mathbb{C}^n$. Prove that $A = B$.

8.52 Prove that if an $n \times n$ matrix P is PD then $aI_n \le P \le bI_n$ and $\frac{1}{b}I_n \le P^{-1} \le \frac{1}{a}I_n$, where $a = \lambda_{\min}(P)$ and $b = \lambda_{\max}(P)$.

8.53 Find a similarity transformation for the simultaneous diagonalization of the diagonalizable matrices A and B given by:

(a) $A = \begin{bmatrix} 2 & 2 \\ 2 & 2 \end{bmatrix}$, $B = \begin{bmatrix} 1 & 2 \\ 2 & 1 \end{bmatrix}$;

(b) $A = \begin{bmatrix} 0 & -2 \\ 1 & 3 \end{bmatrix}$, $B = \begin{bmatrix} -3 & -4 \\ 2 & 3 \end{bmatrix}$;

CHAPTER

9

Vector Norms and Matrix Norms

We have seen in Section 5.4 how the notion of the length of a vector **x** and the distance between two vectors **x** and **y** on the *Euclidean plane* (see Fig. 5.5) can be extended on a *Euclidean* or a *unitary space* by (5.22) and (5.23). The notion of an inner product in these spaces was instrumental in such extension. An obvious question is: how can we introduce these concepts of length and distance for elements in a general vector space which is not necessarily only an inner product space? Can we define the length of a vector or distance between two vectors when these vectors are elements of the *vector space of matrices*? In a general vector space, these concepts are termed respectively *norm* and *metric*. In this chapter, we propose to study such quantitative aspects of vector spaces.

9.1 VECTOR NORMS

From school algebra we know that the magnitude of a real number is measured by its absolute value and that of a complex number by its modulus. For instance, the magnitude of –5 is | –5 | = 5 and that of 3 – $i4$ is | 3 – $i4$ | = $(9 + 16)^{1/2}$ = 5. On many occasions, it becomes necessary to assign a nonnegative number like this to a vector in a vector space by generalizing this concept of magnitude of a real or a complex number.

In an inner product space (see Definition 5.9), the inner product of an element $\mathbf{x} = [x_1\ x_2 \cdots x_n]^T$ is given by

$$\langle \mathbf{x}, \mathbf{x} \rangle = \| \mathbf{x} \|^2 = \begin{cases} (|x_1|^2 + |x_2|^2 + \cdots + |x_n|^2) & \text{in } \mathbb{C}^n, \\ (x_1^2 + x_2^2 + \cdots + x_n^2) & \text{in } \mathbb{R}^n. \end{cases}$$

This definition of inner product gives, in a natural way, a measure of magnitude of these elements in an inner product space. An extension of this measure to vectors **v** in a general vector space V is commonly known as *norm.*

Definition 9.1 *Vector Norm and Normed space.* A real-valued function $\|\cdot\|$ defined on each element **v** of a vector space V is defined as a vector norm or simply a norm on V iff the following axioms are satisfied:

(1) $\|\mathbf{v}\| \geq 0$ and $\|\mathbf{v}\| = 0$ iff $\mathbf{v} = \boldsymbol{\theta}$;

(2) $\|\alpha\mathbf{v}\| = |\alpha|\,\|\mathbf{v}\|$;

(3) $\|\mathbf{u} + \mathbf{v}\| \leq \|\mathbf{u}\| + \|\mathbf{v}\|$. □

Above **u** and **v** are any elements in a vector space V, α any scalar, and $|\alpha|$ the modulus of α. The function $\|\cdot\|$ assigns to each element **v** in V a nonnegative real number. A vector space in which we can ascribe the notion of a norm to its elements is called a *normed space.* In this sense, a Euclidean or a unitary space (see Definition 5.9) is a normed space. Given a vector space, there are numerous ways of expressing the norm.

Example 9.1 Let $\mathbf{x} = [x_1\ x_2\ \ldots\ x_n]^T$ be an element of $\mathbb{C}^n$. Then the following norms, referred to as *vector norms,* are useful for purposes of application:

(a) $\|\mathbf{x}\|_1 = \sum_{i=1}^{n} |x_i|$, (b) $\|\mathbf{x}\|_2 = (\sum_{i=1}^{n} |x_i|^2)^{1/2}$, (Euclidean norm)

(c) $\|\mathbf{x}\|_p = (\sum_{i=1}^{n} |x_i|^p)^{1/2}$, $1 \leq p < \infty$; (d) $\|\mathbf{x}\|_\infty = \max_i |x_i|$. □

Norms (a), (b) and (d) are special cases of norm (c) when $p = 1$, $p = 2$ and $p = \infty$, respectively. To verify the case for $p = \infty$, assume, without loss of generality, that

$$|x_1| = \max_i |x_i|$$

is the maximum of x_i's of multiplicity m, i.e., m number of elements in **x** viz., $x_1, x_2, \ldots, x_m$ have the same maximum modulus or absolute value. Then, putting

$$\left|\frac{x_k}{x_1}\right| = \alpha_{k-m} < 1 \qquad \text{for all } k = m + 1, \ldots, n;$$

$$x_1 = x_2 = \cdots = x_m,$$

we get

$$\|\mathbf{x}\|_p = |x_1|\,(m + \alpha_1^p + \cdots + \alpha_{n-m}^p)^{1/p}.$$

Taking the limit as $p \to \infty$, we obtain

$$\|\mathbf{x}\|_\infty = \lim_{p\to\infty} |x_1|\,(m)^{1/p} = |x_1| = \max_i |x_i|.$$

The vector space with the above norms is denoted by l_p^n for all $1 \leq p \leq \infty$. The fact that all the axioms, (1), (2) and (3) of a norm, are satisfied by vector norms (a)–(d) will depend on Minkowski inequality which will be now derived. It is, however, important to note that if

$p < 1$, then (c) does not define a norm. Indeed, if $\mathbf{x} = [1\ 4]^T$ and $\mathbf{y} = [1\ 9]^T$, then using $p = 1/2$, we have $\| \mathbf{x} \| = 9$, $\| \mathbf{y} \| = 16$, and $\| \mathbf{x} + \mathbf{y} \| = 25.198$, which show that $\| \mathbf{x} + \mathbf{y} \| \le \| \mathbf{x} \| + \| \mathbf{y} \|$ is not satisfied, thus violating axiom (3) of Definition 9.1.

The norm $\| \mathbf{x} \|_1$ is known as *taxicab* or *Manhattan norm*. The name relates to the distance a taxi has to travel in a rectangular street grid to get from the origin to the point $\mathbf{x}$. In general, $\| \mathbf{x} \|_p$ is called *p-norm* while $\| \mathbf{x} \|_\infty$ is *infinity* or *maximum norm*.

To derive Minkowski inequality, we need a result known as Hölder's inequality.

Theorem 9.1 *Hölder's Inequality.* Let $\mathbf{x} = [x_1 \cdots x_n]^T$ and $\mathbf{y} = [y_1 \cdots y_n]^T$ be two elements in $\mathbb{C}^n$ and p and q be any positive real numbers satisfying

$$\frac{1}{p} + \frac{1}{q} = 1. \tag{9.1}$$

Then,

$$\sum_{i=1}^{n} | x_i y_j | \le \left[\sum_{i=1}^{n} | x_i |^p \right]^{1/p} \left[\sum_{i=1}^{n} | y_i |^q \right]^{1/q}. \tag{9.2}$$

Proof We shall prove that

$$\sum_{i=1}^{n} | \alpha_i \beta_i | \le 1 \tag{9.3}$$

for

$$\mathbf{u} = [\alpha_1 \cdots \alpha_n]^T, \qquad \mathbf{v} = [\beta_1 \cdots \beta_n]^T, \qquad \| \mathbf{u} \|_p = \| \mathbf{v} \|_q = 1. \tag{9.4}$$

Then, by putting

$$\alpha_i = \frac{x_i}{\| \mathbf{x} \|_p} \quad \text{and} \quad \beta_i = \frac{y_i}{\| \mathbf{y} \|_q}$$

in (9.3), the desired inequality can be proved.

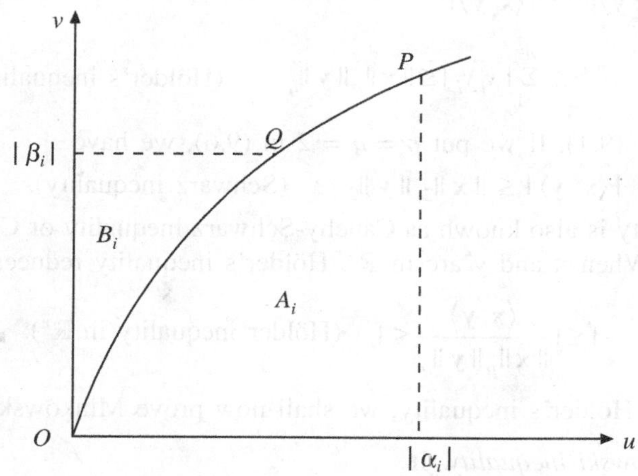

Figure 9.1 Derivation of Hölder inequality.

From (9.1), $p = q(p-1)$ and $q = p(q-1)$; therefore, $v = u^{p-1}$ and $u = v^{q-1}$ represent the same curve. This curve is shown in Figure 9.1. Considering the areas A_i and B_i as shown, we have

$$A_i = \int_0^{|\alpha_i|} u^{p-1}du = \frac{|\alpha_i|^p}{p} \quad \text{and} \quad B_i = \int_0^{|\beta_i|} v^{q-1}dv = \frac{|\beta_i|^q}{q}.$$

Irrespective of the position P relative to Q, we have from Figure 9.1,

$$|\alpha_i\beta_i| = |\alpha_i|\,|\beta_i| \le A_i + B_i = \frac{1}{p}|\alpha_i|^p + \frac{1}{q}|\beta_i|^q.$$

Hence, taking the sum of both sides from $i = 1$ to $i = n$, we get, with the help of (9.4), the equation

$$\sum_{i=1}^{n} |\alpha_i\beta_i| \le \frac{1}{p}\|\mathbf{u}\|_p^p + \frac{1}{q}\|\mathbf{v}\|_q^q = \frac{1}{p} + \frac{1}{q} = 1. \qquad \square$$

For $p = 2$, Hölder's inequality reduces to

$$\sum_{i=1}^{n} |x_iy_i| \le \|\mathbf{x}\|_2\,\|\mathbf{y}\|_2 \quad \text{(Cauchy's inequality).} \tag{9.5}$$

If x_i and y_i are complex numbers, then

$$|x_iy_i| = (x_iy_i\bar{x}_i\bar{y}_i)^{1/2} = (x_i\bar{y}_i\bar{x}_iy_i)^{1/2} = |x_i\bar{y}_i|.$$

Hence, with the help of (5.16), we obtain

$$|\langle\mathbf{x},\mathbf{y}\rangle| = |x_1\bar{y}_1 + \cdots + x_n\bar{y}_n| \le |x_1\bar{y}_1| + \cdots + |x_n\bar{y}_n| = \sum_{i=1}^{n}|x_iy_i|.$$

Therefore, from (9.2) we can write an alternative form of Hölder's inequality as

$$[\langle\mathbf{x},\mathbf{y}\rangle\overline{\langle\mathbf{x},\mathbf{y}\rangle}]^{1/2} = |\langle\mathbf{x},\mathbf{y}\rangle|$$

$$\le \sum_{i=1}^{n}|x_iy_i| \le \|\mathbf{x}\|_p\|\mathbf{y}\|_q \quad \text{(Hölder's inequality in } \mathbb{R}^n\text{),} \tag{9.6}$$

where p and q satisfy (9.1). If we put $p = q = 2$ in (9.6), we have

$$|\langle\mathbf{x},\mathbf{y}\rangle| \le \|\mathbf{x}\|_2\,\|\mathbf{y}\|_2 \quad \text{(Schwarz inequality).} \tag{9.7}$$

The Schwarz inequality is also known as Cauchy-Schwarz inequality or Cauchy-Bunyakovsky-Schwarz inequality. When **x** and **y** are in $\mathbb{R}^n$, Hölder's inequality reduces to

$$-1 \le \left|\frac{\langle\mathbf{x},\mathbf{y}\rangle}{\|\mathbf{x}\|_p\|\mathbf{y}\|_q}\right| \le 1 \quad \text{(Hölder inequality in } \mathbb{R}^n\text{).} \tag{9.8}$$

With the help of Hölder's inequality, we shall now prove Minkowski inequality.

Theorem 9.2 *Minkowski inequality.* If

$$\mathbf{x} = [x_1 \cdots x_n]^T \quad \text{and} \quad \mathbf{y} = [y_1 \cdots y_n]^T$$

are two elements in $\mathbb{C}^n$, then

$$\| \mathbf{x} + \mathbf{y} \|_p \le \| \mathbf{x} \|_p + \| \mathbf{y} \|_p, \quad p \ge 1 \quad \text{(Minkowski inequality).} \tag{9.9}$$

Proof It is easy to verify (9.9) when $p = 1$. Since

$$| x_i + y_i | \le | x_i | + | y_i |,$$

we have

$$| x_i + y_i |^p = | x_i + y_i | \, | x_i + y_i |^{p-1} \le | x_i | \, | x_i + y_i |^{p-1} + | y_i | \, | x_i + y_i |^{p-1}.$$

Taking the sum of both sides from $i = 1$ to $i = n$ and observing that $| \alpha_i | \, | \beta_i | = | \alpha_i \beta_i |$ and from (9.1) $q(p - 1) = p$, we apply Hölder's inequality (9.2) to get

$$\sum_{i=1}^{n} | x_i + y_i |^p \le [\sum_{i=1}^{n} | x_i |^p]^{1/p} [\sum_{i=1}^{n} | x_i + y_i |^{q(p-1)}]^{1/q}$$

$$+ [\sum_{i=1}^{n} | y_i |^p]^{1/p} [\sum_{i=1}^{n} | x_i + y_i |^{q(p-1)}]^{1/q}$$

$$= [(\sum_{i=1}^{n} | x_i |^p)^{1/p} + (\sum_{i=1}^{n} | y_i |^p)^{1/p}] [\sum_{i=1}^{n} | x_i + y_i |^p]^{1/q}$$

On dividing both sides by the extreme right factor and using (9.1), we have finally

$$[\sum_{i=1}^{n} | x_i + y_i |^p]^{1/p} \le [\sum_{i=1}^{n} | x_i |^p]^{1/p} + [\sum_{i=1}^{n} | y_i |^p]^{1/p}.$$ □

Solution of Example 9.1 We shall now show that all the three axioms of a vector norm in Definition 9.1 are satisfied by $\| \mathbf{x} \|_p$. Since the right side of $\| x \|_p^p$ is the sum of positive terms and $\| \mathbf{x} \|_p$ is defined as the positive value of the *pth* root of $\| x \|_p^p$, we have $\| \mathbf{x} \|_p \ge 0$. Also, $\| \mathbf{x} \|_p = 0$ iff x_i's are zero, i.e., iff $\mathbf{x} = \boldsymbol{\theta}$. Axiom (2) is directly verified by changing $\mathbf{x}$ to $\alpha \mathbf{x} = [\alpha x_1 \cdots \alpha x_n]^T$. The axiom (3), i.e., $\| \mathbf{x} + \mathbf{y} \|_p \le \| \mathbf{x} \|_p + \| \mathbf{y} \|_p$, follows from Minkowski inequality for $\mathbf{x}$ and $\mathbf{y}$ in $\mathbb{C}^n$. □

Another important quantitative measurement, that an inner product provides, is the distance $d(\mathbf{x}, \mathbf{y})$, see Definition 5.10, between two vectors $\mathbf{x}$ and $\mathbf{y}$ which, by extending the two-dimensional concept, (see Fig. 5.5), to $\mathbb{C}^n$ is given by

$$d(\mathbf{x}, \mathbf{y}) = \| \mathbf{x} - \mathbf{y} \|_p.$$

The distance function $d(\mathbf{u}, \mathbf{v})$ between vectors $\mathbf{u}$ and $\mathbf{v}$ in a general vector space V is commonly referred to as the *metric.*

Definition 9.2 *Metric.* A metric in a vector space V *is* a real-valued function d of ordered pairs of elements of V which satisfies the following axioms:

(1) $d(\mathbf{u}, \mathbf{v}) \ge 0$ and $d(\mathbf{u}, \mathbf{v}) = 0$ iff $\mathbf{u} = \mathbf{v}$;
(2) $d(\mathbf{u}, \mathbf{v}) = d(\mathbf{v}, \mathbf{u})$ (symmetry);
(3) $d(\mathbf{u}, \mathbf{v}) = d(\mathbf{u}, \mathbf{w}) + d(\mathbf{w}, \mathbf{v})$ (triangle inequality);

for $\mathbf{u}$, $\mathbf{v}$ and $\mathbf{w}$ in V. A vector space for which a metric is defined is called a *metric space.* The function d assigns to each pair $(\mathbf{u}, \mathbf{v})$ of elements a nonnegative real number $d(\mathbf{u}, \mathbf{v})$ which by symmetry does not depend on the order of the elements; $d(\mathbf{u}, \mathbf{v})$ is called the distance between $\mathbf{u}$ and $\mathbf{v}$.

We can define the metric or the distance function of two elements **x** and **y** in $\mathbb{C}^n$ as the norm of the difference of **x** and **y**, i.e.,

$$d(\mathbf{x}, \mathbf{y}) = \| \mathbf{x} - \mathbf{y} \|_p.$$

By Definition 9.1 of a norm, $d(\mathbf{x}, \mathbf{y}) \geq 0$ and $d(\mathbf{x}, \mathbf{y}) = 0$ iff $\mathbf{x} - \mathbf{y} = \boldsymbol{\theta}$, i.e., iff $\mathbf{x} = \mathbf{y}$. Again,

$$d(\mathbf{x}, \mathbf{y}) = [\,| x_1 - y_1 |^p + \cdots + | x_n - y_n |^p]^{1/p} = [\,| y_1 - x_1 |^p + \cdots + | y_n - x_n |^p]^{1/p}$$
$$= : d(\mathbf{y}, \mathbf{x}).$$

Finally, by Minkowski inequality we have

$$d(\mathbf{x}, \mathbf{y}) = \| (\mathbf{x} - \mathbf{z}) + (\mathbf{z} - \mathbf{y}) \|_p \leq \| \mathbf{x} - \mathbf{z} \|_p + \| \mathbf{z} - \mathbf{y} \|_p = d(\mathbf{x}, \mathbf{z}) + d(\mathbf{z}, \mathbf{y}),$$

where **z** is in $\mathbb{C}^n$. So all the axioms (1)–(3) of a metric in Definition 9.2 are satisfied by the function $d(\mathbf{x}, \mathbf{y})$ if we define $d(\mathbf{x}, \mathbf{y}) = \| \mathbf{x} - \mathbf{y} \|_p$. A metric defined as the norm of the difference of two elements, i.e., by $d(\mathbf{x}, \mathbf{y}) := \| \mathbf{x} - \mathbf{y} \|_p$ in general, where $p \geq 1$ is an integer, is referred to as the *metric induced by the norm.* Thus a normed space, that is, a space in which the norm of the elements is defined, is also a metric space in which the distance between the elements can be defined with the help of the norm.

Example 9.2 Find the norm $\| \mathbf{x} \|_p$, $p = 1, 2, \infty$ when $\mathbf{x}^T = [-3, 0\ 4]$.

Solution

$$\| \mathbf{x} \|_1 = 3 + 0 + 4 = 7, \qquad \| \mathbf{x} \|_2 = (9 + 0 + 16)^{1/2} = 5, \qquad \| \mathbf{x} \|_\infty = 4. \qquad \square$$

Example 9.3 The metrics induced by the norms $\| \mathbf{x} \|_1$, $\| \mathbf{x} \|_p$: $1 < p < \infty$, $\| \mathbf{x} \|_\infty$ are respectively,

$$\| \mathbf{x} \|_1: d(\mathbf{x}, \mathbf{y}) = | x_1 - y_1 | + \cdots + | x_n - y_n |,$$
$$\| \mathbf{x} \|_p: d(\mathbf{x}, \mathbf{y}) = [| x_1 - y_1 |^p + \cdots + | x_n - y_n |^p]^{1/p},$$
$$\| \mathbf{x} \|_\infty: d(\mathbf{x}, \mathbf{y}) = \max_i | x_i - y_i |. \qquad \square$$

Example 9.4 Find $d(\mathbf{x}, \mathbf{y})$, where $\mathbf{x} = [1\ 0\ {-2}]^T$ and $\mathbf{y} = [1\ 3\ 2]^T$, d being induced by the norms $\| \mathbf{x} \|_p$, $p = 1, 2, \infty$.

Solution

(a) $d(\mathbf{x}, \mathbf{y}) = \| \mathbf{x} - \mathbf{y} \|_1 = | 1 - 1 | + | 0 - 3 | + | -2 -2 | = 7,$

(b) $d(\mathbf{x}, \mathbf{y}) = \| \mathbf{x} - \mathbf{y} \|_2 = [(1 - 1)^2 + (0 - 3)^2 + (-2 -2)^2]^{1/2} = 5,$

(c) $d(\mathbf{x}, \mathbf{y}) = \| \mathbf{x} - \mathbf{y} \|_\infty = \max_i \{| 1 - 1 |, | 0 - 3 |, | -2 -2 |\} = 4.$ $\square$

9.2 MATRIX NORMS

It is easily verified that the set of all $n \times m$ matrices satisfy all the axioms of a vector space in Definition 5.1 if one defines $n \times m$ null matrix as the zero element in this space. This vector space will be denoted by $\mathbb{R}^{n \times m}$ when the elements in a matrix are real or by $\mathbb{C}^{n \times m}$ when the elements are complex or by $V_{n \times m}$ over a field $\mathbb{F}$, where $\mathbb{F}$ specifies the nature of the elements in a matrix in $V_{n \times m}$. It can be shown that the set of $n \times m$ matrices E_{ij} in which the ijth element is 1 and 0's elsewhere forms a basis of this vector space whose dimension is nm.

Definition 9.3 *Matrix Norm.* Let A be an element in $V_{n \times n}$ over a field $\mathbb{F}$. We shall call a real-valued function $\| \cdot \|$ defined on all square matrices in $V_{n \times n}$ a matrix norm of A on $V_{n \times n}$ and denote by $\| A \|$ iff the following axioms are satisfied:

(1) $\| A \| \geq 0$ and $\| A \| = 0$ iff $A = 0$;

(2) $\| \alpha A \| = |\alpha| \, \| A \|$ for any scalar α in $\mathbb{F}$;

(3) $\| A + B \| \leq \| A \| + \| B \|$;

(4) $\| AB \| \leq \| A \| \, \| B \|$,

where B is another element in $V_{n \times n}$. □

There are numerous ways by which matrix norms can be formed, and these are distinguished from one another by a subscript. If from the context it is clear what norm is referred to, then the subscript is dropped for convenience of notation.

Sometimes by $N(A)$ we also denote a matrix norm; then axioms (1)–(4) can be rewritten by replacing $\| A \|$ with $N(A)$.

Example 9.5 The following are matrix norms on $V_{n \times n}$:

(a) $\| A \|_a = \sum_{i,j} |a_{ij}|$ (sum of *all* elements);

(b) $\| A \|_1 = \max_j \sum_{i,j} |a_{ij}|$;

(c) $\| A \|_\infty = \max_i \sum_{i,j} |a_{ij}|$;

(d) $\| A \|_e = (\sum_{i,j} |a_{ij}|^2)^{1/2} = (\text{trace } (A^* A))^{1/2}$ (Euclidean $\| \cdot \|_e$, Frobenius $\| \cdot \|_F$, or Hilbert-Schmidt norm $\| \cdot \|_{HS}$),

where $A = [a_{ij}]$.

Solution We shall prove all the cases by showing that each one satisfies all the axioms of a matrix norm.

(a) Obviously, $\| A \| \geq 0$ is satisfied. $\| A \| = 0$ iff $|a_{ij}| = 0$, i.e., $a_{ij} = 0$ or iff $A = 0$. Now,

$$\| \alpha A \| = \sum_{i,j} |\alpha a_{ij}| = |\alpha| \sum_{i,j} |a_{ij}| = |\alpha| \, \| A \|.$$

If $B = [b_{ij}]$ is in $V_{n \times n}$, then

$$\| A + B \| = \sum_{i,j} |a_{ij} + b_{ij}| \leq \sum_{i,j} (|a_{ij}| + |b_{ij}|) = \| A \| + \| B \|.$$

Finally,

$$\| AB \| = \sum_{i,j} |(\sum_k a_{ik} b_{kj})| \leq \sum_{i,j} (\sum_k |a_{ik} b_{kj}|) = \sum_{i,j} \sum_k |a_{ik}| \, |b_{kj}|$$

$$\leq \sum_{i,k} |a_{ik}| \sum_{p,j} |b_{pj}| = \| A \| \, \| B \|.$$

All the axioms in Definition 9.3 are satisfied by $\| \cdot \|_a$ and, therefore, (a) is a matrix norm.

(b) Axioms (1)–(3) are easily checked. For axiom (4), by repeated application of the triangle inequality of complex numbers, i.e., $|\alpha_i + \beta_i| \le |\alpha_i| + |\beta_i|$, we have

$$\|AB\| = \max_j \sum_i |\sum_k a_{ik}b_{kj}| \le \max_j \sum_i \sum_k |a_{ik}b_{kj}| = \max_j \sum_k \sum_i |a_{ik}|\,|b_{kj}|$$

$$\le (\max_k \sum_i |a_{ik}|)(\max_j \sum_k |b_{kj}|) = \|A\|\,\|B\|$$

(c) It can be verified as in (b).

(d) Axioms (1) and (2) can be easily verified. To check axiom (3), we observe that

$$|a_{ij} + b_{ij}|^2 = (a_{ij} + b_{ij})(\bar{a}_{ij} + \bar{b}_{ij}) = |a_{ij}|^2 + |b_{ij}|^2 + 2\,\mathrm{Re}\, a_{ij}\bar{b}_{ij}$$

$$\le |a_{ij}|^2 + |b_{ij}|^2 + 2|a_{ij}b_{ij}|.$$

Therefore,

$$\|A+B\|^2 = \sum_{i,j} |a_{ij} + b_{ij}|^2 \le \sum_{i,j} (|a_{ij}|^2 + |b_{ij}|^2 + 2|a_{ij}b_{ij}|)$$

$$\le \|A\|^2 + \|B\|^2 + 2(\sum_{i,j} |a_{ij}|^2)^{1/2} (\sum_{i,j} |b_{ij}|^2)^{1/2}$$

(by Schwarz inequality)

$$= \|A\|^2 + \|B\|^2 + 2\|A\|\,\|B\|.$$

From this inequality we get Axion (3):

$$\|A + B\| \le \|A\| + \|B\|.$$

Finally,

$$\|AB\|^2 = \sum_{i,j} |\sum_k a_{ik}b_{kj}|^2$$

$$\le \sum_{i,j} (\sum_k |a_{ik}|^2)(\sum_k |b_{kj}|^2) \quad \text{(by Schwarz inequality)}$$

$$= \sum_{i,k} |a_{ik}|^2 \sum_{j,k} |b_{kj}|^2 = \|A\|^2\,\|B\|^2. \qquad \square$$

As the norm of a null matrix is zero, the norm of any other matrix gives a measure of the magnitude by which the given matrix deviates, so to say, from the null matrix.

Example 9.6 Compute the matrix norms (a) $\|A\|_a$, (b) $\|A\|_1$, (c) $\|A\|_\infty$, (d) $\|A\|_e$ when

$$A = \begin{bmatrix} 1 & -2 & 3 \\ 0 & -1 & -4 \\ 2 & 1 & 2 \end{bmatrix}$$

The row sums of absolute values are 6, 5 and 5 respectively; and column sums of absolute values are 3, 4 and 9 respectively.

Therefore,

(a) $\| A \|_a = 1 + 2 + 3 + 0 + 1 + 4 + 2 + 1 + 2 = 16,$

(b) $\| A \|_1 = 9,$ (c) $\| A \|_\infty = 6,$

(d) $\| A \|_e = (1 + 4 + 9 + 0 + 1 + 16 + 4 + 1 + 4)^{1/2} = 6.32.$ □

Example 9.7 For any matrix norm, prove the following inequalities:

(a) $\| I \| \geq 1,$ (b) $\| A^{-1} \| \geq 1/\| A \|,$ (c) $\| A^n \| \leq \| A \|^n.$

Solution (a) In axiom (4) of a matrix norm, we put $A = B = I$ to get $\| I \| \leq \| I \|^2$, i.e., $1 \leq \| I \|$, from which follows (a).

(b) From $\| AA^{-1} \| = \| I \| \geq 1$, it follows that $1 \leq \| A^{-1}A \| \leq \| A^{-1} \| \| A \|$ by axiom (4) of Definition 9.3.

(c) By repeated application of axiom (4) of a matrix norm, we have $\| A^n \| = \| A^{n-1}A \| \leq \| A \| \| A^{n-1} \| \leq \| A \|^2 \| A^{n-2} \| \leq \cdots \leq \| A \|^n$. □

9.2.1 Compatible Matrix Norms

Definition 9.4 *Matrix Norm Compatible with a Vector Norm.* Let A be any matrix in $V_{n \times n}$ over a field $\mathbb{F}$ and $\mathbf{x}$ any vector in V_n over the same field $\mathbb{F}$. Then the matrix norm is said to be compatible with the vector norm iff for all $\mathbf{x}$,

$$\| A\mathbf{x} \|_a \leq \| A \|_b \, \| \mathbf{x} \|_a. \tag{9.10}$$

□

We should carefully note that both the norms $\| \mathbf{x} \|_a$ and $\| A\mathbf{x} \|_a$ in (9.10) are formed in a similar fashion in the sense that if $\| \mathbf{x} \|_a$ is the Euclidean norm $\| \mathbf{x} \|_e$, then $\| A\mathbf{x} \|_a$ is the same Euclidean norm $\| A\mathbf{x} \|_e$, while $\| A \|_b$ is any norm compatible with the vector norm $\| \mathbf{x} \|_e$.

We may observe that the vector $A\mathbf{x}$ in V_n is composed of a matrix A in $V_{n \times n}$ multiplied into a vector $\mathbf{x}$ in V_n, and for both of A and $\mathbf{x}$ a norm is defined. With this observation we are motivated by axiom (4) of a matrix norm not to allow the norm of $A\mathbf{x}$ to exceed the product of the norm of A and that of $\mathbf{x}$ and so it is expressed by (9.10).

Example 9.8 Show that the following matrix norms and vector norms are compatible:

(a) $\| A \|_1 = \max\limits_j \sum\limits_i | a_{ij} |$ (column sum), $\quad \| \mathbf{x} \|_1 = \sum\limits_{i=1}^{n} | x_i |;$

(b) $\| A \|_\infty = \max\limits_i \sum\limits_j | a_{ij} |$ (row sum), $\quad \| \mathbf{x} \|_\infty = \max\limits_i | x_i |;$

(c) $\| A \|_e = \left(\sum\limits_{i,j} | a_{ij} |^2 \right)^{1/2},$ $\quad \| \mathbf{x} \|_2 = \left(\sum\limits_{i=1}^{n} | x_i |^2 \right)^{1/2}.$

Solution To prove the claim in this example, we have to show the validity (9.10) in each case. If there is no confusion, we may drop the subscript with the norm.

(a) $\| A\mathbf{x} \|_1 = \sum_{i=1}^{n} |(a_{i1}x_1 + \cdots + a_{in}x_n)| \le \sum_{i=1}^{n}\sum_{j=1}^{n} |a_{ij}||x_j|$ by Minkowski inequality

$$\le (\max_j \sum_{i=1}^{n} |a_{ij}|)(\sum_{j=1}^{n} |x_j|).$$

This gives us the desired result

$$\| A\mathbf{x} \|_1 \le \| A \|_1 \, \| \mathbf{x} \|_1.$$

(b) $\| A\mathbf{x} \|_\infty = \max_i |\sum_{j=1}^{n} a_{ij}x_j| \le \max_i \sum_{j=1}^{n} |a_{ij}||x_j|$ by Minkowski inequality

$$\le (\max_i \sum_{j=1}^{n} |a_{ij}|)(\max_j |x_j|)$$

which leads to the compatibility condition

$$\| A\mathbf{x} \|_\infty \le \| A \|_\infty \, \| \mathbf{x} \|_\infty.$$

(c) See Example 9.11 and Problem 9.10. □

9.2.2 Continuity of Matrix and Vector Norms

If each element of $\mathbf{x} = [x_1\ x_2\ \cdots\ x_n]^T$ tends to a corresponding element of $\mathbf{y} = [y_1\ y_2\ \cdots\ y_n]^T$, i.e., if x_i tends to y_i, where x_i and y_i are the ith component of two vectors $\mathbf{x}$ and $\mathbf{y}$ in V_n respectively, then we say that $\mathbf{y}$ is the *limit of* $\mathbf{x}$ and write $\mathbf{x} \to \mathbf{y}$. Similarly, if $a_{ij} \to b_{ij}$ for each pair of subscripts (i, j), where a_{ij} and b_{ij} are the ijth element of A and B respectively in $V_{n\times n}$, then B is said to be the *limit of* A, and we write $A \to B$. The vector norm $\| \mathbf{x} \|$ is a real-valued function on V_n in the sense that for each element $\mathbf{x}$ in V_n we get a real value for its norm $\| \mathbf{x} \|$. Likewise, the matrix norm $\| A \|$ is a real-valued function on $V_{n\times n}$. It will be shown in Theorem 9.3 *that a vector norm* $\| \mathbf{x} \|$ *is a continuous function of* $\mathbf{x}$ *and a matrix norm* $\| A \|$ *is a continuous function of* A.

Theorem 9.3 (a) Let $\mathbf{x}$ and $\mathbf{y}$ be in V_n. If $\mathbf{x} \to \mathbf{y}$, then the real number $\| \mathbf{x} \| \to$ the real number $\| \mathbf{y} \|$ and $\| \mathbf{x} - \mathbf{y} \| \to 0$.

(b) Let A and B be in $V_{n\times n}$. If $A \to B$, then the real number $\| A \| \to$ the real number $\| B \|$, and $\| A - B \| \to 0$.

Proof (a) We can write

$$\mathbf{x} = \sum_{i=1}^{n} x_i\mathbf{e}_i \quad \text{and} \quad \mathbf{y} = \sum_{i=1}^{n} y_i\mathbf{e}_i,$$

where $\mathbf{e}_i$'s are elementary vectors given by (5.4). Now by Problem 9.1, Minkowski inequality and axiom (2) of a vector norm, we have

$$0 \le |\, \| \mathbf{x} \| - \| \mathbf{y} \| \,| \le \| \mathbf{x} - \mathbf{y} \| = \| \sum_{i=1}^{n} (x_i - y_i)\, \mathbf{e}_i \| \le \sum_{i=1}^{n} |x_i - y_i| \, \| \mathbf{e}_i \|.$$

Hence, $\mathbf{x} \to \mathbf{y}$ implies that $\| \mathbf{x} \| \to \| \mathbf{y} \|$ and $\| \mathbf{x} - \mathbf{y} \| \to 0$.

(b) Denoting by E_{ij} the set of matrices where the ijth element is 1 and the rest are zero (see (5.7)) we have, by the same arguments as above, the relations

$$0 \leq | \, \| A \| - \| B \| \, | \leq \| A - B \| = \| \sum_{i,\, j=1}^{n} (a_{ij} - b_{ij}) \, E_{ij} \|$$

$$\leq \sum_{i,\, j=1}^{n} | a_{ij} - b_{ij} | \, \| E_{ij} \|$$

and so, as $A \to B$, we get $\| A \| \to \| B \|$ and $\| A - B \| \to 0$.

If we define $\max_{i,j} \| E_{ij} \| = c$ and $\delta = \varepsilon / cn^2$ for any $\varepsilon > 0$, then from above, for all pairs (i, j), we have

$$| a_{ij} - b_{ij} | < \delta \text{ implies } | \, \| A \| - \| B \| \, | < \varepsilon.$$

This means that a small variation in the elements of A causes a small variation in the norm of A, i.e., $\| \cdot \|$ is a continuous function of A. □

9.3 INDUCED MATRIX NORMS

We have determined the matrix norm in Example 9.5 in terms of the elements a_{ij} of a square matrix A. We can regard a square matrix A in $V_{n \times n}$ as a transformation from V_n into V_n and with our knowledge of the vector norm of an element $\mathbf{x}$ in V_n we can as well determine the vector norm of $A\mathbf{x}$, and hence the matrix norm. The matrix norm defined in this way is referred to as the *induced matrix norm.*

Definition 9.5 *Induced Matrix Norm.* Let $\mathbf{x}$ be an element in V_n and A a matrix transformation from V_n into V_n. Then the quotient of the vector norms $\| A\mathbf{x} \|$ and $\| \mathbf{x} \|$, where $\| \cdot \|$ is any vector norm, is always positive. The maximum of this quotient for all possible values of $\mathbf{x} \neq \theta$ is defined as the *matrix norm induced by the vector norm* or, briefly, induced matrix norm, and is written as

$$\| A \| = \max_{\mathbf{x} \neq \theta} \frac{\| A\mathbf{x} \|}{\| \mathbf{x} \|}. \tag{9.11}$$

□

Intuitively, one may look upon $\| A \|$ as representing the maximum magnification of a vector $\mathbf{x}$ on multiplication by the matrix A.

Putting $\mathbf{y} = \mathbf{x}/\| \mathbf{x} \|$, where evidently, $\| \mathbf{y} \| = 1$, and using axiom (2) of a vector norm, from (9.11) we have

$$\| A \| = \max_{\substack{\mathbf{x} \neq \theta \\ \|\mathbf{y}\| = 1}} \frac{\| A \, (\| \mathbf{x} \|) \, \mathbf{y} \|}{\| \mathbf{x} \|} = \max_{\|\mathbf{y}\| = 1} \| A\mathbf{y} \|.$$

Therefore, alternatively, we can define the *induced matrix norm* by

$$\| A \| = \max_{\|\mathbf{x}\| = 1} \| A\mathbf{x} \|. \tag{9.12}$$

Definition 9.5 can also be applied to define the *norm of any linear transformation L* acting on a normed linear space V (see Definition 9.1).

The nature of the characteristic $\| \mathbf{x} \|_p = 1$ on $\mathbb{R}^2$ when $p = 1, 2, \infty$ under the norms defined in Example 9.1 is shown in Figure 9.2. Considering $\| \mathbf{x} \|_2 = 1$, if we compute $\| A\mathbf{x} \|_2$

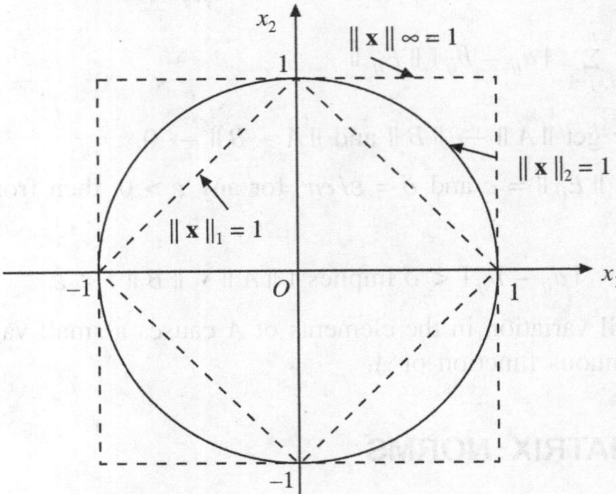

Figure 9.2 Computation of induced matrix norms.

for all possible values of $\mathbf{x}$ lying on the unit circle and then take the maximum value, we get $\| A \|$ induced by the Euclidean norm $\| \mathbf{x} \|_2$. Similar arguments hold for $p = 1$ and $p = \infty$.

The maximum value as defined by (9.12) always exists as a consequence of *Weierstrass theorem*, which states that a continuous function defined on a closed and bounded set of points attains a maximum and a minimum in its domain. The assertion on the existence made above follows from the fact that the set of points defined by $\| \mathbf{x} \| = 1$ constitutes a closed and bounded set and the function $\| A\mathbf{x} \|$ is continuous by Theorem 9.3; hence there exists a unit vector $\mathbf{x}_0$ (i.e., $\| \mathbf{x}_0 \| = 1$) satisfying $\| A \| = \| A\mathbf{x}_0 \|$.

From the result given in the following theorem, the norm of a matrix A induced by the Euclidean norm $\| \mathbf{x} \|_2$ can be easily computed.

Theorem 9.4 *Rayleigh Quotient.* Let H be a Hermitian matrix in $\mathbb{C}^{n \times n}$ and $\mathbf{x}$ be a vector in $\mathbb{C}^n$. Then the quotient

$$R(\mathbf{x}) := \frac{\langle H\mathbf{x}, \mathbf{x} \rangle}{\langle \mathbf{x}, \mathbf{x} \rangle} \tag{9.13}$$

for all $\mathbf{x} \neq \boldsymbol{\theta}$ in $\mathbb{C}^n$ is referred to as the *Rayleigh quotient*. Prove that

$$\max R(\mathbf{x}) = \max_i \; \lambda_i(H) \tag{9.14a}$$

and

$$\min R(\mathbf{x}) = \min_i \; \lambda_i(H), \tag{9.14b}$$

where $\lambda_i(H)$ is the ith eigenvalue of H.

Proof Finding the maximum and the minimum of $R(\mathbf{x})$ for all values of $\mathbf{x} \neq \boldsymbol{\theta}$ is equivalent to determining the maximum and the minimum of $R(\mathbf{x})$ for $\| \mathbf{x} \| = 1$ as observed in (9.12). Since H is Hermitian, by Theorem 7.27 there exists a unitary matrix U such that

$$U^*HU = \text{diag}\ [\lambda_1, \lambda_2, ..., \lambda_n], \qquad U^*U = UU^* = I, \tag{9.15}$$

where λ_i's are the eigenvalues of H. Therefore, using the transformation $\mathbf{x} = U\mathbf{y}$, where $\mathbf{y}$ is any vector in $\mathbb{C}^n$, we have

$$\langle \mathbf{x}, \mathbf{x} \rangle = \langle U\mathbf{y}, U\mathbf{y} \rangle = \langle \mathbf{y}, U^*U\mathbf{y} \rangle = \langle \mathbf{y}, \mathbf{y} \rangle, \tag{9.16}$$

$$\langle H\, \mathbf{x}\ \mathbf{x} \rangle = \langle HU\mathbf{y}, U\mathbf{y} \rangle = \langle U^*HU\mathbf{y}, \mathbf{y} \rangle = \lambda_1 \mid y_1 \mid^2 + \cdots + \lambda_n \mid y_n \mid^2, \tag{9.17}$$

where we have taken the help of Definition 5.9 and (9.15). Combining (9.16) and (9.17), it is easy to check that on $\langle \mathbf{x}, \mathbf{x} \rangle = 1 = \langle \mathbf{y}, \mathbf{y} \rangle$,

$$\langle H\, \mathbf{x}, \mathbf{x} \rangle \leq \lambda_{\max} (\mid y_1 \mid^2 + \cdots + \mid y_n \mid^2) = \lambda_{\max},$$

$$\langle H\, \mathbf{x}, \mathbf{x} \rangle \geq \lambda_{\min} (\mid y_1 \mid^2 + \cdots + \mid y_n \mid^2) = \lambda_{\min},$$

from which

$$\lambda_{\min} \leq \langle H\, \mathbf{x}, \mathbf{x} \rangle \leq \lambda_{\max} \tag{9.18}$$

Now assume, without any loss of generality, that

$$\lambda_{\max} = \lambda_i, \qquad \lambda_{\min} = \lambda_j$$

and since $\mathbf{y}$ is arbitrary, put $\mathbf{y} = \mathbf{e}_i$ and $\mathbf{y} = \mathbf{e}_j$, which are in $\mathbb{C}^n$, successively in (9.17). Then

$$\langle H\mathbf{x}, \mathbf{x} \rangle = \lambda_i = \lambda_{\max} \quad \text{and} \quad \langle H\mathbf{x}, \mathbf{x} \rangle = \lambda_j = \lambda_{\min}$$

and so by (9.18), we have

$$\max \langle H\, \mathbf{x}, \mathbf{x} \rangle = \lambda_{\max} \quad \text{and} \quad \min \langle H\, \mathbf{x}, \mathbf{x} \rangle = \lambda_{\min}.$$

Therefore,

$$\max_{\mathbf{x} \neq \boldsymbol{\theta}} R(\mathbf{x}) = \max_{\|\mathbf{x}\| = 1} \langle H\mathbf{x}, \mathbf{x} \rangle = \lambda_{\max},$$

$$\min_{\mathbf{x} \neq \boldsymbol{\theta}} R(\mathbf{x}) = \min_{\|\mathbf{x}\| = 1} \langle H\mathbf{x}, \mathbf{x} \rangle = \lambda_{\min}. \qquad \square$$

The norm of a matrix A induced by the Euclidean vector norm $\| \mathbf{x} \|_2$ is defined as the *spectral norm.* In view of Theorem 9.4, the spectral norm of A is $[\lambda_{\max} (A^* A)]^{1/2}$, because $\| A\mathbf{x} \|^2 = \mathbf{x}^*A^*A\mathbf{x}$. This spectral norm is the maximum singular value of A (see Section 9.6) and is denoted by $\| A \|_2$ (see Example 9.11).

Example 9.9 Determine the matrix norm $\| A \|$ induced by (a) $\| \mathbf{x} \|_1$, (b) $\| \mathbf{x} \|_2$, and (c) $\| \mathbf{x} \|_\infty$ respectively when

$$A = \begin{bmatrix} 1 & -1 \\ 1 & 1 \end{bmatrix}.$$

Solution Here,

$$A\mathbf{x} = \begin{bmatrix} 1 & -1 \\ 1 & 1 \end{bmatrix} \begin{bmatrix} x_1 \\ x_2 \end{bmatrix} = \begin{bmatrix} x_1 - x_2 \\ x_1 + x_2 \end{bmatrix}.$$

(a) Referring to Figure 9.2, since $|x_1| + |x_2| = 1$ on $\| \mathbf{x} \|_1 = 1$, we have

$$\| A \| = \max_{\|\mathbf{x}\|_1 = 1} \| A\mathbf{x} \|_1 = \max_{\|\mathbf{x}\|_1 = 1} \{| x_1 - x_2 | + | x_1 + x_2 |\} = 1 + 1 = 2$$

which, for example, occurs at $x_1 = 0$ and $x_2 = 1$.

(b) We note that $A^*A = \text{diag}\,[2, 2]$ has an eigenvalue 2 of multiplicity 2. Now by Theorem 9.4,

$$\| A \|_2^2 = \max_{\|\mathbf{x}\|_2 = 1} \| A\mathbf{x} \|_2^2 = \max_{\|\mathbf{x}\|_2 = 1} \langle A\mathbf{x}, A\mathbf{x} \rangle = \max_{\|\mathbf{x}\|_2 = 1} \langle A^* A\mathbf{x}, \mathbf{x} \rangle = 2.$$

Therefore,

$$\| A \|_2 := \max_{\|\mathbf{x}\|_2 = 1} \| A\mathbf{x} \|_2 = \sqrt{2}$$

which is the *spectral norm* of the given matrix A.

(c) Referring to Figure 9.2, since $x_1 = 1$, $x_2 = 1$ is a point on $\| \mathbf{x} \|_\infty = 1$, i.e., on the set of points (x_1, x_2) defined by $\max \{| x_1 |, | x_2 |\} = 1$, we have

$$\| A \| = \max_{\|\mathbf{x}\|_\infty = 1} \| A\mathbf{x} \|_\infty = \max_{\|\mathbf{x}\|_\infty = 1} \{| x_1 - x_2 |, | x_1 + x_2 |\} = 2$$ □

Before proceeding further, the reader should check that he understands clearly the distinction between a matrix norm and an induced matrix norm. In the following theorem, we shall assume that the axioms of a vector norm in Definition 9.1 are true. Hence, we shall show that an induced matrix norm satisfies all the axioms of a matrix norm in Definition 9.3.

Theorem 9.5 Every induced matrix norm is a *compatible matrix norm.* In other words,

(a) $\| A \| = \max_{\|\mathbf{x}\| = 1} \| A\mathbf{x} \|$ is a matrix norm.

(b) The induced matrix norm defined in (a) is compatible with the vector norm $\| \mathbf{x} \|$, i.e., $\| A\mathbf{x} \| \le \| A \| \, \| \mathbf{x} \|$.

Proof At the beginning we shall prove assertion (b)

Let $\mathbf{y}$ be any vector in V^n so that

$$\mathbf{x} = \mathbf{y}/\| \mathbf{y} \| \quad \text{and} \quad \| \mathbf{x} \| = 1.$$

Therefore, with the help of (9.12), we get

$$\| A\mathbf{y} \| = \| A(\| \mathbf{y} \|)\mathbf{x} \| = \| \mathbf{y} \| \, \| A\mathbf{x} \| \le \| \mathbf{y} \| \max_{\|\mathbf{x}\| = 1} \| A\mathbf{x} \| = \| \mathbf{y} \| \, \| A \|$$

which is the compatibility condition in Definition 9.4.

We shall now prove (a), that is, $\| A \|$ is indeed a matrix norm. Since $A\mathbf{x}$ is a vector, in view of Axiom (1) of a vector norm and (9.12), $\| A \| \ge 0$. Also, $A = 0$ implies $\| A \| = 0$. Conversely, $\| A \| = 0$ implies

$$\max_{\|\mathbf{x}\| = 1} \| A\mathbf{x} \| = 0, \quad \text{or} \quad \| A\mathbf{x} \| = 0 \text{ on } \| \mathbf{x} \| = 1,$$

or $A\mathbf{x} = \boldsymbol{\theta}$ on $\| \mathbf{x} \| = 1$ by axiom (1) of a vector norm. Since $A\mathbf{x} = \boldsymbol{\theta}$ for infinitely many values of $\mathbf{x}$, $A = 0$. Axiom (1) of a matrix norm is therefore satisfied.

Again, by Axiom (2) of a vector norm,

$$\| \alpha A \| = \max_{\|\mathbf{x}\|=1} \| \alpha A\mathbf{x} \| = |\alpha| \max_{\|\mathbf{x}\|=1} \| A\mathbf{x} \| = |\alpha| \, \| A \|$$

which verifies Axiom (2) of a matrix norm.

Now, we have, by Axiom (3) of a vector norm,

$$\| A + B \| = \max_{\|\mathbf{x}\|=1} \|(A + B)\mathbf{x}\| = \max_{\|\mathbf{x}\|=1} \| A\mathbf{x} + B\mathbf{x} \|$$

$$\leq \max_{\|\mathbf{x}\|=1} (\| A\mathbf{x} \| + \| B\mathbf{x} \|) = \| A \| + \| B \|$$

which shows that Axiom (3) of a matrix norm is satisfied.

Finally, by Weierstrass theorem, there is a unit vector $\mathbf{x}_0$ so that $\| AB \| = \| AB\mathbf{x}_0 \|$. Therefore, by compatibility condition (b),

$$\| AB \| \leq \| A \| \, \| B\mathbf{x}_0 \| \leq \| A \| \, \| B \| \, \| \mathbf{x}_0 \| = \| A \| \, \| B \|$$

which is Axiom (4) of a matrix norm. □

The following examples take the help of Rayleigh quotient to compute the norm $\| A \|$ induced by the Euclidean norm $\| \mathbf{x} \|_2$.

Example 9.10 Show that if A is a matrix in $V_{n \times n}$ and $\mathbf{x}$ is a vector in $\mathbb{C}^n$, then the induced matrix norm

$$\| A \| = \max_{\|\mathbf{x}\|_2 = 1} (\langle A\mathbf{x}, A\mathbf{x} \rangle)^{1/2} = \max_{\|\mathbf{x}\|_2 = 1} \| A\mathbf{x} \|_2 \tag{9.19}$$

is a matrix norm and

$$\| A \| = \max_i \, [\lambda_i(A^*A)]^{1/2},$$

where $\lambda_i(A^*A)$'s are the eigenvalues of A^*A.

Solution By Theorem 9.4, the maximum in (9.19) exists and is obtained from

$$0 \leq \max_{\|\mathbf{x}\|_2 = 1} \langle A\mathbf{x}, A\mathbf{x} \rangle = \max_{\|\mathbf{x}\|_2 = 1} \langle A^* A\mathbf{x}, \mathbf{x} \rangle = \max_i \lambda_i (A^* A)$$

which is also true if max is replaced by min and so all the eigenvalues of $A^* A$ are nonnegative. Since every induced matrix norm is a matrix norm by Theorem 9.5, so is (9.19). □

Example 9.11 Show that the following matrix norm and vector norm are compatible:

$$\| A \|_2 := \max_i \, [\lambda_i (A^* A)]^{1/2}, \qquad \| \mathbf{x} \|_2 = (\sum_{i=1}^{n} | x_i |^2)^{1/2}.$$

Solution Let $\mathbf{x}$ be any nonzero vector. Then,

$$\| A\mathbf{x} \|_2^2 = \langle A\mathbf{x}, A\mathbf{x} \rangle = \langle A^* A\mathbf{x}, \mathbf{x} \rangle \leq \langle \mathbf{x}, \mathbf{x} \rangle \max_{\mathbf{x} \neq \theta} \frac{\langle A^* A\mathbf{x}, \mathbf{x} \rangle}{\langle \mathbf{x}, \mathbf{x} \rangle}$$

$$= \max_i \lambda_i (A^* A) \langle \mathbf{x}, \mathbf{x} \rangle,$$

where the last step follows from (9.13) and (9.14a). Hence,

$$\| A\mathbf{x} \|_2 \le \| A \|_2 \, \| \mathbf{x} \|_2. \qquad \square$$

Example 9.12 Show that the matrix norms

(a) $\| A \|_1 = \max_j \sum_i | a_{ij} |$,

(b) $\| A \|_2 = \max_i [\lambda_i (A^* A)]^{1/2}$,

(c) $\| A \|_\infty = \max_t \sum_j | a_{ij} |$

are induced respectively by the vector norms

(a) $\| \mathbf{x} \|_1 = \sum_{i=1}^{n} | x_i |$, (b) $\| \mathbf{x} \|_2 = (\sum_{i=1}^{n} | x_i |^2)^{1/2}$, (c) $\| \mathbf{x} \|_\infty = \max_i | x_i |$.

Solution (a) We shall denote by $\| A \|$ the matrix norm induced by the vector norm $\| \mathbf{x} \|_1$, i.e.,

$$\| A \| = \max_{\|\mathbf{x}\|_1 = 1} \| A\mathbf{x} \|_1.$$

Now,

$$\| A\mathbf{x} \|_1 = \sum_{i=1}^{n} | (a_{i1} x_1 + \cdots + a_{in} x_n) | \le \sum_{i=1}^{n} \sum_{j=1}^{n} | a_{ij} x_j | \tag{9.20}$$

$$= \sum_{i=1}^{n} \sum_{j=1}^{n} | a_{ij} | | x_j | \le \sum_{i=1}^{n} \max_j | a_{ij} | \sum_{j=1}^{n} | x_j | = \| A \|_1 \, \| \mathbf{x} \|_1 .$$

Therefore,

$$\| A \| = \max_{\|\mathbf{x}\|_1 = 1} \| A\mathbf{x} \|_1 \le \| A \|_1. \tag{9.21}$$

If $\| A \|_1 = \sum_{i=1}^{n} | a_{im} |$, we select $\mathbf{x} = \mathbf{e}_m$, where $\mathbf{e}_m$ is the mth elementary vector in V_n so that $\| \mathbf{x} \|_1 = 1$. Substituting into (9.20), we get

$$\| A\mathbf{e}_m \|_1 = | a_{1m} | + | a_{2m} | + \cdots + | a_{nm} | = \| A \|_1$$

which implies that the maximum value in (9.21) is attained and is equal to $\| A \|_1$. Therefore, $\| A \| = \| A \|_1$, i.e., the matrix norm $\| A \|_1$ is equal to the norm $\| A \|$ induced by the vector norm $\| \mathbf{x} \|_1$.

(b) This has already been proved in Example 9.10. Note that $\| A \|_e$ is not induced by $\| \mathbf{x} \|_2$ as can be expected.

(c) By the same arguments as in (a), we can show that the induced matrix norm $\| A \|$ is given by

$$\| A \| = \max_{\|\mathbf{x}\|_\infty = 1} \| A\mathbf{x} \|_\infty \le \| A \|_\infty$$

That the maximum is attained is shown by selecting x_j to be

$$x_j = \begin{cases} \dfrac{\bar{a}_{mj}}{|a_{mj}|}, & \text{if } a_{mj} \neq 0; \\ 0, & \text{if } a_{mj} = 0; \end{cases}$$

in the norm

$$\|A\mathbf{x}\|_\infty = |a_{m1}x_1 + a_{m2}x_2 + \cdots + a_{mn}x_n|$$

assuming that in (c) the maximum is attained at $i = m$ for $\|A\|_\infty$. Since $\|\mathbf{x}\|_\infty = 1$ by this selection, we have

$$\|A\| = \|A\mathbf{x}\|_\infty = |a_{m1}| + |a_{m2}| + \cdots + |a_{mn}| = \|A\|_\infty. \qquad \square$$

Among all induced matrix norms the value offered by $\|\cdot\|_2$ is the lowest as shown in Problem 9.20(a) and this norm is most widely used in analysis.

9.3.1 Singularity Index

In the light of Definition 9.5, let us find the minimum value of (9.12) which exists in view of Weierstrass theorem and denote it by m, i.e.,

$$m = \min_{\|\mathbf{x}\|=1} \|A\mathbf{x}\| \tag{9.22}$$

$$= \min_{\mathbf{x}\neq\boldsymbol{\theta}} \frac{\|A\mathbf{x}\|}{\|\mathbf{x}\|}. \tag{9.23}$$

From our knowledge of Rayleigh quotient, we find that if the norm under consideration is the Euclidean norm $(\langle \mathbf{x}, \mathbf{x}\rangle)^{1/2} = \|\mathbf{x}\|_2$ in V_n, then

$$m = \min_i \lambda_i(A^*A)$$

and $m = 0$ if A is singular (see Problem 8.41). So a departure from singularity of a matrix A can be measured by the index m given by (9.22), which we shall call *singularity index* of A with respect to the vector norm $\|\cdot\|$. Our next result is along this direction, which, indeed, shows that if a matrix is singular, the index given by the general expression is zero, and any departure from singularity will be given by a value assumed by m. This value of m depends on the underlying vector norm $\|\cdot\|$.

Theorem 9.6 Let $\|A\|$ be a matrix norm induced by the vector norm $\|\mathbf{x}\|$. Then,

$$m = \begin{cases} 0, & \text{if } \det A = 0; \\ \dfrac{1}{\|A^{-1}\|}, & \text{otherwise.} \end{cases}$$

Proof If $\det A = 0$, then $A\mathbf{x} = \boldsymbol{\theta}$ gives a nonzero solution for $\mathbf{x}$ which if normalized gives $\|\mathbf{x}\| = 1$. Consequently, $m = 0$ if det, $A = 0$.

If det $A \neq 0$, let $A\mathbf{x} = \mathbf{y}$ so that $\mathbf{x} = A^{-1}\mathbf{y}$. Substituting into (9.23), we get

$$m = \min_{\mathbf{y} \neq \boldsymbol{\theta}} \frac{\| \mathbf{y} \|}{\| A^{-1}\mathbf{y} \|} = \left[\max_{\mathbf{y} \neq \boldsymbol{\theta}} \frac{\| A^{-1}\mathbf{y} \|}{\| \mathbf{y} \|} \right]^{-1} = [\| A^{-1} \|]^{-1} \qquad \square$$

Summarizing, we find that a matrix norm $\| A \| = 0$ iff $A = 0$, whereas the singularity index $m(A)$ of a matrix A is 0 iff det $A = 0$. Hence, we can look upon matrix norms as a measure of departure of A from a null matrix, while, the singularity index as a measure of departure of A from a singular matrix.

9.4 EQUIVALENT NORMS

Definition 9.6 *Equivalent Vector Norms.* Let two different ways of forming norms in V_n be denoted by $\| \mathbf{x} \|_a$ and $\| \mathbf{x} \|_b$. Then the norm $\| \mathbf{x} \|_a$ is said to be equivalent to $\| \mathbf{x} \|_b$ iff there exist two positive numbers m and M such that

$$m \| \mathbf{x} \|_a \le \| \mathbf{x} \|_b \le M \| \mathbf{x} \|_a \tag{9.24}$$

for all $\mathbf{x}$ in V_n. The constant m and M must be the same whatever be the element $\mathbf{x}$ in V_n. $\square$

Equivalent norms for matrices in $V_{n \times n}$, i.e., *equivalent matrix norms are similarly* defined.

Theorem 9.7 The vector norm equivalence is (a) reflexive, (b) symmetric, and (c) transitive (see Definition 4.4).

Proof (a) *Reflexivity.* Choose $m \le 1$ and $M \ge 1$ in (9.24). Then,

$$m \| \mathbf{x} \|_a \le \| \mathbf{x} \|_a \le M \| \mathbf{x} \|_a.$$

So, $\| \mathbf{x} \|_a$ is equivalent to $\| \mathbf{x} \|_a$.

(b) *Symmetry.* Let $\| \mathbf{x} \|_a$ be equivalent to $\| \mathbf{x} \|_b$, i.e., let (9.24) be true. Then, we have

$$\frac{1}{M} \| \mathbf{x} \|_b \le \| \mathbf{x} \|_a \le \frac{1}{m} \| \mathbf{x} \|_b.$$

So, $\| \mathbf{x} \|_b$ is equivalent to $\| \mathbf{x} \|_a$.

(c) *Transitivity.* Let $\| \mathbf{x} \|_a$ be equivalent to $\| \mathbf{x} \|_b$ and $\| \mathbf{x} \|_b$ be equivalent to $\| \mathbf{x} \|_c$. Then (9.24) is satisfied and also there exist two positive numbers n and N such that

$$n \| \mathbf{x} \|_b \le \| \mathbf{x} \|_c \le N \| \mathbf{x} \|_b. \tag{9.25}$$

With the help of (9.24) and (9.25), we can write

$$m \| \mathbf{x} \|_a \le \| \mathbf{x} \|_b \le \frac{1}{n} \| \mathbf{x} \|_c \le \frac{N}{n} \| \mathbf{x} \|_b \le \frac{MN}{n} \| \mathbf{x} \|_a .$$

Hence, we have

$$mn \| \mathbf{x} \|_a \le \| \mathbf{x} \|_c \le MN \| \mathbf{x} \|_a$$

which shows that $\| \mathbf{x} \|_a$ is equivalent to $\| \mathbf{x} \|_c$. $\square$

Theorem 9.8 All norms on $\mathbb{C}^n$ are equivalent.

Proof Let us consider the Euclidean norm $\| \mathbf{x} \|_2$. The set of points on the unit circle $\| \mathbf{x} \|_2 = 1$ forms a closed and bounded set. Let $\mathbf{y} = \mathbf{x}/\| \mathbf{x} \|_2$ be any point in $\mathbb{C}^n$. Then, any arbitrary norm $\| \mathbf{y} \|$ on $\mathbb{C}^n$, by Theorem 9.3, is a continuous function of $\mathbf{y}$, and by Weierstrass theorem attains a maximum M and a minimum m on the closed and bounded set of the unit circle $\| \mathbf{y} \|_2 = 1$. The vector $\mathbf{y}$ is so defined, that for all values of $\mathbf{x}$ in $\mathbb{C}^n$, $\| \mathbf{y} \|_2 = 1$. Hence, for all values $\mathbf{x}$ in $\mathbb{C}^n$,

$$m \le \| \mathbf{y} \| = \left\| \frac{\mathbf{x}}{\| \mathbf{x} \|_2} \right\| \le M,$$

where $m > 0$, since $\mathbf{y}$ does not vanish on the circle $\| \mathbf{y} \|_2 = 1$ because of Axiom (1) in Definition 9.1. Thus,

$$m \, \| \mathbf{x} \|_2 \le \| \mathbf{x} \| \le M \, \| \mathbf{x} \|_2.$$

Consequently, $\| \mathbf{x} \|_2$ is equivalent to $\| \mathbf{x} \|$. Since $\| \cdot \|$ is quite arbitrary, $\| \mathbf{x} \|_2$ is, therefore, equivalent to all arbitrary norms. Now, by transitivity, all norms are equivalent. □

Defining a set of matrices in $\mathbb{C}^{n \times n}$ by

$$B = \frac{A}{\| A \|_e},$$

we note that, the set of all matrices B satisfying $\| B \|_e = 1$ forms a closed and bounded set. By the proof of Theorem 9.8, we have the following theorem.

Theorem 9.9 All norms on $\mathbb{C}^{n \times n}$ are equivalent. In other words, if $\| \cdot \|_a$ and $\| \cdot \|_b$ are two different norms on $\mathbb{C}^{n \times n}$, then there exist real numbers m and M such that

$$m \, \| A \|_a \le \| A \|_b \le M \, \| A \|_a$$

for any A in $\mathbb{C}^{n \times n}$. □

It is a consequence of Theorem 9.8 that the norms $\| \mathbf{x} \|_1$, $\| \mathbf{x} \|_2$ and $\| \mathbf{x} \|_\infty$ defined in Example 9.1 are equivalent. Let us look into this in more detail.

Example 9.13 The norms $\| \mathbf{x} \|_p$, $p = 1, 2, \infty$, satisfy the following inequalities:

(a) $\| \mathbf{x} \|_\infty \le \| \mathbf{x} \|_1 \le n \, \| \mathbf{x} \|_\infty$, $\quad \| \mathbf{x} \|_\infty \le \| \mathbf{x} \|_2 \le \sqrt{n} \, \| \mathbf{x} \|_\infty$;

(b) $\| \mathbf{x} \|_2 \le \| \mathbf{x} \|_1 \le \sqrt{n} \, \| \mathbf{x} \|_2$, $\quad \dfrac{1}{\sqrt{n}} \| \mathbf{x} \|_1 \le \| \mathbf{x} \|_2 \le \| \mathbf{x} \|_1$;

(c) $\dfrac{1}{n} \| \mathbf{x} \|_1 \le \| \mathbf{x} \|_\infty \le \| \mathbf{x} \|_1$, $\quad \dfrac{1}{\sqrt{n}} \| \mathbf{x} \|_2 \le \| \mathbf{x} \|_\infty \le \| \mathbf{x} \|_2$.

Solution It follows directly from Definition 9.6 that

$$\max_i |x_i| \le \| \mathbf{x} \|_1 = |x_1| + \cdots + |x_n| \le n \max_i |x_i|,$$

$$\max_i |x_i|^2 \le \| \mathbf{x} \|_2^2 = |x_1|^2 + \cdots + |x_n|^2 \le n \max_i |x_i|^2.$$

Hence, we have (a), from which follows (c). Putting $\mathbf{x} = [x_1 \; x_2 \; \cdots \; x_n]^T$ and $\mathbf{y} = [1 \; 1 \; \cdots \; 1]^T$ in Cauchy's inequality (9.5), we have

$$\| \mathbf{x} \|_1 \le \sqrt{n} \, \| \mathbf{x} \|_2.$$

Also, by definition,

$$\| \mathbf{x} \|_1^2 = (|x_1| + \cdots + |x_n|)^2 \geq (|x_1|^2 + \cdots + |x_n|^2) = \| \mathbf{x} \|_2^2$$

which gives (b). □

The importance of studying equivalent norms is that if a matrix sequence or series is convergent with respect to one norm, it is so with respect to all norms in the equivalent class. This property is true also for continuity and boundedness.

9.5 MATRIX SEQUENCE AND MATRIX SERIES

We know that a sequence

$$\{x_p\}: = \{x_1, x_2, ..., x_p, \dots\}$$

of real numbers is said to be convergent iff there exists a real number x such that x_p is close to x for all sufficiently large p and write

$$x_p \to x, \quad \text{or} \quad \lim_{p\to\infty} x_p = x, \quad \text{or} \quad \lim x_p = x,$$

when x is called the *limit of the sequence* $\{x_p\}$. Alternatively, the sequence $\{x_p\}$ is said to *converge to* x iff given any $\varepsilon > 0$ there exists a positive integer N_ε, generally dependent on ε such that

$$p \geq N_\varepsilon \text{ implies } |x_p - x| < \varepsilon. \tag{9.26}$$

This means that, x_p remains within a prescribed distance ε from x for sufficiently large p, i.e.,

$$p \geq N_\varepsilon \text{ implies } |x| - \varepsilon < |x_p| < |x| + \varepsilon, \tag{9.27}$$

or that every convergent sequence is bounded. Also, every convergent sequence has the following property. Given any $\varepsilon > 0$, there exists a positive integer N_ε, such that

$$p, q \geq N_\varepsilon \text{ implies } |x_p - x_q| < \varepsilon, \tag{9.28}$$

Indeed, in view of (9.26), if $\{x_p\}$ is convergent to x, then for any given $\varepsilon > 0$ there exists a positive integer N_ε such that $p, q \geq N_\varepsilon$ implies

$$|x_p - x| < \varepsilon/2 \quad \text{and} \quad |x_q - x| < \varepsilon/2.$$

Hence,

$$|x_p - x_q| \leq |x_p - x| + |x_q - x| < \frac{\varepsilon}{2} + \frac{\varepsilon}{2} = \varepsilon.$$

A sequence satisfying (9.28) is called a *Cauchy sequence* and in the case of real numbers a sequence is convergent iff it is a Cauchy sequence.

Thus, we have three different ways of characterizing a convergent sequence as in (9.26)–(9.28), of which the last one does not require our knowledge of the limit x and is, therefore, convenient to apply for testing convergence of any sequence.

On the other hand, an infinite series of real numbers

$$a_1 + a_2 + \cdots + a_p + \cdots$$

is said to be *convergent* if the sequence $\{x_p\}$ of partial sums defined by

$$x_p = a_1 + a_2 + \cdots + a_p$$

is convergent. In this case, the limit x of $\{x_p\}$ is called the *sum of the infinite series* and is written as

$$x = a_1 + a_2 + \cdots + a_p + \cdots, \text{ and}$$

$$x = \lim_{p\to\infty} x_p = \lim_{p\to\infty} (a_1 + a_2 + \cdots + a_p).$$

Example 9.14 For a series in geometric progression, i.e., for the GP series

$$1 + \alpha + \alpha^2 + \cdots + \alpha^p + \cdots, \tag{9.29}$$

where α is a nonnegative real number with $\alpha < 1$, we have

$$x_p = 1 + \alpha + \cdots + \alpha^p = (1 - \alpha^{p+1})/(1 - \alpha). \tag{9.30}$$

Therefore, with positive integers k and q with $p = q + k$,

$$x_p - x_q = \alpha^{q+1}(1 - \alpha^k)/(1 - \alpha).$$

Hence, given any $\varepsilon > 0$, we have

$$| x_p - x_q | \le \alpha^{q+1}/(1 - \alpha) < \varepsilon \quad \text{whenever } p, q \ge N_\varepsilon,$$

where N_ε is a positive integer satisfying

$$N_\varepsilon > \frac{\log \varepsilon(1-\alpha)}{\log \alpha} - 1. \tag{9.31}$$

Thus, $\{x_p\}$ is convergent to a positive real number x. Given any $\varepsilon > 0$, from (9.30) we can write

$$\left| x_p - \frac{1}{1-\alpha} \right| = \frac{\alpha^{p+1}}{1-\alpha} < \varepsilon \quad \text{whenever } p \ge N_\varepsilon,$$

where N_ε is a positive integer satisfying (9.31). Hence, $x = (1 - \alpha)^{-1}$ is the limit of the sequence $\{x_p\}$ and thus is the sum of (9.29). □

It might be observed that the above GP series is convergent for any α, real or complex, with $|\alpha| < 1$. However, we shall require only the property of the series as stated in the last example in the following discussion.

Example 9.15 For the infinite series

$$1 + a + \frac{a^2}{2!} + \cdots + \frac{a^p}{p!} + \cdots, \tag{9.32}$$

where a is any nonnegative real number, the partial sum is written as

$$x_p = 1 + a + \frac{a^2}{2!} + \cdots + \frac{a^p}{p!}.$$

Since a is finite, we choose a positive integer q such that $a/q =: \alpha < 1$. Hence, for positive integers k and $p := q + k$, we have successively,

$$a_q := a^q/q!, \quad a_{q+1} \le \alpha a_q, \ldots, a_p \le \alpha^k a_q,$$

Thus,

$$0 \leq \lim_{p\to\infty} a_p = \lim_{p\to\infty} a_{q+k} \leq \lim_{k\to\infty} \alpha^k a_q$$

and hence,

$$\lim_{p\to\infty} a_p = \lim_{p\to\infty} \frac{a^p}{p!} = 0.$$

Now,

$$x_p - x_q = a_{q+1} + \cdots + a_{q+k} \leq a_q \alpha (1 + \alpha + \cdots + \alpha^{k-1}).$$

Since $\alpha < 1$ and as $q \to \infty$, $a_q \to 0$, for any given $\varepsilon > 0$, there exists a positive integer N_ε such that

$$p, q \geq N_\varepsilon \text{ implies } x_p - x_q < \varepsilon.$$

So the infinite series (9.32) is convergent and is represented by exp (a). This series can be shown to be convergent when a is, in general, complex, which result will not be used later. □

Analogous to the sequence and the infinite series of real numbers, we can define the convergence of a matrix sequence

$$\{A_1, A_2, \ldots, A_p, \cdots\} =: \{A_p\}$$

and that of a matrix series

$$A_1 + A_2 + \cdots + A_p + \cdots, \tag{9.33}$$

where A_i's are in $V_{n\times n}$.

Definition 9.7 *Convergence of a Matrix Sequence.* Let the *ij*th element of A_p in $V_{n\times n}$ be denoted by $a_{ij}^{(p)}$, $p = 1, 2, \ldots$. The sequence $\{A_p\}$ is said to *converge to the matrix A* in $V_{n\times n}$ iff there exist elements a_{ij} in V such that

$$\lim_{p\to\infty} a_{ij}^{(p)} = a_{ij}$$

when the matrix $A = [a_{ij}]$ is called the limit of the sequence $\{A_p\}$ and is denoted by

$$A_p \to A, \quad \text{or} \quad \lim_{p\to\infty} A_p = A, \quad \text{or} \quad \lim A_p = A. \tag{9.34}$$

□

Although the above definition for convergence is valid for a matrix A of any order $n \times m$, we shall be mainly concerned with convergence of square matrices for which the norm is defined.

Example 9.16 If

$$A_p = \begin{bmatrix} \left(1+\dfrac{1}{p}\right)^p & (p-1)/p & 2^p/p! \\ p \sin\dfrac{1}{p} & \text{arc tan } p & \text{In}\left(1+\dfrac{1}{p}\right) \end{bmatrix}$$

then

$$\lim_{p\to\infty} A_p = \begin{bmatrix} e & 1 & 0 \\ 1 & \pi/2 & 0 \end{bmatrix}. \qquad \square$$

As a consequence of Theorem 9.3(b), if (9.34) holds, then

$$\lim_{p\to\infty} \| A_p - A \| = 0, \quad \text{or} \quad \lim_{p\to\infty} \| A_p \| = \| A \|.$$

Conversely, since $\| \cdot \|$ is a continuous function and $\| A \| = 0$ implies $A = 0$, from

$$\lim_{p\to\infty} \| A_p - A \| = \| \lim_{p\to\infty} (A_p - A) \| = 0,$$

we have

$$\lim_{p\to\infty} (A_p - A) = 0.$$

The above arguments prove the following theorem:

Theorem 9.10 A matrix sequence $\{A_p\}$ is convergent to A iff $\lim_{p\to\infty} \| A_p - A \| = 0$. $\square$

This theorem can also be stated alternatively as in the next theorem.

Theorem 9.11 A matrix sequence $\{A_p\}$ is convergent to A iff for each $\varepsilon > 0$ there exists a positive integer N_ε such that

$$p \ge N_\varepsilon \text{ implies } \| A_p - A \| < \varepsilon. \qquad \square$$

From this and in view of Problem 9.1, we have the following theorem.

Theorem 9.12 The sequence $\{A_p\}$ is convergent to A iff for any $\varepsilon > 0$ there exists a positive integer N_ε such that $p \ge N_\varepsilon$ implies $-\varepsilon + \| A \| \le \| A_p \| \le \varepsilon + \| A \|$, i.e., every convergent matrix sequence is bounded and so $\| A_p \|$ remains within a prescribed distance ε from $\| A \|$ when p is above some sufficiently large positive number N_ε. $\square$

Alternatively, drawing analogy on the sequence of real numbers, we can obtain the following theorem.

Theorem 9.13 The matrix sequence $\{A_p\}$ is convergent iff given any $\varepsilon > 0$, there exists a positive integer N_ε such that

$$p, q \ge N_\varepsilon \text{ implies } \| A_p - A_q \| < \varepsilon. \qquad \square$$

By this theorem, the sequence $\{A_p\}$ can be tested for convergence without knowing its limit.

To define the convergence of the infinite matrix series (9.33), we define first a sequence of partial sums by

$$B_p = A_1 + A_2 + \cdots + A_p.$$

Then the matrix series (9.33) is said to be convergent iff $\{B_p\}$ is convergent. The limit B of $\{B_p\}$ is then said to be the sum of the infinite matrix series (9.33) and is written as

$$B = A_1 + A_2 + \cdots + A_p + \cdots.$$

In view of Theorems 9.10–9.13 on the convergence of a matrix sequence, we can now state the following theorem on the convergence of an infinite matrix series.

Theorem 9.14 The infinite matrix series (9.33) converges to B iff any of the following conditions holds:

(a) $\lim_{p\to\infty} \| B_p - B \| = 0.$

(b) For every $\varepsilon > 0$ there exists a positive integer N_ε such that
$p \ge N_\varepsilon$ implies $\| B_p - B \| < \varepsilon.$

(c) For every $\varepsilon > 0$ there exists a positive integer N, such that
$p \ge N_\varepsilon$ implies $-\varepsilon + \| B \| \le \| B_p \| \le \varepsilon + \| B \|$, i.e., $\{B_p\}$ is bounded.

(d) For every $\varepsilon > 0$ there exists a positive integer N_ε such that
$p, q \ge N_\varepsilon$ implies $\| B_p - B_q \| < \varepsilon.$ □

By Theorem 9.14(d), the convergence of a matrix series can be ascertained without knowing its limit.

Example 9.17 *Sufficient Condition for Convergence of a Matrix GP Series.* Prove that if $\| \cdot \|$ denotes any matrix norm in $V_{n\times n}$ and if $\alpha = \| A \| < 1$ for a matrix A in $V_{n\times n}$, then

$$I + A + \cdots + A^p + \cdots \tag{9.35}$$

converges to $(I - A)^{-1}$.

Solution We have to show that (a) the matrix GP series (9.35) is convergent and (b) $(I - A)^{-1}$ is the limit of this series. Let

$$B_p = I + A + \cdots + A^p.$$

Therefore, in view of Example 9.14 and axioms (3) and (4) of a matrix norm (see Definition 9.3), we have

$$\| B_p - B_q \| = \| A^{q+1} + \cdots + A^p \| \le \alpha^{q+1}/(1 - \alpha).$$

Hence, given any $\varepsilon > 0$, we can find a positive integer N_ε such that $p, q \ge N_\varepsilon$ implies $\| B_p - B_q \| < \varepsilon$, where N_ε satisfies (9.31). So the series (9.35) converges to an element B in $V_{n\times n}$, i.e., $\lim B_p = B$ by Theorem 9.14. We now observe that

$$B_p - AB_p = I - A^{p+1}.$$

So,

$$0 \le \| I - B_p + AB_p \| = \| A^{p+1} \| \le \alpha^{p+1}$$

which in the limit $p \to \infty$ implies that

$$\lim_{p\to\infty} \| I - B_p + AB_p \| = 0.$$

Hence,

$$\lim(I - B_p + AB_p) = 0$$

and so

$$I = \lim(I - A)B_p = (I - A) \lim B_p = (I - A)B.$$

Similarly,

$$B(I - A) = I. \qquad \square$$

A matrix GP series may be convergent even if the condition $\| A \| < 1$ for some norm is violated. For example, with

$$A = \begin{bmatrix} 0.9 & 2 \\ 0 & 0.7 \end{bmatrix}, \tag{9.36}$$

the matrix GP is convergent even though

$$\| A \|_1 = 2.7, \quad \| A \|_\infty = 2.9, \quad \| A \|_e = 2.3.$$

This is so because the *n.a.s.c. for the convergence of* (9.35) *is that all the eigenvalues of A are less than one in absolute value, see Section* 12.5, *i.e.*, $|\lambda_i(A)| < 1$, $i = 1, 2, \ldots, n$, and the matrix A above has 0.9 and 0.7 as eigenvalues (see Problem 7.7).

Example 9.18 Show that for any matrix A in $V_{n \times n}$, the infinite series

$$I + A + \frac{1}{2!}A^2 + \cdots + \frac{1}{p!}A^p + \cdots \tag{9.37}$$

is convergent. This is defined as exp (A) or e^A.

Solution Let $\| A \| = a$, $A_p = A^p/p!$ and B_p denote the sum of first $p + 1$ terms. Then,

$$B_p - B_q = A_{q+1} + \cdots + A_p,$$

where q is so chosen that $a/q =: \alpha < 1$ and $p = q + k$. As in Example 9.15, we can show that

$$\| B_p - B_q \| \le \frac{a^q}{q!} \alpha (1 + \alpha + \cdots + \alpha^{k-1}).$$

Hence, it follows that a positive integer N_ε exists such that

$$p, q \ge N_\varepsilon \text{ implies } \| B_p - B_q \| < \varepsilon.$$

So, (9.37) is convergent to a matrix B in $V_{n \times n}$ by Theorem 9.14(d). If $\| I \| = 1$ as for matrix norms defined in Example 9.12, it is easy to show that

$$\| e^A \| \le \exp (\| A \|). \qquad \square$$

The convergence of a matrix sequence and a matrix series can be related to the eigenvalues of the matrix. To establish such results, Jordan normal forms play a crucial role. These results will be discussed in Section 12.5.

9.6 GENERALIZED INVERSE OF A MATRIX

As discussed in Section 4.8, when an $m \times n$ matrix A is given, the $n \times m$ matrix B is called the *left inverse* of A if $BA = I_n$, and the $n \times m$ matrix C, the *right inverse of A* if $AC = I_m$. If both of these inverses exist for A, then we have $B = C$ which is a unique inverse of A, and $BA = AB = I$, $m = n$. In this case, A is nonsingular. Thus, it is not difficult to see that, for a left inverse B as well as for a right inverse C of A, we have

$$ABA = A, \qquad BAB = B; \tag{9.38a}$$

$$ACA = A, \qquad CAC = C. \tag{9.38b}$$

Moreover, there are an infinite number of right and left inverses of A. Relations (9.38a) and (9.38b) are obviously satisfied for the unique inverse of a nonsingular A.

The object of this section is to constructively establish that for every matrix A, even though it is not nonsingular, nor right or left invertible, there exists a matrix B satisfying relations of the form (9.38a). In this case, B is called a *generalized* (*pseudo* or *natural*) *inverse of* A and is denoted by $A^{\dagger}$. Under certain conditions to be specified later, this generalized inverse is unique.

The following lemma plays a crucial role in establishing the construction of a generalized inverse.

Lemma 9.1 Let A be an $m \times n$ matrix over $\mathbb{C}$. Then, there exist unitary matrices U and V of order m and n respectively, such that

$$A = UDV^*,$$

where

$$D = \text{diag}\,[\sigma_1, ..., \sigma_r, 0, ..., 0] \text{ if } m = n; \tag{9.39}$$

or

$$D = \left[\begin{array}{ccccc|ccc} \sigma_1 & & & & & 0 & \cdots & 0 \\ & \ddots & & & & \vdots & & \vdots \\ & & \sigma_r & & & \vdots & & \vdots \\ & & & 0 & & \vdots & & \vdots \\ & & & & \ddots & \vdots & & \vdots \\ & & & & 0 & 0 & \cdots & 0 \end{array}\right]_{m \times n} \text{ if } m < n; \tag{9.40}$$

or

$$D = \left[\begin{array}{ccccc} \sigma_1 & & & & \\ & \ddots & & & \\ & & \sigma_r & & \\ & & & 0 & \\ & & & & \ddots \\ & & & & 0 \\ \hline 0 & & \cdots & & 0 \\ \vdots & & & & \vdots \\ 0 & & \cdots & & 0 \end{array}\right]_{m \times n} \text{ if } m > n. \tag{9.41}$$

The latter two are called *rectangular diagonal matrices,* where σ_1, ..., σ_r are real and positive, and $r \le p = \min\,(m, n)$. The diagonal elements of D, i.e., $\sigma_1,\ldots\sigma_r$, $\sigma_{r+1},\ldots,\sigma_p$ are called the *singular values* of A and if they are arranged in non-increasing order $\sigma_1 \ge \cdots \ge \sigma_r \ge \cdots \ge \sigma_p \ge 0$ then D is unique while U and V are not.

Proof By Problem 8.40, the matrix A^*A is positive semi-definite and therefore has nonnegative eigenvalues given by $\sigma_1^2, \sigma_2^2, \ldots, \sigma_r^2, 0, \ldots, 0$, where no σ_i's are zero and $r \le \min\,(m, n)$. As a result of Theorem 7.27, there exists a unitary matrix V of order n such that

$$V^* A^* AV = \text{diag}[\sigma_1^2, \sigma_2^2, \ldots, \sigma_r^2, 0, \ldots, 0]_{n\times n}. \tag{9.42}$$

If we set $AV = [\mathbf{x}_1\ \mathbf{x}_2 \cdots \mathbf{x}_n]$, where $\mathbf{x}_i$'s are elements of $\mathbb{C}^m$, then from (9.42),

$$\mathbf{x}_i^*\mathbf{x}_j = \begin{cases} \sigma_i^2 \neq 0, & i = j = 1,\dots,r; \\ 0, & \text{otherwise}. \end{cases} \tag{9.43}$$

From (9.43) we have $\mathbf{x}_i^*\ \mathbf{x}_i = 0$, and hence $\mathbf{x}_i = \boldsymbol{\theta}$ for $i = r + 1, \dots, n$ by Definition 5.9. Consequently, the vectors $\mathbf{x}_1, \mathbf{x}_2, \dots, \mathbf{x}_r$ form an orthogonal set. We now put

$$\mathbf{u}_i = \sigma_i^{-1}\mathbf{x}_i \quad i = 1, 2, \dots, r;$$

so that $\{\mathbf{u}_1, \mathbf{u}_2, \dots, \mathbf{u}_r\}$ is an orthonormal set of vectors in $\mathbb{C}^m$ and apply Theorems 5.4 and 5.8 to extend this set to an orthonormal basis $\{\mathbf{u}_1, \dots, \mathbf{u}_r, \mathbf{u}_{r+1}, \dots, \mathbf{u}_m\}$ in $\mathbb{C}^m$. The matrix U formed with these vectors as columns is obviously unitary. Therefore,

$$\begin{aligned} AV &= [\mathbf{x}_1, \dots, \mathbf{x}_r, \boldsymbol{\theta}, \dots, \boldsymbol{\theta}]_{m\times n} \\ &= [\sigma_1\mathbf{u}_1, \dots, \sigma_r\mathbf{u}_r, \boldsymbol{\theta}, \dots, \boldsymbol{\theta}] \\ &= UD. \end{aligned}$$

Therefore, $A = UDV^*$, where D is of any form given in the Lemma. □

Singular values and singular vectors

The numbers $\sigma_1, \dots \sigma_r$ together with $\sigma_{r+1} = 0, \dots, \sigma_p = 0$, $p = \min(m, n)$ in Lemma 9.1, which are the positive square roots of the eigenvalues of A^*A, are called the *singular values* of A, and $A = UDV^*$, the *singular value decomposition* (SVD) of A.

From $A = UDV^*$ we can write, $A^*\ AV = VD^T\ D$ and $AA^*\ U = UDD^T$. Therefore,

$$A^*A\mathbf{v}_i = \sigma_i^2\mathbf{v}_i, \text{ and } A\,A^*\mathbf{u}_i = \sigma_i^2\mathbf{u}_i$$

which show that $\sigma_1^2, \dots \sigma_r^2$ are the non-zero eigenvalues of both AA^* and $A^*\ A$ with columns of U and V as the respective eigenvectors. The columns of U are called the *left singular vectors* and those of V, the *right singular vectors* of A because of their location with respect to A^*A in the above expressions as explained in Section 7.2.1. Moreover, from the above relations, it is obvious that the columns of U are formed by the eigenvectors of AA^* and those of V by the eigenvectors of $A^*\ A$. The choice of $A^*\ A$ in place of AA^* to define the singular values is only by convention.

From $A = UDV^*$ and the above arguments, it is evident that rank (A^*A) = rank (AA^*) = rank $A = r$.

Using the structure of D given in (9.39)–(9.41), we can write from $A = UDV^*$,

$$A = \sigma_1\mathbf{u}_1\mathbf{v}_1^* + \cdots + \sigma_r\mathbf{u}_r\mathbf{v}_r^*, \tag{9.43a}$$

where $\mathbf{u}_k$ and $\mathbf{v}_k$ are columns of U and V respectively. This representation of A in terms of its singular vectors is known as the *dyadic expansion* of A, or the *outer product* of the SVD. If we multiply any column of U, say $\mathbf{u}_k$, and the like numbered column $\mathbf{v}_k$ in V by $e^{i\varphi}$ (or by -1 in the real case) then the SVD of A remains unchanged although both U and V are now altered. Thus, in the SVD the matrices U and V are not unique although D is unique as the eigenvalues of A^*A for a non-increasing order are so.

Properties of singular values and singular vectors

Rearranging $A = UDV^*$, the SVD of A can be written in two basic forms, $AV = UD$ and $A^*U = VD^T$. Equating the columns from both sides,

$$A\mathbf{v}_i = \sigma_i\mathbf{u}_i, i = 1,\ldots,r;\ A\mathbf{v}_i = 0, i = r+1,\ldots,n; \tag{9.43b}$$

$$A^*\mathbf{u}_i = \sigma_i\mathbf{v}_i, i = 1,\ldots,r;\ A^*\mathbf{u}_i = 0, i = r+1,\ldots,m, \tag{9.43c}$$

1. It appears from (9.43b) that if we select $\{\mathbf{v}_1,\ldots,\mathbf{v}_n\}$ as a basis in the domain and $\{\mathbf{u}_1,\ldots,\mathbf{u}_m\}$ as a basis in the range space of $A \in \mathbb{C}^{m\times n}$, its representation is a diagonal matrix (9.39), (9.40), or (9.41) according to the theory developed in Section 6.3. These bases are called *singular bases* of A.
2. It follows from (9.43b) that the range space of A is spanned by $\{\mathbf{u}_1,\ldots,\mathbf{u}_r\}$ and the null space by $\{\mathbf{v}_{r+1},\ldots,\mathbf{v}_n\}$. Therefore, rank $A = r$ and null $A = n - r$. Thus the number of non-zero singular values is the rank of a matrix. As the singular values can be computed in a numerically stable way, the best method to determine the rank of a matrix is to count the number of non-zero singular values greater than a judiciously chosen tolerance.
3. We conclude from (9.43c) that the range space of A^* is spanned by $\{\mathbf{v}_1,\ldots,\mathbf{v}_r\}$ and the null space by $\{\mathbf{u}_{r+1},\ldots,\mathbf{u}_m\}$.
4. Define $U_1 = [\mathbf{u}_1\ldots\mathbf{u}_r]$, where the set $\{\mathbf{u}_1,\ldots\mathbf{u}_r\}$ spans $\mathbb{R}(A)$ as shown in 2. By Corollary 6.9, orthogonal projection (OP) on $\mathbb{R}(A)$ is $U_1(U_1^*U_1)^{-1}U_1^* = U_1U_1^*$ since $U_1^*U_1 = I_r$. Define $V_2 = [\mathbf{v}_{r+1}\cdots\mathbf{v}_n]$, where the set $\{\mathbf{v}_{r+1},\ldots,\mathbf{v}_n\}$ spans $\mathbb{N}(A)$ as shown in 2. By Corollary 6.9, projection on $\mathbb{N}(A)$ is $V_2(V_2^*V_2)^{-1}V_2^* = V_2V_2^*$ since $V_2^*V_2 = I_{n-r}$.
5. Define $V_1 = [\mathbf{v}_1\cdots\mathbf{v}_r]$, where the set $\{\mathbf{v}_1,\ldots,\mathbf{v}_r\}$ spans $\mathbb{R}(A^*)$ as shown in 3. By Corollary 6.9, projection on $\mathbb{R}(A^*)$ is $V_1(V_1^*V_1)^{-1}V_1^* = V_1V_1^*$ since $V_1^*V_1 = I_r$. Define $U_2 = [\mathbf{u}_{r+1}\cdots\mathbf{u}_m]$, where the set $\{\mathbf{u}_{r+1},\ldots,\mathbf{u}_m\}$ spans $\mathbb{N}(A^*)$ as shown in 3. By Corollary 6.9, projection on $\mathbb{N}(A^*)$ is $U_2(U_2^*U_2)^{-1}U_2^* = U_2U_2^*$ since $U_2^*U_2 = I_{n-r}$.
6. For a normal matrix A which includes Hermitian matrices, $\sigma_i(A) = |\lambda_i(A)|$. This follows from the relation $A = U$ diag $[\lambda_1, \lambda_2,\ldots\lambda_n]U^*$ for a normal matrix as given in Theorem 7.33, where U is unitary. In particular, for a PD or a PSD matrix, $\sigma_i(A) = \lambda_i(A)$.
7. As explained in Example 9.9(b), the spectral norm of A of order n is defined as $\|A\|_2 = [\lambda_{\max}(A^*A)]^{1/2} = \sigma_{\max}(A) = \sigma_1$. If A is non-singular, then $\|A^{-1}\|_2 = 1/\sigma_n$ and the condition number of A as defined by (13.5) is $k(A) := \|A\|_2\,\|A^{-1}\|_2 = \sigma_1/\sigma_n$.
8. Using the definition in Example 9.5(d), the Euclidean Norm $\|A\|_e$ or Frobenius norm $\|A\|_F$ of a square matrix A is $\|A\|_e^2 = \sigma_1^2 + \sigma_2^2 + \cdots + \sigma_r^2$. Indeed, from (9.42) trace (V^*A^*AV) = trace (VV^*A^*A) = trace (A^*A) = trace of diag $[\sigma_1^2, \sigma_2^2,\ldots,\sigma_r^2, 0,\ldots,0]$, where we have used Definition 1.14 and the property of a trace of the product of two matrices.

Note: Singular values are positive while eigenvalues are not.

If A is symmetric and positive definite, then its eigenvalues are also its singular values, and $U = V$. If A is symmetric with some eigenvalues negative (they will be real, because A is symmetric), then the singular values are the magnitudes of the eigenvalues, and U and V are the same up to multiplication by minus one for the columns corresponding to negative eigenvalues.

On the other hand for a normal matrix A of order n if we choose the eigenvectors as the basis then $A = U\,\text{diag}[\lambda_1,\ldots,\lambda_n]U^*$, where λ_is are the eigenvalues of A implying that eigenvectors are singular bases for a normal matrix.

Summary of SVD Properties

$\mathbb{R}(A) = \mathbb{R}(U_1)$;	$\mathbb{N}(A) = \mathbb{R}(V_2)$;
$\mathbb{R}(A^*) = \mathbb{R}(V_1)$;	$\mathbb{N}(A^*) = \mathbb{R}(U_2)$.
$U_1U_1^* = \text{OP on } \mathbb{R}(A)$;	$V_2V_2^* = \text{OP on } \mathbb{N}(A)$;
$V_1V_1^* = \text{OP on } \mathbb{R}(A^*)$;	$U_2U_2^* = \text{OP on } \mathbb{N}(A^*)$.
$\|A\|_2 = \sigma_1$.	$\|A\|_F^2 = \sigma_1^2 + \sigma_2^2 + \cdots + \sigma_r^2$.

Here, $\mathbb{R}(A)$ represents range space of A, $\mathbb{N}(A)$, the null space of A and OP, the orthogonal projection. We define $U = [U_1\ U_2]$, $U_1 = [\mathbf{u}_1\ldots\mathbf{u}_r]$, $U_2 = [\mathbf{u}_{r+1}\ldots\mathbf{u}_m]$ and $V = [V_1\ V_2]$, $V_1 = [\mathbf{v}_1\ldots\mathbf{v}_r]$, $V_2 = [\mathbf{v}_{r+1}\ldots\mathbf{v}_n]$

SVD as rotation and stretching

If we apply a tangential force to the surface DC of a cubical block, the thin layers of the block parallel to DC will slide over those lying below as shown in Figure 9.3. The tangent of the angle through which two adjacent sides rotate relative to their initial position is termed as shear strain, and the tangential force per unit area is the shear stress. A shear transformation is a linear map that displaces each point in fixed direction, by an amount proportional to its signed distance from a line that is parallel to that direction of stress. An example is the mapping that takes any point D with coordinates (x, y) to the point $F{:}(x + my, y)$, so in Figure 9.3, $m = \tan\theta$, where θ is the angle between AD and AF. In this case, the displacement is horizontal, the fixed line is the x-axis, and the signed distance is the y coordinate. In general, we represent the mapping as

$$A\begin{bmatrix} x \\ y \end{bmatrix} = \begin{bmatrix} x + my \\ y \end{bmatrix}, \text{ where } A = \begin{bmatrix} 1 & m \\ 0 & 1 \end{bmatrix}$$

and m is the factor of proportionality.

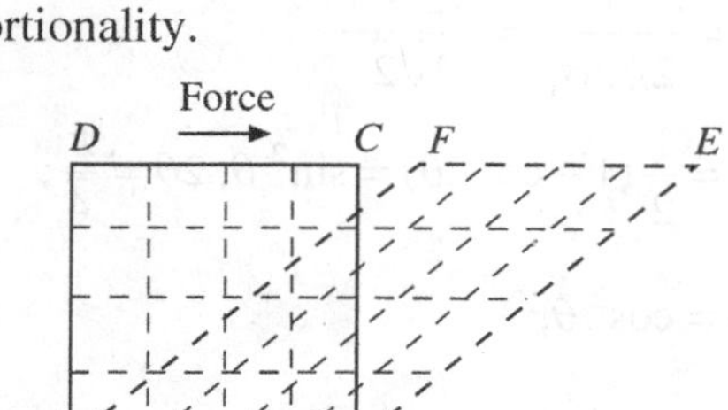

Figure 9.3 Shear in a cubical block for m = 1.2.

The shear will usually distort the shape of a geometric figure, for example turning squares into non-square parallelograms, and circles into ellipses. However, a shearing does preserve the area of geometric figures, the alignment and relative distances of collinear points.

The same definition is used in three-dimensional geometry, except that the distance is measured from a fixed plane. A three-dimensional shearing transformation preserves the volume of solid figures, but changes areas of plane figures (except those that are parallel to the displacement).

In the general n-dimensional Cartesian space $\mathbb{R}^n$, the distance is measured from a fixed hyperplane parallel to the direction of displacement. This geometric transformation is a linear transformation of $\mathbb{R}^n$ that preserves the n-dimensional measure (hyper-volume) of any set.

We will describe below using SVD, how is a circle distorted by shearing transformation to an ellipse with a simple second order example.

Let the shearing transformation with $m = 2$ be given by

$$A = \begin{bmatrix} 1 & 2 \\ 0 & 1 \end{bmatrix}.$$

The matrix $A^T A$ and AA^T are, respectively,

$$A^T A = \begin{bmatrix} 1 & 2 \\ 2 & 5 \end{bmatrix}, \text{ and } AA^T = \begin{bmatrix} 5 & 2 \\ 2 & 1 \end{bmatrix}.$$

For both of them, the characteristic equation is $\lambda^2 - 6\lambda + 1 = 0$ and the eigenvalues, $\sigma_1^2 = (\sqrt{2}+1)^2$ and $\sigma_2^2 = (\sqrt{2}-1)^2$. As a result the singular values of A are $\sigma_1 = \sqrt{2}+1$ and $\sigma_2 = \sqrt{2}-1$. The eigenvectors of $A^T A$ and AA^T are, respectively,

$$\tilde{\mathbf{v}}_1 = \begin{bmatrix} 1 \\ \sigma_1 \end{bmatrix}, \tilde{\mathbf{v}}_2 = \begin{bmatrix} -1 \\ \sigma_2 \end{bmatrix}, \text{ and } \tilde{\mathbf{u}}_1 = \begin{bmatrix} 1 \\ \sigma_2 \end{bmatrix}, \tilde{\mathbf{u}}_2 = \begin{bmatrix} -1 \\ \sigma_1 \end{bmatrix}.$$

Hence, the orthonormal eigenvectors of $A^T A$ and AA^T are, respectively,

$$\mathbf{v}_1 = \begin{bmatrix} \sin\theta \\ \cos\theta \end{bmatrix}, \mathbf{v}_2 = \begin{bmatrix} -\cos\theta \\ \sin\theta \end{bmatrix}, \text{ and } \mathbf{u}_1 = \begin{bmatrix} \cos\theta \\ \sin\theta \end{bmatrix}, \mathbf{u}_2 = \begin{bmatrix} -\sin\theta \\ \cos\theta \end{bmatrix},$$

which follow from $1+\sigma_1^2 = 2\sqrt{2}\sigma_1$, $1+\sigma_2^2 = 2\sqrt{2}\sigma_2$,

$$\frac{1}{1+\sigma_1^2} = \frac{1}{2\sqrt{2}\sigma_1} = \frac{\sqrt{2}-1}{2\sqrt{2}}$$

$$= \frac{1}{2}(1-\cos 2\theta) = \sin^2\theta,\ 2\theta = \frac{\pi}{4};$$

$$\frac{\sigma_1^2}{1+\sigma_1^2} = \cos^2\theta,$$

$$\frac{1}{1+\sigma_2^2} = \frac{1}{2\sqrt{2}\sigma_2} = \frac{\sqrt{2}+1}{2\sqrt{2}}$$

$$= \frac{1}{2}(1+\cos 2\theta) = \cos^2\theta,$$

$$\frac{\sigma_2^2}{1+\sigma_2^2} = \sin^2\theta.$$

So, the SVD of A is $A = U\Sigma V^*$ where $U = [\mathbf{u}_1\ \mathbf{u}_2]$ and $V = [\mathbf{v}_1\ \mathbf{v}_2]$ given by

$$V = \begin{bmatrix} \sin\theta & -\cos\theta \\ \cos\theta & \sin\theta \end{bmatrix}, \text{ and } U = \begin{bmatrix} \cos\theta & -\sin\theta \\ \sin\theta & \cos\theta \end{bmatrix}$$

are unitary, $\Sigma = \mathrm{diag}[\sigma_1, \sigma_2]$, and $A\mathbf{u}_i = \sigma_i\mathbf{v}_i$, $i = 1, 2$.

The vectors $\mathbf{u}_1$ and $\mathbf{u}_2$ are unit vectors making angles θ and $(\pi/2) + \theta$, respectively, and $\mathbf{v}_1$ and $\mathbf{v}_2$ are also unit vectors making angles $(\pi/2) - \theta$ and $\pi - \theta$, respectively with the x-axis as shown in Figure 9.4, where $\theta = \pi/8$ and $\phi = (\pi/2) - \theta$. The transformation A maps $\mathbf{v}_1$ and $\mathbf{v}_2$ lying on a unit circle, respectively, to $\sigma_1\mathbf{u}_1$ and $\sigma_2\mathbf{u}_2$ lying on an ellipse with semi-major axis σ_1, the first singular value and semi-minor axis σ_2, the second singular value of A. The semi-major axis, representing the farthest point $A\mathbf{x}$ on the ellipse, is the *spectral norm* $\| A \|_2$, and the inverse of the semi-minor axis, representing the nearest point $A\mathbf{x}$, is the *spectral norm* $\| A^{-1} \|_2$.

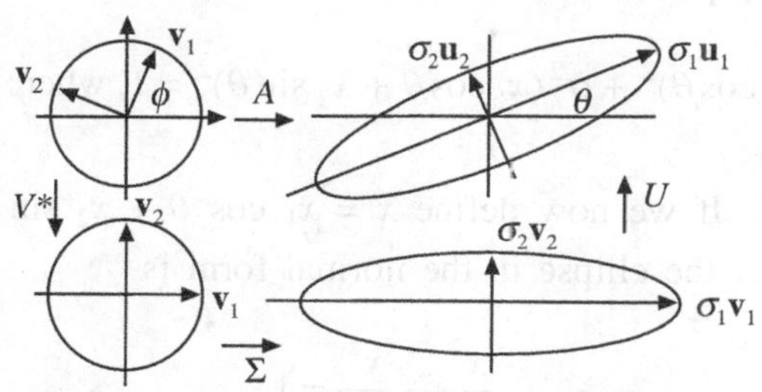

Figure 9.4 Geometric interpretation of SVD and $\| A \|_2$.

Looking into the mapping in terms of the components in the SVD of A, V^* maps the unit circle into another unit circle by rotating it through ϕ, the diagonal matrix Σ stretches $\mathbf{v}_1$ to $\sigma_1\mathbf{v}_1$ and compresses $\mathbf{v}_2$ to $\sigma_2\mathbf{v}_2$ transforming the circle into an ellipse and finally U rotates the ellipse anticlockwise through θ. Thus, the map under A is a composition of three maps defined by V^*, Σ and U.

Any point on the unit circle can be expressed as $(\cos\psi, \sin\psi)$ and the image (u, v) of this point by the shearing transformation matrix A is given by

$$\begin{bmatrix} u \\ v \end{bmatrix} = \begin{bmatrix} 1 & 2 \\ 0 & 1 \end{bmatrix} \begin{bmatrix} \cos\psi \\ \sin\psi \end{bmatrix}.$$

Therefore, $u = \cos\psi + 2\sin\psi$ and $v = \sin\psi$, from which it follows that $u^2 - 4uv + 5v^2 = 1$ is the equation of the desired ellipse which written in a matrix form is

$$[u \quad v]\begin{bmatrix} 1 & -2 \\ -2 & 5 \end{bmatrix}\begin{bmatrix} u \\ v \end{bmatrix} = 1.$$

The symmetric matrix S in the above form has a characteristic equation $\lambda^2 - 6\lambda + 1 = 0$ with eigenvalues $\sigma_1^2(\sqrt{2}+1)^2$ and $\sigma_2^2 = (\sqrt{2}-1)^2$. The eigenvectors of S associated with these eigenvalues are, respectively,

$$\begin{bmatrix} 1 \\ -\sigma_1 \end{bmatrix} \text{ and } \begin{bmatrix} 1 \\ \sigma_2 \end{bmatrix}$$

which after normalization are

$$\begin{bmatrix} \sin\theta \\ -\cos\theta \end{bmatrix} \text{ and } \begin{bmatrix} \cos\theta \\ \sin\theta \end{bmatrix},$$

where $2\theta = \pi/4$ in the way which follow proved before. If we define P with these normalized eigenvectors as columns:

$$P = \begin{bmatrix} \sin\theta & \cos\theta \\ -\cos\theta & \sin\theta \end{bmatrix},$$

then the equation of the ellipse in a normal form is

$$\sigma_1^2(x_1 \sin\theta - x_2 \cos\theta)^2 + \sigma_2^2(x_1\cos\theta + x_2\sin\theta)^2 = 1, \text{ where } \begin{bmatrix} x_1 \\ x_2 \end{bmatrix} = P^T\begin{bmatrix} u \\ v \end{bmatrix},$$

and $P^TSP = \text{diag}[\sigma_1^2, \sigma_2^2]$. If we now define $x = x_1 \cos\theta + x_2 \sin\theta$ and $y = x_1 \sin\theta - x_2 \cos\theta$, then the equation of the ellipse in the normal form is

$$\frac{x^2}{\sigma_1^2} + \frac{y^2}{\sigma_2^2} = 1$$

with x-axis defined by $x_1 \sin\theta - x_2 \cos\theta = 0$ at an angle θ with the horizontal direction and y-axis by $x_1 \cos\theta + x_2 \sin\theta = 0$ at an angle $(\pi/2) + \theta$ as shown in Figure 9.4.

Let $A = U\Sigma V^*$ be an SVD of an $m \times n$ matrix A with rank r. Then the image of the unit sphere $\|\mathbf{x}\| = 1$ in $\mathbb{R}^n$ under A that maps $\mathbf{x}$ to $A\mathbf{x}$ is

(a) the surface of an ellipsoid in $\mathbb{R}^m$ if $r = n$,

(b) a solid ellipsoid in $\mathbb{R}^m$ if $r < n$.

We note that since $\|\mathbf{x}\| = 1$, by Theorem 6.13 $\|V^T\mathbf{x}\| = 1$ as V is orthogonal. Denoting $\mathbf{x} = [x_1 \cdots x_n]^T$, we have,

$$(\mathbf{v}_1^T\mathbf{x})^2 + \cdots + (\mathbf{v}_n^T\mathbf{x})^2 = 1,$$

where $V = [\mathbf{v}_1 \cdots \mathbf{v}_n]$. Post-multiplying the dyadic expansion (9.43a) by $\mathbf{x}$,

$$\begin{aligned}A\mathbf{x} &= \sigma_1\mathbf{u}_1\mathbf{v}_1^T\mathbf{x} + \cdots + \sigma_r\mathbf{u}_r\mathbf{v}_r^T\mathbf{x}\\ &= (\sigma_1\mathbf{v}_1^T\mathbf{x})\mathbf{u}_1 + \cdots + (\sigma_r\mathbf{v}_r^T\mathbf{x})\mathbf{u}_r\\ &= y_1\mathbf{u}_1 + \cdots + y_r\mathbf{u}_r,\ y_i = \sigma_i\mathbf{v}_i^T\mathbf{x},\\ &= U\mathbf{y},\end{aligned}$$

where $\mathbf{y} = [y_1 \cdots y_r]^T$ and $\| A\mathbf{x} \| = \| U\mathbf{y} \| = \|\mathbf{y}\|$ by Theorem 6.13 as U is orthogonal.

To prove (a), we take $r = n$. Then, $n \le m$. Since, $y_i = \sigma_i\mathbf{v}_i^T\mathbf{x},\ i = 1,\ldots,n,$

$$\left(\frac{y_1}{\sigma_1}\right)^2 + \cdots + \left(\frac{y_n}{\sigma_n}\right)^2 = (\mathbf{v}_1^T\mathbf{x})^2 + \cdots + (\mathbf{v}_n^T\mathbf{x})^2 = 1,$$

which shows that the vectors $A\mathbf{x}$ form the surface of an ellipsoid in $\mathbb{R}^n$ as $\| A\mathbf{x} \| = \| \mathbf{y} \|$.
To prove (b), we take $r < n$. Then,

$$\begin{aligned}\left(\frac{y_1}{\sigma_1}\right)^2 + \cdots + \left(\frac{y_r}{\sigma_r}\right)^2 &= (\mathbf{v}_1^T\mathbf{x})^2 + \cdots + (\mathbf{v}_r^T\mathbf{x})^2\\ &< (\mathbf{v}_1^T\mathbf{x})^2 + \cdots + (\mathbf{v}_n^T\mathbf{x})^2 = 1,\end{aligned}$$

an ellipsoid in $\mathbb{R}^n$ of dimension r.
For the SVD given below,

$$A = \begin{bmatrix} 0 & 1 & 1\\ 1 & 0 & 0\end{bmatrix} = \begin{bmatrix} 1 & 0\\ 0 & 1\end{bmatrix}\begin{bmatrix} \sqrt{2} & 0 & 0\\ 0 & 1 & 0\end{bmatrix}\begin{bmatrix} 0 & \frac{1}{\sqrt{2}} & \frac{1}{\sqrt{2}}\\ 1 & 0 & 0\\ 0 & \frac{1}{\sqrt{2}} & -\frac{1}{\sqrt{2}}\end{bmatrix},$$

the matrix A transforms the sphere $x_1^2 + x_2^2 + x_3^2 = 1$ in $\mathbb{R}^3$ to the ellipsoid $\frac{y_1^2}{2} + y_2^2 < 1$ in $\mathbb{R}^2$ noting that $\sigma_1^2 = 2$ and $\sigma_2^2 = 1$.

In a different problem, we consider a quadratic form $q = \mathbf{x}^T A\mathbf{x}$ represented in a standard basis in $\mathbb{R}^n$. If A is PD then it follows from Corollary 7.7 and Theorem 8.19(a) that, $A = VDV^T$, V is orthogonal, where $D = \text{diag}[\lambda_1, \lambda_2,\ldots,\lambda_n]$, and λ_is are positive eigenvalues of A in a non-increasing order. If we now select the columns $\{\mathbf{v}_1, \mathbf{v}_2,\ldots,\mathbf{v}_n\}$ of V as a basis in $\mathbb{R}^n$, then the representation of $\mathbf{x}$ in terms of the new basis is $\mathbf{y} = [y_1\ y_2 \cdots y_n]^T$ such that $\mathbf{x} = V\mathbf{y}$. Consequently,

$$\mathbf{x}^T A\mathbf{x} = \mathbf{x}^T V DV^T \mathbf{x} = \mathbf{y}^T D\mathbf{y}.$$

That is,

$$q = \mathbf{x}^T A\mathbf{x} = \lambda_1 y_1^2 + \lambda_2 y_2^2 - \cdots + \lambda_n y_n^2,$$

which represents the surface of an n-dimensional ellipsoid, a *closed surface* with principal axes along $\mathbf{v}_i$, $i = 1, 2,\ldots, n$. If $q = 1$, the semi-principal axes of this ellipsoid are $1/\sqrt{\lambda_i}$ with $1/\sqrt{\lambda_n}$

as the semi-major axis and $1/\sqrt{\lambda_1}$ as the semi-minor axis, since λ_1 is the largest and λ_n is the smallest eigenvalue. As A is PD, if follows from $A = VDV^T$ that λ_is are also the singular values of A.

Example 9.19 Let

$$A = \begin{bmatrix} 1 & 1-i \\ 1+i & 2 \\ 1 & 1-i \end{bmatrix}. \tag{9.44}$$

Therefore,

$$A^*A = 4\begin{bmatrix} 1 & 1-i \\ 1+i & 2 \end{bmatrix}$$

which has eigenvalues as $\sigma_1^2 = 12, \sigma_2 = 0$ and the corresponding orthogonal set of eigenvectors as

$$[1, \quad 1+i]^T \quad \text{and} \quad [1-i, \quad -1]^T.$$

After normalization, we choose a unitary matrix V with the above eigenvectors as columns.

$$V = \frac{1}{\sqrt{3}}\begin{bmatrix} 1 & 1-i \\ 1+i & -1 \end{bmatrix} \tag{9.45}$$

so that $V^*A^*AV = \text{diag}\,[12, 0]$. Now

$$AV = \sqrt{3}\begin{bmatrix} 1 & 0 \\ 1+i & 0 \\ 1 & 0 \end{bmatrix} =: [\sigma_1\mathbf{u}_1, \mathbf{0}],$$

where $\mathbf{u}_1 = \frac{1}{2}[1\ 1+i\ 1]^T$ in $\mathbb{C}^3$, which we extend to a basis $\{\mathbf{u}_1, \mathbf{e}_2, \mathbf{e}_3\}$ in $\mathbb{C}^3$, where $\mathbf{e}_2$ and $\mathbf{e}_3$ are elementary vectors in $\mathbb{C}^3$. We now apply Theorem 5.7 to construct an orthonormal basis $\{\mathbf{x}_1, \mathbf{x}_2, \mathbf{x}_3\}$ in $\mathbb{C}^3$:

$$\mathbf{x}_1 = \mathbf{u}_1, \quad \mathbf{x}_2 = \mathbf{e}_2 - \frac{\langle \mathbf{e}_2, \mathbf{x}_1 \rangle}{\langle \mathbf{x}_1, \mathbf{x}_1 \rangle}\mathbf{x}_1 = \frac{1}{4}\begin{bmatrix} -1+i \\ 2 \\ -1+i \end{bmatrix},$$

$$\mathbf{x}_3 = \mathbf{e}_3 - \frac{\langle \mathbf{e}_3, \mathbf{x}_1 \rangle}{\langle \mathbf{x}_1, \mathbf{x}_1 \rangle}\mathbf{x}_1 - \frac{\langle \mathbf{e}_3, \mathbf{x}_2 \rangle}{\langle \mathbf{x}_2, \mathbf{x}_2 \rangle}\mathbf{x}_2 = \frac{1}{2}\begin{bmatrix} -1 \\ 0 \\ 1 \end{bmatrix}.$$

After normalization, we choose

$$\left\{ \mathbf{u}_1 = \frac{1}{2}[1\ \ 1+i\ \ 1]^T,\ \mathbf{u}_2 = \frac{\sqrt{2}}{4}[-1+i, 2, -1+i]^T,\ \mathbf{u}_3 = \frac{\sqrt{2}}{2}[-1\ 0\ 1]^T \right\}$$

as the orthonormal basis in $\mathbb{C}^3$. Consequently, we set

$$U = \frac{\sqrt{2}}{2}\begin{bmatrix} 1/\sqrt{2} & (-1+i)/2 & -1 \\ (1+i)/\sqrt{2} & 1 & 0 \\ 1/\sqrt{2} & (-1+i)/2 & 1 \end{bmatrix}. \tag{9.46}$$

It is easy to check that

$$U^*AV = \left[\begin{array}{cc} 2\sqrt{3} & 0 \\ 0 & 0 \\ \hdashline 0 & 0 \end{array}\right] = D \tag{9.47}$$

which is a rectangular diagonal matrix.

Theorem 9.15 Let A be a matrix of order $m \times n$ over $\mathbb{C}$. Then there exists a *unique* matrix $A^\dagger$ of order $n \times m$ satisfying (a) $AA^\dagger A = A$, (b) $A^\dagger AA^\dagger = A^\dagger$, such that (c) $AA^\dagger$ and $A^\dagger A$ are Hermitian.

Proof Suppose, at first, that A is a diagonal matrix having any of the forms (9.39)-(9.41). Define an $n \times m$ matrix $D^\dagger$ by

$$D^\dagger = \text{diag}\ [\sigma_1^{-1}, \sigma_2^{-1}, \ldots, \sigma_r^{-1}, 0\ldots, 0] \qquad \text{if } m = n;$$

$$D^\dagger = \left[\begin{array}{ccc} \sigma_1^{-1} & & & & \\ & \ddots & & & \\ & & \sigma_r^{-1} & & \\ & & & 0 & \\ & & & & \ddots & \\ & & & & & 0 \\ \hline 0 & & \cdots & & & 0 \\ \vdots & & & & & \vdots \\ 0 & & \cdots & & & 0 \end{array}\right]_{n \times m} \qquad \text{if } n > m;$$

or

$$D^\dagger = \left[\begin{array}{cccccc|ccc} \sigma_1^{-1} & & & & & & 0 & \ldots & 0 \\ & \ddots & & & & & & & \\ & & \sigma_r^{-1} & & & & \vdots & & \vdots \\ & & & 0 & & & & & \\ & & & & \ddots & & & & \\ & & & & & 0 & 0 & \ldots & 0 \end{array}\right]_{n \times m} \qquad \text{if } n < m.$$

Then by direct multiplication,

$$D^{\dagger}D = \text{diag}\ [1, \ldots, 1,\ \ 0, \ldots, 0]_{n\times n},\quad DD^{\dagger} = \text{diag}\ [1, \ldots, 1,\ \ 0, \ldots, 0]_{m\times m};$$

which are obviously Hermitian, thereby giving us property (c). Also, it can be easily verified that $DD^{\dagger}D = D$ and $D^{\dagger}DD^{\dagger} = D^{\dagger}$, thus proving property (a) and (b).

Suppose now that A is an arbitrary $m \times n$ matrix. Then, in view of Lemma 9.1, there exist unitary matrices U and V of order m and n respectively such that $A = UDV^*$, where D is a diagonal matrix of the form given in (9.39)–(9.41). We now define an $n \times m$ matrix $A^{\dagger}$ by

$$A^{\dagger} = VD^{\dagger}U^*$$

Then, we get

$$AA^{\dagger} = UDV^*VD^{\dagger}U^* = UDD^{\dagger}U^*;\quad A^{\dagger}A = VD^{\dagger}U^*UDV^* = VD^{\dagger}DV^*$$

which show that $AA^{\dagger}$ and $A^{\dagger}A$ are Hermitian because $DD^{\dagger}$ and $D^{\dagger}D$ are so. Moreover,

$$AA^{\dagger}A = UDD^{\dagger}U^*UDV^* = UDD^{\dagger}DV^* = UDV^* = A,$$

$$A^{\dagger}AA^{\dagger} = VD^{\dagger}DV^*VD^{\dagger}U^* = VD^{\dagger}DD^{\dagger}U^* = VD^{\dagger}U^* = A^{\dagger}$$

which prove (a) and (b) respectively.

To show that $A^{\dagger}$ is a unique inverse of A, we assume that $A_1{}^{\dagger}$ is another inverse of A. We note that by property (c)

$$AA^{\dagger} = (A^{\dagger})^*A^*,\quad A^{\dagger}A = A^*(A^{\dagger})^*,\qquad AA_1^{\dagger} = (A_1^{\dagger})^*A^*,\qquad A_1^{\dagger}A = A^*(A_1^{\dagger})^*,$$

Hence,

$$\begin{aligned} A_1^{\dagger} &= A^{\dagger}AA^{\dagger} = A^*(A^{\dagger})^*A^{\dagger} = \{A^*(A_1^{\dagger})^*A^*\}(A^{\dagger})^*A^{\dagger} \\ &= A_1^{\dagger}AA^{\dagger}AA^{\dagger} = A_1^{\dagger}AA^{\dagger} = A_1^{\dagger}(A^{\dagger})^*A^* \\ &= A_1^{\dagger}(A^{\dagger})^*\{A^*(A_1^{\dagger})^*A^*\} = A_1^{\dagger}AA^{\dagger}AA_1^{\dagger} \\ &= A_1^{\dagger}AA_1^{\dagger} = A_1^{\dagger} \end{aligned}$$

Therefore, $A^{\dagger}$ is a unique generalized or pseudo inverse of A, see (9.38a). □

Corollary 9.1 If $A \in \mathbb{R}^{m\times n}$ (resp $\mathbb{C}^{m\times n}$) is a matrix with linearly independent columns, then the pseudo-inverse of A is the matrix $A^{\dagger}$ defined by

$$A^{\dagger} = (A^T A)^{-1} A^T \quad (\text{resp } A^{\dagger} = (A^* A)^{-1} A^*)$$

and is unique.

Proof The three properties in Theorem 9.15 are satisfied by $A^{\dagger}$ defined in the Corollary.

Remark 9.1 When A is square and nonsingular, premultiplying and postmultiplying both sides of $AA^{\dagger}A = A$ by A^{-1}, we have $A^{\dagger}A = I$ and $AA^{\dagger} = I$ respectively. Therefore, $A^{\dagger} = A^{-1}$.

Example 9.20 To determine the generalized inverse of A given by (9.44), we note that D given by (9.47) has a generalized inverse

$$D^{\dagger} = \begin{bmatrix} 1/2\sqrt{3} & 0 & 0 \\ 0 & 0 & 0 \end{bmatrix}$$

So, the generalized inverse of A is $A^{\dagger} = VD^{\dagger}U^*$, where U and V are given by (9.46) and (9.45), respectively. Therefore,

$$A^{\dagger} = \frac{1}{\sqrt{3}}\begin{bmatrix} 1 & 1-i \\ 1+i & -1 \end{bmatrix}\begin{bmatrix} 1/2\sqrt{3} & 0 & 0 \\ 0 & 0 & 0 \end{bmatrix}\frac{\sqrt{2}}{2}\begin{bmatrix} 1/\sqrt{2} & (1-i)/\sqrt{2} & 1/\sqrt{2} \\ -(1+i)/2 & 1 & -(1+i)/2 \\ -1 & 0 & 1 \end{bmatrix}$$

$$= \frac{1}{\sqrt{6}}\begin{bmatrix} 1 & 1-i \\ 1+i & -1 \end{bmatrix}\begin{bmatrix} 1/2\sqrt{6} & (1-i)/2\sqrt{6} & 1/2\sqrt{6} \\ 0 & 0 & 0 \end{bmatrix}$$

$$= \frac{1}{\sqrt{6}}\begin{bmatrix} 1/2\sqrt{6} & (1-i)/2\sqrt{6} & 1/2\sqrt{6} \\ (1+i)/2\sqrt{6} & 2/2\sqrt{6} & (1+i)/2\sqrt{6} \end{bmatrix}$$

$$= \frac{1}{12}\begin{bmatrix} 1 & 1-i & 1 \\ 1+i & 2 & 1+i \end{bmatrix}. \qquad \square \tag{9.48}$$

The generalized inverse has an important role to play in the solutions of linear simultaneous equations. We shall now consider these equations to be described by (5.35). Considering homogeneous equations we have the following theorem.

Theorem 9.16 Let A and $A^{\dagger}$ be matrices of order $m \times n$ and $n \times m$ respectively over $\mathbb{C}$ such that $AA^{\dagger}A = A$. Then $A\mathbf{x} = \boldsymbol{\theta}$ has a solution iff

$$\mathbf{x} = (I - A^{\dagger}A)\mathbf{y} \tag{9.49}$$

for some vector $\mathbf{y}$ in $\mathbb{C}^n$.

Proof Suppose that $A\mathbf{x} = \boldsymbol{\theta}$. Then, we have

$$\mathbf{x} = \mathbf{x} - A^{\dagger}A\mathbf{x} = (I - A^{\dagger}A)\mathbf{x}.$$

Now, (9.49) follows by selecting $\mathbf{y} = \mathbf{x}$. Conversely, if (9.49) holds, then

$$A\mathbf{x} = A(I - A^{\dagger}A)\mathbf{y} = \boldsymbol{\theta}. \qquad \square$$

In studying the linear simultaneous equations $A\mathbf{x} = \mathbf{b}$ in Section 5.6, we have seen that if A is square and nonsingular, then $\mathbf{x} = B\mathbf{b}$, where $B = A^{-1}$ is a unique solution. Also, if A is right invertible and B is a corresponding right inverse, then $\mathbf{x} = B\mathbf{b}$ gives an infinite number of solutions for $A\mathbf{x} = \mathbf{b}$ for every right inverse B of A. In both the cases above we see that the relation $AB\mathbf{b} = A\mathbf{x} = \mathbf{b}$ is satisfied. We show below that it is true in general and constitutes a necessary and sufficient condition for the existence of solutions of $A\mathbf{x} = \mathbf{b}$ as given in the following theorem.

Theorem 9.17 Let A and $A^{\dagger}$ be two matrices over $\mathbb{C}$ of order $m \times n$ and $n \times m$ respectively satisfying $AA^{\dagger}A = A$, then $A\mathbf{x} = \mathbf{b}$ has a solution iff $AA^{\dagger}\mathbf{b} = \mathbf{b}$. In this case, the most general solution is

$$\mathbf{x} = A^{\dagger}\mathbf{b} + (I - A^{\dagger}A)\mathbf{y}, \tag{9.50}$$

where $\mathbf{y}$ is arbitrary.

Proof Suppose $A = AA^{\dagger}A$. We now assume that $A\mathbf{x} = \mathbf{b}$ has a solution which implies that there is a vector $\mathbf{u}$ such that $A\mathbf{u} = \mathbf{b}$. Consequently,

$$AA^{\dagger}\mathbf{b} = AA^{\dagger}A\mathbf{u} = A\mathbf{u} = \mathbf{b}$$

Conversely, assume that $AA^{\dagger}\mathbf{b} = \mathbf{b}$. If we now set

$$\mathbf{x} = A^{\dagger}\mathbf{b}, \tag{9.51}$$

then $A\mathbf{x} = AA^{\dagger}\mathbf{b} = \mathbf{b}$ which shows that $A^{\dagger}\mathbf{b}$ is a solution of $A\mathbf{x} = \mathbf{b}$. □

Finally, in view of Theorem 5.11, (9.50) follows from (9.51) and Theorem 9.16.

Example 9.21 Find a general solution of

$$\begin{bmatrix} 1 & 1-i \\ 1+i & 2 \\ 1 & 1-i \end{bmatrix} \begin{bmatrix} x_1 \\ x_2 \end{bmatrix} = \begin{bmatrix} i \\ -1+i \\ i \end{bmatrix}.$$

Solution The above equation is of the form $A\mathbf{x} = \mathbf{b}$. The generalized inverse $A^{\dagger}$ of the coefficient matrix A is given by (9.48) which satisfies $AA^{\dagger}A = A$. Since

$$AA^{\dagger}\mathbf{b} = \frac{1}{12}\begin{bmatrix} 1 & 1-i \\ 1+i & 2 \\ 1 & 1-i \end{bmatrix}\begin{bmatrix} 1 & 1-i & 1 \\ 1+i & 2 & 1+i \end{bmatrix}\begin{bmatrix} i \\ -1+i \\ i \end{bmatrix}$$

$$= \frac{1}{4}\begin{bmatrix} 1 & 1-i & 1 \\ 1+i & 2 & 1+i \\ 1 & 1-i & 1 \end{bmatrix}\begin{bmatrix} i \\ -1+i \\ i \end{bmatrix} = \begin{bmatrix} i \\ -1+i \\ i \end{bmatrix} = \mathbf{b},$$

the given equation has a solution. The general solution is

$$\mathbf{x} = A^{\dagger}\mathbf{b} + (I - A^{\dagger}A)\mathbf{y} = \frac{1}{3}\begin{bmatrix} i \\ -1+i \end{bmatrix} + \frac{1}{3}\begin{bmatrix} 2 & -1+i \\ -1+i & 1 \end{bmatrix}\mathbf{y},$$

where $\mathbf{y} = [y_1\ y_2]^T$ is any arbitrary vector. □

9.6.1 Least Squares Solution of Ax = b

We shall now prove that the solution $A^{\dagger}\mathbf{b}$ of $A\mathbf{x} = \mathbf{b}$ in Theorem 9.17 has two important properties: (a) $\| A\mathbf{x} - \mathbf{b} \|$ is minimum if $\mathbf{y} = A^{\dagger}\mathbf{b}$ is a solution and (b) if $\boldsymbol{\eta} \neq \mathbf{y}$ is any other vector for which $\| A\mathbf{x} - \mathbf{b} \|$ is minimum, then $\| \boldsymbol{\eta} \| > \| \mathbf{y} \|$, where $\| \cdot \|$ is the Euclidean norm. That is, $A^{\dagger}\mathbf{b}$ is a least squares minimum norm solution.

Indeed, since under unitary transformation, Euclidean norm is invariant (see Corollary 6.4), for unitary matrices U and V satisfying SVD: $A = UDV^*$, we have

$$\| \mathbf{b} - A\mathbf{x} \| = \| U^*(\mathbf{b} - AVV^*\mathbf{x}) \|.$$

Setting

$$V*\mathbf{x}=\begin{bmatrix}\mathbf{w}\\ \mathbf{z}\end{bmatrix},\qquad U*\mathbf{b}=\begin{bmatrix}\mathbf{a}\\ \mathbf{c}\end{bmatrix},\qquad L:=\begin{bmatrix}\sigma_1 & & \\ & \ddots & \\ & & \sigma_r\end{bmatrix},$$

where $\mathbf{w}$ and $\mathbf{a}$ are in $\mathbb{C}^r$, we have

$$\|\mathbf{b}-A\mathbf{x}\|=\left\|\begin{bmatrix}\mathbf{a}-L\mathbf{w}\\ \mathbf{c}\end{bmatrix}\right\| \tag{9.52}$$

which is minimum if $\mathbf{w} = L^{-1}\mathbf{a}$ and $\mathbf{z}$ arbitrary. Thus, there are an infinite number of vectors minimizing the norm in (a). Setting $\mathbf{z} = \boldsymbol{\theta}$, we get

$$\mathbf{y}=V\begin{bmatrix}L^{-1}\mathbf{a}\\ \boldsymbol{\theta}\end{bmatrix}=V\begin{bmatrix}L^{-1} & 0\\ 0 & 0\end{bmatrix}\begin{bmatrix}\mathbf{a}\\ \mathbf{c}\end{bmatrix}=V\begin{bmatrix}L^{-1} & 0\\ 0 & 0\end{bmatrix}U*\mathbf{b}=A^{\dagger}\mathbf{b},$$

which gives the solution in (a) minimizing (9.52). If

$$\boldsymbol{\eta}=V\begin{bmatrix}L^{-1}\mathbf{a}\\ \mathbf{z}\end{bmatrix},\qquad \mathbf{z}\neq\boldsymbol{\theta}$$

is any other vector minimizing (9.52), then

$$\|\mathbf{y}\| = \|L^{-1}\mathbf{a}\| \quad < \quad \|L^{-1}\mathbf{a}\| + \|\mathbf{z}\| = \|\boldsymbol{\eta}\|$$

which is the assertion in (b).

The solution $\mathbf{y} = A^{\dagger}\mathbf{b}$ is the best approximate solution of $A\mathbf{x} = \mathbf{b}$. Also, $\mathbf{y}$ is known as the least squares solution of the inconsistent, or over-determined system $A\mathbf{x} = \mathbf{b}$ for which $\|Ax - \mathbf{b}\|$ is a minimum and $\|\mathbf{y}\|$ has the least value among all the solutions minimizing $\|Ax - \mathbf{b}\|$, see Section 13.3.3.

Minimizing $\|A\mathbf{x} - \mathbf{b}\|^2$, the least squares minimum norm solution of $A\mathbf{x} - \mathbf{b}$, where $A \in \mathbb{C}^{m\times n}$, is given by $\hat{\mathbf{x}} = A^{\dagger}\mathbf{b}$, where $A^{\dagger}$ is the generalized inverse of A. If A has full column rank, this solution is given by

$$\hat{\mathbf{x}} = (A^*A)^{-1}A^*\mathbf{b}$$

using Corollary 9.1. So

$$A\hat{\mathbf{x}} = A(A^*A)^{-1}A^*\mathbf{b} = P\mathbf{b},$$

where $P := A(A^*A)^{-1}A^*$ is the orthogonal projection on the subspace formed by the columns of A as shown in Figure 9.5. The minimum error $\mathbf{e} = \mathbf{b} - A\hat{\mathbf{x}}$ is exactly the distance of $\mathbf{b}$ from the point $A\hat{\mathbf{x}}$. Moreover, $\mathbf{e}$ is orthogonal to the range space of A:

$$A^*(\mathbf{b} - A\hat{\mathbf{x}}) = A^*(\mathbf{b} - A(A^*A)^{-1}A^*\mathbf{b}) = 0.$$

Thus, the least squares minimum norm solution $\hat{\mathbf{x}}$ is such that $A\hat{\mathbf{x}}$ is the orthogonal projection of **b** on the range space of A.

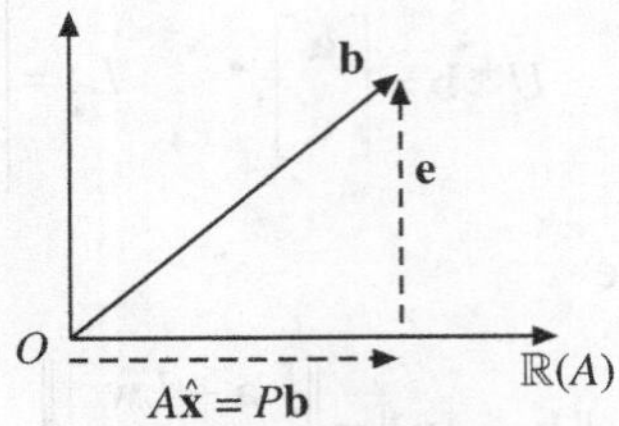

Figure 9.5 Orthogonal projection.

9.7 SOLUTION OF Ax = b WITH MATLAB

To compute the norm of a matrix of order $m \times n$ we use the following commands
norm (*A*): is the largest singular value of A.
norm (*A*, 2): is the same as norm (A).
norm (*A*, 1): is the 1–norm of A.
norm (*A*, inf): is the infinity norm of A.
norm (*A*, 'fro'): is the Frobenius norm of A.

The command

[*U D V*] = *svd* (*A*)

produces a diagonal matrix D of the same dimension as A with nonnegative diagonal elements in decreasing order and unitary matrices U and V, so that $A = UDV^T$.
MATLAB command X = pinv (A) computes a matrix X, the pseudoinverse of A, of the same dimension as A^T, so that

$$A*X*A = A, \qquad X*A*X = X,$$

where $A*X$ and $X*A$ are Hermitian. The computation is based on svd (*A*) and any singular values less than a tolerance are treated as zero. The default tolerance is

max (size (A))∗norm (A)∗eps

where eps = 2.2004 e-016 in MATLAB6p5 release 13, size (*A*) = m if $m > n$ and n if $n \geq m$. However, pinv (*A*, tol) uses the prescribed tolerance tol instead of the default.
To find a *particular solution* of $A\mathbf{x} = \mathbf{b}$, one can use the MATLAB command

pinv (A)∗b

to get an exact solution if it exists. If A is square and nonsingular, x = A\b also gives an exact solution. A particular solution is not unique if A is singular and $A\mathbf{x} = \mathbf{b}$ has a solution. If $A\mathbf{x} = \mathbf{b}$ does not have an exact solution, we get a least squares solution using the above command.

The general solution of $A\mathbf{x} = \mathbf{b}$ is the sum of

1. A null solution of the homogeneous system $A\mathbf{x} = \boldsymbol{\theta}$, and
2. A particular solution of the nonhomogeneous system $A\mathbf{x} = \mathbf{b}$.

To get the solution in step 1, we use the null command by typing null (A). This returns a basis for the solution space to $A\mathbf{x} = \boldsymbol{\theta}$. A linear combination of the basis vectors is a null solution. To find a particular solution we use the pinv (A) command as already outlined.

Example 9.22 Obtain a general solution of $A\mathbf{x} = \mathbf{b}$ if

$$A = \begin{bmatrix} 1 & 1 & 1 \\ 1 & 1 & 1 \\ 1 & 1 & 1 \end{bmatrix}, \qquad \mathbf{b} = \begin{bmatrix} 1 \\ 2 \\ 3 \end{bmatrix}.$$

Solution **Step 1** We have from null (A) the response

$$\begin{matrix} 0.8165 & 0 \\ -0.4082 & -0.7071 \\ -0.4082 & 0.7071 \end{matrix}$$

which are the basis vectors for the null space of A. Therefore, a null solution $\mathbf{x}_0$ is a linear combination of these basis vectors:

$$\mathbf{x}_0 = \alpha \begin{bmatrix} 0.8165 \\ -0.4082 \\ -0.4082 \end{bmatrix} + \beta \begin{bmatrix} 0 \\ -0.7071 \\ 0.7071 \end{bmatrix},$$

where α and β are arbitrary scalars.

Step 2 The command **xls** = pinv (A) $*$ **b** generates the least squares minimum norm particular solution

$$\mathbf{xls} = \begin{bmatrix} 0.6667 \\ 0.6667 \\ 0.6667 \end{bmatrix}.$$

Hence, the general solution is $\mathbf{x} = \mathbf{x}_0 + \mathbf{xls}$.

The SVD of $A = UDV'$, where $D = \text{diag}\ (3, 0, 0)$,

$$U = \begin{bmatrix} -0.5774 & -0.5774 & -0.5774 \\ -0.5774 & 0.7887 & -0.2113 \\ -0.5774 & -0.2113 & 0.7887 \end{bmatrix}, \quad V = \begin{bmatrix} -0.5774 & 0.8165 & 0 \\ -0.5774 & -0.4082 & -0.7071 \\ -0.5774 & -0.4082 & 0.7071 \end{bmatrix}.$$

Referring to Section 9.6.1, since $U\mathbf{b} = [-3.4641 \quad 0.3660 \quad 1.3660]'$, define $\mathbf{a} = -3.4641$, $L = 3$ and so $L^{-1}\mathbf{a} = -1.1547$. A least squares minimum norm solution is $\mathbf{x} = A^{\dagger}\mathbf{b} = [0.6667\ 0.6667\ 0.6667]'$. The general least squares solution is

$$\mathbf{u} = V \begin{bmatrix} L^{-1}\mathbf{a} \\ \mathbf{z} \end{bmatrix}, \quad \text{where } \mathbf{z} = \begin{bmatrix} z_1 \\ z_2 \end{bmatrix} \text{ is arbitrary.}$$

Let $\mathbf{z} = [1\ 1]'$. Then, a least squares solution is $\mathbf{u} = [1.4832 - 0.4487\ 0.9655]'$. It can be verified that $\mathbf{b} - A\mathbf{x} = [-1\ 0\ 1]' = \mathbf{b} - A\mathbf{u}$, $\| \mathbf{x} \| = 1.1547$ and $\| \mathbf{u} \| = 1.8257$. □

PROBLEMS

9.1 Prove that for any matrix and vector norm
(a) $| \|A\| - \|B\| | \le \|A - B\|$, (b) $| \|\mathbf{x}\| - \|\mathbf{y}\| | \le \|\mathbf{x} - \mathbf{y}\|$.

9.2 Show that $\|\cdot\|_p$ defined by $\|\mathbf{x}\|_p = (|x_1|^p + |x_2|^p)^{1/p}$, $0 < p < 1$, where $\mathbf{x} = [x_1\ x_2]^T$ does not represent a norm.

9.3 *Minkowski inequality for infinite sums.* Prove that if $\mathbf{x} = \{x_1, x_2, ..., x_i, \ldots\}$ and $\mathbf{y} = \{y_1, y_2, ..., y_i, \ldots\}$ are two sequences of real or complex numbers such that $\sum_{i=1}^{\infty} |x_i|^p$ and $\sum_{i=1}^{\infty} |y_i|^p$ are convergent, then $\sum_{i=1}^{\infty} |x_i + y_i|^p$ is also convergent and

$$\left(\sum_{i=1}^{\infty} |x_i + y_i|^p\right)^{1/p} \le \left(\sum_{i=1}^{\infty} |x_i|^p\right)^{1/p} + \left(\sum_{i=1}^{\infty} |y_i|^p\right)^{1/p}, \qquad p \ge 1.$$

9.4 State and prove Hölder's inequality for infinite sums.

9.5 Determine $\|\mathbf{x}\|_p$, $p = 1, 2, \infty$, when $\mathbf{x}^T$ is given by
(a) $[1\ \ 1\ \ 1]$, (b) $[i\ \ i\ \ i]$, (c) $[1 + i, 1 - i, 1]$, (d) $[-9, -4, 8]$,

9.6 Compute $\|\mathbf{x}\|_p$, $p = 1, 2, \infty$ for the following n-vectors:
(a) $[1\ \ 1\ \ \cdots\ \ 1]^T$, (b) $\mathbf{e}_i$, an ith elementary vector,
(c) $[1\ \ 2\ \ \cdots\ \ n]^T$, (d) $[-1,\ \ -2,\ \ ....,\ \ -n]^T$.

9.7 (a) Show that $\|\cdot\|$ defined by $\|\mathbf{x}\| = \langle H\mathbf{x}, \mathbf{x}\rangle^{1/2}$ is a vector norm, where H is a PD Hermitian matrix.
(b) If $\|\cdot\|$ is a vector norm in $\mathbb{C}^n$, so also $\|\cdot\|_\alpha$ defined by $\|\mathbf{x}\|_\alpha = \|A\mathbf{x}\|$, where A is a nonsingular matrix in $\mathbb{C}^{n \times n}$.
(c) A Euclidean vector norm in $\mathbb{C}^n$ remains invariant under a unitary transformation.

9.8 Show that
(a) $\|\cdot\|_n$ defined by $\|A\|_n = n \max_{i,j} |a_{ij}|$ is a matrix norm,
(b) $\|A\|_e \le \|A\|_n$,
(c) $\|A\|_n$ is compatible with the vector norms $\|\mathbf{x}\|_p$, $p = 1, 2, \infty$.

9.9 Show that $\max_{i,j} |a_{ij}|$ satisfies axioms (1)–(3) of a matrix norm but not axiom (4).

9.10 Show that:
(a) $\|A\|_e = \text{trace}\,(A^*A) = \sigma_1^2 + \cdots + \sigma_r^2$, where $\sigma_1, ..., \sigma_r$ are nonzero singular values of A.

(b) Euclidean norms $\| A \|_e$ and $\| \mathbf{x} \|_2$ are compatible.

(c) $\| A \|_2 \leq \| A \|_e \leq n^{1/2} \| A \|_2$.

(d) $\| A^*A \|_2 = \| A \|_2^2$.

9.11 If $\| A \|$ is a matrix norm of A induced by $\| \mathbf{x} \|$ and if $N(A)$ is any other norm of A compatible with $\| \mathbf{x} \|$, then $\| A \| \leq N(A)$.

9.12 Prove an alternative statement of Theorem 9.3(b): Any norm of a matrix $A = [a_{ij}]$ depends continuously on the elements a_{ij}, i.e., given any $\varepsilon > 0$, there exists a $\delta(\varepsilon) > 0$ such that for all pairs of subscripts (i, j),

$$| a_{ij} - b_{ij} | < \delta(\varepsilon) \text{ implies } | \, \| A \| - \| B \| \, | < \varepsilon.$$

9.13 Show that the induced norm (a) for an identity matrix I is $\| I \| = 1$, and (b) for a scalar matrix $S_n = \text{diag}\,[\alpha, ..., \alpha]_n$, it is $\| S_n \| = | \alpha |$.

9.14 (a) Let $D = \text{diag}\,[d_1, d_2, ..., d_n]$ be a diagonal matrix. Show that $\| D \|_p = \max_i |d_i|$, $p = 1, 2, \infty$.

(b) Let

$$\hat{D} = \begin{bmatrix} 0 & \cdots & 0 & \hat{d}_1 \\ 0 & \cdots & \hat{d}_2 & 0 \\ \vdots & & \vdots & \vdots \\ \hat{d}_n & \cdots & 0 & 0 \end{bmatrix} \tag{9.53}$$

be an *anti-diagonal matrix.* Show that

$$\| \hat{D} \|_p = \max_i | \hat{d}_i |, \qquad p = 1, 2, \infty.$$

9.15 Determine for the matrix E_{ij} (see (5.7)) the norms
(a) $\| \cdot \|_p$, $p = 1, 2, \infty$, (b) $\| \cdot \|_e$, and (c) $\| \cdot \|_n$ (see Problem 9.8).

9.16 Determine $\| A \|_p$, $p = e, 1, 2, \infty$, when

$$A = \begin{bmatrix} \cos\theta & -\sin\theta & 0 \\ \sin\theta & \cos\theta & 0 \\ 0 & 0 & 1 \end{bmatrix}.$$

9.17 Prove that the Rayleigh quotient for a matrix A has the same value for all vectors $\mathbf{x}$ lying on a straight line through the origin.

9.18 Prove that for any matrix A in $\mathbb{C}^{n \times n}$ the eigenvalues of A^*A are nonnegative, and if 0 is an eigenvalue of A then it is also of A^*A and that $| \det A |^2 = \det (A^*A)$.

9.19 If A is nonsingular and $\sigma_1^2, \sigma_2^2, ..., \sigma_n^2$ are the eigenvalues of A^*A in order of increasing magnitude, then

$$\| A \|_2 = \sigma_n, \quad \| A^{-1} \|_2 = \sigma_1^{-1}, \quad \| A^{-1} \|_2 \le \| A \|_2^{n-1} / | \det A |.$$

9.20 Let A be in $\mathbb{C}^{n \times n}$ and $\lambda_i(A)$ be the ith eigenvalue of A. Then $r(A) = \max_i |\lambda_i(A)|$ is defined as the *spectral radius* of A. Show that

(a) $r(A) \le \| A \|$, $\| A \|$ is any matrix norm in $\mathbb{C}^{n \times n}$ induced by the vector norm $\|\mathbf{x}\|$ in $\mathbb{C}^n$.

(b) $r(A) \le N(A)$, $N(A)$ is any matrix norm in $\mathbb{C}^{n \times n}$.

(c) $r(A) \le \| A \| \le N(A)$, where $N(A)$ is compatible with the vector norm $\| \mathbf{x} \|$ in $\mathbb{C}^n$ and $\| A \|$ the norm induced by $\| \mathbf{x} \|$.

(d) $r(A)$ = spectral norm of A, if A is a normal matrix.

9.21 Show that:

(a) $\| A \|_1 = \| A^* \|_\infty, \quad \| A \|_\infty = \| A^* \|_1$;

(b) $\|A\|_2^2 \le \|A\|_1 \, \|A\|_\infty$.

9.22 Find (i) the matrix norm $\| \cdot \|_2$ and (ii) the singularity index with respect to the vector norm $\| \cdot \|_2$ for (a) an orthogonal matrix P, (b) a unitary matrix U, and (c) $A = [a_{ij}]$, where $a_{ij} = 1$ for all (i, j).

9.23 Find the singularity index with respect to the vector norms $\| \cdot \|_p$, $p = 1, 2, \infty$, (a) for the anti-diagonal matrix given by (9.53), and (b) for the diagonal matrix $D = \text{diag}\,[d_1, ..., d_n]$.

9.24 If $m(\cdot)$ denotes singularity index, then

$$m(AB) \ge m(A)m(B).$$

9.25 Show that the singularity index for an identity matrix and a scalar matrix $S_n = \text{diag}\,[\alpha \cdots \alpha]_n$ are, respectively, 1 and $| \alpha |$.

9.26 (a) If a matrix A in $V_{m \times n}$ is the limit of the sequence $\{A_p\}$, then

$$\lim A_pB = AB \text{ and } \lim CA_p = CA,$$

where B and C are in $V_{n \times k}$ and $V_{r \times m}$ respectively.

(b) If A and T are in $V_{n \times n}$ with T nonsingular and $\lim A_p = A$, then

$$\lim_{p \to \infty} T^{-1}A_pT = T^{-1}AT.$$

(c) If $A^p \to B$, then any eigenvalue λ of A satisfies $| \lambda | \le 1$ with $| \lambda | = 1$ only when λ is the real number 1.

9.27 Prove:

(a) $\lim_{p\to\infty} A^p \to B$ iff $\| A^p - B \| \to 0$.

(b) If $A^p \to B$ then $\| A \|^p \to \| B \|$.

(c) If $\| A \| < 1$ then $\lim_{p\to\infty} A^p = 0$.

9.28 Prove that if $\| A \| < 1$ and $\| I \| = 1$, then

$$\frac{1}{1+\| A \|} \le \| (I+A)^{-1} \| \le \frac{1}{1-\| A \|}$$

9.29 If A is similar to B, then a sufficient condition for convergence of $I + A + A^2 + \cdots$ is that $\| B \| < 1$.

9.30 Show that a GP series in A with

$$A = \begin{bmatrix} 0.9 & 10^6 \\ 0 & 0.7 \end{bmatrix}$$

is convergent.

9.31 Prove that $|\mathbf{x}^T\mathbf{y}| \le \|\mathbf{x}\|_2 \|\mathbf{y}\|_2$, where the equality holds iff **x** and **y** are linearly dependent.

9.32 If P is unitary, show that $\|P\|_2 = 1$.

9.33 Show that if λ is an eigenvalue of A with eigenvector **x**, then $\lambda = \mathbf{x}^*A\mathbf{x}/\mathbf{x}^*\mathbf{x}$.

9.34 Let **x** be an n-vector and A an $n \times n$ matrix. Show that

$$\| A \|_p := \left(\sum_{i,j}^{n} | a_{ij} |^p \right)^{1/p}, \quad p \ge 1$$

is a matrix norm on $\mathbb{C}^{n\times n}$ iff $1 \le p \le 2$. However,

$$\| \mathbf{x} \|_p := \left(\sum_{i=1}^{n} | x_i |^p \right)^{1/p}, \quad 1 \le p < \infty$$

is a vector norm. These norms $\| \mathbf{x} \|_p$ and $\| A \|_p$ are called *Hölder's norm.*

9.35 Prove that a square matrix is unitary iff all its singular values are equal to one.

9.36 Prove that for an n-square matrix A,

$$| \det A | = \sigma_1(A)\sigma_2(A) \cdots \sigma_n(A).$$

9.37 Prove that (i) $\| UA \|_2 = \| A \|_2$, and (ii) $\| AV \|_2 = \| A \|_2$, where U and V are orthogonal or unitary.

9.38 Prove that if σ_i is a singular value of a matrix A of order $m \times n$, then $\exists$ nonzero vectors $\mathbf{u}_i$ and $\mathbf{v}_i$ such that

$$A\mathbf{v}_i = \sigma_i \mathbf{u}_i \text{ and } A^*\mathbf{u}_i = \sigma_i \mathbf{v}_i.$$

9.39 Prove that $\sigma_{\min}(A) - 1 \le \sigma_{\min}(I + A) \le \sigma_{\min}(A) + 1.$

9.40 Prove that:

(i) $2x_1x_2 \le x_1^2 + x_2^2$;

(ii) $(x_1 + x_2 + \cdots + x_n)^2 \le n(x_1^2 + x_2^2 + \cdots + x_n^2)$.

9.41 For a normal matrix $A \in \mathbb{C}^{n \times n}$, $\sigma_i(A) = |\lambda_i(A)|$.

9.42 Prove that for PD or PSD matrices, singular values and eigenvalues are identical.

9.43 If λ_i's are eigenvalues of A and μ_i's are those of A^*A, $i = 1, 2, \ldots, n$, show that

$$\sum_{i=1}^{n} |\lambda_i|^2 \le \sum_{i=1}^{n} \mu_i.$$

The equality holds iff A is normal.

9.44 Two matrices A and B, each of order $m \times n$, are defined to be *unitarily equivalent* if $A = UBV^*$, where U and V are unitary. Prove that two matrices are unitarily equivalent iff they have the same singular values.

9.45 Prove that two $m \times n$ matrices A and B are unitarily equivalent iff A^*A and B^*B are similar.

9.46 Let $A = [2\ 3\ 1;\ 4\ 6\ 2;\ 9\ 10\ 2]$ and $\mathbf{b} = [8\ 27\ 16]'$. Show that a least squares minimum norm solution of $A\mathbf{x} = \mathbf{b}$ is $\mathbf{x} = [-6.4;\ 5.8;\ 7.8]$. Determine that another least squares solution is $\mathbf{u} = [-5.0766;\ 4.1459;\ 10.1155]$, $\| \mathbf{x} \| = 11.6379$ and $\| \mathbf{u} \| = 12.0534$ but $\| A\mathbf{x} - \mathbf{b} \| = \| A\mathbf{u} - \mathbf{b} \| = 4.9193$. Use MATLAB.

9.47 Let $A \in \mathbb{R}^{m \times n}$ be a simple matrix with σ_i and λ_i, $i = 1, 2, \ldots, n$ as the respective singular values and eigenvalues. Show that

$$\sigma_i = u_i^T M \operatorname{diag}(M^{-1}\mathbf{v}_i)\, \lambda,$$

where $\lambda = [\lambda_1\ \lambda_2\ \ldots\ \lambda_n]^T$, matrices U and V are formed with the left singular vectors $\{\mathbf{u}_1, \mathbf{u}_2, \ldots, \mathbf{u}_n\}$ and right singular vectors $\{\mathbf{v}_1, \mathbf{v}_2, \ldots, \mathbf{v}_n\}$ as columns respectively, M is the modal matrix of A, and for a vector $\mathbf{x} = [x_1, x_2, \ldots, x_n]^T$, we define diag $(\mathbf{x})$: = diag $[x_1, x_2, \ldots x_n]$.

9.48 Prove that the singular values of A and $B = QAQ^*$ where Q is unitary are the same.

CHAPTER

10

Normal Forms

We may recall from Section 6.4 that a single linear transformation on a vector space V_n (see Definition 6.1) can be represented by a set of infinitely many matrices, each for a different selection of basis in the vector space. These matrices are called *similar matrices* (see Corollary 6.3 and Definition 6.11). If A and J are two matrices of order n in this set, i.e., if A and J are similar, then there exists a nonsingular matrix T of the same order n such that $J = T^{-1}AT$. For a simple matrix A (see Definition 7.21), one can find a T (see Theorem 7.23) such that J is diagonal having the eigenvalues of A as the diagonal elements, i.e., A is similar to a diagonal matrix. If A is not simple, such nonsingular transformation to a diagonal J is not possible. In that event, how a nonsimple A can be transformed into a quasi-diagonal canonical form, e.g., a Jordan canonical form, will be the major problem to be discussed in this chapter. Our approach will be purely algebraic, i.e., we shall mainly manipulate polynomials in a single variable λ to get the final result, and in the course of deriving quasi-diagonal canonical forms, we shall introduce Hermite and Smith canonical forms for a polynomial matrix $A(\lambda)$ and its invariant polynomials and elementary divisors.

10.1 ELEMENTARY OPERATIONS ON λ-MATRICES

In order to reduce any matrix A to a quasi-diagonal canonical form, we need to perform some kind of elementary operations on the rows and columns of its characteristic matrix $(\lambda I - A)$ which we are going to fix now. Instead of narrowing down our attention at the beginning to a characteristic matrix alone, we shall describe these operations with respect to a larger class of λ-matrices (see Section 7.3). This line of approach, though a bit digressive, would produce some important results on equivalent class of λ-matrices.

The elementary row (resp column) operations for a λ-matrix are defined in a manner similar to that for an ordinary matrix (see Section 4.2). These operations are:

(i) Interchange of any two rows (resp columns).
(ii) Multiplication of any row (resp column) by a nonzero constant k.
(iii) Addition to any row (resp column) of any other row (resp column) multiplied by an arbitrary polynomial $f(\lambda)$.

These operations are the same as those enumerated in Section 4.2, differing only in respect of (iii), where $f(\lambda)$ is taken as a constant, and can be used in a similar fashion to reduce any λ-matrix to a very simple form.

To effect any elementary row (resp column) operations on a λ-matrix $A(\lambda)$ of order $m \times n$, first perform these operations on a unit matrix of order m (resp n) and then premultiply (resp postmultiply) $A(\lambda)$ by the transformed unit matrix. The matrices obtained from the unit matrix by elementary row and column operations are called *elementary λ-matrices* because, evidently, they are λ-matrices; moreover, their determinant is a nonzero constant independent of λ (see Section 2.3). These *elementary λ-matrices* are often referred to as *unimodular.*

Let a λ-matrix be given by

$$A(\lambda) = \begin{bmatrix} \lambda & 1 & 2\lambda \\ \lambda & 3\lambda & 2 \\ 3 & 1 & 2\lambda \end{bmatrix} \tag{10.1}$$

and the matrices

$$\text{(a) } P_1 = \begin{bmatrix} 1 & 0 & 0 \\ 0 & 0 & 1 \\ 0 & 1 & 0 \end{bmatrix}, \quad \text{(b) } P_2 = \begin{bmatrix} 1 & 0 & 0 \\ 0 & 1 & 0 \\ 0 & 0 & k \end{bmatrix},$$

$$\text{(c) } P_3 = \begin{bmatrix} 1 & 0 & \lambda^2 \\ 0 & 1 & 0 \\ 0 & 0 & 1 \end{bmatrix} \tag{10.2}$$

be obtained respectively from a unit matrix of order 3 by row operations (RO), viz.,

RO(a) interchanging the second and third rows

RO(b) multiplying the third row by a nonzero constant k, and

RO(c) adding to the first row λ^2 times the third row.

These three operations can be realized in $A(\lambda)$ by premultiplying it by P_1, P_2 and P_3 respectively:

$$P_1 A(\lambda) = \begin{bmatrix} \lambda & 1 & 2\lambda \\ 3 & 1 & 2\lambda \\ \lambda & 3\lambda & 2 \end{bmatrix},$$

$$P_2A(\lambda) = \begin{bmatrix} \lambda & 1 & 2\lambda \\ \lambda & 3\lambda & 2 \\ 3k & k & 2k\lambda \end{bmatrix},$$

$$P_3A(\lambda) = \begin{bmatrix} \lambda + 3\lambda^2 & 1 + \lambda^2 & 2\lambda + 2\lambda^3 \\ \lambda & 3\lambda & 2 \\ 3 & 1 & 2\lambda \end{bmatrix},$$

where $P_1A(\lambda)$ is the matrix $A(\lambda)$ with the second and third rows interchanged, $P_2A(\lambda)$ is the matrix $A(\lambda)$ with the third row multiplied by k, and $P_3A(\lambda)$ with λ^2 times third row added to the first row. If we obtain the product $P_3P_2P_1A(\lambda)$, the three elementary row operations RO(a), RO(b), RO(c) would be realized in $A(\lambda)$ simultaneously.

Alternatively, the matrices in (10.2) can be regarded as if they were obtained from a unit matrix by the following elementary column operations (CO):

CO(a) interchanging the second and third columns,

CO(b) multiplication of the third column by the nonzero constant k,

CO(c) adding, to the third column, λ^2 times the first column.

If we postmultiply $A(\lambda)$ by P_1, P_2 and P_3 given in (10.2), these operations CO(a), CO(b) and CO(c) will be realized in $A(\lambda)$. Thus,

$$A(\lambda)P_1 = \begin{bmatrix} \lambda & 2\lambda & 1 \\ \lambda & 2 & 3\lambda \\ 3 & 2\lambda & 1 \end{bmatrix},$$

$$A(\lambda)P_2 = \begin{bmatrix} \lambda & 1 & 2k\lambda \\ \lambda & 3\lambda & 2k \\ 3 & 1 & 2k\lambda \end{bmatrix},$$

$$A(\lambda)P_3 = \begin{bmatrix} \lambda & 1 & 2\lambda + \lambda^3 \\ \lambda & 3\lambda & 2 + \lambda^3 \\ 3 & 1 & 2\lambda + 3\lambda^2 \end{bmatrix},$$

where $A(\lambda)P_1$ is the matrix $A(\lambda)$ with second and third columns interchanged, etc. These three column operations can be realized simultaneously in $A(\lambda)$ by performing the product $A(\lambda)P_1P_2P_3$. Furthermore, the product $P_3P_2P_1A(\lambda)P_1P_2P_3 = P(\lambda)A(\lambda)Q(\lambda)$, where $P(\lambda) = P_3P_2P_1$ and $Q(\lambda) = P_1P_2P_3$ are elementary λ-matrices, will give a matrix obtained from $A(\lambda)$, subjecting it successively to the three elementary row and column operations, viz., RO(a), RO(b), RO(c) and CO(a), CO(b), CO(c) mentioned above. Instead of getting $P(\lambda)$ as a product $P_3P_2P_1$, it could be obtained by subjecting a unit matrix to the row operations RO(a)–RO(c) mentioned above in succession. Similarly, $Q(\lambda)$ can be obtained by subjecting a unit matrix successively to the same column operations as those of $A(\lambda)$.

It is important to observe that det $P(\lambda)$ or det $Q(\lambda)$ is a nonzero constant independent of λ. This suggests that $P^{-1}(\lambda)$ = adj $P(\lambda)$/det $P(\lambda)$ is again a λ-matrix, and that these elementary operations affect the value of det $A(\lambda)$ to the extent of multiplying it by a nonzero constant.

In view of (10.2), the matrix $P(\lambda)$ is

$$P(\lambda) = P_3P_2P_1 = \begin{bmatrix} 1 & \lambda^2 k & 0 \\ 0 & 0 & 1 \\ 0 & k & 0 \end{bmatrix},$$

and hence

$$P^{-1}(\lambda) = \begin{bmatrix} 1 & 0 & -\lambda^2 \\ 0 & 0 & 1/k \\ 0 & 1 & 0 \end{bmatrix}.$$

Since det $P(\lambda) = -k$, a nonzero constant independent of λ, it is obvious that the product of any number of elementary λ-matrices (or unimodular matrices), i.e., $P(\lambda)$ in this case, is an elementary λ-matrix (or a unimodular matrix). Moreover, since det $P^{-1}(\lambda) = -1/k$, a nonzero constant independent of λ, the inverse of an elementary λ-matrix (or a unimodular matrix) is again an elementary λ-matrix (or a unimodular matrix). Furthermore, instead of obtaining $P(\lambda)$ from matrix multiplication as above, it can be had from a third order unit matrix by subjecting it to the three elementary row operations mentioned above. Thus, we get successively,

$$\begin{bmatrix} 1 & 0 & 0 \\ 0 & 1 & 0 \\ 0 & 0 & 1 \end{bmatrix} \xrightarrow{\text{RO(a)}} P_1 = \begin{bmatrix} 1 & 0 & 0 \\ 0 & 0 & 1 \\ 0 & 1 & 0 \end{bmatrix} \xrightarrow{\text{RO(b)}} P_2P_1 = \begin{bmatrix} 1 & 0 & 0 \\ 0 & 0 & 1 \\ 0 & k & 0 \end{bmatrix}$$

$$\xrightarrow{\text{RO(c)}} P_3P_2P_1 = \begin{bmatrix} 1 & \lambda^2 k & 0 \\ 0 & 0 & 1 \\ 0 & k & 0 \end{bmatrix} = P(\lambda).$$

Identical operations but with columns can be used to give $Q(\lambda) = P_1P_2P_3$.

10.2 LEFT EQUIVALENCE: COLUMN HERMITE FORMS

Definition 10.1 *Left Equivalence.* Two $m \times n$ λ-matrices $A(\lambda)$ and $B(\lambda)$ are said to be *left equivalent* if one can be obtained from the other by a sequence of elementary row operations. In other words, $A(\lambda)$ is left equivalent to $B(\lambda)$ if there exists a unimodular matrix $U(\lambda)$ of order m such that

$$U(\lambda)A(\lambda) = B(\lambda).$$

Since $U^{-1}(\lambda)$ is unimodular, we have $U^{-1}(\lambda)B(\lambda) = A(\lambda)$, and hence $B(\lambda)$ is also left equivalent to $A(\lambda)$. Thus, $A(\lambda)$ and $B(\lambda)$ are left equivalent if $U(\lambda)A(\lambda) = B(\lambda)$ for a unimodular matrix $U(\lambda)$. □

By elementary row operations, a λ-matrix can be reduced to an upper right triangular λ-matrix known as *column Hermite form.* That is, any λ-matrix is left equivalent to an upper right triangular λ-matrix. Likewise, employing column operations, we can show that λ-matrices are also reduced to a lower triangular form known as *row Hermite form.*

Theorem 10.1 *Column Hermite Form.* Any $m \times n$ λ-matrix $A(\lambda)$ is left equivalent to an upper right triangular λ-matrix $B(\lambda)$, i.e., there always, exists a unimodular matrix $U(\lambda)$ which reduces $A(\lambda)$ to an upper right triangular form $B(\lambda) = U(\lambda)A(\lambda)$, where

$$B(\lambda)=\begin{bmatrix} b_{11}(\lambda) & b_{12}(\lambda) & \cdots & b_{1k}(\lambda) & \cdots & b_{1m}(\lambda) & \cdots & b_{1n}(\lambda) \\ 0 & b_{22}(\lambda) & \cdots & b_{2k}(\lambda) & \cdots & b_{2m}(\lambda) & \cdots & b_{2n}(\lambda) \\ \vdots & \vdots & & \vdots & & \vdots & & \vdots \\ 0 & 0 & \cdots & b_{k-1,k}(\lambda) & \cdots & b_{k-1,m}(\lambda) & \cdots & b_{k-1,n}(\lambda) \\ 0 & 0 & \cdots & b_{kk}(k) & \cdots & b_{km}(\lambda) & \cdots & b_{kn}(\lambda) \\ \vdots & \vdots & & \vdots & & \vdots & & \vdots \\ 0 & 0 & & 0 & & b_{mm}(\lambda) & \cdots & b_{mn}(\lambda) \end{bmatrix}, m \le n; \tag{10.3}$$

or

$$B(\lambda)=\begin{bmatrix} b_{11}(\lambda) & b_{12}(\lambda) & \cdots & b_{1k}(\lambda) & \cdots & b_{1n}(\lambda) \\ 0 & b_{22}(\lambda) & \cdots & b_{2k}(\lambda) & \cdots & b_{2n}(\lambda) \\ \vdots & \vdots & & \vdots & & \vdots \\ 0 & 0 & \cdots & b_{k-1,k}(\lambda) & \cdots & b_{k-1,n}(\lambda) \\ 0 & 0 & \cdots & b_{kk}(\lambda) & \cdots & b_{kn}(\lambda) \\ \vdots & \vdots & & \vdots & & \vdots \\ 0 & 0 & \cdots & 0 & \cdots & b_{nn}(\lambda) \\ 0 & 0 & \cdots & 0 & \cdots & 0 \\ \vdots & \vdots & & \vdots & & \vdots \\ 0 & 0 & \cdots & 0 & \cdots & 0 \end{bmatrix}, \quad m \ge n; \tag{10.4}$$

and for all $k = 2, 3, ..., \min(m, n)$ polynomials $b_{1k}(\lambda), b_{2k}(\lambda), ..., b_{k-1,k}(\lambda)$ are of lower degree than $b_{kk}(\lambda)$ if $b_{kk}(\lambda) \ne 0$ and are identically zero if $b_{kk}(\lambda)$ = a constant $\ne 0$.

Proof (a) Assuming that the elements of the first column are not identically zero, we choose among them a polynomial, say $a_{i1}(\lambda)$, of least degree. By interchanging the ith row with the first row (this is elementary row operation (i)), we bring this element to position (1, 1) and call it $a_{11}(\lambda)$. Since the remaining elements $a_{21}(\lambda), ..., a_{m1}(\lambda)$ in the first column are of a degree greater than or equal to $a_{11}(\lambda)$, they give, on dividing by $a_{11}(\lambda)$ (in view of the Euclidean division algorithm), the relations

$$a_{i1}(\lambda) = a_{11}(\lambda)q_{i1}(\lambda) + r_{i1}(\lambda), \qquad i = 2, ..., m, \tag{10.5}$$

where $q_{i1}(\lambda)$ is the quotient and $r_{i1}(\lambda)$, the remainder, is of lower degree than $a_{11}(\lambda)$. After subtracting $q_{i1}(\lambda)$ times the first row from the ith row for all $i = 2, ..., m$ (this is row operation (iii)), $A(\lambda)$ reduces to

$$\begin{bmatrix} a_{11}(\lambda) & a_{12}(\lambda) & \cdots & a_{1n}(\lambda) \\ r_{21}(\lambda) & & & \\ \vdots & & A_1 & \\ r_{m1}(\lambda) & & & \end{bmatrix},$$

where A_1, an $(m-1) \times (n-1)$ matrix, represents the remaining elements. If $r_{21}(\lambda), ..., r_{m1}(\lambda)$ are not all zero polynomials, then, as they are of lower degree than $a_{11}(\lambda)$, by interchanging again the rows, we bring in the (1, 1) position the polynomial of least degree among these $r_{i1}(\lambda)$'s and repeat the above procedure. Noting that the degree of the polynomial elements is finite and the degree of the polynomial element in the (1, 1) position is decreasing at each step, all the elements in the first column except the one in the (1, 1) position are reduced to zero after a finite number of operations.

(b) We now turn our attention to the element in position (2, 2) and applying step (a) to the rows 2, 3, ..., m, we reduce to zero all the elements in the second column below the (2, 2) position. If the element in the (1, 2) position is not of lower degree than that in the (2, 2) position, then in view of the division algorithm we achieve this by subtracting a suitable multiple of the second row from the first row. In this last step, if the element in position (2, 2) is of degree zero, then the element in position (1, 2) becomes identically zero.

(c) Continuing this process for the remaining diagonal elements, we finally end up with a matrix of the form either (10.3) or (10.4) having the stipulated degree property for the column elements. □

Example 10.1 Reduce to column Hermite form the following λ-matrix:

$$A(\lambda) = \begin{bmatrix} \lambda^3+1 & \lambda^3-\lambda^2+\lambda & \lambda & \lambda^2 \\ \lambda^2+\lambda & \lambda^2 & 1 & \lambda+1 \\ -\lambda^3-\lambda^2 & -\lambda^3 & 1 & -\lambda^2-1 \end{bmatrix}.$$

Solution Concentrating on the elements of the first column, perform the following operations:

(i) Interchange the first and second rows to bring an element of the lowest degree to the (1, 1) position:

$$\begin{bmatrix} \lambda^2+\lambda & \lambda^2 & 1 & \lambda+1 \\ \lambda^3+1 & \lambda^3-\lambda^2+\lambda & \lambda & \lambda^2 \\ -\lambda^3-\lambda^2 & -\lambda^3 & 1 & -\lambda^2-1 \end{bmatrix}.$$

Applying the division algorithm (10.5), we get

$$(\lambda^3+1) = (\lambda-1)(\lambda^2+\lambda)+(\lambda+1),$$

$$-(\lambda^3+\lambda^2) = -\lambda(\lambda^2+\lambda).$$

(ii) Subtract $(\lambda - 1)$ times the first row from the second row and

(iii) Subtract $-\lambda$ times the first row from the third row to get

$$\begin{bmatrix} \lambda^2+\lambda & \lambda^2 & 1 & \lambda+1 \\ \lambda+1 & \lambda & 1 & 1 \\ 0 & 0 & \lambda+1 & \lambda-1 \end{bmatrix}.$$

(iv) Again interchange the first and second rows:

$$\begin{bmatrix} \lambda+1 & \lambda & 1 & 1 \\ \lambda^2+\lambda & \lambda^2 & 1 & \lambda+1 \\ 0 & 0 & \lambda+1 & \lambda-1 \end{bmatrix}.$$

(v) Since $\lambda^2 + \lambda = \lambda(\lambda + 1) + 0$, subtract λ times the first row from the second row to get

$$\begin{bmatrix} \lambda+1 & \lambda & 1 & 1 \\ 0 & 0 & -\lambda+1 & 1 \\ 0 & 0 & \lambda+1 & \lambda-1 \end{bmatrix}.$$

Noticing that the elements in the second column below the first row are zero, we come to the element $(\lambda + 1)$ in the (3, 3) position.

(vi) Add third row to the second row so that the elements in the third column above row three have lower degree than the element in position (3, 3) to get

$$B(\lambda) = \begin{bmatrix} \lambda+1 & \lambda & 1 & 1 \\ 0 & 0 & 2 & \lambda \\ 0 & 0 & \lambda+1 & \lambda-1 \end{bmatrix}, \tag{10.6}$$

which is in column Hermite form. Applying the elementary row operations listed in (i)–(vi) to an identity matrix of order 3, we have

$$U(\lambda) = \begin{bmatrix} 1 & -\lambda+1 & 0 \\ -\lambda & \lambda^2+1 & 1 \\ 0 & \lambda & 1 \end{bmatrix},$$

so that $U(\lambda)A(\lambda) = B(\lambda)$, i.e., $A(\lambda)$ is left equivalent to the column Hermite form $B(\lambda)$.

(vii) Again interchange the second and third rows in (10.6).

(viii) Divide third row by 2.

(ix) Subtract third row from the first row.

(x) Subtract $(1 + \lambda)$ times third row from the second row.

Thus, through steps (vii)–(x) we get, finally, the matrix

$$\tilde{B}(\lambda) = \begin{bmatrix} \lambda+1 & \lambda & 0 & 1-(\lambda/2) \\ 0 & 0 & 0 & (-\lambda^2+\lambda-2)/2 \\ 0 & 0 & 1 & \lambda/2 \end{bmatrix}.$$

Subjecting an identity matrix of order 3 to the operations (i)–(x) [equivalently subjecting $U(\lambda)$ to (vii)–(x)], we obtain

$$\tilde{U}(\lambda) = \frac{1}{2}\begin{bmatrix} 2+\lambda & -\lambda^2-2\lambda+1 & -1 \\ \lambda(1+\lambda) & -\lambda^3-\lambda^2+\lambda-1 & -\lambda+1 \\ -\lambda & \lambda^2+1 & 1 \end{bmatrix},$$

so that

$$\tilde{U}(\lambda)A(\lambda) = \tilde{B}(\lambda),$$

where $\tilde{B}(\lambda)$ is in column Hermite form. Since $B(\lambda)$ and $\tilde{B}(\lambda)$ are different, the column Hermite form of a λ-matrix is, in general, not unique. Observe also that the matrices $U(\lambda)$ and $\tilde{U}(\lambda)$ are unimodular, because det $U(\lambda)$ = 1 and det $\tilde{U}(\lambda)$ = –1/2, which are nonzero constants independent of λ. If, however, $A(\lambda)$ is square and nonsingular, then its column Hermite form can be given uniquely, as shown in Theorem 10.3. □

Definition 10.2 *Rank of a λ-Matrix.* An $m \times n$ λ-matrix $A(\lambda)$ is said to be of rank r if it has at least one submatrix of order $r \le$ main (m, n) which is nonsingular (whose determinant is not identically zero) but has all submatrices of order greater than r which are singular (whose determinants are identically zero). □

For example, the rank of the matrix

$$\begin{bmatrix} \lambda+\lambda^2 & \lambda^3 & \lambda \\ 2+2\lambda & 2\lambda^2 & 2 \end{bmatrix}$$

is one because all submatrices of order 2 are singular. The rank of the matrix $A(\lambda)$ in Example 10.1 is 3.

The following theorem can be easily verified.

Theorem 10.2 Elementary (row or column) operations do not alter the rank of a λ-matrix.

Proof The property, whether or not a matrix or its submatrix is nonsingular, is not affected by elementary operations. This observation proves directly the proposition of the theorem. □

Remark 10.1 The column Hermite form $B(\lambda)$ of an $m \times n$ λ-matrix $A(\lambda)$ having rank r in Theorem 10.1 can be arranged to have (i) last $m - r$ rows identically zero, and (ii) all nonzero diagonal elements $b_{kk}(\lambda)$'s to be monic polynomials.

Remark 10.2 If $B(\lambda)$ and $\tilde{B}(\lambda)$ are two column Hermite forms of the matrix $A(\lambda)$, then there exists unimodular matrices $U(\lambda)$ and $\tilde{U}(\lambda)$, such that

$$U(\lambda)A(\lambda) = B(\lambda), \qquad \tilde{U}(\lambda)A(\lambda) = \tilde{B}(\lambda).$$

It follows easily that

$$\tilde{B}(\lambda) = \tilde{U}(\lambda)U^{-1}(\lambda)B(\lambda) = V(\lambda)B(\lambda), \qquad V(\lambda) = \tilde{U}(\lambda)U^{-1}(\lambda),$$

where $V(\lambda)$ is a unimodular matrix, because the product of any number of unimodular matrices is again unimodular. So two column Hermite forms of the same matrix are related together by a unimodular matrix.

Theorem 10.3 A nonsingular $m \times m$ λ-matrix $A(\lambda)$ has a unique column Hermite form $B(\lambda)$, see (10.3) and (10.4), in which the diagonal elements $b_{kk}(\lambda)$'s are nonzero monic polynomials.

Proof In view of Theorem 10.1 and Remark 10.1, let $B(\lambda)$ and $\tilde{B}(\lambda)$ be two column Hermite forms of the λ-matrix $A(\lambda)$ given respectively by

$$B(\lambda) = \begin{bmatrix} b_{11}(\lambda) & b_{12}(\lambda) & b_{13}(\lambda) & \cdots & b_{1m}(\lambda) \\ & b_{22}(\lambda) & b_{23}(\lambda) & \cdots & b_{2m}(\lambda) \\ & & b_{33}(\lambda) & \cdots & b_{3m}(\lambda) \\ & & & \ddots & \vdots \\ & & & & b_{mm}(\lambda) \end{bmatrix}$$

and

$$\tilde{B}(\lambda) = \begin{bmatrix} \tilde{b}_{11}(\lambda) & \tilde{b}_{12}(\lambda) & \tilde{b}_{13}(\lambda) & \cdots & \tilde{b}_{1m}(\lambda) \\ & \tilde{b}_{22}(\lambda) & \tilde{b}_{23}(\lambda) & \cdots & \tilde{b}_{2m}(\lambda) \\ & & \tilde{b}_{33}(\lambda) & \cdots & \tilde{b}_{3m}(\lambda) \\ & & & \ddots & \vdots \\ & & & & \tilde{b}_{mm}(\lambda) \end{bmatrix}, \tag{10.7}$$

where $b_{kk}(\lambda)$'s and $\tilde{b}_{kk}(\lambda)$'s are monic. By Remark 10.2, there exists a unimodular matrix $V(\lambda)$ of order m such that

$$\tilde{B}(\lambda) = V(\lambda)B(\lambda). \tag{10.8}$$

To prove uniqueness, we have to show that $V(\lambda)$ is an identity matrix. Since rank $A(\lambda) = m$, we observe by Theorem 10.2 that $B(\lambda)$ and $\tilde{B}(\lambda)$ have also rank m. So, $B(\lambda)$ is nonsingular and as it is upper triangular, its inverse $B^{-1}(\lambda)$ too is so (see Problem 3.22). Therefore, the product $\tilde{B}(\lambda)B^{-1}(\lambda) = V(\lambda)$ is also upper triangular (see Problem 1.28).

We now write $V(\lambda)$ explicitly as

$$V(\lambda) = \begin{bmatrix} v_{11}(\lambda) & v_{12}(\lambda) & \cdots & v_{1m}(\lambda) \\ & v_{22}(\lambda) & \cdots & v_{2m}(\lambda) \\ & & \ddots & \vdots \\ & & & v_{mm}(\lambda) \end{bmatrix}. \tag{10.9}$$

Since det $V(\lambda) = v_{11}(\lambda) \cdots v_{mm}(\lambda)$ is a nonzero constant, it follows that each of $v_{kk}(\lambda)$'s is also a nonzero constant. The diagonal elements in the product (10.8) obtained by substituting from (10.7) and (10.9) are

$$\tilde{b}_{kk}(\lambda) = v_{kk}(\lambda) b_{kk}(\lambda).$$

Since $\tilde{b}_{kk}(\lambda)$ and $b_{kk}(\lambda)$ are monic and $v_{kk}(\lambda)$ is a scalar constant, comparing the highest coefficient from both sides, we have

$$v_{kk}(\lambda) = 1 \quad \text{and} \quad \tilde{b}_{kk}(\lambda) = b_{kk}(\lambda) \neq 0$$

for all k = 1, 2, ..., m. Again, from (10.8) we have

$$\tilde{b}_{12}(\lambda) = b_{12}(\lambda) + v_{12}(\lambda) b_{22}(\lambda). \tag{10.10}$$

Since

$$\deg \tilde{b}_{12}(\lambda) < \deg \tilde{b}_{22}(\lambda) = \deg b_{22}(\lambda) \qquad \text{and} \qquad \deg b_{12}(\lambda) < \deg b_{22}(\lambda)$$

by the column degree property of a Hermite form, it follows from (10.10) that $v_{12}(\lambda) = 0$. Next, from

$$\tilde{b}_{13}(\lambda) = b_{13}(\lambda) + v_{13}(\lambda) b_{33}(\lambda),$$

it follows that $v_{13}(\lambda) = 0$ by similar arguments. Thus continuing with the elements along the first row of $\tilde{B}(\lambda)$, we have

$$v_{12}(\lambda) = v_{13}(\lambda) = \cdots = v_{1,m-1}(\lambda) = 0,$$

and so, finally,

$$\tilde{b}_{1m}(\lambda) = b_{1m}(\lambda) + v_{1m}(\lambda) b_{mm}(\lambda).$$

Hence, by similar arguments $v_{1m}(\lambda) = 0$. This procedure can be continued for the remaining rows of $\tilde{B}(\lambda)$ which shows that all the elements above the diagonal of $V(\lambda)$ are zero. Since $V(\lambda)$ is upper triangular with $v_{kk}(\lambda) = 1$, it follows that $V(\lambda) = I_m$, which is an identity matrix. So the column Hermite form of a nonsingular λ-matrix $A(\lambda)$ is unique. □

Example 10.2 Reduce to column Hermite form the following nonsingular λ-matrix:

$$A(\lambda) = \begin{bmatrix} (\lambda-1)^2 & 2\lambda - 1 & (\lambda+1)(\lambda^3 - 1) \\ \lambda^2 - 1 & -\lambda^2 + \lambda + 1 & \lambda(\lambda^3 + 3\lambda^2 + 4\lambda + 4) \\ \lambda - 1 & 1 & \lambda(\lambda^2 + 2\lambda + 2) \end{bmatrix}.$$

Solution We execute successively the following row operations on $A(\lambda)$:

(i) Interchange the first and third rows.
(ii) Subtract $(\lambda + 1)$ times the first row from the second row.
(iii) Subtract $(\lambda - 1)$ times the first row from the third row.
(iv) Interchange the second and third rows.
(v) Add λ times the second row to the third row.
(vi) Subtract $(\lambda + 1)$ times the third row from the first row.

Thus we get the unique column Hermite form $B(\lambda)$ and the requisite unimodular matrix $U(\lambda)$ as

$$B(\lambda) = \begin{bmatrix} \lambda - 1 & 1 & \lambda \\ 0 & \lambda & \lambda - 1 \\ 0 & 0 & \lambda(\lambda+1) \end{bmatrix},$$

$$U(\lambda) = \begin{bmatrix} -\lambda(\lambda+1) & -(\lambda+1) & (\lambda^3+\lambda^2+\lambda+2) \\ 1 & 0 & -\lambda+1 \\ \lambda & 1 & -(\lambda^2+1) \end{bmatrix}.$$ □

10.3 RIGHT EQUIVALENCE: ROW HERMITE FORMS

Instead of applying row operations which we have been doing so far, we shall now apply elementary column operations to reduce a λ-matrix $A(\lambda)$. In this case, $A(\lambda)$ can be transformed into a lower triangular form (having row degree property), to be called the *row Hermite form*, which becomes unique if $A(\lambda)$ is nonsingular. The same logical sequence of arguments as in obtaining column Hermite form is necessary in the present case, the major difference being that now we have to focus our attention on elementary column operations. Therefore, we shall enumerate only the main results without striving to prove them. This can obviously be done only by stretching one's imagination. The row Hermite form of $A(\lambda)$ can be obtained indirectly by determining the column Hermite form $B(\lambda)$ of $A^T(\lambda)$, such that

$$U(\lambda)A^T(\lambda) = B(\lambda),$$

where $U(\lambda)$ is unimodular, and then transpositing to get

$$A(\lambda)U^T(\lambda) = B^T(\lambda)$$

with $B^T(\lambda)$ representing row Hermite form of $A(\lambda)$.

Definition 10.3 *Right Equivalence.* Two $m \times n$ λ-matrices $A(\lambda)$ and $B(\lambda)$ are said to be *right equivalent* if one can be obtained from the other by a sequence of elementary column operations. In other words, $A(\lambda)$ is right equivalent to $B(\lambda)$ if there exists a unimodular matrix $V(\lambda)$ of order n such that $A(\lambda)V(\lambda) = B(\lambda)$. This implies that $B(\lambda)$ is also right equivalent to $A(\lambda)$ as $V^{-1}(\lambda)$ is unimodular. Thus, $A(\lambda)$ and $B(\lambda)$ are right equivalent if $A(\lambda)V(\lambda) = B(\lambda)$ for a unimodular matrix $V(\lambda)$. □

Theorem 10.4 *Row Hermite Form.* Any $m \times n$ λ-matrix $A(\lambda)$ of rank r is right equivalent to a lower left triangular λ-matrix $B(\lambda)$, i.e., there always exists a unimodular matrix $V(\lambda)$ which reduces $V(\lambda)$ to a lower left triangular form $B(\lambda) = A(\lambda)V(\lambda)$, where

$$B(\lambda)=\begin{bmatrix} b_{11}(\lambda) & 0 & \cdots & 0 & 0 & 0 & 0 & \cdots & 0\\ b_{21}(\lambda) & b_{22}(\lambda) & \cdots & 0 & 0 & \cdots 0 & 0 & & 0\\ \vdots & \vdots & & \vdots & & \vdots & \vdots & & \vdots\\ b_{k1}(\lambda) & b_{k2}(\lambda) & \cdots & b_{k,k-1}(\lambda) & b_{kk}(\lambda) & 0 & 0 & & 0\\ \vdots & \vdots & & \vdots & \vdots & \vdots & \vdots & & \vdots\\ b_{m1}(\lambda) & b_{m2}(\lambda) & \cdots & b_{m,k-1}(\lambda) & b_{mk}(\lambda) & \cdots b_{mm}(\lambda) & 0 & \cdots & 0 \end{bmatrix},\quad m\le n;$$

or

$$B(\lambda)=\begin{bmatrix} b_{11}(\lambda) & 0 & \cdots & 0 & 0 & \cdots & 0\\ b_{21}(\lambda) & b_{22}(\lambda) & \cdots & 0 & 0 & \cdots & 0\\ \vdots & \vdots & & \vdots & \vdots & & \vdots\\ b_{k1}(\lambda) & 0 & \cdots & b_{k,k-1}(\lambda) & b_{kk}(\lambda) & \cdots & 0\\ \vdots & \vdots & & \vdots & \vdots & & \vdots\\ b_{n1}(\lambda) & b_{n2}(\lambda) & \cdots & b_{n,k-1}(\lambda) & b_{nk}(A) & \cdots & b_{nn}(\lambda)\\ \vdots & \vdots & & \vdots & \vdots & & \vdots\\ b_{m1}(\lambda) & b_{m2}(\lambda) & \cdots & b_{m,k-1}(\lambda) & b_{mk}(\lambda) & \cdots & b_{mn}(\lambda) \end{bmatrix},\quad m\ge n;$$

where (i) the last $n-r$ columns are identically zero; (ii) the diagonal elements $b_{kk}(\lambda)$'s are monic polynomials; (iii) for all $k = 2, 3, ..., \min(m, n)$, polynomials $b_{k1}(\lambda), b_{k2}(\lambda), ..., b_{k,k-1}(\lambda)$ are of lower degree than $b_{kk}(\lambda)$ if $b_{kk}(\lambda) \neq 0$ and are identically zero if $b_{kk}(\lambda)$ = a constant $\neq 0$; and (iv) $B(\lambda)$ is unique if $m = n = r$. □

Example 10.3 Reduce to row Hermite form the following nonsingular λ-matrix:

$$A(\lambda)=\begin{bmatrix} \lambda & 1 & 2-\lambda\\ \lambda^2+\lambda+1 & 1+2\lambda & -2\lambda^2+3\lambda+2\\ -2\lambda^2-\lambda^2+2\lambda+4 & -2\lambda^2-\lambda+1 & 2\lambda^3-3\lambda^2-2\lambda+3 \end{bmatrix}.$$

Solution We successively perform the following column operations on $A(\lambda)$:

(i) Interchange the first and second columns.

(ii) Subtract λ times the first column from the second column and $(2-\lambda)$ times from the third column.

(iii) Subtract the third column from the second column and $(1-2\lambda)$ times from the first column. Thus, we get the unique row Hermite form $B(\lambda)$ and the requisite unimodular matrix $V(\lambda)$ as

$$B(\lambda)=\begin{bmatrix} 1 & 0 & 0\\ 1+2\lambda & 1-\lambda^2 & 0\\ 0 & 3 & 1+\lambda \end{bmatrix}$$

$$V(\lambda)=\begin{bmatrix} 0 & 1 & 0\\ 2\lambda^2-5\lambda+3 & -2\lambda+2 & \lambda-2\\ 2\lambda-1 & -1 & 1 \end{bmatrix}.$$ □

10.4 EQUIVALENCE OF LAMBDA MATRICES

On a λ-matrix $A(\lambda)$, we shall now perform simultaneously both elementary row and column operations to reduce the number of elements in it as far as practicable. We shall see that the number of nonzero elements in it can always be made equal to its rank r. Moreover, these r number of elements will occupy the first r positions on the main diagonal of the transformed matrix and have a characteristic divisibility property. This reduction to a diagonal form would be instrumental in establishing Smith's canonical form in Section 10.5.

Definition 10.4 *Equivalence of Two λ-Matrices.* Two $m \times n$ λ-matrices $A(\lambda)$ and $B(\lambda)$ are said to be *equivalent* if one can be obtained from the other by a sequence of elementary row and column operations. In other words, $A(\lambda)$ is equivalent to $B(\lambda)$ if there exist unimodular matrices $U(\lambda)$ and $V(\lambda)$ such that

$$U(\lambda)A(\lambda)V(\lambda) = B(\lambda). \tag{10.11}$$

□

If $V(\lambda) = I_n$ in Definition 10.4, we have left equivalence; if $U(\lambda) = I_m$, we have right equivalence of $A(\lambda)$ and $B(\lambda)$. Besides, if $A(\lambda)$ is equivalent to $B(\lambda)$, then also $B(\lambda)$ is equivalent to $A(\lambda)$, because from (10.11) we can write

$$U^{-1}(\lambda)B(\lambda)V^{-1}(\lambda) = A(\lambda)$$

and the inverse of unimodular matrices is also unimodular. So, if (10.11) holds, we say synonymously that $A(\lambda)$ and $B(\lambda)$ are equivalent. It can be easily proved that the equivalence relation defined by (10.11) is reflexive, symmetric and transitive (see Problem 10.5). When $A(\lambda)$ is transformed into $B(\lambda)$ by a sequence of elementary row and column operations, we say that $B(\lambda)$ is obtained from $A(\lambda)$ under *equivalent* or *unimodular* transformation.

Example 10.4 The matrix

$$A(\lambda) = \begin{bmatrix} \lambda^2 - \lambda + 1 & 1 + \lambda^3 \\ 1 + \lambda^3 & (\lambda + 1)(\lambda^3 + 2) \end{bmatrix}$$

is equivalent to the matrix

$$B(\lambda) = \begin{bmatrix} \lambda + 1 & 0 \\ 0 & \lambda^2 - \lambda + 1 \end{bmatrix},$$

because by successive elementary row and column operations we have

$$\underbrace{\begin{bmatrix} 0 & 1 \\ 1 & 0 \end{bmatrix}}_{P_2} \underbrace{\begin{bmatrix} 1 & 0 \\ -(1+\lambda) & 1 \end{bmatrix}}_{P_1} \underbrace{\begin{bmatrix} \lambda^2 - \lambda + 1 & 1 + \lambda^3 \\ 1 + \lambda^3 & (\lambda + 1)(\lambda^3 + 2) \end{bmatrix}}_{A(\lambda)} \underbrace{\begin{bmatrix} 1 & -(1+\lambda) \\ 0 & 1 \end{bmatrix}}_{Q_1} \underbrace{\begin{bmatrix} 0 & 1 \\ 1 & 0 \end{bmatrix}}_{Q_2}$$

$$= \underbrace{\begin{bmatrix} \lambda + 1 & 0 \\ 0 & \lambda^2 - \lambda + 1 \end{bmatrix}}_{B(\lambda)},$$

i.e., $P(\lambda)A(\lambda)Q(\lambda) = B(\lambda)$, where

$$P(\lambda) = P_2P_1 = \begin{bmatrix} -(\lambda+1) & 1 \\ 1 & 0 \end{bmatrix},$$

$$Q(\lambda) = Q_1Q_2 = \begin{bmatrix} -(\lambda+1) & 1 \\ 1 & 0 \end{bmatrix}$$

are unimodular matrices. □

Based on the above notion of equivalence, we are now going to establish a useful lemma, probably the most important lemma in this book, which will later prove to be crucial in establishing the basic invariant structure of λ-matrices. Henceforth, we denote

$$\text{diag}\,[\varepsilon_1(\lambda), \varepsilon_2(\lambda), ..., \varepsilon_r(\lambda), 0, ..., 0]$$

an $m \times n$ matrix, with $r \leq \min\,(m, n)$ number of polynomials $\varepsilon_k(\lambda)$'s in position (k, k), where $k = 1, 2, ..., r$, and the remaining elements all equal to zero.

Lemma 10.1 Any $m \times n$ λ-matrix $A(\lambda)$ of rank r over a field $\mathbb{F}$ is equivalent to a diagonal matrix, i.e., there exist unimodular matrices $P(\lambda)$ and $Q(\lambda)$ such that

$$P(\lambda)A(\lambda)Q(\lambda) = \text{diag}\,[\varepsilon_1(\lambda), \varepsilon_2(\lambda), ..., \varepsilon_r(\lambda), 0, ..., 0],$$

where $\varepsilon_{i+1}(\lambda)$'s are divisible by $\varepsilon_i(\lambda)$, $i = 1, 2, ..., r - 1$, and are monic polynomials over $\mathbb{F}$.

Proof The lemma will be proved constructively in four steps.

1. Since the elements of $A(\lambda)$ are polynomials in λ, we can find out among its elements the element having the least degree. We denote this element $a_{ij}(\lambda)$ and bring it to position (1, 1) by interchanging the ith row and the jth column with the first row and the first column respectively (this is an elementary row as well as column operation (i)). We call now this element $a_{11}(\lambda)$. The remaining elements in the first row $a_{1j}(\lambda)$, $j = 2, ..., n$, and in the first column $a_{i1}(\lambda)$, $i = 2, ..., m$, being of equal or higher degree polynomials than $a_{11}(\lambda)$, on dividing by $a_{11}(\lambda)$, are related respectively by

$$a_{1j}(\lambda) = a_{11}(\lambda)q_{1j}(\lambda) + r_{1j}(\lambda), \qquad a_{i1}(\lambda) = a_{11}(\lambda)q_{i1}(\lambda) + r_{i1}(\lambda),$$

where $i = 2, ..., m$; $j = 2, ..., n$, with $q_{1j}(\lambda)$, $q_{i1}(\lambda)$ as the quotients and $r_{1j}(\lambda)$, $r_{i1}(\lambda)$ as the remainders of lower degree than $a_{11}(\lambda)$. Subtracting now $q_{1j}(\lambda)$ times the first column from the jth column (this is elementary column operation (iii)), we have $r_{1j}(\lambda)$ in the position $(1, j)$ for all $j = 2, ..., n$. Similarly, by subtracting $q_{i1}(\lambda)$ times the first row from the ith row (this is elementary row operation (iii)), we have $r_{i1}(\lambda)$ in position $(i, 1)$, for all $i = 2, ..., m$. If all $r_{i1}(\lambda)$ and $r_{1j}(\lambda)$ are zero, step 1 is complete and we have $A(\lambda)$ reduced to the matrix shown in (10.12). Otherwise, since $r_{1j}(\lambda)$ and $r_{i1}(\lambda)$ are of degree lower than $a_{11}(\lambda)$, by interchanging rows or columns, we can bring the lowest degree polynomial among these $r(\lambda)$'s to the (1, 1) position, and can repeat the above procedure, thus reducing further the degree of the element in the (1, 1) position and that of the remaining elements in the first row and the first column. Observing that the degree of the polynomial elements is finite and the degree of the element in the (1, 1) position is reduced in each step, we obtain, after a finite number of operations, all the elements in the first row as also those in the first column reduced to zero except the one in the (1, 1) position. Calling this nonzero polynomial element still by $a_{11}(\lambda)$, we have now $A(\lambda)$ reduced to the form

$$\begin{bmatrix} a_{11}(\lambda) & 0 & \cdots & 0 \\ 0 & a_{22}(\lambda) & \cdots & a_{2n}(\lambda) \\ \vdots & \vdots & & \vdots \\ 0 & a_{m2}(\lambda) & \cdots & a_{mn}(\lambda) \end{bmatrix}. \tag{10.12}$$

2. After the reduction by the previous step, if $a_{11}(\lambda)$ is not of the least degree, by interchanging rows and columns, we can bring the lowest degree element to the (1, 1) position from among the remaining elements in (10.12) and repeating step 1, transform $A(\lambda)$ to the form (10.12). Still, if $a_{11}(\lambda)$ is not of the least degree, we repeat the above process till $A(\lambda)$ is reduced to the form (10.12) with $a_{11}(\lambda)$ of the lowest degree polynomial in λ among the elements.

3. After steps 1 and 2, all the remaining elements $a_{ij}(\lambda)$, $i = 2, ..., m; j = 2, ..., n$, may not be divisible by $a_{11}(\lambda)$. If $a_{ij}(\lambda)$ is such an element, we add the jth column to the first column and repeat steps 1 and 2, so that the reduced $A(\lambda)$ is of the desired form. This process is repeated until we have $A(\lambda)$ in the form (10.12) with $a_{11}(\lambda)$ of the least degree and, moreover, all the remaining elements are divisible by $a_{11}(\lambda)$. If $a_{11}(\lambda)$ is not monic, we divide the first row by coefficient of the highest term in $a_{11}(\lambda)$ and thereby get $A(\lambda)$ reduced after the above three steps to

$$\begin{bmatrix} \varepsilon_1(\lambda) & 0 & \cdots & 0 \\ 0 & b_{22}(\lambda) & \cdots & b_{2n}(\lambda) \\ \vdots & \vdots & & \vdots \\ 0 & b_{m2}(\lambda) & \cdots & b_{mn}(\lambda) \end{bmatrix}, \tag{10.13}$$

where $\varepsilon_1(\lambda)$ is a monic polynomial and all the remaining elements $b_{ij}(\lambda)$ are divisible by it.

4. The lemma is proved if $b_{ij}(\lambda)$'s in (10.13) are zero. Otherwise, steps 1–3 are applied to the submatrix $[b_{ij}(\lambda)]$, so that $A(\lambda)$ is reduced to

$$\begin{bmatrix} \varepsilon_1(\lambda) & 0 & 0 & \cdots & 0 \\ 0 & \varepsilon_2(\lambda) & 0 & \cdots & 0 \\ 0 & 0 & c_{33}(\lambda) & \cdots & c_{3n}(\lambda) \\ \vdots & \vdots & \vdots & & \vdots \\ 0 & 0 & c_{m3}(\lambda) & \cdots & c_{mn}(\lambda) \end{bmatrix}, \tag{10.14}$$

where $c_{ij}(\lambda)$'s are divisible by the monic polynomial $\varepsilon_2(\lambda)$ which is itself divisible by $\varepsilon_1(\lambda)$. If $c_{ij}(\lambda)$'s are not all zero, the above process is continued until we get

$$\text{diag } [\varepsilon_1(\lambda), \varepsilon_2(\lambda), ..., \varepsilon_k(\lambda), 0, ..., 0]. \tag{10.15}$$

Since (10.15) is obtained from $A(\lambda)$ by elementary row and column operations, there exist unimodular matrices $P(\lambda)$ and $Q(\lambda)$ such that

$$P(\lambda)A(\lambda)Q(\lambda) = \text{diag}\ [\varepsilon_1(\lambda), \varepsilon_2(\lambda), ..., \varepsilon_k(\lambda), 0, ..., 0].$$

As elementary operations do not change the rank of a λ-matrix by Theorem 10.2, we have

$$k = \text{rank}\ A(\lambda) = r \leq \min\ (m, n).$$

Thus, the proof of the lemma is complete. □

Example 10.5 Reduce the matrix

$$A(\lambda) = \begin{bmatrix} 0 & 0 & \lambda(\lambda+1) \\ 2\lambda & -2\lambda & -\lambda \\ 0 & \lambda(\lambda-1) & \lambda^2-\lambda \end{bmatrix}$$

to a diagonal form of Lemma 10.1.

Solution (1) Interchanging the first and second rows and (1′) multiplying the first column by 1/2, we have

$$\begin{bmatrix} \lambda & -2\lambda & -\lambda \\ 0 & 0 & \lambda(\lambda+1) \\ 0 & \lambda(\lambda-1) & \lambda(\lambda-1) \end{bmatrix}.$$

(2′) Adding twice the first column to the second column and once to the third column respectively, we obtain

$$\begin{bmatrix} \lambda & 0 & 0 \\ 0 & 0 & \lambda(\lambda+1) \\ 0 & \lambda(\lambda-1) & \lambda(\lambda-1) \end{bmatrix}.$$

(2) Interchanging the second and third rows and (3′) subtracting the second column from the third column, we get

$$\begin{bmatrix} \lambda & 0 & 0 \\ 0 & \lambda(\lambda-1) & 0 \\ 0 & 0 & \lambda(\lambda+1) \end{bmatrix}.$$

Since $\lambda(\lambda + 1)$ is not divisible by $\lambda(\lambda - 1)$, (4′) adding the third column to the second column, we obtain

$$\begin{bmatrix} \lambda & 0 & 0 \\ 0 & \lambda(\lambda-1) & 0 \\ 0 & \lambda(\lambda+1) & \lambda(\lambda+1) \end{bmatrix}.$$

As $\lambda(\lambda + 1) = \lambda(\lambda - 1)\cdot 1 + 2\lambda$, (3) subtracting the second row from the third and (4) interchanging the second and third rows we get,

$$\begin{bmatrix} \lambda & 0 & 0 \\ 0 & 2\lambda & \lambda(\lambda+1) \\ 0 & \lambda(\lambda-1) & 0 \end{bmatrix}.$$

Again (5) subtracting $(\lambda - 1)/2$ times the second row from the third row, then (5′) multiplying the second column by 1/2 and subtracting $(\lambda + 1)$ times the second column from the third column and multiplying the third column by -2, we have, finally, $A(\lambda)$ reduced to the form

$$\text{diag}\,[\varepsilon_1(\lambda), \varepsilon_2(\lambda), \varepsilon_3(\lambda)] = B(\lambda) = \begin{bmatrix} \lambda & 0 & 0 \\ 0 & \lambda & 0 \\ 0 & 0 & \lambda(\lambda^2-1) \end{bmatrix}.$$

Subjecting a unit matrix of order three to the row operations (1) to (5) and another to the column operations (1′) to (5′) mentioned above, we have, respectively,

$$P(\lambda) = \begin{bmatrix} 0 & 1 & 0 \\ 1 & 0 & -1 \\ (1-\lambda)/2 & 0 & (1+\lambda)/2 \end{bmatrix}, \qquad Q(\lambda) = \begin{bmatrix} 1/2 & 1/4 & (\lambda+3)/2 \\ 0 & 0 & 2 \\ 0 & 1/2 & (1-\lambda) \end{bmatrix}.$$

By direct multiplication, it can be verified that $P(\lambda)A(\lambda)Q(\lambda) = B(\lambda)$, in which $\varepsilon_{i+1}(\lambda)$'s are divisible by $\varepsilon_i(\lambda)$ for $i = 1, 2$.

Example 10.6 Reduce the matrix

$$A(\lambda) = \begin{bmatrix} \lambda(\lambda^4+\lambda^3-4\lambda^2-6\lambda-2) & \lambda(\lambda^3+\lambda^2-4\lambda-5) & -\lambda(\lambda^2+2\lambda+3) \\ \lambda(\lambda^4-\lambda^2+2\lambda+2) & \lambda(\lambda^3-\lambda+3) & 2\lambda(\lambda^2-\lambda+1) \\ \lambda(\lambda+1)^2 & \lambda(\lambda+2) & \lambda(\lambda^2+1) \end{bmatrix}$$

to a diagonal form as in Lemma 10.1.

Solution We interchange (1) the first and third rows and (1′) the first and second columns to bring the lowest degree polynomial element $\lambda(\lambda + 2)$ to the (1, 1) position to get

$$\begin{bmatrix} \lambda(\lambda+2) & \lambda(\lambda+1)^2 & \lambda(\lambda^2+1) \\ \lambda(\lambda^3-\lambda+3) & \lambda(\lambda^4-\lambda^2+2\lambda+2) & 2\lambda(\lambda^2-\lambda+1) \\ \lambda(\lambda^3+\lambda^2-4\lambda-5) & \lambda(\lambda^4+\lambda^3-4\lambda^2-6\lambda-2) & -\lambda(\lambda^2+2\lambda+3) \end{bmatrix}.$$

Now, as

$$\lambda(\lambda^3-\lambda+3) = \lambda(\lambda+2)\cdot(\lambda^2-2\lambda+3)-3\lambda,$$

$$\lambda(\lambda^3+\lambda^2-4\lambda-5) = \lambda(\lambda+2)\cdot(\lambda^2-\lambda-2)-\lambda,$$

$$\lambda(\lambda+1)^2 = \lambda(\lambda+2)\cdot\lambda+\lambda,$$

$$\lambda(\lambda^2+1) = \lambda(\lambda+2)\cdot(\lambda-2)+5\lambda.$$

we have, (2) subtracting $(\lambda^2 - 2\lambda + 3)$ and $(\lambda^2 - \lambda - 2)$ times the first row respectively from the second and the third row and (2′) λ and $(\lambda - 2)$ times the first column respectively from the second and third columns, $A(\lambda)$ reduced to the form

$$\begin{bmatrix} \lambda(\lambda+2) & \lambda & 5\lambda \\ -3\lambda & \lambda(-\lambda^2+\lambda-1) & \lambda(-\lambda^4+2\lambda^3-2\lambda^2+3\lambda-7) \\ -\lambda & -\lambda^3 & \lambda(-\lambda^4+\lambda^3-3) \end{bmatrix}.$$

Again, we bring the lowest degree element λ to the (1, 1) position by (3′) interchanging the first and second columns to get

$$\begin{bmatrix} \lambda & \lambda(\lambda+2) & 5\lambda \\ \lambda(-\lambda^2+\lambda-1) & -3\lambda & \lambda(-\lambda^4+2\lambda^3-2\lambda^2+3\lambda-7) \\ -\lambda^3 & -\lambda & \lambda(-\lambda^4+\lambda^3-3) \end{bmatrix}.$$

The above matrix is reduced further by (3) subtracting $(-\lambda^2 + \lambda - 1)$ and $-\lambda^2$ times the first row from the second and the third row respectively, and (4′) $(\lambda + 2)$ and 5 times the first column respectively from the second and third columns to give

$$\begin{bmatrix} \lambda & 0 & 0 \\ 0 & \lambda(\lambda^3+\lambda^2-\lambda-1) & \lambda(-\lambda^4+2\lambda^3+3\lambda^2-2\lambda-2) \\ 0 & \lambda(\lambda^3+2\lambda^2-1) & \lambda(-\lambda^4+\lambda^3+5\lambda^2-3) \end{bmatrix}.$$

Again (4) subtracting the second row from the third row, and as

$$\lambda(-\lambda^4+2\lambda^3+3\lambda^2-2\lambda-2) = \lambda(\lambda^3+\lambda^2-\lambda-1)(3-\lambda)+(\lambda-\lambda^3)$$

(5′) subtracting $(3 - \lambda)$ times the second column from the third column, we get

$$\begin{bmatrix} \lambda & 0 & 0 \\ 0 & \lambda(\lambda^3+\lambda^2-\lambda-1) & \lambda(1-\lambda^2) \\ 0 & \lambda(\lambda^2+\lambda) & -\lambda(1+\lambda) \end{bmatrix}.$$

Now (5) interchanging the second and third rows and (6′) the second and third columns and then multiplying the second column by –1, we bring $\lambda(1 + \lambda)$ to the (2, 2) position to get

$$\begin{bmatrix} \lambda & 0 & 0 \\ 0 & \lambda(\lambda+1) & \lambda^2(\lambda+1) \\ 0 & \lambda(\lambda^2-1) & \lambda(\lambda^2-1)(\lambda+1) \end{bmatrix}.$$

Finally (6) subtracting $(\lambda - 1)$ times the second row from the third row and (7′) λ times the second column from the third column, we obtain

$$\text{diag}\,[\varepsilon_1(\lambda), \varepsilon_2(\lambda), \varepsilon_3(\lambda)] = B(\lambda) = \begin{bmatrix} \lambda & 0 & 0 \\ 0 & \lambda(\lambda+1) & 0 \\ 0 & 0 & \lambda(\lambda^2-1) \end{bmatrix}$$

as the reduced form of $A(\lambda)$. Taking into consideration (a) the elementary row operations (1) to (6) and (b) the elementary column operations (1′) to (7′), which when applied separately to a unit matrix of order 3, give us

$$\text{(a) } P(\lambda) = \begin{bmatrix} 0 & 0 & 1 \\ 1 & -1 & 4 \\ 1-\lambda & \lambda & 2-3\lambda \end{bmatrix} \text{ and}$$

$$\text{(b) } Q(\lambda) = \begin{bmatrix} 1 & (\lambda^2-\lambda-1) & (-\lambda^3+\lambda^2-2) \\ -\lambda & (-\lambda^3+\lambda^2+\lambda+1) & (\lambda^4-\lambda^3+\lambda+1) \\ 0 & -1 & \lambda \end{bmatrix}$$

respectively. It can be verified that $P(\lambda)A(\lambda)Q(\lambda) = B(\lambda)$, in which $\varepsilon_{i+1}(\lambda)$'s are divisible by $\varepsilon_i(\lambda)$ for $i = 1, 2$. □

10.5 INVARIANT POLYNOMIALS AND SMITH CANONICAL FORMS

Now, we shall introduce invariant polynomials of a λ-matrix and two more lemmas, before establishing formally the famous Smith canonical form. Even though a λ-matrix $A(\lambda)$ is transformed by elementary operations, its invarient polynomials remain unchanged. The Smith canonical form of $A(\lambda)$ is a *quasi-diagonal matrix* (see Definition 3.5) having the invariant polynomials along its diagonal.

Lemma 10.2 The greatest common divisor (GCD, see Definition 7.12) of kth order minors of an $m \times n$ λ-matrix $A(\lambda)$ is invariant under equivalent transformations.

Proof Let the λ-matrix $B(\lambda)$ be obtained from $A(\lambda)$ by a sequence of elementary operations. Since these operations change the minors of kth order in $A(\lambda)$ to the extent of a nonzero scalar multiple, the kth order minors in $B(\lambda)$ differ from those in $A(\lambda)$ by a scalar multiple. As GCDs are monic polynomials, there follows the proposition of the theorem. □

The following example explains the basic fact that the GCD of minors of order k of a λ-matrix is divisible by the GCD of minors of order $k - 1$.

Example 10.7 Consider the following λ-matrix:

$$A(\lambda) = \begin{bmatrix} 1 & 1 & \lambda+1 \\ \lambda+1 & 1 & 1 \\ \lambda-1 & -1 & \lambda-1 \end{bmatrix}.$$

Solution Let $M_{ij}(\lambda)$ denote the minor of the element $a_{ij}(\lambda)$ in $A(\lambda)$. Then we can write

$$\begin{bmatrix} M_{11}(\lambda) & M_{12}(\lambda) & M_{13}(\lambda) \\ M_{21}(\lambda) & M_{22}(\lambda) & M_{23}(\lambda) \\ M_{31}(\lambda) & M_{32}(\lambda) & M_{33}(\lambda) \end{bmatrix} = \begin{bmatrix} \lambda & \lambda(\lambda-1) & -2\lambda \\ 2\lambda & \lambda(1-\lambda) & -\lambda \\ -\lambda & -\lambda(\lambda+2) & -\lambda \end{bmatrix}.$$

Now, a minor of order 3 in $A(\lambda)$, denoted by $M(\lambda)$, which is in fact det $A(\lambda)$ itself, can be expressed as a linear combination of these minors M_{ij} (see Definition 2.4 and Equation (2.10)), viz.,

$$\begin{aligned} M(\lambda) &= 1 \cdot M_{11}(\lambda) - 1 \cdot M_{12}(\lambda) + (1+\lambda)M_{13}(\lambda) \\ &= -(\lambda+1)M_{21}(\lambda) + 1 \cdot M_{22}(\lambda) - 1 \cdot M_{23}(\lambda) \\ &= (\lambda-1)M_{31}(\lambda) + 1 \cdot M_{32}(\lambda) + (\lambda-1)M_{33}(\lambda) \end{aligned}$$

which shows clearly that the GCD $d_2(\lambda)$ of the minors $M_{ij}(\lambda)$'s of order two divides a minor of order three and hence the GCD $d_3(\lambda)$ of these third order minors when $A(\lambda)$ is of higher order. In fact, here we have $d_1(\lambda) = 1$, $d_2(\lambda) = \lambda$, $d_3(\lambda) = \lambda^2$. □

Let us denote by $d_k(\lambda)$ the GCD of kth order minors in an $m \times n$ λ-matrix $A(\lambda)$. Since any kth order minor can be expressed as a linear combination of minors of $(k - 1)$ order, $d_k(\lambda)$ is divisible by $d_{k-1}(\lambda)$ for all $k = 1, ..., n$.

Definition 10.5 *Invariant Polynomials.* Let $d_0(\lambda) := 1$, $d_1(\lambda)$, ..., $d_r(\lambda)$ be the GCDs of minors of order 0, 1, ..., r, respectively, of an $m \times n$ λ-matrix $A(\lambda)$ of rank r. Then the monic polynomials

$$i_1(\lambda) = \frac{d_1(\lambda)}{d_0(\lambda)},\ i_2(\lambda) = \frac{d_2(\lambda)}{d_1(\lambda)}, ..., i_r(\lambda) = \frac{d_r(\lambda)}{d_{r-1}(\lambda)}$$

are called the *invariant polynomials* of $A(\lambda)$. □

It is evident that $d_k(\lambda) = i_1(\lambda)\ i_2(\lambda) \cdots i_k(\lambda)$. In Theorem 10.5 below, it is shown that $i_{k+1}(\lambda)$ is divisible by $i_k(\lambda)$ for all $k = 1, 2, ..., r - 1$.

The attribute *invariant* in invariant polynomials is with respect to elementary row and column operations. In other words, even though a matrix changes under elementary operations, its invariant polynomials do not, as a result of Lemma 10.2. Hence follows trivially the following Lemma.

Lemma 10.3 Two equivalent $m \times n$ λ-matrices have the same invariant polynomials.

Theorem 10.5 *Smith Canonical Form.* Any $m \times n$ λ-matrix $A(\lambda)$ of rank r over a field $\mathbb{F}$ is equivalent to

$$\text{diag}\ [i_1(\lambda), ..., i_r(\lambda), 0, ..., 0],$$

where $i_k(\lambda)$'s are invariant polynomials of $A(\lambda)$ over $\mathbb{F}$. In other words, there exist unimodular matrices $P(\lambda)$ and $Q(\lambda)$ such that

$$P(\lambda)\ A(\lambda)\ Q(\lambda) = \text{diag}\ [i_1(\lambda), ..., i_r(\lambda), 0, ..., 0]$$

and that $i_{k+1}(\lambda)$ is divisible by $i_k(\lambda)$ for all $k = 1, ..., r - 1$. The above diagonal matrix is called Smith canonical (or normal) form of $A(\lambda)$.

Proof By Lemma 10.1, there exist unimodular matrices $P(\lambda)$ and $Q(\lambda)$ such that

$$P(\lambda)\ A(\lambda)\ Q(\lambda) = \text{diag}\ [\varepsilon_1(\lambda), ..., \varepsilon_r(\lambda), 0, ..., 0] = B(\lambda), \text{ say,}$$

where $\varepsilon_{k+1}(\lambda)$ is divisible by $\varepsilon_k(\lambda)$. The GCDs of minors of all orders of $B(\lambda)$ are $d_0(\lambda): = 1$, $d_1(\lambda) = \varepsilon_1(\lambda)$, $d_2(\lambda) = \varepsilon_1(\lambda)\ \varepsilon_2(\lambda), ..., d_r(\lambda) = \varepsilon_1(\lambda) \cdots \varepsilon_r(\lambda)$ and, consequently, its invariant polynomials are

$$i_1(\lambda) = \varepsilon_1(\lambda),\ i_2(\lambda) = \varepsilon_2(\lambda),\ ...,\ i_r(\lambda) = \varepsilon_r(\lambda).$$

The assertion of the theorem follows by noting that, in view of Lemma 10.3, the invariant polynomials of $A(\lambda)$ are the same as those of its equivalent λ-matrix $B(\lambda)$. Though the invariant polynomials are uniquely determined, the unimodular matrices $P(\lambda)$ and $Q(\lambda)$ are not, in general, unique. □

Example 10.8 Find Smith canonical form of $A(\lambda): = [a_{ij}(\lambda)]$ given in Example 10.5.

Solution Let $M_{ij}(\lambda)$ denote the minors of order two (see Definition 2.4) in $A(\lambda)$ given by

$$A(\lambda) = \begin{bmatrix} 0 & 0 & \lambda(\lambda+1) \\ 2\lambda & -2\lambda & -\lambda \\ 0 & \lambda(\lambda-1) & \lambda(\lambda-1) \end{bmatrix}. \tag{10.16}$$

Then,

$$\begin{bmatrix} M_{11}(\lambda) & M_{12}(\lambda) & M_{13}(\lambda) \\ M_{21}(\lambda) & M_{22}(\lambda) & M_{23}(\lambda) \\ M_{31}(\lambda) & M_{32}(\lambda) & M_{33}(\lambda) \end{bmatrix} = \begin{bmatrix} \lambda^2(1-\lambda) & 2\lambda^2(\lambda-1) & 2\lambda^2(\lambda-1) \\ \lambda^2(1-\lambda^2) & 0 & 0 \\ 2\lambda^2(1+\lambda) & -2\lambda^2(1+\lambda) & 0 \end{bmatrix}, \tag{10.17}$$

$$\det A(\lambda) = 2\lambda^3(\lambda^2-1). \tag{10.18}$$

Hence, from (10.16)–(10.18),

$$\begin{aligned} d_0(\lambda) &= 1, \\ d_1(\lambda) &= \text{GCD of } a_{ij}(\lambda)\text{'s} = \lambda, \\ d_2(\lambda) &= \text{GCD of } M_{ij}(\lambda)\text{'s} = \lambda^2, \\ d_3(\lambda) &= \text{GCD of } \det A(\lambda) = \lambda^3(\lambda^2-1). \end{aligned}$$

The last GCD follows from the fact that it is a monic polynomial. Therefore (see Definition 10.5 and Theorem 10.5),

$$\varepsilon_1(\lambda) = 1, \qquad \varepsilon_2(\lambda) = 1, \qquad \varepsilon_3(\lambda) = \lambda(\lambda^2-1),$$

and Smith canonical form becomes

$$\text{diag}\ [\lambda,\ \lambda,\ \lambda(\lambda^2-1)].$$

We can verify that $A(\lambda)$ in Example 10,1 has a Smith canonical form

$$\begin{bmatrix} 1 & 0 & 0 & 0 \\ 0 & 1 & 0 & 0 \\ 0 & 0 & \lambda^2-\lambda+2 & 0 \end{bmatrix}.$$ □

10.6 SIMILARITY AND EQUIVALENCE—FIRST AND SECOND NATURAL NORMAL FORMS: JORDAN CANONICAL FORMS

So far, we were concerned with arbitrary rectangular $m \times n$ matrices under equivalence transformations. An important characteristic in this respect is that, the class of all equivalent λ-matrices have the same invariant polynomials (see Lemma 10.3). In this section, however, we deal with arbitrary square $n \times n$ scalar matrices A in $V_{n\times n}$ and their relationship with the characteristic matrices $(\lambda I - A)$, see (7.7), which are λ-matrices of order n and degree one. An important characteristic of such matrices is that if A and B are two similar matrices in $V_{n\times n}$ (see Definition 6.11) then $(\lambda I - A)$ and $(\lambda I - B)$ are equivalent and vice versa. In other words, the class of similar matrices in $V_{n\times n}$ has a corresponding class of equivalent characteristic matrices. This characteristic leads to the representation by canonical forms (e.g., companion forms, Jordan canonical forms, etc.) of a matrix A in $V_{n\times n}$ via invariant polynomials or Smith normal form of the characteristic matrix $(\lambda I - A)$.

Theorem 10.6 Two matrices A and B in $V_{n\times n}$ are similar iff $(\lambda I - A)$ and $(\lambda I - B)$ are equivalent.

Proof (a) *Necessity.* Suppose that A and B are similar (see Definition 6.11). Then there exists a nonsingular transformation T such that

$$T^{-1}AT = B.$$

Hence, it follows that

$$\lambda I - T^{-1}AT = (\lambda I - B), \quad \text{or} \quad T^{-1}(\lambda I - A)T = (\lambda I - B)$$

which shows that $(\lambda I - A)$ and $(\lambda I - B)$ are equivalent.

(b) *Sufficiency.* Now suppose that $(\lambda I - A)$ and $(\lambda I - B)$ are equivalent. Therefore, there exist unimodular matrices $P(\lambda)$ and $Q(\lambda)$ such that

$$P(\lambda)(\lambda I - A)\, Q(\lambda) = (\lambda I - B);$$

or, since $Q^{-1}(\lambda) = R(\lambda)$ is unimodular,

$$P(\lambda)\,(\lambda I - A) = (\lambda I - B)\, R(\lambda). \tag{10.19}$$

By the remainder theorem, we can write the right division of $R(\lambda)$ by $(\lambda I - A)$ and the left division of $P(\lambda)$ by $(\lambda I - B)$ as (see Theorem 7.11)

$$R(\lambda) = C(\lambda)(\lambda I - A) + R_0, \tag{10.20}$$

$$P(\lambda) = (\lambda I - B)\, D(\lambda) + P_0, \tag{10.21}$$

where (see Definition 7.10) R_0 = right value of $R(\lambda)$ at A and P_0 = left value of $P(\lambda)$ at B; $C(\lambda)$ and $D(\lambda)$ are polynomial matrices of degree less than that of $R(\lambda)$ and $P(\lambda)$ respectively. Substituting (10.20) and (10.21) into (10.19) and rearranging, we get

$$(\lambda I - B)\,[C(\lambda) - D(\lambda)]\,(\lambda I - A) = P_0(\lambda I - A) - (\lambda I - B)R_0.$$

Observing that the left side is at least of degree two, the above identity in λ is valid if

$$C(\lambda) = D(\lambda). \tag{10.22}$$

Consequently,

$$P_0(\lambda I - A) = (\lambda I - B)R_0. \tag{10.23}$$

Hence, equating coefficients of λ, we have

$$P_0A = BR_0 \quad \text{and} \quad P_0 = R_0. \tag{10.24}$$

Thus,

$$R_0A = BR_0. \tag{10.25}$$

Now, as a result of (10.25), A and B are similar if we can prove that R_0 is nonsingular. To show this we write by remainder Theorem 7.11 the right division of $Q(\lambda)$ by $(\lambda I - B)$ as

$$Q(\lambda) = E(\lambda)(\lambda I - B) + Q_0, \tag{10.26}$$

where $Q_0 = Q_r(B)$ is the right value of $Q(\lambda)$ at B (see Definition 7.10), and $E(\lambda)$ is a polynomial matrix of lower degree than $Q(\lambda)$. We substitute (10.22) into (10.20) to get

$$R(\lambda) = D(\lambda)(\lambda I - A) + R_0. \tag{10.27}$$

Since $Q^{-1}(\lambda) = R(\lambda)$, from (10.23), (10.26) and (10.27), we have

$$I = Q(\lambda)\, R(\lambda) = [Q(\lambda)\, D(\lambda) + E(\lambda)P_0](\lambda I - A) + Q_0R_0$$

which, being an identity in λ, yields

$$Q(\lambda)\, D(\lambda) + E(\lambda)P_0 = 0,$$

$$Q_0R_0 = I. \tag{10.28}$$

Equation (10.28) shows that det $R_0 \neq 0$ and by (10.25), A and B are similar. □

Remark 10.3 The sufficiency proof above shows that if A and B are similar, such that

$$T^{-1}AT = B$$

and if we can determine unimodular matrices $P(\lambda)$ and $Q(\lambda)$ satisfying

$$P(\lambda)\ (\lambda I - A)\ Q(\lambda) = (\lambda I - B)$$

then, in view of (10.24), (10.25) and (10.28), the nonsingular matrix T is given by

$$T = R_0^{-1} = Q_0 = Q_r(B) = \text{right value of } Q(\lambda) \text{ at } B,$$

or

$$T^{-1} = R_0 = P_0 = P_l(B) = \text{left value of } P(\lambda) \quad \text{at } B.$$ □

Corollary 10.1 Two regular λ-matrices $A_1\lambda + A_0$ and $B_1\lambda + B_0$, both of order n and degree 1 (see Definition 7.5), are *strictly equivalent.* In other words, if

$$(B_1\lambda + B_0) = P(\lambda)\,(A_1\lambda + A_0)\,Q(\lambda), \tag{10.29}$$

then the unimodular matrices $P(\lambda)$ and $Q(\lambda)$ are independent of λ.

Proof In view of (10.29), where A_1 and B_1 are nonsingular (see Definition 7.5), we have, after some minor rearrangement, the relation

$$(\lambda I - B) = R(\lambda)(\lambda I - A)\ Q(\lambda), \tag{10.30}$$

where

$$R(\lambda) = B_1^{-1}P(\lambda)A_1, \qquad A = -A_1^{-1}A_0, \qquad B = -B_1^{-1}B_0. \tag{10.31}$$

It is evident from (10.30) that $(\lambda I - A)$ and $(\lambda I - B)$ are equivalent, and hence A and B are similar by Theorem 10.6. So there exists a nonsingular matrix T satisfying

$$B = T^{-1}AT, \quad \text{or} \quad (\lambda I - B) = T^{-1}(\lambda I - A)T.$$

Using (10.31), we now have

$$(\lambda I + B_1^{-1}B_0) = T^{-1}(\lambda I + A_1^{-1}A_0)T,$$

i.e.,

$$(B_1\lambda + B_0) = B_1T^{-1}A_1^{-1}(A_1\lambda + A_0)T$$

which proves Corollary 10.1. □

Theorem 10.7 Two $m \times n$ λ-matrices $A(\lambda)$ and $B(\lambda)$ over a field $\mathbb{F}$ are equivalent iff they have the same invariant polynomials.

Proof (a) *Necessity.* It follows by Lemma 10.3.

(b) *Sufficiency.* Suppose that $A(\lambda)$ and $B(\lambda)$ have the same invariant polynomials $i_1(\lambda)$, $i_2(\lambda)$, ..., $i_r(\lambda)$. Then there exist unimodular matrices $P(\lambda)$, $Q(\lambda)$, $R(\lambda)$ and $T(\lambda)$ such that by Theorem 10.5,

$$P(\lambda)\ A(\lambda)\ Q(\lambda) = R(\lambda)\ B(\lambda)\ T(\lambda) = \text{diag}\ [i_1(\lambda), \ldots, i_r(\lambda), 0, \ldots, 0],$$

i.e.,

$$R^{-1}(\lambda)\ P(\lambda)\ A(\lambda)\ Q(\lambda)\ T^{-1}(\lambda) = B(\lambda)$$

Since $R^{-1}(\lambda)\ P(\lambda)$ and $Q(\lambda)\ T^{-1}(\lambda)$ are unimodular, $A(\lambda)$ and $B(\lambda)$ are equivalent. □

As an easy consequence of Theorems 10.6 and 10.7, we have the following theorem.

Theorem 10.8 Two matrices A and B in $V_{n\times n}$ are similar iff $(A - \lambda I)$ and $(B - \lambda I)$ have the same invariant polynomials.

Proof By Theorem 10.6, A and B are similar, iff $(\lambda I - A)$ and $(\lambda I - B)$ are equivalent; by Theorem 10.7, $(\lambda I - A)$ and $(\lambda I - B)$ are equivalent iff $(\lambda I - A)$ and $(\lambda I - B)$ have the same invariant polynomials (see Example 10.10). □

Remark 10.4 If one knows that A and B in $V_{n\times n}$ are similar, i.e., there exists a nonsingular matrix T such that

$$T^{-1}AT = B,$$

then by Theorem 10.8, $(\lambda I - A)$ and $(\lambda I - B)$ have the same invariant polynomials, say, $i_1(\lambda)$, ..., $i_n(\lambda)$. So, in view of Lemma 10.1 and Theorem 10.5, we can determine unimodular matrices $U(\lambda)$, $V(\lambda)$, $R(\lambda)$ and $W(\lambda)$ satisfying

$$U(\lambda)(\lambda I - A)\ V(\lambda) = \text{diag}\ [i_1(\lambda), \ldots, i_n(\lambda)] = R(\lambda)(\lambda I - B)\ W(\lambda).$$

Hence,

$$P(\lambda)(\lambda I - A)\ Q(\lambda) = (\lambda I - B),$$

where $P(\lambda) = R^{-1}(\lambda)\ U(\lambda)$ and $Q(\lambda) = V(\lambda)\ W^{-1}(\lambda)$. Since $Q(\lambda)$ and $P(\lambda)$ are now known, by sufficiency proof of Theorem 10.6 (see Remark 10.3), the nonsingular matrix T can be computed by (see Definition 7.10)

$$T = \text{right value of } Q(\lambda) \quad \text{at } B = Q_r(B), \tag{10.32}$$

or

$$T^{-1} = \text{left value of } P(\lambda) \quad \text{at } B = P_l(B). \tag{10.33}$$

□

Example 10.9 It is known that the matrices A and C given below are similar. Determine a nonsingular matrix T such that $T^{-1}AT = C$.

$$A = \begin{bmatrix} -2 & 1 & 0 \\ -9 & 4 & 1 \\ 12 & -3 & -3 \end{bmatrix}, \qquad C = \begin{bmatrix} 0 & 1 & 0 \\ 0 & 0 & 1 \\ 3 & 2 & -1 \end{bmatrix}.$$

Solution The elementary row and column operations defined by:

(a') interchange the first and second columns, (b') add $(\lambda + 2)$ times the first column to the second column (a) add $(\lambda - 4)$ times the first row to the second row and three times to the third row, (c') interchange the second and third columns (d') add $(\lambda^2 - 2\lambda + 1)$ times the second column to the third column (b) add $(\lambda + 3)$ times the second row to the third row, finally (e') multiply the first and second columns by –1, determine matrices $U(\lambda)$ and $V(\lambda)$ such that

$$U(\lambda)(\lambda I - A)\, V(\lambda) = \text{diag}\,[1, 1, \lambda^3 + \lambda^2 - 2\lambda - 3]$$

Again the elementary row and column operations defined by:

(1′) interchange the first and second columns, (1) add λ times the first row to the second row and –2 times to the third row, (2′) add λ times first column to the second column, (3′) interchange the second and third columns, (4′) add λ^2 times the second column to the third column, (2) add $(\lambda + 1)$ times the second row to the third row, finally (5′) multiply the first and second columns by –1, generate respectively matrices $R(\lambda)$ and $W(\lambda)$ such that

$$R(\lambda)(\lambda I - C)\, W(\lambda) = \text{diag}\,[1, 1, \lambda^3 + \lambda^2 - 2\lambda - 3].$$

In view of Remark 10.4, we find that $V(\lambda)$ and $W(\lambda)$ can respectively be constructed from a third order unit matrix by subjecting it separately to the column operations (a′)–(e′) and (l′)–(5′) respectively. Now, using the notation given in Section 4.2 and considering column operations (l′)–(5′), we have

$$W(\lambda) = K_{12}K_{21}(\lambda)\, K_{23}K_{32}(\lambda^2)\, K_1(-1)K_2(-1) \tag{10.34}$$

and, consequently by (a′)–(e′) and (10.34), see inverse operations defined in Section 4.2,

$$\begin{aligned} Q(\lambda) = V(\lambda)\ W^{-1}(\lambda) &= K_{12}K_{21}(\lambda + 2)\ K_{23}K_{32}(\lambda^2 - 2\lambda + 1)\ K_1(-1)K_2(-1) \\ &\quad K_2(-1)\ K_1(-1)\ K_{32}(-\lambda^2)\ K_{23}K_{21}(-\lambda)K_{12} \end{aligned} \tag{10.35}$$

$$= \begin{bmatrix} 1 & 0 & 0 \\ 2 & 1 & 0 \\ -2\lambda + 1 & 0 & 1 \end{bmatrix},$$

where the last step is accomplished by imposing the elementary column operations defined in (10.35) on a third order unity matrix. Hence, in view of (10.32), the-transformation matrix T is

$$T = Q_r(C) = \begin{bmatrix} 1 & 0 & 0 \\ 2 & 1 & 0 \\ 1 & 0 & 1 \end{bmatrix} + \begin{bmatrix} 0 & 0 & 0 \\ 0 & 0 & 0 \\ -2 & 0 & 0 \end{bmatrix} C = \begin{bmatrix} 1 & 0 & 0 \\ 2 & 1 & 0 \\ 1 & -2 & 1 \end{bmatrix}.$$

Alternatively, we can compute T by (10.33). In this situation, $P(\lambda) = R^{-1}(\lambda)\, U(\lambda)$, where we compute $R(\lambda)$ and $U(\lambda)$ by the row operations defined by (1)–(2) and (a)–(b). Therefore, see discussion on inverse operations in Section 4.2,

$$\begin{aligned} P(\lambda) &= R^{-1}(\lambda)\, U(\lambda) = \{H_{32}(\lambda+1)\, H_{31}(-2)\, H_{21}(\lambda)\}^{-1}\, \{H_{32}(\lambda+3)\, H_{31}(3)\, H_{21}(\lambda-4)\} \\ &= H_{21}(-\lambda)\, H_{31}(2)\, H_{32}(-\lambda-1)\, H_{32}(\lambda+3)\, H_{31}(3)\, H_{21}(\lambda-4) \\ &= \begin{bmatrix} 1 & 0 & 0 \\ -4 & 1 & 0 \\ -3+2\lambda & 2 & 1 \end{bmatrix}. \end{aligned}$$

Hence, by (10.33),

$$\begin{aligned} T^{-1} = P_l(C) &= \begin{bmatrix} 1 & 0 & 0 \\ -4 & 1 & 0 \\ -3 & 2 & 1 \end{bmatrix} + \begin{bmatrix} 0 & 1 & 0 \\ 0 & 0 & 1 \\ 3 & 2 & -1 \end{bmatrix} \begin{bmatrix} 0 & 0 & 0 \\ 0 & 0 & 0 \\ 2 & 0 & 0 \end{bmatrix} \\ &= \begin{bmatrix} 1 & 0 & 0 \\ -2 & 1 & 0 \\ -5 & 2 & 1 \end{bmatrix}. \end{aligned}$$

□

Lemma 10.4 Let C given by

$$C = \begin{bmatrix} 0 & 1 & 0 & \cdots & 0 \\ 0 & 0 & 1 & \cdots & 0 \\ \vdots & \vdots & \vdots & & \vdots \\ 0 & 0 & 0 & \cdots & 1 \\ -a_0 & -a_1 & -a_2 & \cdots & -a_{n-1} \end{bmatrix}$$

be the companion matrix of $\phi(\lambda) = \lambda^n + a_{n-1}\lambda^{n-1} + \cdots + a_1\lambda + a_0$ (see Definition 7.19: a companion matrix is also called a *Frobenius matrix*) in $V_{n\times n}$. Then the invariant polynomials of $(\lambda I - C)$ are

$$\{1, \ldots, 1, \phi(\lambda)\}.$$

Proof We have

$$(\lambda I - C) = \left[\begin{array}{c|ccccc} \lambda & -1 & 0 & \cdots & 0 & 0 \\ 0 & \lambda & -1 & \cdots & 0 & 0 \\ \vdots & \vdots & \vdots & & \vdots & \vdots \\ 0 & 0 & 0 & \cdots & \lambda & -1 \\ \hline a_0 & a_1 & a_2 & \cdots & a_{n-2} & \lambda + a_{n-1} \end{array}\right],$$

where at least one minor of all orders from 1 to $(n-1)$ of the upper right submatrix is 1 or -1. Also, by Example 7.32, $\det(\lambda I - C) = \phi(\lambda)$. Hence the GCDs of minors of all orders are $d_1 = \cdots = d_{n-1} = 1$ and $d_n(\lambda) = \phi(\lambda)$. Therefore, the lemma follows by Definition 10.5. □

Remark 10.5 From Lemma 10.4, we observe that *given* any monic polynomial $\phi(\lambda)$, we can construct a companion matrix C where the last row has all the coefficients of $\phi(\lambda)$ with a negative sign. Furthermore, for this companion matrix C, $(\lambda I - C)$ has the given polynomial as the *only nonzero degree invariant polynomial.* Conversely, *given* any companion matrix C, one can immediately write down the invariant polynomials associated with $(\lambda I - C)$.

Because $(\lambda I - A)$ is nonsingular for any matrix A in $V_{n\times n}$ and no principal minors of any order in $(\lambda I - A)$ are zero, we have the following corollary.

Corollary 10.2 No invariant polynomials of $(\lambda I - A)$, where A is in $V_{n\times n}$, can be identically zero, i.e., $i_k(\lambda) \neq 0$ for any $k = 1, ..., n$. □

Lemma 10.5 Let $i_1(\lambda)$ and $i_2(\lambda)$ be two monic polynomials such that $i_2(\lambda)$ is divisible by $i_1(\lambda)$, and C_1 and C_2 be respectively the associated companion matrices. Then the invariant polynomials of $(\lambda I - C)$, where $C = \text{diag}\,[C_1, C_2]$, are $1, ..., 1, i_1(\lambda), i_2(\lambda)$.

Proof By Lemma 10.4, the invariant polynomials of $(\lambda I - C_1)$ and $(\lambda I - C_2)$ are, respectively,

$$\{1, ..., 1, i_1(\lambda)\} \text{ and } \{1, ..., 1, i_2(\lambda)\}.$$

By Theorem 10.5, there exist unimodular matrices $P(\lambda)$, $Q(\lambda)$, $R(\lambda)$, and $S(\lambda)$ such that

$$P(\lambda)(\lambda I - C_1)Q(\lambda) = \text{diag}\;[1, ..., 1, i_1(\lambda)],$$

$$R(\lambda)(\lambda I - C_2)S(\lambda) = \text{diag}\;[1, ..., 1, i_2(\lambda)].$$

Hence,

$$\begin{bmatrix} P(\lambda) & 0 \\ 0 & R(\lambda) \end{bmatrix} \begin{bmatrix} \lambda I - C_1 & 0 \\ 0 & \lambda I - C_2 \end{bmatrix} \begin{bmatrix} Q(\lambda) & 0 \\ 0 & S(\lambda) \end{bmatrix}$$

$$= \text{diag}\,[1, ..., 1, i_1(\lambda); 1, ..., 1, i_2(\lambda)].$$

Interchanging rows and columns on both sides, we see that $(\lambda I - C) = \text{diag}\,[\lambda I - C_1, \lambda I - C_2]$ is equivalent to $[1, ..., 1; i_1(\lambda), i_2(\lambda)]$. Since $i_2(\lambda)$ is divisible by $i_1(\lambda)$, the assertion follows from Theorem 10.5. □

Example 10.10 The invariant polynomials of $(\lambda I - A)$, where

$$A = \begin{bmatrix} 0 & 1 & 0 & 0 & 0 \\ 0 & -1 & 0 & 0 & 0 \\ -2 & -1 & 0 & 1 & 0 \\ 5 & 2 & 2 & -1 & 1 \\ 7 & -1 & 2 & -2 & 1 \end{bmatrix} \tag{10.36}$$

are$\{1, 1, 1, \lambda(\lambda+1), \lambda(\lambda^2-1)\}$. These can be computed by the procedure given in Lemma 10.1.

The companion matrices associated with the invariant polynomials of nonzero degree, viz.,

(a) $i_1(\lambda) = \lambda(\lambda + 1)$ and (b) $i_2(\lambda) = \lambda(\lambda^2 - 1)$,

are respectively

$$\text{(a)}\ C_1 = \begin{bmatrix} 0 & 1 \\ 0 & -1 \end{bmatrix} \quad \text{and} \quad \text{(b)}\ C_2 = \begin{bmatrix} 0 & 1 & 0 \\ 0 & 0 & 1 \\ 0 & 1 & 0 \end{bmatrix}. \tag{10.37}$$

It is easy to verify that $(\lambda I - C)$, where C is the quasi-diagonal matrix, $C = \text{diag}\,[C_1, C_2]$ has the same set of invariant polynomials as that of $(\lambda I - A)$, which is an assertion to be proved in Theorem 10.9. By Theorem 10.8, A and C are therefore similar. □

Lemma 10.6 Let $i_1(\lambda), ..., i_n(\lambda)$ be the invariant polynomials of $(\lambda I - A)$, where A is in $V_{n \times n}$. Then $\det(\lambda I - A) = i_1(\lambda) \cdots i_n(\lambda)$, implying that the sum of the degrees of $i_k(\lambda)$'s is n.

Proof By Theorem 10.5, there exist unimodular matrices $P(\lambda)$ and $Q(\lambda)$ such that

$$P(\lambda)\,(\lambda I - A)\,Q(\lambda) = \text{diag}\,[i_1(\lambda), ..., i_n(\lambda)].$$

On taking the determinants of both sides, we obtain

$$\det P(\lambda) \det(\lambda I - A) \det Q(\lambda) = i_1(\lambda) \cdots i_n(\lambda).$$

Since $\det P(\lambda)$ and $\det Q(\lambda)$ are independent of λ, and since $\det(\lambda I - A)$ is a monic polynomial and all $i_k(\lambda)$'s are monic, we have

$$\det(\lambda I - A) = i_1(\lambda) \cdots i_n(\lambda).$$

The assertion in the lemma follows by noting that the degree of the characteristic polynomial $\det(\lambda I - A)$ of A is n. □

Remark 10.6 Lemma 10.6 remains valid if $(\lambda I - A)$ is replaced by any $n \times n$ λ-matrix $A(\lambda)$ of full rank.

Theorem 10.9 *First Natural Normal Form.* Let $i_1(\lambda), ..., i_p(\lambda)$ be invariant polynomials of nonzero degree of $\lambda I - A$, where A is in $V_{n \times n}$, $p \le n$. Then A is similar to a quasi-diagonal companion form

$$C = \text{diag}\,[C_1, C_2, ..., C_p]$$

where C_k's are the companion matrices associated with $i_k(\lambda)$'s.

Proof Having deepened our understanding by means of Example 10.10, we observe that, in view of repeated application of Lemma 10.5, the invariant polynomials of $(\lambda I - C)$ are $1, ..., 1, i_1(\lambda), ..., i_p(\lambda)$, where 1's are $(n - p)$ in number. Hence, $(\lambda I - C)$ and $(\lambda I - A)$ have the same set of invariant polynomials and, therefore, by Theorem 10.8, A and C are similar. □

Example 10.11 Let A be given by (10.36). Then one can determine unimodular matrices $U(\lambda)$ and $V(\lambda)$ such that (see Example 10.9)

$$U(\lambda)\,(\lambda I - A)\,V(\lambda) = \text{diag}\,(1, 1, 1, \lambda(\lambda+1), \lambda(\lambda^2 - 1)), \tag{10.38}$$

where

$$U(\lambda) = H_3(-1)\,H_{53}(\lambda-1)\,H_{34}H_2(-1)\,H_{52}(2)H_{42}(\lambda+1)H_{23}H_1(-1)\times$$
$$H_{51}(1)H_{41}(-2)H_{31}(1)H_{21}(\lambda+1),$$
$$V(\lambda) = K_{12}K_{21}(\lambda)K_{24}K_{32}(\lambda)K_{42}(\lambda+2)K_{35}K_{43}(\lambda^2+\lambda-3)K_{53}(\lambda^2+\lambda-2)K_{45}(-1).$$

From Theorem 10.9, we have

$$i_1(\lambda)=\lambda(\lambda+1) \quad \text{and} \quad i_2(\lambda)=\lambda(\lambda^2-1),$$

and the associated companion matrices C_1 and C_2 are given by (10.37). Denoting $C = \text{diag}\,[C_1, C_2]$, we have, by a similar computation,

$$R(\lambda)\,(\lambda I - C)\,W(\lambda) = \text{diag}\,[1, 1, 1, \lambda(\lambda+1), \lambda(\lambda^2-1)], \tag{10.39}$$

where the unimodular matrices $R(\lambda)$ and $W(\lambda)$ are given by

$$R(\lambda) = H_3(-1)H_{53}(\lambda)H_{34}H_2(-1)H_{52}(-1)H_{42}(\lambda)H_{23}H_1(-1)H_{21}(\lambda+1),$$
$$W(\lambda) = K_{12}K_{21}(\lambda)K_{24}K_{32}(\lambda)K_{35}K_{53}(\lambda^2).$$

Now, (10.38) and (10.39) show that $(\lambda I - A)$ and $(\lambda I - C)$ have the same set of invariant polynomials, and so by Theorem 10.8, A and C are similar. Consequently, there exists a nonsingular matrix T such that

$$T^{-1}AT = T^{-1}\begin{bmatrix} 0 & 1 & 0 & 0 & 0 \\ 0 & -1 & 0 & 0 & 0 \\ -2 & -1 & 0 & 1 & 0 \\ 5 & 2 & 2 & -1 & 1 \\ 7 & -1 & 2 & -2 & 1 \end{bmatrix} T$$

$$= \left[\begin{array}{cc|ccc} 0 & 1 & & & \\ 0 & -1 & & & \\ \hline & & 0 & 1 & 0 \\ & & 0 & 0 & 1 \\ & & 0 & 1 & 0 \end{array}\right] = \text{diag}\,[C_1, C_2] = C.$$

Taking cognizance of Remark 10.4, we have

$$Q(\lambda) = V(\lambda)W^{-1}(\lambda) = K_{12}K_{21}(\lambda)K_{24}K_{32}(\lambda)K_{42}(\lambda+2)K_{35}K_{43}(\lambda^2+\lambda-3)$$
$$\times K_{53}(\lambda^2+\lambda-2)K_{45}(-1)K_{53}(-\lambda^2)K_{35}K_{32}(-\lambda)K_{24}K_{21}(-\lambda)K_{12}$$

$$= \begin{bmatrix} 1 & 0 & 0 & 0 & 0 \\ 0 & 1 & 0 & 0 & 0 \\ -1 & 0 & 1 & 0 & 0 \\ 2 & 0 & 0 & 1 & 0 \\ -1 & 0 & \lambda-2 & 0 & 1 \end{bmatrix}.$$

Hence, in view of (10.32), the nonsingular transformation matrix T is

$$T = Q_r(C) = \begin{bmatrix} 1 & 0 & 0 & 0 & 0 \\ 0 & 1 & 0 & 0 & 0 \\ -1 & 0 & 1 & 0 & 0 \\ 2 & 0 & 0 & 1 & 0 \\ -1 & 0 & -2 & 1 & 1 \end{bmatrix}.$$

Also, direct computation gives det $(\lambda I - A)$ as $\lambda^2(\lambda+1)^2(\lambda-1)$ which is the product of invariant polynomials of $(\lambda I - A)$, and proves thereby the assertion of Lemma 10.6. □

It is possible, as we shall now see, to reduce further the matrix A in the sense that the diagonal elements of C are now sub-matrices of order lower than C_1 and C_2. With this object in view, we proceed to define *elementary divisors*. We shall first restrict ourselves to the field $\mathbb{C}$ of complex numbers and then generalize the results obtained to any other field such as the real field $\mathbb{R}$. However, we note that, as a consequence of Lemma 10.6, det $(\lambda I - A)$ and the product of invariant polynomials $i_1(\lambda)$, ..., $i_n(\lambda)$ of $(\lambda I - A)$ contain the same factors over a field $\mathbb{F}$.

Definition 10.6 *Elementary Divisors over a Complex Field* $\mathbb{C}$. Let $\lambda_1, \lambda_2, ..., \lambda_n$ be distinct eigenvalues of a matrix A in $\mathbb{C}^{n\times n}$, and $i_1(\lambda)$, $i_2(\lambda)$, ..., $i_n(\lambda)$ be invariant polynomials of $(\lambda I - A)$. Then

$$\begin{aligned} i_1(\lambda) &= (\lambda-\lambda_1)^{\alpha_{11}}(\lambda-\lambda_2)^{\alpha_{12}}\cdots(\lambda-\lambda_k)^{\alpha_{1k}}\cdots(\lambda-\lambda_n)^{\alpha_{1n}}, \\ i_2(\lambda) &= (\lambda-\lambda_1)^{\alpha_{21}}(\lambda-\lambda_2)^{\alpha_{22}}\cdots(\lambda-\lambda_k)^{\alpha_{2k}}\cdots(\lambda-\lambda_n)^{\alpha_{2n}}, \\ &\vdots \\ i_p(\lambda) &= (\lambda-\lambda_1)^{\alpha_{p1}}(\lambda-\lambda_2)^{\alpha_{p2}}\cdots(\lambda-\lambda_k)^{\alpha_{pk}}\cdots(\lambda-\lambda_n)^{\alpha_{pn}}, \\ &\vdots \\ i_n(\lambda) &= (\lambda-\lambda_1)^{\alpha_{n1}}(\lambda-\lambda_2)^{\alpha_{n2}}\cdots(\lambda-\lambda_k)^{\alpha_{nk}}\cdots(\lambda-\lambda_n)^{\alpha_{nn}}, \end{aligned}$$

where some of α_{ij}'s may be zero, and since $i_{p+1}(\lambda)$ is divisible by $i_p(\lambda)$, for all p = 1, 2, ..., $n-1$, we have

$$\begin{aligned} &0 \le \alpha_{1k} \le \alpha_{2k} \le \cdots \le \alpha_{pk} \le \cdots \le \alpha_{nk} \le n_k, \\ &\alpha_{1k} + \alpha_{2k} + \cdots + \alpha_{nk} = n_k, \\ &n_1 + n_2 + \cdots + n_n = n, \end{aligned}$$

and $0 \le n_k \le n$ is the multiplicity of the eigenvalue λ_k. Each factor $(\lambda - \lambda_k)^{\alpha_{pk}}$ with $\alpha_{pk} > 0$ in the above factorizations of invariant polynomials is defined as the *elementary divisor* of $(\lambda I - A)$ over the complex field $\mathbb{C}$, corresponding to the eigenvalue λ_k. The product $i_1(\lambda) \cdots i_n(\lambda) = d_n(\lambda)$ is the characteristic polynomial (CP) of A, $d_{n-1}(\lambda)$ is the GCD of all minors of order $n-1$ of $(\lambda I - A)$ and $i_n(\lambda) = d_n(\lambda)/d_{n-1}(\lambda)$ is the minimum polynomial (MP) of A (see Definition 10.5 and Theorems 7.14 and 10.5). An elementary divisor $(\lambda - \lambda_k)^{\alpha_{pk}}$ is said to be *linear* if $\alpha_{pk} = 1$; otherwise, it is *nonlinear.* Elementary divisors of any other $m \times n$ polynomial matrix $A(\lambda)$ of rank r over a complex field $\mathbb{C}$ are similarly defined. In this case, the invariant polynomials $i_1(\lambda)$, $i_2(\lambda)$, ..., $i_r(\lambda)$ are resolved into irreducible factors which are the elementary divisors of $A(\lambda)$. □

Example 10.12 The eigenvalues of A as given by (10.36) are $\lambda_1 = 0$, $\lambda_2 = -1$, $\lambda_3 = 1$ obtained in Example 10.11, and the invariant polynomials of $(\lambda I - A)$ are

$$i_1(\lambda) = i_2(\lambda) = i_3(\lambda) = 1, \qquad i_4(\lambda) = \lambda(\lambda+1) \qquad i_5(\lambda) = \lambda(\lambda^2 - 1)$$

which can be factorized as

$$i_1(\lambda) = \lambda^0(\lambda+1)^0\,(\lambda-1)^0,$$
$$i_2(\lambda) = \lambda^0(\lambda+1)^0\,(\lambda-1)^0,$$
$$i_3(\lambda) = \lambda^0(\lambda+1)^0\,(\lambda-1)^0,$$
$$i_4(\lambda) = \lambda(\lambda+1)\,(\lambda-1)^0,$$
$$i_5(\lambda) = \lambda(\lambda+1)\,(\lambda-1).$$

Considering only the factors of nonzero degree, we find the elementary divisors of $(\lambda I - A)$ as λ, λ,$(\lambda + 1)$, $(\lambda + 1)$ and $(\lambda - 1)$ which are all linear. The CP and MP of A are $\phi(\lambda) = \lambda^2(\lambda - 1)(\lambda + 1)^2$ and $\psi(\lambda) = \lambda(\lambda^2 - 1)$ respectively. □

Example 10.13 The elementary divisors of $(\lambda I - D)$, where $D = \text{diag}\,[d_1, d_2, ..., d_n]$ in $V_{n \times n}$, are $(\lambda - d_i)$, $i = 1, 2, ..., n$, which are all linear. □

Definition 10.7 *Elementary Divisors over any Field* $\mathbb{F}$. The irreducible factors of the invariant polynomials $i_1(\lambda)$, ..., $i_n(\lambda)$ are not always linear over all fields. For example, an invariant polynomial $(\lambda^2 + 1)$ is irreducible over the real field $\mathbb{R}$ and cannot be split into simpler factors such as $(\lambda + i)(\lambda - i)$ in $\mathbb{R}$ (see Definition 7.13 and Example 7.24). So over any field $\mathbb{F}$, let $\phi_1(\lambda)$, $\phi_2(\lambda)$, ..., $\phi_n(\lambda)$ be distinct irreducible monic factors of $\det(\lambda I - A)$, where A is in $V_{n \times n}$. Then the invariant polynomials $i_1(\lambda)$, ..., $i_n(\lambda)$ of $(\lambda I - A)$ can be resolved into factors, e.g.,

$$i_1(\lambda) = [\phi_1(\lambda)]^{\alpha_{11}}\,[\phi_2(\lambda)]^{\alpha_{12}} \cdots [\phi_k(\lambda)]^{\alpha_{1k}} \cdots [\phi_n(\lambda)]^{\alpha_{1n}},$$
$$i_2(\lambda) = [\phi_1(\lambda)]^{\alpha_{21}}\,[\phi_2(\lambda)]^{\alpha_{22}} \cdots [\phi_k(\lambda)]^{\alpha_{2k}} \cdots [\phi_n(\lambda)]^{\alpha_{2n}},$$
$$\vdots$$
$$i_p(\lambda) = [\phi_1(\lambda)]^{\alpha_{p1}}\,[\phi_2(\lambda)]^{\alpha_{p2}} \cdots [\phi_k(\lambda)]^{\alpha_{pk}} \cdots [\phi_n(\lambda)]^{\alpha_{pn}},$$
$$\vdots$$
$$i_n(\lambda) = [\phi_1(\lambda)]^{\alpha_{n1}}\,[\phi_2(\lambda)]^{\alpha_{n2}} \cdots [\phi_k(\lambda)]^{\alpha_{nk}} \cdots [\phi_n(\lambda)]^{\alpha_{nn}},$$

where

$$0 \le \alpha_{1k} \le \alpha_{2k} \le \cdots \le \alpha_{pk} \le \cdots \le \alpha_{nk} \le \beta_k$$

with

$$\beta_k = \alpha_{1k} + \alpha_{2k} + \cdots + \alpha_{nk},$$

because $i_{p+1}(\lambda)$ is divisible by $i_p(\lambda)$ for all $p = 1, 2, ..., n - 1$. Also, β_k, the multiplicity of the irreducible factor $\phi_k(\lambda)$, satisfies $0 \le \beta_k \le n$. Each monic factor $[\phi_k(\lambda)]^{\alpha_{pk}}$ with $\alpha_{pk} > 0$ in the above factorization of invariant polynomials is defined as the *elementary divisor* of $(\lambda I - A)$ over $\mathbb{F}$. The elementary divisors of any $m \times n$ λ-matrix $A(\lambda)$ of rank r over $\mathbb{F}$ are similarly defined. □

Example 10.14 The invariant polynomials of $(\lambda I - A)$, where

$$A = \begin{bmatrix} 1 & 0 & 0 & 0 & 0 & 0 & 0 & 0 & 0 \\ 0 & 0 & 0 & 0 & 0 & 1 & 0 & 0 & 0 \\ 0 & 0 & 0 & 0 & 0 & 0 & 0 & 1 & 0 \\ 0 & 0 & 0 & 0 & 1 & 0 & 0 & 0 & 0 \\ 0 & 0 & 0 & -1 & 0 & 0 & 0 & 0 & 0 \\ 0 & -1 & 0 & 0 & 0 & 0 & 0 & 0 & 0 \\ 0 & 0 & -2 & 0 & 0 & 0 & 1 & 0 & 1 \\ 0 & 0 & -5 & 0 & 0 & 0 & 0 & 2 & 4 \\ 0 & 0 & 2 & 0 & 0 & 0 & 4 & -2 & 1 \end{bmatrix} \tag{10.40}$$

are 1, $(\lambda^2 + 1)$, $[(\lambda - 1)^2 + 4]^2\,(\lambda^2 + 1)(\lambda - 1)$. Resolving these polynomials into monic factors over $\mathbb{R}$, we have

$$i_1(\lambda) = (\lambda - 1)^0\,(\lambda^2 + 1)^0\,[(\lambda - 1)^2 + 4]^0,$$

$$i_2(\lambda) = (\lambda - 1)^0\,(\lambda^2 + 1)\,[(\lambda - 1)^2 + 4]^0,$$

$$i_3(\lambda) = (\lambda - 1)\,(\lambda^2 + 1)\,[(\lambda - 1)^2 + 4]^2.$$

Hence, the elementary divisors of $(\lambda I - A)$ are

$$(\lambda - 1),\ (\lambda^2 + 1),\ (\lambda^2 + 1),\ [(\lambda - 1)^2 + 4]^2 \text{ over } \mathbb{R} \tag{10.41}$$

of which all but the first are nonlinear. It is not difficult to see that the elementary divisors of $(\lambda I - A)$ over the complex field $\mathbb{C}$ are

$$(\lambda - 1),\ (\lambda + i),\ (\lambda + i),\ (\lambda - i),\ (\lambda - i),\ (\lambda - 1 + 2i)^2,\ (\lambda - 1 - 2i)^2 \text{ over } \mathbb{C} \tag{10.42}$$

which are different from (10.41). □

Remark 10.7 Example 10.14 corroborates what is true in general, i.e., elementary divisors of $(\lambda I - A)$ over different field $\mathbb{F}$'s, all of which contain the elements of A, are different in general though its invariant polynomials over these fields are the same. Sometimes, a matrix does not have eigenvalues over $\mathbb{F}$, for instance, A in Example 10.14 does not have eigenvalues over $\mathbb{R}$ but elementary divisors under such circumstances do however exist.

The following two lemmas are true in general over any field not necessarily real or complex.

Lemma 10.7 Two matrices A and B in $V_{n\times n}$ are similar iff $(\lambda I - A)$ and $(\lambda I - B)$ have the same elementary divisors.

Proof Let A and B be similar. Then by Theorem 10.8, $(\lambda I - A)$ and $(\lambda I - B)$ have the same invariant polynomials and hence the same elementary divisors.

Conversely, let $(\lambda I - A)$ and $(\lambda I - B)$ have the same elementary divisors. They must therefore have the same invariant polynomials. Hence by Theorem 10.8, A and B are similar. □

Lemma 10.8 The elementary divisors over any field $\mathbb{F}$ of diag $[(\lambda I - A), (\lambda I - B)]$, where A and B are in $V_{n\times n}$ and $V_{m\times m}$, respectively, are the union of the elementary divisors of $(\lambda I - A)$ and $(\lambda I - B)$.

Proof We shall prove the lemma over a complex field $\mathbb{C}$; the proof over any other field $\mathbb{F}$ is exactly the same.

Let the invariant polynomials of $(\lambda I - A)$ and $(\lambda I - B)$ having the common factor $(\lambda - x)$ be, respectively,

$$p_1(\lambda) = (\lambda - x)^{\alpha_1}\ f_1(\lambda),$$
$$\vdots$$
$$p_n(\lambda) = (\lambda - x)^{\alpha_n}\ f_n(\lambda),$$

and

$$q_1(\lambda) = (\lambda - x)^{\beta_1}\ g_1(\lambda),$$
$$\vdots$$
$$q_m(\lambda) = (\lambda - x)^{\beta_m}\ g_m(\lambda),$$

where $0 \le \delta_1 \le \delta_2 \le \cdots \le \alpha_n$, $0 \le \beta_1 \le \beta_2 \le \cdots \le \beta_m$, $f_k(\lambda)$'s and $g_k(\lambda)$'s are not divisible by $(\lambda - x)$. The above two sets of invariant polynomials can be written together in order of non-decreasing powers of $(\lambda - x)$ after renaming $f_i(\lambda)$'s and $g_j(\lambda)$'s by $h_k(\lambda)$'s, $k = 1, 2, ..., m + n = r$, where $h_k(\lambda)$'s are not divisible by $(\lambda - x)$. Therefore, diag $[(\lambda I - A), (\lambda I - B)]$ is equivalent to (see constructive proof of Lemma 10.5)

$$C(\lambda) = \text{diag}\,[(\lambda - x)^{\delta_1}\ h_1(\lambda), (\lambda - x)^{\delta_2}\ h_2(\lambda), ..., (\lambda - x)^{\delta_r}\ h_r(\lambda)],$$

where $0 \le \delta_1 \le \delta_2 \le \cdots \le \delta_r$. It is easy to check that $(\lambda - x)^{\delta_1}, ..., (\lambda - x)^{\delta_r}$ represent the union of the elementary divisors of $(\lambda I - A)$ and $(\lambda I - B)$. The GCDs of minors of $C(\lambda)$ of order r, $r - 1, ..., 1$ are respectively of the form

$$d_r(\lambda) = (\lambda - x)^{\delta_1 + \delta_2 + \cdots + \delta_r}\ \phi_r(\lambda),\ d_{r-1}(\lambda) = (\lambda - x)^{\delta_1 + \delta_2 + \cdots + \delta_{r-1}}\ \phi_{r-1}(\lambda), ...,$$
$$d_1(\lambda) = (\lambda - x)^{\delta_1}\ \phi_1(\lambda),$$

where $d_{k+1}(\lambda)$ is divisible by $d_k(\lambda)$ but $\phi_k(\lambda)$'s are not divisible by $(\lambda - x)$. Consequently, the invariant polynomials of $C(\lambda)$ are of the form

$$i_1(\lambda) = (\lambda - x)^{\delta_1}\ a_1(\lambda),\ ...,\ i_r(\lambda) = (\lambda - x)^{\delta_r}\ a_r(\lambda),$$

where $a_k(\lambda)$'s are not divisible by $(\lambda - x)$. Hence, the elementary divisors of $C(\lambda)$ corresponding to the eigenvalue $\lambda = x$ are

$$(\lambda - x)^{\delta_1} \cdots (\lambda - x)^{\delta_r}$$

which are the same as the union of those of $(\lambda I - A)$ and $(\lambda I - B)$. Similar results can be proved for other distinct eigenvalues. □

Remark 10.8 Lemma 10.8 remains valid if λ-matrices $A(\lambda)$ and $B(\lambda)$ of order $m \times n$ and $p \times q$ replace $(\lambda I - A)$ and $(\lambda I - B)$ respectively.

Example 10.15 The elementary divisors of

$$A(\lambda) = \text{diag}\ [\lambda, \lambda^2(\lambda + 1), \lambda^2(\lambda + 1)^3],$$

$$B(\lambda) = \text{diag}\ [(\lambda + 1), \lambda^3(\lambda^2 - 1)]$$

are

$$\lambda, \lambda^2, \lambda^2, (\lambda + 1), (\lambda + 1)^3$$

and

$$\lambda^3, (\lambda + 1), (\lambda + 1), (\lambda - 1)$$

respectively. The quasi-diagonal matrix

$$C(\lambda) = \begin{bmatrix} A(\lambda) & 0 \\ 0 & B(\lambda) \end{bmatrix} = \text{diag}\ [\lambda, \lambda^2(\lambda + 1), \lambda^2(\lambda + 1)^3, (\lambda + 1), \lambda^3(\lambda^2 - 1)]$$

is equivalent to Smith form

$$\text{diag}\ [1, \lambda(\lambda + 1), \lambda^2(\lambda + 1), \lambda^2(\lambda + 1), \lambda^3(\lambda + 1)^3 (\lambda - 1)],$$

and has, therefore, the elementary divisors

$$\lambda, \lambda^2, \lambda^2, \lambda^3, (\lambda + 1), (\lambda + 1), (\lambda + 1), (\lambda + 1)^3, (\lambda - 1)$$

which are the union of the elementary divisors of $A(\lambda)$ and $B(\lambda)$. □

Theorem 10.10 *Second Natural Normal Form.* Let $e_1(\lambda), e_2(\lambda), ..., e_p(\lambda)$ be the elementary divisors over any field $\mathbb{F}$ of $(\lambda I - A)$, where A is in $V_{n\times n}$ and $p \le n$. Then A is similar to the quasi-diagonal companion form

$$C = \text{diag}\ [C_1, C_2, ..., C_p],$$

where C_k is the *companion matrix* associated with $e_k(\lambda)$, $k = 1, 2, ..., p$ (see Definition 7.16).

Proof By Lemma 10.4, the invariant polynomial of nonzero degree of $(\lambda I - C_k)$ is $e_k(\lambda)$ for all $k = 1, 2, ..., p$. Since, only a single eigenvalue is associated with each $e_k(\lambda)$, $e_k(\lambda)$ is the only elementary divisor of $(\lambda I - C_k)$. By repeated application of Lemma 10.8, we find that $e_1(\lambda)$, $e_2(\lambda)$, ..., $e_p(\lambda)$ are also the elementary divisors of $(\lambda I - C)$. Inasmuch as $(\lambda I - A)$ and $(\lambda I - C)$ have the same elementary divisors, A and C are similar by Lemma 10.7. □

Example 10.16 The second natural normal forms over $\mathbb{R}$ and $\mathbb{C}$ of the matrix A given by (10.40) in Example 10.14 are respectively,

$$\begin{bmatrix} 1 & & & & & & & & \\ & 0 & 1 & & & & & & \\ & -1 & 0 & & & & & & \\ & & & 0 & 1 & & & & \\ & & & -1 & 0 & & & & \\ & & & & & 0 & 1 & 0 & 0 \\ & & & & & 0 & 0 & 1 & 0 \\ & & & & & 0 & 0 & 0 & 1 \\ & & & & & -25 & 20 & -14 & 4 \end{bmatrix}$$

and

$$\begin{bmatrix} 1 & & & & & & & & \\ & -i & & & & & & & \\ & & -i & & & & & & \\ & & & i & & & & & \\ & & & & i & & & & \\ & & & & & 0 & 1 & & \\ & & & & & 3+4i & 2(1-2i) & & \\ & & & & & & & 0 & 1 \\ & & & & & & & 3-4i & 2(1+2i) \end{bmatrix}$$ □

We have seen what are the ingredients necessary to establish Jordan normal (or canonical) form of a matrix in $V_{n \times n}$ over a complex field $\mathbb{C}$ as well as over a real field $\mathbb{R}$. The constructional procedure we have set forth would enable us to find the elementary divisors of $(\lambda I - A)$. *Over the complex field* $\mathbb{C}$, with each elementary divisor of the form $(\lambda - \alpha)^k$, we can associate a special matrix termed the *Jordan block* having this factor as the *only* elementary divisor, as discussed in the following definition.

Definition 10.8 *Jordan Block.* The Jordan block associated with the elementary divisor $(\lambda - \alpha)^k$ of the characteristic matrix $(\lambda I - A)$ of a matrix A in $V_{n \times n}$ over $\mathbb{C}$ is defined as the upper triangular matrix J in $V_{k \times k}$ with α on the main diagonal, 1's on the super-diagonal, and zeros elsewhere:

$$J(\alpha) = \begin{bmatrix} \alpha & 1 & \cdots & 0 \\ 0 & \alpha & \cdots & 0 \\ \vdots & \vdots & & \vdots \\ 0 & 0 & \cdots & 1 \\ 0 & 0 & \cdots & \alpha \end{bmatrix}_k . \tag{10.43}$$

Alternatively, the Jordan block associated with $(\lambda - \alpha)^k$ can be written with 1's on the subdiagonal as

$$J(\alpha) = \begin{bmatrix} \alpha & 0 & \cdots & 0 & 0 \\ 1 & \alpha & \cdots & 0 & 0 \\ 0 & 1 & \cdots & 0 & 0 \\ \vdots & \vdots & & \vdots & \vdots \\ 0 & 0 & \cdots & 1 & \alpha \end{bmatrix}_k \tag{10.44}$$

when we refer to (10.43) as the *upper Jordan block* and (10.44) as the *lower Jordan block.*

Conversely, for the Jordan block represented by (10.43) or (10.44), $(\lambda I - J)$ has only one elementary divisor given by $(\lambda - \alpha)^k$. □

Let $\hat{J}$ be a matrix which is the same as J but 1's replaced with arbitrary nonzero real numbers $\beta_1, \beta_2, \ldots, \beta_{k-1}$. Then, it can be shown that $T^{-1}JT = \hat{J}$, where $T = \text{diag}[t, t\beta_1, t\beta_1\beta_2, \ldots, t\beta_1\beta_2 \cdots \beta_{k-1}]$, $t \neq 0$, an arbitrary real number; and $(\lambda I - J)$ and $(\lambda I - \hat{J})$ have the same elementary divisor. Hence, the unit values along the super-diagonal or sub-diagonal of J have no special significance (see Problem 10.28).

Over the field $\mathbb{R}$ of real numbers an elementary divisor of $(\lambda I - A)$ is either of the form $(\lambda - \alpha)^k$ or of $[(\lambda - \alpha)^2 + \beta^2]^m$. The Jordan block associated with the elementary divisor $(\lambda - \alpha)^k$ is defined as in (10.43) and (10.44) which are now over $\mathbb{R}$. However, the Jordan block pertaining to the elementary divisor $[(\lambda - \alpha)^2 + \beta^2]^m$ is

$$V = \begin{bmatrix} W & I & \cdots & 0 \\ 0 & W & \cdots & 0 \\ \vdots & \vdots & & \vdots \\ 0 & 0 & \cdots & I \\ 0 & 0 & \cdots & W \end{bmatrix}_{2m}, \text{ or } V = \begin{bmatrix} W & 0 & \cdots & 0 & 0 \\ I & W & \cdots & 0 & 0 \\ 0 & I & & 0 & 0 \\ \vdots & \vdots & & \vdots & \vdots \\ 0 & 0 & \cdots & I & W \end{bmatrix}_{2m}, \tag{10.45}$$

where

$$W = \begin{bmatrix} \alpha & \beta \\ -\beta & \alpha \end{bmatrix}, \quad I = \begin{bmatrix} 1 & 0 \\ 0 & 1 \end{bmatrix}, \quad 0 = \begin{bmatrix} 0 & 0 \\ 0 & 0 \end{bmatrix}.$$

It might be observed that the Jordan block V is strikingly similar to J in which α and 1 are now replaced as if by second order matrices W and I. Conversely, the Jordan block V has only one elementary divisor $[(\lambda - \alpha)^2 + \beta^2]^m$, see Problem 10.32.

Theorem 10.11 *Jordan Normal (or Canonical) Form over a Complex Field* $\mathbb{C}$. In $\mathbb{C}^{n\times n}$, every matrix is similar to a quasi-diagonal matrix, called *Jordan matrix J:*

$$J = \text{diag}\,[J_1, J_2, \ldots, J_p]$$

where J_k's are the Jordan blocks associated with the elementary divisor $e_k(\lambda)$ of $(\lambda I - A)$ for $k = 1, 2, \ldots, p$.

Proof The proof of the theorem evidently follows (i) by observing that $e_k(\lambda)$ is the only invariant polynomial of nonzero degree of $(\lambda I - J_k)$ and hence its only elementary divisor, and (ii) by applying the arguments that established Theorem 10.10. □

Similarly, we can establish the following theorem.

Theorem 10.12 *Jordan Normal (or Canonical) Form over a Real Field* $\mathbb{R}$. Let $(\lambda I - A)$ have k elementary divisors of the form $(\lambda - \mu_i)^{m_i}$ and l elementary divisors of the form $[(\lambda - \alpha_j)^2 + \beta_j^2]^{n_j}$, with A as a matrix in $\mathbb{R}^{n\times n}$. Then, A is similar to a quasi-diagonal matrix J, called Jordan normal (or canonical) form and is designated by

$$J = \text{diag}\,[J_1, \ldots, J_k, V_1, \ldots, V_l],$$

where J_i is an $m_i \times m_i$ matrix of the form (10.43) or (10.44) and V_j, is a $2n_j \times 2n_j$ matrix with either of the forms in (10.45). □

Example 10.17 In view of elementary divisors given by (10.41) and (10.42), the Jordan normal (or canonical) forms of the matrix A given by (10.40) over $\mathbb{R}$ and $\mathbb{C}$ are, respectively,

$$J_{\text{over }\mathbb{R}} = \begin{bmatrix} 1 & & & & & & & & \\ & 0 & 1 & & & & & & \\ & -1 & 0 & & & & & & \\ & & & 0 & 1 & & & & \\ & & & -1 & 0 & & & & \\ & & & & & 1 & 2 & 1 & 0 \\ & & & & & -2 & 1 & 0 & 1 \\ & & & & & 0 & 0 & 1 & 2 \\ & & & & & 0 & 0 & -2 & 1 \end{bmatrix}$$

$$J_{\text{over }\mathbb{C}} = \begin{bmatrix} 1 & & & & & & & & \\ & -i & & & & & & & \\ & & -i & & & & & & \\ & & & i & & & & & \\ & & & & i & & & & \\ & & & & & 1-2i & 1 & & \\ & & & & & 0 & 1-2i & & \\ & & & & & & & 1+2i & 1 \\ & & & & & & & 0 & 1+2i \end{bmatrix}$$ □

Example 10.18 Find the Jordan normal form J of the matrix A given by

$$A = \frac{1}{2}\begin{bmatrix} 10 & -24 & -18 & 54 & -12 \\ -1 & 6 & 3 & -6 & 1 \\ 1 & -8 & 0 & 16 & -4 \\ -1 & 0 & 2 & 2 & 0 \\ -1 & -4 & 0 & 6 & 2 \end{bmatrix} \tag{10.46}$$

as also a nonsingular transformation T such that $J = T^{-1}AT$.

Solution By subjecting $(-2)(\lambda I - A)$ to elementary row and column operations as in Lemma 10.1, it is possible to find unimodular matrices $U(\lambda)$ and $V(\lambda)$ such that

$$U(\lambda)(\lambda I - A)V(\lambda) = \text{diag}\,[1, 1, 1, (\lambda - 2)^2, (\lambda - 2)^3],$$

where (see Section 4.2 for notation)

$$U(\lambda) = (-2)\bar{U}(\lambda),$$

$$\bar{U}(\lambda) = H_5\left(-\frac{1}{2}\right)H_{54}(\lambda-1)H_4\left(-\frac{1}{2}\right)H_{53}(3-\lambda)H_{43}(-2)H_{34}H_{45}\left(\frac{(1-\lambda)}{2}\right)$$

$$\times H_{35}(-2)H_2\left(-\frac{1}{8}\right)H_{52}\left(-\frac{3}{2}\right)H_{42}\left(-\frac{(1+\lambda)}{4}\right)H_{32}(-2\lambda+7)$$

$$\times H_{24}H_{51}(1)H_{41}(1)H_{31}(2\lambda-10)H_{21}(1)H_{13},$$

$$V(\lambda) = K_{21}(8)K_{31}(2\lambda)H_{41}(-16)K_{51}(4)K_{32}\left(\frac{(1-\lambda)}{4}\right)K_{42}\left(\frac{(9-\lambda)}{4}\right)$$

$$\times K_{52}\left(-\frac{1}{2}\right)K_{43}(\lambda^2-3)K_{53}(2\lambda-\lambda^2)K_{45}K_{54}(3).$$

The elementary divisors of $(\lambda I - A)$ are, therefore, $(\lambda - 2)^2$ and $(\lambda - 2)^3$ over $\mathbb{R}$ as well as over $\mathbb{C}$ and by Theorem 10.11, A is similar to

$$J = \left[\begin{array}{cc|ccc} 2 & 1 & & & \\ 0 & 2 & & & \\ \hline & & 2 & 1 & 0 \\ & & 0 & 2 & 1 \\ & & 0 & 0 & 2 \end{array}\right]. \tag{10.47}$$

Following the above procedure, one can find unimodular matrices $R(\lambda)$ and $W(\lambda)$ such that

$$R(\lambda)\,(\lambda I - J)W(\lambda) = \text{diag}\,[1, 1, 1, (\lambda - 2)^2, (\lambda - 2)^3],$$

where

$$R(\lambda) = H_3(-1)H_{53}(\lambda - 2)H_{34}H_2(-1)H_{42}(\lambda - 2)H_{23}H_1(-1)H_{21}(\lambda - 2),$$

$$W(\lambda) = K_{12}K_{21}(\lambda - 2)K_{24}K_{32}(\lambda - 2)K_{35}K_{53}((\lambda - 2)^2).$$

In view of Remark 10.4, we define

$$Q(\lambda) = V(\lambda)W^{-1}(\lambda) = K_{21}(8)K_{31}(2\lambda)K_{41}(-16)K_{51}(4)K_{32}\left(\frac{(1-\lambda)}{4}\right)$$

$$\times K_{42}\left(\frac{(9-\lambda)}{4}\right)K_{52}\left(-\frac{1}{2}\right)K_{43}(\lambda^2-3)\,K_{53}(2\lambda-\lambda^2)\,K_{45}K_{54}(3)$$

$$\times K_{53}\,(-(\lambda-2)^2)\,K_{35}K_{32}(2-\lambda)\,K_{24}K_{21}(2-\lambda)K_{12}$$

$$
= \begin{bmatrix} -2\lambda^2+3\lambda+2, & 1, & -6\lambda^2+10\lambda+4, & 8, & 2 \\ \dfrac{(\lambda^3-3\lambda^2+2\lambda-2)}{4}, & 0, & \dfrac{(3\lambda^3-13\lambda^2+12\lambda+4)}{4}, & 1, & \dfrac{(1-\lambda)}{4} \\ 2\lambda-\lambda^2, & 0, & -3\lambda^2+10\lambda-7, & 0, & 1 \\ 0, & 0, & 1, & 0, & 0 \\ 1, & 0, & 3, & 0, & 0 \end{bmatrix}
$$

and hence by (10.32), the required nonsingular transformation T is

$$
T = Q_r(J) = \left[\begin{array}{cc|ccc} 0 & -4 & 0 & -6 & -4 \\ -1/2 & 1/2 & 0 & 0 & 1 \\ 0 & -2 & 1 & -2 & -2 \\ 0 & 0 & 1 & 0 & 0 \\ 1 & 0 & 3 & 0 & 0 \end{array}\right]. \tag{10.48}
$$

□

The transformation matrix T as determined above is not unique. In fact, there are innumerable nonsingular transformations which achieve the same Jordan normal form, e.g., the nonsingular transformation S given by

$$
S = \begin{bmatrix} 0 & 2 & 0 & -6 & 2 \\ 1 & -2 & 0 & 0 & 2 \\ 1 & 0 & 1 & -3 & 2 \\ 1 & -2 & 1 & -1 & 2 \\ 1 & -4 & 3 & -3 & 4 \end{bmatrix}, \qquad S^{-1} = \frac{1}{2}\begin{bmatrix} 0 & -2 & 0 & 6 & -2 \\ -1 & 2 & 3 & -6 & 1 \\ -1 & 0 & 2 & -3 & 1 \\ -1 & 2 & 2 & -5 & 1 \\ -1 & 4 & 3 & -9 & 2 \end{bmatrix}
$$

transforms A to the same Jordan form (10.47), which can be verified by direct multiplication. From the relation

$$
J = T^{-1}AT = S^{-1}AS
$$

we can easily derive

$$
RA = AR, \qquad R = ST^{-1}.
$$

Thus, the class of transformation matrices S for Jordan normal form J of the square matrix A is given by $S = RT$, where R is any nonsingular matrix permutable with A and T is one of the required transformation matrices.

Let us consider the matrices

$$\text{(a) } U = \left[\begin{array}{cc|ccc} \alpha & 0 & 0 & 0 & 0 \\ 0 & \alpha & 0 & 0 & 0 \\ 0 & 0 & 1 & 0 & 0 \\ 0 & 0 & 0 & 1 & 0 \\ 0 & 0 & 0 & 0 & 1 \end{array}\right], \qquad \text{(b) } V = \left[\begin{array}{cc|ccc} 1 & 0 & 0 & 0 & 0 \\ 0 & 1 & 0 & 0 & 0 \\ 0 & 0 & 1 & \beta & 0 \\ 0 & 0 & 0 & 1 & \beta \\ 0 & 0 & 0 & 0 & 1 \end{array}\right],$$

$$\text{(c) } W = \left[\begin{array}{cc|ccc} 1 & 0 & 0 & 0 & 0 \\ 0 & 1 & 0 & 0 & 0 \\ 1 & 0 & 1 & 0 & 0 \\ 0 & 1 & 0 & 1 & 0 \\ 0 & 0 & 0 & 0 & 1 \end{array}\right]$$

obtained from a unit matrix of order 5 which is a column partitioned into blocks as in Jordan form J of (10.47) by

(i) multiplying the first two columns by a nonzero scalar α,

(ii) adding β times the second column to the third column and β times the first column to the second column in the second block, and

(iii) adding the first and the second columns of the larger block to the corresponding columns of the first block, respectively.

It is easily verified that for the matrix A given by (10.46)

(a) $(TU)^{-1}ATU = U^{-1}JU = J$;

(b) $(TV)^{-1}A(TV) = V^{-1}JV = J$;

(c) $(TW)^{-1}A(TW) = W^{-1}JW = J$.

This shows that the Jordan form of A remains unaltered by operations (i)–(iii) for this simple example. Concentrating on TU, TV and TW, we can formulate the following general rules governing the elementary column transformations allowed on T which is column-partitioned as in the case of J, without altering the Jordan normal form of A:

(i) Multiplication of all the columns in T corresponding to a Jordan block in J by the same nonzero scalar α.

(ii) Addition of the same constant multiple of a column to the next column performed on all the columns of any block in T as permissible, starting from the last column.

(iii) Addition to all the columns of a block the corresponding columns of any other larger or same size block in T pertaining to the same eigenvalue.

According to these rules, we add the first column to the second column and multiply the first and second columns by 2 in T given by (10.48) to get

$$T_1 = \left[\begin{array}{cc|ccc} 0 & -8 & 0 & -6 & -4 \\ -1 & 0 & 0 & 0 & 1 \\ 0 & -4 & 1 & -2 & -2 \\ 0 & 0 & 1 & 0 & 0 \\ 2 & 2 & 3 & 0 & 0 \end{array}\right] \qquad (10.49)$$

which transforms A in (10.46) to the Jordan form J in (10.47) as was done by T.

Remark 10.9 Finally, we like to remark that if L is a linear transformation $L: V_n \to V_n$ and A is its matrix representation with respect to the basis, $X = \{\mathbf{x}_1, ..., \mathbf{x}_n\}$ in V_n and J the matrix representation with respect to the basis $\hat{X} = \{\hat{\mathbf{x}}_1, ..., \hat{\mathbf{x}}_n\}$, then $J = T^{-1}AT$, where

$$\hat{\mathbf{x}}_i = t_{1i}\mathbf{x}_1 + t_{2i}\mathbf{x}_2 + \cdots + t_{ni}\mathbf{x}_n, \qquad T = [t_{ij}] \tag{10.50}$$

(see Theorem 6.12 and Corollary 6.3). If L and X are given, one can compute A as in Section 6.3 and then T by the method given in this section. Consequently, the basis vectors in $\hat{X}$, by means of which the matrix representation in Jordan normal form is obtained, can be determined by (10.50).

Example 10.19 A linear transformation $L : \mathbb{R}^5 \to \mathbb{R}^5$ is given by

$$L\mathbf{x} = \frac{1}{2}\begin{bmatrix} 10x_1 \\ -x_1 + 3x_2 + x_5 \\ x_1 \\ -x_1 \\ -x_1 \end{bmatrix} + \begin{bmatrix} -12x_2 - 9x_3 + 27x_4 - 6x_5 \\ 3x_2 \qquad\quad - 3x_4 \\ -4x_2 \qquad\quad + 8x_4 - 2x_5 \\ x_3 + \quad x_4 \\ -2x_2 \qquad\quad + 3x_4 + \ x_5 \end{bmatrix},$$ □

where $\mathbf{x} = [x_1\ x_2\ x_3\ x_4\ x_5]^T$ is in $\mathbb{R}^5$. With respect to a natural basis $X = \{\mathbf{e}_1\ \mathbf{e}_2\ \mathbf{e}_3\ \mathbf{e}_4\ \mathbf{e}_5\}$ in $\mathbb{R}^5$ (see Definition 5.6), the matrix representation of L (see Example 6.15) is given by A in (10.46). Therefore, the columns of T in (10.48) or those of T_1 in (10.49) form the basis for representation of L in Jordan normal form J in (10.47). □

We would like to remark that if an $n \times n$ matrix A is similar to a Jordan canonical form J, i.e., if there exists a nonsingular matrix $T = [\mathbf{y}_1, \mathbf{y}_2, ..., \mathbf{y}_n]$ such that $AT = TJ$, then all the n-vectors $\mathbf{y}_i$'s in T are not eigenvectors of A. For instance, in Example 10.18, $n = 5$, and only $\mathbf{y}_1$ and $\mathbf{y}_3$ are eigenvectors of A since, by multiplication with the help of (10.46)–(10.48), we have $(A - 2I)\mathbf{y}_1 = (A - 2I)\mathbf{y}_3 = \boldsymbol{\theta}$. Also, both the Jordan blocks here are associated with the same eigenvalue 2.

It can be checked that $\mathbf{y}_1 = (A - 2I)\mathbf{y}_2$, $(A - 2I)^2\mathbf{y}_2 = \boldsymbol{\theta}$, and $\mathbf{y}_4 = (\mathbf{y} - 2I)\mathbf{y}_5$, $\mathbf{y}_3 = (A - 2I)^2\mathbf{y}_5$, $(A - 2I)^3\mathbf{y}_5 = \boldsymbol{\theta}$. The vectors $\mathbf{y}_i$'s being the columns of a nonsingular matrix T are *LI*. The subspace, say T_{11}, generated by $\{\mathbf{y}_1, \mathbf{y}_2\}$ is $(A - 2I)$-cyclic (see Definition 11.8), where $A - 2I$ takes the part of L as also the subspace T_{12} generated by $\{\mathbf{y}_3, \mathbf{y}_4, \mathbf{y}_5\}$. So, the vector space $\mathbb{R}^5$ can be decomposed as the direct sum of two $(A - 2I)$ cyclic subspaces T_{11} and I_{12} generated by the sets $\{(A - 2I)\ \mathbf{y}_2, \mathbf{y}_2\}$ and $\{(A - 2I)^2\ \mathbf{y}_5, (A - 2I)\ \mathbf{y}_5, \mathbf{y}_5\}$, and associated with the eigenvalue $\lambda_1 = 2$ (see Definition 11.2). The matrix A, in this case, is defined to be derogatory as either its Jordan form has more than one Jordan block or it is the direct sum of more than one cyclic subspace associated with the same eigenvalue $\lambda_1 = 2$. The characteristic polynomial of A is $\phi(\lambda) = (\lambda - 2)^5$ and is different from the minimum polynomial $\psi(\lambda) = (\lambda - 2)^3$ (see Definition 7.17).

On the other hand, if we take

$$A = \frac{1}{2}\begin{bmatrix} 10 & -24 & -18 & 46 & -12 \\ -1 & 6 & 3 & -5 & 1 \\ 1 & -8 & 0 & 12 & -4 \\ -1 & 0 & 2 & 2 & 0 \\ -1 & -4 & 0 & 6 & 2 \end{bmatrix}$$

then it can be verified that

$$T^{-1}AT = \begin{bmatrix} 2 & 1 & 0 & 0 & 0 \\ 0 & 2 & 1 & 0 & 0 \\ 0 & 0 & 2 & 1 & 0 \\ 0 & 0 & 0 & 2 & 1 \\ 0 & 0 & 0 & 0 & 2 \end{bmatrix} = J,$$

where T is given by (10.48) and

$$T^{-1} = \begin{bmatrix} 0 & 0 & 0 & -3 & 1 \\ 1 & -2 & -3 & 6 & -1 \\ 0 & 0 & 0 & 1 & 0 \\ -1/2 & 0 & 1 & -1 & 0 \\ -1/2 & 2 & 3/2 & -9/2 & 1 \end{bmatrix}.$$

If we write $T = [\mathbf{y}_1\ \mathbf{y}_2\ \mathbf{y}_3\ \mathbf{y}_4\ \mathbf{y}_5]$ as before, then by direct multiplication we find that $\mathbf{y}_4 = (A-2I)\mathbf{y}_5$, $\mathbf{y}_3 = (A-2I)^2\mathbf{y}_5$, $\mathbf{y}_2 = (A-2I)^3\mathbf{y}_5$, $\mathbf{y}_1 = (A-2I)^4\mathbf{y}_5$ and $(A-2I)^5\mathbf{y}_5 = \mathbf{0}$. As T is nonsingular, the set $\{\mathbf{y}_1, \mathbf{y}_2, \mathbf{y}_3, \mathbf{y}_4, \mathbf{y}_5\}$ generates the vector space $\mathbb{R}^5$. In this case, we say that this set, written alternatively as

$$\{(A-2I)^4\mathbf{y}_5,\ (A-2I)^3\mathbf{y}_5,\ (A-2I)^2\mathbf{y}_5,\ (A-2I)\mathbf{y}_5,\ \ \mathbf{y}_5\},$$

is a cyclic basis of $\mathbb{R}^5$ and A is defined to be *nonderogatory*. The CP and MP of A are the same being equal to $\phi(\lambda) = \psi(\lambda) = (\lambda - 2)^5$ (see Definition 7.17), and also the Jordan form of A has only one Jordan block associated with the eigenvalue $\lambda_1 = 2$. Moreover, $\mathbf{y}_1 = [0, -1/2, 0, 0, 1]^T$ is the only eigenvector of A corresponding to the eigenvalue $\lambda_1 = 2$.

This induces us to give the following definition.

Definition 10.9 *Nonderogatory Matrix.* A square matrix A is called *nonderogatory* if the geometric multiplicity k_i of each of its distinct eigenvalues λ_i is 1. □

Let n_i be the (algebraic) multiplicity of an eigenvalue λ_i of A and k_i be its geometric multiplicity (see Definition 7.20). Then, A is *nonderogatory* if $k_i = 1$, $\forall i$; and A is *derogatory* if $k_i > 1$ for at least one i. The matrix A is *simple* if $k_i = n_i$, $\forall i$; and A is *defective* if $k_i < n_i$ for at least one i.

From this definition and that of elementary divisors (see Definition 10.6), it is clear that a nonderogatory matrix has only one invariant polynomial of nonzero degree and hence is similar to a companion matrix of the form (7.73), while a derogatory matrix is similar to a quasi-diagonal companion form or a first natural normal form as in Theorem 10.9. Both of them are, however, similar to the second natural normal form as in Theorem 10.10 or the Jordan normal form in Theorem 10.11. From the fact that a nonderogatory matrix is similar to a companion matrix (7.92), it can be inferred that a matrix is nonderogatory iff its *minimum polynomial* and *characteristic polynomial* are the same (see Problem 10.33).

In place of the term *normal,* the term *canonical* is very often used, and normal forms in Theorems 10.9–10.12 are then called *canonical forms*. The attribute *canonical* asserts that any linear transformation L has only one such matrix representation in any class of these canonical forms or the class of matrices representing L (see Chapter 6) is similar to exactly one such matrix in any canonical form. The first natural normal form is sometimes referred to as *rational canonical form,* the word *rational* testifying to the fact that it can be derived by transformations which are rational in the field of the elements of A.

Example 10.20 Let $(\lambda I - A)$ be a matrix of order 3 having invariant polynomials $i_1(\lambda)$, $i_2(\lambda)$ and $i_3(\lambda)$ where $i_{r+1}(\lambda)$ is divisible by $i_r(\lambda)$; $r = 1, 2$. Then $(\lambda I - A)$ is equivalent to diag $[i_1(\lambda), i_2(\lambda), i_3(\lambda)] =: D(\lambda)$. The CP of A is $\phi(\lambda) = i_1(\lambda)\ i_2(\lambda)\ i_3(\lambda)$ (see Lemma 10.6). Now adj $D(\lambda) =$ diag $[i_2 i_3, i_3 i_1, i_1 i_2]$. Therefore, GCD of minors of order 2 of $(\lambda I - A)$ is $i_1(\lambda)\ i_2(\lambda)$ (see Lemma 10.2), and MP of A is $\phi(\lambda)/i_1(\lambda)\ i_2(\lambda) = i_3(\lambda)$, the invariant polynomial of highest degree of $(\lambda I - A)$.

Similarly, it can be shown that if $(\lambda I - A)$ is of order n having invariant polynomials $i_1(\lambda), \ldots, i_n(\lambda)$; $i_{r+1}(\lambda)$ is divisible by $i_r(\lambda)$, $r = 1, 2, \ldots, n - 1$, then the CP of A is $\phi(\lambda) = i_1(\lambda) \ldots i_n(\lambda)$ and MP $\psi(\lambda) = i_n(\lambda)$, which is the invariant polynomial of highest degree.

In view of Definition 10.8, for a Jordan block $J(\alpha)$ of order k, $(\lambda I - J)$ has only one elementary divisor, viz. $(\lambda - \alpha)^k$, which is the invariant polynomial of highest degree and is therefore its MP. If $J_1(\alpha)$ is a Jordan block of order 2, then $(\lambda I - J_1)$ is equivalent to diag $[1, (\lambda - \alpha)^2]$ and if $J_2(\alpha)$ is a Jordan block of order 3, then $(\lambda I - J_2)$ is equivalent to diag $[1, 1, (1 - \alpha)^3]$. By virtue of Lemma 10.5, if $J(\alpha) :=$ diag $[J_1(\alpha), J_2(\alpha)]$, then $(\lambda I - J)$ is equivalent to diag $[1, 1, 1, (\lambda - \alpha)^2, (1 - \alpha)^3]$ and therefore has a CP $\phi(\lambda) = (\lambda - \alpha)^5$ and MP $\psi(\lambda) = (\lambda - \alpha)^3$. □

Example 10.21 Show that if $A \in \mathbb{C}^{n\times n}$ is idempotent, and $A \neq I$ or $A \neq 0$, then the minimum polynomial of A is $\lambda(\lambda - 1)$ and the characteristic polynomial has the form $(\lambda - 1)^r\ \lambda^p$, where r and p are integers. Prove that A is similar to diag $[I_r \quad 0]$.

Solution Since A is idempotent, $A^2 = A$ (see Problem 1.35), and $\lambda^2 - \lambda = \lambda(\lambda - 1)$ is an annihilating polynomial of A. A lower degree polynomial such as λ or $\lambda - 1$ cannot be an annihilating polynomial, since $A \neq 0$ or $A \neq I$. Moreover, a polynomial such as $A = \alpha I$, $\alpha \neq 1$ (a scalar) is not idempotent. So $\lambda(\lambda - 1)$ is a minimum polynomial of A, and its elementary divisors can only be λ and $\lambda - 1$. Positive integral powers of these factors combined with multiplication will give the characteristic polynomial as stipulated: $\phi(\lambda) = (\lambda - 1)^r\ \lambda^p$, $p + r = n$, with eigenvalues 1 of multiplicity r and 0 of multiplicity $n - r$. Since the elementary divisors are linear, A is similar to a diagonal matrix with 1 occurring r times and 0, $n - r$ times. As a result, there exists a nonsingular matrix T such that

$$A = T \operatorname{diag}[\underbrace{1, \ldots, 1}_{r \text{ times}}, \underbrace{0, \ldots, 0}_{n-r \text{ times}}]_n T^{-1}. \qquad \square$$

PROBLEMS

10.1 Show that "subtraction from any row (resp column) of any other row (resp column), multiplied by an arbitrary polynomial" is a valid elementary operation.

10.2 Prove that (a) the determinant of a unimodular matrix $U(\lambda)$ is always nonzero and independent of λ, and (b) $U^{-1}(\lambda)$ is unimodular.

10.3 A 3×3 λ-matrix $B(\lambda)$ is obtained from a 3×3 λ-matrix $A(\lambda)$ by a sequence of elementary row and column operations such that $P(\lambda)\,A(\lambda)\,Q(\lambda) = B(\lambda)$. Determine $P(\lambda)$ and $Q(\lambda)$ when operations in order of execution are (see Section 4.2):

(a) H_{13}, $K_{21}(\lambda + 1)$, K_{12}, $H_{21}(-\lambda)$, $H_{31}(\lambda - 1)$;

(b) $K_{21}(-1)$, $K_{31}(1 + \lambda)$, $H_{12}(-\lambda - 1)$, H_{12}, $K_{12}(-\lambda)$, $H_2(\lambda^2)$.

10.4 Determine $P(\lambda)$, $Q(\lambda)$ and $B(\lambda)$ such that $P(\lambda)\,A(\lambda)\,Q(\lambda) = B(\lambda)$ when

$$A(\lambda) = \begin{bmatrix} 1 & \lambda & \lambda \\ 1+\lambda^2 & 2 & 0 \\ 0 & 1-\lambda & 1+\lambda \end{bmatrix}$$

is subjected to the following elementary operations in the given order:

$$K_{32}(-1)\ K_{21}(-\lambda)\ H_{21}(-1 - \lambda^2)\ K_{23}K_2(-1/2)\ H_{32}(\lambda)\ K_{32}(-2 + \lambda + \lambda^3).$$

10.5 Prove that each of left equivalent, right equivalent, and equivalent relations satisfies the properties of (i) reflexivity, (ii) symmetry, and (iii) transitivity (see Section 4.3).

10.6 Prove that two row Hermite forms of the same λ-matrix $A(\lambda)$ are related together by a unimodular matrix.

10.7 Determine the column Hermite form of the following λ-matrices:

(a) $\begin{bmatrix} \lambda & 2 \\ \lambda+1 & 3 \end{bmatrix}$, (b) $\begin{bmatrix} \lambda+1 & \lambda-1 \\ \lambda^2 & 1 \end{bmatrix}$, (c) $\begin{bmatrix} \lambda^2+\lambda & \lambda \\ \lambda+1 & 1 \end{bmatrix}$,

(d) $\begin{bmatrix} \lambda^2+\lambda & \lambda^3 & \lambda \\ \lambda+1 & \lambda^2-1 & 1 \end{bmatrix}$,

$$\text{(e)} \begin{bmatrix} (\lambda-1)^2 & 2\lambda-1 & (\lambda+1)(\lambda^3-1) \\ \lambda^2-1 & -\lambda^2+\lambda+1 & \lambda(\lambda^3+3\lambda^2+4\lambda+4) \\ \lambda-1 & 1 & \lambda(\lambda^2+2\lambda+2) \end{bmatrix},$$

$$\text{(f)} \begin{bmatrix} \lambda^2+2 & \lambda & -\lambda^2+2\lambda+1 \\ -\lambda^3+7\lambda+3 & 1+\lambda+\lambda^2 & -\lambda^3+\lambda^2+3\lambda+3 \\ 3\lambda^2+14\lambda+5 & \lambda+4 & -\lambda^2-2\lambda+8 \end{bmatrix},$$

$$\text{(g)} \begin{bmatrix} 2\lambda^2+2\lambda+1 & 4\lambda+2 \\ 2\lambda+1 & 3 \\ 0 & 2 \end{bmatrix}, \quad \text{(h)} \begin{bmatrix} 0 & \lambda(\lambda+1) \\ \lambda^2-1 & \lambda^3+\lambda-1 \\ \lambda+1 & \lambda^2+\lambda+1 \end{bmatrix}.$$

10.8 Determine the row Hermite form of all the λ-matrices given in Problem 10.7.

10.9 Determine the Smith form of the λ-matrices given in Problem 10.7.

10.10 Determine (a) the row and (b) the column Hermite form of the λ-matrix in Example 10.6.

10.11 Determine the Smith form of the λ-matrix given in (a) Example 10.1, (b) Example 10.2, (c) Example 10.3, (d) Example 10.7.

10.12 Determine (a) column Hermite, (b) row Hermite, and (c) Smith forms of the following λ-matrix as also the corresponding transforming matrices:

$$A(\lambda) = \begin{bmatrix} \lambda^2 & \lambda^2-2\lambda & 8\lambda \\ \lambda & -\lambda & -2\lambda \\ \lambda^2 & \lambda^2-2\lambda & \lambda^2+7\lambda \end{bmatrix}.$$

10.13 Prove that the invariant polynomials of $\phi(\lambda)\,(\lambda I - C)^{-1}$ are $1, \phi(\lambda), \ldots, \phi(\lambda)$, where C and $\phi(\lambda)$ are given as in Lemma 10.4.

10.14 Determine which of the following matrices are similar:

$$\text{(a)}\quad A = \begin{bmatrix} 1 & 2 \\ 0 & 1 \end{bmatrix}, \quad B = \begin{bmatrix} 1 & 0 \\ 0 & 1 \end{bmatrix}, \quad C = \begin{bmatrix} -1 & -2 \\ 2 & 3 \end{bmatrix}, \quad D = \begin{bmatrix} 2 & 2 \\ 1 & 1 \end{bmatrix}.$$

$$\text{(b)}\quad A = \begin{bmatrix} -2 & 1 & -1 \\ 3 & -6 & 7 \\ 2 & -3 & 3 \end{bmatrix}, \quad B = \begin{bmatrix} -1 & -1 & 2 \\ 2 & -4 & 4 \\ 1 & -1 & 0 \end{bmatrix},$$

$$C = \begin{bmatrix} -6 & -5 & -7 \\ 2 & 1 & 3 \\ 1 & 1 & 0 \end{bmatrix}, \quad D = \begin{bmatrix} -2 & 0 & 0 \\ 1 & -1 & 0 \\ 1 & 2 & -1 \end{bmatrix}.$$

10.15 In each of the following cases, the matrices A and B are known to be similar. Determine a nonsingular matrix T such that' $T^{-1}AT = B$.

(a) $A = \begin{bmatrix} 1 & 2 \\ 0 & 1 \end{bmatrix}, \quad B = \begin{bmatrix} -1 & -2 \\ 2 & 3 \end{bmatrix}.$

(b) $A = \begin{bmatrix} -2 & 1 & -1 \\ 3 & -6 & 7 \\ 2 & -3 & 3 \end{bmatrix}, \quad B = \begin{bmatrix} -6 & -5 & -7 \\ 2 & 1 & 3 \\ 1 & 1 & 0 \end{bmatrix},$

10.16 Is T unique in the similarity transformation of two matrices A and B given by $T^{-1}AT = B$, where T is nonsingular? Why?

10.17 Give an example to show that two matrices may have the same characteristic polynomial but with different Jordan forms.

10.18 Determine a nonsingular transformation T which transforms A to a Jordan normal form when

(a) $A = \begin{bmatrix} -2 & 1 & -1 \\ 3 & -6 & 7 \\ 2 & -3 & 3 \end{bmatrix}, \quad$ (b) $A = \begin{bmatrix} -6 & -5 & -7 \\ 2 & 1 & 3 \\ 1 & 1 & 0 \end{bmatrix}.$

10.19 Determine (i) the invariant polynomials and (ii) the elementary divisors of $(\lambda I - A)$ when A is given by

(a) $A = \begin{bmatrix} 2 & 1 & 1 & 2 \\ -4 & -1 & 0 & -4 \\ 1 & 1 & 2 & 2 \\ 1 & 0 & -1 & 1 \end{bmatrix}, \quad$ (b) $A = \begin{bmatrix} 3 & 0 & -1 \\ 0 & 3 & 0 \\ 1 & 0 & 1 \end{bmatrix},$

(c) $\begin{bmatrix} 1 & 0 & 0 & 1 & -1 \\ -4 & 1 & -2 & 0 & 0 \\ -4 & 0 & -1 & 0 & 0 \\ 0 & -1 & 1 & 0 & 1 \\ 0 & -1 & 1 & -1 & 2 \end{bmatrix}.$

Hence, find the lower Jordan normal form of A.

10.20 Determine the invariant polynomials and elementary divisors of $(\lambda I - A)$, where

$$A = \begin{bmatrix} 0 & 1 & 0 & 0 & 0 \\ 0 & -1 & 0 & 0 & 0 \\ -2 & -1 & 0 & 1 & 0 \\ -2 & 2 & 0 & 1 & 0 \\ -4 & -1 & 0 & 2 & 0 \end{bmatrix}.$$

Compute (a) the first natural normal form, (b) the second natural normal form, and (c) the Jordan normal form of A.

10.21 A linear transformation L on $\mathbb{R}^3$ is defined by

$$\mathbf{Lx} = \begin{bmatrix} -x_1 - x_2 + x_3 \\ x_1 - 4x_2 + x_3 \\ 2x_1 - 5x_2 \end{bmatrix}, \qquad \mathbf{x} = \begin{bmatrix} x_1 \\ x_2 \\ x_3 \end{bmatrix}.$$

Determine the basis vectors in $\mathbb{R}^3$ in terms of which matrix representation of L is in Jordan normal form.

10.22 The linear transformation on $\mathbb{R}^4$ is defined by

$$\mathbf{Lx} = \begin{bmatrix} 2x_1 - x_2 + x_4 \\ 3x_2 - x_3 \\ x_2 + x_3 \\ -x_2 + 3x_4 \end{bmatrix}.$$

Determine a matrix representation of L which is in (a) first natural normal form, (b) second natural normal form, and (c) Jordan normal form.

10.23 Show that the minimum polynomial of a matrix A in $V_{n\times n}$ is the invariant polynomial of the highest degree of $(\lambda I - A)$. Employ this principle to determine the minimum polynomial of A given in Problem 10.20.

[*Hint:* Characteristic polynomial = $i_1\, i_2\, \ldots\, i_r = d_{r-1}\, i_r$. See Theorem 7.14.]

10.24 Prove that for a derogatory matrix the degree of the minimum polynomial is less than that of the characteristic polynomial.

10.25 Let C be the companion matrix associated with $x^n - c_1x^{n-1} - c_2x^{n-2} - \cdots - c_n = 0$ which has distinct eigenvalues $\lambda_1, \lambda_2, \ldots, \lambda_n$. Prove that in the similarity transformation $T^{-1}CT = \text{diag}\,(\lambda_1, \lambda_2, \ldots, \lambda_n)$,

$$T = \begin{bmatrix} 1 & 1 & \cdots & 1 \\ \lambda_1 & \lambda_2 & \cdots & \lambda_n \\ \vdots & \vdots & & \vdots \\ \lambda_1^{n-1} & \lambda_2^{n-2} & \cdots & \lambda_n^{n-1} \end{bmatrix}.$$

10.26 Determine the rational canonical form (i.e., the first natural normal form) of A given by (10.40).

10.27 A matrix A is simple iff the elementary divisors of $(\lambda I - A)$ are linear, i.e., of degree 1. Use this property to check whether A given in Problem 7.53 is simple or not.

10.28 Find a diagonal similarity transformation T such that $T^{-1}JT = \hat{J}$, where

$$J = \begin{bmatrix} \alpha & 1 & & \\ & \alpha & \ddots & \\ & & \ddots & 1 \\ & & & \alpha \end{bmatrix}_k \qquad \hat{J} = \begin{bmatrix} \alpha & \beta_1 & & \\ & \alpha & \ddots & \\ & & \ddots & \beta_{k-1} \\ & & & \alpha \end{bmatrix}_k$$

[*Hint:* Equating the corresponding elements in $JT = T\hat{J}$, determine the elements t_{ij} of T.]

10.29 Let $x(\lambda)$ and $y(\lambda)$ be two co-prime polynomials in λ and let the polynomials $\alpha(\lambda)$ and $\beta(\lambda)$ be also co-prime. Then prove that in the product

$$\begin{bmatrix} \alpha & -\beta \\ \beta & \alpha \end{bmatrix} \begin{bmatrix} x & -y \\ y & x \end{bmatrix} = \begin{bmatrix} u & -v \\ v & u \end{bmatrix}$$

u and v are also co-prime.

10.30 Show that if

$$W(\lambda) = \begin{bmatrix} \lambda - \alpha & \beta \\ -\beta & \lambda - \alpha \end{bmatrix}$$

is a λ-matrix, where α and β are real numbers, then the row elements or the column elements of W^n are co-prime.

[*Hint:* The solution follows by repeated application of Problem 10.29.]

10.31 Show that

$$(\lambda I - J) = \begin{bmatrix} W & I & & \\ & W & \ddots & \\ & & \ddots & I \\ & & & W \end{bmatrix}_{2m},$$

where W given in Problem 10.30, is column equivalent to

$$\left[\begin{array}{ccccc} 0 & I & 0 & \cdots & 0 \\ 0 & 0 & I & \cdots & 0 \\ \vdots & \vdots & \vdots & \vdots & \vdots \\ 0 & 0 & 0 & \cdots & I \\ \hline (-1)^{m+1}W^m & (-1)^m W^{m-1} & \cdots & -W^2 & W \end{array}\right]_{2m} \qquad (10.51)$$

[*Hint:* See Example 7.32.]

10.32 Prove that the invariant polynomials of $(\lambda I - J)$ given in Problem 10.31 are 1, 1, ..., 1, $[(\lambda - \alpha)^2 + \beta^2]^m$.

10.33 A matrix is nonderogatory as in Definition 10.9 iff its characteristic and minimum polynomials are the same.

10.34 Prove that the minimum polynomial and the characteristic polynomial of a matrix are identical iff its Jordan matrix representation has only one Jordan block associated with each distinct eigenvalue of the matrix.

10.35 Let $\psi(\lambda)$ be the minimum polynomial of a matrix $A \in \mathbb{C}_{n \times n}$ and $f(\lambda)$ be a polynomial over $\mathbb{C}$. Prove that $f(A)$ is nonsingular iff $\psi(\lambda)$ and $f(\lambda)$ are relatively prime.

10.36 A matrix $A \in \mathbb{C}_{n \times n}$ is simple iff all the elementary divisors of $(\lambda I - A)$ are linear.

10.37 A matrix in $V_{n \times n}$ is nonderogatory iff the first $n - 1$ number of invariant polynomials of $(\lambda I - A)$ are equal to 1.

10.38 For a matrix A of order n, let

$$\phi(\lambda) = (\lambda - \lambda_1)^{n_1} (\lambda - \lambda_2)^{n_2} \cdots (\lambda - \lambda_k)^{n_k}$$

be the characteristic polynomial, where $\lambda_1, \lambda_2, ..., \lambda_k$ are the eigenvalues of A of multiplicity $n_1, n_2, ..., n_k$ respectively. Prove that trace $(A) = n_1\lambda_1 + n_2\lambda_2 + \cdots + n_k \lambda_k$.

10.39 A matrix $A \in \mathbb{C}^{n \times n}$ of rank r is idempotent iff r of its eigenvalues are equal to 1 and the remaining $n - r$ are equal to 0.

10.40 Prove that trace of a matrix is invariant under similarity transformation.

[*Hint:* Characteristic polynomial is invariant under an ST, see Theorem 7.1.]

10.41 Prove that if A is idempolent, then rank (A) is equal to its trace.

10.42 Prove that the only nonsingular idempotent matrix is the identity matrix.

10.43 An identity matrix of order n which is symmetric and of rank r is positive definite if $r = n$ and positive semidefinite if $r < n$.

10.44 If A and B are similar, then

(a) A is idempotent, so is B;

(b) A is nilpotent of index k and so is B.

10.45 Show that an idempotent matrix is a simple matrix.

10.46 If P is idempotent and $\mathbb{R}(A)$ denotes the range space of A, then,

(i) $(I - P)$ is idempotent;

(ii) $\mathbb{R}(I - P) = \ker P$; and

(iii) $\ker (I - P) = \mathbb{R}(P)$.

10.47 If an n-square matrix A of rank r is idempotent, then there exists a biorthogonal system of vectors in $\mathbb{R}^n$ or $\mathbb{C}^n$ such that

$$A = \sum_{j=1}^{r} \mathbf{x}_j \mathbf{x}_j^*. \tag{10.52}$$

Conversely, any matrix of the form (10.52) is idempotent.

CHAPTER

11

Linear Transformations and Normal Forms

In Chapter 10, we have given a detailed account on how to obtain a nonsingular transformation T for a given matrix A such that $T^{-1}AT$ is in a quasi-diagonal companion form or in a Jordan normal form. This technique can also be used (as mentioned in Remark 10.9) to quasi-diagonalize a linear transformation L on V_n, i.e., to find a basis in V_n with respect to which any linear transformation L on V_n can be represented by a quasi-diagonal companion matrix or a Jordan normal form. The chain of arguments leading to the above development is based on polynomial matrices which are also known as *λ-matrices*. Taking a different approach, which uses the intrinsic vector space property of the linear transformation L to decompose V_n as the direct sum of L-invariant (cyclic) subspaces and to generate a basis of these subspaces, we set up a powerful mathematical machinery to fully realize the above normal forms along with the relevant nonsingular transformation T. If L is replaced with a square matrix A which is looked upon as a linear transformation on $\mathbb{C}^n$ or $\mathbb{R}^n$, the analsis of this chapter, which is devoted to the linear transformation L, remains unaltered.

11.1 DIRECT SUM OF SUBSPACES

In Section 5.3, we have offered introductory ideas regarding subspaces and direct sum of these spaces. Given two or more subspaces, one may be interested in constructing new sub-spaces by set-theoretic operations. This important problem would be investigated here in some detail.

It is not difficult to show that if S_1 and S_2 are two subspaces of a vector space V then $S_1 \cap S_2$ is also a subspace of V.

Theorem 11.1 The intersection of two subspaces S_1 and S_2 of a vector space V over a field $\mathbb{F}$ is a subspace of V over $\mathbb{F}$.

Proof Put $S = S_1 \cap S_2$ and suppose that $\mathbf{x}, \mathbf{y} \in S$, and α is a scalar in $\mathbb{F}$. Then $\mathbf{x}$ and $\mathbf{y}$ are elements of each of S_1 and S_2, and $\mathbf{x} + \mathbf{y}$ and $\alpha\mathbf{x}$ are elements of each of S_1 and S_2. Hence, $\mathbf{x} + \mathbf{y}$ and $\alpha\mathbf{x} \in S$. By Theorem 5.3, S is a subspace. □

Example 11.1 If in $\mathbb{R}^3$ we have $S_1 = x_1x_2$-plane and $S_2 = x_2x_3$-plane, then $S = S_1 \cap S_2 = x_2$-axis is a subspace of $\mathbb{R}^3$ over $\mathbb{R}$ (see Figure 11.1). □

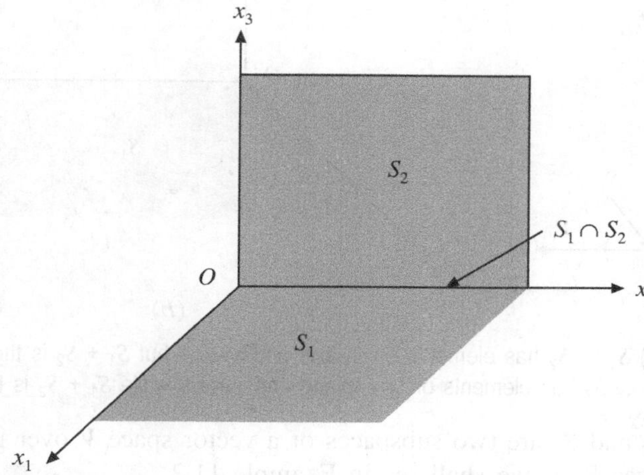

Figure 11.1 Intersection of two subspaces.

Following the argument of Theorem 11.1, we can prove the more general result given in the following corollary.

Corollary 11.1 The intersection of subspaces of a vector space V over a field $\mathbb{F}$ is a subspace of V over $\mathbb{F}$. □

It is, however, easily verified by a counter-example that the set-theoretic union $S_1 \cup S_2$ of two subspaces of V is not a subspace. In fact (see Figure 11.2(a)), in $\mathbb{R}^2$ if $S_1 = \{\mathbf{e}_1\}$ is generated by the elementary vector $\mathbf{e}_1 = (1, 0)$ and $S_2 = \{\mathbf{e}_2\}$ by $\mathbf{e}_2 = (0, 1)$, then $S_1 \cup S_2$ has elements only on x_1-axis as well as x_2-axis. The element $\mathbf{e}_1 + \mathbf{e}_2 = (1, 1)$ is not an element of $S_1 \cup S_2$ which it must be, had $S_1 \cup S_2$ been a vector subspace by Theorem 5.3. Thus, condition (i) of this theorem is not satisfied although it can be easily shown that condition (ii) is satisfied by the above set $S_1 \cup S_2$. So the set constructed as the union of two subspaces is not in general a subspace. But the operation contained in (i) of Theorem 5.3 can be included to construct a subspace from a finite number of subspaces, which is known as the *sum of subspaces.*

Definition 11.1 *Sum of Subspaces.* If S_1 and S_2 are two subspaces of a vector space V over $\mathbb{F}$, then the set

$$S = S_1 + S_2 = \{\mathbf{x} + \mathbf{y} : \mathbf{x} \in S_1, \mathbf{y} \in S_2\}$$

is defined as the sum of S_1 and S_2. The sum of a finite number of subspaces $S_1, S_2, ..., S_p$ is then defined by

$$\begin{aligned} S &= S_1 + S_2 + \cdots + S_k \\ &= \{\mathbf{x}_1 + \mathbf{x}_2 + \cdots + \mathbf{x}_k: \mathbf{x}_1 \in S_1, \mathbf{x}_2 \in S_2, ..., \mathbf{x}_k \in S_k\}. \end{aligned}$$ □

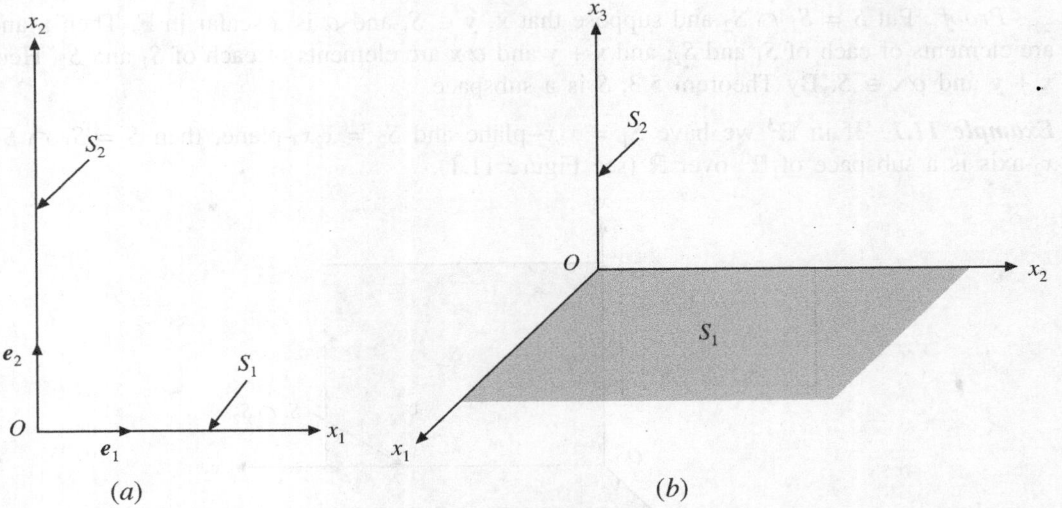

Figure 11.2 (a) $S_1 \cup S_2$ has elements on x_1-axis and x_2-axis but $S_1 + S_2$ is the whole x_1x_2-plane; (b) $S_1 \cup S_2$ has elements on x_1x_2-plane and x_3-axis while $S_1 + S_2$ is the whole $x_1x_2x_3$-space.

Theorem 11.2 If S_1 and S_2 are two subspaces of a vector space V over $\mathbb{F}$, then $S = S_1 + S_2$ is a subspace of V over $\mathbb{F}$, as we shall see in Example 11.2.

Proof Let $\mathbf{x}$ and $\mathbf{y}$ be two elements in S and α be a scalar in $\mathbb{F}$. Then there are elements $\mathbf{x}_1, \mathbf{y}_1 \in S_1$ and $\mathbf{x}_2, \mathbf{y}_2 \in S_2$ such that $\mathbf{x} = \mathbf{x}_1 + \mathbf{x}_2$ and $\mathbf{y} = \mathbf{y}_1 + \mathbf{y}_2$. Therefore,

$$\mathbf{x} + \mathbf{y} = (\mathbf{x}_1 + \mathbf{x}_2) + (\mathbf{y}_1 + \mathbf{y}_2) = (\mathbf{x}_1 + \mathbf{y}_1) + (\mathbf{x}_2 + \mathbf{y}_2)$$

is an element of S because $\mathbf{x}_1 + \mathbf{y}_1 \in S_1$ and $\mathbf{x}_2 + \mathbf{y}_2 \in S_2$. Moreover,

$$\alpha\mathbf{x} = \alpha(\mathbf{x}_1 + \mathbf{x}_2) = \alpha\mathbf{x}_1 + \alpha\mathbf{x}_2$$

is an element of S because $\alpha\mathbf{x}_1 \in S_1$ and $\alpha\mathbf{x}_2 \in S_2$. Therefore, by Theorem 5.3, $S = S_1 + S_2$ is a subspace of V over $\mathbb{F}$. □

The following corollary can be similarly proved.

Corollary 11.2 The sum of a finite number of subspaces of a vector space V over $\mathbb{F}$ is a subspace of V over $\mathbb{F}$. □

Example 11.2 (a) If $V = \mathbb{R}^2$ is a vector space over $\mathbb{R}$, $S_1 = \{(a, 0): a \in \mathbb{R}\} = x_1$-axis and $S_2 = \{(0, b) : b \in \mathbb{R}\} = x_2$-axis are subspaces of V, then

$$S = S_1 + S_2 = \{(a, b) : a, b \in \mathbb{R}\}$$

is the entire x_1x_2-plane and is a vector subspace of V (see Figure 11.2(a)).

(b) If $V = \mathbb{R}^3$ is a vector space over $\mathbb{R}$, $S_1 = \{(a, b, 0) : a, b \in \mathbb{R}\} = x_1x_2$-plane and $S_2 = \{(0, 0, c) : a \in \mathbb{R}\} = x_3$-axis are vector spaces of V, then $S_1 \cup S_2$ is not a vector subspace although $S_1 + S_2 = \{(a, b, c) : a, b, c \in \mathbb{R}\} =$ space $\mathbb{R}^3$ is a vector subspace of $\mathbb{R}^3$ (see Figure 11.2(b)). □

Relating the two subspaces just considered, we have the following important Theorem.

Theorem 11.3 If S_1 and S_2 are subspaces of a vector space V over $\mathbb{F}$, then

$$\dim (S_1 + S_2) = \dim S_1 + \dim S_2 - \dim (S_1 \cap S_2). \tag{11.1}$$

Proof Let $X = \{\mathbf{x}_1, ..., \mathbf{x}_p\}$ be a basis of $S_1 \cap S_2$. Since $S_1 \cap S_2 \subset S_1$, X can be extended to a basis $\{\mathbf{x}_1, ..., \mathbf{x}_p, \mathbf{y}_1, ..., \mathbf{y}_q\}$ of S_1. Similarly X can be extended to a basis $\{\mathbf{x}_1, ..., \mathbf{x}_p, \mathbf{z}_1, ..., \mathbf{z}_r\}$ of S_2. It is obvious that the set

$$\{\mathbf{x}_1, ..., \mathbf{x}_p, \mathbf{y}_1, ..., \mathbf{y}_q, \mathbf{z}_1, ..., \mathbf{z}_r\} \tag{11.2}$$

spans the vector space $S_1 + S_2$. We shall now show that the set (11.2) is linearly independent. For this, suppose that

$$\alpha_1\mathbf{x}_1 + \cdots + \alpha_p\mathbf{x}_p + \beta_1\mathbf{y}_1 + \cdots + \beta_q\mathbf{y}_q + \gamma_1\mathbf{z}_1 + \cdots + \gamma_r\mathbf{z}_r = \boldsymbol{\theta}$$

for scalars α's, β's and γ's in $\mathbb{F}$. Therefore,

$$\begin{aligned}\alpha_1\mathbf{x}_1 + \cdots + \alpha_p\mathbf{x}_p + \beta_1\mathbf{y}_1 + \cdots + \beta_q\mathbf{x}_q &= -(\gamma_1\mathbf{z}_1 + \cdots + \gamma_r\mathbf{z}_r) \\ &= \delta_1\mathbf{x}_1 + \cdots + \delta_p\mathbf{x}_p\end{aligned} \tag{11.3}$$

since the left-hand side of (11.3) is an element of S_1 and the right-hand side is an element of S_2, so both of them being in $S_1 \cap S_2$ are expressed as a linear combination of $\mathbf{x}_i$'s where δ's are in $\mathbb{F}$. Hence,

$$\delta_1\mathbf{x}_1 + \cdots + \delta_p\mathbf{x}_p + \gamma_1\mathbf{z}_1 + \cdots + \gamma_r\mathbf{z}_r = \boldsymbol{\theta}$$

which implies that $\delta_1 = \cdots = \delta_p = \gamma_1 = \cdots = \gamma^r = 0$, because $\{\mathbf{x}_1, ..., \mathbf{x}_p, \mathbf{z}_1, ..., \mathbf{z}_r\}$ are linearly independent. It now follows from (11.3) that $\alpha_1 = \cdots = \alpha_p = \beta_1 = \cdots = \beta_q = 0$ because $\{\mathbf{x}_1, ..., \mathbf{x}_p, \mathbf{y}_1, ..., \mathbf{y}_q\}$ are linearly independent. Consequently, the set (11.2) is linearly independent and, therefore, forms a basis of the vector space $S_1 + S_2$. Thus from (11.2),

$$\begin{aligned}\dim (S_1 + S_2) &= p + q + r = (p + q) + (p + r) - p \\ &= \dim S_1 + \dim S_2 - \dim (S_1 \cap S_2).\end{aligned}$$ □

We have seen above that the sum of two subspaces S_1 and S_2 is a vector space. We now consider a vector space V which is the sum of two subspaces S_1 and S_2 such that

$$S_1 \cap S_2 = \{\boldsymbol{\theta}\}.$$

Theorem 11.4 Let S_1 and S_2 be two subspaces of a vector space V over $\mathbb{F}$. Then any element $\mathbf{v} \in V$ can be represented uniquely in the form

$$\mathbf{v} = \mathbf{x} + \mathbf{y}, \tag{11.4}$$

where $\mathbf{x} \in S_1$ and $\mathbf{y} \in S_2$ iff

$$\text{(a) } V = S_1 + S_2, \quad \text{(b) } S_1 \cap S_2 = \{\boldsymbol{\theta}\}. \tag{11.5}$$

Proof Suppose that any element $\mathbf{v} \in V$ can be represented uniquely by (11.4), where $\mathbf{x} \in S_1$ and $\mathbf{y} \in S_2$. Then obviously (a) is satisfied. If $\mathbf{z} \in S_1 \cap S_2$, then

$$\begin{aligned}\mathbf{z} &= \mathbf{z} + \boldsymbol{\theta}, \quad \mathbf{z} \in S_1 \text{ and } \boldsymbol{\theta} \in S_2; \\ &= \boldsymbol{\theta} + \mathbf{z}, \quad \boldsymbol{\theta} \in S_1 \text{ and } \mathbf{z} \in S_2.\end{aligned}$$

Since any of the above representations is unique, $\mathbf{z} = \boldsymbol{\theta}$, and consequently,

$$S_1 \cap S_2 = \{\boldsymbol{\theta}\}.$$

Conversely, suppose that (a) and (b) are true. In view of (a), let an element $\mathbf{v} \in V$ have the representation $\mathbf{v} = \mathbf{x} + \mathbf{y}$, where $\mathbf{x} \in S_1$ and $\mathbf{y} \in S_2$. Suppose $\mathbf{v}$ can also be represented by $\mathbf{v} = \hat{\mathbf{x}} + \hat{\mathbf{y}}$, where $\hat{\mathbf{x}} \in S_1$ and $\hat{\mathbf{y}} \in S_2$. Hence, $\mathbf{x} + \mathbf{y} = \hat{\mathbf{x}} + \hat{\mathbf{y}}$ or $\mathbf{x} - \hat{\mathbf{x}} = \hat{\mathbf{y}} - \mathbf{y}$. Now, $\mathbf{x} - \hat{\mathbf{x}} \in S_1$ and $\hat{\mathbf{y}} - \mathbf{y} \in S_2$. So by (b) $\mathbf{x} - \hat{\mathbf{x}} = \hat{\mathbf{y}} - \mathbf{y} = \boldsymbol{\theta}$. Therefore, $\mathbf{x} = \hat{\mathbf{x}}$ and $\mathbf{y} = \hat{\mathbf{y}}$ and the representation $\mathbf{v} = \mathbf{x} + \mathbf{y}$ is unique. □

In Definition 5.8, we have called a vector space V as the direct sum of two subspaces S_1 and S_2 if any element $\mathbf{v} \in V$ can be represented uniquely as $\mathbf{v} = \mathbf{x} + \mathbf{y}$, where $\mathbf{x} \in S_1$ and $\mathbf{y} \in S_2$. Alternatively, we have the following definition.

Definition 11.2 *Direct Sum of Two Subspaces.* A vector space V is said to be the direct sum of two subspaces S_1 and S_2 of V if

(a) $V = S_1 + S_2$ and (b) $S_1 \cap S_2 = \{\boldsymbol{\theta}\}$.

Then, V is written as

$$V = S_1 \oplus S_2.$$ □

We have already observed in Definition 11.1 that, if only (a) above is satisfied, then V is called the *sum* of two subspaces S_1 and S_2. The additional constraint (b) makes such representation unique. Combining Definition 11.2 and Theorem 11.3, we have the following theorem.

Theorem 11.5 Let a vector space V be represented as the sum of two subspaces S_1 and S_2 of $V : V = S_1 + S_2$. Then this sum is the direct sum iff

$$\dim V = \dim S_1 + \dim S_2. \tag{11.6}$$

Proof Suppose that $V = S_1 + S_2$ and $S_1 \cap S_2 = \{\boldsymbol{\theta}\}$. Then by Theorem 11.3 we have (11.6). Conversely, let $V = S_1 + S_2$ and (11.6) be satisfied. Then comparing (11.6) with (11.1), we have $S_1 \cap S_2 = \{\boldsymbol{\theta}\}$. □

Example 11.3 (a) If $S_1 = \{(a, 0, b): a, b \in \mathbb{R}\} = x_1x_3$-plane and $S_2 = \{(0, c, d): c,d \in \mathbb{R}\} = x_2x_3$-plane are vector subspaces (see Figure 11.3(a)) of $\mathbb{R}^3$, then $\mathbb{R}^3 = S_1 + S_2$ but $\mathbb{R}^3 \neq S_1 \oplus S_2$ because $S_1 \cap S_2 = \{(0, 0, g) : g \in \mathbb{R}\} = \mathbf{x}_3$-axis $\neq \boldsymbol{\theta}$. However, if $S_3 = \{(h, 0, h) : h \in \mathbb{R}\}$, then $\mathbb{R}^3 = S_2 + S_3$, see Figure 11.3(b), and $S_2 \cap S_3 = (0, 0, 0)$. Hence, $\mathbb{R}^3 = S_2 \oplus S_3$.

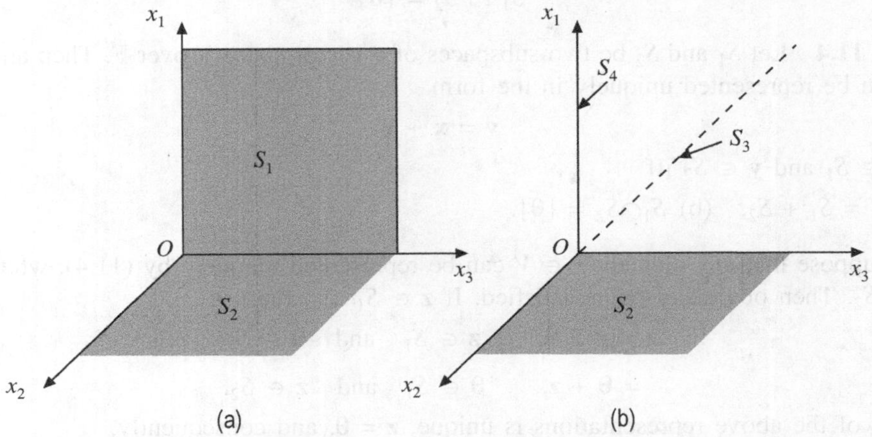

Figure 11.3 (a) Sum of two subspaces S_1 and S_2; (b) sum of three subspaces S_2, S_3 and S_4.

(b) Along with the subspaces S_2 and S_3 given in (a), if we define another subspace $S_4 = \{(f, 0, 0): f \in \mathbb{R}\}$, then (see Figure 11.3(b)) evidently,

$$\mathbb{R}^3 = S_2 + S_3 + S_4, \tag{11.7}$$

$$S_2 \cap S_3 = S_3 \cap S_4 = S_4 \cap S_2 = \{(0, 0, 0)\}. \tag{11.8}$$

However, the representation of every element in $\mathbb{R}^3$ by (11.7) is not unique. Indeed, the element $\mathbf{v} = (0, 0, \alpha) \in \mathbb{R}^3$, $\alpha \in \mathbb{R}$ can be represented as

$$\mathbf{v} = \alpha(\mathbf{v}_3 - \mathbf{v}_4) = \alpha\,\mathbf{v}_2,$$

where

$$\mathbf{v}_3 = (1, 0, 1) \in S_3, \qquad \mathbf{v}_4 = (1, 0, 0) \in S_4, \qquad \mathbf{v}_2 = (0, 0, 1) \in S_2.$$

Example 11.3(b) gives an indication that the generalization of Definition 11.2 to define the direct sum of more than two subspaces is not as straightforward as it looks.

(c) If

$$S_1 = \{(x\ y\ z) : x + 2y + z = 0\},$$

$$S_2 = \{(x\ y\ z) : x + y - z = 0\}$$

are subspaces of $\mathbb{R}^3$, then (i) find a basis of $S_1 \cap S_2$, (ii) determine dim $(S_1 + S_2)$, and (iii) describe S_1, S_2, $S_1 \cap S_2$ and $S_1 + S_2$ geometrically.

Solution (i) The relation $x + 2y + z = 0$ shows that the elements of S_1 are of the form $(\alpha, \beta, -\alpha - 2\beta)$, where α and β are arbitrary and, consequently, S_1 has the set

$$\{(1\ 0\ -1), (0\ 1\ -2)\}$$

as a basis. Similarly, $\{(1\ 0\ 1), (0\ 1\ 1)\}$ is a basis of S_2. Now, if $\mathbf{x} \in S_1 \cap S_2$, then

$$\mathbf{x} = a(1\ 0\ -1) + b(0\ 1\ -2) = c(1\ 0\ 1) + d(0\ 1\ 1)$$

for $a, b, c, d \in \mathbb{R}$. So,

$$a = c, \qquad b = d, \qquad a + 2b + c + d = 0$$

which on solving give $a = c = -3\alpha$, $b = d = 2\alpha$, where α is real. Therefore,

$$\mathbf{x} = \alpha(-3\ 2\ -1)$$

and consequently, $(-3\ 2\ -1)$ is a basis of $S_1 \cap S_2$.

(ii) dim $(S_1 + S_2)$ = dim S_1 + dim S_2 − dim $S_1 \cap S_2 = 2 + 2 - 1 = 3$

(iii) S_1 is a plane in $\mathbb{R}^3$ passing through (0, 0, 0), (1, 0, −1) and (0, 1, −2); S_2 is a plane passing through (0, 0, 0), (1, 0, 1), (0, 1, 1), $S_1 \cap S_2$ is a line joining (0, 0, 0) and (−3, 2, −1) and $S_1 + S_2$ is the whole space $\mathbb{R}^3$. □

(d) Prove that for an $m \times n$ matrix A,

$$V_n = \mathbb{N}(A) \oplus \mathbb{R}(A^T),$$

i.e., the direct sum of null space of A and range space of A^T is the n-dimensional vector space V_n.

Solution To prove the result, let us assume that

$$\mathbf{y} \in \mathbb{N}(A) \cap \mathbb{R}(A^T).$$

Therefore, $\mathbf{y} \in \mathbb{N}(A)$ as well as $\mathbf{y} \in \mathbb{R}(A^T)$. Consequently, $A\mathbf{y} = \boldsymbol{\theta}$ and $A^T\mathbf{u} = \mathbf{y}$ for some $\mathbf{u} \in V_m$.

Hence,

$$\mathbf{u}^T A\mathbf{y} = \mathbf{y}^T\ \mathbf{y} = 0$$

as $A\mathbf{y} = \boldsymbol{\theta}$. The result $\mathbf{y}^T\ \mathbf{y} = 0$ implies that $\mathbf{y} = \boldsymbol{\theta}$ showing that

$$(i)\quad \mathbb{N}(A) \cap \mathbb{R}(A^T) = \boldsymbol{\theta}.$$

This satisfies condition (b) of Definition 11.2 of a direct sum.

In view of Theorem 5.16, let $\{\mathbf{x}_1,\ldots,\mathbf{x}_r\} \in V_n$ be a basis of $\mathbb{R}(A^T)$ and $\{\mathbf{x}_{r+1},\ldots,\mathbf{x}_n\} \in V_n$ be a basis of $\mathbb{N}(A)$. Assume that

$$\alpha_1\mathbf{x}_1 + \cdots + \alpha_r\mathbf{x}_r + \alpha_{r+1}\mathbf{x}_{r+1} + \cdots + \alpha_n\mathbf{x}_n = \boldsymbol{\theta}.$$

Therefore,

$$\alpha_1\mathbf{x}_1 + \cdots + \alpha_r\mathbf{x}_r = -(\alpha_{r+1}\mathbf{x}_{r+1} + \cdots + \alpha_n\mathbf{x}_n).$$

As the LHS is in $\mathbb{N}(A)$ and RHS is in $\mathbb{R}(A^T)$, and $\mathbb{N}(A) \cap \mathbb{R})(A^T) = \boldsymbol{\theta}$, we must have

$$\alpha_1\mathbf{x}_1 + \cdots + \alpha_r\mathbf{x}_r = -(\alpha_{r+1}\mathbf{x}_{r+1} + \cdots + \alpha_n\mathbf{x}_n) = \boldsymbol{\theta}.$$

By definition of basis vectors, $\alpha_i = 0$, $i = 1,\ldots, n$ and so

$$(ii)\quad \text{the set}\quad \{\mathbf{x}_1,\ldots,\mathbf{x}_r,\mathbf{x}_{r+1},\ldots,\mathbf{x}_n\}$$

is linearly independent and spans V_n. Thus condition (a) of Definition 11.2 is satisfied. The conditions in (i) and (ii) prove the assertion. For a complex space we replace A^T by A^*. □

Definition 11.3 *Direct Sum of Subspaces.* A vector space V is said to be the direct sum of the subspaces S_1, S_2, ..., S_k denoted by

$$V = S_1 \oplus S_2 \oplus \cdots \oplus S_k$$

if every vector $\mathbf{v} \in V$ has a unique representation in the form

$$\mathbf{v} = \mathbf{x}_1 + \mathbf{x}_2 + \cdots + \mathbf{x}_k \tag{11.9}$$

where $\mathbf{x}_i \in S_i$, $i = 1, 2, \ldots, k$. □

Theorem 11.6 The vector space V is the direct sum of the subspaces S_1, S_2, ..., S_k iff

(a) $V = S_1 + S_2 + \cdots + S_k$;

(b) $S_i \cap (\sum_{j\neq i} S_j) = \{\boldsymbol{\theta}\}$, $i = 1, 2, \ldots, k$.

Proof Suppose that $\mathbf{v} \in V$ can be represented uniquely by (11.9). Then obviously, (a) is satisfied. Now, let

$$\mathbf{v} \in S_i \cap (\sum_{j\neq i} S_j).$$

Then

$$\mathbf{v} = \mathbf{x}_i + \boldsymbol{\theta} = \boldsymbol{\theta} + \sum_{j\neq i} \mathbf{x}_j,$$

where $\mathbf{x}_i \in S_i$, $i = 1, 2, ..., k$. Since any of the above representations is unique, we have

$$\mathbf{x}_i = \sum_{j \neq i} x_j = \boldsymbol{\theta}$$

which is the condition (b).

Conversely, suppose that (a) and (b) are true. Then, in view of (a), any vector $\mathbf{x} \in V$ can be written as

$$\mathbf{v} = \mathbf{x}_1 + \mathbf{x}_2 + \cdots + \mathbf{x}_k, \tag{11.10}$$

where $\mathbf{x}_i \in S_i$, $i = 1, 2, ..., k$. Suppose that $\mathbf{v}$ is alternatively represented by

$$\mathbf{v} = \mathbf{y}_1 + \mathbf{y}_2 + \cdots + \mathbf{y}_k, \tag{11.11}$$

where $\mathbf{y}_i \in S_i$, $i = 1, 2, ..., k$. Then it follows from (11.10) and (11.11) that

$$\mathbf{x}_i - \mathbf{y}_i = -\sum_{j \neq i} (\mathbf{x}_j - \mathbf{y}_j) = \boldsymbol{\theta},$$

where the last step follows by (b). Hence, $\mathbf{x}_i = \mathbf{y}_i$ for all $i = 1, 2, ..., k$, and the representation (11.10) is unique. □

Lemma 11.1 Prove that if any vector $\mathbf{v} \in V$ can be expressed uniquely in terms of the vectors $\mathbf{x}_1, \mathbf{x}_2, ..., \mathbf{x}_n$ in V, then these vectors form a basis of V.

Proof As any vector $\mathbf{v} \in V$ can be represented uniquely by

$$\mathbf{v} = \alpha_1\mathbf{x}_1 + \alpha_2\mathbf{x}_2 + \cdots + \alpha_n\mathbf{x}_n \tag{11.12}$$

and so none of the vectors $\mathbf{x}_i$'s can be zero. Indeed if $\mathbf{x}_1 = \boldsymbol{\theta}$, then α_1 in the above representation can be arbitrary which is against our assumption that representation (11.12) is unique. Now,

$$\boldsymbol{\theta} = \alpha_1\mathbf{x}_1 + \alpha_2\mathbf{x}_2 + \cdots + \alpha_n\mathbf{x}_n = \boldsymbol{\theta} + \boldsymbol{\theta} + \cdots + \boldsymbol{\theta}$$

which, because of uniqueness of representation, implies that

$$\alpha_i\mathbf{x}_i = \boldsymbol{\theta}, \qquad i = 1, 2, ..., n.$$

Since $\mathbf{x}_i \neq \boldsymbol{\theta}$, $\alpha_i = 0$. Hence, $\mathbf{x}_1, \mathbf{x}_2, ..., \mathbf{x}_n$ are linearly independent and, by Definition 5.5, these vectors form a basis of V. □

Theorem 11.7 (a) If the vector space V is a direct sum of the subspaces S_i, $i = 1, 2, ..., k$, i.e., if

$$V = S_1 \oplus S_2 \oplus \cdots \oplus S_k, \tag{11.13}$$

then the union of arbitrary bases of these subspaces is a basis of the whole space V.

(b) Conversely, if the union of arbitrary bases of subspaces S_i, $i \neq 1, 2, ..., k$, is a basis of the space V, then V can be represented as the direct sum of these subspaces.

Proof Let V be represented by (11.13) and

$$\{\mathbf{x}_{i1}, \mathbf{x}_{i2}, ..., \mathbf{x}_{in_i}\} \tag{11.14}$$

be some basis of S_i, where n_i is the dimension of S_i, $i = 1, 2, ..., k$. From (11.13) it follows that for any vector $\mathbf{v} \in V$,

$$\mathbf{v} = \mathbf{y}_1 + \mathbf{y}_2 + \cdots + \mathbf{y}_k, \tag{11.15}$$

where $\mathbf{y}_i \in S_i$, $i = 1, 2, ..., k$. Hence,

$$\mathbf{v} = \sum_{j=1}^{n_1} \alpha_{1j}\mathbf{x}_{1j} + \sum_{j=1}^{n_2} \alpha_{2j}\mathbf{x}_{2j} + \cdots + \sum_{j=1}^{n_k} \alpha_{kj}\mathbf{x}_{kj}, \tag{11.16}$$

where $(\alpha_{i1}\ \alpha_{i2} \dots \alpha_{in_i})$ is a coordinate of the vector $\mathbf{y}_i$ in respect of the basis (11.14). Since the representation (11.15) is unique, so is (11.16). This means that any vector $\mathbf{v} \in V$ can be represented uniquely in terms of the vectors $\mathbf{x}_{ij}$ $(i = 1, 2, \dots, k; j = 1, 2, \dots, n_i)$ which therefore form a basis of V by Lemma 11.1.

(b) Now, let $\{\mathbf{x}_{i1}, \mathbf{x}_{i2}, \dots, \mathbf{x}_{in_i}\}$, $i = 1, 2, \dots, k$, be bases of S_i's, the union of which is a basis of V. Then for any $\mathbf{v} \in V$, the decomposition (11.16) is unique which implies the uniqueness of the representation (11.15). Consequently, V is a direct sum of the subspaces S_i's by Definition 11.3. □

Corollary 11.3 A vector space V is the sum of subspaces:

$$V = S_1 + S_2 + \cdots + S_k. \tag{11.17}$$

This sum is direct iff

$$\dim V = \dim S_1 + \dim S_2 + \cdots + \dim S_k. \tag{11.18}$$

Proof (a) *Necessity.* Suppose that the sum (11.17) is direct. Now, if

$$\{\mathbf{x}_{i1}\ \mathbf{x}_{i2} \cdots \mathbf{x}_{in_i}\}$$

is a basis of S_i, then by Theorem 11.7,

$$\{\mathbf{x}_{11}\ \mathbf{x}_{12} \cdots \mathbf{x}_{1n_1};\ \mathbf{x}_{21}\ \mathbf{x}_{22} \cdots \mathbf{x}_{2n_2} \cdots;\ \mathbf{x}_{k1}\ \mathbf{x}_{k2} \cdots \mathbf{x}_{kn_k}\}$$

is a basis of V. Hence,

$$\dim V = n_1 + n_2 + \cdots + n_k = \dim S_1 + \dim S_2 + \cdots + \dim S_k.$$

(b) *Sufficiency.* Suppose V is the sum of subspaces given by (11.17). Then by repeated application of Theorem 11.3, we have

$$\dim V = \dim S_1 + \dim S_2 + \cdots + \dim S_k$$

$$- \dim (S_1 \cap \sum_{j=2}^{k} S_j) - \dim (S_2 \cap \sum_{j=3}^{k} S_j) - \cdots - \dim (S_{k-1} \cap S_k)$$

which, when compared with (11.18), gives

$$\dim (S_1 \cap \sum_{j=2}^{k} S_j) = 0$$

If the first subspace in (11.17) is written as S_i, after rearranging we get, similarly,

$$\dim (S_i \cap \sum_{j \neq i} S_j) = 0, \quad i = 1, 2, \dots, k$$

which shows, in view of Theorem 11.6, that the sum (11.17) is direct. □

The direct sum of subspaces defined in Definition 5.8 and in Definition 11.2 are sometimes called *internal direct sum of subspaces.* This is so because of the possibility of another kind of direct sum known as *external direct sum of vector spaces.* Let S_1 and S_2 be two m- and n-dimensional vector spaces whose elements are, respectively,

$$\mathbf{x}_1 = [\mathbf{x}_{11}\ \mathbf{x}_{12} \cdots \mathbf{x}_{1m}]^T \quad \text{and} \quad \mathbf{x}_2 = [\mathbf{x}_{21}\ \mathbf{x}_{22} \cdots \mathbf{x}_{2n}]^T.$$

Then the external direct sum of these spaces S_1 and S_2 designated also by $S_1 \oplus S_2$ is defined as

$$S = S_1 \oplus S_2: = \{\mathbf{x} = \begin{bmatrix} \mathbf{x}_1 \\ \mathbf{x}_2 \end{bmatrix} : \mathbf{x}_1 \in S_1, \mathbf{x}_2 \in S_2\}. \tag{11.19}$$

Obviously,

$$\mathbf{x} : = [x_{11}\ x_{12}\ \cdots\ x_{1m};\ x_{21}\ x_{22}\ \cdots\ x_{2n}]^T$$

and this representation (11.19) is unique.

If we embed the elements of S_1 in the vector space S by considering them to be of the form

$$[x_{11}\ x_{12}\ \cdots\ x_{1m};\ 0\ 0\ \cdots\ 0]^T$$

and, similarly, if those of S_2 are embedded in S having the form

$$[0\ 0\ \cdots\ 0;\ x_{21}\ x_{22}\ \cdots\ x_{2n}]^T,$$

then (11.19) can be looked upon as the internal direct sum of the subspaces S_1 and S_2. This definition of external direct sum can be extended to any finite number of vector spaces. Referring to Example 5.20 and Figure 5.4, let

$$S_1 = \{\mathbf{x}_1 = [x_1\ x_2]^T : x_1, x_2 \in \mathbb{R}\}, \text{ and}$$
$$S_2 = \{\mathbf{x}_2 = \mathbf{x}_3: x_3 \in \mathbb{R}\}.$$

Then, the external direct sum of S_1 and S_2 is

$$S_1 \oplus S_2 = \left\{\begin{bmatrix} \mathbf{x}_1 \\ \mathbf{x}_2 \end{bmatrix} = [x_1\ x_2\ x_3]^T : \mathbf{x}_1 \in S_1, \mathbf{x}_2 \in S_2\right\}$$

which is the space $\mathbb{R}^3$.

11.2 INVARIANT SUBSPACES

We have explained in Example 7.1, that the set of all vectors, which are (a nonzero scalar) k times an eigenvector $\mathbf{x}_1$ of a square matrix A over $\mathbb{R}$, are eigenvectors of A associated with the same eigenvalue λ. This is written mathematically as $A(k\mathbf{x}_1) = \lambda(k\mathbf{x}_1)$. This set of eigenvectors, together with the zero vector $\boldsymbol{\theta}$, forms a vector subspace S which points in the same direction as $\mathbf{x}_1$ (see Figure 7.2). Moreover, if $\mathbf{x}_1 \in S$, then obviously $A\mathbf{x}_1 \in S$. These properties are generally true for any linear transformation.

Theorem 11.8 Let L be a linear transformation on an n-dimensional vector space V over $\mathbb{F}$. Then the set of all eigenvectors of L associated with an eigenvalue λ, together with $\boldsymbol{\theta}$, forms a subspace E_λ of V, which is called the *characteristic space* or *eigenspace* of L associated with λ. Moreover, $L\mathbf{x} \in E_\lambda$, for every $\mathbf{x} \in E_\lambda$.

Proof Suppose that $\mathbf{x}, \mathbf{y} \in E_\lambda$ and that α is a scalar in $\mathbb{F}$. Then $L\mathbf{x} = \lambda\mathbf{x}$ and $L\mathbf{y} = \lambda\mathbf{y}$. Consequently,

$$L(\mathbf{x} + \mathbf{y}) = L\mathbf{x} + L\mathbf{y} = \lambda(\mathbf{x} + \mathbf{y}) \quad \text{and} \quad L(\alpha\,\mathbf{x}) = \alpha L\mathbf{x} = \lambda(\alpha\,\mathbf{x})$$

which show that $\mathbf{x} + \mathbf{y} \in E_\lambda$ and $\alpha\mathbf{x} \in E_\lambda$. Since $E_\lambda \subseteq V$, by Theorem 5.3, E_λ is a subspace of V. We note that $\lambda\mathbf{x} \in E_\lambda$ whenever $\mathbf{x} \in E_\lambda$, which proves that $\lambda\mathbf{x} = L\mathbf{x} \in E_\lambda$ for every $\mathbf{x} \in E_\lambda$. □

We have proved in Example 6.7 that ker L is a subspace of V and in Definition 6.2 we have shown that $L(V)$ = range of L is also a subspace of V. An intrinsic property of these subspaces is contained in the following theorem.

Theorem 11.9 Let L be a linear transformation on an n-dimensional vector space V over $\mathbb{F}$. Then, (a) $L\mathbf{x} \in \ker L$ whenever $\mathbf{x} \in \ker L$, and (b) $L\mathbf{x} \in L(V)$ whenever $\mathbf{x} \in L(V)$.

Proof (a) Suppose that $\mathbf{x} \in \ker L$. Then $L\mathbf{x} = \mathbf{\theta}$ and $L(L\mathbf{x}) = L\mathbf{\theta} = \mathbf{\theta}$ which shows that $L\mathbf{x} \in \ker L$. (b) Suppose that $\mathbf{x} \in L(V) \subseteq V$; then obviously, $L\mathbf{x} \in L(V)$. □

The important intrinsic property that pervades the assertions of Theorems 11.8 and 11.9 is this: if L is a linear transformation on a vector space V and if S (imagine S to be E_λ, $\ker L$ or $L(V)$ in the above theorems) is a subspace of V, then every element $\mathbf{x}$ in S is mapped by L into an element of the same subspace S. In other words, $L(S)$ is contained in S. This notion gives rise to the following definition.

Definition 11.4 *L-invariant Subspace,* Let L be a linear transformation on a vector space V. Then a subspace S *of* V is called invariant under L (or L-invariant) if $L\mathbf{x} \in S$ for every $\mathbf{x} \in S$, or if $L(S) \subseteq S$. □

Example 11.4 Let a linear transformation $L: \mathbb{R}^4 \to \mathbb{R}^4$ be defined as

$$L(x, y, z, u) = (x + y, x - y, z + 2u, 2z - u).$$

Then the subspace

$$S = \{(a, b, 0, 0): a, b \in \mathbb{R}\}$$

is an L-invariant subspace of $\mathbb{R}^4$. Indeed,

$$L(a, b, 0, 0) = (a + b, a - b, 0, 0) \in S.$$ □

Another important notion to be used in the sequel is the restriction of a linear transformation.

Definition 11.5 *Restriction of a Linear Transformation.* Let $L : V \to W$ be a linear transformation and let S be a subspace of V. Then the transformation $L_S : S \to W$ defined by $L_S(\mathbf{v}) = L(\mathbf{v})$ for each $\mathbf{v} \in S$ is defined as the restriction of L to the subspace S (see Figure 11.4). □

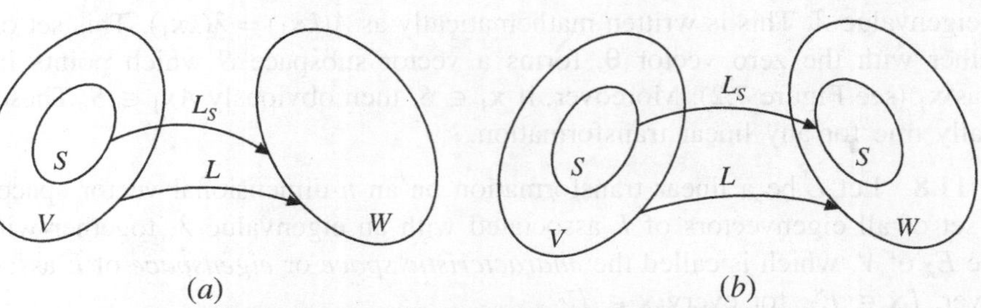

Figure 11.4 (a) L_s is the restriction of L to the subspace S; (b) L_s is the restriction of L to the L-invariant subspace S.

Example 11.5 Let a linear transformation be defined by $L: \mathbb{R}^2 \to \mathbb{R}^2: (x, y) \to (x + y, x - y)$ and a subspace of $\mathbb{R}^2$ by

$$S = \{(a, 0) : a \to \mathbb{R}\}.$$

Then the restriction of L to S is

$$L_s : S \to \mathbb{R}^2 : (a, 0) \to (a, a).$$

See Figure 11.5. □

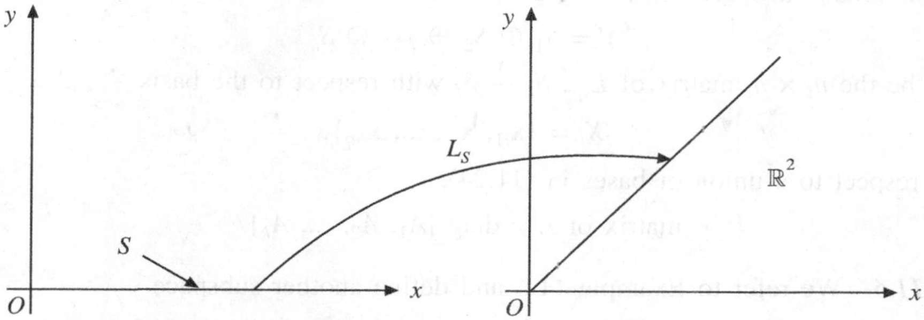

Figure 11.5 Restriction of a linear transformation defined in Example 11.5.

If a subspace S of V is L-invariant, then L_s is a mapping from S into itself. To determine the matrix representation of such L, we choose any basis

$$\{\mathbf{x}_1, \mathbf{x}_2, ..., \mathbf{x}_k\}$$

in S which is extended to a basis

$$\{\mathbf{x}_1, \mathbf{x}_2, ..., \mathbf{x}_k; \mathbf{x}_{k+1}, ..., \mathbf{x}_n\} \tag{11.20}$$

of the whole space V. Hence, we have

$$L_s\mathbf{x}_i = L\mathbf{x}_i = a_{i1}\mathbf{x}_1 + a_{i2}\mathbf{x}_2 + \cdots + a_{ik}\mathbf{x}_k, \qquad i = 1, 2, ..., k; \tag{11.21}$$

$$L\mathbf{x}_j = a_{j1}\mathbf{x}_1 + \cdots + a_{jk}\mathbf{x}_k + \cdots + a_{jn}\mathbf{x}_n, \qquad j = k+1, ..., n; \tag{11.22}$$

from which it follows that the matrix of L with respect to the basis (11.20) has the form

$$\begin{array}{cc} k & n-k \end{array}$$
$$\begin{bmatrix} A & B \\ 0 & C \end{bmatrix}\begin{matrix} k \\ n-k \end{matrix},$$

where the $k \times k$ matrix A represents L_s on the L-invariant subspace S determined by the coefficients in (11.21) and the matrices B and C by those in (11.22). We have just now considered V to be an n-dimensional vector space over $\mathbb{F}$ and L a linear transformation on V, which we shall now assume unless otherwise stated.

Suppose now that S_1 and S_2 are two L-invariant subspaces of V with bases

$$X_1 = \{\mathbf{x}_{11}, \mathbf{x}_{12}, ..., \mathbf{x}_{1n_2}\} \quad \text{and} \quad X_2 = \{\mathbf{x}_{21}, \mathbf{x}_{22}, ..., \mathbf{x}_{2n_2}\} \tag{11.23}$$

respectively. If $V = S_1 \oplus S_2$, then by Theorem 11.7, the union of bases in (11.23) is a basis of V. Consequently, following the chain of arguments given in the foregoing discussion,

$$\text{matrix of } L = \text{diag } [A_1, A_2]$$

is a *quasi-diagonal matrix* with respect to the basis (11.23), where A_i is an $n_i \times n_i$ matrix of

$$L_{s_i}: S_i \to S_i$$

with respect to the basis X_i, $i = 1, 2$. Thus, we can prove the following theorem.

Theorem 11.10 Let L be a linear transformation on V which can be decomposed into a direct sum of L-invariant subspaces S_i, $i = 1, 2, ..., k$, i.e.,

$$V = S_1 \oplus S_2 \oplus \cdots \oplus S_k$$

and let A_i be the $n_i \times n_i$ matrix of $L_{s_i} : S_i \to S_i$ with respect to the basis

$$X_i = \{\mathbf{x}_{i1}, \mathbf{x}_{i2}, ..., \mathbf{x}_{in_i}\}, \tag{11.24}$$

then with respect to a union of bases in (11.24),

$$\text{matrix of } L = \text{diag } [A_1, A_2, ..., A_k]. \qquad \square$$

Example 11.6 We refer to Example 11.4 and define another subspace

$$\hat{S} = \{(0, 0, c, d): c, d \in \mathbb{R}\}$$

which is L-invariant, because $L(0, 0, c, d) = (0, 0, c + 2d, 2c - d)$. Now the vector space $\mathbb{R}^4$ can be expressed as a sum of the subspaces S and $\hat{S}$, since

$$(x, y, z, u) = (x, y, 0, 0) + (0, 0, z, u)$$

and, moreover, $\dim \mathbb{R}^4 = \dim S + \dim \hat{S}$. Therefore, by Corollary 11.3, $\mathbb{R}^4 = S \oplus \hat{S}$. As a result, the matrix of A of L_s: $S \to S$ with respect to the elementary row vectors $\{\mathbf{e}_1, \mathbf{e}_2\}$ in $\mathbb{R}^4$ and the matrix B of $L_{\hat{S}} : \hat{S} \to \hat{S}$ with respect to the elementary row vectors $\{\mathbf{e}_3, \mathbf{e}_4\}$ in $\mathbb{R}^4$ are

$$A = \begin{bmatrix} 1 & 1 \\ 1 & -1 \end{bmatrix} \quad \text{and} \quad B = \begin{bmatrix} 1 & 2 \\ 2 & -1 \end{bmatrix},$$

because $L\mathbf{e}_1 = (1\ 1\ 0\ 0) = 1 \cdot \mathbf{e}_1 + 1 \cdot \mathbf{e}_2$, $L\mathbf{e}_2 = 1 \cdot \mathbf{e}_1 - 1 \cdot \mathbf{e}_2$, $L\mathbf{e}_3 = 1 \cdot \mathbf{e}_3 + 2 \cdot \mathbf{e}_4$ and $L\mathbf{e}_4 = 2 \cdot \mathbf{e}_3 - 1 \cdot \mathbf{e}_4$. Hence, the matrix of L with respect to the basis $\{\mathbf{e}_1, \mathbf{e}_2, \mathbf{e}_3, \mathbf{e}_4\}$ is

$$\left[\begin{array}{cc|cc} 1 & 1 & & \\ 1 & -1 & & \\ \hline & & 1 & 2 \\ & & 2 & -1 \end{array}\right] \qquad \square$$

11.3 ROOT SUBSPACES: QUASI-DIAGONAL FORM

We shall now introduce the concept of *root subspace* of a linear transformation L on a vector space V associated with an eigenvalue λ, and show that this subspace is L-invariant. Then we shall see that if L has k distinct eigenvalues, then V can be decomposed into a direct sum of k number of L-invariant root subspaces. Finally, we shall prove that there always exists a basis in V with respect to which L can be represented by a quasi-diagonal matrix with k diagonal elements.

Let $\lambda_1, \lambda_2, ..., \lambda_k$, $k \leq n$, be the distinct eigenvalues of the linear transformation L on the n-dimensional vector space V over $\mathbb{C}$, each with a multiplicity $n_i \geq 1$, $i = 1, 2, ..., k$; $n_1 + n_2 + \cdots + n_k = n$. Then, the characteristic polynomial of L is, evidently,

$$\phi(\lambda) = (\lambda - \lambda_1)^{n_1} (\lambda - \lambda_2)^{n_2} \cdots (\lambda - \lambda_k)^{n_k} \tag{11.25}$$

and the minimum polynomial of L can be written as

$$\psi(\lambda) = (\lambda - \lambda_1)^{m_1} (\lambda - \lambda_2)^{m_2} \cdots (\lambda - \lambda_k)^{m_k}, \tag{11.26}$$

where $1 \leq m_i \leq n_i$, see Problem 7.32.

Definition 11.6 *Root Space of a Linear Transformation L.* The set S_i of all vectors $\mathbf{x} \in V$ defined by

$$S_i = \{\mathbf{x} : \mathbf{x} \in V, (\lambda_i I_V - L)^{m_i}\, \mathbf{x} = \boldsymbol{\theta}\}, \qquad i = 1, 2, ..., k; \tag{11.27}$$

where m_i is the multiplicity of the eigenvalue λ_i in the minimum polynomial $\psi(\lambda)$ of L, is called the *root space of L* associated with λ_i. In particular, when $m_i = 1$, the root space S_i becomes the *eigenspace* E_i *of L* associated with the eigenvalue λ_i. □

Suppose now, that with respect to the basis $\{\mathbf{x}_1, \mathbf{x}_2, ..., \mathbf{x}_n\}$ in V, A is the matrix of L, then in view of the results given in Section 7.3.4, $(\lambda_i I - A)^{m_i}$ is the matrix of $(\lambda_i I_V - L)^{m_i}$ with respect to the same basis. Hence,

$$(\lambda_i I_V - L)^{m_i}\mathbf{x} = \boldsymbol{\theta}$$

iff

$$(\lambda_i I - A)^{m_i}\mathbf{u} = \boldsymbol{\theta}, \tag{11.28}$$

where

$$\mathbf{x} = \alpha_1\mathbf{x}_1 + \alpha_2\mathbf{x}_2 + \cdots + \alpha_n\mathbf{x}_n \quad \text{and} \quad \mathbf{u} = [\alpha_1\ \alpha_2\ \cdots\ \alpha_n]^T. \tag{11.29}$$

Therefore, root space as a solution of (11.27) can be determined by first finding the matrix A of L with respect to a chosen basis, then solving (11.28) for $\mathbf{u}$, and finally computing $\mathbf{x}$ with the help of (11.29).

The nonzero eigenvector of L associated with λ_i is always an element of S_i and so the root space S_i is not identically the zero subspace. Indeed, if $\mathbf{x} \neq \boldsymbol{\theta}$ is an eigenvector of L associated with λ_i, then $(\lambda_i I_V - L)\mathbf{x} = \boldsymbol{\theta}$, and consequently,

$$(\lambda_i I_V - L)^{m_i}\mathbf{x} = (\lambda_i I_V - L)^{m_i - 1}(\lambda_i I_V - L)\mathbf{x} = \boldsymbol{\theta}, \qquad \mathbf{x} \neq \boldsymbol{\theta}.$$

The following theorem contains two important properties of a root space.

Theorem 11.11 Let L be a linear transformation on a vector space V over $\mathbb{F}$. Then, (a) its root space S_i associated with eigenvalue λ_i is a subspace of V invariant under L, and (b) S_i does not contain any eigenvector associated with an eigenvalue λ_j of L different from λ_i.

Proof (a) Let S_i $(i = 1, 2, ..., k)$ be a root space of L: $V \to V$ associated with λ_i, i.e., S_i is the set of vectors $\mathbf{x} \in V$ satisfying (11.27). Obviously, $S_i \subseteq V$. For any $\mathbf{x}, \mathbf{y} \in S_i$, and arbitrary scalars α and β in $\mathbb{F}$ we have

$$\begin{aligned}(\lambda_i I_V - L)^{m_i}(\alpha\,\mathbf{x} + \beta\,\mathbf{y}) &= \alpha(\lambda_i I_V - L)^{m_i}\mathbf{x} + \beta\,(\lambda_i I_V - L)^{m_i}\mathbf{y} \\ &= \alpha\cdot\boldsymbol{\theta} + \beta\cdot\boldsymbol{\theta} = \boldsymbol{\theta}.\end{aligned}$$

Consequently, $\alpha\mathbf{x} + \beta\mathbf{y} \in S_i$ and by Theorem 5.3, S_i is a subspace of V. Moreover, S_i is L-invariant because $(\lambda_i I_V - L)^{m_i}$ and L are commutative, and for $\mathbf{x} \in S_i$, we have

$$(\lambda_i I_V - L)^{m_i} L\mathbf{x} = L(\lambda_i I_V - L)^{m_i}\mathbf{x} = L\boldsymbol{\theta} = \boldsymbol{\theta},$$

i.e., $L\mathbf{x} \in S_i$. In other words, $L\mathbf{x} \in S_i$ whenever $\mathbf{x} \in S_i$.

(b) Suppose now that $\mathbf{y}$ is an eigenvector of L associated with an eigenvalue λ_j of L and $\lambda_j \neq \lambda_i$. Then,

$$L\mathbf{y} = \lambda_j\mathbf{y}; \qquad \mathbf{y} \neq \boldsymbol{\theta}, \qquad \lambda_j \neq \lambda_i.$$

Hence,

$$(\lambda_i I_V - L)^{m_i}\mathbf{y} = (\lambda_i I_V - L)^{m_i-1}(\lambda_i I_V - L)\mathbf{y} = (\lambda_i - \lambda_j)(\lambda_i I_V - L)^{m_i-1}\mathbf{y}$$
$$= \cdots = (\lambda_i - \lambda_j)^{m_i}\mathbf{y} \neq \boldsymbol{\theta}$$

and consequently, $\mathbf{y} \notin S_i$. □

We recall that L has an eigenvalue λ for the subspace S_i if there exists a nonzero vector $\mathbf{x} \in S_i$, such that $L\mathbf{x} = \lambda\mathbf{x}$. Thus, from part (b) of Theorem 11.11 an easy corollary follows.

Corollary 11.4 The linear transformation L restricted to its root subspace S_i has a unique eigenvalue λ_i $(i = 1, 2, ..., k)$. □

Theorem 11.12 *Primary Decomposition of a Vector Space V over* $\mathbb{C}$. Let L be a linear transformation on an n-dimensional vector space V over $\mathbb{C}$. Then V can be decomposed as the direct sum of invariant root subspaces of L associated with all of its distinct eigenvalues $\lambda_1, \lambda_2, ..., \lambda_k$, $k \leq n$.

Proof Let $\psi(\lambda)$ given by (11.26) be the minimum polynomial of L. We split $1/\psi(\lambda)$ into a partial fraction to get

$$\frac{1}{\psi(\lambda)} = \left[\frac{a_{11}}{\lambda - \lambda_1} + \cdots + \frac{a_{1m_i}}{(\lambda - \lambda_1)^{m_1}}\right] + \cdots + \left[\frac{a_{k1}}{(\lambda - \lambda_k)} + \cdots + \frac{a_{km_k}}{(\lambda - \lambda_k)^{m_k}}\right],$$

where a_{ij}, $i = 1, ..., k$; $j = 1, ... m_i$, are elements of $\mathbb{C}$. Hence,

$$\frac{1}{\psi(\lambda)} = \frac{f_1(\lambda)}{(\lambda - \lambda_1)^{m_1}} + \cdots + \frac{f_k(\lambda)}{(\lambda - \lambda_k)^{m_k}}, \tag{11.30}$$

where $f_i(\lambda)$, $i = 1, ..., k$, are polynomials which are co-prime with $(\lambda - \lambda_i)^{m_i}$, i.e., $f_i(\lambda_i) \neq 0$. We now define

$$\psi_i(\lambda) = \frac{\psi(\lambda)}{(\lambda - \lambda_i)^{m_i}}, \qquad i = 1, 2, ..., k; \tag{11.31}$$

and multiply both sides of (11.30) by $\psi(\lambda)$ to get the identity

$$1 = f_1(\lambda)\psi_1(\lambda) + \cdots + f_k(\lambda)\psi_k(\lambda). \tag{11.32}$$

For any vector $\mathbf{x} \in V$, we define a vector $\mathbf{x}_i \in V$ by

$$\mathbf{x}_i := f_i(L)\psi_i(L)\mathbf{x}, \qquad i = 1, ..., k. \tag{11.33}$$

The operator polynomial associated with (11.32) is

$$I_V = f_1(L)\psi_1(L) + \cdots + f_k(L)\psi_k(L) \tag{11.34}$$

which by virtue of (11.33) gives, for any vector $\mathbf{x} \in V$, the relation

$$\mathbf{x} = f_1(L)\psi_1(L)\mathbf{x} + \cdots + f_k(L)\psi_k(L)\mathbf{x} = \mathbf{x}_1 + \cdots + \mathbf{x}_k. \tag{11.35}$$

This shows that any vector $\mathbf{x} \in V$ can be decomposed as the sum of k number of vectors. Let S_i be a root subspace of L associated with the eigenvalue λ_i, i.e.,

$$S_i = \{\mathbf{x} : \mathbf{x} \in V \text{ and } (\lambda_i I_V - L)^{m_i}\mathbf{x} = \boldsymbol{\theta}\}, \qquad i = 1, \ldots, k.$$

We shall now show that $\mathbf{x}_i$ defined by (11.33) $\in S_i$. Indeed, using the operator polynomial associated with (11.31), we have

$$\begin{aligned}(\lambda_i I_V - L)^{m_i}\mathbf{x}_i &= (\lambda_i I_V - L)^{m_i} f_i(L)\psi_i(L)\mathbf{x}\\ &= f_i(L)(\lambda_i I_V - L)^{m_i}\psi_i(L)\mathbf{x}\\ &= f_i(L)\psi(L)\mathbf{x} = \boldsymbol{\theta},\end{aligned}$$

because $\psi(L) = \Theta$ by definition of a minimum polynomial. So, $\mathbf{x}_i \in S_i$.

We shall next prove that the decomposition (11.35) of $\mathbf{x}$ into its components $\mathbf{x}_i \in S_i$ is unique. To see this, let us suppose that $\mathbf{x}$ has another decompostion

$$\mathbf{x} = \mathbf{y}_1 + \cdots + \mathbf{y}_k; \qquad \mathbf{y}_i \in \mathrm{S}_i, \qquad i = 1, \ldots, k.$$

We now observe that for $\mathbf{z} \in S_i$ and $p \neq i$,

$$\begin{aligned}\psi_p(L)\mathbf{z} &= (\lambda_1 I_V - L)^{m_1} \cdots (\lambda_{p-1} I_V - L)^{m_{p-1}}(\lambda_{p+1} I_V - L)^{m_{p+1}} \cdots (\lambda_k I_V - L)^{m_k}\,\mathbf{z}\\ &= \left\{\prod_{\substack{t\neq p\\ t\neq i}} (\lambda_t I_V - L)^{m_t}\right\}(\lambda_i I_V - L)^{m_i}\mathbf{z} = \boldsymbol{\theta}.\end{aligned}$$

Hence, using the resolution of identity (11.34), we have

$$\begin{aligned}\mathbf{y}_i &= \left\{\sum_{j=1}^{k} f_j(L)\psi_j(L)\right\}\mathbf{y}_i = f_i(L)\psi_i(L)\mathbf{y}_i\\ &= \sum_{j=1}^{k} \{f_i(L)\psi_i(L)\mathbf{y}_j\} = f_i(L)\psi_i(L)\sum_{j=1}^{k}\mathbf{y}_j = f_i(L)\psi_i(L)\mathbf{x}\\ &= \mathbf{x}_i,\end{aligned}$$

where $i = 1, \ldots, k$. Thus any vector $\mathbf{x} \in V$ can be decomposed uniquely as the sum (11.35), where $\mathbf{x}_i \in S_i$, $i = 1, \ldots, k$. Consequently, V is the direct sum of root sub-spaces:

$$V = S_1 \oplus \cdots \oplus S_k, \text{ and} \tag{11.36}$$

$$\dim V = \dim S_1 + \cdots + \dim S_k. \tag{11.37}$$

□

With the help of Theorem 11.12, we shall now prove that the dimension of the root subspace S_i of L associated with an eigenvalue λ_i is equal to its multiplicity n_i.

Theorem 11.13 Let L be a linear transformation on an n-dimensional vector space V over $\mathbb{C}$. For the space V, let L have distinct eigenvalues λ_i with multiplicities n_i and associated root subspaces S_i for all $i = 1, \ldots, k$. Then $\dim S_i = n_i$, $i = 1, \ldots, k$.

Proof The theorem will be proved by the method of induction with respect to the number of eigenvalues of L.

(a) If L has a unique eigenvalue λ_1 for the vector space V, then its characteristic polynomial has the form

$$\phi(\lambda) = (\lambda - \lambda_1)^n$$

which shows that the multiplicity of λ_1 is $n_1 = n$. So by (11.36) and (11.37),

$$V = S_1 \quad \text{and} \quad \dim S_1 = \dim V = n = n_1.$$

(b) Now let L have $(k - 1)$ number of distinct eigenvalues $\lambda_1, ..., \lambda_{k-1}$ for the vector space V. Then its characteristic polynomial has the form

$$\phi(\lambda) = (\lambda - \lambda_1)^{n_1} \cdots (\lambda - \lambda_{k-1})^{n_{k-1}},$$

and by way of induction hypothesis we have

$$\dim S_1 = n_1, ..., \dim S_{k-1} = n_{k-1}.$$

(c) Let us finally assume that L has k distinct eigenvalues $\lambda_1, ..., \lambda_k$ for the n-dimensional vector space V with multiplicities $n_1, ..., n_k$ respectively, and $S_1, ..., S_k$ are the respective root subspaces of L. Then, the characteristic polynomial of L is

$$\phi(\lambda) = (\lambda - \lambda_1)^{n_1} \cdots (\lambda - \lambda_{k-1})^{n_k},$$

where

$$n_1 + \cdots + n_k = n \tag{11.38}$$

and by (11.36),

$$V = S_1 \oplus \cdots \oplus S_{k-1} \oplus S_k \tag{11.39}$$

We now define the subspace $S \subseteq V$ by

$$S = S_1 \oplus \cdots \oplus S_{k-1}.$$

Since S_i's are L-invariant, so is S. Let $L_s : S \to V$ be the restriction of L to S. Then L_s has $k - 1$ distinct eigenvalues for the subspace S in view of Corollary 11.4. So, by our induction hypothesis in (b), we have

$$\dim S_i = n_i; \qquad i = 1, ..., k - 1. \tag{11.40}$$

Consequently, by (11.39) and (11.40),

$$n = \dim V = \sum_{t=1}^{k-1} \dim S_t + \dim S_k = \sum_{t=1}^{k-1} n_t + \dim S_k$$

which, with the help of (11.38), gives

$$\dim S_k = n - \sum_{t=1}^{k-1} n_t = n_k. \qquad \square$$

Corollary 11.5 The linear transformation L on V restricted to its root subspace S_i associated with the eigenvalue λ_i has a characteristic polynomial of the form

$$c_i(\lambda) = (\lambda - \lambda_i)^{n_i}, \tag{11.41}$$

where n_i is the multiplicity of the eigenvalue λ_i.

Proof The assertion (11.41) follows from the fact that by Theorem 11.13, dim $S_i = n_i$, and by Corollary 11.4 the transformation L restricted to S_i has a unique eigenvalue λ_i. □

In Theorem 11.12, we have been able to decompose the vector space V as the direct sum of L-invariant subspaces S_i, each associated with the eigenvalue λ_i of L, and Theorem 11.13 establishes that the dimension of S_i is equal to the multiplicity of λ_i. In respect of this decomposition of V, the matrix representation of any linear transformation L on V becomes very much simplified and can always be represented as a *quasi-diagonal form*. This result is contained in the following theorem whose proof is based on Theorems 11.12, 11.11(a) and 11.10.

Theorem 11.14 Let L be a linear transformation on the n-dimensional vector space V with distinct eigenvalues $\lambda_1, ..., \lambda_k$ of respective multiplicities $n_1, ..., n_k$, where $n_1 + \cdots + n_k = n$. Then there exists a basis in V with respect to which the matrix of L has the *quasi-diagonal form*

$$\text{diag } [A_1, ..., A_k],$$

where A_t is a matrix of order n_i with the characteristic polynomial

$$c_i(\lambda) = (\lambda - \lambda_i)^{ni}.$$ □

The following example will show how to construct the basis in this theorem so that the linear transformation under consideration has a quasi-diagonal form.

Example 11.7 A linear transformation $L: \mathbb{R}^4 \to \mathbb{R}^4$ is defined by

$$L\mathbf{x} = \begin{bmatrix} 3x_1 - 2x_2 + 2x_4 \\ 5x_1 - 6x_2 - x_3 + 2x_4 \\ -23x_1 + 35x_2 + 7x_3 - 5x_4 \\ 3x_1 - 5x_2 - x_3 + x_4 \end{bmatrix}, \quad \mathbf{x} = \begin{bmatrix} x_1 \\ x_2 \\ x_3 \\ x_4 \end{bmatrix}.$$

The matrix A of L with respect to the natural basis $\{\mathbf{e}_1, \mathbf{e}_2, \mathbf{e}_3, \mathbf{e}_4\}$ in $\mathbb{R}^4$ is

$$A = \begin{bmatrix} 3 & -2 & 0 & 2 \\ 5 & -6 & -1 & 2 \\ -23 & 35 & 7 & -5 \\ 3 & -5 & -1 & 1 \end{bmatrix}.$$

The characteristic polynomial of L is

$$\phi(\lambda) = \det(\lambda I - A) = (\lambda - 1)^3(\lambda - 2)$$

and the minimum polynomial (see Section 7.4.1) can be shown to be

$$\psi(\lambda) = (\lambda - 1)^2(\lambda - 2).$$

So, to compute the root space S_1 of L associated with $\lambda_1 = 1$, we need to find the set of independent vectors $\mathbf{x}$ satisfying $(I - A)^2\mathbf{x} = \boldsymbol{\theta}$, i.e.,

$$\begin{bmatrix} 0 & 0 & 0 & 0 \\ 4 & -6 & -1 & 1 \\ -24 & 36 & 6 & -6 \\ 4 & -6 & -1 & 1 \end{bmatrix} \begin{bmatrix} x_1 \\ x_2 \\ x_3 \\ x_4 \end{bmatrix} = \boldsymbol{\theta},$$

which has three linearly independent solution vectors (see Example 5.30) given by

$$\mathbf{x}_{11} = [1\ 0\ 0\ {-4}]^T, \quad \mathbf{x}_{12} = [0\ 1\ 0\ 6]^T, \quad \mathbf{x}_{13} = [0\ 0\ 1\ 1]^T.$$

In view of (11.29), where now $\mathbf{x}_i = \mathbf{e}_i$ (i = 1, ..., 4), n = 4, the above set is a basis for the L-invariant root subspace S_1, which has therefore dimension 3, equal to the multiplicity of eigenvalue 1. The independent vectors $\mathbf{x}$ satisfying $(2I - A)\mathbf{x} = \mathbf{\theta}$ will generate the root subspace S_2 associated with $\lambda_2 = 2$. These are obtained from

$$\begin{bmatrix} -1 & 2 & 0 & -2 \\ -5 & 8 & 1 & -2 \\ 23 & -35 & -5 & 5 \\ -3 & 5 & 1 & 1 \end{bmatrix} \begin{bmatrix} x_1 \\ x_2 \\ x_3 \\ x_4 \end{bmatrix} = \mathbf{\theta}$$

which has one linearly independent solution vector given by

$$\mathbf{x}_{21} = [0\ 1\ {-6}\ 1]^T.$$

As remarked above, in view of (11.29), $\mathbf{x}_{21}$ generates the L-invariant root subspace S_2 which has consequently a dimension l, equal to the multiplicity of eigenvalue 2. As a result of Theorem 11.12, we have

$$\mathbb{R}^4 = S_1 \oplus S_2,$$

where S_1 and S_2 are L-invariant subspaces of $\mathbb{R}^4$. Define

$$T = [\mathbf{x}_{11}\ \mathbf{x}_{12}\ \mathbf{x}_{13} \mid \mathbf{x}_{21}].$$

Then, in view of Remark 6.4, we can write

$$L[\mathbf{x}_{11}\ \mathbf{x}_{12}\ \mathbf{x}_{13} \mid \mathbf{x}_{21}] = AT = [\mathbf{x}_{11}\ \mathbf{x}_{12}\ \mathbf{x}_{13} : \mathbf{x}_{21}]T^{-1}AT,$$

where

$$T^{-1}AT = \begin{bmatrix} 1 & 0 & 0 & 0 \\ -4 & 7 & 1 & -1 \\ 24 & -36 & -5 & 6 \\ 4 & -6 & -1 & 1 \end{bmatrix} \begin{bmatrix} 3 & -2 & 0 & 2 \\ 5 & -6 & -1 & 2 \\ -23 & 35 & 7 & -5 \\ 3 & -5 & -1 & 1 \end{bmatrix} \begin{bmatrix} 1 & 0 & 0 & 0 \\ 0 & 1 & 0 & 1 \\ 0 & 0 & 1 & -6 \\ -4 & 6 & 1 & 1 \end{bmatrix}$$

$$= \left[\begin{array}{ccc|c} -5 & 10 & 2 & 0 \\ -3 & 6 & 1 & 0 \\ -3 & 5 & 2 & 0 \\ \hline 0 & 0 & 0 & 2 \end{array}\right]$$

is the matrix of L in quasi-diagonal form with respect to the basis $\{\mathbf{x}_{11}, \mathbf{x}_{12}, \mathbf{x}_{13}, \mathbf{x}_{21}\}$. □

In Theorem 7.23, a necessary and sufficient condition is given for diagonalization of a square matrix. We shall now establish an alternative criterion for this.

Theorem 11.15 A linear transformation L on an n-dimensional vector space V (resp an $n \times n$ matrix A) is diagonalizable iff its minimum polynomial factorizes into distinct linear factors.

Proof Suppose that L is diagonalizable. So there exists a basis in V such that L is represented by a diagonal matrix

$$D: = \text{diag}\ [d_1, ..., d_1; d_2, ..., d_k],$$

where d_i is of multiplicity n_i, $i = 1, ..., k$, and $n_1 + \cdots + n_k = n$. Hence,

$$\text{adj}\ (\lambda I - D) = \text{diag}\ [\phi_1(\lambda), ..., \phi_1(\lambda); \phi_2(\lambda), ..., \phi_k(\lambda)]$$

in which the multiplicity of $\phi_i(\lambda)$ is n_i, $i = 1, ..., k$, and

$$\phi_1(\lambda) = (\lambda - d_1)^{n_1 - 1}(\lambda - d_2)^{n_2} \cdots (\lambda - d_k)^{n_k} = \phi(\lambda)/(\lambda - d_1)$$

where $\phi(\lambda)$ is expressed by (11.25), i.e.,

$$\phi_i(\lambda) = \phi(\lambda)/(\lambda - d_i), \qquad i = 1..., k;$$

$$D(\lambda): = \text{GCD}\ [\phi_1(\lambda), ..., \phi_k(\lambda)] = (\lambda - d_1)^{n_1-1}(\lambda - d_2)^{n_2-1} \cdots (\lambda - d_k)^{n_k-1}.$$

Therefore, by Theorems 7.14 and 7.15, the minimum polynomial of L is

$$\psi(\lambda) = \phi(\lambda)/D(\lambda) = (\lambda - d_1)(\lambda - d_2) \cdots (\lambda - d_k).$$

Conversely, suppose that the minimum polynomial $\psi(\lambda)$ of L has k distinct factors, i.e.,

$$\psi(\lambda) = (\lambda - \lambda_1) \cdots (\lambda - \lambda_k)$$

where λ_i's are distinct. By (11.36) and Definition 11.6, we have

$$V = S_1 \oplus S_2 \oplus \cdots \oplus S_k = E_1 \oplus E_2 \oplus \cdots \oplus E_k,$$

where E_i is the eigenspace of L associated with λ_i. Let dim $E_i = n_i$ and $\{\mathbf{x}_{i1}, ..., \mathbf{x}_{in_i}\}$ be a basis of E_i. Obviously, dim $V = n = n_1 + n_2 + \cdots + n_k$ and by Theorem 11.7(a), the union of the above bases, i.e., the set of vectors

$$\{\mathbf{x}_{11}, ..., \mathbf{x}_{1n_1}; \cdots ; \mathbf{x}_{k1}, ..., \mathbf{x}_{kn_k}\} \tag{11.42}$$

forms a basis of V. In view of (11.27), where now $m_i = 1$, it is easy to see that

$$L\mathbf{x}_{ij} = \lambda_i \mathbf{x}_{ij}, \qquad i = 1, ..., k; \qquad j = 1, ..., n_i;$$

and hence by definition of the matrix of a linear transformation in Section 6.3, we have

$$\text{matrix of } L = \text{diag}\ [\lambda_1, ..., \lambda_1; \lambda_2, ..., \lambda_k]$$

which is a diagonal matrix with respect to the basis (11.42). □

11.4 DECOMPOSITION OF ROOT-SUBSPACES: JORDAN NORMAL FORM

In Section 11.3, we have studied the decomposition of a vector space V into root subspaces of a given linear transformation L and its matrix representation by quasi-diagonal form. Consequently, we are now eager to know whether or not further decomposition of these root subspaces is feasible, which might expose the finer structure of V under the linear transformation L. We shall show that such refined decomposition of V is possible, in which case the matrix representation of L can be given by Jordan normal form with respect to a suitable basis. An elementary proof of Jordan normal form is included in Section 12.4.

Definition 11.7 *Linear Independence over a Subspace.* Let T and S be two subspaces of a vector space V and $T \subset S$. Then the vectors $\mathbf{x}_1, ..., \mathbf{x}_k$ of S are called linearly independent over the subspace T if (i) they are linearly independent and (ii) the space spanned by these vectors has with T zero intersection, i.e.,

$$\alpha_1\mathbf{x}_1 + \cdots + \alpha_k\mathbf{x}_k \in T$$

iff

$$\alpha_1 = \cdots = \alpha_k = 0. \qquad \square$$

We observe that if the vectors $\mathbf{x}_1, ..., \mathbf{x}_k$ are linearly independent over $T = O$, a null subspace, then they are linearly independent in the usual sense.

Example 11.8 Let S and T be two subspaces of $\mathbf{R}^3$ generated respectively by

$$\{\mathbf{x}_1 = [1\ 1\ 1]^T, \quad \mathbf{x}_2 = [1\ 1\ 0]^T, \quad \mathbf{x}_3 = [1\ 0\ 0]^T\}; \text{ and } \quad \{\mathbf{y} = [0\ 0\ 1]^T\}.$$

By the Gaussian elimination method of Section 5.8.1, we observe that

$$\begin{array}{c} \begin{matrix} \mathbf{y} & \mathbf{x}_1 & \mathbf{x}_2 & \mathbf{x}_3 \end{matrix} \\ \left[\begin{array}{c|ccc} 0 & 1 & 1 & 1 \\ 0 & 1 & 1 & 0 \\ 1 & 1 & 0 & 0 \end{array}\right] \end{array} \sim \begin{array}{c} \begin{matrix} \mathbf{y} & \mathbf{x}_1 & \mathbf{x}_2 & \mathbf{x}_3 \end{matrix} \\ \left[\begin{array}{c|ccc} 0 & 1 & 0 & 1 \\ 0 & 1 & 0 & 0 \\ 1 & 0 & 0 & 0 \end{array}\right] \end{array}$$

which shows that $\{\mathbf{x}_1, \mathbf{x}_3\}$ are linearly independent over T. It is not difficult to see that $\{\mathbf{x}_2, \mathbf{x}_3\}$ are also linearly independent over T, but not $\{\mathbf{x}_1, \mathbf{x}_2\}$. $\square$

Theorem 11.16 Let V be an n-dimensional vector space over $\mathbb{C}$ and let

$$O = U_0 \subset U_1 \subset \cdots \subset U_{m-1} \subset U_m = V$$

be an *ascending chain of subspaces*, each of them being a proper subspace of the succeeding one. Let

$$\mathbf{x}_{i1}, ..., \mathbf{x}_{ir_i}; \qquad i = 1, 2, ..., m; \qquad r_i \geq 1, \tag{11.43}$$

be the maximum number of linearly independent vectors in U_i over U_{i-1}. Then the union of all vectors in (11.43) is a basis of V.

Proof First, we shall prove that the complete aggregate of all vectors (11.43) is a linearly independent set of V. To see this, suppose that

$$\sum_{j=1}^{r_1} \alpha_{1j}\mathbf{x}_{1j} + \cdots + \sum_{j=1}^{r_p} \alpha_{pj}\mathbf{x}_{pj} + \cdots + \sum_{j=1}^{r_m} \alpha_{mj}\mathbf{x}_{mj} = \boldsymbol{\theta}, \tag{11.44}$$

where $\alpha_{ij} = 0$, $i > p$ and $\alpha_{pj} \neq 0$ for at least one value of $j = 1, ..., r_p$. Then,

$$\sum_{j=1}^{r_p} \alpha_{pj}\mathbf{x}_{pj} = -\sum_{j=1}^{r_1} \alpha_{1j}\mathbf{x}_{1j} - \cdots - \sum_{j=1}^{r_{p-1}} \alpha_{p-1,j}\mathbf{x}_{p-1,j}$$

is a nonzero element in the subspace U_{p-1} spanned by the vectors $\mathbf{x}_{p1}, ..., \mathbf{x}_{pr_p}$ which contradicts the definition of these vectors in (11.43). Hence $\alpha_{ij} = 0$, $i = 1, ..., m$; $j = 1, ..., r_i$, in (11.44) and the complete aggregate of the set (11.43) is therefore linearly independent in V.

Second, we shall prove that the set of vectors (11.43) is a basis of V by the method of *mathematical induction*.

(a) Suppose that $m = 1$. Then,

$$O = U_0 \subset U_1 = V$$

and hence the vectors given by (11.43) are the maximum number of linearly independent vectors in V and consequently form its basis.

(b) We now suppose that the theorem is true for a chain of proper subspaces

$$O = U_0 \subset U_1 \subset \cdots \subset U_{m-1}$$

containing $(m-1)$ elements, $m > 1$, different from zero. That is, we assume that the union of all vectors

$$\mathbf{x}_{i1}, \ldots, \mathbf{x}_{ir_i}; \qquad i = 1, 2, \ldots, m-1; \qquad r_i \geq 1, \tag{11.45}$$

is a basis of U_{m-1}.

(c) Suppose, in this final step, that the vectors

$$\mathbf{x}_{ml}, \ldots, \mathbf{x}_{mr_m} \tag{11.46}$$

and a vector $\mathbf{x} \in U_m = V$ are linearly independent. In the opposite case, $\mathbf{x}$ is a linear combination of the vectors (11.46), and we are done. Otherwise, the subspace

$$\{\alpha_1\mathbf{x}_{ml} + \cdots + \alpha_{r_m}\mathbf{x}_{mr_m} + \beta\mathbf{x}\}$$

spanned by the vectors (11.46) and $\mathbf{x}$ has with U_{m-1} nonzero intersection because the vectors in (11.46) are the maximum number of linearly independent vectors in U_m over U_{m-1}. Consequently, we can find a linear combination such that the vector (see Figure 11.6)

$$\mathbf{z} = a_1\mathbf{x}_{ml} + \cdots + a_{r_m}\mathbf{x}_{mr_m} + b\mathbf{x} \in U_{m-1}, \tag{11.47}$$

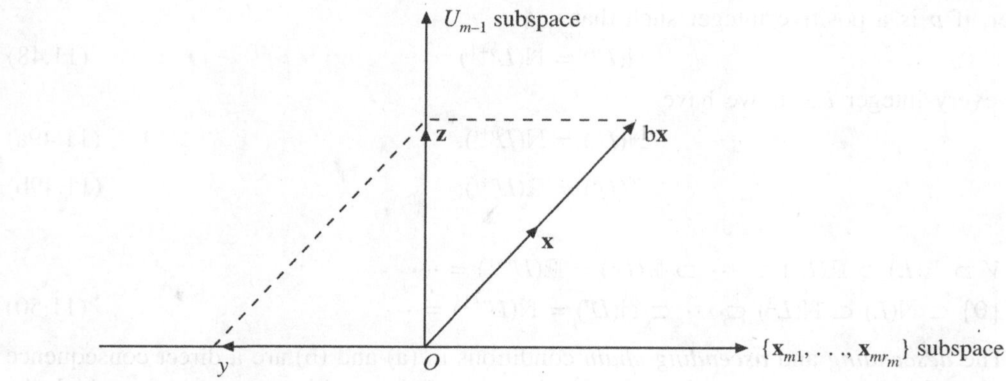

Figure 11.6 $\mathbf{z} = b\mathbf{x} + \mathbf{y}$, where $\mathbf{y}$ is generated by the vectors (11.46).

where obviously $b \neq 0$. But on the basis of induction hypothesis in (b), $\mathbf{z} \in U_{m-1}$ is a linear combination of the vectors in (11.45). Thus, from (11.47) it is evident that any arbitrary vector $\mathbf{x} \in V = U_m$ can be expressed as a linear combination of the vectors given in (11.43). Consequently, the set of vectors given by (11.43) forms a basis of V. □

We recall that the null space and the range space (see Section 6.1) of a linear transforma tion L: $V \to W$ are defined respectively by

$$\mathbb{N}(L) = \{\mathbf{x} : \mathbf{x} \in V \text{ and } L\mathbf{x} = \boldsymbol{\theta}\},$$

$$\mathbb{R}(L) = \{\mathbf{y} : \mathbf{y} \in W \text{ and } \exists \text{ a vector } \mathbf{x} \in V \text{ s.t. } L\mathbf{x} = \mathbf{y}\},$$

and that

$$\text{null } L: = \dim \mathbb{N}(L), \qquad \text{rank } L: = \dim \mathbb{R}(L), \qquad \text{null } L + \text{rank } L = \dim V.$$

The following lemma plays a crucial role in establishing the ascending and the decending chains of proper subspaces associated respectively with the null and the range space of L in Theorem 11.17.

Lemma 11.2 Let $K : U \to V$ and $L : V \to W$ be two linear transformations. Then,

(a) $\mathbb{R}(L{\cdot}K) \subseteq \mathbb{R}(L)$ and rank $(L{\cdot}K) \leq$ rank L,

(b) $\mathbb{N}(L{\cdot}K) \supseteq \mathbb{N}(K)$ and null $(L{\cdot}K) \geq$ null K.

Proof (a) Let $\mathbf{z} \in \mathbb{R}(L{\cdot}K)$. Then there exists a vector $\mathbf{x} \in U$ such that $(L{\cdot}K)\mathbf{x} = L(K\mathbf{x}) = \mathbf{z}$. Now $K\mathbf{x} = \mathbf{y} \in V$. So, $L\mathbf{y} = \mathbf{z}$ implies that $\mathbf{z} \in \mathbb{R}(L)$. The rank inequality follows by Definition 5.7 and Theorem 6.5.

(b)If $\mathbf{x} \in \mathbb{N}(K)$, then $K\mathbf{x} = \boldsymbol{\theta}$ which implies that $(L{\cdot}K)\mathbf{x} = \boldsymbol{\theta}$. So, $\mathbf{x} \in \mathbb{N}(L{\cdot}K)$. Hence, the result in (b) follows. □

Theorem 11.17 If L is a linear transformation on an n-dimensional vector space V, then

(a) $V \supseteq \mathbb{R}(L) \supseteq \mathbb{R}(L^2) \supseteq \cdots \supseteq \mathbb{R}(L^k) \supseteq \cdots,$

(b) $\{\boldsymbol{\theta}\} \subseteq \mathbb{N}(L) \subseteq \mathbb{N}(L^2) \subseteq \cdots \subseteq \mathbb{N}(L^k) \subseteq \cdots.$

Moreover, if p is a positive integer such that

$$\mathbb{N}(L^p) = \mathbb{N}(L^{p+1}) \tag{11.48}$$

then for every integer $i \geq 1$, we have

$$\mathbb{N}(L^p) = \mathbb{N}(L^{p+i}), \tag{11.49a}$$

$$\mathbb{R}(L^p) = \mathbb{R}(L^{p+i}); \tag{11.49b}$$

and

(c) $V \supset \mathbb{R}(L) \supset \mathbb{R}(L^2) \supset \cdots \supset \mathbb{R}(L^p) = \mathbb{R}(L^{p+1}) = \cdots,$

(d) $\{\boldsymbol{\theta}\} \subset \mathbb{N}(L) \subset \mathbb{N}(L^2) \subset \cdots \subset \mathbb{N}(L^p) = \mathbb{N}(L^{p+1}) = \cdots.$ (11.50)

Proof The *descending* and *ascending chain* conditions in (a) and (b) are a direct consequence of Lemma 11.2. Since V is finite dimensional, there is a finite positive integer p for which the equality (11.48) must hold.

Therefore, if $\mathbf{x} \in \mathbb{N}(L^{p+i})$, then $\boldsymbol{\theta} = L^{p+i}\mathbf{x} = L^{p+1}(L^{i-1}\mathbf{x})$ which shows that

$$L^{i-1}\mathbf{x} \in \mathbb{N}(L^{p+1}) = \mathbb{N}(L^p)$$

Hence, if $i > 1$,

$$\boldsymbol{\theta} = L^p(L^{i-1}\mathbf{x}) = L^{p+1}(L^{i-2}\mathbf{x}),$$

and so,

$$L^{i-2}\mathbf{x} \in \mathbb{N}(L^{p+1}) = \mathbb{N}(L^p).$$

In this way it follows that $\mathbf{x} \in \mathbb{N}(L^p)$ whenever $\mathbf{x} \in \mathbb{N}(L^{p+i})$ and, consequently,

$$\mathbb{N}(L^p) \supseteq \mathbb{N}(L^{p+i}).$$

Taking (b) into account, we have the equality (11.49a) and the ascending chain (d). The equality (11.49b) follows by virtue of Theorem 6.5. □

We now propose to study in detail the structure of the ascending chain of null spaces in (11.50) associated with the linear transformation $(\lambda_i I_V - L)$, on the n-dimensional vector space V, where λ_i is an eigenvalue of L.

Definition 11.8 *Cyclic Space: Cyclic Basis.* Let L be a linear transformation on an n-dimensional vector space V over $\mathbb{C}$. Then V is said to be cyclic relative to its linear transformation L or L-cyclic if the chain of vectors

$$\mathbf{x},\ L\mathbf{x},\ \ldots,\ L^{n-1}\mathbf{x}; \qquad \mathbf{x} \neq \boldsymbol{\theta}, \tag{11.51}$$

forms a basis of this space. This basis is called the *cyclic basis* of V. On the other hand, a subspace W of V is L-cyclic if there exists a chain of vectors $\mathbf{x}, L\mathbf{x}, \ldots, L^{k-1}\mathbf{x}$; $\mathbf{x} \neq \boldsymbol{\theta}$, $k \geq 1$, for the basis of W. □

It is obvious that an L-cyclic subspace is also L-invariant (see Problem 11.21).

The decomposition of a vector space into a direct sum of L-cyclic subspaces gives rise to a refined matrix structure of the linear transformation L, known as *Jordan normal form* which we propose to study in the sequel. Before actually taking up this study, let us note the following consequence of an L-cyclic space on the matrix structure of L.

Theorem 11.18 Let L be a linear transformation on an n-dimensional vector space V over $\mathbb{C}$ with a unique eigenvalue λ, and let V be cyclic relative to the transformation

$$K = L - \lambda I_V.$$

Then the matrix of L in cyclic basis

$$\mathbf{x}_n = K\mathbf{x}_{n-1},\ \ldots,\ \mathbf{x}_2 = K\mathbf{x}_1,\ \mathbf{x}_1 \tag{11.52}$$

is represented by *Jordan's form*

$$J(\lambda) = \begin{bmatrix} \lambda & 1 & \ldots & 0 & 0 \\ 0 & \lambda & \ldots & 0 & 0 \\ \vdots & \vdots & & \vdots & \vdots \\ 0 & 0 & \ldots & \lambda & 1 \\ 0 & 0 & \ldots & 0 & \lambda \end{bmatrix} \tag{11.53}$$

corresponding to λ.

Proof The linear transformation L has only one eigenvalue in the n-dimensional vector space V. So, L has a single root subspace S associated with λ and, by Corollary 11.5, it has a characteristic polynomial of the form $\phi(\mu) = (\mu - \lambda)^n$. By the Cayley-Hamilton theorem (see Corollary 7.3), we have

$$(L - \lambda I_V)^n = \Theta$$

and consequently,

$$\mathbf{x}_{n+1} = K\mathbf{x}_n = K^n\mathbf{x}_1 = \boldsymbol{\theta}.$$

From (11.52), we also have

$$L\mathbf{x}_i = (\lambda I_V + K)\mathbf{x}_i = \mathbf{x}_{i+1} + \lambda\mathbf{x}_i; \quad i = 1, ..., n-1.$$

Hence,

$$\begin{aligned} L\mathbf{x}_n &= \lambda\mathbf{x}_n, \\ L\mathbf{x}_{n-1} &= \mathbf{x}_n + \lambda\mathbf{x}_{n-1}, \\ &\vdots \\ L\mathbf{x}_2 &= \mathbf{x}_3 + \lambda\mathbf{x}_2, \\ L\mathbf{x}_1 &= \mathbf{x}_2 + \lambda\mathbf{x}_1. \end{aligned}$$

Therefore, the matrix of L with respect to the *cyclic basis* (11.52) is represented by (11.53). □

We shall next concentrate on a root-subspace S_i of a linear transformation L on V associated with an eigenvalue λ_i. We shall show that S_i can be decomposed into a direct sum of L-cyclic subspaces, the matrices of L associated with these subspaces would later turn out to be the building blocks of Jordan canonical structure of a matrix.

Definition 11.9 We denote by $\mathbb{N}_{ij}$ the null space of the linear transformation $(\lambda_i I_V - L)^j$ on the n-dimensional vector space V associated with λ_i. Thus $\mathbb{N}_{ij}$ is the set of all vectors $\mathbf{x} \in V$ defined by

$$\mathbb{N}_{ij} = \{\mathbf{x} : \mathbf{x} \in V, (\lambda_i I_V - L)^j \mathbf{x} = \boldsymbol{\theta}\}$$

for all $j = 0, 1, ...$. The least value μ_i of J for which $\mathbb{N}_{ij} = \mathbb{N}_{i,\, j+1}$ is called *the index of L* associated with λ_i. In particular, $\mathbb{N}_{i1}$ is the eigenspace E_{λ_i} defined in Theorem 11.8. □

In the light of Definition 11.9, we have as an easy consequence of Theorem 11.17(d).

Corollary 11.6 The null spaces $\mathbb{N}_{ij}$ of the linear transformation $(\lambda_i I_V - L)^j$ on V associated with the eigenvalue λ_i of L are subspaces of V and satisfy the following ascending chain conditions:

$$\{\boldsymbol{\theta}\} = \mathbb{N}_{i0} \subset \mathbb{N}_{i1} \subset \cdots \subset \mathbb{N}_{i\mu_i} = \mathbb{N}_{i,\, \mu_i+1} = ...,$$

$i = 1, 2, ..., k$, where k is the number of distinct eigenvalues of L. □

Physically interpreted, this corollary says that for $j \geq \mu_i$ all the null spaces $\mathbb{N}_{ij}$ become the same. Moreover, as will be proved in Lemma 11.3, μ_i is equal to the multiplicity of eigenvalue λ_i in the minimum polynomial of L.

Lemma 11.3 Let L be a linear transformation on an n-dimensional vector space V. Also, let m_i be the multiplicity of the eigenvalue λ_i in the minimum polynomial $\psi(\lambda)$ of L and μ_i be the index of L associated with λ_i. Then,

$$\mu_i = m_i; \quad i = 1, 2, ..., k,$$

where k is the number of distinct eigenvalues of L.

Proof First, we shall prove that $m_i \geq \mu_i$ by the method of contradiction. For this, suppose that $m_i < \mu_i$. By Corollary 11.6, there exists a vector $\mathbf{z} \neq \boldsymbol{\theta}$ in V such that $\mathbf{z} \in \mathbb{N}_{i,m_i+1}$ but $\mathbf{z} \notin \mathbb{N}_{im_i}$. That is, $(\lambda_i I_V - L)^{m_i+1}\, \mathbf{z} = \boldsymbol{\theta}$ but $\mathbf{y} := (\lambda_i I_V - L)^{m_i}\mathbf{z} \neq \boldsymbol{\theta}$. We observe that $L\mathbf{y} = \lambda_i\mathbf{y}$. Indeed,

$$L\mathbf{y} = -(\lambda_i I_V - L - \lambda_i I_V)(\lambda_i I_V - L)^{m_i}\, \mathbf{z} = \lambda_i\mathbf{y}. \tag{11.54}$$

We define (see (11.31))

$$\psi_i(\lambda) = \psi(\lambda)/(\lambda - \lambda_i)^{m_i},$$

where $\psi_i(\lambda_i) \neq 0$. Now, in view of (11.54) and the property of a minimum polynomial of L, we have

$$\boldsymbol{\theta} = \psi(L)\mathbf{z} = \psi_i(L)\,(L - \lambda_i I_V)^{m_i}\,\mathbf{z} = \psi_i(L)\mathbf{y} = \psi_i(\lambda_i)\mathbf{y},$$

where $\psi_i(\lambda_i) \neq 0$ and $\mathbf{y} \neq \boldsymbol{\theta}$. This contradiction proves that $m_i \geq \mu_i$. Suppose now that $m_i > \mu_i$, and hence

$$\mathrm{N}_{i\mu_i} = \mathrm{N}_{i\mu_i+1} = \cdots = \mathrm{N}_{im_i} =: S_i$$

is the root subspace of L associated with λ_i. Therefore,

$$(\lambda_i I_V - L)^{\mu_i}\,\mathbf{x}_i = \boldsymbol{\theta}, \qquad \forall\, \mathbf{x}_i \in S_i. \tag{11.55}$$

We now define

$$\delta(\lambda) = (\lambda - \lambda_1)^{\mu_1} \cdots (\lambda - \lambda_k)^{\mu_k}.$$

Note that in view of Theorem 11.12, V can be decomposed as the direct sum of root subspaces:

$$V = S_i \oplus S_2 \oplus \cdots \oplus S_k.$$

So, for any vector $\mathbf{x} \in V$, we have the unique decomposition

$$\mathbf{x} = \mathbf{x}_1 + \mathbf{x}_2 + \cdots + \mathbf{x}_k, \tag{11.56}$$

where $\mathbf{x}_i \in S_i$, $i = 1, 2, ..., k$. Consequently, with the help of (11.55) and (11.56), we have

$$\delta(L)\mathbf{x} = (L - \lambda_1 I_V)^{\mu_1} \cdots (L - \lambda_k I_V)^{\mu_k}\,\mathbf{x} = \boldsymbol{\theta}$$

for any $\mathbf{x} \in V$. Hence, $\delta(L) = \Theta$ and $\delta(\lambda)$ is an annihilating polynomial of L. Since $\psi(\lambda)$ is a minimum polynomial of L, $\delta(\lambda)$ must be divisible by $\psi(\lambda)$ (see Theorems 7.15 and 7.13). So $m_i > \mu_i$ is not admissible, and hence $\mu_i = m_i$, because $\delta(\lambda)$ and $\psi(\lambda)$ have the same linear factors. □

Example 11.9 Let $A : \mathbb{R}^4 \to \mathbb{R}^4$ be a linear transformation, where

$$A = \begin{bmatrix} 1 & 1 & 0 & 0 \\ 0 & 1 & 0 & 0 \\ 0 & 0 & 1 & 0 \\ 0 & 0 & 0 & 2 \end{bmatrix}$$

with the minimum polynomial $(\lambda - 1)^2(\lambda - 2)$. For the eigenvalue $\lambda_1 = 1$, we have $m_1 = 2$ and

$$(I - A) = \begin{bmatrix} 0 & -1 & 0 & 0 \\ 0 & 0 & 0 & 0 \\ 0 & 0 & 0 & 0 \\ 0 & 0 & 0 & -1 \end{bmatrix},$$

$$(I - A)^2 = \begin{bmatrix} 0 & 0 & 0 & 0 \\ 0 & 0 & 0 & 0 \\ 0 & 0 & 0 & 0 \\ 0 & 0 & 0 & 1 \end{bmatrix}.$$

Therefore,

$$\mathbb{N}_{11} = \{\mathbf{x} : (I - A)\ \mathbf{x} = \boldsymbol{\theta}\} = \{\mathbf{x} : x_2 = x_4 = 0\},$$
$$\mathbb{N}_{12} = \{\mathbf{x} : (I - A)^2\ \mathbf{x} = \boldsymbol{\theta}\} = \{\mathbf{x} : x_4 = 0\} \supset \mathbb{N}_{11},$$

where $\mathbf{x} = [x_1\ x_2\ x_3\ x_4]^T$. Similarly, for $\lambda_2 = 2$ we have $m_2 = 1$ and

$$\mathbb{N}_{21} = (\mathbf{x} : (2I - A)\ \mathbf{x} = \boldsymbol{\theta}\} = \{\mathbf{x} : x_1 = x_2 = x_3 = 0\}.$$

Let

$$\mathbf{x}_{j1}^{(i)}, \mathbf{x}_{j2}^{(i)}, \ldots, \mathbf{x}_{jr_{ij}}^{(i)}; \qquad j = 1, 2, \ldots, m_i; \tag{11.57}$$

be the set of all vectors in $\mathbb{N}_{ij}$ which are linearly independent (LI) over $\mathbb{N}_{i,\ j-1}$, see Definition 11.7 and Example 11.8. In other words, the vectors in (11.57) are LI and the subspace spanned by them has zero intersection with $\mathbb{N}_{i,\ j-1}$. For this set of vectors, we can write

$$\{\mathbf{x}_{jt}^{(i)} : (\lambda_i I_V - L)^j \mathbf{x}_{jt}^{(i)} = \boldsymbol{\theta} \quad \text{but} \quad (\lambda_i I_V - L)^{j-1} \mathbf{x}_{jt}^{(i)} \neq \boldsymbol{\theta};\ t = 1, 2, \ldots, r_{ij}\},$$

which are associated with the eigenvalue λ_i as indicated by the superscript (i). In short, all $\mathbf{x}_{jt}^{(i)} \in \mathbb{N}_{ij}$ are LI but $\mathbf{x}_{jt}^{(i)} \notin \mathbb{N}_{i,\ j-1}$. Index i or superscript (i) refers to the ith eigenvalue λ_i and j, to the power of $(\lambda_i I_V - L)$ in $(\lambda_i I_V - L)^j$ and t to the number of LI vectors in $\mathbb{N}_{ij}$. The set of vectors in (11.57) in subspace $\mathbb{N}_{ij}$, which are LI over $\mathbb{N}_{i,j-1}$, are defined to be the generalized eigenvectors of order j of the linear transformation L. The vectors $\mathbf{x}_{jt}^{(i)}$ are related to $\mathbf{x}_{j-1,t}^{(i)}$ by

$$\mathbf{x}_{j-1,t}^{(i)} = (\lambda_i I_V - L)\mathbf{x}_{jt}^{(i)},$$

which is proved below.

Example 11.10 Referring to Example 11.9, the basis vectors in space $\mathbb{N}_{11}$ may be selected as

$$\mathbf{x}_{11}^{(1)} = \mathbf{e}_1, \qquad \mathbf{x}_{12}^{(1)} = \mathbf{e}_3 \in \mathbb{N}_{11}, \qquad r_{11} = 2;$$

and those in $\mathbb{N}_{12}$ as

$$\mathbf{e}_1, \quad \mathbf{e}_2, \quad \mathbf{e}_3 \in \mathbb{N}_{12}.$$

The vector $\mathbf{x}_{21}^{(1)} = \mathbf{e}_2 \in \mathbb{N}_{12}$ is the only vector LI over $\mathbb{N}_{11}$, see Definition 11.9 and Example 11.8, so that $r_{12} = 1$, the number of LI eigenvectors associated with $\lambda_1 = 1$ in $\mathbb{N}_{12}$ over $\mathbb{N}_{11}$. □

We write

$$\mathbf{x}_{j-s,t}^{(i)} = (\lambda_i I_V - L)^s \mathbf{x}_{jt}^{(i)}, \qquad t = 1, \ldots, r_{ij}; \qquad s < j. \tag{11.58}$$

Then the vectors $\mathbf{x}_{j-s,t}^{(i)} \in \mathbb{N}_{i,j-s}$ and are LI over $\mathbb{N}_{i,j-s-1}$ for $j > 1$.

The development in the following are general but the particular case for which $s = 1$ is what we need to explain the Jordan matrix representation of the linear transformation L. To prove the above assertion we have,

$$(\lambda_i I_V - L)^{j-s} \mathbf{x}_{j-s,t}^{(i)} = (\lambda_i I_V - L)^j \mathbf{x}_{jt}^{(i)} = \boldsymbol{\theta},$$

which shows that $\mathbf{x}_{j-s,t}^{(i)} \in \mathbb{N}_{i,j-s}$. Moreover, the vectors $\mathbf{x}_{j-s,t}^{(i)}$ in (11.58) are LI, because if

$$\sum_t \alpha_t \mathbf{x}_{j-s,t}^{(i)} = (\lambda_i I_V - L)^s \sum_t \alpha_t \mathbf{x}_{jt}^{(i)} = \boldsymbol{\theta}, \tag{11.59}$$

then
$$\sum_t \alpha_t \mathbf{x}_{jt}^{(i)} \in \mathbb{N}_{is} \subset \mathbb{N}_{i,j-1}, \qquad s < j. \tag{11.60}$$

Noticing that the vectors in (11.57) belong to $\mathbb{N}_{ij}$ and are LI over $\mathbb{N}_{i,j-1}$, it follows from (11.60) that (see also Definition 11.7)

$$\sum_t \alpha_t \mathbf{x}_{jt}^{(i)} = \boldsymbol{\theta}.$$

Hence, $\alpha_1 = \alpha_2 = \cdots = \alpha_{r_{ij}} = 0$.

(i) From (11.59) it now follows that the vectors in (11.58) are LI.

Finally, let

$$\mathbf{z} = \sum_t \alpha_t \mathbf{x}_{j-s,t}^{(i)} \in \mathbb{N}_{i,j-s-1}.$$

Then, by Definition 11.9 and (11.58)

$$\boldsymbol{\theta} = (\lambda_i I_V - L)^{j-s-1} \sum_t \alpha_t \mathbf{x}_{j-s,t}^{(i)} = (\lambda_i I_V - L)^{j-1} \sum_t \alpha_t \mathbf{x}_{jt}^{(i)}.$$

Consequently, $\sum_t \alpha_t \mathbf{x}_{jt}^{(i)} \in \mathbb{N}_{i,j-1}$, but the vectors $\mathbf{x}_{jt}^{(i)}$ are LI over $\mathbb{N}_{i,j-1}$. Hence,

$$\sum_t \alpha_t \mathbf{x}_{jt}^{(i)} = \boldsymbol{\theta} \qquad \text{and} \qquad \alpha_1 = \alpha_2 = \cdots = \alpha_{r_{ij}} = 0.$$

This implies that $\mathbf{z} = \boldsymbol{\theta}$.

(ii) Thus the space spanned by the vectors in (11.58) has zero intersection with $\mathbb{N}_{i,j-s-1}$.

In view of the results (i) and (ii), the vectors $\mathbf{x}_{j-s,t}^{(i)} \in \mathbb{N}_{i,j-s}$ defined by (11.58) are LI over $\mathbb{N}_{i,j-s-1}$ for $j > 1$. Thus the following Lemma is proved.

Lemma 11.4 Let

$$\mathbf{x}_{j1}^{(i)}, \quad \mathbf{x}_{j2}^{(i)}, \quad \ldots, \quad \mathbf{x}_{jr_{ij}}^{(i)}; \qquad j = 1, 2, \ldots, m_i;$$

be the set of vectors in the null space of $\mathbb{N}_{ij}$ of $(\lambda_i I_V - L)^j$ which are LI over the null space $\mathbb{N}_{i,j-1}$ of $(\lambda_i I_V - L)^{j-1}$, where m_i is the multiplicity of λ_i in the minimum polynomial of L. Then the vectors

$$\begin{aligned} \mathbf{x}_{j-s,1}^{(i)} &= (\lambda_i I_V - L)^s \, \mathbf{x}_{j1}^{(i)}, \\ &\vdots \\ \mathbf{x}_{j-s,r_{ij}}^{(i)} &= (\lambda_i I_V - L)^s \, \mathbf{x}_{jr_{ij}}^{(i)}, \qquad s < j \end{aligned}$$

in $\mathbb{N}_{i,j-s}$ are LI over $\mathbb{N}_{i,j-s-1}$. □

We shall now focus our attention on a root subspace S_i of L associated with an eigenvalue λ_i of multiplicity m_i in the minimum polynomial of L. The null spaces in Definition 11.9 are now denoted by $\mathbb{N}_{i1}, \mathbb{N}_{i2}, \ldots, \mathbb{N}_{im_i}$. Before proceeding further, we specialize this Lemma to the case $s = 1$ to get the following Corollary. The set of elements in $\mathbb{N}_{ij}$ that are LI over $\mathbb{N}_{i,\, j-1}$ is designated by $\mathbb{N}_{ij} - \mathbb{N}_{i,\, j-1}$; $j = m_i, m_i - 1, ..., 2, 1$.

Corollary 11.7 Let $\mathbf{x}_{j1}^{(i)}, \mathbf{x}_{j2}^{(i)}, ..., \mathbf{x}_{jr_{ij}}^{(i)}$ be the set of vectors in $\mathbb{N}_{ij}$ LI over $\mathbb{N}_{i,\, j-1}$ with dim $(\mathbb{N}_{ij} - \mathbb{N}_{i,\, j-1}) = r_{ij}$. Then,

$$\begin{aligned} \mathbf{x}_{j-1,1}^{(i)} &= (\lambda_i I_V - L)\, \mathbf{x}_{j1}^{(i)}, \\ \mathbf{x}_{j-1,2}^{(i)} &= (\lambda_i I_V - L)\, \mathbf{x}_{j2}^{(i)}, \\ &\vdots \\ \mathbf{x}_{j-1,r_{ij}}^{(i)} &= (\lambda_i I_V - L)\, \mathbf{x}_{jr_{ij}}^{(i)}, \end{aligned} \tag{11.61}$$

are the vectors in $\mathbb{N}_{i,\, j-1}$ LI over $\mathbb{N}_{i,\, j-2}$. □

If necessary, we add additional vectors to the set (11.61) so that

$$\mathbf{x}_{j-1,1}^{(i)}, ..., \mathbf{x}_{j-1,r_{ij}}^{(i)}; \quad \mathbf{x}_{j-1,r_{ij}+1}^{(i)}, ..., \mathbf{x}_{j-1,r_{i,j-1}}^{(i)}; \quad r_{i,j-1} \geq r_{ij},$$

are the maximum number of vectors in $\mathbb{N}_{i,j-1}$ LI over $\mathbb{N}_{i,j-2}$. The first r_{ij} number of vectors are defined by (11.61). The additional vectors are the number of vectors in $\mathbb{N}_{i,j-1} - \mathbb{N}_{i,j-2}$ depleted by the set of LI vectors in (11.61).

Thus, in every subspace $\mathbb{N}_{ij}$, $1 \leq j \leq m_i$, a maximum number r_{ij} of vectors given by

$$\mathbf{x}_{j1}^{(i)}, ..., \mathbf{x}_{jr_{ij}}^{(i)}; \quad j = m_i, m_i - 1, ..., 2, 1; \quad r_{i,j-1} \geq r_{ij}, \tag{11.62}$$

are LI over $\mathbb{N}_{i,j-1}$, where it is true that

$$\mathbf{x}_{j-1,1}^{(i)} = (\lambda_i I_V - L)\mathbf{x}_{j1}^{(i)},\ \mathbf{x}_{j-1,2}^{(i)} = (\lambda_i I_V - L)\,\mathbf{x}_{j2}^{(i)}, ..., \mathbf{x}_{j-1,r_{ij}}^{(i)} = (\lambda_i I_V - L)\mathbf{x}_{jr_{ij}}^{(i)}; 1 < j \leq m_i;$$

and that

$$\mathbf{x}_{11}^{(i)}, \quad \mathbf{x}_{12}^{(i)}, \quad ..., \quad \mathbf{x}_{1r_{i1}}^{(i)}$$

are the number of LI vectors in $\mathbb{N}_{i1} = E_{\lambda_i}$, the eigenspace of L on V associated with λ_i. By Theorem 11.16, the union of the set of vectors given by (11.61) are LI and *with a change of sign* can be used as in Table 11.1. Thus

(I) $\mathbf{x}_{m_i 1}^{(i)}, \quad ..., \quad \mathbf{x}_{m_i, r_{im_i}}^{(i)}$ are LI in $\mathbb{N}_{im_i} - \mathbb{N}_{i,m_i-1}$;

(II) $\mathbf{x}_{m_i-1,1}^{(i)} = (L - \lambda_i I_V)\,\mathbf{x}_{m_i 1}^{(i)}, \quad ..., \quad \mathbf{x}_{m_i-1,r_{im_i}}^{(i)} = (L - \lambda_i I_V)\,\mathbf{x}_{m_i,r_{im_i}}^{(i)}$, s

$$\mathbf{x}_{m_i-1,r_{im_i}+1}^{(i)}, ..., \mathbf{x}_{m_i-1,\, r_{i,\, m_i-1}}^{(i)} \text{ are LI in } \mathbb{N}_{i,m_i-1} - \mathbb{N}_{i,m_i-2};$$

$$\text{(III)} \qquad \mathbf{x}^{(i)}_{m_i-2,1} = (L-\lambda_i I_V)\mathbf{x}^{(i)}_{m_i-1,1} = (L-\lambda_i I_V)^2\mathbf{x}^{(i)}_{m_i 1}, \ldots,$$

$$\mathbf{x}^{(i)}_{m_i-2,r_{im_i}} = (L-\lambda_i I_V)^2\mathbf{x}^{(i)}_{m_i,r_{im_i}}, \mathbf{x}^{(i)}_{m_i-2,r_{im_i}+1} = (L-\lambda_i I_V)\mathbf{x}^{(i)}_{m_i-1,r_{im_i}+1}, \ldots,$$

$$\mathbf{x}^{(i)}_{m_i-2,r_{i,m_i-1}} = (L-\lambda_i I_V)\mathbf{x}^{(i)}_{m_i-1,r_{i,m_i-1}},$$

$$\mathbf{x}^{(i)}_{m_i-2,r_{i,m_i-1}+1}, \ldots, \mathbf{x}^{(i)}_{m_i-2,r_{i,m_i-2}} \text{ are LI in } \mathbb{N}_{i,m_i-2} - \mathbb{N}_{i,m_i-3},$$

$$\text{(IV)} \qquad \mathbf{x}^{(i)}_{j1} = (L-\lambda_i I_V)^{m_i-j}\mathbf{x}^{(i)}_{m_i 1}, \ldots, \mathbf{x}^{(i)}_{jr_{m_i}} = (L-\lambda_i I_V)^{m_i-j}\mathbf{x}^{(i)}_{m_i,r_{im_i}},$$

$$\mathbf{x}^{(i)}_{j,r_{m_i}+1} = (L-\lambda_i I_V)^{m_i-j-1}\mathbf{x}^{(i)}_{m_i-1,r_{im_i}+1}, \ldots,$$

$$\mathbf{x}^{(i)}_{j,r_{m_i}-1} = (L-\lambda_i I_V)^{m_i-j-1}\mathbf{x}^{(i)}_{m_i-1,r_{i,m_i-1}}, \ldots, \mathbf{x}^{(i)}_{j,r_{i,j+1}+1}, \ldots, \mathbf{x}^{(i)}_{jr_{ij}} \text{ are LI in } \mathbb{N}_{ij} - \mathbb{N}_{i,j-1}$$

$$\text{(V)} \qquad \mathbf{x}^{(i)}_{11} = (L-\lambda_i I_V)^{m_i-1}\mathbf{x}^{(i)}_{m_i 1}, \quad \mathbf{x}^{(i)}_{12} = (L-\lambda_i I_V)^{m_i-1}\mathbf{x}^{(i)}_{m_i 2}, \ldots,$$

$$\mathbf{x}^{(i)}_{1,r_{im_i}} = (L-\lambda_i I_V)^{m_i-1}\mathbf{x}^{(i)}_{m_i,r_{im_i}}, \ldots, \mathbf{x}^{(i)}_{1,r_{im_i}-1} = (L-\lambda_i I_V)^{m_i-2}\mathbf{x}^{(i)}_{m_i-1,r_{i,m_i-1}}, \ldots,$$

$$\mathbf{x}^{(i)}_{1,r_{i,j-1}+1} = (L-\lambda_i I_V)^{j-1}\mathbf{x}^{(i)}_{j,r_{i,j-1}+1}, \ldots, \mathbf{x}^{(i)}_{1r_{ij}} = (L-\lambda_i I_V)^{j-1}\mathbf{x}^{(i)}_{jr_{ij}}, \ldots,$$

$$\mathbf{x}^{(i)}_{1,r_{i3}+1} = (L-\lambda_i I_V)\mathbf{x}^{(i)}_{2,r_{i3}+1}, \ldots, \mathbf{x}^{(i)}_{1r_{i2}} = (L-\lambda_i I_V)\mathbf{x}^{(i)}_{2r_{i2}}, \ldots,$$

$$\mathbf{x}^{(i)}_{1,r_{i2}+1}, \ldots, \mathbf{x}^{(i)}_{1r_{i1}} \text{ are LI in } \mathbb{N}_{i1}.$$

The elements in subspace $\mathbb{N}_{i1}$ are numbered as

$$1, 2, \ldots, r_{i1},$$

which can further be detailed as

$$1, 2, \ldots, r_{im_i}, r_{im_i}+1, \ldots, r_{i,m_i-1}, r_{i,m_i-1}+1, \ldots, r_{i,m_i-2}, r_{i,m_i-2}, +1, \ldots, r_{i,m_i-3},$$
$$r_{i,m_i-3}+1, \ldots, r_{i2}, r_{i2}+1, \ldots, r_{i1}; \qquad 0 < r_{im_i} < r_{i,m_i-1} < \cdots < r_{ij} < \cdots < r_{i2} < r_{i1}.$$

Let the set

$$\begin{aligned}\{\mathbf{x}^{(i)}_{11} &= (L-\lambda_i I_V)^{m_i-1}\mathbf{x}^{(i)}_{m_i 1},\\ &\vdots\\ \mathbf{x}^{(i)}_{j1} &= (L-\lambda_i I_V)^{m_i-j}\mathbf{x}^{(i)}_{m_i 1},\\ &\vdots\\ \mathbf{x}^{(i)}_{m_i-1,1} &= (L-\lambda_i I_V)\,\mathbf{x}^{(i)}_{m_i 1}, \mathbf{x}^{(i)}_{m_i 1}\}\end{aligned} \tag{11.63}$$

formed by the first elements in the list of (I)–(V) (see column 1 of Table 11.1), counted in the reverse order, generate the subspace T_{i1} associated with the eigenvalue λ_i. T_{i1} is a subspace of S_i which is generated by a set of LI vectors contained in (I)–(V). Since $(L-\lambda_i I_V)\,\mathbf{x}^{(i)}_{11} = \boldsymbol{\theta}$

(multiply the first equation of (11.63) by $(L-\lambda_i I_V)$ and use Definition 11.6), the subspace T_{i1} is $(L-\lambda_i I_V)$-cyclic and in view of Theorem 11.18, the matrix of L restricted to T_{i1}, i.e., $L : T_{i1} \to V$ can be represented with respect to the basis (11.63) by

$$J_{i1}(\lambda_i) = \begin{bmatrix} \lambda_i & 1 & \cdots & 0 & 0 \\ 0 & \lambda_i & \cdots & 0 & 0 \\ \vdots & \vdots & & \vdots & \vdots \\ 0 & 0 & \cdots & \lambda_i & 1 \\ 0 & 0 & \cdots & 0 & \lambda_i \end{bmatrix}_{m_i} . \tag{11.64}$$

Continuing in this fashion we observe that the set

$$\left\{ \begin{aligned} \mathbf{x}^{(i)}_{1r_{im_i}} &= (L-\lambda_i I_V)^{m_i-1}\mathbf{x}^{(i)}_{m_i,\, r_{im_i}}, \\ &\vdots \\ \mathbf{x}^{(i)}_{jr_{im_i}} &= (L-\lambda_i I_V)^{m_i-j}\mathbf{x}^{(i)}_{m_i,\, r_{im_i}}, \\ &\vdots \\ \mathbf{x}^{(i)}_{m_i-1,\, r_{im_i}} &= (L-\lambda_i I_V)\ \mathbf{x}^{(i)}_{m_i,\, r_{im_i}},\quad \mathbf{x}^{(i)}_{m_i,\, r_{im_i}} \end{aligned} \right\} \tag{11.65}$$

formed by the r_{im_i}th elements in the list of (I)–(V) (see also Table 11.1), generates the subspace $T_{i,r_{im_i}}$ associated with the eigenvalue λ_i. Since $(L-\lambda_i I_V)\mathbf{x}^{(i)}_{1r_{im_i}} = \boldsymbol{\theta}$, the subspace $T_{i,r_{im_i}}$ is $(L-\lambda_i I_V)$-cyclic and in view of Theorem 11.18, the matrix of L restricted to $T_{i,r_{im_i}}$, i.e., $L : T_{i,r_{im_i}} \to V$ can be represented with respect to the basis (11.65) by $J_{i,r_{im_i}}(\lambda_i)$ which is identical with $J_{i1}(\lambda_i)$ in (11.64). This proves that the matrix representations of L restricted to each of the subspaces $T_{i1}, T_{i2}, \ldots, T_{i,r_{im_i}}$ with respect to the basis elements, respectively in 1, 2, ..., r_{im_i} columns in Table 11.1 are $J_{i1}(\lambda_i), J_{i2}(\lambda_i), \ldots, J_{i,r_{im_i}}(\lambda_i)$, each of which is equal to the Jordan block (11.64).

The subspaces (see Table 11.1),

$$T_{i,r_{i3}+1}, T_{i,r_{i3}+2}, \ldots, T_{i,r_{i2}-1}, T_{i,r_{i2}} \tag{11.66}$$

are generated respectively by

$$\begin{aligned} &\{\mathbf{x}^{(i)}_{1,r_{i3}+1} = (L-\lambda_i I_V)\,\mathbf{x}^{(i)}_{2,r_{i3}+1}, \mathbf{x}^{(i)}_{2,r_{i3}+1}\}, \\ &\{\mathbf{x}^{(i)}_{1,r_{i3}+2} = (L-\lambda_i I_V)\,\mathbf{x}^{(i)}_{2,r_{i3}+2}, \mathbf{x}^{(i)}_{2,r_{i3}+2}\}, \\ &\qquad\vdots \\ &\{\mathbf{x}^{(i)}_{1,r_{i2}-1} = (L-\lambda_i I_V)\,\mathbf{x}^{(i)}_{2,r_{i2}-1}, \mathbf{x}^{(i)}_{2,r_{i3}+1}\}, \\ &\{\mathbf{x}^{(i)}_{1\,r_{i2}} = (L-\lambda_i I_V)\,\mathbf{x}^{(i)}_{2\,r_{i2}}, \mathbf{x}^{(i)}_{2\,r_{i2}}\}, \end{aligned} \tag{11.67}$$

Table 11.1 Crate diagram. The elements in any row are linearly independent elements in $\mathbb{N}_{ij}$ over $\mathbb{N}_{i,j-1}$ and are designated by $\mathbb{N}_{ij} - \mathbb{N}_{i,j-1}$, $j = m_i, m_i - 1, ..., 2, 1$. This representation is called a *crate diagram* in which each independent vector is put as if in a crate.

Column → Row ↓	1	...	r_{im_i}
$\mathbb{N}_{im_i} - \mathbb{N}_{i,m_i-1}$	$\mathbf{x}_{m_i1}^{(i)}$	...	$\mathbf{x}_{m_i,r_{im_i}}^{(i)}$
$\mathbb{N}_{i,m_i-1} - \mathbb{N}_{i,m_i-2}$	$\mathbf{x}_{m_i-1,1}^{(i)} =$ $(L-\lambda_i I_V)\mathbf{x}_{m_i1}^{(i)}$	...	$\mathbf{x}_{m_i-1,r_{im_i}}^{(i)} =$ $(L-\lambda_i I_V)\mathbf{x}_{m_i,r_{im_i}}^{(i)}$
⋮	⋮	⋮	⋮
$\mathbb{N}_{i2} - \mathbb{N}_{i,j-1}$	$\mathbf{x}_{i1}^{(i)} =$ $(L-\lambda_i I_V)^{m_i-j}\mathbf{x}_{m_i1}^{(i)}$	...	$\mathbf{x}_{i,r_{im_i}}^{(i)} =$ $(L-\lambda_i I_V)^{m_i-j}\mathbf{x}_{m_i,r_{im_i}}^{(i)}$
⋮	⋮	⋮	⋮
$\mathbb{N}_{i2} - \mathbb{N}_{i1}$	$\mathbf{x}_{21}^{(i)} =$ $(L-\lambda_i I_V)^{m_i-2}\mathbf{x}_{m_i1}^{(i)}$	...	$\mathbf{x}_{2,r_{im_i}}^{(i)} =$ $(L-\lambda_i I_V)^{m_i-2}\mathbf{x}_{m_i,r_{im_i}}^{(i)}$
$\mathbb{N}_{i1}$	$\mathbf{x}_{11}^{(i)} =$ $(L-\lambda_i I_V)^{m_i-1}\mathbf{x}_{m_i1}^{(i)}$	...	$\mathbf{x}_{1,r_{im_i}}^{(i)} =$ $(L-\lambda_i I_V)^{m_i-1}\mathbf{x}_{m_i,r_{im_i}}^{(i)}$
Subspace	T_{i1}	...	$T_{ir_{im_i}}$
Number of elements	$s_{i1} = m_i$	...	$s_{ir_{im_i}} = m_i$

and are of dimensions

$$s_{i,r_{i3}+1}, s_{i,r_{i3}+2}, ..., s_{i,r_{i2}-1}, s_{i\,r_{i2}}$$

respectively, each equal to 2. Since

$$(L-\lambda_i I_V)\mathbf{x}_{1,r_{i3}+1}^{(i)} = \boldsymbol{\theta}, (L-\lambda_i I_V)\mathbf{x}_{1,r_{i3}+2}^{(i)} = \boldsymbol{\theta}, ..., (L-\lambda_i I_V)\mathbf{x}_{1,r_{i2}-1}^{(i)} = \boldsymbol{\theta}, (L-\lambda_i I_V)\mathbf{x}_{1\,r_{i2}}^{(i)} = \boldsymbol{\theta},$$

the matrix representation of L with respect, to the basis in (11.67) restricted to each of the subspaces in (11.66) is

$$\begin{bmatrix} \lambda_i & 1 \\ 0 & \lambda_i \end{bmatrix} = J_{i,r_{i3}+1}(\lambda_i) = J_{i,r_{i3}+2}(\lambda_i) = \cdots = J_{i,r_{i2}-1}(\lambda_i) = J_{i\,r_{i2}}(\lambda_i).$$

Table 11.1 (Continued)

$r_{im_i}+1$	$\cdots\ r_{i,m_i-1}$	$\cdots\ r_{ij}$
$\mathbf{x}^{(i)}_{m_i-1,\, r_{im_i}+1}$	$\cdots\ \mathbf{x}^{(i)}_{m_i-1,\, r_{i,m_i}-1}$	
$\vdots$	$\cdots\ \vdots$	
$\mathbf{x}^{(i)}_{j,\, r_{i,m_i}+1} =$ $(L-\lambda_i I_V)^{m_i-j-1}\mathbf{x}^{(i)}_{m_i-1,\, r_{im_i}+1}$	$\cdots\ \mathbf{x}^{(i)}_{j,\, r_{i,m_i}-1} =$ $(L-\lambda_i I_V)^{m_i-j-1}\mathbf{x}^{(i)}_{m_i-1,\, r_{i,m_i-1}}$	$\cdots\ \mathbf{x}^{(i)}_{j\, r_{ij}}$
$\vdots$	$\cdots\ \vdots$	$\cdots$
$\mathbf{x}^{(i)}_{2,\, r_{im_i}+1} =$ $(L-\lambda_i I_V)^{m_i-3}\mathbf{x}^{(i)}_{m_i-1,\, r_{im_i}+1}$	$\cdots\ \mathbf{x}^{(i)}_{2,\, r_{i,m_i}-1} =$ $(L-\lambda_i I_V)^{m_i-3}\mathbf{x}^{(i)}_{m_i-1,\, r_{i,m_i-1}}$	$\cdots\ \mathbf{x}^{(i)}_{2\, r_{ij}} =$ $(L-\lambda_i I_V)^{j-2}\mathbf{x}^{(i)}_{j\, r_{ij}}$
$\mathbf{x}^{(i)}_{1,\, r_{im_i}+1} =$ $(L-\lambda_i I_V)^{m_i-2}\mathbf{x}^{(i)}_{m_i-1,\, r_{im_i}+1}$	$\cdots\ \mathbf{x}^{(i)}_{1,\, r_{i,m_i}-1} =$ $(L-\lambda_i I_V)^{m_i-2}\mathbf{x}^{(i)}_{m_i-1,\, r_{i,m_i-1}}$	$\cdots\ \mathbf{x}^{(i)}_{1\, r_{ij}} =$ $(L-\lambda_i I_V)^{j-1}\mathbf{x}^{(i)}_{j\, r_{ij}}$
$T_{i,r_{im_i}+1}$	$\cdots\ T_{i,r_{i,m_i}-1}$	$\cdots\ T_{i\, r_{ij}}$
$s_{i,r_{im_i}+1} = m_i - 1$	$\cdots\ s_{i,r_{i,m_i}-1} = m_i - 1$	$\cdots\ s_{i\, r_{ij}} = j$

Table 11.1 (Continued)

$\cdots$	$r_{i3}+1$	$\cdots$	r_{i2}	$r_{i2}+1$	$\cdots$	r_{i1}
$\cdots$	$\mathbf{x}^{(i)}_{2,r_{i3}+1}$	$\cdots$	$\mathbf{x}^{(i)}_{2,r_{i2}}$			
$\cdots$	$\mathbf{x}^{(i)}_{1,r_{i3}+1} =$ $(L-\lambda_i I_V)\mathbf{x}^{(i)}_{2,r_{i3}+1}$	$\cdots$	$\mathbf{x}^{(i)}_{1 r_{i2}} =$ $(L-\lambda_i I_V)\mathbf{x}^{(i)}_{2,r_{i2}}$	$\mathbf{x}^{(i)}_{1,r_{i2}+1}$	$\cdots$	$\mathbf{x}^{(i)}_{1 r_{i1}}$
$\cdots$	$T_{i,r_{i3}+1}$	$\cdots$	$T_{i\, r_{i2}}$	$T_{i\, r_{i2}+1}$	$\cdots$	$T_{i\, r_{i1}}$
$\cdots$	$s_{i,r_{i3}+1} = 2$	$\cdots$	$s_{i\, r_{i2}} = 2$	$s_{i,r_{i2}+1} = 1$	$\cdots$	$s_{i\, r_{i1}} = 1$

Finally, the subspaces

$$T_{i,r_{i2}+1}, T_{i,r_{i2}+2}, \ldots, T_{i,r_{i1}-1}, T_{ir_{i1}} \tag{11.68}$$

are generated respectively by

$$\{\mathbf{x}^{(i)}_{1,r_{i2}+1}\}, \{\mathbf{x}^{(i)}_{1,r_{i2}+2}\}, \ldots, \{\mathbf{x}^{(i)}_{1,r_{i1}-1}\}, \{\mathbf{x}^{(i)}_{1r_{i1}}\} \tag{11.69}$$

are of dimensions

$$s_{i,r_{i2}+1}, s_{i,r_{i2}+2}, \ldots, s_{i,r_{i1}-1}, s_{ir_{i1}}$$

respectively, each equal to 1. Since $(L-\lambda_i I_V)\mathbf{x}^{(i)}_{1,r_{i2}+1} = \boldsymbol{\theta}$, $(L-\lambda_i I_V)\mathbf{x}^{(i)}_{1,r_{i2}+2} = \boldsymbol{\theta}$, ..., $(L-\lambda_i I_V)\mathbf{x}^{(i)}_{1,r_{i1}-1} = \boldsymbol{\theta}$, $(L-\lambda_i I_V)\mathbf{x}^{(i)}_{1r_{i1}} = \boldsymbol{\theta}$, by virtue of the property of subspace $\mathbb{N}_{i1}$, the matrix representation of L restricted to each of these subspaces in (11.68) with respect, to the basis in (11.69) is

$$J_{i,r_{i2}+1}(\lambda_i) = J_{i,r_{i2}+2}(\lambda_i) = \cdots = J_{i,r_{i1}-1}(\lambda_i) = J_{ir_{i1}}(\lambda_i) = [\lambda_i].$$

Based on the preceding developments, the following theorem is easily established.

Theorem 11.19 *Cyclic Decomposition of a Root Subspace.* Let S_i be the root subspace of a linear transformation L on V over $\mathbb{C}$ associated with the eigenvalue λ_i of multiplicity m_i and n_i in the minimum and the characteristic polynomial respectively of L, $n_i \geq m_i$. Then S_i can be decomposed into a direct sum of r_{i1} number of $(L-\lambda_i I_V)$-cyclic, L-invariant subspaces T_{iq} with dimensions s_{iq}, $q = 1, 2, \ldots, r_{i1}$ such that

$$s_{i1} \geq s_{i2} \geq \cdots \geq s_{ir_{i1}}, \tag{11.70}$$

$$s_{i1} + s_{i2} + \cdots + s_{ir_{i1}} = n_i, \tag{11.71}$$

$$S_i = T_{i1} \oplus T_{i2} \oplus \cdots \oplus T_{ir_{i1}}, \tag{11.72}$$

where $r_{i1} = \dim E_{\lambda_i}$, E_{λ_i} being the eigenspace of L associated with, λ_i, and (i) the first r_{im_i} number of subspaces in (11.72) have each dimension m_i, (ii) the next $r_{i,m_i-1} - r_{im_i}$ number of subspaces have each dimension $m_i - 1$, so on, and (iii) the last $r_{i1} - r_{i2}$ number of subspaces have each dimension 1,

$$r_{ij} = \dim \mathbb{N}_{ij} - \dim \mathbb{N}_{i,j-1}, \qquad j = 1, 2, \ldots, m_i; \tag{11.73}$$

$\mathbb{N}_{ij}$ being the null spaces of the linear transformation $(L-\lambda_i I_V)^j$ and

$$\{\boldsymbol{\theta}\} = \mathbb{N}_{i0} \subset \mathbb{N}_{ij} \subset \cdots \subset \mathbb{N}_{im_i}. \qquad \square$$

It is clear from (11.70), (11.71) and (i)–(iii) that

$$s_{i1} = s_{i2} = \cdots = s_{ir_{im_i}} = m_i; \; s_{i,r_{im_i}+1} = \cdots = s_{ir_{i,m_i-1}} = m_i - 1;$$

$$\cdots s_{i,r_{i2}+1} = \cdots = s_{ir_{i1}} = 1,$$

and if $r_{ij} - r_{i,j+1} = 0$, there is no subspace in S_i, with dimension j ($j = 1, 2, ..., m_i - 1$). Moreover, the dimension of S_i as computed from (i)–(iii) is

$$\dim S_i = m_i r_{im_i} + (m_i - 1)(r_{i,m_i-1} - r_{im_i}) + (m_i - 2)(r_{i,m_i-2} - r_{i,m_i-1}) + \cdots + 1 \cdot (r_{i1} - r_{i2}) = r_{im_i} + r_{i,m_i-1} + \cdots + r_{i1} = n_i.$$

If we premultiply each element in row $\mathbb{N}_{i1}$ of Table 11.1 by $(L - \lambda_i I_V)$ and use Definition 11.9, we have the relation

$$(L - \lambda_i I_V)\mathbf{x}_{1q}^{(i)} = (L - \lambda_i I_V)^j \mathbf{x}_{jq}^{(i)} = \boldsymbol{\theta}, \quad j = m_i, m_i - 1, ..., 2, 1,$$
$$q = r_{i,j+1} + 1, ..., r_{ij}; r_{i,m_i+1} := 0, \tag{11.74}$$

where $\mathbf{x}_{jq}^{(i)} \in \mathbb{N}_{ij}$. So each column of Table 11.1 represents a $(L - \lambda_i I_V)$-cyclic subspace of S_i which is also $(L - \lambda_i I_V)$-invariant. Moreover, since

$$r_{im_i} \le r_{i,m_i-1} \le \cdots \le r_{i2} \le r_{i1},$$

the root subspace S_i can be decomposed as the direct sum of r_{i1} number of $(L - \lambda_i I_V)$-cyclic subspaces T_{iq} defined by

$$T_{iq} = \{\mathbf{x}_{1q}^{(i)} = (L - \lambda_i I_V)^{j-1}\mathbf{x}_{jq}^{(i)}, ..., \mathbf{x}_{j-1,q}^{(i)} = (L - \lambda_i I_V)\mathbf{x}_{jq}^{(i)}, \mathbf{x}_{jq}^{(i)};\}, \tag{11.75}$$
$$j = m_i, m_i - 1, ..., 2, 1; \quad q = 1 + r_{i,j+1}, ..., r_{ij}; \quad r_{i,m_i+1} := 0;$$

Table 11.2 Crate diagram for Example 11.11

Column → Row ↓	1	$r_{14} = r_{13} = 2$	$r_{12} = 3$	4	$r_1 = 5$
$\mathbb{N}_{14} - \mathbb{N}_{13}$	$\mathbf{x}_{41}^{(1)}$	$\mathbf{x}_{42}^{(1)}$			
$\mathbb{N}_{13} - \mathbb{N}_{12}$	$(L - \lambda_1 I_V)\mathbf{x}_{41}^{(1)}$	$(L - \lambda_1 I_V)\mathbf{x}_{42}^{(1)}$			
$\mathbb{N}_{12} - \mathbb{N}_{11}$	$(L - \lambda_1 I_V)^2\mathbf{x}_{41}^{(1)}$	$(L - \lambda_1 I_V)^2\mathbf{x}_{42}^{(1)}$	$\mathbf{x}_{23}^{(1)}$		
$\mathbb{N}_{11}$	$(L - \lambda_1 I_V)^3\mathbf{x}_{41}^{(1)}$	$(L - \lambda_1 I_V)^3\mathbf{x}_{42}^{(1)}$	$(L - \lambda_1 I_V)\mathbf{x}_{23}^{(1)}$	$\mathbf{x}_{14}^{(1)}$	$\mathbf{x}_{15}^{(1)}$
Subspace	T_{11}	T_{12}	T_{13}	T_{14}	T_{15}
Number of elements	$s_{11} = 4$	$s_{12} = 4$	$s_{13} = 2$	$s_{14} = 1$	$s_{15} = 1$

each with a dimension s_{iq}, so that (11.70) to (11.72) are satisfied. The subspace T_{iq}, $q = 1, 2, ..., r_{im_i}, ..., r_{i2}, ..., r_{i1}$ are generated by the LI vectors occupying the qth column of the crate diagram in Table 11.1. Each $(L - \lambda_1 I_V)$ cyclic subspace T_{iq} in S_i is also L-invariant. Indeed, with the help of (11.74) and (11.75), we have

$$L\mathbf{x}_{jq}^{(i)} = \mathbf{x}_{j-1,q}^{(i)} + \lambda_i \mathbf{x}_{jq}^{(i)}, L\mathbf{x}_{j-1,q}^{(i)} = \mathbf{x}_{j-2,q}^{(i)} + \lambda_i \mathbf{x}_{j-1,q}^{(i)}, ..., L\mathbf{x}_{1q}^{(i)} = \lambda_i \mathbf{x}_{1q}^{(i)}, \tag{11.76}$$

which proves the assertion that T_{iq} is L-invariant.

Example 11.11 Let λ_i be the only eigenvalue of L on a 12-dimensional real vector space V and let $m_1 = 4$ be the multiplicity of λ_1 in its minimum polynomial; $n_1 = 12$, that in its characteristic polynomial, dim $\mathbb{N}_{11} = 5$, dim $\mathbb{N}_{12} = 8$, dim $\mathbb{N}_{13} = 10$, and dim $\mathbb{N}_{14} = 12$. Hence, by (11.73)

$$r_{11} = 5, \quad r_{12} = 3, \quad r_{13} = 2, \quad r_{14} = 2,$$

and the crate diagram appears as in Table 11.2.

The elements in the jth column are LI and generate the $(L - \lambda_i I_V)$-cyclic subspace T_{1j}, $j = 1, 2, ..., 5$ of dimension 4, 4, 2, 1 and 1 respectively. Since $r_{13} = r_{14}$, there is no L-invariant subspace of dimension 3 (see Theorem 11.19), and the root subspace S_1 is the direct sum of these subspaces T_{1j}s:

$$S_1 = \mathbb{R}^{12} = T_{11} \oplus T_{12} \oplus T_{13} \oplus T_{14} \oplus T_{15}.$$

As $\mathbf{x}_{41}^{(1)} \in \mathbb{N}_{14}$, we have

$$(L - \lambda_1 I_V)^4 \mathbf{x}_{41}^{(1)} = \boldsymbol{\theta}.$$

So, from Column 1 and row $\mathbb{N}_{11}$ we have $(L - \lambda_1 I_V)^4 \mathbf{x}_{11}^{(1)} = \boldsymbol{\theta}$. Hence, with other elements in Column 1 we can write (see Theorem 11.18),

$$L\mathbf{x}_{11}^{(1)} = \lambda_1 \mathbf{x}_{11}^{(1)}, \qquad L\mathbf{x}_{21}^{(1)} = \mathbf{x}_{11}^{(1)} + \lambda_1 \mathbf{x}_{21}^{(1)},$$

$$L\mathbf{x}_{31}^{(1)} = \mathbf{x}_{21}^{(1)} + \lambda_1 \mathbf{x}_{31}^{(1)}, \qquad L\mathbf{x}_{41}^{(1)} = \mathbf{x}_{31}^{(1)} + \lambda_1 \mathbf{x}_{41}^{(1)}.$$

Consequently, the matrix representation of L with respect to

$$\{\mathbf{x}_{11}^{(1)}, \mathbf{x}_{21}^{(1)}, \mathbf{x}_{31}^{(1)}, \mathbf{x}_{41}^{(1)}\},$$

when restricted to the subspace T_{11}, is the Jordan block

$$J_{11}(\lambda_1) = \begin{bmatrix} \lambda_1 & 1 & 0 & 0 \\ 0 & \lambda_1 & 1 & 0 \\ 0 & 0 & \lambda_1 & 1 \\ 0 & 0 & 0 & \lambda_1 \end{bmatrix}.$$

Similarly, the matrix representation of L, restricted to T_{12}, with respect to basis $\{\mathbf{x}_{12}^{(1)}, \mathbf{x}_{22}^{(1)}, \mathbf{x}_{32}^{(1)}, \mathbf{x}_{42}^{(1)}\}$, is $J_{12}(\lambda_1) = J_{11}(\lambda_1)$; and those restricted to T_{13}, T_{14}, and T_{15} with respect to bases $\{\mathbf{x}_{13}^{(1)}, \mathbf{x}_{23}^{(1)}\}$, $\{\mathbf{x}_{14}^{(1)}\}$, and $\{\mathbf{x}_{15}^{(1)}\}$, are respectively the Jordan blocks

$$J_{13}(\lambda_1) = \begin{bmatrix} \lambda_1 & 1 \\ 0 & \lambda_1 \end{bmatrix}, \qquad J_{14}(\lambda_1) = J_{15}(\lambda_1) = [\lambda_1].$$

By virtue of Theorem 11.10, as the subspaces T_{11}, ..., T_{15} are L-invariant, the matrix representation of L on $\mathbb{R}^{12}$ is the Jordan matrix

$$J(\lambda_1) = \text{diag}\,[J_{11}(\lambda_1), J_{12}(\lambda_1), J_{13}(\lambda_1), J_{14}(\lambda_1), J_{15}(\lambda_1)].$$

□

It is important to note that to decompose the root subspaee S_i of L associated with its eigenvalue λ_i, it is not necessary to evaluate the minimum polynomial of L. If we keep on computing the null spaces of $(L - \lambda_i I_V)^j$, then the least integer j for which

$$\mathbb{N}(L-\lambda_i I_V)^j = \mathbb{N}(L-\lambda_i I_V)^{j+1},$$

i.e. rank $(L-\lambda_i I_V)^j$ = rank $(L-\lambda_i I_V)^{j+1}$, gives $j = m_i$, the multiplicity m_i of λ_i in the minimum polynomial of L in view of Lemma 11.3. Nevertheless, we do need to compute the distinct eigenvalues of L.

Example 11.12 Let $A:\mathbb{R}^4 \to \mathbb{R}^4$ be given as in Example 11.9. Hence, associated with the eigenvalue $\lambda_1 = 1$ we have the null spaces of $(L-\lambda_1 I_V)^j$, $j = 1, 2, ...,$ given by

$$\mathbb{N}_{11} = \{\mathbf{x} : x_2 = x_4 = 0\}, \qquad \mathbb{N}_{12} = \{\mathbf{x} : x_4 = 0\} = \mathbb{N}_{13}.$$

The basis of $\mathbb{N}_{11}$ is $\{\mathbf{e}_1, \mathbf{e}_3\}$ and that of $\mathbb{N}_{12}$ is $\{\mathbf{e}_1, \mathbf{e}_2, \mathbf{e}_3\}$ in $\mathbb{R}_4$. Therefore,

$$m_{11} = 2, \quad r_{11} = \dim \mathbb{N}_{11} = 2, \quad r_{12} = \dim \mathbb{N}_{12} - \dim \mathbb{N}_{11} = 3 - 2 = 1.$$

The crate diagram of LI vectors of root subspace S_1 of L associated with, $\lambda_1 = 1$ is as given in Table 11.3.

Thus, T_{11} is generated by $\{\mathbf{e}_1, \mathbf{e}_2\}$, T_{12} by $\{\mathbf{e}_3\}$, and $S_1 = T_{11} \oplus T_{12}$. Since from Example 11.9, S_2 is generated by $\{\mathbf{e}_4\}$, we have, finally,

$$\mathbb{R}^4 = S_1 \oplus S_2 = T_{11} \oplus T_{12} \oplus S_2.$$ □

Table 11.3 Crate diagram for Example 11.12 for $\lambda_1 = 1$

Column → Row ↓	$r_{12} = 1$	$r_{11} = 2$
$\mathbb{N}_{12} - \mathbb{N}_{11}$	$\mathbf{e}_2$	
$\mathbb{N}_{11}$	$\mathbf{e}_1 = (A - I)\,\mathbf{e}_2$	$\mathbf{e}_3$
Subspace	T_{11}	T_{12}

Theorem 11.20 Let S_i be the root subspace of a linear transformation L on V over $\mathbb{C}$ associated with eigenvalue λ_i. Then there exists a basis with respect to which the matrix of L restricted to S_i has the *quasi-diagonal form*

$$J_i(\lambda_i) = \text{diag}\,[J_{i1}, J_{i2}, \ldots, J_{ir_{i1}}]_{n_i}, \tag{11.77}$$

where J_{iq} is a Jordan block (see (10.43)) having the form

$$J_{iq}(\lambda_i) = \begin{bmatrix} \lambda_i & 1 & \cdots & 0 & 0 \\ 0 & \lambda_i & \cdots & 1 & 0 \\ \vdots & \vdots & \ddots & \vdots & \vdots \\ 0 & 0 & \cdots & \lambda_i & 1 \\ 0 & 0 & \cdots & 0 & \lambda_i \end{bmatrix}_{s_{iq}} \tag{11.78}$$

of order s_{iq}, $q = 1, 2, ..., r_{im_i}, ..., r_{i2}, ..., r_{i1}$ associated with λ_i, and

$$r_{i1} = \dim \mathbb{N}(\lambda_i I_V - L), \qquad r_{iq} = \dim \mathbb{N}_{iq} - \dim \mathbb{N}_{i,q-1}, \qquad \mathbb{N}_{iq} = \mathbb{N}(\lambda_i I_V - L)^q;$$

$$s_{i1} = s_{i2} = \cdots = s_{ir_{im_i}} = m_i, \qquad s_{i,r_{im_i}+1} = \cdots = s_{i,r_{i,m_i-1}} = m_i - 1, \ldots,$$

$$s_{i,\,r_{i3}+1} = \cdots = s_{ir_{i2}} = 2, \qquad s_{i,r_{i2}+1} = \cdots = s_{ir_{i1}} = 1,$$

$$s_{i1} \geq s_{i2} \geq \cdots \geq s_{ir_{i1}}, \quad s_{i1} + s_{i2} + \cdots + s_{ir_{i1}} = n_i;$$

m_i and n_i are multiplicities of λ_i in the minimum polynomial and the characteristic polynomial of L respectively.

Proof In view of Theorem 11.19, S_i can be decomposed into a direct sum of r_{i1} number of $(L - \lambda_i I_V)$-cyclic L-invariant subspaces T_{iq}:

$$S_i = T_{i1} \oplus T_{i2} \oplus \cdots \oplus T_{ir_{i1}}.$$

We choose the set of vectors (11.75) as the basis of subspace T_{iq} and the linear transformation L restricted to T_{iq} has the matrix representation with respect to this basis given by (11.78) (see also (11.76)), in view of Theorem 11.18. The quasi-diagonal form in (11.77) follows from an application of Theorem 11.10. □

Theorem 11.21 Let L be a linear transformation on an n-dimensional vector space V over $\mathbb{C}$ with k distinct eigenvalues λ_i of multiplicity n_i and m_i in its charactenstic and minimum polynomials $\phi(\lambda)$ and $\psi(\lambda)$ respectively, for $i = 1, 2, ..., k$. Then, there exists a basis in V with respect to which the matrix of L has the form

$$J = \operatorname{diag}[J_1, J_2, ..., J_k]_n, \tag{11.79}$$

where

$$J_i = \operatorname{diag}[J_{i1}, J_{i2}, ..., J_{ir_{i1}}]_{n_i}, \qquad i = 1, 2, ..., k; \tag{11.80}$$

$$J_{ij}(\lambda_i) = \begin{bmatrix} \lambda_i & 1 & \cdots & 0 & 0 \\ 0 & \lambda_i & \cdots & 1 & 0 \\ \vdots & \vdots & \ddots & \vdots & \vdots \\ 0 & 0 & \cdots & \lambda_i & 1 \\ 0 & 0 & \cdots & 0 & \lambda_i \end{bmatrix}_{s_{i\,r_{ij}}}, \qquad j = 1, 2, ..., r_{i1}; \tag{11.81}$$

$$r_{i1} = \dim \mathbb{N}(\lambda_i I_V - L),$$

$$\mathbb{N}_{it} = \mathbb{N}(\lambda_i I_V - L)^t, \qquad t = 1, 2, ..., m_i; \tag{11.82}$$

$$r_{it} = \dim \mathbb{N}_{it} - \dim \mathbb{N}_{i,t-1}, \tag{11.83}$$

$$s_{i1} = s_{i2} = \cdots = s_{ir_{im_i}} = m_i,$$

$$s_{i,r_{im_i}+1} = \cdots = s_{i,r_{i,m_i-1}} = m_i - 1, \ldots, s_{i,r_{i2}+1} = \cdots = s_{ir_{i1}} = 1, \tag{11.84}$$

$$s_{i1} \geq S_{i2} \geq \cdots \geq s_{ir_{i1}}, \qquad r_{i1} \geq r_{i2} \geq \cdots \geq r_{im_{i1}};$$

$$n_1 + n_2 + \cdots + n_k = n, \qquad s_{i1} + s_{i2} + \cdots + s_{ir_{i1}} = n_i.$$

Proof Let S_i be the root subspace of L associated with the eigenvalue λ_i. Then, in view of Theorems 11.10, 11.18 and 11.19, we have

$$V = S_1 \oplus S_2 \oplus \cdots \oplus S_k,$$

where

$$S_i = T_{i1} \oplus T_{i2} \oplus \cdots \oplus T_{ir_{i1}}, \qquad i = 1, 2, \ldots, k; \tag{11.85}$$

and the basis of T_{ij} is represented by the vectors in the jth column of Table 11.1 arranged in order counted from the bottom as in (11.75). The root subspares S_i are L-invariant and the subspaces T_{ij}s are $(L - \lambda_i I_V)$-cyclic and also L-invariant. Following Theorems 11.10 and 11.20, if we choose the union of vectors in Table 11.1, or as indicated in (11.75) for $i = 1, 2, \ldots, k$, as a basis for the vector space V, then we have *Jordan form* J as the matrix of L which is stipulated in the theorem. □

Example 11.13 Determine the Jordan form of the linear transformation $L : \mathbb{R}^5 \to \mathbb{R}^5$ defined by

$$\mathbf{L}\mathbf{x} = [x_1 + x_4 - x_5, -4x_1 + x_2 - 2x_3, -4x_1 - x_3, -x_2 + x_3 + x_5,$$
$$-x_2 + x_3 - x_4 + 2x_5]^T, \qquad \text{where } \mathbf{x} = [x_1\ x_2\ x_3\ x_4\ x_5]^T. \tag{11.86}$$

Solution With respect to the standard basis $\{\mathbf{e}_1\ \mathbf{e}_2\ \mathbf{e}_3\ \mathbf{e}_4\ \mathbf{e}_5\}$ in $\mathbb{R}^5$, the matrix A of L is (see Example 6.17)

$$A = \begin{bmatrix} 1 & 0 & 0 & 1 & -1 \\ -4 & 1 & -2 & 0 & 0 \\ -4 & 0 & -1 & 0 & 0 \\ 0 & -1 & 1 & 0 & 1 \\ 0 & -1 & 1 & -1 & 2 \end{bmatrix}.$$

This gives the characteristic polynomial of L as (see Section 7.1.1)

$$\phi(\lambda) = \det(\lambda I - A) = (\lambda + 1)(\lambda - 1)^4$$

and its eigenvalues $\lambda_1 = -1$ and $\lambda_2 = 1$ of multiplicities $n_1 = 1$ and $n_2 = 4$ respectively. From (11.79), the Jordan form of L will therefore have the appearance

$$A = \begin{bmatrix} J_1 & \\ & J_2 \end{bmatrix},$$

when J_1 is of order $n_1 = 1$ and J_2 is of order $n_2 = 4$.

(a) Considering the eigenvalue $\lambda_1 = -1$, and in view of (11.28) which holds even if m_i is an arbitrary integer and Theorem 6.5, we have

$$r_{11} = \dim \mathbb{N}(-I_V - L) = 5 - \text{rank}(-I_V - L) = 5 - \text{rank}(-I_V - A) = 5 - 4 = 1,$$

where $V = \mathbb{R}^5$ and

$$(I + A) = \begin{bmatrix} 2 & 0 & 0 & 1 & -1 \\ -4 & 2 & -2 & 0 & 0 \\ -4 & 0 & 0 & 0 & 0 \\ 0 & -1 & 1 & 1 & 1 \\ 0 & -1 & 1 & -1 & 3 \end{bmatrix}. \tag{11.87}$$

Therefore, by (11.80) $J_1 = J_{11}$ which is of order s_{11}. Since $1 \le m_1 \le n_1 = 1$, we have $m_1 = 1$ and as a result, $s_{11} = m_1 = 1$ by (11.84). Hence, by (11.81).

$$J_1 = J_{11} = [-1].$$

(b) We now consider the eigenvalue $\lambda_2 = 1$ and evaluate successively $(\lambda_2 I - A)$ and $(\lambda_2 I - A)^2$ to get

$$(I - A) = \begin{bmatrix} 0 & 0 & 0 & -1 & 1 \\ 4 & 0 & 2 & 0 & 0 \\ 4 & 0 & 2 & 0 & 0 \\ 0 & 1 & -1 & 1 & -1 \\ 0 & 1 & -1 & 1 & -1 \end{bmatrix},$$

$$(I - A)^2 = \begin{bmatrix} 0 & 0 & 0 & 0 & 0 \\ 8 & 0 & 4 & -4 & 4 \\ 8 & 0 & 4 & -4 & 4 \\ 0 & 0 & 0 & 0 & 0 \\ 0 & 0 & 0 & 0 & 0 \end{bmatrix}. \tag{11.88}$$

Therefore,

$$r_{21} = \dim \mathbb{N}(I_V - L) = 5 - \text{rank}(I - A) = 5 - 3 = 2. \tag{11.89}$$

Since, $\dim \mathbb{N}(I_V - L)^2 = 5 - \text{rank}(I - A)^2 = 5 - 1 = 4$, by (11.83), we have $r_{22} = \dim \mathbb{N}(I_V - L)^2 - r_{21} = 4 - 2 = 2$. Therefore, by Theorem 11.13 and Lemma 11.3, since $n_2 = 4$, we have $\text{rank}(I - A)^2 = \text{rank}(I - A)^3$ and, consequently, $m_2 = 2$. In view of (11.81)

$$J_2 = \begin{bmatrix} J_{21} & \\ & J_{22} \end{bmatrix}_4,$$

where by (11.81) the sub-blocks J_{21} and J_{22} are respectively of order s_{21} and s_{22} which, with the help of (11.84), are given by $s_{21} = s_{22} = 2$. By (11.81)

$$J_{21} = J_{22} = \begin{bmatrix} 1 & 1 \\ 0 & 1 \end{bmatrix}.$$

Finally, therefore, Jordan normal form of L is

$$J = \begin{bmatrix} J_1 & \\ & J_2 \end{bmatrix} = \begin{bmatrix} J_{11} & & \\ & J_{21} & \\ & & J_{22} \end{bmatrix} = \begin{bmatrix} -1 & & & & \\ & 1 & 1 & & \\ & 0 & 1 & & \\ & & & 1 & 1 \\ & & & 0 & 1 \end{bmatrix}. \quad (11.90)$$

□

In the above example, we have computed Jordan normal form of a linear transformation L which is, in fact, based on the computation of r_{it}'s and s_{ij}'s defined by (11.83) and (11.84) respectively. Having determined r_{it}'s, we can form the crate diagram shown in Table 11.4, where number of elements in the first row $r_{21} = 2$ and those in the second row $r_{22} = 2$.

Table 11.4 Crate diagram for Example 11.13 if $\lambda_2 = 1$

$\mathbf{x}_{21}$	$\mathbf{x}_{22}$
$(L - I_V)\mathbf{x}_{21}$	$(L - I_V)\mathbf{x}_{22}$
T_{21}	T_{22}

Hence, s_{21} is computed as the number of elements in the first column and s_{22} that in the second column. If one is interested in computing the basis with respect to which the above Jordan form is defined, one needs to solve (11.82) to determine the basis vectors in the null spaces $\mathbb{N}_{it}$. This will be illustrated in the following example.

Example 11.14 Determine the basis in terms of which the linear transformation (11.86) has Jordan's normal form given by (11.90).

Solution **(a)** Since $\lambda_1 = -1$ is of multiplicity $n_1 = 1$, the root subspace S_1 is of dimension 1 and in (11.85), $S_1 = T_{11} = \mathbb{N}_{11}$, where by Definition 11.9, $\mathbb{N}_{11}$ is the null space of $(\lambda_1 I_V - L) = (-I_V - L)$. Now, solving $(-I - A)\mathbf{u} = \boldsymbol{\theta}$ for $\mathbf{u}$, where $(I + A)$ is given by (11.87), we have

$$\mathbf{u} = [0\ \alpha\ \alpha\ 0\ 0]^T, \qquad \alpha \text{ real.} \quad (11.91)$$

By (11.29), $\mathbf{x}_{11} = \mathbf{u}$ is a basis of T_{11}.

(b) We now consider $\lambda_2 = 1$ which, as shown in Example 11.13, is of multiplicity $n_2 = 4$, index $m_2 = 2$, and from (11.89), $r_{21} = 2$. Hence, the root subspace S_2 in (11.85) is of dimension 4 and can be decomposed into $S_2 = T_{21} \oplus T_{22}$.

Taking cognizance of (11.88), by solving

$$(I - A)\mathbf{u} = \boldsymbol{\theta}, \qquad (I - A)^2\mathbf{u} = \boldsymbol{\theta},$$

we get respectively

$$\mathbf{u} = [\alpha, -2\alpha, -2\alpha, \beta, \beta]^T, \qquad \alpha, \beta \text{ real}; \tag{11.92}$$

$$\mathbf{u} = [-b + c - d, a, 2b, 2c, 2d]^T, \qquad a, b, c, d \text{ real}. \tag{11.93}$$

Since, A is the matrix of L with respect to the natural basis $\{\mathbf{e}_1, \mathbf{e}_2, \mathbf{e}_3, \mathbf{e}_4, \mathbf{e}_5\}$, by (11.29) we can determine from (11.92) that the set

$$\mathbb{N}_{21} : \{[1, -2, -2, 0, 0]^T, [0, 0, 0, 1, 1]^T\}$$

form a basis of $\mathbb{N}_{21}$, and from (11.93), we have

$$\mathbb{N}_{22} : \{[0, 1, 0, 0, 0]^T, [-1, 0, 2, 0, 0]^T, [1, 0, 0, 2, 0]^T, [-1, 0, 0, 0, 2]^T\}$$

as a basis of $\mathbb{N}_{22}$. Now we apply Gaussian elimination method of Section 5.8.1 to determine the number of LI vectors in $\mathbb{N}_{22}$ over $\mathbb{N}_{21}$. By column equivalence,

$$\left[\begin{array}{cc|cccc} 1 & 0 & 0 & -1 & 1 & -1 \\ -2 & 0 & 1 & 0 & 0 & 0 \\ -2 & 0 & 0 & 2 & 0 & 0 \\ 0 & 1 & 0 & 0 & 2 & 0 \\ 0 & 1 & 0 & 0 & 0 & 2 \end{array}\right] \sim \left[\begin{array}{cc|cccc} 1 & 0 & 0 & 0 & 0 & 0 \\ -2 & 0 & 1 & 0 & 0 & 0 \\ -2 & 0 & 0 & 0 & 2 & 0 \\ 0 & 1 & 0 & 0 & 0 & 0 \\ 0 & 1 & 0 & 0 & -2 & 0 \end{array}\right]$$

from which, see Table 11.4, we have

$$\mathbf{x}_{21} = [0, 1, 0, 0, 0]^T, \qquad \mathbf{x}_{22} = [1, 0, 0, 2, 0]^T \tag{11.94}$$

as the set of vectors in $\mathbb{N}_{22}$ which are LI of $\mathbb{N}_{21}$. Since, there are two vectors in the basis of $\mathbb{N}_{21}$, they can be chosen as (see Table 11.4),

$$\begin{aligned} (L - I_V)\mathbf{x}_{21} &= (A - I_V)\mathbf{x}_{21} = [0, 0, 0, -1, -1]^T, \\ (L - I_V)\mathbf{x}_{22} &= (A - I_V)\mathbf{x}_{21} = [2, -4, -4, -2, -2]^T. \end{aligned} \tag{11.95}$$

Taking into consideration (11.91) for $\alpha = 1$, (11.94) and (11.95) and Table 11.4, we choose the set

$$\begin{aligned} &\mathbf{x}_{11} = [0, 1, 1, 0, 0]^T, \\ &B\mathbf{x}_{21} = [0, 0, 0, -1, -1]^T, \qquad \mathbf{x}_{21} = [0, 1, 0, 0, 0]^T, \\ &B\mathbf{x}_{22} = [2, -4, -4, -2, -2]^T, \qquad \mathbf{x}_{22} = [1, 0, 0, 2, 0]^T, \end{aligned} \tag{11.96}$$

where $B = (A - I)$, a basis in $\mathbb{R}^5$, in terms of which the normal form of (11.86) is given by (11.90). Indeed, setting

$$T = [\mathbf{x}_{11} \vdots B\mathbf{x}_{21}, \mathbf{x}_{21} \vdots B\mathbf{x}_{22}, \mathbf{x}_{22}],$$

in view of Remark 6.4, we can write

$$L[\mathbf{x}_{11} \vdots B\mathbf{x}_{21}, \mathbf{x}_{21} \vdots B\mathbf{x}_{22}, \mathbf{x}_{22}] = AT = [\mathbf{x}_{11} \vdots B\mathbf{x}_{21}, \mathbf{x}_{21} \vdots B\mathbf{x}_{22}, \mathbf{x}_{22}]T^{-1}AT,$$

where

$$T^{-1}AT = \begin{bmatrix} 2 & 0 & 1 & -1 & 1 \\ -1 & 0 & 0 & \frac{1}{2} & -\frac{3}{2} \\ 0 & 1 & -1 & 0 & 0 \\ \frac{1}{2} & 0 & 0 & -\frac{1}{4} & \frac{1}{4} \\ 0 & 0 & 0 & \frac{1}{2} & -\frac{1}{2} \end{bmatrix} \begin{bmatrix} 1 & 0 & 0 & 1 & -1 \\ -4 & 1 & -2 & 0 & 0 \\ -4 & 0 & -1 & 0 & 0 \\ 0 & -1 & 1 & 0 & 1 \\ 0 & -1 & 1 & -1 & 2 \end{bmatrix}$$

$$\times \begin{bmatrix} 0 & 0 & 0 & 2 & 1 \\ 1 & 0 & 1 & -4 & 0 \\ 1 & 0 & 0 & -4 & 0 \\ 0 & -1 & 0 & -2 & 2 \\ 0 & -1 & 0 & -2 & 0 \end{bmatrix} = \begin{bmatrix} -1 & & & & \\ & 1 & 1 & & \\ & 0 & 1 & & \\ & & & 1 & 1 \\ & & & 0 & 1 \end{bmatrix}$$

is the matrix of L in Jordan normal form with respect to the basis in (11.96). Denoting the Jordan matrix by J, we have $A = TJT^{-1}$. Decomposing J as

$$J = \begin{bmatrix} -1 & 0 & 0 & 0 & 0 \\ 0 & 1 & 0 & 0 & 0 \\ 0 & 0 & 1 & 0 & 0 \\ 0 & 0 & 0 & 1 & 0 \\ 0 & 0 & 0 & 0 & 1 \end{bmatrix} + \begin{bmatrix} 0 & 0 & 0 & 0 & 0 \\ 0 & 0 & 1 & 0 & 0 \\ 0 & 0 & 0 & 0 & 0 \\ 0 & 0 & 0 & 0 & 1 \\ 0 & 0 & 0 & 0 & 0 \end{bmatrix} = D + E,$$

where D is a diagonal matrix with diagonal elements as all the eigenvalues of A counting multiplicities and E is block diagonal, with each diagonal block having two superdiagonal elements equal to 1 but other elements are equal to 0. It, is easy to check that $DE = ED$. With this representation of J, A can be split into two parts S and $N : A = S + N = T(D + E)T^{-1}$, where

$$S := TDT^{-1} = \begin{bmatrix} 1 & 0 & 0 & 0 & 0 \\ -4 & 1 & -2 & 2 & -2 \\ -4 & 0 & -1 & 2 & -2 \\ 0 & 0 & 0 & 1 & 0 \\ 0 & 0 & 0 & 0 & 1 \end{bmatrix}, \qquad N := TET^{-1} = \begin{bmatrix} 0 & 0 & 0 & 1 & -1 \\ 0 & 0 & 0 & -2 & 2 \\ 0 & 0 & 0 & -2 & 2 \\ 0 & -1 & 1 & -1 & 1 \\ 0 & -1 & 1 & -1 & 1 \end{bmatrix},$$

the latter is nilpotent of index 2 (see Problem 1.34). A simpler way of computing N is $N = A - S$. Since $DE = ED$, it follows that $SN = NS$. □

A matrix A of order n is looked upon as a linear transformation (LT) $A : \mathbb{C}^n \rightarrow \mathbb{C}^n$, so finding the Jordan matrix representation of L in this case becomes : given a matrix A determine a set of basis vectors with respect to which the representation of A is in Jordan form J. In other words, find a nonsingular matrix T with these basis vectors as columns such that $T^{-1}AT = J$. We shall restate the results of Jordan normal form developed for a LT in case the LT is a matrix in an easy-to-understand set-up in what follows.

Definition 11.10 ***Generalized Eigenvectors and Generalized Eigenspace of a Square Matrix A.*** Let A be an n-square matrix over $\mathbb{C}$. Then a vector $\mathbf{x}$ in $\mathbb{C}^n$ is called a generalized eigenvector of order r of matrix A corresponding to the eigenvalue λ if

$$(\lambda I - A)^r \mathbf{x} = \boldsymbol{\theta}, \quad \text{but} \quad (\lambda I - A)^{r-1} \mathbf{x} \neq \boldsymbol{\theta}.$$

If $r = 1$, then the above definition reduces to $(\lambda I - A)\mathbf{x} = \boldsymbol{\theta}$, but $\mathbf{x} \neq \boldsymbol{\theta}$, which is the definition of an eigenvector. □

Let $\mathbf{x}$ be a generalized eigenvector of order r associated with the eigenvalue λ. The set of vectors defined by $\mathbf{x}_r := \mathbf{x}$ and $\mathbf{x}_i := (A - \lambda I)^{r-i}\mathbf{x}_r, i = r, \ldots, 2, 1,$ i.e.

$$\begin{aligned}
\mathbf{x}_r &:= \mathbf{x}, \\
\mathbf{x}_{r-1} &:= (A - \lambda I)\mathbf{x}_r, \\
\mathbf{x}_{r-2} &:= (A - \lambda I)^2\mathbf{x}_r, \\
&\vdots \\
\mathbf{x}_1 &:= (A - \lambda I)^{r-1}\mathbf{x}_r,
\end{aligned}$$

is LI in view of Corollary 11.7. We note that

$$(A - \lambda I)^i \mathbf{x}_i := (A - \lambda I)^i (A - \lambda I)^{r-i} \mathbf{x}_r = (A - \lambda I)^r \mathbf{x}_r = \boldsymbol{\theta}.$$

However,

$$(A - \lambda I)^{i-1} \mathbf{x}_i := (A - \lambda I)^{i-1} (A - \lambda I)^{r-i} \mathbf{x}_r = (A - \lambda I)^r \mathbf{x}_r \neq \boldsymbol{\theta}.$$

This implies that $\mathbf{x}_i$, $1 \le i \le r$ is a generalized eigenvector of order i corresponding to λ. The set of vectors

$$\mathbf{x}_1, \mathbf{x}_2, \ldots, \mathbf{x}_{r-1}, \mathbf{x}_r$$

is called a *chain of generalized eigenvectors* corresponding to the eigenvalue λ, and is LI. There may be more than one generalized eigenvector and consequently more than one chain of generalized eigenvectors of order r. The set of all generalized vectors, all of which are LI, plus the null vector $\boldsymbol{\theta}$ is said to form the *generalized eigenspace of A corresponding to* λ. This definition is applicable to both real and complex eigenvalues.

Let A be a matrix of order n having only real eigenvalues. First, compute all the chains of its generalized eigenvectors according to a prescribed order of real eigenvalues and set them as columns of a matrix T. It can be shown that this matrix is nonsingular and $T^{-1}AT = J_{real}$ is in Jordan canonical form. Now, consider a matrix A of order $2m$ having all complex eigenvalues. Compute all the chains of generalized eigenvectors of A as in the real case. If $\mathbf{u}_r + i\mathbf{v}_r$ is an eigenvector in the set of generalized eigenvectors corresponging to an eigenvalue $\lambda_r = a_r + ib_r$, $r = 1, 2 \ldots, m,$ constitute a matrix T of order $2m$ with successive columns $\mathbf{u}_r$ and $\mathbf{v}_r$, $r = 1, 2, \ldots, m$. Then, $T^{-1}AT = J_{\text{comp}}$ is in Jordan canonical form as in J_{real} except that the diagonal elements are now of the form

$$W_r = \begin{bmatrix} a_r & b_r \\ -b_r & a_r \end{bmatrix}$$

and ones in the super- or sub-diagonals are replaced with an identity matrix of order 2. If the eigenvalues are a combination in the order of real and complex numbers, Jordan matrix representation will be a concatenation of J_{real} and J_{comp} in the form diag $[J_{\text{real}}, J_{\text{comp}}]$.

Theorem 11.22 *Jordan Form of A with all real Eigenvalues.* Let A be a real matrix of order n with k number of distinct real eigenvalues $\lambda_1, \lambda_2, ..., \lambda_k$ of multiplicities $n_1, n_2, ..., n_k$ respectively such that $n_1 + n_2 + \cdots + n_k = n$. Let $T = [\mathbf{x}_1, \mathbf{x}_2, ..., \mathbf{x}_n]$ be a matrix with columns formed by all the chains of generalized eigenvectors of A according to a prescribed order of the eigenvalues. Because of Theorem 11.21, the matrix T thus formed is nonsingular. Then, $T^{-1}AT = J$, where J is described by (11.79)–(11.81) and $A = S + N$ where

$$T^{-1}ST = \text{diag}\,[\underbrace{\lambda_1, \ldots, \lambda_1}_{n_1 \text{ times}}, \ldots, \underbrace{\lambda_k, \ldots, \lambda_k}_{n_k \text{ times}}],$$

$N = A - S$ is nilpotent and $SN = NS$. □

Example 11.15 Find a nonsingular matrix T such that $T^{-1}AT = J$ is in the Jordan form where A is a real matrix, and decompose it as $A = S + N$ where $T^{-1}ST = D$, a diagonal matrix and $T^{-1}NT = E$, a nilpotent matrix if

$$A = \begin{bmatrix} 4 & 1 & 0 & 0 \\ -6 & 0 & 1 & 0 \\ 4 & 0 & 0 & 1 \\ -1 & 0 & 0 & 0 \end{bmatrix}.$$

Solution Since A is in a companion form, its characteristic polynomial is (see Example 7.32 and Problem 7.44), $\phi(\lambda) = \lambda^4 - 4\lambda^3 + 6\lambda^2 - 4\lambda + 1 = (\lambda - 1)^4$. Therefore, A has only one distinct eigenvalue $\lambda_1 = 1$ of multiplicity 4. We now have

$$(A - I) = \begin{bmatrix} 3 & 1 & 0 & 0 \\ -6 & -1 & 1 & 0 \\ 4 & 0 & -1 & 1 \\ -1 & 0 & 0 & -1 \end{bmatrix}, \qquad (A - I)^2 = \begin{bmatrix} 3 & 2 & 1 & 0 \\ -8 & -5 & -2 & 1 \\ 7 & 4 & 1 & -2 \\ -2 & -1 & 0 & 1 \end{bmatrix}, \tag{11.97}$$

$$(A - I)^3 = \begin{bmatrix} 1 & 1 & 1 & 1 \\ -3 & -3 & -3 & -3 \\ 3 & 3 & 3 & 3 \\ -1 & -1 & -1 & -1 \end{bmatrix}, \qquad (A - I)^4 = \begin{bmatrix} 0 & 0 & 0 & 0 \\ 0 & 0 & 0 & 0 \\ 0 & 0 & 0 & 0 \\ 0 & 0 & 0 & 0 \end{bmatrix}$$

The solutions of $(A - I)\mathbf{x} = \boldsymbol{\theta}$, $(A - I)^2\mathbf{x} = \boldsymbol{\theta}$, $(A - I)^3\mathbf{x} = \boldsymbol{\theta}$, and $(A - I)^4\mathbf{x} = \boldsymbol{\theta}$, generate the null spaces $\mathbb{N}_{1j}$, $j = 1, 2, 3, 4$ and are respectively

$$\mathbb{N}_{11} : \{[1, -3, 3, -1]^T\};$$

$$\mathbb{N}_{12} : \{[1, -3, 3, -1]^T, [1, -2, 1, 0]^T\};$$

$$\mathbb{N}_{13} : \{[1, -3, 3, -1]^T, [1, -2, 1, 0]^T, [1, -1, 0, 0]^T\};$$

$$\mathbb{N}_{14} : \{[1, -3, 3, -1]^T, [1, -2, 1, 0]^T, [1, -1, 0, 0]^T, [0, 0, 0, 1]^T\}.$$

We may use MATLAB command null $((A - \lambda I)^j)$ to generate the basis vectors for the above null spaces. The generalized eigenvector of order 4 corresponding to $\lambda_1 = 1$ is $\mathbf{x}_{41}^{(1)} = [0\ 0\ 0\ 1]^T$ and a single chain of generalized eigenvectors is

$$= \{(A - I)^3\mathbf{x}_{41}^{(1)}, (A - I)^2\mathbf{x}_{41}^{(1)}, (A - I)\mathbf{x}_{41}^{(1)}, \mathbf{x}_{41}^{(1)}\}$$
$$= \{[1, -3, 3, -1]^T, [0, 1, -2, 1]^T, [0, 0, 1, -1]^T, [0, 0, 0, 1]^T\}.$$

The nonsingular matrix T is formed with these vectors as columns, so that T and $J = T^{-1}AT = D + E$ are respectively

$$T = \begin{bmatrix} 1 & 0 & 0 & 0 \\ -3 & 1 & 0 & 0 \\ 3 & -2 & 1 & 0 \\ -1 & 1 & -1 & 1 \end{bmatrix}, \quad J = \begin{bmatrix} 1 & 1 & 0 & 0 \\ 0 & 1 & 1 & 0 \\ 0 & 0 & 1 & 1 \\ 0 & 0 & 0 & 1 \end{bmatrix} = \begin{bmatrix} 1 & 0 & 0 & 0 \\ 0 & 1 & 0 & 0 \\ 0 & 0 & 1 & 0 \\ 0 & 0 & 0 & 1 \end{bmatrix} + \begin{bmatrix} 0 & 1 & 0 & 0 \\ 0 & 0 & 1 & 0 \\ 0 & 0 & 0 & 1 \\ 0 & 0 & 0 & 0 \end{bmatrix}.$$

Hence, $S = TDT^{-1} = T(I)T^{-1} = I$ and $N = A - S = A - I$ as given in (11.97). □

The computation of a Jordan matrix representation for a third order matrix A is illustrated in Example 12.17.

Example 11.16 Find the Jordan canonical form J of

$$A = \left[\begin{array}{ccccccc|c} 1 & 1 & 1 & 1 & 1 & 1 & 1 & 1 \\ 0 & 0 & 0 & 0 & 0 & 0 & 0 & 1 \\ 0 & 0 & 0 & 0 & 0 & 0 & 0 & -1 \\ 0 & -1 & 1 & 0 & 0 & 0 & 0 & 1 \\ 0 & 0 & 0 & 1 & 1 & 0 & 0 & 0 \\ 0 & 1 & 1 & 1 & 1 & 1 & 0 & 1 \\ 0 & -1 & -1 & -1 & -1 & 0 & 1 & -1 \\ 0 & 0 & 0 & 0 & 0 & 0 & 0 & 0 \end{array}\right]$$

and a nonsingular matrix T such that $J = T^{-1}AT$. Decompose A as $A = S + N$ were $T^{-1}ST$ is a diagonal matrix with diagonal elements as the eigenvalues of A and $T^{-1}NT$ is a nilpotent matrix. Determine the index of nilpoteney.

Solution Partitioning the matrix properly, the characteristic polynomial of A is $\phi(\lambda) = \det(\lambda I - A)$:

$$\phi(\lambda) = (\lambda - 1)\begin{vmatrix} \lambda & 0 & 0 \\ 0 & \lambda & 0 \\ 1 & -1 & \lambda \end{vmatrix}\begin{vmatrix} \lambda-1 & 0 & 0 \\ -1 & \lambda-1 & 0 \\ 1 & 0 & \lambda-1 \end{vmatrix}\lambda = \lambda^4(\lambda-1)^4.$$

Hence, the eigenvalues of A are $\lambda_1 = 0$ and $\lambda_2 = 1$, each of multiplicity 4, so that $n_1 = n_2 = 4$.

(a) *Null Space* $\mathbb{N}_{11}$ Considering the eigenvalue $\lambda_1 = 0$, we have $\dim \mathbb{N}(\lambda_1 I - A) = 8 - \text{rank}(A) = 8 - 6 = 2$. Therefore, the number of LI eigenvectors of A associated with $\lambda_1 = 0$ is $r_{11} = 2$ and hence (see (11.80)), the Jordan block J_1 associated with λ_1 has two sub-blocks: $J_1 = \text{diag}[J_{11}, J_{12}]_4$, where J_{11} and J_{12} are still unknown. The LI eigenvectors associated with $\lambda_1 = 0$ are the solutions of $(\lambda_1 I - A)\mathbf{x} = -A\mathbf{x} = \boldsymbol{\theta}$ and are given by

$$\mathbf{x} = [2\alpha, -\alpha, -\alpha, \beta, -\beta, 2\alpha, -2\alpha, 0]^T, \quad \alpha, \beta \text{ real.}$$

Putting $\alpha = 0$, $\beta = 1$ and $\alpha = 1$, $\beta = 0$, we find that the vectors

$$\mathbf{x}_{11}^{(1)} = [0, 0, 0, 1, -1, 0, 0, 0]^T, \quad \mathbf{x}_{12}^{(1)} = [2, -1, -1, 0, 0, 2, -2, 0]^T$$

form a basis of the null space $\mathbb{N}_{11}$ of $(\lambda_1 I - A)$.

(b) *Null Space* $\mathbb{N}_{12}$ *and* $\mathbb{N}_{13}$ Since $\lambda_1 = 0$, matrices $(\lambda_1 I - A)^2 = A^2$ and $(\lambda_1 I - A)^3 = -A^3$, where

$$A^2 = \begin{bmatrix} 1 & 0 & 2 & 2 & 2 & 2 & 2 & 2 \\ 0 & 0 & 0 & 0 & 0 & 0 & 0 & 0 \\ 0 & 0 & 0 & 0 & 0 & 0 & 0 & 0 \\ 0 & 0 & 0 & 0 & 0 & 0 & 0 & -2 \\ 0 & -1 & 1 & 1 & 1 & 0 & 0 & 1 \\ 0 & 0 & 2 & 2 & 2 & 1 & 0 & 2 \\ 0 & 0 & -2 & -2 & -2 & 0 & 1 & -2 \\ 0 & 0 & 0 & 0 & 0 & 0 & 0 & 0 \end{bmatrix},$$

$$A^3 = \begin{bmatrix} 1 & -1 & 3 & 3 & 3 & 3 & 3 & 1 \\ 0 & 0 & 0 & 0 & 0 & 0 & 0 & 0 \\ 0 & 0 & 0 & 0 & 0 & 0 & 0 & 0 \\ 0 & 0 & 0 & 0 & 0 & 0 & 0 & 0 \\ 0 & -1 & 1 & 1 & 1 & 0 & 0 & -1 \\ 0 & -1 & 3 & 3 & 3 & 1 & 0 & 1 \\ 0 & 1 & -3 & -3 & -3 & 0 & 1 & -1 \\ 0 & 0 & 0 & 0 & 0 & 0 & 0 & 0 \end{bmatrix}.$$

The solutions of $(\lambda_1 I - A)^2\mathbf{x} = A^2\mathbf{x} = \boldsymbol{\theta}$ and $(\lambda_1 I - A)^3\mathbf{x} = -A^3\mathbf{x} = \boldsymbol{\theta}$ are respectively:

$$\mathbf{x} = [2\alpha, -\alpha, \beta, \gamma, -(\alpha+\beta+\gamma), 2\alpha, -2\alpha, 0]^T, \quad \alpha, \beta, \gamma \text{ real;}$$

$$\mathbf{x} = [2a, -1-2d, b, c, -(a+b+c+d), 2a, -2a, d]^T, \quad a, b, c, d \text{ real.}$$

Hence, putting $\alpha = 0$, $\beta = 0$, $\gamma = 1$; $\alpha = 1$, $\beta = -1$, $\gamma = 0$ and $\alpha = 0$, $\beta = 1$, $\gamma = 0$, the null space $\mathbb{N}_{12}$ is generated by the vectors

$$\mathbb{N}_{12} : \{[0, 0, 0, 1, -1, 0, 0, 0]^T, [2, -1, -1, 0, 0, 2, -2. 0]^T,$$
$$\mathbf{x}_{21}^{(1)} = [0, 0, 1, 0, -1, 0, 0, 0]^T\}.$$

Again, putting $a = 0$, $b = 0$, $c = 1$, $d = 0$; $a = 1$, $b = -1$, $c = 0$, $d = 0$; $a = 0$, $b = 1$, $c = 0$, $d = 0$; and $a = 0$, $b = 1$, $c = 0$, $d = -1$, we have

$$\mathbb{N}_{13} : \{[0, 0, 0, 1, -1, 0, 0, 0]^T, \quad [2, -1, -1, 0, 0, 2, -2, 0]^T,$$
$$[0, 0, 1, 0, -1, 0, 0, 0]^T, \ \mathbf{x}_{31}^{(1)} = [0, 2, 1, 0, 0, 0, 0, -1]^T\}.$$

By computing the null space of A^4, we can determine that $\mathbb{N}_{14} = \mathbb{N}_{13}$ and so we do not need null spaces beyond $\mathbb{N}_{13}$. If we know the multiplicity n_1 of λ_1, we can stop at $\mathbb{N}_{13}$ by observing that dim $\mathbb{N}_{13} = n_1 = 4$. Examining the null spaces $\mathbb{N}_{11}$, $\mathbb{N}_{12}$ and $\mathbb{N}_{13}$ we conclude that there is only one vector $\mathbf{x}_{31}^{(1)}$ in $\mathbb{N}_{13}$ which is LI over $\mathbb{N}_{12}$ and is therefore a generalized eigenvector of order 3 associated with $\lambda_1 = 0$.

Considering similarly, determine that $\mathbf{x}_{21}^{(1)}$ is a generalized eigenvector of order 2. Therefore, $r_{11} = 2$ number of $(A - \lambda_1 I)$-cyclic and A-invariant subspaces T_{11} and T_{12} for $\lambda_1 = 0$ are generated respectively by

$$\begin{aligned} &T_{11} : A^2\mathbf{x}_{31}^{(1)}, \quad A\mathbf{x}_{31}^{(1)}, \quad \mathbf{x}_{31}^{(1)}; \\ &T_{12} : \qquad\qquad\qquad\quad \mathbf{x}_{12}^{(1)}. \end{aligned} \tag{11.98}$$

The other vectors $\mathbf{x}_{11}^{(1)}$ and $\mathbf{x}_{21}^{(1)}$ do not, in general, participate in forming the basis of S_1. The matrix representation of A restricted to the root subspace $S_1 = T_{11} \oplus T_{12}$ with respect to the basis in (11.98) is

$$J_1 = \begin{bmatrix} 0 & 1 & 0 & 0 \\ 0 & 0 & 1 & 0 \\ 0 & 0 & 0 & 0 \\ 0 & 0 & 0 & 0 \end{bmatrix} = \begin{bmatrix} J_{11} & 0 \\ 0 & J_{12} \end{bmatrix}.$$

(c) *Null Space* $\mathbb{N}_{21}$ Considering the eigenvalue $\lambda_2 = 1$, we have dim $\mathbb{N}(I - A) = 8 - \text{rank}(I - A) = 8 - 6 = 2$. Therefore, the number of LI eigenvectors of A associated with $\lambda_2 = 1$ is $r_{21} = 2$ and hence, the Jordan block J_2 associated with λ_2 has two sub-blocks: $J_2 = \text{diag}[J_{21}\ J_{22}]_4$, where J_{21} and J_{22} are still unknown. The LI eigenvectors associated with $\lambda_2 = 1$ are the solutions of $(\lambda_2 I - A)\mathbf{x} = (I - A)\mathbf{x} = \boldsymbol{\theta}$ and are given by

$$\mathbf{x} = [\alpha, 0, 0, 0, 0, \beta, -\beta, 0]^T; \qquad \alpha, \beta, \text{ real.}$$

Putting $\alpha = 1$, $\beta = 0$ and $\alpha = 0$, $\beta = 1$, we find that the vectors

$$\mathbf{x}_{11}^{(2)} = [1, 0, 0, 0, 0, 0, 0, 0]^T, \qquad \mathbf{x}_{12}^{(2)} = [0, 0, 0, 0, 0, 1, -1, 0]^T$$

form a basis of the null space $\mathbb{N}_{21}$ of $(I - A)$.

(d) *Null Space* $\mathbb{N}_{22}$ The solutions of $(\lambda_2 I - A)^2\mathbf{x} = (I - A)^2\mathbf{x} = \boldsymbol{\theta}$ are

$$\mathbf{x} = [a, 0, 0, 0, b, c, d, 0]^T; \qquad a, b, c, d \text{ real.}$$

Hence, putting $a = 1$, $b = c = d = 0$; $a = b = 0$, $c = 1$, $d = -1$; $a = b = d = 0$, $c = 1$ and $a = c = d = 0$, $b = 1$ we find the null space $\mathbb{N}_{22}$ is generated by the set of vectors

$$\mathbf{e}_1, \qquad [0, 0, 0, 0, 0, 1, -1, 0]^T, \qquad \mathbf{e}_5, \qquad \mathbf{e}_6.$$

Examining the null spaces $\mathbb{N}_{21}$ and $\mathbb{N}_{22}$ we conclude that vectors $\mathbf{x}_{21}^{(2)} := \mathbf{e}_6$ and $\mathbf{x}_{22}^{(2)} := \mathbf{e}_5$ in $\mathbb{N}_{22}$ are LI over $\mathbb{N}_{21}$ and are generalized eigenvectors of order 2 associated with $\lambda_2 = 1$. Since $n_2 = 4$

and there are four vectors in null space $\mathbb{N}_{22}$, we need not compute null spaces beyond $\mathbb{N}_{22}$. Therefore, $r_{21} = 2$ number of $(A - \lambda_2 I) = (A - I)$-cyclic and A-invariant subspaces T_{21} and T_{22} for $\lambda_2 = 1$ are generated respectively by

$$T_{21} : (A - I)\mathbf{x}_{21}^{(2)},\quad \mathbf{x}_{21}^{(2)};$$
$$T_{22} : (A - I)\mathbf{x}_{22}^{(2)},\quad \mathbf{x}_{22}^{(2)}. \tag{11.99}$$

The other vectors $\mathbf{x}_{11}^{(2)}$ and $\mathbf{x}_{12}^{(2)}$ do not, in general, take part in forming the basis of S_2. The matrix representation of A restricted to the root subspace $S_2 = T_{21} \oplus T_{22}$ with respect to the basis in (11.99) is

$$J_2 = \begin{bmatrix} 1 & 1 & 0 & 0 \\ 0 & 1 & 0 & 0 \\ 0 & 0 & 1 & 1 \\ 0 & 0 & 0 & 1 \end{bmatrix} = \begin{bmatrix} J_{21} & 0 \\ 0 & J_{22} \end{bmatrix}.$$

Let us define $T = [A^2\mathbf{x}_{31}^{(1)}, A\mathbf{x}_{31}^{(1)}, \mathbf{x}_{31}^{(1)}, \mathbf{x}_{12}^{(1)}, (A - I)\mathbf{x}_{21}^{(2)}, \mathbf{x}_{21}^{(2)}, (A - I)\mathbf{x}_{22}^{(2)}, \mathbf{x}_{22}^{(2)}]$, so that

$$T = \begin{bmatrix} 0 & 2 & 0 & 2 & 1 & 0 & 1 & 0 \\ 0 & -1 & 2 & -1 & 0 & 0 & 0 & 0 \\ 0 & 1 & 1 & -1 & 0 & 0 & 0 & 0 \\ 2 & -2 & 0 & 0 & 0 & 0 & 0 & 0 \\ -2 & 0 & 0 & 0 & 0 & 0 & 0 & 1 \\ 0 & 2 & 0 & 2 & 0 & 1 & 1 & 0 \\ 0 & -2 & 0 & -2 & 0 & 0 & -1 & 0 \\ 0 & 0 & -1 & 0 & 0 & 0 & 0 & 0 \end{bmatrix}$$

and

$$J = T^{-1}AT = \left[\begin{array}{ccc|c|cc|cc} 0 & 1 & 0 & 0 & 0 & 0 & 0 & 0 \\ 0 & 0 & 1 & 0 & 0 & 0 & 0 & 0 \\ 0 & 0 & 0 & 0 & 0 & 0 & 0 & 0 \\ \hline 0 & 0 & 0 & 0 & 0 & 0 & 0 & 0 \\ \hline 0 & 0 & 0 & 0 & 1 & 1 & 0 & 0 \\ 0 & 0 & 0 & 0 & 0 & 1 & 0 & 0 \\ \hline 0 & 0 & 0 & 0 & 0 & 0 & 1 & 1 \\ 0 & 0 & 0 & 0 & 0 & 0 & 0 & 1 \end{array}\right] = \begin{bmatrix} J_1 & 0 \\ 0 & J_2 \end{bmatrix}$$

Let us define a diagonal matrix $D = \text{diag}\ [0\ 0\ 0\ 0\ 1\ 1\ 1\ 1]^T$ with the eigenvalues of A and so $E = J - D$, a nilpotent matrix of index 3. As E is block diagonal, this index can be easily evaluated. We decompose A as $A = S + N$ where $S = TDT^{-1}$ and $N = TET^{-1} = A - S$, since $J = D + E = T^{-1}(S + N)T$. Therefore,

$$S=\begin{bmatrix}1&2&0&0&0&0&0&4\\0&0&0&0&0&0&0&0\\0&0&0&0&0&0&0&0\\0&0&0&0&0&0&0&0\\0&-1&1&1&1&0&0&-1\\0&2&0&0&0&1&0&4\\0&-2&0&0&0&0&1&-4\\0&0&0&0&0&0&0&0\end{bmatrix},\quad N=\begin{bmatrix}0&-1&1&1&1&1&1&-3\\0&0&0&0&0&0&0&1\\0&0&0&0&0&0&0&-1\\0&-1&1&0&0&0&0&1\\0&1&-1&0&0&0&0&1\\0&-1&1&1&1&0&0&-3\\0&1&-1&-1&-1&0&0&3\\0&0&0&0&0&0&0&0\end{bmatrix},$$

so that $T^{-1}ST = D$ and $T^{-1}NT = E$, a nilpotent of order 3. □

Theorem 11.23 *Jordan Form of A with all Complex Eigenvalues.* Let A be a real matrix of order $2n$ with k number of distinct complex eigenvalues $\lambda_1, \lambda_2, \ldots, \lambda_k$ of multiplicities $n_1, n_2, \ldots, n_k$ respeetively, such that $n_1 + n_2 + \cdots + n_k = n$ with correspondmg n number of complex conjugate parts. Let $\mathbf{w}_r = \mathbf{u}_r + i\mathbf{v}_r$; $\mathbf{u}_r, \mathbf{v}_r \in \mathbb{R}^{2n}$, $i = \sqrt{-1}$ be a generalized complex eigenvector associated with eigenvalue $\lambda_r = a_r + ib_r$. Then $\bar{\mathbf{w}}_r = \mathbf{u}_r - i\mathbf{v}_r$ is also a generalized eigenvector associated with eigenvalue $\bar{\lambda}_r = a_r - ib_r$. Let $T = [\mathbf{u}_1, \mathbf{v}_1, \mathbf{u}_2, \mathbf{v}_2, \ldots, \mathbf{u}_n, \mathbf{v}_n]$ be a matrix with columns formed by all the chains of generalized eigenvectors of A according to a prescribed order of the eigenvalues. The matrix T thus formed, is nonsingular. Then, $T^{-1}AT = J = D + E$, where J is described by (11.79)–(11.81) in which the eigenvalue λ_r is replaced with W_r defined by (see (10.45) and Example 10.17)

$$W_r=\begin{bmatrix}a_r&b_r\\-b_r&a_r\end{bmatrix}$$

and 1 in the superdiagonal, by an identity matrix of order 2. Moreover, $A = S + N$ where

$$T^{-1}ST=\text{diag}[\underbrace{W_1,\ldots,W_1}_{n_1\text{ times}},\ \ldots,\ \underbrace{W_k,\ldots,W_k}_{n_k\text{ times}}],$$

$N = A - S$ is nilpotent and $SN = NS$.

Example 11.17 Find a nonsingular matrix T such that $T^{-1}AT$ is in Jordan form, if

$$A=\begin{bmatrix}0&-1&0&0\\1&0&0&0\\0&0&0&-1\\2&0&1&0\end{bmatrix}.$$

Solution The matrix A is in block triangular form and its characteristic polynomial is $\phi(\lambda) = (\lambda^2 + 1)^2$. Consequently, the eigenvalues are $\lambda_1 = i$, $\bar{\lambda}_i = -i$, each of multiplicity 2. Now, solutions of

$$(A-\lambda_1 I)\mathbf{x}_1=\begin{bmatrix}i&-1&0&0\\1&-i&0&0\\0&0&-i&-1\\2&0&1&-i\end{bmatrix}\mathbf{x}_1=\begin{bmatrix}0\\0\\0\\0\end{bmatrix},$$

$$(A-\lambda_1 I)^2\mathbf{x}_2 = \begin{bmatrix} -2 & 2i & 0 & 0 \\ -2i & -2 & 0 & 0 \\ -2 & 0 & -2 & 2i \\ -4i & -2 & -2i & -2 \end{bmatrix}\mathbf{x}_2 = \begin{bmatrix} 0 \\ 0 \\ 0 \\ 0 \end{bmatrix}$$

are $\mathbf{x}_1 = [0\ 0\ i\ 1]^T$; and $\mathbf{x}_2 = [0\ 0\ i\ 1]^T$ and $[i\ 1\ 0\ 1]^T$. So, $\mathbf{x}_1$ is an eigenvector and $\mathbf{x}_2 = [i\ 1\ 0\ 1]^T$ is a generalized eigenvector of order 2, as $(A-\lambda_1 I)\ [i\ 1\ 0\ 1]^T \neq \boldsymbol{\theta}$, associated with the eigenvalue λ_1. Therefore, $\mathbf{w}_1 := (A-\lambda_1 I)\mathbf{x}_2 = [0,\ 0,\ -1,\ i]^T = \mathbf{u}_1 + i\mathbf{v}_1$ and $\mathbf{w}_2 := [i\ 1\ 0\ 1]^T = \mathbf{u}_2 + i\mathbf{v}_2$ form a chain of generalized eigenvectors of A associated with λ_1. Consequently, we have, $T = [\mathbf{u}_1, \mathbf{v}_1, \mathbf{u}_2, \mathbf{v}_2]$ and $T^{-1}AT$ is

$$T^{-1}AT = \begin{bmatrix} 0 & 0 & -1 & 0 \\ 0 & -1 & 0 & 1 \\ 0 & 1 & 0 & 0 \\ 1 & 0 & 0 & 0 \end{bmatrix}\begin{bmatrix} 0 & -1 & 0 & 0 \\ 1 & 0 & 0 & 0 \\ 0 & 0 & 0 & -1 \\ 2 & 0 & 1 & 0 \end{bmatrix}\begin{bmatrix} 0 & 0 & 0 & 1 \\ 0 & 0 & 1 & 0 \\ -1 & 0 & 0 & 0 \\ 0 & 1 & 1 & 0 \end{bmatrix}$$

$$= \begin{bmatrix} 0 & 1 & 1 & 0 \\ -1 & 0 & 0 & 1 \\ 0 & 0 & 0 & 1 \\ 0 & 0 & -1 & 0 \end{bmatrix}$$

which is in a Jordan form for a matrix having complex eigenvalues. Also, $S = TDT^{-1}$, so that

$$S = \begin{bmatrix} 0 & 0 & 0 & 1 \\ 0 & 0 & 1 & 0 \\ -1 & 0 & 0 & 0 \\ 0 & 1 & 1 & 0 \end{bmatrix}\begin{bmatrix} 0 & 1 & 0 & 0 \\ -1 & 0 & 0 & 0 \\ 0 & 0 & 0 & 1 \\ 0 & 0 & -1 & 0 \end{bmatrix}\begin{bmatrix} 0 & 0 & -1 & 0 \\ 0 & -1 & 0 & 1 \\ 0 & 1 & 0 & 0 \\ 1 & 0 & 0 & 0 \end{bmatrix}$$

$$= \begin{bmatrix} 0 & -1 & 0 & 0 \\ 1 & 0 & 0 & 0 \\ 0 & 1 & 0 & -1 \\ 1 & 0 & 1 & 0 \end{bmatrix},$$

where $D = \text{diag}\ [W_1, W_1]$ is a block diagonal matrix with W_1, a matrix of order 2, having 0 diagonal elements and 1, –1 as antidiagonal elements. Also,

$$N = A - S = \begin{bmatrix} 0 & 0 & 0 & 0 \\ 0 & 0 & 0 & 0 \\ 0 & -1 & 0 & 0 \\ 1 & 0 & 0 & 0 \end{bmatrix}.$$

Theorem 11.24 *Jordan Form of A with a Combination of Real and Complex Eigenvalues.* Let A be a real matrix of order n with p number of distinct real eigenvalues $\lambda_1, \lambda_2, ..., \lambda_p$ of multiplicities $n_1, n_2, ..., n_p$ respectively such that $n_1 + n_2 + \cdots + n_p = q$; and let $\{\mathbf{x}_1, \mathbf{x}_2, ..., \mathbf{x}_q\}$ be the set representing the chains of generalized eigenvectors corresponding to these eigenvalues. Moreover, let k be the number of distinct complex eigenvalues $\lambda_{p+1}, \lambda_{p+2}, ..., \lambda_{p+k}$ of multiplicities $n_{p+1}, n_{p+2}, ..., n_{p+k}$ respectively, such that

$n_{p+1} + n_{p+2} + \cdots + n_{p+k} = m$ with corresponding m number of complex conjugate parts. Let $\mathbf{w}_r = \mathbf{u}_r + i\mathbf{v}_r$; $\mathbf{u}_r$, $\mathbf{v}_r \in \mathbb{R}^n$, $i = \sqrt{-1}$ be a generalized complex eigenvector associated with eigenvalue $\lambda_r = a_r + ib_r$. Then $\overline{\mathbf{w}}_r = \mathbf{u}_r - i\mathbf{v}_r$, is also a generalized eigenvector associated with eigenvalue $\overline{\lambda}_r = a_r - ib_r$, $r = p + 1, p + 2, \ldots, p + k$. Let $T = [\mathbf{x}_1, \mathbf{x}_2, \ldots, \mathbf{x}_r; \mathbf{u}_1, \mathbf{v}_1, \mathbf{u}_2, \mathbf{v}_2, \ldots, \mathbf{u}_m, \mathbf{v}_m]$ be a matrix with columns formed by all the chains of generalized eigenvectors of A according to a prescribed order of the, eigenvalues. The matrix T thus formed is nonsingular. Then, $A = S + N$ and $T^{-1}AT = J = D + E$, where $D = T^{-1}ST$ is a quasi-diagonal matrix, E is an idempotent matrix and J is described by (11.79)–(11.81) in which, the eigenvalue λ_r is replaced with W_r defined by (see (10.45) and Example 10.17),

$$W_r = \begin{bmatrix} a_r & b_r \\ -b_r & a_r \end{bmatrix}$$

and 1 in the superdiagonal by an identity matrix of order 2, $r = p+1, p+2, \ldots, p+k$. Moreover,

$$T^{-1}ST = \text{diag}[\underbrace{\lambda_1, \ldots, \lambda_1}_{n_1 \text{ times}}, \ldots, \underbrace{\lambda_p, \ldots, \lambda_p}_{n_p \text{ times}};\ \underbrace{W_{p+1}, \ldots, W_{p+1}}_{n_{p+1} \text{ times}}, \ldots, \underbrace{W_{p+k}, \ldots, W_{p+k}}_{n_{p+k} \text{ times}}],$$

$N = A - S$ is nilpotent and $SN = NS$.

11.5 JORDAN FORMS WITH MATLAB

In MATLAB, jordan (A) computes the Jordan canonical or normal form J of A. [T, J] = jordan (A) also computes the similarity transformation, T, so that $T^{-1}AT = J$. The columns of T are the generalized eigenvectors.

Example 11.18 Let

$$A = \begin{bmatrix} 1 & -1 & 0 & 0 \\ 1 & 1 & 0 & 0 \\ 0 & 0 & 3 & -2 \\ 0 & 0 & 1 & 1 \end{bmatrix}.$$

```
>>[T, J] = jordan (A)
```

responds with

$$T = \begin{bmatrix} 0 & 0 & 0.5 & 0.5 \\ 0 & 0 & 0.0-0.5i & 0.0+0.5i \\ 0.5-0.5i & 0.5+0.5i & 0 & 0 \\ 0.0-0.5i & 0.0+0.5i & 0 & 0 \end{bmatrix}$$

J = diag [2 + i, 2 − i, 1 + i, 1 − i]

Set $$P = \begin{bmatrix} 0 & 0 & 0.5 & 0 \\ 0 & 0 & 0 & -0.5 \\ 0.5 & -0.5 & 0 & 0 \\ 0 & -0.5 & 0 & 0 \end{bmatrix}, \quad \text{then } P^{-1}AP = \begin{bmatrix} 2 & 1 & 0 & 0 \\ -1 & 2 & 0 & 0 \\ 0 & 0 & 1 & 1 \\ 0 & 0 & -1 & 1 \end{bmatrix}.$$

PROBLEMS

11.1 Test if $\mathbb{R}^3$ is a direct sum of subspaces S_1 and S_2 when

(a) $S_1 = \{[a, b, 0] : a, b \in \mathbb{R}\}$, $S_2 = \{[0, c, d] : c, d \in \mathbb{R}\}$;

(b) $S_1 = \{[a, b, 0] : a, b \in \mathbb{R}\}$, $S_2 = \{[c, c, c] : c \in \mathbb{R}\}$;

(c) $S_1 = \{[a, 0, 0] : a \in \mathbb{R}\}$, $S_2 = \{[0, b, 0] : b \in \mathbb{R}\}$;

(d) $S_1 = \{[a, b, 0] : a, b \in \mathbb{R}\}$, $S_2 = \{[0, 0, c] : c \in \mathbb{R}\}$.

11.2 Test if $\mathbb{R}^3$ is a sum of subspaces S_1 and S_2 when these are given as in Problem 11.1.

11.3 (a) Show that the following sets are subspaces of $\mathbb{R}^3$ whose elements are row vectors, where $a, b, c \in \mathbb{R}$:

$S_1 = \{[a, b, c] : a + b - 2c = 0\}$, $S_2 = \{[a, b, c] : a = b\}$,
$S_3 = \{[a, b, c] : a + 2c = 0, b - 4c = 0\}$.

(b) Determine (i) $S_1 \cap S_2$, (ii) $S_2 \cap S_3$, and (iii) $S_3 \cap S_1$.

(c) Determine the basis of the subspaces in (b).

(d) Express $\mathbb{R}^3$ as the direct sum of subspaces given in (a).

11.4 (a) Show that the following sets are subspaces of $\mathbb{R}^3$, whose elements are row vectors, where $a, b, c \in \mathbb{R}$:

$S_1 = \{[a, b, c] : c = 0\}$, $S_2 = \{[a, b, c] : b = 0\}$,
$S_3 = \{[a, b, c] : b = c = 0\}$, $S_4 = \{[a, b, c]\}$.

(b) Determine (i) $S_1 \cap S_2 \cap S_3$, (ii) $S_1 \cap S_3 \cap S_4$, (iii) $S_2 \cap S_3 \cap S_4$, (iv) $S_1 \cap S_2 \cap S_4$, and give a geometrical picture of the resultant subspace.

(c) Determine a basis of the subspaces in (b).

11.5 Apply Theorem 11.3 to determine the dimension of the subspaces

(a) $S_1 + S_2$, (b) $S_2 + S_3$, (c) $S_3 + S_1$,

when S_i's are given as in Problem 11.3(a).

11.6 Test if $\mathbb{R}^4$ is the direct sum of the following subspaces by applying Theorem 11.6 where $a, b, c, d \in \mathbb{R}$:

$S_1 = \{[a, b, c, d] : a + b = 0\}$, $S_2 = \{[a, b, c, d] : a - b = 0\}$,
$S_3 = \{[a, b, c, d] : c + d = 0\}$, $S_4 = \{[a, b, c, d] : c - d = 0\}$.

11.7 Prove that

(a) $\mathbb{R}^3 = S_1 \oplus S_2 \oplus S_3$,

where

$S_1 = \{[a\ 0\ 0] : a \in \mathbb{R}\}$, $S_2 = \{[0\ b\ 0] : b \in \mathbb{R}\}$, $S_3 = \{[0\ 0\ c] : c \in \mathbb{R}\}$.

(b) $\mathbb{R}^4 = S_1 \oplus S_2 \oplus S_3 \oplus S_4$,

where

$S_i = \{[a_1, a_2, a_3, a_4] : a_i \neq 0; a_j = 0, j \neq i\}$

for $i, j = 1, ..., 4$ and $a_i \in \mathbb{R}$.

(c) $\mathbb{R}^3 = S_1 \oplus S_2 \oplus S_3$,

where

$S_1 = \{[a\ b\ c] : a = b = c\}$, $S_2 = \{[a\ b\ c\ ] : a = b = -c\}$.

$S_3 = \{[a\ b\ c] : a = -b = c\}$; $a, b, c \in \mathbb{R}$.

(d) $\mathbb{R}^4 = S_1 \oplus S_2 \oplus S_3$,

where

$S_1 = \{[a\ b\ c\ d] : c = d = 0\}$, $S_2 = \{[a\ b\ c\ d] : a = b = 0, c = d\}$,

$S_3 = \{[a\ b\ c\ d] : a = b = 0, c = -d\}$; $a, b, c, d \in \mathbb{R}$,

11.8 Let

$S_1 = \{A \in V_{n\times n} : a_{ij} = 0 \text{ whenever } i > j\}$,

$S_2 = \{A \in V_{n\times n} : a_{ij} = 0 \text{ whenever } i \le j\}$,

i.e., S_1 is the subspace of upper triangular matrices and S_2 is that of lower triangular matrices with zero diagonal elements. Show that S_1 and S_2 form a subspace of $V_{n\times n}$ and $V_{n\times n} = S_1 \oplus S_2$.

11.9 Let

$V = \{A \in V_{n\times n} : a_{ij} = 0 \text{ whenever } i < j\}$

be the vector space of lower triangular matrices;

$S_1 = \{A \in V_{n\times n} : a_{ij} = 0 \text{ whenever } i \ne j\}$,

the subspace of diagonal matrices; and

$S_2 = \{A \in V_{n\times n} : a_{ij} = 0 \text{ whenever } i \le j\}$,

the subspace of lower triangular matrices with zero diagonal elements. Show that $V = S_1 \oplus S_2$.

11.10 Let S_1 be the set of all symmetric and S_2 skew-symmetric matrices in $V_{n\times n}$. Prove that S_1 and S_2 are subspaces in $V_{n\times n}$ and that $V_{n\times n} = S_1 \oplus S_2$.

11.11 Let S_1 be the set of all Hermitian; and S_2, the skew-Hermitian matrices in $\mathbb{C}^{n\times n}$. Show that S_1 and S_2 are subspaces of $\mathbb{C}^{n\times n}$ and that $\mathbb{C}^{n\times n} = S_1 \oplus S_2$.

11.12 Let S_1 and S_2 be the subspaces of $\mathbb{P}_2(x)$ generated by $\{1, x + x^2\}$ and $\{1 + x + x^2, x\}$ respectively. Determine (a) a basis for $S_1 \cap S_2$ and (b) the dimension of $S_1 + S_2$.

11.13 Let L be a linear transformation on $\mathbb{R}^3$ defined by

$L[x\ y\ z]^T = [-3x + 3y - 2z, -7x + 6y - 3z, x - y + 2z]^T$.

(a) Show that the subspace S generated by $\{\mathbf{x} = [1, 2, 1]^T, \mathbf{y} = [1\ 1\ -1]^T\}$ is L-invariant.

(b) Determine the matrix of L restricted to S with respect to the basis $\{\mathbf{x}, \mathbf{y}\}$.

11.14 Let L be a linear transformation on $\mathbb{R}^4$ defined by

$L[w\ x\ y\ z]^T = [-3x + y + 2z, -2w + x - y + 2z, -2w + x - y + 2z, -2w - 3x + y + 4z]^T$.

(a) Show that the subspace S generated by

$\{\mathbf{x} = [1\ 1\ 1\ 1]^T, \mathbf{y} = [1\ 0\ 0\ 1]^T, \mathbf{z} = [-1\ 1\ 1\ 0]^T\}$

is L-invariant.

(b) Determine the matrix of L restricted to S with respect to the basis $\{\mathbf{x}, \mathbf{y}, \mathbf{z}\}$.

11.15 Let L be a linear transformation on $\mathbb{P}(x)$ defined by $Lp(x) = x(dp(x)/dx)$ and $S = \mathbb{P}_2(x)$, a subspace of $\mathbb{P}(x)$. Show that S is L-invariant.

11.16 Let S_1 and S_2 be two subspaces in $\mathbb{R}^4$ generated respectively by

$\{\mathbf{w} = [-1, -1, -1, -1]^T, \mathbf{x} = [0\ 1\ 2\ 0]^T\}$

and

$\{\mathbf{y} = [1\ 0\ 0\ 1]^T, \mathbf{z} = [-1\ 1\ 10]^T\}$.

Show that $\mathbb{R}^4 = S_1 \oplus S_2$ and find the matrix of L defined in Problem 11.14 with respect to the basis $\{\mathbf{w}\ \mathbf{x}\ \mathbf{y}\ \mathbf{z}\}$.

11.17 Determine the minimum polynomial of the following linear transformation L on $\mathbb{R}^4$ and then find the root subspace associated with each eigenvalue:

(a) $L\mathbf{x} = [b\ c\ d\ 0]^T$,

(b) $L\mathbf{x} = [\alpha\ \alpha\ \alpha\ \alpha]^T$; $\quad \alpha = a + b + c + d$,

(c) $L\mathbf{x} = [b\ c\ d, -a + 4b - 6c + 4d]^T$,

where $\mathbf{x} = [a\ b\ c\ d]^T$.

11.18 Determine the root subspaces of the following linear transformation L on $\mathbb{R}^4$ associated with its eigenvalues, and hence find a basis to represent L in a quasi-diagonal form:

(a) $L\mathbf{x} = [4a - b - 5c - 2d, a - c, a - c - d, -b + c + 2d]^T$,

(b) $L\mathbf{x} = [-2a + 4b + 4c - 3d, -a + b - d, c, -a]^T$.

11.19 Write, by inspection, the characteristic polynomial and minimum polynomial of the following matrices A over $\mathbb{R}$ (see Theorem 11.15 and Problem 7.31a):

(a) $\begin{bmatrix} a & & & \\ & a & & \\ & & a & \\ & & & a \end{bmatrix}$, (b) $\begin{bmatrix} a & 1 & & \\ & a & & \\ & & a & \\ & & & a \end{bmatrix}$, (c) $\begin{bmatrix} a & 1 & & \\ & a & 1 & \\ & & a & \\ & & & a \end{bmatrix}$,

(d) $\begin{bmatrix} a & 1 & & \\ & a & & \\ & & a & 1 \\ & & & a \end{bmatrix}$.

11.20 Find the root subspace determined by each of the matrices given in Problem 11.19 and express it as the direct sum of $(A - aI)$-cyclic spaces.

11.21 Prove that an L-cyclic subspace S of V is L-invariant.

11.22 Let $\lambda_1, \lambda_2, \ldots, \lambda_k$ be the distinct eigenvalues of the linear transformation L on an n-dimensional vector space V and let the largest Jordan block associated with λ_i in the Jordan canonical form of L be of order μ_i. Then using Problem 7.31, show that the minimum polynomial of L is

$$(\lambda - \lambda_1)^{\mu_1} (\lambda - \lambda_2)^{\mu_2} \cdots (\lambda - \lambda_k)^{\mu_k}.$$

11.23 If k is the least positive integer such that $(\lambda I_V - L)^k \mathbf{x} = \boldsymbol{\theta}$ for a vector $\mathbf{x} \neq \boldsymbol{\theta}$ in V, where λ is the eigenvalue of the linear transformation L on V, then prove by mathematical induction that the set of vectors $(\lambda I_V - L)^{k-1} \mathbf{x}, \ldots, (\lambda I_V - L)\mathbf{x}, \mathbf{x}$ is linearly independent.

11.24 Determine a basis to represent the matrix of L, which is as given in Problem 11.18, in Jordan normal form.

11.25 Determine a nonsingular matrix T such that $T^{-1}AT$ is in Jordan form, where A is given by (10.40).

11.26 Find a nonsingular matrix T such that $T^{-1}AT = J$ is in Jordan form, where A is a real matrix, and decompose it as $A = S + N$, where $T^{-1}ST = D$, a diagonal or a quasidiagonal matrix, and $T^{-1}NT = E$, a nilpotent matrix, if A is

(i) $A = \begin{bmatrix} -2 & 0 & -2 \\ -1 & 1 & -2 \\ 0 & 1 & -2 \end{bmatrix}$, (ii) $A = \begin{bmatrix} -6 & 31 & -14 \\ -1 & 6 & -2 \\ 0 & 2 & 1 \end{bmatrix}$,

(iii) $A = \begin{bmatrix} 5 & -3 & -2 \\ 8 & -5 & -4 \\ -4 & 3 & 3 \end{bmatrix}$, (iv) $A = \begin{bmatrix} 2 & 0 & 0 \\ 1 & 2 & 0 \\ 4 & -1 & 3 \end{bmatrix}$,

(v) $A = \begin{bmatrix} 1 & 2 & 3 \\ 0 & 1 & 2 \\ 0 & 0 & 1 \end{bmatrix}$, (vi) $A = \begin{bmatrix} 2 & 0 & 0 \\ 1 & 2 & 0 \\ 1 & 2 & 2 \end{bmatrix}$,

(vii) $A = \begin{bmatrix} 1 & -1 & 1 \\ 0 & 2 & 1 \\ 0 & 0 & 2 \end{bmatrix}$, (viii) $A = \begin{bmatrix} 2 & 1 & 0 \\ 0 & 2 & 0 \\ 0 & -1 & 2 \end{bmatrix}$,

(ix) $A = \begin{bmatrix} 1 & 0 & 0 \\ 0 & 1 & 1 \\ 0 & 0 & -1 \end{bmatrix}$, (x) $A = \begin{bmatrix} -1 & 1 & -2 \\ 0 & -1 & 4 \\ 0 & 0 & 1 \end{bmatrix}$,

(xi) $A = \begin{bmatrix} 0 & 1 & 0 & 0 \\ 0 & 0 & 1 & 0 \\ 0 & 0 & 0 & 1 \\ -1 & 4 & -6 & 4 \end{bmatrix}$, (xii) $A = \begin{bmatrix} 0 & -2 & -1 & -1 \\ 1 & 2 & 1 & 0 \\ 0 & 1 & 1 & 0 \\ 0 & 0 & 0 & 1 \end{bmatrix}$,

(xiii) $A = \begin{bmatrix} 0 & -1 & 0 & 0 \\ 1 & 0 & 0 & 0 \\ 0 & 0 & 0 & -1 \\ 2 & 0 & 1 & 0 \end{bmatrix}$, (xiv) $A = \begin{bmatrix} 1 & 1 & 1 & 1 \\ 2 & 2 & 2 & 2 \\ 3 & 3 & 3 & 3 \\ 4 & 4 & 4 & 4 \end{bmatrix}$,

(xv) $A = \begin{bmatrix} 1 & 1 & 1 & 0 \\ 0 & 2 & 0 & 0 \\ 0 & 0 & 1 & 1 \\ 0 & 0 & 0 & 2 \end{bmatrix}$, (xvi) $A = \begin{bmatrix} 0 & -1 & -2 & -1 \\ 1 & 2 & 1 & 1 \\ 0 & 0 & 1 & 0 \\ 0 & 0 & 1 & 1 \end{bmatrix}$,

(xvii) $A = \begin{bmatrix} 0 & -2 & -1 & -1 \\ 1 & 2 & 1 & 1 \\ 0 & 1 & 1 & 0 \\ 0 & 0 & 0 & 1 \end{bmatrix}$, (xviii) $A = \begin{bmatrix} 2 & 1 & 4 & 0 \\ 0 & 2 & 1 & -1 \\ 0 & 0 & 2 & 1 \\ 0 & 0 & 0 & 2 \end{bmatrix}$.

11.27 Let A be a matrix of order 5. It has either (i) one LI eigenvector, (ii) two LI eigenvectors, (iii) three LI eigenvectors, (iv) four LI eigenvectors or (v) five LI eigenvectors corresponding to an eigenvalue μ of multiplicity 5. In each case determine the structure of Jordan form of A. Compute the CP and MP in each case.

CHAPTER

12

Function of a Matrix

We recall the definition of a matrix polynomial given in Section 7.3.1. That is, given a scalar polynomial

$$p(\lambda) = a_n\lambda^n + a_{n-1}\lambda^{n-1} + \cdots + a_1\lambda + a_0, \tag{12.1}$$

the associated matrix polynomial of degree n in the variable A of $\mathbb{C}^{n\times n}$ is defined by

$$p(A) = a_nA^n + a_{n-1}A^{n-1} + \cdots + a_1A + a_0I,$$

replacing λ^k by A^k, $k = 0, 1, ..., n$, in (12.1). If the minimum polynomial of A, viz.,

$$\psi(\lambda) = \lambda^m + b_{m-1}\lambda^{m-1} + \cdots + b_1\lambda + b_0 \tag{12.2}$$

is of degree $m \le n$, then $\psi(A) = 0$ and $f(A)$ can be represented by a matrix polynomial of degree $m - 1$ (see Example 7.27(b)).

Now, an important question is: Given any *arbitrary function* $f(\lambda)$, which is not necessarily a polynomial, how to define $f(A)$, which is to be called the function of a matrix? Suppose that $f(\lambda) = \exp(\lambda)$. Can we define $f(A) = \exp(A)$ in the same way as for a polynomial? If yes, can we compute $f(A)$ using the power series expression

$$\exp(\lambda) = 1 + \lambda + \frac{\lambda^2}{2} + \cdots$$

as in the case of a matrix polynomial?

In principle, in the same way as in a polynomial case, we can represent

$$\exp(A) = c_{m-1}A^{m-1} + \cdots + c_1A + c_0I,$$

where c_i's are unknown scalars, by using the fact that A satisfies the minimum polynomial $\psi(\lambda)$

of A given by (12.2) (see Example 7.27(b)). But, till now, we have not developed any theory to determine these unknown scalars.

This chapter is, therefore, concerned with introducing a meaningful definition of the function of a matrix which is computationally viable. In fact, any function of a matrix $f(A)$ is defined as being equal to a matrix polynomial $p(A)$ when the associated scalar function $f(\lambda)$ and scalar polynomial $p(\lambda)$ and all their derivatives up to order $m_i - 1$ have the same values at the eigenvalues λ_i of A, $i = 1, 2, ..., k$, where m_i is the multiplicity of λ_i in the minimum polynomial $\psi(\lambda)$ of A.

12.1 DEFINITION AND EVALUATION OF THE FUNCTION OF A MATRIX

We recall the definition of a minimum polynomial $\psi(\lambda)$ of an n-square matrix A over the complex field $\mathbb{C}$ (see Definition 7.15, Theorem 7.14, and Problem 10.23) and write it in the factored form as

$$\psi(\lambda) = (\lambda - \lambda_1)^{m_1} (\lambda - \lambda_2)^{m_2} \cdots (\lambda - \lambda_k)^{m_k}, \tag{12.3}$$

where $m_1 + m_2 + \cdots + m_k = m = \deg$ of $\psi(\lambda)$ and m_i is called the *index of the eigenvalue.* λ_i.

First, we consider the case when $\psi(\lambda)$ contains linear factors, i.e., $m_1 = m_2 = \cdots = m_k = 1$ and

$$\psi(\lambda) = (\lambda - \lambda_1)(\lambda - \lambda_2) \cdots (\lambda - \lambda_m),$$

where λ_i's are distinct real or complex numbers. We now split $1/\psi(\lambda)$ into a partial fraction to get

$$\frac{1}{\psi(\lambda)} = \frac{c_1}{\lambda - \lambda_1} + \frac{c_2}{\lambda - \lambda_2} + \dots + \frac{c_m}{\lambda - \lambda_m}, \tag{12.4}$$

where

$$c_i = \frac{1}{\prod\limits_{\substack{j=1 \\ j \neq i}}^{m} (\lambda_i - \lambda_j)}. \tag{12.5}$$

Multiplying both sides of (12.4) by $\psi(\lambda)$ and then inserting (12.5) therein, we obtain

$$1 = \sum_{i=1}^{m} \frac{\psi(\lambda) c_i}{(\lambda - \lambda_i)} = l_1(\lambda) + l_2(\lambda) + \dots + l_m(\lambda),$$

where

$$l_i(\lambda) = \frac{\psi(\lambda) c_i}{\lambda - \lambda_i} = \prod_{\substack{j=1 \\ j \neq i}}^{m} \left(\frac{\lambda - \lambda_j}{\lambda_i - \lambda_j} \right). \tag{12.6}$$

The polynomials $l_i(\lambda)$ have the following important properties:

$$l_i(\lambda_j) = \delta_{ij}; \quad i, j = 1, 2, ..., m; \tag{12.7}$$

where δ_{ij} is the *Kronecker delta.* Now, the polynomial of degree at most $m - 1$ defined by

$$L(\lambda) = \sum_{i=1}^{m} f_i l_i(\lambda) \tag{12.8}$$

is known as the *Lagrange interpolating polynomial* which takes the specified values f_k at $\lambda = \lambda_k$ (see Figure 12.1 which is drawn when f_k's and λ_k's are real and λ_k's are arranged in increasing order).

Given m distinct points $\lambda_1, \lambda_2, ..., \lambda_m$, real or complex, and a set of real or complex values $f_1, f_2, ..., f_m$, there exists a unique polynomial $p(\lambda)$ of degree less than m such that $p(\lambda_i) = f_i$, $i = 1,2, ..., m$. We claim that $p(\lambda) = L(\lambda)$ given by (12.8) is such a polynomial. Indeed,

$$p(\lambda_k) = \sum_{i=1}^{m} f_i l_i(\lambda_k) = f_k l_k(\lambda_k) = f_k.$$

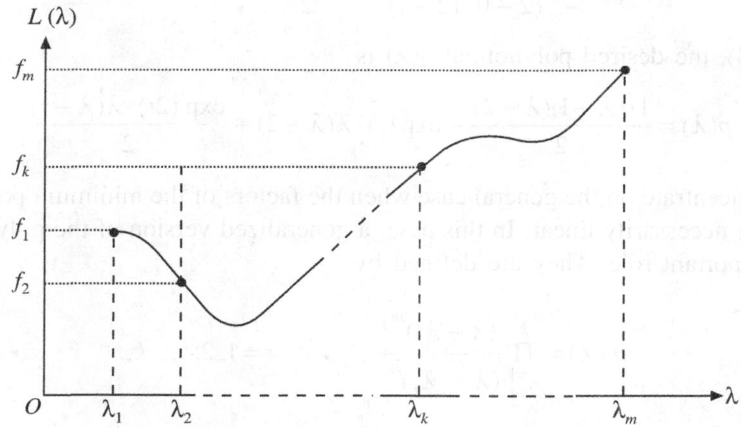

Figure 12.1 Lagrange interpolating polynomial.

The polynomial $p(\lambda)$ given by (12.8) is unique because, if $q(\lambda)$ is another polynomial with the same properties, then $p(\lambda) - q(\lambda)$ is a polynomial of degree at most $m - 1$ which vanishes at m points $\lambda_1, \lambda_2, ..., \lambda_m$. Hence, $p(\lambda) - q(\lambda) = 0$ (see Problem 12.1), i.e., $p(\lambda) = q(\lambda)$.

Example 12.1 Let

$$A = \begin{bmatrix} 0 & 1 & 0 \\ 0 & 0 & 1 \\ 0 & -2 & 3 \end{bmatrix} \tag{12.9}$$

whose characteristic polynomial and minimum polynomial are the same and are given by

$$\psi(\lambda) = \det(\lambda I - A) = \lambda(\lambda - 1)(\lambda - 2). \tag{12.10}$$

Therefore, we have from (12.10),

$$\lambda_1 = 0, \quad m_1 = 1; \quad \lambda_2 = 1, \quad m_2 = 1; \quad \lambda_3 = 2, \quad m_3 = 1. \tag{12.11}$$

It is desired that we compute a polynomial $p(\lambda)$, whose values at $\lambda = 0, 1, 2$ are respectively those of exp (λt), where t is real, at the same locations. So, in (12.8), we have

$$f_1 = \exp(\lambda t)|_{\lambda=0} = 1, \qquad f_2 = \exp(\lambda t)|_{\lambda=1} = \exp(t), \qquad f_3 = \exp(\lambda t)|_{\lambda=2} = \exp(2t).$$

In view of (12.6), substituting from (12.11), we get

$$l_1(\lambda) = \frac{(\lambda-1)(\lambda-2)}{(0-1)(0-2)} = \frac{(\lambda-1)(\lambda-2)}{2},$$

$$l_2(\lambda) = \frac{\lambda(\lambda-2)}{(1-0)(1-2)} = -\lambda(\lambda-2),$$

$$l_3(\lambda) = \frac{\lambda(\lambda-1)}{(2-0)(2-1)} = \frac{\lambda(\lambda-1)}{2}.$$

Hence, by (12.8), the desired polynomial $p(\lambda)$ is

$$p(\lambda) = \frac{1\cdot(\lambda-1)(\lambda-2)}{2} - \exp(t)\cdot\lambda(\lambda-2) + \frac{\exp(2t)\cdot\lambda(\lambda-1)}{2}. \qquad \square$$

We now concentrate on the general case when the factors of the minimum polynomial $\psi(\lambda)$ in (12.3) are not necessarily linear. In this case, a generalized version of the polynomials (12.6) will play an important role. They are defined by

$$h_i(\lambda) = \prod_{\substack{j=1\\ j\neq i}}^{k} \frac{(\lambda-\lambda_j)^{m_j}}{(\lambda_i-\lambda_j)^{m_j}}, \qquad i = 1, 2, \ldots, k, \tag{12.12}$$

which have the important property

$$h_i(\lambda_j) = \delta_{ij}. \tag{12.13}$$

We now expand $1/h_i(\lambda)$ by *Taylor's series* about $\lambda = \lambda_i$:

$$1/h_i(\lambda) = a_{i0}(\lambda_i) + a_{i1}(\lambda_i)(\lambda-\lambda_i) + \cdots + a_{i,m_i-1}(\lambda_i)(\lambda-\lambda_i)^{m_i-1} + \cdots, \tag{12.14}$$

where

$$a_{ij}(\lambda) = \frac{1}{j!}\frac{d^j}{d\lambda^j}\left[\frac{1}{h_i(\lambda)}\right]. \tag{12.15}$$

Because of (12.13), $a_{i0}(\lambda_i) = 1/h_i(\lambda_i) = 1$. With the help of the first $m_i - j + 1$ terms in (12.14), we now define a polynomial

$$p_{i,m_i-j}(\lambda) = 1 + a_{i1}(\lambda_i)(\lambda-\lambda_i) + \cdots + a_{i,m_i-j}(\lambda_i)(\lambda-\lambda_i)^{m_i-j}. \tag{12.16}$$

We then employ (12.12) and (12.16) to define finally the polynomials

$$\phi_{ij}(\lambda) = \frac{(\lambda - \lambda_i)^{j-1}}{(j-1)!} h_i(\lambda) p_{i,m_i-j}(\lambda), \tag{12.17}$$

$$i = 1, 2, \ldots, k; \qquad j = 1, 2, \ldots, m_i$$

which are at most of degree $j - 1 + m - m_i + m_i - j = m - 1$. To investigate the interpolating properties of the polynomials (12.17), we recall the *Leibniz rule* for the l-th derivative of the product of two polynomials $q(\lambda)$ and $r(\lambda)$:

$$\begin{aligned}[q(\lambda)r(\lambda)]^{(l)} &= q^{(l)}(\lambda)r(\lambda) + {}^{l}C_1 q^{(l-1)}(\lambda)r^{(1)}(\lambda) + \cdots + {}^{l}C_{l-j+1} q^{(j-1)}(\lambda) r^{(l-j+1)}(\lambda) \\ &\quad + {}^{l}C_{l-j+2} q^{(j-2)}(\lambda) r^{(l-j+2)}(\lambda) + \cdots + q(\lambda) r^{(l)}(\lambda),\end{aligned} \tag{12.18}$$

where ${}^{l}C_i$'s are *binomial coefficients*. Obviously, $\phi_{i1}(\lambda_i) = 1$. If we set $q(\lambda) = (\lambda - \lambda_i)^{j-1}/(j-1)!$ and $r(\lambda) = h_i(\lambda)p_{i,m_i-j}(\lambda)$ in (12.17), then we have

$$[q(\lambda_i)]^{(l)} = \frac{d^l}{d\lambda^l}\left[\frac{(\lambda - \lambda_i)^{j-1}}{(j-1)!}\right]_{\lambda=\lambda_i} = \begin{cases} 0, & l = 0, 1, \ldots, (j-2), j, \ldots, m_i - 1; \\ 1, & l = j - 1. \end{cases}$$

Consequently, by applying (12.18), it follows from (12.17) that

$$\phi_{ij}^{(l)}(\lambda_i) = \begin{cases} 0, & l = 0, 1, \ldots, j-2; \\ 1, & l = j - 1; \end{cases} \tag{12.19}$$

and

$$\begin{aligned}\phi_{ij}^{(l)}(\lambda_i) &= {}^{l}C_{l-j+1} \left.\frac{d^{l-j+1} r(\lambda)}{d\lambda^{l-j+1}}\right|_{\lambda=\lambda_i}, \qquad l = j, \ldots, m_i - 1 \\ &= {}^{l}C_{l-j+1}\, [h_i^{(l-j+1)}(\lambda_i) + {}^{l-j+1}C_1 h_i^{(l-j)}(\lambda_i) a_{i1}(\lambda_i) + {}^{l-j+1}C_2 h_i^{(l-j-1)}(\lambda_i) 2!\, a_{i2}(\lambda_i) \\ &\quad + \cdots + h_i(\lambda_i)(l-j+1)!\, a_{i,l-j+1}(\lambda_i)] \\ &= {}^{l}C_{l-j+1} \left\{ h_i^{(l-j+1)}(\lambda_i) + {}^{l-j+1}C_1 h_i^{(l-j)}(\lambda_i) \left[\frac{1}{h_i(\lambda)}\right]^{(1)}_{\lambda=\lambda_i} \right. \\ &\quad \left. + {}^{l-j+1}C_2 h_i^{(l-j-1)}(\lambda_i) \left[\frac{1}{h_i(\lambda)}\right]^{(2)}_{\lambda=\lambda_i} + \cdots + h_i(\lambda_i) \left[\frac{1}{h_i(\lambda)}\right]^{(l-j+1)}_{\lambda=\lambda_i} \right\} \\ &= {}^{l}C_{l-j+1} \frac{d^{l-j+1}}{d\lambda^{l-j+1}} \left[h_i(\lambda) \cdot \frac{1}{h_i(\lambda)} \right]_{\lambda=\lambda_i} = 0, \quad l = j, \ldots, m_i - 1,\end{aligned} \tag{12.20}$$

where we have employed (12.18) and (12.16) in the second step and (12.15) in the third step. Moreover, it is evident from (12.17) and (12.12) that

$$\phi_{ij}^{(l)}(\lambda_s) = 0 \quad \text{for} \quad s \neq i, \quad l = 0, 1, \ldots, m_s - 1. \tag{12.21}$$

A geometrical interpretation of the properties of $\phi_{ij}(\lambda)$ as given by (12.19)–(12.21) shows that at the point $\lambda = \lambda_i$ which refers to the first subscript i of ϕ_{ij}

$$\phi_{i1}(\lambda) = \phi_{i2}^{(1)}(\lambda) = \cdots = \phi_{im_i}^{(m_i-1)}(\lambda) = 1 \quad \text{at } \lambda = \lambda_i$$

for all $i = 1, 2, \ldots, k$. However,

$$\phi_{j1}(\lambda) = \phi_{j2}^{(1)}(\lambda) = \cdots = \phi_{jm_i}^{(m_i-1)}(\lambda) = 0 \quad \text{at } \lambda = \lambda_i, \qquad i \neq j.$$

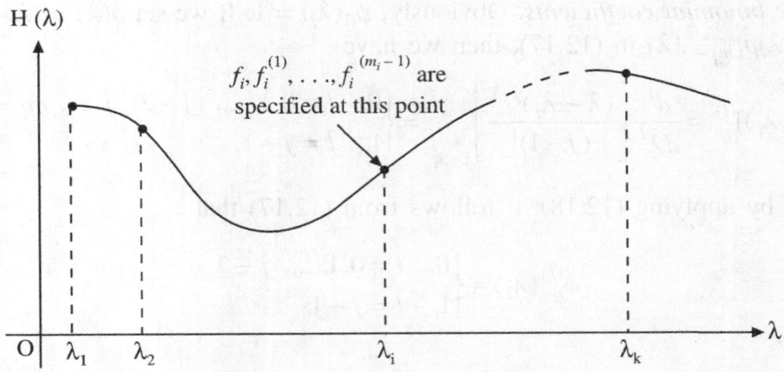

Figure 12.2 Interpolating with Hermite polynomials.

We now define a polynomial of degree at most $(m - 1)$ by

$$H(\lambda) = \sum_{i=1}^{k} \sum_{j=1}^{m_i} f_i^{(j-1)} \phi_{ij}(\lambda) \tag{12.22}$$

which is known as *Hermite interpolating polynomial*. This polynomial assumes at $\lambda = \lambda_i$, the specified value f_i, and $m_i - 1$ number of its consecutive derivatives (see Figure 12.2):

$$H^{(1)}(\lambda_i) = f_i^{(1)}, H^{(2)}(\lambda_i) = f_i^{(2)}, \ldots, H^{(m_i-1)}(\lambda_i) = f_i^{(m_i-1)}$$

which can be easily verified by the properties of ϕ_{ij}'s given in (12.19) and (12.20). The Hermite interpolating polynomial passing through k number of points and satisfying $(m_i - 1)$ number of specified derivatives at each point $\lambda = \lambda_i$ is unique. Indeed, if $G(\lambda)$ is another polynomial of degree $(m - 1)$ satisfying these properties, then $p(\lambda) := H(\lambda) - G(\lambda)$ is at most of degree $(m - 1)$ and can be represented by (12.35a) such that

$$p(\lambda_i) = 0, \qquad p^{(1)}(\lambda_i) = 0, \ldots, p^{(m_i-1)}(\lambda_i) = 0; \qquad i = 1, 2, \ldots, k.$$

Following the notation given in (12.36a)–(12.36e), we have

$$A^T\mathbf{x} = \mathbf{\theta},$$

where A^T is a nonsingular matrix. Consequently, $\mathbf{x} = \mathbf{\theta}$ and $p(\lambda)$ is a zero polynomial. It follows therefore that $H(\lambda) = G(\lambda)$, i.e., $H(\lambda)$ is a unique interpolating polynomial satisfying the above properties.

Definition 12.1 Let $f(\lambda)$ be any arbitrary function and A be an $n \times n$ matrix over $\mathbb{C}$, with distinct eigenvalues λ_i of index m_i, $i = 1, 2, ..., k$. If the m numbers

$$f(\lambda_i), f^{(1)}(\lambda_i), ..., f^{(m_i-1)}(\lambda_i), \qquad i = 1, 2, ..., k \tag{12.23}$$

exist, then we say that *the function $f(\lambda)$ is defined on the spectrum of A* and we call these m numbers, the *values of the function $f(\lambda)$ on the spectrum of A*. □

We shall now prove an important theorem concerning two polynomials.

Theorem 12.1 Let $g(\lambda)$ and $h(\lambda)$ be two polynomials defined on the spectrum of an $n \times n$ matrix A over $\mathbb{C}$. Then, $g(A) = h(A)$ iff $g(\lambda)$ and $h(\lambda)$ have the same values on the spectrum of A.

Proof Suppose that $g(A) = h(A)$. Then, $g(A) - h(A) = 0$ which shows that $g(\lambda) - h(\lambda)$ is an annihilating polynomial of A and is consequently divisible by the minimum polynomial of $\psi(\lambda)$ of A. Therefore,

$$g(\lambda) - h(\lambda) = \psi(\lambda)q(\lambda)$$

for some polynomial $q(\lambda)$. In view of (12.3), $g(\lambda) - h(\lambda)$ vanishes on the spectrum of A, i.e.,

$$g(\lambda_i) = h(\lambda_i),\ g^{(1)}(\lambda_i) = h^{(1)}(\lambda_i),\ ...,\ g^{(m_i-1)}(\lambda_i) = h^{(m_i-1)}(\lambda_i), i = 1, 2, ..., k. \tag{12.24}$$

Conversely, if $g(\lambda)$ and $h(\lambda)$ have the same values on the spectrum of A, then (12.24) is satisfied. As a result, $g(\lambda) - h(\lambda)$ is divisible by $(\lambda - \lambda_i)^{m_i}$ for all $i = 1, 2,..., k$, i.e., $g(\lambda) - h(\lambda)$ is divisible by the minimum polynomial $\psi(\lambda)$ of A. So, $g(\lambda) - h(\lambda)$ is an annihilating polynomial of A satisfying $g(A) - h(A) = 0$. □

Theorem 12.1 attests to the fact that if two polynomials $g(\lambda)$ and $h(\lambda)$ have the same values on the spectrum of an $n \times n$ matrix A, then they give rise to the same matrix polynomial, i.e., $g(A) = h(A)$. In such cases, if $g(\lambda)$ is of higher degree than $h(\lambda)$, one can employ $h(A)$ to evaluate $g(A)$. On the other hand, if $f(\lambda)$ is *any arbitrary function* whose values on the spectrum of A are given by (12.23), then there exists a polynomial $H(\lambda)$ of least possible degree provided by (12.22), which has the same values on the spectrum of A as those of $f(\lambda)$. This significant observation is instrumental in offering the following definition of $f(A)$ *for any arbitrary function* $f(\lambda)$.

Definition 12.2 *Function of a Square Matrix A.* Let $f(\lambda)$ be *an arbitrary function* which has the same values on the spectrum of an $n \times n$ matrix A as those of a polynomial $p(\lambda)$. Then, $f(A)$, the function of a square matrix A associated with $f(\lambda)$, is defined as $p(A)$. □

Constructive procedures to evaluate the function of a square matrix are given in Examples 12.3 an 12.4.

Example 12.2 Let us consider the companion matrix

$$A=\begin{bmatrix}0 & 1 & 0\\ 0 & 0 & 1\\ 0 & 0 & 1\end{bmatrix} \tag{12.25}$$

whose minimum polynomial is $\psi(\lambda) = \lambda^2(\lambda - 1)$. Compute $f(A)$, if $f(\lambda) = 1/(\lambda^2 + 1)$.
Solution In this case,

$$\lambda_1 = 0, \quad m_1 = 2; \quad \lambda_2 = 1, \quad m_2 = 1; \quad k = 2.$$

The associated polynomials defined by (12.12) are

$$h_1(\lambda) = \frac{(\lambda - \lambda_2)}{(\lambda_1 - \lambda_2)} = -(\lambda - 1); \qquad h_2(\lambda) = \frac{(\lambda - \lambda_1)^2}{(\lambda_2 - \lambda_1)^2} = \lambda^2.$$

Consequently, by (12.16),

$$\frac{1}{h_1(\lambda)} = (1-\lambda)^{-1} = 1 + \lambda + \lambda^2 + \cdots,$$

$$p_{11}(\lambda) = 1 + \lambda, \; p_{10}(\lambda) = 1,$$

$$\frac{1}{h_2(\lambda)} = \lambda^{-2} = [1 + (\lambda - 1)]^{-2} = 1 - 2(\lambda - 1) + 3(\lambda - 1)^2 - \cdots,$$

$$p_{20}(\lambda) = 1,$$

$$\phi_{11}(\lambda) = h_1(\lambda)p_{11}(\lambda) = (1-\lambda)(1+\lambda) = 1 - \lambda^2,$$

$$\phi_{12}(\lambda) = (\lambda - \lambda_1)h_1(\lambda)p_{10}(\lambda) = \lambda(1-\lambda),$$

$$\phi_{21}(\lambda) = h_2(\lambda)p_{20}(\lambda) = \lambda^2$$

$$f_1 = f(\lambda_1) = 1, \quad f_1^{(1)} = f^{(1)}(\lambda_1) = 0, \quad f_2 = f(\lambda_2) = \frac{1}{2}.$$

Hence, inserting these values in (12.22), we obtain

$$H(\lambda) = f_1\phi_{11}(\lambda) + f_1^{(1)}\phi_{12}(\lambda) + f_2\phi_{21}(\lambda) = (2 - \lambda^2)/2.$$

Finally, by Definition 12.2, we have

$$f(A) = H(A) = \frac{1}{2}(-A^2 + 2I) = \frac{1}{2}\begin{bmatrix}2 & 0 & -1\\ 0 & 2 & -1\\ 0 & 0 & 1\end{bmatrix}.$$

Computing $f(A) = (A^2 + I)^{-1}$, we also get the matrix as given above. □

It is important to note that, since $g(\lambda) = 1/\lambda(\lambda + 1)$ is not defined at $\lambda = 0$, the associated matrix function $g(A)$ does not exist for A given by (12.25).

12.2 SPECTRAL RESOLUTION $f(A)$ WHEN A IS ARBITRARY

In Theorem 7.35, we have established that when A is a simple matrix, any matrix polynomial $f(A)$ can be expressed in terms of the spectrum $f(\lambda_i)$'s of $f(A)$, where λ_i's are the eigenvalues of A. We shall now generalize this result in two ways: (i) when A is *not necessarily simple* and (ii) when $f(A)$ is *an arbitrary* function of a matrix. In fact, we shall prove the following result (cf. Theorem 7.35).

Theorem 12.2 *Spectral Resolution of $f(A)$.* If A is a matrix in $\mathbb{C}^{n\times n}$ and the function $f(\lambda)$ is defined on the spectrum of A with values

$$f(\lambda_i) = f_i,\ f^{(1)}(\lambda_i) = f_i^{(1)},\ \ldots,\ f^{(m_i-1)}(\lambda_i) = f_i^{(m_i-1)} \tag{12.26}$$

for all $i = 1, \ldots, k$, where λ_i is an eigenvalue of A with index m_i, then there exist matrices Z_{ij} in $\mathbb{C}^{n\times n}$ which depend only on the properties of A and are independent of $f(\lambda)$ such that

$$f(A) = \sum_{i=1}^{k} \sum_{j=1}^{m_i} f_i^{(j-1)} Z_{ij}. \tag{12.27}$$

The matrices Z_{ij} are called *constituent matrices* of A and commute with A.

Proof The polynomial $H(\lambda)$ defined by (12.22) and $f(\lambda)$ have the same values on the spectrum of A. Therefore, by Definition 12.2,

$$f(A) = H(A)$$

which gives (12.27), where

$$Z_{ij} = \phi_{ij}(A) \tag{12.28}$$

and $\phi_{ij}(\lambda)$ is defined by (12.17). Since $\phi_{ij}(\lambda)$ is defined with the help of the minimum polynomial $\psi(\lambda)$ of A, Z_{ij}'s depend only on the properties of A, and obviously they commute with A. ☐

Lemma 12.1 Let $p(\lambda)$ and $q(\lambda)$ be two arbitrary polynomials. Then, $p(A) = q(A)$ iff $p(\lambda) - q(\lambda)$ is divisible by the minimum polynomial $\psi(\lambda)$ of A.

Proof Suppose that $p(A) = q(A)$. Then, $p(A) - q(A) = 0$ which shows that $p(\lambda) - q(\lambda)$ is an annihilating polynomial of A and so is divisible by $\psi(\lambda)$. Conversely, if $p(\lambda) - q(\lambda)$ is divisible by $\psi(\lambda)$, then $p(\lambda) - q(\lambda) = q(\lambda)\psi(\lambda)$ for some polynomial $q(\lambda)$. Since $\psi(A) = 0$, it follows from above that $p(A) - q(A) = 0$, i.e., $p(A) = q(A)$. ☐

Theorem 12.3 The constituent matrices Z_{ij} ($i = 1, 2, \ldots, k$; $j = 1, 2, \ldots, m_i$) of A in Theorem 12.2 have the following important properties:

(i) *Resolution of identity* $\sum_{i=1}^{k} Z_{i1} = I$.

(ii) $Z_{i1}^2 = Z_{i1}$. (12.29)

(iii) $Z_{i1}Z_{it} = Z_{it}Z_{i1} = Z_{it}$; $t = 1, 2, \ldots, m_i$.

(iv) $Z_{ij}Z_{lt} = Z_{lt}Z_{ij} = 0$, $i \neq l$.

Proof (i) If we consider $f(\lambda) = 1$, then all its derivatives in (12.26) are equal to zero and, consequently, from (12.27) we get the required result. This formula is called the *resolution of identity*.

(ii) It is a particular case of (iii) proved below.

(iii) In view of properties of $\phi_{ij}(\lambda)$ given by (12.19) to (12.21), we have

$$\phi_{i1}(\lambda_i) = 1, \qquad \phi_{i1}^{(1)}(\lambda_i) = \phi_{i1}^{(2)}(\lambda_i) = \cdots = \phi_{i1}^{(m_i-1)}(\lambda_i) = 0$$

which means that $\phi_{i1}(\lambda) - 1$ has $(\lambda - \lambda_i)^{m_i}$ as the factor. Since $\phi_{it}(\lambda)$, as given by (12.12) and (12.17), has all the factors of the minimum polynomial $\psi(\lambda)$ of A except $(\lambda - \lambda_i)^{m_i}$, it is evident that $\phi_{it}(\lambda)\{\phi_{i1}(\lambda) - 1\}$ is divisible by $\psi(\lambda)$. Hence by Lemma 12.1,

$$\phi_{it}(A)\{\phi_{i1}(A) - I\} = 0$$

which, together with (12.28), proves (iii).

(iv) For $i \neq l$, the polynomial

$$\phi_{ij}(\lambda)\phi_{lt}(\lambda)$$

has $\psi(\lambda)$ as a factor in view of (12.12) and (12.17). This proves the assertion by Lemma 12.1. □

We have already observed that, a matrix A, for which $A = A^2$, is said to be *idempotent* (see Section 7.7, Remark 7.1). The matrices Z_{i1}, $i = 1, 2, ..., k$, are idempotent as is evident from (ii) of Theorem 12.3. Once these idempotent elements of the constituent matrices of A are known, the rest of the constituent matrices can be determined as a result of the following theorem.

Theorem 12.4 The constituent matrices Z_{ij} of A, for all $i = 1, ..., k$; $j = 1, ..., m_i$, have the following property:

$$Z_{ij} = \frac{1}{(j-1)!}(A - \lambda_i I)^{j-1} Z_{i1} = \frac{1}{(j-1)!} Z_{i1}(A - \lambda_i I)^{j-1}. \tag{12.30}$$

Proof Employing (12.16) and (12.17), we can write

$$\frac{(\lambda - \lambda_i)^{j-1}}{(j-1)!}\phi_{i1}(\lambda) - \phi_{ij}(\lambda) = \frac{(\lambda - \lambda_i)^{j-1}}{(j-1)!} h_i(\lambda)[a_{i,m_i-j+1}(\lambda_i)(\lambda - \lambda_i)^{m_i-j+1}$$

$$+ \cdots + a_{i,m_i-1}(\lambda_i)(\lambda - \lambda_i)^{m_i-1}]$$

which has obviously $\psi(\lambda)$ as a factor. Hence, by Lemma 12.1 and (12.28), relation (12.30) follows. □

Corollary 12.1 The constituent matrices of A satisfy the relations

$$Z_{i1}, Z_{i2}, Z_{i3} = \frac{1}{2!} Z_{i2}^2, \ldots, Z_{im_i} = \frac{1}{(m_i - 1)!} Z_{i2}^{m_i-1}, Z_{i2}^{m_i} = 0 \tag{12.31}$$

for all $i = 1, 2, ..., k$.

Proof We have, by (12.30),

$$Z_{i2} = (A - \lambda_i I)Z_{i1} = Z_{i1}(A - \lambda_i I),$$

$$Z_{i3} = \frac{1}{2!}(A - \lambda_i I)^2 Z_{i1} = \frac{1}{2!}(A - \lambda_i I)Z_{i2}.$$

Premultiplying both sides of the last equation by Z_{i1} and taking cognizance of (iii) in Theorem 12.3, we get

$$Z_{i3} = \frac{1}{2!} Z_{i2}^2$$

etc. Finally, from (12.30),

$$Z_{im_i} = \frac{1}{(m_i - 1)!}(A - \lambda_i I)^{m_i - 1} Z_{i1} = \frac{1}{(m_i - 1)!}(A - \lambda_i I)^{m_i - 2} Z_{i2}.$$

Therefore,

$$Z_{im_i} = Z_{i1} Z_{im_i} = \frac{1}{(m_i - 1)!}(A - \lambda_i I)^{m_i - 3} Z_{i2}^2 = \cdots = \frac{1}{(m_i - 1)!} Z_{i2}^{m_i - 1}.$$

Moreover, by (12.17),

$$\phi_{i2}(\lambda) = (\lambda - \lambda_i) h_i(\lambda) p_{i,m_i - 2}(\lambda)$$

which shows that $\phi_{i2}^{m_i}(\lambda)$ contains $\psi(\lambda)$, the minimum polynomial of A, as a factor since $h_i(\lambda)$ has all the factors of $\psi(\lambda)$ except those pertaining to $(\lambda - \lambda_i)$. Hence,

$$Z_{i2}^{m_i} = \phi_{i2}^{m_i}(A) = 0.$$ □

Remark 12.1 A matrix A is called *nilpotent* if there exists a positive integer k such that $A^k = 0$. The lowest exponent k for which $A^k = 0$ is called the *degree of nilpotency* of A. For example, the matrix

$$A = \begin{bmatrix} 0 & 1 & 2 \\ 0 & 0 & 1 \\ 0 & 0 & 0 \end{bmatrix}$$

is nilpotent of degree 3. The elements Z_{i2} in the constituent matrices of A is nilpotent of degree m_i, $i = 1, 2, \ldots, k$. We shall call Z_{i1} the idempotent and Z_{i2} the nilpotent corresponding to λ_i.

Theorem 12.5 *Partial Fraction Decomposition of the Resolvent.* The resolvent $(\lambda I - A)^{-1}$ of the matrix A in $\mathbb{C}^{n\times n}$ can be decomposed into a partial fraction of the form

$$R(\lambda, A) = (\lambda I - A)^{-1} = \sum_{i=1}^{k} \left\{ \frac{Z_{i1}}{(\lambda - \lambda_i)} + \frac{Z_{i2}}{(\lambda - \lambda_i)^2} + \cdots + \frac{Z_{i2}^{m_i - 1}}{(\lambda - \lambda_i)^{m_i}} \right\}, \tag{12.32}$$

where k represents the number of distinct eigenvalues of A and

(i) $Z_{i1} + Z_{i2} + \cdots + Z_{ik} = I_n$,

(ii) $Z_{i1}Z_{j1} = \delta_{ij} Z_{i1}$, $\quad Z_{i1}Z_{i2} = Z_{i2}Z_{i1} = Z_{i2}$;

(iii) $Z_{i1}Z_{j2} = 0$, $\quad Z_{i2}Z_{j2} = 0$, $\quad$ if $i \neq j$.

Proof Here we define the function $f(\mu)$ by

$$f(\mu) = (\lambda - \mu)^{-1},$$

where $(j-1)$th derivative at $\mu = \lambda_i$ is

$$f_i^{(j-1)} := f^{(j-1)}(\lambda_i) = \frac{(j-1)!}{(\lambda - \lambda_i)^j}, \qquad j = 1, 2, \ldots, m_i.$$

This, along with (12.30) and (12.31) when put in (12.27), gives the desired result. □

The criterion that defines a function $f(A)$ of a given square matrix A is provided explicitly by Definition 12.2. Making sure that $f(A)$ is defined for a given matrix A, the next problem one encounters is, how to compute such $f(A)$ systematically. The obvious procedure would be to employ the defining expression for $f(A)$ given by (12.27). This will be illustrated by the following example.

Example 12.3 *Computation of $f(A)$ Using Hermite Interpolation Polynomials* Compute exp (At), where t is a real number and

$$A = \begin{bmatrix} 8 & 9 & -9 & 0 \\ 0 & 2 & 0 & 2 \\ 4 & 6 & -4 & 0 \\ 0 & 0 & 0 & 3 \end{bmatrix}. \tag{12.33}$$

Solution At first, we reduce $(\lambda I - A)$ to Smith canonical form following the constructive procedure given in Lemma 10.1 to get

$$(\lambda I - A) = \begin{bmatrix} \lambda - 8 & -9 & 9 & 0 \\ 0 & \lambda - 2 & 0 & -2 \\ -4 & -6 & \lambda + 4 & 0 \\ 0 & 0 & 0 & \lambda - 3 \end{bmatrix} \sim \begin{bmatrix} 1 & 0 & 0 & 0 \\ 0 & 1 & 0 & 0 \\ 0 & 0 & (\lambda - 2) & 0 \\ 0 & 0 & 0 & (\lambda - 2)^2(\lambda - 3) \end{bmatrix}.$$

Hence, in view of Problem 10.23, the minimum polynomial of A is

$$\psi(\lambda) = (\lambda - 2)^2(\lambda - 3).$$

Here, $\lambda_1 = 2$, $\lambda_2 = 3$, $m = 3$, $m_1 = 2$, $m_2 = 1$. Using (12.12), we now construct $h_1(\lambda)$ and $h_2(\lambda)$ to get

$$h_1(\lambda) = -(\lambda - 3) \qquad \text{and} \qquad h_2(\lambda) = (\lambda - 2)^2.$$

Expanding the reciprocal of the above polynomials by Taylor's series about $\lambda = 2$ and $\lambda = 3$ respectively, we have

$$\frac{1}{h_1(\lambda)} = \frac{1}{[1-(\lambda-2)]} = 1 + (\lambda-2) + (\lambda-2)^2 + \cdots,$$

$$p_{11}(\lambda) = \lambda - 1, \quad p_{10}(\lambda) = 1,$$

$$\frac{1}{h_2(\lambda)} = \frac{1}{[1+(\lambda-3)]^2} = 1 - 2(\lambda-3) + 3(\lambda-3)^2 - 4(\lambda-3)^3 + \ldots$$

$$p_{20}(\lambda) = 1.$$

Consequently, from (12.17),

$$\phi_{11}(\lambda) = h_1(\lambda)p_{11}(\lambda) = -(\lambda-3)(\lambda-1),$$

$$\phi_{12}(\lambda) = (\lambda-2)h_1(\lambda)p_{10}(\lambda) = -(\lambda-2)(\lambda-3),$$

$$\phi_{21}(\lambda) = h_2(\lambda)p_{20}(\lambda) = (\lambda-2)^2.$$

By (12.28), we now define

$$Z_{11} = \phi_{11}(A) = (A-3I)(I-A), \qquad Z_{12} = \phi_{12}(A) = (A-3I)(2I-A),$$

$$Z_{21} = \phi_{21}(A) = (A-2I)^2.$$

□ Inserting A from (12.33) into the above expressions, the constituent matrices of A are obtained as

$$Z_{11} = \begin{bmatrix} 5 & 9 & -9 & 0 \\ 0 & -1 & 0 & 2 \\ 4 & 6 & -7 & 0 \\ 0 & 0 & 0 & 0 \end{bmatrix} \begin{bmatrix} -7 & -9 & 9 & 0 \\ 0 & -1 & 0 & -2 \\ -4 & -6 & 5 & 0 \\ 0 & 0 & 0 & -2 \end{bmatrix} = \begin{bmatrix} 1 & 0 & 0 & -18 \\ 0 & 1 & 0 & -2 \\ 0 & 0 & 1 & -12 \\ 0 & 0 & 0 & 0 \end{bmatrix},$$

$$Z_{12} = \begin{bmatrix} 5 & 9 & -9 & 0 \\ 0 & -1 & 0 & 2 \\ 4 & 6 & -7 & 0 \\ 0 & 0 & 0 & 0 \end{bmatrix} \begin{bmatrix} -6 & -9 & 9 & 0 \\ 0 & 0 & 0 & -2 \\ -4 & -6 & 6 & 0 \\ 0 & 0 & 0 & -1 \end{bmatrix} = \begin{bmatrix} 6 & 9 & -9 & -18 \\ 0 & 0 & 0 & 0 \\ 4 & 6 & -6 & -12 \\ 0 & 0 & 0 & 0 \end{bmatrix},$$

$$Z_{21} = \begin{bmatrix} 6 & 9 & -9 & 0 \\ 0 & 0 & 0 & 2 \\ 4 & 6 & -6 & 0 \\ 0 & 0 & 0 & 1 \end{bmatrix}^2 = \begin{bmatrix} 0 & 0 & 0 & 18 \\ 0 & 0 & 0 & 2 \\ 0 & 0 & 0 & 12 \\ 0 & 0 & 0 & 1 \end{bmatrix}.$$

Hence, by (12.27) $f(A)$ is given by

$$f(A) = f_1 Z_{11} + f_1^{(1)} Z_{12} + f_2 Z_{21}. \tag{12.34}$$

To determine exp (At), we consider the scalar function $f(\lambda) = \exp(\lambda t)$ which is defined at $\lambda_1 = 2$ and $\lambda_2 = 3$. Also,

$$f_1 = f(\lambda_1) = \exp(2t), \quad f_1^{(1)} = f^{(1)}(\lambda_1) = t\exp(2t), \quad f_2 = f(\lambda_2) = \exp(3t),$$

where $f^{(1)}(\lambda)$ is the derivative of $f(\lambda)$ with respect to λ. Finally, inserting these values in (12.34), we have

$$\exp(At) = \exp(2t)\begin{bmatrix} 1 & 0 & 0 & -18 \\ 0 & 1 & 0 & -2 \\ 0 & 0 & 1 & -12 \\ 0 & 0 & 0 & 0 \end{bmatrix} + t\exp(2t)\begin{bmatrix} 6 & 9 & -9 & -18 \\ 0 & 0 & 0 & 0 \\ 4 & 6 & -6 & -12 \\ 0 & 0 & 0 & 0 \end{bmatrix}$$

$$+\exp(3t)\begin{bmatrix} 0 & 0 & 0 & 18 \\ 0 & 0 & 0 & 2 \\ 0 & 0 & 0 & 12 \\ 0 & 0 & 0 & 1 \end{bmatrix}$$

$$= \begin{bmatrix} (1+6t)\exp(2t) & 9t\exp(2t) & -9t\exp(2t) & -18(1+t)\exp(2t)+18\exp(3t) \\ 0 & \exp(2t) & 0 & -2\exp(2t)+2\exp(3t) \\ 4t\exp(2t) & 6t\exp(2t) & (1-6t)\exp(2t) & -12(1+t)\exp(2t)+12\exp(3t) \\ 0 & 0 & 0 & \exp(3t) \end{bmatrix}. \quad \square$$

In the above example, we could employ (12.30) to compute Z_{12} after Z_{11} has been determined. In general, one can determine Z_{11}, Z_{21}, ..., Z_{k1} with the help of (12.17) and (12.28) and then take the advantage of the recursive relations (12.30) to construct Z_{i2}, ..., Z_{im_i}, $i = 1, 2, ..., k$.

12.2.1 Computation of *f*(*A*) Using Vandermonde Matrix

Alternatively, we can compute $f(A)$ by first constructing a scalar polynomial $p(\lambda)$ having on the spectrum of A the same values as those of $f(\lambda)$, and then $f(A)$ is given by $p(A)$. This is, in fact, Definition 12.2 for the function of a matrix. This polynomial $p(\lambda)$ can be taken as the Hermite interpolation polynomial $H(\lambda)$ given directly by (12.22). Otherwise, we can obtain it by a different process which we shall discuss now. We have already noted that, this polynomial $p(\lambda)$ having properties mentioned above is unique (see Problem 12.3 and the discussion following (12.22)) and of degree $m - 1$, and can be written as

$$p(\lambda) = \alpha_0 + \alpha_1\lambda + \cdots + \alpha_{m-2}\lambda^{m-2} + \alpha_{m-1}\lambda^{m-1}, \tag{12.35a}$$

where m is the degree of minimum polynomial of A.

The coefficients α_i's are unknown and should be so determined as to satisfy

$$p(\lambda_i) = f(\lambda_i), \quad p^{(1)}(\lambda_i) = f^{(1)}(\lambda_i), \quad ..., \quad p^{(m_i-1)}(\lambda_i) = f_i^{(m_i-1)}(\lambda_i), \tag{12.35b}$$

where $f(\lambda)$ is a scalar function corresponding to the given arbitrary function $f(A)$ of a known matrix A and λ_i's are its *distinct* eigenvalues of index m_i. Setting the values of (12.35b) in (12.35a), we have

$$\begin{aligned}
p(\lambda_i) &= \alpha_0 + \alpha_1\lambda_i + \cdots + \alpha_{m_i-1}\lambda_i^{m_i-1} + \cdots + \alpha_{m-2}\lambda_i^{m-2} + \alpha_{m-1}\lambda_i^{m-1},\\
p^{(1)}(\lambda_i) &= 0 + \alpha_1 + \cdots + (m_i-1)\alpha_{m_i-1}\lambda_i^{m_i-2} + \cdots + (m-2)\,\alpha_{m-2}\lambda_i^{m-3}\\
&\qquad + (m-1)\,\alpha_{m-1}\lambda_i^{m-2},\\
&\vdots\\
p^{(m_i-1)}(\lambda_i) &= 0 + 0 + \cdots + (m_i-1)!\,\alpha_{m_i-1} + \cdots + \frac{(m-2)!\,\alpha_{m-2}\lambda_i^{m-m_i-1}}{(m-m_i-1)!}\\
&\qquad + \frac{(m-1)!\,\alpha_{m-1}\lambda_i^{m-m_i}}{(m-m_i)!}.
\end{aligned} \tag{12.36a}$$

Let us now define

$$A_i = \begin{bmatrix}
1 & 0 & \cdots & 0\\
\lambda_i & 1 & \cdots & 0\\
\lambda_i^2 & 2\lambda_i & \cdots & 0\\
\vdots & \vdots & & \vdots\\
\lambda_i^{m_i-1} & (m_i-1)\lambda_i^{m_i-2} & \cdots & (m_i-1)!\\
\vdots & \vdots & & \vdots\\
\lambda_i^{m-2} & (m-2)\lambda_i^{m-3} & \cdots & \dfrac{(m-2)!\,\lambda_i^{m-m_i-1}}{(m-m_i-1)!}\\
\lambda_i^{m-1} & (m-1)\lambda_i^{m-2} & \cdots & \dfrac{(m-1)!\,\lambda_i^{m-m_i}}{(m-m_i)!}
\end{bmatrix}_{m\times m_i}, \tag{12.36b}$$

$$\mathbf{b} = \left[\, p(\lambda_1)\, p^{(1)}(\lambda_1) \cdots p^{(m_1-1)}(\lambda_1);\, p(\lambda_2) \cdots p^{(m_k-1)}(\lambda_k)\right]^T, \tag{12.36c}$$

$$\mathbf{x} = \left[\,\alpha_0\ \alpha_1 \cdots \alpha_{m_i-1} \cdots \alpha_{m-1}\right]^T, \tag{12.36d}$$

$$A = \left[A_1\ A_2 \cdots A_k\right]_{m\times m}, \tag{12.36e}$$

where $\quad i = 1, 2, ..., k, \qquad m_i \le m; \qquad m_1 + m_2 + \cdots + m_k = m.$

The elements of column 2 of A_i are thought of as the first derivative of the components in column 1, ..., the elements in column $j + 1$ are the first derivative of the components in column

j. The matrix A is called the *generalized* or *confluent Vandermonde matrix*. Obviously, A_i is of full rank m_i. Since λ_i's are distinct, by considering the columns of A, it is not hard to see that rank $A = m$, i.e., A is nonsingular. Equations in (12.36) can now be written as

$$\mathbf{b}^T = \mathbf{x}^T A, \qquad \text{or} \qquad A^T\mathbf{x} = \mathbf{b}.$$

Hence,

$$\mathbf{x} = (A^{-1})^T \mathbf{b} \tag{12.37}$$

which determines the coefficients of the Hermite polynomial (12.35). We now illustrate this procedure for the computation of $f(A)$ by an example.

Example 12.4 *Computation of f(A) Using Vandermonde Matrix.* Let us compute $f(A) = \exp(At)$ when A is given by (12.33) as in Example 12.3.

Solution Since the degree of minimum polynomial of A is $m = 3$, we find that $p(\lambda)$ is given by

$$p(\lambda) = \alpha_0 + \alpha_1\lambda + \alpha_2\lambda^2. \tag{12.38}$$

The eigenvalues of A are $\lambda_1 = 2$ of index $m_1 = 2$ and $\lambda_2 = 3$ of index $m_2 = 1$. Now $f(\lambda) = \exp(\lambda t)$, and for $f(\lambda)$ and $p(\lambda)$ to have the same values on the spectrum of A, it is necessary that

$$\exp(2t) = \alpha_0 + 2\alpha_1 + 4\alpha_2, \qquad t\exp(2t) = \alpha_1 + 4\alpha_2,$$
$$\exp(3t) = \alpha_0 + 3\alpha_1 + 9\alpha_2.$$

Hence,

$$\begin{bmatrix}\alpha_0\\ \alpha_1\\ \alpha_2\end{bmatrix} = \begin{bmatrix}1 & 2 & 4\\ 0 & 1 & 4\\ 1 & 3 & 9\end{bmatrix}^{-1}\begin{bmatrix}\exp(2t)\\ 2\exp(2t)\\ \exp(3t)\end{bmatrix} = \begin{bmatrix}-3 & -6 & 4\\ 4 & 5 & -4\\ -1 & -1 & 1\end{bmatrix}\begin{bmatrix}\exp(2t)\\ 2\exp(2t)\\ \exp(3t)\end{bmatrix}$$

$$= \begin{bmatrix}-3(1+2t)\exp(2t) + 4\exp(3t)\\ (4+5t)\exp(2t) - 4\exp(3t)\\ -(1+t)\exp(2t) + \exp(3t)\end{bmatrix}.$$

Therefore,

$$\begin{aligned} f(A) = p(A) &= \alpha_0 I + \alpha_1 A + \alpha_2 A^2\\ &= [-3(1+2t)\exp(2t) + 4\exp(3t)]I + [(4+5t)\exp(2t) - 4\exp(3t)]A\\ &\quad + [-(1+t)\exp(2t) + \exp(3t)]A^2 \end{aligned}$$

$$= [-3(1+2t)\exp(2t) + 4\exp(3t)]\begin{bmatrix}1 & 0 & 0 & 0\\ 0 & 1 & 0 & 0\\ 0 & 0 & 1 & 0\\ 0 & 0 & 0 & 1\end{bmatrix}$$

$$+[(4+5t)\exp(2t)-4\exp(3t)]\begin{bmatrix}8 & 9 & -9 & 0\\ 0 & 2 & 0 & 2\\ 4 & 6 & -4 & 0\\ 0 & 0 & 0 & 3\end{bmatrix}$$

$$+[-(1+t)\exp(2t)+\exp(3t)]\begin{bmatrix}28 & 36 & -36 & 18\\ 0 & 4 & 0 & 10\\ 16 & 24 & -20 & -12\\ 0 & 0 & 0 & 9\end{bmatrix}$$

$$=\exp(2t)\begin{bmatrix}(1+6t) & 9t & -9t & -18(1+t)\\ 0 & 1 & 0 & -2\\ 4t & 6t & 1-6t & -12(1+t)\\ 0 & 0 & 0 & 0\end{bmatrix}+\exp(3t)\begin{bmatrix}0 & 0 & 0 & 18\\ 0 & 0 & 0 & 2\\ 0 & 0 & 0 & 12\\ 0 & 0 & 0 & 1\end{bmatrix}$$

which is identical with what we obtained in Example 12.3. □

We conclude this section by establishing, with the help of the following lemma, an important property of constituent matrices Z_{ij}, i.e., they are linearly independent elements in $\mathbb{C}^{n\times n}$.

Lemma 12.2 The m number of polynomials $\phi_{ij}(\lambda)$'s described by (12.17) are linearly independent elements of $\mathbb{P}_{m-1}(\lambda)$.

Proof The polynomials $\phi_{ij}(\lambda)$'s can be written explicitly as

$$\phi_{ij}(\lambda)=\frac{(\lambda-\lambda_i)^{j-1}}{(j-1)!}\left[\prod_{\substack{j=1\\ j\neq i}}^{k}\frac{(\lambda-\lambda_j)^{m_j}}{(\lambda_i-\lambda_j)^{m_j}}\right][1+a_{i1}(\lambda_i)(\lambda-\lambda_i)+\cdots$$
$$+a_{im_i-j}(\lambda_i)(\lambda-\lambda_i)^{m_i-j}], \quad (12.39)$$

where $i = 1, 2, ..., k$; $j = 1, 2, ..., m_i$. We shall now prove that $\phi_{i1}(\lambda), \phi_{i2}(\lambda), ..., \phi_{im_i}(\lambda)$ are linearly independent elements of $\mathbb{P}_{m-1}(\lambda)$. Suppose that for scalars $\alpha_1, \alpha_2, ..., \alpha_{m_i}$,

$$\alpha_1\phi_{i1}(\lambda)+\alpha_2\phi_{i2}(\lambda)+\cdots+\alpha_{m_i}\phi_{im_i}(\lambda)=0. \quad (12.40)$$

By (12.19), $\phi_{i1}(\lambda_i) = 1$ and $\phi_{i2}(\lambda_i) = \cdots = \phi_{im_i}(\lambda_i) = 0$ which, when inserted in (12.40) at $\lambda = \lambda_i$, gives $\alpha_1 = 0$. Differentiating (12.40) once and inserting from (12.19), we have $\alpha_2 = 0$. In this way, all α_i's in (12.40) are zero, and, consequently, $\phi_{i1}(\lambda), \phi_{i2}(\lambda), ..., \phi_{im_i}(\lambda)$ are linearly independent. Again, if

$$c_1\phi_{i1}(\lambda)+c_2\phi_{j1}(\lambda)=0, \qquad i\neq j \quad (12.41)$$

for scalars c_1 and c_2, then in view of (12.19) and (12.21) at $\lambda = \lambda_i$, we have $\phi_{i1}(\lambda_i) = 1$ but $\phi_{j1}(\lambda_i) = 0$, which when put in (12.41) gives $c_1 = 0$. Similarly for $\lambda = \lambda_j$, from (12.41), $c_2 = 0$. This shows that $\phi_{i1}(\lambda)$ and $\phi_{j1}(\lambda)$, $i \neq j$ are linearly independent. It follows therefore that $\phi_{i1}(\lambda)$'s defined by (12.39) are linearly independent. □

Theorem 12.6 If A is a matrix in $\mathbb{C}^{n \times n}$, then its *constituent matrices* Z_{ij} defined by (12.28) are linearly independent elements of $\mathbb{C}^{n \times n}$.

Proof By definition, $Z_{ij} = \phi_{ij}(A)$, where $\phi_{ij}(\lambda)$'s are polynomials of degree less than or equal to $(m - 1)$ given by (12.39). Suppose now that for scalars α_{ij} in $\mathbb{C}$,

$$\sum_{i=1}^{k} \sum_{j=1}^{m_i} \alpha_{ij} \phi_{ij}(A) = 0 \tag{12.42}$$

which is a null matrix. We now define a polynomial $p(\lambda)$ in $\mathbb{P}_{m-1}(\lambda)$ by

$$p(\lambda) = \sum_{i=1}^{k} \sum_{j=1}^{m_i} \alpha_{ij} \phi_{ij}(\lambda) \tag{12.43}$$

which has a maximum degree $m - 1$. Hence by (12.42), $p(A) = 0$ which shows that A has an annihilating polynomial $p(\lambda)$ of degree less than m, the degree of minimum polynomial of A. This is contradictory by the definition of a minimum polynomial. Consequently, $p(\lambda) = 0$, and by Lemma 12.2, from (12.43), we have $\alpha_{ij} = 0$ for $i = 1, ..., k; j = 1, ..., m_i$. It now follows from (12.42) that the matrices $\phi_{ij}(A)$ are linearly independent in $\mathbb{C}^{n \times n}$. □

12.3 SQUARE ROOT OF A MATRIX A, sin A, cos A, ln A

A natural generalization of a positive number is a positive definite (PD) matrix. The square root of a positive number is well defined, so it is interesting to know if a PD matrix possesses a PD square root. In the following a PD (or PSD) matrix will refer to a real symmetric or a Hermitian matrix, that is PD (or PSD).

Definition 12.3 *Square Root of a Matrix.* Let A be a PD (or PSD) matrix. If there exists a matrix A_s such that $A_s^2 = A$, then A_s is defined to be the square root of A and is denoted by $A_s := A^{1/2}$. Also represented as $f(A) = A^{1/2}$, a square root is a function of a matrix A. □

Example 12.5 The matrices

$$A_s = \frac{1}{2}\begin{bmatrix} 3 & -1 \\ -1 & 3 \end{bmatrix} \quad \text{and} \quad H_s = \frac{1}{\sqrt{2}}\begin{bmatrix} 1 & i \\ -i & 1 \end{bmatrix}$$

are respectively the square roots of

$$A = \frac{1}{2}\begin{bmatrix} 5 & -3 \\ -3 & 5 \end{bmatrix} \quad \text{and} \quad H = \begin{bmatrix} 1 & i \\ -i & 1 \end{bmatrix},$$

because, by direct multiplication, it can be verified that $A_s^2 = A$ and $H_s^2 = H$. In this example, A and A_s are both PD, while H and H_s are both PSD, each with rank equal to 1. On the other hand, $-A_s$ and $-H_s$ are also square roots of A and H respectively, however, they are respectively ND and NSD. This is analogous to a real number such as 4 having two square roots such as 2 and –2. In this text, by a square root of a PD (resp PSD) matrix A, we will refer to its PD (resp PSD) square root and denote it by $A^{1/2}$. □

Let H be a PD Hermitian matrix of order n. Then, by corollary 8.12, all of its eigenvalues $\lambda_1, \lambda_2, \ldots, \lambda_n$ are positive. By Theorem 7.27, there exists a unitary transformation U, $UU^* = U^*U = I$, such that

$$H = U \text{ diag } [\lambda_1, \lambda_2, \ldots, \lambda_n] \; U^*.$$

Define:

(1) $H_s = UD_sU^*$, where $D_s = \text{diag}\left[\sqrt{\lambda_1}, \sqrt{\lambda_2}, \ldots, \sqrt{\lambda_n}\right]$.

Then, $H^{1/2} := H_s$ is the PD square root of H. Indeed,

$$H_s^2 = UD_sU^*UD_sU^* = UD_s^2U^* = H.$$

The representation $H_s = UD_sU^*$ shows that, the eigenvalues of H_s are the diagonal elements of D_s which are all positive. Hence,

(2) rank (H) = rank (H_s).

For PSD matrices, the definition of a PSD square root (1), and rank condition (2) are also true. If rank $(H) = r$, then $H^{1/2}$ has r number of positive eigenvalues as it follows from the similarity relation:

$$H^{1/2} = UD_sU^*, \; U^*U = I.$$

(3) In general, for each real number $\alpha > 0$, we define

$$H^{\alpha} = UD_{\alpha}U^*, \; D_{\alpha} = \text{diag}[\lambda_1^{\alpha}, \lambda_2^{\alpha}, \ldots, \lambda_n^{\alpha}].$$

If $\alpha = \frac{1}{2}$, we get the square root of H, while if $\alpha = \frac{1}{3}$, we get the cube root of H, etc.

Theorem 12.7 (a) Let H be a PD Hermitian matrix. The PD square root H_s of H is unique and is given by $H_s := U D_sU^*$, where U is a unitary matrix such that $H = UDU^*$, D is a diagonal matrix containing the eigenvalues $\{\lambda_1, \lambda_2, \ldots, \lambda_n\}$ of H and $D_s = \text{diag}\left[\sqrt{\lambda_1}, \sqrt{\lambda_2}, \ldots, \sqrt{\lambda_n}\right]$.

(b) The matrices $(AA^*)^{1/2}$ and $(A^*A)^{1/2}$ have the same number of identical positive eigenvalues.

Proof (a) Let H_s be a square root of H given by (1). To prove the uniqueness, suppose H_{sr} is another square root of H such that $H = H_{sr}^2$. By Theorem 7.18, eigenvalues of H_{sr} are the positive square roots of those of H. Since $HH_{sr} = H_{sr}^3 = H_{sr}H$, H commute with H_{sr} and the same unitary transformation U diagonalizes both H and H_{sr} such that $H = UDU^*$ and $H_{sr} = UD_sU^*$ in view of Theorem 8.30. Consequently, by definition of a square root,

$$H_s := UD_sU^* = H_{sr}.$$

Similarly, we can prove that the square root of a positive semi-definite Hermitian matrix is unique.

(b) For an $m \times n$ matrix A, $m \neq n$, n-square matrix A^*A and the n-square matrix AA^* are PSD (see Problem 8.40). Therefore, both the matrices AA^* and A^*A have PSD square root. We prove the assertion in the theorem for the matrices AA^* and A^*A without the square root. As A^*A

is Hermitian, by Theorem 7.31, there exists a set of orthonormal eigenvectors $\mathbf{x}_1, \mathbf{x}_2, \ldots, \mathbf{x}_n$ associated with the eigenvalues $\lambda_1, \lambda_2, \ldots, \lambda_n$ such that

$$A^*A\mathbf{x}_i = \lambda_i \mathbf{x}_i, \quad \langle \mathbf{x}_i, \mathbf{x}_j \rangle = \delta_{ij} \text{ and } \langle A^*A\mathbf{x}_i, \mathbf{x}_j \rangle = \langle \lambda_i \mathbf{x}_i, \mathbf{x}_j \rangle = \lambda_i \delta_{ij}, 1 \le i \le n.$$

Hence, in view of Definition 5.9, for $i = j$,

$$\langle A\mathbf{x}_i, A\mathbf{x}_i \rangle = \lambda_i$$

which leads to

$$A\mathbf{x}_i = \boldsymbol{\theta} \Leftrightarrow \lambda_i = 0.$$

Since,

$$AA^*(A\mathbf{x}_i) = A(A^*A\mathbf{x}_i) = A(\lambda_i \mathbf{x}_i) = \lambda_i (A\mathbf{x}_i),$$

we have $A\mathbf{x}_i$ as an eigenvector of AA^* corresponding to the eigenvalue λ_i if $\lambda_i \neq 0$. Thus, if λ_i is a non-zero eigenvalue of A^*A with corresponding eigenvector $\mathbf{x}_i$, then λ_i is also an eigenvalue of AA^* with corresponding eigenvector $A\mathbf{x}_i$. Denoting by spec(A) the spectrum of a matrix A, we have proved that if

$$\lambda_i \in \text{spec}\,(A^*A),\ \lambda_i \neq 0 \Rightarrow \lambda_i \in \text{spec}\,(AA^*).$$

Interchanging the roles of A and A^* to start with, we can similarly show that

$$\lambda_i \in \text{spec}\,(AA^*),\ \lambda_i \neq 0 \Rightarrow \lambda_i \in \text{spec}\,(A^*A).$$

Combining these two results, we have the assertion (b) in the theorem.

Example 12.6 Determine $H^{1/2}$ if

$$H = \begin{bmatrix} 2 & i & 0 \\ -i & 1 & -i \\ 0 & i & 2 \end{bmatrix}.$$

Solution (a) *Using the Definition* As developed in Example 8.12, for the unitary matrix U given by

$$U = \frac{1}{\sqrt{6}} \begin{bmatrix} \sqrt{3} & \sqrt{2}i & -i \\ 0 & \sqrt{2} & 2 \\ -\sqrt{3} & \sqrt{2}i & -i \end{bmatrix},$$

we have $H = U$ diag [2, 3, 0] U^*. Therefore,

$$H^{1/2} = U\,\text{diag}\,[\sqrt{2}, \sqrt{3}, 0]\,U^* = \frac{1}{\sqrt{3}} \begin{bmatrix} \frac{\sqrt{6}}{2}+1 & i & -\frac{\sqrt{6}}{2}+1 \\ -i & 1 & -i \\ -\frac{\sqrt{6}}{2}+1 & i & \frac{\sqrt{6}}{2}+1 \end{bmatrix}.$$

Here, both H and $H^{1/2}$ are PSD Hermitian matrices.

(b) *Using Vandermonde Matrix* The eigenvalues of H are 2, 3 and 0 which are distinct, so the minimum polynomial of H is of degree 3. Hence, $p(\lambda) = \alpha_0 + \alpha_1\lambda + \alpha_2\lambda^2$ (see Example 12.4). To compute $f(H) = H^{1/2}$, we have $f(\lambda) = \sqrt{\lambda}$. For $p(\lambda)$ and $f(\lambda)$ to have the same eigenvalues on the spectrum of H,

$$\sqrt{2} = \alpha_0 + 2\alpha_1 + 4\alpha_2, \qquad \sqrt{3} = \alpha_0 + 3\alpha_1 + 9\alpha_2, \qquad \alpha_0 = 0.$$

Simplifying, $\alpha_0 = 0, \alpha_1 = \frac{3}{2}\sqrt{2} - \frac{2}{3}\sqrt{3}, \alpha_2 = \frac{1}{3}\sqrt{3} - \frac{1}{2}\sqrt{2}$. Therefore, $H^{1/2} = \alpha_0 I + \alpha_1 H + \alpha_2 H^2$ gives the same value as in (a).

(c) *Using Lagrange Interpolation Polynomials.* The minimum polynomial of H is $\psi(\lambda) = \lambda(\lambda - 2)(\lambda - 3)$. In this case, in (12.3), $m_1 = m_2 = m_3 = 1$; $\lambda_1 = 0$, $\lambda_2 = 2$ and $\lambda_3 = 3$. The Lagrange polynomials are

$$l_1(\lambda) = \frac{(\lambda - \lambda_2)(\lambda - \lambda_3)}{(\lambda_1 - \lambda_2)(\lambda_1 - \lambda_3)} = \frac{1}{6}(\lambda - 2)(\lambda - 3),$$

$$l_2(\lambda) = \frac{(\lambda - \lambda_1)(\lambda - \lambda_3)}{(\lambda_2 - \lambda_1)(\lambda_2 - \lambda_3)} = -\frac{1}{2}\lambda(\lambda - 3),$$

$$l_3(\lambda) = \frac{(\lambda - \lambda_1)(\lambda - \lambda_2)}{(\lambda_3 - \lambda_1)(\lambda_3 - \lambda_2)} = \frac{1}{3}\lambda(\lambda - 2).$$

To compute $f(H) = H^{1/2}$, we define $f(\lambda) = \sqrt{\lambda}$ so that $f_1 = f(\lambda_1) = 0$, $f_2 = f(\lambda_2) = \sqrt{2}$, $f_3 = f(\lambda_3) = \sqrt{3}$. Therefore, using (12.8)

$$f(\lambda) = -\frac{\sqrt{2}}{2}\lambda(\lambda - 3) + \frac{\sqrt{3}}{3}\lambda(\lambda - 2),$$

and

$$H^{1/2} = f(H) = \left(\frac{3}{2}\sqrt{2} - \frac{2}{3}\sqrt{3}\right)H + \left(\frac{1}{3}\sqrt{3} - \frac{1}{2}\sqrt{2}\right)H^2$$

which gives $H^{1/2}$ as in (b). □

Example 12.7 Compute (a) sin A and (b) cos A if

$$A = \frac{\pi}{2}\begin{bmatrix} 2 & 0 & 0 \\ 0 & 1 & 1 \\ 0 & 0 & 1 \end{bmatrix}.$$

Solution The eigenvalues of the triangular matrix A are its diagonal elements. It has one simple eigenvalue at $\lambda_1 = \pi$ and the other at $\lambda_2 = \pi/2$ of multiplicity 2. We can show that the characteristic polynomial and the minimum polynomial are the same for A, and of degree 3. So, in Vandermonde matrix method, the polynomial $p(\lambda)$ is assumed as $p(\lambda) = \alpha_0 + \alpha_1\lambda + \alpha_2\lambda^2$.

(a) For $p(\lambda)$ and $f(\lambda) = \sin \lambda$ to have the same values on the spectrum of A we have, using (12.36a)

$$0 = \alpha_0 + \alpha_1 \pi + \alpha_2 \pi^2, \qquad 1 = \alpha_0 + \frac{\alpha_1 \pi}{2} + \frac{\alpha_2 \pi^2}{4}, \qquad 0 = \alpha_1 + \alpha_2 \pi.$$

Solving we get $\alpha_0 = 0$, $\alpha_1 = 4/\pi$ and $\alpha_2 = -4/\pi^2$.

Therefore,

$$\sin A = \frac{4}{\pi} A - \frac{4}{\pi^2} A^2 = \begin{bmatrix} 0 & 0 & 0 \\ 0 & 1 & 0 \\ 0 & 0 & 1 \end{bmatrix}.$$

(b) In this case $f(\lambda) = \cos \lambda$ and $p(\lambda) = \alpha_0 + \alpha_1 \lambda + \alpha_2 \lambda^2$, so, we have using (12.36a) at $\lambda_1 = \pi$ and $\lambda_2 = \pi/2$ of multiplicity 2,

$$-1 = \alpha_0 + \alpha_1 \pi + \alpha_2 \pi^2, \qquad 0 = \alpha_0 + \frac{\alpha_1 \pi}{2} + \frac{\alpha_2 \pi^2}{4}, \qquad -1 = \alpha_1 + \alpha_2 \pi.$$

Solving we get, $\alpha_0 = \pi - 1$, $\alpha_1 = -3 + (4/\pi)$ and $\alpha_2 = (2\pi - 4)/\pi^2$. Therefore,

$$\cos A = (\pi - 1)I + (4 - 3\pi)\frac{1}{\pi} A + (2\pi - 4)\frac{1}{\pi^2} A^2$$

$$= \begin{bmatrix} -1 & 0 & 0 \\ 0 & 0 & -\pi/2 \\ 0 & 0 & 0 \end{bmatrix}$$

□

All solutions X of the matrix equation $\exp(X) = A$ are called *natural logarithms* of A and are denoted by $\ln A$. If ξ_i is an eigenvalue of X and λ_i that of A, then $\lambda_i = \exp(\xi_i)$ (see Theorem 7.18). So, if $\exp(X) = A$ has a solution, all eigenvalues of A are different from zero and A is nonsingular. The condition $\det A \neq 0$ is necessary for the existence of a solution of $\exp(X) = A$, and it can be shown that this condition is also sufficient.

Example 12.8 Determine In A if

$$A = \begin{bmatrix} 1 & 0 & 0 \\ 0 & 1 & 1 \\ 0 & 0 & 1 \end{bmatrix}.$$

Solution The minimum polynomial of A is $\phi(\lambda) = (\lambda - 1)^2$, and $\lambda_1 = 1$ is of multiplicity, $m = 2$. So, $p(\lambda) = \alpha_0 + \alpha_1 \lambda$ and $f(\lambda) = \ln \lambda$. Using (12.36a), we have $0 = \alpha_0 + \alpha_1$ and $1 = \alpha_1$, giving us $\alpha_0 = -1$ and $\alpha_1 = 1$. The desired value of $\ln A = \alpha_0 I + \alpha_1 A = A - I$ is:

$$\ln A = \begin{bmatrix} 0 & 0 & 0 \\ 0 & 0 & 1 \\ 0 & 0 & 0 \end{bmatrix}. \qquad \square$$

We can use Hermite interpolation polynomial and a generalization of Cauchy's integral formula (12.62) to evaluate sin A, cos A and ln A.

Polar decomposition

The polar representation of any complex number z is $z = re^{i\theta}$, $r \geq 0$, where $r = \sqrt{\bar{z}z} = \sqrt{z\bar{z}}$ and the number $q = e^{i\theta}$ has the property $\bar{q}q = q\bar{q} = 1$. This representation is generalized to $m \times n$ matrices A, $m \geq n$ as $A = QS$, where Q plays the role of $q = e^{i\theta}$, a rotation, satisfying $Q^*Q = I_n$ and S plays the role of r, a dilation, having the property $S = \sqrt{A^*A} \geq 0$, a positive semi-definite matrix. Thus, every matrix can be written as the product of a rotation and a dilation analogous to the polar representation of a complex number.

If A is a square matrix, then by SVD, $A = U\,DV^* = (UV^*)(V\,DV^*) = QS$, where $Q := UV^*$ is unitary and $S := V\,DV^*$ is Hermitian. Alternatively, $A = (UDU^*)(UV^*) = SQ$, where $S := UDU^*$ is Hermitian and $Q := UV^*$ is unitary. But for a non-square matrix A, we have the following result.

Theorem 12.8 *Polar Decomposition* (a) Let $A \in \mathbb{C}^{m\times n}$, $m \geq n$. Then, $\exists$ a matrix $Q \in \mathbb{C}^{m\times n}$ and a unique Hermitian positive semi-definite matrix $S \in \mathbb{C}^{n\times n}$ such that

$$A = QS, \quad Q^*Q = I_n, \text{ and } A^*A = S^2,$$

S is the square root of A^*A. If rank $A = n$, then S is positive definite and Q is uniquely determined.

(b) Let $A \in \mathbb{C}^{m\times n}$, $m \leq n$. Then, $\exists$ a matrix $Q \in \mathbb{C}^{m\times n}$ and a unique Hermitian positive semi-definite matrix $S \in \mathbb{C}^{m\times m}$ such that

$$A = SQ, \quad QQ^* = I_m, \text{ and } AA^* = S^2,$$

S is the square root of AA^*. If rank $A = n$, then S is positive definite and Q is uniquely determined.

Proof (a) Let $A \in \mathbb{C}^{m\times n}$, $m \geq n$ have the SVD

$$A = U\begin{bmatrix} \Sigma \\ 0 \end{bmatrix}V^*,$$

where $U \in \mathbb{C}^{m\times m}$ and $V = \in \mathbb{C}^{n\times n}$ are unitary and

$$\Sigma = \text{diag}[\sigma_1, \sigma_2, \ldots, \sigma_n], \text{ and } \sigma_1 \geq \sigma_2 \geq \cdots \geq \sigma_n \geq 0$$

are the singular values of A. Partitioning as

$$U = \underset{\quad n \qquad\quad m-n}{m[U_1 \;\; \vdots \;\; U_2]}, \quad U_1^*U_1 = I_n,$$

we have,

$$A = U_1\Sigma V^* = (U_1V^*)(V\Sigma V^*),$$

and setting $Q = U_1V^*$ and $S = V\Sigma V^*$ we get the polar decomposition $A = QS$.

Conversely, given the polar decomposition $A = QS \in \mathbb{C}^{m\times n}$, from a spectral decomposition $S = V\Sigma V^*$, $V^*V = I$, one can construct the singular value decomposition

$$A = (QV)\Sigma V^* = UDV^*,$$

where the $m \times n$ matrix $U_1 := QV$ with orthonormal columns is enlarged to an $m \times m$ unitary matrix $U = [U_1\ U_2]$ by appending $m - n$ number of orthonormal columns and Σ to an $m \times n$ matrix

$$D = \begin{bmatrix} \Sigma \\ 0 \end{bmatrix}$$

by appending $m - n$ zero rows.

If rank $A = n$, then rank (A) = rank $(A^*A) = n$ and S is unique and PD. In this case, $Q = S^{-1}A$ is unique.

(b) The proof is similar to (a) if we consider A^* in place of A in the above proof. Let $A \in \mathbb{C}^{m\times n}$, $m \le n$ have the SVD

$$A = U\,[\Sigma\ \ 0]\,V^*,$$

where $U \in \mathbb{C}^{m\times m}$ and $V = \in \mathbb{C}^{n\times n}$ are unitary and $\Sigma = \text{diag}\,[\sigma_1, \sigma_2, \ldots, \sigma_m]$, and $\sigma_1 \ge \sigma_2 \ge \cdots \ge \sigma_m \ge 0$ are the singular values of A. Partitioning V as

$$V = n[\underset{m}{V_1} \ \vdots\ \underset{n-m}{V_2}],\ V_1^*V_1 = I_m,$$

we have, $A = U\Sigma V_1^* = (U\Sigma U^*)(UV_1^*)$, and setting $S = U\Sigma U^*$ and $Q = UV_1^*$ we get the polar decomposition $A = SQ$. To find the polar decomposition of A = (1/15)[10 14 19; 20 –2 8], we note that its SVD computed by MATLAB is

$$U = \frac{1}{5}\begin{bmatrix} 4 & 3 \\ 3 & -4 \end{bmatrix},\quad D = \begin{bmatrix} 2 & 0 & 0 \\ 0 & 1 & 0 \end{bmatrix},\quad V = \frac{1}{3}\begin{bmatrix} 2 & -2 & 1 \\ 1 & 2 & 2 \\ 2 & 1 & -2 \end{bmatrix}$$

We define

$$V_1 = \frac{1}{3}\begin{bmatrix} 2 & -2 \\ 1 & 2 \\ 2 & 1 \end{bmatrix},\quad \Sigma = \begin{bmatrix} 2 & 0 \\ 0 & 1 \end{bmatrix},$$

so that the polar decomposition of A is $A = SQ$, where

$$S = U\Sigma U^T = \frac{1}{25}\begin{bmatrix} 41 & 12 \\ 12 & 34 \end{bmatrix}, \text{and } Q = UV_1^T = \frac{1}{15}\begin{bmatrix} 2 & 10 & 11 \\ 14 & -5 & 2 \end{bmatrix}. \qquad \square$$

12.4 AN ELEMENTARY PROOF OF JORDAN NORMAL FORM

The decomposition of the resolvent of A in Theorem 12.5 can be employed to give an alternative proof of Jordan normal form of an $n \times n$ matrix A over $\mathbb{C}$. This will be taken up in what follows. To start with, we observe that in view of (i) in Theorem 12.3, the idempotents Z_{i1}'s in the constituent matrices of A can be used for the resolution of identity I in $V: = \mathbb{C}^{n \times n}$ given by

$$Z_{11} + Z_{21} + \cdots + Z_{k1} = I. \tag{12.44}$$

The idempotents Z_{i1}'s are $n \times n$ matrices and, therefore, act on the elements $\mathbf{x}$ in the n-dimensional vector space V to give elements $\mathbf{x}_i: = Z_{i1}\mathbf{x}$ in V. If we look upon Z_{i1} as a linear transformation on V, i.e., $Z_{i1}: V \to V$, then $X_i = Z_{i1}(V)$ is a subspace of V (see the discussion following Definition 6.2) which is the range of Z_{i1} and defines a projection of V onto a linear subspace X_i of V. Thus, for any vector $\mathbf{x}$ in V, from (12.44), we have

$$\mathbf{x} = I\mathbf{x} = Z_{11}\mathbf{x} + Z_{21}\mathbf{x} + \cdots + Z_{k1}\mathbf{x} = \mathbf{x}_1 + \mathbf{x}_2 + \cdots + \mathbf{x}_k, \tag{12.45}$$

where $\mathbf{x}_i \in X_i$. Moreover, if

$$\mathbf{x} \in X_i \cap \sum_{j \neq i} X_j, \tag{12.46}$$

then, there are elements $\mathbf{y}$ and $\mathbf{y}_j$ in V such that

$$\mathbf{x} = Z_{i1}\mathbf{y} = (\sum_{j \neq i} Z_{j1}\mathbf{y}_j).$$

Hence, as a result of properties (ii) and (iv) in Theorem 12.3, we have

$$Z_{i1}\mathbf{x} = Z_{i1}\mathbf{y} = \mathbf{x} = Z_{i1}(\sum_{j \neq i} Z_{j1}\mathbf{y}_j) = \boldsymbol{\theta}$$

which shows that any vector belonging to the intersection defined by (12.46) is a zero vector for all $i = 1, 2, ..., k$. In view of Theorem 11.6, this fact together with (12.45) suggests that V is the direct sum of the subspaces X_i's i.e.,

$$V = X_1 \oplus X_2 \oplus \cdots \oplus X_k. \tag{12.47}$$

We shall now demonstrate that these subspaces X_i's are the same as the root subspaces S_i's of A associated with its eigenvalues λ_i's as in Definition 11.6. Indeed, we have from (12.30) and (12.31),

$$Z_{i2} = (A - \lambda_i I)Z_{i1} \quad \text{and} \quad Z_{i2}^{m_i} = 0.$$

By property (ii) in Theorem 12.3, these equations yield

$$(\lambda_i I - A)^{m_i} Z_{i1} = 0.$$

This shows that, each column of Z_{i1} belongs to the null space of $(\lambda_i I - A)^{m_i}$. Consequently,

$$X_i = \text{range space of } Z_{i1} \subseteq S_i \tag{12.48}$$

by Definition 11.6, where S_i is the *root subspace* of A associated with its eigenvalue λ_i. The assertion made above follows by comparing (12.47) with (see (11.36))

$$V = S_1 \oplus S_2 \oplus \cdots \oplus S_k \tag{12.49}$$

on the basis of (12.48). By (12.48),

$$(\dim S_i - \dim X_i) \geq 0. \tag{12.50}$$

Because of Corollary 11.3, from (12.47) and (12.49),

$$(\dim S_1 - \dim X_1) + (\dim S_2 - \dim X_2) + \cdots + (\dim S_k - \dim X_k) = 0.$$

It follows from this equation that

$$\dim S_i = \dim X_i \quad \text{and} \quad S_i = X_i,$$

since (12.50) has to be satisfied. Thus, we have established the result included in the following theorem.

Theorem 12.9 The idempotent component Z_{i1} of a matrix A in $V: = \mathbb{C}^{n \times n}$ project V onto the root subspace $S_i \subseteq V$ of A associated with its eigenvalue λ_i for all $i = 1, 2, ..., k$ with $\dim S_i = m_i$, where m_i is the index of A associated with λ_i. □

Let $\mathbb{N}_j$ denote the null space of the linear transformation $(\mu I - A)^j$ for all $j = 0, 1, ..., p$, where μ is an eigenvalue of A of index p and multiplicity ν. Because of Corollary 11.6, we have the following ascending chain conditions:

$$\{\mathbf{\theta}\} = \mathbb{N}_0 \subset \mathbb{N}_1 \subset \cdots \subset \mathbb{N}_p = S,$$

where S is the root subspace of A associated with μ. To construct a basis of S, let

$$\{\mathbf{x}_1, ..., \mathbf{x}_{r(p)}\} \tag{12.51}$$

be a basis of $\mathbb{N}_p$ over $\mathbb{N}_{p-1}$ (see Definition 11.7). The number of independent elements in $\mathbb{N}_p$ is a function of p, and is denoted by $r(p)$. We denote $C: = (A - \mu I)$ and claim that the vectors $\{C\mathbf{x}_1, ..., C\mathbf{x}_{r(p)}\}$ are linearly independent elements of $\mathbb{N}_{p-1}$ over $\mathbb{N}_{p-2}$. Obviously, these elements are in $\mathbb{N}_{p-1}$, (see Table 12.1):

$$\mathbf{\theta} = C^p \mathbf{x}_i = C^{p-1}(C\mathbf{x}_i), \qquad i = 1, 2, \ldots, r(p).$$

Suppose now that

$$\mathbf{x} = \alpha_1 C\mathbf{x}_1 + \cdots + \alpha_{r(p)} C\mathbf{x}_{r(p)} \in \mathbb{N}_{p-2}, \tag{12.52}$$

where the scalars α_i's are in $\mathbb{C}$. Premultiplying both sides of (12.52) by $\mathbb{C}^{p-2}$, we have

$$\mathbf{\theta} = \alpha_1 C^{p-1}\mathbf{x}_1 + \cdots + \alpha_{r(p)} C^{p-1}\mathbf{x}_{r(p)} = C^{p-1}(\alpha_1 \mathbf{x}_1 + \cdots + \alpha_{r(p)}\mathbf{x}_{r(p)})$$

which shows that $\alpha_1\mathbf{x}_1 + \cdots + \alpha_{r(p)}\mathbf{x}_{r(p)} \in \mathbb{N}_{p-1}$. In view of Definition 11.7, $\alpha_1 = \cdots = \alpha_{r(p)} = 0$. Taking note of (12.52), we now conclude that $C\mathbf{x}_1, ..., C\mathbf{x}_{r(p)}$ are linearly independent elements in $\mathbb{N}_{p-1}$ over $\mathbb{N}_{p-2}$. We now extend the above set to a basis of $\mathbb{N}_{p-1}$ over $\mathbb{N}_{p-2}$ by adjoining additional elements $\mathbf{x}_{r(p)+1}, ..., \mathbf{x}_{r(p-1)}$. We repeat this procedure and taking cognizance of Theorem 11.16, we show in Table 12.1 the set of vectors for the basis of S.

Table 12.1 Selection of a basis for Jordan normal form

	1	. . .	$r(p)$	$r(p)+1$	. . .	$r(p-1)$	. . .	↓ $r(j+2)+1$	. . .	↓ $r(j+1)$	. . .	$r(2)+1$	. . .	$r(1)$
$\mathbb{N}_p$	$\mathbf{x}_1$	. . .	$\mathbf{x}_{r(p)}$											
$\mathbb{N}_{p-1}$	$C\mathbf{x}_1$	. . .	$C\mathbf{x}_{r(p)}$	$\mathbf{x}_{r(p)+1}$	. . .	$\mathbf{x}_{r(p-1)}$								
⋮	⋮		⋮	⋮		⋮								
$\mathbb{N}_{j+1}$	$C^{p-j-1}\mathbf{x}_1$	. . .	$C^{p-j-1}\mathbf{x}_{r(p)}$	$C^{p-j-2}\mathbf{x}_{r(p)+1}$	. . .	$C^{p-j-2}\mathbf{x}_{r(p-1)}$	. . .	$\mathbf{x}_{r(j+2)+1}$	. . .	$\mathbf{x}_{r(j+1)}$				
⋮	⋮		⋮	⋮		⋮								
$\mathbb{N}_1$	$C^{p-1}\mathbf{x}_1$	. . .	$C^{p-1}\mathbf{x}_{r(p)}$	$C^{p-2}\mathbf{x}_{r(p)+1}$	. . .	$C^{p-2}\mathbf{x}_{r(p-1)}$	. . .	$C^j\mathbf{x}_{r(j+1)+1}$	. . .	$C^j\mathbf{x}_{r(j+1)}$	. . .	$\mathbf{x}_{r(2)+1}$	. . .	$\mathbf{x}_{r(1)}$
	T_1	. . .	$T_{r(p)}$	$T_{r(p)+1}$	. . .	$T_{r(p-1)}$	. . .	$T_{r(j+2)+1}$	. . .	$T_{r(j+1)}$	. . .	$T_{r(2)+1}$	. . .	$T_{r(1)}$

Any independent vectors in space $\mathbb{N}_1$ when premultiplied by C results in a zero vector, e.g.,

$$C \cdot C^j \mathbf{x}_{r(j+1)} = C^{j+1} \mathbf{x}_{r(j+1)} = \boldsymbol{\theta},$$

since $\mathbf{x}_{r(j+1)} \in \mathbb{N}_{j+1}$. This fact and that the elements of any column in Table 12.1 are linearly independent show that the listed vectors in Table 12.1 form C-cyclic and C-invariant subspaces $T_1, T_2, ..., T_{r(1)}$ of S. Thus,

$$S = T_1 \oplus T_2 \oplus \cdots \oplus T_{r(1)}, \tag{12.53}$$

where $r(1)$ = dimension of eigenspace of A associated with μ. Concentrating on any of the columns *marked by arrows* in Table 12.1, we see that these elements are of the form

$$X_i := \{\mathbf{x}, C\mathbf{x}, ..., C^j\mathbf{x}\} \tag{12.54}$$

which generate, say, the subspace T_i.

We shall now show that, this subspace T_i is A-invariant. Indeed, if $C^l\mathbf{x}$, $0 \le l \le j$, is any element of T_i, then as C commutes with A,

$$AC^l\mathbf{x} = C^l A\mathbf{x} = C^l(A - \mu I + \mu I)\mathbf{x} = C^{l+1}\mathbf{x} + \mu C^l\mathbf{x}, \qquad l = 0, 1, ..., j \tag{12.55}$$

which is a linear combination of basis vectors in T_i, observing that $C^{j+1}\mathbf{x} = \boldsymbol{\theta}$. We, therefore, find that the root subspace S is decomposed into a direct sum of A-invariant and *C-cyclic subspaces* of S determined by (12.53). If we concentrate on any T_i of these subspaces generated by the *cyclic basis* in (12.54), then the matrix of A restricted to T_i, in view of (12.55), with respect to this cyclic basis, is of order $s_i := (j + 1)$ and is given by the Jordan block

$$J_i = \begin{bmatrix} \mu & 1 & 0 & \cdots & 0 \\ 0 & \mu & 1 & \cdots & 0 \\ \vdots & \vdots & \vdots & & \vdots \\ 0 & 0 & 0 & \cdots & \mu \end{bmatrix}_{s_i}. \tag{12.56}$$

Consequently, the matrix of A restricted to S given by (12.53) with respect to the basis

$$\{X_1, X_2, ..., X_{r(1)}\},$$

where X_i is given by (12.54), is (see Theorem 11.10)

$$\text{diag}\,[J_1, J_2, ..., J_r] \tag{12.57}$$

with J_i as a Jordan block associated with μ of order s_i given by (12.56), $r(1) = r$,

$$s_1 \ge s_2 \ge \cdots \ge s_r, \qquad s_1 + s_2 + \cdots + s_r = v$$

$$s_1 = s_2 = \cdots = s_{r(p)} = p; \quad s_{r(p)+1} = \cdots = s_{r(p-1)} = p - 1; \cdots;$$

$$s_{r(2)+1} = \cdots = s_{r(1)} = 1.$$

Finally, focussing our attention on representation of V by (12.49) and any root subspace $S_i = S$ by (12.53), the matrix of A on V with respect to a basis which is a union of cyclic bases given in Table 12.1 for all eigenvalues can be described by (see Theorem 11.10)

$$\text{diag } [A_1, A_2, \ldots, A_k],$$

where A_i is the matrix restricted to $S_i = S$ associated with the eigenvalue $\lambda_i = \mu$ formed by the Jordan blocks as in (12.57), $v = n_i$ the multiplicity of λ_i in characteristic equation of A, and $p = m_i$ is the index of λ_i.

12.5 INTEGRAL REPRESENTATION OF $f(A)$

In this section, we shall study the representation of a function $f(A)$ of a matrix A, by using results from the complex variable theory. A function $f(z)$ of a complex variable z is called *analytic* at point $z = a$ if it is single-valued and differentiable at this point. It is analytic in a domain D if it is analytic at every point of D. Points at which $f(z)$ is not analytic are called poles or singular points. We now give some results from the complex variable theory.

Cauchy's Integral Formula. Let $f(z)$ be analytic in a domain D including the point $z = a$ which is bounded by the curve C. Then,

$$f(a) = \frac{1}{2\pi i} \int_C \frac{f(z)}{z-a} dz, \qquad i = \sqrt{-1}, \tag{12.58}$$

where the integral is to be evaluated in an *anticlockwise direction* along C. □

This allows us to calculate the value of $f(z)$ at an interior point of D from the values of $f(z)$ on its boundary C. The pth derivative of (12.58) can be shown to be

$$f^{(p)}(a) = \frac{p!}{2\pi i} \int_C \frac{f(z)}{(z-a)^{p+1}} dz, \qquad i = \sqrt{-1}. \tag{12.59}$$

In particular, if $f(z) = 1$, then from (12.58) and (12.59), we have

$$\int_C \frac{dz}{(z-a)^p} dz = \begin{cases} 2\pi i, & \text{if } p = 1, \quad i = \sqrt{-1}; \\ 0, & \text{otherwise}; \end{cases} \tag{12.60}$$

which is also true for nonpositive values of p as is evident from the following theorem.

Cauchy's Residue Theorem If a function $f(z)$ is analytic in a domain D but for a finite number of poles and is continuous on its boundary C, then

$$\int_C f(z)\, dz = 2\pi i \,\{\text{sum of residues of } f(z) \text{ at its poles in } D\}. \qquad □$$

The resolvent of a square matrix A as defined in Definition 7.16 is

$$R(z, A) = (zI - A)^{-1}$$

if z is not an eigenvalue of A. We now consider the scalar function associated with this resolvent, viz.,

$$f(\mu) = (z - \mu)^{-1}.$$

Differentiating $(j - 1)$ times with respect to μ, at $\mu = \lambda_r$ we have

$$f_r^{(j-1)} := f^{(j-1)}(\lambda_r) = (j-1)!/(z-\lambda_r)^j$$

which when substituted in (12.27) gives

$$f(A) = (zI - A)^{-1} = R(z, A)$$

$$= \sum_{r=1}^{k} \left[\frac{1}{(z-\lambda_r)} Z_{r1} + \frac{1!}{(z-\lambda_r)^2} Z_{r2} + \cdots + \frac{(j-1)!}{(z-\lambda_r)^j} Z_{rj} + \cdots + \frac{(m_r-1)!}{(z-\lambda_r)^{m_r}} Z_{rm_r} \right], \quad (12.61)$$

where λ_r's, $r = 1, 2, \ldots, k$, are distinct eigenvalues of A of index m_i located in a domain D bounded by the closed curve C. Suppose that $h(z)$ is a complex-valued function of z which is continuous on C and analytic in D. We now multiply both sides of (12.61) by $h(z)$ and integrate on the closed curve C to get (see Section 1.5)

$$\int_C (zI - A)^{-1} h(z)\, dz = \sum_{r=1}^{k} \int_C \left[\frac{1}{(z-\lambda_r)} Z_{r1} + \frac{1!}{(z-\lambda_r)^2} Z_{r2} + \cdots \right.$$

$$\left. + \frac{(j-1)!}{(z-\lambda_r)^j} Z_{rj} + \cdots + \frac{(m_r-1)!}{(z-\lambda_r)^{m_r}} Z_{rm_r} \right] h(z)\, dz$$

$$= 2\pi i \sum_{r=1}^{k} [h(\lambda_r) Z_{r1} + h^{(1)}(\lambda_r) Z_{r2} + \cdots + h^{(j-1)}(\lambda_r) Z_{rj}$$

$$+ \cdots + h^{(m_r-1)}(\lambda_r) Z_{rm_r}]$$

$$= 2\pi i h(A),$$

where in the second step we have used (12.58) and (12.59) and in third step (12.27). Hence,

$$h(A) = \frac{1}{2\pi i} \int_C (zI - A)^{-1} h(z)\, dz \quad (12.62)$$

which is a generalization of Cauchy's integral formula to the matrix case and is an alternative representation of an arbitrary function $h(A)$ of a matrix A when the associated scalar function $h(\lambda)$ is defined on the spectrum of A.

It is also possible to obtain integral representation of the constituent matrices Z_{pj}'s of A. To this effect, we note that, if C_p is a closed curve enclosing only the eigenvalue λ_i, then integrating both sides of (12.61) around C_p, we have

$$\int_{C_p} R(z, A)\, dz = \int_{C_p} \frac{1}{z-\lambda_p} Z_{p1}\, dz = 2\pi i\, Z_{p1},$$

because other terms are zero if $r \neq p$ by Cauchy's residue theorem or by (12.60) when $r = p$. Hence,

$$Z_{p1} = \frac{1}{2\pi i} \int_{C_p} R(z, A)\, dz, \qquad p = 1, 2, ..., k.$$

In view of Cauchy's residue theorem, it follows that

$$\int_{C_p} \frac{(z-\lambda_p)^l}{(z-\lambda_r)^j}\, dz = \begin{cases} 2\pi i, & \text{if } r = p, l = j-1; \\ 0, & \text{otherwise}. \end{cases}$$

Therefore, multiplying both sides of (12.61) by $(z - \lambda_p)^{j-1}$, and then integrating on a closed contour C_p, the only term on the right hand side which is nonzero corresponds to $r = p$ associated with Z_{rj}. So,

$$\begin{aligned} \int_{C_p} (z-\lambda_p)^{j-1} R(z, A)\, dz &= \int_{C_p} \frac{(j-1)!(z-\lambda_p)^{j-1}}{(z-\lambda_p)^j} Z_{pj}\, dz \\ &= (j-1)!\, 2\pi i Z_{pj}, \quad p = 1, ..., k; \quad j = 1, ..., m_p. \end{aligned}$$

Hence,

$$Z_{pj} = \frac{1}{2\pi i (j-1)!} \int_{C_p} (z-\lambda_p)^{j-1} R(z, A)\, dz, \tag{12.63}$$

where $p = 1, ..., k$ and $j = 1, ..., m_p$.

Example 12.9 A matrix $A = \begin{bmatrix} 0 & 1 & 0 \\ 0 & 0 & 1 \\ 0 & 0 & 1 \end{bmatrix}$

is given. Compute the *constituent matrices* of A and exp (At).

Solution For A, we have $\phi(\lambda) = \psi(\lambda) = \lambda^2(\lambda - 1)$, as shown in Example 12.2, with eigenvalues $\lambda_1 = 0$, and $\lambda_2 = 1$ of multiplicity $m_1 = 2$, and $m_2 = 0$ respectively. The resolvent of the square matrix A is

$$\begin{aligned} R(z, A) = (zI - A)^{-1} &= \begin{bmatrix} z & -1 & 0 \\ 0 & z & -1 \\ 0 & 0 & z-1 \end{bmatrix}^{-1} \\ &= \frac{1}{z^2(z-1)} \begin{bmatrix} z(z-1) & (z-1) & 1 \\ 0 & z(z-1) & z \\ 0 & 0 & z^2 \end{bmatrix}. \end{aligned}$$

We now consider the scalar function $h(z) = \exp(zt)$, and consequently, (12.62) becomes

$$h(A) = \exp(At) = \frac{1}{2\pi i}\int_C \begin{bmatrix} \frac{1}{z} & \frac{1}{z^2} & \frac{1}{z^2(z-1)} \\ 0 & \frac{1}{z} & \frac{1}{z(z-1)} \\ 0 & 0 & \frac{1}{(z-1)} \end{bmatrix} \exp(zt)\, dz,$$

where the contour encloses the eigenvalues 0 and 1 of A. Hence, by applying Cauchy's residue theorem, we have

$$\exp(At) = h(A) = \begin{bmatrix} 1 & t & \exp t - 1 - t \\ 0 & 1 & \exp t - 1 \\ 0 & 0 & \exp t \end{bmatrix}.$$

For the constituent matrices of A, we apply (12.63) to get

$$Z_{11} = \frac{1}{2\pi i}\int_{C_1} R(z, A)\, dz = \frac{1}{2\pi i}\int_{C_1} \begin{bmatrix} \frac{1}{z} & \frac{1}{z^2} & \frac{1}{z^2(z-1)} \\ 0 & \frac{1}{z} & \frac{1}{z(z-1)} \\ 0 & 0 & \frac{1}{(z-1)} \end{bmatrix} dz, \qquad (12.64)$$

where C_1 encloses $\lambda_1 = 0$. Therefore, by Cauchy's residue theorem,

$$Z_{11} = \begin{bmatrix} 1 & 0 & -1 \\ 0 & 1 & -1 \\ 0 & 0 & 0 \end{bmatrix}.$$

Similarly, Z_{21} is given by (12.64) in which C_1 is replaced by C_2 which encloses now only $\lambda_2 = 1$. Therefore,

$$Z_{21} = \frac{1}{2\pi i}\int_{C_2} \begin{bmatrix} \frac{1}{z} & \frac{1}{z^2} & -\frac{1}{z} - \frac{1}{z^2} + \frac{1}{(z-1)} \\ 0 & \frac{1}{z} & \frac{1}{(z-1)} - \frac{1}{z} \\ 0 & 0 & \frac{1}{(z-1)} \end{bmatrix} dz = \begin{bmatrix} 0 & 0 & 1 \\ 0 & 0 & 1 \\ 0 & 0 & 1 \end{bmatrix}.$$

Finally,

$$Z_{12} = \frac{1}{2\pi i}\int_{C_2} z\begin{bmatrix} \frac{1}{z} & \frac{1}{z^2} & -\frac{1}{z}-\frac{1}{z^2}+\frac{1}{(z-1)} \\ 0 & \frac{1}{z} & \frac{1}{(z-1)}-\frac{1}{z} \\ 0 & 0 & \frac{1}{(z-1)} \end{bmatrix} dz = \begin{bmatrix} 0 & 1 & -1 \\ 0 & 0 & 0 \\ 0 & 0 & 0 \end{bmatrix}. \qquad \square$$

12.6 FURTHER DISCUSSION ON MATRIX SEQUENCE AND MATRIX SERIES

We have given introductory ideas on matrix sequence and matrix series in Section 9.6. We now propose to apply Jordan normal forms and spectral representation of an arbitrary function to establish new results on the convergence of a matrix sequence and a matrix power series.

Let J be a Jordan block of order k corresponding to the eigenvalue λ of a matrix A over $\mathbb{C}$. Then, (see (10.43)),

$$J = \lambda I_k + E_k,$$

where E_k is a matrix of order k with 1's in the *superdiagonal* and zeros elsewhere. We can check by direct computation that

$$(E_k)^p = 0, \qquad p \geq k.$$

Therefore,

$$J^p = (\lambda I_k + E_k)^p = \lambda^p I_k + {}^pC_1\lambda^{p-1}E_k + \cdots + {}^pC_{k-1}\lambda^{p-k+1}(E_k)^{k-1}, \qquad p \geq k. \quad (12.65)$$

Lemma 12.3 Prove that $\lim\limits_{p\to\infty} J^p$ exists iff (i) $|\lambda| < 1$, and (ii) $\lambda = 1$, $k = 1$.

Proof We know that if λ is a complex number, then

$$\lim_{p\to\infty} \lambda^p \text{ exists iff } \lambda = 1, \quad \text{or} \quad |\lambda| < 1.$$

Note that, when λ is complex and $\lambda = i$, $\lim \lambda^p$ does not exist. We now write (12.65) explicitly to get

$$J^p = \begin{bmatrix} \lambda^p & {}^pC_1\lambda^{p-1} & {}^pC_2\lambda^{p-2} & \cdots & {}^pC_{k-1}\lambda^{p-k+1} \\ 0 & \lambda^p & {}^pC_1\lambda^{p-1} & \cdots & {}^pC_{k-2}\lambda^{p-k+2} \\ 0 & 0 & \lambda^p & \cdots & {}^pC_{k-3}\lambda^{p-k+3} \\ \vdots & \vdots & \vdots & \ddots & \vdots \\ 0 & 0 & 0 & \cdots & \lambda^p \end{bmatrix}. \qquad (12.66)$$

Suppose that lim J^p exists, but (i) is not true. Then,

$$\lim_{p\to\infty} {}^pC_{k-1}\lambda^{p-k+1} \to \infty$$

and if (ii) is not true, i.e., if $\lambda = 1$ but $k \neq 1$, then

$$\lim_{p\to\infty} {}^pC_{k-1} \to \infty.$$

Conversely, if (i) and (ii) are true, then obviously lim J^p exists. We note that lim $J^p = 0$ for condition (i) and lim $J^p = [1]$ for condition (ii). □

Theorem 12.10 Let A be an $n \times n$ matrix over $\mathbb{C}$. Then, lim A^p exists iff the following conditions hold:

(a) if λ is an eigenvalue of A then $|\lambda| < 1$,

(b) if λ is an eigenvalue of A such that $|\lambda| = 1$ then λ is the real number 1,

(c) if 1 is the eigenvalue of A then the associated elementary divisors are linear.

Proof Let T be a nonsingular transformation in $\mathbb{C}^{n\times n}$ such that

$$T^{-1}A\,T = \text{diag}\,[J_1, J_2, \ldots, J_r], \tag{12.67}$$

$$A^p = T\,\text{diag}\left[J_1^p, J_2^p, \ldots, J_r^p\right]T^{-1}. \tag{12.68}$$

Suppose that lim A^p exists. In view of Problem 9.26(b), lim J_i^p exists for $i = 1, 2, \ldots, r$, which by Lemma 12.3 implies that conditions (a), (b) and (c) are satisfied. The condition (c) implies that $T^{-1}AT$ is diagonal (see Problem 10.27). Conversely, if conditions (a), (b) and (c) hold, then obviously lim A^p exists in view of Lemma 12.3. □

An important consequence of the above theorem which is easy to verify is the following corollary.

Corollary 12.2 If A is a matrix in $\mathbb{C}^{n\times n}$, then lim $A^p = 0$ iff $|\lambda| < 1$ for all eigenvalues λ of A. □

The convergence of matrix series of the form (9.33) is defined in Section 9.5. In respect of applications, matrix *power series* plays a dominant role. In this case, in matrix series (9.33), $A_i = \alpha_i A^i$, $i = 0, 1, \ldots,$ where α_i's are scalars.

Definition 12.4 *Matrix Power Series.* Let $f(x)$ represent the scalar power series in x described by

$$f(x) = \alpha_0 + \alpha_1 x + \cdots + \alpha_k x^k + \cdots$$

which is defined for values of x for which it is convergent. A matrix power series associated with $f(x)$ is obtained by replacing x by a square matrix variable A and is given by (see Section 7.3.1)

$$f(A) = \alpha_0 I + \alpha_1 A + \cdots + \alpha_k A^k + \cdots$$

which is defined for all matrices A for which it is convergent. □

We shall now prove an important theorem for the convergence of a matrix power series.

Theorem 12.11 If the scalar power series

$$f(x) = \alpha_0 + \alpha_1 x + \cdots + \alpha_k x^k + \cdots$$

is convergent for all $|x| < r$, then the associated matrix power series

$$f(A) = \alpha_0 I + \alpha_1 A + \cdots + \alpha_k A^k + \cdots \tag{12.69}$$

converges, if for every eigenvalue λ of the n-square matrix A, $|\lambda| < r$.

Proof Let T be a nonsingular transformation in $\mathbb{C}^{n\times n}$ such that $T^{-1}AT$ and A^p are represented by (12.67) and (12.68), respectively. Considering the partial sum of (12.69), we have

$$f_p(A) = \alpha_0 I + \alpha_1 A + \cdots + \alpha_p A^p = T \sum_{j=0}^{p} \alpha_j \operatorname{diag}\left[J_1^j, J_2^j, \ldots, J_r^j\right] T^{-1}.$$

If we set $J_i = J$, where J is a Jordan block of order k, then for $p \geq k$

$$\sum_{j=0}^{p} \alpha_j J^j = \begin{bmatrix} f_p(\lambda) & f_p^{(1)}(\lambda) & \cdots & \frac{1}{l!} f_p^{(l)}(\lambda) & \cdots & \frac{1}{(k-1)!} f_p^{(k-1)}(\lambda) \\ & f_p(\lambda) & \cdots & \frac{1}{(l-1)!} f_p^{(l-1)}(\lambda) & \cdots & \frac{1}{(k-2)!} f_p^{(k-2)}(\lambda) \\ & & \ddots & \vdots & & \vdots \\ & & & f_p(\lambda) & \cdots & \frac{1}{(k-l-1)!} f_p^{(k-l-1)}(\lambda) \\ & & & & \ddots & \vdots \\ & & & & & f_p(\lambda) \end{bmatrix} \tag{12.70}$$

This is because, the term in $(1, l+1)$th position of (12.70) can be written by using (12.66) as

$$\sum_{j=l}^{p} \alpha_j \, {}^{j}C_l \lambda^{j-l} = \alpha_l + \alpha_{l+1} \, {}^{l+1}C_l \lambda + \cdots + \alpha_p \, {}^{p}C_l \lambda^{p-l}.$$

On the other hand, we differentiate l-times the polynomial

$$f_p(x) = \alpha_0 + \alpha_1 x + \cdots + \alpha_l x^l + \alpha_{l+1} x^{l+1} + \cdots + \alpha_p x^p$$

with respect to x to get

$$\begin{aligned} f_p^{(l)}(x) &= \alpha_l \, l! + \alpha_{l+1} \frac{(l+1)!}{1!} x + \alpha_{l+2} \frac{(l+2)!}{2!} x^2 + \cdots + \alpha_p \frac{p!}{(p-l)!} x^{p-l} \\ &= l! \sum_{j=l}^{p} \alpha_j \, {}^{j}C_l x^{j-l}. \end{aligned}$$

Suppose, now, that the sequence $\{f_p(x)\}$ is convergent for $|x| < r$. Then, all its derivatives are also convergent (see Problem 12.12). This shows that, if the eigenvalues of A are limited by $|\lambda| < r$, the matrix (12.70) and, consequently, $f_p(A)$ are convergent (see Problem 9.26(b)). □

Example 12.10 We know that the power (i.e., Taylor) series representation of elementary functions such as exp x, sin x, sinh x, cos x, and cosh x are valid *for all complex values* of x. In view of Theorem 12.11, we can therefore write for all square matrices A,

(a) $\exp A = \sum_{i=0}^{\infty} \frac{1}{i!} A^i$, (b) $\sin A = \sum_{i=0}^{\infty} \frac{1}{(2i+1)!} (-1)^i A^{2i+1}$,

(c) $\sinh A = \sum_{i=0}^{\infty} \frac{1}{(2i+1)!} A^{2i+1}$, (d) $\cos A = \sum_{i=0}^{\infty} \frac{1}{(2i)!} (-1)^i A^{2i}$,

(e) $\cosh A = \sum_{i=0}^{\infty} \frac{1}{(2i)!} A^{2i}$. □

The representation of an arbitrary matrix function by constituent matrices helps us formulate a necessary and sufficient condition for the convergence of a matrix sequence and a matrix series.

We now consider a sequence of arbitrary matrix functions

$$f_1(A), f_2(A), \ldots, f_p(A), \ldots. \tag{12.71}$$

By Theorem 12.2, we can represent $f_p(A)$ by

$$f_p(A) = \sum_{i=1}^{k} \sum_{j=1}^{m_i} f_p^{(j-1)}(\lambda_i) Z_{ij}, \tag{12.72}$$

where Z_{ij} are the constituent matrices of A. We shall now show that m scalar sequences

$$\{f_p^{(j-1)}(\lambda_i)\}, \qquad i = 1, 2, \ldots, k; \qquad j = 1, 2, \ldots, m_i; \tag{12.73}$$

and $m_1 + m_2 + \cdots + m_k = m$ determines the convergence of the matrix sequence (12.71).

Theorem 12.12 Let the sequence of scalar functions $f_1(\lambda), f_2(\lambda), \ldots, f_p(\lambda), \ldots$ be defined on the spectrum of an n-square matrix A over $\mathbb{C}$. Then, the matrix sequence (12.71) converges iff the m number of scalar sequences (12.73) are convergent.

Proof (a) *Necessity.* For an $n \times n$ matrix $f(A) = [\mathbf{a}_1, \mathbf{a}_2, \ldots, \mathbf{a}_n]$, where $\mathbf{a}_i$'s are n-vectors, we define an associated n^2-vector $\mathbf{x}$ by

$$\mathbf{x} = \left[\mathbf{a}_1^T, \mathbf{a}_2^T, \ldots, \mathbf{a}_n^T\right]^T.$$

Let $\mathbf{y}$ be an m-vector defined by

$$\mathbf{y} = \left[f_p(\lambda_1), f_p^{(1)}(\lambda_1), \ldots, f_p^{(m_1-1)}(\lambda_1); f_p(\lambda_2), \ldots, f_p^{(m_k-1)}(\lambda_k)\right]^T,$$

$\mathbf{x}_p$ be the n^2-vector associated with $f_p(A)$, and $\mathbf{x}_{ij}$ the n^2-vectors associated with Z_{ij}. Then, (12.72) can be represented (try for a third order system !) as

$$\mathbf{x}_p = [\mathbf{x}_{11}, \mathbf{x}_{12}, \ldots, \mathbf{x}_{1m_1}; \mathbf{x}_{21} \ldots \mathbf{x}_{km_k}]\mathbf{y},$$

where the coefficient matrix has rank m (see Problem 12.13), and can, therefore, be solved uniquely for $\mathbf{y}$ in terms of the elements in $\mathbf{x}_p$, which are also the elements of $f_p(A) = \left[a_{ij}^{(p)}\right]$. In other words, the solution of (12.72) for $f_p^{(j-1)}$ can be written as

$$f_p^{(j-1)}(\lambda_i) = \sum_{i,j=1}^{n} \alpha_{ij} a_{ij}^{(p)}. \tag{12.74}$$

Suppose that $\lim f_p(A) = f(A)$, where $f(A) = [a_{ij}]$, i.e.,

$$\lim a_{ij}^{(p)} = a_{ij} \qquad \text{for all} \quad i, j = 1, 2, \ldots, n$$

which, when incorporated into (12.74), shows that the m scalar sequences (12.73) converge, i.e.,

$$\lim \{f_p^{(j-1)}(\lambda_i)\} = \{f^{(j-1)}(\lambda_i)\} \tag{12.75}$$

for all $i = 1, 2, \ldots, k$; $j = 1, 2, \ldots, m_i$.

(b) *Sufficiency.* Suppose that the limit in (12.75) exists. Then, in the limit $p \to \infty$, (12.72) becomes

$$\lim f_p(A) = \lim \sum_{i=1}^{k} \sum_{j=1}^{m_i} f_p^{(j-1)}(\lambda_i) Z_{ij} = \sum_{i=1}^{k} \sum_{j=1}^{m_i} f^{(j-1)}(\lambda_i) Z_{ij} = f(A)$$

which shows that, the sequence (12.71) is convergent. □

Example 12.11 Prove that

$$f(A) = I + A + A^2 + \cdots$$

is convergent iff $|\lambda_i| < 1$ for all eigenvalues of A and that $f(A) = (I - A)^{-1}$.

Solution We define

$$f_p(A) = I + A + A^2 + \cdots + A^p.$$

By Theorem 12.12, the sequence $\{f_p(A)\}$ is convergent iff the m scalar sequences defined by (12.73), viz.,

$$f_p(\lambda_i) = 1 + \lambda_i + \lambda_i^2 + \cdots + \lambda_i^p, \; f_p^{(1)}(\lambda_i), \; \ldots, \; f_p^{(m_i-1)}(\lambda_i), \tag{12.76}$$

where $i = 1, 2, \ldots, k$, are convergent. The sequences (12.76) are convergent iff $f_p(\lambda_i)$ is convergent (see Example 9.14 and Problem 12.12), i.e., iff $|\lambda_i| < 1$, $i = 1, 2, \ldots, k$. Since $f_p(\lambda_i)$ converges to $f(\lambda_i) = (1 - \lambda_i)^{-1}$, we have

$$\lim f_p(A) = f(A) = (I - A)^{-1}.$$ □

12.7 SOLUTION OF VECTOR-MATRIX DIFFERENTIAL EQUATIONS

A set of differential equations in the variables $x_1, x_2, ..., x_n$ can be described by

$$\dot{\mathbf{x}}(t) = A\mathbf{x}(t), \qquad \mathbf{x}(0) = \mathbf{x}_0, \tag{12.77}$$

where $\mathbf{x} = [x_1\ x_2 \cdots x_n]^T$ and $A = [a_{ij}]$ is a constant matrix of order n. Such differential equations occur in many branches of physical sciences, one of which is illustrated by (7.3) and Example 7.4. For a scalar differential equation (DE) $\dot{x}(t) = ax(t)$, $x(0) = x_0$, the solution is $x(t) = x_0 \exp at$, $t \geq 0$. Analogously, the solution of (12.77) is assumed to be of the form

$$\mathbf{x}(t) = e^{At}\,\mathbf{x}_0, \qquad t \geq 0, \tag{12.78}$$

where (see Example 9.18),

$$e^{At} = I + At + \frac{1}{2!}A^2t^2 + \cdots + \frac{1}{n!}A^nt^n + \cdots \tag{12.79}$$

has the following properties:

Property 1 $e^{At} = I$ at $t = 0$.

Property 2 $e^{At}\,e^{Bt} = e^{(A+B)t}$, if $AB = BA$.

Property 3 $(e^{At})^{-1} = e^{-At}$,

Property 4 $\dfrac{d}{dt}e^{At} = Ae^{At} = e^{At}A$.

Property 5 $e^{T^{-1}(At)T} = T^{-1}e^{At}T$, where T is nonsingular.

Differentiating both sides of (12.78) with respect to t and using Property 4 (see Section 1.5), we have

$$\dot{\mathbf{x}}(t) = Ae^{At}\,\mathbf{x}_0 = A\mathbf{x}(t)$$

which shows that (12.78) satisfies DE (12.77) with the initial condition

$$\mathbf{x}(t)\big|_{t=0} = e^{At}\,\mathbf{x}_0\big|_{t=0} = I\mathbf{x}_0 = \mathbf{x}_0.$$

Example 12.12 Determine the solution of $\dot{\mathbf{x}} = A\mathbf{x}$ if

$$A = \begin{bmatrix} 1 & 0 \\ 0 & 2 \end{bmatrix}, \qquad \mathbf{x}_0 = \begin{bmatrix} 2 \\ 1 \end{bmatrix}.$$

Solution We have

$$e^{At} = I + t\begin{bmatrix} 1 & 0 \\ 0 & 2 \end{bmatrix} + \frac{t^2}{2!}\begin{bmatrix} 1 & 0 \\ 0 & 2 \end{bmatrix}^2 + \cdots$$

$$= \begin{bmatrix} 1+t+\frac{1}{2!}t^2+\cdots & 0 \\ 0 & 1+2t+\frac{1}{2!}(2t)^2+\cdots \end{bmatrix}$$

$$= \begin{bmatrix} e^t & 0 \\ 0 & e^{2t} \end{bmatrix}.$$

Therefore, the required solution is

$$\mathbf{x}(t) = \begin{bmatrix} e^t & 0 \\ 0 & e^{2t} \end{bmatrix}\begin{bmatrix} 2 \\ 1 \end{bmatrix} = \begin{bmatrix} 2e^t \\ e^{2t} \end{bmatrix}.$$ □

Example 12.13 Compute e^{Dt} if $D = \text{diag}\left[\lambda_1, \lambda_2, \ldots, \lambda_n\right]$. Hence, determine the solution of $\dot{\mathbf{x}} = D\mathbf{x}, \mathbf{x}(0) = [1\ 1\ \cdots\ 1]^T$.

Solution We know that $D^k = \text{diag}\,[\lambda_1^k, \lambda_2^k, \ldots, \lambda_n^k]$, $k = 1, 2, \ldots$. So, in series representation of e^{Dt}, only the diagonal elements are nonzero. The ith diagonal element is

$$1 + \lambda_i t + \frac{1}{2!}(\lambda_i t)^2 + \cdots = e^{\lambda_i t}, \quad i = 1, 2, \ldots, n.$$

Hence,

$$e^{Dt} = \text{diag}\left[e^{\lambda_1 t}, e^{\lambda_2 t}, \ldots, e^{\lambda_n t}\right].$$

If $\lambda_1 = \lambda_2 = \cdots = \lambda_n = \lambda$, then $e^{Dt} = e^{\lambda t I} = e^{\lambda t} I$. The solution of the given DE, therefore, is

$$\mathbf{x}(t) = e^{Dt}\,\mathbf{x}(0) = \left[e^{\lambda_1 t}\ e^{\lambda_2 t}\ \cdots\ e^{\lambda_n t}\right]^T.$$ □

Example 12.14 Compute e^{Jt} where J is a Jordan block of order n associated with eigenvalue λ (see (11.53)). Hence, determine the solution of $\dot{\mathbf{x}} = J\mathbf{x}, \mathbf{x}(0) = [1\ 1\ \cdots\ 1]^T$.

Solution To understand the computation, let us first consider a Jordan block of order 3 associated with λ, so that $J = \lambda I + N$ where

$$N = \begin{bmatrix} 0 & 1 & 0 \\ 0 & 0 & 1 \\ 0 & 0 & 0 \end{bmatrix}, \qquad N^2 = \begin{bmatrix} 0 & 0 & 1 \\ 0 & 0 & 0 \\ 0 & 0 & 0 \end{bmatrix},$$

N^k = a null matrix of order 3 for $k = 3, 4, \ldots$. So, using (12.79)

$$e^{Jt} = e^{(\lambda I + N)t} = e^{\lambda t I}\, e^{Nt} = e^{\lambda t} I\left(I + Nt + \frac{1}{2}N^2 t^2\right)$$

$$= e^{\lambda t}\begin{bmatrix} 1 & t & \frac{1}{2}t^2 \\ 0 & 1 & t \\ 0 & 0 & 1 \end{bmatrix}.$$

Hence, for a Jordan block of order n associated with λ,

$$e^{Jt} = e^{\lambda t}\begin{bmatrix} 1 & t & \frac{t^2}{2} & \cdots & \frac{t^{n-1}}{(n-1)!} \\ 0 & 1 & t & \cdots & \frac{t^{n-2}}{(n-2)!} \\ \vdots & \vdots & \vdots & \cdots & \vdots \\ 0 & 0 & 0 & \cdots & t \\ 0 & 0 & 0 & \cdots & 1 \end{bmatrix}.$$

The solution of the given DE is, therefore,

$$\mathbf{x}(t) = e^{Jt}\ \mathbf{x}(0) = \begin{bmatrix} \sum_{i=1}^{n} \frac{t^{i-1}}{(i-1)!} \\ \vdots \\ 1+t \\ 1 \end{bmatrix}.$$

□

Example 12.15 Compute e^{At} if

$$A = \begin{bmatrix} a & -b \\ b & a \end{bmatrix}.$$

Hence, solve $\dot{\mathbf{x}} = A\mathbf{x}$, $\mathbf{x}(0) = [1 \quad -1]^T$.

Solution We can write $A = aI + b\,P$, where

$$P = \begin{bmatrix} 0 & -1 \\ 1 & 0 \end{bmatrix}, \quad P^2 = -I, \quad P^3 = -P, \quad P^4 = I,$$

so that $P^k = P^{k+4}$, k 1, 2, ... Therefore, using (12.79)

$$e^{btP} = I + bt\,P + \frac{1}{2}(bt)^2\,P^2 + \frac{1}{3!}(bt)^3\,P^3 + \cdots$$

so that in the matrix $e^{btP} = [p_{ij}]$, the elements

$$p_{11} = p_{22} = 1 - \frac{1}{2}(bt)^2 + \frac{1}{4!}(bt)^4 - \cdots = \cos bt,$$

$$p_{21} = -p_{12} = bt - \frac{1}{3!}(bt)^3 + \cdots = \sin bt.$$

Consequently,

$$e^{At} = e^{(aI + bP)t} = e^{atI}\ e^{bt\,P}$$

$$= e^{at} I \begin{bmatrix} p_{11} & p_{12} \\ p_{21} & p_{22} \end{bmatrix} = e^{at} \begin{bmatrix} \cos bt & -\sin bt \\ \sin bt & \cos bt \end{bmatrix}.$$

Hence, the solution of the given DE is

$$\mathbf{x}(t) = e^{At}\ \mathbf{x}(0) = e^{at} \begin{bmatrix} \cos bt + \sin bt \\ \sin bt - \cos bt \end{bmatrix}$$ □

Example 12.16 Determine e^{Jt} if $J = \text{diag}\,[J_1, J_2, ..., J_k]$ where J_i is a Jordan block of order m_i, $i = 1, 2, ..., k$ associated with λ. Hence, solve $\dot{\mathbf{x}} = J\mathbf{x}$, $\mathbf{x}(0) = [1\ \ 1\ \ 1\ \ 1]$, if

$$J = \begin{bmatrix} 2 & 1 & 0 & 0 \\ 0 & 2 & 0 & 0 \\ 0 & 0 & 2 & 0 \\ 0 & 0 & 0 & 3 \end{bmatrix}$$

Solution Using the series representation (12.79) we have

$$e^{Jt} = I + Jt + \frac{1}{2!}(Jt)^2 + \frac{1}{3!}(Jt)^3 + \cdots$$

$$= \text{diag}\Big[I + tJ_1 + \frac{1}{2!}t^2 J_1^2 + \frac{1}{3!}t^3\ J_1^3 + \cdots,$$

$$I + tJ_2 + \frac{1}{2!}t^2\ J_2^2 + \frac{1}{3!}t^3\ J_2^3 + \cdots,$$

$$I + tJ_k + \frac{1}{2!}t^2\ J_k^2 + \frac{1}{3!}t^3\ J_k^3 + \cdots\Big]$$

$$= \text{diag}\left[e^{J_1 t}, e^{J_2 t}, ..., e^{J_k t}\right].$$

Setting

$$J_1 = \begin{bmatrix} 2 & 1 \\ 0 & 2 \end{bmatrix}, \quad J_2 = [2], \quad J_3 = [3],$$

the solution of the given DE is (see Example 12.12),

$$\mathbf{x}(t) = e^{Jt}\ \mathbf{x}(0) = \text{diag}\left[e^{J_1 t}, e^{J_2 t}, e^{J_3 t},\right] \mathbf{x}(0)$$

$$= \begin{bmatrix} e^{2t}\begin{bmatrix} 1 & t \\ 0 & 1 \end{bmatrix} & & \\ & e^{2t} & \\ & & e^{3t} \end{bmatrix} \mathbf{x}(0)$$

$$= \left[(1+t)\,e^{2t}, e^{2t}, e^{2t}, e^{3t}\right]^T.$$ □

Example 12.17 Determine the Jordan normal form of

$$A = \begin{bmatrix} 0 & -3 & -1 \\ 0 & 3 & 0 \\ 4 & 5 & 4 \end{bmatrix}$$

and hence, compute e^{At}. Solve $\dot{\mathbf{x}} = A\mathbf{x}$, $\mathbf{x}(0) = [0\ 1\ 3]^T$.

Solution The characteristic polynomial of A is

$$\begin{aligned} \phi(\lambda) &= \lambda^3 - (\text{trace } A)\,\lambda^2 + (\text{sum of all principal minors of order 2})\lambda - \det A \\ &= \lambda^3 - (0+3+4)\,\lambda^2 + (0+4+12)\,\lambda - 12 \\ &= (\lambda-2)^2\,(\lambda-3) \end{aligned}$$

with eigenvalues $\lambda_1 = 2$ of multiplicity 2 and $\lambda_2 = 3$, a simple one. The set of generalized eigenvectors of A associated with $\lambda_1 = 2$ are obtained from solving $(\lambda I - A)^k\ \mathbf{x} = \boldsymbol{\theta}$ for $\mathbf{x}$, $k = 1, 2$:

$$\begin{bmatrix} 2 & 3 & 1 \\ 0 & -1 & 0 \\ -4 & -5 & -2 \end{bmatrix}\begin{bmatrix} x_1 \\ x_2 \\ x_3 \end{bmatrix} = \begin{bmatrix} 0 \\ 0 \\ 0 \end{bmatrix}, \qquad \begin{bmatrix} 0 & -2 & 0 \\ 0 & 1 & 0 \\ 0 & 3 & 0 \end{bmatrix}\begin{bmatrix} x_1 \\ x_2 \\ x_3 \end{bmatrix} = \begin{bmatrix} 0 \\ 0 \\ 0 \end{bmatrix}.$$

The solution gives

$$[1,\quad 0,\quad -2]^T \in \mathbb{N}_{11} \quad \text{and} \quad \left\{[1,\quad 0,\quad 0]^T, [0,\quad 0,\quad 1]^T\right\} \in \mathbb{N}_{12}.$$

Since, by column transformations,

$$\left[\begin{array}{c|cc} 1 & 1 & 0 \\ 0 & 0 & 0 \\ -2 & 0 & 1 \end{array}\right] \sim \left[\begin{array}{c|cc} 1 & 0 & 0 \\ 0 & 0 & 0 \\ -2 & 2 & 1 \end{array}\right] \sim \left[\begin{array}{c|cc} 1 & 0 & 0 \\ 0 & 0 & 0 \\ -2 & 2 & 0 \end{array}\right],$$

the eigenvector $[1\quad 0\quad 0]^T \in \mathbb{N}_{12}$ is LI over $\mathbb{N}_{11}$. The set of generalized eigenvectors associated with $\lambda_1 = 2$ is

$$\mathbf{x}_1 = [1\quad 0\quad 0]^T, \qquad \mathbf{x}_2 = (A - 2I)\mathbf{x}_1 = [-2,\quad 0,\quad 4]^T.$$

The eigenvector associated with $\lambda_2 = 3$ is a solution of $(\lambda_2 I - A)\,\mathbf{x} = \boldsymbol{\theta}$:

$$\begin{bmatrix} 3 & 3 & 1 \\ 0 & 0 & 0 \\ -4 & -5 & -1 \end{bmatrix}\begin{bmatrix} x_1 \\ x_2 \\ x_3 \end{bmatrix} = \begin{bmatrix} 0 \\ 0 \\ 0 \end{bmatrix}$$

which gives $\mathbf{x}_3 = [-2,\ 1,\ 3]^T$. With respect to the basis $\{(A - 2I)\,\mathbf{x}_1, \mathbf{x}_1, \mathbf{x}_3\}$, the matrix representation of A is $J = Q^{-1}AQ$, where

$$Q: = \left[(A-2I)\,\mathbf{x}_1, \mathbf{x}_1, \mathbf{x}_3\right], \text{ so that}$$

$$J = \begin{bmatrix} -2 & 1 & -2 \\ 0 & 0 & 1 \\ 4 & 0 & 3 \end{bmatrix}^{-1} \begin{bmatrix} 0 & -3 & -1 \\ 0 & 3 & 0 \\ 4 & 5 & 4 \end{bmatrix} \begin{bmatrix} -2 & 1 & -2 \\ 0 & 0 & 1 \\ 4 & 0 & 3 \end{bmatrix}$$

$$= \left[\begin{array}{cc|c} 2 & 1 & 0 \\ 0 & 2 & 0 \\ \hline 0 & 0 & 3 \end{array}\right], \qquad Q = \begin{bmatrix} -2 & 1 & -2 \\ 0 & 0 & 1 \\ 4 & 0 & 3 \end{bmatrix}$$

Now,

$$A = QJQ^{-1}, A^2 = (QJQ^{-1})(QJQ^{-1}) = QJ^2Q^{-1}, \ldots, A^k = QJ^kQ^{-1}.$$

Using the series representation (12.79)

$$e^{At} = Qe^{Jt}\,Q^{-1} = Q\begin{bmatrix} e^{2t}\begin{bmatrix} 1 & t \\ 0 & 1 \end{bmatrix} & \begin{matrix} 0 \\ 0 \end{matrix} \\ \begin{matrix} 0 & 0 \end{matrix} & e^{3t} \end{bmatrix} Q^{-1}$$

$$= \begin{bmatrix} (1-2t)\,e^{2t} & (2-t)\,e^{2t} - 2e^{3t} & -te^{2t} \\ 0 & e^{3t} & 0 \\ 4t\,e^{2t} & (-3+2t)\,e^{2t} + 3e^{3t} & (1+2t)\,e^{2t} \end{bmatrix}. \tag{12.80}$$

The solution of the given DE, therefore, is

$$\mathbf{x}(t) = e^{At}\,\mathbf{x}(0) = \begin{bmatrix} 2(1-2t)\,e^{2t} - 2e^{3t} \\ e^{3t} \\ 8t\,e^{2t} + 3e^{3t} \end{bmatrix}.$$

Alternatively, if we use Vandermonde matrix method (see (12.36a)), we define $p(\lambda) = \alpha_0 + \alpha_1\lambda + \alpha_2\lambda^2$ and $f(\lambda) = e^{\lambda t}$. On the spectrum of A we have,

$$\alpha_0 + 2\alpha_1 + 4\alpha_2 = e^{2t}, \quad \alpha_1 + 4\alpha_2 = te^{2t}, \quad \alpha_0 + 3\alpha_1 + 9\alpha_2 = e^{3t}. \tag{12.81}$$

Solving these equations, we get

$$\alpha_2 = e^{3t} - (1+t)\,e^{2t}, \quad \alpha_1 = (4+5t)\,e^{2t} - 4e^{3t},$$

$$\alpha_0 = -3(1+2t)\,e^{2t} + 4e^{3t}. \tag{12.82}$$

Hence, simplifying with (12.81) and (12.82), the matrix

$$e^{At} = (\alpha_0 I + \alpha_1 A + \alpha_2 A^2)$$

$$= \alpha_0 \begin{bmatrix} 1 & 0 & 0 \\ 0 & 1 & 0 \\ 0 & 0 & 1 \end{bmatrix} + \alpha_1 \begin{bmatrix} 0 & -3 & -1 \\ 0 & 3 & 0 \\ 4 & 5 & 4 \end{bmatrix} + \alpha_2 \begin{bmatrix} -4 & -14 & -4 \\ 0 & 9 & 0 \\ 16 & 23 & 12 \end{bmatrix}$$

gives the same expression (12.80) as before. □

12.8 SOLUTION OF VECTOR-MATRIX DIFFERENCE EQUATIONS

A set of difference equations in the variables $x_1(k)$, $x_2(k)$, ..., $x_n(k)$ can be described by

$$\mathbf{x}(k+1) = A\,\mathbf{x}(k), \qquad \mathbf{x}(0) = \mathbf{x}_0, \tag{12.83}$$

where $\mathbf{x}(k) = [x_1(k), x_2(k), ..., x_n(k)]^T$ and $A = [a_{ij}]$ is a constant matrix of order n. The vector variables $\mathbf{x}(k)$ and $\mathbf{x}(k+1)$ at two time instants are related by the transformation matrix A. If A is a simple matrix, then there exists a nonsingular matrix P such that

$$P^{-1}AP = \text{diag}\,[\lambda_1, \lambda_2, ..., \lambda_n] = : D,$$

where $\lambda_1, \lambda_2, ..., \lambda_n$ are the eigenvalues of A, real or complex. Therefore,

$$A^2 = (PDP^{-1})(PDP^{-1}) = PD^2P^{-1}, \;..., \; A^k = PD^kP^{-1}.$$

Putting $k = 0, 1, ...$ in (12.83), we get successively,

$$\mathbf{x}(1) = A\mathbf{x}_0, \qquad \mathbf{x}(2) = A\mathbf{x}(1) = A^2\mathbf{x}_0, \;...,$$

$$\mathbf{x}(k) = A^k\mathbf{x}_0 = PD^kP^{-1}\,\mathbf{x}_0. \tag{12.84}$$

If $\mathbf{v}_1, \mathbf{v}_2, ..., \mathbf{v}_n$ are columns of P and $[a_1, a_2, ..., a_n]^T : = P^{-1}\mathbf{x}_0$ then

$$\mathbf{x}(k) = [\mathbf{v}_1, \mathbf{v}_2, ..., \mathbf{v}_n]\ \text{diag}\ [\lambda_1^k, \lambda_2^k, ..., \lambda_n^k]\,P^{-1}\mathbf{x}_0$$

$$= a_1\,\lambda_1^k\,\mathbf{v}_1 + a_2\,\lambda_2^k\,\mathbf{v}_2 + \cdots + a_n\,\lambda_n^k\,\mathbf{v}_n. \tag{12.85}$$

If $|\lambda_i| < 1$ for all i, then vector $\mathbf{x}(k)$ tends to a zero vector if k tends to infinity, as λ_i^k tends to zero in (12.85) for all i. On the other hand, if there exists at least one eigenvalue λ_i with $|\lambda_i| > 1$, then λ_i^k in (12.85) tends to infinity, making at least one element of $\mathbf{x}(k)$ infinitely large. However, if the maximum absolute value of the eigenvalues is 1, then it is evident from (12.85) that $\mathbf{x}(k)$ remains a finite vector. To evaluate the growth of λ^k, if λ is complex, represent it as $\lambda = r\,e^{i\theta}$, $i = \sqrt{-1}$, where $r = |\lambda|$. The above discussion prompts us to furnish the following definition.

Definition 12.5 The *discrete-time system* described by a linear difference equation (12.83) is said to be

(1) stable if $|\lambda_i| < 1, \forall i$;
(2) unstable if $|\lambda_i| > 1$ for at least one value of i; and
(3) neutrally stable if the maximum absolute value of the eigenvalues is 1. □

Example 12.18 Study the response of the system $\mathbf{x}(k+1) = A\mathbf{x}(k)$, $\mathbf{x}(0) = [a \;\; b]^T$ if

$$A = \begin{bmatrix} 0 & 1 \\ -1 & 0 \end{bmatrix}.$$

Solution Computing the various powers of A, we have using (12.84) for $k = 0, 1, 2, \ldots$

$$\mathbf{x}(4k) = \mathbf{x}_0,\ \mathbf{x}(4k+1) = \begin{bmatrix} 0 & 1 \\ -1 & 0 \end{bmatrix}\begin{bmatrix} a \\ b \end{bmatrix} = \begin{bmatrix} b \\ -a \end{bmatrix},$$

$$\mathbf{x}(4k+2) = \begin{bmatrix} -1 & 0 \\ 0 & -1 \end{bmatrix}\mathbf{x}_0 = -\mathbf{x}_0,$$

$$\mathbf{x}(4k+3) = \begin{bmatrix} 0 & -1 \\ 1 & 0 \end{bmatrix}\mathbf{x}_0 = \begin{bmatrix} -b \\ a \end{bmatrix}$$

since,

$$A = \begin{bmatrix} 0 & 1 \\ -1 & 0 \end{bmatrix} = A^5 = A^{4k+1}, \qquad A^2 = -I = A^{4k+2},$$

$$A^3 = -A = A^{4k+3}, \qquad A^4 = I = A^{4k}.$$

The response $\mathbf{x}(k)$ does not tend to a zero vector but its elements remain finite. So the system is neutrally stable. The eigenvalues of A are $\pm\sqrt{-1}$, which corresponds to such kind of stability. □

Example 12.19 Study the response of the system $\mathbf{x}(k+1) = A\mathbf{x}(k)$, $\mathbf{x}(0) = [1 \quad 0]^T$ if

$$A = \begin{bmatrix} 0 & 1 \\ 1 & 1 \end{bmatrix}.$$

Solution Since A is in companion form, its characteristic polynomial is (see Example 7.32),

$$\phi(\lambda) = \lambda^2 - \lambda - 1,$$

and hence, the eigenvalues of A are $\lambda_1 = \frac{1}{2}(1+\sqrt{5})$ and $\lambda_2 = \frac{1}{2}(1-\sqrt{5})$. If we construct P as

$$P = \begin{bmatrix} 1 & 1 \\ \lambda_1 & \lambda_2 \end{bmatrix} = \begin{bmatrix} 1 & 1 \\ \frac{1}{2}(1+\sqrt{5}) & \frac{1}{2}(1-\sqrt{5}) \end{bmatrix},$$

then $P^{-1}AP = \text{diag}\ [\lambda_1, \quad \lambda_2]$, and

$$\mathbf{x}(k) = PD^kP^{-1}\ \mathbf{x}_0 = \frac{\lambda_1\lambda_2}{\lambda_2 - \lambda_1}\begin{bmatrix} \lambda_1^{k-1} - \lambda_2^{k-1} \\ \lambda_1^k \quad - \lambda_2^k \end{bmatrix}.$$

As k tends to infinity, $\lambda_1^{k-1} = (1.618)^{k-1}$ also tends to infinity. So, both the elements of $\mathbf{x}(k)$ tend to infinity and the system response is unbounded. The reason is that one of the eigenvalues $\lambda_1 = 1.618$ is greater than 1 and the given discrete-time system is unstable according to Definition 12.5. □

12.9 MATLAB COMPUTATION OF MATRIX FUNCTION

MATLAB commands expm(A), logm(A), and sqrtm(A) compute the exponential, natural logarithm, and squre root of a matrix A. In computing with expm, MATLAB, uses a Padé expansion after scaling A for more reliable computation. sqrtm(A) produces a complex matrix if A has nonpositive eigenvalues. MATLAB command funm(A, fun) evaluates the matrix function specified by fun. For example, funm(A, @ sin) and funm(A, @ cos) computes sin

A and $\cos A$ respectively. The matrix A = magic(3) is

$$A = \begin{bmatrix} 8 & 1 & 6 \\ 3 & 5 & 7 \\ 4 & 9 & 2 \end{bmatrix}$$

and has eigenvalues 15, $\pm$4.8990 for which

$$\sin A = \begin{bmatrix} -0.3850 & 1.0191 & 0.0162 \\ 0.6179 & 0.2168 & -0.1844 \\ 0.4173 & -0.5856 & 0.8185 \end{bmatrix},$$

$$\cos A = -\begin{bmatrix} 0.1296 & 0.3151 & 0.3151 \\ 0.3151 & 0.1296 & 0.3151 \\ 0.3151 & 0.3151 & 0.1296 \end{bmatrix}.$$

However, sqrtm(A) generates a complex matrix.

PROBLEMS

12.1 Prove that if $p(\lambda) = a_{m-1}\lambda^{m-1} + \cdots + a_1\lambda + a_0 = 0$ at m distinct values of λ, then $a_0 = a_1 = \cdots = a_{m-1} = 0$.

12.2 Prove that if $f(\lambda) = a_{m-1}\lambda^{m-1} + \cdots + a_1\lambda + a_0 = 0$ on the spectrum of A (as defined in Definition 12.1) when $m_1 + \cdots + m_k = m$, then $a_0 = a_1 = \cdots = a_{m-1} = 0$.

12.3 Prove that the *Hermite* interpolating polynomial $H(\lambda)$ defined by (12.22) is unique.

12.4 Determine the polynomial $p(\lambda)$ which passes through the points –1, 0, 1 and has the following values at these points:

(a) –1, 0, 1, (b) 1, 2, 3, and (c) same as $\exp \lambda t$.

12.5 Determine the polynomial $p(\lambda)$ which passes through the roots of the minimum polynomial of A = diag [1 1 2 2] and has the same values as those of $(\lambda - 2)/(\lambda - 3)$ at these points.

12.6 Compute (a) $\exp (At)$, (b) $\sin At$, (c) $\cos At$, (d) $\cosh At$, and (e) $\sinh At$ when A is given by (12.9). Can you compute $(I - A)^{-1}$ for the same A?

12.7 Compute (a) A^8 and (b) $(I - A + A^2)^{-1}$ if A is given by (12.9).

12.8 Repeat Problem 12.6 when A is given by (12.25).

12.9 Prove that an $n \times n$ matrix A is simple iff the minimum polynomial of A contains linear factors (use Problems 10.23 and 10.27), i.e., $m_1 = m_2 = \cdots = m_k = 1$ in (12.3). Hence, show that

$$f(A) = \sum_{i=1}^{k} f(\lambda_i) Z_{i1}$$

for a simple matrix (cf. Theorem 7.35).

12.10 Prove that if A is an $n \times n$ matrix over $\mathbb{C}$ then

(a) $A = \sum_{i=1}^{k} (\lambda_i Z_{i1} + Z_{i2}),$

(b) $A^p = \sum_{i=1}^{k} \sum_{j=1}^{m_i} \frac{p!}{(p-j+1)!} \lambda^{p-j+1} Z_{ij},$

(c) $\exp(At) = \sum_{i=1}^{k} \sum_{j=1}^{m_i} t^{j-1} \exp(\lambda_i t) Z_{ij}.$

12.11 Determine the constituent matrices of A when it is given by

(a) $\begin{bmatrix} 3 & 0 & 2 \\ 1 & 1 & 1 \\ -1 & 0 & 0 \end{bmatrix},$ (b) $\begin{bmatrix} 4 & 1 & 0 & -3 \\ -6 & -1 & 1 & 4 \\ 4 & 1 & 0 & -3 \\ 3 & 1 & 0 & -2 \end{bmatrix},$

using (i) definition given by (12.28), and (ii) Cauchy's Residue Theorem expressed by (12.63).

12.12 If the infinite series

$$f(x) = \alpha_0 + \alpha_1 x + \cdots + \alpha_p x^p + \cdots$$

is convergent, then prove that all its derivatives are also convergent.

12.13 We construct an n^2-vector $\mathbf{x}$ associated with an n-square matrix $A = [\mathbf{a}_1, \mathbf{a}_2, ..., \mathbf{a}_n]$, where $\mathbf{a}_i$'s are n-vectors such that $\mathbf{x} = \left[\mathbf{a}_1^T, \mathbf{a}_2^T, \ldots, \mathbf{a}_n^T\right]^T$. Prove that, the matrices $A_1, A_2, ..., A_p$ in $\mathbb{C}^{n \times n}$, $p \leq n$, are linearly independent iff the associated n^2-vectors $\mathbf{x}_1, \mathbf{x}_2, ..., \mathbf{x}_p$ corresponding to $A_1, A_2, ..., A_p$ as defined above, are linearly independent.

12.14 Prove Theorem 12.11 with the help of Theorem 12.12.

12.15 Solve $\dot{\mathbf{x}}(t) = A\mathbf{x}(t)$ if

(i) $A = \begin{bmatrix} 2 & 3 \\ 0 & 2 \end{bmatrix},$ $\mathbf{x}(0) = \begin{bmatrix} 2 \\ -1 \end{bmatrix};$

(ii) $A = \begin{bmatrix} 1 & 1 \\ 0 & 3 \end{bmatrix},$ $\mathbf{x}(0) = \begin{bmatrix} 1 \\ -1 \end{bmatrix};$

(iii) $A = \begin{bmatrix} 0 & 2 \\ -2 & 0 \end{bmatrix},$ $\mathbf{x}(0) = \begin{bmatrix} 1 \\ 1 \end{bmatrix};$

(iv) $A = \begin{bmatrix} 1 & \pi \\ -\pi & 1 \end{bmatrix},$ $\mathbf{x}(0) = \begin{bmatrix} 1 \\ -1 \end{bmatrix};$

(v) $A = \begin{bmatrix} 0 & 1 \\ -1 & -2 \end{bmatrix}$, $\mathbf{x}(0) = \begin{bmatrix} -1 \\ 1 \end{bmatrix}$;

(vi) $A = \begin{bmatrix} 2 & 1 & -1 \\ -3 & -1 & 1 \\ 0 & 3 & -4 \end{bmatrix}$, $\mathbf{x}(0) = \begin{bmatrix} -1 \\ -1 \\ 1 \end{bmatrix}$;

(vii) $A = \begin{bmatrix} 2 & 1 & 0 \\ 0 & 2 & 0 \\ 0 & 0 & 3 \end{bmatrix}$, $\mathbf{x}(0) = \begin{bmatrix} 1 \\ 1 \\ 1 \end{bmatrix}$;

(viii) $A = \begin{bmatrix} 4 & -3 & -1 \\ 1 & 0 & -1 \\ -1 & 2 & 3 \end{bmatrix}$, $\mathbf{x}(0) = \begin{bmatrix} 1 \\ 2 \\ -3 \end{bmatrix}$;

(ix) $A = \begin{bmatrix} -1 & 0 & 2 \\ 2 & 1 & -2 \\ -2 & 0 & 3 \end{bmatrix}$, $\mathbf{x}(0) = \begin{bmatrix} 1 \\ 0 \\ 2 \end{bmatrix}$;

(x) $A = \begin{bmatrix} 8 & -1 & 4 \\ 24 & -6 & 8 \\ -12 & 2 & -6 \end{bmatrix}$, $\mathbf{x}(0) = \begin{bmatrix} -1 \\ 2 \\ -1 \end{bmatrix}$.

12.16 Find the response $\mathbf{x}(k)$ for the discrete system $\mathbf{x}(k+1) = A\mathbf{x}(k)$ if

(i) $A = \begin{bmatrix} 1 & 1 \\ 1 & 0 \end{bmatrix}$, $\mathbf{x}(0) = \begin{bmatrix} 0 \\ 1 \end{bmatrix}$;

(ii) $A = \begin{bmatrix} 0 & 1 \\ -2 & -3 \end{bmatrix}$, $\mathbf{x}(0) = \begin{bmatrix} 1 \\ -1 \end{bmatrix}$;

(iii) $A = \begin{bmatrix} 0 & 1 \\ -1 & -2 \end{bmatrix}$, $\mathbf{x}(0) = \begin{bmatrix} 1 \\ -1 \end{bmatrix}$;

(iv) $A = \begin{bmatrix} -1 & 1 & 0 \\ 0 & -1 & 0 \\ 0 & 0 & -2 \end{bmatrix}$, $\mathbf{x}(0) = \begin{bmatrix} 0 \\ 1 \\ 1 \end{bmatrix}$;

(v) $A = \begin{bmatrix} 2 & 1 & 0 \\ 0 & 2 & 1 \\ 0 & 0 & 2 \end{bmatrix}$, $\mathbf{x}(0) = \begin{bmatrix} 0 \\ 1 \\ 0 \end{bmatrix}$;

(vi) $A = \begin{bmatrix} 0 & 1 & 0 \\ 0 & 0 & 1 \\ 1 & -3 & 3 \end{bmatrix}, \quad \mathbf{x}(0) = \begin{bmatrix} 1 \\ -1 \\ 0 \end{bmatrix};$

(vii) $A = \begin{bmatrix} 1 & 0 & 0 & 0 \\ 0 & 2 & 1 & 0 \\ 0 & 0 & 2 & 1 \\ 0 & 0 & 0 & 2 \end{bmatrix}, \quad \mathbf{x}(0) = \begin{bmatrix} 0 \\ 1 \\ 0 \\ 1 \end{bmatrix};$

(viii) $A = \begin{bmatrix} 0 & -3 & 1 & 2 \\ -2 & 1 & -1 & 2 \\ -2 & 1 & -1 & 2 \\ -2 & -3 & 1 & 4 \end{bmatrix}, \quad \mathbf{x}(0) = \begin{bmatrix} 1 \\ -1 \\ 0 \\ 0 \end{bmatrix}.$

12.17 Let x_n be a sequence with $x_0 = 1$, $x_1 = 2$, $x_2 = 0$ and $x_n = 2x_{n-1} + x_{n-2} - 2x_{n-3}$, $n \geq 3$. Find the 100th term.

12.18 Prove that a confluent Vandermonde matrix is nonsingular.

12.19 Find the polar decomposition of $A = \frac{1}{15}[10\ \ 20;\ 14\ -2;\ 9\ \ 8]$.

CHAPTER

13

Numerical Linear Algebra

Today, digital computers are invariably used in any computation involving matrices. These machines operate in *finite arithmetic* and so no real number can be described with a very high degree of precision. Because of this, well-entrenched algorithms to solve the linear simultaneous equations or to determine the eigenvalues and eigenvectors of a matrix may completely fail if they are not numerically viable in relation to the constraints of finite arithmetic. This chapter aims at highlighting the numerical instability on a finite-word-length digital computer associated with these problems and suggests reliable algorithms to get around them. Consequently, it emphasizes the need to develop *QR*-factorization of a matrix for linear simultaneous equations and *QR*-algorithm to solve the eigenvalue problem.

13.1 BASIC CONCEPTS OF FINITE ARITHMETIC

Any real number can be expressed in the form

$$r = \pm\, 0.a_1a_2 \, . \, . \, . \; a_m \times 10^n, \tag{13.1}$$

where a_1, ..., a_m are integers between 0 and 9:

$$0 < a_1 \le 9;\; 0 \le a_i \le 9, \qquad i = 2, \ldots, m;$$

and n is an integer, positive or negative, e.g.,

$$-99 \le n \le 99.$$

The real number represented by (13.1) is called m-digit *floating point number* which, in short, is expressed as $r = a.10^n$, where $a = \pm\, 0.a_1a_2 \, . \, . \, . \; a_m$. The fraction a is called the *mantissa,* the integer n the *characteristic* or *exponent,* and the number of digits m in a the *precision* of the floating point number. The arithmetic involving *m*-digit floating point numbers is called *m-digit arithmetic*.

Taking the help of a hypothetical computer with the characteristic

$$-5 \le n \le 5 \tag{13.2}$$

and precision 4, we shall illustrate some fundamental concepts.

Example 13.1 The real number 1/320 has the floating-point representation given by

$$1/320 = 0.003125 = 0.3125 \times 10^{-2} \tag{13.3}$$

which has 0.3125 as the mantissa, $n = -2$ as the characteristic, and $m = 4$ as the precision. In other words, (13.3) is a 4-digit floating point number. □

The characteristic or exponent n in a floating-point representation determines the range of numbers which a computing machine can store or handle. Considering that n is given by (13.2), no number greater than 10^5 and less than 10^{-6} can be accommodated in our hypothetical computer, which can deal with numbers ranging from 10^{-6} to 10^5. In such computers, the division of 0.3121×10^5 by 0.1000×10^{-4} results in 0.3121×10^{10}, which the computer fails to compute as it is greater than the largest number it can accommodate. Numerical computation such as this, whose value is above the maximum computer capacity, is termed *overflow*. On the other hand, computations like the division of 0.1000×10^{-4} by 0.3121×10^5, resulting in 0.3204×10^{-9} which is below the smallest number the computer can handle, causes *underflow*. Algorithms should be deemed efficient if they can always avoid the occurrence of such overflows and underflows.

It is evident from the foregoing discussion that a computer working on *m*-digit arithmetic has a finite precision and operates with numbers lying in a finite range. If it encounters a number greater than *m*-digits within its range, it cannot store all the digits without shortening up to m digits either by truncation or rounding the number. Both numerical errors resulting from this process of truncation or rounding are known as *roundoff* errors.

Example 13.2 Let the computer work on 4-digit arithmetic with the characteristic $-5 \le n \le 5$. If $a = 0.5312$, $b = 0.3121$, $c = 0.2163$, then $ab = 0.16578752$, $bc = 0.6750723 \times 10^{-1}$, $(ab)c = 0.35859840576 \times 10^{-1} = a(bc)$, which the computer in 4-digit arithmetic computes by truncation as $ab \simeq 0.1657$, $bc \simeq 0.6750 \times 10^{-1}$, $(ab)c \simeq (0.1657)(0.2163) \simeq 0.3584 \times 10^{-1}$, $a(bc) = (0.5312)(0.6750 \times 10^{-1}) \simeq 0.3585 \times 10^{-1}$. Therefore, the roundoff errors due to truncation in computing ab and bc are 0.8752×10^{-4} and 0.7230×10^{-5}, respectively. It follows from above that the associative law of scalar multiplication is not valid in finite arithmetic. On the other hand, rounding off ab and bc to 4 significant digits, we have $ab \simeq 0.1658$ and $bc \simeq 0.6751 \times 10^{-1}$ in which roundoff errors are -0.1248×10^{-4} and -0.2770×10^{-5} respectively. □

The notion of *relative error* is much more significant than that of *absolute error*. If a denotes a real number and a_c the resulting number stored in a computer, then we define absolute error $= a_c - a$ and relative error $= (a_c - a)/a$ of the number a. Thus, if 0.1 is an error in the quantity 0.1, it is not negligible; it is negligible if the quantity is 10^4. The relative error in the former case is 1 while it is 10^{-5} in the latter.

13.2 CONDITIONING AND NUMERICAL STABILITY

There are two concepts which are of fundamental importance in numerical linear algebra: (i) conditioning of a numerical problem, and (ii) its numerical stability.

Any numerical problem is deemed to be *ill-conditioned* if small changes in the data cause a large *relative* change in the computed solution. On the other hand, if the computed solution is not sensitive to small changes in the data of the problem, then we say that the problem is *well-conditioned.* Given the specific nature of a problem, it is possible in some simple cases to give a mathematical estimate of conditioning of a problem in terms of the *condition number*. This concept will be explained in terms of three numerical problems:

(1) Inverse of a nonsingular matrix A.

(2) System of linear equations $A\mathbf{x} = \mathbf{b}$.

(3) Eigenvalues and eigenvectors of a square matrix A.

In the above problems, A and $\mathbf{b}$ may be determined from (i) physical measurements which are subject to errors inherent in instruments and observation or (ii) an algebraic expression by computers where accurate evaluation is not possible due to their finite precision and finite range. (iii) Even though exact data for A and $\mathbf{b}$ are used, in the process of computation the machine alters the matrix elements by roundoff. It therefore follows that true matrices for which computation is done in (i) and (ii) are not A and $\mathbf{b}$ but $A + E$ and $\mathbf{b} + \mathbf{d}$, where E and $\mathbf{d}$ are error matrices whose elements are bounded by $| e_{ij} | \le \varepsilon$ and $| d_i | \le \delta$ while in (iii) the solution is obtained not for A and $\mathbf{b}$ but for $A + E$ and $\mathbf{b} + \mathbf{d}$. This leads us to the question as to how small changes in the elements of A and $\mathbf{b}$ propagate to the solution of the problem. If the change in the solution is small, the result of the problem obtained by the numerical computation in a digital computer is acceptable and the problem is regarded as *well-conditioned.*

13.2.1 Inverse of a Perturbed Matrix: Condition Number

In what follows, the following lemma will play a crucial role. It is assumed that $\| I \| = 1$, which is true in 1-, 2- and ∞-norms.

Lemma 13.1 If $\| A \| < 1$, then $(I - A)$ is nonsingular and

(a) $\|(I-A)^{-1}\| \le 1/(1-\| A \|)$, if $\| I \| = 1$;

(b) $\| I-(I-A)^{-1} \| \le \| A \|/(1-\| A \|)$, if $\| I \| = 1$;

Proof It follows from Example 9.17 that if $\| A \| < 1$, then $(I - A)$ is nonsingular and $(I - A)^{-1} = I + A + A^2 + \cdots$.

Therefore, (a) follows from the result (see Definition 9.3)

$$\|(I-A)^{-1}\| \le 1+\|A\|+\|A\|^2+\cdots.$$

To prove (b), we have $I - (I - A)^{-1} = -A(I - A)^{-1}$, and hence

$$\|I-(I-A)^{-1}\| \;\le\; \|A\|\,\|(I-A)^{-1}\| \;\le\; \frac{\|A\|}{(1-\|A\|)} \qquad \square$$

Theorem 13.1 The relative error in the inverse of $(A + E)$ if E is a perturbation in A is

$$\alpha_I = \frac{\|A^{-1}-(A+E)^{-1}\|}{\|A^{-1}\|} \le \frac{\|A^{-1}E\|}{\left(1-\|A^{-1}E\|\right)}, \text{ if } \|A^{-1}E\|<1. \tag{13.4}$$

Proof We have $(A + E) = A(I + A^{-1}E)$, and hence if $\| A^{-1}E \| < 1$, then

$$A^{-1} - (A + E)^{-1} = [I - (I + A^{-1}E)^{-1}]A^{-1}$$

By Lemma 13.1, it therefore follows that

$$\|A^{-1}-(A+E)^{-1}\| \;\le\; \|A^{-1}\|\frac{\|A^{-1}E\|}{\left(1-\|A^{-1}E\|\right)}$$

which is the same as inequality (13.4). $\square$

Remark 13.1 If $\| A^{-1} \| \| E \| < 1$, a pessimistic estimate of the relative error in $(A + E)^{-1}$ which follows from (13.4) is

$$\alpha_I \le k(A)\mu[1-k(A)\mu]^{-1}, \tag{13.5}$$

where

$$\mu:= \text{relative change in } A = \frac{\|E\|}{\|A\|};$$

$$k(A):= \text{condition number of } A = \|A\|\,\|A^{-1}\|.$$

The condition number $k(A)$ determines whether the relative error α_I is large or not.

Example 13.3 The matrix

$$A=\begin{bmatrix}1 & 1\\ 1 & 1.01\end{bmatrix} \tag{13.6}$$

is *ill-conditioned* with respect to *inversion.* To see this, let the elements in A be changed by a small amount so that $(A + E)$ is the perturbed matrix, where

$$E=\begin{bmatrix}-0.001 & 0.001\\ 0.001 & -0.010\end{bmatrix}. \tag{13.7}$$

Now,

$$A^{-1}=\begin{bmatrix}101 & -100\\ -100 & 100\end{bmatrix}, \qquad (A+E)^{-1}=\begin{bmatrix}-333.2 & 333.5\\ 333.5 & -332.8\end{bmatrix}. \tag{13.8}$$

If the perturbed matrix $A + E$ is used to estimate A^{-1}, the error is

$$A^{-1} - (A+E)^{-1} = \begin{bmatrix} 434.2 & -433.5 \\ -433.5 & 432.5 \end{bmatrix}$$

which shows that $(A + E)^{-1}$ is nowhere near A^{-1}, and hence the matrix A given in (13.6) is ill-conditioned with respect to inversion. Using ∞-norm (see Example 9.6), we have

$$\alpha = \| A^{-1} - (A + E)^{-1} \| = 867.7, \qquad \beta = \| A^{-1} \| = 201,$$

and hence the mathematical estimate of the relative error is $\alpha/\beta = 867.7/201 = 4.316$.

On the other hand, if

$$A = \begin{bmatrix} 1 & 1 \\ 1 & 2 \end{bmatrix} \tag{13.9}$$

and E is the same as (13.7), then

$$A^{-1} = \begin{bmatrix} 2 & -1 \\ -1 & 1 \end{bmatrix}, \qquad (A+E)^{-1} = \begin{bmatrix} 2.018 & -1.015 \\ -1.015 & 1.013 \end{bmatrix}. \tag{13.10}$$

The error now becomes

$$A^{-1} - (A+E)^{-1} = \begin{bmatrix} -0.018 & 0.015 \\ 0.015 & -0.013 \end{bmatrix}$$

which shows that a small change E in A produces a small change in its inverse $(A + E)^{-1}$. The relative error in respect to ∞-norm (see Example 9.6) is

$$\frac{\| A^{-1} - (A+E)^{-1} \|}{\| A^{-1} \|} = \frac{0.033}{3} = 0.011$$

and this inversion problem is well-conditioned. □

13.2.2 Perturbed Linear Equations: Condition Number

We consider a set of linear simultaneous equations

$$A\mathbf{x} = \mathbf{b}, \tag{13.11}$$

where A is nonsingular. If $\mathbf{b}$ is perturbed to $\mathbf{b} + \mathbf{k}$, then the new solution $\mathbf{x} + \mathbf{h}$ satisfies

$$A(\mathbf{x} + \mathbf{h}) = \mathbf{b} + \mathbf{k}, \tag{13.12}$$

Subtracting (13.11) from (13.12), we have $\mathbf{h} = A^{-1}\mathbf{k}$. Hence, from (13.11) it follows that

$$\| \mathbf{h} \| \le \| A^{-1} \| \, \| \mathbf{k} \|, \qquad \| \mathbf{b} \| \le \| A \| \, \| \mathbf{x} \|. \tag{13.13}$$

Therefore, the relative error in the solution **x** is

$$\varepsilon := \frac{\|\mathbf{h}\|}{\|\mathbf{x}\|} \le \|A\|\|A^{-1}\| \frac{\|\mathbf{k}\|}{\|\mathbf{b}\|} = k(A)\eta, \tag{13.14}$$

where η is the relative change in **b**.

On the other hand, if only A is perturbed to $A + E$, then the new solution $\mathbf{x} + \mathbf{h}$ satisfies

$$(A + E)(\mathbf{x} + \mathbf{h}) = \mathbf{b}. \tag{13.15}$$

Subtracting (13.11) from (13.15) and rearranging, we get

$$\mathbf{h} = -(I + A^{-1}E)^{-1}A^{-1}E\mathbf{x}.$$

Hence, by Lemma 13.1,

$$\frac{\|\mathbf{h}\|}{\|\mathbf{x}\|} \le \frac{\|A^{-1}E\|}{\left(1 - \|A^{-1}E\|\right)^{-1}}, \quad \text{if } \|A^{-1}E\| < 1.$$

If $\|A^{-1}\|\,\|E\| < 1$, a pessimistic estimate of the above result is

$$\varepsilon := \frac{\|\mathbf{h}\|}{\|\mathbf{x}\|} \le k(A)\mu[1 - k(A)\mu]^{-1} \tag{13.16}$$

which is the same as in the inverse of A.

In the above problems, if the condition number $k(A) = \|A\|\,\|A^{-1}\|$ is large [see (13.5), (13.14) and (13.16)], the relative error is large and the solution corresponding to a small perturbation in A or **b** is far away from the solution of unperturbed equation. In such cases, i.e., when $k(A)$ is large, the problems are known as *ill-conditioned*.

Example 13.4 The solution of linear simultaneous equations $A\mathbf{x} = \mathbf{b}$, where A is given by (13.9) and $\mathbf{b} = [2\ 3]^T$ is $\mathbf{x} = [1\ 1]^T$. If A is changed to $A + E$, where E is given by (13.7), we have to solve $(A + E)\tilde{\mathbf{x}} = \mathbf{b}$. This gives $\tilde{\mathbf{x}} = [0.991,\ 1.009]^T$, thereby showing that a small change in A also causes a small change in the solution **x**.

On the other hand, if A is given by (13.6) while **b** remains the same, the solution is $\mathbf{x} = [-98,\ 100]^T$. However, in this case, a change of A by the amount E given in (13.7) results in the solution $\hat{\mathbf{x}} = [334.22,\ -331.56]^T$ which shows a large deviation from $[-98\ \ 100]^T$. As a small change in A produces a large change in the solution, the problem here is very *ill-conditioned*. □

13.2.3 Eigenvalues and Eigenvectors of a Square Matrix

Let A be a simple matrix. Then in view of Theorem 7.23, there exists a nonsingular matrix P with eigenvectors of A as columns such that

$$P^{-1}AP = \operatorname{diag}[\lambda_1\ \lambda_2 \cdots \lambda_n] =: D, \tag{13.17}$$

where λ_i's are eigenvalues of A not necessarily distinct. Let λ be an eigenvalue of $A + E$, where E represents a perturbation in A, different from λ_i's. Then, $(\lambda I - A)$ is nonsingular and $\lambda I - A - E$ singular [see (7.8)]. Taking into consideration (13.17), we have

$$P^{-1}(\lambda I - A - E)P = [(\lambda I - D) - P^{-1}EP] = (\lambda I - D)[I - (\lambda I - D)^{-1}F],$$

where $(\lambda I - D)$ is nonsingular and $F: = P^{-1}EP$. Because of Corollary 2.1 and Corollary 5.14, $[I - (\lambda I - D)^{-1} F]$ is singular, which implies, in view of Example 9.17, that

$$1 \le \| (\lambda I - D)^{-1} F \| \le \| (\lambda I - D)^{-1} \| \, \| P^{-1} \| \, \| E \| \, \| P \|. \tag{13.18}$$

Now, in the $\| \cdot \|_p$ norm, where $p = 1, 2, \infty$,

$$\| (\lambda I - D)^{-1} \| = \max_i | (\lambda - \lambda_i)^{-1} | = \frac{1}{\min_i | \lambda - \lambda_i |}.$$

Consequently, from (13.18), we have

$$\min_i | \lambda - \lambda_i | \le k(P) \, \| E \|, \tag{13.19}$$

where

$$k(P) := \| P^{-1} \| \, \| P \| \tag{13.20}$$

is the *condition number* of the matrix P. Since $\| A \|_2 \le \| A \|_e$ [see Problem 9.10(c)], the result (13.19) is also true in $\| \cdot \|_e$ norm and in the case when $\lambda = \lambda_i$ for some i. This shows that the eigenvalue of the perturbed matrix $A + E$ lies within a disc of radius $k(P) \, \| E \|$ around the eigenvalues λ_i of A if the minimum is attained at λ_i.

As the scalar multiple of an eigenvector is again an eigenvector of a matrix A, the diagonalizing matrix P in (13.17) is not unique. However, in case of $\| \cdot \|_p$ norms, $p = 1, 2, \infty$, it follows from

$$k(P) = \| P^{-1} \| \, \| P \| \ge \| P^{-1} P \| = 1$$

that the minimum value of $k(P)$ is 1.

On the other hand, if A is normal (in particular Hermitian or unitary), then the matrix P in (13.17) is unitary and (see Theorem 7.26 and Problem 9.32) the condition number is

$$k(P) = \| P^* \|_2 \, \| P \|_2 = 1 \tag{13.21}$$

in the $\| \cdot \|_2$ norm. Thus, in view of (13.19) and (13.21), the eigenvalues of a normal matrix are always well-conditioned as their perturbed values are within a distance of $\| E \|_2$ of λ_i's. As a consequence, the *singular values* of a matrix are well-conditioned (see Section 9.6).

The effect of perturbation on a multiple eigenvalue of a nonsimple matrix A as also on the eigenvectors cannot be given an elementary treatment as above and is therefore outside the scope of this book. Nonetheless, we shall explain some basic concepts with the help of examples.

Example 13.5 Let

$$A=\begin{bmatrix}1 & 0\\ 0 & 1-10^{-8}\end{bmatrix},\qquad E=\begin{bmatrix}0 & 0\\ 10^{-10} & 0\end{bmatrix}.$$

The matrix A is a simple matrix having the eigenvalues 1 and $1-10^{-8}$ with the corresponding eigenvectors $\mathbf{x}_1=[1\ 0]^T$ and $\mathbf{x}_2=[0\ 1]^T$, respectively. However, the matrix $A+E$ has the same eigenvalues with the corresponding eigenvectors $\mathbf{y}_1=[100,1]^T$ and $\mathbf{y}_2=[0\ 1]^T$. Since the change in the eigenvector $\mathbf{x}_1$ with a small perturbation E in A, i.e., $\|\mathbf{y}_1-\mathbf{x}_1\|_1/\|E\|_1=100/10^{-10}=10^{12}$ is large, the matrix A is ill-conditioned with respect to eigenvectors although well-conditioned in relation to eigenvalues. □

Example 13.6 Let

$$A=\begin{bmatrix}0 & 1 & 0\\ 0 & 0 & 1\\ 0 & 0 & 0\end{bmatrix},\qquad E=\begin{bmatrix}0 & 0 & 0\\ 0 & 0 & 0\\ 10^{-6} & 0 & 0\end{bmatrix}.$$

The eigenvalues of A and $A+E$ are respectively $\lambda=0,0,0$; $\mu=10^{-2}$, $10^{-2}\exp(i2\pi/3)$, $10^{-2}\exp(-i2\pi/3)$, and the eigenvectors associated with 0 and 10^{-2} are $\mathbf{x}=[1\ 0\ 0]^T$ and $\mathbf{y}=[10^4\ 10^2\ 1]^T$. Hence, $|\lambda-\mu|/\|E\|_\infty=10^{-2}/10^{-6}=10^4$ and $\|\mathbf{x}-\mathbf{y}\|_\infty/\|E\|_\infty=10^4/10^{-6}=10^{10}$. This shows that A is ill-conditioned with respect to both eigenvalue λ and eigenvector $\mathbf{x}$. □

Numerical methods applied to an ill-conditioned problem are likely to give rise to inaccurate solutions. An algorithm or a numerical procedure will *be called numerically unstable* if it produces poor solutions to a well-conditioned problem. Because of roundoff errors, the *Gaussian algorithms* in finite arithmetic is a numerically unstable procedure, as shown in the following example.

Example 13.7 To solve the linear equations

$$\begin{bmatrix}-0.1\times10^{-4} & 1\\ 1 & 1\end{bmatrix}\begin{bmatrix}x_1\\ x_2\end{bmatrix}=\begin{bmatrix}1\\ 3\end{bmatrix}\tag{13.22}$$

by Gaussian elimination in 4-digit arithmetic, add $1/0.1\times10^{-4}=10^5$ times the first row to the second row of the augmented matrix to get

$$\left[\begin{array}{cc|c}-0.1\times10^{-4} & 1 & 1\\ 0 & 0.1\times10^{6} & 0.1\times10^{6}\end{array}\right].\tag{13.23}$$

The element -0.1×10^{-4} used to reduce 1 to 0 in the (2, 1) position is known *as the pivot element.* From (13.23), $x_1=0$, $x_2=1$ (see Example 5.30) which are different from the correct solution: $x_1=2$, $x_2=1$. This is because when we add 10^5 times the first row to the second row we make the approximation $1 + 10^5 \simeq 10^5$ and $3 + 10^5 \simeq 10^5$ due to rounding off in 4-digit arithmetic, thereby losing 1 and 3 in the process. This happens while choosing too small a number in absolute value as a pivot element. Since $\| A \|_1 \| A^{-1} \|_1 = 2 \times 2 = 4$, the system (13.22) is well-conditioned, which therefore shows that Gaussian algorithm is numerically unstable. □

To get around this difficulty, a *partial pivoting strategy* is adopted in which by interchanging rows, the element having the maximum absolute value is brought to the (1, 1) position to act as the pivot element. So with the equations interchanged, the system (13.22) becomes

$$\left[\begin{array}{cc|c} 1 & 1 & 3 \\ -0.1\times10^{-4} & 1 & 1 \end{array}\right] \sim \left[\begin{array}{cc|c} 1 & 1 & 3 \\ 0 & 1 & 1 \end{array}\right],$$

where the second matrix is obtained from the first by adding 10^{-5} times the first row to its second row. Hence, we get $x_1 = 2$, $x_2 = 1$ which is the correct solution.

Sometimes to improve accuracy, a *complete pivoting* strategy is used, by which an element with the maximum absolute value is brought to the (1, 1) position of the submatrix involved in the elimination process to act as the pivot element. Thus, if the pivot element is *too small in absolute value* and a large multiplier is used for reduction, the Gaussian algorithm is numerically unstable and partial or complete pivoting strategy is used to improve the accuracy of solution.

Since wide disparity in the sizes of the elements of $[A\ \mathbf{b}]$ is responsible for an inaccurate solution, sometimes by *scaling* its rows and columns, $[A\ \mathbf{b}]$ is balanced. After scaling, if the matrix $[A\ \mathbf{b}]$ is such that the element of the largest magnitude in each row and column is between 0.10 and 1.0, then it is said to be *equibrated.* The matrix $[A\ \mathbf{b}]$ in (13.22) is not equibrated.

Example 13.8 To solve the system

$$\begin{bmatrix} 1 & 1 & 1 \\ 0.4 & -10^{-5} & 10^{-5} \\ 0.2 & -2\times10^{-5} & 3\times10^{-5} \end{bmatrix} \begin{bmatrix} x_1 \\ x_2 \\ x_3 \end{bmatrix} = \begin{bmatrix} 2 \\ 0.6 \\ -0.1 \end{bmatrix} \tag{13.24}$$

in 4-digit arithmetic, we use the largest element 1 in (1, 1) position for complete pivoting to get, after one step of Gaussian elimination, the matrix

$$\left[\begin{array}{ccc|c} 1 & 1 & 1 & 2 \\ 0 & -0.4 & -0.4 & -0.2 \\ 0 & -0.2 & -0.2 & -0.5 \end{array}\right]. \tag{13.25}$$

The equations described by (13.25) are inconsistent, on the basis of which one may think that (13.24) does not have any solution. This is not true.

We now *scale* the second and third equations given in (13.24), multiplying each by 10^5 to get

$$\left[\begin{array}{ccc|c} 1 & 1 & 1 & 2 \\ 4\times10^4 & -1 & 1 & 6\times10^4 \\ 2\times10^4 & -2 & 3 & -1\times10^4 \end{array}\right]$$

which is then subjected to Gaussian elimination using partial pivoting strategy. This gives us successively,

$$\left[\begin{array}{ccc|c} 4\times10^4 & -1 & 1 & 6\times10^4 \\ 1 & 1 & 1 & 2 \\ 2\times10^4 & -2 & 3 & -1\times10^4 \end{array}\right] \sim \left[\begin{array}{ccc|c} 4\times10^4 & -1 & 1 & 6\times10^4 \\ 0 & 1 & 1 & 0.5 \\ 0 & -1.5 & 2.5 & -4\times10^4 \end{array}\right]$$

$$\sim \left[\begin{array}{ccc|c} 4\times10^4 & -1 & 1 & 6\times10^4 \\ 0 & -1.5 & 2.5 & -4\times10^4 \\ 0 & 1 & 1 & 0.5 \end{array}\right] \sim \left[\begin{array}{ccc|c} 4\times10^4 & -1 & 1 & 6\times10^4 \\ 0 & -1.5 & 2.5 & -4\times10^4 \\ 0 & 0 & a & -a\times10^4 \end{array}\right]$$

where $a = 2.667$. Hence, $x_1 = 2$, $x_2 = 10^4$, $x_3 = -10^4$, which is the correct solution. □

It may be observed that the system (13.24) is ill-conditioned, because $\| A \|_1 \, \| A^{-1} \|_1 = 1.6 \times 5 \times 10^4 = 8 \times 10^4$ and that Gaussian elimination with pivot strategy does not give the correct solution before the scaling operation. In spite of intensive theoretical investigation, there is no satisfactory algorithm for scaling a general matrix [183].

13.3 ORTHOGONAL TRANSFORMATIONS

In solving the linear, simultaneous equations $A\mathbf{x} = \mathbf{b}$ by Gaussian elimination, we apply elementary row transformations to the matrix A which is equivalent to premultiplying A by a nonsingular matrix P not necessarily orthogonal. Then we solve the triangular system of equations $PA\mathbf{x} = P\mathbf{b}$ by back substitution. The numerical instability to this artifice may be traced to the fact that under P the norm of a vector $\mathbf{v}$ may increase, i.e., $\| P\mathbf{v} \|$ may be greater than $\| \mathbf{v} \|$. In Example 13.7, P transforming (13.22) to (13.23) is $\begin{bmatrix} 1 & 0 \\ 10^5 & 1 \end{bmatrix}$ which testifies to the above fact.

A remedy to this may be sought in applying orthogonal transformation R in place of P, since under it the norm of a vector remains invariant, i..e., $\| \mathbf{v} \| = \| R\mathbf{v} \|$ (see Section 6.5). To improve numerical stability, orthogonal transformations are therefore recommended. In this section, Jacobi and Householder transformations which are orthogonal in nature will be introduced and used for triangularization of an arbitrary matrix and tridiagonalization of a symmetric matrix.

13.3.1 Householder Transformation

Definition 13.1 *Householder Transformation.* If $\mathbf{w}$ is a unit vector in $\mathbb{C}^n$, then a matrix P in $\mathbb{C}^{n\times n}$ of the form

$$P = I - 2\mathbf{w}\mathbf{w}^*, \qquad \mathbf{w}^*\mathbf{w} = 1 \tag{13.26}$$

is called a *Householder transformation, elementary Hermitian matrix* or an *elementary reflector.* □

(i) *Properties of Householder transformation.* It is evident from (13.26) that P has the following properties:

Property I P is Hermitian: $P^* = P$.

Property II P is unitary: $P^*P = PP^* = I$.

Property III P is involutory: $P^2 = I$.

Property IV Invariance of norm: If $P\mathbf{x} = \mathbf{y}$, then $\| \mathbf{x} \|_2 = \| \mathbf{y} \|_2$.

If $\mathbf{w}$ is real, then P is orthogonal and symmetric. In what follows, we deal with transformations P in $\mathbb{R}^{n\times n}$. Under transformation P, an n-vector $\mathbf{x}$ can be reduced to a vector, all of whose elements except for the first row are equal to zero. This result, which is crucial from the viewpoint of application, is contained in the following theorem.

Theorem 13.2 Given a nonzero vector $\mathbf{x} = [x_1, ..., x_k]^T$ in $\mathbb{R}^k$, there exists a Householder transformation $P_k = I - 2\mathbf{w}\mathbf{w}^T$ of order k such that

$$P_k\mathbf{x} = \alpha\,\mathbf{e}_1, \tag{13.27}$$

where α is a real number, $\mathbf{e}_1 = [1\ 0\ .\ .\ .\ 0]^T$ is an elementary vector in $\mathbb{R}^k$, and I is an identity matrix of order k.

Proof We use Example 9.1(b) to define the positive number m by

$$m: = \| \mathbf{x} \|_2 = \sqrt{(x_1^2 + \cdots + x_k^2)}. \tag{13.28}$$

Since $P_k^T P_k = I$, it now follows from (13.27) that

$$m^2 = \|\mathbf{x}\|_2^2 = \mathbf{x}^T P_k^T P_k \mathbf{x} = \alpha^2 \mathbf{e}_1^T \mathbf{e}_1 = \alpha^2.$$

Hence,

$$\alpha = \pm m = \pm\|\mathbf{x}\|_2. \tag{13.29}$$

Using the expression given by the right-hand side of (13.26) for P_k where $\mathbf{w} = [w_1, ..., w_k]^T$ in $\mathbb{R}^k$, and equating the components of both sides of (13.27), we have

$$x_1 - 2w_1(\mathbf{w}^T\mathbf{x}) = \alpha = \pm m,$$

$$x_i - 2w_i(\mathbf{w}^T\mathbf{x}) = 0, \qquad i = 2, . . ., k.$$

To determine the unknown $\mathbf{w}$, by setting q: $= \mathbf{w}^T\mathbf{x}$, we have

$$2w_1 q = x_1 \mp m; \qquad 2w_i q = x_i, \qquad i = 2, . . ., k. \tag{13.30}$$

Observing that $\mathbf{w}^T\mathbf{w} = w_1^2 + \cdots + w_k^2 = 1$, by squaring and adding the equations in (13.30), we have

$$4q^2 = (x_1 \mp m)^2 + x_2^2 + \cdots + x_k^2 = 2(m^2 \mp x_1 m).$$

Hence,

$$2q^2 = m(m \mp x_1) \tag{13.31}$$

is known. It therefore follows from (13.30) that

$$\mathbf{w} = (2q)^{-1}\mathbf{u}, \qquad \mathbf{u} = [x_1 \mp m, x_2, \ldots, x_k]^T; \tag{13.32}$$

and from (13.26) that $P_k = I - (2q^2)^{-1}\mathbf{u}\mathbf{u}^T$. Among the two values admissible for $2q^2$ in (13.31), the larger one is selected for numerical stability and so,

$$2q^2 = m[m + (\text{sgn } x_1)x_1],$$

where sgn x = 1, 1, –1 according as $x > 0$, $x = 0$, $x < 0$. The assertion in the theorem is now proved with

$$\alpha = \pm\, m = -(\mp m) = -(\text{sgn } x_1)\,\|\mathbf{x}\|_2\,, \text{ and}$$

$$P_k\mathbf{x} = -(\text{sgn } x_1)\,\|\mathbf{x}\|_2\,\mathbf{e}_1.$$

□

(ii) *Algorithm for Householder transformation.* Given a nonzero vector $\mathbf{x} = [x_1, \ldots, x_n]^T$, compute

(a) $m = \sqrt{(x_1^2 + \cdots + x_n^2)}$, (b) $2q^2 = m^2 + mx_1 \text{ sgn } x_1$,

(c) $\mathbf{u} = [x_1 + m \text{ sgn } x_1, x_2, \ldots, x_n]^T$, (d) $P_n = I - \dfrac{\mathbf{u}\mathbf{u}^T}{2q^2}$. (13.33)

Then,

$$P_n\mathbf{x} = \mathbf{x} - (\mathbf{u}^T\mathbf{x})\mathbf{u}/2q^2 = -\text{sgn } x_1\ \|\mathbf{x}\|_2\ \mathbf{e}_1.$$

The total number of multiplications required to compute **u** is n and only one square root, and to determine $P_n\mathbf{x}$ additional $2n + 1$ multiplications.

(iii) *Orthogonal reduction* to a *triangular form and QR-decomposition* by *Householder transformation.* To reduce any $m \times n$ matrix $A \in \mathbb{R}^{m \times n}$:

$$A = \begin{bmatrix} a_{11} & a_{12} & \cdots & a_{1n} \\ a_{21} & a_{22} & \cdots & a_{2n} \\ \vdots & \vdots & & \vdots \\ a_{m1} & a_{m2} & \cdots & a_{mn} \end{bmatrix}$$

to an upper triangular form, we set $\mathbf{x}^T = [a_{11}\ a_{21}\ \cdots\ a_{m1}]$. Let P_m be a matrix of order m such that $P_m\mathbf{x} = \alpha_{11}\mathbf{e}_1$, where $\alpha_{11} = -\text{sgn } a_{11}\ \|\mathbf{x}\|_2$ and $\mathbf{e}_1 \in \mathbb{R}^m$. This results in

$$P_mA = \begin{bmatrix} \alpha_{11} & \alpha_{12} & \cdots & \alpha_{1n} \\ 0 & b_{22} & \cdots & b_{2n} \\ \vdots & \vdots & & \vdots \\ 0 & b_{m-1,2} & \cdots & b_{m-1,n} \\ 0 & b_{m2} & \cdots & b_{mn} \end{bmatrix}.$$

Next we set $\mathbf{x}^T = [b_{22}\ \cdots\ b_{m2}]$ and find a matrix P_{m-1} of order $m - 1$ such that $P_{m-1}\,\mathbf{x} = \alpha_{22}\mathbf{e}_1$, where $\alpha_{22} = -\text{sgn } b_{22}\ \|\mathbf{x}\|_2$ and $\mathbf{e}_1 \in \mathbb{R}^{m-1}$. Hence,

$$\begin{bmatrix} 1 & 0 \\ 0 & P_{m-1} \end{bmatrix} P_m A = \begin{bmatrix} \alpha_{11} & \alpha_{12} & \alpha_{13} & \cdots & \alpha_{1n} \\ 0 & \alpha_{22} & \alpha_{23} & \cdots & \alpha_{2n} \\ 0 & 0 & c_{33} & \cdots & c_{3n} \\ \vdots & \vdots & \vdots & & \vdots \\ 0 & 0 & c_{m3} & \cdots & c_{mn} \end{bmatrix}.$$

Setting $\mathbf{x}^T = [c_{33} \cdots c_{m3}]$, we can proceed to reduce it to $P_{m-2}\,\mathbf{x} = \alpha_{33}\mathbf{e}_1$, where $\alpha_{33} = -\text{sgn}\; c_{33}\; \| \mathbf{x} \|_2$ and $\mathbf{e}_1 \in \mathbb{R}^{m-2}$. Thus, if

$$U_{k+1} := \begin{bmatrix} I_k & 0 \\ 0 & P_{m-k} \end{bmatrix}, \qquad k = 0, 1, \ldots, r-1,$$

where P_{m-k} is a Householder transformation of order $(m - k)$, then

$$Q^T A = U_r \cdots U_1 A = \begin{bmatrix} \alpha_{11} & \alpha_{12} & \cdots & \alpha_{1n} \\ 0 & \alpha_{22} & \cdots & \alpha_{2n} \\ \vdots & \vdots & & \vdots \\ 0 & 0 & \cdots & \alpha_{nn} \\ 0 & 0 & \cdots & 0 \\ \vdots & \vdots & & \vdots \\ 0 & 0 & \cdots & 0 \end{bmatrix}, \qquad m > n;$$

$$= \begin{bmatrix} \alpha_{11} & \alpha_{12} & \cdots & \alpha_{1m} & \cdots & \alpha_{1n} \\ 0 & \alpha_{22} & \cdots & \alpha_{2m} & \cdots & \alpha_{2n} \\ \vdots & \vdots & & \vdots & & \vdots \\ 0 & 0 & \cdots & \alpha_{mm} & \cdots & \alpha_{mn} \end{bmatrix}, \qquad m < n; \tag{13.34}$$

where $r = \min\,(m - 1, n)$ and $Q^T = U_r \cdots U_1$ is an orthogonal matrix. At the kth step, if $\mathbf{x}$ is a zero vector, we set $U_k = I_m$ and pass on to the next step. Hence, $A = QR$ is called the *QR-decomposition* (or factorization) of an arbitrary matrix A, where Q is orthogonal and R is upper triangular. If we determine an orthogonal matrix U such that $UA^T = L^T$ is in upper triangular form, then $A = LU$, where L is a lower triangular matrix.

For an n-square matrix A, the total numbers of multiplications and square roots required to triangularize A by the Householder transformation are

$$\frac{2}{3}n^3 + \frac{3}{2}n^2 + \frac{11}{6}n - 4 \approx \frac{2}{3}n^3$$

if n is very large and $(n - 1)$ respectively.

Example 13.9 Determine the *QR*-decomposition of A:

$$A = \begin{bmatrix} 0 & 3 & 50 \\ 3 & 5 & 25 \\ 4 & 0 & 25 \end{bmatrix}.$$

Solution First, we find an orthogonal matrix P_3 so that $P_3\mathbf{x}$, where $\mathbf{x} = [0\ 3\ 4]^T$, the first column of A, is of the form $\alpha\,\mathbf{e}_1$, where $\mathbf{e}_1 \in \mathbb{R}^3$. Using (13.33), $m = \sqrt{(0 + 9 + 16)} = 5$, $\mathbf{u} = [5\ 3\ 4]^T$, $2q^2 = 25$ and

$$U_1 := P_3 = I_3 - \frac{\mathbf{u}\mathbf{u}^T}{2q^2} = \frac{1}{25}\begin{bmatrix} 0 & -15 & -20 \\ -15 & 16 & -12 \\ -20 & -12 & 9 \end{bmatrix},$$

$$U_1A = P_3A = \begin{bmatrix} -5 & -3 & -35 \\ 0 & 7/5 & -26 \\ 0 & -24/5 & -43 \end{bmatrix}.$$

We shall now determine P_2 so that $P_2\mathbf{x}$, where $\mathbf{x} = [7/5,\ -24/5]^T$ is of the form $\alpha\,\mathbf{e}_1$, with $\mathbf{e}_1 \in \mathbb{R}^2$. Again using (13.33), $m = 5$, $2q^2 = 32$, $\mathbf{u}^T = [32, -24]/5$, and

$$P_2 = I_2 - \frac{\mathbf{u}\mathbf{u}^T}{2q^2} = \frac{1}{25}\begin{bmatrix} -7 & 24 \\ 24 & 7 \end{bmatrix}.$$

Define

$$U_2 = \frac{1}{25}\begin{bmatrix} 25 & 0 & 0 \\ 0 & -7 & 24 \\ 0 & 24 & 7 \end{bmatrix}$$

so that

$$U_2U_1A = -\begin{bmatrix} 5 & 3 & 35 \\ 0 & 5 & 34 \\ 0 & 0 & 37 \end{bmatrix} =: R, \qquad U_2U_1 = -\frac{1}{25}\begin{bmatrix} 0 & 15 & 20 \\ 15 & 16 & -12 \\ 20 & -12 & 9 \end{bmatrix}$$

and $A = QR$, where $Q = U_1U_2$ is the QR-decomposition of A. □

(iv) *Orthogonal reduction to tridiagonal and Hessenberg forms by Householder Transformation.* We now show how to reduce a real symmetric matrix to a tridiagonal form and a real arbitrary matrix to a Hessenberg form. Complex matrices may be treated similarly (see Problem 13.9).

Let $A = [a_{ij}]$, $a_{ij} = a_{ji}$ be a symmetric matrix of order n given by

$$A = \left[\begin{array}{c|cccc} a_{11} & a_{12} & a_{13} & \cdots & a_{1n} \\ \hline a_{21} & a_{22} & a_{23} & \cdots & a_{2n} \\ a_{31} & a_{32} & a_{33} & \cdots & a_{3n} \\ \vdots & \vdots & \vdots & & \vdots \\ a_{n1} & a_{n2} & a_{n3} & \cdots & a_{nn} \end{array}\right] = \begin{bmatrix} a_{11} & \mathbf{x}^T \\ \mathbf{x} & A_{n-1} \end{bmatrix},$$

where

$$\mathbf{x} = \begin{bmatrix} a_{21} \\ a_{31} \\ \vdots \\ a_{n1} \end{bmatrix}, \qquad A_{n-1} = \begin{bmatrix} a_{22} & a_{23} & \cdots & a_{2n} \\ a_{32} & a_{33} & \cdots & a_{3n} \\ \vdots & \vdots & & \vdots \\ a_{n2} & a_{n3} & \cdots & a_{nn} \end{bmatrix}.$$

Employing Theorem 13.2, we can construct a matrix P_{n-1} such that

$$P_{n-1}\,\mathbf{x} = c_{21}\mathbf{e}_1, \quad \text{where } c_{21} = -\text{sgn } a_{21} \,\|\, \mathbf{x} \,\|_2 \tag{13.35}$$

is a scalar, $\mathbf{e}_1 \in \mathbb{R}^{n-1}$ and P_{n-1} is given by (13.33). Consequently, with $c_{11} = a_{11}$ and $c_{12} = c_{21}$, we have

$$\begin{bmatrix} 1 & 0 \\ 0 & P_{n-1} \end{bmatrix} \begin{bmatrix} a_{11} & \mathbf{x}^T \\ \mathbf{x} & A_{n-1} \end{bmatrix} \begin{bmatrix} 1 & 0 \\ 0 & P_{n-1} \end{bmatrix} = \begin{bmatrix} a_{11} & \mathbf{x}^T P_{n-1} \\ P_{n-1}\mathbf{x} & P_{n-1}A_{n-1}P_{n-1} \end{bmatrix} \tag{13.36}$$

$$= \left[\begin{array}{c|cccc} c_{11} & c_{12} & 0 & \cdots & 0 \\ \hline c_{21} & b_{11} & b_{12} & \cdots & b_{1,\,n-1} \\ 0 & b_{21} & b_{22} & \cdots & b_{2,\,n-1} \\ \vdots & \vdots & \vdots & & \vdots \\ 0 & b_{n-1,\,1} & b_{n-1,\,2} & \cdots & b_{n-1,\,n-1} \end{array}\right]. \tag{13.37}$$

We now subject the symmetric matrix $B = [b_{ij}] := P_{n-1}A_{n-1}P_{n-1}$ to the transformation P_{n-2} given in (13.33) so that $P_{n-2}\,\mathbf{y} = c_{32}\mathbf{e}_1$, where $\mathbf{y} = [b_{21}, \ldots, b_{n-1,\,1}]^T$, $\mathbf{e}_1 \in \mathbb{R}^{n-2}$, and $c_{23} = c_{32} = -\text{sgn } b_{21} \,\|\, \mathbf{y} \,\|$. Hence, if $c_{22} = b_{11}$, we have

$$\begin{bmatrix} 1 & 0 \\ 0 & P_{n-2} \end{bmatrix} \begin{bmatrix} b_{11} & \mathbf{y}^T \\ \mathbf{y} & B_{n-2} \end{bmatrix} \begin{bmatrix} 1 & 0 \\ 0 & P_{n-2} \end{bmatrix} = \begin{bmatrix} b_{11} & \mathbf{y}^T P_{n-2} \\ P_{n-2}\mathbf{y} & P_{n-2}B_{n-2}P_{n-2} \end{bmatrix}$$

$$= \begin{bmatrix} b_{11} & c_{23} & 0 & \cdots & 0 \\ c_{32} & d_{11} & d_{12} & \cdots & d_{1,\,n-2} \\ 0 & d_{21} & d_{22} & \cdots & d_{2,\,n-2} \\ \vdots & \vdots & \vdots & & \vdots \\ 0 & d_{n-2,\,1} & d_{n-2,\,2} & \cdots & d_{n-2,\,n-2} \end{bmatrix};$$

$$\begin{bmatrix} I_2 & 0 \\ 0 & P_{n-2} \end{bmatrix} \begin{bmatrix} 1 & 0 \\ 0 & P_{n-1} \end{bmatrix} A \begin{bmatrix} 1 & 0 \\ 0 & P_{n-1} \end{bmatrix} \begin{bmatrix} I_2 & 0 \\ 0 & P_{n-2} \end{bmatrix}$$

$$= \begin{bmatrix} c_{11} & c_{12} & 0 & \cdots & 0 \\ c_{21} & c_{22} & c_{23} & \cdots & 0 \\ 0 & c_{32} & d_{11} & \cdots & d_{1,\,n-2} \\ \vdots & \vdots & \vdots & & \vdots \\ 0 & 0 & d_{n-2,\,1} & \cdots & d_{n-2,\,n-2} \end{bmatrix}.$$

In this way, we can define a sequence of matrices $P_{n-1}, P_{n-2}, \ldots, P_2$ so that the orthogonal matrix $P := U_{n-2} \cdots U_2U_1$, where

$$U_k = \begin{bmatrix} I_k & 0 \\ 0 & P_{n-k} \end{bmatrix}, \qquad k = 1, 2, \ldots, n-2$$

transforms A to the tridiagonal form

$$PAP^T = \begin{bmatrix} c_{11} & c_{12} & 0 & \cdots & 0 & 0 \\ c_{21} & c_{22} & c_{23} & \cdots & 0 & 0 \\ 0 & c_{32} & c_{33} & \cdots & 0 & 0 \\ \vdots & \vdots & \vdots & \ddots & \vdots & \vdots \\ 0 & 0 & 0 & \cdots & c_{n-1,\,n-1} & c_{n-1,\,n} \\ 0 & 0 & 0 & \cdots & c_{n,\,n-1} & c_{nn} \end{bmatrix}, \tag{13.38}$$

where $c_{i,i+1} = c_{i+1,i}$, $i = 1, 2, ..., n - 1$ and P is an orthogonal transformation.

If the matrix A is not symmetric, then in view of (13.35), the matrix in (13.37) will have $(n - 2)$ zeros in the first column and not necessarily in the first row. So, finally, in place of (13.38) we shall have

$$PAP^T = \begin{bmatrix} c_{11} & c_{12} & \cdots & c_{1,\,n-1} & c_{1n} \\ c_{21} & c_{22} & \cdots & c_{2,\,n-1} & c_{2n} \\ 0 & c_{32} & \cdots & c_{3,\,n-1} & c_{3n} \\ \vdots & \vdots & \ddots & \vdots & \vdots \\ 0 & 0 & \cdots & c_{n,\,n-1} & c_{nn} \end{bmatrix} \tag{13.39}$$

which is defined as an *upper Hessenberg matrix.* One requires $2n^2 + 5n - 7 \approx 2n^2$ multiplications and $(n - 2)$ square roots to get a tridiagonal or an upper Hessenberg form by the Householder transformation.

Definition 13.2 *Hessenberg Matrix.* A square matrix $C = [c_{ij}]$ is defined as *upper Hessenberg* if

$$c_{ij} = 0, \qquad i \geq j + 2;$$

and an *unreduced upper Hessenberg* if in addition $c_{i+1,\,i} \neq 0$, i.e., no subdiagonal elements are zero. It is called a *lower Hessenberg* if $c_{ij} = 0$, $j \geq i + 2$, and *unreduced lower Hessenberg* if in addition $c_{i,\,i+1} \neq 0$, i.e., no superdiagonal elements are zero. Hence, if C is upper Hessenberg, then C^T is lower Hessenberg. A square matrix C is *tridiagonal* if it is both upper and lower Hessenberg. □

To reduce a matrix A to a lower Hessenberg form, we may follow the above procedure to determine an orthogonal transformation P so that PA^TP^T is upper Hessenberg. Consequently, its transpose PAP^T is lower Hessenberg.

Example 13.10 Show that the matrix

$$A = \begin{bmatrix} 1 & 1 & 2 & 1 \\ -3 & 0 & 1 & -1 \\ 0 & 1 & -1 & 1 \\ 4 & 0 & 1 & -1 \end{bmatrix}$$

is orthogonally similar to an upper Hessenberg form.

Solution In view of (13.33), to obtain the Householder transformation for the vector $[-3\ 0\ 4]^T$, we have $m = 5$, $\mathbf{u} = [-8\ 0\ 4]^T$, $2q^2 = 25 + 3 \times 5 = 40$, and so

$$P_3 := I_3 - (2q^2)^{-1}\mathbf{u}\mathbf{u}^T = \begin{bmatrix} -0.6 & 0 & 0.8 \\ 0.0 & 1 & 0.0 \\ 0.8 & 0 & 0.6 \end{bmatrix}.$$

Defining

$$U_1 = \begin{bmatrix} 1 & 0 & 0 & 0 \\ 0 & -0.6 & 0 & 0.8 \\ 0 & 0.0 & 1 & 0.0 \\ 0 & 0.8 & 0 & 0.6 \end{bmatrix}$$

we have

$$U_1 A U_1^T = \begin{bmatrix} 1 & 0.2 & 2.0 & 1.4 \\ 5 & -0.16 & 0.2 & -0.12 \\ 0 & 0.2 & -1.0 & 1.4 \\ 0 & -1.12 & 1.4 & -0.84 \end{bmatrix}.$$

Again, to obtain the Householder transformation for the vector $[0.2, -1.12]^T$, we have $m = 1.138$, $2q^2 = 1.522$, and $\mathbf{u} = [1.338, -1.12]^T$ which are correct to three decimal places, and hence

$$P_2 = I - (2q^2)^{-1}\mathbf{u}\mathbf{u}^T = \begin{bmatrix} -0.176 & 0.985 \\ 0.985 & 0.176 \end{bmatrix}.$$

Define

$$U_2 = \begin{bmatrix} 1 & 0 & 0 & 0 \\ 0 & 1 & 0 & 0 \\ 0 & 0 & -0.176 & 0.985 \\ 0 & 0 & 0.985 & 0.176 \end{bmatrix}.$$

Then

$$PAP^T = \begin{bmatrix} 1 & 0.200 & 1.027 & 2.216 \\ 5 & -0.160 & -0.153 & 0.176 \\ 0 & -1.138 & -1.332 & 1.343 \\ 0 & 0.000 & 1.343 & -0.511 \end{bmatrix}, \qquad P = U_2 U_1.$$

□

13.3.2 Plane Rotation

Definition 13.3 *Plane Rotation.* An nth order unitary matrix Q, all of whose elements are the same as the identity matrix I_n except the elements

$$q_{jj} = \exp(i\alpha)\cos\theta, \qquad q_{jk} = \exp(i\beta)\sin\theta,$$

$$q_{kj} = -\exp(-i\beta)\sin\theta, \quad q_{kk} = \exp(-i\alpha)\cos\theta, \qquad i = \sqrt{-1},$$

where α, β and θ are real, is called a *plane rotation,* an *elementary unitary matrix* or *Jacobi transformation* in the (j, k)-plane. □

We can verify that $QQ^* = Q^*Q = I_n$. So, $Q^{-1} = Q^*$. The choice $\alpha = \beta = 0$ gives an orthogonal matrix having the structure

$$Q = \begin{array}{c} \\ \\ \\ j \\ \\ k \\ \\ \\ \\ \end{array}\overset{\begin{array}{ccc} \qquad\qquad\qquad j & \qquad\qquad k & \end{array}}{\begin{bmatrix} 1 & \cdots & 0 & 0 & \cdots & 0 & 0 & \cdots & 0 \\ \vdots & & \vdots & \vdots & & \vdots & \vdots & & \vdots \\ 0 & \cdots & 1 & 0 & \cdots & 0 & 0 & \cdots & 0 \\ 0 & \cdots & 0 & c & \cdots & s & 0 & \cdots & 0 \\ \vdots & & \vdots & \vdots & & \vdots & \vdots & & \vdots \\ 0 & \cdots & 0 & -s & \cdots & c & 0 & \cdots & 0 \\ 0 & \cdots & 0 & 0 & \cdots & 0 & 1 & \cdots & 0 \\ \vdots & & \vdots & \vdots & & \vdots & \vdots & & \vdots \\ 0 & \cdots & 0 & 0 & \cdots & 0 & 0 & \cdots & 1 \end{bmatrix}}, \quad c^2 + s^2 = 1, \tag{13.40}$$

where $c = \cos\theta$ and $s = \sin\theta$, which will be useful for our purpose. The significance of plane rotation is obvious if one considers $n = 2$ (see Section 6.5).

Because of the structure of Q, in the product $Q\mathbf{x} = \mathbf{y}$, $\mathbf{y}$ differs from $\mathbf{x}$ only in the elements of rows j and k of $\mathbf{x}$, i.e.,

$$y_m = x_m \ (m \neq j, k); \qquad y_j = cx_j + sx_k; \qquad y_k = -sx_j + cx_k.$$

Solving $c^2 + s^2 = 1$ and $-sx_j + cx_k = 0$, we get

$$c = \frac{x_j}{\alpha}, \qquad s = \frac{x_k}{\alpha}, \qquad \alpha = \sqrt{(x_j^2 + x_k^2)}. \tag{13.41}$$

Equations (13.41) define a plane rotation Q in the plane (j, k) for which the kth row y_k of $Q\mathbf{x} = \mathbf{y}$ is zero, i.e.,

$$y_j = \alpha, \qquad y_k = 0. \tag{13.42}$$

This needs four multiplications and one square root.

For an $n \times m$ matrix A, the product $B{:} = QA$ differs from A only in the jth and kth rows. The elements in this product are

$$b_{jr} = ca_{jr} + sa_{kr}; \qquad b_{kr} = -sa_{jr} + ca_{kr};$$

$$b_{ir} = a_{ir}; \qquad i = 1, \ldots, n; \qquad i \neq j, k; \qquad r = 1, \ldots, m.$$

It can be shown that to form the product QA, where all the elements in the first column except one are zero and A is an $n \times n$ matrix, $4n(n-1)$ multiplications and $(n-1)$ square roots are needed. Similarly, in the product AQ, where A is an $m \times n$ matrix, only the elements in the jth and kth columns are different from those of A.

(i) *Orthogonal reduction to a triangular form and QR-decomposition by plane rotation.* In view of the above property, we can successively use plane rotations in the planes (1, 2),

(1, 3), ..., (1, n) to annihilate the elements of the first column of the $n \times m$ matrix A in the positions (2, 1), (3, 1), ..., (n, 1). Similarly, by plane rotations in the planes (2, 3), (2, 4), ..., (2, n), elements of the second column of A in the positions (3, 2), (4, 2), ..., (n, 2) are annihilated without altering the elements in the first column. In this way, any $n \times m$ matrix A can be reduced to an upper triangular form R given in (13.34), which is now of order $n \times m$. If Q_{jk} denotes a plane rotation in the (j, k)-plane, then $Q_{n-1,\,n} \cdots Q_{13}Q_{12}A = R$, $n \le m$. Hence, we have

$$A = QR, \qquad Q^T = Q_{n-1,\,n} \cdots Q_{13}Q_{12}, \qquad n \le m, \tag{13.43}$$

where the matrix Q is orthogonal, since each Q_{jk} is orthogonal and R is upper triangular. This is known as the QR-decomposition of A by *Givens method* or *plane rotation* and requires $n(n-1)/2$ plane rotations.

To triangularize an n-square matrix by plane rotations, $(4/3)\, n(n^2 - 1) \approx (4/3)n^3$ multiplications and $n(n-1)/2 \approx n^2/2$ square roots are needed, if n is very large.

Theorem 13.3 Any nonsingular matrix A can be decomposed uniquely in the form

$$A = QR, \qquad Q^TQ = I, \tag{13.44}$$

where R is nonsingular and upper triangular with positive diagonal elements.

Proof The matrix A can be decomposed in the form (13.44) by the Householder transformation [see (13.34)] or by plane rotation [see (13.43)]. To prove the uniqueness, let there exist another factorization of A:

$$A = Q_1R_1, \qquad Q_1^T Q_1 = I, \tag{13.45}$$

where R_1 is nonsingular and upper triangular with positive diagonal elements. From (13.44) and (13.45), we have

$$QR = Q_1R_1, \tag{13.46}$$

$$A^TA = R_1^T R_1 = R^T R,$$

i.e.,

$$R_1R^{-1} = (RR_1^{-1})^T, \tag{13.47}$$

where the left-hand side is upper triangular while the right-hand side is lower triangular, and so both of them must be diagonal. If δ_i's and d_i's are diagonal elements of R_1 and R respectively, then equating diagonal elements of both sides of (13.47), we have

$$\frac{\delta_i}{d_i} = \frac{d_i}{\delta_i},$$

which shows that $R_1R^{-1} = I$, i.e., $R_1 = R$. The fact that $Q_1 = Q$, follows from (13.46). □

Example 13.11 Solve the simultaneous equations $A\mathbf{x} = \mathbf{b}$, where

$$A = \begin{bmatrix} 0 & 3 & 50 \\ 3 & 5 & 25 \\ 4 & 0 & 25 \end{bmatrix}, \qquad \mathbf{b} = \begin{bmatrix} 53 \\ 36 \\ 33 \end{bmatrix}. \tag{13.48}$$

Solution Using plane rotation, we shall reduce A to an upper triangular matrix. For rotation in the (1, 2)-plane, from (13.41) we have $\alpha = 3$, $c = 0$, and $s = 1$, and hence

$$Q_{12} = \begin{bmatrix} 0 & 1 & 0 \\ -1 & 0 & 0 \\ 0 & 0 & 1 \end{bmatrix}, \qquad Q_{12}[A\ \mathbf{b}] = \left[\begin{array}{ccc|c} 3 & 5 & 25 & 36 \\ 0 & -3 & -50 & -53 \\ 4 & 0 & 25 & 33 \end{array}\right].$$

Now, for rotation in the (1, 3)-plane, $\alpha = \sqrt{(9+16)} = 5$, $c = 3/5$ and $s = 4/5$, and therefore,

$$Q_{13} = \frac{1}{5}\begin{bmatrix} 3 & 0 & 4 \\ 0 & 5 & 0 \\ -4 & 0 & 3 \end{bmatrix}, \qquad Q_{13}Q_{12}[A\ \mathbf{b}] = \left[\begin{array}{ccc|c} 5 & 3 & 35 & 48 \\ 0 & -3 & -50 & -53 \\ 0 & -4 & -5 & -9 \end{array}\right].$$

Finally, for rotation in the (2, 3)-plane, $\alpha = \sqrt{(9+16)} = 5$, $c = -3/5$ and $s = -4/5$. Consequently,

$$Q_{23} = \frac{1}{5}\begin{bmatrix} 5 & 0 & 0 \\ 0 & -3 & -4 \\ 0 & 4 & -3 \end{bmatrix}, \qquad Q^T[A\ \mathbf{b}] = \left[\begin{array}{ccc|c} 5 & 3 & 35 & 48 \\ 0 & 5 & 34 & 39 \\ 0 & 0 & -37 & -37 \end{array}\right], \tag{13.49}$$

where $Q^T = Q_{23}Q_{13}Q_{12}$. From the third equation in (13.49), $x_3 = 1$, and so by back substitution, $x_2 = 1$ and $x_1 = 2$. In view of (13.43),

$$Q^T = Q_{23}Q_{13}Q_{12} = \frac{1}{25}\begin{bmatrix} 0 & 15 & 20 \\ 15 & 16 & -12 \\ -20 & 12 & -9 \end{bmatrix}$$

and, therefore, the QR-decomposition of A is

$$A = \frac{1}{25}\begin{bmatrix} 0 & 15 & -20 \\ 15 & 16 & 12 \\ 20 & -12 & -9 \end{bmatrix}\begin{bmatrix} 5 & 3 & 35 \\ 0 & 5 & 34 \\ 0 & 0 & -37 \end{bmatrix}.$$

□

(ii) *Orthogonal reduction to tridiagonal and Hessenberg forms by plane rotation.* Using plane rotations in the planes (2, 3), ..., (2, n) to an n-square matrix A, we have

$$Q_{2n}\cdots Q_{23}A = \begin{bmatrix} a_{11} & a_{12} & a_{13} & \cdots & a_{1n} \\ b_{21} & b_{22} & b_{23} & \cdots & b_{2n} \\ 0 & b_{32} & b_{33} & \cdots & b_{3n} \\ \vdots & \vdots & \vdots & & \vdots \\ 0 & b_{n2} & b_{n3} & \cdots & b_{nn} \end{bmatrix}$$

and if A is symmetric, then

$$Q_{2n}\cdots Q_{23}AQ_{23}^T\cdots Q_{2n}^T = \begin{bmatrix} a_{11} & b_{12} & 0 & \cdots & 0 \\ b_{21} & c_{22} & c_{23} & \cdots & c_{2n} \\ 0 & c_{32} & c_{33} & \cdots & c_{3n} \\ \vdots & \vdots & \vdots & & \vdots \\ 0 & c_{n2} & c_{n3} & \cdots & c_{nn} \end{bmatrix}$$

which is obviously symmetric. Introducing further rotations in the planes (3, 4), ..., (3, n); ..., (n – 1, n), we can determine an orthogonal transformation Q such that QAQ^T is tridiagonal [see (13.38)] if A is symmetric; otherwise it is in upper Hessenberg form [see (13.39)].

13.3.3 Least Squares Solution of Ax = b

Let A be an $n \times m$ matrix with all columns linearly independent, implying that $m \le n$. Our problem is to determine $\mathbf{x}$ such that $\mathbf{r} = \mathbf{b} - A\mathbf{x}$ has a minimum Euclidean norm denoted by $\|\cdot\|$. Let the QR-decomposition of A be written as

$$QA = \begin{bmatrix} R \\ 0 \end{bmatrix},$$

where R is nonsingular and upper triangular. We set $Q\mathbf{b} = [\mathbf{a}^T \ \mathbf{c}^T]^T$, where $\mathbf{a} \in \mathbb{R}^m$. Consequently,

$$Q\mathbf{r} = Q(\mathbf{b} - A\mathbf{x}) = \begin{bmatrix} \mathbf{a} - R\mathbf{x} \\ \mathbf{c} \end{bmatrix}. \tag{13.50}$$

Observing that Q is orthogonal, the Euclidean norm of $\mathbf{r}$ is

$$\|\mathbf{r}\|^2 = \|Q\mathbf{r}\|^2 = \|\mathbf{a} - R\mathbf{x}\|^2 + \|\mathbf{c}\|^2$$

which is minimum if $R\mathbf{x} = \mathbf{a}$ and can be solved for $\mathbf{x}$ by back substitution as it is a triangular system of equations. The minimum residual in the sense of Euclidean norm from (13.50) is $\mathbf{r} = Q^T[\boldsymbol{\theta}^T \ \ \mathbf{c}^T]^T$. See Section 9.6.1 for the general case.

Example 13.12 Obtain the least squares solution of $A\mathbf{x} = \mathbf{b}$, where

$$A = \begin{bmatrix} 0 & 1 \\ -3 & 0 \\ 0 & 2 \\ 4 & 10/3 \end{bmatrix}, \quad \mathbf{x} = \begin{bmatrix} x_1 \\ x_2 \end{bmatrix}, \quad \mathbf{b} = \begin{bmatrix} 1 \\ 1 \\ 1 \\ 1 \end{bmatrix}. \tag{13.51}$$

Solution Since rank A = 2 and rank $[A \ \mathbf{b}]$ = 3, equations (13.51) are inconsistent (see Theorem 5.10). So we shall determine $\mathbf{x}$ such that the Euclidean norm of $\mathbf{r} = \mathbf{b} - A\mathbf{x}$ is minimum.

To do this, at first a Householder transformation U_1 is determined so that $U_1[0 \ -3 \ 0 \ 4]^T = \alpha\,\mathbf{e}_1$, α is a scalar. Applying (13.33), we have m = 5, $2q^2 = 25$, $\mathbf{u}^T = [5 \ -3 \ 0 \ 4]$. Hence,

$$U_1 = \begin{bmatrix} 0 & 0.6 & 0 & -0.8 \\ 0.6 & 0.64 & 0 & 0.48 \\ 0 & 0 & 1 & 0 \\ -0.8 & 0.48 & 0 & 0.36 \end{bmatrix},$$

$$U_1[A \,\vdots\, \mathbf{b}] = \left[\begin{array}{cc|c} -5 & -8/3 & -0.2 \\ 0 & 2.2 & 1.72 \\ 0 & 2.0 & 1 \\ 0 & 0.4 & 0.04 \end{array}\right].$$

Again, we apply (13.33) to the vector $[2.2,\ 2,\ 0.4]^T$ to get $m = 3$, $2q^2 = 15.6$, and $\mathbf{u}^T = [5.2,\ 2.0,\ 0.4]$. Define

$$U_2 = \begin{bmatrix} 1 & 0 \\ 0 & I_3 - (2q^2)^{-1}\mathbf{u}\mathbf{u}^T \end{bmatrix} = \begin{bmatrix} 1 & 0 & 0 & 0 \\ 0 & -0.733 & -0.667 & -0.133 \\ 0 & -0.667 & 0.744 & -0.051 \\ 0 & -0.133 & -0.051 & 0.990 \end{bmatrix}.$$

Then

$$U_2U_1[A\ \mathbf{b}] = -\begin{bmatrix} 5 & 8/3 & 0.2 \\ 0 & 3 & 1.933 \\ 0 & 0 & 0.405 \\ 0 & 0 & 0.241 \end{bmatrix},$$

$$R = -\begin{bmatrix} 5 & 8/3 \\ 0 & 3 \end{bmatrix}, \qquad \mathbf{a} = -[0.2 \quad 1.933]^T, \quad \mathbf{c} = -[0.405 \quad 0.241]^T.$$

Hence solving $R\mathbf{x} = \mathbf{a}$, we get

$$x_1 = -0.3034, \qquad x_2 = 0.644,$$

$$\mathbf{r} = U_1U_2[0\ 0\ \mathbf{c}^T]^T = [0.356,\ 0.089,\ -0.289,\ 0.066]^T.$$

The computation of $(A^TA)^{-1}A^T\mathbf{b}$ gives us the same values for x_1 and x_2. □

13.4 NUMERICAL EVALUATION OF EIGENVALUES AND EIGENVECTORS

The various properties of eigenvalues and eigenvectors are established in Chapter 7. For large matrices, it is highly impractical to calculate the characteristic polynomial of a matrix, not to speak of evaluation of the eigenvalues and eigenvectors. Some of the computational aspects of these problems are taken care of in this section.

13.4.1 Gerschgorin Method

Sometimes, a rough estimate regarding the bounds of eigenvalues is required, which forms the starting point for numerically evaluating them from the characteristic polynomial. This can be done by Gerschgorin method.

To understand this, from (7.6) we have

$$(\lambda - a_{ii})x_i = \sum_{j=1,\, j\neq i}^{n} a_{ij}x_j, \qquad i = 1, 2, \ldots, n. \tag{13.52}$$

Let x_k, the kth component of $\mathbf{x}$, be the largest in absolute value:

$$\left|\frac{x_j}{x_k}\right| \le 1, \qquad j = 1, 2, \ldots, n. \tag{13.53}$$

Then using the triangle inequality and (13.53), from (13.52) at $i = k$, we have

$$|\lambda - a_{kk}| = \left| \sum_{j=1,\, j\neq k}^{n} a_{kj} \left(\frac{x_j}{x_k}\right) \right| \leq \sum_{j=1,\, j\neq k}^{n} |a_{kj}| =: r_k \tag{13.54}$$

which implies that for each eigenvalue there is some k for which (13.54) is true. But it is not known *a priori* which k to choose for each eigenvalue. For each k, the set of λ's which satisfy (13.54) is a disc with centre at a_{kk} and radius r_k. If each of these discs is denoted by D_k, i.e.,

$$D_k := \{\lambda : |\lambda - a_{kk}| \leq r_k\}, \qquad k = 1, 2, \ldots, n,$$

then each eigenvalue of A lies in at least one of the D_k's in the complex z-plane. If D is the union of these discs D_k's, then all eigenvalues of A lie inside D. Since A^T has the same eigenvalues as A, the eigenvalues of A also lie in the set of discs $\tilde{D}_1, \tilde{D}_2, \ldots, \tilde{D}_n$, where

$$\tilde{D}_k := \{\lambda : |\lambda - a_{kk}| \leq \tilde{r}_k\}, \qquad k = 1, 2, \ldots, n,$$

where

$$\tilde{r}_k := \sum_{i=1,\, i\neq k}^{n} |a_{ik}|$$

is the kth column sum of A. If $\tilde{D}$ is the union of these discs $\tilde{D}_k$'s, then all eigenvalues of A also lie inside $\tilde{D}$. Cousequently, eigenvalues of A lie inside $D \cap \tilde{D}$. The set of discs D_k and $\tilde{D}_k$ are referred to as the *Gerschgorin discs*.

Example 13.13 (a) The eigenvalues of the matrix

$$A = \begin{bmatrix} 1 & 0 & -1 \\ 1 & -2 & 1 \\ 2 & -1 & -1 \end{bmatrix} \tag{13.55}$$

are $\lambda_1 = 0$, $\lambda_2 = -1 + i$, $\lambda_3 = -1 - i$. The set D is the union of Gerschgorin discs:

$$D_1 = \{\lambda : |\lambda - 1| \leq 1\}, \qquad D_2 = \{\lambda : |\lambda + 2| \leq 2\}, \qquad D_3 = \{\lambda : |\lambda + 1| \leq 3\}$$

which are shown in Figure 13.1 along with the eigenvalues. Since A^T has the same eigenvalues as A, we also get a different set $\tilde{D}$ of Gerschgorin discs: $\tilde{D}_1 = \{\lambda : |\lambda - 1| \leq 3\}$, $\tilde{D}_2 = \{\lambda : |\lambda + 2| \leq 1\}$, and $\tilde{D}_3 = \{\lambda : |\lambda + 1| \leq 2\}$, so that the eigenvalues of A lie in the intersection of D and $\tilde{D}$. □

A recipe. In eigenvalue problems, we shall repeatedly use the following recipe: "If **x** is an eigenvector of the matrix A associated with λ, so also is any nonzero multiple 'α' of **x**, i.e., $A\mathbf{x} = \lambda\mathbf{x}$ implies and is implied by $A(\alpha\mathbf{x}) = \lambda(\alpha\,\mathbf{x})$". So one should not be confused if from $A(\lambda^k\alpha)\mathbf{x} = (\lambda^{k+1}\alpha)\mathbf{x}$ we state that **x** is an eigenvector of A corresponding to the eigenvalue λ (see Example 7.1).

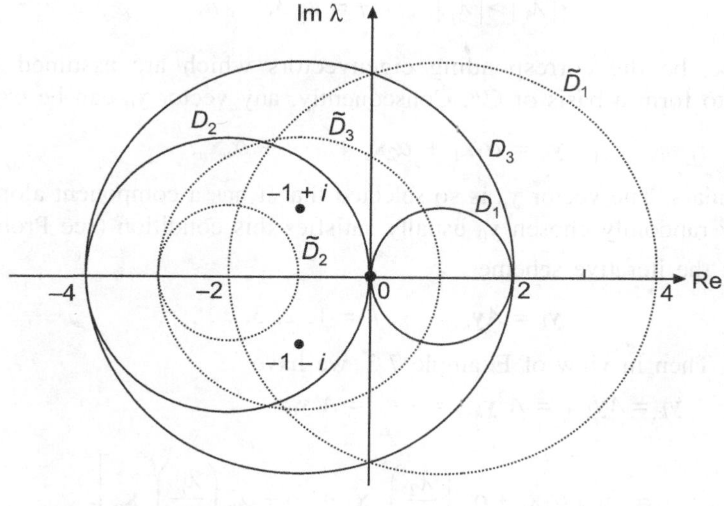

Figure 13.1 Eigenvalues of A lie inside D, the union of Gerschgorin discs of A and also inside $\tilde{D}$, the union of Gerschgorin discs of A^T, and hence in $D \cap \tilde{D}$.

(b) The largest eigenvalue of a stochastic matrix (i.e. a matrix whose entries are positive and whose rows add up to 1) is 1.

Solution Say A is an $n \times n$ row stochastic matrix. Now:

$$A\begin{bmatrix}1\\1\\\vdots\\1\end{bmatrix} = \begin{bmatrix}\sum_{i=1}^{n} a_{1i}\\ \sum_{i=1}^{n} a_{2i}\\ \vdots \\ \sum_{i=1}^{n} a_{ni}\end{bmatrix} = \begin{bmatrix}1\\1\\\vdots\\1\end{bmatrix}$$

Thus the eigenvalue 1 is attained. To show that this is the largest eigenvalue you can use the Gershgorin circle theorem. Using (13.54), for the diagonal element a_{kk} in kth row

$$|\lambda - a_{kk}| \le \sum_{i \ne k} |a_{ki}| = \sum_{i \ne k} a_{ki} = 1 - a_{kk},$$

since all $a_{ki} \ge 0$. Therefore,

$$|\lambda| = |\lambda - a_{kk} + a_{kk}| \le |\lambda - a_{kk}| + a_{kk} \le 1 - a_{kk} + a_{kk} = 1.$$

Thus for all rows, $|\lambda| \le 1$ implying that all eigenvalues λ_i satisfy $|\lambda_i| \le 1$. □

13.4.2 The Power Method

A simple method to compute the eigenvalue, which is the largest in absolute value, and the associated eigenvector of a matrix A, is the power method. Let λ_1 be such an eigenvalue to be termed a *dominant eigenvalue,* i.e.,

$$|\lambda_1| > |\lambda_j|, \qquad j = 2, 3, \ldots, n, \tag{13.56}$$

and $\mathbf{x}_1$, $\mathbf{x}_2$, ..., $\mathbf{x}_n$ be the corresponding eigenvectors which are assumed to be linearly independent and to form a basis of $\mathbb{C}^n$. Consequently, any vector $\mathbf{y}_0$ can be expressed as

$$\mathbf{y}_0 = \alpha_1\mathbf{x}_1 + \alpha_2\mathbf{x}_2 + \cdots + \alpha_n\mathbf{x}_n, \tag{13.57}$$

where α_i's are scalars. The vector $\mathbf{y}_0$ is so selected that it has a component along $\mathbf{x}_1$, in which case $\alpha_1 \neq 0$. Any randomly chosen $\mathbf{y}_0$ usually satisfies this condition (see Problem 13.16).

Let us define the iterative scheme

$$\mathbf{y}_k = A\mathbf{y}_{k-1}, \qquad k = 1, 2, 3, \ldots \tag{13.58}$$

with $\mathbf{y}_0$ arbitrary. Then in view of Example 7.2, we have

$$\mathbf{y}_k = A\mathbf{y}_{k-1} = A^2\mathbf{y}_{k-2} = \cdots = A^k\mathbf{y}_0$$

$$= \lambda_1^k\left[\alpha_1\mathbf{x}_1 + \alpha_2\left(\frac{\lambda_2}{\lambda_1}\right)^k \mathbf{x}_2 + \cdots + \alpha_n\left(\frac{\lambda_n}{\lambda_1}\right)^k \mathbf{x}_n\right]. \tag{13.59}$$

Now, because of (13.56), see also Problem 13.15, $(\lambda_i/\lambda_1)^k$ tends to zero as $k \to \infty$ for $i = 2$, ..., n, and hence, for sufficiently large k, $\mathbf{y}_k \approx \lambda_1^k\alpha_1\mathbf{x}_1$ and $\mathbf{y}_k$ given by (13.58) is an approximation to the dominant eigenvector $\mathbf{x}_1$. Each component of $\mathbf{y}_k$ computed recursively may become unbounded if $|\lambda_1| > 1$ or zero if $|\lambda_1| < 1$, as is evident from (13.59). Therefore, the modified iterative scheme is

$$\mathbf{y}_k = \beta_k^{-1} A\mathbf{y}_{k-1}, \qquad k = 1, 2, \ldots, \tag{13.60}$$

where β_k is a scaling factor which may be the absolute value of the largest or any nonzero component of $A\mathbf{y}_{k-1}$. This ensures that at least one component of $\mathbf{y}_k$ is one. Once the dominant eigenvector $\mathbf{x}_1$ is determined for the given A, the corresponding eigenvalue follows from the relation $A\mathbf{x}_1 = \lambda_1\mathbf{x}_1$.

To enhance the convergence of power method, a device known as *shift of origin* is used. We know that if μ is a scalar then $(A - \mu I)\mathbf{x}_i = (\lambda_i - \mu)\mathbf{x}_i$. If (13.57) is premultiplied by $(A - \mu I)$, then in place of $|\lambda_i/\lambda_1|$, the convergence of (13.59) depends on $|(\lambda_i - \mu)/(\lambda_1 - \mu)|$ which can be made much smaller than $|\lambda_i/\lambda_1|$ by choosing μ properly. For example, taking $\lambda_1 = 3$, $\lambda_2 = 2$, $\lambda_3 = 1$, with $\mu = 1$, we have $(\lambda_2 - \mu)/(\lambda_1 - \mu) = 0.5$ in place of $\lambda_2/\lambda_1 = 2/3$ and $(\lambda_3 - \mu)/(\lambda_1 - \mu) = 0$ in place of $\lambda_3/\lambda_1 = 1/3$ and so there is an improved rate of convergence.

Example 13.14 The eigenvalues of the matrix

$$A = \begin{bmatrix} 3 & 1 & 0 \\ 13 & 0 & 1 \\ -15 & 0 & 0 \end{bmatrix}$$

are 5, –3, and 1 with the corresponding eigenvectors $[1 \quad 2 \quad -3]^T$, $[1 \quad -6 \quad 5]^T$, and $[1 \quad -2 \quad -15]^T$. Choosing an initial vector $\mathbf{y}_0 = [1 \quad 1 \quad -1]^T$ and using (13.60), we obtain

$$A\mathbf{y}_0 = \begin{bmatrix} 3 & 1 & 0 \\ 13 & 0 & 1 \\ -15 & 0 & 0 \end{bmatrix} \begin{bmatrix} 1 \\ 1 \\ -1 \end{bmatrix} = \begin{bmatrix} 4 \\ 12 \\ -15 \end{bmatrix}, \qquad \mathbf{y}_1 = \begin{bmatrix} 1 \\ 3 \\ -3.75 \end{bmatrix},$$

where we have used β_k as the first component of $A\mathbf{y}_{k-1}$. The successive values of $\mathbf{y}_2, \mathbf{y}_3, \ldots$ given in Table 13.1 converge in 11 iterations. Since $A\mathbf{y}_{11} = 5\mathbf{y}_{11}$, the eigenvalue $\lambda_1 = 5$. A different choice of $\mathbf{y}_0$ will lead to different number of iterations for convergence. Using *shift*

Table 13.1 Eigenvector computation by power method in Example 13.14

k	$\mathbf{y}_k^T$			k	$\mathbf{y}_k^T$			k	$\mathbf{y}_k^T$		
1	1	3	–3.75	5	1	2.11	–3.11	9	1	2.01	–3.01
2	1	1.54	–2.50	6	1	1.93	–2.93	10	1	1.99	–2.99
3	1	2.31	–3.30	7	1	2.04	–3.04	11	1	2.00	–3.00
4	1	1.82	–2.82	8	1	1.97	–2.97	12	1	2	–3

of origin with $\mu = -3$ and the same $\mathbf{y}_0$, we have $(A + 3I)\mathbf{y}_0 = [7 \quad 15 \quad -18]$, $\mathbf{y}_1 = [1, 2.14, -2.57]$. In this way, after seven iterations, we get the correct eigenvector. An ad hoc choice of μ can be made as given by J.H. Wilkinson [205]. □

13.4.3 Method of Deflation

Once an eigenvalue and an eigenvector of an n-square matrix A are found, a *deflated* $(n-1) \times (n-1)$ matrix A_1 is to be determined which contains other eigenvalues of A and whose eigenvectors are simply related to those of A.

Suppose that $\mathbf{x} = [1 \quad \mathbf{z}^T]^T$ is an eigenvector of A associated with λ. Then,

$$A\mathbf{x} = \begin{bmatrix} a_{11} & \mathbf{a}^T \\ \mathbf{c} & B \end{bmatrix} \begin{bmatrix} 1 \\ \mathbf{z} \end{bmatrix} = \lambda \begin{bmatrix} 1 \\ \mathbf{z} \end{bmatrix} \tag{13.61}$$

gives

$$a_{11} + \mathbf{a}^T\mathbf{z} = \lambda, \qquad \mathbf{c} + B\mathbf{z} = \lambda\mathbf{z}.$$

Define

$$P = \begin{bmatrix} 1 & 0 \\ \mathbf{z} & I \end{bmatrix} \tag{13.62}$$

so that

$$P^{-1} = \begin{bmatrix} 1 & 0 \\ -\mathbf{z} & I \end{bmatrix}.$$

Hence, taking (13.61) into consideration, we have

$$P^{-1}AP = \begin{bmatrix} \lambda & \mathbf{a}^T \\ 0 & B - \mathbf{z}\mathbf{a}^T \end{bmatrix}, \tag{13.63}$$

where $A_1 = B - \mathbf{z}\mathbf{a}^T$ is the *deflated* matrix containing the remaining eigenvalues of A. Let $\mu \neq \lambda$ be an eigenvalue of A_1 and $\mathbf{u}$ be the corresponding eigenvector so that $A_1\mathbf{u} = (B - \mathbf{z}\mathbf{a}^T)\mathbf{u} = \mu\mathbf{u}$. Then, $\mathbf{y} = [1\ \mathbf{v}^T]^T$, where $\mathbf{v} = (\mu - \lambda)\mathbf{u}/\mathbf{a}^T\mathbf{u}$, is an eigenvector of $P^{-1}AP$; and $P\mathbf{y}$ that of A (see Theorem 7.1) associated with μ. Indeed, with the help of (13.63), we have

$$P^{-1}AP\begin{bmatrix} 1 \\ \dfrac{(\mu-\lambda)\mathbf{u}}{\mathbf{a}^T\mathbf{u}} \end{bmatrix} = \mu\begin{bmatrix} 1 \\ \dfrac{(\mu-\lambda)\mathbf{u}}{\mathbf{a}^T\mathbf{u}} \end{bmatrix}, \qquad P\mathbf{y} = \begin{bmatrix} 1 \\ \mathbf{z}+\mathbf{v} \end{bmatrix}$$

while $[0\ \mathbf{u}^T]^T$ is an eigenvector of $P^{-1}AP$ if $\mathbf{a}^T\mathbf{u} = 0$. Hence, if μ and $\mathbf{u}$ are determined from the *deflated* matrix A_1 by the power method, the corresponding eigenvector of A is given by $\mathbf{y}$. By repeated application of the above technique, all the eigenvalues and eigenvectors of A can be determined provided they are separated:

$$|\lambda_1| > |\lambda_2| > \cdots > |\lambda_n|.$$

Example 13.15 Continuing with Example 13.14, we have $\lambda = 5$, $\mathbf{z}^T = [2, -3]$, $\mathbf{a}^T = [1\quad 0]$. Hence, by (13.62),

$$P = \begin{bmatrix} 1 & 0 & 0 \\ 2 & 1 & 0 \\ -3 & 0 & 1 \end{bmatrix}, \qquad B = \begin{bmatrix} 0 & 1 \\ 0 & 0 \end{bmatrix}, \qquad A_1 = B - \mathbf{z}\mathbf{a}^T = \begin{bmatrix} -2 & 1 \\ 3 & 0 \end{bmatrix}.$$

By the power method the dominant eigenvalue of A_1 is $\mu = -3$ and the eigenvector $\mathbf{u}^T = [1, -1]$. Hence, $\mathbf{v} = (\mu - \lambda)\mathbf{u}/\mathbf{a}^T\mathbf{u} = [-8, 8]^T$ and the corresponding eigenvector of A is

$$P\mathbf{y} = \begin{bmatrix} 1 \\ \mathbf{z}+\mathbf{v} \end{bmatrix} = \begin{bmatrix} 1 \\ -6 \\ 5 \end{bmatrix}.$$

□

The power method has a very slow convergence rate and can be extended to the case when A has dominant, multiple or complex conjugate eigenvalues for which the reader is referred to J.H. Wilkinson [205].

13.4.4 Inverse Iteration

In matrix eigenproblems, one of the most powerful methods is known as *inverse iteration* which has a fast rate of convergence and enables one to determine all nonrepeated eigenvalues and eigenvectors. Apart from this, as we shall see later, this method will provide an explanation of remarkable numerical properties of *QR*-algorithm which can determine the complete set of eigenvalues and eigenvectors of any arbitrary matrix.

It follows from Theorem 7.18 that if λ_i's are the eigenvalues of A with the corresponding linearly independent eigenvectors $\mathbf{x}_i$'s and if μ_0 is a scalar different from λ_i's, then the eigenvalues of $(\mu_0 I - A)^{-1}$ are $(\mu_0 - \lambda_i)^{-1}$ with the same eigenvectors $\mathbf{x}_i$'s, i.e., $(\mu_0 I - A)^{-1}\mathbf{x}_i = (\mu_0 - \lambda_i)^{-1}\mathbf{x}_i$. If we choose an initial vector

$$\mathbf{y}_0 = \alpha_1\mathbf{x}_1 + \alpha_2\mathbf{x}_2 + \cdots + \alpha_n\mathbf{x}_n, \tag{13.64}$$

then premultiplying both sides of (13.64) by $(\mu_0 I - A)^{-1}$, we have

$$\mathbf{y}_1 := (\mu_0 I - A)^{-1}\mathbf{y}_0 \tag{13.65}$$

$$= \alpha_1(\mu_0 - \lambda_1)^{-1}\mathbf{x}_1 + \alpha_2(\mu_0 - \lambda_2)^{-1}\mathbf{x}_2 + \cdots + \alpha_n(\mu_0 - \lambda_n)^{-1}\mathbf{x}_n. \tag{13.66}$$

If μ_0 is chosen much closer to λ_1 than to other λ_i's, then the first term in the sum of (13.66) will predominate provided α_1 is not too small. This implies that with almost any initial vector $\mathbf{y}_0$, after one operation indicated in (13.65), $\mathbf{y}_1$ will be a good approximation of the eigenvector $\mathbf{x}_1$. In practice, the process may have to be repeated a few times to improve accuracy:

$$\mathbf{y}_k = (\mu_0 I - A)^{-1}\mathbf{y}_{k-1}, \qquad k = 1, 2, \ldots \tag{13.67}$$

and all other nonrepeated eigenvalues λ_i's and eigenvectors of A can be determined by selecting μ_0 close to λ_i.

The successful application of inverse iteration thus depends on two factors: (i) a judicious guess on μ_0 nearer an eigenvalue and (ii) computation of $\mathbf{y}_k$ by (13.67) which involves a matrix inversion. As for a choice of μ_0, we use

$$\mu_k = \frac{\mathbf{y}_k^T A \mathbf{y}_k}{\mathbf{y}_k^T \mathbf{y}_k}, \qquad k = 0, 1, \ldots \tag{13.68}$$

which is called the *Rayleigh quotient* (see Theorem 9.4). We note that if $\mathbf{y}_k = \mathbf{x}$ is an eigenvector of A associated with λ, i.e., if $A\mathbf{x} = \lambda\mathbf{x}$, then $\mu_k = \lambda$, and so the Rayleigh quotient gives an exact value for the eigenvalue of A. Thus for an initial choice of $\mathbf{y}_0$, we calculate μ_0 from (13.68) and use it to get an approximate eigenvector $\mathbf{y}_1$ from (13.65). In the kth iterative step if μ_k is calculated from (13.68), then a new eigenvector $\mathbf{y}_{k+1}$ approximating $\mathbf{x}_1$ is generated from

$$\mathbf{y}_{k+1} = (\mu_k I - A)^{-1}\mathbf{y}_k. \tag{13.69}$$

The inverse iteration (13.69) is obtained as the solution of the matrix equation

$$(\mu_k I - A)\,\beta_{k+1}^{-1}\,\mathbf{y}_{k+1} = \mathbf{y}_k \tag{13.70}$$

for the choice of μ_k obtained from (13.68) by Gaussian elimination with pivoting strategy or using orthogonal transformations (see Example 13.12) where β_{k+1} is a scaling factor used for the same reason as explained in connection with (13.60).

Example 13.16 Let us consider the eigenvalue problem in Example 13.14 with the initial vector $\mathbf{y}_0 = [1\ 1, -1]$. (i) By the Rayleigh quotient $\mathbf{y}_0^T A\mathbf{y}_0 / \mathbf{y}_0^T\mathbf{y}_0 = 31/3$ and so setting $\mu_0 = 10$, and (ii) by using (13.70), we get

$$\mathbf{y}_1 = \beta_1^{-1}(\mu_0 I - A)^{-1}\mathbf{y}_0 = \beta_1^{-1}\begin{bmatrix} 7 & -1 & 0 \\ -13 & 10 & -1 \\ 15 & 0 & 10 \end{bmatrix}^{-1}\begin{bmatrix} 1 \\ 1 \\ -1 \end{bmatrix} = \begin{bmatrix} 1 \\ 1.66 \\ -2.05 \end{bmatrix},$$

where $\beta_1 = 0.18$. Steps (i) and (ii) are repeated following (13.68) and (13.70), with β_{k+1} equal to the first component of $(\mu_k I - A)^{-1}\mathbf{y}_k$. The computation is shown in the Table 13.2, which in comparison with Table 13.2 shows that inverse iteration converges much faster than the power method.

Table 13.2 Eigenvector computation by inverse iteration in Example 13.16

k		$\mathbf{y}_k^T$		β_k	μ_k
1	1	1.66	−2.05	0.18	6.73
2	1	1.95	−2.76	0.56	5.33
3	1	1.99	−2.98	3.03	5.03
4	1	2.00	−3.00	33.35	5.00

13.4.5 Jacobi and Givens Method

In the foregoing discussion, we have emphasised how a dominant eigenvalue can be computed by power method and nonrepeated eigenvalues along with eigenvectors by inverse iteration. We shall now turn to those algorithms which are designed to compute the complete set of eigenvalues and eigenvectors of any matrix. The basis of these algorithms is to apply a sequence of orthogonal similarity transformations to A until its transformed form gives the eigenvalues with little effort. Two existence theorems, which we now discuss, are crucial for this purpose.

Theorem 13.4 For every n-square Hermitian (resp symmetric) matrix, there exists a unitary (resp orthogonal) matrix U (resp Q) such that U^*AU(resp Q^TAQ) is $D = \text{diag}(\lambda_1, \lambda_2, ..., \lambda_n)$, where λ_i's are real but not necessarily distinct. □

For proof, see Theorem 7.27 and Corollary 7.7.

Theorem 13.5 *Real Schur Form.* Let $a_1 \pm ib_1, ..., a_m \pm ib_m, \lambda_{2m+1}, ..., \lambda_n$ be the eigenvalues of an n-square real matrix A. There exists an orthogonal similarity transformation Q such that

$$S := Q^TAQ = \begin{bmatrix} B_1 & & & & & \\ & \cdot & & & & \\ & & B_m & & C & \\ & & & \lambda_{2m+1} & & \\ & & & & \cdot & \\ & 0 & & & & \lambda_n \end{bmatrix} \tag{13.71}$$

is in quasi-upper triangular form, where B_r is a 2 × 2 matrix with eigenvalues $a_r \pm ib_r$, $r = 1, 2, ..., m$.

Proof Consequent on Example 7.9, if $\mathbf{x} \pm i\mathbf{y}$ are eigenvectors corresponding to the eigenvalues $a \pm ib$: $= a_1 \pm ib_1$ of A, then **x** and **y** are linearly independent and so we have [see (7.23)]

$$A[\mathbf{x}, \mathbf{y}] = [\mathbf{x} \;\; \mathbf{y}]\begin{bmatrix} a & b \\ -b & a \end{bmatrix} = [\mathbf{x} \;\; \mathbf{y}]E, \; E = \begin{bmatrix} a & b \\ -b & a \end{bmatrix}. \tag{13.72}$$

Now there exists an orthogonal matrix of Householder transformation, or plane rotation H such that

$$H[\mathbf{x} \;\; \mathbf{y}] = \begin{bmatrix} U \\ \hline 0 \end{bmatrix}, \tag{13.73}$$

where U is a 2 × 2 nonsingular matrix. We set

$$HAH^T = \begin{bmatrix} B_1 & C_1 \\ D_1 & A_2 \end{bmatrix}$$

and using (13.72) and (13.73) we get

$$\begin{bmatrix} B_1 & C_1 \\ D_1 & A_2 \end{bmatrix}\begin{bmatrix} U \\ \hline 0 \end{bmatrix} = \begin{bmatrix} U \\ \hline 0 \end{bmatrix}E. \tag{13.74}$$

Therefore, $B_1U = UE$ and $D_1U = 0$. So, B_1 and E are similar, the former having the eigenvalues $a_1 \pm ib_1$ and $D_1 = 0$. Consequently,

$$HAH^T = \begin{bmatrix} B_1 & C_1 \\ 0 & A_2 \end{bmatrix}.$$

Treating A_2 similarly, we get (13.71) by an induction procedure. If there are repeated roots, we can apply Theorem 11.20 to find a set of independent vectors and repeat (13.73) and (13.74). □

Example 13.17 Transform the following matrix to a real Schur form:

$$A = \begin{bmatrix} 1 & 0 & 0 & 0 \\ 1 & -1 & -3 & 1 \\ 0 & 1 & 1 & 1 \\ 0 & 0 & -2 & 3 \end{bmatrix}.$$

Solution The characteristic equation of A using Theorem 7.16 is $\lambda^4 - 4\lambda^3 + 7\lambda^2 - 6\lambda + 2 = 0$ with roots at 1, 1, $1 \pm i$. Since rank $(A - I) = 3$ and rank $(A - I)^2 = 2$, we solve $(A - I)^2\mathbf{y} = \boldsymbol{\theta}$ to get $\mathbf{y} = [-1\ 1\ 0\ 1]^T$ such that $\mathbf{x} = (A - I)\mathbf{y} = 2[0, -1, 1, 1]^T \neq \boldsymbol{\theta}$. Taking cognizance of Theorem 11.20, we have

$$A[\mathbf{x} \;\; \mathbf{y}] = [\mathbf{x} \;\; \mathbf{y}]\begin{bmatrix} 1 & 1 \\ 0 & 1 \end{bmatrix}.$$

Considering the vector **x**, we have the Householder transformation $P = I - (1/12)[2\sqrt{3}, -2, 2, 2]^T [2\sqrt{3}, -2, 2, 2]$ for which $P\mathbf{x} = -2\sqrt{3}\mathbf{e}_1$ and $\mathbf{z} = P\mathbf{y} = [0, 1 - a, a, 1 + a]^T$, where $1/a = \sqrt{3}$. We now determine the Householder transformation P_1 for which $P_1\mathbf{z} = -\sqrt{3}\mathbf{e}_2$, where

$$P_1 = I - (\sqrt{3} + 2)^{-1} [0, 1 - a + a^{-1}, a, 1 + a]^T [0, 1 - a + a^{-1}, a, 1 + a].$$

It can be verified that

$$Q = P_1 P = \frac{1}{\sqrt{3}} \begin{bmatrix} 0 & 1 & -1 & -1 \\ 1 & -1 & 0 & -1 \\ -1 & 0 & 1 & -1 \\ -1 & -1 & -1 & 0 \end{bmatrix}, \qquad Q^T AQ = \begin{bmatrix} 1 & 2 & 0 & 1 \\ 0 & 1 & 3 & -2 \\ 0 & 0 & 2 & -1 \\ 0 & 0 & 2 & 0 \end{bmatrix}$$

which is in real Schur form.

Now in $Q^T AQ$, the bottom corner matrix

$$B := \begin{bmatrix} 2 & -1 \\ 2 & 0 \end{bmatrix}$$

has eigenvalues $1 + i$ and $1 - i$ with corresponding eigenvectors

$$\begin{bmatrix} 1 \\ 1-i \end{bmatrix} \text{ and } \begin{bmatrix} 1 \\ 1+i \end{bmatrix}.$$

Normalizing by Gram-Schmidt process, the corresponding orthonormal eigenvectors are

$$\frac{1}{\sqrt{3}} \begin{bmatrix} 1 \\ 1-i \end{bmatrix} \text{ and } \frac{1}{\sqrt{3}} \begin{bmatrix} 1-i \\ i \end{bmatrix}.$$

The unitary matrix formed with these eigenvectors as columns is

$$U = \frac{1}{\sqrt{3}} \begin{bmatrix} 1 & 1-i \\ 1-i & i \end{bmatrix} \text{ and } C := U^* BU = \begin{bmatrix} 1+i & 2-i \\ 0 & 1-i \end{bmatrix}.$$

Defining

$$V = Q \begin{bmatrix} I & 0 \\ 0 & U \end{bmatrix},$$

by direct computation

$$V^* AV = Q \begin{bmatrix} 1 & 2 & a(1-i) & ai \\ 0 & 1 & a(1+2i) & a(3-5i) \\ 0 & 0 & 1+i & 2-i \\ 0 & 0 & 0 & 1-i \end{bmatrix}, a = \frac{1}{\sqrt{3}},$$

which is the Schur decomposition as established in Corollary 7.8. □

We shall now show how a symmetric (resp Hermitian) matrix can be reduced to a diagonal form by a sequence of orthogonal transformations. Applying plane rotation in the *jk*-plane to a symmetric $n \times n$ matrix A, we have [see (13.40)] $Q^T AQ = B$ with elements

$$\begin{aligned}
b_{pq} &= a_{pq}, \qquad p \neq j, \qquad q \neq k; \\
b_{jj} &= c^2 a_{jj} - 2cs a_{jk} + s^2 a_{kk}, \\
b_{jk} &= cs(a_{jj} - a_{kk}) + a_{jk}(c^2 - s^2) = b_{kj}, \\
b_{kk} &= s^2 a_{jj} + 2cs a_{jk} + c^2 a_{kk}, \\
b_{jq} &= c a_{jq} + s a_{kq}, \qquad q \neq j,\ k; \\
b_{kq} &= s a_{jq} + c a_{kq}, \qquad q \neq j,\ k; \\
b_{pj} &= c a_{pj} - s a_{pk}, \qquad p \neq j,\ k; \\
b_{qk} &= s a_{pj} + c a_{pk}, \qquad p \neq j,\ k.
\end{aligned} \tag{13.75}$$

Setting θ such that $b_{jk} = 0$, we have

$$b_{jq}^2 + b_{kq}^2 = a_{jq}^2 + a_{kq}^2, \qquad q \neq j,\ k; \tag{13.76}$$

$$b_{pj}^2 + b_{qk}^2 = a_{pj}^2 + a_{pk}^2, \qquad p \neq j,\ k; \tag{13.77}$$

$$b_{jj}^2 + b_{kk}^2 + b_{jk}^2 + b_{kj}^2 = a_{jj}^2 + a_{kk}^2 + a_{jk}^2 + a_{kj}^2, \tag{13.78}$$

when $\tan 2\theta = 2a_{jk}/(a_{kk} - a_{jj})$.

If we use the Euclidean norm as in Example 9.5 and denote the sum of squares of the off-diagonal elements of A and B by a and b respectively, then it follows from (13.75)-(13.78) that $\| A \| = \| B \|$, but $b < a$. Since by (13.78), $b = a - 2a_{jk}^2$, the off-diagonal sum of squares decreases by $2a_{jk}^2$ while the diagonal sum of squares increases by the same amount. In this way, if the given symmetric matrix A is subjected to a set of orthogonal similarity transformations $Q_1, \ldots, Q_m$, then $A_m := Q_m^T \cdots Q_1^T A Q_1 \cdots Q_m = Q_m^T A_{m-1} Q_m; A_0 := A$. At each stage we select Q_m to reduce to zero one off-diagonal element (possibly the largest requiring a time consuming search) of A_{m-1}. If a_m denotes the off-diagonal sum of squares of A_m, then $0 \leq a_m < \cdots < a_1 < a_0 := a$ and as $m \to \infty$, $a_m \to 0$ and $A_m \to D$, a diagonal matrix containing the eigenvalues of A.

The above method established by C.G.J. Jacobi in 1846 goes by his name. The demerits of this method are: (i) an off-diagonal element made zero at one stage becomes nonzero at a later stage, and (ii) the number of iterations required is very large, making the process a slow one. The criterion to limit the number of iterations is to make the off-diagonal norm a_m less than some prescribed tolerance.

To get around difficulties (i) and (ii), W.J. Givens in 1954 suggested to transform the symmetric matrix A, instead of a diagonal form, to a tridiagonal form, requiring only a finite number of operations, known as *Givens method* (see Section 13.3.2). This tridiagonal form can also be obtained by the Householder transformation as suggested by A.S. Householder in 1958, needing $(n - 2)$ square roots and only about half the number of multiplications required by Givens method.

Sturm sequences and bisection. The theory of Sturm sequence for root computation is now applied to the symmetric tridiagonal matrix

$$T = \begin{bmatrix} \alpha_1 & \beta_1 & 0 & \cdots & 0 & 0 \\ \beta_1 & \alpha_2 & \beta_2 & \cdots & 0 & 0 \\ 0 & \beta_2 & \alpha_3 & \cdots & 0 & 0 \\ \vdots & \vdots & \vdots & & \vdots & \vdots \\ 0 & 0 & 0 & \cdots & \alpha_{n-1} & \beta_{n-1} \\ 0 & 0 & 0 & \cdots & \beta_{n-1} & \alpha_n \end{bmatrix}.$$

If any of the β_i's is zero, then T consists of tridiagonal submatrices of lower order, for which the eigenvalues can be computed separately. So, assume now that no β_i is zero. The leading principal minors $p_r(\lambda)$ of order r of $T - \lambda I$ are given by the recursive relations

$$p_0(\lambda) = 1,\ p_1(\lambda) = \alpha_1 - \lambda,\ ...,\ p_r(\lambda) = (\alpha_r - \lambda)p_{r-1}(\lambda) - \beta_{r-1}^2 p_{r-2}(\lambda),$$

where $r = 2, 3, ..., n$ with $p_n(\lambda)$ as the characteristic polynomial. If none of the β_i's are zero, it can be shown that, zeros λ_i's of $p_r(\lambda)$ are strictly separate from zeros μ_i's of $p_{r+1}(\lambda)$, i.e., $\mu_1 > \lambda_1 > \mu_2 > \lambda_2 > \ldots > \mu_r > \lambda_r > \mu_{r+1}$. For some real number λ, we compute $p_0(\lambda)$, $p_1(\lambda)$, ..., $p_n(\lambda)$ and denote by $s(\lambda)$ the number of agreements in sign between successive members of the sequence of real numbers $\{p_r(\lambda)\}$. For example, the sequence 1, 1, 0, –1, –1 when $\lambda = 1$ gives + + + – – and $s(1) = 3$. When $p_r(\lambda)$ is zero, its sign is taken to be that of $p_{r-1}(\lambda)$. It can be shown that the set $p_r(\lambda)$ satisfies the following property of a Sturm sequence.

Theorem 13.6 *Property of Sturm Sequence.* Let the sequence $p_0(\lambda)$, $p_1(\lambda)$, ..., $p_n(\lambda)$ be evaluated at some real number λ. The number of eigenvalues of T strictly greater than λ is equal to the number of agreements in sign $s(\lambda)$ of successive members of the above sequence. In fact, the sequence of leading principal minors of $(A - \lambda I)$, where A is any symmetric matrix, possesses this property. □

For several values of λ, a systematic search is made to determine the intervals in which the eigenvalues, are located. If, for example, $s(\alpha) = k$ and $s(\mu) = k + 1$, $\mu < \alpha$, then one eigenvalue λ of T lies in the interval $(\mu, \alpha]$. Setting $\beta = (\alpha + \mu)/2$, if $s(\beta) = k$, then λ lies in $(\mu, \beta]$, otherwise in $(\beta, \alpha]$. This *method of bisection,* if repeated, allows a tight interval for λ. Once the eigenvalues are known, the eigenvectors are determined by inverse iteration, and not by solving $(T - \lambda I)\mathbf{x} = 0$ which is generally unstable.

Example 13.18 Let us consider the tridiagonal matrix

$$T = \begin{bmatrix} 0 & 1 & 0 & 0 \\ 1 & 0 & 1 & 0 \\ 0 & 1 & 0 & 1 \\ 0 & 0 & 1 & 0 \end{bmatrix}. \tag{13.79}$$

If $\lambda = 0$, we have $p_0 = 1$, $p_1 = 0$, $p_2 = -1$, $p_3 = 0$, and $p_4 = 1$ with the sequence of signs + + – – +, and so $s(0) = 2$ and the two eigenvalues are greater than 0. Again for $\lambda = 1$, $p_0 = 1$, $p_1 = -1$, $p_2 = 0$, $p_3 = 1$, $p_4 = -1$ with the sequence of signs + – – + –, and so $s(1) = 1$ and one eigenvalue is greater than 1. Hence one eigenvalue μ lies in the interval $0 < \mu \le 1$. Now choose $\lambda = (0 + 1)/2 = 0.5$ for which $s(0.5) = 2$ and $0.5 < \mu \le 1$. Proceeding in this way, we ultimately find that $\mu = 0.62$ as one of the eigenvalues. Others are ± 1.61 and –0.62. For eigenvectors of T, see Problem 13.22. □

13.4.6 *LR*-factorization and *LR*-algorithm

Let L_i be defined as a unit matrix except for its ith column given by

$$L_i = \begin{bmatrix} 1 & \cdots & 0 & 0 & \cdots & 0 \\ \vdots & & \vdots & \vdots & & \vdots \\ 0 & \cdots & 1 & 0 & \cdots & 0 \\ 0 & \cdots & -r_{i+1,i} & 1 & \cdots & 0 \\ \vdots & & \vdots & \vdots & & \vdots \\ 0 & \cdots & -r_{ni} & 0 & \cdots & 1 \end{bmatrix}$$

which is nonsingular, and lower triangular with all diagonal elements equal to 1. A lower triangular matrix with all diagonal elements equal to 1 is called *unit lower triangular*. To compute L_i^{-1}, we reverse the signs of r_{ji}'s in L_i. Premultiplication of any n-square matrix A by L_i has the effect of subtracting r_{ji} times the ith row from the jth row for $j > i$. If we choose $r_{j1} = a_{j1}/a_{11}$, then the matrix $A_1 := L_1A$ has zeros in the positions 2 to n of the first column. If $a_{22}^{(1)}$, the (2, 2) element of A_1 is nonzero, then choosing $r_{j2} = a_{j2}^{(1)} / a_{22}^{(1)}$ in L_2, the elements 3 to n of the second column of $A_2 = L_2A_1$ can be reduced to zero. If all the *pivot elements* $a_{j+1,j+1}^{(j)}$ of $A_j = L_j L_{j-1} \cdots L_1A$ are nonzero for $j = 0, 1, \ldots, n - 2$, the above process known as *Gaussian elimination* is well defined, and we have A_{n-1} in an upper triangular form R, i.e., $L_{n-1} \cdots L_1A = R$. Hence,

$$A = LR, \tag{13.80}$$

where $L = (L_{n-1} \cdots L_1)^{-1}$, is a *unit* (i.e., with unit diagonal elements) *lower triangular matrix*, and (13.80) is known as the *LR*-decomposition of a nonsingular matrix A. Concentrating on the leading submatrices of A and noting that the above premultiplication by L_i's are nothing but elementary row operations applied to these submatrices, all pivot elements are leading minors of A and can be nonzero iff all leading principal minors of A of orders from 1 to $(n - 1)$ are nonzero. If any of these submatrices is singular, then the *LR*-decomposition of A fails. The number of multiplications needed for triangularization by Gaussian elimination is $n(n^2 - 1)/3$.

Let D be any nonsingular diagonal matrix. Define $L_1 := LD^{-1}$ and $R_1 := DR$ so that $A = LR = L_1R_1$, and hence the *LR*-factorization of A is not unique. If, however, one insists on a unit lower triangular L and L_1, then it follows that $L^{-1}L_1 = RR_1^{-1}$, where the right-hand side is upper triangular while the left-hand side is lower triangular. Hence, $L^{-1}L_1 = I$, and the decomposition $A = LR$, where L is unit lower triangular or R unit upper triangular, is unique.

Let $A = LR'$ be the *LR*-decomposition of A and let D be a diagonal matrix whose elements are the diagonal elements of R'. Then $A = LDR$, where $R' = DR$ and L and R are unit lower and unit upper triangular respectively.

The *LR*-decomposition in (13.80), where L is unit lower triangular, is known as *Doolittle* decomposition; if R is unit upper triangular and L is lower triangular, it is known as *Crout decomposition* of A. To understand how the latter is obtained, define R_i, a unit matrix except for its ith row:

$$R_i = \begin{bmatrix} 1 & \cdots & 0 & 0 & \cdots & 0 \\ \vdots & & \vdots & \vdots & & \vdots \\ 0 & \cdots & 1 & -r_{i,i+1} & \cdots & -r_{in} \\ 0 & \cdots & 0 & 1 & \cdots & 0 \\ \vdots & & \vdots & \vdots & & \vdots \\ 0 & \cdots & 0 & 0 & \cdots & 1 \end{bmatrix}$$

which is nonsingular, unit upper triangular in form. The inverse of R_i, is obtained by reversing the negative signs of r_{ij} in R_i. Postmultiplying any n-square matrix A by R_i has the effect of subtracting r_{ij} times the ith column from the jth column for $j > i$. If we choose $r_{1j} = a_{1j}/a_{11}$, then the matrix $A_1 := AR_1$ has zeros in the positions 2 to n of the first row. As with row operations, we can define the matrices $R_1, ..., R_{n-1}$ such that $A = LR$, where $R = (R_1 \cdots R_{n-1})^{-1}$, is unit upper triangular. This decomposition is known as *Crout decomposition* and is possible iff all leading principal minors of all orders from 1 to $n - 1$ of A are nonzero.

Let A be symmetric. Then, $A = LDR = R^TDL^T$, i.e., $L^{-1}R^T = DR(DL^T)^{-1}$, where the left-hand side is unit lower and the right-hand side unit upper triangular in form, and hence each side must be equal to an identity matrix. Therefore, $L = R^T$ and $A = LDL^T$.

It is known [see Theorem 8.19(c)] that a symmetric matrix A is positive semidefinite iff the eigenvalues $\lambda_1, ..., \lambda_n$ of A are nonnegative. In view of Corollary 7.7, there exists an orthogonal matrix Q such that $Q^TAQ = \text{diag}\,[\lambda_1, ..., \lambda_n]$. Define $D := \text{diag}\,[\sqrt{\lambda_1}, ..., \sqrt{\lambda_n}]$ and $P := DQ^T$. Then $A = P^TP$. Now, let H be a Householder transformation, such that $L^T := HP$ is upper triangular. This gives the factorization of a positive definite symmetric matrix $A = LL^T$, which is known as the *Cholesky* decomposition, where L is lower triangular.

The development of LR-algorithm by Rutishauser in 1958 relying on the LR-decomposition of a matrix is one of the landmarks in the history of algebraic eigenvalue problems. The most powerful QR-algorithm for such problems set forth by J.G.F. Francis (1961-62) and V.N. Kublanovskaya (1961) has genesis in this LR-algorithm which we shall discuss very briefly. This will prepare the ground for a thorough understanding of the QR-algorithm in Section 13.4.7.

Given a matrix $A = A_0$, we construct a sequence of matrices A_k, $k = 0, 1, ...,$ as

$$A_k = L_kR_k, \qquad A_{k+1} = R_kL_k \tag{13.81}$$

assuming the LR-decomposition of A_k to exist. The factorization of A_k as the product L_kR_k and multiplication of the factors in the reverse order to construct A_{k+1} successively is known as the LR-algorithrm or *Rutishauser method*. From (13.81), it follows easily that

$$A_{k+1} = L_k^{-1} A_k L_k = L_k^{-1} \cdots L_0^{-1} A L_0 \cdots L_k. \tag{13.82}$$

Hence, denoting $L^{(k)} := L_0 \cdots L_k$ and $R^{(k)} := R_k \cdots R_0$, we have

$$\begin{aligned} L^{(k+1)}R^{(k+1)} &= L_0 \cdots L_k(L_{k+1}R_{k+1})R_k \cdots R_0 \\ &= L_0 \cdots L_kA_{k+1}R_k \cdots R_0 = AL_0 \cdots L_kR_k \cdots R_0 = AL^{(k)}R^{(k)} \end{aligned}$$

and, consequently, $L^{(k+1)}R^{(k+1)} = A^{k+2}$ is obtained recursively. It can be shown that under certain restrictions, the *LR*-algorithm (13.81) transforms a real matrix into a real Schur form (13.71), i.e., $A_k \to S$ as $k \to \infty$.

For rapid convergence, we introduce the *LR*-algorithm with origin shift μ_k defined by

$$A_k - \mu_k I = L_k R_k, \qquad A_{k+1} - \mu_k I = R_k L_k, \tag{13.83}$$

where μ_k is a suitably chosen scalar. It can be verified that

$$A_{k+1} = L_k^{-1} A_k L_k, \qquad L^{(k+1)}R^{(k+1)} = (A - \mu_0 I)(A - \mu_1 I) \cdots (A - \mu_{k+1} I)$$

and the modified sequence also converges to S. The *LR*-factorization is also called *LU-factorization* in which the upper triangular matrix R is denoted by U.

The shortcomings of the *LR*-algorithm are: (i) the triangular factorization of all matrices, even nonsingular, does not exist; (ii) the Gaussian elimination used to get this factorization is numerically unstable. It is known that even when interchanges are allowed, the stability of triangular decomposition cannot be guaranteed unconditionally.

13.4.7 *QR*-algorithm

In the *LR*-algorithm, a sequence of similarity transformations is applied to the given matrix by means of *triangular matrices* (see (13.81)), for which the process turns out to be numerically unstable. To circumvent this, *orthogonal matrices* are used in the *QR*-algorithm which, with other attractive features, develops into the most powerful technique of eigenvalue computation of an arbitrary matrix.

To explain this algorithm, we write the *QR*-decomposition of the matrix $(A_k - \mu_k I)$, where A_k and the scalar μ_k are given in the form

$$(A_k - \mu_k I) = Q_k R_k, \qquad Q_k Q_k^T = I \tag{13.84}$$

with R_k as upper triangular. By Theorem 13.3, the factorization (13.84) with R_k having positive diagonal elements is unique if $(A_k - \mu_k I)$ is nonsingular. Moreover, we define

$$(A_{k+1} - \mu_k I) = R_k Q_k. \tag{13.85}$$

It now follows from (13.84) and (13.85) that

$$A_{k+1} = Q_k^T (A_k - \mu_k I) Q_k + \mu_k I = Q_k^T A_k Q_k \tag{13.86}$$

which shows that A_k is orthogonally similar to A_{k+1}. It will be shown later how μ_k, called the *origin shift*, is determined from the elements of A_k.

Thus a *QR*-algorithm with an origin shift μ_k generates a sequence of orthogonally similar matrices $A_0, A_1, \ldots$, employing the recursive relations (13.84) and (13.85) when $A_0 := A$ is given and μ_k is determined from the parameters of A_k.

Noting that $Q_k^{-1} = Q_k^T$, we get from (13.84),

$$Q_k = (A_k^T - \mu_k I)^{-1} R_k^T. \tag{13.87}$$

Let us denote the last column of the lower triangular matrix R_k^T by $\rho[0\ldots 0\,1]^T = \rho\mathbf{e}_n$, where ρ is a scalar and $Q_k = [\mathbf{q}_1^{(k)}\ \mathbf{q}_2^{(k)} \cdots \mathbf{q}_n^{(k)}]$. Then from (13.87),

$$\mathbf{q}_n^{(k)} = \rho(A_k^T - \mu_k I)^{-1}\mathbf{e}_n \tag{13.88}$$

which shows that $\mathbf{q}_n^{(k)}$ is an approximate eigenvector of A_k^T obtained after applying one step of inverse iteration with a shift μ_k to the vector $\mathbf{e}_n$. If we set

$$Q_k = [K\ \ \mathbf{q}], \qquad \mathbf{q} := \mathbf{q}_n^{(k)} \quad \text{and} \quad \mathbf{q}^T\mathbf{q} = 1, \tag{13.89}$$

where K represents the first $(n - 1)$ columns of the orthogonal matrix Q_k, then from (13.86),

$$A_{k+1} = \begin{bmatrix} K^T \\ \mathbf{q}^T \end{bmatrix} A_k [K\ \ \mathbf{q}] = \begin{bmatrix} B_{k+1} & \mathbf{b}_{k+1} \\ \mathbf{c}_{k+1}^T & \alpha_{k+1} \end{bmatrix}, \tag{13.90}$$

where $B_{k+1}: = K^T A_k K$, $\mathbf{b}_{k+1}: = K^T A_k \mathbf{q}$, $\mathbf{c}_{k+1}^T := \mathbf{q}^T A_k K$ and $\alpha_{k+1} = \mathbf{q}^T A_k \mathbf{q}$. If $\mathbf{q}$ were an exact eigenvector of A_k^T, then α_{k+1} would have been the eigenvalue associated with $\mathbf{q}$ and $\mathbf{c}_{k+1} = \boldsymbol{\theta}$ because of relations (13.84) and (13.89).

Assuming $\mathbf{x}$ to be an approximate eigenvector corresponding to the approximate eigenvalue μ of a matrix A, we define the residual of the approximate eigenvector $\mathbf{x}$ by

$$\mathbf{r}: = A\mathbf{x} - \mu\mathbf{x} \tag{13.91}$$

which is zero if $\mathbf{x}$ is an exact eigenvector and μ the corresponding eigenvalue, i.e., if $A\mathbf{x} = \mu\mathbf{x}$. Otherwise, by varying μ, (13.91) can be made to attain a minimum value. Now denoting the Euclidean norm [see Example 9.1(b)] by $\|\cdot\|$, from (13.91) we have

$$\|\mathbf{r}\|^2 = (A\mathbf{x} - \mu\mathbf{x})^T (A\mathbf{x} - \mu\mathbf{x})$$

$$= \mathbf{x}^T A^T A\mathbf{x} - \frac{\mathbf{x}^T A^T \mathbf{x}\mathbf{x}^T A\mathbf{x}}{\mathbf{x}^T\mathbf{x}} + \frac{(\mu\mathbf{x}^T\mathbf{x} - \mathbf{x}^T A^T\mathbf{x})(\mu\mathbf{x}^T\mathbf{x} - \mathbf{x}^T A\mathbf{x})}{\mathbf{x}^T\mathbf{x}}$$

which is minimum when $\mu = \mathbf{x}^T A\mathbf{x}/\mathbf{x}^T\mathbf{x}$. Consequently, the smallest residual of $\mathbf{q}$, an approximate eigenvector of A_k^T, occurs if $\mu = \mathbf{q}^T A_k^T \mathbf{q}$ and is given by

$$\|A_k^T\mathbf{q} - \mu\mathbf{q}\| = \|[K\ \ \mathbf{q}]^T (A_k^T\mathbf{q} - \mu\mathbf{q})\| = \|K^T A_k^T \mathbf{q}\| = \|\mathbf{c}_{k+1}\|, \tag{13.92}$$

where the first step follows from relations (13.84) and (13.89).

Analysing (13.88), we see that $\mathbf{e}_n$ is taken as the initial approximate eigenvector of A_k^T which has a representation similar to (13.90), now in terms of B_k, $\mathbf{b}_k$, $\mathbf{c}_k$ and α_k. Consequently, the residual of $\mathbf{e}_n$ as suggested by (13.92) has the smallest value $\|\mathbf{c}_k\|$. If μ_k is close to an eigenvalue of A_k giving $\mathbf{q}$ a better approximate eigenvector than $\mathbf{e}_n$, then $\|\mathbf{c}_{k+1}\|$ will be smaller than $\|\mathbf{c}_k\|$ and approach zero if the above process is repeated with $k = 0, 1, 2, \ldots$ at a fast rate as k tends to infinity. If the origin shift is properly selected, the number of steps k required is not large. Thus, when $\mathbf{c}_k$ associated with A_k approaches zero as k tends to infinity, α_k becomes an eigenvalue of A (see Theorem 7.1). The other eigenvalues can be determined successively by subjecting B_k to the QR-algorithm.

As with inverse iteration, we shall set the origin shift μ_k by the Rayleigh quotient. Since $\mathbf{e}_n$ is our starting vector in (13.88), it follows from (13.90) that

$$\mu_k = \frac{\mathbf{e}_n^T A_k^T \mathbf{e}_n}{\mathbf{e}_n^T \mathbf{e}_n} = \alpha_k. \tag{13.93}$$

On the other hand, for a symmetric A, the eigenvalue of the second order submatrix at the right bottom corner of A_k, i.e., the eigenvalue of

$$\begin{bmatrix} a_{n-1,n-1} & a_{n-1,n} \\ a_{n,n-1} & a_{nn} \end{bmatrix}, \qquad a_{nn} := \alpha_k \tag{13.94}$$

which is nearer to a_{nn}, is taken to be the origin shift μ_k. This choice gives a good convergence of eigenvalues.

To reduce excessively high number of operations needed to obtain the QR-decomposition of $(A_k - \mu_k I)$ at each step, the given matrix A is transformed at the beginning orthogonally into an upper Hessenberg form or if A is symmetric to a tridiagonal form (see Section 13.3). It might be recalled that a triangularization of an arbitrary matrix needs about $(2/3)n^3$ [resp $(4/3)n^3$] multiplications by Householder [respectively Givens] method while they are about $2n^2$ in both these methods if A is in upper Hessenberg (resp tridiagonal) form. Moreover, if A is in Hessenberg or tridiagonal form, so also $Q^TAQ = A_1$. In other words, a Hessenberg or a tridiagonal form is invariant under QR-transformation. To see this, let us introduce rotations $Q_{12}, Q_{23}, \ldots, Q_{n-1,n}$ in planes (1, 2), (2, 3), ..., $(n-1, n)$ to reduce the upper Hessenberg matrix in (13.39) to the upper triangular form

$$Q^TA = R = \begin{bmatrix} b_{11} & b_{12} & \cdots & b_{1,n-1} & b_{1n} \\ & b_{22} & \cdots & b_{2,n-1} & b_{2n} \\ & & & \vdots & \\ & 0 & & b_{n-1,n-1} & b_{n-1,n} \\ & & & & b_{nn} \end{bmatrix}, \tag{13.95}$$

where $Q^T := Q_{n-1,n} \cdots Q_{23}Q_{12}$. Now, in the product RQ_{12}^T, the first two columns are linear combinations of those in R, so a nonzero entry will appear in the (2, 1) position. Similarly, $RQ_{12}^TQ_{23}^T$ will have another nonzero element in the (3, 2) position until in $RQ_{12}^T \cdots Q_{n-1,n}^T = Q^TAQ$ nonzero subdiagonal elements are introduced, making it an upper Hessenberg form.

On the other hand, if A is in tridiagonal form as in (13.38), with the same kind of plane rotations as above we have, in place of (13.95) and $A_1 := Q^TAQ$, the matrices

$$Q^TA = \begin{bmatrix} c_1 & d_2 & e_3 & & & 0 \\ & c_2 & d_3 & \cdot & & \\ & & c_3 & \cdot & \cdot & e_n \\ & & & \cdot & \cdot & d_n \\ 0 & & & & & c_n \end{bmatrix}$$

and

$$A_1 = \begin{bmatrix} \alpha_1 & \beta_2 & \gamma_3 & & 0 \\ \beta_2 & \alpha_2 & \beta_3 & & \\ & \beta_3 & \alpha_3 & & \gamma_n \\ & & & & \beta_n \\ 0 & & & \beta_n & \alpha_n \end{bmatrix} \tag{13.96}$$

respectively. Since A_1 is symmetric, $\gamma_3 = \cdots = \gamma_n = 0$, and hence it is in tridiagonal form. In course of iteration, if a subdiagonal element is zero such as β_3 in (13.96), then we have two unreduced tridiagonal or upper Hessenberg submatrices, each of which should be subjected separately to QR-transformation.

Steps in explicitly shifted QR-algorithm:

Step I. Given an n-square matrix A, determine an orthogonal transformation U such that U^TAU is in upper Hessenberg form, or in tridiagonal form if A is symmetric. Set A equal to U^TAU.

Step II. Compute the origin shift μ for A determined in Step I by the Rayleigh quotient (13.93) or by (13.94).

Step III. Determine the QR-factorization of $(A - \mu I)$, i.e., $Q^T(A - \mu I) = R$, $Q^T = Q_{n-1,n} \cdots Q_{23}Q_{12}$ by plane rotations.

Step IV. Initiate the next QR-iteration by forming, $Q^T(A - \mu I)Q = RQ = RQ_{12}^TQ_{23}^T \cdots Q_{n-1,n}^T$.

Step V. Restore the origin shift: $A_1 = Q^TAQ = RQ + \mu I$.

Step VI. Go to Step I with $A = A_1$ and $U = UQ$.

At the end of the QR algorithm, the eigenvalues of A, if real, appear along the diagonal of A_k which is now a triangular matrix. For complex eigenvalues, the reader is referred to double-shifted QR-algorithm.

Example 13.19 Apply the QR-algorithm to find the eigenvalues of the tridiagonal matrix

$$A = \begin{bmatrix} 2 & 1 & 0 \\ 1 & 2 & 1 \\ 0 & 1 & 2 \end{bmatrix} \tag{13.97}$$

Solution The eigenvalues of the second order submatrix at the right bottom corner of A are 1 and 3, both of which are at the same distance from $a_{33} = 2$. We choose $\mu = 3$ as the origin shift, and so

$$A - \mu I = \begin{bmatrix} -1 & 1 & 0 \\ 1 & -1 & 1 \\ 0 & 1 & -1 \end{bmatrix}$$

whose QR-factorization is now determined by plane rotation. For rotation in the (1, 2) plane from (13.41), we have $\alpha = \sqrt{2}$, $c = -1/\sqrt{2} = -s$, and hence

$$Q_{12} = \begin{bmatrix} -1/\sqrt{2} & 1/\sqrt{2} & 0 \\ -1/\sqrt{2} & -1/\sqrt{2} & 0 \\ 0 & 0 & 1 \end{bmatrix}, \qquad Q_{12}(A - \mu I) = \begin{bmatrix} \sqrt{2} & -\sqrt{2} & 1/\sqrt{2} \\ 0 & 0 & -1/\sqrt{2} \\ 0 & 1 & -1 \end{bmatrix}.$$

Again for rotation in the (2, 3) plane applied to the last matrix we have $\alpha = 1$, $c = 0$, $s = 1$, and therefore

$$Q_{23} = \begin{bmatrix} 1 & 0 & 0 \\ 0 & 0 & 1 \\ 0 & -1 & 0 \end{bmatrix}, \qquad Q_{23}Q_{12}(A - \mu I) = \begin{bmatrix} \sqrt{2} & -\sqrt{2} & 1/\sqrt{2} \\ 0 & 1 & -1 \\ 0 & 0 & 1/\sqrt{2} \end{bmatrix}.$$

Define $Q_0^T = Q_{23}Q_{13}$. Then

$$Q_0^T(A - \mu I)Q_0 = \begin{bmatrix} -2 & 1/\sqrt{2} & 0 \\ 1/\sqrt{2} & -1 & 1/\sqrt{2} \\ 0 & 1/\sqrt{2} & 0 \end{bmatrix}.$$

Finally, restoring the shift, we get

$$A_1 := Q_0^T A Q_0 = \begin{bmatrix} 1 & 1/\sqrt{2} & 0 \\ 1/\sqrt{2} & 2 & 1/\sqrt{2} \\ 0 & 1/\sqrt{2} & 3 \end{bmatrix}$$

where

$$Q_0 = \begin{bmatrix} -1/\sqrt{2} & 1/\sqrt{2} & 0 \\ 0 & 0 & 1 \\ 1/\sqrt{2} & 1/\sqrt{2} & 0 \end{bmatrix}.$$

Let the above process be repeated with A_1 whose right bottom corner submatrix of order 2 has eigenvalues 3.3660 and 1.6340. We choose $\mu = 3.3660$ which is nearer $a_{33} = 3$. Computing correct to 4 decimal places, we have successively

$$(A_1 - \mu I) = \begin{bmatrix} -2.3660 & 0.7071 & 0 \\ 0.7071 & -1.3660 & 0.7071 \\ 0 & 0.7071 & -0.3660 \end{bmatrix},$$

$$Q_{12} = \begin{bmatrix} -0.9581 & 0.2863 & 0 \\ -0.2863 & -0.9581 & 0 \\ 0 & 0 & 1 \end{bmatrix},$$

$$Q_{12}(A_1 - \mu I) = \begin{bmatrix} 2.4693 & -1.0686 & 0.2024 \\ 0 & 1.1063 & -0.6775 \\ 0 & 0.7071 & -0.3660 \end{bmatrix},$$

$$Q_{23} = \begin{bmatrix} 1 & 0 & 0 \\ 0 & 0.8426 & 0.5385 \\ 0 & -0.5385 & 0.8426 \end{bmatrix},$$

$$Q_{23}Q_{12}(A_1 - \mu I)Q_{12}^T Q_{23}^T = \begin{bmatrix} -2.6718 & 0.3761 & 0 \\ 0.3759 & -1.4735 & 0.0303 \\ 0 & 0.0304 & 0.0475 \end{bmatrix},$$

and finally, restoring the origin shift with $Q_1^T = Q_{23}Q_{12}$, we have

$$A_2 := Q_1^T A_1 Q_1 = \begin{bmatrix} 0.6942 & 9.3761 & 0 \\ 0.3759 & 1.8925 & 0.0303 \\ 0 & 0.0304 & 3.4135 \end{bmatrix}.$$

If the elements in positions (3, 2) and (2, 3) are considered insignificant, then 3.4135 is one of the eigenvalues of A. The other eigenvalues obtained from the left upper corner submatrix of A_2 are 2.0007 and 0.5860. In fact, the true eigenvalues of A are 2, 0.5858 and 3.4142, and so in this example, the QR-algorithm gives a good approximation only in two passes. □

13.4.8 Implicitly Shifted *QR*-algorithm

In an explicitly shifted QR-algorithm, the origin shift is carried along with the computation in each pass and is finally restored to initiate the next pass. This transport of origin shift is dispensed with in the implicitly shifted algorithm. For reasons already explained, following (13.95) we shall, without any loss of generality, assume that the given matrix A is in unreduced upper Hessenberg or tridiagonal form.

Given A in the above form and an origin shift μ determined from A by (13.93) or (13.94), we write the QR-factorization of $(A - \mu I)$ as

$$(A - \mu I) = QR, \qquad Q^TQ = I, \tag{13.98}$$

R is upper triangular with positive diagonal elements and $A_1 := Q^TAQ$ is in upper Hessenberg or tridiagonal form as discussed in the text following (13.95). Instead of determining this decomposition, particularly Q, by the explicitly shifted method given in Step III of the last algorithm, we now follow a different procedure. We shall prove that Q determined by the new method is the same as that obtained in the explicity shifted method. If $Q = [\mathbf{q}_1\ \mathbf{q}_2\ \ldots\ \mathbf{q}_n]$ and the first columns of $(A - \mu I)$ and R are denoted by $\mathbf{a} := [a_{11} - \mu, a_{21}, 0, ..., 0]^T$ and $\mathbf{r} = \rho\mathbf{e}_1$, respectively, where ρ is a positive real number, then it follows from (13.98) that

$$\mathbf{a} = (A - \mu I)\mathbf{e}_1 = QR\mathbf{e}_1 = \rho\, \mathbf{q}_1. \tag{13.99}$$

In view of Householder transformation or plane rotation, it is possible to select an orthogonal matrix P satisfyirig $P^TP = I$ such that

$$P\mathbf{a} = \|\mathbf{a}\|\,\mathbf{e}_1, \qquad \text{i.e., } P^T\mathbf{e}_1 = \frac{\mathbf{a}}{\|\mathbf{a}\|}, \tag{13.100}$$

where $\|\cdot\|$ denotes Euclidean norm. Since from (13.98) $\mathbf{q}_1^T\mathbf{q}_1 = 1$, so from (13.99), $\rho = \|\mathbf{a}\|$. Hence, by (13.100)

$$\mathbf{q}_1 = \rho^{-1}\mathbf{a} = \frac{\mathbf{a}}{\|\mathbf{a}\|} = \text{the first column of } P^T. \tag{13.101}$$

Let $U_1, U_2, ..., U_{n-2}$ be the Householder transformations such that (see Section 13.3.1)

$$U_{n-2}\cdots U_2U_1PAP^T(U_{n-2}\cdots U_2U_1)^T$$

is in upper Hessenberg form. The first row in the product

$$C^T := U_{n-2}\cdots U_2U_1P \tag{13.102}$$

is the same as that of P, as a result of the structure of U_i's, see the text preceding (13.38), i.e., C and P^T have the same first column. It now follows from (13.101) that C and Q also have the same first column. As a result of like first columns, $C = Q$ as ensured by the following lemma, and so Q in (13.98) can be constructed as the product of matrices shown in (13.102). Thus, Q need not be computed from $(A - \mu I)$ inserting a origin shift μ but from A as with (13.102). However, only P is to be determined from the first column of $(A - \mu I)$. So, the QR-iterations to construct a sequence of orthogonally similar matrices $\{A_{k+1} = Q_k^TA_kQ_k\}$, see (13.86), with Q_k determined not directly from $(A_k - \mu_kI)$ involving an origin shift, are called implicitly shifted QR-iterations. In place of Householder transformations, one can as well use plane rotations, see (13.43).

Lemma 13.2 Let A, H and Q be three matrices in $\mathbb{C}^{n\times n}$ with H unreduced upper Hessenberg having positive subdiagonal elements and Q unitary such that $QH = AQ$. If A and the first column of Q are given, then H and Q are uniquely determined.

Proof Let $Q = [\mathbf{q}_1, \mathbf{q}_2, ..., \mathbf{q}_n]$ be represented by its columns $\mathbf{q}_i$'s where $\mathbf{q}_1$ is known. To prove the lemma by mathematical induction, we assume that the first k columns $\mathbf{q}_1, \mathbf{q}_2, ..., \mathbf{q}_k$ of Q and $(k - 1)$ columns $\mathbf{h}_1, \mathbf{h}_1, ..., \mathbf{h}_{k-1}$ of H are determined at the $(k - 1)$ step. To compute $\mathbf{h}_k$:

$$\mathbf{h}_k := [h_{1k}\ \cdots\ h_{k+1,k}\ 0\ \cdots\ 0]^T \tag{13.103}$$

and $\mathbf{q}_{k+1}$, from $QH = AQ$ since H is upper Hessenberg, we get

$$h_{1k}\mathbf{q}_1 + h_{2k}\mathbf{q}_2 + \cdots + h_{k+1,k}\,\mathbf{q}_{k+1} = A\mathbf{q}_k. \tag{13.104}$$

Because of orthogonality of Q, for all $i, j = 1, 2, ..., k$,

$$\mathbf{q}_i^* \mathbf{q}_j = \begin{cases} 0, & i \neq j; \\ 1, & i = j. \end{cases}$$

Hence, premultiplying (13.104) by $\mathbf{q}_i^*$, we get

$$h_{ik} = \mathbf{q}_i^* A \mathbf{q}_k, \qquad i = 1, 2, ..., k.$$

So, all the elements in $\mathbf{h}_k$ given by (13.103) are determined except $h_{k+1,k}$, because $\mathbf{q}_{k+1}$ is still unknown by induction hypothesis. Now, from (13.104), since H is unreduced and so $h_{k+1,k} \neq 0$, we have

$$\mathbf{q}_{k+1} = \frac{\mathbf{v}}{h_{k+1,k}}, \qquad \mathbf{v} := [A\mathbf{q}_k - \sum_{i=1}^{k} h_{ik}\mathbf{q}_i] \tag{13.105}$$

from which, as $\mathbf{q}_{k+1}^* \mathbf{q}_{k+1} = 1$ it follows that

$$\mathbf{v}^*\mathbf{v} = h_{k+1,k}^2.$$

As $\mathbf{v}$ is known, the positive number $h_{k+1,k}$ is uniquely determined and $\mathbf{q}_{k+1}$ from (13.105). The subdiagonal elements of H should be positive for H and Q to be unique. □

The key steps of the modified algorithm are now given.

Steps in implicitly shifted QR-algorithm:

Step I. Given an arbitrary (resp symmetric) matrix A, transform it orthogonally by U to an upper Hessenberg (resp tridiagonal) form with positive subdiagonal elements. Set A equal to UAU^T.

Step II. Compute the origin shift μ using (13.93) or (13.94).

Step III. If $\mathbf{a}$ is the first column of $(A - \mu I)$, determine the Householder transformation or plane rotation P such that $P\mathbf{a} = \|\mathbf{a}\|\mathbf{e}_1$ and compute PAP^T.

Step.IV. Determine Householder transformations or plane rotations $U_1, U_2, ..., U_{n-2}$ to make $U_{n-2} \ldots U_n U_1 PAP^T (U_{n-2} \ldots U_2 U_1)^T$ in Hessenberg (resp tridiagonal) form.

This process, if repeated, implies subjecting A to successively QR-transformations which will culminate in zero elements in the last row of the transformed A but for the element a_{nn}. This a_{nn} gives one of the eigenvalues of A.

Example 13.20 Use the implicitly shifted QR-algorithm to find the eigenvalues of the tridiagonal matrix A given by (13.97).

Solution Proceeding as in Example 13.19, we get $(A - \mu I)$ given therein. So $\mathbf{a} = [-1\ 1\ 0]^T$ and the orthogonal matrix

$$P = \begin{bmatrix} -1/\sqrt{2} & 1/\sqrt{2} & 0 \\ -1/\sqrt{2} & -1/\sqrt{2} & 0 \\ 0 & 0 & 1 \end{bmatrix} \text{ with } PAP^T = \begin{bmatrix} 1 & 0 & 1/\sqrt{2} \\ 0 & 3 & -1/\sqrt{2} \\ 1/\sqrt{2} & -1/\sqrt{2} & 2 \end{bmatrix}$$

determined with rotation in the (1, 2)-plane of $\mathbf{a}$ gives $P\mathbf{a} = \sqrt{2}\mathbf{e}_1$. To transform PAP^T to a tridiagonal form, we use plane rotation in the (2, 3)-plane of its first column to get

$$U_1 = \begin{bmatrix} 1 & 0 & 0 \\ 0 & 0 & 1 \\ 0 & -1 & 0 \end{bmatrix}, \qquad A_1 := U_1 PAP^T U_1^T = \begin{bmatrix} 1 & 1/\sqrt{2} & 0 \\ 1/\sqrt{2} & 2 & 1/\sqrt{2} \\ 0 & 1/\sqrt{2} & 3 \end{bmatrix}.$$

To start the next pass, we have the eigenvalues of the right corner submatrix of order 2 of A_1 as 3.3660 and 1.6340. We choose μ = 3.366 and note that the first column of $(A_1 - \mu I)$ is $\mathbf{a} = [-2.366, 0.7071, 0]^T$. The orthogonal matrix

$$P = \begin{bmatrix} -0.9581 & 0.2863 & 0 \\ -0.2863 & -0.9581 & 0 \\ 0 & 0 & 1 \end{bmatrix} \quad \text{with} \quad \bar{A} = \begin{bmatrix} 0.6940 & 0.3169 & 0.2024 \\ 0.3169 & 2.3058 & -0.6774 \\ 0.2024 & -0.6775 & 3 \end{bmatrix}$$

determined with rotation in the (1, 2)-plane of $\mathbf{a}$ gives $P\mathbf{a} = 2.4694\mathbf{e}_1$ where $\bar{A} := PA_1P^T$. To transform $\bar{A}$ to a tridiagonal form, we use plane rotation in the (2, 3)-plane of its first column to get

$$U_1 = \begin{bmatrix} 1 & 0 & 0 \\ 0 & 0.8428 & 0.5383 \\ 0 & -0.5383 & 0.8428 \end{bmatrix}, \qquad A_2 = \begin{bmatrix} 0.6940 & 0.3760 & 0 \\ 0.3760 & 1.8924 & 0.0301 \\ 0 & 0.0300 & 3.4137 \end{bmatrix},$$

where $A_2 := U_1PA_1P^TU^T$, and the same conclusion follows as in Example 13.19. □

13.4.9 Double-shifted *QR*-algorithm

It is evident from (13.88) that for the convergence of the last diagonal element of A_k to an eigenvalue of A, the origin shift μ_k must be very close to that eigenvalue. This algorithm becomes virtually useless if we are concerned with a pair of complex eigenvalues and the origin shift is a real quantity. So to move arbitrarily close to a complex eigenvalue, the origin shift must be complex in which case the *QR*-factorization of the complex matrix $(A_k - \mu_k I)$ must be carried out in complex arithmetic, although originally the matrix A is real.

To get around this difficulty of pursuing complex arithmetic, a pair of conjugate origin shifts is introduced with which the real character of the *QR*-algorithm can still be preserved. The sequence of matrices A_k now tends to a block upper triangular form with 1 × 1 or 2 × 2 blocks of real numbers on the diagonal, the former corresponding to real eigenvalues and the latter to a pair of complex conjugate eigenvalues. This technique is known as the *double-shifted QR-algorithm.*

To understand how to apply two origin shifts at a single stroke, consider an unreduced real upper Hessenberg matrix A, the eigenvalues of the 2 × 2 real matrix at its bottom right corner (see (13.94)), are λ and μ such that

$$\lambda + \mu = a_{n-1,n-1} + a_{nn}, \qquad \lambda\mu = a_{nn}a_{n-1,n-1} - a_{n-1,n}a_{n,n-1}. \tag{13.106}$$

We now perform two steps of QR-algorithm with shifts λ and μ to get

$$A - \lambda I = Q_1R_1, \qquad A_1 - \lambda I = R_1Q_1; \tag{13.107}$$

$$A_1 - \mu I = Q_2R_2, \qquad A_2 - \mu I = R_2Q_2; \tag{13.108}$$

from which, in view of (13.86), it follows that

$$A_1 = Q_1^T AQ_1, \qquad A_2 = Q_2^T A_1Q_2 = (Q_1Q_2)^T A(Q_1Q_2), \tag{13.109}$$

$$Q_1Q_2R_2R_1 = Q_1(A_1 - \mu I)R_1 = Q_1(Q_1^T AQ_1 - \mu I)R_1 = (A - \mu I)Q_1R_1$$

$$= (A - \mu I)\ (A - \lambda I). \tag{13.110}$$

Hence, setting $Q = Q_1Q_2$ and $R = R_2R_1$, we have

$$QR = (A - \lambda I)\ (A - \mu I) =: B \tag{13.111}$$

$$A_2 = Q^TAQ \tag{13.112}$$

in which the QR-product is given in terms of a real matrix B even though λ and μ, being conjugate pairs, are complex numbers.

Since A is upper Hessenberg, the first three elements in the first column of $(A - \lambda I) \times (A - \mu I)$ are nonzero. These elements are

$$\alpha_1 = a_{11}^2 - (\lambda + \mu) + \lambda\mu + a_{12}a_{21}, \qquad \alpha_2 = a_{21}[a_{11} + a_{22} - (\lambda + \mu)],$$

$$\alpha_3 = a_{21}a_{32}.$$

Substituting from (13.106), these become

$$\alpha_1 = (a_{11} - a_{n-1,n-1})\,(a_{11} - a_{nn}) + (a_{12}a_{21} - a_{n-1,n}\ a_{n,n-1}),$$

$$\alpha_2 = a_{21}(a_{11} + a_{22} - a_{n-1,\,n-1} - a_{nn}), \qquad \alpha_3 = a_{21}a_{32}. \tag{13.113}$$

We now determine an orthogonal matrix P by Householder transformation or plane rotation such that (13.100) is satisfied, where $\mathbf{a} = [\alpha_1\ \alpha_2\ \alpha_3\ 0\ \cdots\ 0]^T$. By (13.101), the first columns of P and Q given in (13.111) and (13.112) are the same. As in a single shift case, we can determine Householder transformations $U_1, U_2, ..., U_{n-2}$ and define $C^T := U_{n-2} \cdots U_2U_1P$ in which C has the same first column as Q. Thus by Lemma 13.2, $C = Q$.

This double-shifted algorithm for complex eigenvalues of A is performed above in real arithmetic and is then repeated to A_2 given by (13.112), etc. The use of explicitly shifted QR-decomposition may not give the above results and cause troubles, see J.H. Wilkinson [205].

Example 13.21 Use double-shifted QR-algorithm to determine the eigenvalues of the upper Hessenberg matrix

$$A = \begin{bmatrix} 1 & 0 & 0 & 0 \\ 1 & -1 & -3 & 1 \\ 0 & 1 & 1 & 1 \\ 0 & 0 & -2 & 3 \end{bmatrix} \tag{13.114}$$

Solution Only one iteration will be shown. The eigenvalues of the bottom right corner submatrix of A are $2 \pm i$. Using (13.113), $\alpha_1 = 2$, $\alpha_2 = -4$, $\alpha_3 = 1$. Using (13.33), we compute the Householder transformation

$$P = \begin{bmatrix} -0.437 & 0.873 & -0.218 & 0 \\ 0.873 & 0.470 & 0.133 & 0 \\ -0.218 & 0.133 & 0.967 & 0 \\ 0 & 0 & 0 & 1 \end{bmatrix}$$

so that $P\mathbf{a} = 4.583\mathbf{e}_1$, where $\mathbf{a} = [2, -4, 1, 0]^T$, and

$$PAP^T = \begin{bmatrix} -0.525 & -0.509 & -2.984 & 0.655 \\ -0.603 & 0.844 & -1.572 & 0.603 \\ 0.641 & 0.393 & 0.678 & 1.100 \\ 0.436 & -0.266 & -1.934 & 3.000 \end{bmatrix}. \tag{13.115}$$

We shall now compute U_1 and U_2 to transform (13.115) orthogonally to an upper Hessenberg form. We use (13.33) to find the Householder transformation for the vector $[-0.603, 0.641, 0.436]^T$ to get

$$U_1 = \begin{bmatrix} 1 & 0 & 0 & 0 \\ 0 & -0.614 & 0.652 & 0.444 \\ 0 & 0.652 & 0.736 & -0.179 \\ 0 & 0.444 & -0.179 & 0.878 \end{bmatrix},$$

$$U_1PAP^TU_1^T = \begin{bmatrix} -0.525 & -1.342 & -2.645 & 0.883 \\ 0.982 & 1.337 & -0.144 & 1.207 \\ 0 & -0.366 & 0.327 & 1.011 \\ 0 & -0.484 & -2.290 & 2.857 \end{bmatrix}.$$

Using (13.33), we now find the Householder transformation for the vector $[-0.366, -0.484]^T$ to obtain

$$U_2 = \begin{bmatrix} 1 & 0 & 0 & 0 \\ 0 & 1 & 0 & 0 \\ 0 & 0 & -0.605 & -0.798 \\ 0 & 0 & -0.798 & 0.603 \end{bmatrix},$$

$$A_2 := U_2U_1PAP^TU_1^TU_2^T = \begin{bmatrix} -0.525 & -1.342 & 0.896 & 2.643 \\ 0.982 & 1.337 & -0.876 & 0.842 \\ 0 & 0.608 & 1.322 & -3.045 \\ 0 & 0 & 0.262 & 1.863 \end{bmatrix}.$$

The above computation has to be repeated to A_2 and so on. The eigenvalues of A can be shown to be $1, 1, 1 \pm i$. □

13.5 DETERMINATION OF EIGENVECTORS VIA *QR*-ALGORITHM

Given a real (resp symmetric) matrix A, we determine first an orthogonal matrix P such that $A_0 := P^TAP$ is in upper Hessenberg (resp tridiagonal) form. By applying the QR-algorithm, the orthogonal matrix Q is determined, giving thereby

$$S := Q^TA_0Q = Q^TP^TAPQ \tag{13.116}$$

in real Schur or quasi-triangular (resp diagonal) form (see Theorems 13.4 and 13.5). It is known that if $\mathbf{x}_i$ is an eigenvector of S, then $PQ\mathbf{x}_i$ is that of A associated with λ_i. Indeed, we have $S\mathbf{x}_i = \lambda_i\mathbf{x}_i$ and from (13.116),

$$A(PQ\mathbf{x}_i) = \lambda_i(PQ\mathbf{x}_i). \tag{13.117}$$

For a symmetric A, S is diagonal whose eigenvectors are elementary vectors $\mathbf{x}_i = \mathbf{e}_i$, see Problem 7.5. In view of (13.117), the columns of PQ form the eigenvectors corresponding to the eigenvalues λ_i of A which are now the diagonal elements of S.

For a general real matrix A, S is quasi-triangular, the eigenvector $\mathbf{x}_r$ corresponding to a real eigenvalue λ_r of S is obtained by setting $x_r = 1$ and solving the system of equations $(S - \lambda_rI)\mathbf{x} = \boldsymbol{\theta}$, each 2×2 block generating a system of two equations. On the other hand, to determine a pair of complex eigenvectors $\mathbf{x}$ and $\bar{\mathbf{x}}$ corresponding to complex eigenvalues λ and $\bar{\lambda}$ associated with a 2×2 block in S with diagonal elements a_{r-1} and a_r, we determine only $\mathbf{x}$ from $(S - \lambda I)\mathbf{x} = \boldsymbol{\theta}$ with $x_r = 1$, after finding an equation for x_{r-1} and noting that each 2×2 block generates a system of two equations.

Example 13.22 Determine the eigenvectors of the upper Hessenberg matrix A given by (13.114).

Solution The orthogonal matrix transforming A to the quasi-triangular form [see Example 13.17]

$$S = \left[\begin{array}{cc|cc} 1 & 2 & 0 & 1 \\ 0 & 1 & 3 & -2 \\ \hline 0 & 0 & 2 & -1 \\ 0 & 0 & 2 & 0 \end{array}\right]$$

is

$$Q = \frac{1}{\sqrt{3}}\begin{bmatrix} 0 & 1 & -1 & -1 \\ 1 & -1 & 0 & -1 \\ -1 & 0 & 1 & -1 \\ -1 & -1 & -1 & 0 \end{bmatrix}.$$

The eigenvalues of S, and consequently of A, are 1, 1, $1 \pm i$. To get the eigenvector of S associated with 1, we solve $(S - I)\,\mathbf{x} = \boldsymbol{\theta}$, i.e., $2x_2 + x_4 = 0$, $3x_3 - 2x_4 = 0$, $x_3 - x_4 = 0$, $2x_3 - x_4 = 0$, which gives $\mathbf{x} = [1\ 0\ 0\ 0]^T$ as the desired eigenvector. For the complex eigenvalue $1 + i$, we solve $[S - (1 + i)I]\,\mathbf{x} = \boldsymbol{\theta}$ with $x_4 = 1$. This gives us the eigenvector as $\mathbf{y} = (1/2)[2 - 4i, 3 + i, -1 + i, 2]^T$ associated with $1 + i$ and that for $1 - i$ is $\bar{\mathbf{y}}$. Hence, $Q\mathbf{x}$, $Q\mathbf{y}$ and $Q\bar{\mathbf{y}}$ are the three eigenvectors of A. □

13.6 COMPUTATION WITH MATLAB

The *QR*-factorization expresses any rectangular $m \times n$ matrix A as a product of an orthogonal or a unitary matrix Q and an upper triangular matrix R. The command

```
>> A = rand (4, 3)
A =
    0.1389    0.2722    0.4451
    0.2028    0.1988    0.9318
    0.1987    0.0153    0.4660
    0.6038    0.7468    0.4186
```

generates a 4×3 matrix A with random entries, chosen from a uniform distribution on the interval (0.0, 1.0). In the QR-factorization, we get successively matrices Q and R, using the command

```
>> [Q, R] = qr (A)
Q =
   -0.2038    0.4540    0.4621   -0.7340
   -0.2975   -0.1343    0.8001    0.5033
   -0.2916   -0.8526    0.0209   -0.4332
   -0.8860    0.2213   -0.3819    0.1424
>> R
R =
   -0.6815   -0.7807   -0.8747
         0    0.2491   -0.2278
         0         0    0.8011
         0         0         0
```

One may check that $QQ' = I$.

Schur decomposition expresses a square matrix A as $A = QSQ'$, where S is in real Schur, (i.e. a quasi-upper triangular) form. This decomposition is shown for a sixth order matrix with randomly generated entries:

```
>> A = rand (6)
A =
    0.8983    0.2731    0.4091    0.5972    0.8121    0.1662
    0.7546    0.6262    0.4740    0.1614    0.6101    0.8332
    0.7911    0.5369    0.9090    0.8295    0.7015    0.8386
    0.8150    0.0595    0.5962    0.9561    0.0922    0.4516
    0.6700    0.0890    0.3290    0.5955    0.4249    0.9566
    0.2009    0.2713    0.4782    0.0287    0.3756    0.1472
>> [Q, S] = schur [A]
```

```
Q =
    0.4088   -0.4384   -0.3317    0.5883    0.0408    0.4277
    0.4173    0.5786    0.2605    0.5461   -0.0276   -0.3524
    0.5862    0.2036    0.1251   -0.4676   -0.4420    0.4306
    0.3877   -0.5565    0.5782   -0.1608    0.2997   -0.2999
    0.3484   -0.1017   -0.6575   -0.2409   -0.1061   -0.6056
    0.2087    0.3341   -0.1994   -0.2305    0.8373    0.2255
S =
    3.1069   -0.4404   -0.5178   -0.0292    0.9464    0.1888
    0          0.7573   -0.3083    0.0243   -0.0122   -0.0817
    0          0          0.3353    0.0551   -0.1055    0.1669
    0          0         -0.3564    0.3353   -0.1545   -0.1805
    0          0          0          0        -0.2866    0.0448
    0          0          0          0        -0.5713   -0.2866
>> eig (A)
ans =
    3.1070
    0.7574
   -0.2866 + 0.1601i
   -0.2866 - 0.1601i
    0.3353 + 0.1402i
    0.3353 - 0.1402i
```

The Cholesky factorization expresses a symmetric matrix as the product of a triangular matrix and its transpose $A = R'R$, where R is an upper triangular matrix. Not all symmetric matrices can be factored in this way; the matrices that have such a factorization are said to be *positive definite.* This implies that all the diagonal elements of A are positive and that the off-diagonal elements are *not too big*. The pascal matrices provide an interesting example.

```
>> A = pascal (6)
A =
    1     1     1     1     1     1
    1     2     3     4     5     6
    1     3     6    10    15    21
    1     4    10    20    35    56
    1     5    15    35    70   126
    1     6    21    56   126   252
```

The elements of A are binomial coefficients. Each element is the sum of its north and west neighbours. The Cholesky factorization is

```
>> R = chol (A)
R =
     1     1     1     1     1     1
     0     1     2     3     4     5
     0     0     1     3     6    10
     0     0     0     1     4    10
     0     0     0     0     1     5
     0     0     0     0     0     1
```

The elements are again binomial coefficients. The fact that $R'R$ is equal to A demonstrates an identity involving sums of products of binomial coefficients. The Cholesky factorization also applies to complex matrices. Any complex matrix which has a Cholesky factorization satisfies $A' = A$ and is said to be *Hermitian positive definite.* The Cholesky factorization allows the linear system $A\mathbf{x} = \mathbf{b}$ to be replaced by $R'R\mathbf{x} = \mathbf{b}$.

Because the backslash operator recognizes triangular systems, this can be solved in MATLAB quickly with x=R\(R'\b). If A is n-by-n, the computational complexity of chol (A) is $O(n^3)$, but the complexity of the subsequent backslash solutions is only $O(n^2)$.

LU – factorization, or Gaussian elimination, expresses any square matrix A as the product of a permutation of a lower triangular matrix and an upper triangular matrix $A = LU$, where L is a permutation of a lower triangular matrix with ones on its diagonal and U is an upper triangular matrix. Partial pivoting ensures that the elements of L are bounded by one in magnitude and that the elements of U are not much larger than those of A. For example, if B=magic(3),

```
>> [L,U] = lu(B)
L =
    1.0000         0         0
    0.3750    0.5441    1.0000
    0.5000    1.0000         0
U =
    8.0000    1.0000    6.0000
         0    8.5000   -1.0000
         0         0    5.2941
```

The LU-factorization of A allows the linear system $A\mathbf{x} = \mathbf{b}$ to be solved quickly with x = U/(L/b). Determinants and inverses are computed from the LU – factorization using det (A) = det (L)*det (U) and inv (A) = inv (U)*inv (L). You can also compute the determinants using det (A) = prod (diag (U)), though the signs of the determinants may be reversed.

13.7 SVD IN IMAGE PROCESSING AND DATA ANALYSIS

An important application of singular value decomposition (SVD) described in Section 9.6 occurs in the image processing and the data analysis. A crucial concept in this respect is the numerical rank of a matrix.

13.7.1 Numerical Rank of a Matrix

As established in Theorem 5.9, the number of independent columns or rows of a matrix A is equal to its rank. A classical method to evaluate the rank of A is to convert it to row reduced echelon form (rref) by the Gaussian elimination as described in Section 5.9. However, in floating-point arithmetic, this is not a numerically viable method due to round-off and truncation errors in computation. We know by Lemma 9.1 that there exist orthogonal matrices $U \in \mathbb{R}^{m\times m}$ and $V \in \mathbb{R}^{n\times n}$, such that any matrix $A \in \mathbb{R}^{m\times n}$ can be represented as $A = U\Sigma V^T$, where Σ contains the singular values of A along the diagonal, other elements being zero. As the singular values of a matrix are well-conditioned which is analogous to the case of eigenvalues of a symmetric matrix [see (13.2.3)], by counting the non-zero singular values we can determine the rank of A. But we have to know when to accept a computed singular value to be zero. The standard practice is: *accept a singular value to be zero if it is less than or equal to* $\delta = 10^{-m} \| A \|_{\infty}$, *where m is the precision of the floating point numbers in* A. A matrix A is defined to have a numerical rank r if the computed singular values $\hat{\sigma}_i$, $i = 1, 2, \ldots, n$ satisfy

$$\hat{\sigma}_1 \geq \cdots \geq \hat{\sigma}_r > \delta \geq \hat{\sigma}_{r+1} \geq \cdots \geq \hat{\sigma}_n.$$

13.7.2 Image Compression Using SVD

A black-and-white video image or photograph is digitized by subdividing it into tiny rectangular cells called picture elements abbreviated as pixels while transmitting it from, say, Mars mission down to earth. Each pixel is represented by a single floating point number between 0 and 1 that records the average light intensity in that rectangular cell known as gray levels. The value 0 corresponds to black while the value 1 to white. If each photograph is divided into 256 × 256 pixels, it would have to send 65, 536 numbers stored in a matrix A of order 256 × 256 for each picture. This amounts to a great deal of time and extensive storage of pixels. However, to get around these difficulties, the matrix A of order $m \times n$ as above is approximated in practice, by a matrix of much lower rank using the SVD technique. As established in Section 9.6, every $A \in \mathbb{R}^{m\times n}$ of rank r can be represented by a dyadic representation of the form

$$A = \sigma_1 \mathbf{u}_1 \mathbf{v}_1^T + \sigma_2 \mathbf{u}_2 \mathbf{v}_2^T + \cdots + \sigma_r \mathbf{u}_r \mathbf{v}_r^T, \tag{13.118}$$

where $\sigma_1 \geq \sigma_2 \geq \cdots \geq \sigma_r > 0$ are non-zero singular values of A, $\mathbf{u}_i \in \mathbb{R}^m$ and $\mathbf{v}_i \in \mathbb{R}^n$ are the corresponding left and right singular vectors of A. Some of the non-zero singular values of A are appreciably small and can be neglected in storing the information in A. If $k \leq r$ singular values are retained, then A is compressed to

$$A_c = \sigma_1 \mathbf{u}_1 \mathbf{v}_1^T + \sigma_2 \mathbf{u}_2 \mathbf{v}_2^T + \cdots + \sigma_k \mathbf{u}_k \mathbf{v}_k^T, \tag{13.119}$$

which is also of order $m \times n$. So, for image transmission we need to store σ_i, $\mathbf{u}_i$ and $\mathbf{v}_i$, $i = 1, 2, \ldots, k$ requiring only $n_{A_c} = (m + n + 1)k$ memory units instead of $n_A = mn$ affecting a considerable saving in storage space. The ratio $n_A / n_{A_c} = mn / (m + n + 1)k$ is an important factor in image processing and is called *compression ratio*. For the above 256 × 256 pixel image, if

20 terms are retained, we have to provide 10260 memory units for the storage space instead of 65,536 with a compression ratio 65,536/10,260 = 6.4 giving a great saving. While recovering the image in the receiving side, if one dyadic term is included, one can recognize nothing, but as more and more terms are being included, all of a sudden the image becomes strikingly impressive. This process is known as image compression as the image represented by a very large matrix is converted to a lower-order approximation of A still preserving the excellent image quality.

To process a colour picture, each pixel is composed of three primary colour components – a red, a green, and a blue. In storing colour pictures, each pixel is represented by three floating point numbers corresponding to three colour intensities, so three matrices each of dimension $m \times n$ are needed to capture a picture and by using the SVD technique a considerable saving in storage for the colour picture can be achieved.

A black and white picture 'lena_bw.gif' of Swedish girl Lena Söderberg, the first lady of the internet, is downloaded from the online image repository and is stored in a matrix A of size (704 × 704 uint 8) and with the colour code matrix X of size 256 × 3. The compression of the image with SVD is studied using MATLAB with the code listed below.

```
[A,X]=imread('lena_bw.gif');% stores image in (A uint8) with
%colour code in X
figure
subimage(A,X)
title('Lena:Original Image')
A=double(A); % converts uint8 to double precision
[U,D,V]=svd(A);
k=60;
Ak=U(:,1:k)*D(1:k,1:k)*V(:,1:k)';
figure
subimage(Ak,X)
title('Lena: SVD compressed Image, k=60')
```

The 704 singular values of A varies from $\sigma_{max} = 7.2001 \times 10^4$ to $\sigma_{min} = 0.0056$. With $k = 10$ the SVD compressed image is distorted but it progressively improves at $k = 15$ and excellent with $k = 30$ for which $\sigma_{30} = 1.5935 \times 10^3$. In Figure 13.2, the original image along with the SVD compressed images for $k = 15$, $\sigma_{15} = 3.3119 \times 10^3$ and $k = 60$, $\sigma_{60} = 7.1723 \times 10^2$ are shown.

Original $k = 15$ $k = 60$

Figure 13.2 Image compression with SVD.

There are two ways to represent the number that corresponds to the brightness of the pixel. One known as the double class or data type assigns a floating number, i.e., a number with decimals, between 0 and 1 to each pixel. The value 0 corresponds to black and the value 1 corresponds to white. The other class is called uint8 which assigns an integer between 0 and 255 to represent the brightness of a pixel, i.e., each pixel can be represented by a single 8-bit byte. Uint stands for unsigned integer. The value 0 corresponds to black and 255 to white. The class uint8 only requires roughly 1/8 of the storage compared to the class double. *Many mathematical functions can only be applied to the double class.*

When decomposing a grayscale image, only one step is needed. Because the grayscale image is only one layer thick, it is much easier to manage. Colour images, however, are much more common, and are also much more difficult to decompose. Because true colour images are composed of a red, green, and blue layer for each pixel, we cannot simply compute the SVD of the original matrix. In order to get around this, the three layers of the image are to be separated and stored in different matrices. In this fashion the matrices can be decomposed individually and then later stacked back on top of one another to create a new SVD true colour image as outlined in the following MATLAB code.

```
L=imread('lena_images.jpg');
imshow(L)
L1=L(:,:,1);
L2=L(:,:,2);
L3=L(:,:,3);
I1=im2double(L1); I2=im2double(L2); I3=im2double(L3);
[u1,s1,v1]=svd(I1);
[u2,s2,v2]=svd(I2);
[u3,s3,v3]=svd(I3);
k=15;
C1=u1(:,1:k)*s1(1:k,1:k)*v1(:,1:k)';
C2=u2(:,1:k)*s2(1:k,1:k)*v2(:,1:k)';
C3=u3(:,1:k)*s3(1:k,1:k)*v3(:,1:k)';
R1=im2uint8(C1);
R2=im2uint8(C2);
R3=im2uint8(C3);
Q(:,:,1)=R1;
Q(:,:,2)=R2;
Q(:,:,3)=R3;
figure
imshow(Q,[])
```

Downloading the coloured image 'lena_images.jpg' of size (225 × 225 uint8) from the internet, and running the above MATLAB program, even with k = 15 one gets an excellent reconstructed image with the SVD compression. The above program will also work if the extension of colour images is other than jpg.

Comparison and errors in image compression

(1) Two of the error metrics used to compare the various image compression techniques are the Mean Square Error (MSE) between the original and the compressed image and Peak Signal to Noise Ratio (PSNR). The mathematical formula for the two are:

$$\text{MSE} = \frac{1}{mn}\sum_{1}^{m}\sum_{1}^{n}[A(i,j) - A_c(i,j)]^2, \tag{13.120}$$

$$\text{PSNR} = 20\log_{10}(255/\sqrt{\text{MSE}}). \tag{13.121}$$

where $A(i, j)$ is the original image (13.118) and $A_c(i, j)$ is the image after compression as given by (13.119). The MSE is the cumulative squared error between the compressed and the original image, whereas PSNR is a measure of the peak error.

(2) If we want to compress an image, we usually want to remove as much data as possible without losing too much energy of the image. The Frobenius norm of a matrix A is an l_2 norm and is related to the concept of energy and for this reason the Frobenius matrix norm is sometimes useful in image processing. The difference in energy between the original image and the compressed image is one possible way to design an error measure when compressing images.

(3) In some cases the energy error measure might be small, while the visual error is terrible. So another error measure is the relative (Frobenius) error E_{rel} defined as

$$E_{rel} = \frac{\| A - A_c \|_F}{\| A \|_F}$$

(4) To measure the efficiency in compression, let n_{A_c} be the memory units needed to store a compressed image and n_A, an original image. Then the ratio n_A/n_{A_c} is an important factor in image processing and is called *compression ratio*.

13.7.3 Principal Component Analysis

In many physical sciences such as neuroscience, photo-science, meteorology, oceanography, and credit scoring models for consumer lending problems in banks, the number of variables to be measured may be contaminated with noise and may also contain redundant data. In the analysis of such statistical data, we use an orthogonal transformation to convert a set of observations of possibly correlated variables into a set of values of linearly uncorrelated variables called *principal components*. The number of principal components is less than or equal to the number of original variables. Under this transformation, the first principal component has the largest possible variance, which accounts for as much of the variability in the data as possible. Each succeeding component in turn has the next highest variance possible under the constraint that it is orthogonal to the preceding components.

This procedure of computing the principal components of a large set of data contaminated with noise and containing redundancy for statistical analysis is known as *principal component analysis* (PCA). PCA can be done by eigenvalue decomposition of a data covariance matrix or singular value decomposition of a data matrix. In *regression analysis*, we we explore statistical data to establish a relationship between two or more variables so that the information about one of them can be obtained by knowing the values of others. It is the usual practice, especially when there are strong correlations between different possible explanatory variables, to reduce them to a few principal components and then run the regression against them, a method called principal component regression.

The principal components are termed factors in Exploratory Factor Analysis (EFA), where we determine the nature of and the number of latent variables that account for observed variation and covariation among set of observed indicators (items or variables). In contrast, in PCA we select a fewer principal components that summarize their variances to reduce the redundancy in the data set.

PCA by eigenvalue decomposition of a data covariance matrix

We define the number of variables as the dimension of the data set. Each data sample is, therefore, a vector in an m-dimensional vector space to be spanned by an *orthonormal* standard basis. Let $X \in \mathbb{R}^{m\times n}$ represent n *measurements* or *samples* of the data set and let $P \in \mathbb{R}^{m\times m}$ be an orthogonal matrix. We now define a new matrix $Y \in \mathbb{R}^{m\times n}$ by

$$X = PY, \quad PP^T = P^T P = I. \tag{13.122}$$

If $\mathbf{x}_i$, $\mathbf{y}_i$, $\mathbf{p}_i \in \mathbb{R}^m$ represent the columns of X, Y and P, respectively, then

$$\mathbf{y}_i = \begin{bmatrix} \langle \mathbf{p}_1, \mathbf{x}_i \rangle \\ \vdots \\ \langle \mathbf{p}_m, \mathbf{x}_i \rangle \end{bmatrix} \tag{13.123}$$

implying that the components of $\mathbf{y}_i$ are the orthogonal projections of $\mathbf{x}_i$ on the m orthogonal coordinates defined by the columns of P, see Definition 5.14. These m orthogonal column vectors $\mathbf{p}_i \in \mathbb{R}^m$ of P are called the *principal components* of the data set X. The principal components are the eigenvectors of the covariance matrix, XX^T as proved below, and the $\mathbf{p}_i$s in order of the size of eigenvalues of XX^T, tell us how *significant principal* each principal component is.

Let us now consider two data sets:

$$\mathbf{w} = [w_1, w_2, \ldots, w_n]; \quad \mathbf{z} = [z_1, z_2, \ldots, z_n]$$

recording two different variables measured n times and represented as row vectors with zero means as the relevant means have been subtracted from them. The covariance of these two variables is

$$\sigma_{wz}^2 = \frac{1}{n-1}\mathbf{w}\mathbf{z}^T,$$

where the coefficient $1/(n-1)$ is a normalization factor for an unbiased estimator. If $\sigma_{wz}^2 = 0$, then $\mathbf{w}$ and $\mathbf{z}$ are entirely uncorrelated, and

$$\sigma_w^2 = \frac{1}{n-1}\mathbf{w}\mathbf{w}^T, \quad \sigma_z^2 = \frac{1}{n-1}\mathbf{z}\mathbf{z}^T$$

are the *variances* of $\mathbf{w}$ and $\mathbf{z}$, respectively, as the samples in $\mathbf{w}$ and $\mathbf{z}$ are given above their means.

Let the ith row, $i = 1, 2,\ldots, m$, of the measurement matrix X, correspond to the ith variable measured n times. The m elements in each column $\mathbf{x}_i$ of X represent the ith measurement of m variables and the n columns of X, the measurement of m variables n times. The mean of the ith variable in the data matrix X, $i = 1, 2,\ldots, m$ is assumed to be zero and this is ensured by subtracting the mean of the measured values from each component of the ith row of X. It is assumed henceforth that the data sets are in this *mean deviation form*. We now define the covariance matrix S_X:

$$S_X = \frac{1}{n-1}XX^T = \frac{1}{n-1}\begin{bmatrix} \mathbf{x}_1^T\mathbf{x}_1 & \mathbf{x}_1^T\mathbf{x}_2 & \cdots & \mathbf{x}_1^T\mathbf{x}_m \\ \mathbf{x}_2^T\mathbf{x}_1 & \mathbf{x}_2^T\mathbf{x}_2 & \cdots & \mathbf{x}_2^T\mathbf{x}_m \\ \vdots & \vdots & \ddots & \vdots \\ \mathbf{x}_m^T\mathbf{x}_1 & \mathbf{x}_m^T\mathbf{x}_2 & \cdots & \mathbf{x}_m^T\mathbf{x}_m \end{bmatrix},$$

where the ith diagonal element represents the variance of the ith variable and ijth off-diagonal terms are the covariances of ith and jth variables. Thus, S_X captures all the correlations between all possible pairs of variables. A large covariance corresponds to the fact that this (i, j) pair of variables are highly correlated and one of them is redundant; and zero covariance corresponds to entirely uncorrelated variables. We now choose a transformation of X defined by (13.118), i.e., $X = PY$, so that the covariance matrix in terms of Y is

$$S_Y = \frac{1}{n-1}YY^T = \frac{1}{n-1}P^T(XX^T)P.$$

Since XX^T is real, symmetric and positive semi-definite, by Corollary 7.7 and Theorem 8.19 there exists an orthogonal matrix P such that $P^T(XX^T)P = D$ is diagonal, where the columns of P are the eigenvectors of XX^T and the diagonal elements of D are the eigenvalues of XX^T. If $\sigma_1^2 \geq \sigma_2^2 \geq \cdots \geq \sigma_r^2 > 0 = \cdots = 0$ are the eigenvalues of XX^T arranged in a decreasing order of magnitude, then by properly manipulating the columns of P, we can write

$$(n-1)\,S_Y = YY^T$$

$$= \begin{bmatrix} \mathbf{y}_1^T\mathbf{y}_1 & \cdots & \mathbf{y}_1^T\mathbf{y}_r & \mathbf{y}_1^T\mathbf{y}_{r+1} & \cdots & \mathbf{y}_1^T\mathbf{y}_m \\ \vdots & \ddots & \vdots & \vdots & \cdots & \vdots \\ \mathbf{y}_r^T\mathbf{y}_1 & \cdots & \mathbf{y}_r^T\mathbf{y}_r & \mathbf{y}_r^T\mathbf{y}_{r+1} & \cdots & \mathbf{y}_r^T\mathbf{y}_m \\ \mathbf{y}_{r+1}^T\mathbf{y}_1 & \cdots & \mathbf{y}_{r+1}^T\mathbf{y}_r & \mathbf{y}_{r+1}^T\mathbf{y}_{r+1} & \cdots & \mathbf{y}_{r+1}^T\mathbf{y}_m \\ \vdots & \cdots & \vdots & \vdots & \cdots & \vdots \\ \mathbf{y}_m^T\mathbf{y}_1 & \cdots & \mathbf{y}_m^T\mathbf{y}_r & \mathbf{y}_m^T\mathbf{y}_{r+1} & \cdots & \mathbf{y}_m^T\mathbf{y}_m \end{bmatrix} = \begin{bmatrix} \sigma_1^2 & \cdots & 0 & 0 & \cdots & 0 \\ \vdots & \ddots & \vdots & \vdots & \cdots & \vdots \\ 0 & \cdots & \sigma_r^2 & 0 & \cdots & 0 \\ 0 & \cdots & 0 & 0 & \cdots & 0 \\ \vdots & \cdots & \vdots & \vdots & \ddots & \vdots \\ 0 & \cdots & 0 & 0 & \cdots & 0 \end{bmatrix},$$

where r is the rank of XX^T. In view of (13.119), the components of $\mathbf{y}_i$, the ith column of Y, are along the directions of the eigenvectors of XX^T which are the columns of P and the covariances of the transformed matrix Y are zero. The columns $\mathbf{p}_i$s of P which are the eigenvectors of XX^T with decreasing variances σ_is are defined to be the *principal components* of X. The ith diagonal value of S_Y is the variance $\sigma_i^2/(n-1)$ of X along $\mathbf{p}_i$, and the covariances along the principal directions are zero. The technique of computing Y and the variances of X along the principal directions is known as *principal component analysis* (PCA).

PCA by SVD from covariance matrix of X

Let us define a new matrix $Z \in \mathbb{R}^{n \times m}$:

$$Z = \frac{1}{\sqrt{n-1}} X^T,$$

where each column of Z has a zero mean. It can be easily verified that

$$Z^T Z = S_X.$$

By construction $Z^T Z$ equals the *covariance* matrix of X. Since $Z^T Z = \frac{1}{n-1} XX^T$, the eigenvalues of $Z^T Z$ are $\frac{1}{n-1}\{\sigma_1^2, \sigma_2^2, \ldots, \sigma_m^2\}$, where $\{\sigma_1^2, \sigma_2^2, \ldots, \sigma_m^2\}$ are the eigenvalues of XX^T as defined before. By Lemma 9.1, $Z = U\Sigma V^T$ is the SVD of Z, where the columns of V contain the eigenvectors of $Z^T Z = \frac{1}{n-1} XX^T$ and Σ is a diagonal matrix in one of the forms (9.39)–(9.41) with non-zero diagonal elements $\sigma_1/\sqrt{n-1}, \sigma_2/\sqrt{n-1}, \cdots, \sigma_r/\sqrt{n-1}$. Note that $Z^T Z$ and XX^T have the same eigenvectors but the eigenvalues differ by a multiplying factor of $(n - 1)$. Consequently, the columns of V are also the eigenvectors of XX^T and construct the columns of P defined before which is formed by the eigenvectors of XX^T. This is, therefore, an alternative method to construct the principal components of X via SVD.

Incidentally, since $Z^T Z = (U\Sigma V^T)^T (U\Sigma V^T) = V \operatorname{diag}[\sigma_1^2/(n-1), \sigma_2^2/(n-1), \ldots, \sigma_r^2/(n-1), 0, \ldots, 0]V^T$, an SVD of $Z^T Z$, the principal components of X can also be constructed from $\operatorname{cov} Z = S_X$, a covariance matrix of X.

PCA method of data analysis

PCA is a process of identifying patterns in statistical data, and expressing the data in such a way as to highlight their similarities and differences. According to the patterns, the data can be compressed by reducing the number of dimensions without much loss of information as in image processing. Basic assumptions are: Linearity, so we can change the frame of representation by a change of basis. The data are Gaussian as mean and variance are sufficient statistics that entirely describe a probability distribution. The data have a high SNR, signal to noise ratio.

To illustrate PCA for data analysis, we consider the population of freshmen in the whole of Japan, and plan to study the dependence of college GPA (grade point average) on high school GPA. To explain the technique, only ten observations are taken although many more belong to the population.

High School GPA	6.75	7.75	5.25	8.00	6.00	8.50	6.50	5.00	7.75	6.25
College GPA	5.50	7.00	6.00	9.50	4.75	8.75	7.75	3.50	8.50	7.50

The diversity of college GPA values among students of the same high school GPA value may be attributed to uncertain factors such as motivation, physical and mental health, extracurricular activities, parents' guidance, etc. The following are the major steps.

Step 1: Subtract the mean

Having the data matrix,

$$A = \begin{bmatrix} 6.75 & 7.75 & 5.25 & 8.00 & 6.00 & 8.50 & 6.50 & 5.00 & 7.75 & 6.25 \\ 5.50 & 7.00 & 6.00 & 9.50 & 4.75 & 8.75 & 7.75 & 3.50 & 8.50 & 7.50 \end{bmatrix},$$

we subtract the mean of each row, 6.7750 and 6.8750 from the corresponding row elements, and get,

$$X = \begin{bmatrix} -0.0250 & 0.9750 & -1.5250 & 1.2250 & -0.7750 & 1.7250 & -0.2750 & -1.7750 & 0.9750 & -0.5250 \\ -1.3750 & 0.1250 & -0.8750 & 2.6250 & -2.1250 & 1.8750 & 0.8750 & -3.3750 & 1.6250 & 0.6250 \end{bmatrix}.$$

The data distribution and that above the mean are shown in Figures 13.3 and 13.5, respectively.

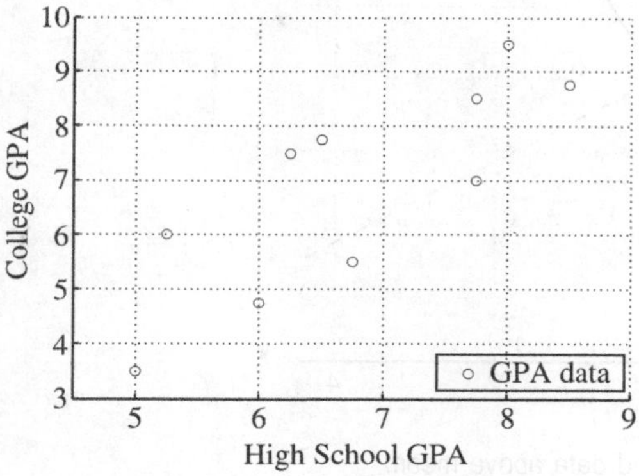

Figure 13.3 Data distribution.

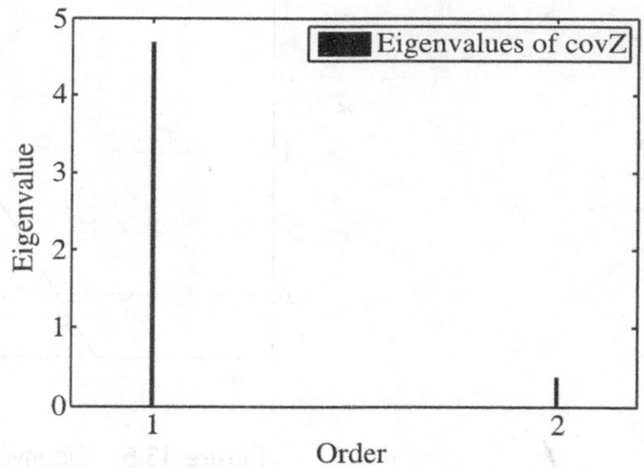

Figure 13.4 Eigenvalues of covZ.

Step 2: Calculate the covariance matrix and its SVD

The covariance of $Z = \frac{1}{\sqrt{n-1}} X^T$ is

$$\text{cov}Z = \begin{bmatrix} 1.4229 & 1.8437 \\ 1.8437 & 3.6424 \end{bmatrix}.$$

Since the off-diagonal elements of covZ are positive, both the variables College GPA and High School GPA increase together. The SVD of covZ is USV^T where

$$U = V = \begin{bmatrix} -0.4921 & -0.8705 \\ -0.8705 & 0.4921 \end{bmatrix}, S = \begin{bmatrix} 4.6846 & 0 \\ 0 & 0.3807 \end{bmatrix}.$$

The singular values of covZ are $s_1^2 = 4.6846$ and $s_2^2 = 0.3807$. As $Z^T Z$ is PSD, the eigenvalues of covZ are also $s_1^2 = 4.6846$ and $s_2^2 = 0.3807$ shown in Figure 13.4 with the columns of V as the corresponding *orthogonal eigenvectors*. The columns of V are called the *principal components* of the data set X. From the plot of these eigenvectors in Figure 13.5, the first eigenvector associated with the maximum eigenvalue goes through the middle of the data showing an important pattern that a line along this direction is the best fit for the data. The second eigenvector exhibits another pattern, less important, that all the points follow the main line but off to the side of the main line by some amount. So the eigenvectors of the cov$Z$ enable us to extract the first few of the lines that characterize the data, and discard the rest of them. This is based on the fact that the eigenvectors with eigenvalues that are much much higher than others contain more information about the data distribution and so we might be interested in keeping them.

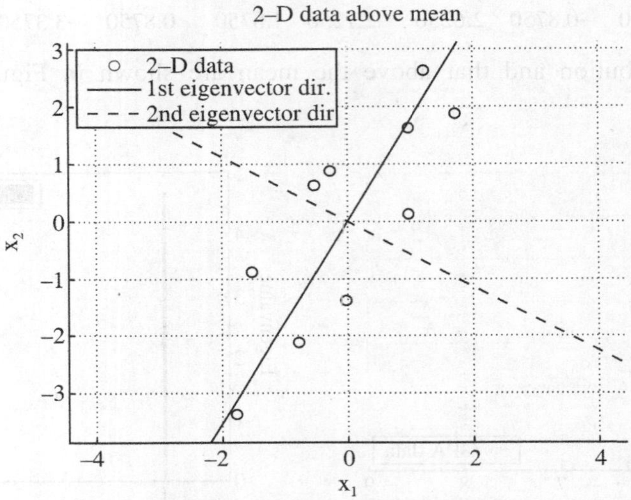

Figure 13.5 Eigenvectors and data above mean.

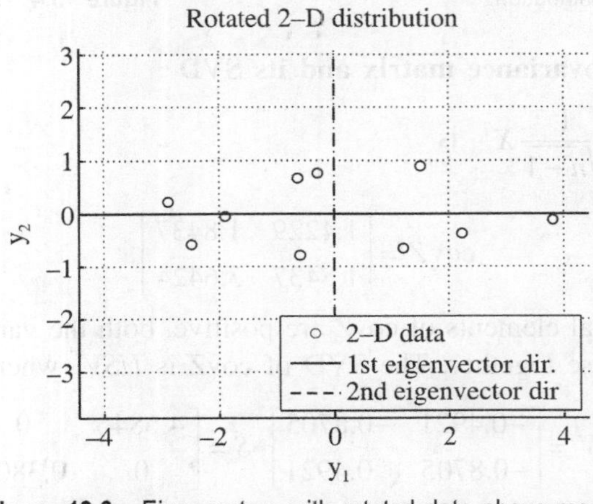

Figure 13.6 Eigenvectors with rotated data above mean.

Step 3: Construct the feature vector

In the general $m \times n$ data set X, as the eigenvalues in S are arranged in decreasing order, the variance of X along the first eigenvector is the largest where the data is most spread, along the second eigenvector is the second largest and so on. As along the first eigenvector the variance is the largest, it is the most important, i.e., most principal in PCA, the other eigenvectors come successively in order of significance. If we decide to discard the principal components of lesser significance, the matrix formed with the eigenvectors which we find to preserve the most important features of the data set is called a *feature vector*. This is just a fancy name for a matrix of vectors.

In the present case, if we decide to keep only the first principal component, then the feature vector is

$$\mathbf{f} = \begin{bmatrix} -0.4921 \\ -0.8705 \end{bmatrix}.$$

In general, for an $m \times m$ cov$Z$, if we keep the first $k \le m$ columns from V, then the feature vector is

$$F = [\mathbf{v}_1\ \mathbf{v}_2 \cdots \mathbf{v}_k]$$

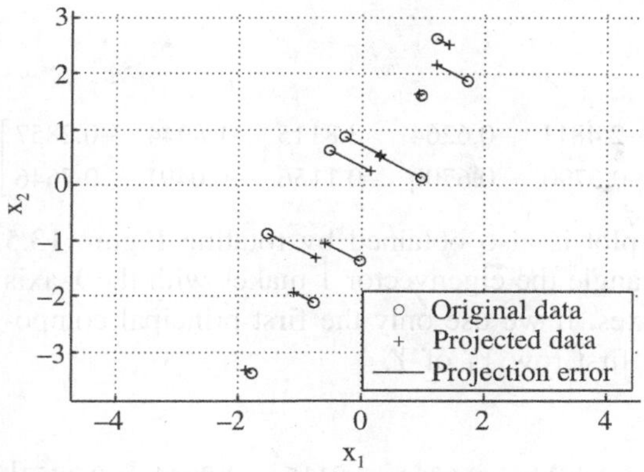

Figure 13.7 Projection on the primary eigenvector.

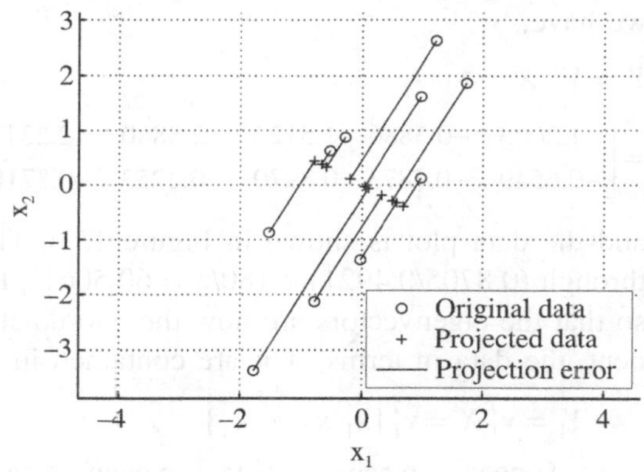

Figure 13.8 Projection on the secondary eigenvector.

Step 4: Data with reduced dimension

If we select the columns $\{\mathbf{v}_1, \mathbf{v}_2, \ldots, \mathbf{v}_m\}$ of V as the basis of $\mathbb{R}^m$ and a standard basis in $\mathbb{R}^n$,

$$[\mathbf{x}_1\ \mathbf{x}_2 \cdots \mathbf{x}_n][\mathbf{e}_1\ \mathbf{e}_2 \cdots \mathbf{e}_n] = [\mathbf{v}_1\ \mathbf{v}_2 \cdots \mathbf{v}_m][\mathbf{y}_1\ \mathbf{y}_2 \cdots \mathbf{y}_n],$$

then the representation of X is $Y = [\mathbf{y}_1\ \mathbf{y}_2 \cdots \mathbf{y}_n]$ and

$$YY^T = \Sigma,$$

a diagonal matrix noting that $X = VY$. Using the relations, $Z^T Z = VSV^T$, $Z = \frac{1}{\sqrt{n-1}} X^T$ and $X = VY$,

$$\begin{aligned} \frac{1}{n-1} YY^T &= \frac{1}{n-1} V^T XX^T V \\ &= V^T Z^T ZV = S \end{aligned}$$

$$= \frac{1}{n-1}\operatorname{diag}[\sigma_1^2, \sigma_2^2, \ldots, \sigma_m^2],$$

where σ_i^2 s are the eigenvalues of XX^T. Therefore,

$$Y = V^T X = \begin{bmatrix} \mathbf{v}_1^T\mathbf{x}_1 & \mathbf{v}_1^T\mathbf{x}_2 & \cdots & \mathbf{v}_1^T\mathbf{x}_n \\ \mathbf{v}_2^T\mathbf{x}_1 & \mathbf{v}_2^T\mathbf{x}_2 & \cdots & \mathbf{v}_2^T\mathbf{x}_n \\ \vdots & \vdots & \ddots & \vdots \\ \mathbf{v}_m^T\mathbf{x}_1 & \mathbf{v}_m^T\mathbf{x}_2 & \cdots & \mathbf{v}_m^T\mathbf{x}_n \end{bmatrix},$$

where the ith row of Y is the projection of $\{\mathbf{x}_1, \mathbf{x}_2, \ldots, \mathbf{x}_n\}$ on the eigenvector $\mathbf{v}_i$. The data matrix Y is now in terms of basis $\{\mathbf{v}_1, \mathbf{v}_2, \ldots, \mathbf{v}_m\}$ in place of X that is in terms of a standard basis. If we plan to approximate the data by the first k eigenvectors, then we retain the first k rows of Y and neglect the remaining ones getting data in a reduced dimension. In the present 2-D case, we have,

$Y = V^T X$

$$= \begin{bmatrix} 1.2093 & -0.5886 & 1.5122 & -2.8880 & 2.2313 & -2.4811 & -0.6264 & 3.8115 & -1.8944 & -0.2857 \\ -0.6549 & -0.7873 & 0.8970 & 0.2253 & -0.3710 & -0.5790 & 0.6700 & -0.1156 & -0.0491 & 0.7646 \end{bmatrix},$$

and the data plot is shown in Figure 13.6. This plot is also obtained by rotating Figure 13.5 through $(0.8705/0.4921) * 180/\pi = 60.5061°$, the angle the eigenvector 1 makes with the x-axis so that the eigenvectors are now the coordinate axes. If we use only the first principal component, the data in terms of it are contained in the first row Y_1 of Y,

$$Y_1 = \mathbf{v}_1^T X = \mathbf{v}_1^T[\mathbf{x}_1\ \mathbf{x}_2 \cdots \mathbf{x}_n]$$

$$= [1.2093\quad -0.5886\quad 1.5122\quad -2.8880\quad 2.2313\quad -2.4811\quad -0.6264\quad 3.8115\quad -1.8944\quad -0.2857],$$

so the approximate data *above mean* in terms of old variables along the first principal component are

$$\tilde{X}_1 = \mathbf{v}_1 Y_1$$

$$= \begin{bmatrix} -0.5951 & 0.2897 & -0.7441 & 1.4212 & -1.0980 & 1.2210 & 0.3082 & -1.8756 & 0.9322 & 0.1406 \\ -1.0527 & 0.5124 & -1.3164 & 2.5141 & -1.9424 & 2.1599 & 0.5453 & -3.3181 & 1.6492 & 0.2487 \end{bmatrix}.$$

The error in this approximation is calculated as

$$\text{error1} = \sum_{i=1}^{n} \operatorname{norm}([\tilde{X}_1(1,i) - X(1,i); \tilde{X}_2(2,i) - X(2,i)], 2) = 5.1138,$$

which is taken as the sum of error of each data point. Adding the mean 6.7750 and 6.8750 to the corresponding row elements of $\tilde{X}_1$, the original data along the first principal eigenvector are

$$\tilde{A}_1 = \begin{bmatrix} 6.1799 & 7.0647 & 6.0309 & 8.1962 & 5.6770 & 7.9960 & 7.0832 & 4.8994 & 7.7072 & 6.9156 \\ 5.8223 & 7.3874 & 5.5586 & 9.3891 & 4.9326 & 9.0349 & 7.4203 & 3.5569 & 8.5242 & 7.1237 \end{bmatrix}.$$

Figure 13.9 shows the plot of college GPA against high school GPA which captures essential feature that college GPA increases with high school GPA although we lost the variation along the other principal component which is of less importance to us.

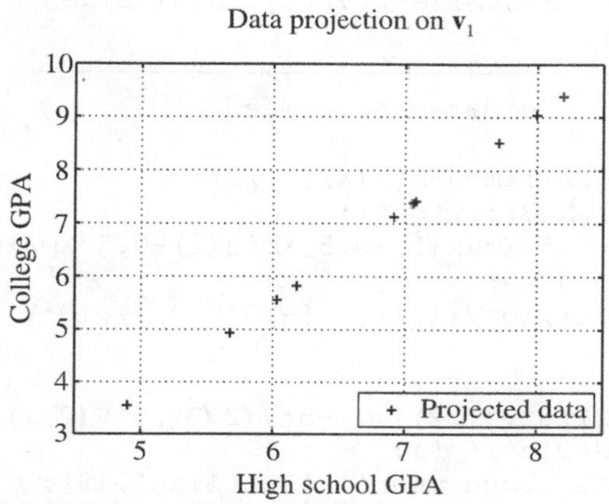

Figure 13.9 Data projection on first principal component.

The projection on the second principal component $\mathbf{v}_2$ of the data contained in X is $\tilde{X}_1 = \mathbf{v}_1 Y_1$,

$$\tilde{X}_2 = \begin{bmatrix} 0.5701 & 0.6853 & -0.7809 & -0.1962 & 0.3230 & 0.5040 & -0.5832 & 0.1006 & 0.0428 & -0.6656 \\ -0.3223 & -0.3874 & 0.4414 & 0.1109 & -0.1826 & -0.2849 & 0.3297 & -0.0569 & -0.0242 & 0.3763 \end{bmatrix},$$

where Y_2 is the second row of Y. The error in this approximation is calculated as

$$\text{error2} = \sum_{i=1}^{n} \text{norm}([\tilde{X}_2(1,i) - X(1,i); \tilde{X}_2(2,i) - X(2,i)], 2) = 17.5286.$$

The MATLAB script for the above computations is given as follows:

```
clear all; close all; clc;
A=[6.75 7.75 5.25 8.00 6.00 8.50 6.50 5.00 7.75 6.25; 5.50 7.00 6.00
9.50 4.75 8.75 7.75 3.50 8.50 7.50];
[m,n]=size(A);
MAX=max(A,[ ],2);MIN=min(A,[ ],2);
figure
scatter(A(1,:),A(2,:),'bo');
axis([MIN(1)-0.5 MAX(1)+0.5 MIN(2)-0.5 MAX(2)+0.5]);
grid
title('GPA Data Distribution', 'FontSize',18);
xlabel('High  School  GPA','FontSize',18);  ylabel('College  GPA',
'FontSize',18);
```

```
Amean = mean(A,2); X = A - repmat(Amean,1,n); % subtract row mean to
%obtain X
Z=1/sqrt(n-1)*X'; covZ=Z'*Z; % Singular value decomposition
```

```
[U,S,V] = svd(covZ);
variances=sqrt(diag(S).*diag(S)); figure;
bar(variances(1:2),0.01) % scree plot of variances
title('Eigenvalues', 'FontSize',18);
xlabel('Order','FontSize',18); ylabel('Eigenvalue',
'FontSize',18);
```

```
% Plotting X, the data above mean
figure
Xmax=max(X,[ ],2);Xmin=min(X,[ ],2);
scatter(X(1,:),X(2,:),'bo');
axis([Xmin(1)-0.5 Xmax(1)+0.5 Xmin(2)-0.5 Xmax(2)+0.5]);
axis equal; hold on;
line([-10*V(1,1),10*V(1,1)], [-10*V(2,1),10*V(2,1)], 'Color',
[1 0 0]);
hold on;
line([-5*V(1,2),5*V(1,2)], [-5*V(2,2),5*V(2,2)], 'Color',
[0 0 0],'LineStyle','-');
title('2-D data above mean', 'FontSize',18);
xlabel('x_1','FontSize',18); ylabel('x_2', 'FontSize',18);
grid
```

```
% Rotate data matrix using principal axes
Y=V'*X;
% Plot rotated data
figure;
scatter(Y(1,:), Y(2,:), 'ok');
axis([Xmin(1)-0.5 Xmax(1)+0.5 Xmin(2)-0.5 Xmax(2)+0.5]);
axis equal;
title('Rotated 2-D data distribution', 'FontSize',18);
xlabel('y_1','FontSize',18); ylabel('y_2', 'FontSize',18);
% SVD for rotated x matrix
[U1,S1,V1] = svd(cov(Y'));
% Plot rotated principal directions
hold on;
line([-10*V1(1,1),10*V1(1,1)], [-10*V1(2,1), 10*V1(2,1)], 'Color',
[1 0 0]);
hold on;
line([-5*V1(1,2), 5*V1(1,2)], [-5*V1(2,2), 5*V1(2,2)], 'Color',
[0 0 0],'LineStyle','-');
legend('2-D data','1st eigenvector dir.','2nd eigenvector dir',4);
grid
```

```
% Reproject rotated data into original space: primary eigenvector
X1tilde=V(:,1)*Y(1,:);
figure
MAX=max(X,[ ],2);MIN=min(X,[ ],2);
scatter(X(1,:),X(2,:),'ok'); hold on;
scatter(X1tilde(1,:),X1tilde(2,:),'+r'); hold on;
axis([MIN(1)-0.5 MAX(1)+0.5 MIN(2)-0.5 MAX(2)+0.5])
```

```
error1 = 0;
for i=1:n;
    line([X1tilde(1,i),X(1,i)], [X1tilde(2,i),X(2,i)]);
    error1 = error1 + norm([X1tilde(1,i)-X(1,i); X1tilde(2,i)
    X(2,i)],2);
    hold on;
end
title('Projection on the primary eigenvector', 'FontSize',18);
xlabel('x_1','FontSize',18); ylabel('x_2', 'FontSize',18);
axis equal;grid
```

```
% Reproject rotated data into original space: secondary eigenvector
X2tilde=V(:,2)*Y(2,:);
figure; scatter(X(1,:),X(2,:),'ok'); hold on;
scatter(X2tilde(1,:),X2tilde(2,:),'+r'); hold on;
axis([MIN(1)-0.5 MAX(1)+0.5 MIN(2)-0.5 MAX(2)+0.5])
error2 = 0;
for i=1:n
    line([X2tilde(1,i),X(1,i)], [X2tilde(2,i),X(2,i)]);
    error2 = error2 + norm([X2tilde(1,i)-X(1,i); X2tilde(2,i)-
    X(2,i)],2);
    hold on;
end
title('Projection on the secondary eigenvector', 'FontSize',18);
xlabel('x_1','FontSize',18); ylabel('x_2', 'FontSize',18);
axis equal;
grid
display(error1)
display(error2)
```

```
% Projection of original data on first principal component v_1
A1tilde=X1tilde+repmat(Amean,1,n);
figure
plot(A1tilde(1,:),A1tilde(2,:), '+k')
title('Data Projection on v_1', 'FontSize',18);
xlabel('High School GPA','FontSize',18); ylabel('College GPA',
'FontSize',18);
grid
```

The command repmat(A,m,n) repeats matrix A row-wise m times and column-wise n times. For example, B=repmat([1 3;2 4],2,3) is a 4×6 matrix. The above program is written for a 2-D data but can be extended to m-D data as illustrated in the following example of an image compression.

PCA method of image compression

1. We take the same black and white picture 'lena_bw.gif' studied with SVD and store in a matrix A of size 704×704 uint8) and with the colour code matrix map of size 256×3.
2. It is then converted to double precision, A=double(A).

3. Then, Amean, an $m \times 1$ vector containing the means of the corresponding rows of $m \times n$ matrix A are calculated. The mean of ith row is subtracted from each element of the corresponding row of A and the result is stored in X = A - repmat(Amean,1,n).
4. Successively, `Z=1/sqrt(n-1)*X'`, `covZ=Z'*Z` and `[U,S,V] = svd(covZ)` are computed.
5. A scree plot is used in principal components analysis to display the eigenvalues associated with a component or factor in descending order versus the number of the component or factor to visualize which components or factors explain most of the variability in the data. The ideal pattern in a scree plot is a steep curve, followed by a bend and then a flat or horizontal line. Retain those components or factors in the steep curve before the first point that starts the flat line trend. The word scree means slope of loose rock debris at the base of a steep incline or cliff. The lower part of Figure 13.10 strikingly resembles such a view.
6. The 704 singular values of covZ varies from $s_1^2 = 5.3351 \times 10^5$ to $s_{740}^2 = 2.3689 \times 10^{-10}$. With $k = 8$ the PCA compressed image is distorted but it progressively improves at $k = 19$ and excellent with $k = 30$. In Figure 13.11, the original image along with the PCA compressed images for $k = 8$, $s_8^2 = 4.3503 \times 10^4$, $k = 19$, $s_{19}^2 = 8.9423 \times 10^3$ and $k = 30$, $s_{30}^2 = 3.6104 \times 10^3$ are shown.

The following MATLAB program carries out the above steps and the results are displayed in Figures 13.10 and 13.11. This program can be extended to colour image compression by considering the red, green and blue layers separately as discussed in respect to SVD.

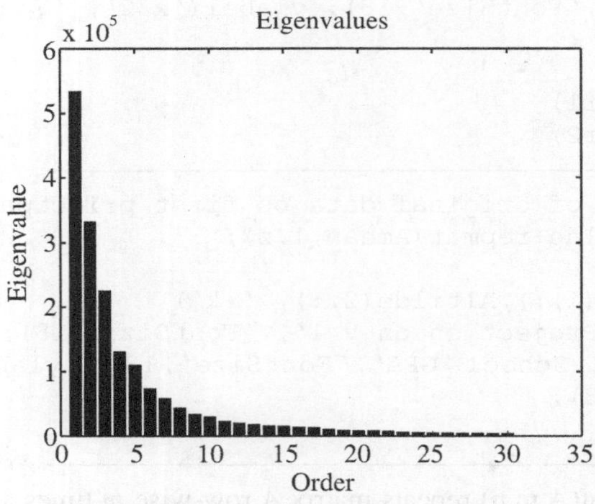

Figure 13.10 Scree plot of Lena with PCA.

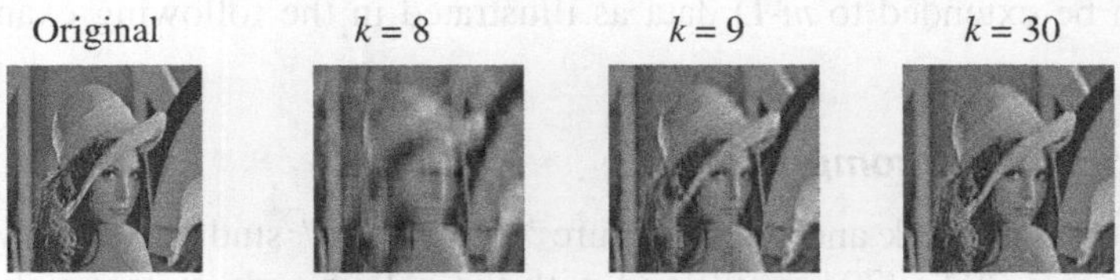

Figure 13.11 Image compression with PCA.

```
clear all; close all; clc;
[A,map] = imread('lena_bw.gif'); % load image into MATLAB
A=double(A); % convert to double precision
figure;
image(A),colormap(map); % display image
axis off, axis equal
[m n]=size(A);
Amean = mean(A,2); % compute row mean which is a vector of size m
%by 1
X = A - repmat(Amean,1,n); % subtract row mean to obtain X
Z=1/sqrt(n-1)*X'; % create matrix, Z
covZ=Z'*Z; % covariance matrix of Z
% Singular value decomposition
[U,S,V] = svd(covZ);
variances=diag(S); % compute variances
figure;
bar(variances(1:30)) % scree plot of variances
   title('Eigenvalues', 'FontSize',15);
xlabel('Order','FontSize',15); ylabel('Eigenvalue',
'FontSize',15);
% Extract three among first 30 principal components
nEigs = 8:11:30;
for iter=1:size(nEigs,2)  % dimension of cols. in (nEigs):1 for
row, 2 for col
   PCs = nEigs(iter);  % select number of principal components
   V1=V(:,1:PCs);
   Y=V1'*X; % project data onto PCs
   XX=V1*Y; % convert back to original basis
   XX=XX+repmat(Amean,1,n); % add the row means back on
   figure;
   image(XX),colormap(map),axis off; % display results
end
```

PROBLEMS

13.1 Consider the problem of solving $A\mathbf{x} = \mathbf{b}$ with 4-digit arithmetic, where

$$A = \begin{bmatrix} 10^{-4} & 1 \\ 1 & -1 \end{bmatrix}, \qquad \mathbf{b} = \begin{bmatrix} 1 \\ 0 \end{bmatrix}.$$

Show that the above problem is well-conditioned. Use Gaussian elimination with and without pivot strategy to solve it and compare the results.

13.2 For the Problem 13.1, consider

$$A = \begin{bmatrix} 51 & 50 \\ 50 & a \end{bmatrix}, \qquad \mathbf{b} = \begin{bmatrix} 2 \\ 2 \end{bmatrix}.$$

Explain graphically, by plotting the equations on x_1x_2-plane, why this problem is ill-conditioned if $a = 49$ while well-conditioned if $a = -49$. Confirm this by solving the equation with (a) a small change in **b** and (b) a small change in A.

13.3 Let $A = [a_{ij}]$ be an n-square companion matrix with $a_{i,\, i+1} = 1$ and other elements equal to zero. Show that the eigenvalues and eigenvectors of A are ill-conditioned by considering a perturbation 2^{-n} to a_{n1}.

13.4 Prove that under Householder transformation the norm of a vector is invariant.

13.5 Show that if P and Q are Householder transformations, then PQ is not so.

13.6 Show how to use algorithm (13.33) to represent $A = QR$ uniquely, where A is an $n \times m$ matrix with full column rank, Q has orthonomal columns and R is upper triangular with positive diagonal elements.

13.7 Explain why the QR-decomposition of the same A in Examples 13.9 and 13.11 are different. Obtain a unique decomposition for the same.

13.8 If $\mathbf{x}$ is a vector in $\mathbb{C}^k$, then show that there exists an elementary Hermitian matrix $P = I - 2\mathbf{w}\mathbf{w}^*$ such that $P\mathbf{x} = \alpha\, \mathbf{e}_1$, where α is a scalar.

13.9 Show how a complex matrix can be reduced to (a) a triangular and (b) a Hessenberg form, while a Hermitian matrix to (c) a tridiagonal form.

13.10 Obtain by Householder transformation an orthogonal matrix Q such that Q^TA is upper triangular, where

$$A = \begin{bmatrix} 0 & 0 \\ -1 & 2 \\ 2 & -1 \\ 2 & 0 \end{bmatrix}.$$

13.11 Find an orthogonal similarity transformation Q such that QAQ^T is upper Hessenberg by (a) Householder transformation and (b) plane rotation when

$$A = \begin{bmatrix} 0 & 1 & 2 & 1 \\ 1 & 0 & 1 & 1 \\ 2 & 7 & -1 & 1 \\ -2 & -8 & 1 & 0 \end{bmatrix}.$$

13.12 Find an orthogonal similarity transformation Q and amatrix B such that $B = QAQ^T$ is tridiagonal by (a) Householder transformation and (b) plane rotation when

$$A = \begin{bmatrix} 0 & 1 & 2 & -2 \\ 1 & 0 & -8 & -2 \\ 2 & -8 & -1 & 1 \\ -2 & -2 & 1 & 0 \end{bmatrix}.$$

13.13 Obtain the least squares solution and the minimum residual of $A\mathbf{x} = \mathbf{b}$ if

$$A = \begin{bmatrix} 1 & 2 \\ -1 & -5 \\ 1 & 5 \\ -1 & -2 \end{bmatrix}, \quad \mathbf{b} = \begin{bmatrix} 6 \\ 12 \\ -6 \\ 6 \end{bmatrix}.$$

13.14 Obtain bounds on the eigenvalues and graph the Gerschgorin discs for each of the following matrices:

$$\text{(a)} \begin{bmatrix} 2 & 0 & 1 \\ 0 & 0 & 1 \\ 1 & 0 & -2 \end{bmatrix}, \quad \text{(b)} \begin{bmatrix} -1 & 1 & 0 \\ 1 & 5 & -1 \\ 0 & -1 & 2 \end{bmatrix}, \quad \text{(c)} \begin{bmatrix} 2+i2 & 1-i & 0 \\ 0 & 2-i2 & 1+i \\ 1 & 0 & -1 \end{bmatrix}.$$

13.15 If $\lambda = (a + ib)/(c + id)$ and $c^2 + d^2 > a^2 + b^2$, show that $|\lambda|^k \to 0$ as $k \to \infty$.

13.16 Show that if $\mathbf{y}_0$ in (13.57) is selected as $[1, -1]^T$, then it has no component along the dominant eigenvector of A with $a_{22} = 2\,a_{11} = 2$, $a_{12} = 1$, $a_{21} = 0$. Explain why.

13.17 If the n-square matrix A has a dominant eigenvalue of multiplicity m with m linearly independent eigenvectors, test if the power method will converge to λ and its eigenvectors.

13.18 Find the dominant eigenvalue and eigenvector by the power method:

$$\text{(a)} \begin{bmatrix} -1 & 1 & 0 \\ 1 & 5 & -1 \\ 0 & -1 & 2 \end{bmatrix}, \quad \text{(b)} \begin{bmatrix} 23 & 15 & 0 \\ -24 & -15 & 0 \\ 3 & 2 & 1 \end{bmatrix}, \quad \text{(c)} \begin{bmatrix} -1 & -1 & 0 \\ 12 & 12 & 3 \\ -21 & -21 & -5 \end{bmatrix}.$$

Hence, determine other eigenvalues and eigenvectors by the method of deflation.

13.19 By inverse iteration, determine the complete set of eigenvalues and eigenvectors of the matrices given in Problem 13.18.

13.20 By orthogonal similarity transformation, reduce the following matrix to a real Schur form:

$$\begin{bmatrix} -1 & 1 & 0 & -4 \\ -1 & 2 & -1 & -8 \\ -1 & 2 & -1 & -4 \\ 2 & -1 & 0 & 0 \end{bmatrix}.$$

13.21 Perform one step of Jacobi transformation to each of the following symmetric matrices to show a reduction of the off-diagonal sum of squares:

$$\text{(a)} \begin{bmatrix} 0 & 1 & 0 \\ 1 & 0 & 1 \\ 0 & 1 & 0 \end{bmatrix}, \quad \text{(b)} \begin{bmatrix} 1 & 0 & 2 \\ 0 & 1 & 0 \\ 2 & 0 & 1 \end{bmatrix}, \quad \text{(c)} \begin{bmatrix} 1 & 1 & 0 \\ 1 & 1 & 1 \\ 0 & 1 & 1 \end{bmatrix}.$$

13.22 Use the method of bisection and inverse iteration to compute the complete set of eigenvalues and eigenvectors of A given by (13.79) and by the three matrices in Problem 13.21.

13.23 Obtain a complete set of eigenvalues and eigenvectors of the following matrices by QR-algorithm:

$$\text{(a)} \begin{bmatrix} 0 & 1 & -2 & 1 \\ 3 & 2 & -7 & -7 \\ 0 & 1 & -1 & -3 \\ 0 & 0 & -1 & -2 \end{bmatrix}, \quad \text{(b)} \begin{bmatrix} 0 & 3 & 0 & 0 \\ 3 & -4 & 4 & 0 \\ 0 & 4 & -5 & 1 \\ 0 & 0 & 1 & 8 \end{bmatrix}, \quad \text{(c)} \begin{bmatrix} 1 & 1 & 0 & 0 \\ 1 & 1 & 1 & 0 \\ 0 & 1 & 1 & 1 \\ 0 & 0 & 1 & 1 \end{bmatrix}.$$

13.24 Solve Problem 13.23 by LR-algorithm.

13.25 Let

$$A = \begin{bmatrix} a_{11} & a_{12} & a_{13} & \cdots & a_{1n-1} & a_{1n} \\ 1 & a_{22} & a_{23} & \cdots & a_{2n-1} & a_{2n} \\ 0 & 1 & a_{33} & \cdots & a_{3n-1} & a_{3n} \\ \vdots & \vdots & \vdots & & & \vdots \\ 0 & 0 & 0 & \cdots & a_{n-1,n-1} & a_{n-1,n} \\ 0 & 0 & 0 & \cdots & 1 & a_{nn} \end{bmatrix}$$

be the unreduced upper Hessenberg matrix. Show that its determinant Δ can be computed by

$$\Delta = a_{11}A_1 + a_{12}A_2 + \cdots + a_{1n}A_n,$$

where A_i's are the cofactors of the first row of A determined recursively by $A_n = (-1)^{n+1}$, $A_{n-1} = (-1)^n a_{nn}$, $A_{n-2} = -(a_{n-1,\,n-1} A_{n-1} + a_{n-1,\,n} A_n)$, ..., $A_{i-1} = -(a_{i1}A_i + a_{i,\,i+1}A_{i+1} + \cdots + a_{in}A_n)$, where $i = 2, 3, ..., n-1$.

13.26 If $\| A \|_2$ denotes the spectral norm of an n-squrae matrix A (see Theorem 9.4), then prove that the condition number of A: $k(A) = \sigma_1/\sigma_n$, where σ_1 and σ_n are the largest and the smallest singular value of A. If A is normal then $k(A) = |\lambda_1|/|\lambda_n|$, where $|\lambda_1|$ and $|\lambda_n|$ are the largest and the smallest modulus of eigenvalues of A.

13.27 Let $A \in \mathbb{C}^{n\times n}$ and $A = QR$ be the QR-decomposition of A. Prove that A and R have the same singular values.

13.28 *Hadamard's inequality*. Let A be an $n \times n$ nonsingular matrix. Then

$$|\det A| \le \Pi_{j=1}^{n} \| A\mathbf{e}_j \|,$$

with the equality if and only if the columns of A are orthogonal, where $\mathbf{e}_j$ is the jth unit vector.

13.29 A matrix A is said to be *strictly diagonally dominant* if $|a_{kk}| > \Sigma_{j\ne k} |a_{kj}|$, $\forall\, k$. Using the bounds of eigenvalues by Gerschgorin method prove that a strictly diagonally dominant Hermitian matrix A with positive diagonal entries is PD.

References

[1] Adby, P.R., *Applied Circuit Theory: Matrix and Computer Methods*, Ellis Horwood, Sussex, 1980.

[2] Agnew, J. and R.C. Knapp, *Linear Algebra with Applications*, 3rd ed., Brooks-Cole, London, 1988.

[3] Aitken, A.C., *Determinants and Matrices*, 6th ed., Greenwood Press, London, 1983.

[4] Anton, H., *Elementary Linear Algebra*, 5th ed., Wiley, New York, 1987.

[5] Anton, H. and C. Rorres, *Elementary Linear Algebra with Applications*, 3rd ed., Wiley, New York, 1987.

[6] Ayres, F., *Theory and Problems of Matrices*, McGraw-Hill, New York, 1982.

[7] Balsievsky, A., *Applied Matrix Algebra in the Statistical Sciences*, North-Holland, Amsterdam, 1983.

[8] Banchoff, T. and J. Wermer, *Linear Algebra through Geometry*, Springer-Verlag, Berlin, 1983.

[9] Barnett, Raymond A. and Michael R. Ziegler, *Linear Algebra: An Introduction with Applications*, Macmillan, New York, 1986.

[10] Barnett, S., *Matrices in Control Theory*, Van Nostrand Reinhold, London, 1971.

[11] Barnett, S., *Matrix Methods for Engineers and Scientists*, McGraw-Hill, New York, 1982.

[12] Barnett, S. and C. Storey, *Matrix Methods in Stability Theory*, Nelson, London, 1970.

[13] Belitsku, G.R. and Y.I. Lyubich, *Matrix Norms and their Applications*, Birkhauser, Boston, 1988.

[14] Bell, E.T., *The Development of Mathematics*, McGraw-Hill, New York, 1945.

[15] Bell, W.W., *Matrices for Scientists and Engineers*, Van Nostrand Reinhold, London, 1975.

[16] Bellman, R., *Introduction to Matrix Analysis*, 2nd ed., McGraw-Hill, New York, 1970.

[17] Ben-Israel, A. and T.N.E. Greville, *Generalized Inverses: Theory and Applications*, Wiley, New York, 1974.

[18] Bentley, D.L. and K.L. Cook, *Linear Algebra with Differential Equations*, Holt, Rinehart and Winston, New York, 1973.

[19] Berman, A. and R.J. Plemmons, *Nonnegative Matrices in the Mathematical Sciences*, Academic Press, London, 1979.

[20] Bhattacharyya, P.B., S.K. Jain and S.R. Nagpaul, *First Course in Linear Algebra*, Wiley Eastern, New Delhi, 1983.

[21] Bhirud, L.L., *Matrix Operations on the Computer*, Oxford and IBH, New Delhi, 1975.

[22] Bishop, R.E.D., G.M.L. Gladwell and S. Michaelson, *Matrix Analysis of Vibrations*, 2nd ed., Cambridge University Press, London, 1979.
[23] Bloch, N.J. and J.G. Michaels, *Linear Algebra*, McGraw-Hill, New York, 1978.
[24] Bloom, D.M., *Linear Algebra and Geometry*, Cambridge University Press, London, 1979.
[25] Blyth, T.S. and E.F. Robertson, *Essential Student Algebra*, Chapman and Hall, London, 1986.
[26] Boullion, T.L. and P.L. Odell, *Generalized Inverse Matrices*, Wiley Interscience, New York, 1971.
[27] Braae, R., *Matrix Algebra for Electrical Engineers*, Pitman, London, 1963.
[28] Brand, T. and A. Sherlock, *Matrices: Pure and Applied*, Arnold, London, 1970.
[29] Branfield, J.R. and H.W. Bell, *Matrices and their Applications*, Macmillan, London, 1970.
[30] Brickell, R., *Matrices and Vector Spaces*, Allen and Unwin, London, 1972.
[31] Brinkmann, H.W. and E.A. Klotz, *Linear Algebra and Analytic Geometry*, Addison-Wesley, Reading (Massachussetts), 1971.
[32] Brisley, W.A., *A Basis for Linear Algebra*, Wiley, New York, 1973.
[33] Bronson, R., *Matrix Methods: An Introduction*, Academic Press, London, 1970.
[34] Brouwer, W., *Matrix Methods in Optical Instrument Design, Benjamin*, New York, 1964.
[35] Campbell, H.G., *Introduction to Matrices, Vectors and Linear Programming*, 2nd ed., Prentice-Hall, Englewood Cliffs (New Jersey), 1975.
[36] Campbell, H.G., *Linear Algebra with Applications*, 2nd ed., Prentice-Hall, Englewood Cliffs (New Jersey), 1980.
[37] Chatterjee, B.C., *Linear Algebra*, Das Gupta and Co., Calcutta, 1967.
[38] Cohen, E.L. and R.N. Kesarwani, *Linear Algebra*, Kendall Hunt, Dubuque (Indiana), 1984.
[39] Coulson, A.E., *Introduction to Matrices*, Longman, London, 1965.
[40] Crouch, T., *Matrix Methods Applied to Engineering Rigid Body Mechanics*, Pergamon Press, Oxford, 1980.
[41] Cullen, C.G., *Matrices and Linear Transformation*, 2nd ed., Addison-Wesley, Reading (Massachussetts), 1972.
[42] Cullen, C.G., *Linear Algebra and Differential Equations*, Scott F., Sheffield, 1988.
[43] Curtis, C.W., *Linear Algebra: An Introductory Approach*, 4th ed., Springer-Verlag, Berlin, 1986.
[44] Curtis, M.L., *Matrix Groups*, Springer-Verlag, Berlin, 1979.
[45] Daniel, J.W., *Elementary Linear Algebra and its Applications*, Prentice-Hall, Englewood Cliffs (New Jersey), 1981.
[46] Davis, S.M. and P.R. Lawrence, *Matrix*, Addison-Wesley, Reading (Massachussetts), 1977.
[47] Deif, A.S., *Advanced Matrix Theory for Scientists and Engineers*, Abacus Press, Kent, 1982.
[48] Dickson, L.E., *Linear Algebra*, Hafner, New York, 1969.
[49] Dieudonne, J., *Linear Algebra and Geometry*, Kenhaw, London, 1983.
[50] Dixon, C., *Linear Algebra*, Van Nostrand Reinhold, New York, 1971.
[51] Dorf, R.C., *Matrix Algebra: A Programmed Introduction*, Wiley, New York, 1969.
[52] Dorfman, R., P. Samuelson and R. Solow, *Linear Programming and Economic Analysis*, McGraw-Hill Kogakusha, Tokyo, 1958.
[53] Edwards, C.H., Jr., and David E. Penney, *Elementary Linear Algebra*, Prentice-Hall, Englewood Cliffs (New Jersey), 1988.
[54] Edelen D.G. and A.D. Kydoniefs, *Introduction to Linear Algebra for Science and Engineering*, Elsevier, Amsterdam, 1976.
[55] Efimov, N.V. and E.R. Rozendorn, *Linear Algebra and Multidimensional Geometry*, Mir Publishers, Moscow, 1975.
[56] Eisenschitz, R.K., *Matrix Algebra for Physicists*, Plenum, London, 1966.
[57] Eves, Howard, *Elementary Matrix Theory*, Dover, New York, 1980.
[58] Faddeeva, V.N., *Computational Methods of Linear Algebra*, Dover, New York, 1959.

[59] Faddeev, D.K. and V.N. Faddeeva, *Computational Methods in Linear Algebra*, Freeman, San Francisco, 1963.

[60] Farkas, I. and M. Farkas, *Introduction to Linear Algebra*, Heyden, Sussex (UK), 1975.

[61] W.L. Ferrar, *Finite Matrices*, Oxford University Press, London, 1951.

[62] Finkbeiner, D.T., II, *Introduction to Matrices and Linear Transformations*, 3rd ed., Freeman, Oxford, 1978.

[63] Fletcher, T.J., *Linear Algebra through its Applications*, Van Nostrand Reinhold, London, 1972.

[64] Florey, F.G., *Elementary Linear Algebra with Applications*, Prentice-Hall, Englewood Cliffs (New Jersey), 1979.

[65] Forsythe, G.E. and C.B. Moler, *Computer Solution of Linear Algebraic Systems*, Prentice-Hall, Englewood Cliffs (New Jersey), 1967.

[66] Fox, L., *Introduction to Numerical Linear Algebra*, Oxford University Press, London, 1974.

[67] Fraleigh, B. John and Raymond A. Beauregard, *Linear Algebra*, Addison-Wesley, Reading (Massachussetts), 1987.

[68] Frazer, R.A., W. J. Duncan and A.R. Collar, *Elementary Matrices and Some Applications to Dynamics and Differential Equations*, Cambridge University Press, London, 1938.

[69] Frederiksen, N. and H. Gulliksen, *Contributions to Mathematical Psychology*, Holt, Rinehart and Winston, New York, 1964.

[70] Freeman, G.F. and N.R. Grabois, *Linear Algebra and Multivariable Calculus*, McGraw-Hill, New York, 1970.

[71] Friedberg, S.H., A.J. Insel and L.E. Spence, *Linear Algebra*, 2nd ed., Prentice-Hall, Englewood Cliffs (New Jersey), 1989.

[72] Fuller, L.E., *Basic Matrix Theory*, Prentice-Hall, Englewood Cliffs (New Jersey), 1962.

[73] Gantmacher, F.R., *The Theory of Matrices*, Vols. I and II, Chelsea, New York, 1959.

[74] Gantmacher, F.R., *Applications of the Theory of Matrices*, Interscience, New York, 1959.

[75] Garbow, B.S., *Matrix Eigensystem Routines—Eispack Guide*, Springer-Verlag, Berlin, 1988.

[76] Gelbaum, B.R., *Linear Algebra: Basics, Practice and Theory*, Elsevier, New York, 1988.

[77] Gel'fand, I.M., *Lectures on Linear Algebra*, Interscience, New York, 1961.

[78] Gere, J.M. and Weaver, W., Jr., *Matrix Algebra for Engineers*, Brooks-Cole, Pacific Grove (California), 1982.

[79] Gerrard, A. and I. Burch, *Introduction to Matrix Methods in Optics*, Wiley, New York, 1975.

[80] Gillett, P.W., *Introduction to Linear Algebra*, Houghton Mifflin, New York, 1975.

[81] Gohberg, I. et al., *Matrix Polynomials*, Academic Press, New York, 1982.

[82] Goult, R.J., *Applied Linear Algebra*, Ellis Horwood, Chichester (UK), 1978.

[83] Golub, G.H. and C.F. Van Loan, *Matrix Computations*, 2nd ed., Johns Hopkins University Press, Baltimore, (Maryland), 1989.

[84] Goult, R.J., *Computational Methods in Linear Algebra*, Stanley Thornes, Chester Springs, Pennsylvania, 1975.

[85] Graham, A., *Matrix Theory and Applications for Engineers and Mathematicians*, Halsted Press, Wiley, New York, 1980.

[86] Graybill, F.A., *Matrices with Applications in Statistics*, Wadsworth, Belmont (California), 1983.

[87] Greub, W.H., *Linear Algebra*, Vol. 23, 4th ed., Springer-Verlag, Berlin, 1982.

[88] Griffel, D., *Linear Algebra and its Applications*, Vols. 1 and 2, Wiley, New York, 1988, 1989.

[89] Grossman, S., *Elementary Linear Algebra*, Wadsworth, Belmont (California), 1980.

[90] Groetsch, Charles and J. Thomas King, *Matrix Methods and Applications: An Introduction to Linear Algebra*, Prentice-Hall, Englewood Cliffs (New Jersey), 1988.

[91] Halmos, P.R., *Finite Dimension Vector Spaces*, Van Nostrand Reinhold, London, 1958.

[92] Hammer, A.G., *Elementary Matrix Algebra for Psychologists and Social Scientists*, Pergamon Press, New York, 1971.

[93] Hartfiel and Hobbs, *Elementary Linear Algebra*, PWS-Kent Pub., Boston (Massachussetts), 1987.
[94] Helzer, G., *Applied Linear Algebra with Applications*, Little, Brown and Company, Boston (Massachussetts), 1983.
[95] Herstein, Israel N. and David J. Winter, *Matrix Theory and Linear Algebra*, Macmillan, London, 1988.
[96] Heylings, M.R., *Matrices and Transformations*, Schofield and Sims, Huddersfield (UK), 1984.
[97] Hill, Jr., R.O., *Elementary Linear Algebra*, Academic Press, London, 1986.
[98] Hoffman, K. and R. Kunze, *Linear Algebra*, 2nd ed., Prentice-Hall, Englewood Cliffs (New Jersey), 1971.
[99] Hohn, F.E., *Elementary Matrix Algebra*, 3rd ed., Macmillan, London, 1973.
[100] Horn, R.A. and C.R. Johnson, *Matrix Analysis*, Cambridge University Press, London, 1985.
[101] Hadley, G., *Linear Algebra*, Addison-Wesley, Reading (Massachussetts), 1961.
[102] Horst, P., *Matrix Algebra for Social Scientists*, Irvington, New York, 1982.
[103] Householder, A.S., *Theory of Matrices in Numerical Analysis*, Dover, New York, 1975.
[104] Irkamov, H.D., *Linear Algebra Problems Book*, Mir Publishers, Moscow, 1983.
[105] Jacob, William, *Linear Algebra*, W.H. Freeman, San Francisco, 1989.
[106] Jennings, A., *Matrix Computation for Engineers and Scientists*, Wiley, New York, 1980.
[107] Johnson, Robert S. and Thomas O. Vinson, Jr., *Elementary Linear Algebra*, Hartcourt Brace Jovanovich, San Diego (California), 1987.
[108] Johnson, L.W. and R.D. Riess, *Introduction to Linear Algebra*, Addison-Wesley, Reading (Massachussetts), 1981.
[109] Kaplansky, I., *Linear Algebra and Geometry: A Second Course*, Chelsea, New York, 1974.
[110] Knight, B. and R. Adams, *Matrices and Vectors*, Allen and Unwin, London, 1975.
[111] Kolman, B., *Elementary Linear Algebra*, 4th ed., Macmillan, London, 1986.
[112] Kolman, B., *Introductory Linear Algebra with Applications*, 3rd ed., Macmillan, London, 1984.
[113] Konvisser, M.W., *Elementary Linear Algebra with Applications*, Ardsley, New York, 1987.
[114] Kraus, Allan D., *Matrices for Engineers*, Hemisphere Publisher, New York, 1988.
[115] Krishnamurthy, V., V.P. Mainra and J.L. Arora, *An Introduction to Linear Algebra,* Affiliated East-West Press, New Delhi, 1976.
[116] Kumpel, P.G. and J.A. Thorpe, *Linear Algebra with Applications to Differential Equations*, Holt, Rinehart and Winston, New York, 1983.
[117] Lancaster, P. and M. Tismenestky, *The Theory of Matrices: With Applications*, 2nd ed., Academic Press, London, 1985.
[118] Lang, S.A., *Linear Algebra*, 3rd ed., Spring-Verlag, Berlin, 1987.
[119] Larson, Ronald E. and Brace H. Edwards, *Elementary Linear Algebra*, Heath, Lexington (Massachussetts), 1988.
[120] Leon, S.J., *Linear Algebra with Applications*, 3rd ed., Macmillan, London, 1990.
[121] Lewis, W.E. and D.G. Pryce, *The Application of Matrix Theory to Electrical Engineering*, E. and F.N. Spon London, 1965.
[122] Lightstone, A.H., *Linear Algebra*, Irvington, New York, 1969.
[123] Lipschutz, S., *Linear Algebra* (Schaum), McGraw-Hill, New York, 1968.
[124] Lowenthal, Franklin, *Linear Algebra with Linear Differential Equations*, Wiley, New York, 1975.
[125] Mac Duffee, C.C., *The Theory of Matrices*, 2nd ed., Chelsea, New York, 1956.
[126] Magid, A.R., *Applied Matrix Models: A Second Course in Linear Algebra with Computer Applications*, Wiley, New York, 1985.
[127] Mansfield, E., *Linear Algebra with Geometric Applications*, Marcel Dekker, New York, 1976.
[128] Marcus, M. and H. Minc, *Introduction to Linear Algebra*, Macmillan, New York, 1965.
[129] Martin, A.D. and V.J. Mizel, *Introduction to Linear Algebra*, McGraw-Hill, New York, 1966.
[130] Maxwell, E.A., *Algebraic Structure and Matrices*, Cambridge University Press, London, 1965.

[131] Mills, G., *Introduction to Linear Algebra*, Allen and Unwin, London, 1969.
[132] Mirsky, L., *An Introduction to Linear Algebra*, Dover, New York, 1982.
[133] Mishina, A.P. and I.V. Proskuryakov, *Higher Algebra*, Pergamon Press, Oxford, 1965.
[134] Moore, J.T., *Elements of Linear Algebra and Matrix Theory*, McGraw-Hill, New York, 1968.
[135] Moore, J.T., *Elementary Linear and Matrix Algebra*, McGraw-Hill, New York, 1972.
[136] Morris, A.O., *Linear Algebra: An Introduction*, 2nd ed., Van Nostrand Reinhold, London, 1982.
[137] Muir Thomas, *The Theory of Determinants in the Historical Order of Development*, Vols. I-IV, Dover, New York, 1906, 1911, 1920, 1923.
[138] Muir, T. and W.H. Metzler, *A Treatise on the Theory of Determinants*, Longmans, London, 1933.
[139] Munkres, J.R., *Elementary Linear Algebra*, Addison-Wesley, Reading (Massachussetts), 1964.
[140] Murdoch, *Linear Algebra for Undergraduates*, Wiley, New York, 1957.
[141] Namboodiri, N.K., *Matrix Algebra*, Sage, Newbury (California), 1984.
[142] Nering, E.D., *Linear Algebra and Matrix Theory*, 2nd ed., Wiley, New York, 1970.
[143] Noble, B. and J.W. Daniel, *Applied Linear Algebra*, 3rd ed., Prentice-Hall, Englewood Cliffs (New Jersey), 1988.
[144] O'Nan, M., *Linear Algebra*, 2nd ed., Hartcourt Brace Jovanovich, California, 1977.
[145] O'Neil, P.V., *Introduction to Linear Algebra*, Wadsworth, Belmont (California), 1979.
[146] Ortega, James M., *Matrix Theory: A Second Course*, Plenum Publishers, New York, 1987.
[147] Paige, L.J. and J. Dean Swift, *Elements of Linear Algebra*, Blaisdell, New York, 1961.
[148] Painter, R.J. and R.C. Yantis, *Elementary Matrix Algebra with Applications*, 2nd ed., Prindle, Weber and Schmidt, Boston (Massachussetts), 1977.
[149] Parlett, B.N., *The Symmetric Eigenvalue Problem*, Prentice-Hall, Englewood Cliffs (New Jersey), 1980.
[150] Pease, M.C., (Ed.), *Methods of Matrix Algebra*, Academic Press, London, 1965.
[151] Pelletier, D., *Linear Algebra: A Comprehensive Introduction*, Prentice-Hall, Englewood Cliffs (New Jersey), 1986.
[152] Perry, William L., *Elementary Linear Algebra*, McGraw-Hill, New York, 1988.
[153] Pettofrezzo, J., *Matrices and Transformations*, Dover, New York, 1978.
[154] Pillis, John de, *Linear Algebra*, Holt, Rinehart and Winston, New York, 1969.
[155] Pipes, L.A., *Matrix Methods for Engineering*, Prentice-Hall, Englewood Cliffs (New Jersey), 1963.
[156] Proskuryakov, I.V., *Problems in Linear Algebra*, Central Books, London, 1979.
[157] Pullman, N.J., (Ed.), *Matrix Theory and its Applications*, Marcel Dekker, New York, 1976.
[158] Rabenstein, A.L., *Elementary Differential Equations with Linear Algebra*, 3rd ed., Academic Press, London, 1982.
[159] Ralston, A. and P. Rabinowitz, *A First Course in Numerical Analysis*, 2nd ed., McGraw-Hill, New York, 1978.
[160] Rao, C.R. and S.K. Mitra, *Generalized Inverse of Matrices and its Applications*, Wiley, New York, 1970.
[161] Ramey, R.L. and E.J. White, *Matrices and Computers in Electronic Circuit Analysis,* McGraw-Hill, New York, 1971.
[162] Reiner, I., *Introduction to Matrix Theory and Linear Algebra*, Holt, Rinehart and Winston, New York, 1971.
[163] Rice, J.R., *Matrix Computations and Mathematical Software*, McGraw-Hill, New York, 1981.
[164] Roberts, W.A., *Linear Algebra,* Benjamin-Cummings, Redwood City (California), 1985.
[165] Rogers, A., *Matrix Methods in Urban and Regional Analysis*, Holden-Day, San Francisco, 1971.
[166] Roman, S., *Introduction to Linear Algebra with Applications*, Holt, Rinehart and Winston, New York, 1985.
[167] Rothenberg, R.L, *Linear Algebra with Computer Applications*, Wiley, New York, 1983.
[168] Rudman, Jack, *Linear Algebra*, National Learning Corporation, New York, 1988.

[169] Samelson, H., *Introduction to Linear Algebra*, Wiley, New York, 1974.
[170] Satake, I., *Linear Algebra,* Marcel Dekker, New York, 1975.
[171] Schneider, D.M., *Linear Algebra: A Concrete Approach*, Macmillan, London, 1982.
[172] Schneider, H. and G.P. Barker, *Matrices and Linear Algebra*, 2nd ed., McGraw-Hill, New York, 1972.
[173] Searle, S.R. and Hausman, W.H., *Matrix Algebra for Business and Economics*, Wiley, New York, 1970.
[174] Shields, P.C., *Elementary Linear Algebra*, 3rd ed., Worth, New York, 1980.
[175] Shilov, G.E., *Linear Spaces*, Prentice-Hall, Englewood Cliffs (New Jersey), 1961.
[176] Shilov, G.E., *Linear Algebra*, Dover, New York, 1977.
[177] Smirnov, V.I., *Linear Algebra and Group Theory*, McGraw-Hill, New York, 1961.
[178] Smith, B.T., et al., *Matrix Eigensystem Routines—Eispack Guide*, Springer-Verlag, Berlin, 1988.
[179] Smith, D.E., *History of Mathematics*, Vol. II, Dover, New York, 1958.
[180] Smith, L., *Linear Algebra*, 2nd ed., Springer-Verlag, Berlin, 1985.
[181] Stecher, Michael, *Linear Algebra*, Harper & Row, New York, 1987.
[182] Steinberg, D.I., *Computational Matrix Algebra*, McGraw-Hill, New York, 1974.
[183] Stewart, G.W., *Introduction to Matrix Computations*, Academic Press, New York, 1973.
[184] Stewart, P.M., *Introduction to Linear Algebra*, Van Nostrand Reinhold, London, 1963.
[185] Stoll, R.R., *Linear Algebra and Matrix Theory*, McGraw-Hill, New York, 1952.
[186] Stoll, R. and E.T. Wong, *Linear Algebra*, Academic Press, New York, 1968.
[187] Strang, W.G., *Linear Algebra and its Applications*, 3rd ed., Harcourt Brace Jovanovich, California, 1988.
[188] Thompson, R.C. and A. Yaqub, *Introduction to Linear Algebra*, Scott, Foresman and Company, Glenview (Illinois), 1970.
[189] Thorpe, J.A. and P.O. Kumpel, *Elementary Linear Algebra*, Holt, Rinehart and Winston, New York, 1984.
[190] Thorpe, J.A. and P.O. Sunny, *Elementary Linear Algebra*, Holt, Rinehart and Winston, New York, 1984.
[191] Todd, J., *Basic Numerical Mathematics*, Vol. 1: *Numerical Algebra*, Academic Press, New York, 1981.
[192] Tropper, A.M., *Matrix Analysis for Electrical Engineering Students*, George G. Harrap, London, 1962.
[193] Turnbull, H.W., *Theory of Determinants, Matrices and Invariants*, 2nd ed., Blackie, Glasgow, 1948.
[194] Varga, R.S., *Matrix Iterative Analysis*, Prentice-Hall, Englewood Cliffs (New Jersey), 1962.
[195] Venit, S. and W. Bishop, *Elementary Linear Algebra*, 2nd ed., Prindle, Weber and Schmidt, Boston (Massachussetts), 1985.
[196] Voyevodin, V.V., *Linear Algebra*, Mir Publishers, Moscow, 1983.
[197] Weiss, A. Von, *Matrix Analysis for Electrical Engineers*, Van Nostrand Reinhold, London, 1964.
[198] Whitelaw, T.A., *Introduction to Linear Algebra*, Blackie, Glasgow, 1983.
[199] Zelinsky, Daniel, *A First Course in Linear Algebra*, 2nd ed., Academic Press, London, 1973.
[200] Yaqub, A. and H.G. Moore, *Elementary Linear Algebra with Applications*, Addison-Wesley, Reading (Massachussetts), 1980.
[201] Yaari, M.E., *Linear Algebra for Social Sciences*, Prentice-Hall, Englewood Cliffs (New Jersey), 1971.
[202] Wilde Carroll, O., *Linear Algebra*, Addison-Wesley, Reading (Massachussetts), 1987.
[203] Williams, G., *Elementary Linear Algebra with Applications*, Auyn and Bacon, London, 1984.
[204] Willams, G., *Course in Linear Algebra*, Gordon and Beach, London, 1972.
[205] Wilkinson, J.H., *The Algebraic Eigenvalue Problem*, Oxford University Press, London, 1965.
[206] Wilkinson, J.H. and C. Reinsch, *Handbook for Automatic Computation*, Vol. II: *Linear Algebra*, Springer-Verlag, Berlin, 1971.

List of Corollaries, Definitions, Examples, Lemmas, Remarks, Theorems

Answers

CHAPTER 1

1.1 (a) $PA = \begin{bmatrix} 1 & 0 & 0 \\ 0 & 2 & 2 \\ 0 & 4 & -4 \end{bmatrix}$; (b) $AQ = \begin{bmatrix} 1 & 0 & 0 \\ 0 & 1 & 2 \\ 0 & 1 & -2 \end{bmatrix}$;

(c) $PQ = I$; (d), (e), (f) $PAQ = \text{diag}\,[1, 2, 4]$; (g), (h), (i) 7.

1.2 (a) $AB = \begin{bmatrix} 3 & 0 \\ 7 & 3 \end{bmatrix}$; (b) $BA = \begin{bmatrix} 1 & 0 & 2 \\ 11 & 3 & 10 \\ 1 & 0 & 2 \end{bmatrix}$; (c), (d) 6.

1.3 (a) $\begin{bmatrix} 2 & \lambda^2 \\ \lambda & 2+\lambda \\ (1+\lambda)^3 & -1 \end{bmatrix}$; (b) $\begin{bmatrix} 1 & 10i \\ i-1 & 1 \end{bmatrix}$.

1.5 (a) $x = -3$, $y = 0$; (b) $x = y = \pm 2$.

1.6 (a) $P = \begin{bmatrix} 1 & 0 & 0 \\ -a_{21} & 1 & 0 \\ -a_{31} + a_{32}a_{21} & -a_{32} & 1 \end{bmatrix}$; (b) $P = \begin{bmatrix} 1 & -a_{12} & a_{23}a_{12} - a_{13} \\ 0 & 1 & -a_{23} \\ 0 & 0 & 1 \end{bmatrix}$.

1.7 $\mathbf{x}\mathbf{y}^T = x_1y_1 + x_2y_2 + \cdots + x_ny_n$,

$$\mathbf{x}^T\mathbf{y} = \begin{bmatrix} x_1y_1 & x_1y_2 & \cdots & x_1y_n \\ x_2y_1 & x_2y_2 & \cdots & x_2y_n \\ \vdots & \vdots & & \vdots \\ x_ny_1 & x_ny_2 & \cdots & x_ny_n \end{bmatrix}.$$

1.14 $(A + B + \cdots + C)^T = A^T + B^T + \cdots + C^T$; $(AB \cdots C)^T = C^T \cdots B^T A^T$.

1.20 (a) $[a_{i1}\ a_{i2} \cdots a_{in}]$; (b) $[a_{1i}\ a_{2i} \cdots a_{ni}]'$; (c) a_{ii}; (d) $0 : i \neq j$; $1 : i = j$; (e) 1 in the ijth position, rest are zero; (f) a_{ij}.

1.22 (a) E_{ij} if $i = j$; a null matrix if $i \neq j$; (b) E_{kl} if $j = k$; a null matrix if $j \neq k$; (c) $\mathbf{e}_j$; (d) a zero-vector; (e) $\mathbf{e}_j^T$ if $k = i$; a zero-vector if $k \neq i$.

1.23 (a) $\dot{A} = \begin{bmatrix} -\sin t & \cos t \\ -\cos t & -\sin t \end{bmatrix}$, $\dot{B} = \begin{bmatrix} -\sin t & \cos t \\ \cos t & \sin t \end{bmatrix}$,

$$\dot{P} = \begin{bmatrix} -ae^{-at} & 1 \\ -t^{-2} & ae^{at} \end{bmatrix}, \quad \dot{Q} = \begin{bmatrix} ae^{at} & -1 \\ t^{-2} & -ae^{-at} \end{bmatrix};$$

(b) $D(AB) = D(PQ)$ = a null matrix;

(c) $(DA)B = \begin{bmatrix} 0 & -1 \\ -1 & 0 \end{bmatrix}, \quad B(DA) = (DB)A = \begin{bmatrix} -\sin 2t & \cos 2t \\ \cos 2t & \sin 2t \end{bmatrix};$

(d) $D(AB)^T$ = a null matrix,

$$D(PQ^T) = \begin{bmatrix} -2t & (1 + t^{-2} - at + at^{-1})\, e^{-at} \\ -e^{at}(1 + e^{-2} - at^{-1} + at) & 2t^{-3} \end{bmatrix}.$$

1.24 (a) $\begin{bmatrix} 0 & 2 \\ -2 & 0 \end{bmatrix}$; (b) $\begin{bmatrix} \frac{1}{a}(e^{-a/2} - e^{-a}) & \frac{3}{8} \\ \ln 2 & \frac{1}{a}\left(e^{a} - e^{a/2}\right) \end{bmatrix}$.

1.34 (a) 2; (b) – (d) n.

CHAPTER 2

2.1 (a) 9; (b) 12; (c) 21; (d) $n(n-1)/2$; (e) $n(n-1)/2$; (f) $n(n+1)/2$; (g) $n(3n+1)/2$; (h) $3n(n-1)/2$.

2.2 (a) $i = 3, k = 7$; (b) $i = 1, k = 5$; (c) $i = 2, k = 4$; (d) $i = 6, k = 2$; (e) $i = 3, k = 7$; (f) $i = 2, k = 6$.

2.3 (a) $i = 2, k = 4$; (b) $i = 5, k = 4$; (c) $i = 3, k = 4$; (d) $i = 4, k = 3$; (e) $i = 4, k = 7$; (f) $i = 4, k = 7$.

2.4 (a) $c(123) = 0$, + ; (b) $c(231) = 2$, + ; (c) $c(312) = 2$, + ; (d) $c(321) = 3$, – ; (e) $c(132) = 1$, – ; (f) $c(213) = 1$, – .

2.5 (a) $r(231) = 2$, + ; (b) $r(312) = 2$, + ; (c) $r(321) = 3$, – ; (d) $r(132) = 1$, – ; (e) $r(213) = 1$, – ; (f) $r(123) = 0$, + .

2.6 The following figure shows the column expansion of a third order determinant.

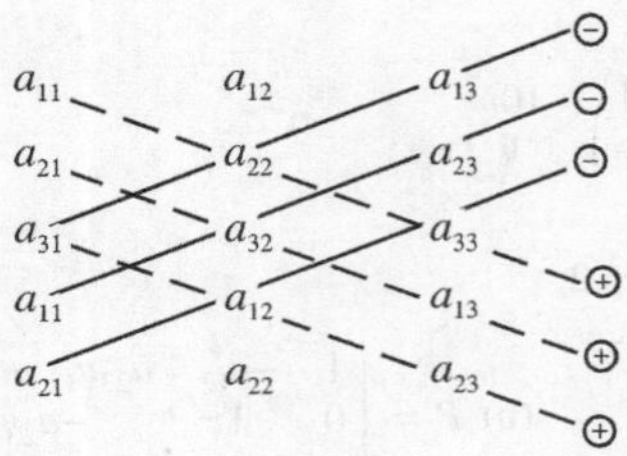

2.7 (a) $c(12 \ldots n) = 0$, + ; (b) $c(n, n-1, \ldots, 2\ 1) = n(n-1)/2$, sign depends on n.

2.8 (a) Yes, – ; (b) No; (c) Yes, – ; (d) Yes, + ; (e) No; (f) Yes, + .

2.9 (a) 1; (b) $\cos(\alpha - \beta)$; (c) 1; (d) 0; (e) 1; (f) $-\omega^2$; (g) –1.

2.11 (a) –18; (b) $3\,abc - a^3 - b^3 - c^3$; (c) –3; (d) $2\,(i + 1)$; (e) 70.

2.16 (a) 0; (b) $3\sqrt{3}/4$; (c) 27; (d) 120; (e) 32.

2.24 (a) $d_n = 2d_{n-1}\cos\theta - d_{n-2}$, $d_n = \det A$.

2.27 (a) $\det A_1 \det A_2 \ldots \det A_n$.

CHAPTER 3

3.1 (a) ns; (b) s; (c) ns; (d) s; (e) ns; (f) s.

3.2 (a) $\begin{bmatrix} 8 & -12 & 20 \\ -4 & 3 & -5 \\ 0 & 2 & -6 \end{bmatrix}$; (b) $\begin{bmatrix} -2 & 4 & -2 \\ -2 & 4 & -2 \\ 6 & -12 & 6 \end{bmatrix}$;

(c) $\begin{bmatrix} 1 & -2 & 2+2i \\ -i & 2i & 1+i \\ 1-2i & -1+i & 1+3i \end{bmatrix}$; (d) $\begin{bmatrix} 0 & 0 & 0 \\ 0 & \sigma^2-1 & \sigma-\sigma^3 \\ 0 & \sigma-\sigma^3 & \sigma^4-\sigma^2 \end{bmatrix}$;

(e) $\begin{bmatrix} 1 & -\sigma & 0 \\ 0 & \sigma & -1 \\ -1 & 1 & 1 \end{bmatrix}$; (f) $\begin{bmatrix} 1-\omega^2 & \omega-1 & \omega-1 \\ 0 & 0 & 0 \\ 1-\omega & \omega-\omega^2 & \omega-\omega^2 \end{bmatrix}$.

3.3 Same as in Problem 3.2.

3.5 (a) $-\dfrac{1}{8}\begin{bmatrix} 8 & -12 & 20 \\ -4 & 3 & -5 \\ 0 & 2 & -6 \end{bmatrix}$; (c) $\dfrac{1}{10}\begin{bmatrix} -1-3i & 2+6i & 4-8i \\ -3+i & 6-2i & 2-4i \\ -7-i & 4+2i & 8-6i \end{bmatrix}$;

(e) Same as in Problem 3.2(e).

3.6 (a) $x = 6, y = 3, z = -4$; (b) $x = -4, y = 7, z = -3$;
(c) $x = 2, y = -2, z = 9$; (d) $x = z = 1, y = 0$.

3.10 (a) $\begin{bmatrix} 1 & 0 \\ 0 & 1 \\ 1 & 2 \\ 6 & 11 \end{bmatrix}$; (b) $\begin{bmatrix} 15 & 19 & 13 \\ 5 & 10 & 11 \\ 5 & 10 & 9 \end{bmatrix}$.

3.13 (a) $\dfrac{2}{3}\begin{bmatrix} \cos\theta & \cos(\theta-120°) & \cos(\theta+120°) \\ -\sin\theta & -\sin(\theta-120°) & -\sin(\theta+120°) \\ \dfrac{1}{2} & \dfrac{1}{2} & \dfrac{1}{2} \end{bmatrix}$;

(b) $\dfrac{1}{15}\begin{bmatrix} 0 & -6 & 3 \\ -10 & 11 & 2 \\ 5 & 2 & -1 \end{bmatrix}$, (c) $\dfrac{1}{55}\begin{bmatrix} -5 & 20 & -5 & 5 \\ 20 & 8 & -13 & 2 \\ -5 & -13 & 28 & -17 \\ 5 & 2 & -17 & 28 \end{bmatrix}$;

(d) $\dfrac{1}{4}\begin{bmatrix} 8 & 3 & 2 & -1 \\ -8 & -2 & -4 & 2 \\ 12 & 9 & 10 & -3 \\ -4 & -3 & -2 & 1 \end{bmatrix}$; (e) $\dfrac{1}{37}\begin{bmatrix} -1 & 3 & 2 & 5 \\ 3 & 28 & -6 & -15 \\ 2 & -6 & 33 & -10 \\ 5 & -15 & -10 & 12 \end{bmatrix}$;

(f) $\begin{bmatrix} a^{-1} & -a^{-1}b^{-1} & -2a^{-1}c^{-1} & 2a^{-1}d^{-1}(b^{-1}+c^{-1}) \\ 0 & b^{-1} & 0 & -2b^{-1}d^{-1} \\ 0 & 0 & c^{-1} & -c^{-1}d^{-1} \\ 0 & 0 & 0 & d^{-1} \end{bmatrix}$; (g) $\dfrac{1}{4}\begin{bmatrix} 4 & -4 & 1 \\ -4 & 8 & -3 \\ 0 & 0 & 1 \end{bmatrix}$;

(h) $\dfrac{1}{2}\begin{bmatrix} 2 & -4 & 0 \\ -2 & 6 & 0 \\ -3 & 5 & 1 \end{bmatrix}$; (i) $\begin{bmatrix} -16 & 11 & -10 & 12 \\ -5 & 3 & -3 & 4 \\ -1 & 2 & -2 & 2 \\ 1 & -1 & 1 & -1 \end{bmatrix}$.

3.20 Any column of A^{-1} is a scalar multiple of the corresponding column of B^{-1}.

3.23 (a) $A^{-1} = iI$ or $-iI$; (b) $A^{-1} = \dfrac{1}{2}(1-i\sqrt{3})I$ or $\dfrac{1}{2}(1+i\sqrt{3})I$.

3.33 (a) $\begin{bmatrix} 2 & -1 \\ -1 & 1 \end{bmatrix}$, (b) $\begin{bmatrix} 3 & -2 & 1 \\ -2 & 2 & -1 \\ 1 & -1 & 1 \end{bmatrix}$, (c) $\begin{bmatrix} 4 & -3 & 2 & -1 \\ -3 & 3 & -2 & 1 \\ 2 & -2 & 2 & -1 \\ -1 & 1 & -1 & 1 \end{bmatrix}$,

(d) $\begin{bmatrix} n & -(n-1) & n-2 & \cdots & -2 & 1 \\ -(n-1) & n-1 & -(n-2) & \cdots & 2 & -1 \\ n-2 & -(n-2) & n-2 & \cdots & -2 & 1 \\ \vdots & \vdots & \vdots & & \vdots & \vdots \\ 1 & -1 & 1 & \cdots & -1 & 1 \end{bmatrix}$, n odd; $\begin{bmatrix} n & -(n-1) & n-2 & \cdots & 2 & -1 \\ -(n-1) & n-1 & -(n-2) & \cdots & -2 & 1 \\ n-2 & -(n-2) & n-2 & \cdots & 2 & -1 \\ \vdots & \vdots & \vdots & & \vdots & \vdots \\ -1 & 1 & -1 & \cdots & 1 & -1 \end{bmatrix}$, n even.

CHAPTER 4

4.1 (a) $\begin{bmatrix} 3 & 3 & 1 \\ 5 & 3 & 2 \\ 1 & 2 & -1 \end{bmatrix}$, $\begin{bmatrix} 3 & 3 \\ 5 & 3 \end{bmatrix}$, $\begin{bmatrix} 3 & 1 \\ 5 & 2 \end{bmatrix}$, $\begin{bmatrix} 3 & 1 \\ 3 & 2 \end{bmatrix}$, $\begin{bmatrix} 5 & 3 \\ 1 & 2 \end{bmatrix}$, $\begin{bmatrix} 5 & 2 \\ 1 & -1 \end{bmatrix}$, $\begin{bmatrix} 3 & 2 \\ 2 & -1 \end{bmatrix}$, $\begin{bmatrix} 3 & 3 \\ 1 & 2 \end{bmatrix}$, $\begin{bmatrix} 3 & 1 \\ 1 & -1 \end{bmatrix}$, $\begin{bmatrix} 3 & 1 \\ 2 & -1 \end{bmatrix}$,

[3], [3], [1], [5], [3], [2], [1], [2], [–1].

4.2 (a) 3; (b) 2; (c) 2; (d) 2.

4.3 (a) $\begin{bmatrix} 1 & 0 & 0 \\ 0 & 1 & 0 \\ -1 & 0 & 1 \end{bmatrix}\begin{bmatrix} 1 & 0 & 0 \\ 0 & 0 & 1 \\ 0 & 1 & 0 \end{bmatrix}$; (b) $\begin{bmatrix} 1 & 0 & 0 \\ 0 & 1 & 0 \\ -2 & 0 & 1 \end{bmatrix}\begin{bmatrix} 1 & 0 & 0 \\ -2 & 1 & 0 \\ 0 & 0 & 1 \end{bmatrix}$;

(c) $\begin{bmatrix} 1 & 0 & 0 & 0 \\ 0 & 0 & 1 & 0 \\ 0 & 1 & 0 & 0 \\ 0 & 0 & 0 & 1 \end{bmatrix}\begin{bmatrix} 1 & 0 & 0 & 0 \\ 1 & 1 & 0 & 0 \\ 0 & 0 & 1 & 0 \\ 0 & 0 & 0 & 1 \end{bmatrix}\begin{bmatrix} 1 & 0 & 0 & 0 \\ 0 & 1 & 0 & 0 \\ -2 & 0 & 1 & 0 \\ 0 & 0 & 0 & 1 \end{bmatrix}\begin{bmatrix} 1 & 0 & 0 & 0 \\ 0 & 1 & 0 & 0 \\ 0 & 0 & 1 & 0 \\ 0 & 0 & -2 & 1 \end{bmatrix}\begin{bmatrix} 0 & 0 & 1 & 0 \\ 0 & 1 & 0 & 0 \\ 1 & 0 & 0 & 0 \\ 0 & 0 & 0 & 1 \end{bmatrix}$;

(d) $\begin{bmatrix} 1 & 0 & 0 & 0 \\ 0 & 1 & 0 & 0 \\ 0 & 0 & 1 & 0 \\ 0 & -2 & 0 & 1 \end{bmatrix}\begin{bmatrix} 1 & 0 & 0 & 0 \\ 0 & 1 & 0 & 0 \\ -1 & 0 & 1 & 0 \\ 0 & 0 & 0 & 1 \end{bmatrix}\begin{bmatrix} 1 & -1 & 0 & 0 \\ 0 & 1 & 0 & 0 \\ 0 & 0 & 1 & 0 \\ 0 & 0 & 0 & 1 \end{bmatrix}\begin{bmatrix} 0 & 1 & 0 & 0 \\ 1 & 0 & 0 & 0 \\ 0 & 0 & 1 & 0 \\ 0 & 0 & 0 & 1 \end{bmatrix}$.

4.4 (a) $\begin{bmatrix} 1 & 0 & -1 \\ 0 & 1 & 0 \\ 0 & 0 & 1 \end{bmatrix}\begin{bmatrix} 1 & 0 & 0 \\ 0 & 1 & 0 \\ 0 & -1 & 1 \end{bmatrix}\begin{bmatrix} 1 & 0 & 0 \\ -1 & 1 & 0 \\ 0 & 0 & 1 \end{bmatrix}$; (b) $\begin{bmatrix} 1 & -1 & 0 & 0 \\ 0 & 1 & 0 & 0 \\ 0 & 0 & 1 & 0 \\ 0 & 0 & 0 & 1 \end{bmatrix}\begin{bmatrix} 2 & 0 & 0 & 0 \\ 0 & 1 & 0 & 0 \\ 0 & 0 & 1 & 0 \\ 0 & 0 & 0 & 1 \end{bmatrix}$;

(c) $\begin{bmatrix} 1 & 0 & -1 & 0 \\ 0 & 1 & 0 & 0 \\ 0 & 0 & 1 & 0 \\ 0 & 0 & 0 & 1 \end{bmatrix}\begin{bmatrix} 1 & 0 & 0 & 0 \\ 0 & 1 & 0 & 0 \\ 0 & 0 & 2 & 0 \\ 0 & 0 & 0 & 1 \end{bmatrix}$.

4.5 (a) $\begin{bmatrix} 1 & 0 & 0 \\ 0 & 0 & 1 \\ 1 & 1 & 0 \end{bmatrix}$; (b) $\begin{bmatrix} 2 & 0 & 0 \\ 0 & 1 & 0 \\ 0 & -2 & 1 \end{bmatrix}$; (c) $\begin{bmatrix} 1 & 0 & 0 \\ 0 & 0 & 2 \\ -2 & -1 & 2 \end{bmatrix}$.

4.6 (a) $\begin{bmatrix} 1 & 0 & 0 \\ 0 & 0 & 1 \\ -1 & 1 & 0 \end{bmatrix}$; (b) $\begin{bmatrix} 1/2 & 0 & 0 \\ 0 & 1 & 0 \\ 0 & 1/2 & 1 \end{bmatrix}$; (c) $\begin{bmatrix} 1 & 0 & 0 \\ 0 & 0 & -1/2 \\ 2 & 1 & 1 \end{bmatrix}$.

4.7 (a) $\begin{bmatrix} 0 & 0 & 1 \\ -1/3 & 0 & 1 \\ 1 & -3/7 & -6/7 \end{bmatrix}$; $\begin{bmatrix} 1 & -2 & -5/3 \\ 0 & 1 & 4/3 \\ 0 & 0 & 1 \end{bmatrix}$; (b) $\begin{bmatrix} 1 & 0 & 0 \\ 2/7 & -1/7 & 0 \\ -11/7 & 2/7 & 1 \end{bmatrix}$; $\begin{bmatrix} 1 & -2 & -2 & -3 \\ 0 & 1 & 1 & 1 \\ 0 & 0 & 1 & 0 \\ 0 & 0 & 0 & 1 \end{bmatrix}$;

(c) $\begin{bmatrix} 0 & 1 & 0 & 0 \\ 0 & -1 & 1 & 0 \\ 1 & 3 & -5 & 0 \\ 0 & 5 & -8 & 1 \end{bmatrix}$; $\begin{bmatrix} 1 & 1 & -2 & 1 \\ 0 & 1 & 1 & -1 \\ 0 & 0 & 1 & 0 \\ 0 & 0 & 0 & 1 \end{bmatrix}$; (d) $\begin{bmatrix} 0 & 0 & 1 & 0 \\ 0 & -1/10 & 3/10 & 0 \\ 1 & -1/2 & -1/2 & 0 \\ 0 & -17/10 & 1/10 & 1 \end{bmatrix}$; $\begin{bmatrix} 1 & -4 \\ 0 & 1 \end{bmatrix}$.

4.8 (a) See Problem 3.13.

4.9 (a) $\begin{bmatrix}1 & 0\\1 & 1\end{bmatrix}\begin{bmatrix}1 & 1\\0 & 1\end{bmatrix}$; (b) $\begin{bmatrix}3 & 0\\0 & 1\end{bmatrix}\begin{bmatrix}1 & 0\\4 & 1\end{bmatrix}\begin{bmatrix}1 & 0\\0 & -6\end{bmatrix}\begin{bmatrix}1 & 2\\0 & 1\end{bmatrix}$;

(c) $\begin{bmatrix}0 & 0 & 1\\0 & 1 & 0\\1 & 0 & 0\end{bmatrix}\begin{bmatrix}1 & 0 & 0\\0 & 1 & 0\\2 & 0 & 1\end{bmatrix}\begin{bmatrix}1 & 0 & 0\\0 & 1 & 0\\0 & 0 & -3\end{bmatrix}\begin{bmatrix}1 & 0 & 0\\0 & 1 & 2\\0 & 0 & 1\end{bmatrix}\begin{bmatrix}1 & 0 & 3\\0 & 1 & 0\\0 & 0 & 1\end{bmatrix}$;

(d) $\begin{bmatrix}1 & 0 & 0\\3 & 1 & 0\\0 & 0 & 1\end{bmatrix}\begin{bmatrix}1 & 0 & 0\\0 & 1 & 0\\1 & 0 & 1\end{bmatrix}\begin{bmatrix}1 & 0 & 0\\0 & 1 & 0\\0 & 1 & 1\end{bmatrix}\begin{bmatrix}1 & 0 & 0\\0 & 1 & 0\\0 & 0 & 2\end{bmatrix}\begin{bmatrix}1 & 0 & 0\\0 & 1 & 1\\0 & 0 & 1\end{bmatrix}\begin{bmatrix}1 & 0 & 0\\0 & -3 & 0\\0 & 0 & 1\end{bmatrix}\begin{bmatrix}1 & 3 & 0\\0 & 1 & 0\\0 & 0 & 1\end{bmatrix}\begin{bmatrix}0 & 1 & 0\\1 & 0 & 0\\0 & 0 & 1\end{bmatrix}$.

4.10 (a) $\begin{bmatrix}1 & 3 & 0 & -3\\0 & 0 & 1 & 2\\0 & 0 & 0 & 0\end{bmatrix}$; (b) $\begin{bmatrix}1 & 0 & -3-2i\\0 & 1 & 2-i\\0 & 0 & 0\end{bmatrix}$; (c) $\begin{bmatrix}1 & 3 & 1 & 0\\0 & 0 & 0 & 1\\0 & 0 & 0 & 0\end{bmatrix}$.

4.11 (a) Left inverse, $\begin{bmatrix}\alpha & 2+4\alpha & -1-3\alpha\\\beta & -3+4\beta & 2-3\beta\end{bmatrix}$; (b) Left inverse, $\begin{bmatrix}2+3.5\alpha & \alpha & -1-2.5\alpha\\-2.5+3.5\beta & \beta & 1.5-2.5\beta\end{bmatrix}$;

(c) Inverse does not exist; (d) $\begin{bmatrix}1-2\alpha & -1/2-2\gamma\\\alpha & \gamma\\0 & 1/2\\\beta & \delta\end{bmatrix}$, right inverse.

CHAPTER 5

5.5 (a) For all t; (b) $t \neq (2n+1)\pi/4$, n integer; (c) for all t.

5.8 Other vectors in the basis are (a) $\mathbf{e}_2$, $\mathbf{e}_3$; (b) [1 –1 1]′; (c) [1, –i, 0]′.

5.9 (a) $-\mathbf{x}_1 + \mathbf{x}_2 + 4\mathbf{x}_3$, (–1, 1, 4); (b) $(3/2)\,\mathbf{x}_1 - \mathbf{x}_2 - (7/2)\,\mathbf{x}_3$, (3/2, –1, –7/2);
(c) $6\mathbf{x}_1 - 5\mathbf{x}_2 - 6\mathbf{x}_3$, (6, –5, –6); (d) $5\mathbf{x}_1 - 4\mathbf{x}_2 + (2i - 9)\,\mathbf{x}_3$, (5, –4, $2i$ – 9);
(e) $-(1/2)\,\mathbf{x}_1 + (3/2)\,\mathbf{x}_3$, (–1/2, 0, 3/2).

5.12 (a) 1; (b) 2; (c) 2; (d) 3; (e) 3.

5.13 (i) (a) 6; (b) 6; (c) $\sqrt{10}$, 2; (d) $12i$; (e) $\sqrt{2}$.
(ii) (a) $-1 + 2i$; (b) $-1 - 2i$; (c) $\sqrt{3}, \sqrt{3}$; (d) $4 + 2i$; (e) $2\sqrt{2}$.
(iii) (a) $2i$; (b) $-2i$; (c) $\sqrt{6}$, 2; (d) $2(-1 + i)$; (e) $\sqrt{10}$,

5.14 (i) [1, –1, 3]′; (ii) [1 + i, –1 + i, 0]′; (iii) [2, 1, –5]′.

5.15 (a) (2/3)**x**, –(1/3)[1, –4, 1]′;
(b) (3/2)**x**, (1/2)[–1, –2, 4, 5]′;
(c) (1/2)[1 + i, 1, –i]′, (1/2)[1 – i, 1 + 2i, i]′;
(d) [0, 0, 0, 0]′, [1 + i, 2 + i, –5, –6]′;
(e) [0, 0, 0, 0]′, [2, –3, 0 2]′.

5.16 (a) (1/3)[2, 1, 2]′, $(1/3\sqrt{2})$[–1, 4, –1]′;
(b) (1/2)[i – 1, i, 1]′, $(1/2\sqrt{2})$[1 – i, 1 + 2i, i]′;
(c) $(1/\sqrt{6})$ [1, 2, 0, 1]′, $(1/\sqrt{46})$[–1, –2, 4, 5]′, $(1/\sqrt{2990})$[44, –27, –15, 10]′;
(d) $(1/\sqrt{6})$[1, 2, –1, 0]′, $(1/\sqrt{66})$[1, 2, 5, –6]′;
(e) (1/3)[1, 2 + i, i, 1 – i]′, $(1/\sqrt{68})$[1 + i, 2 + i, –5, –6]′;
(f) Dependent.

5.18 (a) [1, 0, 0]′, [0, 1, 0]′, [0, 0, 1]′;
(b) $(1/\sqrt{2})$[1, 0, 1]′, $(1/\sqrt{6})$[1, 2, –1]′, $(1/\sqrt{3})$[–1, 1, 1]′;
(c) $(1/\sqrt{3})$[1, 1, 1]′, $(1/\sqrt{2})$[1, 0, –1]′, $(1/\sqrt{6})$[1, –2, 1]′;
(d) $(1/\sqrt{6})$[1, 2, 1]′, $(1/\sqrt{3})$[1, –1, 1]′, $(1/\sqrt{2})$[1, 0, –1]′;
(e) $(1/\sqrt{2})$[i, 0, 1]′, $(1/\sqrt{6})$[1, –2i, i]′, $(1/\sqrt{15})$[2 –i, 1 + 2i, 1 + 2i]′;
(f) (1/3)[2i, 1, –2i]′, $(1/\sqrt{5})$[1, 2i, 0]′, $((1-4i)/\sqrt{765})$[4, –2i, 5]′.

5.19 (a) [1 0 0 0]′, [0 1 0 0]′, [0 0 1 0]′, [0 0 0 1]′;
(b) [1 –1 1 –1]′, [3 3 –1 –1]′, [–1 1 1 –1]′, [1 1 3 3]′;

(c) [2 –1 –1 0]′, [1 2 0 1]′, [–2 –5 1 12]′, [5 –2 12 –1]′;
(d) [1 2 0 0]′, [2 –1 5 0]′, [2 –1 –1 6]′, [–2 1 1 1]′.

5.20 (a) $(1/\sqrt{3})$ [1, 1, 1]′, $(1/\sqrt{2})$ [–1, 1, 0]′, $(1/\sqrt{6})$ [–1, –1, 2]′;
(b) $(1/\sqrt{3})$ [i + 1, 1, 0]′, $(1/\sqrt{15})$ [i – 1, –2i, 3]′, $(1/\sqrt{5})$ [1, –1, + i, 1 + i]′;
(c) $(1/\sqrt{6})$ [2, 1, –1]′, $(1/\sqrt{10})$ [0, 2+i, 2+i]′, $(1/\sqrt{3})$ [1, –1, 1]′.

5.21 (a) 2, 4, –4; (b) 2, – 2, 9; (c) inconsistent; (d) $\alpha + 1, -2\alpha + 1, \alpha$, α is scalar; (e) –2, 3, –1; (f) $2 - \alpha, 2 + \alpha, \alpha$; α is scalar; (g) 1, –1, 4; (h) 1, –3, 0; (i) inconsistent; (j) inconsistent; (k) 1, 0, –3, 0; (l) $4a - 3b - 3, a, b, 2 - 3a + 2b$, a and b are scalars; (m) 27/19, 77/19, 22/19, –18/19; (n) 1/3, –13/3, 19/3; (o) –3/2, 1, 3/2, –3.

5.23 (a) 1, –2; (b) 0; (c) 1, –1.

5.26 (a) 2, 2; (b) 3, 1; (c) 2, 2.

5.27 (a) [1 3 1]′, [2 0 –1]′, [4 1 0 –3]′, [–3 0 1 2]′; (b) [1 2 1]′, [2 –3 4]′, [1 7 –2]′, [3 1 0 1]′;
(c) [2 1 1 3]′, [3 –1 0 5]′, [–2 1 1 0]′, [1 –1 0 1]′.

5.33 (a) $\mathbf{x}_3 = 3\mathbf{x}_1 + \mathbf{x}_2$, $\mathbf{x}_4 = -4\mathbf{x}_1 + 3\mathbf{x}_2$; (b) $\mathbf{x}_3 = 2\mathbf{x}_1 - \mathbf{x}_2$, $\mathbf{x}_4 = 3\mathbf{x}_1 - 2\mathbf{x}_2$;
(c) $\mathbf{x}_3 = \mathbf{x}_1 - \mathbf{x}_2$, $\mathbf{x}_4 = 2\mathbf{x}_1 - 3\mathbf{x}_2$; (d) $\mathbf{x}_3 = -2\mathbf{x}_1 + \mathbf{x}_2$, $\mathbf{x}_4 = 3\mathbf{x}_1 - \mathbf{x}_2$.

5.34 (a) $\mathbf{x}_2 = -\mathbf{x}_1$, $\mathbf{x}_3 = 2\mathbf{x}_1$; (b) $\mathbf{x}_3 = -\mathbf{x}_1 + \mathbf{x}_2$;
(c) $(-2 + 3i)\,\mathbf{x}_1 - (1 + 2i)\,\mathbf{x}_2 + i\mathbf{x}_3 = \boldsymbol{\theta}$; (d) $\mathbf{x}_1 + \mathbf{x}_2 = \boldsymbol{\theta}$.

5.37 $\mathbf{x}_1, \mathbf{x}_2, \mathbf{x}_3$ do not form a basis of $\mathbb{R}^3$.

5.40 For $n = 3$; $\begin{bmatrix} 0 & 1 & 0 \\ 1 & 0 & 0 \\ 0 & 0 & 0 \end{bmatrix}, \begin{bmatrix} 0 & 0 & 1 \\ 0 & 0 & 0 \\ 1 & 0 & 0 \end{bmatrix}, \begin{bmatrix} 0 & 0 & 0 \\ 0 & 0 & 1 \\ 0 & 1 & 0 \end{bmatrix}$, E_{11}, E_{22}, E_{33}.

5.42 (a) E_{ij}, $i, j = 1, 2, 3$ and $i \le j$ for upper triangular matrices.

5.43 (–1, 2, 0, 1).

5.44 (0 2 1).

5.45 (a) (–1, 3, –1); (b) (–1, 2, 0); (c) (2/3, 1, –2/3); (d) (4/3, –2/3, 1);
(e) (3(–1 + 2i)/5, –2(1 + 3i)/5, (3 – i)/5); (f) (0, 0, 1).

CHAPTER 6

6.4 (a) Linear. The image of $\mathbb{R}^3$ is the xy-plane.
(b) Not linear. The image of $\mathbb{R}^3$ is the straight line $y = 1$ on the xy-plane.
(c) Not linear. The image of any point shifts by 1 unit parallel to the x-axis.
(d) Linear. The image of any point is a reflection on the yz-plane.
(e) Linear.

6.6 (a) (i) The point (0, 1, 1); (ii) The line parallel to the x-axis and passing through (0, 1, 1).
(b) (i) The straight line passing through (0, 0, 1) and parallel to the y-axis;
(ii) The plane parallel to the xy-plane and passing through (0, 0, 1);
(c) (i) A plane generated by (1 0 –1) and (–1 1 0) or the plane passing through the lines $x = -z$, $y = 0$ and $x = -y$, $z = 0$;
(ii) The plane perpendicular to the vector (1 1 1) passing through the origin;
(d) (i) The y-axis; (ii) The xy-plane. The codomain is $\mathbb{R}^3$.

6.7 (a) $L(x\ y\ z) = x\ (2\ 1)$;
(b) $L(x\ y) = (1/5)\ (-y + 4x, 5x)$;
(c) $L[x\ y\ z]^T = (1/6)\ [2x + 8y - 4z, 5x - y - z, -4x - 4y + 8z]^T$;
(d) $L[x\ y\ z\ w]^T = [x + y + z + 2w, 2x + y + 2z + 2w, 3x + 2y + 2z + 2w]^T$;
(e) $L[x\ y\ z]^T = [-x - 2y + z, x + y, x]^T$;
(f) $L(x\ y\ z) = (x - y + z, x + y + z, x - y + 2z)$;
(g) $L[x\ y]^T = [4x + y, (-4x + y)/3]^T$;
(h) $L(x\ y\ z) = (5y + 3z, x + 6y + 4z)$;
(i) $L(x\ y\ z) = (y\ x + y - z)$; (b), (c), (e), (f), (g).

6.9 (a) $[-22\ 13\ 6\ -13]^T$, $[-30\ 17\ 8\ -17]^T$, $[5\ -3\ -1\ 3]^T$; $[-129\ 74\ 35\ -74]^T$, (c) $[0\ 0\ 0\ 0]^T$.

6.10 (a) The straight line $x = y$ and $z = 1$; 1;
(b) The straight line $x = y = z$; 1;

(c) The point $x = y = z = 1$; 0;
(d) The plane orthogonal to the vector (1 1 –1) and through the origin; 2;
(e) A straight line $x = y = -z = \alpha$; 1;
(f) $\boldsymbol{\theta}$; 0.

6.11 (a) $(\pi(2n + 1)/2, m\pi)$, m and n are integers; (b) 0; (c) 1;
(d) $(0, (2n + 1)\pi/2)$, n is any integer; (e) (1, 0), (0, 1).

6.12 (a) a; (b) $at + b$; (c) $at^{n-1} + \cdots + bt + c$;
(d) $a \exp(-\alpha t) + b \exp(-\beta t)$, where a, b, and c are arbitrary constants.

6.13 ker L = all $n \times n$ lower triangular matrices with zero diagonal elements and nullity $L = n(n - 1)/2$.

6.16 (a) Neither onto nor one-to-one; (b) onto and one-to-one; (c) onto and one-to-one;
(d) onto but not one-to-one; (e) onto but not one-to-one; (f) one-to-one but not onto.

6.18 $L[x\ y\ z\ w]^T = [3(-10x - 9y + 2z)/2, (-7x - 7y + 2z + w), (x - y/2 + z + 2w), 3(3x + 2y + w)]^T$, assuming $\{\mathbf{x}_1, \mathbf{x}_2, \mathbf{e}_3, \mathbf{e}_4\}$ to be an extended basis in $\mathbb{R}^4$.

6.19 $L[x\ y\ z\ w]^T = [(-(x/2) + y) \cos \theta - (x/2) + z) \sin \theta, (-(x/2) + y) \sin \theta + ((x/2) + z) \cos \theta, x + w]^T$, assuming $\{\mathbf{x}_1, \mathbf{e}_2, \mathbf{e}_3, \mathbf{e}_4\}$ to be an extended basis in $\mathbb{R}^4$.

6.20 $L[x\ y\ z\ w]^T = [z\ w\ 0\ 0]^T$ if $\{\mathbf{e}_1, \mathbf{e}_2\}$ forms a basis of null space which is extended to a basis $\{\mathbf{e}_1, \mathbf{e}_2, \mathbf{e}_3, \mathbf{e}_4\}$ in $\mathbb{R}^4$.

6.22 $L[x\ y\ z\ w]^T = [0\ 0\ z\ w]^T$, assuming $\{\mathbf{e}_1, \mathbf{e}_2, \mathbf{e}_3, \mathbf{e}_4\}$ is an extended basis in $\mathbb{R}^4$ of that $\{\mathbf{e}_1, \mathbf{e}_2\}$ of null space.

6.23 (a) (i) $\mathbf{e}_2^T, \mathbf{e}_3^T$, (ii) (2, 1);
(b) (i) Zero space, (ii) {(4 5), (–1 0)};
(c) (i) Zero space, (ii) $[2\ 5\ -4]^T$, $[8\ -1\ -4]^T$, $[-4\ -1\ 8]^T$;
(d) (i) $\mathbf{x} = [2\ -2\ -2\ 1]^T$,
(ii) $[1\ 2\ 3]^T$, $[1\ 1\ 2]^T$, $[1\ 2\ 2]^T$, assuming $\{\mathbf{e}_1, \mathbf{e}_2, \mathbf{e}_3, \mathbf{x}\}$ to be a basis in $\mathbb{R}^4$;
(e) (i) Zero space; (ii) $[-1\ 1\ 1]^T$, $[-2\ 1\ 0]^T$, $[1\ 0\ 0]^T$;
(f) (i) Zero space; (ii) (1 1 1), (–1 1 –1), (1 1 2);
(g) (i) Zero space; (ii) $[4\ -4/3]^T$, $[1\ 1/3]^T$;
(h) (i) $\mathbf{x}$ = (–2 –3 5); (ii) (0 1), (5 6), assuming $\{\mathbf{e}_1, \mathbf{e}_2, \mathbf{x}\}$ to be basis of $\mathbb{R}^3$;
(i) (i) $\mathbf{x}$ = (1 0 1); (ii) (0 1), (1 1), assuming $\mathbf{x}, \mathbf{e}_1, \mathbf{e}_2$ to be basis of $\mathbb{R}^3$.

Remark: When null space is the zero space, the elementary vectors are assumed to form the extended basis.

6.24 (a) $(y, 0)$, $(z - x, x + y - z, y)$; (b) $(y - z, x - z, x - y)$, $(-x, y)$; (c) (x, y), (x, y); (d) $\mathbf{x}^T\mathbf{a}$, not defined.

6.25 (a) $(y - z - x, 0, x + y + z)$; (b) $2(x - z, y - x, z - y)$.

6.26 (a) $K[x\ y]^T = (1/3)\ [5x + y, -3x]^T$; (b) $K[x\ y\ z]^T = [z, y - z, x + 2y - z]^T$;
(c) $K[x\ y\ z]^T = (1/2)\ [x - y + z, x + y + z, x - y + 2z]^T$.

6.27 (a) $L[x\ y\ z]^T = [z\ y]^T$; $\mathbf{y} = [u\ v]^T$ is the image of $\mathbf{x} = [w\ v\ u]^T$, where w = arbitrary; $L = M \bullet K$ is the composition of two transformations, one projection of $\mathbb{R}^2$ (yz-plane) followed by reflection about $y = z$ (see parts (a) and (b) of the following figure);

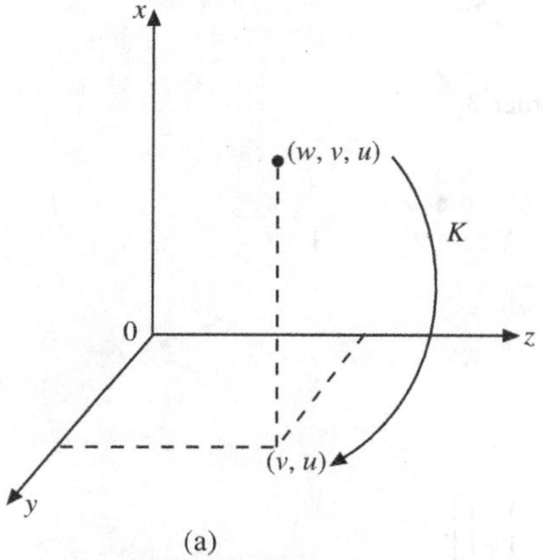

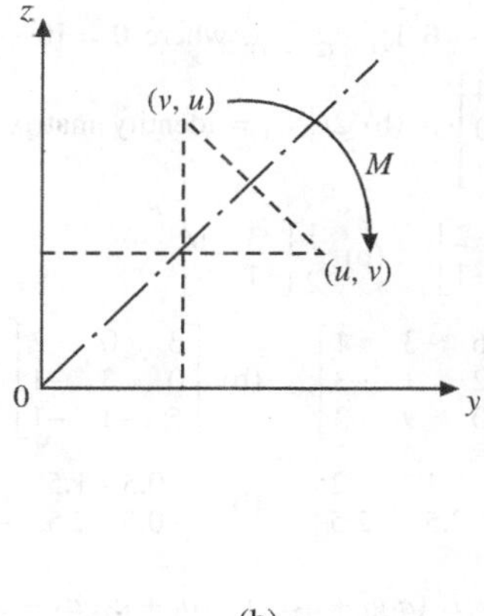

(a) K is a projection on $\mathbb{R}^2$; (b) M is a reflection about $y = z$.

(b) $K[u\ v]^T = [\alpha\ v\ u]^T$, α being arbitrary.

6.28 (a) $K[t\ u\ v]^T = [v - 2t + 2\alpha, -u + 2t - 2\alpha, u - v + t - 2\alpha, \alpha]^T$, α being arbitrary;
(b) $K[u\ v]^T = (1/5)\ [-2\alpha + 5v - 6u, u - 3\alpha, 5\alpha]^T$; (c) $K(u\ v) = (-u + v + \alpha, u, \alpha)$.

6.29 (a) Infinite number of left inverses exist; $K(x, y, z) = (x, x - y)$;
(b) L is one-to-one and onto; a unique left inverse exists; $K(x, y, z) = -0.5\ (y + z, x + z, x + y)$;
(c) $K(x, y, z) = (y, (x - y + z)/2, (x - y - z)/2)$.

6.30 (a) $L[x\ y\ z]^T = [x + y + z \quad x \quad y + z]^T$; (b) $L[x\ y\ z]^T = [x + y + z \quad 0 \quad x + y + z]^T$;
(c) $L[x\ y\ z]^T = [x + y + z \quad x + y + z \quad x]^T$.

6.32 (a) All operators of the form $L_\lambda\ (\lambda a_0 + a_1 t + \cdots + a_n t^n) = a_1 + a_2 t + \cdots + a_n t^{n-1}$ are left inverses of K.

6.34 (a) $\begin{bmatrix} -1 & -1 & 1 \\ 1 & -4 & 1 \\ 2 & -5 & 0 \end{bmatrix}$; (b) $\begin{bmatrix} -2 & 1 & -1 \\ 3 & -6 & 7 \\ 2 & -3 & 3 \end{bmatrix}$; (c) $\begin{bmatrix} 7 & -2 & 6 \\ 3 & -2 & 2 \\ -12 & 3 & -10 \end{bmatrix}$.

6.35 diag (1, –1, 1, –1).

6.36 diag (1, –1).

6.37 The kth column of A is obtained after carrying out the given permutation of the elements in the elementary vector $\mathbf{e}_k$.

6.38 $[\boldsymbol{\theta}\ \Delta]^T$, where $\boldsymbol{\theta} = [0\ 0\ \cdots\ 0]^T$ and $\Delta = \text{diag}\ [1\ (1/2)\ \cdots\ 1/(n + 1)]$.

6.39 (a) $[\boldsymbol{\theta}_{n+1}\ I_{n+1}]^T$, where $\boldsymbol{\theta}_{n+1} = [0\ 0\ \cdots\ 0]^T_{1\times(n+1)}$; (b) Yes; (c) No.

6.40 $\begin{bmatrix} 0 & 1 & -2 & 3 & \cdots & n(-1)^{n+1} \\ 0 & 0 & 2 & -3 & \cdots & n(-1)^n \\ 0 & 0 & 0 & 3 & \cdots & n(-1)^{n-1} \\ \vdots & \vdots & \vdots & \vdots & \ddots & \vdots \\ 0 & 0 & 0 & 0 & \cdots & n \end{bmatrix}_{n\times(n+1)}$.

6.41 $\begin{bmatrix} 1 & 1 & 1 & \cdots & 1 \\ 0 & 1 & 2 & \cdots & {}^nc_1 \\ 0 & 0 & 1 & \cdots & {}^nc_2 \\ \vdots & \vdots & \vdots & \ddots & \vdots \\ 0 & 0 & 0 & \cdots & {}^nc_n \end{bmatrix}$.

6.42 $\begin{bmatrix} 1 & 0 & -1 & 0 \\ 0 & 1 & 0 & -1 \\ -1 & 0 & 1 & 0 \\ 0 & -1 & 0 & 1 \end{bmatrix}$.

6.43 $\begin{bmatrix} 1 & 0 & -1 & 0 \\ 0 & 1 & 0 & -1 \\ 1 & 0 & 1 & 0 \\ 0 & 1 & 0 & 1 \end{bmatrix}$.

6.44 $L\mathbf{x} = [59x_1 + 16x_2 - 99x_3, 18x_1 + 5x_2 - 30x_3, 37x_1 + 10x_2 - 62x_3]^T$, $\mathbf{x} = [x_1, x_2, x_3, x_4]^T$.

6.45 $Lp(t) = 2(2a_0 + a_1 + a_2 + a_3) + 2(2a_0 + a_1)t + 2(2a_0 + 2a_1 + a_2)t^2 + (2a_0 + 3a_1 + 2a_2 + a_3)t^3$,
$p(t) = a_0 + a_1 t + a_2 t^2 + a_3 t^3$.

6.46 $\begin{bmatrix} 1 & 0 & 0 & 1 \\ 1 & 1 & 0 & 0 \\ 0 & 1 & 1 & 0 \end{bmatrix}$.

6.47 $[1\ 1\ \cdots\ 1]_{1\times n}$.

6.48 $[1\ \boldsymbol{\theta}\ 1\ \boldsymbol{\theta}\ \cdots\ \boldsymbol{\theta}\ 1]_{1\times(2n-1)}$, where $\boldsymbol{\theta} = [0\ \cdots\ 0]_{1\times n}$.

6.49 (a) $\begin{bmatrix} 0 & 4 \\ 1 & 0 \\ 1 & 1 \end{bmatrix}$; (b) $2I_3$, I_3 = identity matrix of order 3.

6.50 (a) $\begin{bmatrix} 5 & -2 \\ 3 & -1 \end{bmatrix}$; (b) $\frac{1}{2}\begin{bmatrix} 3 & 1 \\ 11 & 5 \end{bmatrix}$

6.51 (a) $\frac{1}{2}\begin{bmatrix} 6 & -3 & 1 \\ -2 & 1 & -3 \\ 0 & 9 & 3 \end{bmatrix}$; (b) $\begin{bmatrix} 3 & 0 & -2 \\ 0 & 3 & -1 \\ 5 & -1 & -1 \end{bmatrix}$.

6.52 (a) $\begin{bmatrix} -2 & 1 & 2 \\ 1.5 & 2.5 & -2.5 \end{bmatrix}$; (b) $\begin{bmatrix} -0.5 & 1.5 & 1.5 \\ -0.5 & 2.5 & -1.5 \end{bmatrix}$.

6.60 (a) $L^T g = L^T (a_1 g_1 + a_2 g_2) = [a_1 + a_2, a_2 - a_1]$; $\begin{bmatrix} 1 & 1 \\ -1 & 1 \end{bmatrix}$;

(b) $L^T g = L^T (a_1 g_1 + a_2 g_2) = [a_2, -2a_1 + 5a_2]$; $\begin{bmatrix} -4 & 13 \\ -2 & 6 \end{bmatrix}$;

(c) $L^T g = L^T(a_1g_1 + a_2g_2 + a_3g_3) = [-9a_1 + 3a_2 + 5a_3, -4a_1 + 3a_3]$; $\begin{bmatrix} -22 & 6 & 13 \\ -13 & 3 & 8 \end{bmatrix}$.

6.61 $L^T g(x\ y\ z) = 3(y - z)$.

6.62 $L^T g(\mathbf{x}) = (a_1g_1 + a_2g_2 + a_3g_3 + a_4g_4)(L\mathbf{x}) = a_1(-2a + b + c)/3$

$$+a_2(2a - b + 2c)/6 + a_3(2a - b + 2c)/6 + a_4(a + b - 2c)/3;$$

$$A^T = \frac{1}{6}\begin{bmatrix} -2 & 1 & 1 & 4 \\ 4 & 1 & 1 & -2 \\ 2 & 2 & 2 & -4 \end{bmatrix}.$$

6.66 (a) $P = \dfrac{1}{2}\begin{bmatrix} 1 & 1 \\ 1 & 1 \end{bmatrix}$, orthogonal; (b) $P = \dfrac{1}{a^2 + b^2}\begin{bmatrix} a^2 & ab \\ ab & b^2 \end{bmatrix}$, orthogonal;

(c) $P = \dfrac{1}{2}\begin{bmatrix} 1 & 0 \\ 1 & 0 \end{bmatrix}$, oblique (d) $P = \dfrac{\sqrt{3}}{15}\begin{bmatrix} 2a & 2b \\ a & b \end{bmatrix}$, $a = 1 + 2\sqrt{3}$, orthogonal $b = -2 + \sqrt{3}$.

6.67 Orthogonal

$$P = \frac{1}{3}\begin{bmatrix} 1 & 1 & 1 \\ 1 & 1 & 1 \\ 1 & 1 & 1 \end{bmatrix}.$$

6.68 Orthogonal

$$P = \frac{1}{3}\begin{bmatrix} 2 & -1 & -1 \\ -1 & 2 & -1 \\ -1 & -1 & 2 \end{bmatrix}.$$

6.69 Oblique,

$$P = \begin{bmatrix} 0 & -1 & -1 \\ 1 & 2 & 1 \\ -1 & -1 & 0 \end{bmatrix}.$$

CHAPTER 7

7.6 (a) 0, any vector $\mathbf{x} \neq \boldsymbol{\theta}$; (b) 0, $[\alpha\ 0\ \dots\ 0]^T$, α arbitrary;
(c) 1, –1, ..., $(-1)^{n-1}$; for $\lambda = 1$ and $n = 2m$ even, $\mathbf{x} = [\alpha_1, \dots, \alpha_m, \alpha_m, \dots, \alpha_1]^T$; $n = 2m + 1$ odd, $\mathbf{x} = [\alpha_1 \dots, \alpha_m, \beta, \alpha_m \dots \alpha_1]^T$; for $\lambda = -1$ and $n = 2m$ even $\mathbf{x} = [\alpha_1 \dots, \alpha_m, -\alpha_m, \dots, -\alpha_1]^T$; $n = 2m + 1$ odd $\mathbf{x} = [\alpha_1 \dots \alpha_m, 0, -\alpha_m, \dots, -\alpha_1]^T$, where α_i's and β are arbitrary.

7.7 (a) $a_{11}, \dots, a_{nn}$; (b) $b_{11}, b_{22}, \dots, b_{nn}$. The diagonal elements are distinct or if λ_i is an eigenvalue of multiplicity n_i, then rank $(A - \lambda_i I) = n - n_i$.

7.8 (a) –1, $\alpha[-3, 1]^T$; 4, $\beta[2\ 1]^T$; (b) 1, 1; $\alpha[1, -1]^T$;
(c) 0, $\alpha[1\ 0\ -1]^T$; $2\sqrt{2}$, $\beta[1\ \sqrt{2}\ 1]^T$; $-2\sqrt{2}$, $\gamma[1, -\sqrt{2}, 1]^T$; (d) 0, $\alpha[1\ 0, -1]^T$; 2, $\beta[0\ 1\ 0]^T$; 4, $\gamma[1\ 0\ 1]^T$;
(e) 2, $\alpha[2\ 1\ 0]^T$; i, $\beta[1\ 1 + i, -2 + i]^T$; $-i$, $\gamma[1, 1 - i, -2 - i]^T$; (f) 2, $\alpha[-2, 3, -1]^T$.

7.9 (a) $[e^{-t} + 2e^{3t}, -e^{-t} + 2e^{3t}]^T$; (b) $e^t[1 + 2t, 2 + 2t]^T$; (c) $(4e^{-t} - e^{-4t})/3$.

7.10 (a) 0, $\alpha[1\ 1\ 1]^T$; $(3 \pm \sqrt{3}i)/2$, $\beta[1, -(1 \pm \sqrt{3}i)/2, -(1 \mp \sqrt{3}i)/2]^T$;
(b) $\cos\theta \pm i\sin\theta$, $\alpha[1, \mp i]^T$ if $\sin\theta \neq 0$ and $[\beta\ \gamma]$ if $\sin\theta = 0$;
(c) –1, $\alpha(1\ 1\ 1)$; 2, 2, $(\beta, \gamma, -\beta - \gamma)$, (d) $1 \pm i$, $\alpha(1, \mp i)$.

7.11 (a) 1 α; 2, βx; 3, γx^2; (b) –1, α; –2, $\beta(1 - x)$; –3, $\gamma(1 - x)^2$;

(c) $1, 1, \begin{bmatrix} \alpha & \beta \\ -\beta & \alpha \end{bmatrix}; -1, -1, \begin{bmatrix} \gamma & \delta \\ \delta & -\gamma \end{bmatrix}$.

7.12 No, because L is not a linear transformation from V on itself.

7.13 1, 1, $\alpha[1\ 0\ 0]^T$; –3, $\beta[1, -2, 2]^T$.

7.14 (a) $\begin{bmatrix} 1 & 1 \\ 1 & -1 \end{bmatrix}$, diag [1, – 1]; (b) $\begin{bmatrix} 1 & 0 & -1 \\ 0 & 1 & 0 \\ 1 & 0 & 1 \end{bmatrix}$, diag [1, 1, – 1];

(c) $\begin{bmatrix} 1 & 1 & 1 \\ 0 & \sqrt{2} & -\sqrt{2} \\ -1 & 1 & 1 \end{bmatrix}$, diag $[0, 2\sqrt{2}, -2\sqrt{2}]$;

(d) $\begin{bmatrix} 1 & 0 & 1 \\ 0 & 1 & 0 \\ -1 & 0 & 1 \end{bmatrix}$, diag [0, 2, 4]; (e) $\begin{bmatrix} 2 & 1 & 1 \\ 1 & 1+i & 1-i \\ 0 & -2+i & -2-i \end{bmatrix}$, diag $[2, i, -i]$.

7.17 (a) $\begin{bmatrix} 1 & 1 & 0 \\ 1 & 1 & 2 \\ 1 & -3 & 4 \end{bmatrix}$; (b) $\begin{bmatrix} 2 & 1 & 1 \\ 1 & 1 & 1 \\ 0 & -2 & 1 \end{bmatrix}$.

7.18 (a) $\{1, 1 - x, (1 - x)^2\}$; (b) $\{1, x, x^2\}$; (c) {(1, 1, 1), (1, 0, –1), (0, 1, –1)};

(d) $\begin{bmatrix} 1 & 0 \\ 0 & 1 \end{bmatrix}, \begin{bmatrix} 0 & 1 \\ -1 & 0 \end{bmatrix}, \begin{bmatrix} 1 & 0 \\ 0 & -1 \end{bmatrix}, \begin{bmatrix} 0 & 1 \\ 1 & 0 \end{bmatrix}$.

7.19 (a) $\{[-3, 1]^T, [2, 1]^T\}$, $\left\{\frac{1}{5}[-1, 2]^T, \frac{1}{5}[1, 3]^T\right\}$;

(b) $\{[1\ 0\ {-1}]^T, [0\ 1\ 0]^T, [1\ 0\ 1]^T\}$, $\{\frac{1}{2}[1\ 0\ {-1}]^T, [0\ 2\ 0]^T, \frac{1}{2}[1\ 0\ 1]^T\}$;

(c) $\{[1, 2, -1]^T, [-2, -5, 3]^T, [1\ 2\ 0]^T\}$, $\{[6, -3, -1]^T, [2, -1, 0]^T, [-1, 1, 1]^T\}$.

7.20 (a) $\frac{1}{5}\begin{bmatrix} 3+2\cdot4^{20}, & -6+6\cdot4^{20} \\ -1+4^{20}, & 2+3\cdot4^{20} \end{bmatrix}$; (b) $\frac{1}{2}\begin{bmatrix} 4^{20} & 0 & 4^{20} \\ 0 & 2^{21} & 0 \\ 4^{20} & 0 & 4^{20} \end{bmatrix}$; (c) $\begin{bmatrix} -4-20^{20}, & 2+2^{20}, & 2^{20} \\ -10-2^{21}, & 5+2^{21}, & 2^{21} \\ 6, & -3, & 0 \end{bmatrix}$.

7.21 $\frac{1}{5}\begin{bmatrix} 3+2\cdot4^{30} & -3+3\cdot4^{30} \\ -2+2\cdot4^{30} & 2+3\cdot4^{30} \end{bmatrix}$.

7.22 (a) $x + 2$; $x + 1$; (b) $x - 1$; 0; (c) 0; $f(x)$; (d) $3x^2 + 2x + 1$; 2.

7.23 (A) For $A = \begin{bmatrix} 1 & 6 \\ 1 & 2 \end{bmatrix}$

(a) $f(A) = \begin{bmatrix} 44 & 126 \\ 21 & 65 \end{bmatrix}$, $d(A) = \begin{bmatrix} 8 & 18 \\ 3 & 11 \end{bmatrix}$, $q(A) = \begin{bmatrix} 3 & 6 \\ 1 & 4 \end{bmatrix}$, $r(A) = \begin{bmatrix} 2 & 6 \\ 1 & 3 \end{bmatrix}$.

(b) $f(A) = \begin{bmatrix} 24 & 78 \\ 13 & 37 \end{bmatrix}$, $d(A) = \begin{bmatrix} 9 & 24 \\ 4 & 13 \end{bmatrix}$, $q(A) = \begin{bmatrix} 0 & 5 \\ 1 & 1 \end{bmatrix}$, $r(A) = \begin{bmatrix} 0 & 0 \\ 0 & 0 \end{bmatrix}$.

(c) $f(A) = \begin{bmatrix} 441 & 1296 \\ 216 & 657 \end{bmatrix}$, $d(A) = \begin{bmatrix} 2121 & 6390 \\ 1065 & 3186 \end{bmatrix}$, $q(A) = \begin{bmatrix} 0 & 0 \\ 0 & 0 \end{bmatrix}$, $r(A) = f(A)$.

(d) $f(A) = \begin{bmatrix} 206 & 618 \\ 103 & 309 \end{bmatrix}$, $d(A) = \begin{bmatrix} 3 & 12 \\ 2 & 5 \end{bmatrix}$, $q(A) = \begin{bmatrix} 24 & 66 \\ 11 & 35 \end{bmatrix}$, $r(A) = \begin{bmatrix} 2 & 0 \\ 0 & 2 \end{bmatrix}$.

(B) For $A = \begin{bmatrix} 2 & 0 & 2 \\ 0 & 2 & 0 \\ 2 & 0 & 2 \end{bmatrix}$

(a) $f(A)=\begin{bmatrix}55 & 0 & 52\\ 0 & 23 & 0\\ 52 & 0 & 55\end{bmatrix}$, $d(A)=\begin{bmatrix}9 & 0 & 8\\ 0 & 5 & 0\\ 8 & 0 & 9\end{bmatrix}$, $q(A)=\begin{bmatrix}4 & 0 & 2\\ 0 & 4 & 0\\ 2 & 0 & 4\end{bmatrix}$, $r(A)=\begin{bmatrix}3 & 0 & 2\\ 0 & 3 & 0\\ 2 & 0 & 3\end{bmatrix}$.

(b) $f(A)=\begin{bmatrix}31 & 0 & 32\\ 0 & 7 & 0\\ 32 & 0 & 31\end{bmatrix}$, $d(A)=\begin{bmatrix}11 & 0 & 10\\ 0 & 7 & 0\\ 10 & 0 & 11\end{bmatrix}$, $q(A)=\begin{bmatrix}1 & 0 & 2\\ 0 & 1 & 0\\ 2 & 0 & 1\end{bmatrix}$, $r(A)=\begin{bmatrix}0 & 0 & 0\\ 0 & 0 & 0\\ 0 & 0 & 0\end{bmatrix}$.

(c) $f(A)=\begin{bmatrix}545 & 0 & 544\\ 0 & 81 & 0\\ 544 & 0 & 545\end{bmatrix}$, $r(A)=f(A)$, $d(A)=\begin{bmatrix}2658 & 0 & 2658\\ 0 & 186 & 0\\ 2658 & 0 & 2658\end{bmatrix}$, $q(A)=\begin{bmatrix}0 & 0 & 0\\ 0 & 0 & 0\\ 0 & 0 & 0\end{bmatrix}$.

(d) $f(A)=\begin{bmatrix}259 & 0 & 256\\ 0 & 87 & 0\\ 256 & 0 & 259\end{bmatrix}$, $d(A)=\begin{bmatrix}5 & 0 & 4\\ 0 & 5 & 0\\ 4 & 0 & 5\end{bmatrix}$, $q(A)=\begin{bmatrix}29 & 0 & 28\\ 0 & 16 & 0\\ 28 & 0 & 29\end{bmatrix}$, $r(A)=\begin{bmatrix}2 & 0 & 0\\ 0 & 2 & 0\\ 0 & 0 & 2\end{bmatrix}$.

(C) For $A=\begin{bmatrix}-6 & 4 & 2\\ -14 & 9 & 4\\ 6 & -3 & 0\end{bmatrix}$

(a) $f(A)=\begin{bmatrix}37 & 30 & 20\\ -90 & 68 & 40\\ 30 & -15 & 3\end{bmatrix}$, $d(A)=\begin{bmatrix}-7 & 6 & 4\\ -18 & 14 & 8\\ 6 & -3 & 1\end{bmatrix}$, $q(A)=\begin{bmatrix}-4 & 4 & 2\\ -14 & 11 & 4\\ 6 & -3 & 2\end{bmatrix}$, $r(A)=\begin{bmatrix}-5 & 4 & 2\\ -14 & 10 & 4\\ 6 & -3 & 1\end{bmatrix}$.

(b) $f(A)=\begin{bmatrix}-13 & 10 & 8\\ -26 & 20 & 16\\ 6 & -3 & -1\end{bmatrix}$, $d(A)=\begin{bmatrix}-13 & 10 & 6\\ -32 & 23 & 12\\ 12 & -6 & 1\end{bmatrix}$, $q(A)=\begin{bmatrix}-7 & 4 & 2\\ -14 & 8 & 4\\ 6 & -3 & -1\end{bmatrix}$, $r(A)=\begin{bmatrix}0 & 0 & 0\\ 0 & 0 & 0\\ 0 & 0 & 0\end{bmatrix}$.

(c) $f(A)=\begin{bmatrix}-111 & 96 & 80\\ -240 & 201 & 160\\ 48 & -24 & 1\end{bmatrix}$, $r(A)=f(A)$, $d(A)=\begin{bmatrix}-222 & 204 & 186\\ -462 & 417 & 372\\ 54 & -27 & 0\end{bmatrix}$, $q(A)=\begin{bmatrix}0 & 0 & 0\\ 0 & 0 & 0\\ 0 & 0 & 0\end{bmatrix}$.

(d) $f(A)=\begin{bmatrix}-149 & 118 & 84\\ -338 & 256 & 168\\ 102 & -51 & 3\end{bmatrix}$, $d(A)=\begin{bmatrix}-11 & 8 & 4\\ -28 & 19 & 8\\ 12 & -6 & 1\end{bmatrix}$, $q(A)=\begin{bmatrix}-35 & 26 & 16\\ -82 & 58 & 32\\ 30 & -15 & 1\end{bmatrix}$, $r(A)=\begin{bmatrix}2 & 0 & 2\\ 0 & 2 & 0\\ 0 & 0 & 2\end{bmatrix}$.

7.25 (a) $A_r(B)=\begin{bmatrix}0 & 10\\ 0 & 15\end{bmatrix}$ and $A_l(B)=\begin{bmatrix}0 & 2\\ -3 & 15\end{bmatrix}$. (b) $A_r(B)=\begin{bmatrix}-2 & 2\\ -17 & 1\end{bmatrix}$ and $A_l(B)=\begin{bmatrix}6 & 10\\ -6 & -7\end{bmatrix}$.

(c) $A_r(B)=\begin{bmatrix}-2 & 2\\ -9 & 9\end{bmatrix}$ and $A_l(B)=\begin{bmatrix}5 & -3\\ -4 & 2\end{bmatrix}$.

7.26 (i) (a) $Q(\lambda)=\begin{bmatrix}2 & -2\\ -1 & 2\end{bmatrix}\lambda+\begin{bmatrix}-9 & 16\\ 7 & -11\end{bmatrix}$, $R(\lambda)=\begin{bmatrix}10 & -42\\ -4 & 29\end{bmatrix}$.

(b) $Q(\lambda)=\begin{bmatrix}2 & -1\\ -2 & 2\end{bmatrix}\lambda+\begin{bmatrix}-10 & 7\\ 16 & -10\end{bmatrix}$, $R(\lambda)=\begin{bmatrix}27 & -18\\ -29 & 20\end{bmatrix}$.

(ii) (a) $Q(\lambda)=\begin{bmatrix}1 & 0\\ 0 & 0\end{bmatrix}\lambda^2+\begin{bmatrix}0 & 0\\ 1 & 0\end{bmatrix}$, $R(\lambda)=\begin{bmatrix}0 & 0\\ 0 & 0\end{bmatrix}$.

(b) $Q(\lambda)=\begin{bmatrix}1 & 0\\ 0 & 0\end{bmatrix}\lambda^2+\begin{bmatrix}0 & 1\\ -1 & 0\end{bmatrix}\lambda+\begin{bmatrix}1 & 0\\ 1 & -1\end{bmatrix}$, $R(\lambda)=\begin{bmatrix}-1 & 1\\ -1 & 1\end{bmatrix}$.

7.27 (i) $Q(\lambda)=\begin{bmatrix}2\lambda+4 & 3\\ 3 & \lambda+2\end{bmatrix}$; $R=\begin{bmatrix}12 & 6\\ 11 & 5\end{bmatrix}$.

(ii) $Q(\lambda)=\begin{bmatrix}\lambda^2+2\lambda+6 & \lambda+4\\ 1 & 0\end{bmatrix}$; $R=\begin{bmatrix}16 & 10\\ 2 & 2\end{bmatrix}$.

7.28 (c) $16\begin{bmatrix}0&1&0\\1&0&1\\0&1&0\end{bmatrix}, 32\begin{bmatrix}1&0&1\\0&2&0\\1&0&1\end{bmatrix}$; (d) $8\begin{bmatrix}4&0&4\\0&1&0\\4&0&4\end{bmatrix}, 16\begin{bmatrix}8&0&8\\0&1&0\\8&0&8\end{bmatrix}$; (e) $\begin{bmatrix}10&-4&3\\5&-2&1\\-1&2&0\end{bmatrix}, \begin{bmatrix}19&-6&6\\9&-2&3\\0&0&1\end{bmatrix}$;

(f) $\begin{bmatrix}-88&-72&-24\\156&128&48\\-60&-48&-16\end{bmatrix}, \begin{bmatrix}-272&-224&-96\\464&384&176\\-176&-144&-64\end{bmatrix}$.

7.29 (a) $\dfrac{1}{4}\begin{bmatrix}4&-4&1\\-4&8&-3\\0&0&1\end{bmatrix}$; (b) $\dfrac{1}{37}\begin{bmatrix}-1&3&2&5\\3&28&-6&-15\\2&-6&33&-10\\5&-15&-10&12\end{bmatrix}$.

7.30 (a) $(\lambda-1)^2$; (b) $(\lambda-a)$; (c) $(\lambda-2)^2$; (d) $\lambda^3-3\lambda-2$; (e) $(\lambda-2)^2(\lambda-4)$.
7.31 (b) $(\lambda+1)^2(\lambda-1)(\lambda+2)(\lambda-4)$; $(\lambda^2-1)(\lambda+2)(\lambda-4)$.
7.33 (a) $(\lambda-2)^3$, $(\lambda-2)^2$; (b) $(\lambda-1)(\lambda-2)(\lambda-3)$, $(\lambda-1)(\lambda-2)(\lambda-3)$; (c) $(\lambda^2-1)^2$, (λ^2-1).

7.34 (a) $\begin{bmatrix}\lambda-1&1&0\\0&\lambda-1&0\\0&0&\lambda-1\end{bmatrix}$; (b) I_3; (c) $\begin{bmatrix}\lambda-2&1&0&0\\0&\lambda-2&0&0\\0&0&\lambda-2&1\\0&0&0&\lambda-2\end{bmatrix}$;

(d) $\begin{bmatrix}(\lambda-2)(\lambda+1)&0&0&3(\lambda+1)\\-2(\lambda-2)&(\lambda^2-4)&(\lambda-2)&3(\lambda+1)\\2(\lambda-2)&-(\lambda-2)&\lambda(\lambda-2)&3(\lambda+1)\\0&0&0&(\lambda+1)^2\end{bmatrix}$;

(e) $\begin{bmatrix}(\lambda+2)(\lambda-3)&-3(3\lambda-10)&-(5\lambda-14)&-3(3\lambda-10)\\(\lambda+2)&(\lambda+2)(\lambda-5)&-(\lambda+2)&3(\lambda-6)\\2(\lambda-4)&-6(\lambda-4)&(\lambda-4)^2&-6(\lambda-4)\\-4&-4(\lambda-5)&4&\lambda^2-10\lambda+28\end{bmatrix}$.

7.35 (c) $\begin{bmatrix}39&48&0\\32&55&0\\0&0&7\end{bmatrix}$; (d) $\begin{bmatrix}26216&39321&0\\26214&39323&0\\0&0&2\end{bmatrix}$.

7.36 $\begin{bmatrix}329&-328\\-328&329\end{bmatrix}$.

7.39 (a) $\lambda^2(\lambda-3)$; (b) $\lambda^3-(a+b+c)\lambda^2-b(a-c)^2$; (c) $\lambda^4-60\lambda^2-192\lambda-112$.
7.40 (a) 4, $\alpha[0\ 0\ 1]^T$; 2, $\beta[1, -1, 0]^T$; 142, $\gamma[1, 1.5, 0]^T$; (b) 2, 2, $[\alpha, \beta, -\alpha-\beta]^T$; 62, $\gamma[1\ 1\ 1]^T$.

7.47 (a) -1, $[1, -1]^T$; -2, $[1, -2]^T$; $V=\begin{bmatrix}1&1\\-1&-2\end{bmatrix}$; $\lambda^2+3\lambda+2$.

(b) 0, $[1\ 0\ 0]^T$; 1, $[1\ 1\ 1]^T$; -1, $[1, -1, 1]^T$; $V=\begin{bmatrix}1&1&1\\0&1&-1\\0&1&1\end{bmatrix}$; $\lambda^3-\lambda$.

(c) 1, $[1\ 1\ 1]^T$; -1, $[1, -1, 1]^T$; 2, $[1\ 2\ 4]^T$; $V=\begin{bmatrix}1&1&1\\1&-1&-2\\1&1&4\end{bmatrix}$; $\lambda^3-2\lambda^2-\lambda+2$.

7.48 Eigenvalues of A_n are: $\lambda_k = 2\left(c - \cos\dfrac{k\pi}{n+1}\right)$, $k = 1, 2, \ldots, n$.
Eigenvectors of A_n are obtained from the recursive relations: $x_1 = 1$, $x_2 = -\alpha_k$, $x_p + \alpha_k x_{p+1} + x_{p+2} = 0$, $p = 1, 2, \ldots, n-2$, where $\alpha_k = 2c - \lambda_k = 2\cos(k\pi/(n+1))$.

7.49 $-\dfrac{1}{400}[A^4 + 2A^2 - 103\,I]$.

7.52 (a) –1, –1, –1, 2; rank $(-I - A) = 2$; (b) 2, 2, 2, 4; rank $(2I - A) = 2$.

7.53 $R = \begin{bmatrix} 0.8321 & 0.3548 & -0.4264 & 0 \\ -0.5547 & 0.5321 & -0.6396 & 0 \\ 0 & 0.7687 & 0.6396 & 0 \\ 0 & 0 & 0 & 1 \end{bmatrix}$, $R^T AR = \begin{bmatrix} 2 & 0 & -11.7080 & -1.1094 \\ 0 & 2 & -12.2034 & 1.0643 \\ 0 & 0 & 2 & -1.2792 \\ 0 & 0 & 0 & 3 \end{bmatrix}$

7.54 (a) 2, $[1\ 0\ 1]^T$; $1 \pm \sqrt{3}$, $[i,\ 1\pm\sqrt{3},\ -i]^T$; $U = \begin{bmatrix} c & d & f \\ 0 & e & g \\ c & -d & -f \end{bmatrix}$ where $c = 1/\sqrt{2}$, $d = i/a$, $f = i/b$,

$e = (1 + \sqrt{3})/a$, $g = (1 - \sqrt{3})/b$, $a^2 = 6 + 2\sqrt{3}$, $b^2 = 6 - 2\sqrt{3}$;

(b) –1, –1, $\mathbf{x}_1 = [1, -1\ 0\ 0]^T$, $\mathbf{x}_2 = [0\ 0\ 1\ 1]^T$; 5, 5, $\mathbf{x}_3 = [1\ 1\ 0\ 0]^T$, $\mathbf{x}_4 = [0\ 0\ 1\ -1]^T$; $U = \dfrac{1}{\sqrt{2}}[\mathbf{x}_1\ \mathbf{x}_2\ \mathbf{x}_3\ \mathbf{x}_4]$;

(c) 3, $\mathbf{x}_1 = [2\ 2\ -1]^T$; 6, $\mathbf{x}_2 = [2\ -1\ 2]^T$; 9, $\mathbf{x}_3 = [-1\ 2\ 2]^T$; $U = \dfrac{1}{3}[\mathbf{x}_1\ \mathbf{x}_2\ \mathbf{x}_3]$;

(d) 2, $\mathbf{x}_1 = [-1\ 1\ 1\ 1]^T$; 2, $\mathbf{x}_2 = [1\ -1\ 1\ 1]^T$; 4, $\mathbf{x}_3 = [1\ 1\ -1\ 1]^T$; 4, $\mathbf{x}_4 = [1\ 1\ 1\ -1]^T$; $U = \dfrac{1}{2}[\mathbf{x}_1\ \mathbf{x}_2\ \mathbf{x}_3\ \mathbf{x}_4]$;

(e) 2, $\mathbf{x}_1 = [1\ 1\ 0]^T$; 4, $\mathbf{x}_2 = [1\ -1\ 0]^T$; 4, $\mathbf{x}_3 = [0\ 0\ \sqrt{2}]^T$; $U = \dfrac{1}{\sqrt{2}}[\mathbf{x}_1\ \mathbf{x}_2\ \mathbf{x}_3]^T$.

7.56 (a) 4, 4; 4, 4; (b) –2, –3; 0, –2, –3.

7.59 $P = UQR$, where $U = \begin{bmatrix} 0 & -2a & a \\ 0 & a & 2a \\ 1 & 0 & 0 \end{bmatrix}$, $Q = \text{diag}\,[1, 1/2, 1/3]$, $R = \begin{bmatrix} 3b & 4d & 30ac \\ 10ab & -6ad & -9c \\ 0 & 87ad & -2c \end{bmatrix}$,

where $a = 1/\sqrt{5}, b = 1/\sqrt{29}, c = 1/\sqrt{265}$, and $d = 1/\sqrt{1537}$; $P^*BP = \text{diag}\,[0, 0, 53/36]$.

7.60 (a) $\dfrac{1}{\sqrt{3}}\begin{bmatrix} 1 & 1 & 1 \\ -1 & \alpha & \bar{\alpha} \\ 1 & -\bar{\alpha} & -\alpha \end{bmatrix}$, where $\alpha = (1 + i\sqrt{3})/2$; (b) $\dfrac{1}{\sqrt{6}}\begin{bmatrix} 2 & 1 & 1 \\ 1 & \alpha & \bar{\alpha} \\ -1 & -\bar{\alpha} & -\alpha \end{bmatrix}$, where $\alpha = (-2 + i\sqrt{6})/2$;

(c) $\begin{bmatrix} 0 & c & d \\ e & a & b \\ -e & a & b \end{bmatrix}$,

where $a = -(1+\sqrt{3})c/2$, $b = (-1+\sqrt{3})d/2$, $c = (3+\sqrt{3})^{-1/2}$, $d = (3-\sqrt{3})^{-1/2}$, $e = 1/\sqrt{2}$.

7.61 (a) $A = 0\cdot Q_1 + 4Q_2$, where $Q_1 = \dfrac{1}{4}\begin{bmatrix} 3 & -1 & -1 & -1 \\ -1 & 3 & -1 & -1 \\ -1 & -1 & 3 & -1 \\ -1 & -1 & -1 & 3 \end{bmatrix}$, $Q_2 = \dfrac{1}{4}\begin{bmatrix} 1 & 1 & 1 & 1 \\ 1 & 1 & 1 & 1 \\ 1 & 1 & 1 & 1 \\ 1 & 1 & 1 & 1 \end{bmatrix}$;

(b) $A = 2Q_1 + 4Q_2$, where $Q_1 = \dfrac{1}{2}\begin{bmatrix} -1 & 3 & 3 & 3 \\ -3 & 5 & 3 & 3 \\ 0 & 0 & 2 & 0 \\ 2 & -2 & -2 & 0 \end{bmatrix}$, $Q_2 = \dfrac{1}{2}\begin{bmatrix} 3 & -3 & -3 & -3 \\ 3 & -3 & -3 & -3 \\ 0 & 0 & 0 & 0 \\ -2 & 2 & 2 & 2 \end{bmatrix}$.

7.64 (a) diag [1, 1, −1], $P=\begin{bmatrix}1 & 7a & -2\\ 0 & -5a & 1\\ 0 & a & 0\end{bmatrix}$, $a=1/3\sqrt{2}$;

(b) diag [1, −1, −1], $P=\begin{bmatrix}a & -a & -1/2b\\ a & a & -b\\ 0 & 0 & b/\sqrt{2}\end{bmatrix}$, $a=1/\sqrt{2}, b=1/\sqrt{3}$;

(c) diag [1, 1, 1, −1], $P=\begin{bmatrix}0 & a/3 & -b & b\\ 1 & a/4 & -b & -b\\ 0 & a/4 & b & b\\ 0 & a/6 & 0 & 0\end{bmatrix}$, $a=\sqrt{3}, b=1/2$.

7.65 (a) diag [1, −1], $U=\begin{bmatrix}a & -\bar{b}\\ a & b\end{bmatrix}$, $a=1/\sqrt{2}, b=(1-i)/2$; (b) diag [1, −1 0], $U=\begin{bmatrix}a & ia & 1\\ ia & a & -1\\ 0 & 0 & 1\end{bmatrix}$, $a=1/\sqrt{2}$;

(c) diag [1, 1, −1], $U=\begin{bmatrix}1 & -ia & -3a-i\sqrt{2}\\ 0 & 0 & \sqrt{2}\\ 0 & a & -ia\end{bmatrix}$, $a=1/\sqrt{2}$.

7.67 (a) 4, 0; (b) 10, 32.

CHAPTER 8

8.1 (a) $\begin{bmatrix}2 & 1 & -2\\ -2 & 3 & 0\end{bmatrix}$; (b) $\begin{bmatrix}1 & 2 & 1\\ 3 & -2 & 0\\ 0 & 1 & -3\end{bmatrix}$; (c) $\begin{bmatrix}-2 & -1\\ 2 & 0\\ -1 & 3\end{bmatrix}$; (d) $\begin{bmatrix}3 & 1 & 0\\ 1 & -2 & -4\\ 0 & -4 & 3\end{bmatrix}$. Only (d) is symmetric.

8.2 The matrices P and Q are not unique.

(a) $P^T=\begin{bmatrix}1/2 & 0\\ 1/4 & 1/4\end{bmatrix}$; $Q=\begin{bmatrix}1 & -1/2 & 3/4\\ 0 & 1 & 1/2\\ 0 & 0 & 1\end{bmatrix}$;

(b) $P^T=\begin{bmatrix}1 & 0 & -2\\ 0 & 0 & 1\\ -3 & 1 & 8\end{bmatrix}$; $Q=\begin{bmatrix}1 & 0 & 7/27\\ 0 & 1 & -3/27\\ 0 & 0 & -1/27\end{bmatrix}$;

(c) $P^T=\begin{bmatrix}0 & 1/2 & 1/3\\ 0 & 1/6 & 1/3\\ -1/7 & -1/6 & -1/21\end{bmatrix}$; $Q=I_3$;

(d) $P^T=\begin{bmatrix}0 & 1 & 0\\ 0 & 0 & 1\\ 1/23 & -3/23 & -4/23\end{bmatrix}$; $Q=\begin{bmatrix}1 & 4/3 & 22/3\\ 0 & 0 & 1\\ 0 & 1/3 & 4/3\end{bmatrix}$;

8.5 (a) $B=\begin{bmatrix}a & 0 & a\\ a & 0 & -a\\ 0 & 1 & 0\end{bmatrix}$, $a=1/\sqrt{2}$; diag [3, 3, −1]; (b) $B=\begin{bmatrix}0 & 1 & 0\\ a & 0 & -a\\ a & 0 & a\end{bmatrix}$, $a=1/\sqrt{2}$; diag [1, 4, 4];

(c) $B=\begin{bmatrix}3a & 0 & 2a\\ 0 & 1 & 0\\ 2a & 0 & -3a\end{bmatrix}$, $a=1/\sqrt{3}$; diag [16, −2, −10].

8.6 (a) $\frac{1}{3}\begin{bmatrix} 2 & 2 & 1 \\ 2 & -1 & -2 \\ -1 & 2 & -2 \end{bmatrix}$; (b) $\frac{1}{\sqrt{3}}\begin{bmatrix} 1 & 1 & 1 \\ -1 & \alpha & \bar{\alpha} \\ 1 & -\bar{\alpha} & -\alpha \end{bmatrix}$; $\alpha = (1+\sqrt{3}\,i)/2$.

8.8 (a) $2x_1^2 - 3x_2^2 + 3x_3^2 - 2x_1x_2 + 6x_1x_3$; (b) $A = \begin{bmatrix} 2 & -1 & 3 \\ -1 & -3 & 0 \\ 3 & 0 & 3 \end{bmatrix}$; (c) 6; (d) Yes; (e) $\begin{bmatrix} 1 & 2 \\ 4 & 2 \end{bmatrix}\begin{bmatrix} 1 & 3 \\ 3 & 2 \end{bmatrix}$.

8.9 (i) (a) $x_1^2 + 5x_2^2 + 3x_3^2 - 8x_1x_3 + 8x_2x_3$; (b) $x_1^2 + x_2^2 + x_3^2$;

(ii) (a) $3y_1^2 - 3y_2^2 + 9y_3^2$; (b) $x_1^2 - x_2^2 + x_3^2$.

8.10 (a) $P = \frac{1}{3}\begin{bmatrix} 2 & 2 & -1 \\ 2 & -1 & 2 \\ -1 & 2 & 2 \end{bmatrix}$, $D = \text{diag}\,[3, 6, 9]$; (b) $P = \frac{1}{2}\begin{bmatrix} -1 & 1 & 1 & 1 \\ 1 & -1 & 1 & 1 \\ 1 & 1 & -1 & 1 \\ 1 & 1 & 1 & -1 \end{bmatrix}$, $D = \text{diag}\,[2, 2, 4, 4]$;

(c) $P = \frac{1}{\sqrt{2}}\begin{bmatrix} 1 & 1 & 0 \\ 1 & -1 & 0 \\ 0 & 0 & \sqrt{2} \end{bmatrix}$, $D = \text{diag}[2\ \ 4\ \ 4]$.

8.11 (a) $P = \begin{bmatrix} 5 & -2 & 0 \\ 0 & 13/5 & 1 \\ 0 & 0 & 1 \end{bmatrix}$, $\frac{1}{5}y_1^2 + \frac{10}{13}y_2^2 + \frac{81}{13}y_3^2$; (b) $P = \begin{bmatrix} 3 & 1 & 0 & 0 \\ 0 & 1 & 0 & 0 \\ 0 & 0 & 3 & -1 \\ 0 & 0 & 0 & 1 \end{bmatrix}$, $\frac{1}{3}y_1^2 + \frac{8}{3}y_2^2 + \frac{1}{3}y_3^2 + \frac{8}{3}y_4^2$;

(c) $P = \begin{bmatrix} 3 & -1 & 0 \\ 0 & 1 & 0 \\ 0 & 0 & 1 \end{bmatrix}$, $\frac{1}{3}y_1^2 + \frac{8}{3}y_2^2 + 4y_3^2$.

8.12 (a) $\begin{bmatrix} 1 & 1 & 2 \\ -1 & 1 & 0 \\ 0 & 0 & 1 \end{bmatrix}$, $(1/4)(z_1^2 - z_2^2) - z_3^2$; (b) $\begin{bmatrix} 2 & -1 & 3 \\ 0 & 1 & 1 \\ 0 & 1 & -1 \end{bmatrix}$, $z_1^2 + (3/2)(z_2^2 - z_3^2)$; (c) $\begin{bmatrix} 1 & 2 & 1 \\ 1 & 0 & 1 \\ 0 & 0 & 1 \end{bmatrix}$, $(1/4)(z_1^2 - z_2^2)$.

8.14 (a) $5y_1^2 + \frac{26}{5}y_2^2 + \frac{81}{13}y_3^2$, $\frac{1}{5}z_1^2 + \frac{1}{130}z_2^2 + \frac{1}{4212}z_3^2$;

(b) $3y_1^2 + \frac{8}{3}y_2^2 + 3y_3^2 + \frac{8}{3}y_4^2$, $\frac{1}{3}z_1^2 + \frac{1}{24}z_2^2 + \frac{1}{192}z_3^2 + \frac{1}{1536}z_4^2$;

(c) $3y_1^2 + \frac{8}{3}y_2^2 + 4y_3^2$, $\frac{1}{3}z_1^2 + \frac{1}{24}z_2^2 + \frac{1}{256}z_3^2$.

For remaining answers, see Problem 8.11.

8.16 (a) 3, 1, –1; (b) 3, 2, 1; (c) 2, 2, 0.

8.17 (a) $\begin{bmatrix} a & a & 1 \\ -a & a & 0 \\ 0 & 0 & 1 \end{bmatrix}$, $a = 1/2$; (b) $\begin{bmatrix} 2 & -1 & 3 \\ 0 & a & a \\ 0 & a & -a \end{bmatrix}$, $a = \sqrt{3/2}$; (c) $\begin{bmatrix} a & 1 & a \\ a & 0 & a \\ 0 & 0 & 1 \end{bmatrix}$, $a = 1/2$.

8.18 p and q are equivalent over the complex field $\mathbb{C}$. The transformation $y_1 = x_1$, $iy_2 = x_2$, and $iy_3 = x_3$ reduces p to q.

8.19 (a) No; (b) Yes.

8.21 (a) Yes; (b) Yes; (c) No.

8.22 (a) $(\bar{y}_1y_1 - 11\bar{y}_2y_2)/2$, $y_1 = 2x_1 - 3ix_2$, $y_2 = x_2$; $(\bar{y}_1y_1 - \bar{y}_2y_2)/2$, $y_1 = x_1 + (2-i)x_2$, $y_2 = -x_1 + (2-i)x_2$;

(b) $\bar{y}_1y_1 - 3\bar{y}_2y_2$, $y_1 = x_1 + (1+2i)x_2$, $y_2 = x_2$; $(\bar{y}_1y_1 - \bar{y}_2y_2)/2$, $y_1 = x_1 + ix_2$, $y_2 = -x_1 + ix_2$;

(c) $\bar{y}_1y_1 - \bar{y}_2y_2$, $\sqrt{2}y_1 = x_1 + ix_2 + (i-1)x_3$, $\sqrt{2}y_2 = -x_1 + ix_2 + (i+1)x_3$

$\bar{y}_1y_1 - \bar{y}_2y_2 + \bar{y}_3y_3$, $y_1 = x_1 + (1+i)x_2$, $y_2 = x_2$, $y_3 = \sqrt{2}x_3$.

8.24 (a) Indefinite; (b) indefinite; (c) indefinite; (d) PSD; (e) PD; (f) NSD; (g) ND; (h) NSD.

8.25 (a) Indefinite; (b) NSD; (c) PSD; (d) PD; (e) ND.

8.26 (i) (e); (ii) (g).

8.27 (i) (b); (ii) (d).

8.30 (a) $a < -4.1131$; (b) $a < -2$.

8.33 PSD.

8.46 (a) 1/2, 1/2, 1/3; $\alpha[1\ 0\ 1]^T$, $\beta[1\ 1\ 1]^T$, $\gamma[1\ 0\ -1]^T$, $P = \dfrac{1}{12}\begin{bmatrix} 3 & 1 & 2\sqrt{3} \\ 0 & 4 & 0 \\ 3 & 1 & -2\sqrt{3} \end{bmatrix}$, $P^*BP = \text{diag}\,[1/2, 1/2, 1/3]$;

(b) 9, 9, −9/10; $\alpha[0\ 1\ -1]^T$, $\beta[2\ -1\ 0]^T$, $\gamma[1\ 2\ 2]^T$, $P = \dfrac{1}{3\sqrt{10}}\begin{bmatrix} 0 & 4\sqrt{5} & 1 \\ 3\sqrt{5} & -\sqrt{5} & 2 \\ -3\sqrt{5} & -\sqrt{5} & 2 \end{bmatrix}$, $P^*BP = \text{diag}\,[9, 9, -9/10]$;

(c) 1, 1, −1; $\alpha[2\ -1\ 0]^T$, $\beta[-6\ 3\ 5]^T$, $\gamma[3\ -1\ -2]^T$, $P = \begin{bmatrix} 2/\sqrt{5} & -6/\sqrt{5} & 3 \\ -1/\sqrt{5} & 3/\sqrt{5} & -1 \\ 0 & \sqrt{5} & -2 \end{bmatrix}$.

8.48 (a) $P = \dfrac{1}{2\sqrt{3}}\begin{bmatrix} 2 & 2 & 2 \\ 2 & -1 & -1 \\ 0 & -\sqrt{2} & \sqrt{2} \end{bmatrix}$, $a = 0, b = \sqrt{2}, c = -\sqrt{2}$;

(b) $P = \dfrac{1}{\sqrt{15}}\begin{bmatrix} \sqrt{5} & \sqrt{15} & -1 \\ \sqrt{5} & 0 & 2 \\ 0 & -\sqrt{15}/2 & -3/2 \end{bmatrix}$, $a = 0, b = -1, c = 3/5$.

8.49 (a) $\left\{[1\,0\,0]^T, [1\,1\,0]^T, [-1\,0\,1]^T\right\}$;

(b) $\left\{\left[1/\sqrt{2}, 1/\sqrt{2}, 0\right]^T, \left[-1/\sqrt{2}, 1/\sqrt{2}, 0\right]^T, \left[-1/\sqrt{2}, -1/\sqrt{2}, 1/\sqrt{2}\right]^T\right\}$;

(c) $\left\{\left[1/\sqrt{2}, 0, 0\right]^T, \left[-1/\sqrt{14}, 2/\sqrt{14}, 0\right]^T, \left[3/2\sqrt{91}, 1/2\sqrt{91}, 7/2\sqrt{91}\right]^T\right\}$.

8.50 (a) $\left\{\left[1/\sqrt{2}, i/\sqrt{2}, 0\right]^T, \left[i/\sqrt{2}, 1/\sqrt{2}, 0\right]^T, [1, -1, 1]^T\right\}$;

(b) $\left\{[1\,0\,0]^T, [0\,0\,1]^T, \left[-(1+2i)/\sqrt{3}, 1/\sqrt{3}, 0\right]^T\right\}$;

(c) $\left\{[1\,0\,0]^T, [-i\,1\,0]^T, [-3/2, -i, 1/2]^T\right\}$.

8.53 (a) $P = \begin{bmatrix} 1 & 1 \\ 1 & -1 \end{bmatrix}$,

(b) $P = \begin{bmatrix} 2 & -1 \\ -1 & 1 \end{bmatrix}$,

CHAPTER 9

9.1 (a) $\|A\| = \|A - B + B\| \le \|A - B\| + \|B\|$.

9.2 The triangle inequality, for example, is not satisfied by such a norm. Take $\mathbf{x} = [4, 9]^T$, $\mathbf{y} = [1, 4]^T$. Then for $p = 1/2$, $\|\mathbf{x}\|_p = 25$, $\|\mathbf{y}\|_p = 9$ and $\|\mathbf{x}+\mathbf{y}\|_p = 18 + 2\sqrt{65} > 34 = \|\mathbf{x}\|_p + \|\mathbf{y}\|_p$.

9.5 (a) $3, \sqrt{3}, 1$; (b) $3, \sqrt{3}, 1$; (c) $1 + 2\sqrt{2}, \sqrt{5}, \sqrt{2}$; (d) $21, \sqrt{161}, 9$.

9.6 (a) $n, \sqrt{n}, 1$; (b) 1, 1, 1; (c) $n(n + 1)/2$, $\{n(n + 1)(2n + 1)/6\}^{1/2}$, n; (d) same as (c).

9.7 (a) By Theorem 7.27, $S^*HS = \text{diag}\,[\lambda_1, ..., \lambda_n]$, where λ_i's are all positive and $S^*S = I$. Now define $S\mathbf{u} = \mathbf{x}$, $S\mathbf{v} = \mathbf{y}$; then $\|\mathbf{x}\|^2 = \mathbf{x}^*H\mathbf{x} = \mathbf{u}^*S^*HS\mathbf{u} = \sum_i \lambda_i\, |u_i|^2$, etc., and

$$(\mathbf{x}+\mathbf{y})^*H(\mathbf{x}+\mathbf{y}) = \Sigma\lambda_i\,(u_i + v_i)(\bar{u}_i + \bar{v}_i)$$

$$= \|\mathbf{x}\|^2 + \|\mathbf{y}\|^2 + \sum_{i=1}^{n} \lambda_i\,(u_i\bar{v}_i + \bar{u}_i v_i)$$

$$\le \|\mathbf{x}\|^2 + \|\mathbf{y}\|^2 + 2\sum_{i=1}^{n}(\lambda_i\,|u_i|^2)^{1/2}\,(\sum_{i=1}^{n}\lambda_i\,|v_i|^2)^{1/2}$$

$$= (\|\mathbf{x}\| + \|\mathbf{y}\|)^2.$$

9.8 (a) $\|A\|_n = n \max_{i,j} |a_{ij}| = 0$ implies $a_{ij} = 0$ for all (i, j)

$$\|\alpha A\|_n = n \max_{i,j} |\alpha a_{ij}| = |\alpha|\, n \max_{i,j} |a_{ij}| = |\alpha|\, \|A\|_n$$

$$\|A + B\|_n = n \max_{i,j} |a_{ij} + b_{ij}| \le n \max_{i,j} \left(|a_{ij}| + |b_{ij}|\right) = \|A\|_n + \|B\|_n$$

$$\|AB\|_n = n \max_{i,j} |a_{i1} + b_{1j} + \cdots + a_{in}\, b_{nj}| \le n \max_{i,j} \left(|a_{i1}|\,|b_{1j}| + \cdots + |a_{in}|\,|b_{nj}|\right)$$

$$\le n \max_{i,k} |a_{ik}| \max_j \left(|b_{1j}| + \cdots + |b_{nj}|\right) \le n \max_{i,k} |a_{ik}|\, n \max_{k,j} |b_{kj}|$$

$$= \|A\|_n\, \|B\|_n;$$

(b) $\|A\|_e^2 = \sum_{i,j} |a_{ij}|^2 \le n^2 \max_{i,j} |a_{ij}|^2 = \|A\|_n^2$;

(c) $\|A\mathbf{x}\|_1 = \sum_{i=1}^{n} |\sum_{j=1}^{n} a_{ij}x_j| \le \sum_{i=1}^{n}\sum_{j=1}^{n} |a_{ij}|\,|x_j| \le \sum_{i=1}^{n} \max_j |a_{ij}| \sum_{j=1}^{n} |x_j|$

$$\le n \max_{i,j} |a_{ij}|\, \|\mathbf{x}\|_1 = \|A\|_n\, \|\mathbf{x}\|_1$$

$\|A\mathbf{x}\|_2^2 = \sum_{i=1}^{n} |\sum_{j=1}^{n} a_{ij}x_j|^2 \le \sum_{i=1}^{n} (\sum_{j=1}^{n} |a_{ij}|^2)(\sum_{j=1}^{n} |x_j|^2)$ by Schwartz's inequality

$$= \|A\|_e^2\, \|\mathbf{x}\|_2^2 \le \|A\|_n^2\, \|\mathbf{x}\|_2 \text{ by (b)}$$

$$\|A\mathbf{x}\|_\infty = \max_i |\sum_{j=1}^{n} a_{ij}x_j| \le \max_i \sum_{j=1}^{n} |a_{ij}|\,|x_j| \le \max_j |x_j| \max_i \sum_{j=1}^{n} |a_{ij}|$$

$$\le \max_j |x_j|\, n \max_{i,j} |a_{ij}| = \|A\|_n\, \|\mathbf{x}\|_\infty.$$

9.9 $\| AB \| = \max_{i,j} |\sum_{j=1}^{n} a_{ik} b_{kj}| \le \max_{i,j} \sum_{k=1}^{n} |a_{ik}| |b_{kj}| \le \max_{i,k} |a_{ik}| \max_{j} \sum_{k=1}^{n} |b_{kj}| = \| A \| \| B \|_1$.

9.10 (b) See Problem 9.8(c);

(c) $\| A \|_e^2 = \text{trace } A^*A = \lambda_1 + \lambda_2 + \cdots + \lambda_n \ge \max_i \lambda_i = \| A \|_2^2$, λ_i's are eigenvalues of A^*A.

Also, $\| A \|_e^2 = \lambda_1 + \lambda_2 + \cdots + \lambda_n \le n \max_i \lambda_i = n \| A \|_2^2$. See also Corollary 7.4.

9.11 $\max_{\|\mathbf{x}\|=1} \| A\mathbf{x} \| \le \max_{\|\mathbf{x}\|=1} N(A) \| \mathbf{x} \|$, i.e., $\| A \| \le N(A)$.

9.15 (a) 1, 1, 1; (b) 1; (c) n.

9.16 $\sqrt{3}$, $\sqrt{2}$, 1, $\sqrt{2}$.

9.17 Let $\mathbf{y} = \alpha\mathbf{x}$, α a scalar in (9.13). Then $R(\mathbf{x}) = R(\mathbf{y})$.

9.18 A^*A is a PSD matrix and by Theorem 8.19(c), its eigenvalues are non-negative, $\det (A^*A) = \det A^* \det A = \det \bar{A} \det A = |\det A|^2$, see Property III of Section 2.3.

9.19 $\| A \|_2^2 = \max_{\|\mathbf{x}\|=1} \| A\mathbf{x} \|_2 = \max_i \lambda_i (A^*A) = \lambda_n^2$

Let $A^{-1}\mathbf{x} = \mathbf{y}$, $A\mathbf{y} \ne \boldsymbol{\theta}$ implies $\mathbf{y} \ne \boldsymbol{\theta}$. So, $\max_{\mathbf{x} \ne \boldsymbol{\theta}} \frac{\| A^{-1}\mathbf{x} \|}{\| \mathbf{x} \|} = \max_{A\mathbf{y} \ne \boldsymbol{\theta}} \frac{\| \mathbf{y} \|}{\| A\mathbf{y} \|} = 1/\max_{\mathbf{y} \ne \boldsymbol{\theta}} \frac{\| A\mathbf{y} \|}{\| \mathbf{y} \|}$.

Hence, for $\| \cdot \|_2$ norm $\| A^{-1} \|_2 = 1/\lambda_1$ by Theorem 9.4. Moreover,

$$\| A^{-1} \|_2 = \frac{1}{\lambda_1} \le \frac{\lambda_n^{n-1}}{\lambda_1 \lambda_2 \dots \lambda_n} = \frac{\| A \|_2^{n-1}}{|\det A|} \text{ by Problem 9.18.}$$

9.20 (a) For any eigenvalue λ and associated eigenvector $\mathbf{x}$, we have $\lambda\mathbf{x} = A\mathbf{x}$, whence $|\lambda| \| \mathbf{x} \| \le \| A \| \| \mathbf{x} \|$ (by Theorem 9.5), which is true for any eigenvalue λ.

(b) Let $X = [\mathbf{x}, \boldsymbol{\theta}, \dots, \boldsymbol{\theta}]$. Therefore, $AX = \lambda X$. By axioms (2) and (4) of a matrix norm, we have $|\lambda| N(X) \le N(A) N(X)$, whence $r(A) \le N(A)$.

(c) It follows from Problem 9.11 and part (b) of the current problem.

9.21 (a) $\| A \|_1 = \max_j \sum_{i=1}^{n} |a_{ij}| = \max_j \sum_{i=1}^{n} |\bar{a}_{ij}| = \| A^* \|_\infty$;

(b) $r(A^*A) = \| A \|_2^2 \le \| A^*A \|_1$ by Problem 9.20(a)

$$\le \| A^* \|_1 \| A \|_1 = \| A \|_\infty \| A \|_1.$$

9.22 (a), (b) (i) 1, (ii) 1; (c) $\because \det (\lambda I - A^*A) = \lambda^{n-1} (\lambda - n^2)$, (i) n, (ii) 0.

9.23 Apply Theorem 9.6 and Example 9.12.

(a) $\min_i |\hat{d}_i|$ for $p = 1, 2, \infty$; (b) $\min_i |d_i|$ for $p = 1, 2, \infty$.

9.24 If either of A and B is singular, then the inequality is obvious. When both are non-singular,

$$m(AB) = \frac{1}{\| (AB)^{-1} \|} \ge \frac{1}{\| A^{-1} \| \| B^{-1} \|} = m(A)\, m(B).$$

9.26 (a) $\lim A_p B = [\lim_{p\to\infty} \sum_{k=1}^{n} a_{ik}^{(p)} b_{kj}] = [\sum_{k=1}^{n} \lim_{p\to\infty} a_{ik}^{(p)} b_{kj}]$;

(c) We know that $\lim \lambda^p$ exists iff $\lambda = 1$ or $|\lambda| < 1$. Note when $\lambda = i$, $\lim \lambda^p$ does not exist. Use $A\mathbf{x} = \lambda\mathbf{x}$ and $A^p\mathbf{x} = \lambda^p\mathbf{x}$ and (a) to prove (c) by contradiction.

9.27 (a) Use Theorem 9.10; (b) Use Theorem 9.3(b); (c) If $\| A \| < 1$, then $\lim \| A \|^p = 0$. By Example 9.7(c), $0 \le \| A^p \| \le \| A \|^p$ and hence $\lim \| A^p \| = 0$. By (a), $\lim A^p = 0$.

9.28 We have $I = (I + A)^{-1} (I + A)$ and $(I + A)^{-1} = I - (I + A)^{-1}A$; whence $1 \le \| (I + A)^{-1} \| \, \| (I + A) \| \le \| (I + A)^{-1} \| (1 + \| A \|)$ and $\| (I + A)^{-1} \| \le 1 + \| (I + A)^{-1} \| \, \| A \|$.

9.30 Here $\| A \|_2 < 1$.

9.32 Use Theorem 7.26 and Example 7.2.

CHAPTER 10

10.3 (a) $\begin{bmatrix} 0 & 0 & 1 \\ 0 & 1 & -\lambda \\ 1 & 0 & \lambda-1 \end{bmatrix}, \begin{bmatrix} \lambda+1 & 1 & 0 \\ 1 & 0 & 0 \\ 0 & 0 & 1 \end{bmatrix}$; (b) $\begin{bmatrix} 0 & 1 & 0 \\ \lambda^2 & -\lambda^3-\lambda^2 & 0 \\ 0 & 0 & 1 \end{bmatrix}, \begin{bmatrix} 1+\lambda & -1 & 1+\lambda \\ -\lambda & 1 & 0 \\ 0 & 0 & 1 \end{bmatrix}$;

(c) $\begin{bmatrix} 1 & 0 & 1+\lambda^2 \\ 0 & 1 & \lambda+\lambda^2 \\ -\lambda & 0 & 1-\lambda-\lambda^2 \end{bmatrix}, \begin{bmatrix} 1 & -\lambda & 0 \\ 0 & 1 & 0 \\ 0 & 0 & \lambda \end{bmatrix}$. [Order P, Q].

10.4 $\begin{bmatrix} 1 & 0 & 0 \\ -1-\lambda^2, & 1 & 0 \\ -\lambda-\lambda^3, & \lambda & 1 \end{bmatrix}, \begin{bmatrix} 1 & 0 & -\lambda \\ 0 & 1/2 & (\lambda+\lambda^3)/2 \\ 0 & -1/2 & (2-\lambda-\lambda^3)/2 \end{bmatrix}, \begin{bmatrix} 1 & 0 & 0 \\ 0 & 1 & 0 \\ 0 & 0 & (1-\lambda)(\lambda^3+\lambda^2+2\lambda+1) \end{bmatrix}$. [Order P, Q, B].

10.7 (a) $\begin{bmatrix} 1 & 1 \\ 0 & \lambda-2 \end{bmatrix}$; (b) $\begin{bmatrix} 1 & -\lambda^2+2\lambda \\ 0 & \lambda^3-\lambda^2-\lambda-1 \end{bmatrix}$; (c) $\begin{bmatrix} \lambda+1 & 1 \\ 0 & 0 \end{bmatrix}$;

(d) $\begin{bmatrix} \lambda+1 & -1 & 1 \\ 0 & \lambda & 0 \end{bmatrix}$; (g) $\begin{bmatrix} 1 & 0 \\ 0 & 1 \\ 0 & 0 \end{bmatrix}$; (h) $\begin{bmatrix} \lambda+1 & 1 \\ 0 & \lambda \\ 0 & 0 \end{bmatrix}$.

10.8 (a) $\begin{bmatrix} 1 & 0 \\ 3/2 & \lambda-2 \end{bmatrix}$; (b) $\begin{bmatrix} 1 & 0 \\ (\lambda^2-1)/2 & \lambda^3-\lambda^2-\lambda-1 \end{bmatrix}$; (c) $\begin{bmatrix} \lambda & 0 \\ 1 & 0 \end{bmatrix}$;

(d) $\begin{bmatrix} \lambda & 0 & 0 \\ 0 & 1 & 0 \end{bmatrix}$; (e) $\begin{bmatrix} 1 & 0 & 0 \\ 0 & 1 & 0 \\ a & b & \lambda^2(\lambda^2-1) \end{bmatrix}$,

where $a = -(35\lambda^3 - \lambda^2 - 48\lambda)/24$, $b = -(25\lambda^3 + 13\lambda^2 - 24\lambda - 24)/24$;

(f) $V(\lambda) = \begin{bmatrix} 0 & -1 & 1 \\ \lambda-2 & -2\lambda^3+2\lambda^2+9\lambda-2 & 2\lambda^3-3\lambda^2-7\lambda+3 \\ 1 & -2\lambda^2-2\lambda+2 & 2\lambda^2+\lambda-2 \end{bmatrix}$,

$B(\lambda) = \begin{bmatrix} 1 & 0 & 0 \\ 1+2\lambda & 1-\lambda^2 & 0 \\ 0 & 3 & 1+\lambda \end{bmatrix}$; (g) $\begin{bmatrix} 1 & 0 \\ \lambda+(1/2) & \lambda^2+\lambda-(1/2) \\ -(2\lambda+1) & -(2\lambda^2+2\lambda+1) \end{bmatrix}$; (h) $\begin{bmatrix} \lambda(\lambda+1) & 0 \\ 2\lambda-1 & \lambda^2-1 \\ 1 & \lambda+1 \end{bmatrix}$.

10.9 (a) diag $[1, \lambda - 2]$; (b) diag $[1, \lambda^3 - \lambda^2 - \lambda - 1]$; (c) diag $[1, 0]$; (d) diag $[1, \lambda, 0]$; (e) diag $[1, 1, \lambda^2(\lambda^2 - 1)]$; (f) diag $[1, 1, (\lambda + 1)(\lambda^2 - 1)]$; (g) diag $[1, 1, 0]$; (h) diag $[1, \lambda(\lambda + 1), 0]$.

10.10 (a) Not given; (b) $\begin{bmatrix} \lambda(\lambda+1) & 0 & 2\lambda(\lambda+1) \\ 0 & \lambda & -\lambda(\lambda+1) \\ 0 & 0 & \lambda(\lambda^2-1) \end{bmatrix}$.

10.11 (a) [diag[1, 1, $\lambda^2 - \lambda + 2$] $0_{3\times1}$]; (b) diag [1, 1, $\lambda^2(\lambda^2 - 1)$];
(c) diag[1, 1, $(1 - \lambda^2)(1 + \lambda)$]; (d) diag [1, λ, λ].

10.12 (a) $U(\lambda) = \begin{bmatrix} 0 & 1 & 0 \\ 3/2 & -\lambda/2 & -1 \\ -1 & 0 & 1 \end{bmatrix}$, $B(\lambda) = \begin{bmatrix} \lambda & -\lambda & -2\lambda \\ 0 & \lambda^2 - \lambda & 5\lambda \\ 0 & 0 & \lambda^2 - \lambda \end{bmatrix}$;

(b) $B(\lambda) = \begin{bmatrix} \lambda & 0 & 0 \\ -\lambda/4 & \lambda & 0 \\ \lambda(\lambda+7)/8 & \lambda(1-\lambda)/10 & \lambda(\lambda-1)^2 \end{bmatrix}$, $V(\lambda) = \begin{bmatrix} 0 & 2/5 & \lambda - 6 \\ 0 & -2/5 & -\lambda - 4 \\ 1/8 & -1/10 & \lambda - 1 \end{bmatrix}$;

(c) $U(\lambda) = \begin{bmatrix} 0 & 1 & 0 \\ 1 & -\lambda & 0 \\ -(\lambda+9)/10 & \lambda(\lambda-1)/10 & 1 \end{bmatrix}$, $B(\lambda) = \begin{bmatrix} \lambda & 0 & 0 \\ 0 & \lambda & 0 \\ 0 & 0 & \lambda(\lambda-1)^2 \end{bmatrix}$, $V(\lambda) = \begin{bmatrix} 1 & 0.1 & \lambda - 6 \\ 0 & -0.1 & -\lambda - 4 \\ 0 & 0.1 & \lambda - 1 \end{bmatrix}$.

10.14 (a) A and C; (b) A and C.

10.15 (a) $\begin{bmatrix} -1 & 0 \\ 1 & 1 \end{bmatrix}$; (b) $\begin{bmatrix} 0 & 1 & -2 \\ 1 & 3 & 1 \\ 1 & 2 & 2 \end{bmatrix}$.

10.16 No.

10.17 $\begin{bmatrix} \lambda & 1 \\ 0 & \lambda \end{bmatrix}$, $\begin{bmatrix} \lambda & 0 \\ 0 & \lambda \end{bmatrix}$.

10.19 (a) (i) 1, 1, $(\lambda - 1)^2$, $(\lambda - 1)^2$; (b) (i) 1, 1, $(\lambda - 3)(\lambda - 2)^2$,
(c) (i) 1, 1, 1, $(\lambda - 1)^2$, $(\lambda - 1)^2(\lambda + 1)$.

10.20 1, 1, λ, λ, $\lambda(\lambda^2 - 1)$.

10.22 Inv. poly: 1, 1, $(\lambda - 2)$, $(\lambda - 2)^2(\lambda - 3)$.

10.27 Not simple.

10.28 $T = \text{diag}\ [x, \beta_1 x, \beta_1\beta_2 x, \ldots, \beta_1\beta_2 \ldots \beta_{k-1}x]$, x = arbitrary $\neq 0$.

CHAPTER 11

11.1 (a), (c) No; (b), (d) Yes.

11.2 (a), (b), (d) Yes; (c) No.

11.3 (b) $S_1 \cap S_2 = \{(a, b, c); a = b = c\}$, $S_2 \cap S_3 = \{(0, 0, 0)\}$, $S_3 \cap S_1 = \{(a, b, c); a + 2c = 0, b = 4c\}$;
(c) (i) (1 1 1), (ii) (0 0 0), (iii) (–2, 4, 1); (d) $\mathbb{R}^3 = S_2 \oplus S_3$.

11.4 (b) (i), (ii), (iii), (iv) $\{(a, b, c) : b = c = 0\}$; (c) (1, 0, 0).

11.5 (a) 3; (b) 3; (c) 2.

11.6 No.

11.12 (a) $1 + x + x^2$; (b) 3.

11.13 (b) diag [1, 2].

11.14 (b) diag [0, 2, 2].

11.16 $\begin{bmatrix} 0 & 1 & 0 & 0 \\ 0 & 0 & 0 & 0 \\ 0 & 0 & 2 & 0 \\ 0 & 0 & 0 & 2 \end{bmatrix}$.

11.17 (a) λ^4, $\mathbb{R}^4$; (b) $\lambda(\lambda - 4)$, $0 : \{[1, -1, 0, 0]^T, [1, 0, -1, 0]^T, [1, 0, 0, -1]^T\}$; $4: \{[1\ 1\ 1\ 1]^T\}$; (c) $(\lambda - 1)^4$, $1 : \mathbb{R}^4$.

11.18 (a) $\{\mathbf{x}_{11} = [2\ 1\ 1\ 0]^T, \mathbf{x}_{12} = [1\ 1\ 0\ 1]^T, \mathbf{x}_{13} = [2\ 0\ 1\ 0]^T, \mathbf{x}_{21} = [3\ 1\ 1\ 0]^T\}$,

$$\left[\begin{array}{ccc|c} 1 & 0 & 0 & 0 \\ 0 & 1 & 1 & 0 \\ 0 & 0 & 1 & 0 \\ \hline 0 & 0 & 0 & 2 \end{array}\right].$$

(b) $\{[0\ 1\ -1\ 0]^T, [1\ 0\ 0\ -1]^T; [-1\ -1\ 0\ -1]^T, [1\ 0\ 0\ 0]^T\}$, $\begin{bmatrix} 1 & 0 & 0 & 0 \\ 0 & 1 & 0 & 0 \\ 0 & 0 & -1 & 1 \\ 0 & 0 & 0 & -1 \end{bmatrix}$.

11.19 (a) $(\lambda - a)$; (b) $(\lambda - a)^2$; (c) $(\lambda - a)^3$; (d) $(\lambda - a)^2$.

11.20 (a) $\{\mathbf{e}_1\} \oplus \{\mathbf{e}_2\} \oplus \{\mathbf{e}_3\} \oplus \{\mathbf{e}_4\}$; (b) $\{\mathbf{e}_1, \mathbf{e}_2\} \oplus \{\mathbf{e}_3\} \oplus \{\mathbf{e}_4\}$; (c) $\{\mathbf{e}_1, \mathbf{e}_2, \mathbf{e}_3\} \oplus \{\mathbf{e}_4\}$; (d) $\{\mathbf{e}_1, \mathbf{e}_2\} \oplus \{\mathbf{e}_3 \oplus \mathbf{e}_4\}$.

11.24 See Problem 11.18.

11.25 $$\left[\begin{array}{ccccc|cccc} 1 & 0 & 0 & 0 & 0 & 0 & 0 & 0 & 0 \\ 0 & 1 & 0 & 1 & 0 & 0 & 0 & 0 & 0 \\ 0 & 0 & 0 & 0 & 0 & 1 & 0 & 1 & 0 \\ 0 & 0 & 1 & 0 & 1 & 0 & 0 & 0 & 0 \\ 0 & 0 & i & 0 & -i & 0 & 0 & 0 & 0 \\ 0 & i & 0 & -i & 0 & 0 & 0 & 0 & 0 \\ \hline 0 & 0 & 0 & 0 & 0 & i & 0 & -i & 0 \\ 0 & 0 & 0 & 0 & 0 & 1+2i & 1 & 1-2i & 1 \\ 0 & 0 & 0 & 0 & 0 & 0 & i & 0 & -i \end{array}\right], \quad J = \begin{bmatrix} 1 & & & & & & & & \\ & i & & & & & & & \\ & & i & & & & & & \\ & & & -i & & & & & \\ & & & & -i & & & & \\ & & & & & 1+2i & 1 & & \\ & & & & & 0 & 1+2i & & \\ & & & & & & & 1-2i & 1 \\ & & & & & & & 0 & 1-2i \end{bmatrix}.$$

CHAPTER 12

12.4 (a) λ; (b) $\lambda + 2$; (c) $\lambda^2\ [e^{-t} - 2 + e^t]/2 + \lambda\ [e^t - e^{-t}]/2 + 1$.

12.5 $1 - 0.5\lambda$.

12.6 (a) $Z_1 + Z_2\, e^t + Z_3\, e^{2t}$, (b) $Z_2 \sin t + Z_3 \sin 2t$, (c) $Z_1 + Z_2 \cos t + Z_3 \cos 2t$, (d) $Z_1 + Z_2 \cosh t + Z_3 \cosh 2t$, (e) $Z_2 \sinh t + Z_3 \sinh 2t$,

where $Z_1 = \begin{bmatrix} 1 & -1.5 & 0.5 \\ 0 & 0 & 0 \\ 0 & 0 & 0 \end{bmatrix}$, $Z_2 = \begin{bmatrix} 0 & 2 & -1 \\ 0 & 2 & -1 \\ 0 & 2 & -1 \end{bmatrix}$, $Z_3 = \begin{bmatrix} 0 & -0.5 & 0.5 \\ 0 & -1 & 1 \\ 0 & -2 & 2 \end{bmatrix}$.

No.

12.7 (a) $\begin{bmatrix} 0 & -126 & 127 \\ 0 & -254 & 255 \\ 0 & -510 & 511 \end{bmatrix}$; (b) $\begin{bmatrix} 1 & 1/3 & -1/3 \\ 0 & 5/3 & -2/3 \\ 0 & 4/3 & -1/3 \end{bmatrix}$.

12.8 (a) $Z_{11} + tZ_{12} + e^t\, Z_{21}$; (b) $tZ_{12} + \sin t\, Z_{21}$;

(c) $Z_{11} + \cos t\, Z_{21}$; (d) $Z_{11} + \cosh t\, Z_{21}$; (e) $t\,Z_{12} + \sinh t\, Z_{21}$,

where $Z_{11} = \begin{bmatrix} 1 & 0 & -1 \\ 0 & 1 & -1 \\ 0 & 0 & 0 \end{bmatrix}$, $Z_{12} = \begin{bmatrix} 0 & 1 & -1 \\ 0 & 0 & 0 \\ 0 & 0 & 0 \end{bmatrix}$, $Z_{21} = \begin{bmatrix} 0 & 0 & 1 \\ 0 & 0 & 1 \\ 0 & 0 & 1 \end{bmatrix}$. No.

12.11 (a) $Z_{11} = \begin{bmatrix} -1 & 0 & -2 \\ -1 & 1 & -1 \\ 1 & 0 & 2 \end{bmatrix}$, $Z_{21} = \begin{bmatrix} 2 & 0 & 2 \\ 1 & 0 & 1 \\ -1 & 0 & -1 \end{bmatrix}$;

(b) $Z_{11} = \begin{bmatrix} -1 & 0 & 0 & 2 \\ 3 & 1 & 0 & -3 \\ -2 & 0 & 1 & 2 \\ -1 & 0 & 0 & 2 \end{bmatrix}$, $Z_{12} = \begin{bmatrix} 2 & 1 & 0 & -1 \\ -3 & -1 & 1 & 1 \\ 2 & 1 & 0 & -1 \\ 2 & 1 & 0 & -1 \end{bmatrix}$, $Z_{13} = \frac{1}{2}\begin{bmatrix} -1 & 0 & 1 & 0 \\ 1 & 0 & -1 & 0 \\ -1 & 0 & 1 & 0 \\ -1 & 0 & 1 & 0 \end{bmatrix}$, $Z_{21} = \begin{bmatrix} 2 & 0 & 0 & -2 \\ -3 & 0 & 0 & 3 \\ 2 & 0 & 0 & -2 \\ 1 & 0 & 0 & -1 \end{bmatrix}$.

12.15 (i) $\begin{bmatrix} e^{2t}\,(2-3t) \\ -e^{2t} \end{bmatrix}$ (ii) $\begin{bmatrix} (3e^{t} - e^{3t}) \\ 2 \\ -e^{3t} \end{bmatrix}$ (iii) $\begin{bmatrix} \cos 2t + \sin 2t \\ \cos 2t - \sin 2t \end{bmatrix}$ (iv) $e^{t}\begin{bmatrix} \cos \pi t - \sin \pi t \\ -\cos \pi t - \sin \pi t \end{bmatrix}$ (v) $e^{-t}\begin{bmatrix} -1 \\ 1 \end{bmatrix}$

(vi) $e^{-t}\begin{bmatrix} -1-5t+2t^2 \\ -1+4t \\ 1-15t+6t^2 \end{bmatrix}$ (vii) $\begin{bmatrix} (1+t)\,e^{2t} \\ e^{2t} \\ e^{3t} \end{bmatrix}$ (viii) $\begin{bmatrix} 3e^{2t} + te^{2t} - 2e^{3t} \\ 3e^{2t} + te^{2t} - e^{3t} \\ -4e^{2t} - te^{2t} + e^{3t} \end{bmatrix}$ (ix) $\begin{bmatrix} 1+2t \\ -2t \\ 2+2t \end{bmatrix}$ (x) $\begin{bmatrix} -3e^{2t} + 2e^{-4t} \\ -6e^{2t} + 8e^{-4t} \\ 3e^{2t} - 4e^{-4t} \end{bmatrix}$

12.16 (i) $\frac{1}{\sqrt{5}}\begin{bmatrix} \lambda_1\lambda_2^{h+1} - \lambda_1^{k+1}\lambda_2 \\ \lambda_1\lambda_2^{k} - \lambda_1^{k}\lambda_2 \end{bmatrix}$, $\lambda_1, \lambda_2 = \frac{1}{2}(1 \pm \sqrt{5})$; (ii) $\begin{bmatrix} (-1)^k \\ -(-1)^k \end{bmatrix}$; (iii) $(-1)^k\begin{bmatrix} 1 \\ -1 \end{bmatrix}$; (iv) $\begin{bmatrix} k(-1)^{k-1} \\ (-1)^k \\ (-2)^k \end{bmatrix}$;

(v) $2^{k-1}\begin{bmatrix} k \\ 2 \\ 0 \end{bmatrix}$; (vi) $\begin{bmatrix} (3k-1)(k-2)/2 \\ (k-1)(3k+2)/2 \\ k(3k+5)/2 \end{bmatrix}$; (vii) $\begin{bmatrix} 0 \\ 2^{k-3}(k^2-k+8) \\ k2^{k-1} \\ 2^k \end{bmatrix}$; (viii) $[0\ 0\ 0\ 0]^T$.

12.17 $x_k = 2 - \frac{2}{3}(-1)^k - \frac{1}{3}2^k$.

12.19 $Q = \frac{1}{15}\begin{bmatrix} 2 & 14 \\ 10 & -5 \\ 11 & 2 \end{bmatrix}$, $S = \frac{1}{25}\begin{bmatrix} 41 & 12 \\ 12 & 34 \end{bmatrix}$.

CHAPTER 13

13.1 $x_1 = x_2 = 0.1000 \times 10$.

13.10 $Q^T = \begin{bmatrix} 0 & 0.3333 & -0.6667 & -0.6667 \\ 0 & -0.8666 & 0.0619 & -0.4952 \\ -0.6556 & 0.2804 & 0.5609 & -0.4207 \\ 0.7551 & -0.2435 & -0.4870 & 0.3652 \end{bmatrix}$.

13.11 (a) $Q = \begin{bmatrix} 1 & 0 & 0 & 0 \\ 0 & (-1/3) & (-2/3) & 2/3 \\ 0 & (-2/3) & (-1/3) & (-2/3) \\ 0 & 2/3 & (-2/3) & (-1/3) \end{bmatrix}$; $B = \begin{bmatrix} 0 & -1 & -2 & -1 \\ -3 & 2 & 7 & -7 \\ 0 & -1 & -1 & 3 \\ 0 & 0 & 1 & -2 \end{bmatrix}$;

(b) $Q = \begin{bmatrix} 1 & 0 & 0 & 0 \\ 0 & 1/3 & 2/3 & (-2/3) \\ 0 & (-2/3) & (-1/3) & (-2/3) \\ 0 & (-2/3) & 2/3 & 1/3 \end{bmatrix}$; $B = \begin{bmatrix} 0 & 1 & -2 & 1 \\ 3 & 2 & -7 & -7 \\ 0 & 1 & -1 & -3 \\ 0 & 0 & -1 & -2 \end{bmatrix}$.

13.12 (a) $Q=\begin{bmatrix}1 & 0 & 0 & 0\\ 0 & (-1/3) & (-2/3) & 2/3\\ 0 & 2/3 & 1/3 & 2/3\\ 0 & (-2/3) & 2/3 & 1/3\end{bmatrix}$; $B=\begin{bmatrix}0 & -3 & 0 & 0\\ -3 & -4 & 4 & 0\\ 0 & 4 & -5 & -1\\ 0 & 0 & -1 & 8\end{bmatrix}$;

(b) Q same as in 13.11(b); $B=\begin{bmatrix}0 & 3 & 0 & 0\\ 3 & -4 & 4 & 0\\ 0 & 4 & -5 & 1\\ 0 & 0 & 1 & 8\end{bmatrix}$.

13.13 $\mathbf{x} = [6, -3]^T$, $[6\ 3\ 3\ 6]^T$.

13.18 (a) 5.45, $[1, 6.44, -1.87]^T$; (b) 5, $[20, -24, 3]^T$; (c) 4, $[3, -15, 28]^T$.

13.20 The eigenvalues are $-(1 \pm i)$, $1 \pm i$ and the eigenvector associated with $-(1 + i)$ is $[-i2,\ 2 - i4,\ 2,\ 1 - i]^T$.

$$Q=\begin{bmatrix}0 & (-\sqrt{3}/3) & (3-35\sqrt{3})/78 & (-15-7\sqrt{3})/78\\ (-2/3) & (-\sqrt{3}/3) & (2+20\sqrt{3})/78 & (-10+4\sqrt{3})/78\\ (-2/3) & \sqrt{3}/3 & (5-15\sqrt{3})/78 & (-25-3\sqrt{3})/78\\ (-1/3) & 0 & (-14-10\sqrt{3})/78 & (70-2\sqrt{3})/78\end{bmatrix}.$$

Index